tables of BULLET PERFORMANCE

Philip Mannes

1980
WOLFE PUBLISHING CO., INC.
PRESCOTT, ARIZONA

Manufactured in the United States of America

ISBN 0-935632-05-0 (Soft Bound)
0-935632-06-9 (Hard Bound)

Wolfe Publishing Co., Inc.

P.O. Box 30-30
Prescott, AZ 86302

Introduction

These tables compiled by Philip Mannes list exterior-ballistics data in a form that requires no separate computation by the user. You may want to make notes as you use these tables, but you will need no arithmetic, no calculator, no slide rule to find ballistics data. The only information you need, to start with, is the ballistic coefficient and muzzle velocity of your bullet. Ballistic coefficients for commercial bullets have been published and are available from the manufacturers of those bullets. Typical muzzle velocities, as well as ballistic coefficients, are listed in current manuals for handloaders. More-precise velocities may be measured by the shooter using a chronograph.

Trajectory

The first section of this book covers bullets with ballistic coefficients between 0.09 and 0.70 traveling at initial velocities of 800 to 1,900 feet per second. The second section covers ballistic coefficients from 0.10 to 0.70 and muzzle velocities from 2,000 to 4,000 feet per second. Trajectory data in the first or low-velocity section are for ranges out to 500 yards; the high-velocity section lists trajectory data for ranges out to 1,000 yards.

If you want to find, for example, data for a .30-06 firing a 150-grain Sierra bullet at 2,900 feet per second, you already have half the information you need to start with. The other half is Sierra's listed ballistic coefficient for this bullet: 0.397 (we'll use 0.40). Thumbing through this book, we find several tables for a ballistic coefficient of 0.40, among them a table for a bullet with an initial velocity of 2,900 feet per second.

All tables are for typical shooting conditions and scope mounting: a crosswind blowing at ten miles per hour and a scope mounted 1.5 inches above the axis of the bore. The data for remaining velocities, trajectories, drift, and times of flight for a variety of distances are tabulated in ten columns.

The first column, headed *range,* is the distance of the bullet from the muzzle, in yards. The second column lists the remaining *velocity* at each distance shown in the first column. The third column lists the midrange *height* of the trajectory above the line of sight when the gun is sighted-in at each distance in the first column.

The next five columns, headed *50 yd* to *250 yd* in the first section, and *100 yd* out to *300 yd* in the last section, list the height of the trajectory above or below the line of sight, at each distance in the *range* column, when the gun is zeroed at the distance designated at the head of any of these five trajectory columns.

The last two columns list the downwind *drift* of the bullet at each range distance, in that ten-mile-per-hour crosswind already mentioned; and the *time* of flight, in seconds, from the muzzle to any distance listed in the first column.

In this table, we find that the 150-grain Sierra with a ballistic coefficient of 0.40 and an initial velocity of 2,900 feet per second has a remaining velocity (according to the second column) of 2,666 at 100 yards, 2,444 at 200 yards, and ultimately to 1,139 feet per second at 1,000 yards. The next column shows that with the gun zeroed at 200 yards, the bullet would strike 1.7 inches high at the midrange distance (just past 100 yards). With the rifle zeroed at 300 yards, the bullet would strike 5.3 inches high at a distance just beyond 150 yards. ("Midrange" distance, because of the increasing drop of the bullet, is always a bit farther downrange than exactly half of the zeroed range; this is the point where the bullet is at the peak of its arc.)

In the trajectory column headed *200 yd*, we see that with the rifle zeroed at this distance, the path of the bullet is 1.7 inches above the line of sight at 100 yards, crosses the line of sight at 200 yards, and is 7.6 inches below the line of sight at 300 yards. The column headed *300 yd* shows that the trajectory of the bullet fired from a rifle zeroed at 300 yards is 4.2 inches high at 100 yards and crosses the line of sight at 300 yards.

The trajectory tables thus make it possible to sight-in a rifle at a known, convenient range — say 100 yards — and then use data from the tables to determine where the point of impact will be at other ranges.

A deer hunter who can tolerate an error of three to four inches above or below the point of aim can use the tables to determine the best range at which to zero his rifle, as well as where the bullet should strike in relation to the point of aim if he test-fires his rifle at 100 yards.

He refers first to the *height* column. The midrange height for a 200-yard zero is 1.7 inches; and for a 300-yard zero, it is 5.3 inches. So 250 yards is the sight-in distance he decides to try. From the *250 yd* column, he finds that his bullet will strike 2.9 inches high at 100 yards, 2.4 inches high at 200 yards, and 4.0 inches low at 300 yards — he knows that at any distance up to 300 yards, he can hold dead-on and be certain that his error won't be more than four inches.

The column headed *drift* shows how many inches off-course the bullet will be blown by a crosswind of ten miles per hour. At 300 yards, the bullet would strike 7.7 inches from the point of aim. Since drift is directly proportional to the speed of the crosswind, a five-mile-per-hour crosswind will cause half as much drift — 3.9 inches — and a 20-mile-per-hour crosswind will cause twice as much — 15.4 inches.

The column headed *time* shows the number of seconds required for the bullet to travel any listed distance from the muzzle. This figure can be used to estimate the lead required to hit a moving target at a particular distance. When a bullet requires, for example, 0.35 second to reach a target at 300 yards, a deer running broadside to the hunter at 20 miles per hour — or about 30 feet per second — requires a lead of 10.5 feet at 300 yards (30 fps times 0.35 second equals 10.5 feet).

The methods used by bullet manufacturers to compute or estimate ballistic coefficients are not standardized, so one manufacturer's bullet may have a somewhat different coefficient than an essentially similar bullet made by another manufacturer. At first glance, this would seem to introduce a serious source of error into the use of these tables. But as it turns out, this error is much smaller than the normal spread of typical five-shot groups fired from accurate hunting rifles, from benchrests, by very skillful shooters.

Let's compare the performance of a couple of typical .257 bullets, both starting at 3,000 feet per second and both having similar weights and shapes but somewhat different ballistic coefficients, as these performances are reported in the bullet manufacturers' manuals and as they are predicted in these tables. Speer's ballistic coefficient for their 120-grain flat-base spitzer is 0.465; Sierra's coefficient for their 117-grain flat-base spitzer is 0.404. The

Speer manual shows a 300-yard correction figure of 6.7 inches for their bullet, and Sierra's handbook lists 7.01 inches for theirs, with both bullets starting at 3,000 feet per second from rifles zeroed at 200 yards.

Speer's coefficient, 0.465, falls halfway between the 0.46 and 0.47 listed in these tables. The table under 0.46 and a muzzle velocity of 3,000 feet per second lists a drop of 6.7 inches, and the table for this velocity and a coefficient of 0.47 also lists a drop of 6.7 inches, which agrees with Speer's figure.

Looking up the Sierra bullet's coefficient of 0.404 and its initial velocity of 3,000 feet per second, we find that our tables predict that the bullet will strike 7.0 inches below the point of aim at 300 yards, and Sierra's ballistics tables predict that it will strike 7.01 inches low — exactly the same. If these two randomly chosen examples are good indications, we can conclude that the differences in the manufacturers' methods of deriving their ballistic coefficients — and consequently the differences in their figures — are not great enough to lessen the usefulness of the tables in this book.

Energy

Our trajectory tables do not list the diameter or weight of any bullet — only ballistic coefficients, velocities, and trajectory data. But remaining *energy* at a known range is often of interest to shooters, so a separate set of tables (following this introduction) lists energies of bullets in a range of weights from ten to 500 grains, at velocities from 500 to 5,000 feet per second.

To find the energy of any bullet, look first for the column headed by that bullet's weight in grains, then look down that column to the energy figure on a line with the bullet's velocity, at the left, to the nearest hundred feet per second. For example, the 150-grain Sierra fired from a .30-06 at a muzzle velocity of 2,900 feet per second has 2,800 foot-pounds of energy at the muzzle, and out at 200 yards — where we've already found that the remaining velocity is 2,441 feet per second, or near an even 2,400 feet per second — its remaining energy is 1,918 foot-pounds.

Keep in mind that the best prediction of performance is only an approximation, not a narrowly precise figure that leaves no margin for error or deviation. The muzzle velocity you start with, for example, is an average figure; the ten rounds used to chronograph that load may have produced ten different velocities — and not one of them exactly the figure you now use as an average. In turn, any prediction based on this approximation, even if it were done by an extremely precise method, would be only another approximation.

But this caution applies almost entirely to figures on paper, not so much to actual performance in the field. Out there where the bullet is or will be, a hundredth of an inch is nothing, and a tenth of an inch isn't much. When your bullet gets 'way out there at several hundred yards, even an inch or two is negligible — simply because it is still far below the usual and normal round-to-round variation in point of impact of the most accurate bullets fired from the most accurate sporting rifles from the firmest of rests by the most skillful riflemen.

Within the bounds of normal variation, then, you will find that the tables in this book are entirely adequate for use in practical estimation of downrange performance. — Ken Howell

Philip Mannes is a registered mechanical and controls engineer who lives in Torrance, California. His specific engineering interests are in the areas of nonlinear dynamics, evaluation of materials for space and weapons systems, firearms design, and interior and exterior ballistics. He is an avid experimenter, handloader, and hunter of twenty-five years' experience. This book is a product of his knowledge of computers, his vast experience, and a truly fertile mind.— Dave Wolfe

ENERGY (FOOT-POUNDS)

feet per second	bullet weight (grains)									
	10	20	30	40	50	60	70	80	90	100
500	5	11	16	22	27	33	38	44	49	55
600	7	15	23	31	39	47	55	63	71	79
700	10	21	32	43	54	65	76	87	97	108
800	14	28	42	56	71	85	99	113	127	142
900	17	35	53	71	89	107	125	143	161	179
1000	22	44	66	88	111	133	155	177	199	222
1100	26	53	80	107	134	161	188	214	241	268
1200	31	63	95	127	159	191	223	255	287	319
1300	37	75	112	150	187	225	262	300	337	375
1400	43	87	130	174	217	261	304	348	391	435
1500	49	99	149	199	249	299	349	399	449	499
1600	56	113	170	227	284	341	397	454	511	568
1700	64	128	192	256	320	384	449	513	577	641
1800	71	143	215	287	359	431	503	575	647	719
1900	80	160	240	320	400	480	561	641	721	801
2000	88	177	266	355	444	532	621	710	799	888
2100	97	195	293	391	489	587	685	783	881	979
2200	107	214	322	429	537	644	752	859	967	1074
2300	117	234	352	469	587	704	822	939	1056	1174
2400	127	255	383	511	639	767	895	1023	1150	1278
2500	138	277	416	555	693	832	971	1110	1248	1387
2600	150	300	450	600	750	900	1050	1200	1350	1500
2700	161	323	485	647	809	971	1132	1294	1456	1618
2800	174	348	522	696	870	1044	1218	1392	1566	1740
2900	186	373	560	746	933	1120	1306	1493	1680	1867
3000	199	399	599	799	999	1198	1398	1598	1798	1998
3100	213	426	640	853	1066	1280	1493	1706	1920	2133
3200	227	454	682	909	1136	1364	1591	1818	2046	2273
3300	241	483	725	967	1208	1450	1692	1934	2175	2417
3400	256	513	769	1026	1283	1539	1796	2053	2309	2566
3500	271	543	815	1087	1359	1631	1903	2175	2447	2719
3600	287	575	863	1150	1438	1726	2014	2301	2589	2877
3700	303	607	911	1215	1519	1823	2127	2431	2735	3039
3800	320	641	961	1282	1602	1923	2244	2564	2885	3205
3900	337	675	1013	1350	1688	2026	2363	2701	3039	3376
4000	355	710	1065	1420	1776	2131	2486	2841	3196	3552
4100	373	746	1119	1492	1865	2239	2612	2985	3358	3731
4200	391	783	1174	1566	1958	2349	2741	3132	3524	3916
4300	410	820	1231	1641	2052	2462	2873	3283	3694	4104
4400	429	859	1289	1719	2149	2578	3008	3438	3868	4298
4500	449	899	1348	1798	2247	2697	3146	3596	4046	4495
4600	469	939	1409	1879	2348	2818	3288	3758	4227	4697
4700	490	980	1471	1961	2452	2942	3432	3923	4413	4904
4800	511	1023	1534	2046	2557	3069	3580	4092	4603	5115
4900	533	1066	1599	2132	2665	3198	3731	4264	4797	5330
5000	555	1110	1665	2220	2775	3330	3885	4440	4995	5550

ENERGY (FOOT-POUNDS)

feet per second	bullet weight (grains)									
	110	120	130	140	150	160	170	180	190	200
500	61	66	72	77	83	88	94	99	105	111
600	87	95	103	111	119	127	135	143	151	159
700	119	130	141	152	163	174	184	195	206	217
800	156	170	184	198	213	227	241	255	269	284
900	197	215	233	251	269	287	305	323	341	359
1000	244	266	288	310	333	355	377	399	421	444
1100	295	322	349	376	402	429	456	483	510	537
1200	351	383	415	447	479	511	543	575	607	639
1300	412	450	487	525	562	600	637	675	712	750
1400	478	522	565	609	652	696	739	783	826	870
1500	549	599	649	699	749	799	849	899	949	999
1600	625	682	738	795	852	909	966	1023	1079	1136
1700	705	769	834	898	962	1026	1090	1154	1219	1283
1800	791	863	935	1007	1078	1150	1222	1294	1366	1438
1900	881	961	1041	1122	1202	1282	1362	1442	1522	1602
2000	976	1065	1154	1243	1332	1420	1509	1598	1687	1776
2100	1076	1174	1272	1370	1468	1566	1664	1762	1860	1958
2200	1181	1289	1396	1504	1611	1719	1826	1934	2041	2149
2300	1291	1409	1526	1644	1761	1879	1996	2113	2231	2348
2400	1406	1534	1662	1790	1918	2046	2173	2301	2429	2557
2500	1526	1665	1803	1942	2081	2220	2358	2497	2636	2775
2600	1650	1800	1950	2101	2251	2401	2551	2701	2851	3001
2700	1780	1942	2103	2265	2427	2589	2751	2913	3075	3236
2800	1914	2088	2262	2436	2610	2784	2958	3132	3307	3481
2900	2053	2240	2427	2613	2800	2987	3174	3360	3547	3734
3000	2197	2397	2597	2797	2997	3196	3396	3596	3796	3996
3100	2346	2560	2773	2986	3200	3413	3626	3840	4053	4266
3200	2500	2728	2955	3182	3410	3637	3864	4092	4319	4546
3300	2659	2901	3142	3384	3626	3868	4110	4351	4593	4835
3400	2823	3079	3336	3592	3849	4106	4362	4619	4876	5132
3500	2991	3263	3535	3807	4079	4351	4623	4895	5167	5439
3600	3164	3452	3740	4028	4315	4603	4891	5178	5466	5754
3700	3343	3647	3951	4254	4558	4862	5166	5470	5774	6078
3800	3526	3846	4167	4488	4808	5129	5449	5770	6090	6411
3900	3714	4052	4389	4727	5065	5402	5740	6078	6415	6753
4000	3907	4262	4617	4972	5328	5683	6038	6393	6749	7104
4100	4105	4478	4851	5224	5597	5971	6344	6717	7090	7463
4200	4307	4699	5091	5482	5874	6265	6657	7049	7440	7832
4300	4515	4925	5336	5746	6157	6567	6978	7388	7799	8209
4400	4727	5157	5587	6017	6447	6876	7306	7736	8166	8596
4500	4945	5394	5844	6293	6743	7193	7642	8092	8541	8991
4600	5167	5637	6106	6576	7046	7516	7986	8455	8925	9395
4700	5394	5884	6375	6865	7356	7846	8337	8827	9317	9808
4800	5626	6138	6649	7161	7672	8184	8695	9207	9718	10230
4900	5863	6396	6929	7462	7995	8528	9061	9594	10127	10660
5000	6105	6660	7215	7770	8325	8880	9435	9990	10545	11100

ENERGY (FOOT-POUNDS)

feet per second	bullet weight (grains) 210	220	230	240	250	260	270	280	290	300
500	116	122	127	133	138	144	149	155	160	166
600	167	175	183	191	199	207	215	223	231	239
700	228	239	250	261	271	282	293	304	315	326
800	298	312	326	341	355	369	383	397	412	426
900	377	395	413	431	449	467	485	503	521	539
1000	466	488	510	532	555	577	599	621	643	666
1100	564	590	617	644	671	698	725	752	779	805
1200	671	703	735	767	799	831	863	895	927	959
1300	787	825	862	900	937	975	1013	1050	1088	1125
1400	913	957	1000	1044	1087	1131	1174	1218	1261	1305
1500	1048	1098	1148	1198	1248	1298	1348	1398	1448	1498
1600	1193	1250	1307	1364	1420	1477	1534	1591	1648	1705
1700	1347	1411	1475	1539	1604	1668	1732	1796	1860	1924
1800	1510	1582	1654	1726	1798	1870	1942	2014	2085	2157
1900	1683	1763	1843	1923	2003	2083	2163	2244	2324	2404
2000	1864	1953	2042	2131	2220	2308	2397	2486	2575	2664
2100	2056	2153	2251	2349	2447	2545	2643	2741	2839	2937
2200	2256	2363	2471	2578	2686	2793	2901	3008	3116	3223
2300	2466	2583	2701	2818	2936	3053	3170	3288	3405	3523
2400	2685	2813	2941	3069	3196	3324	3452	3580	3708	3836
2500	2913	3052	3191	3330	3468	3607	3746	3885	4023	4162
2600	3151	3301	3451	3601	3751	3901	4052	4202	4352	4502
2700	3398	3560	3722	3884	4046	4207	4369	4531	4693	4855
2800	3655	3829	4003	4177	4351	4525	4699	4873	5047	5221
2900	3920	4107	4294	4480	4667	4854	5041	5227	5414	5601
3000	4195	4395	4595	4795	4995	5194	5394	5594	5794	5994
3100	4480	4693	4907	5120	5333	5547	5760	5973	6187	6400
3200	4774	5001	5228	5456	5683	5910	6138	6365	6592	6820
3300	5077	5318	5560	5802	6044	6285	6527	6769	7011	7252
3400	5389	5646	5902	6159	6416	6672	6929	7185	7442	7699
3500	5711	5983	6255	6527	6798	7070	7342	7614	7886	8158
3600	6042	6329	6617	6905	7193	7480	7768	8056	8343	8631
3700	6382	6686	6990	7294	7598	7902	8206	8509	8813	9117
3800	6732	7052	7373	7693	8014	8335	8655	8976	9296	9617
3900	7091	7428	7766	8104	8441	8779	9117	9454	9792	10130
4000	7459	7814	8169	8525	8880	9235	9590	9945	10301	10656
4100	7837	8210	8583	8956	9329	9703	10076	10449	10822	11195
4200	8224	8615	9007	9398	9790	10182	10573	10965	11356	11748
4300	8620	9030	9441	9851	10262	10672	11083	11493	11904	12314
4400	9025	9455	9885	10315	10745	11174	11604	12034	12464	12894
4500	9440	9890	10339	10789	11239	11688	12138	12587	13037	13486
4600	9865	10334	10804	11274	11744	12213	12683	13153	13623	14093
4700	10298	10789	11279	11769	12260	12750	13241	13731	14221	14712
4800	10741	11253	11764	12276	12787	13299	13810	14322	14833	15345
4900	11193	11726	12259	12792	13325	13859	14392	14925	15458	15991
5000	11655	12210	12765	13320	13875	14430	14985	15540	16095	16650

ENERGY (FOOT-POUNDS)

feet per second	310	320	330	340	350	360	370	380	390	400
	bullet weight (grains)									
500	172	177	183	188	194	199	205	210	216	222
600	247	255	263	271	279	287	295	303	311	319
700	337	348	358	369	380	391	402	413	424	435
800	440	454	468	483	497	511	525	539	554	568
900	557	575	593	611	629	647	665	683	701	719
1000	688	710	732	754	777	799	821	843	865	888
1100	832	859	886	913	940	967	993	1020	1047	1074
1200	991	1023	1054	1086	1118	1150	1182	1214	1246	1278
1300	1163	1200	1238	1275	1313	1350	1388	1425	1463	1500
1400	1348	1392	1435	1479	1522	1566	1609	1653	1697	1740
1500	1548	1598	1648	1698	1748	1798	1848	1898	1948	1998
1600	1761	1818	1875	1932	1989	2046	2102	2159	2216	2273
1700	1988	2053	2117	2181	2245	2309	2373	2438	2502	2566
1800	2229	2301	2373	2445	2517	2589	2661	2733	2805	2877
1900	2484	2564	2644	2724	2805	2885	2965	3045	3125	3205
2000	2752	2841	2930	3019	3108	3196	3285	3374	3463	3552
2100	3035	3132	3230	3328	3426	3524	3622	3720	3818	3916
2200	3330	3438	3545	3653	3760	3868	3975	4083	4190	4298
2300	3640	3758	3875	3993	4110	4227	4345	4462	4580	4697
2400	3964	4092	4219	4347	4475	4603	4731	4859	4987	5115
2500	4301	4440	4578	4717	4856	4995	5133	5272	5411	5550
2600	4652	4802	4952	5102	5252	5402	5552	5702	5852	6003
2700	5017	5178	5340	5502	5664	5826	5988	6150	6311	6473
2800	5395	5569	5743	5917	6091	6265	6439	6614	6788	6962
2900	5787	5974	6161	6348	6534	6721	6908	7094	7281	7468
3000	6193	6393	6593	6793	6993	7193	7392	7592	7792	7992
3100	6613	6827	7040	7253	7467	7680	7893	8107	8320	8533
3200	7047	7274	7502	7729	7956	8184	8411	8638	8866	9093
3300	7494	7736	7978	8220	8461	8703	8945	9187	9428	9670
3400	7955	8212	8469	8725	8982	9239	9495	9752	10008	10265
3500	8430	8702	8974	9246	9518	9790	10062	10334	10606	10878
3600	8919	9207	9494	9782	10070	10357	10645	10933	11221	11508
3700	9421	9725	10029	10333	10637	10941	11245	11549	11853	12157
3800	9937	10258	10579	10899	11220	11540	11861	12181	12502	12823
3900	10467	10805	11143	11480	11818	12156	12493	12831	13169	13506
4000	11011	11366	11721	12077	12432	12787	13142	13498	13853	14208
4100	11569	11942	12315	12688	13061	13434	13808	14181	14554	14927
4200	12140	12531	12923	13315	13706	14098	14489	14881	15273	15664
4300	12725	13135	13546	13956	14367	14777	15188	15598	16009	16419
4400	13323	13753	14183	14613	15043	15473	15902	16332	16762	17192
4500	13936	14386	14835	15285	15734	16184	16633	17083	17533	17982
4600	14562	15032	15502	15972	16441	16911	17381	17851	18320	18790
4700	15202	15693	16183	16674	17164	17654	18145	18635	19126	19616
4800	15856	16368	16879	17391	17902	18414	18925	19437	19948	20460
4900	16524	17057	17590	18123	18656	19189	19722	20255	20788	21321
5000	17205	17760	18315	18870	19425	19980	20535	21090	21645	22200

ENERGY (FOOT-POUNDS)

feet per second	410	420	430	440	450	460	470	480	490	500
	bullet weight (grains)									
500	227	233	238	244	249	255	260	266	271	277
600	327	335	343	351	359	367	375	383	391	399
700	446	456	467	478	489	500	511	522	533	543
800	582	596	610	625	639	653	667	682	696	710
900	737	755	773	791	809	827	845	863	881	899
1000	910	932	954	976	999	1021	1043	1065	1087	1110
1100	1101	1128	1155	1181	1208	1235	1262	1289	1316	1343
1200	1310	1342	1374	1406	1438	1470	1502	1534	1566	1598
1300	1538	1575	1613	1650	1688	1725	1763	1800	1838	1875
1400	1784	1827	1871	1914	1958	2001	2045	2088	2132	2175
1500	2048	2097	2147	2197	2247	2297	2347	2397	2447	2497
1600	2330	2387	2443	2500	2557	2614	2671	2728	2784	2841
1700	2630	2694	2758	2823	2887	2951	3015	3079	3143	3208
1800	2949	3021	3093	3164	3236	3308	3380	3452	3524	3596
1900	3285	3366	3446	3526	3606	3686	3766	3846	3927	4007
2000	3640	3729	3818	3907	3996	4084	4173	4262	4351	4440
2100	4014	4112	4209	4307	4405	4503	4601	4699	4797	4895
2200	4405	4512	4620	4727	4835	4942	5050	5157	5265	5372
2300	4815	4932	5049	5167	5284	5402	5519	5637	5754	5872
2400	5242	5370	5498	5626	5754	5882	6010	6138	6265	6393
2500	5688	5827	5966	6105	6243	6382	6521	6660	6798	6937
2600	6153	6303	6453	6603	6753	6903	7053	7203	7353	7503
2700	6635	6797	6959	7121	7282	7444	7606	7768	7930	8092
2800	7136	7310	7484	7658	7832	8006	8180	8354	8528	8702
2900	7655	7841	8028	8215	8401	8588	8775	8961	9148	9335
3000	8192	8391	8591	8791	8991	9191	9390	9590	9790	9990
3100	8747	8960	9174	9387	9600	9814	10027	10240	10454	10667
3200	9320	9548	9775	10002	10230	10457	10684	10912	11139	11366
3300	9912	10154	10395	10637	10879	11121	11362	11604	11846	12088
3400	10522	10778	11035	11292	11548	11805	12062	12318	12575	12832
3500	11150	11422	11694	11966	12238	12510	12782	13054	13325	13597
3600	11796	12084	12372	12659	12947	13235	13522	13810	14098	14386
3700	12461	12764	13068	13372	13676	13980	14284	14588	14892	15196
3800	13143	13464	13784	14105	14426	14746	15067	15387	15708	16028
3900	13844	14182	14519	14857	15195	15532	15870	16208	16545	16883
4000	14563	14918	15274	15629	15984	16339	16694	17050	17405	17760
4100	15300	15674	16047	16420	16793	17166	17540	17913	18286	18659
4200	16056	16448	16839	17231	17622	18014	18406	18797	19189	19581
4300	16830	17240	17651	18061	18472	18882	19293	19703	20114	20524
4400	17622	18051	18481	18911	19341	19771	20200	20630	21060	21490
4500	18432	18881	19331	19780	20230	20679	21129	21579	22028	22478
4600	19260	19730	20199	20669	21139	21609	22079	22548	23018	23488
4700	20106	20597	21087	21578	22068	22559	23049	23539	24030	24520
4800	20971	21483	21994	22506	23017	23529	24040	24552	25063	25575
4900	21854	22387	22920	23453	23986	24519	25052	25585	26118	26651
5000	22755	23310	23865	24420	24975	25530	26085	26640	27195	27750

BALLISTIC COEFFICIENT .09, MUZZLE VELOCITY FT/SEC 800.

RANGE	VELOCITY	HEIGHT	50 YD	100 YD	150 YD	200 YD	250 YD	DRIFT	TIME
0.	800.	0.0	-1.5	-1.5	-1.5	-1.5	-1.5	0.0	0.0000
50.	733.	1.1	0.0	7.4	16.3	26.7	38.5	1.5	.1960
100.	670.	7.3	-14.8	0.0	17.9	38.6	62.2	6.2	.4101
150.	610.	19.2	-49.0	-26.8	0.0	31.0	66.5	14.5	.6448
200.	554.	38.4	-106.6	-77.1	-41.3	0.0	47.3	26.9	.9029
250.	500.	67.0	-192.5	-155.5	-110.8	-59.2	0.0	44.0	1.1877
300.	450.	107.9	-313.3	-269.0	-215.3	-153.3	-82.4	66.8	1.5043
350.	403.	164.6	-476.5	-424.8	-362.2	-289.9	-207.1	95.7	1.8561
400.	360.	242.2	-692.9	-633.8	-562.2	-479.6	-384.9	131.9	2.2496
450.	319.	347.4	-977.5	-911.1	-830.5	-737.6	-631.1	177.0	2.6933
500.	281.	489.2	-1348.8	-1274.9	-1185.4	-1082.2	-963.8	232.3	3.1948

BALLISTIC COEFFICIENT .09, MUZZLE VELOCITY FT/SEC 900.

RANGE	VELOCITY	HEIGHT	50 YD	100 YD	150 YD	200 YD	250 YD	DRIFT	TIME
0.	900.	0.0	-1.5	-1.5	-1.5	-1.5	-1.5	0.0	0.0000
50.	826.	.7	0.0	5.7	12.6	20.6	29.8	1.3	.1743
100.	758.	5.6	-11.3	0.0	13.9	29.9	48.3	5.4	.3640
150.	693.	14.9	-37.8	-20.9	0.0	24.0	51.5	12.5	.5711
200.	632.	29.8	-82.5	-59.9	-32.0	0.0	36.6	23.1	.7979
250.	574.	51.9	-148.9	-120.6	-85.8	-45.8	0.0	37.6	1.0470
300.	520.	83.0	-241.4	-207.5	-165.7	-117.6	-62.7	56.5	1.3213
350.	469.	126.1	-366.3	-326.7	-278.0	-221.9	-157.9	80.8	1.6259
400.	420.	184.3	-530.1	-484.9	-429.2	-365.1	-291.9	110.9	1.9634
450.	375.	262.5	-744.1	-693.3	-630.6	-558.5	-476.1	148.2	2.3418
500.	333.	366.6	-1020.5	-964.0	-894.3	-814.2	-722.7	193.6	2.7664

BALLISTIC COEFFICIENT .09, MUZZLE VELOCITY FT/SEC 1000.

RANGE	VELOCITY	HEIGHT	50 YD	100 YD	150 YD	200 YD	250 YD	DRIFT	TIME
0.	1000.	0.0	-1.5	-1.5	-1.5	-1.5	-1.5	0.0	0.0000
50.	909.	.4	0.0	4.5	10.2	16.7	24.2	1.3	.1575
100.	833.	4.5	-9.1	0.0	11.4	24.4	39.3	5.3	.3301
150.	765.	12.1	-30.6	-17.0	0.0	19.6	41.9	12.0	.5179
200.	700.	24.3	-67.0	-48.8	-26.1	0.0	29.8	21.6	.7228
250.	639.	42.3	-120.9	-98.3	-69.9	-37.2	0.0	34.7	.9472
300.	581.	67.6	-196.0	-168.8	-134.8	-95.6	-50.9	51.7	1.1935
350.	526.	102.3	-296.8	-265.1	-225.4	-179.6	-127.5	73.1	1.4654
400.	474.	149.0	-428.8	-392.5	-347.1	-294.9	-235.3	99.6	1.7661
450.	426.	211.0	-599.3	-558.6	-507.5	-448.7	-381.7	132.0	2.1002
500.	380.	292.7	-817.6	-772.3	-715.5	-650.2	-575.8	171.2	2.4725

BALLISTIC COEFFICIENT .09, MUZZLE VELOCITY FT/SEC 1100.

RANGE	VELOCITY	HEIGHT	50 YD	100 YD	150 YD	200 YD	250 YD	DRIFT	TIME
0.	1100.	0.0	-1.5	-1.5	-1.5	-1.5	-1.5	0.0	0.0000
50.	986.	.3	0.0	3.7	8.5	14.1	20.5	1.5	.1449
100.	898.	3.7	-7.4	0.0	9.6	20.8	33.5	5.5	.3042
150.	824.	10.2	-25.6	-14.5	0.0	16.7	35.8	12.2	.4786
200.	756.	20.7	-56.5	-41.6	-22.3	0.0	25.4	21.7	.6687
250.	692.	36.1	-102.3	-83.8	-59.7	-31.8	0.0	34.2	.8764
300.	631.	57.7	-166.0	-143.8	-114.9	-81.4	-43.3	50.2	1.1036
350.	573.	87.2	-251.3	-225.3	-191.6	-152.5	-108.0	70.2	1.3533
400.	518.	126.5	-362.5	-332.8	-294.3	-249.6	-198.8	94.6	1.6284
450.	467.	178.7	-506.1	-472.7	-429.3	-379.1	-321.9	124.3	1.9336
500.	419.	247.0	-688.8	-651.8	-603.5	-547.7	-484.2	159.9	2.2721

BALLISTIC COEFFICIENT .09, MUZZLE VELOCITY FT/SEC 1200.

RANGE	VELOCITY	HEIGHT	50 YD	100 YD	150 YD	200 YD	250 YD	DRIFT	TIME
0.	1200.	0.0	-1.5	-1.5	-1.5	-1.5	-1.5	0.0	0.0000
50.	1042.	.1	0.0	3.2	7.5	12.5	18.2	1.8	.1354
100.	944.	3.2	-6.5	0.0	8.6	18.6	30.0	6.5	.2870
150.	863.	9.1	-22.5	-12.8	0.0	15.0	32.1	13.8	.4531
200.	792.	18.6	-50.1	-37.2	-20.1	0.0	22.8	23.7	.6347
250.	726.	32.5	-91.1	-75.0	-53.6	-28.5	0.0	36.5	.8325
300.	663.	52.0	-148.1	-128.7	-103.1	-73.0	-38.8	52.6	1.0488
350.	604.	78.6	-224.4	-201.7	-171.8	-136.7	-96.8	72.3	1.2859
400.	548.	114.1	-324.0	-298.1	-263.9	-223.7	-178.2	96.3	1.5471
450.	495.	161.0	-451.9	-422.8	-384.3	-339.2	-287.9	125.0	1.8354
500.	445.	222.2	-615.0	-582.7	-539.9	-489.7	-432.8	159.3	2.1553

BALLISTIC COEFFICIENT .09, MUZZLE VELOCITY FT/SEC 1300.

RANGE	VELOCITY	HEIGHT	50 YD	100 YD	150 YD	200 YD	250 YD	DRIFT	TIME
0.	1300.	0.0	-1.5	-1.5	-1.5	-1.5	-1.5	0.0	0.0000
50.	1092.	.0	0.0	2.9	6.8	11.3	16.6	2.0	.1270
100.	981.	2.8	-5.7	0.0	7.9	17.0	27.5	7.3	.2724
150.	893.	8.3	-20.3	-11.8	0.0	13.7	29.4	15.4	.4334
200.	820.	17.0	-45.3	-33.9	-18.2	0.0	21.0	25.8	.6080
250.	752.	29.9	-82.9	-68.7	-49.0	-26.2	0.0	39.1	.7992
300.	688.	48.0	-135.2	-118.0	-94.5	-67.1	-35.7	55.5	1.0078
350.	627.	72.6	-205.0	-185.0	-157.6	-125.6	-88.9	75.4	1.2362
400.	570.	105.4	-296.2	-273.4	-242.0	-205.5	-163.5	99.3	1.4873
450.	516.	148.6	-413.3	-387.6	-352.3	-311.3	-264.1	127.7	1.7641
500.	465.	205.1	-562.5	-534.0	-494.7	-449.1	-396.7	161.4	2.0707

BALLISTIC COEFFICIENT .09, MUZZLE VELOCITY FT/SEC 1400.

RANGE	VELOCITY	HEIGHT	50 YD	100 YD	150 YD	200 YD	250 YD	DRIFT	TIME
0.	1400.	0.0	-1.5	-1.5	-1.5	-1.5	-1.5	0.0	0.0000
50.	1148.	-.1	0.0	2.5	6.1	10.3	15.2	2.1	.1191
100.	1016.	2.5	-5.0	0.0	7.1	15.6	25.4	7.8	.2589
150.	922.	7.5	-18.3	-10.7	0.0	12.7	27.3	16.3	.4140
200.	844.	15.6	-41.3	-31.3	-17.0	0.0	19.5	27.4	.5843
250.	775.	27.7	-76.0	-63.4	-45.6	-24.3	0.0	41.2	.7697
300.	710.	44.6	-124.3	-109.2	-87.8	-62.3	-33.1	57.9	.9721
350.	648.	67.6	-189.0	-171.3	-146.4	-116.7	-82.6	78.0	1.1933
400.	589.	98.3	-273.4	-253.3	-224.7	-190.8	-151.8	101.9	1.4362
450.	534.	138.6	-381.9	-359.2	-327.1	-288.9	-245.1	130.1	1.7037
500.	482.	191.1	-519.7	-494.5	-458.9	-416.4	-367.8	163.3	1.9993

BALLISTIC COEFFICIENT .09, MUZZLE VELOCITY FT/SEC 1500.

RANGE	VELOCITY	HEIGHT	50 YD	100 YD	150 YD	200 YD	250 YD	DRIFT	TIME
0.	1500.	0.0	-1.5	-1.5	-1.5	-1.5	-1.5	0.0	0.0000
50.	1212.	-.1	0.0	2.2	5.5	9.5	14.0	2.1	.1119
100.	1047.	2.2	-4.5	0.0	6.5	14.4	23.5	8.3	.2469
150.	948.	6.8	-16.5	-9.8	0.0	11.8	25.5	17.1	.3970
200.	866.	14.4	-37.8	-28.9	-15.8	0.0	18.2	28.6	.5627
250.	796.	25.8	-70.0	-58.8	-42.5	-22.8	0.0	42.8	.7434
300.	729.	41.7	-115.0	-101.6	-82.0	-58.3	-31.0	59.9	.9404
350.	666.	63.4	-175.4	-159.7	-136.8	-109.2	-77.4	80.2	1.1557
400.	607.	92.2	-254.2	-236.3	-210.1	-178.6	-142.2	104.1	1.3918
450.	550.	130.2	-355.5	-335.4	-306.0	-270.4	-229.5	132.3	1.6516
500.	497.	179.6	-484.3	-461.9	-429.3	-389.8	-344.3	165.1	1.9383

BALLISTIC COEFFICIENT .09, MUZZLE VELOCITY FT/SEC 1600.

RANGE	VELOCITY	HEIGHT	50 YD	100 YD	150 YD	200 YD	250 YD	DRIFT	TIME
0.	1600.	0.0	-1.5	-1.5	-1.5	-1.5	-1.5	0.0	0.0000
50.	1281.	-.2	0.0	1.9	5.0	8.6	12.9	2.0	.1052
100.	1082.	1.9	-3.8	0.0	6.1	13.4	22.0	8.2	.2339
150.	974.	6.2	-14.9	-9.1	0.0	11.0	23.9	17.4	.3802
200.	888.	13.3	-34.5	-26.8	-14.7	0.0	17.1	29.3	.5417
250.	815.	24.0	-64.5	-54.9	-39.8	-21.4	0.0	43.9	.7182
300.	748.	39.0	-106.6	-95.1	-76.9	-54.8	-29.2	61.2	.9104
350.	684.	59.5	-163.1	-149.6	-128.4	-102.6	-72.7	81.7	1.1202
400.	623.	86.7	-236.9	-221.5	-197.3	-167.8	-133.7	105.6	1.3501
450.	566.	122.6	-331.9	-314.6	-287.3	-254.2	-215.8	133.6	1.6030
500.	512.	169.3	-452.9	-433.7	-403.3	-366.5	-323.8	166.2	1.8820

BALLISTIC COEFFICIENT .09, MUZZLE VELOCITY FT/SEC 1700.

RANGE	VELOCITY	HEIGHT	50 YD	100 YD	150 YD	200 YD	250 YD	DRIFT	TIME
0.	1700.	0.0	-1.5	-1.5	-1.5	-1.5	-1.5	0.0	0.0000
50.	1355.	-.3	0.0	1.6	4.4	7.9	11.9	1.9	.0992
100.	1122.	1.6	-3.3	0.0	5.6	12.5	20.5	8.0	.2217
150.	1000.	5.6	-13.3	-8.4	0.0	10.3	22.4	17.5	.3639
200.	910.	12.3	-31.5	-24.9	-13.8	0.0	16.1	29.6	.5214
250.	834.	22.4	-59.4	-51.3	-37.3	-20.1	0.0	44.5	.6939
300.	765.	36.6	-98.9	-89.0	-72.3	-51.6	-27.5	62.0	.8817
350.	700.	55.9	-151.8	-140.4	-120.8	-96.7	-68.6	82.5	1.0865
400.	639.	81.7	-221.2	-208.1	-185.8	-158.2	-126.1	106.5	1.3108
450.	581.	115.6	-310.5	-295.8	-270.6	-239.6	-203.5	134.3	1.5571
500.	526.	159.8	-424.3	-408.0	-380.0	-345.6	-305.4	166.6	1.8289

BALLISTIC COEFFICIENT .09, MUZZLE VELOCITY FT/SEC 1800.

RANGE	VELOCITY	HEIGHT	50 YD	100 YD	150 YD	200 YD	250 YD	DRIFT	TIME
0.	1800.	0.0	-1.5	-1.5	-1.5	-1.5	-1.5	0.0	0.0000
50.	1434.	-.3	0.0	1.4	4.0	7.2	11.0	1.8	.0935
100.	1168.	1.4	-2.8	0.0	5.1	11.6	19.2	7.6	.2101
150.	1026.	5.1	-11.9	-7.7	0.0	9.6	21.1	17.3	.3484
200.	931.	11.3	-28.7	-23.1	-12.8	0.0	15.3	29.6	.5017
250.	852.	20.9	-54.9	-48.0	-35.1	-19.1	0.0	44.8	.6713
300.	782.	34.3	-91.7	-83.4	-67.9	-48.7	-25.8	62.3	.8540
350.	717.	52.6	-141.6	-131.9	-113.9	-91.4	-64.7	82.9	1.0544
400.	654.	77.1	-207.0	-195.9	-175.3	-149.7	-119.1	106.8	1.2737
450.	595.	109.3	-291.2	-278.7	-255.5	-226.7	-192.3	134.5	1.5141
500.	540.	151.1	-398.3	-384.4	-358.7	-326.7	-288.5	166.4	1.7786

BALLISTIC COEFFICIENT .09, MUZZLE VELOCITY FT/SEC 1900.

RANGE	VELOCITY	HEIGHT	50 YD	100 YD	150 YD	200 YD	250 YD	DRIFT	TIME
0.	1900.	0.0	-1.5	-1.5	-1.5	-1.5	-1.5	0.0	0.0000
50.	1517.	-.4	0.0	1.2	3.5	6.5	10.1	1.7	.0885
100.	1223.	1.2	-2.3	0.0	4.6	10.7	17.8	7.3	.1992
150.	1053.	4.6	-10.5	-7.0	0.0	9.0	19.7	16.8	.3326
200.	953.	10.4	-26.0	-21.4	-12.1	0.0	14.3	29.4	.4826
250.	870.	19.4	-50.4	-44.5	-32.9	-17.8	0.0	44.5	.6476
300.	799.	32.1	-85.1	-78.1	-64.1	-46.0	-24.7	62.3	.8275
350.	732.	49.6	-132.2	-123.9	-107.7	-86.6	-61.6	82.9	1.0237
400.	669.	72.8	-193.9	-184.5	-165.9	-141.8	-113.3	106.8	1.2381
450.	609.	103.4	-273.5	-262.9	-242.0	-214.9	-182.8	134.2	1.4731
500.	553.	143.2	-374.9	-363.2	-339.9	-309.8	-274.1	165.8	1.7317

BALLISTIC COEFFICIENT .10, MUZZLE VELOCITY FT/SEC 800.

RANGE	VELOCITY	HEIGHT	50 YD	100 YD	150 YD	200 YD	250 YD	DRIFT	TIME
0.	800.	0.0	-1.5	-1.5	-1.5	-1.5	-1.5	0.0	0.0000
50.	740.	1.1	0.0	7.2	15.9	25.8	36.9	1.3	.1950
100.	682.	7.2	-14.5	0.0	17.4	37.1	59.4	5.5	.4062
150.	628.	18.6	-47.7	-26.0	0.0	29.6	63.1	12.9	.6356
200.	576.	36.9	-103.1	-74.1	-39.4	0.0	44.7	23.8	.8851
250.	527.	63.6	-184.7	-148.5	-105.1	-55.8	0.0	38.7	1.1577
300.	480.	100.9	-297.2	-253.8	-201.8	-142.6	-75.6	58.2	1.4555
350.	436.	152.0	-447.9	-397.3	-336.5	-267.5	-189.3	83.0	1.7841
400.	394.	220.4	-644.1	-586.2	-516.8	-438.0	-348.6	113.8	2.1463
450.	355.	310.7	-896.1	-831.0	-752.9	-664.2	-563.6	151.3	2.5472
500.	319.	429.1	-1216.8	-1144.4	-1057.6	-959.0	-847.3	196.7	2.9925

BALLISTIC COEFFICIENT .10, MUZZLE VELOCITY FT/SEC 900.

RANGE	VELOCITY	HEIGHT	50 YD	100 YD	150 YD	200 YD	250 YD	DRIFT	TIME
0.	900.	0.0	-1.5	-1.5	-1.5	-1.5	-1.5	0.0	0.0000
50.	833.	.7	0.0	5.5	12.3	20.0	28.6	1.2	.1736
100.	771.	5.5	-11.1	0.0	13.5	28.8	46.1	4.9	.3609
150.	712.	14.5	-36.9	-20.2	0.0	23.0	48.9	11.2	.5634
200.	656.	28.7	-79.8	-57.6	-30.6	0.0	34.5	20.5	.7829
250.	603.	49.3	-142.9	-115.2	-81.5	-43.1	0.0	33.1	1.0214
300.	552.	78.1	-230.0	-196.7	-156.2	-110.3	-58.5	49.6	1.2818
350.	504.	117.0	-345.3	-306.5	-259.3	-205.6	-145.2	70.3	1.5661
400.	459.	168.5	-494.5	-450.1	-396.2	-334.9	-265.9	95.8	1.8779
450.	416.	236.2	-685.4	-635.5	-574.8	-505.8	-428.2	127.1	2.2220
500.	375.	324.2	-926.6	-871.2	-803.7	-727.1	-640.8	164.6	2.6020

BALLISTIC COEFFICIENT .10, MUZZLE VELOCITY FT/SEC 1000.

RANGE	VELOCITY	HEIGHT	50 YD	100 YD	150 YD	200 YD	250 YD	DRIFT	TIME
0.	1000.	0.0	-1.5	-1.5	-1.5	-1.5	-1.5	0.0	0.0000
50.	918.	.4	0.0	4.4	9.9	16.2	23.3	1.2	.1567
100.	848.	4.4	-8.9	0.0	11.0	23.6	37.7	4.8	.3271
150.	785.	11.8	-29.8	-16.5	0.0	18.8	40.0	10.7	.5110
200.	726.	23.4	-64.9	-47.1	-25.1	0.0	28.2	19.3	.7098
250.	669.	40.3	-116.3	-94.1	-66.6	-35.2	0.0	30.8	.9252
300.	615.	63.7	-187.0	-160.4	-127.3	-89.7	-47.4	45.6	1.1590
350.	564.	95.2	-280.4	-249.4	-210.8	-166.9	-117.6	64.0	1.4138
400.	515.	136.7	-401.0	-365.5	-321.5	-271.3	-214.9	86.7	1.6923
450.	469.	190.8	-554.3	-514.3	-464.7	-408.3	-344.9	114.0	1.9978
500.	426.	260.6	-746.8	-702.4	-647.3	-584.6	-514.1	146.7	2.3335

BALLISTIC COEFFICIENT .10, MUZZLE VELOCITY FT/SEC 1100.

RANGE	VELOCITY	HEIGHT	50 YD	100 YD	150 YD	200 YD	250 YD	DRIFT	TIME
0.	1100.	0.0	-1.5	-1.5	-1.5	-1.5	-1.5	0.0	0.0000
50.	996.	.2	0.0	3.6	8.3	13.7	19.7	1.3	.1438
100.	914.	3.6	-7.3	0.0	9.4	20.1	32.1	5.0	.3010
150.	845.	9.9	-24.9	-14.0	0.0	16.1	34.1	11.1	.4719
200.	783.	19.9	-54.7	-40.1	-21.4	0.0	24.0	19.5	.6563
250.	723.	34.4	-98.4	-80.2	-56.8	-30.1	0.0	30.6	.8558
300.	667.	54.4	-158.5	-136.7	-108.6	-76.5	-40.4	44.7	1.0719
350.	613.	81.2	-237.8	-212.3	-179.6	-142.1	-100.0	62.0	1.3067
400.	562.	116.4	-339.7	-310.6	-273.2	-230.3	-182.3	82.9	1.5622
450.	513.	162.1	-469.1	-436.4	-394.2	-346.1	-292.0	108.2	1.8420
500.	467.	220.8	-631.0	-594.7	-547.9	-494.3	-434.2	138.1	2.1485

BALLISTIC COEFFICIENT .10, MUZZLE VELOCITY FT/SEC 1200.

RANGE	VELOCITY	HEIGHT	50 YD	100 YD	150 YD	200 YD	250 YD	DRIFT	TIME
0.	1200.	0.0	-1.5	-1.5	-1.5	-1.5	-1.5	0.0	0.0000
50.	1054.	.1	0.0	3.1	7.3	12.1	17.5	1.6	.1343
100.	962.	3.1	-6.3	0.0	8.3	17.9	28.7	5.9	.2837
150.	885.	8.8	-21.9	-12.4	0.0	14.4	30.6	12.5	.4463
200.	820.	17.8	-48.4	-35.8	-19.2	0.0	21.6	21.5	.6223
250.	759.	30.9	-87.5	-71.7	-51.0	-27.0	0.0	33.0	.8126
300.	700.	49.0	-141.2	-122.3	-97.4	-68.7	-36.2	47.2	1.0184
350.	645.	73.3	-212.2	-190.1	-161.1	-127.5	-89.7	64.6	1.2418
400.	592.	105.1	-303.5	-278.3	-245.1	-206.8	-163.6	85.3	1.4847
450.	542.	146.1	-418.8	-390.5	-353.2	-310.0	-261.4	109.9	1.7493
500.	495.	198.9	-563.4	-531.9	-490.5	-442.5	-388.5	138.9	2.0394

BALLISTIC COEFFICIENT .10, MUZZLE VELOCITY FT/SEC 1300.

RANGE	VELOCITY	HEIGHT	50 YD	100 YD	150 YD	200 YD	250 YD	DRIFT	TIME
0.	1300.	0.0	-1.5	-1.5	-1.5	-1.5	-1.5	0.0	0.0000
50.	1107.	.0	0.0	2.8	6.5	10.9	15.9	1.9	.1260
100.	1001.	2.7	-5.5	0.0	7.5	16.3	26.2	6.7	.2690
150.	918.	7.9	-19.6	-11.3	0.0	13.2	28.1	14.0	.4255
200.	848.	16.3	-43.7	-32.6	-17.6	0.0	19.9	23.6	.5958
250.	786.	28.4	-79.4	-65.6	-46.8	-24.8	0.0	35.7	.7796
300.	726.	45.2	-128.6	-112.0	-89.5	-63.1	-33.3	50.3	.9782
350.	669.	67.6	-193.6	-174.3	-148.0	-117.2	-82.5	67.9	1.1935
400.	615.	97.0	-277.2	-255.1	-225.0	-189.8	-150.1	88.7	1.4271
450.	564.	135.0	-382.9	-358.0	-324.2	-284.6	-240.0	113.2	1.6818
500.	516.	183.7	-515.2	-487.5	-450.0	-406.0	-356.4	141.9	1.9601

BALLISTIC COEFFICIENT .10, MUZZLE VELOCITY FT/SEC 1400.

RANGE	VELOCITY	HEIGHT	50 YD	100 YD	150 YD	200 YD	250 YD	DRIFT	TIME
0.	1400.	0.0	-1.5	-1.5	-1.5	-1.5	-1.5	0.0	0.0000
50.	1167.	-.1	0.0	2.4	5.9	9.9	14.5	1.9	.1179
100.	1037.	2.4	-4.8	0.0	6.9	15.0	24.2	7.2	.2551
150.	948.	7.2	-17.6	-10.3	0.0	12.2	26.0	15.0	.4066
200.	874.	15.0	-39.8	-30.1	-16.3	0.0	18.3	25.3	.5723
250.	809.	26.3	-72.6	-60.5	-43.3	-22.9	0.0	37.7	.7501
300.	749.	41.9	-118.1	-103.5	-82.9	-58.4	-31.0	52.8	.9428
350.	691.	62.9	-178.2	-161.2	-137.2	-108.6	-76.6	70.7	1.1515
400.	636.	90.4	-255.6	-236.2	-208.6	-176.0	-139.4	91.6	1.3777
450.	584.	125.9	-353.5	-331.6	-300.7	-263.9	-222.8	116.1	1.6241
500.	534.	171.3	-476.0	-451.7	-417.3	-376.5	-330.8	144.6	1.8930

BALLISTIC COEFFICIENT .10, MUZZLE VELOCITY FT/SEC 1500.

RANGE	VELOCITY	HEIGHT	50 YD	100 YD	150 YD	200 YD	250 YD	DRIFT	TIME
0.	1500.	0.0	-1.5	-1.5	-1.5	-1.5	-1.5	0.0	0.0000
50.	1235.	-.2	0.0	2.1	5.3	9.0	13.3	1.9	.1109
100.	1072.	2.1	-4.2	0.0	6.3	13.9	22.4	7.4	.2421
150.	976.	6.5	-15.8	-9.5	0.0	11.3	24.2	15.7	.3891
200.	897.	13.8	-36.1	-27.7	-15.1	0.0	17.2	26.4	.5497
250.	830.	24.4	-66.6	-56.1	-40.3	-21.5	0.0	39.3	.7235
300.	769.	39.1	-109.0	-96.3	-77.4	-54.8	-29.0	54.8	.9113
350.	710.	58.9	-165.0	-150.3	-128.2	-101.8	-71.7	72.9	1.1145
400.	654.	84.8	-237.2	-220.4	-195.1	-165.0	-130.6	94.1	1.3346
450.	601.	118.1	-328.5	-309.6	-281.1	-247.2	-208.6	118.6	1.5737
500.	550.	160.9	-443.0	-422.0	-390.3	-352.7	-309.8	147.0	1.8351

BALLISTIC COEFFICIENT .10, MUZZLE VELOCITY FT/SEC 1600.

RANGE	VELOCITY	HEIGHT	50 YD	100 YD	150 YD	200 YD	250 YD	DRIFT	TIME
0.	1600.	0.0	-1.5	-1.5	-1.5	-1.5	-1.5	0.0	0.0000
50.	1308.	-.2	0.0	1.8	4.7	8.2	12.2	1.8	.1040
100.	1111.	1.8	-3.6	0.0	5.8	12.8	20.9	7.4	.2293
150.	1004.	5.9	-14.2	-8.7	0.0	10.4	22.5	16.0	.3721
200.	921.	12.6	-32.8	-25.6	-13.9	0.0	16.1	26.9	.5280
250.	850.	22.6	-61.2	-52.1	-37.6	-20.2	0.0	40.3	.6979
300.	788.	36.5	-100.7	-89.8	-72.4	-51.5	-27.3	56.1	.8812
350.	728.	55.2	-153.1	-140.5	-120.1	-95.7	-67.5	74.5	1.0794
400.	671.	79.6	-220.8	-206.3	-183.0	-155.2	-122.9	95.8	1.2941
450.	617.	111.2	-306.4	-290.1	-263.9	-232.5	-196.2	120.3	1.5270
500.	566.	151.5	-413.6	-395.5	-366.4	-331.6	-291.3	148.5	1.7811

BALLISTIC COEFFICIENT .10, MUZZLE VELOCITY FT/SEC 1700.

RANGE	VELOCITY	HEIGHT	50 YD	100 YD	150 YD	200 YD	250 YD	DRIFT	TIME
0.	1700.	0.0	-1.5	-1.5	-1.5	-1.5	-1.5	0.0	0.0000
50.	1385.	-.3	0.0	1.5	4.2	7.5	11.2	1.7	.0980
100.	1157.	1.5	-3.1	0.0	5.3	11.8	19.4	7.2	.2172
150.	1031.	5.3	-12.6	-8.0	0.0	9.7	21.0	16.0	.3558
200.	944.	11.6	-29.8	-23.7	-13.0	0.0	15.1	27.2	.5077
250.	870.	21.0	-56.1	-48.4	-35.1	-18.8	0.0	40.8	.6731
300.	806.	34.1	-93.1	-83.8	-67.8	-48.3	-25.7	56.8	.8523
350.	745.	51.8	-142.2	-131.4	-112.7	-90.0	-63.6	75.4	1.0459
400.	688.	74.9	-205.7	-193.4	-172.0	-146.1	-115.9	96.7	1.2555
450.	633.	104.8	-286.3	-272.4	-248.3	-219.1	-185.2	121.2	1.4829
500.	581.	142.9	-386.9	-371.5	-344.8	-312.3	-274.7	149.2	1.7301

BALLISTIC COEFFICIENT .10, MUZZLE VELOCITY FT/SEC 1800.

RANGE	VELOCITY	HEIGHT	50 YD	100 YD	150 YD	200 YD	250 YD	DRIFT	TIME
0.	1800.	0.0	-1.5	-1.5	-1.5	-1.5	-1.5	0.0	0.0000
50.	1467.	-.3	0.0	1.3	3.7	6.8	10.3	1.6	.0924
100.	1212.	1.3	-2.6	0.0	4.8	10.9	18.0	6.8	.2055
150.	1060.	4.8	-11.1	-7.2	0.0	9.2	19.8	15.6	.3386
200.	967.	10.7	-27.0	-21.8	-12.3	0.0	14.1	27.2	.4876
250.	890.	19.5	-51.4	-44.9	-33.0	-17.6	0.0	40.9	.6490
300.	824.	31.9	-86.0	-78.2	-63.8	-45.4	-24.3	57.1	.8243
350.	762.	48.6	-132.1	-123.0	-106.3	-84.8	-60.1	75.7	1.0137
400.	704.	70.5	-191.9	-181.5	-162.4	-137.9	-109.6	97.1	1.2186
450.	648.	98.9	-267.8	-256.1	-234.6	-207.0	-175.3	121.6	1.4408
500.	595.	135.1	-362.8	-349.8	-325.9	-295.3	-260.0	149.4	1.6823

BALLISTIC COEFFICIENT .10, MUZZLE VELOCITY FT/SEC 1900.

RANGE	VELOCITY	HEIGHT	50 YD	100 YD	150 YD	200 YD	250 YD	DRIFT	TIME
0.	1900.	0.0	-1.5	-1.5	-1.5	-1.5	-1.5	0.0	0.0000
50.	1552.	-.4	0.0	1.1	3.2	6.1	9.4	1.5	.0875
100.	1273.	1.1	-2.2	0.0	4.3	10.0	16.7	6.5	.1946
150.	1092.	4.3	-9.7	-6.5	0.0	8.6	18.5	15.1	.3229
200.	990.	9.8	-24.4	-20.1	-11.5	0.0	13.2	26.8	.4683
250.	910.	18.0	-47.1	-41.6	-30.8	-16.5	0.0	40.7	.6257
300.	841.	29.8	-79.4	-72.9	-59.9	-42.8	-22.9	57.0	.7974
350.	779.	45.6	-122.8	-115.2	-100.1	-80.1	-57.0	75.7	.9828
400.	719.	66.5	-179.2	-170.5	-153.2	-130.3	-103.9	97.1	1.1832
450.	663.	93.4	-250.8	-241.1	-221.7	-195.9	-166.1	121.5	1.4006
500.	609.	127.8	-340.7	-329.9	-308.3	-279.7	-246.6	149.1	1.6368

BALLISTIC COEFFICIENT .11, MUZZLE VELOCITY FT/SEC 800.

RANGE	VELOCITY	HEIGHT	50 YD	100 YD	150 YD	200 YD	250 YD	DRIFT	TIME
0.	800.	0.0	-1.5	-1.5	-1.5	-1.5	-1.5	0.0	0.0000
50.	745.	1.1	0.0	7.1	15.5	25.1	35.7	1.2	.1944
100.	693.	7.1	-14.2	0.0	16.9	35.9	57.2	5.0	.4034
150.	642.	18.2	-46.6	-25.3	0.0	28.6	60.5	11.5	.6281
200.	594.	35.7	-100.3	-71.8	-38.1	0.0	42.5	21.3	.8711
250.	549.	61.0	-178.5	-142.9	-100.8	-53.2	0.0	34.6	1.1339
300.	505.	95.9	-285.4	-242.7	-192.2	-135.1	-71.3	51.7	1.4189
350.	464.	142.7	-426.3	-376.4	-317.5	-250.8	-176.4	73.3	1.7288
400.	424.	204.4	-607.5	-550.6	-483.2	-407.0	-321.9	99.8	2.0672
450.	387.	284.5	-837.2	-773.1	-697.3	-611.6	-515.9	132.1	2.4379
500.	352.	387.7	-1124.9	-1053.7	-969.4	-874.2	-767.9	170.7	2.8448

BALLISTIC COEFFICIENT .11, MUZZLE VELOCITY FT/SEC 900.

RANGE	VELOCITY	HEIGHT	50 YD	100 YD	150 YD	200 YD	250 YD	DRIFT	TIME
0.	900.	0.0	-1.5	-1.5	-1.5	-1.5	-1.5	0.0	0.0000
50.	838.	.7	0.0	5.5	12.0	19.4	27.7	1.1	.1729
100.	782.	5.4	-10.9	0.0	13.2	28.0	44.5	4.4	.3582
150.	728.	14.2	-36.1	-19.8	0.0	22.2	46.9	10.1	.5572
200.	676.	27.8	-77.7	-55.9	-29.6	0.0	33.0	18.4	.7710
250.	627.	47.4	-138.4	-111.1	-78.2	-41.2	0.0	29.6	1.0016
300.	579.	74.3	-221.0	-188.3	-148.8	-104.4	-55.0	44.1	1.2505
350.	534.	110.2	-329.7	-291.5	-245.4	-193.6	-135.9	62.3	1.5204
400.	491.	157.1	-468.5	-424.9	-372.2	-313.1	-247.1	84.5	1.8135
450.	451.	217.5	-643.2	-594.1	-534.8	-468.3	-394.1	111.3	2.1324
500.	412.	294.7	-860.6	-806.1	-740.2	-666.3	-583.8	143.3	2.4808

BALLISTIC COEFFICIENT .11, MUZZLE VELOCITY FT/SEC 1000.

RANGE	VELOCITY	HEIGHT	50 YD	100 YD	150 YD	200 YD	250 YD	DRIFT	TIME
0.	1000.	0.0	-1.5	-1.5	-1.5	-1.5	-1.5	0.0	0.0000
50.	925.	.4	0.0	4.4	9.7	15.8	22.5	1.1	.1562
100.	859.	4.3	-8.7	0.0	10.7	22.9	36.3	4.3	.3246
150.	802.	11.5	-29.2	-16.1	0.0	18.2	38.4	9.7	.5053
200.	747.	22.7	-63.2	-45.8	-24.3	0.0	26.9	17.5	.6992
250.	694.	38.8	-112.6	-90.9	-64.0	-33.7	0.0	27.7	.9076
300.	644.	60.7	-180.0	-153.8	-121.6	-85.2	-44.8	40.8	1.1318
350.	596.	89.9	-268.1	-237.6	-200.0	-157.5	-110.4	57.0	1.3740
400.	550.	127.8	-380.5	-345.7	-302.7	-254.1	-200.3	76.8	1.6362
450.	507.	176.3	-521.3	-482.1	-433.7	-379.1	-318.5	100.4	1.9204
500.	465.	237.8	-695.6	-652.1	-598.3	-537.7	-470.3	128.4	2.2295

BALLISTIC COEFFICIENT .11, MUZZLE VELOCITY FT/SEC 1100.

RANGE	VELOCITY	HEIGHT	50 YD	100 YD	150 YD	200 YD	250 YD	DRIFT	TIME
0.	1100.	0.0	-1.5	-1.5	-1.5	-1.5	-1.5	0.0	0.0000
50.	1004.	.2	0.0	3.5	8.1	13.3	19.0	1.3	.1435
100.	928.	3.5	-7.1	0.0	9.1	19.5	31.0	4.6	.2986
150.	863.	9.7	-24.3	-13.7	0.0	15.5	32.8	10.1	.4664
200.	805.	19.3	-53.1	-38.9	-20.7	0.0	23.0	17.8	.6464
250.	750.	33.1	-95.2	-77.4	-54.7	-28.8	0.0	27.8	.8396
300.	697.	51.9	-152.4	-131.1	-103.8	-72.7	-38.2	40.3	1.0470
350.	647.	76.7	-227.2	-202.4	-170.5	-134.3	-94.0	55.6	1.2705
400.	599.	108.9	-322.3	-294.0	-257.5	-216.1	-170.1	74.0	1.5114
450.	553.	150.1	-441.6	-409.6	-368.7	-322.1	-270.3	96.0	1.7726
500.	509.	202.1	-588.8	-553.3	-507.8	-456.0	-398.5	121.8	2.0556

BALLISTIC COEFFICIENT .11, MUZZLE VELOCITY FT/SEC 1200.

RANGE	VELOCITY	HEIGHT	50 YD	100 YD	150 YD	200 YD	250 YD	DRIFT	TIME
0.	1200.	0.0	-1.5	-1.5	-1.5	-1.5	-1.5	0.0	0.0000
50.	1063.	.1	0.0	3.1	7.1	11.7	16.9	1.6	.1340
100.	977.	3.0	-6.1	0.0	8.1	17.3	27.7	5.5	.2810
150.	905.	8.6	-21.3	-12.1	0.0	13.9	29.4	11.5	.4406
200.	843.	17.3	-46.9	-34.7	-18.5	0.0	20.7	19.8	.6125
250.	786.	29.7	-84.5	-69.2	-49.0	-25.9	0.0	30.3	.7969
300.	732.	46.7	-135.7	-117.4	-93.2	-65.4	-34.3	43.1	.9947
350.	680.	69.2	-202.6	-181.2	-153.0	-120.5	-84.3	58.5	1.2071
400.	630.	98.3	-287.9	-263.4	-231.2	-194.1	-152.7	76.8	1.4366
450.	583.	135.4	-394.3	-366.8	-330.5	-288.8	-242.2	98.4	1.6840
500.	538.	182.2	-525.7	-495.1	-454.9	-408.5	-356.8	123.6	1.9520

BALLISTIC COEFFICIENT .11, MUZZLE VELOCITY FT/SEC 1300.

RANGE	VELOCITY	HEIGHT	50 YD	100 YD	150 YD	200 YD	250 YD	DRIFT	TIME
0.	1300.	0.0	-1.5	-1.5	-1.5	-1.5	-1.5	0.0	0.0000
50.	1119.	.0	0.0	2.7	6.3	10.6	15.3	1.7	.1251
100.	1017.	2.6	-5.4	0.0	7.3	15.9	25.3	6.2	.2661
150.	939.	7.7	-19.0	-10.9	0.0	12.8	26.9	12.9	.4197
200.	872.	15.8	-42.5	-31.7	-17.1	0.0	18.8	22.0	.5866
250.	814.	27.2	-76.6	-63.1	-44.9	-23.5	0.0	32.9	.7637
300.	758.	43.0	-123.4	-107.3	-85.4	-59.7	-31.5	46.2	.9546
350.	705.	63.8	-184.8	-166.0	-140.5	-110.5	-77.6	62.0	1.1599
400.	654.	90.8	-262.9	-241.4	-212.2	-178.0	-140.4	80.6	1.3808
450.	606.	125.1	-360.5	-336.2	-303.4	-264.9	-222.6	102.2	1.6191
500.	560.	168.3	-480.7	-453.8	-417.3	-374.5	-327.5	127.2	1.8764

BALLISTIC COEFFICIENT .11, MUZZLE VELOCITY FT/SEC 1400.

RANGE	VELOCITY	HEIGHT	50 YD	100 YD	150 YD	200 YD	250 YD	DRIFT	TIME
0.	1400.	0.0	-1.5	-1.5	-1.5	-1.5	-1.5	0.0	0.0000
50.	1184.	-.1	0.0	2.3	5.7	9.6	14.0	1.7	.1169
100.	1055.	2.3	-4.7	0.0	6.7	14.4	23.2	6.6	.2519
150.	970.	7.0	-17.1	-10.1	0.0	11.5	24.7	14.0	.4012
200.	899.	14.4	-38.3	-28.9	-15.4	0.0	17.6	23.3	.5609
250.	838.	25.1	-69.8	58.1	-41.2	-22.0	0.0	34.9	.7339
300.	781.	39.8	-113.1	-99.0	-78.8	-55.7	-29.3	48.6	.9193
350.	727.	59.3	-169.8	-153.4	-129.8	-102.9	-72.1	64.8	1.1184
400.	676.	84.5	-242.1	-223.3	-196.4	-165.6	-130.4	83.6	1.3324
450.	626.	116.5	-332.4	-311.3	-281.0	-246.4	-206.8	105.4	1.5631
500.	579.	156.9	-443.8	-420.3	-386.6	-348.2	-304.2	130.4	1.8121

BALLISTIC COEFFICIENT .11, MUZZLE VELOCITY FT/SEC 1500.

RANGE	VELOCITY	HEIGHT	50 YD	100 YD	150 YD	200 YD	250 YD	DRIFT	TIME
0.	1500.	0.0	-1.5	-1.5	-1.5	-1.5	-1.5	0.0	0.0000
50.	1255.	-.2	0.0	2.0	5.1	8.7	12.8	1.7	.1098
100.	1094.	2.0	-4.0	0.0	6.1	13.3	21.5	6.8	.2386
150.	1000.	6.3	-15.2	-9.1	0.0	10.8	23.1	14.5	.3824
200.	925.	13.2	-34.7	-26.6	-14.4	0.0	16.4	24.4	.5385
250.	859.	23.2	-63.9	-53.8	-38.5	-20.5	0.0	36.4	.7070
300.	802.	37.1	-104.1	-92.0	-73.7	-52.0	-27.4	50.6	.8877
350.	747.	55.4	-156.9	-142.8	-121.4	-96.2	-67.5	67.2	1.0816
400.	694.	79.1	-224.3	-208.1	-183.8	-154.9	-122.1	86.2	1.2900
450.	644.	109.3	-308.6	-290.4	-262.9	-230.5	-193.6	108.1	1.5142
500.	596.	147.3	-412.7	-392.5	-362.0	-325.9	-284.9	133.1	1.7564

BALLISTIC COEFFICIENT .11, MUZZLE VELOCITY FT/SEC 1600.

RANGE	VELOCITY	HEIGHT	50 YD	100 YD	150 YD	200 YD	250 YD	DRIFT	TIME
0.	1600.	0.0	-1.5	-1.5	-1.5	-1.5	-1.5	0.0	0.0000
50.	1330.	-.2	0.0	1.7	4.5	7.9	11.7	1.6	.1031
100.	1138.	1.7	-3.5	0.0	5.6	12.3	19.9	6.7	.2257
150.	1029.	5.7	-13.6	-8.4	0.0	10.0	21.4	14.8	.3654
200.	949.	12.1	-31.4	-24.5	-13.3	0.0	15.3	24.9	.5167
250.	881.	21.5	-58.4	-49.8	-35.7	-19.1	0.0	37.3	.6808
300.	821.	34.5	-95.9	-85.5	-68.7	-48.7	-25.8	51.9	.8573
350.	766.	51.8	-145.2	-133.1	-113.5	-90.2	-63.5	68.7	1.0466
400.	712.	74.2	-208.3	-194.5	-172.1	-145.5	-114.9	88.0	1.2499
450.	661.	102.7	-287.2	-271.6	-246.4	-216.5	-182.1	109.9	1.4683
500.	612.	138.7	-384.9	-367.6	-339.6	-306.3	-268.1	135.0	1.7043

BALLISTIC COEFFICIENT .11, MUZZLE VELOCITY FT/SEC 1700.

RANGE	VELOCITY	HEIGHT	50 YD	100 YD	150 YD	200 YD	250 YD	DRIFT	TIME
0.	1700.	0.0	-1.5	-1.5	-1.5	-1.5	-1.5	0.0	0.0000
50.	1410.	-.3	0.0	1.5	4.0	7.1	10.7	1.5	.0970
100.	1192.	1.4	-2.9	0.0	5.0	11.3	18.4	6.5	.2132
150.	1059.	5.0	-11.9	-7.5	0.0	9.4	20.1	14.6	.3475
200.	974.	11.0	-28.4	-22.5	-12.5	0.0	14.3	25.1	.4955
250.	902.	19.9	-53.3	-46.1	-33.5	-17.9	0.0	37.7	.6556
300.	840.	32.2	-88.2	-79.5	-64.4	-45.7	-24.2	52.5	.8280
350.	784.	48.5	-134.5	-124.3	-106.7	-84.8	-59.8	69.6	1.0130
400.	729.	69.7	-193.7	-182.0	-161.9	-136.9	-108.3	89.0	1.2116
450.	678.	96.7	-267.8	-254.7	-232.0	-203.9	-171.8	111.0	1.4248
500.	628.	130.7	-359.5	-345.0	-319.8	-288.6	-252.9	136.0	1.6550

BALLISTIC COEFFICIENT .11, MUZZLE VELOCITY FT/SEC 1800.

RANGE	VELOCITY	HEIGHT	50 YD	100 YD	150 YD	200 YD	250 YD	DRIFT	TIME
0.	1800.	0.0	-1.5	-1.5	-1.5	-1.5	-1.5	0.0	0.0000
50.	1495.	-.3	0.0	1.2	3.5	6.4	9.7	1.4	.0915
100.	1252.	1.2	-2.4	0.0	4.5	10.3	17.0	6.1	.2016
150.	1092.	4.5	-10.4	-6.8	0.0	8.7	18.8	14.2	.3307
200.	999.	10.1	-25.5	-20.6	-11.6	0.0	13.4	24.9	.4747
250.	923.	18.4	-48.7	-42.6	-31.3	-16.8	0.0	37.7	.6311
300.	858.	30.0	-81.3	-73.9	-60.4	-43.0	-22.9	52.8	.7999
350.	801.	45.4	-124.6	-116.0	-100.2	-79.9	-56.4	69.9	.9807
400.	746.	65.5	-180.2	-170.4	-152.3	-129.2	-102.3	89.4	1.1748
450.	694.	91.1	-250.0	-239.0	-218.7	-192.6	-162.4	111.5	1.3834
500.	643.	123.4	-336.4	-324.1	-301.5	-272.6	-239.0	136.3	1.6079

BALLISTIC COEFFICIENT .11, MUZZLE VELOCITY FT/SEC 1900.

RANGE	VELOCITY	HEIGHT	50 YD	100 YD	150 YD	200 YD	250 YD	DRIFT	TIME
0.	1900.	0.0	-1.5	-1.5	-1.5	-1.5	-1.5	0.0	0.0000
50.	1581.	-.4	0.0	1.0	3.0	5.7	8.9	1.4	.0867
100.	1316.	1.0	-2.0	0.0	4.0	9.4	15.7	5.8	.1909
150.	1129.	4.0	-9.1	-6.1	0.0	8.1	17.5	13.7	.3146
200.	1023.	9.2	-22.9	-18.8	-10.8	0.0	12.6	24.5	.4549
250.	944.	17.0	-44.4	-39.3	-29.2	-15.8	0.0	37.5	.6078
300.	877.	27.9	-74.8	-68.7	-56.6	-40.5	-21.5	52.6	.7727
350.	818.	42.5	-115.4	-108.3	-94.1	-75.3	-53.2	69.9	.9495
400.	762.	61.6	-167.7	-159.6	-143.5	-121.9	-96.7	89.4	1.1395
450.	709.	85.9	-233.6	-224.5	-206.3	-182.1	-153.7	111.5	1.3438
500.	658.	116.6	-315.1	-305.0	-284.8	-258.0	-226.4	136.2	1.5633

BALLISTIC COEFFICIENT .12, MUZZLE VELOCITY FT/SEC 800.

RANGE	VELOCITY	HEIGHT	50 YD	100 YD	150 YD	200 YD	250 YD	DRIFT	TIME
0.	800.	0.0	-1.5	-1.5	-1.5	-1.5	-1.5	0.0	0.0000
50.	750.	1.1	0.0	7.0	15.3	24.5	34.7	1.1	.1939
100.	701.	7.0	-14.0	0.0	16.6	35.0	55.4	4.5	.4007
150.	655.	17.8	-45.8	-24.8	0.0	27.7	58.3	10.5	.6223
200.	610.	34.7	-98.0	-70.0	-36.9	0.0	40.8	19.3	.8597
250.	568.	58.9	-173.5	-138.5	-97.1	-51.0	0.0	31.2	1.1147
300.	527.	91.9	-276.0	-234.0	-184.3	-129.0	-67.8	46.5	1.3892
350.	488.	135.6	-409.7	-360.7	-302.8	-238.3	-166.8	65.6	1.6854
400.	450.	192.4	-579.9	-523.9	-457.7	-383.9	-302.3	89.0	2.0058
450.	415.	265.0	-792.6	-729.6	-655.1	-572.1	-480.2	117.1	2.3528
500.	381.	356.9	-1055.0	-985.0	-902.2	-810.0	-708.0	150.4	2.7295

BALLISTIC COEFFICIENT .12, MUZZLE VELOCITY FT/SEC 900.

RANGE	VELOCITY	HEIGHT	50 YD	100 YD	150 YD	200 YD	250 YD	DRIFT	TIME
0.	900.	0.0	-1.5	-1.5	-1.5	-1.5	-1.5	0.0	0.0000
50.	843.	.7	0.0	5.4	11.8	19.0	26.9	1.0	.1724
100.	791.	5.3	-10.8	0.0	12.8	27.3	43.1	4.0	.3562
150.	741.	13.9	-35.4	-19.3	0.0	21.6	45.4	9.1	.5520
200.	693.	27.1	-76.1	-54.5	-28.8	0.0	31.7	16.7	.7615
250.	647.	45.9	-134.7	-107.8	-75.7	-39.7	0.0	26.8	.9856
300.	603.	71.4	-214.1	-181.7	-143.2	-100.0	-52.4	39.7	1.2257
350.	560.	104.9	-317.1	-279.4	-234.4	-184.0	-128.5	55.8	1.4837
400.	520.	148.2	-447.8	-404.7	-353.3	-295.7	-232.2	75.4	1.7617
450.	481.	203.3	-610.5	-562.0	-504.2	-439.4	-368.0	98.9	2.0618
500.	444.	272.7	-810.9	-757.0	-692.8	-620.8	-541.4	126.8	2.3869

BALLISTIC COEFFICIENT .12, MUZZLE VELOCITY FT/SEC 1000.

RANGE	VELOCITY	HEIGHT	50 YD	100 YD	150 YD	200 YD	250 YD	DRIFT	TIME
0.	1000.	0.0	-1.5	-1.5	-1.5	-1.5	-1.5	0.0	0.0000
50.	930.	.4	0.0	4.3	9.6	15.4	21.9	1.0	.1557
100.	870.	4.2	-8.6	0.0	10.5	22.3	35.3	4.0	.3226
150.	816.	11.3	-28.7	-15.8	0.0	17.7	37.1	8.9	.5007
200.	765.	22.1	-61.8	-44.6	-23.6	0.0	26.0	15.9	.6906
250.	716.	37.6	-109.7	-88.2	-61.9	-32.4	0.0	25.2	.8933
300.	669.	58.4	-174.4	-148.7	-117.1	-81.8	-42.8	37.0	1.1102
350.	624.	85.7	-258.3	-228.3	-191.5	-150.2	-104.8	51.4	1.3423
400.	581.	120.8	-364.3	-330.0	-287.9	-240.7	-188.8	68.9	1.5914
450.	539.	165.2	-495.9	-457.2	-409.9	-356.9	-298.5	89.7	1.8596
500.	499.	220.9	-657.0	-614.1	-561.5	-502.6	-437.7	114.2	2.1487

BALLISTIC COEFFICIENT .12, MUZZLE VELOCITY FT/SEC 1100.

RANGE	VELOCITY	HEIGHT	50 YD	100 YD	150 YD	200 YD	250 YD	DRIFT	TIME
0.	1100.	0.0	-1.5	-1.5	-1.5	-1.5	-1.5	0.0	0.0000
50.	1011.	.2	0.0	3.5	7.9	12.9	18.5	1.3	.1435
100.	940.	3.5	-6.9	0.0	8.9	19.0	30.1	4.2	.2965
150.	878.	9.5	-23.7	-13.4	0.0	15.1	31.7	9.3	.4619
200.	824.	18.8	-51.7	-37.9	-20.1	0.0	22.2	16.3	.6382
250.	773.	32.0	-92.4	-75.2	-52.9	-27.8	0.0	25.4	.8263
300.	723.	49.9	-147.4	-126.7	-99.9	-69.8	-36.5	36.7	1.0269
350.	676.	73.2	-218.8	-194.6	-163.4	-128.2	-89.4	50.5	1.2415
400.	631.	103.2	-308.8	-281.2	-245.5	-205.4	-160.9	67.0	1.4715
450.	587.	140.9	-420.4	-389.3	-349.1	-304.0	-254.0	86.4	1.7181
500.	545.	188.1	-556.6	-522.1	-477.5	-427.3	-371.8	109.1	1.9833

BALLISTIC COEFFICIENT .12, MUZZLE VELOCITY FT/SEC 1200.

RANGE	VELOCITY	HEIGHT	50 YD	100 YD	150 YD	200 YD	250 YD	DRIFT	TIME
0.	1200.	0.0	-1.5	-1.5	-1.5	-1.5	-1.5	0.0	0.0000
50.	1072.	.1	0.0	3.1	6.9	11.4	16.4	1.4	.1330
100.	990.	3.0	-6.1	0.0	7.8	16.8	26.7	5.2	.2798
150.	922.	8.4	-20.8	-11.6	0.0	13.5	28.5	10.7	.4358
200.	863.	16.8	-45.8	-33.5	-18.0	0.0	20.0	18.3	.6042
250.	809.	28.7	-82.2	-66.9	-47.4	-25.0	0.0	28.0	.7839
300.	759.	44.9	-131.4	-113.0	-89.7	-62.7	-32.8	39.6	.9751
350.	710.	66.1	-195.4	-173.9	-146.7	-115.3	-80.3	53.6	1.1798
400.	663.	93.1	-275.9	-251.4	-220.4	-184.4	-144.5	70.1	1.3983
450.	618.	127.2	-375.7	-348.1	-313.1	-272.7	-227.7	89.3	1.6325
500.	575.	169.7	-497.7	-467.1	-428.3	-383.3	-333.4	111.7	1.8844

BALLISTIC COEFFICIENT .12, MUZZLE VELOCITY FT/SEC 1300.

RANGE	VELOCITY	HEIGHT	50 YD	100 YD	150 YD	200 YD	250 YD	DRIFT	TIME
0.	1300.	0.0	-1.5	-1.5	-1.5	-1.5	-1.5	0.0	0.0000
50.	1131.	-.0	0.0	2.7	6.2	10.3	14.8	1.6	.1243
100.	1031.	2.6	-5.3	0.0	7.0	15.3	24.4	5.9	.2646
150.	958.	7.5	-18.6	-10.5	0.0	12.4	26.0	12.1	.4147
200.	893.	15.3	-41.3	-30.6	-16.5	0.0	18.1	20.5	.5778
250.	837.	26.3	-74.2	-60.9	-43.3	-22.6	0.0	30.5	.7505
300.	786.	41.3	-119.3	-103.3	-82.2	-57.4	-30.2	42.8	.9355
350.	736.	60.8	-177.8	-159.1	-134.5	-105.6	-73.9	57.2	1.1328
400.	688.	85.9	-251.7	-230.3	-202.2	-169.1	-132.9	74.0	1.3437
450.	642.	117.4	-343.0	-319.0	-287.4	-250.2	-209.4	93.4	1.5692
500.	598.	156.7	-454.7	-428.0	-392.9	-351.5	-306.2	115.7	1.8114

BALLISTIC COEFFICIENT .12, MUZZLE VELOCITY FT/SEC 1400.

RANGE	VELOCITY	HEIGHT	50 YD	100 YD	150 YD	200 YD	250 YD	DRIFT	TIME
0.	1400.	0.0	-1.5	-1.5	-1.5	-1.5	-1.5	0.0	0.0000
50.	1200.	-.1	0.0	2.3	5.6	9.3	13.5	1.6	.1162
100.	1072.	2.2	-4.5	0.0	6.6	14.0	22.4	6.1	.2491
150.	990.	6.8	-16.7	-9.8	0.0	11.1	23.8	13.1	.3959
200.	922.	13.9	-37.1	-28.0	-14.9	0.0	16.9	21.7	.5520
250.	862.	24.2	-67.5	-56.1	-39.7	-21.1	0.0	32.5	.7204
300.	809.	38.1	-109.1	-95.4	-75.7	-53.4	-28.1	45.3	.9001
350.	759.	56.4	-163.0	-147.1	-124.2	-98.2	-68.6	60.1	1.0914
400.	710.	79.9	-231.4	-213.2	-187.0	-157.3	-123.5	77.3	1.2961
450.	663.	109.4	-316.0	-295.5	-266.0	-232.6	-194.5	96.9	1.5146
500.	618.	146.1	-419.3	-396.6	-363.8	-326.6	-284.4	119.2	1.7489

BALLISTIC COEFFICIENT .12, MUZZLE VELOCITY FT/SEC 1500.

RANGE	VELOCITY	HEIGHT	50 YD	100 YD	150 YD	200 YD	250 YD	DRIFT	TIME
0.	1500.	0.0	-1.5	-1.5	-1.5	-1.5	-1.5	0.0	0.0000
50.	1273.	-.2	0.0	1.9	4.9	8.4	12.3	1.6	.1091
100.	1114.	1.9	-3.9	0.0	5.9	12.9	20.7	6.3	.2356
150.	1021.	6.1	-14.7	-8.9	0.0	10.4	22.2	13.5	.3767
200.	948.	12.7	-33.5	-25.7	-13.9	0.0	15.7	22.8	.5293
250.	885.	22.3	-61.5	-51.8	-37.0	-19.6	0.0	34.0	.6931
300.	830.	35.4	-100.1	-88.4	-70.7	-49.8	-26.2	47.2	.8682
350.	779.	52.6	-150.3	-136.7	-116.0	-91.6	-64.1	62.4	1.0546
400.	729.	74.7	-214.0	-198.5	-174.8	-147.0	-115.6	79.9	1.2539
450.	682.	102.5	-292.8	-275.3	-248.8	-217.4	-182.1	99.7	1.4665
500.	636.	137.1	-389.4	-370.0	-340.4	-305.6	-266.3	122.2	1.6945

BALLISTIC COEFFICIENT .12, MUZZLE VELOCITY FT/SEC 1600.

RANGE	VELOCITY	HEIGHT	50 YD	100 YD	150 YD	200 YD	250 YD	DRIFT	TIME
0.	1600.	0.0	-1.5	-1.5	-1.5	-1.5	-1.5	0.0	0.0000
50.	1350.	-.2	0.0	1.6	4.3	7.6	11.2	1.5	.1023
100.	1164.	1.6	-3.3	0.0	5.4	11.8	19.1	6.1	.2223
150.	1051.	5.4	-13.0	-8.0	0.0	9.7	20.7	13.6	.3586
200.	974.	11.6	-30.2	-23.6	-12.9	0.0	14.6	23.2	.5070
250.	908.	20.6	-56.1	-47.8	-34.5	-18.3	0.0	34.8	.6665
300.	850.	32.9	-91.9	-82.0	-65.9	-46.5	-24.6	48.4	.8375
350.	798.	49.1	-138.8	-127.2	-108.5	-85.9	-60.2	63.9	1.0195
400.	748.	70.0	-198.3	-185.1	-163.7	-137.9	-108.6	81.6	1.2139
450.	699.	96.2	-272.1	-257.3	-233.2	-204.1	-171.2	101.6	1.4212
500.	653.	128.9	-362.7	-346.2	-319.4	-287.1	-250.5	124.3	1.6435

BALLISTIC COEFFICIENT .12, MUZZLE VELOCITY FT/SEC 1700.

RANGE	VELOCITY	HEIGHT	50 YD	100 YD	150 YD	200 YD	250 YD	DRIFT	TIME
0.	1700.	0.0	-1.5	-1.5	-1.5	-1.5	-1.5	0.0	0.0000
50.	1432.	-.3	0.0	1.4	3.8	6.8	10.2	1.4	.0963
100.	1223.	1.4	-2.8	0.0	4.8	10.8	17.6	5.9	.2099
150.	1085.	4.8	-11.4	-7.2	0.0	9.0	19.3	13.4	.3409
200.	1000.	10.6	-27.1	-21.6	-12.0	0.0	13.7	23.3	.4852
250.	931.	19.0	-51.0	-44.1	-32.1	-17.1	0.0	35.1	.6409
300.	870.	30.6	-84.3	-76.0	-61.6	-43.6	-23.0	49.0	.8078
350.	816.	45.9	-128.1	-118.4	-101.6	-80.6	-56.7	64.8	.9857
400.	765.	65.6	-183.9	-172.8	-153.6	-129.6	-102.2	82.7	1.1756
450.	716.	90.4	-253.2	-240.7	-219.1	-192.2	-161.3	102.8	1.3782
500.	669.	121.4	-338.2	-324.4	-300.3	-270.4	-236.1	125.4	1.5951

BALLISTIC COEFFICIENT .12, MUZZLE VELOCITY FT/SEC 1800.

RANGE	VELOCITY	HEIGHT	50 YD	100 YD	150 YD	200 YD	250 YD	DRIFT	TIME
0.	1800.	0.0	-1.5	-1.5	-1.5	-1.5	-1.5	0.0	0.0000
50.	1519.	-.4	0.0	1.2	3.3	6.1	9.3	1.3	.0908
100.	1287.	1.1	-2.3	0.0	4.3	9.8	16.2	5.6	.1984
150.	1123.	4.3	-9.9	-6.4	0.0	8.3	17.9	13.0	.3238
200.	1026.	9.6	-24.3	-19.7	-11.1	0.0	12.7	23.1	.4645
250.	953.	17.5	-46.3	-40.5	-29.8	-15.9	0.0	35.0	.6158
300.	890.	28.4	-77.3	-70.3	-57.5	-40.8	-21.7	49.1	.7788
350.	834.	42.8	-118.3	-110.2	-95.2	-75.7	-53.4	65.1	.9531
400.	782.	61.5	-170.6	-161.3	-144.2	-121.9	-96.4	83.1	1.1387
450.	733.	85.1	-235.8	-225.4	-206.2	-181.1	-152.4	103.3	1.3370
500.	685.	114.4	-315.8	-304.2	-282.8	-255.0	-223.1	125.9	1.5487

BALLISTIC COEFFICIENT .12, MUZZLE VELOCITY FT/SEC 1900.

RANGE	VELOCITY	HEIGHT	50 YD	100 YD	150 YD	200 YD	250 YD	DRIFT	TIME
0.	1900.	0.0	-1.5	-1.5	-1.5	-1.5	-1.5	0.0	0.0000
50.	1606.	-.4	0.0	1.0	2.9	5.4	8.4	1.2	.0860
100.	1355.	.9	-1.9	0.0	3.8	8.9	14.9	5.3	.1879
150.	1167.	3.8	-8.6	-5.7	0.0	7.6	16.6	12.4	.3075
200.	1053.	8.7	-21.6	-17.7	-10.1	0.0	12.0	22.5	.4434
250.	976.	16.0	-42.0	-37.2	-27.7	-15.0	0.0	34.6	.5916
300.	910.	26.3	-70.8	-65.0	-53.6	-38.4	-20.4	48.8	.7509
350.	852.	40.0	-109.2	-102.4	-89.1	-71.4	-50.4	64.9	.9216
400.	799.	57.7	-158.3	-150.7	-135.4	-115.2	-91.1	83.0	1.1033
450.	749.	80.1	-219.8	-211.2	-194.1	-171.3	-144.2	103.3	1.2975
500.	700.	107.9	-295.2	-285.6	-266.6	-241.2	-211.2	125.8	1.5044

BALLISTIC COEFFICIENT .13, MUZZLE VELOCITY FT/SEC 800.

RANGE	VELOCITY	HEIGHT	50 YD	100 YD	150 YD	200 YD	250 YD	DRIFT	TIME
0.	800.	0.0	-1.5	-1.5	-1.5	-1.5	-1.5	0.0	0.0000
50.	753.	1.0	0.0	6.9	15.0	24.0	33.9	1.0	.1934
100.	709.	6.9	-13.9	0.0	16.2	34.2	53.9	4.2	.3988
150.	665.	17.5	-45.1	-24.3	0.0	26.9	56.5	9.6	.6173
200.	624.	33.9	-96.1	-68.3	-35.9	0.0	39.5	17.6	.8501
250.	584.	57.2	-169.4	-134.7	-94.2	-49.4	0.0	28.4	1.0988
300.	545.	88.7	-268.3	-226.7	-178.1	-124.2	-65.0	42.2	1.3649
350.	508.	129.9	-396.4	-347.9	-291.1	-228.3	-159.2	59.4	1.6501
400.	473.	182.9	-557.9	-502.4	-437.5	-365.8	-286.8	80.3	1.9560
450.	439.	249.7	-757.4	-695.0	-622.0	-541.3	-452.4	105.1	2.2845
500.	407.	333.9	-1003.2	-933.8	-852.7	-763.0	-664.3	134.7	2.6404

BALLISTIC COEFFICIENT .13, MUZZLE VELOCITY FT/SEC 900.

RANGE	VELOCITY	HEIGHT	50 YD	100 YD	150 YD	200 YD	250 YD	DRIFT	TIME
0.	900.	0.0	-1.5	-1.5	-1.5	-1.5	-1.5	0.0	0.0000
50.	847.	.7	0.0	5.3	11.6	18.7	26.3	1.0	.1721
100.	799.	5.3	-10.6	0.0	12.7	26.7	42.1	3.7	.3543
150.	753.	13.7	-34.9	-19.0	0.0	21.0	44.1	8.4	.5479
200.	708.	26.5	-74.6	-53.4	-28.1	0.0	30.7	15.3	.7536
250.	664.	44.6	-131.7	-105.2	-73.5	-38.4	0.0	24.5	.9723
300.	623.	69.0	-208.3	-176.5	-138.5	-96.4	-50.3	36.1	1.2054
350.	583.	100.8	-307.4	-270.2	-225.9	-176.8	-123.0	50.6	1.4544
400.	544.	141.4	-431.9	-389.5	-338.8	-282.6	-221.2	68.2	1.7209
450.	508.	192.5	-585.7	-537.9	-480.9	-417.8	-348.7	89.2	2.0066
500.	472.	256.0	-772.9	-719.8	-656.4	-586.3	-509.5	113.8	2.3130

BALLISTIC COEFFICIENT .13, MUZZLE VELOCITY FT/SEC 1000.

RANGE	VELOCITY	HEIGHT	50 YD	100 YD	150 YD	200 YD	250 YD	DRIFT	TIME
0.	1000.	0.0	-1.5	-1.5	-1.5	-1.5	-1.5	0.0	0.0000
50.	935.	.4	0.0	4.2	9.4	15.1	21.5	.9	.1553
100.	878.	4.2	-8.5	0.0	10.3	21.8	34.4	3.7	.3210
150.	828.	11.1	-28.2	-15.5	0.0	17.2	36.1	8.2	.4969
200.	781.	21.7	-60.6	-43.6	-23.0	0.0	25.2	14.7	.6833
250.	735.	36.6	-107.3	-86.0	-60.2	-31.5	0.0	23.2	.8816
300.	690.	56.5	-169.9	-144.5	-113.5	-79.0	-41.2	33.9	1.0923
350.	648.	82.5	-250.6	-220.9	-184.8	-144.6	-100.5	46.9	1.3167
400.	607.	115.5	-351.8	-317.9	-276.5	-230.6	-180.2	62.6	1.5559
450.	568.	156.8	-476.4	-438.2	-391.7	-340.0	-283.3	81.2	1.8116
500.	530.	208.0	-627.7	-585.3	-533.6	-476.2	-413.2	103.0	2.0853

BALLISTIC COEFFICIENT .13, MUZZLE VELOCITY FT/SEC 1100.

RANGE	VELOCITY	HEIGHT	50 YD	100 YD	150 YD	200 YD	250 YD	DRIFT	TIME
0.	1100.	0.0	-1.5	-1.5	-1.5	-1.5	-1.5	0.0	0.0000
50.	1017.	.2	0.0	3.5	7.8	12.7	18.1	1.0	.1422
100.	951.	3.4	-6.9	0.0	8.7	18.5	29.3	3.9	.2948
150.	892.	9.3	-23.5	-13.1	0.0	14.7	30.9	8.6	.4579
200.	840.	18.4	-50.9	-37.0	-19.6	0.0	21.6	15.1	.6311
250.	792.	31.2	-90.6	-73.3	-51.5	-27.0	0.0	23.5	.8153
300.	746.	48.3	-143.8	-123.1	-96.9	-67.6	-35.2	33.8	1.0104
350.	701.	70.4	-212.4	-188.2	-157.7	-123.4	-85.6	46.3	1.2177
400.	658.	98.6	-298.4	-270.8	-235.9	-196.7	-153.5	61.2	1.4386
450.	617.	133.8	-404.1	-373.0	-333.8	-289.7	-241.1	78.6	1.6741
500.	577.	177.2	-532.2	-497.7	-454.1	-405.1	-351.1	98.9	1.9256

BALLISTIC COEFFICIENT .13, MUZZLE VELOCITY FT/SEC 1200.

RANGE	VELOCITY	HEIGHT	50 YD	100 YD	150 YD	200 YD	250 YD	DRIFT	TIME
0.	1200.	0.0	-1.5	-1.5	-1.5	-1.5	-1.5	0.0	0.0000
50.	1079.	.1	0.0	3.0	6.8	11.2	16.0	1.3	.1324
100.	1002.	2.9	-6.0	0.0	7.7	16.4	26.1	4.8	.2771
150.	937.	8.2	-20.5	-11.5	0.0	13.1	27.6	10.0	.4319
200.	880.	16.4	-44.8	-32.9	-17.5	0.0	19.4	17.1	.5970
250.	829.	27.9	-80.2	-65.3	-46.1	-24.2	0.0	26.0	.7729
300.	782.	43.4	-127.8	-109.9	-86.9	-60.6	-31.6	36.8	.9590
350.	736.	63.5	-189.3	-168.4	-141.5	-110.9	-77.0	49.6	1.1569
400.	692.	89.0	-266.3	-242.5	-211.7	-176.8	-138.1	64.6	1.3673
450.	649.	120.8	-361.0	-334.2	-299.6	-260.2	-216.7	82.1	1.5912
500.	608.	160.0	-475.6	-445.8	-407.4	-363.7	-315.3	102.1	1.8301

BALLISTIC COEFFICIENT .13, MUZZLE VELOCITY FT/SEC 1300.

RANGE	VELOCITY	HEIGHT	50 YD	100 YD	150 YD	200 YD	250 YD	DRIFT	TIME
0.	1300.	0.0	-1.5	-1.5	-1.5	-1.5	-1.5	0.0	0.0000
50.	1141.	-.0	0.0	2.6	6.1	10.0	14.5	1.5	.1236
100.	1044.	2.5	-5.2	0.0	6.9	14.9	23.7	5.5	.2621
150.	974.	7.3	-18.2	-10.4	0.0	11.9	25.2	11.3	.4105
200.	912.	14.8	-40.2	-29.7	-15.9	0.0	17.7	19.0	.5697
250.	858.	25.5	-72.4	-59.4	-42.1	-22.2	0.0	28.6	.7396
300.	809.	39.8	-115.9	-100.3	-79.5	-55.6	-29.0	40.0	.9194
350.	762.	58.4	-172.1	-153.8	-129.6	-101.8	-70.7	53.3	1.1103
400.	717.	82.0	-242.6	-221.8	-194.1	-162.3	-126.8	68.7	1.3132
450.	674.	111.5	-329.3	-305.9	-274.8	-239.0	-199.1	86.4	1.5291
500.	632.	147.8	-434.4	-408.3	-373.8	-334.0	-289.6	106.5	1.7591

BALLISTIC COEFFICIENT .13, MUZZLE VELOCITY FT/SEC 1400.

RANGE	VELOCITY	HEIGHT	50 YD	100 YD	150 YD	200 YD	250 YD	DRIFT	TIME
0.	1400.	0.0	-1.5	-1.5	-1.5	-1.5	-1.5	0.0	0.0000
50.	1213.	-.1	0.0	2.2	5.4	9.0	13.1	1.5	.1156
100.	1087.	2.2	-4.4	0.0	6.4	13.6	21.8	5.7	.2467
150.	1007.	6.6	-16.2	-9.6	0.0	10.9	23.1	12.3	.3911
200.	942.	13.5	-36.1	-27.3	-14.5	0.0	16.3	20.4	.5446
250.	884.	23.4	-65.5	-54.5	-38.5	-20.3	0.0	30.5	.7089
300.	833.	36.7	-105.6	-92.3	-73.2	-51.4	-27.0	42.4	.8837
350.	786.	54.1	-157.5	-142.0	-119.7	-94.3	-65.8	56.2	1.0692
400.	740.	76.1	-222.6	-204.9	-179.4	-150.3	-117.8	71.9	1.2657
450.	695.	103.7	-302.8	-283.0	-254.2	-221.6	-184.9	89.9	1.4752
500.	653.	137.7	-400.1	-378.0	-346.1	-309.8	-269.1	110.3	1.6981

BALLISTIC COEFFICIENT .13, MUZZLE VELOCITY FT/SEC 1500.

RANGE	VELOCITY	HEIGHT	50 YD	100 YD	150 YD	200 YD	250 YD	DRIFT	TIME
0.	1500.	0.0	-1.5	-1.5	-1.5	-1.5	-1.5	0.0	0.0000
50.	1288.	-.2	0.0	1.9	4.7	8.2	11.9	1.5	.1083
100.	1133.	1.9	-3.8	0.0	5.7	12.6	20.1	5.8	.2330
150.	1039.	5.9	-14.2	-8.6	0.0	10.3	21.5	12.6	.3716
200.	969.	12.3	-32.7	-25.1	-13.7	0.0	15.0	21.5	.5222
250.	909.	21.5	-59.6	-50.2	-35.8	-18.7	0.0	31.9	.6811
300.	855.	34.1	-96.9	-85.6	-68.4	-47.9	-25.4	44.4	.8523
350.	806.	50.4	-144.9	-131.7	-111.7	-87.7	-61.5	58.5	1.0322
400.	759.	71.1	-205.5	-190.4	-167.5	-140.2	-110.2	74.6	1.2237
450.	714.	97.1	-280.3	-263.4	-237.6	-206.8	-173.1	92.9	1.4277
500.	671.	129.1	-371.1	-352.2	-323.6	-289.4	-251.9	113.4	1.6445

BALLISTIC COEFFICIENT .13, MUZZLE VELOCITY FT/SEC 1600.

RANGE	VELOCITY	HEIGHT	50 YD	100 YD	150 YD	200 YD	250 YD	DRIFT	TIME
0.	1600.	0.0	-1.5	-1.5	-1.5	-1.5	-1.5	0.0	0.0000
50.	1367.	-.3	0.0	1.6	4.2	7.3	10.8	1.4	.1016
100.	1188.	1.6	-3.2	0.0	5.2	11.4	18.5	5.7	.2196
150.	1072.	5.2	-12.5	-7.7	0.0	9.4	20.0	12.6	.3531
200.	997.	11.2	-29.2	-22.8	-12.5	0.0	14.1	21.7	.4985
250.	932.	19.8	-54.1	-46.2	-33.3	-17.6	0.0	32.6	.6542
300.	876.	31.6	-88.6	-79.1	-63.6	-44.8	-23.7	45.5	.8209
350.	826.	46.9	-133.4	-122.2	-104.2	-82.3	-57.6	59.9	.9967
400.	778.	66.5	-190.0	-177.3	-156.7	-131.7	-103.4	76.3	1.1838
450.	733.	91.0	-260.0	-245.7	-222.5	-194.3	-162.6	94.8	1.3827
500.	688.	121.2	-344.9	-329.0	-303.2	-272.0	-236.7	115.5	1.5940

BALLISTIC COEFFICIENT .13, MUZZLE VELOCITY FT/SEC 1700.

RANGE	VELOCITY	HEIGHT	50 YD	100 YD	150 YD	200 YD	250 YD	DRIFT	TIME
0.	1700.	0.0	-1.5	-1.5	-1.5	-1.5	-1.5	0.0	0.0000
50.	1451.	-.3	0.0	1.3	3.6	6.5	9.8	1.3	.0956
100.	1251.	1.3	-2.7	0.0	4.6	10.4	16.9	5.4	.2073
150.	1110.	4.6	-10.9	-6.9	0.0	8.7	18.5	12.4	.3352
200.	1024.	10.2	-26.1	-20.8	-11.6	0.0	13.1	21.8	.4766
250.	956.	18.2	-49.0	-42.4	-30.9	-16.4	0.0	32.9	.6279
300.	897.	29.2	-81.0	-73.0	-59.2	-41.8	-22.1	45.9	.7904
350.	845.	43.7	-122.7	-113.4	-97.3	-77.1	-54.1	60.7	.9625
400.	796.	62.2	-175.8	-165.1	-146.7	-123.6	-97.3	77.4	1.1454
450.	750.	85.4	-241.4	-229.4	-208.7	-182.6	-153.1	96.0	1.3396
500.	705.	114.0	-321.0	-307.7	-284.7	-255.8	-223.0	116.8	1.5459

BALLISTIC COEFFICIENT .13, MUZZLE VELOCITY FT/SEC 1800.

RANGE	VELOCITY	HEIGHT	50 YD	100 YD	150 YD	200 YD	250 YD	DRIFT	TIME
0.	1800.	0.0	-1.5	-1.5	-1.5	-1.5	-1.5	0.0	0.0000
50.	1539.	-.4	0.0	1.1	3.2	5.8	8.9	1.2	.0902
100.	1318.	1.1	-2.2	0.0	4.1	9.4	15.5	5.1	.1958
150.	1153.	4.1	-9.5	-6.2	0.0	7.9	17.1	12.1	.3185
200.	1052.	9.2	-23.2	-18.7	-10.5	0.0	12.3	21.3	.4545
250.	980.	16.7	-44.3	-38.8	-28.5	-15.4	0.0	32.7	.6024
300.	918.	27.0	-73.8	-67.2	-54.8	-39.1	-20.7	45.9	.7606
350.	863.	40.7	-112.9	-105.2	-90.7	-72.4	-50.9	60.9	.9293
400.	814.	58.2	-162.6	-153.8	-137.3	-116.3	-91.7	77.7	1.1084
450.	767.	80.2	-224.2	-214.3	-195.7	-172.1	-144.4	96.5	1.2983
500.	722.	107.2	-299.1	-288.1	-267.4	-241.2	-210.5	117.3	1.4998

BALLISTIC COEFFICIENT .13, MUZZLE VELOCITY FT/SEC 1900.

RANGE	VELOCITY	HEIGHT	50 YD	100 YD	150 YD	200 YD	250 YD	DRIFT	TIME
0.	1900.	0.0	-1.5	-1.5	-1.5	-1.5	-1.5	0.0	0.0000
50.	1627.	-.4	0.0	.9	2.7	5.1	8.0	1.1	.0853
100.	1389.	.9	-1.8	0.0	3.6	8.4	14.2	4.8	.1852
150.	1205.	3.6	-8.2	-5.4	0.0	7.2	15.9	11.4	.3015
200.	1082.	8.3	-20.5	-16.9	-9.6	0.0	11.5	20.8	.4337
250.	1004.	15.3	-40.0	-35.5	-26.4	-14.4	0.0	32.3	.5781
300.	939.	25.0	-67.4	-61.9	-51.1	-36.6	-19.3	45.5	.7323
350.	882.	37.9	-103.8	-97.5	-84.8	-67.9	-47.8	60.6	.8972
400.	831.	54.5	-150.5	-143.2	-128.7	-109.4	-86.4	77.6	1.0727
450.	783.	75.3	-208.4	-200.2	-183.9	-162.2	-136.3	96.4	1.2585
500.	737.	101.0	-279.1	-270.0	-251.8	-227.8	-199.0	117.3	1.4559

BALLISTIC COEFFICIENT .14, MUZZLE VELOCITY FT/SEC 800.

RANGE	VELOCITY	HEIGHT	50 YD	100 YD	150 YD	200 YD	250 YD	DRIFT	TIME
0.	800.	0.0	-1.5	-1.5	-1.5	-1.5	-1.5	0.0	0.0000
50.	757.	1.0	0.0	6.9	14.9	23.6	33.2	1.0	.1929
100.	715.	6.8	-13.7	0.0	16.0	33.5	52.7	3.9	.3970
150.	674.	17.3	-44.6	-24.0	0.0	26.3	55.1	8.9	.6131
200.	635.	33.3	-94.6	-67.1	-35.1	0.0	38.3	16.2	.8422
250.	598.	55.8	-166.1	-131.7	-91.8	-47.9	0.0	26.1	1.0855
300.	562.	86.0	-261.9	-220.7	-172.8	-120.1	-62.6	38.6	1.3444
350.	527.	125.3	-385.5	-337.4	-281.5	-220.0	-153.0	54.2	1.6207
400.	493.	175.3	-540.2	-485.2	-421.4	-351.1	-274.5	73.1	1.9153
450.	461.	237.8	-729.8	-667.9	-596.0	-517.0	-430.8	95.4	2.2293
500.	430.	315.7	-961.5	-892.8	-812.9	-725.1	-629.3	121.9	2.5675

BALLISTIC COEFFICIENT .14, MUZZLE VELOCITY FT/SEC 900.

RANGE	VELOCITY	HEIGHT	50 YD	100 YD	150 YD	200 YD	250 YD	DRIFT	TIME
0.	900.	0.0	-1.5	-1.5	-1.5	-1.5	-1.5	0.0	0.0000
50.	851.	.7	0.0	5.3	11.5	18.3	25.8	.9	.1718
100.	806.	5.2	-10.5	0.0	12.4	26.2	41.1	3.5	.3529
150.	762.	13.5	-34.5	-18.7	0.0	20.6	43.0	7.8	.5442
200.	720.	26.0	-73.4	-52.3	-27.4	0.0	29.9	14.1	.7466
250.	680.	43.6	-129.1	-102.8	-71.6	-37.3	0.0	22.5	.9609
300.	640.	67.0	-203.6	-172.0	-134.7	-93.5	-48.7	33.1	1.1883
350.	603.	97.4	-299.3	-262.5	-218.9	-170.9	-118.6	46.3	1.4300
400.	566.	135.9	-418.9	-376.8	-327.0	-272.1	-212.4	62.3	1.6871
450.	531.	183.8	-565.4	-518.0	-462.0	-400.3	-333.1	81.1	1.9608
500.	497.	242.8	-742.2	-689.5	-627.3	-558.7	-484.0	103.1	2.2524

BALLISTIC COEFFICIENT .14, MUZZLE VELOCITY FT/SEC 1000.

RANGE	VELOCITY	HEIGHT	50 YD	100 YD	150 YD	200 YD	250 YD	DRIFT	TIME
0.	1000.	0.0	-1.5	-1.5	-1.5	-1.5	-1.5	0.0	0.0000
50.	940.	.4	0.0	4.2	9.3	14.9	21.1	.9	.1548
100.	886.	4.1	-8.4	0.0	10.2	21.5	33.7	3.4	.3194
150.	838.	10.9	-27.8	-15.3	0.0	16.9	35.3	7.6	.4934
200.	794.	21.3	-59.7	-42.9	-22.6	0.0	24.5	13.6	.6774
250.	751.	35.7	-105.3	-84.3	-58.9	-30.6	0.0	21.4	.8718
300.	709.	55.0	-166.2	-141.0	-110.5	-76.6	-39.9	31.2	1.0774
350.	669.	79.8	-244.2	-214.9	-179.3	-139.8	-96.9	43.2	1.2953
400.	630.	111.1	-341.5	-308.0	-267.3	-222.1	-173.1	57.4	1.5263
450.	593.	149.9	-460.5	-422.8	-377.0	-326.2	-271.0	74.2	1.7719
500.	557.	197.6	-603.7	-561.8	-511.0	-454.5	-393.2	93.8	2.0329

BALLISTIC COEFFICIENT .14, MUZZLE VELOCITY FT/SEC 1100.

RANGE	VELOCITY	HEIGHT	50 YD	100 YD	150 YD	200 YD	250 YD	DRIFT	TIME
0.	1100.	0.0	-1.5	-1.5	-1.5	-1.5	-1.5	0.0	0.0000
50.	1022.	.2	0.0	3.4	7.7	12.5	17.7	1.0	.1420
100.	959.	3.4	-6.8	0.0	8.6	18.3	28.7	3.6	.2932
150.	904.	9.2	-23.1	-12.9	0.0	14.6	30.2	8.0	.4544
200.	854.	18.1	-50.2	-36.6	-19.5	0.0	20.8	14.2	.6264
250.	809.	30.4	-88.7	-71.7	-50.3	-26.0	0.0	21.8	.8059
300.	765.	46.9	-140.5	-120.1	-94.4	-65.2	-34.0	31.4	.9965
350.	723.	68.1	-206.9	-183.1	-153.1	-119.0	-82.7	42.9	1.1981
400.	683.	94.9	-289.6	-262.4	-228.1	-189.2	-147.6	56.4	1.4115
450.	643.	128.0	-390.6	-360.0	-321.4	-277.6	-230.9	72.3	1.6380
500.	605.	168.6	-512.3	-478.4	-435.5	-386.8	-334.9	90.6	1.8785

BALLISTIC COEFFICIENT .14, MUZZLE VELOCITY FT/SEC 1200.

RANGE	VELOCITY	HEIGHT	50 YD	100 YD	150 YD	200 YD	250 YD	DRIFT	TIME
0.	1200.	0.0	-1.5	-1.5	-1.5	-1.5	-1.5	0.0	0.0000
50.	1086.	.1	0.0	3.0	6.7	11.0	15.7	1.2	.1320
100.	1012.	2.9	-6.0	0.0	7.5	16.1	25.4	4.7	.2765
150.	950.	8.1	-20.1	-11.2	0.0	12.9	26.9	9.4	.4284
200.	895.	16.1	-44.1	-32.1	-17.2	0.0	18.7	16.1	.5917
250.	847.	27.2	-78.4	-63.5	-44.9	-23.4	0.0	24.4	.7635
300.	802.	42.2	-124.7	-106.8	-84.5	-58.7	-30.6	34.4	.9454
350.	759.	61.4	-184.1	-163.3	-137.2	-107.0	-74.3	46.2	1.1376
400.	717.	85.6	-258.2	-234.4	-204.6	-170.1	-132.8	60.1	1.3412
450.	676.	115.6	-348.7	-321.9	-288.3	-249.6	-207.5	76.0	1.5567
500.	637.	152.2	-457.5	-427.7	-390.4	-347.4	-300.7	94.2	1.7852

BALLISTIC COEFFICIENT .14, MUZZLE VELOCITY FT/SEC 1300.

RANGE	VELOCITY	HEIGHT	50 YD	100 YD	150 YD	200 YD	250 YD	DRIFT	TIME
0.	1300.	0.0	-1.5	-1.5	-1.5	-1.5	-1.5	0.0	0.0000
50.	1150.	-.0	0.0	2.5	6.0	9.8	14.1	1.4	.1231
100.	1056.	2.5	-5.1	0.0	6.9	14.6	23.2	5.1	.2597
150.	988.	7.2	-17.9	-10.4	0.0	11.5	24.5	10.8	.4076
200.	929.	14.5	-39.3	-29.2	-15.4	0.0	17.3	17.9	.5634
250.	877.	24.8	-70.7	-58.1	-40.8	-21.6	0.0	27.0	.7302
300.	830.	38.6	-112.9	-97.7	-77.0	-53.9	-28.0	37.6	.9057
350.	786.	56.4	-167.2	-149.6	-125.4	-98.4	-68.2	49.9	1.0914
400.	743.	78.8	-235.0	-214.8	-187.1	-156.4	-121.8	64.2	1.2877
450.	702.	106.6	-317.8	-295.1	-264.0	-229.4	-190.5	80.4	1.4955
500.	662.	140.5	-417.6	-392.3	-357.8	-319.3	-276.1	98.9	1.7157

BALLISTIC COEFFICIENT .14, MUZZLE VELOCITY FT/SEC 1400.

RANGE	VELOCITY	HEIGHT	50 YD	100 YD	150 YD	200 YD	250 YD	DRIFT	TIME
0.	1400.	0.0	-1.5	-1.5	-1.5	-1.5	-1.5	0.0	0.0000
50.	1225.	-.1	0.0	2.2	5.3	8.8	12.8	1.4	.1148
100.	1101.	2.1	-4.3	0.0	6.2	13.3	21.2	5.3	.2446
150.	1023.	6.4	-15.8	-9.3	0.0	10.6	22.5	11.5	.3865
200.	960.	13.1	-35.2	-26.6	-14.2	0.0	15.8	19.2	.5376
250.	904.	22.7	-63.8	-53.0	-37.5	-19.8	0.0	28.7	.6988
300.	854.	35.6	-102.9	-90.0	-71.4	-50.1	-26.3	40.1	.8706
350.	809.	52.2	-152.8	-137.7	-116.0	-91.2	-63.5	52.8	1.0501
400.	766.	73.1	-215.4	-198.2	-173.3	-145.0	-113.3	67.5	1.2407
450.	724.	99.1	-292.0	-272.6	-244.7	-212.8	-177.1	84.1	1.4421
500.	683.	130.8	-384.3	-362.7	-331.7	-296.2	-256.6	102.8	1.6555

BALLISTIC COEFFICIENT .14, MUZZLE VELOCITY FT/SEC 1500.

RANGE	VELOCITY	HEIGHT	50 YD	100 YD	150 YD	200 YD	250 YD	DRIFT	TIME
0.	1500.	0.0	-1.5	-1.5	-1.5	-1.5	-1.5	0.0	0.0000
50.	1301.	-.2	0.0	1.8	4.6	7.9	11.6	1.3	.1076
100.	1151.	1.8	-3.7	0.0	5.6	12.2	19.5	5.4	.2307
150.	1056.	5.7	-13.9	-8.4	0.0	10.0	20.9	11.8	.3671
200.	988.	12.0	-31.8	-24.5	-13.3	0.0	14.5	20.2	.5150
250.	930.	20.8	-57.9	-48.7	-34.8	-18.2	0.0	30.1	.6708
300.	877.	32.9	-93.8	-82.8	-66.1	-46.2	-24.3	41.8	.8374
350.	830.	48.5	-140.2	-127.4	-107.9	-84.6	-59.2	55.1	1.0129
400.	786.	68.2	-198.5	-183.8	-161.6	-134.9	-105.9	70.2	1.1986
450.	743.	92.6	-269.8	-253.3	-228.3	-198.3	-165.6	87.1	1.3948
500.	702.	122.5	-355.9	-337.5	-309.7	-276.4	-240.1	106.0	1.6025

BALLISTIC COEFFICIENT .14, MUZZLE VELOCITY FT/SEC 1600.

RANGE	VELOCITY	HEIGHT	50 YD	100 YD	150 YD	200 YD	250 YD	DRIFT	TIME
0.	1600.	0.0	-1.5	-1.5	-1.5	-1.5	-1.5	0.0	0.0000
50.	1382.	-.3	0.0	1.5	4.0	7.1	10.5	1.3	.1010
100.	1211.	1.5	-3.1	0.0	5.0	11.1	17.9	5.2	.2173
150.	1092.	5.1	-12.1	-7.5	0.0	9.1	19.3	11.8	.3483
200.	1017.	10.8	-28.3	-22.1	-12.1	0.0	13.7	20.4	.4910
250.	954.	19.1	-52.4	-44.7	-32.2	-17.1	0.0	30.7	.6433
300.	899.	30.4	-85.5	-76.3	-61.3	-43.1	-22.6	42.7	.8053
350.	850.	45.1	-128.7	-118.0	-100.5	-79.3	-55.4	56.5	.9770
400.	805.	63.7	-183.1	-170.8	-150.8	-126.6	-99.3	71.9	1.1583
450.	762.	86.7	-249.7	-235.9	-213.4	-186.1	-155.4	89.1	1.3498
500.	720.	114.9	-330.2	-314.8	-289.8	-259.5	-225.4	108.2	1.5523

BALLISTIC COEFFICIENT .14, MUZZLE VELOCITY FT/SEC 1700.

RANGE	VELOCITY	HEIGHT	50 YD	100 YD	150 YD	200 YD	250 YD	DRIFT	TIME
0.	1700.	0.0	-1.5	-1.5	-1.5	-1.5	-1.5	0.0	0.0000
50.	1468.	-.3	0.0	1.3	3.5	6.3	9.5	1.2	.0951
100.	1276.	1.3	-2.6	0.0	4.4	10.0	16.3	5.0	.2050
150.	1134.	4.5	-10.5	-6.6	0.0	8.4	17.9	11.5	.3301
200.	1045.	9.8	-25.2	-20.1	-11.2	0.0	12.6	20.4	.4689
250.	979.	17.5	-47.3	-40.8	-29.8	-15.8	0.0	30.9	.6166
300.	921.	28.0	-77.9	-70.2	-56.9	-40.1	-21.2	43.1	.7745
350.	870.	41.9	-118.1	-109.1	-93.6	-74.0	-51.9	57.2	.9424
400.	824.	59.4	-168.9	-158.6	-140.9	-118.5	-93.2	72.8	1.1195
450.	780.	81.2	-231.3	-219.7	-199.8	-174.6	-146.2	90.2	1.3066
500.	737.	107.9	-306.8	-294.0	-271.9	-243.8	-212.3	109.5	1.5046

BALLISTIC COEFFICIENT .14, MUZZLE VELOCITY FT/SEC 1800.

RANGE	VELOCITY	HEIGHT	50 YD	100 YD	150 YD	200 YD	250 YD	DRIFT	TIME
0.	1800.	0.0	-1.5	-1.5	-1.5	-1.5	-1.5	0.0	0.0000
50.	1557.	-.4	0.0	1.1	3.0	5.5	8.5	1.1	.0897
100.	1346.	1.0	-2.1	0.0	3.9	9.0	14.9	4.7	.1935
150.	1183.	3.9	-9.1	-5.9	0.0	7.6	16.5	11.0	.3126
200.	1076.	8.8	-22.2	-17.9	-10.1	0.0	11.9	19.8	.4461
250.	1004.	16.0	-42.6	-37.3	-27.5	-14.9	0.0	30.7	.5910
300.	943.	25.9	-71.0	-64.6	-52.9	-37.7	-19.9	43.2	.7452
350.	890.	38.9	-108.3	-100.9	-87.2	-69.5	-48.6	57.3	.9086
400.	842.	55.4	-155.7	-147.2	-131.6	-111.3	-87.5	73.1	1.0819
450.	797.	76.1	-214.4	-204.8	-187.2	-164.4	-137.6	90.7	1.2652
500.	754.	101.4	-285.4	-274.7	-255.2	-229.9	-200.1	110.1	1.4588

BALLISTIC COEFFICIENT .14, MUZZLE VELOCITY FT/SEC 1900.

RANGE	VELOCITY	HEIGHT	50 YD	100 YD	150 YD	200 YD	250 YD	DRIFT	TIME
0.	1900.	0.0	-1.5	-1.5	-1.5	-1.5	-1.5	0.0	0.0000
50.	1646.	-.4	0.0	.9	2.6	4.9	7.7	1.0	.0849
100.	1420.	.9	-1.7	0.0	3.5	8.0	13.6	4.4	.1830
150.	1240.	3.5	-7.8	-5.2	0.0	6.8	15.2	10.5	.2964
200.	1111.	7.9	-19.6	-16.1	-9.1	0.0	11.1	19.2	.4248
250.	1029.	14.6	-38.3	-34.0	-25.3	-13.9	0.0	30.2	.5662
300.	966.	23.9	-64.6	-59.4	-48.9	-35.3	-18.6	42.7	.7165
350.	910.	36.1	-99.3	-93.2	-81.0	-65.0	-45.6	56.9	.8760
400.	859.	51.8	-143.7	-136.7	-122.8	-104.6	-82.3	72.9	1.0459
450.	814.	71.3	-198.8	-190.9	-175.3	-154.8	-129.8	90.6	1.2253
500.	770.	95.3	-265.6	-256.9	-239.6	-216.7	-189.0	110.1	1.4149

BALLISTIC COEFFICIENT .15, MUZZLE VELOCITY FT/SEC 800.

RANGE	VELOCITY	HEIGHT	50 YD	100 YD	150 YD	200 YD	250 YD	DRIFT	TIME
0.	800.	0.0	-1.5	-1.5	-1.5	-1.5	-1.5	0.0	0.0000
50.	760.	1.0	0.0	6.8	14.7	23.3	32.7	.9	.1924
100.	720.	6.7	-13.6	0.0	15.8	33.0	51.8	3.6	.3953
150.	682.	17.1	-44.1	-23.6	0.0	25.9	54.0	8.2	.6093
200.	646.	32.8	-93.3	-66.1	-34.5	0.0	37.4	15.1	.8355
250.	610.	54.7	-163.4	-129.4	-90.0	-46.8	0.0	24.1	1.0747
300.	576.	83.9	-256.8	-216.0	-168.7	-116.9	-60.7	35.7	1.3276
350.	543.	121.5	-376.4	-328.7	-273.5	-213.1	-147.5	49.9	1.5959
400.	511.	169.1	-525.7	-471.2	-408.1	-339.1	-264.2	67.1	1.8813
450.	480.	228.1	-707.0	-645.8	-574.8	-497.1	-412.8	87.3	2.1833
500.	450.	301.0	-927.6	-859.6	-780.7	-694.4	-600.7	111.3	2.5072

BALLISTIC COEFFICIENT .15, MUZZLE VELOCITY FT/SEC 900.

RANGE	VELOCITY	HEIGHT	50 YD	100 YD	150 YD	200 YD	250 YD	DRIFT	TIME
0.	900.	0.0	-1.5	-1.5	-1.5	-1.5	-1.5	0.0	0.0000
50.	854.	.7	0.0	5.2	11.4	18.1	25.4	.8	.1714
100.	812.	5.2	-10.4	0.0	12.3	25.8	40.4	3.2	.3517
150.	771.	13.3	-34.1	-18.5	0.0	20.2	42.1	7.3	.5413
200.	731.	25.6	-72.5	-51.6	-27.0	0.0	29.2	13.1	.7412
250.	693.	42.7	-127.1	-101.0	-70.2	-36.5	0.0	20.9	.9519
300.	656.	65.4	-199.8	-168.5	-131.5	-91.1	-47.3	30.7	1.1743
350.	620.	94.6	-292.4	-255.9	-212.8	-165.6	-114.5	42.7	1.4091
400.	586.	131.3	-408.1	-366.3	-317.0	-263.1	-204.7	57.3	1.6586
450.	552.	176.7	-548.7	-501.7	-446.3	-385.6	-319.9	74.4	1.9227
500.	520.	232.0	-717.1	-664.9	-603.3	-535.9	-462.9	94.2	2.2021

BALLISTIC COEFFICIENT .15, MUZZLE VELOCITY FT/SEC 1000.

RANGE	VELOCITY	HEIGHT	50 YD	100 YD	150 YD	200 YD	250 YD	DRIFT	TIME
0.	1000.	0.0	-1.5	-1.5	-1.5	-1.5	-1.5	0.0	0.0000
50.	943.	.4	0.0	4.1	9.1	14.7	20.7	.9	.1550
100.	893.	4.1	-8.2	0.0	10.1	21.1	33.1	3.2	.3181
150.	848.	10.8	-27.4	-15.1	0.0	16.6	34.6	7.2	.4906
200.	806.	20.9	-58.7	-42.3	-22.1	0.0	24.0	12.7	.6722
250.	765.	35.0	-103.4	-82.8	-57.6	-29.9	0.0	19.9	.8632
300.	726.	53.7	-162.8	-138.1	-107.9	-74.7	-38.7	29.0	1.0646
350.	688.	77.5	-238.6	-209.8	-174.6	-135.8	-93.9	40.0	1.2771
400.	651.	107.5	-332.7	-299.7	-259.5	-215.2	-167.3	53.1	1.5015
450.	615.	144.3	-446.9	-409.9	-364.6	-314.8	-260.9	68.4	1.7385
500.	581.	189.2	-583.7	-542.5	-492.2	-436.9	-377.0	86.1	1.9892

BALLISTIC COEFFICIENT .15, MUZZLE VELOCITY FT/SEC 1100.

RANGE	VELOCITY	HEIGHT	50 YD	100 YD	150 YD	200 YD	250 YD	DRIFT	TIME
0.	1100.	0.0	-1.5	-1.5	-1.5	-1.5	-1.5	0.0	0.0000
50.	1027.	.2	0.0	3.4	7.6	12.3	17.4	1.0	.1421
100.	967.	3.4	-6.8	0.0	8.4	17.8	28.0	3.5	.2929
150.	914.	9.0	-22.7	-12.5	0.0	14.2	29.5	7.5	.4516
200.	867.	17.7	-49.1	-35.6	-18.9	0.0	20.5	13.2	.6203
250.	824.	29.8	-87.0	-70.1	-49.2	-25.6	0.0	20.4	.7977
300.	783.	45.8	-137.5	-117.2	-92.2	-63.9	-33.1	29.3	.9845
350.	743.	66.2	-202.0	-178.3	-149.1	-116.1	-80.2	39.9	1.1812
400.	704.	91.8	-282.1	-255.0	-221.6	-183.8	-142.9	52.4	1.3888
450.	667.	123.3	-379.3	-348.9	-311.3	-268.8	-222.7	67.0	1.6079
500.	631.	161.6	-495.7	-461.9	-420.1	-372.9	-321.7	83.7	1.8393

BALLISTIC COEFFICIENT .15, MUZZLE VELOCITY FT/SEC 1200.

RANGE	VELOCITY	HEIGHT	50 YD	100 YD	150 YD	200 YD	250 YD	DRIFT	TIME
0.	1200.	0.0	-1.5	-1.5	-1.5	-1.5	-1.5	0.0	0.0000
50.	1092.	.1	0.0	2.9	6.6	10.8	15.4	1.2	.1316
100.	1021.	2.9	-5.8	0.0	7.5	15.8	25.0	4.2	.2740
150.	962.	7.9	-19.9	-11.2	0.0	12.5	26.3	8.9	.4255
200.	910.	15.7	-43.2	-31.6	-16.7	0.0	18.3	15.1	.5858
250.	863.	26.6	-76.9	-62.5	-43.8	-22.9	0.0	22.9	.7552
300.	820.	41.1	-122.1	-104.7	-82.4	-57.3	-29.8	32.3	.9335
350.	779.	59.6	-179.8	-159.6	-133.5	-104.2	-72.1	43.4	1.1213
400.	739.	82.8	-251.4	-228.3	-198.5	-165.1	-128.4	56.2	1.3191
450.	700.	111.3	-338.5	-312.5	-278.9	-241.3	-200.0	70.9	1.5276
500.	663.	145.9	-442.8	-413.9	-376.6	-334.8	-289.0	87.6	1.7479

BALLISTIC COEFFICIENT .15, MUZZLE VELOCITY FT/SEC 1300.

RANGE	VELOCITY	HEIGHT	50 YD	100 YD	150 YD	200 YD	250 YD	DRIFT	TIME
0.	1300.	0.0	-1.5	-1.5	-1.5	-1.5	-1.5	0.0	0.0000
50.	1158.	-.0	0.0	2.5	5.8	9.7	13.9	1.3	.1228
100.	1065.	2.5	-5.1	0.0	6.6	14.3	22.7	5.0	.2591
150.	1001.	7.1	-17.5	-9.9	0.0	11.5	24.1	10.1	.4035
200.	944.	14.2	-38.6	-28.5	-15.3	0.0	16.8	17.0	.5584
250.	893.	24.3	-69.3	-56.7	-40.2	-21.0	0.0	25.6	.7223
300.	848.	37.6	-110.3	-95.1	-75.3	-52.3	-27.0	35.4	.8937
350.	806.	54.7	-163.0	-145.4	-122.2	-95.4	-66.0	47.1	1.0752
400.	766.	76.2	-228.6	-208.4	-181.9	-151.3	-117.6	60.4	1.2661
450.	726.	102.6	-308.3	-285.6	-255.8	-221.4	-183.5	75.5	1.4673
500.	688.	134.7	-403.8	-378.6	-345.5	-307.2	-265.1	92.5	1.6797

BALLISTIC COEFFICIENT .15, MUZZLE VELOCITY FT/SEC 1400.

RANGE	VELOCITY	HEIGHT	50 YD	100 YD	150 YD	200 YD	250 YD	DRIFT	TIME
0.	1400.	0.0	-1.5	-1.5	-1.5	-1.5	-1.5	0.0	0.0000
50.	1235.	-.1	0.0	2.1	5.1	8.6	12.5	1.3	.1145
100.	1114.	2.1	-4.2	0.0	6.0	13.0	20.7	5.0	.2428
150.	1037.	6.3	-15.4	-9.1	0.0	10.5	22.0	10.8	.3826
200.	976.	12.8	-34.5	-26.0	-13.9	0.0	15.4	18.2	.5319
250.	922.	22.1	-62.3	-51.8	-36.7	-19.2	0.0	27.2	.6900
300.	874.	34.6	-100.4	-87.7	-69.6	-48.7	-25.6	38.0	.8585
350.	830.	50.5	-148.7	-133.9	-112.7	-88.3	-61.4	49.9	1.0336
400.	789.	70.6	-209.2	-192.3	-168.2	-140.3	-109.5	63.7	1.2190
450.	749.	95.3	-282.9	-263.9	-236.7	-205.3	-170.7	79.2	1.4142
500.	710.	125.2	-371.2	-350.1	-319.9	-285.0	-246.6	96.6	1.6201

BALLISTIC COEFFICIENT .15, MUZZLE VELOCITY FT/SEC 1500.

RANGE	VELOCITY	HEIGHT	50 YD	100 YD	150 YD	200 YD	250 YD	DRIFT	TIME
0.	1500.	0.0	-1.5	-1.5	-1.5	-1.5	-1.5	0.0	0.0000
50.	1313.	-.2	0.0	1.8	4.5	7.7	11.3	1.3	.1072
100.	1168.	1.8	-3.6	0.0	5.4	11.9	19.0	5.0	.2286
150.	1072.	5.6	-13.5	-8.1	0.0	9.7	20.3	11.1	.3632
200.	1005.	11.7	-30.9	-23.8	-12.9	0.0	14.2	19.1	.5085
250.	948.	20.3	-56.4	-47.5	-33.9	-17.7	0.0	28.4	.6616
300.	897.	31.9	-91.2	-80.5	-64.2	-44.8	-23.5	39.5	.8246
350.	852.	46.9	-136.2	-123.7	-104.7	-82.0	-57.2	52.1	.9961
400.	809.	65.7	-192.4	-178.1	-156.4	-130.5	-102.2	66.3	1.1768
450.	769.	88.9	-261.0	-244.9	-220.5	-191.4	-159.5	82.2	1.3670
500.	729.	117.2	-343.3	-325.5	-298.3	-266.0	-230.5	99.9	1.5674

BALLISTIC COEFFICIENT .15, MUZZLE VELOCITY FT/SEC 1600.

RANGE	VELOCITY	HEIGHT	50 YD	100 YD	150 YD	200 YD	250 YD	DRIFT	TIME
0.	1600.	0.0	-1.5	-1.5	-1.5	-1.5	-1.5	0.0	0.0000
50.	1395.	-.3	0.0	1.5	3.9	6.9	10.2	1.2	.1006
100.	1231.	1.5	-3.0	0.0	4.8	10.7	17.4	4.9	.2152
150.	1111.	4.9	-11.7	-7.3	0.0	8.8	18.8	11.0	.3440
200.	1035.	10.5	-27.4	-21.4	-11.8	0.0	13.3	19.2	.4842
250.	974.	18.5	-50.9	-43.4	-31.3	-16.6	0.0	29.0	.6337
300.	921.	29.4	-82.9	-73.9	-59.4	-41.8	-21.8	40.4	.7920
350.	872.	43.6	-125.1	-114.6	-97.7	-77.1	-53.8	53.6	.9611
400.	829.	61.2	-177.0	-165.1	-145.7	-122.2	-95.6	68.0	1.1362
450.	788.	83.1	-241.1	-227.6	-205.9	-179.4	-149.5	84.1	1.3219
500.	748.	109.8	-318.0	-303.1	-278.9	-249.5	-216.3	102.1	1.5174

BALLISTIC COEFFICIENT .15, MUZZLE VELOCITY FT/SEC 1700.

RANGE	VELOCITY	HEIGHT	50 YD	100 YD	150 YD	200 YD	250 YD	DRIFT	TIME
0.	1700.	0.0	-1.5	-1.5	-1.5	-1.5	-1.5	0.0	0.0000
50.	1482.	-.3	0.0	1.2	3.4	6.1	9.2	1.1	.0945
100.	1299.	1.2	-2.5	0.0	4.3	9.7	15.8	4.7	.2029
150.	1157.	4.3	-10.2	-6.4	0.0	8.1	17.3	10.8	.3259
200.	1065.	9.5	-24.4	-19.4	-10.8	0.0	12.2	19.2	.4620
250.	1000.	16.9	-45.8	-39.5	-28.8	-15.2	0.0	29.1	.6066
300.	944.	27.1	-75.5	-68.0	-55.1	-38.9	-20.6	40.8	.7615
350.	893.	40.3	-114.1	-105.4	-90.3	-71.4	-50.0	54.0	.9245
400.	848.	57.0	-162.9	-152.9	-135.7	-114.1	-89.7	68.8	1.0970
450.	806.	77.7	-222.8	-211.6	-192.3	-167.9	-140.5	85.3	1.2785
500.	765.	102.9	-294.9	-282.4	-261.0	-233.9	-203.4	103.3	1.4695

BALLISTIC COEFFICIENT .15, MUZZLE VELOCITY FT/SEC 1800.

RANGE	VELOCITY	HEIGHT	50 YD	100 YD	150 YD	200 YD	250 YD	DRIFT	TIME
0.	1800.	0.0	-1.5	-1.5	-1.5	-1.5	-1.5	0.0	0.0000
50.	1572.	-.4	0.0	1.0	2.9	5.3	8.2	1.1	.0893
100.	1371.	1.0	-2.0	0.0	3.8	8.6	14.4	4.4	.1915
150.	1212.	3.8	-8.8	-5.7	0.0	7.2	15.9	10.3	.3083
200.	1099.	8.5	-21.3	-17.2	-9.7	0.0	11.5	18.5	.4385
250.	1026.	15.4	-41.1	-36.0	-26.5	-14.4	0.0	28.9	.5806
300.	967.	24.9	-68.5	-62.4	-51.0	-36.5	-19.2	40.7	.7315
350.	914.	37.3	-104.2	-97.1	-83.8	-66.9	-46.7	54.0	.8903
400.	867.	53.1	-149.8	-141.6	-126.4	-107.1	-84.1	69.1	1.0590
450.	824.	72.6	-205.9	-196.7	-179.6	-157.9	-131.9	85.6	1.2365
500.	782.	96.5	-273.6	-263.3	-244.4	-220.2	-191.4	103.8	1.4234

BALLISTIC COEFFICIENT .15, MUZZLE VELOCITY FT/SEC 1900.

RANGE	VELOCITY	HEIGHT	50 YD	100 YD	150 YD	200 YD	250 YD	DRIFT	TIME
0.	1900.	0.0	-1.5	-1.5	-1.5	-1.5	-1.5	0.0	0.0000
50.	1662.	-.4	0.0	.8	2.5	4.7	7.3	1.0	.0845
100.	1448.	.8	-1.7	0.0	3.3	7.7	13.0	4.1	.1813
150.	1273.	3.3	-7.5	-5.0	0.0	6.5	14.4	9.7	.2919
200.	1139.	7.6	-18.8	-15.4	-8.7	0.0	10.5	17.8	.4171
250.	1053.	14.0	-36.6	-32.4	-24.1	-13.1	0.0	28.1	.5543
300.	990.	22.9	-62.1	-57.1	-47.1	-34.0	-18.2	40.3	.7024
350.	935.	34.5	-95.3	-89.4	-77.7	-62.4	-44.0	53.6	.8573
400.	885.	49.4	-137.8	-131.0	-117.7	-100.2	-79.2	68.8	1.0222
450.	841.	67.9	-190.3	-182.7	-167.7	-148.1	-124.4	85.4	1.1960
500.	799.	90.5	-254.0	-245.6	-228.9	-207.1	-180.8	103.8	1.3791

BALLISTIC COEFFICIENT .16, MUZZLE VELOCITY FT/SEC 800.

RANGE	VELOCITY	HEIGHT	50 YD	100 YD	150 YD	200 YD	250 YD	DRIFT	TIME
0.	800.	0.0	-1.5	-1.5	-1.5	-1.5	-1.5	0.0	0.0000
50.	762.	1.0	0.0	6.8	14.6	23.0	32.2	.8	.1922
100.	725.	6.7	-13.5	0.0	15.6	32.5	50.8	3.4	.3942
150.	689.	16.9	-43.7	-23.4	0.0	25.4	52.8	7.7	.6065
200.	655.	32.3	-92.1	-65.1	-33.9	0.0	36.5	14.0	.8297
250.	621.	53.6	-160.8	-127.0	-88.0	-45.6	0.0	22.4	1.0645
300.	589.	82.0	-252.4	-211.8	-165.0	-114.2	-59.4	33.1	1.3132
350.	557.	118.3	-368.6	-321.3	-266.7	-207.4	-143.5	46.2	1.5749
400.	527.	163.9	-513.2	-459.0	-396.7	-328.9	-255.9	62.0	1.8523
450.	497.	220.1	-688.5	-627.6	-557.5	-481.2	-399.1	80.6	2.1452
500.	469.	288.9	-899.4	-831.7	-753.8	-669.0	-577.8	102.3	2.4565

BALLISTIC COEFFICIENT .16, MUZZLE VELOCITY FT/SEC 900.

RANGE	VELOCITY	HEIGHT	50 YD	100 YD	150 YD	200 YD	250 YD	DRIFT	TIME
0.	900.	0.0	-1.5	-1.5	-1.5	-1.5	-1.5	0.0	0.0000
50.	857.	.7	0.0	5.2	11.3	17.9	25.1	.8	.1711
100.	817.	5.1	-10.4	0.0	12.2	25.4	39.8	3.0	.3504
150.	778.	13.2	-33.8	-18.2	0.0	19.9	41.4	6.8	.5385
200.	741.	25.2	-71.6	-50.8	-26.5	0.0	28.7	12.2	.7359
250.	705.	42.0	-125.3	-99.4	-69.0	-35.8	0.0	19.4	.9437
300.	670.	64.1	-196.5	-165.5	-129.0	-89.2	-46.2	28.6	1.1622
350.	636.	92.3	-286.9	-250.7	-208.1	-161.7	-111.5	39.7	1.3920
400.	603.	127.4	-398.8	-357.4	-308.8	-255.7	-198.3	53.0	1.6343
450.	571.	170.8	-534.8	-488.2	-433.5	-373.8	-309.3	68.7	1.8905
500.	540.	223.2	-696.6	-644.8	-584.0	-517.7	-446.0	86.9	2.1601

BALLISTIC COEFFICIENT .16, MUZZLE VELOCITY FT/SEC 1000.

RANGE	VELOCITY	HEIGHT	50 YD	100 YD	150 YD	200 YD	250 YD	DRIFT	TIME
0.	1000.	0.0	-1.5	-1.5	-1.5	-1.5	-1.5	0.0	0.0000
50.	947.	.4	0.0	4.1	9.0	14.5	20.3	.9	.1552
100.	899.	4.1	-8.1	0.0	9.9	20.8	32.6	3.0	.3170
150.	856.	10.7	-27.1	-14.9	0.0	16.4	34.0	6.7	.4881
200.	816.	20.6	-57.9	-41.7	-21.8	0.0	23.5	11.9	.6676
250.	778.	34.4	-101.7	-81.4	-56.6	-29.3	0.0	18.6	.8559
300.	740.	52.5	-159.8	-135.5	-105.7	-73.0	-37.7	27.0	1.0535
350.	704.	75.6	-233.8	-205.4	-170.6	-132.5	-91.4	37.2	1.2615
400.	669.	104.4	-325.2	-292.7	-253.0	-209.4	-162.4	49.3	1.4803
450.	635.	139.7	-435.6	-399.1	-354.4	-305.3	-252.5	63.4	1.7103
500.	602.	182.3	-567.3	-526.8	-477.1	-422.6	-363.9	79.7	1.9528

BALLISTIC COEFFICIENT .16, MUZZLE VELOCITY FT/SEC 1100.

RANGE	VELOCITY	HEIGHT	50 YD	100 YD	150 YD	200 YD	250 YD	DRIFT	TIME
0.	1100.	0.0	-1.5	-1.5	-1.5	-1.5	-1.5	0.0	0.0000
50.	1031.	.2	0.0	3.3	7.5	12.1	17.1	1.0	.1422
100.	975.	3.3	-6.6	0.0	8.3	17.6	27.6	3.2	.2909
150.	924.	8.9	-22.4	-12.5	0.0	13.9	29.0	7.0	.4490
200.	878.	17.5	-48.4	-35.2	-18.6	0.0	20.0	12.4	.6159
250.	837.	29.3	-85.5	-69.1	-48.3	-25.0	0.0	19.2	.7906
300.	798.	44.8	-135.0	-115.3	-90.3	-62.4	-32.4	27.5	.9742
350.	760.	64.6	-197.8	-174.8	-145.7	-113.2	-78.1	37.3	1.1667
400.	723.	89.2	-275.6	-249.3	-216.0	-178.8	-138.8	49.0	1.3692
450.	688.	119.4	-369.7	-340.1	-302.7	-260.8	-215.8	62.5	1.5821
500.	653.	155.8	-481.8	-448.9	-407.3	-360.8	-310.8	77.9	1.8061

BALLISTIC COEFFICIENT .16, MUZZLE VELOCITY FT/SEC 1200.

RANGE	VELOCITY	HEIGHT	50 YD	100 YD	150 YD	200 YD	250 YD	DRIFT	TIME
0.	1200.	0.0	-1.5	-1.5	-1.5	-1.5	-1.5	0.0	0.0000
50.	1097.	.1	0.0	2.9	6.5	10.6	15.2	1.1	.1312
100.	1028.	2.8	-5.8	0.0	7.3	15.5	24.5	4.1	.2735
150.	973.	7.8	-19.6	-10.9	0.0	12.3	25.9	8.4	.4227
200.	922.	15.5	-42.5	-30.9	-16.4	0.0	18.1	14.3	.5811
250.	877.	26.2	-75.8	-61.3	-43.1	-22.6	0.0	21.8	.7487
300.	836.	40.2	-119.8	-102.4	-80.7	-56.0	-28.9	30.5	.9234
350.	797.	58.1	-176.2	-155.9	-130.5	-101.7	-70.1	40.9	1.1073
400.	759.	80.4	-245.7	-222.5	-193.5	-160.6	-124.4	52.8	1.3002
450.	722.	107.7	-330.0	-303.9	-271.2	-234.3	-193.5	66.5	1.5029
500.	686.	140.6	-430.5	-401.5	-365.2	-324.1	-278.9	82.1	1.7163

BALLISTIC COEFFICIENT .16, MUZZLE VELOCITY FT/SEC 1300.

RANGE	VELOCITY	HEIGHT	50 YD	100 YD	150 YD	200 YD	250 YD	DRIFT	TIME
0.	1300.	0.0	-1.5	-1.5	-1.5	-1.5	-1.5	0.0	0.0000
50.	1166.	-.0	0.0	2.4	5.8	9.5	13.6	1.2	.1222
100.	1075.	2.4	-4.9	0.0	6.7	14.1	22.2	4.5	.2565
150.	1012.	7.0	-17.4	-10.1	0.0	11.0	23.3	9.8	.4019
200.	958.	13.9	-37.9	-28.1	-14.7	0.0	16.4	16.1	.5529
250.	909.	23.7	-67.9	-55.6	-38.8	-20.5	0.0	24.1	.7138
300.	865.	36.7	-108.1	-93.4	-73.2	-51.2	-26.6	33.6	.8832
350.	824.	53.3	-159.5	-142.3	-118.8	-93.1	-64.5	44.6	1.0609
400.	786.	73.9	-223.2	-203.6	-176.7	-147.3	-114.6	57.1	1.2474
450.	748.	99.2	-300.3	-278.3	-248.0	-215.0	-178.1	71.2	1.4431
500.	712.	129.7	-392.3	-367.8	-334.1	-297.5	-256.5	87.1	1.6487

BALLISTIC COEFFICIENT .16, MUZZLE VELOCITY FT/SEC 1400.

RANGE	VELOCITY	HEIGHT	50 YD	100 YD	150 YD	200 YD	250 YD	DRIFT	TIME
0.	1400.	0.0	-1.5	-1.5	-1.5	-1.5	-1.5	0.0	0.0000
50.	1244.	-.1	0.0	2.1	5.1	8.5	12.2	1.2	.1139
100.	1126.	2.0	-4.1	0.0	6.1	12.9	20.3	4.7	.2411
150.	1049.	6.2	-15.3	-9.1	0.0	10.2	21.3	10.4	.3808
200.	990.	12.6	-34.0	-25.7	-13.6	0.0	14.8	17.5	.5279
250.	938.	21.6	-61.1	-50.7	-35.6	-18.5	0.0	25.8	.6823
300.	891.	33.6	-97.8	-85.4	-67.3	-46.8	-24.6	35.8	.8465
350.	849.	49.1	-145.1	-130.6	-109.4	-85.6	-59.6	47.4	1.0191
400.	809.	68.4	-203.8	-187.3	-163.1	-135.8	-106.1	60.4	1.2002
450.	771.	92.0	-275.1	-256.4	-229.2	-198.5	-165.2	74.9	1.3901
500.	734.	120.5	-360.1	-339.3	-309.1	-275.0	-237.9	91.2	1.5894

BALLISTIC COEFFICIENT .16, MUZZLE VELOCITY FT/SEC 1500.

RANGE	VELOCITY	HEIGHT	50 YD	100 YD	150 YD	200 YD	250 YD	DRIFT	TIME
0.	1500.	0.0	-1.5	-1.5	-1.5	-1.5	-1.5	0.0	0.0000
50.	1323.	-.2	0.0	1.7	4.4	7.5	11.0	1.2	.1067
100.	1183.	1.7	-3.5	0.0	5.3	11.6	18.6	4.7	.2268
150.	1087.	5.5	-13.2	-8.0	0.0	9.4	19.9	10.5	.3596
200.	1021.	11.4	-30.1	-23.2	-12.6	0.0	14.0	18.0	.5023
250.	965.	19.8	-55.2	-46.5	-33.2	-17.5	0.0	27.1	.6541
300.	916.	31.0	-88.9	-78.4	-62.5	-43.7	-22.6	37.5	.8129
350.	871.	45.6	-133.1	-120.9	-102.3	-80.4	-55.8	49.8	.9829
400.	830.	63.6	-187.1	-173.1	-151.9	-126.8	-98.7	62.9	1.1576
450.	791.	85.8	-253.3	-237.6	-213.8	-185.5	-154.0	77.9	1.3427
500.	754.	112.6	-332.5	-315.0	-288.5	-257.1	-222.1	94.5	1.5369

BALLISTIC COEFFICIENT .16, MUZZLE VELOCITY FT/SEC 1600.

RANGE	VELOCITY	HEIGHT	50 YD	100 YD	150 YD	200 YD	250 YD	DRIFT	TIME
0.	1600.	0.0	-1.5	-1.5	-1.5	-1.5	-1.5	0.0	0.0000
50.	1406.	-.3	0.0	1.5	3.8	6.7	10.0	1.1	.1000
100.	1249.	1.4	-2.9	0.0	4.7	10.4	17.0	4.6	.2134
150.	1129.	4.8	-11.4	-7.0	0.0	8.6	18.5	10.4	.3401
200.	1051.	10.2	-26.7	-20.8	-11.5	0.0	13.2	18.1	.4781
250.	992.	18.1	-49.8	-42.5	-30.8	-16.5	0.0	27.8	.6265
300.	940.	28.5	-80.7	-71.9	-57.8	-40.6	-20.9	38.4	.7804
350.	893.	42.1	-121.1	-110.9	-94.4	-74.4	-51.4	50.7	.9443
400.	850.	59.1	-171.8	-160.1	-141.3	-118.4	-92.0	64.5	1.1166
450.	811.	80.0	-233.6	-220.4	-199.3	-173.5	-143.9	79.8	1.2974
500.	772.	105.4	-307.5	-292.9	-269.4	-240.8	-207.9	96.7	1.4869

BALLISTIC COEFFICIENT .16, MUZZLE VELOCITY FT/SEC 1700.

RANGE	VELOCITY	HEIGHT	50 YD	100 YD	150 YD	200 YD	250 YD	DRIFT	TIME
0.	1700.	0.0	-1.5	-1.5	-1.5	-1.5	-1.5	0.0	0.0000
50.	1495.	-.3	0.0	1.2	3.3	5.9	8.9	1.0	.0941
100.	1319.	1.2	-2.4	0.0	4.1	9.3	15.3	4.3	.2011
150.	1180.	4.2	-9.8	-6.2	0.0	7.8	16.8	10.0	.3216
200.	1085.	9.2	-23.5	-18.7	-10.4	0.0	12.0	17.9	.4546
250.	1019.	16.4	-44.4	-38.3	-27.9	-15.0	0.0	27.5	.5974
300.	964.	26.2	-73.2	-65.9	-53.5	-37.9	-20.0	38.7	.7493
350.	915.	38.9	-110.5	-102.0	-87.5	-69.4	-48.4	51.2	.9086
400.	870.	54.9	-157.6	-147.9	-131.4	-110.6	-86.7	65.3	1.0770
450.	829.	74.7	-215.3	-204.4	-185.8	-162.4	-135.5	80.9	1.2536
500.	791.	98.6	-284.6	-272.5	-251.8	-225.8	-195.9	98.0	1.4390

BALLISTIC COEFFICIENT .16, MUZZLE VELOCITY FT/SEC 1800.

RANGE	VELOCITY	HEIGHT	50 YD	100 YD	150 YD	200 YD	250 YD	DRIFT	TIME
0.	1800.	0.0	-1.5	-1.5	-1.5	-1.5	-1.5	0.0	0.0000
50.	1586.	-.4	0.0	1.0	2.8	5.2	7.9	1.0	.0889
100.	1394.	1.0	-2.0	0.0	3.7	8.3	13.9	4.1	.1898
150.	1239.	3.7	-8.5	-5.5	0.0	7.0	15.4	9.5	.3042
200.	1123.	8.2	-20.6	-16.6	-9.3	0.0	11.2	17.3	.4317
250.	1046.	14.9	-39.7	-34.8	-25.6	-14.0	0.0	27.2	.5715
300.	988.	24.1	-66.2	-60.3	-49.3	-35.3	-18.6	38.5	.7190
350.	937.	35.9	-100.7	-93.8	-80.9	-64.6	-45.1	51.2	.8740
400.	890.	51.0	-144.5	-136.6	-122.0	-103.3	-81.0	65.4	1.0384
450.	847.	69.7	-198.5	-189.6	-173.1	-152.1	-126.9	81.2	1.2113
500.	808.	92.3	-263.4	-253.5	-235.2	-211.9	-183.9	98.4	1.3927

BALLISTIC COEFFICIENT .16, MUZZLE VELOCITY FT/SEC 1900.

RANGE	VELOCITY	HEIGHT	50 YD	100 YD	150 YD	200 YD	250 YD	DRIFT	TIME
0.	1900.	0.0	-1.5	-1.5	-1.5	-1.5	-1.5	0.0	0.0000
50.	1676.	-.4	0.0	.8	2.4	4.5	7.0	.9	.0841
100.	1474.	.8	-1.6	0.0	3.2	7.4	12.5	3.8	.1796
150.	1302.	3.2	-7.3	-4.8	0.0	6.3	13.8	9.0	.2880
200.	1167.	7.3	-18.0	-14.8	-8.4	0.0	10.1	16.6	.4100
250.	1076.	13.5	-35.2	-31.1	-23.1	-12.6	0.0	26.3	.5442
300.	1013.	22.1	-59.8	-55.0	-45.3	-32.8	-17.6	38.0	.6895
350.	958.	33.1	-91.7	-86.0	-74.7	-60.1	-42.4	50.6	.8403
400.	910.	47.4	-132.6	-126.1	-113.2	-96.5	-76.3	65.1	1.0012
450.	865.	65.0	-183.0	-175.7	-161.2	-142.4	-119.7	80.9	1.1704
500.	825.	86.5	-244.0	-235.9	-219.7	-198.8	-173.6	98.3	1.3480

BALLISTIC COEFFICIENT .17, MUZZLE VELOCITY FT/SEC 800.

RANGE	VELOCITY	HEIGHT	50 YD	100 YD	150 YD	200 YD	250 YD	DRIFT	TIME
0.	800.	0.0	-1.5	-1.5	-1.5	-1.5	-1.5	0.0	0.0000
50.	764.	1.0	0.0	6.7	14.4	22.8	31.8	.8	.1920
100.	730.	6.7	-13.4	0.0	15.4	32.1	50.1	3.2	.3931
150.	696.	16.7	-43.2	-23.1	0.0	25.1	52.1	7.2	.6036
200.	663.	31.9	-91.1	-64.2	-33.4	0.0	36.0	13.1	.8245
250.	631.	52.8	-158.8	-125.2	-86.8	-45.0	0.0	21.0	1.0567
300.	600.	80.4	-248.2	-207.8	-161.7	-111.5	-57.5	30.8	1.3000
350.	570.	115.7	-362.2	-315.2	-261.3	-202.8	-139.8	43.1	1.5573
400.	541.	159.4	-502.2	-448.4	-386.9	-320.0	-248.0	57.5	1.8268
450.	513.	213.5	-672.7	-612.2	-542.9	-467.7	-386.7	74.8	2.1125
500.	485.	278.8	-875.4	-808.2	-731.3	-647.7	-557.7	94.7	2.4132

BALLISTIC COEFFICIENT .17, MUZZLE VELOCITY FT/SEC 900.

RANGE	VELOCITY	HEIGHT	50 YD	100 YD	150 YD	200 YD	250 YD	DRIFT	TIME
0.	900.	0.0	-1.5	-1.5	-1.5	-1.5	-1.5	0.0	0.0000
50.	859.	.7	0.0	5.1	11.2	17.7	24.8	.7	.1708
100.	821.	5.1	-10.3	0.0	12.1	25.2	39.2	2.8	.3493
150.	785.	13.1	-33.5	-18.1	0.0	19.7	40.7	6.4	.5363
200.	750.	25.0	-70.9	-50.3	-26.2	0.0	28.1	11.5	.7319
250.	715.	41.4	-123.8	-98.0	-67.9	-35.1	0.0	18.2	.9367
300.	682.	62.9	-193.5	-162.7	-126.5	-87.1	-45.0	26.6	1.1513
350.	650.	90.3	-282.2	-246.2	-203.9	-158.1	-108.9	37.0	1.3770
400.	618.	124.2	-391.0	-349.8	-301.6	-249.2	-193.0	49.3	1.6134
450.	588.	165.8	-523.0	-476.7	-422.4	-363.4	-300.2	63.9	1.8628
500.	558.	215.9	-679.4	-627.9	-567.6	-502.1	-431.9	80.6	2.1245

BALLISTIC COEFFICIENT .17, MUZZLE VELOCITY FT/SEC 1000.

RANGE	VELOCITY	HEIGHT	50 YD	100 YD	150 YD	200 YD	250 YD	DRIFT	TIME
0.	1000.	0.0	-1.5	-1.5	-1.5	-1.5	-1.5	0.0	0.0000
50.	950.	.4	0.0	4.1	9.0	14.4	20.1	.7	.1541
100.	904.	4.0	-8.1	0.0	9.8	20.6	32.2	2.8	.3159
150.	863.	10.6	-27.0	-14.8	0.0	16.1	33.5	6.3	.4858
200.	825.	20.4	-57.5	-41.2	-21.5	0.0	23.1	11.2	.6636
250.	789.	33.9	-100.7	-80.4	-55.8	-28.9	0.0	17.5	.8497
300.	753.	51.6	-157.9	-133.4	-103.9	-71.7	-37.0	25.4	1.0443
350.	719.	74.0	-230.1	-201.6	-167.2	-129.5	-89.1	34.8	1.2479
400.	685.	101.8	-319.3	-286.7	-247.4	-204.3	-158.1	46.1	1.4619
450.	653.	135.7	-426.7	-390.1	-345.8	-297.4	-245.4	59.2	1.6863
500.	621.	176.5	-554.0	-513.2	-464.1	-410.3	-352.5	74.2	1.9214

BALLISTIC COEFFICIENT .17, MUZZLE VELOCITY FT/SEC 1100.

RANGE	VELOCITY	HEIGHT	50 YD	100 YD	150 YD	200 YD	250 YD	DRIFT	TIME
0.	1100.	0.0	-1.5	-1.5	-1.5	-1.5	-1.5	0.0	0.0000
50.	1035.	.2	0.0	3.3	7.4	12.0	16.9	.8	.1409
100.	981.	3.3	-6.6	0.0	8.2	17.4	27.2	3.0	.2900
150.	933.	8.8	-22.3	-12.4	0.0	13.7	28.5	6.7	.4469
200.	889.	17.2	-48.0	-34.7	-18.2	0.0	19.7	11.7	.6117
250.	849.	28.8	-84.7	-68.1	-47.5	-24.7	0.0	18.1	.7846
300.	812.	44.0	-133.3	-113.4	-88.7	-61.3	-31.7	25.9	.9654
350.	776.	63.2	-194.9	-171.7	-142.8	-110.9	-76.4	35.2	1.1544
400.	740.	87.0	-270.6	-244.1	-211.1	-174.6	-135.2	46.0	1.3521
450.	706.	116.0	-362.2	-332.3	-295.2	-254.2	-209.8	58.5	1.5599
500.	673.	150.9	-470.7	-437.5	-396.3	-350.7	-301.4	72.8	1.7775

BALLISTIC COEFFICIENT .17, MUZZLE VELOCITY FT/SEC 1200.

RANGE	VELOCITY	HEIGHT	50 YD	100 YD	150 YD	200 YD	250 YD	DRIFT	TIME
0.	1200.	0.0	-1.5	-1.5	-1.5	-1.5	-1.5	0.0	0.0000
50.	1102.	.1	0.0	2.8	6.5	10.5	14.9	1.0	.1308
100.	1036.	2.8	-5.6	0.0	7.3	15.4	24.2	3.8	.2715
150.	982.	7.7	-19.4	-10.9	0.0	12.1	25.3	8.0	.4206
200.	934.	15.2	-42.0	-30.7	-16.2	0.0	17.6	13.6	.5771
250.	890.	25.7	-74.5	-60.4	-42.2	-22.0	0.0	20.6	.7418
300.	850.	39.4	-117.9	-100.9	-79.1	-54.8	-28.4	28.9	.9145
350.	812.	56.8	-173.0	-153.2	-127.7	-99.4	-68.6	38.7	1.0950
400.	776.	78.4	-240.8	-218.2	-189.1	-156.8	-121.6	50.0	1.2838
450.	741.	104.6	-322.5	-297.1	-264.4	-228.0	-188.4	62.7	1.4814
500.	707.	136.2	-419.9	-391.6	-355.2	-314.8	-270.8	77.3	1.6890

BALLISTIC COEFFICIENT .17, MUZZLE VELOCITY FT/SEC 1300.

RANGE	VELOCITY	HEIGHT	50 YD	100 YD	150 YD	200 YD	250 YD	DRIFT	TIME
0.	1300.	0.0	-1.5	-1.5	-1.5	-1.5	-1.5	0.0	0.0000
50.	1172.	-.0	0.0	2.4	5.6	9.4	13.3	1.3	.1226
100.	1084.	2.4	-4.8	0.0	6.5	14.0	21.9	4.3	.2552
150.	1022.	6.9	-16.9	-9.8	0.0	11.2	23.0	9.2	.3982
200.	970.	13.8	-37.5	-27.9	-14.9	0.0	15.8	15.6	.5499
250.	923.	23.3	-66.5	-54.6	-38.3	-19.7	0.0	22.9	.7071
300.	880.	35.9	-105.8	-91.6	-71.9	-49.6	-26.0	31.9	.8738
350.	841.	52.0	-156.0	-139.3	-116.4	-90.4	-62.8	42.3	1.0482
400.	804.	72.0	-218.0	-199.0	-172.9	-143.1	-111.6	54.2	1.2309
450.	768.	96.3	-292.9	-271.5	-242.1	-208.7	-173.2	67.5	1.4219
500.	733.	125.5	-381.9	-358.1	-325.4	-288.2	-248.8	82.4	1.6219

BALLISTIC COEFFICIENT .17, MUZZLE VELOCITY FT/SEC 1400.

RANGE	VELOCITY	HEIGHT	50 YD	100 YD	150 YD	200 YD	250 YD	DRIFT	TIME
0.	1400.	0.0	-1.5	-1.5	-1.5	-1.5	-1.5	0.0	0.0000
50.	1252.	-.1	0.0	2.1	4.9	8.3	12.0	1.1	.1135
100.	1137.	2.0	-4.1	0.0	5.8	12.5	19.9	4.5	.2399
150.	1060.	6.1	-14.8	-8.7	0.0	10.1	21.1	9.7	.3764
200.	1003.	12.3	-33.3	-25.1	-13.5	0.0	14.7	16.5	.5224
250.	953.	21.1	-59.9	-49.7	-35.2	-18.3	0.0	24.6	.6754
300.	907.	32.9	-95.8	-83.5	-66.2	-45.9	-23.9	34.1	.8368
350.	866.	47.8	-141.9	-127.6	-107.3	-83.7	-58.1	45.1	1.0062
400.	828.	66.5	-199.1	-182.7	-159.5	-132.5	-103.2	57.4	1.1835
450.	791.	89.2	-268.3	-249.8	-223.7	-193.4	-160.4	71.2	1.3689
500.	756.	116.5	-350.4	-329.9	-300.9	-267.2	-230.6	86.5	1.5628

BALLISTIC COEFFICIENT .17, MUZZLE VELOCITY FT/SEC 1500.

RANGE	VELOCITY	HEIGHT	50 YD	100 YD	150 YD	200 YD	250 YD	DRIFT	TIME
0.	1500.	0.0	-1.5	-1.5	-1.5	-1.5	-1.5	0.0	0.0000
50.	1332.	-.2	0.0	1.7	4.3	7.4	10.8	1.1	.1064
100.	1198.	1.7	-3.4	0.0	5.2	11.3	18.1	4.5	.2254
150.	1101.	5.3	-12.9	-7.8	0.0	9.2	19.4	9.9	.3563
200.	1035.	11.1	-29.5	-22.6	-12.3	0.0	13.6	17.1	.4971
250.	981.	19.3	-53.9	-45.3	-32.4	-17.1	0.0	25.7	.6462
300.	933.	30.2	-86.9	-76.6	-61.1	-42.7	-22.2	35.7	.8030
350.	889.	44.2	-129.4	-117.4	-99.3	-77.8	-53.9	47.1	.9678
400.	849.	61.7	-182.3	-168.6	-147.9	-123.4	-96.1	59.9	1.1406
450.	812.	83.1	-246.6	-231.2	-207.9	-180.3	-149.6	74.2	1.3213
500.	776.	108.7	-323.0	-305.9	-280.0	-249.3	-215.2	89.8	1.5103

BALLISTIC COEFFICIENT .17, MUZZLE VELOCITY FT/SEC 1600.

RANGE	VELOCITY	HEIGHT	50 YD	100 YD	150 YD	200 YD	250 YD	DRIFT	TIME
0.	1600.	0.0	-1.5	-1.5	-1.5	-1.5	-1.5	0.0	0.0000
50.	1417.	-.3	0.0	1.4	3.7	6.5	9.7	1.0	.0996
100.	1266.	1.4	-2.9	0.0	4.6	10.2	16.6	4.3	.2118
150.	1147.	4.7	-11.1	-6.9	0.0	8.4	18.0	9.7	.3366
200.	1067.	10.0	-26.0	-20.3	-11.2	0.0	12.8	17.2	.4725
250.	1009.	17.6	-48.5	-41.4	-29.9	-16.0	0.0	26.3	.6183
300.	958.	27.7	-78.6	-70.1	-56.3	-39.6	-20.4	36.5	.7698
350.	912.	40.8	-117.9	-107.9	-91.9	-72.4	-50.0	48.3	.9305
400.	870.	57.2	-167.1	-155.6	-137.3	-115.0	-89.4	61.5	1.0992
450.	831.	77.4	-226.9	-214.0	-193.4	-168.3	-139.5	76.0	1.2756
500.	795.	101.6	-298.3	-284.0	-261.1	-233.2	-201.2	92.0	1.4602

BALLISTIC COEFFICIENT .17, MUZZLE VELOCITY FT/SEC 1700.

RANGE	VELOCITY	HEIGHT	50 YD	100 YD	150 YD	200 YD	250 YD	DRIFT	TIME
0.	1700.	0.0	-1.5	-1.5	-1.5	-1.5	-1.5	0.0	0.0000
50.	1507.	-.3	0.0	1.2	3.2	5.7	8.6	1.0	.0938
100.	1338.	1.2	-2.4	0.0	4.0	9.1	14.9	4.1	.1997
150.	1202.	4.1	-9.6	-6.0	0.0	7.5	16.3	9.4	.3181
200.	1104.	8.9	-22.8	-18.1	-10.0	0.0	11.6	16.9	.4487
250.	1037.	15.9	-43.1	-37.2	-27.1	-14.5	0.0	26.0	.5891
300.	983.	25.4	-71.1	-64.0	-51.9	-36.8	-19.4	36.8	.7382
350.	935.	37.7	-107.4	-99.1	-85.0	-67.4	-47.0	48.7	.8945
400.	891.	53.1	-152.9	-143.5	-127.3	-107.2	-84.0	62.2	1.0590
450.	850.	72.0	-208.7	-198.0	-179.9	-157.3	-131.1	77.0	1.2315
500.	813.	94.9	-275.5	-263.6	-243.5	-218.4	-189.3	93.2	1.4119

BALLISTIC COEFFICIENT .17, MUZZLE VELOCITY FT/SEC 1800.

RANGE	VELOCITY	HEIGHT	50 YD	100 YD	150 YD	200 YD	250 YD	DRIFT	TIME
0.	1800.	0.0	-1.5	-1.5	-1.5	-1.5	-1.5	0.0	0.0000
50.	1598.	-.4	0.0	1.0	2.7	5.0	7.7	.9	.0886
100.	1415.	1.0	-1.9	0.0	3.6	8.1	13.4	3.8	.1883
150.	1264.	3.6	-8.2	-5.3	0.0	6.7	14.7	8.9	.3007
200.	1146.	7.9	-20.0	-16.1	-9.0	0.0	10.7	16.2	.4256
250.	1066.	14.4	-38.3	-33.5	-24.6	-13.3	0.0	25.5	.5617
300.	1008.	23.3	-64.2	-58.4	-47.7	-34.2	-18.2	36.5	.7075
350.	957.	34.7	-97.5	-90.8	-78.3	-62.6	-43.9	48.5	.8592
400.	911.	49.2	-139.9	-132.2	-117.9	-100.0	-78.6	62.2	1.0199
450.	869.	67.1	-191.9	-183.2	-167.2	-147.0	-123.0	77.2	1.1887
500.	831.	88.7	-254.4	-244.8	-227.0	-204.5	-177.9	93.6	1.3652

BALLISTIC COEFFICIENT .17, MUZZLE VELOCITY FT/SEC 1900.

RANGE	VELOCITY	HEIGHT	50 YD	100 YD	150 YD	200 YD	250 YD	DRIFT	TIME
0.	1900.	0.0	-1.5	-1.5	-1.5	-1.5	-1.5	0.0	0.0000
50.	1689.	-.4	0.0	.8	2.4	4.4	6.8	.8	.0838
100.	1496.	.8	-1.6	0.0	3.1	7.2	12.0	3.6	.1781
150.	1329.	3.1	-7.1	-4.7	0.0	6.1	13.3	8.4	.2847
200.	1196.	7.1	-17.5	-14.3	-8.1	0.0	9.7	15.5	.4041
250.	1099.	13.0	-33.9	-30.0	-22.2	-12.1	0.0	24.7	.5351
300.	1034.	21.2	-57.5	-52.8	-43.3	-31.2	-16.7	35.6	.6761
350.	980.	31.9	-88.5	-83.0	-72.1	-57.9	-41.0	48.0	.8252
400.	932.	45.6	-128.0	-121.7	-109.1	-93.0	-73.6	61.7	.9823
450.	888.	62.4	-176.5	-169.4	-155.3	-137.1	-115.4	76.9	1.1473
500.	848.	82.9	-235.1	-227.3	-211.6	-191.4	-167.2	93.4	1.3202

BALLISTIC COEFFICIENT .18, MUZZLE VELOCITY FT/SEC 800.

RANGE	VELOCITY	HEIGHT	50 YD	100 YD	150 YD	200 YD	250 YD	DRIFT	TIME
0.	800.	0.0	-1.5	-1.5	-1.5	-1.5	-1.5	0.0	0.0000
50.	766.	1.0	0.0	6.7	14.3	22.6	31.4	.8	.1919
100.	733.	6.6	-13.3	0.0	15.2	31.8	49.4	3.0	.3921
150.	701.	16.6	-42.9	-22.8	0.0	24.9	51.2	6.8	.6010
200.	670.	31.5	-90.3	-63.6	-33.1	0.0	35.2	12.4	.8203
250.	640.	52.1	-156.8	-123.4	-85.4	-44.0	0.0	19.6	1.0489
300.	610.	79.1	-245.0	-205.0	-159.3	-109.6	-56.9	29.0	1.2896
350.	582.	113.2	-356.2	-309.5	-256.2	-198.2	-136.6	40.2	1.5410
400.	554.	155.8	-493.5	-440.1	-379.2	-312.9	-242.6	53.8	1.8059
450.	527.	207.7	-658.8	-598.7	-530.2	-455.7	-376.5	69.7	2.0838
500.	500.	270.1	-854.6	-787.9	-711.8	-628.9	-541.0	88.1	2.3753

BALLISTIC COEFFICIENT .18, MUZZLE VELOCITY FT/SEC 900.

RANGE	VELOCITY	HEIGHT	50 YD	100 YD	150 YD	200 YD	250 YD	DRIFT	TIME
0.	900.	0.0	-1.5	-1.5	-1.5	-1.5	-1.5	0.0	0.0000
50.	861.	.6	0.0	5.1	11.1	17.5	24.5	.7	.1706
100.	826.	5.1	-10.2	0.0	12.0	24.9	38.7	2.7	.3486
150.	791.	13.0	-33.3	-17.9	0.0	19.4	40.1	6.0	.5344
200.	758.	24.7	-70.2	-49.7	-25.8	0.0	27.7	10.8	.7279
250.	725.	40.8	-122.4	-96.8	-66.9	-34.6	0.0	17.1	.9305
300.	693.	61.9	-191.1	-160.4	-124.5	-85.8	-44.3	25.0	1.1423
350.	662.	88.5	-277.8	-241.9	-200.1	-154.9	-106.5	34.6	1.3635
400.	632.	121.5	-384.5	-343.5	-295.7	-244.1	-188.7	46.2	1.5957
450.	603.	161.5	-512.5	-466.4	-412.6	-354.6	-292.3	59.6	1.8386
500.	574.	209.7	-664.8	-613.5	-553.8	-489.3	-420.0	75.2	2.0939

BALLISTIC COEFFICIENT .18, MUZZLE VELOCITY FT/SEC 1000.

RANGE	VELOCITY	HEIGHT	50 YD	100 YD	150 YD	200 YD	250 YD	DRIFT	TIME
0.	1000.	0.0	-1.5	-1.5	-1.5	-1.5	-1.5	0.0	0.0000
50.	953.	.4	0.0	4.0	8.9	14.2	19.9	.7	.1538
100.	909.	4.0	-8.1	0.0	9.7	20.4	31.7	2.7	.3151
150.	870.	10.5	-26.8	-14.6	0.0	15.9	32.9	6.0	.4840
200.	833.	20.2	-56.9	-40.7	-21.2	0.0	22.7	10.6	.6601
250.	799.	33.4	-99.5	-79.3	-54.9	-28.4	0.0	16.5	.8438
300.	765.	50.8	-155.8	-131.5	-102.3	-70.4	-36.4	23.9	1.0359
350.	732.	72.6	-226.8	-198.5	-164.4	-127.2	-87.5	32.8	1.2365
400.	700.	99.6	-313.7	-281.3	-242.3	-199.9	-154.5	43.2	1.4456
450.	669.	132.4	-418.6	-382.2	-338.3	-290.6	-239.5	55.5	1.6653
500.	639.	171.5	-542.2	-501.7	-453.0	-399.9	-343.1	69.4	1.8944

BALLISTIC COEFFICIENT .18, MUZZLE VELOCITY FT/SEC 1100.

RANGE	VELOCITY	HEIGHT	50 YD	100 YD	150 YD	200 YD	250 YD	DRIFT	TIME
0.	1100.	0.0	-1.5	-1.5	-1.5	-1.5	-1.5	0.0	0.0000
50.	1038.	.2	0.0	3.3	7.4	11.9	16.7	.7	.1406
100.	986.	3.3	-6.7	0.0	8.1	17.1	26.8	3.0	.2899
150.	940.	8.7	-22.1	-12.1	0.0	13.5	28.1	6.3	.4448
200.	898.	17.0	-47.5	-34.2	-18.1	0.0	19.4	11.1	.6084
250.	859.	28.4	-83.6	-66.9	-46.8	-24.2	0.0	17.1	.7790
300.	824.	43.2	-131.4	-111.4	-87.2	-60.1	-31.1	24.5	.9572
350.	789.	62.0	-191.9	-168.5	-140.3	-108.8	-74.9	33.2	1.1434
400.	756.	85.1	-266.0	-239.3	-207.1	-171.0	-132.3	43.4	1.3375
450.	723.	113.1	-355.1	-325.1	-288.8	-248.2	-204.7	55.1	1.5404
500.	692.	146.7	-460.7	-427.3	-387.0	-341.9	-293.5	68.5	1.7527

BALLISTIC COEFFICIENT .18, MUZZLE VELOCITY FT/SEC 1200.

RANGE	VELOCITY	HEIGHT	50 YD	100 YD	150 YD	200 YD	250 YD	DRIFT	TIME
0.	1200.	0.0	-1.5	-1.5	-1.5	-1.5	-1.5	0.0	0.0000
50.	1106.	.1	0.0	2.8	6.5	10.4	14.7	1.0	.1306
100.	1042.	2.8	-5.6	0.0	7.3	15.2	23.8	3.7	.2708
150.	990.	7.7	-19.4	-10.9	0.0	11.8	24.7	7.9	.4196
200.	944.	15.1	-41.6	-30.4	-15.8	0.0	17.1	13.0	.5740
250.	901.	25.2	-73.4	-59.4	-41.2	-21.4	0.0	19.5	.7359
300.	863.	38.7	-116.0	-99.2	-77.3	-53.6	-27.9	27.5	.9063
350.	827.	55.7	-170.0	-150.4	-124.9	-97.2	-67.2	36.8	1.0840
400.	792.	76.6	-236.4	-214.0	-184.8	-153.2	-119.0	47.4	1.2695
450.	759.	101.9	-316.1	-290.8	-258.0	-222.5	-183.9	59.4	1.4627
500.	726.	132.3	-410.6	-382.5	-346.1	-306.6	-263.8	73.1	1.6651

BALLISTIC COEFFICIENT .18, MUZZLE VELOCITY FT/SEC 1300.

RANGE	VELOCITY	HEIGHT	50 YD	100 YD	150 YD	200 YD	250 YD	DRIFT	TIME
0.	1300.	0.0	-1.5	-1.5	-1.5	-1.5	-1.5	0.0	0.0000
50.	1178.	-.0	0.0	2.4	5.6	9.2	13.1	1.1	.1216
100.	1092.	2.3	-4.8	0.0	6.5	13.7	21.5	4.1	.2540
150.	1031.	6.8	-16.9	-9.8	0.0	10.7	22.5	8.9	.3968
200.	981.	13.5	-36.8	-27.3	-14.2	0.0	15.7	14.7	.5448
250.	935.	22.9	-65.7	-53.8	-37.5	-19.7	0.0	21.9	.7014
300.	893.	35.3	-104.6	-90.3	-70.7	-49.4	-25.8	30.7	.8667
350.	855.	50.9	-153.4	-136.8	-113.9	-89.0	-61.4	40.4	1.0372
400.	820.	70.2	-213.9	-194.9	-168.7	-140.2	-108.8	51.6	1.2161
450.	786.	93.8	-287.0	-265.6	-236.2	-204.1	-168.8	64.2	1.4033
500.	752.	121.9	-373.5	-349.7	-317.0	-281.4	-242.0	78.3	1.5984

BALLISTIC COEFFICIENT .18, MUZZLE VELOCITY FT/SEC 1400.

RANGE	VELOCITY	HEIGHT	50 YD	100 YD	150 YD	200 YD	250 YD	DRIFT	TIME
0.	1400.	0.0	-1.5	-1.5	-1.5	-1.5	-1.5	0.0	0.0000
50.	1260.	-.1	0.0	2.0	4.9	8.2	11.8	1.0	.1131
100.	1148.	2.0	-4.0	0.0	5.7	12.3	19.6	4.2	.2382
150.	1072.	6.0	-14.6	-8.6	0.0	9.9	20.8	9.2	.3737
200.	1016.	12.1	-32.7	-24.7	-13.2	0.0	14.6	15.7	.5177
250.	966.	20.8	-59.1	-49.1	-34.7	-18.2	0.0	23.6	.6700
300.	922.	32.2	-94.1	-82.0	-64.8	-45.0	-23.1	32.6	.8280
350.	881.	46.7	-139.1	-125.1	-105.0	-81.9	-56.4	43.0	.9945
400.	844.	64.8	-194.9	-178.8	-155.9	-129.5	-100.3	54.8	1.1685
450.	809.	86.8	-262.2	-244.2	-218.4	-188.7	-155.9	67.9	1.3502
500.	775.	113.0	-342.0	-322.0	-293.3	-260.3	-223.8	82.4	1.5395

BALLISTIC COEFFICIENT .18, MUZZLE VELOCITY FT/SEC 1500.

RANGE	VELOCITY	HEIGHT	50 YD	100 YD	150 YD	200 YD	250 YD	DRIFT	TIME
0.	1500.	0.0	-1.5	-1.5	-1.5	-1.5	-1.5	0.0	0.0000
50.	1341.	-.2	0.0	1.7	4.2	7.3	10.6	1.1	.1060
100.	1212.	1.7	-3.4	0.0	5.1	11.2	17.8	4.2	.2239
150.	1114.	5.2	-12.7	-7.6	0.0	9.2	19.0	9.4	.3534
200.	1047.	11.0	-29.1	-22.4	-12.2	0.0	13.1	16.5	.4938
250.	996.	18.9	-52.8	-44.4	-31.7	-16.4	0.0	24.6	.6395
300.	948.	29.5	-85.1	-75.0	-59.8	-41.4	-21.7	34.1	.7940
350.	906.	43.1	-126.6	-114.8	-97.0	-75.6	-52.6	45.0	.9558
400.	866.	60.0	-178.1	-164.7	-144.4	-119.9	-93.6	57.3	1.1254
450.	830.	80.7	-240.6	-225.5	-202.6	-175.1	-145.5	70.8	1.3023
500.	796.	105.4	-314.8	-298.0	-272.6	-242.0	-209.2	85.7	1.4868

BALLISTIC COEFFICIENT .18, MUZZLE VELOCITY FT/SEC 1600.

RANGE	VELOCITY	HEIGHT	50 YD	100 YD	150 YD	200 YD	250 YD	DRIFT	TIME
0.	1600.	0.0	-1.5	-1.5	-1.5	-1.5	-1.5	0.0	0.0000
50.	1426.	-.3	0.0	1.4	3.6	6.4	9.5	1.0	.0993
100.	1281.	1.4	-2.8	0.0	4.5	9.9	16.1	4.0	.2104
150.	1164.	4.6	-10.9	-6.7	0.0	8.2	17.5	9.2	.3335
200.	1082.	9.8	-25.5	-19.9	-10.9	0.0	12.3	16.3	.4679
250.	1024.	17.1	-47.3	-40.3	-29.1	-15.4	0.0	25.0	.6107
300.	974.	27.0	-76.8	-68.5	-55.0	-38.6	-20.1	34.8	.7605
350.	929.	39.7	-115.1	-105.3	-89.7	-70.5	-48.9	46.1	.9182
400.	888.	55.6	-162.8	-151.7	-133.8	-111.9	-87.2	58.7	1.0834
450.	850.	75.0	-220.9	-208.3	-188.2	-163.6	-135.8	72.6	1.2562
500.	815.	98.3	-290.1	-276.1	-253.8	-226.4	-195.5	87.8	1.4364

BALLISTIC COEFFICIENT .18, MUZZLE VELOCITY FT/SEC 1700.

RANGE	VELOCITY	HEIGHT	50 YD	100 YD	150 YD	200 YD	250 YD	DRIFT	TIME
0.	1700.	0.0	-1.5	-1.5	-1.5	-1.5	-1.5	0.0	0.0000
50.	1517.	-.3	0.0	1.2	3.1	5.6	8.4	.9	.0935
100.	1355.	1.1	-2.3	0.0	3.9	8.8	14.5	3.9	.1984
150.	1223.	4.0	-9.4	-5.9	0.0	7.3	15.8	8.8	.3149
200.	1122.	8.7	-22.2	-17.6	-9.8	0.0	11.3	15.9	.4434
250.	1053.	15.5	-41.9	-36.1	-26.3	-14.1	0.0	24.7	.5816
300.	1000.	24.7	-69.2	-62.2	-50.4	-35.8	-18.8	34.9	.7279
350.	953.	36.6	-104.5	-96.4	-82.6	-65.6	-45.8	46.5	.8816
400.	910.	51.4	-148.7	-139.4	-123.8	-104.2	-81.6	59.3	1.0428
450.	870.	69.7	-202.8	-192.3	-174.7	-152.7	-127.3	73.5	1.2116
500.	834.	91.7	-267.4	-255.8	-236.2	-211.8	-183.5	88.9	1.3877

BALLISTIC COEFFICIENT .18, MUZZLE VELOCITY FT/SEC 1800.

RANGE	VELOCITY	HEIGHT	50 YD	100 YD	150 YD	200 YD	250 YD	DRIFT	TIME
0.	1800.	0.0	-1.5	-1.5	-1.5	-1.5	-1.5	0.0	0.0000
50.	1608.	-.4	0.0	.9	2.7	4.9	7.4	.9	.0882
100.	1434.	.9	-1.9	0.0	3.5	7.8	13.0	3.6	.1871
150.	1287.	3.5	-8.0	-5.2	0.0	6.5	14.3	8.4	.2976
200.	1168.	7.7	-19.4	-15.6	-8.7	0.0	10.3	15.3	.4202
250.	1085.	14.0	-37.1	-32.4	-23.8	-12.9	0.0	24.1	.5537
300.	1026.	22.6	-62.2	-56.6	-46.2	-33.1	-17.6	34.6	.6968
350.	977.	33.6	-94.7	-88.1	-76.0	-60.7	-42.7	46.2	.8459
400.	931.	47.6	-135.7	-128.2	-114.3	-96.9	-76.3	59.3	1.0034
450.	890.	64.8	-186.0	-177.5	-162.0	-142.4	-119.2	73.6	1.1682
500.	852.	85.5	-246.4	-237.0	-219.7	-197.9	-172.1	89.3	1.3407

BALLISTIC COEFFICIENT .18, MUZZLE VELOCITY FT/SEC 1900.

RANGE	VELOCITY	HEIGHT	50 YD	100 YD	150 YD	200 YD	250 YD	DRIFT	TIME
0.	1900.	0.0	-1.5	-1.5	-1.5	-1.5	-1.5	0.0	0.0000
50.	1700.	-.4	0.0	.8	2.3	4.2	6.6	.8	.0835
100.	1517.	.8	-1.5	0.0	3.1	6.9	11.6	3.4	.1769
150.	1355.	3.1	-6.9	-4.6	0.0	5.8	12.8	7.9	.2819
200.	1223.	6.9	-16.9	-13.9	-7.8	0.0	9.3	14.5	.3984
250.	1122.	12.6	-32.8	-29.0	-21.4	-11.7	0.0	23.3	.5269
300.	1053.	20.5	-55.6	-51.0	-41.8	-30.1	-16.1	33.7	.6651
350.	1000.	30.9	-85.7	-80.3	-69.6	-56.0	-39.7	45.5	.8114
400.	953.	44.0	-123.8	-117.7	-105.4	-89.9	-71.3	58.7	.9651
450.	910.	60.1	-170.7	-163.8	-150.0	-132.5	-111.5	73.2	1.1263
500.	870.	79.8	-227.2	-219.5	-204.2	-184.8	-161.5	89.0	1.2952

BALLISTIC COEFFICIENT .19, MUZZLE VELOCITY FT/SEC 800.

RANGE	VELOCITY	HEIGHT	50 YD	100 YD	150 YD	200 YD	250 YD	DRIFT	TIME
0.	800.	0.0	-1.5	-1.5	-1.5	-1.5	-1.5	0.0	0.0000
50.	768.	1.0	0.0	6.6	14.2	22.4	31.1	.7	.1917
100.	737.	6.6	-13.2	0.0	15.2	31.5	48.9	2.8	.3910
150.	706.	16.5	-42.6	-22.7	0.0	24.4	50.6	6.5	.5992
200.	677.	31.2	-89.4	-62.9	-32.6	0.0	34.9	11.6	.8161
250.	648.	51.5	-155.4	-122.3	-84.4	-43.6	0.0	18.6	1.0429
300.	619.	77.8	-241.8	-202.1	-156.6	-107.7	-55.3	27.2	1.2793
350.	592.	111.3	-351.4	-305.1	-252.0	-195.0	-133.9	37.9	1.5277
400.	565.	152.5	-485.5	-432.5	-371.8	-306.6	-236.9	50.5	1.7869
450.	539.	202.6	-646.4	-586.8	-518.5	-445.2	-366.7	65.2	2.0582
500.	514.	262.9	-837.8	-771.5	-695.7	-614.2	-527.0	82.5	2.3437

BALLISTIC COEFFICIENT .19, MUZZLE VELOCITY FT/SEC 900.

RANGE	VELOCITY	HEIGHT	50 YD	100 YD	150 YD	200 YD	250 YD	DRIFT	TIME
0.	900.	0.0	-1.5	-1.5	-1.5	-1.5	-1.5	0.0	0.0000
50.	863.	.6	0.0	5.1	11.0	17.4	24.2	.7	.1705
100.	829.	5.1	-10.2	0.0	11.8	24.6	38.3	2.6	.3479
150.	797.	12.9	-33.0	-17.7	0.0	19.2	39.7	5.7	.5324
200.	765.	24.5	-69.6	-49.2	-25.6	0.0	27.3	10.2	.7247
250.	733.	40.3	-121.2	-95.7	-66.1	-34.1	0.0	16.2	.9251
300.	703.	61.0	-188.8	-158.2	-122.7	-84.3	-43.4	23.6	1.1339
350.	673.	87.0	-274.1	-238.5	-197.1	-152.3	-104.5	32.6	1.3522
400.	645.	119.0	-378.5	-337.7	-290.4	-239.2	-184.6	43.4	1.5797
450.	616.	157.8	-503.6	-457.8	-404.5	-347.0	-285.6	55.9	1.8177
500.	589.	204.3	-651.9	-601.0	-541.8	-477.8	-409.6	70.5	2.0670

BALLISTIC COEFFICIENT .19, MUZZLE VELOCITY FT/SEC 1000.

RANGE	VELOCITY	HEIGHT	50 YD	100 YD	150 YD	200 YD	250 YD	DRIFT	TIME
0.	1000.	0.0	-1.5	-1.5	-1.5	-1.5	-1.5	0.0	0.0000
50.	955.	.4	0.0	4.0	8.9	14.1	19.7	.6	.1536
100.	914.	4.0	-8.0	0.0	9.8	20.1	31.4	2.5	.3143
150.	876.	10.5	-26.7	-14.7	0.0	15.5	32.4	5.8	.4831
200.	841.	20.0	-56.4	-40.3	-20.7	0.0	22.5	10.0	.6568
250.	808.	33.0	-98.6	-78.5	-54.1	-28.1	0.0	15.7	.8391
300.	776.	50.0	-154.0	-129.8	-100.5	-69.4	-35.7	22.6	1.0285
350.	744.	71.4	-223.7	-195.6	-161.4	-125.1	-85.7	31.0	1.2260
400.	714.	97.7	-309.2	-277.0	-237.9	-196.4	-151.4	40.8	1.4320
450.	684.	129.4	-411.4	-375.2	-331.2	-284.6	-233.9	52.2	1.6466
500.	655.	167.3	-532.2	-492.0	-443.1	-391.3	-335.0	65.3	1.8711

BALLISTIC COEFFICIENT .19, MUZZLE VELOCITY FT/SEC 1100.

RANGE	VELOCITY	HEIGHT	50 YD	100 YD	150 YD	200 YD	250 YD	DRIFT	TIME
0.	1100.	0.0	-1.5	-1.5	-1.5	-1.5	-1.5	0.0	0.0000
50.	1040.	.2	0.0	3.4	7.4	11.7	16.5	.7	.1404
100.	991.	3.3	-6.7	0.0	8.0	16.8	26.4	3.0	.2898
150.	947.	8.7	-22.1	-12.0	0.0	13.1	27.5	6.2	.4442
200.	907.	16.8	-47.0	-33.6	-17.5	0.0	19.2	10.5	.6049
250.	869.	28.0	-82.7	-65.9	-45.8	-23.9	0.0	16.3	.7742
300.	835.	42.6	-129.7	-109.6	-85.5	-59.3	-30.5	23.2	.9502
350.	802.	60.9	-189.1	-165.6	-137.5	-106.9	-73.4	31.5	1.1334
400.	770.	83.5	-262.0	-235.1	-203.0	-168.0	-129.7	41.1	1.3245
450.	739.	110.6	-348.9	-318.7	-282.6	-243.2	-200.1	52.1	1.5232
500.	708.	143.1	-451.9	-418.3	-378.2	-334.4	-286.5	64.7	1.7310

BALLISTIC COEFFICIENT .19, MUZZLE VELOCITY FT/SEC 1200.

RANGE	VELOCITY	HEIGHT	50 YD	100 YD	150 YD	200 YD	250 YD	DRIFT	TIME
0.	1200.	0.0	-1.5	-1.5	-1.5	-1.5	-1.5	0.0	0.0000
50.	1110.	.1	0.0	2.8	6.3	10.3	14.5	.9	.1303
100.	1048.	2.8	-5.7	0.0	7.0	14.8	23.3	3.7	.2710
150.	998.	7.6	-19.0	-10.4	0.0	11.8	24.6	7.3	.4163
200.	953.	14.9	-41.0	-29.7	-15.8	0.0	17.0	12.3	.5701
250.	912.	24.9	-72.6	-58.4	-41.0	-21.3	0.0	18.7	.7310
300.	874.	38.2	-114.8	-97.8	-76.9	-53.3	-27.7	26.5	.9004
350.	840.	54.6	-167.4	-147.5	-123.1	-95.5	-65.8	35.0	1.0740
400.	807.	75.0	-232.5	-209.8	-182.0	-150.5	-116.4	45.2	1.2566
450.	775.	99.7	-310.5	-284.9	-253.6	-218.2	-179.9	56.6	1.4463
500.	743.	129.0	-402.5	-374.0	-339.3	-299.9	-257.3	69.3	1.6440

BALLISTIC COEFFICIENT .19, MUZZLE VELOCITY FT/SEC 1300.

RANGE	VELOCITY	HEIGHT	50 YD	100 YD	150 YD	200 YD	250 YD	DRIFT	TIME
0.	1300.	0.0	-1.5	-1.5	-1.5	-1.5	-1.5	0.0	0.0000
50.	1184.	-.0	0.0	2.4	5.5	9.2	13.0	1.0	.1211
100.	1100.	2.3	-4.7	0.0	6.3	13.6	21.3	3.9	.2529
150.	1040.	6.7	-16.6	-9.5	0.0	10.9	22.5	8.3	.3933
200.	991.	13.4	-36.7	-27.3	-14.6	0.0	15.4	14.3	.5428
250.	947.	22.6	-65.1	-53.3	-37.5	-19.2	0.0	21.2	.6972
300.	906.	34.6	-102.7	-88.6	-69.6	-47.7	-24.6	29.2	.8580
350.	869.	49.9	-151.0	-134.5	-112.3	-86.8	-59.8	38.6	1.0272
400.	835.	68.7	-210.3	-191.4	-166.1	-136.9	-106.1	49.3	1.2033
450.	802.	91.5	-281.5	-260.3	-231.8	-199.0	-164.4	61.3	1.3866
500.	770.	118.7	-365.9	-342.3	-310.6	-274.2	-235.7	74.6	1.5777

BALLISTIC COEFFICIENT .19, MUZZLE VELOCITY FT/SEC 1400.

RANGE	VELOCITY	HEIGHT	50 YD	100 YD	150 YD	200 YD	250 YD	DRIFT	TIME
0.	1400.	0.0	-1.5	-1.5	-1.5	-1.5	-1.5	0.0	0.0000
50.	1267.	-.1	0.0	2.0	4.8	8.1	11.6	1.0	.1128
100.	1158.	2.0	-4.0	0.0	5.6	12.2	19.2	4.0	.2372
150.	1082.	5.9	-14.4	-8.5	0.0	9.8	20.3	8.8	.3716
200.	1026.	12.0	-32.3	-24.4	-13.1	0.0	14.1	15.1	.5146
250.	979.	20.4	-57.9	-48.0	-33.9	-17.6	0.0	22.5	.6635
300.	936.	31.6	-92.5	-80.6	-63.7	-44.1	-23.0	31.2	.8203
350.	896.	45.8	-136.8	-122.9	-103.2	-80.3	-55.7	41.4	.9849
400.	859.	63.3	-191.1	-175.2	-152.6	-126.5	-98.4	52.5	1.1552
450.	826.	84.6	-256.8	-238.9	-213.5	-184.1	-152.5	65.0	1.3333
500.	793.	110.0	-334.6	-314.7	-286.5	-253.8	-218.7	78.7	1.5188

BALLISTIC COEFFICIENT .19, MUZZLE VELOCITY FT/SEC 1500.

RANGE	VELOCITY	HEIGHT	50 YD	100 YD	150 YD	200 YD	250 YD	DRIFT	TIME
0.	1500.	0.0	-1.5	-1.5	-1.5	-1.5	-1.5	0.0	0.0000
50.	1348.	-.2	0.0	1.6	4.1	7.1	10.4	1.0	.1057
100.	1224.	1.6	-3.3	0.0	5.0	10.9	17.5	4.0	.2226
150.	1127.	5.2	-12.4	-7.5	0.0	8.8	18.8	8.9	.3507
200.	1060.	10.7	-28.3	-21.7	-11.8	0.0	13.3	15.5	.4881
250.	1008.	18.6	-52.1	-43.8	-31.4	-16.7	0.0	23.7	.6345
300.	963.	28.9	-83.5	-73.7	-58.7	-41.0	-21.1	32.7	.7859
350.	921.	42.1	-124.0	-112.4	-95.0	-74.4	-51.1	43.1	.9448
400.	882.	58.5	-174.3	-161.1	-141.2	-117.6	-91.0	54.8	1.1115
450.	847.	78.5	-235.2	-220.4	-197.9	-171.5	-141.5	67.8	1.2852
500.	814.	102.4	-307.4	-291.0	-266.0	-236.6	-203.3	82.0	1.4659

BALLISTIC COEFFICIENT .19, MUZZLE VELOCITY FT/SEC 1600.

RANGE	VELOCITY	HEIGHT	50 YD	100 YD	150 YD	200 YD	250 YD	DRIFT	TIME
0.	1600.	0.0	-1.5	-1.5	-1.5	-1.5	-1.5	0.0	0.0000
50.	1435.	-.3	0.0	1.4	3.6	6.2	9.2	.9	.0990
100.	1295.	1.4	-2.8	0.0	4.4	9.7	15.7	3.8	.2093
150.	1180.	4.5	-10.7	-6.5	0.0	8.0	17.0	8.7	.3307
200.	1097.	9.5	-24.9	-19.4	-10.6	0.0	12.0	15.5	.4629
250.	1038.	16.7	-46.1	-39.3	-28.3	-15.1	0.0	23.7	.6036
300.	989.	26.5	-75.5	-67.2	-54.1	-38.2	-20.1	33.5	.7531
350.	945.	38.8	-112.8	-103.1	-87.9	-69.3	-48.2	44.3	.9078
400.	905.	54.1	-159.0	-148.0	-130.6	-109.3	-85.2	56.2	1.0691
450.	868.	72.9	-215.6	-203.2	-183.5	-159.6	-132.5	69.5	1.2387
500.	833.	95.4	-282.8	-269.0	-247.2	-220.6	-190.5	84.0	1.4150

BALLISTIC COEFFICIENT .19, MUZZLE VELOCITY FT/SEC 1700.

RANGE	VELOCITY	HEIGHT	50 YD	100 YD	150 YD	200 YD	250 YD	DRIFT	TIME
0.	1700.	0.0	-1.5	-1.5	-1.5	-1.5	-1.5	0.0	0.0000
50.	1526.	-.3	0.0	1.1	3.1	5.4	8.2	.9	.0932
100.	1370.	1.1	-2.3	0.0	3.9	8.6	14.1	3.6	.1971
150.	1242.	3.9	-9.2	-5.8	0.0	7.1	15.4	8.4	.3122
200.	1140.	8.5	-21.7	-17.2	-9.5	0.0	11.0	15.1	.4385
250.	1069.	15.1	-40.9	-35.3	-25.6	-13.8	0.0	23.5	.5747
300.	1016.	24.0	-67.5	-60.7	-49.1	-34.9	-18.4	33.3	.7187
350.	969.	35.7	-102.3	-94.4	-80.9	-64.3	-45.0	44.6	.8713
400.	927.	50.0	-145.0	-135.9	-120.5	-101.6	-79.5	56.7	1.0281
450.	888.	67.6	-197.4	-187.3	-169.9	-148.6	-123.8	70.3	1.1935
500.	852.	88.8	-260.1	-248.8	-229.6	-205.9	-178.3	85.1	1.3660

BALLISTIC COEFFICIENT .19, MUZZLE VELOCITY FT/SEC 1800.

RANGE	VELOCITY	HEIGHT	50 YD	100 YD	150 YD	200 YD	250 YD	DRIFT	TIME
0.	1800.	0.0	-1.5	-1.5	-1.5	-1.5	-1.5	0.0	0.0000
50.	1618.	-.4	0.0	.9	2.6	4.7	7.2	.8	.0879
100.	1451.	.9	-1.8	0.0	3.4	7.6	12.6	3.4	.1859
150.	1308.	3.4	-7.8	-5.1	0.0	6.4	13.8	7.9	.2949
200.	1191.	7.5	-18.9	-15.2	-8.5	0.0	10.0	14.4	.4153
250.	1104.	13.6	-36.1	-31.5	-23.1	-12.5	0.0	22.8	.5464
300.	1043.	21.9	-60.5	-54.9	-44.8	-32.1	-17.1	32.9	.6869
350.	994.	32.6	-92.1	-85.6	-73.8	-59.0	-41.5	44.1	.8338
400.	950.	46.1	-132.0	-124.6	-111.1	-94.1	-74.2	56.6	.9882
450.	909.	62.7	-180.7	-172.4	-157.2	-138.2	-115.7	70.3	1.1496
500.	871.	82.9	-240.0	-230.7	-213.9	-192.7	-167.7	85.7	1.3203

BALLISTIC COEFFICIENT .19, MUZZLE VELOCITY FT/SEC 1900.

RANGE	VELOCITY	HEIGHT	50 YD	100 YD	150 YD	200 YD	250 YD	DRIFT	TIME
0.	1900.	0.0	-1.5	-1.5	-1.5	-1.5	-1.5	0.0	0.0000
50.	1710.	-.4	0.0	.7	2.2	4.1	6.4	.7	.0832
100.	1535.	.7	-1.5	0.0	3.0	6.8	11.2	3.2	.1759
150.	1378.	3.0	-6.7	-4.5	0.0	5.7	12.4	7.4	.2791
200.	1249.	6.7	-16.5	-13.5	-7.6	0.0	9.0	13.7	.3936
250.	1144.	12.2	-31.9	-28.1	-20.7	-11.2	0.0	21.9	.5193
300.	1073.	19.8	-53.8	-49.4	-40.4	-29.1	-15.6	31.9	.6550
350.	1019.	29.9	-83.1	-77.8	-67.4	-54.2	-38.5	43.3	.7987
400.	972.	42.5	-120.0	-114.1	-102.2	-87.0	-69.1	56.0	.9495
450.	930.	58.1	-165.4	-158.7	-145.3	-128.3	-108.1	69.8	1.1074
500.	890.	77.0	-220.0	-212.5	-197.7	-178.8	-156.3	85.0	1.2723

BALLISTIC COEFFICIENT .20, MUZZLE VELOCITY FT/SEC 800.

RANGE	VELOCITY	HEIGHT	50 YD	100 YD	150 YD	200 YD	250 YD	DRIFT	TIME
0.	800.	0.0	-1.5	-1.5	-1.5	-1.5	-1.5	0.0	0.0000
50.	770.	1.0	0.0	6.6	14.1	22.2	30.8	.7	.1915
100.	740.	6.5	-13.1	0.0	15.1	31.2	48.4	2.6	.3899
150.	711.	16.4	-42.4	-22.7	0.0	24.1	50.0	6.1	.5973
200.	682.	30.9	-88.7	-62.4	-32.2	0.0	34.5	11.0	.8124
250.	655.	50.9	-153.9	-121.1	-83.3	-43.1	0.0	17.5	1.0372
300.	628.	76.8	-239.4	-200.0	-154.7	-106.4	-54.7	25.7	1.2712
350.	602.	109.4	-346.7	-300.7	-247.9	-191.5	-131.2	35.6	1.5149
400.	576.	149.7	-478.4	-425.9	-365.5	-301.1	-232.2	47.5	1.7701
450.	551.	198.4	-636.4	-577.2	-509.3	-436.8	-359.3	61.5	2.0369
500.	527.	256.6	-822.5	-756.8	-681.3	-600.8	-514.7	77.5	2.3153

BALLISTIC COEFFICIENT .20, MUZZLE VELOCITY FT/SEC 900.

RANGE	VELOCITY	HEIGHT	50 YD	100 YD	150 YD	200 YD	250 YD	DRIFT	TIME
0.	900.	0.0	-1.5	-1.5	-1.5	-1.5	-1.5	0.0	0.0000
50.	865.	.6	0.0	5.1	10.9	17.3	24.0	.7	.1704
100.	833.	5.0	-10.1	0.0	11.7	24.4	37.8	2.4	.3472
150.	801.	12.8	-32.8	-17.6	0.0	19.0	39.2	5.4	.5307
200.	771.	24.3	-69.1	-48.9	-25.4	0.0	26.8	9.7	.7218
250.	741.	39.9	-119.9	-94.6	-65.3	-33.5	0.0	15.2	.9199
300.	712.	60.2	-186.9	-156.5	-121.3	-83.2	-42.9	22.3	1.1268
350.	684.	85.7	-270.7	-235.3	-194.2	-149.7	-102.8	30.8	1.3417
400.	656.	116.9	-373.3	-332.7	-285.8	-235.0	-181.4	40.9	1.5658
450.	629.	154.7	-496.0	-450.4	-397.6	-340.5	-280.1	52.7	1.7995
500.	603.	199.6	-640.3	-589.6	-531.0	-467.5	-400.4	66.2	2.0428

BALLISTIC COEFFICIENT .20, MUZZLE VELOCITY FT/SEC 1000.

RANGE	VELOCITY	HEIGHT	50 YD	100 YD	150 YD	200 YD	250 YD	DRIFT	TIME
0.	1000.	0.0	-1.5	-1.5	-1.5	-1.5	-1.5	0.0	0.0000
50.	957.	.4	0.0	4.0	8.8	14.0	19.5	.6	.1534
100.	918.	4.0	-8.0	0.0	9.6	20.0	31.1	2.4	.3134
150.	881.	10.3	-26.4	-14.4	0.0	15.6	32.2	5.3	.4804
200.	848.	19.8	-56.0	-40.0	-20.8	0.0	22.1	9.5	.6542
250.	816.	32.7	-97.7	-77.7	-53.7	-27.7	0.0	14.9	.8345
300.	785.	49.4	-152.4	-128.4	-99.6	-68.4	-35.2	21.5	1.0220
350.	755.	70.3	-221.1	-193.1	-159.6	-123.1	-84.4	29.4	1.2168
400.	726.	96.0	-305.0	-273.0	-234.7	-193.0	-148.7	38.6	1.4195
450.	697.	126.8	-405.2	-369.2	-326.1	-279.2	-229.4	49.3	1.6304
500.	669.	163.6	-523.3	-483.3	-435.5	-383.3	-328.0	61.7	1.8504

BALLISTIC COEFFICIENT .20, MUZZLE VELOCITY FT/SEC 1100.

RANGE	VELOCITY	HEIGHT	50 YD	100 YD	150 YD	200 YD	250 YD	DRIFT	TIME
0.	1100.	0.0	-1.5	-1.5	-1.5	-1.5	-1.5	0.0	0.0000
50.	1043.	.2	0.0	3.2	7.2	11.6	16.3	.8	.1408
100.	996.	3.2	-6.4	0.0	8.0	16.8	26.3	2.6	.2875
150.	954.	8.6	-21.7	-12.0	0.0	13.2	27.4	5.7	.4414
200.	914.	16.6	-46.5	-33.6	-17.6	0.0	18.9	10.0	.6021
250.	878.	27.7	-81.7	-65.6	-45.6	-23.7	0.0	15.5	.7699
300.	845.	42.0	-128.1	-108.7	-84.7	-58.4	-30.0	22.1	.9438
350.	813.	60.0	-186.6	-164.0	-136.0	-105.3	-72.1	30.0	1.1248
400.	783.	82.0	-257.9	-232.2	-200.1	-165.0	-127.1	39.0	1.3127
450.	753.	108.5	-343.4	-314.4	-278.4	-238.9	-196.3	49.5	1.5083
500.	723.	139.9	-443.7	-411.5	-371.5	-327.6	-280.3	61.2	1.7115

BALLISTIC COEFFICIENT .20, MUZZLE VELOCITY FT/SEC 1200.

RANGE	VELOCITY	HEIGHT	50 YD	100 YD	150 YD	200 YD	250 YD	DRIFT	TIME
0.	1200.	0.0	-1.5	-1.5	-1.5	-1.5	-1.5	0.0	0.0000
50.	1114.	.1	0.0	2.7	6.3	10.2	14.3	.9	.1301
100.	1054.	2.7	-5.5	0.0	7.1	14.8	23.2	3.3	.2687
150.	1005.	7.5	-18.9	-10.7	0.0	11.6	24.1	7.1	.4153
200.	962.	14.7	-40.7	-29.7	-15.4	0.0	16.7	11.8	.5673
250.	922.	24.6	-71.7	-58.0	-40.2	-20.9	0.0	17.8	.7263
300.	885.	37.5	-113.0	-96.5	-75.2	-52.0	-26.9	25.1	.8926
350.	852.	53.7	-165.2	-146.0	-121.0	-94.0	-64.8	33.5	1.0655
400.	820.	73.6	-228.9	-206.9	-178.5	-147.6	-114.2	43.1	1.2447
450.	789.	97.6	-305.5	-280.8	-248.8	-214.0	-176.4	54.0	1.4316
500.	759.	126.0	-395.3	-367.8	-332.2	-293.6	-251.8	66.0	1.6252

BALLISTIC COEFFICIENT .20, MUZZLE VELOCITY FT/SEC 1300.

RANGE	VELOCITY	HEIGHT	50 YD	100 YD	150 YD	200 YD	250 YD	DRIFT	TIME
0.	1300.	0.0	-1.5	-1.5	-1.5	-1.5	-1.5	0.0	0.0000
50.	1190.	-.0	0.0	2.3	5.5	9.0	12.8	1.0	.1209
100.	1107.	2.3	-4.7	0.0	6.4	13.3	20.9	3.7	.2520
150.	1048.	6.7	-16.6	-9.6	0.0	10.4	21.8	8.2	.3929
200.	1001.	13.1	-36.0	-26.6	-13.8	0.0	15.2	13.5	.5380
250.	958.	22.2	-64.0	-52.3	-36.3	-19.0	0.0	20.1	.6911
300.	918.	34.0	-101.2	-87.2	-68.0	-47.3	-24.5	27.9	.8511
350.	882.	49.0	-148.6	-132.3	-109.8	-85.7	-59.1	37.0	1.0179
400.	848.	67.4	-206.9	-188.2	-162.6	-135.0	-104.5	47.3	1.1916
450.	817.	89.6	-276.6	-255.7	-226.8	-195.7	-161.5	58.7	1.3718
500.	786.	115.9	-359.0	-335.7	-303.6	-269.1	-231.1	71.3	1.5592

BALLISTIC COEFFICIENT .20, MUZZLE VELOCITY FT/SEC 1400.

RANGE	VELOCITY	HEIGHT	50 YD	100 YD	150 YD	200 YD	250 YD	DRIFT	TIME
0.	1400.	0.0	-1.5	-1.5	-1.5	-1.5	-1.5	0.0	0.0000
50.	1273.	-.1	0.0	1.9	4.7	7.9	11.5	1.0	.1125
100.	1167.	1.9	-3.9	0.0	5.6	11.9	19.1	3.8	.2358
150.	1092.	5.8	-14.2	-8.3	0.0	9.6	20.3	8.4	.3690
200.	1037.	11.7	-31.7	-23.9	-12.8	0.0	14.3	14.4	.5102
250.	990.	20.2	-57.4	-47.7	-33.8	-17.8	0.0	21.9	.6599
300.	948.	31.0	-91.1	-79.4	-62.7	-43.5	-22.1	30.0	.8132
350.	910.	44.9	-134.3	-120.7	-101.2	-78.9	-53.9	39.6	.9747
400.	874.	62.1	-188.2	-172.6	-150.4	-124.9	-96.3	50.6	1.1447
450.	841.	82.6	-251.8	-234.3	-209.3	-180.5	-148.4	62.2	1.3180
500.	809.	107.3	-327.9	-308.4	-280.6	-248.7	-213.0	75.5	1.5002

BALLISTIC COEFFICIENT .20, MUZZLE VELOCITY FT/SEC 1500.

RANGE	VELOCITY	HEIGHT	50 YD	100 YD	150 YD	200 YD	250 YD	DRIFT	TIME
0.	1500.	0.0	-1.5	-1.5	-1.5	-1.5	-1.5	0.0	0.0000
50.	1355.	-.2	0.0	1.6	4.1	7.0	10.2	1.0	.1056
100.	1235.	1.6	-3.3	0.0	4.9	10.7	17.1	3.8	.2217
150.	1139.	5.1	-12.2	-7.3	0.0	8.7	18.3	8.5	.3484
200.	1072.	10.5	-27.8	-21.3	-11.6	0.0	12.9	14.8	.4842
250.	1021.	18.2	-50.9	-42.8	-30.6	-16.1	0.0	22.5	.6279
300.	976.	28.3	-82.0	-72.2	-57.5	-40.2	-20.9	31.4	.7782
350.	935.	41.2	-121.7	-110.3	-93.2	-73.0	-50.4	41.4	.9353
400.	897.	57.3	-171.0	-158.0	-138.5	-115.3	-89.6	52.7	1.0995
450.	863.	76.6	-230.3	-215.6	-193.7	-167.6	-138.6	65.0	1.2696
500.	830.	99.8	-300.8	-284.5	-260.1	-231.2	-199.0	78.7	1.4470

BALLISTIC COEFFICIENT .20, MUZZLE VELOCITY FT/SEC 1600.

RANGE	VELOCITY	HEIGHT	50 YD	100 YD	150 YD	200 YD	250 YD	DRIFT	TIME
0.	1600.	0.0	-1.5	-1.5	-1.5	-1.5	-1.5	0.0	0.0000
50.	1443.	-.3	0.0	1.3	3.5	6.1	9.0	.9	.0988
100.	1308.	1.3	-2.7	0.0	4.3	9.5	15.4	3.6	.2080
150.	1196.	4.4	-10.5	-6.5	0.0	7.8	16.6	8.3	.3284
200.	1111.	9.3	-24.4	-19.0	-10.4	0.0	11.8	14.7	.4587
250.	1051.	16.4	-45.2	-38.5	-27.7	-14.7	0.0	22.7	.5976
300.	1004.	25.8	-73.7	-65.7	-52.7	-37.2	-19.5	32.0	.7443
350.	960.	37.8	-110.1	-100.7	-85.6	-67.5	-46.8	42.3	.8965
400.	921.	52.8	-155.5	-144.8	-127.5	-106.8	-83.2	53.9	1.0560
450.	884.	71.0	-210.7	-198.6	-179.1	-155.8	-129.3	66.7	1.2226
500.	850.	92.8	-276.2	-262.8	-241.2	-215.3	-185.8	80.7	1.3958

BALLISTIC COEFFICIENT .20, MUZZLE VELOCITY FT/SEC 1700.

RANGE	VELOCITY	HEIGHT	50 YD	100 YD	150 YD	200 YD	250 YD	DRIFT	TIME
0.	1700.	0.0	-1.5	-1.5	-1.5	-1.5	-1.5	0.0	0.0000
50.	1535.	-.3	0.0	1.1	3.0	5.3	8.0	.8	.0930
100.	1385.	1.1	-2.2	0.0	3.8	8.4	13.8	3.4	.1959
150.	1260.	3.8	-9.0	-5.7	0.0	7.0	15.0	7.9	.3096
200.	1157.	8.3	-21.3	-16.9	-9.3	0.0	10.7	14.4	.4345
250.	1085.	14.7	-40.0	-34.4	-25.0	-13.3	0.0	22.4	.5682
300.	1031.	23.6	-66.2	-59.6	-48.3	-34.3	-18.3	32.1	.7116
350.	985.	34.7	-99.8	-92.0	-78.8	-62.5	-43.8	42.6	.8600
400.	944.	48.7	-141.7	-132.9	-117.8	-99.2	-77.8	54.5	1.0153
450.	905.	65.7	-192.6	-182.6	-165.7	-144.7	-120.7	67.4	1.1770
500.	870.	86.2	-253.6	-242.5	-223.7	-200.4	-173.7	81.6	1.3463

BALLISTIC COEFFICIENT .20, MUZZLE VELOCITY FT/SEC 1800.

RANGE	VELOCITY	HEIGHT	50 YD	100 YD	150 YD	200 YD	250 YD	DRIFT	TIME
0.	1800.	0.0	-1.5	-1.5	-1.5	-1.5	-1.5	0.0	0.0000
50.	1627.	-.4	0.0	.9	2.6	4.6	7.0	.8	.0877
100.	1467.	.9	-1.8	0.0	3.3	7.5	12.3	3.2	.1848
150.	1328.	3.4	-7.7	-5.0	0.0	6.2	13.4	7.5	.2924
200.	1212.	7.4	-18.5	-14.9	-8.3	0.0	9.6	13.7	.4111
250.	1123.	13.2	-35.2	-30.7	-22.4	-12.1	0.0	21.7	.5397
300.	1060.	21.3	-58.7	-53.3	-43.4	-30.9	-16.5	31.2	.6773
350.	1011.	31.8	-90.1	-83.8	-72.2	-57.7	-40.8	42.3	.8240
400.	967.	44.9	-128.8	-121.6	-108.4	-91.8	-72.5	54.3	.9753
450.	927.	60.8	-175.9	-167.8	-152.9	-134.3	-112.6	67.4	1.1327
500.	890.	80.1	-232.7	-223.7	-207.1	-186.4	-162.3	81.8	1.2980

BALLISTIC COEFFICIENT .20, MUZZLE VELOCITY FT/SEC 1900.

RANGE	VELOCITY	HEIGHT	50 YD	100 YD	150 YD	200 YD	250 YD	DRIFT	TIME
0.	1900.	0.0	-1.5	-1.5	-1.5	-1.5	-1.5	0.0	0.0000
50.	1719.	-.4	0.0	.7	2.2	4.0	6.2	.7	.0829
100.	1552.	.7	-1.5	0.0	2.9	6.6	10.9	3.0	.1749
150.	1400.	2.9	-6.6	-4.4	0.0	5.5	12.0	7.0	.2767
200.	1273.	6.5	-16.1	-13.2	-7.4	0.0	8.7	12.9	.3892
250.	1167.	11.9	-31.0	-27.3	-20.1	-10.9	0.0	20.7	.5125
300.	1092.	19.3	-52.3	-47.9	-39.2	-28.1	-15.1	30.3	.6457
350.	1037.	29.0	-80.7	-75.5	-65.4	-52.5	-37.3	41.2	.7869
400.	990.	41.4	-117.1	-111.2	-99.6	-84.8	-67.4	53.7	.9366
450.	948.	56.3	-160.7	-154.1	-141.0	-124.5	-104.9	66.8	1.0900
500.	910.	74.4	-213.6	-206.2	-191.7	-173.3	-151.6	81.3	1.2515

BALLISTIC COEFFICIENT .21, MUZZLE VELOCITY FT/SEC 800.

RANGE	VELOCITY	HEIGHT	50 YD	100 YD	150 YD	200 YD	250 YD	DRIFT	TIME
0.	800.	0.0	-1.5	-1.5	-1.5	-1.5	-1.5	0.0	0.0000
50.	771.	1.0	0.0	6.6	14.0	22.0	30.5	.7	.1914
100.	743.	6.5	-13.1	0.0	15.0	31.0	47.9	2.5	.3894
150.	715.	16.3	-42.1	-22.4	0.0	24.0	49.4	5.8	.5955
200.	688.	30.7	-88.2	-61.9	-32.0	0.0	33.8	10.5	.8096
250.	661.	50.3	-152.5	-119.7	-82.3	-42.3	0.0	16.6	1.0317
300.	635.	75.9	-237.0	-197.6	-152.8	-104.7	-54.0	24.3	1.2633
350.	610.	107.9	-343.1	-297.2	-244.8	-188.8	-129.6	33.8	1.5045
400.	586.	147.2	-472.3	-419.9	-360.1	-296.0	-228.3	45.0	1.7554
450.	562.	194.4	-626.6	-567.6	-500.3	-428.2	-352.0	57.9	2.0165
500.	538.	250.9	-808.7	-743.1	-668.4	-588.3	-503.7	73.0	2.2896

BALLISTIC COEFFICIENT .21, MUZZLE VELOCITY FT/SEC 900.

RANGE	VELOCITY	HEIGHT	50 YD	100 YD	150 YD	200 YD	250 YD	DRIFT	TIME
0.	900.	0.0	-1.5	-1.5	-1.5	-1.5	-1.5	0.0	0.0000
50.	867.	.6	0.0	5.0	10.9	17.2	23.8	.6	.1703
100.	836.	5.0	-10.1	0.0	11.7	24.2	37.6	2.3	.3465
150.	806.	12.7	-32.6	-17.5	0.0	18.8	38.8	5.2	.5294
200.	777.	24.1	-68.6	-48.5	-25.1	0.0	26.7	9.2	.7189
250.	748.	39.5	-119.1	-93.9	-64.7	-33.3	0.0	14.5	.9159
300.	720.	59.4	-184.9	-154.7	-119.6	-82.0	-42.0	21.1	1.1199
350.	693.	84.5	-267.9	-232.6	-191.7	-147.8	-101.1	29.2	1.3327
400.	666.	115.1	-368.7	-328.4	-281.7	-231.5	-178.2	38.7	1.5535
450.	640.	151.8	-488.7	-443.4	-390.7	-334.3	-274.3	49.7	1.7824
500.	615.	195.5	-630.4	-580.1	-521.6	-458.9	-392.3	62.5	2.0219

BALLISTIC COEFFICIENT .21, MUZZLE VELOCITY FT/SEC 1000.

RANGE	VELOCITY	HEIGHT	50 YD	100 YD	150 YD	200 YD	250 YD	DRIFT	TIME
0.	1000.	0.0	-1.5	-1.5	-1.5	-1.5	-1.5	0.0	0.0000
50.	959.	.4	0.0	4.0	8.7	13.9	19.4	.5	.1531
100.	921.	3.9	-8.0	0.0	9.5	19.8	30.8	2.2	.3127
150.	886.	10.3	-26.2	-14.3	0.0	15.5	31.9	5.1	.4791
200.	854.	19.6	-55.6	-39.7	-20.6	0.0	21.9	9.1	.6516
250.	823.	32.4	-96.9	-76.9	-53.1	-27.3	0.0	14.2	.8304
300.	794.	48.8	-151.0	-127.1	-98.5	-67.6	-34.7	20.4	1.0161
350.	765.	69.4	-218.7	-190.9	-157.5	-121.4	-83.1	27.9	1.2085
400.	737.	94.4	-301.2	-269.4	-231.2	-190.0	-146.3	36.7	1.4083
450.	709.	124.6	-399.8	-364.0	-321.1	-274.7	-225.5	46.8	1.6161
500.	682.	160.2	-515.2	-475.4	-427.7	-376.1	-321.5	58.3	1.8314

BALLISTIC COEFFICIENT .21, MUZZLE VELOCITY FT/SEC 1100.

RANGE	VELOCITY	HEIGHT	50 YD	100 YD	150 YD	200 YD	250 YD	DRIFT	TIME
0.	1100.	0.0	-1.5	-1.5	-1.5	-1.5	-1.5	0.0	0.0000
50.	1045.	.2	0.0	3.2	7.1	11.5	16.2	.8	.1412
100.	1001.	3.2	-6.3	0.0	7.9	16.6	26.0	2.5	.2869
150.	959.	8.5	-21.4	-11.9	0.0	13.0	27.1	5.4	.4398
200.	922.	16.5	-45.9	-33.2	-17.4	0.0	18.7	9.5	.5994
250.	887.	27.4	-80.8	-64.9	-45.1	-23.3	0.0	14.7	.7656
300.	854.	41.5	-126.6	-107.5	-83.8	-57.7	-29.7	21.1	.9381
350.	824.	59.1	-184.2	-161.9	-134.2	-103.8	-71.1	28.6	1.1168
400.	794.	80.7	-254.5	-229.1	-197.4	-162.6	-125.3	37.2	1.3024
450.	765.	106.5	-338.3	-309.7	-274.1	-234.9	-192.9	47.1	1.4947
500.	737.	137.0	-436.6	-404.8	-365.2	-321.7	-275.1	58.2	1.6943

BALLISTIC COEFFICIENT .21, MUZZLE VELOCITY FT/SEC 1200.

RANGE	VELOCITY	HEIGHT	50 YD	100 YD	150 YD	200 YD	250 YD	DRIFT	TIME
0.	1200.	0.0	-1.5	-1.5	-1.5	-1.5	-1.5	0.0	0.0000
50.	1117.	.1	0.0	2.7	6.3	10.1	14.2	.9	.1304
100.	1058.	2.7	-5.4	0.0	7.2	14.8	23.0	3.1	.2678
150.	1012.	7.5	-18.8	-10.7	0.0	11.5	23.7	7.0	.4147
200.	970.	14.6	-40.5	-29.7	-15.4	0.0	16.3	11.6	.5661
250.	931.	24.3	-70.9	-57.4	-39.5	-20.3	0.0	17.1	.7224
300.	895.	37.1	-111.8	-95.6	-74.2	-51.1	-26.7	24.2	.8875
350.	863.	52.9	-162.9	-144.0	-118.9	-92.0	-63.6	32.1	1.0573
400.	832.	72.4	-225.7	-204.1	-175.5	-144.7	-112.2	41.3	1.2346
450.	802.	95.8	-300.6	-276.4	-244.1	-209.5	-173.0	51.6	1.4181
500.	773.	123.5	-388.8	-361.9	-326.1	-287.6	-247.0	63.2	1.6088

BALLISTIC COEFFICIENT .21, MUZZLE VELOCITY FT/SEC 1300.

RANGE	VELOCITY	HEIGHT	50 YD	100 YD	150 YD	200 YD	250 YD	DRIFT	TIME
0.	1300.	0.0	-1.5	-1.5	-1.5	-1.5	-1.5	0.0	0.0000
50.	1194.	-.0	0.0	2.3	5.4	8.9	12.7	1.0	.1210
100.	1113.	2.3	-4.6	0.0	6.2	13.3	20.8	3.6	.2510
150.	1056.	6.5	-16.2	-9.3	0.0	10.6	21.9	7.6	.3895
200.	1009.	13.1	-35.8	-26.6	-14.2	0.0	15.0	13.2	.5364
250.	967.	22.0	-63.5	-52.0	-36.5	-18.8	0.0	19.6	.6880
300.	929.	33.5	-99.9	-86.2	-67.6	-46.3	-23.8	26.9	.8451
350.	893.	48.3	-146.9	-130.9	-109.2	-84.4	-58.1	35.8	1.0112
400.	861.	66.2	-203.5	-185.2	-160.4	-132.0	-102.0	45.4	1.1808
450.	830.	87.8	-272.1	-251.5	-223.6	-191.7	-157.9	56.3	1.3585
500.	800.	113.4	-352.6	-329.6	-298.7	-263.2	-225.7	68.4	1.5423

BALLISTIC COEFFICIENT .21, MUZZLE VELOCITY FT/SEC 1400.

RANGE	VELOCITY	HEIGHT	50 YD	100 YD	150 YD	200 YD	250 YD	DRIFT	TIME
0.	1400.	0.0	-1.5	-1.5	-1.5	-1.5	-1.5	0.0	0.0000
50.	1278.	-.1	0.0	1.9	4.7	7.9	11.3	.9	.1124
100.	1176.	1.9	-3.9	0.0	5.4	11.8	18.6	3.7	.2353
150.	1101.	5.7	-14.0	-8.2	0.0	9.6	19.8	8.0	.3669
200.	1046.	11.6	-31.5	-23.7	-12.8	0.0	13.6	14.0	.5081
250.	1001.	19.8	-56.3	-46.6	-33.0	-17.0	0.0	20.8	.6537
300.	960.	30.5	-89.6	-78.0	-61.7	-42.5	-22.1	28.8	.8064
350.	922.	44.0	-132.1	-118.6	-99.5	-77.1	-53.3	38.0	.9660
400.	887.	60.8	-184.6	-169.0	-147.3	-121.7	-94.5	48.4	1.1321
450.	855.	80.9	-247.5	-230.0	-205.6	-176.7	-146.1	59.9	1.3045
500.	824.	104.8	-321.7	-302.3	-275.1	-243.1	-209.1	72.5	1.4832

BALLISTIC COEFFICIENT .21, MUZZLE VELOCITY FT/SEC 1500.

RANGE	VELOCITY	HEIGHT	50 YD	100 YD	150 YD	200 YD	250 YD	DRIFT	TIME
0.	1500.	0.0	-1.5	-1.5	-1.5	-1.5	-1.5	0.0	0.0000
50.	1361.	-.2	0.0	1.6	4.0	6.9	10.1	.9	.1052
100.	1246.	1.6	-3.2	0.0	4.8	10.5	17.0	3.6	.2205
150.	1151.	5.0	-12.0	-7.2	0.0	8.5	18.2	8.1	.3460
200.	1083.	10.3	-27.4	-21.0	-11.4	0.0	12.9	14.2	.4806
250.	1032.	18.0	-50.4	-42.4	-30.4	-16.2	0.0	21.9	.6242
300.	988.	27.9	-81.0	-71.3	-56.9	-39.9	-20.5	30.4	.7726
350.	948.	40.4	-119.6	-108.4	-91.6	-71.7	-49.1	39.8	.9263
400.	911.	56.0	-167.8	-155.0	-135.7	-113.0	-87.2	50.6	1.0876
450.	877.	75.0	-226.2	-211.8	-190.1	-164.6	-135.5	62.7	1.2561
500.	845.	97.4	-294.8	-278.7	-254.7	-226.3	-194.0	75.6	1.4297

BALLISTIC COEFFICIENT .21, MUZZLE VELOCITY FT/SEC 1600.

RANGE	VELOCITY	HEIGHT	50 YD	100 YD	150 YD	200 YD	250 YD	DRIFT	TIME
0.	1600.	0.0	-1.5	-1.5	-1.5	-1.5	-1.5	0.0	0.0000
50.	1450.	-.3	0.0	1.3	3.4	6.0	8.9	.8	.0985
100.	1319.	1.3	-2.7	0.0	4.2	9.3	15.2	3.4	.2070
150.	1211.	4.3	-10.3	-6.3	0.0	7.6	16.4	7.9	.3259
200.	1125.	9.2	-23.9	-18.6	-10.2	0.0	11.7	14.0	.4547
250.	1064.	16.1	-44.5	-37.9	-27.3	-14.6	0.0	21.9	.5930
300.	1017.	25.3	-72.2	-64.3	-51.6	-36.3	-18.8	30.6	.7364
350.	974.	37.0	-108.1	-98.8	-84.0	-66.2	-45.7	40.6	.8872
400.	935.	51.6	-152.5	-141.9	-125.0	-104.7	-81.3	51.8	1.0444
450.	899.	69.3	-206.3	-194.4	-175.3	-152.5	-126.1	64.1	1.2079
500.	866.	90.4	-270.2	-257.0	-235.8	-210.4	-181.2	77.5	1.3781

BALLISTIC COEFFICIENT .21, MUZZLE VELOCITY FT/SEC 1700.

RANGE	VELOCITY	HEIGHT	50 YD	100 YD	150 YD	200 YD	250 YD	DRIFT	TIME
0.	1700.	0.0	-1.5	-1.5	-1.5	-1.5	-1.5	0.0	0.0000
50.	1542.	-.3	0.0	1.1	2.9	5.2	7.8	.8	.0927
100.	1398.	1.1	-2.2	0.0	3.7	8.3	13.5	3.3	.1950
150.	1276.	3.8	-8.8	-5.6	0.0	6.8	14.6	7.5	.3075
200.	1174.	8.2	-20.9	-16.5	-9.1	0.0	10.4	13.7	.4308
250.	1100.	14.4	-39.1	-33.7	-24.4	-13.0	0.0	21.3	.5624
300.	1045.	23.0	-64.7	-58.1	-47.0	-33.3	-17.7	30.6	.7034
350.	1000.	33.9	-97.4	-89.7	-76.7	-60.8	-42.6	40.7	.8492
400.	959.	47.5	-138.3	-129.6	-114.8	-96.5	-75.8	52.1	1.0022
450.	921.	64.0	-188.2	-178.3	-161.6	-141.1	-117.8	64.7	1.1618
500.	887.	83.9	-247.6	-236.7	-218.1	-195.3	-169.4	78.4	1.3280

BALLISTIC COEFFICIENT .21, MUZZLE VELOCITY FT/SEC 1800.

RANGE	VELOCITY	HEIGHT	50 YD	100 YD	150 YD	200 YD	250 YD	DRIFT	TIME
0.	1800.	0.0	-1.5	-1.5	-1.5	-1.5	-1.5	0.0	0.0000
50.	1635.	-.4	0.0	.9	2.5	4.5	6.9	.7	.0875
100.	1482.	.9	-1.8	0.0	3.3	7.3	12.0	3.0	.1838
150.	1346.	3.3	-7.5	-4.9	0.0	6.1	13.1	7.1	.2903
200.	1233.	7.2	-18.1	-14.6	-8.1	0.0	9.4	13.0	.4071
250.	1141.	12.9	-34.4	-29.9	-21.8	-11.7	0.0	20.6	.5335
300.	1076.	20.7	-57.3	-52.0	-42.2	-30.1	-16.1	29.8	.6691
350.	1026.	31.0	-87.7	-81.5	-70.1	-56.0	-39.6	40.4	.8129
400.	983.	43.7	-125.5	-118.5	-105.4	-89.3	-70.6	52.0	.9619
450.	943.	59.2	-171.8	-163.8	-149.2	-131.0	-109.9	64.7	1.1178
500.	907.	77.8	-226.8	-217.9	-201.6	-181.4	-158.0	78.5	1.2793

BALLISTIC COEFFICIENT .21, MUZZLE VELOCITY FT/SEC 1900.

RANGE	VELOCITY	HEIGHT	50 YD	100 YD	150 YD	200 YD	250 YD	DRIFT	TIME
0.	1900.	0.0	-1.5	-1.5	-1.5	-1.5	-1.5	0.0	0.0000
50.	1727.	-.4	0.0	.7	2.1	3.9	6.0	.7	.0828
100.	1567.	.7	-1.4	0.0	2.8	6.4	10.7	2.9	.1741
150.	1420.	2.9	-6.4	-4.3	0.0	5.4	11.7	6.6	.2745
200.	1295.	6.4	-15.8	-12.9	-7.2	0.0	8.4	12.3	.3855
250.	1190.	11.6	-30.2	-26.6	-19.5	-10.5	0.0	19.6	.5064
300.	1111.	18.7	-50.9	-46.6	-38.1	-27.3	-14.6	28.8	.6372
350.	1053.	28.2	-78.4	-73.4	-63.5	-50.9	-36.1	39.3	.7760
400.	1007.	40.1	-113.7	-108.0	-96.6	-82.2	-65.4	51.3	.9228
450.	966.	54.7	-156.6	-150.1	-137.3	-121.1	-102.2	64.1	1.0747
500.	928.	72.2	-207.7	-200.5	-186.3	-168.3	-147.2	78.0	1.2324

BALLISTIC COEFFICIENT .22, MUZZLE VELOCITY FT/SEC 800.

RANGE	VELOCITY	HEIGHT	50 YD	100 YD	150 YD	200 YD	250 YD	DRIFT	TIME
0.	800.	0.0	-1.5	-1.5	-1.5	-1.5	-1.5	0.0	0.0000
50.	772.	1.0	0.0	6.5	13.9	21.9	30.3	.7	.1912
100.	745.	6.5	-13.1	0.0	14.8	30.7	47.5	2.4	.3889
150.	719.	16.2	-41.8	-22.2	0.0	23.9	49.1	5.5	.5937
200.	693.	30.5	-87.6	-61.5	-31.9	0.0	33.6	10.0	.8067
250.	667.	49.9	-151.5	-118.8	-81.8	-42.0	0.0	15.8	1.0275
300.	642.	75.0	-234.8	-195.5	-151.1	-103.4	-53.0	23.1	1.2562
350.	618.	106.4	-339.3	-293.5	-241.7	-185.9	-127.1	32.0	1.4941
400.	594.	144.9	-466.8	-414.5	-355.3	-291.6	-224.4	42.6	1.7421
450.	571.	191.2	-618.9	-560.0	-493.4	-421.7	-346.1	55.0	1.9998
500.	549.	246.1	-797.3	-731.9	-657.9	-578.3	-494.3	69.1	2.2678

BALLISTIC COEFFICIENT .22, MUZZLE VELOCITY FT/SEC 900.

RANGE	VELOCITY	HEIGHT	50 YD	100 YD	150 YD	200 YD	250 YD	DRIFT	TIME
0.	900.	0.0	-1.5	-1.5	-1.5	-1.5	-1.5	0.0	0.0000
50.	868.	.6	0.0	5.0	10.8	17.0	23.6	.6	.1702
100.	838.	5.0	-10.0	0.0	11.7	24.1	37.3	2.2	.3458
150.	810.	12.6	-32.5	-17.5	0.0	18.6	38.4	5.0	.5282
200.	782.	23.9	-68.2	-48.1	-24.8	0.0	26.4	8.8	.7164
250.	755.	39.2	-118.2	-93.2	-64.0	-33.0	0.0	13.8	.9119
300.	728.	58.9	-183.5	-153.5	-118.5	-81.3	-41.7	20.1	1.1144
350.	702.	83.4	-265.1	-230.0	-189.2	-145.8	-99.6	27.7	1.3241
400.	676.	113.4	-364.5	-324.5	-277.8	-228.2	-175.4	36.7	1.5421
450.	651.	149.4	-483.0	-438.0	-385.5	-329.6	-270.3	47.3	1.7685
500.	627.	191.9	-621.6	-571.6	-513.3	-451.3	-385.3	59.2	2.0032

BALLISTIC COEFFICIENT .22, MUZZLE VELOCITY FT/SEC 1000.

RANGE	VELOCITY	HEIGHT	50 YD	100 YD	150 YD	200 YD	250 YD	DRIFT	TIME
0.	1000.	0.0	-1.5	-1.5	-1.5	-1.5	-1.5	0.0	0.0000
50.	961.	.4	0.0	4.0	8.7	13.8	19.2	.5	.1531
100.	925.	3.9	-7.9	0.0	9.5	19.7	30.5	2.2	.3123
150.	891.	10.2	-26.1	-14.2	0.0	15.3	31.6	4.9	.4778
200.	859.	19.5	-55.2	-39.3	-20.4	0.0	21.7	8.7	.6492
250.	830.	32.1	-96.1	-76.3	-52.7	-27.1	0.0	13.5	.8269
300.	802.	48.3	-149.5	-125.7	-97.4	-66.7	-34.2	19.5	1.0106
350.	774.	68.5	-216.5	-188.8	-155.7	-120.0	-82.0	26.6	1.2012
400.	747.	93.1	-297.9	-266.2	-228.4	-187.5	-144.1	34.9	1.3985
450.	720.	122.5	-394.6	-358.9	-316.3	-270.4	-221.5	44.5	1.6026
500.	694.	157.4	-508.2	-468.6	-421.3	-370.2	-316.0	55.5	1.8151

BALLISTIC COEFFICIENT .22, MUZZLE VELOCITY FT/SEC 1100.

RANGE	VELOCITY	HEIGHT	50 YD	100 YD	150 YD	200 YD	250 YD	DRIFT	TIME
0.	1100.	0.0	-1.5	-1.5	-1.5	-1.5	-1.5	0.0	0.0000
50.	1047.	.2	0.0	3.2	7.1	11.4	16.1	.9	.1416
100.	1004.	3.2	-6.3	0.0	7.9	16.4	25.8	2.5	.2869
150.	965.	8.5	-21.3	-11.8	0.0	12.8	26.8	5.3	.4394
200.	928.	16.4	-45.5	-32.9	-17.1	0.0	18.7	9.1	.5973
250.	894.	27.2	-80.3	-64.4	-44.7	-23.4	0.0	14.3	.7632
300.	863.	41.0	-125.1	-106.2	-82.5	-56.9	-28.8	20.2	.9327
350.	833.	58.4	-182.0	-159.9	-132.2	-102.3	-69.7	27.3	1.1097
400.	805.	79.5	-251.2	-225.9	-194.4	-160.2	-122.8	35.5	1.2929
450.	777.	104.7	-333.6	-305.1	-269.6	-231.2	-189.1	44.9	1.4824
500.	750.	134.6	-430.3	-398.7	-359.2	-316.5	-269.8	55.5	1.6792

BALLISTIC COEFFICIENT .22, MUZZLE VELOCITY FT/SEC 1200.

RANGE	VELOCITY	HEIGHT	50 YD	100 YD	150 YD	200 YD	250 YD	DRIFT	TIME
0.	1200.	0.0	-1.5	-1.5	-1.5	-1.5	-1.5	0.0	0.0000
50.	1120.	.1	0.0	2.7	6.2	10.0	14.1	.8	.1296
100.	1063.	2.7	-5.5	0.0	6.8	14.5	22.6	3.2	.2680
150.	1018.	7.4	-18.5	-10.3	0.0	11.4	23.7	6.4	.4115
200.	977.	14.4	-39.9	-28.9	-15.2	0.0	16.3	10.9	.5620
250.	940.	24.0	-70.4	-56.6	-39.5	-20.4	0.0	16.4	.7185
300.	905.	36.5	-110.5	-94.0	-73.5	-50.6	-26.1	23.1	.8812
350.	873.	52.4	-161.9	-142.6	-118.7	-92.0	-63.4	31.2	1.0520
400.	843.	71.3	-223.0	-201.0	-173.6	-143.1	-110.4	39.6	1.2249
450.	814.	94.2	-296.9	-272.1	-241.3	-207.0	-170.2	49.5	1.4062
500.	786.	121.2	-383.4	-355.9	-321.7	-283.5	-242.7	60.5	1.5938

BALLISTIC COEFFICIENT .22, MUZZLE VELOCITY FT/SEC 1300.

RANGE	VELOCITY	HEIGHT	50 YD	100 YD	150 YD	200 YD	250 YD	DRIFT	TIME
0.	1300.	0.0	-1.5	-1.5	-1.5	-1.5	-1.5	0.0	0.0000
50.	1199.	-.1	0.0	2.3	5.4	8.8	12.5	.9	.1205
100.	1119.	2.3	-4.6	0.0	6.2	13.0	20.4	3.4	.2502
150.	1062.	6.5	-16.1	-9.3	0.0	10.3	21.4	7.4	.3884
200.	1017.	12.8	-35.2	-26.0	-13.7	0.0	14.8	12.5	.5323
250.	977.	21.6	-62.5	-51.1	-35.7	-18.6	0.0	18.6	.6828
300.	939.	33.1	-98.8	-85.1	-66.5	-46.0	-23.7	25.9	.8394
350.	905.	47.5	-144.6	-128.7	-107.0	-83.1	-57.1	34.2	1.0021
400.	872.	65.3	-201.7	-183.4	-158.7	-131.3	-101.6	44.0	1.1732
450.	842.	86.2	-268.1	-247.5	-219.7	-188.9	-155.5	54.1	1.3461
500.	814.	111.2	-347.2	-324.4	-293.5	-259.3	-222.2	65.8	1.5275

BALLISTIC COEFFICIENT .22, MUZZLE VELOCITY FT/SEC 1400.

RANGE	VELOCITY	HEIGHT	50 YD	100 YD	150 YD	200 YD	250 YD	DRIFT	TIME
0.	1400.	0.0	-1.5	-1.5	-1.5	-1.5	-1.5	0.0	0.0000
50.	1283.	-.1	0.0	1.9	4.6	7.7	11.2	.9	.1121
100.	1184.	1.9	-3.8	0.0	5.4	11.6	18.6	3.4	.2339
150.	1110.	5.6	-13.8	-8.1	0.0	9.3	19.7	7.7	.3651
200.	1055.	11.4	-30.9	-23.3	-12.4	0.0	13.9	13.2	.5038
250.	1011.	19.6	-55.9	-46.4	-32.9	-17.4	0.0	20.3	.6508
300.	970.	30.2	-89.0	-77.6	-61.3	-42.7	-21.9	28.1	.8025
350.	934.	43.3	-130.4	-117.1	-98.1	-76.4	-52.1	36.7	.9583
400.	899.	59.7	-181.7	-166.5	-144.8	-120.0	-92.2	46.6	1.1219
450.	868.	79.4	-243.5	-226.4	-202.1	-174.1	-142.9	57.7	1.2920
500.	838.	102.7	-316.2	-297.2	-270.1	-239.1	-204.3	69.7	1.4677

BALLISTIC COEFFICIENT .22, MUZZLE VELOCITY FT/SEC 1500.

RANGE	VELOCITY	HEIGHT	50 YD	100 YD	150 YD	200 YD	250 YD	DRIFT	TIME
0.	1500.	0.0	-1.5	-1.5	-1.5	-1.5	-1.5	0.0	0.0000
50.	1367.	-.2	0.0	1.6	4.0	6.8	9.9	.9	.1050
100.	1255.	1.6	-3.2	0.0	4.7	10.3	16.6	3.5	.2197
150.	1162.	4.9	-11.9	-7.1	0.0	8.4	17.8	7.7	.3439
200.	1094.	10.2	-27.0	-20.7	-11.2	0.0	12.5	13.6	.4773
250.	1043.	17.6	-49.4	-41.5	-29.7	-15.7	0.0	20.8	.6183
300.	1000.	27.3	-79.3	-69.8	-55.6	-38.8	-20.0	29.0	.7648
350.	961.	39.7	-117.7	-106.6	-90.0	-70.4	-48.5	38.3	.9178
400.	925.	54.9	-165.0	-152.3	-133.4	-111.0	-85.9	48.8	1.0771
450.	891.	73.4	-222.0	-207.7	-186.4	-161.2	-133.0	60.3	1.2426
500.	859.	95.2	-289.3	-273.4	-249.8	-221.8	-190.5	72.9	1.4140

BALLISTIC COEFFICIENT .22, MUZZLE VELOCITY FT/SEC 1600.

RANGE	VELOCITY	HEIGHT	50 YD	100 YD	150 YD	200 YD	250 YD	DRIFT	TIME
0.	1600.	0.0	-1.5	-1.5	-1.5	-1.5	-1.5	0.0	0.0000
50.	1456.	-.3	0.0	1.3	3.4	5.9	8.7	.8	.0983
100.	1330.	1.3	-2.6	0.0	4.2	9.2	14.8	3.3	.2061
150.	1224.	4.3	-10.2	-6.2	0.0	7.5	16.0	7.5	.3237
200.	1138.	9.0	-23.6	-18.3	-10.0	0.0	11.3	13.5	.4514
250.	1076.	15.8	-43.5	-37.0	-26.6	-14.1	0.0	20.8	.5868
300.	1029.	24.9	-71.2	-63.4	-50.9	-35.9	-19.0	29.6	.7308
350.	987.	36.4	-106.4	-97.3	-82.7	-65.2	-45.5	39.3	.8797
400.	949.	50.5	-149.7	-139.2	-122.6	-102.6	-80.0	49.9	1.0335
450.	914.	67.7	-202.3	-190.5	-171.8	-149.3	-123.9	61.7	1.1945
500.	881.	88.3	-264.7	-251.6	-230.8	-205.8	-177.6	74.7	1.3617

BALLISTIC COEFFICIENT .22, MUZZLE VELOCITY FT/SEC 1700.

RANGE	VELOCITY	HEIGHT	50 YD	100 YD	150 YD	200 YD	250 YD	DRIFT	TIME
0.	1700.	0.0	-1.5	-1.5	-1.5	-1.5	-1.5	0.0	0.0000
50.	1549.	-.3	0.0	1.1	2.9	5.1	7.7	.8	.0926
100.	1410.	1.1	-2.1	0.0	3.6	8.1	13.2	3.1	.1940
150.	1292.	3.7	-8.7	-5.5	0.0	6.6	14.3	7.1	.3053
200.	1192.	8.0	-20.4	-16.2	-8.9	0.0	10.2	12.9	.4264
250.	1115.	14.1	-38.3	-33.0	-23.9	-12.8	0.0	20.4	.5569
300.	1059.	22.4	-63.1	-56.7	-45.7	-32.4	-17.1	29.1	.6950
350.	1014.	33.1	-95.4	-87.9	-75.1	-59.6	-41.7	39.1	.8400
400.	974.	46.4	-135.5	-127.0	-112.4	-94.7	-74.2	50.2	.9910
450.	937.	62.5	-184.2	-174.6	-158.2	-138.3	-115.3	62.3	1.1481
500.	902.	81.8	-242.1	-231.4	-213.1	-191.0	-165.4	75.5	1.3112

BALLISTIC COEFFICIENT .22, MUZZLE VELOCITY FT/SEC 1800.

RANGE	VELOCITY	HEIGHT	50 YD	100 YD	150 YD	200 YD	250 YD	DRIFT	TIME
0.	1800.	0.0	-1.5	-1.5	-1.5	-1.5	-1.5	0.0	0.0000
50.	1642.	-.4	0.0	.9	2.5	4.4	6.7	.7	.0873
100.	1495.	.9	-1.7	0.0	3.2	7.1	11.7	2.9	.1830
150.	1363.	3.2	-7.4	-4.8	0.0	5.9	12.8	6.7	.2883
200.	1252.	7.1	-17.7	-14.3	-7.9	0.0	9.2	12.3	.4032
250.	1159.	12.6	-33.6	-29.3	-21.3	-11.5	0.0	19.6	.5280
300.	1092.	20.3	-56.0	-50.8	-41.2	-29.4	-15.6	28.4	.6615
350.	1041.	30.2	-85.5	-79.4	-68.2	-54.4	-38.4	38.6	.8024
400.	999.	42.5	-122.5	-115.5	-102.7	-87.0	-68.6	49.8	.9495
450.	959.	57.6	-167.5	-159.7	-145.3	-127.6	-106.9	62.1	1.1026
500.	923.	75.7	-221.3	-212.6	-196.6	-177.0	-154.0	75.5	1.2621

BALLISTIC COEFFICIENT .22, MUZZLE VELOCITY FT/SEC 1900.

RANGE	VELOCITY	HEIGHT	50 YD	100 YD	150 YD	200 YD	250 YD	DRIFT	TIME
0.	1900.	0.0	-1.5	-1.5	-1.5	-1.5	-1.5	0.0	0.0000
50.	1735.	-.4	0.0	.7	2.1	3.9	5.9	.6	.0826
100.	1581.	.7	-1.4	0.0	2.8	6.3	10.4	2.7	.1733
150.	1439.	2.8	-6.3	-4.2	0.0	5.3	11.4	6.3	.2728
200.	1316.	6.2	-15.4	-12.6	-7.0	0.0	8.2	11.6	.3819
250.	1212.	11.3	-29.5	-26.0	-19.0	-10.3	0.0	18.6	.5007
300.	1129.	18.3	-49.6	-45.4	-37.0	-26.5	-14.1	27.4	.6292
350.	1070.	27.4	-76.4	-71.5	-61.7	-49.4	-35.0	37.5	.7658
400.	1023.	39.0	-110.6	-105.0	-93.8	-79.8	-63.4	49.0	.9098
450.	982.	53.1	-152.4	-146.0	-133.5	-117.7	-99.2	61.4	1.0594
500.	944.	70.2	-202.6	-195.5	-181.6	-164.0	-143.5	75.0	1.2156

BALLISTIC COEFFICIENT .23, MUZZLE VELOCITY FT/SEC 800.

RANGE	VELOCITY	HEIGHT	50 YD	100 YD	150 YD	200 YD	250 YD	DRIFT	TIME
0.	800.	0.0	-1.5	-1.5	-1.5	-1.5	-1.5	0.0	0.0000
50.	774.	1.0	0.0	6.5	13.9	21.8	30.1	.6	.1911
100.	748.	6.5	-13.1	0.0	14.7	30.5	47.1	2.4	.3884
150.	722.	16.1	-41.6	-22.1	0.0	23.6	48.6	5.2	.5923
200.	697.	30.3	-87.0	-60.9	-31.5	0.0	33.3	9.5	.8039
250.	673.	49.5	-150.5	-117.8	-81.1	-41.7	0.0	15.1	1.0232
300.	649.	74.3	-233.1	-194.0	-149.8	-102.6	-52.5	22.1	1.2504
350.	626.	105.2	-336.3	-290.6	-239.1	-184.0	-125.6	30.5	1.4856
400.	603.	142.9	-461.7	-409.5	-350.6	-287.6	-220.9	40.4	1.7298
450.	580.	188.0	-610.8	-552.1	-485.9	-415.0	-340.0	52.0	1.9831
500.	558.	241.6	-786.1	-720.9	-647.3	-568.5	-485.2	65.5	2.2469

BALLISTIC COEFFICIENT .23, MUZZLE VELOCITY FT/SEC 900.

RANGE	VELOCITY	HEIGHT	50 YD	100 YD	150 YD	200 YD	250 YD	DRIFT	TIME
0.	900.	0.0	-1.5	-1.5	-1.5	-1.5	-1.5	0.0	0.0000
50.	870.	.6	0.0	5.0	10.8	17.0	23.5	.6	.1701
100.	841.	5.0	-10.0	0.0	11.6	24.0	37.0	2.1	.3452
150.	814.	12.6	-32.3	-17.4	0.0	18.5	38.0	4.7	.5269
200.	787.	23.8	-67.8	-47.9	-24.7	0.0	26.0	8.4	.7144
250.	760.	38.8	-117.3	-92.4	-63.4	-32.5	0.0	13.1	.9080
300.	735.	58.3	-182.0	-152.1	-117.4	-80.3	-41.3	19.2	1.1090
350.	710.	82.5	-262.9	-228.0	-187.4	-144.2	-98.7	26.5	1.3170
400.	685.	111.9	-360.8	-321.0	-274.6	-225.1	-173.1	35.0	1.5320
450.	661.	147.0	-477.1	-432.3	-380.1	-324.5	-266.0	44.8	1.7547
500.	637.	188.6	-613.5	-563.7	-505.8	-443.9	-379.0	56.2	1.9861

BALLISTIC COEFFICIENT .23, MUZZLE VELOCITY FT/SEC 1000.

RANGE	VELOCITY	HEIGHT	50 YD	100 YD	150 YD	200 YD	250 YD	DRIFT	TIME
0.	1000.	0.0	-1.5	-1.5	-1.5	-1.5	-1.5	0.0	0.0000
50.	962.	.4	0.0	3.9	8.7	13.7	19.1	.6	.1533
100.	928.	3.9	-7.9	0.0	9.5	19.5	30.2	2.1	.3119
150.	895.	10.2	-26.1	-14.3	0.0	15.0	31.1	4.9	.4777
200.	865.	19.4	-54.8	-39.1	-20.0	0.0	21.4	8.3	.6472
250.	837.	31.8	-95.3	-75.6	-51.8	-26.8	0.0	12.9	.8234
300.	809.	47.8	-148.3	-124.7	-96.1	-66.1	-34.0	18.7	1.0060
350.	783.	67.7	-214.4	-186.8	-153.5	-118.5	-81.0	25.4	1.1943
400.	756.	91.9	-294.7	-263.2	-225.1	-185.1	-142.2	33.3	1.3893
450.	731.	120.8	-390.2	-354.7	-311.9	-266.9	-218.6	42.5	1.5913
500.	706.	154.8	-501.7	-462.3	-414.7	-364.7	-311.1	52.8	1.8002

BALLISTIC COEFFICIENT .23, MUZZLE VELOCITY FT/SEC 1100.

RANGE	VELOCITY	HEIGHT	50 YD	100 YD	150 YD	200 YD	250 YD	DRIFT	TIME
0.	1100.	0.0	-1.5	-1.5	-1.5	-1.5	-1.5	0.0	0.0000
50.	1050.	.2	0.0	3.3	7.2	11.4	15.9	.6	.1398
100.	1008.	3.2	-6.5	0.0	7.9	16.3	25.3	2.5	.2870
150.	970.	8.5	-21.6	-11.8	0.0	12.5	26.2	5.3	.4392
200.	935.	16.3	-45.6	-32.5	-16.7	0.0	18.2	8.7	.5951
250.	902.	26.9	-79.7	-63.3	-43.6	-22.7	0.0	13.5	.7583
300.	871.	40.8	-125.4	-105.8	-82.1	-57.1	-29.8	19.8	.9305
350.	842.	57.6	-180.8	-157.9	-130.3	-101.0	-69.2	26.1	1.1030
400.	815.	78.4	-249.2	-223.1	-191.5	-158.1	-121.8	34.0	1.2843
450.	788.	103.2	-330.6	-301.2	-265.7	-228.0	-187.2	43.0	1.4716
500.	762.	132.3	-425.4	-392.8	-353.3	-311.5	-266.1	53.0	1.6649

BALLISTIC COEFFICIENT .23, MUZZLE VELOCITY FT/SEC 1200.

RANGE	VELOCITY	HEIGHT	50 YD	100 YD	150 YD	200 YD	250 YD	DRIFT	TIME
0.	1200.	0.0	-1.5	-1.5	-1.5	-1.5	-1.5	0.0	0.0000
50.	1123.	.1	0.0	2.7	6.2	9.9	14.0	.8	.1294
100.	1068.	2.7	-5.4	0.0	6.9	14.5	22.5	2.9	.2666
150.	1023.	7.3	-18.5	-10.4	0.0	11.4	23.4	6.3	.4107
200.	984.	14.3	-39.8	-29.0	-15.2	0.0	16.1	10.6	.5604
250.	948.	23.8	-69.8	-56.3	-39.0	-20.1	0.0	15.9	.7151
300.	914.	36.1	-109.4	-93.3	-72.5	-49.8	-25.7	22.2	.8762
350.	882.	51.5	-159.4	-140.6	-116.3	-89.8	-61.7	29.6	1.0433
400.	853.	70.3	-220.5	-198.9	-171.2	-140.9	-108.8	38.1	1.2165
450.	825.	92.7	-293.1	-268.9	-237.7	-203.6	-167.5	47.6	1.3952
500.	798.	119.1	-378.1	-351.2	-316.5	-278.6	-238.5	58.1	1.5799

BALLISTIC COEFFICIENT .23, MUZZLE VELOCITY FT/SEC 1300.

RANGE	VELOCITY	HEIGHT	50 YD	100 YD	150 YD	200 YD	250 YD	DRIFT	TIME
0.	1300.	0.0	-1.5	-1.5	-1.5	-1.5	-1.5	0.0	0.0000
50.	1203.	-.1	0.0	2.3	5.3	8.7	12.4	.8	.1202
100.	1125.	2.2	-4.5	0.0	6.1	13.0	20.3	3.3	.2494
150.	1069.	6.4	-15.9	-9.1	0.0	10.3	21.4	7.1	.3863
200.	1025.	12.8	-35.0	-25.9	-13.8	0.0	14.7	12.1	.5304
250.	985.	21.4	-62.1	-50.8	-35.6	-18.4	0.0	18.1	.6800
300.	949.	32.7	-97.8	-84.2	-65.9	-45.3	-23.2	25.0	.8343
350.	915.	46.8	-142.9	-127.1	-105.8	-81.7	-56.0	33.0	.9952
400.	883.	64.1	-198.3	-180.2	-155.9	-128.3	-98.9	42.1	1.1622
450.	854.	84.8	-264.5	-244.2	-216.8	-185.8	-152.7	52.2	1.3351
500.	826.	109.2	-342.2	-319.6	-289.2	-254.7	-217.9	63.3	1.5137

BALLISTIC COEFFICIENT .23, MUZZLE VELOCITY FT/SEC 1400.

RANGE	VELOCITY	HEIGHT	50 YD	100 YD	150 YD	200 YD	250 YD	DRIFT	TIME
0.	1400.	0.0	-1.5	-1.5	-1.5	-1.5	-1.5	0.0	0.0000
50.	1288.	-.1	0.0	1.9	4.6	7.7	11.0	.8	.1119
100.	1192.	1.9	-3.8	0.0	5.3	11.5	18.1	3.4	.2335
150.	1118.	5.6	-13.7	-8.0	0.0	9.3	19.2	7.4	.3637
200.	1063.	11.3	-30.6	-23.0	-12.3	0.0	13.3	12.9	.5018
250.	1020.	19.2	-54.9	-45.4	-32.0	-16.6	0.0	19.2	.6451
300.	981.	29.6	-87.4	-76.0	-60.0	-41.4	-21.6	26.8	.7952
350.	945.	42.7	-128.9	-115.6	-96.9	-75.3	-52.1	35.5	.9519
400.	911.	58.7	-179.2	-164.0	-142.6	-117.9	-91.4	45.0	1.1127
450.	880.	77.9	-239.8	-222.7	-198.6	-170.8	-141.0	55.6	1.2802
500.	851.	100.7	-311.3	-292.3	-265.6	-234.7	-201.5	67.3	1.4539

BALLISTIC COEFFICIENT .23, MUZZLE VELOCITY FT/SEC 1500.

RANGE	VELOCITY	HEIGHT	50 YD	100 YD	150 YD	200 YD	250 YD	DRIFT	TIME
0.	1500.	0.0	-1.5	-1.5	-1.5	-1.5	-1.5	0.0	0.0000
50.	1373.	-.2	0.0	1.6	3.9	6.7	9.7	.8	.1048
100.	1265.	1.5	-3.1	0.0	4.8	10.2	16.3	3.3	.2187
150.	1172.	4.9	-11.8	-7.1	0.0	8.1	17.3	7.6	.3431
200.	1104.	10.0	-26.6	-20.4	-10.9	0.0	12.2	13.0	.4741
250.	1053.	17.3	-48.6	-40.8	-28.9	-15.3	0.0	20.0	.6134
300.	1010.	27.0	-78.6	-69.3	-55.0	-38.7	-20.3	28.3	.7606
350.	972.	39.0	-116.0	-105.0	-88.3	-69.3	-47.9	37.0	.9103
400.	937.	53.9	-162.4	-149.9	-130.9	-109.2	-84.7	47.1	1.0675
450.	904.	71.9	-218.3	-204.2	-182.8	-158.3	-130.8	58.2	1.2304
500.	873.	93.5	-285.2	-269.6	-245.8	-218.6	-188.0	70.7	1.4016

BALLISTIC COEFFICIENT .23, MUZZLE VELOCITY FT/SEC 1600.

RANGE	VELOCITY	HEIGHT	50 YD	100 YD	150 YD	200 YD	250 YD	DRIFT	TIME
0.	1600.	0.0	-1.5	-1.5	-1.5	-1.5	-1.5	0.0	0.0000
50.	1462.	-.3	0.0	1.3	3.3	5.8	8.6	.8	.0981
100.	1340.	1.3	-2.6	0.0	4.1	9.0	14.5	3.1	.2053
150.	1237.	4.2	-10.0	-6.2	0.0	7.3	15.7	7.2	.3221
200.	1151.	8.9	-23.1	-18.0	-9.8	0.0	11.1	12.8	.4478
250.	1088.	15.5	-42.8	-36.4	-26.1	-13.9	0.0	19.9	.5820
300.	1040.	24.3	-69.7	-61.9	-49.6	-35.0	-18.3	28.3	.7230
350.	1000.	35.6	-104.3	-95.2	-80.8	-63.7	-44.3	37.7	.8702
400.	962.	49.5	-147.1	-136.8	-120.3	-100.8	-78.6	48.1	1.0235
450.	927.	66.3	-198.7	-187.0	-168.5	-146.6	-121.6	59.6	1.1822
500.	895.	86.5	-260.3	-247.3	-226.8	-202.4	-174.6	72.3	1.3482

BALLISTIC COEFFICIENT .23, MUZZLE VELOCITY FT/SEC 1700.

RANGE	VELOCITY	HEIGHT	50 YD	100 YD	150 YD	200 YD	250 YD	DRIFT	TIME
0.	1700.	0.0	-1.5	-1.5	-1.5	-1.5	-1.5	0.0	0.0000
50.	1555.	-.3	0.0	1.1	2.8	5.0	7.5	.7	.0924
100.	1421.	1.0	-2.1	0.0	3.6	7.9	12.9	2.9	.1932
150.	1306.	3.7	-8.5	-5.4	0.0	6.5	14.0	6.8	.3034
200.	1208.	7.8	-20.1	-15.9	-8.7	0.0	10.0	12.3	.4230
250.	1129.	13.9	-37.6	-32.3	-23.4	-12.5	0.0	19.5	.5518
300.	1072.	22.0	-61.9	-55.5	-44.8	-31.7	-16.8	28.0	.6883
350.	1027.	32.5	-93.7	-86.4	-73.8	-58.6	-41.1	37.8	.8324
400.	987.	45.5	-133.2	-124.8	-110.5	-93.1	-73.1	48.5	.9817
450.	951.	61.1	-180.5	-171.1	-154.9	-135.4	-112.9	60.0	1.1353
500.	917.	79.8	-237.1	-226.6	-208.6	-186.9	-161.9	72.8	1.2958

BALLISTIC COEFFICIENT .23, MUZZLE VELOCITY FT/SEC 1800.

RANGE	VELOCITY	HEIGHT	50 YD	100 YD	150 YD	200 YD	250 YD	DRIFT	TIME
0.	1800.	0.0	-1.5	-1.5	-1.5	-1.5	-1.5	0.0	0.0000
50.	1649.	-.4	0.0	.9	2.4	4.4	6.6	.7	.0871
100.	1508.	.8	-1.7	0.0	3.1	7.0	11.5	2.7	.1822
150.	1379.	3.2	-7.3	-4.7	0.0	5.8	12.5	6.4	.2864
200.	1270.	6.9	-17.4	-14.0	-7.7	0.0	9.0	11.7	.3998
250.	1177.	12.4	-33.0	-28.7	-20.9	-11.2	0.0	18.7	.5230
300.	1107.	19.8	-54.8	-49.6	-40.2	-28.6	-15.2	27.2	.6543
350.	1055.	29.4	-83.5	-77.5	-66.6	-53.0	-37.3	36.9	.7932
400.	1013.	41.7	-120.4	-113.5	-101.0	-85.5	-67.5	48.2	.9404
450.	974.	56.2	-163.9	-156.2	-142.0	-124.7	-104.5	59.8	1.0895
500.	939.	73.8	-216.4	-207.8	-192.1	-172.8	-150.4	72.7	1.2463

BALLISTIC COEFFICIENT .23, MUZZLE VELOCITY FT/SEC 1900.

RANGE	VELOCITY	HEIGHT	50 YD	100 YD	150 YD	200 YD	250 YD	DRIFT	TIME
0.	1900.	0.0	-1.5	-1.5	-1.5	-1.5	-1.5	0.0	0.0000
50.	1742.	-.4	0.0	.7	2.1	3.8	5.8	.6	.0825
100.	1594.	.7	-1.4	0.0	2.7	6.2	10.2	2.6	.1726
150.	1457.	2.8	-6.2	-4.1	0.0	5.2	11.2	6.0	.2710
200.	1336.	6.1	-15.1	-12.4	-6.9	0.0	8.0	11.1	.3787
250.	1233.	11.1	-29.0	-25.5	-18.6	-10.1	0.0	17.8	.4959
300.	1148.	17.8	-48.4	-44.3	-36.0	-25.7	-13.7	26.1	.6218
350.	1086.	26.7	-74.5	-69.7	-60.1	-48.0	-34.0	35.9	.7564
400.	1039.	37.9	-107.7	-102.2	-91.2	-77.5	-61.4	46.8	.8976
450.	998.	51.7	-148.5	-142.3	-130.0	-114.5	-96.4	58.9	1.0451
500.	961.	68.2	-197.3	-190.4	-176.6	-159.4	-139.3	71.9	1.1983

BALLISTIC COEFFICIENT .24, MUZZLE VELOCITY FT/SEC 800.

RANGE	VELOCITY	HEIGHT	50 YD	100 YD	150 YD	200 YD	250 YD	DRIFT	TIME
0.	800.	0.0	-1.5	-1.5	-1.5	-1.5	-1.5	0.0	0.0000
50.	775.	1.0	0.0	6.5	13.8	21.6	29.9	.6	.1909
100.	750.	6.5	-13.0	0.0	14.7	30.3	46.7	2.3	.3879
150.	725.	16.0	-41.5	-22.0	0.0	23.4	48.1	5.1	.5913
200.	701.	30.1	-86.5	-60.5	-31.2	0.0	33.0	9.0	.8013
250.	678.	49.1	-149.4	-116.8	-80.2	-41.2	0.0	14.3	1.0190
300.	655.	73.6	-231.4	-192.3	-148.3	-101.6	-52.1	21.0	1.2446
350.	632.	104.1	-333.5	-288.0	-236.6	-182.1	-124.4	29.1	1.4778
400.	610.	141.2	-457.5	-405.5	-346.8	-284.4	-218.5	38.6	1.7194
450.	589.	185.5	-604.9	-546.3	-480.3	-410.1	-336.0	49.7	1.9699
500.	568.	237.8	-777.2	-712.1	-638.7	-560.8	-478.4	62.4	2.2294

BALLISTIC COEFFICIENT .24, MUZZLE VELOCITY FT/SEC 900.

RANGE	VELOCITY	HEIGHT	50 YD	100 YD	150 YD	200 YD	250 YD	DRIFT	TIME
0.	900.	0.0	-1.5	-1.5	-1.5	-1.5	-1.5	0.0	0.0000
50.	871.	.6	0.0	5.0	10.7	16.9	23.3	.6	.1700
100.	843.	5.0	-9.9	0.0	11.5	23.8	36.7	2.0	.3449
150.	817.	12.5	-32.2	-17.2	0.0	18.5	37.8	4.5	.5256
200.	791.	23.6	-67.5	-47.6	-24.6	0.0	25.8	8.1	.7125
250.	766.	38.6	-116.7	-91.8	-63.1	-32.3	0.0	12.6	.9051
300.	741.	57.7	-180.6	-150.8	-116.3	-79.3	-40.6	18.3	1.1039
350.	717.	81.6	-260.6	-225.8	-185.5	-142.5	-97.2	25.2	1.3099
400.	693.	110.6	-357.5	-317.8	-271.8	-222.5	-170.8	33.4	1.5230
450.	670.	145.1	-472.5	-427.8	-376.0	-320.6	-262.5	42.8	1.7434
500.	647.	185.8	-606.7	-557.0	-499.5	-437.9	-373.3	53.6	1.9712

BALLISTIC COEFFICIENT .24, MUZZLE VELOCITY FT/SEC 1000.

RANGE	VELOCITY	HEIGHT	50 YD	100 YD	150 YD	200 YD	250 YD	DRIFT	TIME
0.	1000.	0.0	-1.5	-1.5	-1.5	-1.5	-1.5	0.0	0.0000
50.	964.	.4	0.0	3.9	8.6	13.6	18.9	.6	.1535
100.	930.	3.9	-7.8	0.0	9.3	19.4	30.0	2.0	.3114
150.	899.	10.1	-25.7	-14.0	0.0	15.1	31.0	4.5	.4755
200.	870.	19.2	-54.4	-38.8	-20.2	0.0	21.2	8.0	.6453
250.	842.	31.5	-94.5	-75.0	-51.7	-26.5	0.0	12.4	.8203
300.	816.	47.4	-147.0	-123.6	-95.6	-65.4	-33.6	17.8	1.0014
350.	790.	67.0	-212.6	-185.2	-152.6	-117.3	-80.2	24.4	1.1884
400.	765.	90.8	-291.8	-260.6	-223.3	-182.9	-140.5	31.9	1.3812
450.	740.	119.1	-385.7	-350.5	-308.6	-263.2	-215.5	40.5	1.5802
500.	716.	152.5	-495.6	-456.6	-410.0	-359.6	-306.6	50.4	1.7866

BALLISTIC COEFFICIENT .24, MUZZLE VELOCITY FT/SEC 1100.

RANGE	VELOCITY	HEIGHT	50 YD	100 YD	150 YD	200 YD	250 YD	DRIFT	TIME
0.	1100.	0.0	-1.5	-1.5	-1.5	-1.5	-1.5	0.0	0.0000
50.	1052.	.2	0.0	3.3	7.1	11.3	15.8	.6	.1396
100.	1011.	3.2	-6.6	0.0	7.6	16.1	25.1	2.5	.2871
150.	975.	8.4	-21.2	-11.4	0.0	12.7	26.2	4.8	.4364
200.	940.	16.1	-45.3	-32.1	-16.9	0.0	18.0	8.4	.5931
250.	908.	26.6	-79.1	-62.7	-43.7	-22.6	0.0	13.0	.7555
300.	878.	40.2	-123.7	-104.0	-81.2	-55.8	-28.7	18.6	.9239
350.	850.	57.0	-179.2	-156.2	-129.7	-100.0	-68.4	25.1	1.0973
400.	824.	77.4	-246.7	-220.4	-190.0	-156.2	-120.1	32.6	1.2763
450.	798.	101.7	-326.9	-297.3	-263.1	-225.0	-184.4	41.2	1.4613
500.	773.	130.3	-420.6	-387.8	-349.8	-307.4	-262.3	50.9	1.6526

BALLISTIC COEFFICIENT .24, MUZZLE VELOCITY FT/SEC 1200.

RANGE	VELOCITY	HEIGHT	50 YD	100 YD	150 YD	200 YD	250 YD	DRIFT	TIME
0.	1200.	0.0	-1.5	-1.5	-1.5	-1.5	-1.5	0.0	0.0000
50.	1126.	.1	0.0	2.7	6.2	9.9	13.8	.8	.1293
100.	1072.	2.6	-5.4	0.0	7.0	14.5	22.3	2.8	.2660
150.	1028.	7.3	-18.5	-10.4	0.0	11.3	23.1	6.2	.4103
200.	990.	14.3	-39.7	-29.0	-15.1	0.0	15.6	10.5	.5595
250.	955.	23.6	-69.2	-55.8	-38.4	-19.5	0.0	15.3	.7118
300.	922.	35.7	-108.4	-92.4	-71.5	-48.8	-25.4	21.4	.8716
350.	891.	50.9	-157.9	-139.1	-114.8	-88.4	-61.0	28.6	1.0373
400.	863.	69.3	-218.0	-196.6	-168.8	-138.6	-107.3	36.7	1.2084
450.	836.	91.3	-289.6	-265.6	-234.3	-200.3	-165.1	45.8	1.3851
500.	809.	117.2	-373.5	-346.7	-312.0	-274.2	-235.1	55.9	1.5677

BALLISTIC COEFFICIENT .24, MUZZLE VELOCITY FT/SEC 1300.

RANGE	VELOCITY	HEIGHT	50 YD	100 YD	150 YD	200 YD	250 YD	DRIFT	TIME
0.	1300.	0.0	-1.5	-1.5	-1.5	-1.5	-1.5	0.0	0.0000
50.	1207.	-.1	0.0	2.2	5.3	8.7	12.4	.8	.1200
100.	1131.	2.2	-4.5	0.0	6.0	12.9	20.3	3.1	.2487
150.	1075.	6.4	-15.8	-9.0	0.0	10.4	21.4	6.8	.3848
200.	1031.	12.7	-34.9	-25.9	-13.8	0.0	14.6	11.9	.5291
250.	993.	21.3	-61.9	-50.6	-35.6	-18.3	0.0	17.8	.6779
300.	958.	32.3	-96.7	-83.2	-65.2	-44.4	-22.4	24.1	.8293
350.	925.	46.2	-141.4	-125.7	-104.6	-80.4	-54.8	31.9	.9889
400.	893.	63.4	-196.6	-178.6	-154.5	-126.9	-97.6	40.9	1.1557
450.	865.	83.5	-261.2	-240.9	-213.8	-182.7	-149.8	50.4	1.3248
500.	837.	107.4	-337.5	-315.0	-284.9	-250.3	-213.7	61.1	1.5009

BALLISTIC COEFFICIENT .24, MUZZLE VELOCITY FT/SEC 1400.

RANGE	VELOCITY	HEIGHT	50 YD	100 YD	150 YD	200 YD	250 YD	DRIFT	TIME
0.	1400.	0.0	-1.5	-1.5	-1.5	-1.5	-1.5	0.0	0.0000
50.	1292.	-.2	0.0	1.9	4.5	7.5	10.9	.8	.1116
100.	1200.	1.8	-3.7	0.0	5.3	11.3	18.1	3.2	.2323
150.	1126.	5.5	-13.5	-7.9	0.0	9.1	19.2	7.1	.3616
200.	1072.	11.2	-30.2	-22.7	-12.1	0.0	13.5	12.3	.4983
250.	1028.	19.1	-54.5	-45.2	-32.0	-16.9	0.0	18.8	.6426
300.	990.	29.3	-86.8	-75.7	-59.8	-41.6	-21.4	26.2	.7919
350.	955.	42.0	-127.0	-113.9	-95.4	-74.2	-50.6	34.2	.9442
400.	922.	57.8	-176.7	-161.8	-140.7	-116.4	-89.5	43.4	1.1040
450.	891.	76.6	-236.5	-219.7	-195.9	-168.6	-138.3	53.8	1.2698
500.	862.	98.9	-306.6	-287.9	-261.5	-231.2	-197.5	65.0	1.4408

BALLISTIC COEFFICIENT .24, MUZZLE VELOCITY FT/SEC 1500.

RANGE	VELOCITY	HEIGHT	50 YD	100 YD	150 YD	200 YD	250 YD	DRIFT	TIME
0.	1500.	0.0	-1.5	-1.5	-1.5	-1.5	-1.5	0.0	0.0000
50.	1377.	-.2	0.0	1.6	3.9	6.6	9.6	.8	.1046
100.	1273.	1.5	-3.1	0.0	4.6	10.0	16.1	3.2	.2182
150.	1183.	4.8	-11.6	-6.9	0.0	8.1	17.3	7.1	.3403
200.	1114.	9.9	-26.3	-20.1	-10.9	0.0	12.2	12.5	.4712
250.	1062.	17.1	-48.0	-40.3	-28.8	-15.2	0.0	19.3	.6098
300.	1021.	26.5	-77.1	-67.8	-54.0	-37.7	-19.5	27.0	.7534
350.	983.	38.5	-114.4	-103.5	-87.5	-68.5	-47.2	35.9	.9037
400.	948.	53.0	-160.0	-147.6	-129.2	-107.5	-83.2	45.5	1.0586
450.	916.	70.6	-214.9	-200.9	-180.2	-155.8	-128.4	56.2	1.2194
500.	885.	91.5	-279.7	-264.1	-241.2	-214.0	-183.6	68.0	1.3862

BALLISTIC COEFFICIENT .24, MUZZLE VELOCITY FT/SEC 1600.

RANGE	VELOCITY	HEIGHT	50 YD	100 YD	150 YD	200 YD	250 YD	DRIFT	TIME
0.	1600.	0.0	-1.5	-1.5	-1.5	-1.5	-1.5	0.0	0.0000
50.	1468.	-.3	0.0	1.3	3.3	5.7	8.4	.7	.0978
100.	1350.	1.3	-2.6	0.0	4.0	8.8	14.3	3.0	.2046
150.	1249.	4.2	-9.9	-6.1	0.0	7.2	15.4	6.8	.3202
200.	1164.	8.7	-22.8	-17.7	-9.6	0.0	10.9	12.3	.4447
250.	1100.	15.3	-42.2	-35.8	-25.7	-13.6	0.0	19.1	.5775
300.	1051.	23.9	-68.6	-60.9	-48.8	-34.4	-18.0	27.2	.7171
350.	1011.	35.1	-103.1	-94.1	-80.0	-63.2	-44.1	36.7	.8646
400.	974.	48.6	-144.6	-134.4	-118.2	-99.0	-77.2	46.5	1.0140
450.	940.	65.0	-195.2	-183.7	-165.5	-143.9	-119.3	57.5	1.1706
500.	908.	84.6	-255.2	-242.4	-222.2	-198.1	-170.8	69.6	1.3331

BALLISTIC COEFFICIENT .24, MUZZLE VELOCITY FT/SEC 1700.

RANGE	VELOCITY	HEIGHT	50 YD	100 YD	150 YD	200 YD	250 YD	DRIFT	TIME
0.	1700.	0.0	-1.5	-1.5	-1.5	-1.5	-1.5	0.0	0.0000
50.	1561.	-.3	0.0	1.0	2.8	4.9	7.4	.7	.0923
100.	1432.	1.0	-2.1	0.0	3.5	7.8	12.7	2.8	.1926
150.	1319.	3.6	-8.4	-5.3	0.0	6.4	13.7	6.5	.3017
200.	1223.	7.7	-19.8	-15.6	-8.5	0.0	9.8	11.8	.4199
250.	1143.	13.6	-36.9	-31.7	-22.9	-12.2	0.0	18.6	.5470
300.	1085.	21.6	-60.7	-54.5	-43.9	-31.1	-16.4	26.8	.6819
350.	1039.	31.8	-91.7	-84.4	-72.0	-57.1	-40.0	36.2	.8232
400.	1000.	44.5	-130.3	-122.0	-107.9	-90.8	-71.3	46.6	.9705
450.	964.	59.9	-177.3	-168.0	-152.1	-132.9	-110.9	58.1	1.1240
500.	931.	78.1	-232.6	-222.2	-204.5	-183.2	-158.7	70.3	1.2817

BALLISTIC COEFFICIENT .24, MUZZLE VELOCITY FT/SEC 1800.

RANGE	VELOCITY	HEIGHT	50 YD	100 YD	150 YD	200 YD	250 YD	DRIFT	TIME
0.	1800.	0.0	-1.5	-1.5	-1.5	-1.5	-1.5	0.0	0.0000
50.	1655.	-.4	0.0	.8	2.4	4.3	6.5	.6	.0870
100.	1519.	.8	-1.7	0.0	3.1	6.9	11.3	2.6	.1816
150.	1394.	3.1	-7.2	-4.6	0.0	5.7	12.3	6.1	.2847
200.	1287.	6.8	-17.1	-13.8	-7.6	0.0	8.8	11.2	.3968
250.	1195.	12.1	-32.4	-28.2	-20.5	-11.0	0.0	17.9	.5182
300.	1123.	19.4	-53.6	-48.6	-39.3	-27.9	-14.8	26.0	.6476
350.	1069.	28.8	-81.8	-75.9	-65.1	-51.8	-36.4	35.4	.7847
400.	1026.	40.7	-117.5	-110.8	-98.4	-83.3	-65.7	46.2	.9290
450.	988.	55.1	-160.9	-153.3	-139.4	-122.4	-102.6	57.8	1.0785
500.	953.	72.1	-211.8	-203.3	-187.9	-168.9	-147.0	70.1	1.2316

BALLISTIC COEFFICIENT .24, MUZZLE VELOCITY FT/SEC 1900.

RANGE	VELOCITY	HEIGHT	50 YD	100 YD	150 YD	200 YD	250 YD	DRIFT	TIME
0.	1900.	0.0	-1.5	-1.5	-1.5	-1.5	-1.5	0.0	0.0000
50.	1748.	-.4	0.0	.7	2.0	3.7	5.7	.6	.0823
100.	1606.	.7	-1.4	0.0	2.7	6.1	10.0	2.5	.1719
150.	1474.	2.7	-6.1	-4.1	0.0	5.1	11.0	5.7	.2694
200.	1355.	6.0	-14.9	-12.2	-6.8	0.0	7.8	10.6	.3759
250.	1253.	10.8	-28.4	-25.0	-18.3	-9.8	0.0	17.0	.4911
300.	1167.	17.4	-47.4	-43.3	-35.2	-25.1	-13.3	24.9	.6150
350.	1102.	26.1	-72.8	-68.1	-58.6	-46.8	-33.0	34.3	.7475
400.	1053.	37.0	-105.3	-99.8	-89.0	-75.5	-59.8	44.9	.8868
450.	1013.	50.6	-145.8	-139.6	-127.5	-112.2	-94.6	57.0	1.0343
500.	976.	66.4	-192.8	-185.9	-172.4	-155.5	-135.9	69.3	1.1831

BALLISTIC COEFFICIENT .25, MUZZLE VELOCITY FT/SEC 800.

RANGE	VELOCITY	HEIGHT	50 YD	100 YD	150 YD	200 YD	250 YD	DRIFT	TIME
0.	800.	0.0	-1.5	-1.5	-1.5	-1.5	-1.5	0.0	0.0000
50.	776.	1.0	0.0	6.5	13.8	21.6	29.7	.6	.1907
100.	752.	6.5	-13.0	0.0	14.6	30.1	46.4	2.2	.3874
150.	728.	16.0	-41.4	-21.9	0.0	23.3	47.7	4.9	.5902
200.	705.	29.9	-86.2	-60.3	-31.0	0.0	32.6	8.7	.7995
250.	682.	48.8	-148.5	-116.1	-79.5	-40.7	0.0	13.7	1.0155
300.	660.	72.9	-229.6	-190.6	-146.8	-100.2	-51.3	20.0	1.2389
350.	639.	103.0	-330.7	-285.2	-234.1	-179.7	-122.7	27.7	1.4699
400.	617.	139.5	-453.2	-401.2	-342.8	-280.7	-215.5	36.8	1.7091
450.	596.	183.0	-598.6	-540.2	-474.4	-404.6	-331.2	47.4	1.9566
500.	576.	234.3	-768.4	-703.4	-630.4	-552.8	-471.3	59.4	2.2127

BALLISTIC COEFFICIENT .25, MUZZLE VELOCITY FT/SEC 900.

RANGE	VELOCITY	HEIGHT	50 YD	100 YD	150 YD	200 YD	250 YD	DRIFT	TIME
0.	900.	0.0	-1.5	-1.5	-1.5	-1.5	-1.5	0.0	0.0000
50.	872.	.6	0.0	5.0	10.7	16.8	23.2	.6	.1699
100.	845.	4.9	-9.9	0.0	11.4	23.6	36.5	2.0	.3445
150.	820.	12.4	-32.0	-17.1	0.0	18.4	37.6	4.3	.5244
200.	795.	23.5	-67.2	-47.3	-24.5	0.0	25.7	7.7	.7105
250.	771.	38.3	-116.1	-91.2	-62.7	-32.1	0.0	12.1	.9022
300.	747.	57.3	-179.6	-149.8	-115.6	-78.8	-40.3	17.6	1.0999
350.	724.	80.8	-258.6	-223.9	-184.0	-141.1	-96.2	24.1	1.3038
400.	701.	109.3	-354.2	-314.5	-268.9	-219.9	-168.5	31.8	1.5142
450.	678.	143.2	-467.7	-423.0	-371.7	-316.6	-258.8	40.8	1.7320
500.	656.	183.1	-600.2	-550.5	-493.5	-432.3	-368.1	51.1	1.9572

BALLISTIC COEFFICIENT .25, MUZZLE VELOCITY FT/SEC 1000.

RANGE	VELOCITY	HEIGHT	50 YD	100 YD	150 YD	200 YD	250 YD	DRIFT	TIME
0.	1000.	0.0	-1.5	-1.5	-1.5	-1.5	-1.5	0.0	0.0000
50.	965.	.4	0.0	3.9	8.5	13.6	18.8	.7	.1538
100.	933.	3.9	-7.7	0.0	9.3	19.5	29.8	1.9	.3110
150.	903.	10.0	-25.5	-13.9	0.0	15.3	30.9	4.3	.4742
200.	874.	19.2	-54.4	-39.0	-20.4	0.0	20.7	7.9	.6451
250.	848.	31.3	-94.0	-74.6	-51.5	-25.9	0.0	11.9	.8177
300.	822.	47.0	-145.9	-122.6	-94.9	-64.2	-33.1	17.1	.9972
350.	797.	66.4	-210.7	-183.6	-151.2	-115.4	-79.2	23.3	1.1826
400.	773.	89.8	-289.2	-258.2	-221.2	-180.3	-138.9	30.6	1.3738
450.	749.	117.7	-382.1	-347.3	-305.6	-259.6	-213.0	38.9	1.5710
500.	726.	150.4	-490.3	-451.6	-405.3	-354.2	-302.4	48.3	1.7744

BALLISTIC COEFFICIENT .25, MUZZLE VELOCITY FT/SEC 1100.

RANGE	VELOCITY	HEIGHT	50 YD	100 YD	150 YD	200 YD	250 YD	DRIFT	TIME
0.	1100.	0.0	-1.5	-1.5	-1.5	-1.5	-1.5	0.0	0.0000
50.	1053.	.2	0.0	3.2	7.0	11.3	15.7	.6	.1395
100.	1015.	3.1	-6.4	0.0	7.7	16.3	25.1	2.1	.2848
150.	979.	8.3	-21.1	-11.6	0.0	12.8	26.0	4.6	.4353
200.	945.	16.1	-45.3	-32.6	-17.1	0.0	17.6	8.3	.5925
250.	914.	26.4	-78.6	-62.7	-43.4	-22.0	0.0	12.5	.7526
300.	885.	39.8	-122.6	-103.6	-80.4	-54.7	-28.3	17.8	.9195
350.	858.	56.4	-177.6	-155.4	-128.3	-98.4	-67.6	24.1	1.0915
400.	832.	76.6	-244.5	-219.1	-188.1	-154.0	-118.7	31.4	1.2692
450.	807.	100.5	-323.7	-295.1	-260.3	-221.8	-182.2	39.6	1.4523
500.	783.	128.5	-415.9	-384.1	-345.5	-302.7	-258.7	48.8	1.6408

BALLISTIC COEFFICIENT .25, MUZZLE VELOCITY FT/SEC 1200.

RANGE	VELOCITY	HEIGHT	50 YD	100 YD	150 YD	200 YD	250 YD	DRIFT	TIME
0.	1200.	0.0	-1.5	-1.5	-1.5	-1.5	-1.5	0.0	0.0000
50.	1129.	.1	0.0	2.7	6.1	9.8	13.7	.7	.1291
100.	1076.	2.6	-5.3	0.0	6.8	14.2	22.2	2.7	.2654
150.	1034.	7.2	-18.2	-10.2	0.0	11.2	23.1	5.8	.4078
200.	997.	14.1	-39.1	-28.5	-14.9	0.0	15.9	9.8	.5557
250.	962.	23.4	-68.7	-55.4	-38.5	-19.9	0.0	14.8	.7091
300.	930.	35.4	-107.6	-91.6	-71.3	-49.0	-25.1	20.7	.8677
350.	900.	50.3	-156.3	-137.7	-114.0	-87.9	-60.1	27.5	1.0314
400.	872.	68.5	-215.9	-194.6	-167.5	-137.7	-105.9	35.4	1.2012
450.	845.	90.1	-286.5	-262.6	-232.0	-198.6	-162.8	44.2	1.3759
500.	820.	115.4	-368.9	-342.3	-308.4	-271.2	-231.4	53.8	1.5559

BALLISTIC COEFFICIENT .25, MUZZLE VELOCITY FT/SEC 1300.

RANGE	VELOCITY	HEIGHT	50 YD	100 YD	150 YD	200 YD	250 YD	DRIFT	TIME
0.	1300.	0.0	-1.5	-1.5	-1.5	-1.5	-1.5	0.0	0.0000
50.	1210.	-.1	0.0	2.3	5.2	8.6	12.2	.8	.1198
100.	1136.	2.2	-4.5	0.0	5.9	12.6	19.8	3.2	.2488
150.	1081.	6.3	-15.7	-8.9	0.0	10.0	20.8	6.6	.3837
200.	1038.	12.5	-34.2	-25.2	-13.3	0.0	14.4	11.2	.5251
250.	1001.	21.0	-60.8	-49.5	-34.7	-18.0	0.0	16.8	.6725
300.	966.	32.0	-96.1	-82.5	-64.8	-44.8	-23.2	23.6	.8261
350.	934.	45.6	-140.0	-124.1	-103.4	-80.1	-54.9	30.9	.9831
400.	903.	62.3	-193.7	-175.6	-151.9	-125.3	-96.5	39.3	1.1463
450.	875.	82.5	-258.7	-238.3	-211.7	-181.7	-149.3	49.0	1.3168
500.	848.	105.7	-333.3	-310.7	-281.1	-247.8	-211.7	59.1	1.4895

BALLISTIC COEFFICIENT .25, MUZZLE VELOCITY FT/SEC 1400.

RANGE	VELOCITY	HEIGHT	50 YD	100 YD	150 YD	200 YD	250 YD	DRIFT	TIME
0.	1400.	0.0	-1.5	-1.5	-1.5	-1.5	-1.5	0.0	0.0000
50.	1296.	-.2	0.0	1.8	4.5	7.4	10.7	.8	.1116
100.	1207.	1.8	-3.7	0.0	5.2	11.2	17.8	3.0	.2316
150.	1134.	5.5	-13.4	-7.8	0.0	9.0	18.8	6.8	.3600
200.	1079.	11.0	-29.8	-22.4	-12.0	0.0	13.1	11.8	.4957
250.	1037.	18.8	-53.6	-44.4	-31.3	-16.4	0.0	17.9	.6377
300.	999.	28.8	-85.3	-74.3	-58.6	-40.6	-21.0	25.0	.7852
350.	965.	41.6	-125.7	-112.8	-94.5	-73.6	-50.6	33.2	.9388
400.	932.	56.9	-174.6	-159.9	-139.0	-115.0	-88.8	42.1	1.0963
450.	902.	75.4	-233.2	-216.6	-193.1	-166.2	-136.7	52.0	1.2597
500.	874.	97.5	-303.2	-284.8	-258.6	-228.7	-195.9	63.3	1.4308

BALLISTIC COEFFICIENT .25, MUZZLE VELOCITY FT/SEC 1500.

RANGE	VELOCITY	HEIGHT	50 YD	100 YD	150 YD	200 YD	250 YD	DRIFT	TIME
0.	1500.	0.0	-1.5	-1.5	-1.5	-1.5	-1.5	0.0	0.0000
50.	1382.	-.2	0.0	1.5	3.8	6.5	9.5	.8	.1043
100.	1281.	1.5	-3.1	0.0	4.6	9.9	15.9	3.0	.2172
150.	1193.	4.8	-11.5	-6.9	0.0	8.0	16.9	6.9	.3392
200.	1124.	9.8	-26.0	-19.8	-10.6	0.0	11.9	12.0	.4685
250.	1072.	16.8	-47.3	-39.7	-28.1	-14.9	0.0	18.5	.6053
300.	1030.	26.2	-76.5	-67.3	-53.5	-37.6	-19.7	26.4	.7498
350.	993.	38.0	-113.5	-102.7	-86.6	-68.0	-47.2	35.0	.8989
400.	959.	52.2	-157.7	-145.5	-127.0	-105.8	-82.0	44.0	1.0500
450.	927.	69.4	-211.9	-198.0	-177.3	-153.4	-126.7	54.4	1.2093
500.	897.	89.9	-275.7	-260.4	-237.4	-210.8	-181.1	65.9	1.3743

BALLISTIC COEFFICIENT .25, MUZZLE VELOCITY FT/SEC 1600.

RANGE	VELOCITY	HEIGHT	50 YD	100 YD	150 YD	200 YD	250 YD	DRIFT	TIME
0.	1600.	0.0	-1.5	-1.5	-1.5	-1.5	-1.5	0.0	0.0000
50.	1473.	-.3	0.0	1.3	3.3	5.6	8.3	.7	.0976
100.	1358.	1.2	-2.5	0.0	4.0	8.8	14.1	2.9	.2039
150.	1260.	4.1	-9.8	-6.0	0.0	7.2	15.2	6.5	.3184
200.	1176.	8.6	-22.6	-17.5	-9.6	0.0	10.6	11.9	.4425
250.	1111.	15.0	-41.6	-35.2	-25.3	-13.3	0.0	18.4	.5734
300.	1062.	23.6	-67.6	-60.0	-48.1	-33.7	-17.8	26.3	.7120
350.	1022.	34.4	-101.1	-92.2	-78.3	-61.5	-42.9	35.1	.8559
400.	986.	47.8	-142.7	-132.6	-116.7	-97.5	-76.2	45.1	1.0062
450.	952.	63.8	-192.2	-180.7	-162.8	-141.3	-117.4	55.7	1.1600
500.	921.	82.9	-250.8	-238.1	-218.2	-194.3	-167.7	67.3	1.3200

BALLISTIC COEFFICIENT .25, MUZZLE VELOCITY FT/SEC 1700.

RANGE	VELOCITY	HEIGHT	50 YD	100 YD	150 YD	200 YD	250 YD	DRIFT	TIME
0.	1700.	0.0	-1.5	-1.5	-1.5	-1.5	-1.5	0.0	0.0000
50.	1567.	-.3	0.0	1.0	2.8	4.9	7.3	.7	.0921
100.	1442.	1.0	-2.1	0.0	3.5	7.7	12.5	2.7	.1919
150.	1332.	3.6	-8.3	-5.2	0.0	6.3	13.5	6.2	.3002
200.	1237.	7.6	-19.5	-15.4	-8.4	0.0	9.6	11.3	.4173
250.	1157.	13.4	-36.4	-31.3	-22.5	-12.0	0.0	17.9	.5431
300.	1097.	21.2	-59.7	-53.5	-43.0	-30.4	-16.0	25.8	.6759
350.	1051.	31.2	-90.1	-82.9	-70.7	-55.9	-39.1	34.9	.8157
400.	1012.	43.8	-128.7	-120.4	-106.5	-89.7	-70.4	45.3	.9633
450.	977.	58.6	-173.9	-164.6	-149.0	-130.0	-108.4	56.0	1.1121
500.	944.	76.6	-228.6	-218.3	-200.9	-179.8	-155.8	68.1	1.2692

BALLISTIC COEFFICIENT .25, MUZZLE VELOCITY FT/SEC 1800.

RANGE	VELOCITY	HEIGHT	50 YD	100 YD	150 YD	200 YD	250 YD	DRIFT	TIME
0.	1800.	0.0	-1.5	-1.5	-1.5	-1.5	-1.5	0.0	0.0000
50.	1660.	-.4	0.0	.8	2.4	4.2	6.4	.6	.0868
100.	1529.	.8	-1.7	0.0	3.0	6.8	11.1	2.5	.1810
150.	1408.	3.1	-7.1	-4.6	0.0	5.6	12.0	5.8	.2832
200.	1303.	6.7	-16.9	-13.5	-7.4	0.0	8.6	10.7	.3940
250.	1212.	11.9	-31.9	-27.7	-20.1	-10.8	0.0	17.1	.5139
300.	1138.	19.0	-52.7	-47.7	-38.6	-27.4	-14.5	25.0	.6418
350.	1082.	28.3	-80.3	-74.5	-63.8	-50.8	-35.7	34.2	.7774
400.	1039.	39.7	-114.8	-108.2	-96.0	-81.1	-63.9	44.3	.9181
450.	1002.	53.8	-157.3	-149.8	-136.1	-119.3	-99.9	55.6	1.0656
500.	967.	70.6	-208.0	-199.7	-184.4	-165.8	-144.3	67.9	1.2191

BALLISTIC COEFFICIENT .25, MUZZLE VELOCITY FT/SEC 1900.

RANGE	VELOCITY	HEIGHT	50 YD	100 YD	150 YD	200 YD	250 YD	DRIFT	TIME
0.	1900.	0.0	-1.5	-1.5	-1.5	-1.5	-1.5	0.0	0.0000
50.	1754.	-.4	0.0	.7	2.0	3.7	5.6	.6	.0821
100.	1617.	.7	-1.3	0.0	2.7	6.0	9.8	2.3	.1712
150.	1489.	2.7	-6.0	-4.0	0.0	5.0	10.7	5.5	.2679
200.	1372.	5.9	-14.6	-12.0	-6.6	0.0	7.7	10.1	.3730
250.	1273.	10.6	-27.9	-24.5	-17.9	-9.6	0.0	16.2	.4866
300.	1186.	17.0	-46.5	-42.5	-34.4	-24.5	-13.0	23.8	.6088
350.	1118.	25.5	-71.3	-66.7	-57.3	-45.7	-32.3	32.9	.7396
400.	1068.	36.1	-102.9	-97.6	-86.9	-73.6	-58.3	43.1	.8766
450.	1027.	49.3	-142.2	-136.2	-124.2	-109.3	-92.0	54.7	1.0211
500.	990.	65.0	-189.2	-182.5	-169.1	-152.6	-133.4	67.1	1.1707

BALLISTIC COEFFICIENT .26, MUZZLE VELOCITY FT/SEC 800.

RANGE	VELOCITY	HEIGHT	50 YD	100 YD	150 YD	200 YD	250 YD	DRIFT	TIME
0.	800.	0.0	-1.5	-1.5	-1.5	-1.5	-1.5	0.0	0.0000
50.	777.	1.0	0.0	6.5	13.8	21.5	29.6	.5	.1906
100.	753.	6.4	-13.0	0.0	14.6	30.0	46.2	2.1	.3869
150.	731.	15.9	-41.3	-21.8	0.0	23.2	47.5	4.7	.5891
200.	709.	29.8	-85.9	-60.0	-30.9	0.0	32.4	8.4	.7977
250.	687.	48.5	-147.9	-115.5	-79.1	-40.5	0.0	13.2	1.0127
300.	665.	72.4	-228.4	-189.6	-145.9	-99.5	-50.9	19.3	1.2346
350.	644.	102.1	-328.5	-283.2	-232.2	-178.2	-121.4	26.6	1.4636
400.	624.	138.0	-449.6	-397.8	-339.5	-277.7	-212.9	35.2	1.7001
450.	604.	180.8	-593.1	-534.8	-469.3	-399.8	-326.9	45.3	1.9448
500.	584.	231.1	-760.5	-695.7	-622.9	-545.7	-464.7	56.8	2.1976

BALLISTIC COEFFICIENT .26, MUZZLE VELOCITY FT/SEC 900.

RANGE	VELOCITY	HEIGHT	50 YD	100 YD	150 YD	200 YD	250 YD	DRIFT	TIME
0.	900.	0.0	-1.5	-1.5	-1.5	-1.5	-1.5	0.0	0.0000
50.	873.	.6	0.0	5.0	10.6	16.7	23.1	.5	.1698
100.	847.	4.9	-9.9	0.0	11.4	23.5	36.3	1.9	.3442
150.	823.	12.4	-31.9	-17.0	0.0	18.2	37.3	4.2	.5237
200.	799.	23.4	-66.8	-47.0	-24.2	0.0	25.5	7.4	.7086
250.	776.	38.1	-115.4	-90.6	-62.2	-31.9	0.0	11.6	.8994
300.	753.	56.9	-178.5	-148.8	-114.7	-78.3	-40.0	16.9	1.0959
350.	730.	80.2	-257.0	-222.3	-182.5	-140.1	-95.4	23.2	1.2984
400.	708.	108.3	-351.8	-312.1	-266.6	-218.2	-167.1	30.6	1.5072
450.	686.	141.7	-463.8	-419.2	-368.0	-313.5	-256.1	39.1	1.7224
500.	664.	180.8	-594.4	-544.8	-488.0	-427.4	-363.5	48.9	1.9447

BALLISTIC COEFFICIENT .26, MUZZLE VELOCITY FT/SEC 1000.

RANGE	VELOCITY	HEIGHT	50 YD	100 YD	150 YD	200 YD	250 YD	DRIFT	TIME
0.	1000.	0.0	-1.5	-1.5	-1.5	-1.5	-1.5	0.0	0.0000
50.	966.	.4	0.0	3.8	8.5	13.4	18.7	.7	.1540
100.	935.	3.9	-7.7	0.0	9.2	19.2	29.7	1.9	.3106
150.	906.	10.0	-25.4	-13.8	0.0	15.0	30.7	4.1	.4734
200.	878.	19.0	-53.8	-38.4	-20.0	0.0	20.9	7.4	.6420
250.	853.	31.1	-93.3	-74.2	-51.1	-26.1	0.0	11.5	.8152
300.	828.	46.6	-144.9	-121.9	-94.2	-64.3	-32.9	16.5	.9937
350.	804.	65.8	-209.1	-182.2	-149.9	-115.0	-78.4	22.4	1.1775
400.	781.	88.9	-286.6	-255.9	-219.0	-179.0	-137.2	29.3	1.3666
450.	757.	116.3	-378.4	-343.9	-302.3	-257.4	-210.4	37.3	1.5618
500.	735.	148.5	-485.4	-447.0	-400.8	-350.9	-298.7	46.3	1.7632

BALLISTIC COEFFICIENT .26, MUZZLE VELOCITY FT/SEC 1100.

RANGE	VELOCITY	HEIGHT	50 YD	100 YD	150 YD	200 YD	250 YD	DRIFT	TIME
0.	1100.	0.0	-1.5	-1.5	-1.5	-1.5	-1.5	0.0	0.0000
50.	1055.	.2	0.0	3.2	7.0	11.2	15.6	.5	.1394
100.	1017.	3.1	-6.3	0.0	7.8	16.1	24.9	2.0	.2843
150.	983.	8.3	-21.1	-11.6	0.0	12.5	25.7	4.5	.4349
200.	951.	15.9	-44.8	-32.1	-16.6	0.0	17.7	7.8	.5897
250.	920.	26.2	-78.1	-62.3	-42.8	-22.1	0.0	12.0	.7498
300.	892.	39.5	-121.8	-102.8	-79.5	-54.6	-28.1	17.2	.9157
350.	865.	55.9	-176.3	-154.2	-127.0	-97.9	-67.0	23.2	1.0866
400.	840.	75.7	-242.2	-216.9	-185.9	-152.6	-117.3	30.1	1.2622
450.	816.	99.3	-320.6	-292.2	-257.2	-219.8	-180.1	38.1	1.4437
500.	792.	126.9	-411.9	-380.2	-341.4	-299.9	-255.7	47.0	1.6305

BALLISTIC COEFFICIENT .26, MUZZLE VELOCITY FT/SEC 1200.

RANGE	VELOCITY	HEIGHT	50 YD	100 YD	150 YD	200 YD	250 YD	DRIFT	TIME
0.	1200.	0.0	-1.5	-1.5	-1.5	-1.5	-1.5	0.0	0.0000
50.	1131.	.0	0.0	2.6	6.0	9.7	13.7	.7	.1290
100.	1079.	2.6	-5.3	0.0	6.7	14.2	22.2	2.6	.2648
150.	1038.	7.2	-18.0	-10.1	0.0	11.1	23.1	5.6	.4066
200.	1002.	14.0	-38.9	-28.3	-14.9	0.0	16.0	9.5	.5542
250.	968.	23.3	-68.6	-55.4	-38.6	-20.0	0.0	14.6	.7079
300.	937.	35.1	-106.7	-90.9	-70.7	-48.4	-24.4	20.0	.8637
350.	908.	49.8	-155.1	-136.6	-113.0	-87.0	-59.0	26.6	1.0264
400.	880.	67.7	-213.7	-192.6	-165.6	-135.9	-103.9	34.2	1.1941
450.	854.	89.0	-283.6	-259.9	-229.6	-196.1	-160.1	42.7	1.3675
500.	829.	113.9	-365.1	-338.7	-305.0	-267.9	-227.9	52.1	1.5457

BALLISTIC COEFFICIENT .26, MUZZLE VELOCITY FT/SEC 1300.

RANGE	VELOCITY	HEIGHT	50 YD	100 YD	150 YD	200 YD	250 YD	DRIFT	TIME
0.	1300.	0.0	-1.5	-1.5	-1.5	-1.5	-1.5	0.0	0.0000
50.	1213.	-.1	0.0	2.2	5.2	8.5	12.1	.8	.1200
100.	1141.	2.2	-4.4	0.0	5.9	12.7	19.8	2.9	.2473
150.	1087.	6.3	-15.5	-8.9	0.0	10.1	20.8	6.3	.3822
200.	1044.	12.4	-34.1	-25.3	-13.4	0.0	14.3	11.0	.5241
250.	1008.	20.9	-60.5	-49.6	-34.7	-17.9	0.0	16.5	.6709
300.	974.	31.6	-94.9	-81.7	-63.9	-43.7	-22.3	22.6	.8210
350.	942.	45.2	-138.7	-123.3	-102.5	-79.0	-53.9	30.0	.9780
400.	912.	61.6	-191.7	-174.1	-150.4	-123.5	-94.8	38.1	1.1394
450.	885.	81.2	-255.0	-235.3	-208.5	-178.3	-146.1	47.2	1.3064
500.	858.	104.2	-329.1	-307.2	-277.5	-243.9	-208.0	57.2	1.4786

BALLISTIC COEFFICIENT .26, MUZZLE VELOCITY FT/SEC 1400.

RANGE	VELOCITY	HEIGHT	50 YD	100 YD	150 YD	200 YD	250 YD	DRIFT	TIME
0.	1400.	0.0	-1.5	-1.5	-1.5	-1.5	-1.5	0.0	0.0000
50.	1300.	-.2	0.0	1.8	4.4	7.4	10.7	.7	.1112
100.	1213.	1.8	-3.7	0.0	5.1	11.1	17.6	3.0	.2312
150.	1141.	5.4	-13.3	-7.7	0.0	8.9	18.7	6.5	.3585
200.	1087.	10.9	-29.6	-22.2	-11.9	0.0	13.1	11.4	.4935
250.	1044.	18.6	-53.3	-44.1	-31.2	-16.4	0.0	17.5	.6354
300.	1007.	28.6	-84.9	-73.8	-58.4	-40.5	-20.9	24.5	.7821
350.	974.	41.0	-124.2	-111.2	-93.2	-72.4	-49.5	32.1	.9323
400.	942.	56.2	-172.8	-158.0	-137.4	-113.7	-87.4	40.9	1.0893
450.	912.	74.3	-230.5	-213.9	-190.7	-164.0	-134.5	50.4	1.2507
500.	884.	95.7	-298.5	-280.0	-254.3	-224.6	-191.8	61.0	1.4178

BALLISTIC COEFFICIENT .26, MUZZLE VELOCITY FT/SEC 1500.

RANGE	VELOCITY	HEIGHT	50 YD	100 YD	150 YD	200 YD	250 YD	DRIFT	TIME
0.	1500.	0.0	-1.5	-1.5	-1.5	-1.5	-1.5	0.0	0.0000
50.	1386.	-.2	0.0	1.5	3.8	6.4	9.3	.7	.1042
100.	1288.	1.5	-3.0	0.0	4.5	9.8	15.7	2.9	.2166
150.	1203.	4.7	-11.3	-6.8	0.0	7.9	16.7	6.6	.3372
200.	1133.	9.7	-25.7	-19.6	-10.5	0.0	11.7	11.6	.4659
250.	1081.	16.6	-46.7	-39.1	-27.9	-14.7	0.0	17.9	.6017
300.	1039.	25.8	-75.1	-66.0	-52.4	-36.6	-19.0	25.2	.7431
350.	1003.	37.3	-111.5	-100.8	-85.0	-66.6	-46.0	33.6	.8908
400.	969.	51.6	-156.4	-144.2	-126.1	-105.1	-81.6	43.0	1.0443
450.	938.	68.3	-208.9	-195.3	-174.9	-151.2	-124.8	52.8	1.1997
500.	909.	88.3	-271.5	-256.3	-233.7	-207.4	-178.0	63.8	1.3623

BALLISTIC COEFFICIENT .26, MUZZLE VELOCITY FT/SEC 1600.

RANGE	VELOCITY	HEIGHT	50 YD	100 YD	150 YD	200 YD	250 YD	DRIFT	TIME
0.	1600.	0.0	-1.5	-1.5	-1.5	-1.5	-1.5	0.0	0.0000
50.	1478.	-.3	0.0	1.3	3.2	5.6	8.2	.6	.0974
100.	1367.	1.2	-2.5	0.0	3.9	8.6	13.9	2.8	.2032
150.	1271.	4.1	-9.7	-5.9	0.0	7.0	14.9	6.3	.3170
200.	1188.	8.5	-22.2	-17.2	-9.3	0.0	10.5	11.3	.4393
250.	1122.	14.8	-41.0	-34.7	-24.8	-13.2	0.0	17.7	.5693
300.	1072.	23.2	-66.6	-59.0	-47.2	-33.2	-17.4	25.3	.7062
350.	1032.	33.8	-99.5	-90.7	-76.9	-60.6	-42.2	33.9	.8490
400.	997.	47.0	-140.3	-130.2	-114.5	-95.8	-74.7	43.5	.9969
450.	963.	62.8	-189.4	-178.1	-160.4	-139.4	-115.7	54.0	1.1506
500.	932.	81.4	-247.0	-234.5	-214.8	-191.4	-165.1	65.3	1.3084

BALLISTIC COEFFICIENT .26, MUZZLE VELOCITY FT/SEC 1700.

RANGE	VELOCITY	HEIGHT	50 YD	100 YD	150 YD	200 YD	250 YD	DRIFT	TIME
0.	1700.	0.0	-1.5	-1.5	-1.5	-1.5	-1.5	0.0	0.0000
50.	1572.	-.3	0.0	1.0	2.7	4.8	7.2	.7	.0920
100.	1451.	1.0	-2.0	0.0	3.4	7.6	12.3	2.6	.1913
150.	1344.	3.5	-8.2	-5.2	0.0	6.2	13.2	6.0	.2988
200.	1251.	7.5	-19.2	-15.2	-8.3	0.0	9.4	10.8	.4145
250.	1171.	13.2	-35.8	-30.7	-22.1	-11.7	0.0	17.1	.5386
300.	1110.	20.8	-58.7	-52.6	-42.3	-29.9	-15.8	24.8	.6704
350.	1062.	30.7	-88.7	-81.6	-69.5	-55.1	-38.6	33.7	.8093
400.	1024.	42.9	-126.0	-117.9	-104.1	-87.6	-68.8	43.5	.9532
450.	989.	57.7	-171.5	-162.4	-146.9	-128.3	-107.1	54.4	1.1034
500.	956.	75.0	-224.3	-214.1	-196.9	-176.2	-152.7	65.7	1.2558

BALLISTIC COEFFICIENT .26, MUZZLE VELOCITY FT/SEC 1800.

RANGE	VELOCITY	HEIGHT	50 YD	100 YD	150 YD	200 YD	250 YD	DRIFT	TIME
0.	1800.	0.0	-1.5	-1.5	-1.5	-1.5	-1.5	0.0	0.0000
50.	1665.	-.4	0.0	.8	2.3	4.2	6.3	.6	.0867
100.	1539.	.8	-1.6	0.0	3.0	6.7	10.9	2.4	.1803
150.	1421.	3.1	-7.0	-4.5	0.0	5.5	11.8	5.6	.2817
200.	1318.	6.6	-16.6	-13.4	-7.3	0.0	8.4	10.2	.3915
250.	1229.	11.7	-31.3	-27.2	-19.7	-10.5	0.0	16.3	.5095
300.	1153.	18.7	-52.0	-47.0	-38.0	-27.0	-14.4	24.1	.6370
350.	1096.	27.7	-78.7	-72.9	-62.4	-49.5	-34.8	32.7	.7693
400.	1052.	38.9	-112.7	-106.1	-94.1	-79.4	-62.6	42.7	.9090
450.	1015.	52.6	-154.2	-146.8	-133.3	-116.8	-97.9	53.6	1.0545
500.	980.	68.9	-203.5	-195.3	-180.3	-161.9	-140.9	65.4	1.2048

BALLISTIC COEFFICIENT .26, MUZZLE VELOCITY FT/SEC 1900.

RANGE	VELOCITY	HEIGHT	50 YD	100 YD	150 YD	200 YD	250 YD	DRIFT	TIME
0.	1900.	0.0	-1.5	-1.5	-1.5	-1.5	-1.5	0.0	0.0000
50.	1759.	-.4	0.0	.7	2.0	3.6	5.5	.5	.0820
100.	1627.	.6	-1.3	0.0	2.6	5.9	9.6	2.3	.1707
150.	1503.	2.7	-5.9	-4.0	0.0	4.9	10.5	5.2	.2666
200.	1389.	5.8	-14.4	-11.8	-6.5	0.0	7.5	9.6	.3705
250.	1291.	10.4	-27.4	-24.1	-17.5	-9.4	0.0	15.5	.4826
300.	1205.	16.7	-45.6	-41.7	-33.8	-24.0	-12.7	22.8	.6030
350.	1135.	24.9	-69.8	-65.2	-56.0	-44.6	-31.4	31.5	.7315
400.	1082.	35.4	-100.9	-95.6	-85.0	-72.0	-57.0	41.5	.8675
450.	1040.	48.1	-138.8	-132.9	-121.0	-106.4	-89.5	52.4	1.0084
500.	1004.	63.4	-184.8	-178.1	-165.0	-148.7	-129.9	64.5	1.1561

BALLISTIC COEFFICIENT .27, MUZZLE VELOCITY FT/SEC 800.

RANGE	VELOCITY	HEIGHT	50 YD	100 YD	150 YD	200 YD	250 YD	DRIFT	TIME
0.	800.	0.0	-1.5	-1.5	-1.5	-1.5	-1.5	0.0	0.0000
50.	777.	1.0	0.0	6.5	13.7	21.4	29.4	.5	.1904
100.	755.	6.4	-12.9	0.0	14.5	29.9	46.0	2.0	.3864
150.	733.	15.9	-41.1	-21.8	0.0	23.1	47.2	4.5	.5881
200.	712.	29.7	-85.6	-59.7	-30.7	0.0	32.2	8.1	.7958
250.	691.	48.2	-147.2	-114.9	-78.7	-40.3	0.0	12.7	1.0098
300.	670.	71.9	-227.2	-188.4	-144.9	-98.8	-50.5	18.6	1.2304
350.	650.	101.3	-326.6	-281.3	-230.6	-176.8	-120.4	25.6	1.4578
400.	630.	136.7	-446.5	-394.8	-336.8	-275.3	-210.9	33.8	1.6923
450.	610.	178.9	-588.4	-530.3	-465.0	-395.9	-323.4	43.4	1.9344
500.	591.	228.3	-753.8	-689.2	-616.7	-539.8	-459.3	54.4	2.1843

BALLISTIC COEFFICIENT .27, MUZZLE VELOCITY FT/SEC 900.

RANGE	VELOCITY	HEIGHT	50 YD	100 YD	150 YD	200 YD	250 YD	DRIFT	TIME
0.	900.	0.0	-1.5	-1.5	-1.5	-1.5	-1.5	0.0	0.0000
50.	874.	.6	0.0	4.9	10.6	16.6	22.9	.5	.1697
100.	849.	4.9	-9.9	0.0	11.3	23.4	36.0	1.9	.3439
150.	826.	12.4	-31.8	-17.0	0.0	18.1	37.0	4.0	.5229
200.	803.	23.3	-66.6	-46.8	-24.1	0.0	25.2	7.1	.7071
250.	780.	37.8	-114.7	-90.0	-61.7	-31.5	0.0	11.1	.8965
300.	758.	56.5	-177.4	-147.7	-113.7	-77.6	-39.7	16.2	1.0919
350.	736.	79.5	-255.2	-220.6	-180.9	-138.8	-94.6	22.2	1.2930
400.	714.	107.3	-349.2	-309.6	-264.3	-216.0	-165.6	29.4	1.5001
450.	693.	140.2	-460.1	-415.6	-364.6	-310.4	-253.6	37.6	1.7134
500.	672.	178.7	-589.2	-539.7	-483.0	-422.7	-359.7	46.9	1.9333

BALLISTIC COEFFICIENT .27, MUZZLE VELOCITY FT/SEC 1000.

RANGE	VELOCITY	HEIGHT	50 YD	100 YD	150 YD	200 YD	250 YD	DRIFT	TIME
0.	1000.	0.0	-1.5	-1.5	-1.5	-1.5	-1.5	0.0	0.0000
50.	968.	.4	0.0	3.8	8.4	13.3	18.5	.7	.1542
100.	938.	3.9	-7.6	0.0	9.2	19.1	29.5	1.8	.3101
150.	909.	10.0	-25.2	-13.8	0.0	14.8	30.4	4.0	.4726
200.	883.	18.9	-53.3	-38.1	-19.7	0.0	20.8	7.1	.6401
250.	857.	30.9	-92.7	-73.7	-50.7	-26.0	0.0	11.0	.8126
300.	833.	46.3	-143.9	-121.1	-93.5	-63.9	-32.7	15.9	.9902
350.	810.	65.3	-207.6	-181.0	-148.8	-114.3	-77.9	21.6	1.1729
400.	787.	88.1	-284.5	-254.1	-217.3	-177.8	-136.2	28.3	1.3607
450.	765.	115.1	-375.2	-341.0	-299.6	-255.2	-208.4	35.9	1.5538
500.	743.	146.7	-480.7	-442.7	-396.7	-347.3	-295.3	44.5	1.7527

BALLISTIC COEFFICIENT .27, MUZZLE VELOCITY FT/SEC 1100.

RANGE	VELOCITY	HEIGHT	50 YD	100 YD	150 YD	200 YD	250 YD	DRIFT	TIME
0.	1100.	0.0	-1.5	-1.5	-1.5	-1.5	-1.5	0.0	0.0000
50.	1056.	.2	0.0	3.1	7.1	11.1	15.5	.5	.1392
100.	1020.	3.1	-6.3	0.0	7.8	16.0	24.8	2.0	.2839
150.	986.	8.3	-21.2	-11.7	0.0	12.2	25.4	4.5	.4348
200.	955.	15.8	-44.6	-32.0	-16.3	0.0	17.6	7.5	.5880
250.	926.	26.1	-77.7	-62.0	-42.4	-22.0	0.0	11.6	.7476
300.	898.	39.2	-121.1	-102.2	-78.7	-54.3	-27.9	16.6	.9126
350.	872.	55.4	-175.1	-153.1	-125.7	-97.1	-66.3	22.4	1.0820
400.	847.	75.0	-240.5	-215.3	-184.0	-151.4	-116.2	29.1	1.2565
450.	824.	98.2	-317.8	-289.5	-254.3	-217.6	-178.0	36.7	1.4359
500.	801.	125.3	-407.9	-376.4	-337.3	-296.5	-252.5	45.2	1.6205

BALLISTIC COEFFICIENT .27, MUZZLE VELOCITY FT/SEC 1200.

RANGE	VELOCITY	HEIGHT	50 YD	100 YD	150 YD	200 YD	250 YD	DRIFT	TIME
0.	1200.	0.0	-1.5	-1.5	-1.5	-1.5	-1.5	0.0	0.0000
50.	1133.	.0	0.0	2.7	6.0	9.7	13.6	.7	.1288
100.	1082.	2.6	-5.3	0.0	6.7	14.1	21.8	2.6	.2650
150.	1042.	7.2	-18.0	-10.0	0.0	11.1	22.6	5.5	.4062
200.	1007.	14.0	-38.9	-28.2	-14.9	0.0	15.3	9.4	.5536
250.	975.	23.0	-67.8	-54.4	-37.7	-19.2	0.0	13.9	.7037
300.	944.	34.8	-106.2	-90.2	-70.2	-47.9	-24.9	19.5	.8610
350.	915.	49.3	-153.8	-135.1	-111.8	-85.8	-58.9	25.8	1.0215
400.	888.	67.0	-212.0	-190.6	-163.9	-134.2	-103.5	33.1	1.1881
450.	863.	88.0	-280.9	-256.9	-226.8	-193.4	-158.9	41.3	1.3594
500.	838.	112.4	-361.2	-334.6	-301.2	-264.0	-225.7	50.3	1.5357

BALLISTIC COEFFICIENT .27, MUZZLE VELOCITY FT/SEC 1300.

RANGE	VELOCITY	HEIGHT	50 YD	100 YD	150 YD	200 YD	250 YD	DRIFT	TIME
0.	1300.	0.0	-1.5	-1.5	-1.5	-1.5	-1.5	0.0	0.0000
50.	1216.	-.1	0.0	2.2	5.1	8.4	12.0	.7	.1196
100.	1146.	2.2	-4.4	0.0	5.9	12.5	19.5	2.8	.2467
150.	1092.	6.2	-15.4	-8.8	0.0	9.9	20.5	6.1	.3811
200.	1050.	12.3	-33.7	-24.9	-13.1	0.0	14.1	10.5	.5212
250.	1014.	20.6	-59.8	-48.8	-34.1	-17.7	0.0	15.8	.6668
300.	981.	31.3	-94.2	-81.0	-63.4	-43.6	-22.4	22.0	.8172
350.	950.	44.7	-137.3	-122.0	-101.4	-78.4	-53.6	29.0	.9725
400.	921.	60.8	-189.8	-172.3	-148.7	-122.4	-94.1	36.9	1.1326
450.	893.	80.4	-253.3	-233.5	-207.0	-177.4	-145.6	46.1	1.3001
500.	868.	102.8	-325.7	-303.8	-274.3	-241.5	-206.1	55.4	1.4688

BALLISTIC COEFFICIENT .27, MUZZLE VELOCITY FT/SEC 1400.

RANGE	VELOCITY	HEIGHT	50 YD	100 YD	150 YD	200 YD	250 YD	DRIFT	TIME
0.	1400.	0.0	-1.5	-1.5	-1.5	-1.5	-1.5	0.0	0.0000
50.	1303.	-.2	0.0	1.8	4.4	7.3	10.5	.7	.1111
100.	1219.	1.8	-3.6	0.0	5.1	11.0	17.4	2.8	.2302
150.	1148.	5.4	-13.2	-7.7	0.0	8.8	18.4	6.3	.3573
200.	1094.	10.8	-29.3	-22.0	-11.7	0.0	12.8	11.0	.4913
250.	1052.	18.4	-52.6	-43.5	-30.7	-16.0	0.0	16.8	.6313
300.	1016.	28.2	-83.7	-72.8	-57.4	-39.7	-20.5	23.5	.7766
350.	982.	40.5	-123.0	-110.3	-92.3	-71.7	-49.3	31.2	.9272
400.	951.	55.4	-170.7	-156.2	-135.7	-112.2	-86.6	39.6	1.0820
450.	922.	73.3	-227.8	-211.4	-188.3	-161.9	-133.1	48.9	1.2420
500.	894.	94.5	-295.5	-277.3	-251.7	-222.3	-190.3	59.4	1.4089

BALLISTIC COEFFICIENT .27, MUZZLE VELOCITY FT/SEC 1500.

RANGE	VELOCITY	HEIGHT	50 YD	100 YD	150 YD	200 YD	250 YD	DRIFT	TIME
0.	1500.	0.0	-1.5	-1.5	-1.5	-1.5	-1.5	0.0	0.0000
50.	1390.	-.2	0.0	1.5	3.7	6.3	9.2	.7	.1040
100.	1295.	1.5	-3.0	0.0	4.5	9.7	15.5	2.8	.2161
150.	1212.	4.7	-11.2	-6.7	0.0	7.8	16.5	6.3	.3358
200.	1142.	9.6	-25.4	-19.3	-10.4	0.0	11.6	11.2	.4635
250.	1090.	16.4	-46.2	-38.6	-27.5	-14.5	0.0	17.3	.5982
300.	1047.	25.6	-74.7	-65.7	-52.3	-36.7	-19.3	24.8	.7407
350.	1012.	37.0	-110.6	-100.0	-84.4	-66.2	-45.9	32.9	.8867
400.	979.	50.7	-153.8	-141.7	-123.8	-103.0	-79.8	41.4	1.0351
450.	948.	67.3	-206.3	-192.7	-172.6	-149.2	-123.1	51.2	1.1909
500.	919.	86.9	-267.8	-252.6	-230.3	-204.3	-175.3	61.8	1.3512

BALLISTIC COEFFICIENT .27, MUZZLE VELOCITY FT/SEC 1600.

RANGE	VELOCITY	HEIGHT	50 YD	100 YD	150 YD	200 YD	250 YD	DRIFT	TIME
0.	1600.	0.0	-1.5	-1.5	-1.5	-1.5	-1.5	0.0	0.0000
50.	1482.	-.3	0.0	1.2	3.2	5.5	8.1	.6	.0973
100.	1374.	1.2	-2.5	0.0	3.9	8.5	13.7	2.7	.2027
150.	1281.	4.0	-9.6	-5.8	0.0	6.9	14.7	6.1	.3157
200.	1200.	8.4	-22.0	-17.0	-9.2	0.0	10.4	10.9	.4368
250.	1133.	14.6	-40.4	-34.2	-24.5	-13.0	0.0	17.1	.5657
300.	1082.	22.9	-65.8	-58.3	-46.6	-32.8	-17.2	24.5	.7018
350.	1042.	33.4	-98.2	-89.5	-75.9	-59.7	-41.6	32.9	.8430
400.	1007.	46.3	-138.7	-128.7	-113.1	-94.7	-74.0	42.3	.9905
450.	974.	61.7	-186.5	-175.3	-157.8	-137.0	-113.7	52.3	1.1407
500.	944.	80.1	-243.7	-231.2	-211.8	-188.7	-162.8	63.4	1.2980

BALLISTIC COEFFICIENT .27, MUZZLE VELOCITY FT/SEC 1700.

RANGE	VELOCITY	HEIGHT	50 YD	100 YD	150 YD	200 YD	250 YD	DRIFT	TIME
0.	1700.	0.0	-1.5	-1.5	-1.5	-1.5	-1.5	0.0	0.0000
50.	1576.	-.3	0.0	1.0	2.7	4.7	7.1	.6	.0919
100.	1460.	1.0	-2.0	0.0	3.4	7.5	12.1	2.5	.1907
150.	1355.	3.5	-8.1	-5.1	0.0	6.1	13.0	5.8	.2976
200.	1264.	7.4	-19.0	-15.0	-8.1	0.0	9.2	10.4	.4120
250.	1185.	13.0	-35.3	-30.2	-21.7	-11.6	0.0	16.5	.5348
300.	1122.	20.5	-57.8	-51.7	-41.5	-29.3	-15.4	23.9	.6651
350.	1074.	30.1	-87.1	-80.1	-68.1	-53.9	-37.7	32.4	.8019
400.	1035.	42.1	-123.8	-115.8	-102.1	-85.9	-67.4	42.0	.9443
450.	1000.	56.5	-168.1	-159.1	-143.7	-125.5	-104.6	52.4	1.0918
500.	968.	73.8	-221.2	-211.2	-194.1	-173.8	-150.7	64.0	1.2459

BALLISTIC COEFFICIENT .27, MUZZLE VELOCITY FT/SEC 1800.

RANGE	VELOCITY	HEIGHT	50 YD	100 YD	150 YD	200 YD	250 YD	DRIFT	TIME
0.	1800.	0.0	-1.5	-1.5	-1.5	-1.5	-1.5	0.0	0.0000
50.	1670.	-.4	0.0	.8	2.3	4.1	6.2	.6	.0866
100.	1548.	.8	-1.6	0.0	3.0	6.6	10.7	2.3	.1799
150.	1434.	3.0	-6.9	-4.5	0.0	5.4	11.6	5.4	.2806
200.	1332.	6.5	-16.4	-13.2	-7.2	0.0	8.3	9.8	.3892
250.	1245.	11.5	-30.9	-26.8	-19.3	-10.3	0.0	15.7	.5058
300.	1168.	18.3	-50.9	-46.0	-37.1	-26.2	-13.8	22.9	.6303
350.	1110.	27.1	-77.3	-71.6	-61.2	-48.5	-34.1	31.5	.7623
400.	1064.	38.3	-111.0	-104.4	-92.6	-78.1	-61.6	41.3	.9016
450.	1026.	51.7	-151.7	-144.4	-131.0	-114.8	-96.2	51.9	1.0451
500.	992.	67.8	-200.7	-192.6	-177.7	-159.6	-139.0	63.7	1.1952

BALLISTIC COEFFICIENT .27, MUZZLE VELOCITY FT/SEC 1900.

RANGE	VELOCITY	HEIGHT	50 YD	100 YD	150 YD	200 YD	250 YD	DRIFT	TIME
0.	1900.	0.0	-1.5	-1.5	-1.5	-1.5	-1.5	0.0	0.0000
50.	1764.	-.4	0.0	.7	2.0	3.6	5.4	.5	.0819
100.	1637.	.6	-1.3	0.0	2.6	5.8	9.5	2.2	.1702
150.	1517.	2.6	-5.9	-3.9	0.0	4.8	10.3	5.0	.2654
200.	1405.	5.8	-14.2	-11.6	-6.4	0.0	7.4	9.2	.3682
250.	1308.	10.3	-27.0	-23.7	-17.2	-9.2	0.0	14.8	.4789
300.	1223.	16.4	-44.8	-40.9	-33.1	-23.5	-12.4	21.8	.5976
350.	1151.	24.4	-68.5	-63.9	-54.8	-43.6	-30.7	30.2	.7243
400.	1096.	34.6	-98.7	-93.5	-83.1	-70.3	-55.5	39.9	.8580
450.	1053.	47.0	-136.0	-130.1	-118.4	-104.0	-87.4	50.5	.9977
500.	1017.	61.9	-180.7	-174.2	-161.2	-145.2	-126.8	62.2	1.1428

BALLISTIC COEFFICIENT .28, MUZZLE VELOCITY FT/SEC 800.

RANGE	VELOCITY	HEIGHT	50 YD	100 YD	150 YD	200 YD	250 YD	DRIFT	TIME
0.	800.	0.0	-1.5	-1.5	-1.5	-1.5	-1.5	0.0	0.0000
50.	778.	1.0	0.0	6.4	13.7	21.3	29.3	.5	.1902
100.	757.	6.4	-12.9	0.0	14.4	29.7	45.7	1.9	.3859
150.	736.	15.8	-41.0	-21.7	0.0	22.9	46.9	4.3	.5870
200.	715.	29.5	-85.2	-59.5	-30.6	0.0	32.0	7.7	.7940
250.	695.	47.9	-146.5	-114.3	-78.2	-40.0	0.0	12.2	1.0070
300.	674.	71.4	-225.9	-187.3	-144.0	-98.1	-50.1	17.8	1.2262
350.	655.	100.5	-324.5	-279.4	-228.9	-175.3	-119.3	24.6	1.4520
400.	635.	135.4	-443.2	-391.7	-333.9	-272.8	-208.8	32.5	1.6844
450.	616.	176.9	-583.6	-525.6	-460.6	-391.8	-319.8	41.6	1.9240
500.	598.	225.5	-746.7	-682.3	-610.1	-533.7	-453.7	52.1	2.1710

BALLISTIC COEFFICIENT .28, MUZZLE VELOCITY FT/SEC 900.

RANGE	VELOCITY	HEIGHT	50 YD	100 YD	150 YD	200 YD	250 YD	DRIFT	TIME
0.	900.	0.0	-1.5	-1.5	-1.5	-1.5	-1.5	0.0	0.0000
50.	875.	.6	0.0	4.9	10.6	16.6	22.9	.5	.1696
100.	851.	4.9	-9.9	0.0	11.3	23.3	35.9	1.8	.3436
150.	828.	12.3	-31.8	-17.0	0.0	18.0	36.8	3.9	.5222
200.	806.	23.2	-66.4	-46.6	-24.0	0.0	25.1	6.9	.7059
250.	784.	37.7	-114.4	-89.7	-61.4	-31.4	0.0	10.8	.8945
300.	762.	56.1	-176.5	-146.9	-113.0	-76.9	-39.3	15.6	1.0884
350.	741.	78.9	-253.6	-219.0	-179.5	-137.4	-93.5	21.3	1.2879
400.	720.	106.3	-346.6	-307.1	-261.9	-213.8	-163.6	28.1	1.4932
450.	700.	138.7	-456.3	-411.9	-361.0	-307.0	-250.5	36.0	1.7044
500.	680.	176.5	-583.7	-534.4	-477.8	-417.8	-355.0	44.9	1.9219

BALLISTIC COEFFICIENT .28, MUZZLE VELOCITY FT/SEC 1000.

RANGE	VELOCITY	HEIGHT	50 YD	100 YD	150 YD	200 YD	250 YD	DRIFT	TIME
0.	1000.	0.0	-1.5	-1.5	-1.5	-1.5	-1.5	0.0	0.0000
50.	969.	.4	0.0	3.8	8.4	13.3	18.4	.8	.1544
100.	940.	3.9	-7.5	0.0	9.2	19.0	29.3	1.7	.3097
150.	912.	9.9	-25.1	-13.8	0.0	14.8	30.2	3.8	.4718
200.	886.	18.8	-53.1	-38.0	-19.7	0.0	20.6	6.8	.6388
250.	862.	30.8	-92.1	-73.3	-50.4	-25.8	0.0	10.6	.8103
300.	838.	46.0	-142.9	-120.3	-92.8	-63.3	-32.3	15.3	.9867
350.	816.	64.8	-206.2	-179.8	-147.7	-113.3	-77.2	20.8	1.1683
400.	794.	87.4	-282.4	-252.3	-215.6	-176.2	-135.0	27.3	1.3548
450.	772.	114.0	-372.3	-338.4	-297.1	-252.9	-206.5	34.6	1.5465
500.	751.	145.2	-476.7	-439.1	-393.2	-344.0	-292.4	42.9	1.7435

BALLISTIC COEFFICIENT .28, MUZZLE VELOCITY FT/SEC 1100.

RANGE	VELOCITY	HEIGHT	50 YD	100 YD	150 YD	200 YD	250 YD	DRIFT	TIME
0.	1100.	0.0	-1.5	-1.5	-1.5	-1.5	-1.5	0.0	0.0000
50.	1058.	.2	0.0	3.2	7.1	11.1	15.5	.5	.1391
100.	1022.	3.1	-6.3	0.0	7.8	15.8	24.6	2.0	.2840
150.	990.	8.3	-21.2	-11.7	0.0	12.0	25.2	4.5	.4348
200.	959.	15.8	-44.3	-31.6	-16.0	0.0	17.6	7.2	.5864
250.	931.	25.9	-77.3	-61.5	-42.0	-21.9	0.0	11.2	.7455
300.	904.	38.9	-120.2	-101.2	-77.8	-53.7	-27.4	16.0	.9089
350.	878.	55.0	-174.0	-151.9	-124.5	-96.5	-65.8	21.7	1.0779
400.	854.	74.3	-238.7	-213.4	-182.1	-150.1	-115.0	28.1	1.2507
450.	831.	97.2	-315.4	-286.9	-251.7	-215.7	-176.2	35.5	1.4288
500.	809.	123.9	-404.5	-372.8	-333.7	-293.7	-249.8	43.7	1.6118

BALLISTIC COEFFICIENT .28, MUZZLE VELOCITY FT/SEC 1200.

RANGE	VELOCITY	HEIGHT	50 YD	100 YD	150 YD	200 YD	250 YD	DRIFT	TIME
0.	1200.	0.0	-1.5	-1.5	-1.5	-1.5	-1.5	0.0	0.0000
50.	1135.	.1	0.0	2.6	6.0	9.7	13.4	.8	.1298
100.	1086.	2.6	-5.1	0.0	6.8	14.2	21.7	2.5	.2639
150.	1046.	7.2	-17.9	-10.2	0.0	11.0	22.3	5.5	.4064
200.	1012.	13.9	-38.6	-28.4	-14.7	0.0	15.0	9.3	.5529
250.	980.	22.9	-67.1	-54.2	-37.2	-18.8	0.0	13.4	.7012
300.	950.	34.5	-105.0	-89.5	-69.1	-47.0	-24.5	18.8	.8569
350.	922.	48.9	-152.2	-134.3	-110.4	-84.6	-58.3	25.0	1.0168
400.	895.	66.5	-210.3	-189.7	-162.4	-133.0	-103.0	32.3	1.1834
450.	871.	87.0	-278.0	-254.9	-224.2	-191.1	-157.3	40.0	1.3523
500.	847.	111.2	-357.4	-331.7	-297.6	-260.9	-223.3	48.7	1.5269

BALLISTIC COEFFICIENT .28, MUZZLE VELOCITY FT/SEC 1300.

RANGE	VELOCITY	HEIGHT	50 YD	100 YD	150 YD	200 YD	250 YD	DRIFT	TIME
0.	1300.	0.0	-1.5	-1.5	-1.5	-1.5	-1.5	0.0	0.0000
50.	1219.	-.1	0.0	2.2	5.1	8.4	11.9	.7	.1193
100.	1150.	2.2	-4.4	0.0	5.8	12.4	19.4	2.7	.2462
150.	1097.	6.2	-15.3	-8.8	0.0	9.8	20.3	5.9	.3799
200.	1056.	12.2	-33.5	-24.7	-13.0	0.0	14.0	10.2	.5193
250.	1020.	20.4	-59.4	-48.4	-33.8	-17.5	0.0	15.3	.6641
300.	988.	31.1	-93.9	-80.8	-63.3	-43.7	-22.7	21.6	.8152
350.	958.	44.2	-136.2	-120.8	-100.4	-77.6	-53.0	28.1	.9675
400.	929.	60.2	-188.3	-170.8	-147.4	-121.3	-93.3	35.9	1.1268
450.	902.	79.2	-250.0	-230.3	-204.0	-174.7	-143.1	44.3	1.2904
500.	877.	101.6	-322.8	-300.9	-271.7	-239.1	-204.0	54.0	1.4604

BALLISTIC COEFFICIENT .28, MUZZLE VELOCITY FT/SEC 1400.

RANGE	VELOCITY	HEIGHT	50 YD	100 YD	150 YD	200 YD	250 YD	DRIFT	TIME
0.	1400.	0.0	-1.5	-1.5	-1.5	-1.5	-1.5	0.0	0.0000
50.	1307.	-.2	0.0	1.8	4.4	7.3	10.4	.7	.1110
100.	1225.	1.8	-3.6	0.0	5.2	10.9	17.2	2.7	.2297
150.	1154.	5.4	-13.2	-7.8	0.0	8.6	18.1	6.3	.3572
200.	1101.	10.7	-29.0	-21.8	-11.4	0.0	12.7	10.7	.4892
250.	1058.	18.2	-52.1	-43.1	-30.1	-15.8	0.0	16.3	.6282
300.	1023.	27.9	-83.0	-72.1	-56.6	-39.5	-20.5	22.9	.7731
350.	990.	40.2	-122.3	-109.7	-91.5	-71.5	-49.4	30.6	.9238
400.	960.	54.7	-168.8	-154.4	-133.7	-110.8	-85.5	38.4	1.0752
450.	931.	72.4	-225.4	-209.2	-185.8	-160.2	-131.7	47.5	1.2343
500.	904.	93.0	-291.4	-273.3	-247.4	-218.9	-187.2	57.4	1.3976

BALLISTIC COEFFICIENT .28, MUZZLE VELOCITY FT/SEC 1500.

RANGE	VELOCITY	HEIGHT	50 YD	100 YD	150 YD	200 YD	250 YD	DRIFT	TIME
0.	1500.	0.0	-1.5	-1.5	-1.5	-1.5	-1.5	0.0	0.0000
50.	1394.	-.2	0.0	1.5	3.7	6.3	9.1	.7	.1039
100.	1301.	1.5	-3.0	0.0	4.4	9.6	15.3	2.7	.2153
150.	1220.	4.6	-11.1	-6.6	0.0	7.7	16.3	6.1	.3344
200.	1151.	9.5	-25.1	-19.2	-10.3	0.0	11.4	10.8	.4613
250.	1098.	16.2	-45.7	-38.2	-27.2	-14.3	0.0	16.7	.5949
300.	1056.	25.1	-73.4	-64.4	-51.2	-35.7	-18.5	23.6	.7343
350.	1021.	36.3	-108.7	-98.3	-82.8	-64.7	-44.8	31.5	.8790
400.	988.	50.2	-152.6	-140.7	-122.9	-102.3	-79.5	40.5	1.0301
450.	958.	66.3	-203.7	-190.2	-170.3	-147.1	-121.4	49.7	1.1823
500.	930.	85.6	-264.5	-249.6	-227.5	-201.7	-173.1	60.1	1.3416

BALLISTIC COEFFICIENT .28, MUZZLE VELOCITY FT/SEC 1600.

RANGE	VELOCITY	HEIGHT	50 YD	100 YD	150 YD	200 YD	250 YD	DRIFT	TIME
0.	1600.	0.0	-1.5	-1.5	-1.5	-1.5	-1.5	0.0	0.0000
50.	1486.	-.3	0.0	1.2	3.2	5.4	8.0	.6	.0971
100.	1382.	1.2	-2.5	0.0	3.9	8.4	13.5	2.5	.2019
150.	1290.	4.0	-9.5	-5.8	0.0	6.8	14.5	5.8	.3144
200.	1211.	8.3	-21.7	-16.8	-9.1	0.0	10.2	10.5	.4345
250.	1143.	14.4	-39.9	-33.8	-24.1	-12.7	0.0	16.4	.5621
300.	1092.	22.5	-64.8	-57.4	-45.8	-32.2	-16.9	23.6	.6966
350.	1051.	32.8	-96.8	-88.2	-74.6	-58.7	-40.9	31.8	.8367
400.	1017.	45.5	-136.4	-126.6	-111.1	-92.9	-72.5	40.8	.9819
450.	984.	60.8	-184.3	-173.2	-155.8	-135.3	-112.4	50.9	1.1328
500.	954.	78.7	-239.9	-227.6	-208.3	-185.5	-160.0	61.4	1.2866

BALLISTIC COEFFICIENT .28, MUZZLE VELOCITY FT/SEC 1700.

RANGE	VELOCITY	HEIGHT	50 YD	100 YD	150 YD	200 YD	250 YD	DRIFT	TIME
0.	1700.	0.0	-1.5	-1.5	-1.5	-1.5	-1.5	0.0	0.0000
50.	1580.	-.3	0.0	1.0	2.7	4.7	7.0	.6	.0917
100.	1468.	1.0	-2.0	0.0	3.4	7.4	12.0	2.4	.1901
150.	1365.	3.5	-8.0	-5.1	0.0	6.0	12.9	5.5	.2962
200.	1276.	7.3	-18.8	-14.8	-8.1	0.0	9.1	10.0	.4100
250.	1198.	12.8	-34.8	-29.9	-21.4	-11.4	0.0	15.9	.5314
300.	1134.	20.2	-56.9	-51.0	-40.9	-28.8	-15.1	23.0	.6601
350.	1085.	29.6	-85.8	-78.8	-67.0	-52.9	-37.0	31.3	.7955
400.	1045.	41.5	-122.3	-114.3	-100.8	-84.7	-66.5	40.8	.9379
450.	1011.	55.7	-166.2	-157.3	-142.1	-124.0	-103.5	51.1	1.0847
500.	979.	72.3	-217.1	-207.1	-190.3	-170.1	-147.4	61.8	1.2332

BALLISTIC COEFFICIENT .28, MUZZLE VELOCITY FT/SEC 1800.

RANGE	VELOCITY	HEIGHT	50 YD	100 YD	150 YD	200 YD	250 YD	DRIFT	TIME
0.	1800.	0.0	-1.5	-1.5	-1.5	-1.5	-1.5	0.0	0.0000
50.	1675.	-.4	0.0	.8	2.3	4.1	6.1	.5	.0864
100.	1557.	.8	-1.6	0.0	2.9	6.5	10.5	2.2	.1794
150.	1445.	3.0	-6.8	-4.4	0.0	5.3	11.4	5.2	.2795
200.	1346.	6.4	-16.2	-13.0	-7.1	0.0	8.1	9.5	.3870
250.	1259.	11.4	-30.4	-26.4	-19.0	-10.1	0.0	15.0	.5022
300.	1183.	18.0	-50.1	-45.2	-36.4	-25.7	-13.6	22.0	.6252
350.	1123.	26.7	-76.0	-70.3	-60.0	-47.6	-33.4	30.3	.7556
400.	1076.	37.5	-108.7	-102.2	-90.5	-76.2	-60.0	39.7	.8921
450.	1038.	50.6	-148.7	-141.4	-128.2	-112.1	-93.9	50.0	1.0341
500.	1004.	66.3	-196.6	-188.5	-173.8	-156.0	-135.8	61.4	1.1821

BALLISTIC COEFFICIENT .28, MUZZLE VELOCITY FT/SEC 1900.

RANGE	VELOCITY	HEIGHT	50 YD	100 YD	150 YD	200 YD	250 YD	DRIFT	TIME
0.	1900.	0.0	-1.5	-1.5	-1.5	-1.5	-1.5	0.0	0.0000
50.	1769.	-.4	0.0	.6	1.9	3.5	5.3	.5	.0818
100.	1646.	.6	-1.3	0.0	2.6	5.7	9.4	2.1	.1698
150.	1529.	2.6	-5.8	-3.9	0.0	4.7	10.2	4.8	.2643
200.	1420.	5.7	-14.0	-11.4	-6.3	0.0	7.3	8.8	.3660
250.	1324.	10.1	-26.6	-23.4	-16.9	-9.1	0.0	14.2	.4756
300.	1240.	16.1	-44.1	-40.3	-32.5	-23.1	-12.2	21.0	.5928
350.	1167.	24.0	-67.3	-62.8	-53.8	-42.7	-30.0	29.0	.7176
400.	1111.	33.9	-96.9	-91.7	-81.4	-68.8	-54.3	38.4	.8495
450.	1066.	46.1	-133.3	-127.5	-115.9	-101.8	-85.5	48.7	.9874
500.	1029.	60.8	-177.8	-171.3	-158.4	-142.7	-124.6	60.4	1.1324

BALLISTIC COEFFICIENT .29, MUZZLE VELOCITY FT/SEC 800.

RANGE	VELOCITY	HEIGHT	50 YD	100 YD	150 YD	200 YD	250 YD	DRIFT	TIME
0.	800.	0.0	-1.5	-1.5	-1.5	-1.5	-1.5	0.0	0.0000
50.	779.	1.0	0.0	6.4	13.6	21.2	29.2	.5	.1901
100.	758.	6.4	-12.8	0.0	14.4	29.6	45.5	1.8	.3853
150.	738.	15.7	-40.8	-21.6	0.0	22.8	46.7	4.1	.5860
200.	718.	29.4	-84.9	-59.2	-30.4	0.0	31.8	7.4	.7922
250.	698.	47.6	-145.8	-113.7	-77.8	-39.7	0.0	11.7	1.0041
300.	679.	70.9	-224.6	-186.1	-142.9	-97.3	-49.6	17.1	1.2220
350.	659.	99.6	-322.4	-277.4	-227.1	-173.9	-118.2	23.5	1.4462
400.	641.	134.2	-440.1	-388.7	-331.2	-270.3	-206.8	31.1	1.6768
450.	622.	175.2	-579.2	-521.4	-456.7	-388.2	-316.7	40.0	1.9146
500.	604.	223.1	-741.0	-676.8	-604.8	-528.8	-449.3	50.1	2.1597

BALLISTIC COEFFICIENT .29, MUZZLE VELOCITY FT/SEC 900.

RANGE	VELOCITY	HEIGHT	50 YD	100 YD	150 YD	200 YD	250 YD	DRIFT	TIME
0.	900.	0.0	-1.5	-1.5	-1.5	-1.5	-1.5	0.0	0.0000
50.	876.	.6	0.0	4.9	10.6	16.5	22.8	.5	.1695
100.	853.	4.9	-9.9	0.0	11.3	23.2	35.7	1.7	.3432
150.	830.	12.3	-31.7	-16.9	0.0	17.9	36.7	3.8	.5215
200.	809.	23.1	-66.2	-46.5	-23.9	0.0	25.0	6.7	.7046
250.	788.	37.5	-114.0	-89.3	-61.1	-31.2	0.0	10.4	.8926
300.	767.	55.8	-175.8	-146.2	-112.4	-76.5	-39.0	15.1	1.0856
350.	746.	78.4	-252.4	-217.9	-178.5	-136.6	-92.9	20.6	1.2839
400.	726.	105.5	-344.7	-305.3	-260.2	-212.4	-162.4	27.2	1.4877
450.	706.	137.5	-453.5	-409.2	-358.5	-304.6	-248.4	34.7	1.6973
500.	687.	174.9	-579.7	-530.4	-474.1	-414.3	-351.8	43.3	1.9128

BALLISTIC COEFFICIENT .29, MUZZLE VELOCITY FT/SEC 1000.

RANGE	VELOCITY	HEIGHT	50 YD	100 YD	150 YD	200 YD	250 YD	DRIFT	TIME
0.	1000.	0.0	-1.5	-1.5	-1.5	-1.5	-1.5	0.0	0.0000
50.	970.	.4	0.0	3.8	8.3	13.2	18.3	.8	.1547
100.	942.	3.9	-7.5	0.0	9.1	18.9	29.1	1.7	.3098
150.	915.	9.9	-24.9	-13.6	0.0	14.7	30.1	3.7	.4710
200.	890.	18.8	-52.8	-37.8	-19.6	0.0	20.5	6.6	.6375
250.	866.	30.6	-91.7	-72.9	-50.2	-25.7	0.0	10.3	.8085
300.	843.	45.7	-142.1	-119.5	-92.3	-62.9	-32.1	14.8	.9839
350.	821.	64.3	-204.7	-178.4	-146.6	-112.3	-76.4	20.1	1.1640
400.	800.	86.6	-280.2	-250.1	-213.8	-174.6	-133.5	26.2	1.3490
450.	779.	113.0	-369.3	-335.4	-294.6	-250.5	-204.2	33.3	1.5391
500.	758.	143.6	-472.6	-435.0	-389.6	-340.6	-289.2	41.2	1.7343

BALLISTIC COEFFICIENT .29, MUZZLE VELOCITY FT/SEC 1100.

RANGE	VELOCITY	HEIGHT	50 YD	100 YD	150 YD	200 YD	250 YD	DRIFT	TIME
0.	1100.	0.0	-1.5	-1.5	-1.5	-1.5	-1.5	0.0	0.0000
50.	1059.	.2	0.0	3.2	7.1	11.1	15.4	.5	.1390
100.	1025.	3.1	-6.4	0.0	7.8	15.8	24.4	2.0	.2841
150.	993.	8.3	-21.3	-11.7	0.0	12.0	24.9	4.5	.4347
200.	963.	15.7	-44.3	-31.6	-16.0	0.0	17.2	7.1	.5860
250.	936.	25.8	-76.9	-61.0	-41.5	-21.5	0.0	10.8	.7434
300.	909.	38.6	-119.6	-100.5	-77.1	-53.1	-27.3	15.5	.9060
350.	884.	54.5	-172.8	-150.5	-123.1	-95.2	-65.0	20.9	1.0733
400.	861.	73.7	-236.9	-211.5	-180.2	-148.3	-113.8	27.1	1.2452
450.	838.	96.3	-312.8	-284.2	-249.0	-213.1	-174.3	34.2	1.4217
500.	817.	122.6	-401.0	-369.2	-330.2	-290.2	-247.2	42.2	1.6032

BALLISTIC COEFFICIENT .29, MUZZLE VELOCITY FT/SEC 1200.

RANGE	VELOCITY	HEIGHT	50 YD	100 YD	150 YD	200 YD	250 YD	DRIFT	TIME
0.	1200.	0.0	-1.5	-1.5	-1.5	-1.5	-1.5	0.0	0.0000
50.	1137.	.1	0.0	2.6	5.9	9.5	13.4	.7	.1292
100.	1089.	2.6	-5.2	0.0	6.6	13.9	21.7	2.4	.2635
150.	1050.	7.1	-17.7	-9.9	0.0	10.9	22.6	5.1	.4039
200.	1017.	13.7	-38.1	-27.8	-14.5	0.0	15.6	8.7	.5492
250.	985.	22.8	-67.1	-54.2	-37.6	-19.5	0.0	13.3	.7003
300.	956.	34.2	-104.4	-88.9	-69.0	-47.2	-23.8	18.2	.8535
350.	929.	48.5	-151.4	-133.4	-110.2	-84.8	-57.5	24.3	1.0129
400.	903.	65.7	-208.3	-187.6	-161.1	-132.1	-100.9	31.1	1.1765
450.	878.	86.2	-276.1	-252.8	-223.0	-190.4	-155.3	38.9	1.3457
500.	855.	109.9	-354.5	-328.6	-295.5	-259.3	-220.2	47.3	1.5185

BALLISTIC COEFFICIENT .29, MUZZLE VELOCITY FT/SEC 1300.

RANGE	VELOCITY	HEIGHT	50 YD	100 YD	150 YD	200 YD	250 YD	DRIFT	TIME
0.	1300.	0.0	-1.5	-1.5	-1.5	-1.5	-1.5	0.0	0.0000
50.	1222.	-.1	0.0	2.2	5.1	8.3	11.8	.7	.1191
100.	1154.	2.2	-4.5	0.0	5.7	12.2	19.2	2.9	.2470
150.	1102.	6.1	-15.2	-8.5	0.0	9.7	20.3	5.8	.3789
200.	1061.	12.1	-33.3	-24.3	-13.0	0.0	14.1	9.9	.5178
250.	1026.	20.3	-59.2	-48.1	-33.8	-17.6	0.0	15.1	.6627
300.	995.	30.8	-92.8	-79.4	-62.3	-42.9	-21.7	20.7	.8102
350.	965.	43.9	-135.5	-119.8	-99.9	-77.2	-52.5	27.6	.9642
400.	937.	59.6	-186.7	-168.8	-146.0	-120.1	-91.9	34.9	1.1211
450.	910.	78.3	-247.8	-227.7	-202.1	-172.9	-141.2	43.1	1.2835
500.	885.	100.3	-319.2	-296.9	-268.4	-236.0	-200.7	52.2	1.4507

BALLISTIC COEFFICIENT .29, MUZZLE VELOCITY FT/SEC 1400.

RANGE	VELOCITY	HEIGHT	50 YD	100 YD	150 YD	200 YD	250 YD	DRIFT	TIME
0.	1400.	0.0	-1.5	-1.5	-1.5	-1.5	-1.5	0.0	0.0000
50.	1310.	-.2	0.0	1.8	4.3	7.2	10.4	.7	.1109
100.	1230.	1.8	-3.6	0.0	5.0	10.8	17.2	2.6	.2292
150.	1161.	5.3	-12.9	-7.6	0.0	8.6	18.3	5.9	.3548
200.	1108.	10.7	-28.8	-21.6	-11.5	0.0	12.8	10.3	.4874
250.	1065.	18.1	-52.0	-43.0	-30.5	-16.0	0.0	16.1	.6272
300.	1030.	27.8	-82.5	-71.8	-56.7	-39.4	-20.1	22.5	.7706
350.	998.	39.6	-120.5	-107.9	-90.3	-70.1	-47.6	29.4	.9168
400.	968.	54.3	-167.9	-153.6	-133.4	-110.4	-84.7	37.7	1.0713
450.	940.	71.5	-223.0	-206.8	-184.2	-158.2	-129.4	46.2	1.2267
500.	913.	91.8	-288.2	-270.3	-245.1	-216.3	-184.2	55.8	1.3886

BALLISTIC COEFFICIENT .29, MUZZLE VELOCITY FT/SEC 1500.

RANGE	VELOCITY	HEIGHT	50 YD	100 YD	150 YD	200 YD	250 YD	DRIFT	TIME
0.	1500.	0.0	-1.5	-1.5	-1.5	-1.5	-1.5	0.0	0.0000
50.	1397.	-.2	0.0	1.5	3.7	6.2	9.0	.7	.1039
100.	1307.	1.5	-3.0	0.0	4.4	9.5	15.1	2.6	.2148
150.	1228.	4.6	-11.0	-6.6	0.0	7.6	16.1	5.9	.3334
200.	1159.	9.4	-24.9	-19.0	-10.2	0.0	11.3	10.4	.4594
250.	1106.	16.1	-45.2	-37.8	-26.8	-14.1	0.0	16.2	.5919
300.	1064.	24.9	-72.9	-64.0	-50.8	-35.6	-18.6	23.2	.7316
350.	1029.	36.0	-107.9	-97.6	-82.2	-64.4	-44.6	30.9	.8753
400.	997.	49.4	-150.2	-138.4	-120.8	-100.5	-77.9	39.1	1.0219
450.	967.	65.7	-201.9	-188.6	-168.9	-145.9	-120.5	48.6	1.1762
500.	939.	84.4	-261.2	-246.5	-224.5	-199.0	-170.8	58.4	1.3320

BALLISTIC COEFFICIENT .29, MUZZLE VELOCITY FT/SEC 1600.

RANGE	VELOCITY	HEIGHT	50 YD	100 YD	150 YD	200 YD	250 YD	DRIFT	TIME
0.	1600.	0.0	-1.5	-1.5	-1.5	-1.5	-1.5	0.0	0.0000
50.	1490.	-.3	0.0	1.2	3.1	5.4	7.9	.6	.0970
100.	1388.	1.2	-2.4	0.0	3.8	8.3	13.5	2.5	.2014
150.	1299.	4.0	-9.4	-5.7	0.0	6.7	14.4	5.6	.3131
200.	1221.	8.2	-21.5	-16.6	-9.0	0.0	10.3	10.1	.4323
250.	1153.	14.3	-39.7	-33.6	-24.1	-12.9	0.0	16.1	.5603
300.	1102.	22.2	-64.0	-56.7	-45.2	-31.7	-16.3	22.8	.6921
350.	1060.	32.4	-95.5	-87.0	-73.6	-57.9	-39.9	30.7	.8310
400.	1026.	45.0	-134.9	-125.2	-109.9	-91.9	-71.4	39.8	.9760
450.	994.	59.9	-181.5	-170.6	-153.3	-133.1	-110.0	49.3	1.1236
500.	964.	77.6	-237.0	-224.9	-205.7	-183.3	-157.6	59.9	1.2776

BALLISTIC COEFFICIENT .29, MUZZLE VELOCITY FT/SEC 1700.

RANGE	VELOCITY	HEIGHT	50 YD	100 YD	150 YD	200 YD	250 YD	DRIFT	TIME
0.	1700.	0.0	-1.5	-1.5	-1.5	-1.5	-1.5	0.0	0.0000
50.	1584.	-.3	0.0	1.0	2.7	4.6	6.9	.6	.0916
100.	1475.	1.0	-2.0	0.0	3.3	7.3	11.8	2.3	.1896
150.	1375.	3.4	-8.0	-5.0	0.0	5.9	12.7	5.4	.2951
200.	1288.	7.2	-18.6	-14.6	-7.9	0.0	9.0	9.6	.4078
250.	1211.	12.6	-34.4	-29.5	-21.1	-11.2	0.0	15.3	.5280
300.	1146.	19.9	-56.2	-50.3	-40.2	-28.3	-14.9	22.2	.6555
350.	1096.	29.2	-84.6	-77.7	-66.0	-52.1	-36.4	30.3	.7896
400.	1055.	40.7	-120.0	-112.2	-98.8	-82.9	-65.0	39.3	.9291
450.	1021.	54.6	-163.1	-154.3	-139.2	-121.4	-101.2	49.3	1.0740
500.	990.	71.3	-214.7	-204.9	-188.1	-168.3	-145.9	60.3	1.2251

BALLISTIC COEFFICIENT .29, MUZZLE VELOCITY FT/SEC 1800.

RANGE	VELOCITY	HEIGHT	50 YD	100 YD	150 YD	200 YD	250 YD	DRIFT	TIME
0.	1800.	0.0	-1.5	-1.5	-1.5	-1.5	-1.5	0.0	0.0000
50.	1679.	-.4	0.0	.8	2.3	4.0	6.0	.5	.0863
100.	1565.	.8	-1.6	0.0	2.9	6.4	10.4	2.2	.1790
150.	1457.	3.0	-6.8	-4.4	0.0	5.3	11.3	5.0	.2783
200.	1359.	6.4	-16.1	-12.8	-7.0	0.0	8.0	9.1	.3851
250.	1273.	11.2	-30.1	-26.1	-18.8	-10.0	0.0	14.5	.4993
300.	1198.	17.7	-49.4	-44.6	-35.9	-25.3	-13.3	21.2	.6207
350.	1135.	26.3	-75.0	-69.4	-59.2	-46.9	-32.9	29.4	.7503
400.	1088.	36.8	-106.9	-100.5	-88.9	-74.8	-58.8	38.3	.8844
450.	1049.	49.9	-147.0	-139.8	-126.7	-110.9	-92.9	48.8	1.0275
500.	1016.	65.0	-193.0	-185.0	-170.5	-152.9	-132.8	59.3	1.1704

BALLISTIC COEFFICIENT .29, MUZZLE VELOCITY FT/SEC 1900.

RANGE	VELOCITY	HEIGHT	50 YD	100 YD	150 YD	200 YD	250 YD	DRIFT	TIME
0.	1900.	0.0	-1.5	-1.5	-1.5	-1.5	-1.5	0.0	0.0000
50.	1774.	-.4	0.0	.6	1.9	3.5	5.3	.5	.0818
100.	1654.	.6	-1.3	0.0	2.5	5.7	9.2	2.0	.1694
150.	1541.	2.6	-5.7	-3.8	0.0	4.7	10.0	4.7	.2633
200.	1435.	5.6	-13.9	-11.3	-6.2	0.0	7.1	8.5	.3643
250.	1340.	10.0	-26.3	-23.1	-16.7	-8.9	0.0	13.7	.4725
300.	1257.	15.9	-43.5	-39.7	-32.0	-22.7	-12.0	20.2	.5882
350.	1183.	23.5	-66.2	-61.7	-52.8	-41.9	-29.4	27.9	.7112
400.	1125.	33.2	-95.1	-90.0	-79.8	-67.4	-53.1	36.9	.8415
450.	1079.	45.1	-130.8	-125.1	-113.6	-99.6	-83.6	47.0	.9776
500.	1042.	59.4	-173.9	-167.6	-154.8	-139.2	-121.4	58.1	1.1197

BALLISTIC COEFFICIENT .30, MUZZLE VELOCITY FT/SEC 800.

RANGE	VELOCITY	HEIGHT	50 YD	100 YD	150 YD	200 YD	250 YD	DRIFT	TIME
0.	800.	0.0	-1.5	-1.5	-1.5	-1.5	-1.5	0.0	0.0000
50.	780.	1.0	0.0	6.4	13.6	21.1	29.0	.4	.1899
100.	760.	6.4	-12.8	0.0	14.3	29.5	45.3	1.7	.3848
150.	740.	15.7	-40.7	-21.5	0.0	22.7	46.4	3.9	.5849
200.	720.	29.2	-84.6	-59.0	-30.3	0.0	31.6	7.1	.7905
250.	701.	47.4	-145.2	-113.2	-77.4	-39.5	0.0	11.3	1.0017
300.	682.	70.5	-223.7	-185.3	-142.3	-96.8	-49.4	16.5	1.2186
350.	664.	99.0	-321.0	-276.1	-226.0	-173.0	-117.6	22.8	1.4418
400.	646.	133.3	-437.9	-386.7	-329.4	-268.8	-205.5	30.1	1.6710
450.	628.	173.8	-575.8	-518.2	-453.7	-385.5	-314.3	38.6	1.9067
500.	610.	221.0	-735.8	-671.8	-600.1	-524.4	-445.3	48.3	2.1493

BALLISTIC COEFFICIENT .30, MUZZLE VELOCITY FT/SEC 900.

RANGE	VELOCITY	HEIGHT	50 YD	100 YD	150 YD	200 YD	250 YD	DRIFT	TIME
0.	900.	0.0	-1.5	-1.5	-1.5	-1.5	-1.5	0.0	0.0000
50.	876.	.6	0.0	4.9	10.5	16.5	22.7	.5	.1693
100.	854.	4.9	-9.8	0.0	11.2	23.1	35.6	1.7	.3429
150.	833.	12.3	-31.6	-16.8	0.0	17.9	36.5	3.7	.5208
200.	812.	23.0	-66.0	-46.3	-23.8	0.0	24.9	6.5	.7034
250.	791.	37.3	-113.5	-88.9	-60.9	-31.1	0.0	10.1	.8906
300.	771.	55.5	-175.0	-145.5	-111.8	-76.1	-38.8	14.6	1.0827
350.	751.	77.9	-251.2	-216.7	-177.4	-135.7	-92.2	19.9	1.2799
400.	731.	104.7	-342.8	-303.4	-258.5	-210.9	-161.1	26.2	1.4823
450.	712.	136.4	-450.7	-406.4	-355.8	-302.2	-246.3	33.5	1.6903
500.	693.	173.2	-575.6	-526.4	-470.3	-410.7	-348.5	41.7	1.9038

BALLISTIC COEFFICIENT .30, MUZZLE VELOCITY FT/SEC 1000.

RANGE	VELOCITY	HEIGHT	50 YD	100 YD	150 YD	200 YD	250 YD	DRIFT	TIME
0.	1000.	0.0	-1.5	-1.5	-1.5	-1.5	-1.5	0.0	0.0000
50.	971.	.4	0.0	3.8	8.3	13.1	18.2	.9	.1549
100.	943.	3.9	-7.5	0.0	9.0	18.7	29.0	1.8	.3100
150.	918.	9.9	-24.8	-13.5	0.0	14.6	30.0	3.5	.4702
200.	893.	18.7	-52.5	-37.5	-19.5	0.0	20.5	6.4	.6362
250.	870.	30.5	-91.2	-72.4	-50.0	-25.6	0.0	10.0	.8066
300.	848.	45.5	-141.4	-118.8	-91.9	-62.6	-31.9	14.3	.9813
350.	826.	63.9	-203.6	-177.3	-145.9	-111.7	-75.9	19.4	1.1605
400.	806.	86.0	-278.6	-248.5	-212.6	-173.6	-132.6	25.4	1.3444
450.	785.	112.0	-366.9	-333.0	-292.6	-248.7	-202.7	32.2	1.5330
500.	765.	142.3	-469.2	-431.5	-386.6	-337.9	-286.7	39.9	1.7265

BALLISTIC COEFFICIENT .30, MUZZLE VELOCITY FT/SEC 1100.

RANGE	VELOCITY	HEIGHT	50 YD	100 YD	150 YD	200 YD	250 YD	DRIFT	TIME
0.	1100.	0.0	-1.5	-1.5	-1.5	-1.5	-1.5	0.0	0.0000
50.	1060.	.2	0.0	3.2	6.9	11.1	15.3	.5	.1389
100.	1027.	3.1	-6.4	0.0	7.4	15.8	24.2	2.0	.2842
150.	996.	8.2	-20.7	-11.2	0.0	12.5	25.2	3.9	.4313
200.	967.	15.7	-44.4	-31.6	-16.7	0.0	16.9	7.1	.5857
250.	940.	25.6	-76.5	-60.6	-41.9	-21.1	0.0	10.5	.7413
300.	914.	38.4	-118.9	-99.7	-77.4	-52.4	-27.1	14.9	.9031
350.	890.	54.2	-171.7	-149.4	-123.3	-94.1	-64.6	20.3	1.0696
400.	867.	73.1	-235.5	-210.0	-180.2	-146.8	-113.1	26.3	1.2405
450.	845.	95.4	-310.7	-281.9	-248.4	-210.9	-172.9	33.2	1.4157
500.	824.	121.4	-397.9	-366.0	-328.8	-287.0	-244.9	40.8	1.5954

BALLISTIC COEFFICIENT .30, MUZZLE VELOCITY FT/SEC 1200.

RANGE	VELOCITY	HEIGHT	50 YD	100 YD	150 YD	200 YD	250 YD	DRIFT	TIME
0.	1200.	0.0	-1.5	-1.5	-1.5	-1.5	-1.5	0.0	0.0000
50.	1139.	.0	0.0	2.6	5.9	9.5	13.4	.6	.1286
100.	1092.	2.6	-5.2	0.0	6.6	13.8	21.7	2.3	.2631
150.	1054.	7.0	-17.7	-9.9	0.0	10.8	22.6	4.9	.4030
200.	1021.	13.7	-38.0	-27.6	-14.4	0.0	15.7	8.4	.5479
250.	990.	22.7	-67.1	-54.2	-37.7	-19.6	0.0	13.1	.6994
300.	962.	34.0	-104.0	-88.4	-68.6	-47.0	-23.4	17.8	.8510
350.	935.	48.1	-150.6	-132.4	-109.4	-84.1	-56.6	23.6	1.0090
400.	910.	65.1	-207.0	-186.3	-159.9	-131.0	-99.6	30.2	1.1716
450.	885.	85.3	-273.9	-250.6	-220.9	-188.4	-153.1	37.6	1.3388
500.	863.	108.8	-351.6	-325.7	-292.8	-256.7	-217.4	45.8	1.5105

BALLISTIC COEFFICIENT .30, MUZZLE VELOCITY FT/SEC 1300.

RANGE	VELOCITY	HEIGHT	50 YD	100 YD	150 YD	200 YD	250 YD	DRIFT	TIME
0.	1300.	0.0	-1.5	-1.5	-1.5	-1.5	-1.5	0.0	0.0000
50.	1224.	-.1	0.0	2.2	5.1	8.4	11.8	.6	.1190
100.	1158.	2.1	-4.4	0.0	5.7	12.4	19.3	2.6	.2456
150.	1107.	6.1	-15.2	-8.6	0.0	9.9	20.3	5.6	.3779
200.	1065.	12.1	-33.4	-24.7	-13.2	0.0	13.8	10.0	.5182
250.	1031.	20.2	-59.0	-48.1	-33.8	-17.3	0.0	14.9	.6614
300.	1001.	30.5	-92.1	-79.0	-61.8	-42.0	-21.2	20.2	.8070
350.	971.	43.6	-135.0	-119.7	-99.6	-76.5	-52.3	27.1	.9618
400.	944.	59.1	-185.5	-168.1	-145.1	-118.7	-91.0	34.1	1.1168
450.	918.	77.5	-245.5	-225.9	-200.1	-170.3	-139.2	41.9	1.2766
500.	893.	99.4	-317.2	-295.4	-266.7	-233.6	-199.1	51.2	1.4446

BALLISTIC COEFFICIENT .30, MUZZLE VELOCITY FT/SEC 1400.

RANGE	VELOCITY	HEIGHT	50 YD	100 YD	150 YD	200 YD	250 YD	DRIFT	TIME
0.	1400.	0.0	-1.5	-1.5	-1.5	-1.5	-1.5	0.0	0.0000
50.	1312.	-.2	0.0	1.8	4.3	7.1	10.2	.6	.1107
100.	1235.	1.8	-3.6	0.0	5.0	10.7	16.9	2.6	.2290
150.	1167.	5.3	-12.8	-7.5	0.0	8.6	17.9	5.7	.3538
200.	1114.	10.6	-28.5	-21.4	-11.4	0.0	12.4	10.0	.4855
250.	1072.	17.9	-51.2	-42.2	-29.8	-15.5	0.0	15.3	.6229
300.	1037.	27.4	-81.3	-70.6	-55.6	-38.5	-19.9	21.5	.7652
350.	1005.	39.3	-119.7	-107.2	-89.8	-69.8	-48.0	28.8	.9134
400.	976.	53.6	-165.7	-151.4	-131.5	-108.6	-83.8	36.4	1.0638
450.	948.	70.7	-220.9	-204.7	-182.4	-156.7	-128.7	45.0	1.2199
500.	922.	90.7	-285.2	-267.2	-242.4	-213.8	-182.8	54.3	1.3800

BALLISTIC COEFFICIENT .30, MUZZLE VELOCITY FT/SEC 1500.

RANGE	VELOCITY	HEIGHT	50 YD	100 YD	150 YD	200 YD	250 YD	DRIFT	TIME
0.	1500.	0.0	-1.5	-1.5	-1.5	-1.5	-1.5	0.0	0.0000
50.	1400.	-.2	0.0	1.5	3.7	6.2	9.0	.6	.1036
100.	1313.	1.5	-2.9	0.0	4.4	9.4	15.0	2.5	.2143
150.	1235.	4.6	-11.0	-6.6	0.0	7.5	15.9	5.7	.3326
200.	1168.	9.3	-24.7	-18.8	-10.0	0.0	11.2	10.1	.4573
250.	1114.	15.9	-44.8	-37.4	-26.5	-14.0	0.0	15.7	.5890
300.	1072.	24.6	-71.9	-63.0	-49.9	-34.8	-18.1	22.2	.7264
350.	1037.	35.5	-106.3	-96.0	-80.7	-63.2	-43.6	29.7	.8687
400.	1005.	48.9	-149.0	-137.3	-119.7	-99.7	-77.4	38.2	1.0169
450.	976.	64.6	-199.2	-185.9	-166.2	-143.6	-118.5	47.0	1.1672
500.	948.	83.3	-258.4	-243.6	-221.7	-196.7	-168.8	56.9	1.3233

BALLISTIC COEFFICIENT .30, MUZZLE VELOCITY FT/SEC 1600.

RANGE	VELOCITY	HEIGHT	50 YD	100 YD	150 YD	200 YD	250 YD	DRIFT	TIME
0.	1600.	0.0	-1.5	-1.5	-1.5	-1.5	-1.5	0.0	0.0000
50.	1494.	-.3	0.0	1.2	3.1	5.3	7.8	.5	.0969
100.	1395.	1.2	-2.4	0.0	3.8	8.2	13.2	2.4	.2011
150.	1308.	3.9	-9.3	-5.7	0.0	6.7	14.1	5.4	.3121
200.	1231.	8.1	-21.3	-16.5	-8.9	0.0	9.9	9.8	.4304
250.	1164.	14.1	-39.1	-33.0	-23.5	-12.4	0.0	15.3	.5559
300.	1111.	22.0	-63.2	-56.0	-44.6	-31.3	-16.4	22.1	.6880
350.	1069.	32.0	-94.4	-85.9	-72.6	-57.0	-39.7	29.8	.8257
400.	1035.	44.3	-132.9	-123.2	-108.1	-90.3	-70.4	38.4	.9684
450.	1004.	59.1	-179.5	-168.5	-151.5	-131.5	-109.2	48.0	1.1164
500.	974.	76.4	-233.7	-221.5	-202.6	-180.3	-155.6	58.1	1.2674

BALLISTIC COEFFICIENT .30, MUZZLE VELOCITY FT/SEC 1700.

RANGE	VELOCITY	HEIGHT	50 YD	100 YD	150 YD	200 YD	250 YD	DRIFT	TIME
0.	1700.	0.0	-1.5	-1.5	-1.5	-1.5	-1.5	0.0	0.0000
50.	1588.	-.3	0.0	1.0	2.6	4.6	6.8	.6	.0915
100.	1482.	1.0	-1.9	0.0	3.3	7.2	11.6	2.2	.1891
150.	1385.	3.4	-7.9	-5.0	0.0	5.9	12.5	5.1	.2939
200.	1299.	7.2	-18.4	-14.5	-7.9	0.0	8.8	9.3	.4058
250.	1223.	12.5	-34.0	-29.1	-20.9	-11.0	0.0	14.7	.5248
300.	1157.	19.6	-55.6	-49.7	-39.8	-28.0	-14.8	21.5	.6517
350.	1106.	28.7	-83.4	-76.6	-65.0	-51.3	-35.8	29.3	.7839
400.	1065.	40.2	-118.9	-111.1	-97.9	-82.2	-64.5	38.4	.9240
450.	1031.	53.9	-161.4	-152.6	-137.8	-120.1	-100.2	48.1	1.0674
500.	1000.	69.9	-210.7	-201.0	-184.5	-164.8	-142.8	58.2	1.2131

BALLISTIC COEFFICIENT .30, MUZZLE VELOCITY FT/SEC 1800.

RANGE	VELOCITY	HEIGHT	50 YD	100 YD	150 YD	200 YD	250 YD	DRIFT	TIME
0.	1800.	0.0	-1.5	-1.5	-1.5	-1.5	-1.5	0.0	0.0000
50.	1683.	-.4	0.0	.8	2.2	4.0	5.9	.5	.0862
100.	1572.	.8	-1.6	0.0	2.9	6.3	10.3	2.1	.1786
150.	1467.	2.9	-6.7	-4.3	0.0	5.2	11.1	4.8	.2772
200.	1371.	6.3	-15.9	-12.7	-6.9	0.0	7.9	8.8	.3831
250.	1287.	11.1	-29.7	-25.7	-18.5	-9.8	0.0	14.0	.4960
300.	1212.	17.5	-48.8	-44.0	-35.4	-25.0	-13.2	20.5	.6166
350.	1149.	25.8	-73.7	-68.1	-58.1	-45.9	-32.1	28.2	.7435
400.	1099.	36.2	-105.2	-98.8	-87.4	-73.5	-57.7	37.0	.8770
450.	1060.	48.8	-143.7	-136.6	-123.7	-108.0	-90.3	46.8	1.0159
500.	1026.	64.0	-190.3	-182.3	-168.0	-150.6	-130.9	57.7	1.1613

BALLISTIC COEFFICIENT .30, MUZZLE VELOCITY FT/SEC 1900.

RANGE	VELOCITY	HEIGHT	50 YD	100 YD	150 YD	200 YD	250 YD	DRIFT	TIME
0.	1900.	0.0	-1.5	-1.5	-1.5	-1.5	-1.5	0.0	0.0000
50.	1778.	-.4	0.0	.6	1.9	3.4	5.2	.5	.0817
100.	1662.	.6	-1.3	0.0	2.5	5.6	9.1	1.9	.1690
150.	1552.	2.6	-5.7	-3.8	0.0	4.6	9.9	4.5	.2624
200.	1448.	5.6	-13.7	-11.2	-6.2	0.0	7.0	8.2	.3625
250.	1355.	9.8	-26.0	-22.8	-16.5	-8.8	0.0	13.2	.4698
300.	1273.	15.6	-42.9	-39.1	-31.5	-22.3	-11.7	19.4	.5839
350.	1200.	23.1	-65.2	-60.8	-51.9	-41.1	-28.8	26.9	.7054
400.	1139.	32.6	-93.6	-88.5	-78.4	-66.1	-52.0	35.7	.8341
450.	1092.	44.3	-128.6	-122.9	-111.5	-97.7	-81.8	45.4	.9686
500.	1053.	58.2	-170.7	-164.4	-151.7	-136.3	-118.7	56.2	1.1085

BALLISTIC COEFFICIENT .31, MUZZLE VELOCITY FT/SEC 800.

RANGE	VELOCITY	HEIGHT	50 YD	100 YD	150 YD	200 YD	250 YD	DRIFT	TIME
0.	800.	0.0	-1.5	-1.5	-1.5	-1.5	-1.5	0.0	0.0000
50.	780.	1.0	0.0	6.4	13.5	21.1	29.0	.4	.1899
100.	761.	6.4	-12.8	0.0	14.3	29.4	45.1	1.7	.3847
150.	742.	15.6	-40.6	-21.4	0.0	22.6	46.3	3.8	.5844
200.	723.	29.2	-84.4	-58.8	-30.2	0.0	31.5	6.9	.7894
250.	704.	47.2	-144.8	-112.8	-77.1	-39.4	0.0	11.0	.9998
300.	686.	70.2	-222.9	-184.5	-141.6	-96.3	-49.1	16.0	1.2158
350.	668.	98.5	-319.5	-274.7	-224.7	-171.9	-116.8	22.0	1.4376
400.	650.	132.4	-435.6	-384.4	-327.3	-266.9	-203.9	29.1	1.6652
450.	633.	172.3	-572.1	-514.5	-450.2	-382.3	-311.4	37.2	1.8989
500.	616.	218.8	-730.3	-666.3	-594.9	-519.4	-440.6	46.4	2.1389

BALLISTIC COEFFICIENT .31, MUZZLE VELOCITY FT/SEC 900.

RANGE	VELOCITY	HEIGHT	50 YD	100 YD	150 YD	200 YD	250 YD	DRIFT	TIME
0.	900.	0.0	-1.5	-1.5	-1.5	-1.5	-1.5	0.0	0.0000
50.	877.	.6	0.0	4.9	10.5	16.4	22.6	.5	.1692
100.	855.	4.9	-9.8	0.0	11.2	23.1	35.4	1.6	.3426
150.	835.	12.2	-31.5	-16.8	0.0	17.8	36.4	3.5	.5201
200.	814.	22.9	-65.7	-46.1	-23.7	0.0	24.7	6.2	.7021
250.	794.	37.2	-113.1	-88.6	-60.6	-30.9	0.0	9.7	.8887
300.	775.	55.2	-174.3	-144.8	-111.2	-75.6	-38.5	14.0	1.0798
350.	755.	77.4	-249.9	-215.5	-176.4	-134.8	-91.5	19.2	1.2759
400.	736.	104.0	-340.8	-301.5	-256.7	-209.3	-159.8	25.3	1.4769
450.	718.	135.2	-447.7	-403.5	-353.1	-299.8	-244.1	32.2	1.6832
500.	699.	171.6	-571.4	-522.3	-466.3	-407.0	-345.2	40.2	1.8948

BALLISTIC COEFFICIENT .31, MUZZLE VELOCITY FT/SEC 1000.

RANGE	VELOCITY	HEIGHT	50 YD	100 YD	150 YD	200 YD	250 YD	DRIFT	TIME
0.	1000.	0.0	-1.5	-1.5	-1.5	-1.5	-1.5	0.0	0.0000
50.	971.	.4	0.0	3.8	8.2	13.1	18.2	.9	.1551
100.	945.	3.9	-7.5	0.0	8.9	18.7	28.8	1.8	.3102
150.	920.	9.8	-24.6	-13.3	0.0	14.8	29.9	3.4	.4693
200.	896.	18.7	-52.5	-37.5	-19.7	0.0	20.1	6.4	.6362
250.	873.	30.3	-90.8	-71.9	-49.8	-25.1	0.0	9.6	.8048
300.	852.	45.2	-140.7	-118.0	-91.5	-61.8	-31.7	13.9	.9787
350.	831.	63.5	-202.5	-176.1	-145.1	-110.6	-75.4	18.8	1.1570
400.	811.	85.4	-277.0	-246.8	-211.3	-171.9	-131.7	24.6	1.3398
450.	791.	111.2	-364.6	-330.6	-290.7	-246.3	-201.2	31.2	1.5271
500.	772.	141.1	-465.9	-428.2	-383.9	-334.6	-284.4	38.6	1.7191

BALLISTIC COEFFICIENT .31, MUZZLE VELOCITY FT/SEC 1100.

RANGE	VELOCITY	HEIGHT	50 YD	100 YD	150 YD	200 YD	250 YD	DRIFT	TIME
0.	1100.	0.0	-1.5	-1.5	-1.5	-1.5	-1.5	0.0	0.0000
50.	1061.	.2	0.0	3.2	6.9	11.1	15.3	.5	.1394
100.	1029.	3.1	-6.4	0.0	7.4	15.8	24.2	2.0	.2843
150.	999.	8.1	-20.6	-11.0	0.0	12.6	25.3	3.8	.4305
200.	971.	15.7	-44.3	-31.6	-16.8	0.0	16.8	7.1	.5855
250.	944.	25.6	-76.4	-60.5	-42.1	-21.0	0.0	10.4	.7408
300.	919.	38.1	-118.1	-99.0	-76.9	-51.6	-26.4	14.4	.9002
350.	895.	53.9	-171.1	-148.8	-123.1	-93.6	-64.1	19.9	1.0674
400.	873.	72.6	-233.8	-208.4	-179.0	-145.3	-111.6	25.5	1.2359
450.	851.	94.7	-308.4	-279.8	-246.7	-208.8	-170.9	32.2	1.4099
500.	831.	120.3	-394.9	-363.1	-326.3	-284.2	-242.1	39.5	1.5883

BALLISTIC COEFFICIENT .31, MUZZLE VELOCITY FT/SEC 1200.

RANGE	VELOCITY	HEIGHT	50 YD	100 YD	150 YD	200 YD	250 YD	DRIFT	TIME
0.	1200.	0.0	-1.5	-1.5	-1.5	-1.5	-1.5	0.0	0.0000
50.	1141.	.0	0.0	2.6	5.9	9.5	13.3	.6	.1282
100.	1095.	2.6	-5.2	0.0	6.6	13.8	21.3	2.2	.2628
150.	1057.	7.0	-17.6	-9.8	0.0	10.9	22.2	4.8	.4022
200.	1025.	13.6	-38.0	-27.7	-14.5	0.0	15.0	8.4	.5475
250.	995.	22.4	-66.3	-53.3	-36.9	-18.8	0.0	12.3	.6952
300.	967.	33.9	-103.9	-88.3	-68.7	-46.9	-24.4	17.5	.8497
350.	941.	47.8	-149.8	-131.6	-108.7	-83.2	-57.0	23.0	1.0054
400.	916.	64.6	-205.7	-184.9	-158.7	-129.6	-99.6	29.3	1.1667
450.	892.	84.5	-272.0	-248.6	-219.2	-186.4	-152.7	36.6	1.3328
500.	870.	107.7	-349.1	-323.2	-290.4	-254.0	-216.5	44.6	1.5033

BALLISTIC COEFFICIENT .31, MUZZLE VELOCITY FT/SEC 1300.

RANGE	VELOCITY	HEIGHT	50 YD	100 YD	150 YD	200 YD	250 YD	DRIFT	TIME
0.	1300.	0.0	-1.5	-1.5	-1.5	-1.5	-1.5	0.0	0.0000
50.	1226.	-.1	0.0	2.1	5.0	8.2	11.6	.6	.1190
100.	1162.	2.1	-4.3	0.0	5.7	12.1	19.0	2.5	.2448
150.	1111.	6.1	-15.1	-8.6	0.0	9.6	19.8	5.4	.3770
200.	1071.	12.0	-32.8	-24.2	-12.8	0.0	13.7	9.3	.5146
250.	1037.	20.0	-58.1	-47.4	-33.0	-17.1	0.0	14.1	.6569
300.	1006.	30.4	-91.9	-79.0	-61.7	-42.6	-22.1	19.9	.8054
350.	978.	43.0	-133.1	-118.1	-98.0	-75.6	-51.7	25.9	.9551
400.	951.	58.5	-183.7	-166.5	-143.5	-118.0	-90.7	33.0	1.1108
450.	926.	76.7	-243.6	-224.3	-198.4	-169.7	-139.0	40.9	1.2706
500.	901.	98.0	-313.3	-291.8	-263.1	-231.2	-197.0	49.4	1.4346

BALLISTIC COEFFICIENT .31, MUZZLE VELOCITY FT/SEC 1400.

RANGE	VELOCITY	HEIGHT	50 YD	100 YD	150 YD	200 YD	250 YD	DRIFT	TIME
0.	1400.	0.0	-1.5	-1.5	-1.5	-1.5	-1.5	0.0	0.0000
50.	1315.	-.2	0.0	1.8	4.3	7.1	10.1	.6	.1108
100.	1240.	1.8	-3.5	0.0	5.1	10.6	16.8	2.5	.2283
150.	1173.	5.3	-12.9	-7.6	0.0	8.3	17.5	5.7	.3540
200.	1120.	10.5	-28.3	-21.2	-11.1	0.0	12.3	9.7	.4837
250.	1078.	17.7	-50.7	-41.9	-29.2	-15.4	0.0	14.9	.6203
300.	1043.	27.2	-80.8	-70.2	-55.0	-38.4	-19.9	21.1	.7627
350.	1012.	39.0	-119.1	-106.7	-89.0	-69.6	-48.1	28.3	.9106
400.	983.	53.1	-164.5	-150.3	-130.1	-107.9	-83.3	35.6	1.0592
450.	956.	69.9	-218.7	-202.8	-180.0	-155.0	-127.4	43.8	1.2130
500.	930.	89.6	-282.5	-264.8	-239.5	-211.8	-181.0	53.0	1.3723

BALLISTIC COEFFICIENT .31, MUZZLE VELOCITY FT/SEC 1500.

RANGE	VELOCITY	HEIGHT	50 YD	100 YD	150 YD	200 YD	250 YD	DRIFT	TIME
0.	1500.	0.0	-1.5	-1.5	-1.5	-1.5	-1.5	0.0	0.0000
50.	1403.	-.2	0.0	1.5	3.6	6.1	8.9	.6	.1035
100.	1318.	1.4	-2.9	0.0	4.3	9.4	14.8	2.5	.2139
150.	1242.	4.5	-10.9	-6.5	0.0	7.6	15.7	5.5	.3313
200.	1175.	9.2	-24.6	-18.7	-10.1	0.0	10.9	9.9	.4563
250.	1122.	15.7	-44.4	-37.0	-26.2	-13.6	0.0	15.2	.5862
300.	1079.	24.3	-71.1	-62.3	-49.3	-34.2	-17.9	21.6	.7225
350.	1044.	35.2	-105.6	-95.4	-80.2	-62.6	-43.5	29.1	.8652
400.	1013.	48.5	-148.0	-136.3	-119.0	-98.8	-77.1	37.4	1.0127
450.	984.	64.0	-197.4	-184.2	-164.8	-142.1	-117.6	46.0	1.1612
500.	957.	82.2	-255.4	-240.8	-219.2	-193.9	-166.7	55.4	1.3146

BALLISTIC COEFFICIENT .31, MUZZLE VELOCITY FT/SEC 1600.

RANGE	VELOCITY	HEIGHT	50 YD	100 YD	150 YD	200 YD	250 YD	DRIFT	TIME
0.	1600.	0.0	-1.5	-1.5	-1.5	-1.5	-1.5	0.0	0.0000
50.	1497.	-.3	0.0	1.2	3.1	5.3	7.8	.5	.0967
100.	1401.	1.2	-2.4	0.0	3.8	8.2	13.2	2.3	.2004
150.	1316.	3.9	-9.3	-5.7	0.0	6.6	14.1	5.3	.3112
200.	1240.	8.1	-21.1	-16.3	-8.8	0.0	10.0	9.4	.4286
250.	1173.	14.0	-38.9	-32.9	-23.4	-12.5	0.0	15.0	.5542
300.	1120.	21.7	-62.5	-55.3	-44.0	-30.8	-15.8	21.4	.6839
350.	1078.	31.6	-93.2	-84.8	-71.6	-56.2	-38.8	28.9	.8205
400.	1043.	43.8	-131.6	-122.0	-106.9	-89.3	-69.4	37.5	.9630
450.	1012.	58.5	-178.0	-167.2	-150.2	-130.4	-108.0	47.0	1.1108
500.	983.	75.4	-231.1	-219.1	-200.2	-178.2	-153.3	56.6	1.2594

BALLISTIC COEFFICIENT .31, MUZZLE VELOCITY FT/SEC 1700.

RANGE	VELOCITY	HEIGHT	50 YD	100 YD	150 YD	200 YD	250 YD	DRIFT	TIME
0.	1700.	0.0	-1.5	-1.5	-1.5	-1.5	-1.5	0.0	0.0000
50.	1592.	-.3	0.0	1.0	2.6	4.5	6.7	.6	.0914
100.	1489.	1.0	-1.9	0.0	3.3	7.2	11.6	2.1	.1886
150.	1394.	3.4	-7.8	-4.9	0.0	5.8	12.4	5.0	.2929
200.	1309.	7.1	-18.2	-14.3	-7.8	0.0	8.8	9.0	.4040
250	1235.	12.4	-33.7	-28.9	-20.7	-11.0	0.0	14.3	.5224
300.	1169.	19.3	-54.8	-49.0	-39.2	-27.5	-14.3	20.7	.6471
350.	1117.	28.3	-82.3	-75.6	-64.1	-50.5	-35.1	28.3	.7785
400.	1075.	39.5	-116.7	-109.0	-95.9	-80.3	-62.8	36.9	.9155
450.	1041.	52.9	-158.4	-149.8	-135.0	-117.5	-97.8	46.3	1.0574
500.	1010.	69.1	-208.6	-199.0	-182.6	-163.2	-141.2	57.0	1.2060

BALLISTIC COEFFICIENT .31, MUZZLE VELOCITY FT/SEC 1800.

RANGE	VELOCITY	HEIGHT	50 YD	100 YD	150 YD	200 YD	250 YD	DRIFT	TIME
0.	1800.	0.0	-1.5	-1.5	-1.5	-1.5	-1.5	0.0	0.0000
50.	1687.	-.4	0.0	.8	2.2	3.9	5.9	.5	.0861
100.	1579.	.8	-1.6	0.0	2.8	6.3	10.2	2.0	.1782
150.	1477.	2.9	-6.6	-4.3	0.0	5.1	11.0	4.6	.2762
200.	1383.	6.2	-15.7	-12.5	-6.9	0.0	7.8	8.4	.3813
250.	1300.	10.9	-29.3	-25.4	-18.3	-9.7	0.0	13.5	.4931
300.	1226.	17.2	-48.1	-43.4	-34.8	-24.6	-12.9	19.7	.6121
350.	1162.	25.4	-72.6	-67.1	-57.2	-45.2	-31.6	27.2	.7379
400.	1111.	35.6	-103.7	-97.4	-86.0	-72.3	-56.8	35.8	.8702
450.	1071.	48.0	-141.6	-134.5	-121.7	-106.3	-88.8	45.4	1.0077
500.	1037.	62.7	-186.8	-178.9	-164.7	-147.6	-128.2	55.8	1.1501

BALLISTIC COEFFICIENT .31, MUZZLE VELOCITY FT/SEC 1900.

RANGE	VELOCITY	HEIGHT	50 YD	100 YD	150 YD	200 YD	250 YD	DRIFT	TIME
0.	1900.	0.0	-1.5	-1.5	-1.5	-1.5	-1.5	0.0	0.0000
50.	1782.	-.4	0.0	.6	1.9	3.4	5.1	.5	.0816
100.	1669.	.6	-1.2	0.0	2.5	5.5	9.0	1.9	.1686
150.	1562.	2.5	-5.6	-3.8	0.0	4.5	9.7	4.4	.2616
200.	1461.	5.5	-13.6	-11.1	-6.1	0.0	6.9	7.9	.3608
250.	1369.	9.7	-25.6	-22.5	-16.2	-8.6	0.0	12.7	.4669
300.	1288.	15.4	-42.3	-38.6	-31.0	-21.9	-11.6	18.7	.5799
350.	1215.	22.8	-64.3	-59.9	-51.1	-40.5	-28.4	26.0	.7001
400.	1153.	32.1	-92.0	-87.0	-77.0	-64.9	-51.0	34.4	.8268
450.	1104.	43.5	-126.4	-120.8	-109.5	-95.8	-80.3	43.9	.9599
500.	1064.	57.3	-168.3	-162.1	-149.5	-134.4	-117.1	54.7	1.1000

BALLISTIC COEFFICIENT .32, MUZZLE VELOCITY FT/SEC 800.

RANGE	VELOCITY	HEIGHT	50 YD	100 YD	150 YD	200 YD	250 YD	DRIFT	TIME
0.	800.	0.0	-1.5	-1.5	-1.5	-1.5	-1.5	0.0	0.0000
50.	781.	1.0	0.0	6.4	13.5	21.0	28.9	.4	.1900
100.	762.	6.3	-12.8	0.0	14.3	29.3	45.0	1.7	.3844
150.	744.	15.6	-40.5	-21.4	0.0	22.6	46.1	3.8	.5838
200.	725.	29.1	-84.1	-58.6	-30.1	0.0	31.3	6.7	.7883
250.	707.	47.1	-144.3	-112.5	-76.8	-39.2	0.0	10.7	.9980
300.	689.	69.9	-222.0	-183.7	-140.9	-95.8	-48.8	15.5	1.2129
350.	672.	97.9	-317.9	-273.3	-223.3	-170.7	-115.9	21.3	1.4334
400.	655.	131.4	-433.1	-382.1	-325.0	-264.8	-202.2	28.1	1.6594
450.	638.	170.9	-568.3	-510.9	-446.6	-379.0	-308.5	35.8	1.8910
500.	621.	216.8	-725.0	-661.2	-589.8	-514.6	-436.3	44.7	2.1291

BALLISTIC COEFFICIENT .32, MUZZLE VELOCITY FT/SEC 900.

RANGE	VELOCITY	HEIGHT	50 YD	100 YD	150 YD	200 YD	250 YD	DRIFT	TIME
0.	900.	0.0	-1.5	-1.5	-1.5	-1.5	-1.5	0.0	0.0000
50.	878.	.6	0.0	4.9	10.5	16.4	22.5	.4	.1691
100.	857.	4.9	-9.8	0.0	11.2	23.0	35.3	1.6	.3422
150.	837.	12.2	-31.4	-16.7	0.0	17.7	36.2	3.4	.5194
200.	817.	22.8	-65.5	-45.9	-23.6	0.0	24.6	6.0	.7008
250.	798.	37.0	-112.7	-88.2	-60.3	-30.8	0.0	9.4	.8867
300.	778.	54.9	-173.5	-144.1	-110.6	-75.2	-38.3	13.5	1.0770
350.	760.	76.9	-248.6	-214.3	-175.3	-133.9	-90.8	18.5	1.2719
400.	741.	103.2	-338.9	-299.7	-255.1	-207.9	-158.6	24.4	1.4719
450.	723.	134.3	-445.2	-401.1	-350.9	-297.8	-242.4	31.2	1.6771
500.	705.	170.3	-568.2	-519.2	-463.4	-404.4	-342.8	38.9	1.8875

BALLISTIC COEFFICIENT .32, MUZZLE VELOCITY FT/SEC 1000.

RANGE	VELOCITY	HEIGHT	50 YD	100 YD	150 YD	200 YD	250 YD	DRIFT	TIME
0.	1000.	0.0	-1.5	-1.5	-1.5	-1.5	-1.5	0.0	0.0000
50.	973.	.4	0.0	4.0	8.3	13.2	18.2	.4	.1523
100.	947.	3.9	-7.9	0.0	8.8	18.5	28.6	1.8	.3105
150.	922.	9.8	-25.0	-13.2	0.0	14.5	29.7	3.3	.4689
200.	899.	18.5	-52.7	-36.9	-19.3	0.0	20.2	6.0	.6339
250.	877.	30.2	-91.2	-71.4	-49.4	-25.2	0.0	9.3	.8029
300.	856.	45.0	-141.0	-117.3	-90.9	-61.9	-31.6	13.4	.9762
350.	836.	63.1	-202.6	-174.9	-144.1	-110.3	-74.9	18.2	1.1535
400.	816.	84.8	-276.7	-245.0	-209.9	-171.2	-130.8	23.8	1.3352
450.	797.	110.3	-363.7	-328.2	-288.6	-245.1	-199.6	30.1	1.5213
500.	778.	139.9	-464.4	-424.8	-380.9	-332.5	-282.0	37.3	1.7117

BALLISTIC COEFFICIENT .32, MUZZLE VELOCITY FT/SEC 1100.

RANGE	VELOCITY	HEIGHT	50 YD	100 YD	150 YD	200 YD	250 YD	DRIFT	TIME
0.	1100.	0.0	-1.5	-1.5	-1.5	-1.5	-1.5	0.0	0.0000
50.	1062.	.2	0.0	3.2	6.8	10.9	15.1	.6	.1398
100.	1031.	3.1	-6.3	0.0	7.3	15.4	23.9	2.1	.2845
150.	1002.	8.1	-20.5	-11.0	0.0	12.1	24.9	3.7	.4303
200.	975.	15.5	-43.5	-30.8	-16.1	0.0	17.1	6.4	.5819
250.	949.	25.4	-75.7	-59.8	-41.5	-21.3	0.0	9.9	.7380
300.	924.	38.0	-117.4	-98.4	-76.4	-52.2	-26.6	14.0	.8980
350.	901.	53.4	-169.3	-147.2	-121.5	-93.3	-63.4	19.0	1.0622
400.	878.	72.1	-232.4	-207.1	-177.7	-145.5	-111.4	24.8	1.2318
450.	857.	93.9	-306.1	-277.6	-244.6	-208.3	-169.9	31.1	1.4042
500.	837.	119.3	-391.8	-360.1	-323.4	-283.1	-240.5	38.3	1.5813

BALLISTIC COEFFICIENT .32, MUZZLE VELOCITY FT/SEC 1200.

RANGE	VELOCITY	HEIGHT	50 YD	100 YD	150 YD	200 YD	250 YD	DRIFT	TIME
0.	1200.	0.0	-1.5	-1.5	-1.5	-1.5	-1.5	0.0	0.0000
50.	1142.	.0	0.0	2.6	5.8	9.5	13.2	.6	.1282
100.	1097.	2.6	-5.2	0.0	6.5	13.8	21.2	2.2	.2624
150.	1060.	7.0	-17.5	-9.8	0.0	11.0	22.0	4.6	.4013
200.	1028.	13.6	-38.0	-27.7	-14.7	0.0	14.7	8.3	.5470
250.	1000.	22.3	-65.9	-53.0	-36.7	-18.4	0.0	12.0	.6931
300.	973.	33.6	-102.9	-87.4	-67.8	-45.8	-23.8	16.8	.8455
350.	946.	47.6	-149.5	-131.4	-108.6	-82.9	-57.2	22.6	1.0036
400.	922.	64.1	-204.3	-183.6	-157.6	-128.3	-98.8	28.5	1.1621
450.	899.	83.8	-270.1	-246.9	-217.6	-184.6	-151.5	35.6	1.3272
500.	877.	106.7	-346.4	-320.5	-288.0	-251.3	-214.5	43.3	1.4962

BALLISTIC COEFFICIENT .32, MUZZLE VELOCITY FT/SEC 1300.

RANGE	VELOCITY	HEIGHT	50 YD	100 YD	150 YD	200 YD	250 YD	DRIFT	TIME
0.	1300.	0.0	-1.5	-1.5	-1.5	-1.5	-1.5	0.0	0.0000
50.	1229.	-.1	0.0	2.1	5.0	8.2	11.6	.6	.1189
100.	1166.	2.1	-4.3	0.0	5.7	12.0	18.9	2.4	.2444
150.	1116.	6.0	-15.0	-8.6	0.0	9.5	19.8	5.3	.3761
200.	1075.	11.9	-32.6	-24.1	-12.7	0.0	13.7	9.1	.5131
250.	1042.	19.9	-57.9	-47.2	-32.9	-17.1	0.0	13.8	.6553
300.	1012.	30.3	-91.6	-78.8	-61.6	-42.6	-22.1	19.6	.8037
350.	984.	42.8	-132.5	-117.6	-97.6	-75.4	-51.5	25.5	.9524
400.	958.	57.9	-182.3	-165.2	-142.3	-117.0	-89.6	32.2	1.1058
450.	933.	76.0	-241.8	-222.5	-196.8	-168.3	-137.6	39.8	1.2648
500.	909.	97.1	-310.8	-289.5	-260.9	-229.2	-195.1	48.2	1.4277

BALLISTIC COEFFICIENT .32, MUZZLE VELOCITY FT/SEC 1400.

RANGE	VELOCITY	HEIGHT	50 YD	100 YD	150 YD	200 YD	250 YD	DRIFT	TIME
0.	1400.	0.0	-1.5	-1.5	-1.5	-1.5	-1.5	0.0	0.0000
50.	1317.	-.2	0.0	1.8	4.2	7.0	10.1	.6	.1106
100.	1244.	1.7	-3.5	0.0	4.9	10.5	16.6	2.4	.2279
150.	1179.	5.2	-12.7	-7.4	0.0	8.4	17.5	5.4	.3519
200.	1126.	10.4	-28.1	-21.1	-11.2	0.0	12.2	9.4	.4821
250.	1084.	17.6	-50.4	-41.6	-29.2	-15.2	0.0	14.5	.6179
300.	1049.	27.1	-80.7	-70.2	-55.4	-38.6	-20.3	20.9	.7616
350.	1019.	38.5	-117.4	-105.0	-87.8	-68.2	-46.8	27.1	.9039
400.	990.	52.8	-163.7	-149.6	-129.9	-107.5	-83.1	35.0	1.0558
450.	964.	69.3	-217.1	-201.3	-179.1	-153.9	-126.4	42.8	1.2076
500.	938.	88.6	-279.9	-262.3	-237.6	-209.6	-179.1	51.6	1.3647

BALLISTIC COEFFICIENT .32, MUZZLE VELOCITY FT/SEC 1500.

RANGE	VELOCITY	HEIGHT	50 YD	100 YD	150 YD	200 YD	250 YD	DRIFT	TIME
0.	1500.	0.0	-1.5	-1.5	-1.5	-1.5	-1.5	0.0	0.0000
50.	1406.	-.2	0.0	1.5	3.6	6.1	8.8	.6	.1034
100.	1323.	1.4	-2.9	0.0	4.3	9.2	14.7	2.4	.2135
150.	1249.	4.5	-10.8	-6.4	0.0	7.4	15.6	5.3	.3303
200.	1183.	9.1	-24.3	-18.5	-9.9	0.0	10.9	9.5	.4537
250.	1129.	15.6	-44.0	-36.7	-26.0	-13.7	0.0	14.7	.5837
300.	1087.	24.1	-70.5	-61.8	-48.9	-34.1	-17.7	21.0	.7192
350.	1051.	34.7	-104.2	-94.1	-79.0	-61.8	-42.7	28.1	.8596
400.	1021.	47.7	-145.7	-134.1	-116.9	-97.2	-75.3	36.0	1.0046
450.	992.	63.4	-196.1	-183.1	-163.7	-141.5	-117.0	45.1	1.1565
500.	965.	81.4	-253.4	-238.8	-217.4	-192.7	-165.4	54.2	1.3081

BALLISTIC COEFFICIENT .32, MUZZLE VELOCITY FT/SEC 1600.

RANGE	VELOCITY	HEIGHT	50 YD	100 YD	150 YD	200 YD	250 YD	DRIFT	TIME
0.	1600.	0.0	-1.5	-1.5	-1.5	-1.5	-1.5	0.0	0.0000
50.	1500.	-.3	0.0	1.2	3.1	5.2	7.7	.5	.0966
100.	1406.	1.2	-2.4	0.0	3.7	8.1	12.9	2.2	.2000
150.	1323.	3.9	-9.2	-5.6	0.0	6.5	13.8	5.1	.3101
200.	1249.	8.0	-21.0	-16.2	-8.7	0.0	9.7	9.1	.4269
250.	1183.	13.8	-38.3	-32.3	-23.0	-12.1	0.0	14.4	.5503
300.	1129.	21.5	-61.9	-54.7	-43.5	-30.4	-15.9	20.7	.6803
350.	1087.	31.2	-92.2	-83.9	-70.8	-55.6	-38.6	28.1	.8158
400.	1051.	43.1	-129.8	-120.3	-105.3	-87.9	-68.5	36.3	.9562
450.	1021.	57.5	-175.0	-164.3	-147.5	-127.9	-106.1	45.3	1.1012
500.	992.	74.6	-229.2	-217.3	-198.6	-176.8	-152.6	55.5	1.2531

BALLISTIC COEFFICIENT .32, MUZZLE VELOCITY FT/SEC 1700.

RANGE	VELOCITY	HEIGHT	50 YD	100 YD	150 YD	200 YD	250 YD	DRIFT	TIME
0.	1700.	0.0	-1.5	-1.5	-1.5	-1.5	-1.5	0.0	0.0000
50.	1595.	-.3	0.0	1.0	2.6	4.5	6.7	.5	.0913
100.	1495.	1.0	-1.9	0.0	3.3	7.1	11.4	2.1	.1882
150.	1402.	3.3	-7.7	-4.9	0.0	5.8	12.2	4.8	.2919
200.	1319.	7.0	-18.0	-14.2	-7.7	0.0	8.6	8.7	.4023
250.	1246.	12.2	-33.3	-28.5	-20.4	-10.8	0.0	13.8	.5195
300.	1180.	19.1	-54.1	-48.4	-38.6	-27.1	-14.2	20.0	.6431
350.	1127.	28.0	-81.3	-74.6	-63.2	-49.8	-34.7	27.4	.7734
400.	1085.	38.9	-115.2	-107.5	-94.5	-79.2	-61.9	35.8	.9092
450.	1050.	52.2	-156.3	-147.7	-133.1	-115.8	-96.4	45.0	1.0498
500.	1019.	67.8	-205.0	-195.4	-179.2	-160.0	-138.4	55.0	1.1949

BALLISTIC COEFFICIENT .32, MUZZLE VELOCITY FT/SEC 1800.

RANGE	VELOCITY	HEIGHT	50 YD	100 YD	150 YD	200 YD	250 YD	DRIFT	TIME
0.	1800.	0.0	-1.5	-1.5	-1.5	-1.5	-1.5	0.0	0.0000
50.	1690.	-.4	0.0	.8	2.2	3.9	5.8	.5	.0860
100.	1586.	.8	-1.6	0.0	2.8	6.2	10.0	2.0	.1778
150.	1486.	2.9	-6.6	-4.2	0.0	5.1	10.8	4.5	.2753
200.	1394.	6.2	-15.6	-12.4	-6.8	0.0	7.7	8.1	.3796
250.	1312.	10.8	-29.0	-25.1	-18.1	-9.6	0.0	13.0	.4906
300.	1239.	17.0	-47.6	-42.9	-34.4	-24.2	-12.7	19.1	.6084
350.	1174.	25.1	-71.9	-66.5	-56.6	-44.7	-31.3	26.5	.7338
400.	1123.	35.0	-102.2	-95.9	-84.7	-71.1	-55.7	34.6	.8635
450.	1081.	47.3	-139.6	-132.5	-119.9	-104.6	-87.3	44.0	1.0000
500.	1046.	62.0	-184.8	-177.0	-162.9	-145.9	-126.7	54.5	1.1429

BALLISTIC COEFFICIENT .32, MUZZLE VELOCITY FT/SEC 1900.

RANGE	VELOCITY	HEIGHT	50 YD	100 YD	150 YD	200 YD	250 YD	DRIFT	TIME
0.	1900.	0.0	-1.5	-1.5	-1.5	-1.5	-1.5	0.0	0.0000
50.	1785.	-.4	0.0	.6	1.9	3.4	5.1	.5	.0815
100.	1676.	.6	-1.2	0.0	2.5	5.5	8.9	1.8	.1682
150.	1572.	2.5	-5.6	-3.7	0.0	4.5	9.6	4.2	.2608
200.	1474.	5.4	-13.4	-11.0	-6.0	0.0	6.8	7.6	.3591
250.	1383.	9.6	-25.3	-22.2	-16.0	-8.5	0.0	12.2	.4643
300.	1302.	15.2	-41.8	-38.1	-30.6	-21.6	-11.4	18.0	.5761
350.	1230.	22.4	-63.4	-59.0	-50.3	-39.8	-27.9	25.0	.6948
400.	1167.	31.5	-90.6	-85.7	-75.7	-63.8	-50.1	33.2	.8201
450.	1117.	42.7	-124.3	-118.8	-107.6	-94.1	-78.8	42.4	.9516
500.	1076.	56.1	-164.9	-158.7	-146.3	-131.3	-114.3	52.6	1.0884

BALLISTIC COEFFICIENT .33, MUZZLE VELOCITY FT/SEC 800.

RANGE	VELOCITY	HEIGHT	50 YD	100 YD	150 YD	200 YD	250 YD	DRIFT	TIME
0.	800.	0.0	-1.5	-1.5	-1.5	-1.5	-1.5	0.0	0.0000
50.	781.	1.0	0.0	6.4	13.5	21.0	28.8	.4	.1898
100.	763.	6.3	-12.8	0.0	14.2	29.2	44.8	1.6	.3842
150.	745.	15.6	-40.5	-21.4	0.0	22.5	45.9	3.7	.5833
200.	727.	29.0	-84.0	-58.4	-30.0	0.0	31.2	6.6	.7873
250.	710.	46.9	-143.9	-112.0	-76.4	-39.0	0.0	10.3	.9962
300.	693.	69.5	-221.1	-182.9	-140.1	-95.2	-48.4	15.0	1.2101
350.	676.	97.3	-316.4	-271.8	-221.9	-169.5	-114.9	20.5	1.4292
400.	659.	130.5	-430.7	-379.6	-322.7	-262.8	-200.4	27.0	1.6537
450.	642.	169.7	-565.2	-507.8	-443.7	-376.3	-306.2	34.6	1.8843
500.	626.	215.2	-721.2	-657.4	-586.2	-511.2	-433.3	43.3	2.1212

BALLISTIC COEFFICIENT .33, MUZZLE VELOCITY FT/SEC 900.

RANGE	VELOCITY	HEIGHT	50 YD	100 YD	150 YD	200 YD	250 YD	DRIFT	TIME
0.	900.	0.0	-1.5	-1.5	-1.5	-1.5	-1.5	0.0	0.0000
50.	878.	.6	0.0	4.9	10.4	16.3	22.5	.4	.1690
100.	858.	4.9	-9.8	0.0	11.1	22.9	35.1	1.5	.3419
150.	838.	12.2	-31.3	-16.7	0.0	17.6	36.0	3.3	.5186
200.	819.	22.7	-65.3	-45.7	-23.5	0.0	24.5	5.8	.6996
250.	800.	36.8	-112.3	-87.9	-60.1	-30.7	0.0	9.1	.8849
300.	782.	54.7	-172.9	-143.5	-110.2	-74.9	-38.1	13.1	1.0746
350.	764.	76.5	-247.7	-213.5	-174.6	-133.5	-90.5	18.0	1.2689
400.	746.	102.7	-337.6	-298.5	-254.0	-207.0	-157.9	23.7	1.4679
450.	728.	133.4	-443.1	-399.1	-349.1	-296.2	-241.0	30.2	1.6717
500.	710.	169.0	-565.0	-516.2	-460.6	-401.8	-340.5	37.6	1.8804

BALLISTIC COEFFICIENT .33, MUZZLE VELOCITY FT/SEC 1000.

RANGE	VELOCITY	HEIGHT	50 YD	100 YD	150 YD	200 YD	250 YD	DRIFT	TIME
0.	1000.	0.0	-1.5	-1.5	-1.5	-1.5	-1.5	0.0	0.0000
50.	974.	.4	0.0	3.8	8.3	13.1	18.2	.4	.1522
100.	948.	3.8	-7.7	0.0	9.0	18.6	28.6	1.5	.3084
150.	925.	9.8	-25.0	-13.5	0.0	14.4	29.5	3.2	.4685
200.	902.	18.5	-52.5	-37.1	-19.2	0.0	20.1	5.7	.6326
250.	880.	30.1	-90.8	-71.6	-49.1	-25.1	0.0	9.0	.8011
300.	859.	44.8	-140.4	-117.3	-90.4	-61.6	-31.4	13.0	.9738
350.	840.	62.7	-201.5	-174.6	-143.2	-109.6	-74.4	17.6	1.1500
400.	821.	84.3	-275.1	-244.4	-208.5	-170.1	-129.9	23.0	1.3308
450.	802.	109.5	-361.7	-327.1	-286.7	-243.5	-198.3	29.2	1.5159
500.	783.	138.8	-461.7	-423.3	-378.3	-330.4	-280.1	36.1	1.7052

BALLISTIC COEFFICIENT .33, MUZZLE VELOCITY FT/SEC 1100.

RANGE	VELOCITY	HEIGHT	50 YD	100 YD	150 YD	200 YD	250 YD	DRIFT	TIME
0.	1100.	0.0	-1.5	-1.5	-1.5	-1.5	-1.5	0.0	0.0000
50.	1063.	.2	0.0	3.0	6.8	10.8	15.0	.7	.1402
100.	1033.	3.1	-6.0	0.0	7.6	15.6	24.0	1.6	.2821
150.	1004.	8.1	-20.5	-11.4	0.0	11.9	24.7	3.8	.4304
200.	978.	15.4	-43.2	-31.1	-15.9	0.0	17.0	6.2	.5807
250.	952.	25.3	-75.2	-60.1	-41.1	-21.2	0.0	9.6	.7363
300.	928.	37.8	-116.8	-98.7	-75.9	-52.0	-26.5	13.7	.8959
350.	905.	53.1	-168.4	-147.3	-120.6	-92.8	-63.1	18.4	1.0594
400.	884.	71.5	-230.7	-206.5	-176.1	-144.3	-110.4	24.0	1.2272
450.	863.	93.2	-304.1	-276.9	-242.6	-206.9	-168.7	30.2	1.3991
500.	843.	118.3	-389.0	-358.8	-320.7	-281.0	-238.5	37.2	1.5749

BALLISTIC COEFFICIENT .33, MUZZLE VELOCITY FT/SEC 1200.

RANGE	VELOCITY	HEIGHT	50 YD	100 YD	150 YD	200 YD	250 YD	DRIFT	TIME
0.	1200.	0.0	-1.5	-1.5	-1.5	-1.5	-1.5	0.0	0.0000
50.	1144.	.0	0.0	2.6	5.9	9.4	13.2	.5	.1281
100.	1099.	2.5	-5.1	0.0	6.6	13.6	21.2	2.1	.2620
150.	1063.	7.0	-17.7	-9.9	0.0	10.5	21.9	4.8	.4020
200.	1032.	13.5	-37.5	-27.2	-14.0	0.0	15.2	7.8	.5442
250.	1004.	22.3	-65.9	-53.0	-36.4	-19.0	0.0	11.9	.6924
300.	977.	33.4	-102.3	-86.9	-67.0	-46.0	-23.3	16.4	.8430
350.	952.	47.1	-148.0	-130.0	-106.8	-82.4	-55.8	21.7	.9986
400.	928.	63.6	-203.2	-182.6	-156.1	-128.2	-97.8	27.8	1.1582
450.	905.	83.1	-268.3	-245.1	-215.3	-183.9	-149.7	34.6	1.3217
500.	883.	105.8	-343.9	-318.2	-285.1	-250.1	-212.2	42.2	1.4896

BALLISTIC COEFFICIENT .33, MUZZLE VELOCITY FT/SEC 1300.

RANGE	VELOCITY	HEIGHT	50 YD	100 YD	150 YD	200 YD	250 YD	DRIFT	TIME
0.	1300.	0.0	-1.5	-1.5	-1.5	-1.5	-1.5	0.0	0.0000
50.	1231.	-.1	0.0	2.1	5.0	8.1	11.6	.6	.1188
100.	1169.	2.1	-4.3	0.0	5.7	12.0	18.9	2.3	.2440
150.	1119.	6.0	-14.9	-8.5	0.0	9.4	19.9	5.1	.3753
200.	1080.	11.8	-32.4	-23.9	-12.6	0.0	13.9	8.8	.5116
250.	1046.	19.8	-57.9	-47.3	-33.1	-17.4	0.0	13.7	.6550
300.	1017.	29.8	-90.3	-77.5	-60.5	-41.6	-20.8	18.7	.7984
350.	990.	42.6	-132.1	-117.3	-97.4	-75.4	-51.0	25.1	.9504
400.	964.	57.6	-181.4	-164.4	-141.6	-116.5	-88.7	31.6	1.1024
450.	939.	75.3	-239.9	-220.7	-195.2	-166.9	-135.6	38.8	1.2591
500.	916.	96.1	-308.3	-287.0	-258.6	-227.3	-192.5	47.0	1.4208

BALLISTIC COEFFICIENT .33, MUZZLE VELOCITY FT/SEC 1400.

RANGE	VELOCITY	HEIGHT	50 YD	100 YD	150 YD	200 YD	250 YD	DRIFT	TIME
0.	1400.	0.0	-1.5	-1.5	-1.5	-1.5	-1.5	0.0	0.0000
50.	1320.	-.2	0.0	1.8	4.2	7.0	10.0	.6	.1104
100.	1248.	1.7	-3.5	0.0	4.9	10.5	16.5	2.3	.2274
150.	1184.	5.2	-12.6	-7.3	0.0	8.3	17.4	5.2	.3508
200.	1132.	10.3	-27.9	-20.9	-11.1	0.0	12.1	9.2	.4806
250.	1090.	17.5	-50.1	-41.3	-29.0	-15.1	0.0	14.1	.6158
300.	1055.	26.7	-79.4	-68.9	-54.2	-37.5	-19.3	19.9	.7557
350.	1025.	38.2	-116.8	-104.5	-87.4	-67.9	-46.7	26.6	.9011
400.	997.	52.0	-161.6	-147.5	-127.9	-105.7	-81.5	33.7	1.0485
450.	970.	68.8	-216.2	-200.4	-178.3	-153.3	-126.0	42.1	1.2037
500.	946.	87.9	-278.2	-260.6	-236.2	-208.3	-178.1	50.7	1.3592

BALLISTIC COEFFICIENT .33, MUZZLE VELOCITY FT/SEC 1500.

RANGE	VELOCITY	HEIGHT	50 YD	100 YD	150 YD	200 YD	250 YD	DRIFT	TIME
0.	1500.	0.0	-1.5	-1.5	-1.5	-1.5	-1.5	0.0	0.0000
50.	1409.	-.2	0.0	1.4	3.6	6.0	8.8	.6	.1033
100.	1328.	1.4	-2.9	0.0	4.3	9.2	14.6	2.3	.2131
150.	1255.	4.5	-10.8	-6.4	0.0	7.3	15.5	5.2	.3295
200.	1191.	9.1	-24.1	-18.3	-9.8	0.0	10.9	9.2	.4522
250.	1136.	15.5	-43.8	-36.6	-25.9	-13.7	0.0	14.4	.5821
300.	1094.	23.8	-69.9	-61.2	-48.4	-33.7	-17.3	20.4	.7159
350.	1058.	34.4	-103.2	-93.1	-78.1	-61.0	-41.9	27.3	.8552
400.	1028.	47.3	-144.8	-133.2	-116.2	-96.6	-74.8	35.4	1.0009
450.	1000.	62.4	-193.1	-180.1	-160.9	-138.9	-114.3	43.5	1.1471
500.	974.	80.3	-250.4	-235.9	-214.5	-190.1	-162.8	52.7	1.2994

BALLISTIC COEFFICIENT .33, MUZZLE VELOCITY FT/SEC 1600.

RANGE	VELOCITY	HEIGHT	50 YD	100 YD	150 YD	200 YD	250 YD	DRIFT	TIME
0.	1600.	0.0	-1.5	-1.5	-1.5	-1.5	-1.5	0.0	0.0000
50.	1503.	-.3	0.0	1.2	3.0	5.2	7.6	.5	.0965
100.	1412.	1.2	-2.4	0.0	3.7	8.0	12.9	2.1	.1996
150.	1330.	3.8	-9.1	-5.6	0.0	6.5	13.7	4.9	.3092
200.	1258.	7.9	-20.8	-16.1	-8.6	0.0	9.7	8.8	.4252
250.	1192.	13.7	-38.1	-32.2	-22.9	-12.1	0.0	14.0	.5485
300.	1138.	21.3	-61.3	-54.2	-43.1	-30.2	-15.7	20.2	.6771
350.	1095.	30.8	-91.2	-83.0	-70.0	-54.9	-37.9	27.3	.8112
400.	1059.	42.6	-128.3	-118.8	-104.0	-86.7	-67.4	35.3	.9503
450.	1029.	56.9	-173.7	-163.0	-146.3	-126.9	-105.1	44.4	1.0962
500.	1001.	73.3	-225.5	-213.7	-195.1	-173.5	-149.3	53.6	1.2423

BALLISTIC COEFFICIENT .33, MUZZLE VELOCITY FT/SEC 1700.

RANGE	VELOCITY	HEIGHT	50 YD	100 YD	150 YD	200 YD	250 YD	DRIFT	TIME
0.	1700.	0.0	-1.5	-1.5	-1.5	-1.5	-1.5	0.0	0.0000
50.	1598.	-.4	0.0	.9	2.6	4.5	6.6	.5	.0912
100.	1501.	.9	-1.9	0.0	3.2	7.0	11.3	2.0	.1878
150.	1410.	3.3	-7.7	-4.8	0.0	5.7	12.1	4.6	.2910
200.	1329.	7.0	-17.9	-14.1	-7.6	0.0	8.5	8.4	.4008
250.	1256.	12.1	-33.0	-28.3	-20.2	-10.7	0.0	13.3	.5170
300.	1192.	18.9	-53.6	-47.9	-38.2	-26.8	-14.0	19.4	.6396
350.	1137.	27.7	-80.5	-73.9	-62.6	-49.2	-34.3	26.7	.7693
400.	1094.	38.4	-113.8	-106.2	-93.3	-78.0	-61.0	34.7	.9032
450.	1059.	51.4	-154.3	-145.7	-131.2	-114.1	-94.9	43.7	1.0425
500.	1028.	67.0	-203.0	-193.6	-177.4	-158.4	-137.0	53.8	1.1882

BALLISTIC COEFFICIENT .33, MUZZLE VELOCITY FT/SEC 1800.

RANGE	VELOCITY	HEIGHT	50 YD	100 YD	150 YD	200 YD	250 YD	DRIFT	TIME
0.	1800.	0.0	-1.5	-1.5	-1.5	-1.5	-1.5	0.0	0.0000
50.	1693.	-.4	0.0	.8	2.2	3.9	5.7	.5	.0859
100.	1592.	.8	-1.6	0.0	2.8	6.2	9.9	1.9	.1775
150.	1495.	2.9	-6.5	-4.2	0.0	5.0	10.7	4.3	.2745
200.	1405.	6.1	-15.4	-12.3	-6.7	0.0	7.6	7.9	.3781
250.	1324.	10.7	-28.7	-24.9	-17.9	-9.5	0.0	12.6	.4882
300.	1252.	16.8	-47.0	-42.4	-34.0	-23.9	-12.5	18.4	.6048
350.	1188.	24.7	-70.8	-65.4	-55.6	-43.8	-30.6	25.4	.7279
400.	1134.	34.5	-100.8	-94.6	-83.4	-70.0	-54.8	33.5	.8572
450.	1092.	46.5	-137.6	-130.6	-118.0	-102.9	-85.8	42.6	.9922
500.	1057.	60.7	-181.3	-173.6	-159.6	-142.8	-123.9	52.5	1.1318

BALLISTIC COEFFICIENT .33, MUZZLE VELOCITY FT/SEC 1900.

RANGE	VELOCITY	HEIGHT	50 YD	100 YD	150 YD	200 YD	250 YD	DRIFT	TIME
0.	1900.	0.0	-1.5	-1.5	-1.5	-1.5	-1.5	0.0	0.0000
50.	1789.	-.4	0.0	.6	1.8	3.3	5.0	.4	.0815
100.	1682.	.6	-1.2	0.0	2.5	5.4	8.8	1.8	.1679
150.	1581.	2.5	-5.5	-3.7	0.0	4.4	9.5	4.1	.2600
200.	1485.	5.4	-13.3	-10.9	-5.9	0.0	6.8	7.4	.3576
250.	1396.	9.5	-25.1	-22.0	-15.8	-8.5	0.0	11.9	.4621
300.	1316.	15.0	-41.3	-37.7	-30.2	-21.4	-11.2	17.4	.5728
350.	1245.	22.1	-62.6	-58.3	-49.6	-39.3	-27.4	24.2	.6900
400.	1181.	31.0	-89.3	-84.4	-74.5	-62.7	-49.2	32.0	.8137
450.	1129.	42.0	-122.5	-117.0	-105.8	-92.5	-77.3	41.1	.9438
500.	1088.	55.2	-162.3	-156.2	-143.8	-129.0	-112.1	51.0	1.0792

BALLISTIC COEFFICIENT .34, MUZZLE VELOCITY FT/SEC 800.

RANGE	VELOCITY	HEIGHT	50 YD	100 YD	150 YD	200 YD	250 YD	DRIFT	TIME
0.	800.	0.0	-1.5	-1.5	-1.5	-1.5	-1.5	0.0	0.0000
50.	782.	1.0	0.0	6.4	13.5	20.9	28.7	.4	.1898
100.	764.	6.3	-12.7	0.0	14.2	29.1	44.6	1.6	.3841
150.	747.	15.6	-40.4	-21.3	0.0	22.4	45.6	3.6	.5828
200.	730.	28.9	-83.8	-58.3	-29.8	0.0	31.0	6.4	.7862
250.	712.	46.7	-143.4	-111.6	-76.1	-38.8	0.0	10.0	.9943
300.	696.	69.2	-220.2	-182.0	-139.4	-94.6	-48.1	14.5	1.2072
350.	679.	96.7	-314.8	-270.2	-220.5	-168.2	-114.0	19.8	1.4249
400.	663.	129.8	-428.8	-377.8	-321.0	-261.3	-199.3	26.2	1.6490
450.	647.	168.6	-562.6	-505.3	-441.4	-374.2	-304.4	33.6	1.8785
500.	631.	213.6	-717.1	-653.4	-582.4	-507.7	-430.2	42.0	2.1134

BALLISTIC COEFFICIENT .34, MUZZLE VELOCITY FT/SEC 900.

RANGE	VELOCITY	HEIGHT	50 YD	100 YD	150 YD	200 YD	250 YD	DRIFT	TIME
0.	900.	0.0	-1.5	-1.5	-1.5	-1.5	-1.5	0.0	0.0000
50.	879.	.6	0.0	4.9	10.4	16.3	22.4	.5	.1693
100.	859.	4.9	-9.7	0.0	11.1	22.8	35.1	1.5	.3416
150.	840.	12.1	-31.2	-16.6	0.0	17.6	36.0	3.2	.5179
200.	821.	22.7	-65.1	-45.7	-23.5	0.0	24.5	5.6	.6987
250.	803.	36.7	-111.9	-87.7	-60.0	-30.6	0.0	8.9	.8836
300.	785.	54.5	-172.3	-143.1	-109.9	-74.7	-38.0	12.8	1.0726
350.	767.	76.2	-246.7	-212.8	-174.0	-132.9	-90.0	17.5	1.2660
400.	750.	102.1	-336.0	-297.2	-252.9	-205.9	-156.9	23.0	1.4639
450.	733.	132.5	-440.7	-397.0	-347.2	-294.3	-239.3	29.3	1.6663
500.	715.	167.7	-561.5	-513.0	-457.6	-398.9	-337.7	36.4	1.8734

BALLISTIC COEFFICIENT .34, MUZZLE VELOCITY FT/SEC 1000.

RANGE	VELOCITY	HEIGHT	50 YD	100 YD	150 YD	200 YD	250 YD	DRIFT	TIME
0.	1000.	0.0	-1.5	-1.5	-1.5	-1.5	-1.5	0.0	0.0000
50.	974.	.4	0.0	3.8	8.3	13.1	18.1	.4	.1522
100.	950.	3.8	-7.7	0.0	9.0	18.5	28.5	1.4	.3082
150.	927.	9.8	-25.0	-13.5	0.0	14.3	29.4	3.2	.4680
200.	904.	18.4	-52.4	-37.1	-19.1	0.0	20.0	5.6	.6318
250.	883.	30.0	-90.5	-71.4	-48.9	-25.0	0.0	8.8	.7998
300.	863.	44.6	-139.9	-116.9	-89.9	-61.3	-31.2	12.6	.9717
350.	844.	62.4	-200.8	-174.0	-142.6	-109.1	-74.1	17.1	1.1474
400.	825.	83.8	-274.0	-243.4	-207.5	-169.2	-129.2	22.4	1.3273
450.	807.	108.9	-360.0	-325.5	-285.1	-242.1	-197.1	28.4	1.5113
500.	789.	137.9	-459.3	-420.9	-376.0	-328.3	-278.2	35.1	1.6994

BALLISTIC COEFFICIENT .34, MUZZLE VELOCITY FT/SEC 1100.

RANGE	VELOCITY	HEIGHT	50 YD	100 YD	150 YD	200 YD	250 YD	DRIFT	TIME
0.	1100.	0.0	-1.5	-1.5	-1.5	-1.5	-1.5	0.0	0.0000
50.	1064.	.2	0.0	3.0	6.8	10.7	15.0	.7	.1406
100.	1035.	3.1	-6.0	0.0	7.7	15.5	23.9	1.6	.2818
150.	1007.	8.1	-20.4	-11.5	0.0	11.8	24.4	3.8	.4305
200.	981.	15.4	-43.0	-31.0	-15.7	0.0	16.8	6.1	.5799
250.	956.	25.2	-74.8	-59.8	-40.7	-21.0	0.0	9.3	.7346
300.	933.	37.6	-116.2	-98.3	-75.3	-51.7	-26.5	13.3	.8938
350.	910.	52.8	-167.5	-146.6	-119.8	-92.2	-62.8	17.9	1.0565
400.	889.	71.1	-229.4	-205.5	-174.8	-143.4	-109.7	23.3	1.2235
450.	868.	92.6	-302.2	-275.4	-240.9	-205.5	-167.7	29.4	1.3944
500.	849.	117.4	-386.5	-356.7	-318.4	-279.0	-237.0	36.2	1.5691

BALLISTIC COEFFICIENT .34, MUZZLE VELOCITY FT/SEC 1200.

RANGE	VELOCITY	HEIGHT	50 YD	100 YD	150 YD	200 YD	250 YD	DRIFT	TIME
0.	1200.	0.0	-1.5	-1.5	-1.5	-1.5	-1.5	0.0	0.0000
50.	1145.	.0	0.0	2.6	5.8	9.3	13.2	.5	.1280
100.	1102.	2.5	-5.1	0.0	6.5	13.5	21.2	2.1	.2617
150.	1066.	6.9	-17.4	-9.7	0.0	10.6	22.1	4.4	.4001
200.	1036.	13.4	-37.3	-27.1	-14.1	0.0	15.3	7.6	.5430
250.	1008.	22.2	-65.8	-53.0	-36.8	-19.2	0.0	11.8	.6918
300.	982.	33.2	-102.0	-86.6	-67.1	-45.9	-23.0	16.0	.8411
350.	957.	46.8	-147.2	-129.2	-106.5	-81.8	-55.0	21.2	.9953
400.	934.	63.2	-202.1	-181.5	-155.6	-127.4	-96.7	27.1	1.1543
450.	911.	82.5	-266.6	-243.6	-214.4	-182.6	-148.1	33.8	1.3168
500.	890.	104.9	-341.7	-316.1	-283.6	-248.3	-210.0	41.1	1.4836

BALLISTIC COEFFICIENT .34, MUZZLE VELOCITY FT/SEC 1300.

RANGE	VELOCITY	HEIGHT	50 YD	100 YD	150 YD	200 YD	250 YD	DRIFT	TIME
0.	1300.	0.0	-1.5	-1.5	-1.5	-1.5	-1.5	0.0	0.0000
50.	1232.	-.1	0.0	2.2	4.9	8.0	11.4	.7	.1192
100.	1172.	2.1	-4.3	0.0	5.5	11.8	18.5	2.5	.2452
150.	1124.	6.0	-14.8	-8.2	0.0	9.4	19.5	5.0	.3744
200.	1084.	11.8	-32.2	-23.5	-12.5	0.0	13.4	8.6	.5104
250.	1051.	19.6	-57.0	-46.2	-32.4	-16.8	0.0	13.0	.6510
300.	1022.	29.7	-89.8	-76.7	-60.3	-41.5	-21.3	18.3	.7963
350.	996.	42.1	-130.4	-115.2	-96.0	-74.1	-50.6	24.1	.9447
400.	970.	57.3	-180.7	-163.3	-141.3	-116.3	-89.5	31.1	1.0999
450.	946.	74.9	-238.8	-219.3	-194.5	-166.3	-136.2	38.2	1.2556
500.	923.	95.3	-305.8	-284.1	-256.6	-225.3	-191.8	45.8	1.4143

BALLISTIC COEFFICIENT .34, MUZZLE VELOCITY FT/SEC 1400.

RANGE	VELOCITY	HEIGHT	50 YD	100 YD	150 YD	200 YD	250 YD	DRIFT	TIME
0.	1400.	0.0	-1.5	-1.5	-1.5	-1.5	-1.5	0.0	0.0000
50.	1322.	-.2	0.0	1.7	4.2	7.0	9.9	.6	.1103
100.	1252.	1.7	-3.5	0.0	4.9	10.5	16.4	2.2	.2270
150.	1190.	5.1	-12.5	-7.3	0.0	8.4	17.3	5.0	.3500
200.	1137.	10.3	-27.9	-20.9	-11.2	0.0	11.9	9.0	.4798
250.	1095.	17.3	-49.7	-41.0	-28.8	-14.9	0.0	13.7	.6136
300.	1060.	26.5	-78.8	-68.4	-53.7	-37.0	-19.2	19.4	.7528
350.	1031.	38.0	-116.3	-104.1	-87.0	-67.5	-46.7	26.2	.8987
400.	1003.	51.6	-160.6	-146.6	-127.1	-104.8	-81.0	33.0	1.0447
450.	978.	67.8	-213.3	-197.6	-175.7	-150.6	-123.8	40.7	1.1954
500.	953.	86.8	-275.1	-257.7	-233.3	-205.4	-175.7	49.2	1.3509

BALLISTIC COEFFICIENT .34, MUZZLE VELOCITY FT/SEC 1500.

RANGE	VELOCITY	HEIGHT	50 YD	100 YD	150 YD	200 YD	250 YD	DRIFT	TIME
0.	1500.	0.0	-1.5	-1.5	-1.5	-1.5	-1.5	0.0	0.0000
50.	1412.	-.2	0.0	1.4	3.6	6.0	8.7	.6	.1032
100.	1332.	1.4	-2.9	0.0	4.2	9.1	14.4	2.2	.2127
150.	1262.	4.4	-10.7	-6.3	0.0	7.3	15.3	5.0	.3283
200.	1198.	9.0	-24.0	-18.2	-9.8	0.0	10.6	8.9	.4508
250.	1144.	15.3	-43.3	-36.1	-25.5	-13.3	0.0	13.9	.5788
300.	1101.	23.6	-69.2	-60.6	-47.9	-33.3	-17.3	19.8	.7126
350.	1065.	34.2	-102.9	-92.8	-78.1	-61.0	-42.4	27.0	.8533
400.	1035.	46.7	-142.9	-131.4	-114.5	-95.0	-73.7	34.2	.9942
450.	1007.	62.0	-192.0	-179.1	-160.1	-138.1	-114.2	42.8	1.1430
500.	981.	79.4	-248.0	-233.6	-212.5	-188.1	-161.5	51.5	1.2924

BALLISTIC COEFFICIENT .34, MUZZLE VELOCITY FT/SEC 1600.

RANGE	VELOCITY	HEIGHT	50 YD	100 YD	150 YD	200 YD	250 YD	DRIFT	TIME
0.	1600.	0.0	-1.5	-1.5	-1.5	-1.5	-1.5	0.0	0.0000
50.	1506.	-.3	0.0	1.2	3.0	5.2	7.5	.5	.0965
100.	1417.	1.2	-2.4	0.0	3.7	8.0	12.7	2.1	.1993
150.	1337.	3.8	-9.1	-5.5	0.0	6.4	13.5	4.8	.3085
200.	1266.	7.9	-20.6	-15.9	-8.5	0.0	9.5	8.6	.4236
250.	1202.	13.5	-37.6	-31.7	-22.5	-11.9	0.0	13.5	.5454
300.	1147.	21.0	-60.6	-53.6	-42.5	-29.7	-15.5	19.5	.6733
350.	1103.	30.5	-90.3	-82.1	-69.1	-54.2	-37.6	26.5	.8068
400.	1067.	42.1	-127.0	-117.6	-102.8	-85.8	-66.8	34.3	.9451
450.	1037.	56.0	-171.1	-160.5	-143.9	-124.7	-103.4	42.9	1.0877
500.	1009.	72.7	-223.8	-212.1	-193.6	-172.3	-148.6	52.7	1.2367

BALLISTIC COEFFICIENT .34, MUZZLE VELOCITY FT/SEC 1700.

RANGE	VELOCITY	HEIGHT	50 YD	100 YD	150 YD	200 YD	250 YD	DRIFT	TIME
0.	1700.	0.0	-1.5	-1.5	-1.5	-1.5	-1.5	0.0	0.0000
50.	1601.	-.4	0.0	.9	2.5	4.4	6.5	.5	.0911
100.	1507.	.9	-1.9	0.0	3.2	7.0	11.2	1.9	.1875
150.	1418.	3.3	-7.6	-4.8	0.0	5.7	12.0	4.5	.2902
200.	1338.	6.9	-17.7	-14.0	-7.6	0.0	8.4	8.2	.3994
250.	1267.	12.0	-32.7	-28.0	-19.9	-10.5	0.0	12.9	.5145
300.	1202.	18.7	-53.1	-47.4	-37.8	-26.4	-13.8	18.8	.6362
350.	1147.	27.3	-79.4	-72.8	-61.6	-48.4	-33.7	25.8	.7640
400.	1104.	37.9	-112.4	-104.9	-92.1	-76.9	-60.2	33.7	.8975
450.	1068.	50.7	-152.4	-143.9	-129.5	-112.5	-93.6	42.5	1.0357
500.	1037.	65.9	-199.8	-190.4	-174.4	-155.4	-134.5	52.1	1.1783

BALLISTIC COEFFICIENT .34, MUZZLE VELOCITY FT/SEC 1800.

RANGE	VELOCITY	HEIGHT	50 YD	100 YD	150 YD	200 YD	250 YD	DRIFT	TIME
0.	1800.	0.0	-1.5	-1.5	-1.5	-1.5	-1.5	0.0	0.0000
50.	1696.	-.4	0.0	.8	2.2	3.8	5.7	.4	.0858
100.	1598.	.8	-1.5	0.0	2.8	6.1	9.9	1.8	.1771
150.	1504.	2.8	-6.5	-4.2	0.0	5.0	10.6	4.2	.2737
200.	1415.	6.1	-15.3	-12.2	-6.7	0.0	7.5	7.6	.3766
250.	1335.	10.6	-28.5	-24.7	-17.7	-9.4	0.0	12.2	.4861
300.	1264.	16.6	-46.5	-41.9	-33.6	-23.6	-12.3	17.8	.6013
350.	1200.	24.4	-70.0	-64.6	-54.9	-43.3	-30.1	24.6	.7233
400.	1146.	34.0	-99.5	-93.3	-82.2	-68.9	-53.9	32.5	.8513
450.	1102.	45.8	-135.7	-128.7	-116.2	-101.3	-84.3	41.3	.9849
500.	1066.	59.8	-178.8	-171.1	-157.2	-140.6	-121.8	51.0	1.1233

BALLISTIC COEFFICIENT .34, MUZZLE VELOCITY FT/SEC 1900.

RANGE	VELOCITY	HEIGHT	50 YD	100 YD	150 YD	200 YD	250 YD	DRIFT	TIME
0.	1900.	0.0	-1.5	-1.5	-1.5	-1.5	-1.5	0.0	0.0000
50.	1792.	-.4	0.0	.6	1.8	3.3	5.0	.4	.0814
100.	1689.	.6	-1.2	0.0	2.5	5.4	8.7	1.7	.1675
150.	1590.	2.5	-5.5	-3.7	0.0	4.4	9.4	3.9	.2593
200.	1496.	5.3	-13.2	-10.8	-5.9	0.0	6.7	7.1	.3562
250.	1408.	9.4	-24.8	-21.8	-15.7	-8.3	0.0	11.4	.4597
300.	1329.	14.8	-40.9	-37.2	-29.8	-21.1	-11.1	16.9	.5694
350.	1259.	21.8	-61.8	-57.5	-48.9	-38.7	-27.0	23.4	.6854
400.	1196.	30.6	-88.3	-83.4	-73.6	-61.9	-48.6	31.1	.8082
450.	1142.	41.3	-120.6	-115.2	-104.1	-90.9	-76.0	39.7	.9362
500.	1099.	54.2	-159.8	-153.8	-141.5	-126.8	-110.2	49.4	1.0703

BALLISTIC COEFFICIENT .35, MUZZLE VELOCITY FT/SEC 800.

RANGE	VELOCITY	HEIGHT	50 YD	100 YD	150 YD	200 YD	250 YD	DRIFT	TIME
0.	800.	0.0	-1.5	-1.5	-1.5	-1.5	-1.5	0.0	0.0000
50.	783.	1.0	0.0	6.4	13.4	20.9	28.6	.4	.1898
100.	765.	6.3	-12.7	0.0	14.2	29.0	44.5	1.6	.3839
150.	748.	15.5	-40.3	-21.3	0.0	22.3	45.4	3.5	.5823
200.	731.	28.8	-83.5	-58.1	-29.7	0.0	30.8	6.2	.7852
250.	715.	46.5	-143.0	-111.1	-75.7	-38.5	0.0	9.7	.9925
300.	699.	68.9	-219.3	-181.1	-138.6	-94.0	-47.7	14.0	1.2044
350.	682.	96.3	-313.7	-269.1	-219.5	-167.5	-113.6	19.2	1.4217
400.	667.	129.1	-427.2	-376.2	-319.6	-260.1	-198.4	25.5	1.6448
450.	651.	167.6	-560.0	-502.7	-438.9	-372.0	-302.7	32.6	1.8727
500.	635.	212.0	-713.0	-649.3	-578.5	-504.1	-427.1	40.6	2.1055

BALLISTIC COEFFICIENT .35, MUZZLE VELOCITY FT/SEC 900.

RANGE	VELOCITY	HEIGHT	50 YD	100 YD	150 YD	200 YD	250 YD	DRIFT	TIME
0.	900.	0.0	-1.5	-1.5	-1.5	-1.5	-1.5	0.0	0.0000
50.	880.	.6	0.0	4.9	10.4	16.3	22.4	.4	.1690
100.	860.	4.8	-9.7	0.0	11.1	22.8	35.0	1.4	.3413
150.	842.	12.1	-31.2	-16.6	0.0	17.6	35.8	3.1	.5176
200.	824.	22.6	-65.0	-45.6	-23.4	0.0	24.4	5.5	.6980
250.	806.	36.6	-111.8	-87.5	-59.7	-30.5	0.0	8.6	.8824
300.	788.	54.3	-171.9	-142.7	-109.5	-74.3	-37.8	12.4	1.0707
350.	771.	75.8	-246.0	-212.0	-173.2	-132.2	-89.5	17.0	1.2631
400.	754.	101.5	-334.7	-295.8	-251.5	-204.6	-155.9	22.3	1.4598
450.	737.	131.7	-438.6	-394.9	-345.0	-292.3	-237.4	28.3	1.6609
500.	720.	166.5	-558.5	-509.9	-454.4	-395.9	-334.9	35.2	1.8665

BALLISTIC COEFFICIENT .35, MUZZLE VELOCITY FT/SEC 1000.

RANGE	VELOCITY	HEIGHT	50 YD	100 YD	150 YD	200 YD	250 YD	DRIFT	TIME
0.	1000.	0.0	-1.5	-1.5	-1.5	-1.5	-1.5	0.0	0.0000
50.	975.	.4	0.0	3.8	8.3	13.1	18.1	.4	.1521
100.	951.	3.8	-7.7	0.0	9.0	18.5	28.5	1.4	.3079
150.	929.	9.7	-24.9	-13.4	0.0	14.3	29.3	3.1	.4676
200.	907.	18.4	-52.3	-37.0	-19.0	0.0	20.0	5.4	.6310
250.	886.	29.9	-90.3	-71.2	-48.8	-24.9	0.0	8.5	.7985
300.	867.	44.4	-139.4	-116.5	-89.6	-61.0	-31.1	12.3	.9698
350.	848.	62.2	-200.1	-173.3	-141.9	-108.6	-73.7	16.7	1.1448
400.	829.	83.4	-272.9	-242.3	-206.4	-168.3	-128.4	21.8	1.3238
450.	812.	108.2	-358.3	-323.9	-283.5	-240.7	-195.8	27.6	1.5067
500.	794.	136.9	-456.8	-418.5	-373.7	-326.1	-276.2	34.1	1.6935

BALLISTIC COEFFICIENT .35, MUZZLE VELOCITY FT/SEC 1100.

RANGE	VELOCITY	HEIGHT	50 YD	100 YD	150 YD	200 YD	250 YD	DRIFT	TIME
0.	1100.	0.0	-1.5	-1.5	-1.5	-1.5	-1.5	0.0	0.0000
50.	1065.	.2	0.0	3.0	6.8	10.7	14.9	.8	.1410
100.	1036.	3.1	-5.9	0.0	7.7	15.6	23.8	1.5	.2815
150.	1009.	8.1	-20.4	-11.6	0.0	11.8	24.2	3.8	.4305
200.	984.	15.4	-42.9	-31.1	-15.7	0.0	16.5	6.1	.5799
250.	959.	25.0	-74.3	-59.6	-40.3	-20.6	0.0	9.0	.7330
300.	936.	37.4	-115.5	-97.8	-74.7	-51.1	-26.4	12.9	.8916
350.	914.	52.5	-166.5	-145.9	-118.9	-91.4	-62.5	17.4	1.0536
400.	894.	70.7	-228.0	-204.4	-173.6	-142.1	-109.1	22.7	1.2197
450.	873.	92.0	-300.4	-273.8	-239.1	-203.7	-166.6	28.6	1.3898
500.	854.	116.6	-384.0	-354.5	-316.0	-276.6	-235.4	35.2	1.5634

BALLISTIC COEFFICIENT .35, MUZZLE VELOCITY FT/SEC 1200.

RANGE	VELOCITY	HEIGHT	50 YD	100 YD	150 YD	200 YD	250 YD	DRIFT	TIME
0.	1200.	0.0	-1.5	-1.5	-1.5	-1.5	-1.5	0.0	0.0000
50.	1147.	.0	0.0	2.6	5.8	9.3	13.2	.5	.1280
100.	1104.	2.5	-5.1	0.0	6.5	13.5	21.2	2.0	.2614
150.	1069.	6.9	-17.4	-9.7	0.0	10.5	22.1	4.3	.3995
200.	1039.	13.3	-37.2	-26.9	-14.0	0.0	15.5	7.3	.5418
250.	1012.	22.2	-65.8	-53.0	-36.9	-19.3	0.0	11.6	.6912
300.	986.	33.1	-101.9	-86.5	-67.1	-46.1	-22.9	15.9	.8402
350.	962.	46.6	-146.6	-128.7	-106.1	-81.5	-54.5	20.7	.9928
400.	939.	62.8	-200.9	-180.4	-154.6	-126.5	-95.6	26.5	1.1503
450.	917.	81.9	-265.0	-242.0	-212.9	-181.4	-146.6	32.9	1.3119
500.	895.	104.3	-340.2	-314.7	-282.4	-247.3	-208.7	40.3	1.4792

BALLISTIC COEFFICIENT .35, MUZZLE VELOCITY FT/SEC 1300.

RANGE	VELOCITY	HEIGHT	50 YD	100 YD	150 YD	200 YD	250 YD	DRIFT	TIME
0.	1300.	0.0	-1.5	-1.5	-1.5	-1.5	-1.5	0.0	0.0000
50.	1234.	-.1	0.0	2.1	4.9	8.0	11.3	.6	.1190
100.	1175.	2.1	-4.3	0.0	5.5	11.8	18.4	2.4	.2442
150.	1127.	6.0	-14.7	-8.3	0.0	9.3	19.3	4.9	.3737
200.	1088.	11.7	-32.1	-23.5	-12.5	0.0	13.3	8.4	.5092
250.	1056.	19.5	-56.7	-46.0	-32.2	-16.6	0.0	12.7	.6492
300.	1027.	29.6	-89.6	-76.8	-60.2	-41.5	-21.5	18.1	.7950
350.	1001.	41.8	-129.7	-114.7	-95.4	-73.5	-50.3	23.6	.9415
400.	976.	56.6	-178.6	-161.6	-139.4	-114.5	-87.9	30.0	1.0933
450.	952.	74.1	-236.5	-217.3	-192.4	-164.4	-134.5	37.1	1.2490
500.	929.	94.5	-303.8	-282.5	-254.8	-223.7	-190.4	44.8	1.4086

BALLISTIC COEFFICIENT .35, MUZZLE VELOCITY FT/SEC 1400.

RANGE	VELOCITY	HEIGHT	50 YD	100 YD	150 YD	200 YD	250 YD	DRIFT	TIME
0.	1400.	0.0	-1.5	-1.5	-1.5	-1.5	-1.5	0.0	0.0000
50.	1324.	-.2	0.0	1.7	4.2	6.9	9.9	.5	.1103
100.	1256.	1.7	-3.5	0.0	4.9	10.3	16.3	2.2	.2268
150.	1195.	5.1	-12.5	-7.3	0.0	8.1	17.1	5.0	.3497
200.	1143.	10.2	-27.6	-20.6	-10.9	0.0	11.9	8.6	.4776
250.	1101.	17.2	-49.4	-40.6	-28.5	-14.9	0.0	13.3	.6115
300.	1066.	26.3	-78.3	-67.8	-53.2	-36.9	-19.0	18.9	.7500
350.	1037.	37.5	-114.7	-102.5	-85.5	-66.5	-45.6	25.1	.8928
400.	1009.	51.4	-159.9	-146.0	-126.5	-104.8	-80.9	32.5	1.0419
450.	984.	67.4	-212.1	-196.4	-174.5	-150.1	-123.3	39.9	1.1912
500.	960.	86.0	-272.7	-255.3	-230.9	-203.8	-174.0	48.0	1.3440

BALLISTIC COEFFICIENT .35, MUZZLE VELOCITY FT/SEC 1500.

RANGE	VELOCITY	HEIGHT	50 YD	100 YD	150 YD	200 YD	250 YD	DRIFT	TIME
0.	1500.	0.0	-1.5	-1.5	-1.5	-1.5	-1.5	0.0	0.0000
50.	1414.	-.2	0.0	1.4	3.5	5.9	8.6	.5	.1031
100.	1337.	1.4	-2.9	0.0	4.2	9.0	14.3	2.2	.2125
150.	1267.	4.4	-10.6	-6.3	0.0	7.2	15.2	4.9	.3276
200.	1205.	8.9	-23.8	-18.0	-9.6	0.0	10.6	8.7	.4492
250.	1151.	15.2	-42.9	-35.7	-25.3	-13.2	0.0	13.5	.5767
300.	1108.	23.4	-68.7	-60.1	-47.5	-33.1	-17.2	19.3	.7097
350.	1072.	33.7	-101.5	-91.4	-76.7	-59.9	-41.4	25.9	.8474
400.	1041.	46.3	-141.8	-130.3	-113.5	-94.2	-73.1	33.4	.9898
450.	1014.	61.2	-189.7	-176.7	-157.9	-136.2	-112.4	41.5	1.1357
500.	988.	78.8	-246.6	-232.2	-211.3	-187.2	-160.7	50.6	1.2876

BALLISTIC COEFFICIENT .35, MUZZLE VELOCITY FT/SEC 1600.

RANGE	VELOCITY	HEIGHT	50 YD	100 YD	150 YD	200 YD	250 YD	DRIFT	TIME
0.	1600.	0.0	-1.5	-1.5	-1.5	-1.5	-1.5	0.0	0.0000
50.	1508.	-.3	0.0	1.2	3.0	5.1	7.5	.5	.0964
100.	1422.	1.1	-2.3	0.0	3.7	7.9	12.6	2.0	.1988
150.	1344.	3.8	-9.0	-5.5	0.0	6.4	13.4	4.6	.3076
200.	1273.	7.8	-20.5	-15.9	-8.5	0.0	9.3	8.4	.4226
250.	1211.	13.4	-37.3	-31.5	-22.3	-11.7	0.0	13.1	.5431
300.	1155.	20.9	-60.4	-53.4	-42.4	-29.6	-15.6	19.2	.6713
350.	1111.	30.2	-89.4	-81.3	-68.4	-53.5	-37.2	25.8	.8027
400.	1075.	41.7	-125.7	-116.4	-101.7	-84.6	-66.0	33.4	.9399
450.	1044.	55.5	-169.8	-159.3	-142.8	-123.6	-102.6	42.1	1.0829
500.	1017.	71.6	-220.6	-209.0	-190.7	-169.3	-146.0	51.0	1.2274

BALLISTIC COEFFICIENT .35, MUZZLE VELOCITY FT/SEC 1700.

RANGE	VELOCITY	HEIGHT	50 YD	100 YD	150 YD	200 YD	250 YD	DRIFT	TIME
0.	1700.	0.0	-1.5	-1.5	-1.5	-1.5	-1.5	0.0	0.0000
50.	1604.	-.4	0.0	.9	2.5	4.4	6.5	.5	.0910
100.	1512.	.9	-1.9	0.0	3.2	6.9	11.1	1.9	.1873
150.	1425.	3.3	-7.6	-4.8	0.0	5.6	11.9	4.4	.2895
200.	1347.	6.9	-17.6	-13.8	-7.5	0.0	8.3	7.9	.3979
250.	1276.	11.9	-32.4	-27.7	-19.8	-10.4	0.0	12.6	.5125
300.	1213.	18.5	-52.6	-47.0	-37.5	-26.2	-13.7	18.3	.6334
350.	1157.	27.0	-78.8	-72.2	-61.1	-47.9	-33.3	25.1	.7603
400.	1113.	37.4	-111.2	-103.7	-91.0	-76.0	-59.3	32.8	.8921
450.	1076.	50.1	-150.6	-142.2	-127.9	-111.0	-92.2	41.4	1.0291
500.	1045.	65.2	-198.1	-188.7	-172.8	-154.1	-133.2	51.0	1.1724

BALLISTIC COEFFICIENT .35, MUZZLE VELOCITY FT/SEC 1800.

RANGE	VELOCITY	HEIGHT	50 YD	100 YD	150 YD	200 YD	250 YD	DRIFT	TIME
0.	1800.	0.0	-1.5	-1.5	-1.5	-1.5	-1.5	0.0	0.0000
50.	1699.	-.4	0.0	.8	2.1	3.8	5.6	.4	.0857
100.	1603.	.7	-1.5	0.0	2.8	6.1	9.8	1.8	.1768
150.	1511.	2.8	-6.4	-4.1	0.0	4.9	10.5	4.1	.2730
200.	1424.	6.0	-15.2	-12.1	-6.6	0.0	7.4	7.4	.3753
250.	1346.	10.5	-28.2	-24.4	-17.5	-9.3	0.0	11.8	.4838
300.	1276.	16.4	-46.1	-41.5	-33.2	-23.4	-12.2	17.3	.5985
350.	1212.	24.1	-69.4	-64.0	-54.3	-42.8	-29.8	23.9	.7194
400.	1157.	33.6	-98.5	-92.4	-81.4	-68.2	-53.3	31.6	.8465
450.	1113.	45.2	-134.0	-127.1	-114.6	-99.8	-83.1	40.1	.9781
500.	1076.	58.9	-176.4	-168.8	-155.0	-138.5	-120.0	49.6	1.1152

BALLISTIC COEFFICIENT .35, MUZZLE VELOCITY FT/SEC 1900.

RANGE	VELOCITY	HEIGHT	50 YD	100 YD	150 YD	200 YD	250 YD	DRIFT	TIME
0.	1900.	0.0	-1.5	-1.5	-1.5	-1.5	-1.5	0.0	0.0000
50.	1795.	-.4	0.0	.6	1.8	3.3	4.9	.4	.0813
100.	1694.	.6	-1.2	0.0	2.4	5.3	8.6	1.6	.1672
150.	1598.	2.5	-5.5	-3.7	0.0	4.4	9.3	3.8	.2585
200.	1507.	5.3	-13.1	-10.7	-5.8	0.0	6.6	6.9	.3550
250.	1420.	9.3	-24.6	-21.6	-15.5	-8.2	0.0	11.1	.4575
300.	1342.	14.6	-40.4	-36.8	-29.5	-20.8	-10.9	16.3	.5664
350.	1273.	21.5	-61.1	-56.9	-48.3	-38.2	-26.7	22.6	.6812
400.	1210.	30.1	-87.1	-82.3	-72.5	-60.9	-47.8	30.0	.8022
450.	1154.	40.8	-119.4	-114.0	-103.1	-90.0	-75.2	38.7	.9307
500.	1111.	53.4	-157.6	-151.5	-139.3	-124.8	-108.4	48.0	1.0619

BALLISTIC COEFFICIENT .36, MUZZLE VELOCITY FT/SEC 800.

RANGE	VELOCITY	HEIGHT	50 YD	100 YD	150 YD	200 YD	250 YD	DRIFT	TIME
0.	800.	0.0	-1.5	-1.5	-1.5	-1.5	-1.5	0.0	0.0000
50.	783.	1.0	0.0	6.4	13.4	20.8	28.5	.4	.1898
100.	766.	6.3	-12.7	0.0	14.1	28.9	44.3	1.5	.3837
150.	750.	15.5	-40.3	-21.2	0.0	22.2	45.2	3.4	.5818
200.	733.	28.8	-83.3	-57.9	-29.6	0.0	30.7	6.0	.7841
250.	717.	46.4	-142.5	-110.7	-75.3	-38.3	0.0	9.4	.9907
300.	701.	68.6	-218.6	-180.4	-138.0	-93.6	-47.6	13.6	1.2020
350.	686.	95.9	-312.7	-268.2	-218.8	-167.0	-113.3	18.7	1.4189
400.	670.	128.4	-425.4	-374.6	-318.0	-258.8	-197.5	24.7	1.6406
450.	655.	166.5	-557.3	-500.0	-436.4	-369.8	-300.8	31.6	1.8669
500.	640.	210.5	-708.8	-645.2	-574.5	-500.5	-423.9	39.2	2.0977

BALLISTIC COEFFICIENT .36, MUZZLE VELOCITY FT/SEC 900.

RANGE	VELOCITY	HEIGHT	50 YD	100 YD	150 YD	200 YD	250 YD	DRIFT	TIME
0.	900.	0.0	-1.5	-1.5	-1.5	-1.5	-1.5	0.0	0.0000
50.	880.	.6	0.0	4.9	10.4	16.2	22.3	.4	.1688
100.	861.	4.8	-9.7	0.0	11.1	22.7	34.9	1.4	.3412
150.	843.	12.1	-31.2	-16.6	0.0	17.5	35.7	3.0	.5173
200.	826.	22.6	-65.0	-45.5	-23.4	0.0	24.3	5.4	.6973
250.	808.	36.5	-111.5	-87.2	-59.5	-30.3	0.0	8.4	.8811
300.	791.	54.1	-171.4	-142.2	-109.0	-74.0	-37.6	12.1	1.0687
350.	774.	75.5	-245.1	-211.0	-172.3	-131.4	-89.0	16.5	1.2603
400.	758.	101.0	-333.2	-294.3	-250.0	-203.3	-154.8	21.6	1.4558
450.	741.	130.9	-436.6	-392.8	-343.0	-290.4	-235.8	27.4	1.6559
500.	725.	165.5	-556.2	-507.5	-452.2	-393.8	-333.1	34.2	1.8610

BALLISTIC COEFFICIENT .36, MUZZLE VELOCITY FT/SEC 1000.

RANGE	VELOCITY	HEIGHT	50 YD	100 YD	150 YD	200 YD	250 YD	DRIFT	TIME
0.	1000.	0.0	-1.5	-1.5	-1.5	-1.5	-1.5	0.0	0.0000
50.	976.	.4	0.0	3.8	8.3	13.0	18.0	.4	.1521
100.	953.	3.8	-7.6	0.0	9.0	18.4	28.4	1.4	.3077
150.	930.	9.7	-24.9	-13.4	0.0	14.2	29.1	3.0	.4671
200.	909.	18.3	-52.2	-36.9	-19.0	0.0	19.9	5.3	.6301
250.	889.	29.8	-90.0	-71.0	-48.6	-24.8	0.0	8.3	.7972
300.	870.	44.2	-139.0	-116.1	-89.2	-60.7	-30.9	12.0	.9679
350.	851.	61.9	-199.4	-172.6	-141.3	-108.1	-73.3	16.2	1.1423
400.	833.	82.9	-271.7	-241.2	-205.4	-167.4	-127.6	21.2	1.3203
450.	816.	107.6	-356.5	-322.2	-281.9	-239.2	-194.5	26.8	1.5021
500.	799.	136.0	-454.3	-416.1	-371.3	-323.9	-274.2	33.0	1.6877

BALLISTIC COEFFICIENT .36, MUZZLE VELOCITY FT/SEC 1100.

RANGE	VELOCITY	HEIGHT	50 YD	100 YD	150 YD	200 YD	250 YD	DRIFT	TIME
0.	1100.	0.0	-1.5	-1.5	-1.5	-1.5	-1.5	0.0	0.0000
50.	1067.	.2	0.0	3.1	6.9	10.9	15.0	.4	.1386
100.	1038.	3.0	-6.1	0.0	7.8	15.6	23.8	1.5	.2812
150.	1011.	8.1	-20.8	-11.6	0.0	11.8	24.1	3.8	.4306
200.	986.	15.4	-43.5	-31.2	-15.7	0.0	16.5	6.0	.5798
250.	963.	25.0	-74.9	-59.6	-40.2	-20.6	0.0	8.9	.7325
300.	940.	37.2	-115.8	-97.4	-74.2	-50.6	-25.9	12.6	.8896
350.	919.	52.2	-166.6	-145.1	-118.0	-90.5	-61.7	16.9	1.0507
400.	898.	70.3	-228.1	-203.6	-172.6	-141.1	-108.2	22.2	1.2168
450.	878.	91.3	-299.8	-272.2	-237.3	-201.9	-164.9	27.8	1.3851
500.	859.	115.8	-383.1	-352.4	-313.6	-274.3	-233.2	34.2	1.5580

BALLISTIC COEFFICIENT .36, MUZZLE VELOCITY FT/SEC 1200.

RANGE	VELOCITY	HEIGHT	50 YD	100 YD	150 YD	200 YD	250 YD	DRIFT	TIME
0.	1200.	0.0	-1.5	-1.5	-1.5	-1.5	-1.5	0.0	0.0000
50.	1148.	.0	0.0	2.6	5.8	9.3	13.0	.5	.1279
100.	1106.	2.5	-5.1	0.0	6.4	13.5	20.8	2.0	.2612
150.	1072.	6.9	-17.3	-9.6	0.0	10.6	21.6	4.2	.3989
200.	1042.	13.3	-37.2	-27.0	-14.1	0.0	14.7	7.3	.5416
250.	1016.	21.9	-64.8	-52.1	-36.0	-18.4	0.0	10.9	.6869
300.	990.	33.1	-101.8	-86.4	-67.2	-46.0	-24.0	15.7	.8393
350.	966.	46.4	-146.4	-128.5	-106.0	-81.3	-55.6	20.5	.9915
400.	944.	62.5	-200.3	-179.9	-154.2	-126.0	-96.6	26.1	1.1480
450.	922.	81.3	-263.5	-240.5	-211.6	-179.8	-146.8	32.1	1.3074
500.	901.	103.2	-337.2	-311.7	-279.6	-244.3	-207.6	39.0	1.4719

BALLISTIC COEFFICIENT .36, MUZZLE VELOCITY FT/SEC 1300.

RANGE	VELOCITY	HEIGHT	50 YD	100 YD	150 YD	200 YD	250 YD	DRIFT	TIME
0.	1300.	0.0	-1.5	-1.5	-1.5	-1.5	-1.5	0.0	0.0000
50.	1236.	-.1	0.0	2.1	4.9	8.0	11.3	.6	.1189
100.	1178.	2.1	-4.2	0.0	5.6	11.8	18.4	2.2	.2432
150.	1131.	5.9	-14.6	-8.4	0.0	9.3	19.2	4.7	.3730
200.	1092.	11.6	-31.9	-23.5	-12.4	0.0	13.2	8.2	.5081
250.	1060.	19.4	-56.4	-45.9	-32.0	-16.5	0.0	12.4	.6473
300.	1031.	29.5	-89.4	-76.8	-60.1	-41.5	-21.7	17.8	.7937
350.	1006.	41.6	-129.4	-114.7	-95.2	-73.5	-50.4	23.3	.9399
400.	981.	56.2	-177.6	-160.9	-138.6	-113.8	-87.4	29.3	1.0897
450.	958.	73.5	-234.9	-216.1	-191.0	-163.1	-133.4	36.2	1.2440
500.	935.	93.7	-301.8	-280.8	-253.0	-222.0	-189.0	43.8	1.4028

BALLISTIC COEFFICIENT .36, MUZZLE VELOCITY FT/SEC 1400.

RANGE	VELOCITY	HEIGHT	50 YD	100 YD	150 YD	200 YD	250 YD	DRIFT	TIME
0.	1400.	0.0	-1.5	-1.5	-1.5	-1.5	-1.5	0.0	0.0000
50.	1326.	-.2	0.0	1.7	4.1	6.9	9.8	.5	.1102
100.	1260.	1.7	-3.4	0.0	4.8	10.3	16.2	2.1	.2261
150.	1200.	5.1	-12.4	-7.2	0.0	8.2	17.0	4.8	.3485
200.	1148.	10.1	-27.4	-20.5	-10.9	0.0	11.8	8.4	.4764
250.	1106.	17.1	-49.1	-40.5	-28.4	-14.8	0.0	13.0	.6097
300.	1072.	26.1	-77.8	-67.4	-52.9	-36.6	-18.9	18.4	.7474
350.	1042.	37.3	-114.1	-102.1	-85.1	-66.1	-45.4	24.6	.8900
400.	1016.	50.7	-158.0	-144.2	-124.9	-103.1	-79.5	31.4	1.0355
450.	990.	67.0	-211.2	-195.7	-174.0	-149.5	-122.9	39.3	1.1878
500.	966.	85.4	-271.6	-254.4	-230.2	-203.0	-173.4	47.3	1.3401

BALLISTIC COEFFICIENT .36, MUZZLE VELOCITY FT/SEC 1500.

RANGE	VELOCITY	HEIGHT	50 YD	100 YD	150 YD	200 YD	250 YD	DRIFT	TIME
0.	1500.	0.0	-1.5	-1.5	-1.5	-1.5	-1.5	0.0	0.0000
50.	1416.	-.2	0.0	1.4	3.5	5.9	8.5	.6	.1031
100.	1341.	1.4	-2.8	0.0	4.2	9.0	14.3	2.1	.2120
150.	1273.	4.4	-10.6	-6.3	0.0	7.1	15.1	4.8	.3273
200.	1212.	8.9	-23.6	-17.9	-9.5	0.0	10.6	8.4	.4477
250.	1157.	15.1	-42.7	-35.6	-25.1	-13.2	0.0	13.2	.5752
300.	1114.	23.2	-68.2	-59.6	-47.0	-32.7	-16.9	18.8	.7068
350.	1078.	33.4	-100.6	-90.6	-75.9	-59.3	-40.7	25.3	.8436
400.	1047.	46.1	-141.3	-130.0	-113.1	-94.1	-73.0	33.0	.9876
450.	1021.	60.6	-188.0	-175.2	-156.3	-134.9	-111.1	40.5	1.1302
500.	996.	77.8	-243.5	-229.3	-208.2	-184.5	-158.0	49.1	1.2790

BALLISTIC COEFFICIENT .36, MUZZLE VELOCITY FT/SEC 1600.

RANGE	VELOCITY	HEIGHT	50 YD	100 YD	150 YD	200 YD	250 YD	DRIFT	TIME
0.	1600.	0.0	-1.5	-1.5	-1.5	-1.5	-1.5	0.0	0.0000
50.	1511.	-.3	0.0	1.2	3.0	5.1	7.4	.5	.0964
100.	1426.	1.1	-2.3	0.0	3.6	7.8	12.5	1.9	.1986
150.	1350.	3.8	-8.9	-5.5	0.0	6.3	13.3	4.5	.3069
200.	1281.	7.8	-20.3	-15.7	-8.4	0.0	9.3	8.1	.4209
250.	1219.	13.3	-37.0	-31.2	-22.1	-11.6	0.0	12.7	.5409
300.	1164.	20.6	-59.6	-52.6	-41.7	-29.1	-15.2	18.4	.6670
350.	1119.	29.9	-88.7	-80.5	-67.8	-53.1	-36.8	25.1	.7990
400.	1082.	41.3	-124.7	-115.4	-100.8	-84.0	-65.5	32.7	.9358
450.	1051.	54.8	-167.6	-157.2	-140.8	-121.9	-101.0	40.8	1.0757
500.	1024.	70.8	-218.7	-207.1	-188.9	-167.9	-144.7	50.0	1.2213

BALLISTIC COEFFICIENT .36, MUZZLE VELOCITY FT/SEC 1700.

RANGE	VELOCITY	HEIGHT	50 YD	100 YD	150 YD	200 YD	250 YD	DRIFT	TIME
0.	1700.	0.0	-1.5	-1.5	-1.5	-1.5	-1.5	0.0	0.0000
50.	1606.	-.4	0.0	.9	2.5	4.4	6.4	.5	.0909
100.	1517.	.9	-1.9	0.0	3.2	6.9	11.0	1.8	.1870
150.	1432.	3.3	-7.5	-4.7	0.0	5.6	11.7	4.2	.2888
200.	1355.	6.8	-17.5	-13.8	-7.5	0.0	8.2	7.7	.3968
250.	1286.	11.7	-32.2	-27.5	-19.6	-10.2	0.0	12.2	.5102
300.	1223.	18.3	-52.1	-46.4	-37.0	-25.8	-13.5	17.7	.6298
350.	1167.	26.7	-77.8	-71.3	-60.2	-47.2	-32.8	24.3	.7556
400.	1122.	37.0	-109.9	-102.5	-89.8	-74.9	-58.5	31.8	.8867
450.	1085.	49.5	-148.9	-140.5	-126.3	-109.5	-91.1	40.2	1.0228
500.	1053.	64.2	-195.2	-185.8	-170.0	-151.3	-130.8	49.4	1.1632

BALLISTIC COEFFICIENT .36, MUZZLE VELOCITY FT/SEC 1800.

RANGE	VELOCITY	HEIGHT	50 YD	100 YD	150 YD	200 YD	250 YD	DRIFT	TIME
0.	1800.	0.0	-1.5	-1.5	-1.5	-1.5	-1.5	0.0	0.0000
50.	1702.	-.4	0.0	.8	2.1	3.8	5.6	.4	.0856
100.	1608.	.7	-1.5	0.0	2.7	6.0	9.7	1.7	.1764
150.	1519.	2.8	-6.4	-4.1	0.0	4.9	10.4	3.9	.2724
200.	1434.	6.0	-15.1	-12.0	-6.5	0.0	7.3	7.2	.3741
250.	1356.	10.4	-28.0	-24.2	-17.4	-9.2	0.0	11.5	.4819
300.	1287.	16.3	-45.6	-41.1	-32.8	-23.0	-12.0	16.8	.5952
350.	1224.	23.8	-68.5	-63.2	-53.6	-42.1	-29.3	23.1	.7147
400.	1168.	33.1	-97.2	-91.1	-80.1	-67.0	-52.3	30.6	.8404
450.	1123.	44.5	-132.3	-125.4	-113.1	-98.3	-81.8	39.0	.9714
500.	1085.	58.1	-174.2	-166.6	-152.8	-136.5	-118.1	48.2	1.1074

BALLISTIC COEFFICIENT .36, MUZZLE VELOCITY FT/SEC 1900.

RANGE	VELOCITY	HEIGHT	50 YD	100 YD	150 YD	200 YD	250 YD	DRIFT	TIME
0.	1900.	0.0	-1.5	-1.5	-1.5	-1.5	-1.5	0.0	0.0000
50.	1798.	-.4	0.0	.6	1.8	3.3	4.9	.4	.0812
100.	1700.	.6	-1.2	0.0	2.4	5.3	8.6	1.6	.1669
150.	1606.	2.4	-5.4	-3.6	0.0	4.3	9.2	3.7	.2579
200.	1517.	5.3	-13.0	-10.6	-5.8	0.0	6.5	6.7	.3539
250.	1432.	9.2	-24.4	-21.4	-15.4	-8.1	0.0	10.7	.4558
300.	1355.	14.5	-40.1	-36.5	-29.3	-20.6	-10.8	15.9	.5638
350.	1285.	21.3	-60.4	-56.3	-47.8	-37.7	-26.3	21.9	.6772
400.	1223.	29.7	-86.0	-81.3	-71.6	-60.0	-47.0	29.1	.7968
450.	1167.	40.1	-117.5	-112.1	-101.2	-88.2	-73.5	37.3	.9226
500.	1122.	52.5	-155.3	-149.4	-137.2	-122.8	-106.5	46.5	1.0537

BALLISTIC COEFFICIENT .37, MUZZLE VELOCITY FT/SEC 800.

RANGE	VELOCITY	HEIGHT	50 YD	100 YD	150 YD	200 YD	250 YD	DRIFT	TIME
0.	800.	0.0	-1.5	-1.5	-1.5	-1.5	-1.5	0.0	0.0000
50.	783.	1.0	0.0	6.3	13.4	20.8	28.4	.4	.1898
100.	767.	6.3	-12.7	0.0	14.1	28.8	44.1	1.5	.3836
150.	751.	15.5	-40.2	-21.1	0.0	22.1	45.0	3.3	.5813
200.	735.	28.7	-83.1	-57.7	-29.5	0.0	30.5	5.8	.7831
250.	719.	46.2	-141.9	-110.2	-75.0	-38.1	0.0	9.0	.9888
300.	704.	68.4	-218.0	-179.9	-137.7	-93.4	-47.7	13.2	1.2002
350.	689.	95.5	-311.7	-267.3	-218.0	-166.4	-113.0	18.2	1.4160
400.	673.	127.8	-423.7	-372.9	-316.5	-257.5	-196.6	24.0	1.6363
450.	658.	165.5	-554.4	-497.3	-433.9	-367.5	-299.0	30.6	1.8611
500.	644.	209.3	-705.8	-642.3	-571.8	-498.1	-421.9	38.1	2.0917

BALLISTIC COEFFICIENT .37, MUZZLE VELOCITY FT/SEC 900.

RANGE	VELOCITY	HEIGHT	50 YD	100 YD	150 YD	200 YD	250 YD	DRIFT	TIME
0.	900.	0.0	-1.5	-1.5	-1.5	-1.5	-1.5	0.0	0.0000
50.	881.	.6	0.0	4.9	10.4	16.2	22.2	.4	.1688
100.	862.	4.8	-9.7	0.0	11.0	22.7	34.8	1.4	.3411
150.	845.	12.1	-31.2	-16.6	0.0	17.5	35.6	3.0	.5170
200.	828.	22.5	-64.8	-45.4	-23.3	0.0	24.2	5.3	.6965
250.	811.	36.4	-111.2	-86.9	-59.3	-30.2	0.0	8.2	.8798
300.	794.	53.9	-170.9	-141.7	-108.5	-73.6	-37.4	11.8	1.0668
350.	777.	75.1	-244.2	-210.1	-171.4	-130.7	-88.4	16.0	1.2574
400.	761.	100.5	-332.0	-293.1	-248.9	-202.3	-154.0	21.0	1.4524
450.	745.	130.2	-435.0	-391.2	-341.5	-289.1	-234.8	26.7	1.6518
500.	729.	164.5	-553.7	-505.1	-449.9	-391.6	-331.3	33.3	1.8556

BALLISTIC COEFFICIENT .37, MUZZLE VELOCITY FT/SEC 1000.

RANGE	VELOCITY	HEIGHT	50 YD	100 YD	150 YD	200 YD	250 YD	DRIFT	TIME
0.	1000.	0.0	-1.5	-1.5	-1.5	-1.5	-1.5	0.0	0.0000
50.	976.	.4	0.0	3.8	8.3	13.0	18.0	.4	.1520
100.	954.	3.8	-7.6	0.0	8.9	18.4	28.3	1.3	.3075
150.	932.	9.7	-24.8	-13.4	0.0	14.2	29.0	2.9	.4667
200.	912.	18.3	-52.0	-36.8	-18.9	0.0	19.8	5.2	.6293
250.	892.	29.7	-89.8	-70.7	-48.4	-24.7	0.0	8.1	.7959
300.	873.	44.0	-138.5	-115.7	-88.8	-60.5	-30.8	11.6	.9661
350.	855.	61.6	-198.6	-171.9	-140.7	-107.6	-72.9	15.8	1.1397
400.	837.	82.5	-270.5	-240.0	-204.3	-166.4	-126.9	20.6	1.3168
450.	820.	106.9	-354.8	-320.5	-280.3	-237.7	-193.2	26.0	1.4976
500.	803.	135.2	-452.3	-414.2	-369.5	-322.2	-272.7	32.2	1.6828

BALLISTIC COEFFICIENT .37, MUZZLE VELOCITY FT/SEC 1100.

RANGE	VELOCITY	HEIGHT	50 YD	100 YD	150 YD	200 YD	250 YD	DRIFT	TIME
0.	1100.	0.0	-1.5	-1.5	-1.5	-1.5	-1.5	0.0	0.0000
50.	1067.	.2	0.0	3.1	7.0	10.9	15.0	.4	.1385
100.	1039.	3.0	-6.1	0.0	7.8	15.7	23.9	1.4	.2809
150.	1013.	8.2	-20.9	-11.7	0.0	11.8	24.1	3.8	.4307
200.	989.	15.4	-43.5	-31.3	-15.7	0.0	16.4	6.0	.5797
250.	966.	25.0	-75.0	-59.7	-40.2	-20.6	0.0	8.9	.7323
300.	944.	37.2	-115.8	-97.5	-74.1	-50.5	-25.8	12.5	.8890
350.	923.	52.0	-166.0	-144.6	-117.3	-89.8	-61.0	16.5	1.0484
400.	902.	69.8	-226.8	-202.4	-171.2	-139.7	-106.8	21.4	1.2128
450.	883.	90.8	-298.4	-270.9	-235.8	-200.4	-163.4	27.1	1.3811
500.	865.	115.0	-381.1	-350.6	-311.6	-272.3	-231.1	33.3	1.5530

BALLISTIC COEFFICIENT .37, MUZZLE VELOCITY FT/SEC 1200.

RANGE	VELOCITY	HEIGHT	50 YD	100 YD	150 YD	200 YD	250 YD	DRIFT	TIME
0.	1200.	0.0	-1.5	-1.5	-1.5	-1.5	-1.5	0.0	0.0000
50.	1149.	.0	0.0	2.5	5.7	9.3	12.9	.5	.1278
100.	1108.	2.5	-5.1	0.0	6.4	13.6	20.7	1.9	.2609
150.	1074.	6.9	-17.2	-9.6	0.0	10.7	21.5	4.1	.3983
200.	1045.	13.3	-37.3	-27.1	-14.3	0.0	14.4	7.4	.5418
250.	1019.	21.8	-64.6	-51.8	-35.8	-17.9	0.0	10.6	.6854
300.	995.	32.7	-100.5	-85.3	-66.1	-44.6	-23.1	14.9	.8346
350.	971.	46.3	-146.2	-128.4	-106.0	-81.0	-55.9	20.3	.9903
400.	949.	62.0	-198.9	-178.6	-153.0	-124.4	-95.7	25.3	1.1436
450.	927.	80.8	-262.2	-239.3	-210.5	-178.3	-146.0	31.4	1.3034
500.	907.	102.5	-335.5	-310.0	-278.0	-242.2	-206.4	38.2	1.4669

BALLISTIC COEFFICIENT .37, MUZZLE VELOCITY FT/SEC 1300.

RANGE	VELOCITY	HEIGHT	50 YD	100 YD	150 YD	200 YD	250 YD	DRIFT	TIME
0.	1300.	0.0	-1.5	-1.5	-1.5	-1.5	-1.5	0.0	0.0000
50.	1238.	-.1	0.0	2.1	4.9	7.9	11.3	.6	.1187
100.	1181.	2.1	-4.2	0.0	5.6	11.7	18.5	2.1	.2425
150.	1134.	5.9	-14.6	-8.4	0.0	9.3	19.3	4.6	.3723
200.	1096.	11.6	-31.8	-23.5	-12.3	0.0	13.5	8.0	.5069
250.	1063.	19.4	-56.6	-46.2	-32.2	-16.8	0.0	12.4	.6475
300.	1036.	29.1	-88.2	-75.7	-59.0	-40.5	-20.3	17.0	.7888
350.	1010.	41.5	-129.1	-114.6	-95.1	-73.5	-49.9	23.0	.9382
400.	986.	56.0	-177.3	-160.6	-138.3	-113.7	-86.8	29.0	1.0876
450.	963.	73.1	-234.0	-215.2	-190.2	-162.4	-132.1	35.6	1.2405
500.	941.	93.0	-300.0	-279.2	-251.4	-220.5	-186.9	42.9	1.3976

BALLISTIC COEFFICIENT .37, MUZZLE VELOCITY FT/SEC 1400.

RANGE	VELOCITY	HEIGHT	50 YD	100 YD	150 YD	200 YD	250 YD	DRIFT	TIME
0.	1400.	0.0	-1.5	-1.5	-1.5	-1.5	-1.5	0.0	0.0000
50.	1328.	-.2	0.0	1.7	4.1	6.8	9.8	.5	.1101
100.	1263.	1.7	-3.4	0.0	4.8	10.2	16.1	2.0	.2259
150.	1204.	5.1	-12.3	-7.2	0.0	8.1	16.9	4.6	.3477
200.	1153.	10.1	-27.3	-20.4	-10.8	0.0	11.8	8.2	.4751
250.	1111.	17.0	-48.8	-40.2	-28.2	-14.7	0.0	12.7	.6078
300.	1077.	25.9	-77.2	-66.9	-52.5	-36.3	-18.7	18.0	.7448
350.	1047.	37.2	-114.0	-102.0	-85.1	-66.2	-45.7	24.4	.8889
400.	1021.	50.3	-156.9	-143.2	-124.0	-102.4	-78.9	30.7	1.0316
450.	996.	66.1	-208.6	-193.2	-171.6	-147.3	-120.8	38.0	1.1802
500.	973.	84.5	-268.8	-251.7	-227.7	-200.7	-171.3	46.0	1.3326

BALLISTIC COEFFICIENT .37, MUZZLE VELOCITY FT/SEC 1500.

RANGE	VELOCITY	HEIGHT	50 YD	100 YD	150 YD	200 YD	250 YD	DRIFT	TIME
0.	1500.	0.0	-1.5	-1.5	-1.5	-1.5	-1.5	0.0	0.0000
50.	1418.	-.2	0.0	1.4	3.5	5.9	8.5	.5	.1030
100.	1345.	1.4	-2.8	0.0	4.2	8.9	14.1	2.1	.2118
150.	1278.	4.4	-10.5	-6.2	0.0	7.1	14.9	4.6	.3263
200.	1218.	8.8	-23.5	-17.8	-9.5	0.0	10.4	8.2	.4465
250.	1164.	15.0	-42.3	-35.2	-24.8	-13.0	0.0	12.8	.5725
300.	1120.	23.0	-67.6	-59.1	-46.6	-32.4	-16.8	18.3	.7040
350.	1085.	33.1	-99.8	-89.9	-75.3	-58.7	-40.6	24.7	.8401
400.	1054.	45.4	-139.3	-127.9	-111.3	-92.3	-71.6	31.8	.9805
450.	1027.	60.2	-187.1	-174.3	-155.6	-134.2	-110.9	39.9	1.1265
500.	1002.	77.1	-241.7	-227.4	-206.6	-182.9	-157.0	48.1	1.2732

BALLISTIC COEFFICIENT .37, MUZZLE VELOCITY FT/SEC 1600.

RANGE	VELOCITY	HEIGHT	50 YD	100 YD	150 YD	200 YD	250 YD	DRIFT	TIME
0.	1600.	0.0	-1.5	-1.5	-1.5	-1.5	-1.5	0.0	0.0000
50.	1513.	-.3	0.0	1.2	3.0	5.0	7.4	.4	.0963
100.	1431.	1.1	-2.3	0.0	3.6	7.8	12.4	1.9	.1983
150.	1356.	3.8	-8.9	-5.4	0.0	6.2	13.1	4.4	.3063
200.	1288.	7.7	-20.2	-15.6	-8.3	0.0	9.2	7.8	.4196
250.	1227.	13.2	-36.8	-31.0	-21.9	-11.5	0.0	12.4	.5390
300.	1172.	20.4	-59.1	-52.2	-41.3	-28.8	-15.0	17.9	.6642
350.	1127.	29.6	-87.8	-79.7	-67.0	-52.4	-36.3	24.4	.7949
400.	1090.	40.8	-123.3	-114.1	-99.5	-82.9	-64.5	31.7	.9304
450.	1058.	54.2	-165.9	-155.5	-139.2	-120.5	-99.8	39.8	1.0699
500.	1031.	70.3	-217.2	-205.7	-187.5	-166.7	-143.7	49.1	1.2163

BALLISTIC COEFFICIENT .37, MUZZLE VELOCITY FT/SEC 1700.

RANGE	VELOCITY	HEIGHT	50 YD	100 YD	150 YD	200 YD	250 YD	DRIFT	TIME
0.	1700.	0.0	-1.5	-1.5	-1.5	-1.5	-1.5	0.0	0.0000
50.	1609.	-.4	0.0	.9	2.5	4.3	6.4	.5	.0909
100.	1522.	.9	-1.9	0.0	3.1	6.8	10.9	1.8	.1867
150.	1439.	3.2	-7.5	-4.7	0.0	5.5	11.7	4.1	.2883
200.	1363.	6.8	-17.4	-13.6	-7.3	0.0	8.2	7.5	.3953
250.	1294.	11.7	-32.0	-27.3	-19.4	-10.2	0.0	11.8	.5085
300.	1233.	18.2	-51.8	-46.2	-36.7	-25.7	-13.4	17.3	.6276
350.	1177.	26.4	-77.2	-70.7	-59.7	-46.8	-32.5	23.7	.7523
400.	1131.	36.6	-108.8	-101.4	-88.8	-74.1	-57.7	31.0	.8818
450.	1093.	48.9	-147.4	-139.0	-124.8	-108.3	-89.8	39.2	1.0169
500.	1061.	63.5	-193.1	-183.8	-168.0	-149.7	-129.2	48.3	1.1565

BALLISTIC COEFFICIENT .37, MUZZLE VELOCITY FT/SEC 1800.

RANGE	VELOCITY	HEIGHT	50 YD	100 YD	150 YD	200 YD	250 YD	DRIFT	TIME
0.	1800.	0.0	-1.5	-1.5	-1.5	-1.5	-1.5	0.0	0.0000
50.	1705.	-.4	0.0	.8	2.1	3.7	5.6	.4	.0856
100.	1613.	.7	-1.5	0.0	2.7	6.0	9.6	1.7	.1761
150.	1526.	2.8	-6.4	-4.1	0.0	4.9	10.3	3.8	.2718
200.	1443.	5.9	-15.0	-12.0	-6.5	0.0	7.2	7.0	.3730
250.	1366.	10.3	-27.8	-24.0	-17.2	-9.0	0.0	11.1	.4798
300.	1298.	16.1	-45.2	-40.7	-32.5	-22.8	-11.9	16.3	.5925
350.	1236.	23.5	-67.9	-62.7	-53.1	-41.7	-29.1	22.5	.7113
400.	1179.	32.7	-96.1	-90.1	-79.2	-66.2	-51.7	29.7	.8353
450.	1133.	44.0	-130.7	-123.9	-111.6	-97.0	-80.7	37.9	.9652
500.	1095.	57.3	-172.1	-164.5	-150.8	-134.6	-116.5	46.9	1.1000

BALLISTIC COEFFICIENT .37, MUZZLE VELOCITY FT/SEC 1900.

RANGE	VELOCITY	HEIGHT	50 YD	100 YD	150 YD	200 YD	250 YD	DRIFT	TIME
0.	1900.	0.0	-1.5	-1.5	-1.5	-1.5	-1.5	0.0	0.0000
50.	1800.	-.4	0.0	.6	1.8	3.2	4.8	.4	.0811
100.	1705.	.6	-1.2	0.0	2.4	5.3	8.5	1.5	.1666
150.	1614.	2.4	-5.4	-3.6	0.0	4.3	9.2	3.6	.2572
200.	1526.	5.2	-12.9	-10.6	-5.8	0.0	6.5	6.5	.3528
250.	1443.	9.1	-24.2	-21.3	-15.3	-8.1	0.0	10.4	.4540
300.	1367.	14.3	-39.7	-36.1	-28.9	-20.3	-10.6	15.3	.5608
350.	1298.	21.0	-59.8	-55.7	-47.3	-37.2	-25.9	21.3	.6734
400.	1236.	29.4	-85.2	-80.5	-70.8	-59.3	-46.4	28.3	.7923
450.	1180.	39.5	-116.0	-110.7	-99.9	-86.9	-72.4	36.2	.9162
500.	1133.	51.8	-153.3	-147.4	-135.3	-121.0	-104.8	45.2	1.0461

BALLISTIC COEFFICIENT .38, MUZZLE VELOCITY FT/SEC 800.

RANGE	VELOCITY	HEIGHT	50 YD	100 YD	150 YD	200 YD	250 YD	DRIFT	TIME
0.	800.	0.0	-1.5	-1.5	-1.5	-1.5	-1.5	0.0	0.0000
50.	784.	1.0	0.0	6.3	13.4	20.7	28.3	.4	.1898
100.	768.	6.3	-12.7	0.0	14.1	28.7	44.0	1.5	.3834
150.	752.	15.4	-40.1	-21.1	0.0	22.0	44.9	3.2	.5808
200.	737.	28.6	-82.8	-57.5	-29.4	0.0	30.5	5.6	.7820
250.	721.	46.1	-141.6	-109.9	-74.8	-38.1	0.0	8.8	.9875
300.	706.	68.2	-217.5	-179.5	-137.3	-93.2	-47.6	12.9	1.1983
350.	691.	95.1	-310.7	-266.3	-217.1	-165.8	-112.5	17.7	1.4132
400.	677.	127.1	-421.9	-371.1	-314.9	-256.2	-195.3	23.3	1.6321
450.	662.	164.6	-552.2	-495.1	-431.9	-365.8	-297.3	29.7	1.8562
500.	648.	208.1	-702.9	-639.5	-569.2	-495.8	-419.7	37.1	2.0859

BALLISTIC COEFFICIENT .38, MUZZLE VELOCITY FT/SEC 900.

RANGE	VELOCITY	HEIGHT	50 YD	100 YD	150 YD	200 YD	250 YD	DRIFT	TIME
0.	900.	0.0	-1.5	-1.5	-1.5	-1.5	-1.5	0.0	0.0000
50.	881.	.6	0.0	4.9	10.4	16.2	22.2	.4	.1688
100.	863.	4.8	-9.7	0.0	11.0	22.6	34.7	1.3	.3410
150.	846.	12.1	-31.1	-16.5	0.0	17.4	35.4	2.9	.5166
200.	829.	22.5	-64.7	-45.3	-23.2	0.0	24.1	5.1	.6958
250.	813.	36.3	-111.0	-86.6	-59.1	-30.1	0.0	8.0	.8786
300.	797.	53.7	-170.3	-141.1	-108.1	-73.2	-37.2	11.4	1.0648
350.	780.	74.8	-243.3	-209.2	-170.7	-130.0	-87.9	15.5	1.2548
400.	765.	100.1	-330.9	-292.0	-247.9	-201.5	-153.4	20.4	1.4493
450.	749.	129.6	-433.4	-389.6	-340.0	-287.8	-233.7	26.0	1.6478
500.	733.	163.6	-551.2	-502.6	-447.5	-389.5	-329.3	32.3	1.8502

BALLISTIC COEFFICIENT .38, MUZZLE VELOCITY FT/SEC 1000.

RANGE	VELOCITY	HEIGHT	50 YD	100 YD	150 YD	200 YD	250 YD	DRIFT	TIME
0.	1000.	0.0	-1.5	-1.5	-1.5	-1.5	-1.5	0.0	0.0000
50.	977.	.4	0.0	3.8	8.3	13.0	18.0	.3	.1519
100.	955.	3.8	-7.6	0.0	8.9	18.3	28.5	1.3	.3072
150.	934.	9.7	-24.8	-13.4	0.0	14.1	29.3	2.9	.4663
200.	914.	18.2	-51.9	-36.7	-18.8	0.0	20.3	5.0	.6285
250.	894.	29.7	-90.2	-71.2	-48.9	-25.3	0.0	8.3	.7969
300.	876.	43.9	-138.0	-115.2	-88.5	-60.2	-29.8	11.3	.9642
350.	858.	61.3	-197.8	-171.2	-140.0	-107.0	-71.6	15.3	1.1372
400.	841.	82.1	-269.4	-239.0	-203.3	-165.6	-125.1	20.0	1.3135
450.	824.	106.4	-353.5	-319.3	-279.1	-236.7	-191.2	25.4	1.4941
500.	808.	134.4	-450.4	-412.4	-367.8	-320.6	-270.0	31.4	1.6782

BALLISTIC COEFFICIENT .38, MUZZLE VELOCITY FT/SEC 1100.

RANGE	VELOCITY	HEIGHT	50 YD	100 YD	150 YD	200 YD	250 YD	DRIFT	TIME
0.	1100.	0.0	-1.5	-1.5	-1.5	-1.5	-1.5	0.0	0.0000
50.	1068.	.2	0.0	3.1	6.8	10.9	15.0	.4	.1385
100.	1040.	3.0	-6.1	0.0	7.4	15.7	23.9	1.4	.2808
150.	1016.	8.0	-20.3	-11.1	0.0	12.4	24.7	3.2	.4270
200.	991.	15.4	-43.6	-31.4	-16.5	0.0	16.4	6.0	.5797
250.	969.	25.0	-75.0	-59.7	-41.2	-20.5	0.0	8.8	.7321
300.	947.	37.1	-115.8	-97.5	-75.2	-50.4	-25.8	12.4	.8885
350.	927.	51.8	-165.4	-144.1	-118.1	-89.2	-60.4	16.1	1.0462
400.	907.	69.5	-225.9	-201.5	-171.8	-138.8	-105.9	20.9	1.2099
450.	888.	90.3	-297.1	-269.6	-236.2	-199.1	-162.1	26.4	1.3773
500.	869.	114.3	-379.3	-348.8	-311.6	-270.4	-229.3	32.5	1.5483

BALLISTIC COEFFICIENT .38, MUZZLE VELOCITY FT/SEC 1200.

RANGE	VELOCITY	HEIGHT	50 YD	100 YD	150 YD	200 YD	250 YD	DRIFT	TIME
0.	1200.	0.0	-1.5	-1.5	-1.5	-1.5	-1.5	0.0	0.0000
50.	1151.	.0	0.0	2.5	5.7	9.3	12.9	.5	.1278
100.	1110.	2.5	-5.1	0.0	6.4	13.6	20.7	1.9	.2607
150.	1077.	6.8	-17.2	-9.6	0.0	10.9	21.5	4.0	.3978
200.	1048.	13.4	-37.4	-27.2	-14.5	0.0	14.2	7.4	.5421
250.	1022.	21.8	-64.5	-51.8	-35.8	-17.7	0.0	10.5	.6847
300.	998.	32.5	-100.1	-84.9	-65.7	-44.0	-22.7	14.5	.8326
350.	975.	45.8	-144.5	-126.7	-104.4	-79.1	-54.3	19.3	.9847
400.	953.	61.7	-198.0	-177.7	-152.2	-123.2	-94.8	24.7	1.1403
450.	932.	80.3	-260.9	-238.1	-209.4	-176.8	-144.9	30.7	1.2995
500.	912.	101.9	-333.6	-308.3	-276.4	-240.2	-204.7	37.3	1.4620

BALLISTIC COEFFICIENT .38, MUZZLE VELOCITY FT/SEC 1300.

RANGE	VELOCITY	HEIGHT	50 YD	100 YD	150 YD	200 YD	250 YD	DRIFT	TIME
0.	1300.	0.0	-1.5	-1.5	-1.5	-1.5	-1.5	0.0	0.0000
50.	1239.	-.1	0.0	2.1	4.9	7.9	11.2	.6	.1185
100.	1184.	2.1	-4.2	0.0	5.6	11.7	18.2	2.0	.2423
150.	1137.	5.9	-14.6	-8.4	0.0	9.1	18.9	4.6	.3723
200.	1100.	11.5	-31.7	-23.3	-12.1	0.0	13.1	7.8	.5058
250.	1068.	19.2	-55.9	-45.5	-31.5	-16.4	0.0	11.9	.6443
300.	1040.	28.9	-87.8	-75.3	-58.5	-40.3	-20.6	16.6	.7865
350.	1015.	41.0	-127.6	-113.1	-93.5	-72.3	-49.3	22.0	.9329
400.	991.	55.8	-176.8	-160.2	-137.8	-113.5	-87.3	28.6	1.0855
450.	968.	72.8	-233.4	-214.7	-189.5	-162.2	-132.7	35.1	1.2380
500.	947.	92.6	-299.2	-278.4	-250.4	-220.0	-187.3	42.3	1.3944

BALLISTIC COEFFICIENT .38, MUZZLE VELOCITY FT/SEC 1400.

RANGE	VELOCITY	HEIGHT	50 YD	100 YD	150 YD	200 YD	250 YD	DRIFT	TIME
0.	1400.	0.0	-1.5	-1.5	-1.5	-1.5	-1.5	0.0	0.0000
50.	1330.	-.2	0.0	1.7	4.1	6.8	9.7	.5	.1101
100.	1267.	1.7	-3.4	0.0	4.8	10.2	16.0	2.0	.2256
150.	1209.	5.0	-12.3	-7.2	0.0	8.1	16.8	4.5	.3470
200.	1158.	10.1	-27.2	-20.4	-10.8	0.0	11.6	8.1	.4745
250.	1116.	16.9	-48.5	-39.9	-28.0	-14.5	0.0	12.4	.6060
300.	1082.	25.8	-76.9	-66.7	-52.3	-36.1	-18.7	17.6	.7431
350.	1053.	36.7	-112.4	-100.5	-83.8	-64.8	-44.6	23.4	.8831
400.	1026.	50.1	-156.4	-142.7	-123.6	-101.9	-78.8	30.3	1.0292
450.	1002.	65.6	-207.4	-192.0	-170.5	-146.1	-120.1	37.3	1.1761
500.	979.	83.8	-266.8	-249.8	-225.9	-198.8	-169.9	45.0	1.3269

BALLISTIC COEFFICIENT .38, MUZZLE VELOCITY FT/SEC 1500.

RANGE	VELOCITY	HEIGHT	50 YD	100 YD	150 YD	200 YD	250 YD	DRIFT	TIME
0.	1500.	0.0	-1.5	-1.5	-1.5	-1.5	-1.5	0.0	0.0000
50.	1420.	-.2	0.0	1.4	3.5	5.8	8.4	.5	.1028
100.	1348.	1.4	-2.8	0.0	4.1	8.8	14.0	2.0	.2115
150.	1283.	4.3	-10.4	-6.2	0.0	7.1	14.8	4.5	.3255
200.	1224.	8.8	-23.3	-17.7	-9.4	0.0	10.3	7.9	.4452
250.	1171.	14.9	-42.1	-35.0	-24.7	-12.9	0.0	12.4	.5707
300.	1127.	22.9	-67.2	-58.7	-46.3	-32.2	-16.7	17.8	.7014
350.	1091.	32.9	-99.1	-89.2	-74.8	-58.3	-40.2	24.1	.8368
400.	1060.	45.0	-138.2	-126.8	-110.3	-91.5	-70.8	31.0	.9761
450.	1033.	59.4	-185.0	-172.2	-153.6	-132.4	-109.2	38.7	1.1198
500.	1008.	76.6	-240.4	-226.3	-205.6	-182.1	-156.3	47.4	1.2691

BALLISTIC COEFFICIENT .38, MUZZLE VELOCITY FT/SEC 1600.

RANGE	VELOCITY	HEIGHT	50 YD	100 YD	150 YD	200 YD	250 YD	DRIFT	TIME
0.	1600.	0.0	-1.5	-1.5	-1.5	-1.5	-1.5	0.0	0.0000
50.	1515.	-.3	0.0	1.2	2.9	5.0	7.3	.4	.0962
100.	1435.	1.1	-2.3	0.0	3.6	7.8	12.3	1.9	.1981
150.	1361.	3.7	-8.8	-5.4	0.0	6.2	13.1	4.3	.3054
200.	1295.	7.7	-20.1	-15.5	-8.3	0.0	9.2	7.7	.4186
250.	1235.	13.1	-36.6	-30.8	-21.9	-11.5	0.0	12.1	.5375
300.	1180.	20.2	-58.6	-51.7	-40.9	-28.4	-14.7	17.4	.6613
350.	1134.	29.3	-87.0	-79.0	-66.4	-51.8	-35.8	23.8	.7912
400.	1097.	40.4	-122.2	-113.0	-98.6	-82.0	-63.6	30.9	.9258
450.	1065.	53.9	-165.3	-154.9	-138.8	-120.0	-99.4	39.3	1.0670
500.	1038.	69.2	-214.1	-202.5	-184.6	-163.8	-140.9	47.5	1.2071

BALLISTIC COEFFICIENT .38, MUZZLE VELOCITY FT/SEC 1700.

RANGE	VELOCITY	HEIGHT	50 YD	100 YD	150 YD	200 YD	250 YD	DRIFT	TIME
0.	1700.	0.0	-1.5	-1.5	-1.5	-1.5	-1.5	0.0	0.0000
50.	1611.	-.4	0.0	.9	2.5	4.3	6.3	.4	.0908
100.	1526.	.9	-1.9	0.0	3.1	6.8	10.8	1.8	.1864
150.	1445.	3.2	-7.5	-4.7	0.0	5.5	11.5	4.0	.2876
200.	1370.	6.7	-17.3	-13.5	-7.3	0.0	8.1	7.2	.3941
250.	1303.	11.6	-31.7	-27.0	-19.2	-10.1	0.0	11.5	.5063
300.	1242.	18.0	-51.2	-45.7	-36.3	-25.4	-13.3	16.7	.6244
350.	1187.	26.1	-76.4	-69.9	-59.0	-46.2	-32.1	22.9	.7480
400.	1140.	36.2	-107.7	-100.3	-87.8	-73.2	-57.1	30.1	.8770
450.	1101.	48.3	-145.8	-137.4	-123.4	-107.0	-88.8	38.2	1.0110
500.	1069.	62.7	-190.9	-181.6	-166.0	-147.8	-127.6	47.0	1.1493

BALLISTIC COEFFICIENT .38, MUZZLE VELOCITY FT/SEC 1800.

RANGE	VELOCITY	HEIGHT	50 YD	100 YD	150 YD	200 YD	250 YD	DRIFT	TIME
0.	1800.	0.0	-1.5	-1.5	-1.5	-1.5	-1.5	0.0	0.0000
50.	1707.	-.4	0.0	.7	2.1	3.7	5.5	.4	.0855
100.	1618.	.7	-1.5	0.0	2.7	5.9	9.5	1.6	.1758
150.	1533.	2.8	-6.3	-4.1	0.0	4.8	10.2	3.7	.2711
200.	1451.	5.9	-14.9	-11.9	-6.4	0.0	7.2	6.8	.3719
250.	1376.	10.2	-27.6	-23.8	-17.0	-9.0	0.0	10.8	.4781
300.	1308.	15.9	-44.8	-40.4	-32.2	-22.5	-11.7	15.8	.5897
350.	1247.	23.3	-67.2	-62.0	-52.5	-41.2	-28.6	21.8	.7074
400.	1191.	32.4	-95.2	-89.2	-78.3	-65.4	-51.0	28.8	.8306
450.	1143.	43.4	-129.2	-122.5	-110.2	-95.7	-79.6	36.8	.9591
500.	1104.	56.6	-170.0	-162.5	-148.9	-132.8	-114.8	45.7	1.0928

BALLISTIC COEFFICIENT .38, MUZZLE VELOCITY FT/SEC 1900.

RANGE	VELOCITY	HEIGHT	50 YD	100 YD	150 YD	200 YD	250 YD	DRIFT	TIME
0.	1900.	0.0	-1.5	-1.5	-1.5	-1.5	-1.5	0.0	0.0000
50.	1803.	-.4	0.0	.6	1.8	3.2	4.8	.4	.0811
100.	1710.	.6	-1.2	0.0	2.4	5.2	8.4	1.5	.1664
150.	1621.	2.4	-5.3	-3.6	0.0	4.3	9.1	3.5	.2565
200.	1535.	5.2	-12.8	-10.5	-5.7	0.0	6.4	6.3	.3517
250.	1454.	9.1	-24.0	-21.1	-15.1	-8.0	0.0	10.1	.4523
300.	1378.	14.2	-39.3	-35.8	-28.7	-20.1	-10.5	14.9	.5582
350.	1310.	20.8	-59.2	-55.1	-46.8	-36.8	-25.6	20.6	.6698
400.	1249.	29.0	-84.2	-79.5	-70.0	-58.5	-45.8	27.4	.7872
450.	1192.	39.1	-114.9	-109.6	-98.9	-86.0	-71.7	35.3	.9110
500.	1144.	51.0	-151.3	-145.4	-133.5	-119.2	-103.2	43.9	1.0387

BALLISTIC COEFFICIENT .39, MUZZLE VELOCITY FT/SEC 800.

RANGE	VELOCITY	HEIGHT	50 YD	100 YD	150 YD	200 YD	250 YD	DRIFT	TIME
0.	800.	0.0	-1.5	-1.5	-1.5	-1.5	-1.5	0.0	0.0000
50.	784.	1.0	0.0	6.3	13.3	20.6	28.3	.4	.1898
100.	769.	6.3	-12.7	0.0	14.0	28.6	43.9	1.5	.3833
150.	753.	15.4	-40.0	-21.0	0.0	21.9	44.8	3.1	.5803
200.	738.	28.5	-82.6	-57.3	-29.2	0.0	30.5	5.4	.7809
250.	723.	46.0	-141.4	-109.7	-74.7	-38.1	0.0	8.6	.9865
300.	709.	68.0	-217.0	-179.0	-136.9	-93.1	-47.3	12.6	1.1965
350.	694.	94.7	-309.7	-265.3	-216.3	-165.1	-111.8	17.2	1.4103
400.	680.	126.5	-420.0	-369.3	-313.3	-254.8	-193.8	22.5	1.6279
450.	665.	163.9	-550.3	-493.3	-430.3	-364.5	-295.9	28.9	1.8520
500.	651.	206.9	-699.9	-636.6	-566.5	-493.5	-417.2	36.1	2.0801

BALLISTIC COEFFICIENT .39, MUZZLE VELOCITY FT/SEC 900.

RANGE	VELOCITY	HEIGHT	50 YD	100 YD	150 YD	200 YD	250 YD	DRIFT	TIME
0.	900.	0.0	-1.5	-1.5	-1.5	-1.5	-1.5	0.0	0.0000
50.	882.	.6	0.0	4.9	10.4	16.1	22.1	.4	.1688
100.	864.	4.8	-9.7	0.0	11.0	22.6	34.5	1.3	.3409
150.	847.	12.0	-31.1	-16.5	0.0	17.3	35.3	2.9	.5163
200.	831.	22.4	-64.6	-45.1	-23.1	0.0	23.9	5.0	.6951
250.	815.	36.2	-110.6	-86.4	-58.8	-29.9	0.0	7.7	.8773
300.	799.	53.5	-169.7	-140.6	-107.6	-72.9	-37.0	11.1	1.0629
350.	783.	74.6	-242.7	-208.7	-170.1	-129.7	-87.8	15.1	1.2527
400.	768.	99.7	-329.9	-291.0	-247.0	-200.7	-152.8	19.9	1.4465
450.	753.	129.0	-431.7	-388.0	-338.5	-286.4	-232.6	25.3	1.6438
500.	737.	162.6	-548.7	-500.1	-445.1	-387.2	-327.4	31.4	1.8448

BALLISTIC COEFFICIENT .39, MUZZLE VELOCITY FT/SEC 1000.

RANGE	VELOCITY	HEIGHT	50 YD	100 YD	150 YD	200 YD	250 YD	DRIFT	TIME
0.	1000.	0.0	-1.5	-1.5	-1.5	-1.5	-1.5	0.0	0.0000
50.	978.	.4	0.0	3.8	8.2	12.9	17.9	.3	.1519
100.	956.	3.8	-7.6	0.0	8.9	18.3	28.3	1.2	.3070
150.	935.	9.7	-24.7	-13.4	0.0	14.1	29.0	2.8	.4658
200.	916.	18.2	-51.8	-36.6	-18.8	0.0	19.9	4.9	.6277
250.	897.	29.6	-89.6	-70.7	-48.4	-24.9	0.0	7.9	.7947
300.	879.	43.7	-137.6	-114.8	-88.1	-59.9	-30.0	11.0	.9624
350.	861.	61.1	-197.2	-170.6	-139.4	-106.6	-71.7	14.9	1.1348
400.	844.	81.7	-268.6	-238.2	-202.6	-165.1	-125.2	19.5	1.3110
450.	828.	105.9	-352.2	-318.1	-278.0	-235.8	-190.9	24.7	1.4906
500.	812.	133.7	-448.5	-410.5	-366.0	-319.0	-269.2	30.6	1.6736

BALLISTIC COEFFICIENT .39, MUZZLE VELOCITY FT/SEC 1100.

RANGE	VELOCITY	HEIGHT	50 YD	100 YD	150 YD	200 YD	250 YD	DRIFT	TIME
0.	1100.	0.0	-1.5	-1.5	-1.5	-1.5	-1.5	0.0	0.0000
50.	1069.	.2	0.0	3.1	6.7	10.9	15.0	.4	.1385
100.	1042.	3.0	-6.2	0.0	7.3	15.6	23.9	1.5	.2812
150.	1017.	8.0	-20.2	-11.0	0.0	12.5	24.8	3.1	.4265
200.	994.	15.4	-43.6	-31.3	-16.6	0.0	16.4	6.0	.5796
250.	971.	25.0	-75.1	-59.7	-41.3	-20.5	0.0	8.8	.7319
300.	951.	36.8	-114.7	-96.2	-74.2	-49.3	-24.6	11.7	.8845
350.	930.	51.6	-164.9	-143.3	-117.6	-88.6	-59.8	15.8	1.0441
400.	911.	69.2	-225.1	-200.4	-171.1	-137.8	-105.0	20.4	1.2070
450.	892.	89.8	-295.8	-268.1	-235.1	-197.7	-160.7	25.8	1.3736
500.	874.	113.6	-377.5	-346.7	-310.0	-268.4	-227.4	31.7	1.5437

BALLISTIC COEFFICIENT .39, MUZZLE VELOCITY FT/SEC 1200.

RANGE	VELOCITY	HEIGHT	50 YD	100 YD	150 YD	200 YD	250 YD	DRIFT	TIME
0.	1200.	0.0	-1.5	-1.5	-1.5	-1.5	-1.5	0.0	0.0000
50.	1152.	.0	0.0	2.5	5.7	9.2	12.9	.5	.1277
100.	1112.	2.5	-5.1	0.0	6.3	13.3	20.7	1.8	.2604
150.	1079.	6.8	-17.1	-9.5	0.0	10.4	21.6	3.9	.3972
200.	1051.	13.2	-36.7	-26.6	-13.9	0.0	14.9	6.7	.5382
250.	1026.	21.7	-64.5	-51.8	-35.9	-18.6	0.0	10.4	.6842
300.	1002.	32.4	-99.9	-84.7	-65.6	-44.8	-22.5	14.3	.8313
350.	979.	45.6	-143.9	-126.1	-103.9	-79.7	-53.6	18.9	.9822
400.	958.	61.3	-197.0	-176.7	-151.3	-123.6	-93.8	24.1	1.1370
450.	937.	79.8	-259.6	-236.8	-208.3	-177.1	-143.6	30.0	1.2956
500.	917.	101.2	-331.8	-306.5	-274.7	-240.1	-202.9	36.5	1.4571

BALLISTIC COEFFICIENT .39, MUZZLE VELOCITY FT/SEC 1300.

RANGE	VELOCITY	HEIGHT	50 YD	100 YD	150 YD	200 YD	250 YD	DRIFT	TIME
0.	1300.	0.0	-1.5	-1.5	-1.5	-1.5	-1.5	0.0	0.0000
50.	1241.	-.1	0.0	2.1	4.8	7.9	11.1	.5	.1184
100.	1187.	2.1	-4.2	0.0	5.5	11.6	18.1	2.0	.2420
150.	1141.	5.9	-14.5	-8.3	0.0	9.2	18.9	4.4	.3709
200.	1103.	11.5	-31.5	-23.2	-12.2	0.0	13.0	7.6	.5048
250.	1072.	19.1	-55.7	-45.3	-31.6	-16.3	0.0	11.6	.6428
300.	1044.	28.9	-87.8	-75.4	-58.9	-40.5	-21.0	16.5	.7862
350.	1020.	40.8	-126.9	-112.4	-93.2	-71.7	-49.0	21.5	.9301
400.	996.	55.1	-174.8	-158.1	-136.1	-111.7	-85.6	27.5	1.0791
450.	974.	72.0	-231.1	-212.4	-187.7	-160.1	-130.8	34.0	1.2315
500.	953.	91.6	-296.4	-275.6	-248.1	-217.5	-185.0	41.1	1.3872

BALLISTIC COEFFICIENT .39, MUZZLE VELOCITY FT/SEC 1400.

RANGE	VELOCITY	HEIGHT	50 YD	100 YD	150 YD	200 YD	250 YD	DRIFT	TIME
0.	1400.	0.0	-1.5	-1.5	-1.5	-1.5	-1.5	0.0	0.0000
50.	1332.	-.2	0.0	1.7	4.1	6.7	9.6	.5	.1100
100.	1270.	1.7	-3.4	0.0	4.8	10.1	15.9	1.9	.2254
150.	1213.	5.0	-12.3	-7.2	0.0	7.9	16.6	4.5	.3468
200.	1163.	10.0	-27.0	-20.2	-10.6	0.0	11.6	7.8	.4727
250.	1121.	16.8	-48.2	-39.7	-27.7	-14.5	0.0	12.1	.6042
300.	1087.	25.6	-76.3	-66.1	-51.7	-35.9	-18.5	17.1	.7402
350.	1057.	36.4	-111.7	-99.8	-83.0	-64.5	-44.2	22.9	.8801
400.	1031.	49.8	-155.8	-142.1	-123.0	-101.8	-78.7	29.8	1.0267
450.	1007.	65.3	-206.6	-191.3	-169.7	-145.9	-119.9	36.8	1.1732
500.	985.	83.3	-265.7	-248.7	-224.7	-198.2	-169.3	44.3	1.3231

BALLISTIC COEFFICIENT .39, MUZZLE VELOCITY FT/SEC 1500.

RANGE	VELOCITY	HEIGHT	50 YD	100 YD	150 YD	200 YD	250 YD	DRIFT	TIME
0.	1500.	0.0	-1.5	-1.5	-1.5	-1.5	-1.5	0.0	0.0000
50.	1422.	-.2	0.0	1.4	3.5	5.8	8.4	.5	.1028
100.	1352.	1.4	-2.8	0.0	4.1	8.8	14.0	2.0	.2112
150.	1288.	4.3	-10.4	-6.2	0.0	7.0	14.8	4.4	.3248
200.	1230.	8.7	-23.2	-17.6	-9.4	0.0	10.3	7.8	.4441
250.	1177.	14.8	-42.0	-34.9	-24.6	-12.9	0.0	12.2	.5695
300.	1133.	22.7	-66.7	-58.3	-45.9	-31.9	-16.4	17.4	.6989
350.	1097.	32.6	-98.4	-88.5	-74.1	-57.7	-39.6	23.5	.8335
400.	1065.	44.9	-138.1	-126.8	-110.4	-91.6	-71.0	30.8	.9751
450.	1039.	58.9	-183.4	-170.7	-152.2	-131.1	-107.9	37.8	1.1147
500.	1015.	75.6	-237.6	-223.5	-203.0	-179.5	-153.7	46.0	1.2612

BALLISTIC COEFFICIENT .39, MUZZLE VELOCITY FT/SEC 1600.

RANGE	VELOCITY	HEIGHT	50 YD	100 YD	150 YD	200 YD	250 YD	DRIFT	TIME
0.	1600.	0.0	-1.5	-1.5	-1.5	-1.5	-1.5	0.0	0.0000
50.	1518.	-.3	0.0	1.1	2.9	5.0	7.3	.4	.0962
100.	1439.	1.1	-2.3	0.0	3.6	7.7	12.2	1.8	.1979
150.	1367.	3.7	-8.8	-5.3	0.0	6.2	13.0	4.1	.3048
200.	1301.	7.6	-19.9	-15.3	-8.2	0.0	9.1	7.4	.4171
250.	1242.	13.0	-36.3	-30.5	-21.6	-11.4	0.0	11.7	.5354
300.	1188.	20.1	-58.2	-51.3	-40.6	-28.3	-14.7	17.0	.6589
350.	1142.	29.0	-86.3	-78.3	-65.8	-51.4	-35.5	23.1	.7876
400.	1104.	40.0	-121.1	-111.9	-97.7	-81.3	-63.1	30.2	.9215
450.	1072.	53.1	-162.9	-152.6	-136.6	-118.1	-97.6	37.9	1.0594
500.	1045.	68.7	-212.8	-201.4	-183.5	-163.0	-140.3	46.7	1.2029

BALLISTIC COEFFICIENT .39, MUZZLE VELOCITY FT/SEC 1700.

RANGE	VELOCITY	HEIGHT	50 YD	100 YD	150 YD	200 YD	250 YD	DRIFT	TIME
0.	1700.	0.0	-1.5	-1.5	-1.5	-1.5	-1.5	0.0	0.0000
50.	1614.	-.4	0.0	.9	2.5	4.3	6.3	.4	.0907
100.	1531.	.9	-1.8	0.0	3.1	6.7	10.7	1.7	.1862
150.	1451.	3.2	-7.4	-4.7	0.0	5.4	11.4	3.9	.2869
200.	1378.	6.7	-17.2	-13.5	-7.2	0.0	8.0	7.1	.3930
250.	1311.	11.5	-31.5	-26.8	-19.1	-10.0	0.0	11.2	.5045
300.	1251.	17.8	-50.8	-45.3	-36.0	-25.1	-13.1	16.3	.6218
350.	1196.	25.9	-75.9	-69.4	-58.5	-45.8	-31.8	22.4	.7449
400.	1149.	35.8	-106.7	-99.4	-86.9	-72.4	-56.4	29.3	.8725
450.	1110.	47.8	-144.4	-136.1	-122.1	-105.8	-87.8	37.2	1.0056
500.	1077.	61.9	-188.9	-179.6	-164.1	-146.0	-126.0	45.8	1.1427

BALLISTIC COEFFICIENT .39, MUZZLE VELOCITY FT/SEC 1800.

RANGE	VELOCITY	HEIGHT	50 YD	100 YD	150 YD	200 YD	250 YD	DRIFT	TIME
0.	1800.	0.0	-1.5	-1.5	-1.5	-1.5	-1.5	0.0	0.0000
50.	1709.	-.4	0.0	.7	2.1	3.7	5.5	.4	.0854
100.	1622.	.7	-1.5	0.0	2.7	5.9	9.5	1.6	.1756
150.	1539.	2.8	-6.3	-4.1	0.0	4.8	10.1	3.6	.2705
200.	1459.	5.8	-14.8	-11.8	-6.4	0.0	7.1	6.6	.3707
250.	1385.	10.1	-27.4	-23.6	-16.9	-8.9	0.0	10.5	.4762
300.	1318.	15.8	-44.5	-40.0	-31.9	-22.3	-11.7	15.4	.5873
350.	1257.	23.0	-66.6	-61.4	-52.0	-40.8	-28.3	21.2	.7039
400.	1202.	32.0	-94.2	-88.3	-77.5	-64.7	-50.5	28.0	.8260
450.	1153.	43.1	-128.5	-121.8	-109.6	-95.2	-79.2	36.2	.9555
500.	1114.	55.9	-168.1	-160.7	-147.1	-131.1	-113.4	44.5	1.0860

BALLISTIC COEFFICIENT .39, MUZZLE VELOCITY FT/SEC 1900.

RANGE	VELOCITY	HEIGHT	50 YD	100 YD	150 YD	200 YD	250 YD	DRIFT	TIME
0.	1900.	0.0	-1.5	-1.5	-1.5	-1.5	-1.5	0.0	0.0000
50.	1805.	-.4	0.0	.6	1.8	3.2	4.8	.4	.0810
100.	1715.	.6	-1.2	0.0	2.4	5.2	8.4	1.4	.1661
150.	1627.	2.4	-5.3	-3.6	0.0	4.3	9.0	3.4	.2560
200.	1544.	5.2	-12.8	-10.4	-5.7	0.0	6.3	6.2	.3507
250.	1464.	9.0	-23.8	-20.9	-15.0	-7.9	0.0	9.8	.4505
300.	1389.	14.1	-39.0	-35.5	-28.4	-19.9	-10.4	14.4	.5557
350.	1322.	20.6	-58.7	-54.6	-46.3	-36.4	-25.3	20.0	.6664
400.	1261.	28.6	-83.3	-78.6	-69.1	-57.8	-45.1	26.6	.7826
450.	1205.	38.5	-113.4	-108.2	-97.5	-84.7	-70.5	34.1	.9046
500.	1155.	50.5	-149.9	-144.1	-132.2	-118.0	-102.2	42.9	1.0330

BALLISTIC COEFFICIENT .40, MUZZLE VELOCITY FT/SEC 800.

RANGE	VELOCITY	HEIGHT	50 YD	100 YD	150 YD	200 YD	250 YD	DRIFT	TIME
0.	800.	0.0	-1.5	-1.5	-1.5	-1.5	-1.5	0.0	0.0000
50.	785.	1.0	0.0	6.3	13.3	20.6	28.2	.4	.1898
100.	770.	6.3	-12.6	0.0	14.0	28.5	43.8	1.4	.3831
150.	755.	15.4	-39.9	-21.0	0.0	21.8	44.7	3.0	.5798
200.	740.	28.4	-82.4	-57.1	-29.1	0.0	30.5	5.3	.7799
250.	725.	45.9	-141.1	-109.5	-74.6	-38.2	0.0	8.4	.9854
300.	711.	67.8	-216.4	-178.4	-136.5	-92.9	-47.0	12.3	1.1947
350.	697.	94.3	-308.6	-264.3	-215.4	-164.5	-111.0	16.7	1.4075
400.	682.	126.0	-418.8	-368.2	-312.3	-254.1	-193.0	22.0	1.6248
450.	669.	163.1	-548.4	-491.5	-428.6	-363.1	-294.4	28.2	1.8477
500.	655.	205.8	-696.9	-633.7	-563.8	-491.0	-414.7	35.1	2.0743

BALLISTIC COEFFICIENT .40, MUZZLE VELOCITY FT/SEC 900.

RANGE	VELOCITY	HEIGHT	50 YD	100 YD	150 YD	200 YD	250 YD	DRIFT	TIME
0.	900.	0.0	-1.5	-1.5	-1.5	-1.5	-1.5	0.0	0.0000
50.	882.	.6	0.0	4.9	10.3	16.1	22.1	.4	.1688
100.	865.	4.8	-9.7	0.0	11.0	22.5	34.4	1.3	.3408
150.	849.	12.0	-31.0	-16.5	0.0	17.3	35.2	2.8	.5160
200.	833.	22.4	-64.4	-45.0	-23.1	0.0	23.8	4.9	.6944
250.	817.	36.1	-110.3	-86.1	-58.6	-29.8	0.0	7.5	.8761
300.	801.	53.3	-169.3	-140.2	-107.2	-72.6	-36.9	10.8	1.0613
350.	786.	74.3	-242.1	-208.1	-169.6	-129.3	-87.6	14.8	1.2508
400.	771.	99.3	-328.8	-290.0	-246.1	-200.0	-152.3	19.4	1.4436
450.	756.	128.3	-430.0	-386.3	-336.9	-285.0	-231.4	24.6	1.6398
500.	741.	161.7	-546.3	-497.8	-442.9	-385.2	-325.6	30.5	1.8398

BALLISTIC COEFFICIENT .40, MUZZLE VELOCITY FT/SEC 1000.

RANGE	VELOCITY	HEIGHT	50 YD	100 YD	150 YD	200 YD	250 YD	DRIFT	TIME
0.	1000.	0.0	-1.5	-1.5	-1.5	-1.5	-1.5	0.0	0.0000
50.	978.	.4	0.0	3.8	8.2	12.9	17.8	.3	.1518
100.	957.	3.8	-7.6	0.0	8.9	18.2	28.1	1.2	.3067
150.	937.	9.6	-24.7	-13.3	0.0	14.0	28.8	2.7	.4654
200.	918.	18.1	-51.6	-36.5	-18.7	0.0	19.6	4.7	.6269
250.	899.	29.4	-89.1	-70.2	-47.9	-24.5	0.0	7.5	.7924
300.	881.	43.6	-137.2	-114.5	-87.8	-59.7	-30.3	10.7	.9607
350.	864.	60.9	-196.7	-170.2	-139.1	-106.3	-71.9	14.6	1.1330
400.	848.	81.4	-267.8	-237.5	-201.9	-164.5	-125.2	19.1	1.3084
450.	832.	105.4	-350.9	-316.9	-276.8	-234.7	-190.6	24.1	1.4871
500.	816.	133.0	-446.5	-408.7	-364.2	-317.4	-268.3	29.7	1.6690

BALLISTIC COEFFICIENT .40, MUZZLE VELOCITY FT/SEC 1100.

RANGE	VELOCITY	HEIGHT	50 YD	100 YD	150 YD	200 YD	250 YD	DRIFT	TIME
0.	1100.	0.0	-1.5	-1.5	-1.5	-1.5	-1.5	0.0	0.0000
50.	1070.	.2	0.0	3.1	6.7	10.7	14.8	.4	.1384
100.	1043.	3.1	-6.2	0.0	7.3	15.1	23.4	1.6	.2816
150.	1019.	8.0	-20.2	-10.9	0.0	11.8	24.2	3.0	.4260
200.	996.	15.1	-42.7	-30.3	-15.8	0.0	16.5	5.2	.5751
250.	975.	24.6	-73.9	-58.4	-40.3	-20.6	0.0	8.0	.7273
300.	954.	36.7	-114.3	-95.7	-73.9	-50.3	-25.6	11.4	.8829
350.	934.	51.4	-164.3	-142.6	-117.2	-89.6	-60.8	15.4	1.0420
400.	914.	68.8	-224.2	-199.3	-170.3	-138.8	-105.9	19.9	1.2041
450.	896.	89.6	-295.3	-267.3	-234.7	-199.2	-162.2	25.4	1.3716
500.	878.	112.9	-375.6	-344.5	-308.3	-268.9	-227.7	30.9	1.5390

BALLISTIC COEFFICIENT .40, MUZZLE VELOCITY FT/SEC 1200.

RANGE	VELOCITY	HEIGHT	50 YD	100 YD	150 YD	200 YD	250 YD	DRIFT	TIME
0.	1200.	0.0	-1.5	-1.5	-1.5	-1.5	-1.5	0.0	0.0000
50.	1153.	.0	0.0	2.5	5.7	9.1	12.9	.5	.1276
100.	1114.	2.5	-5.1	0.0	6.4	13.2	20.7	1.8	.2602
150.	1081.	6.8	-17.2	-9.6	0.0	10.3	21.5	3.9	.3973
200.	1054.	13.1	-36.6	-26.5	-13.7	0.0	15.0	6.6	.5374
250.	1028.	21.7	-64.4	-51.8	-35.8	-18.7	0.0	10.3	.6838
300.	1005.	32.4	-99.8	-84.7	-65.5	-45.0	-22.5	14.2	.8307
350.	983.	45.4	-143.7	-126.0	-103.6	-79.7	-53.5	18.7	.9810
400.	962.	61.0	-196.4	-176.2	-150.6	-123.2	-93.3	23.7	1.1346
450.	942.	79.4	-258.6	-235.9	-207.2	-176.3	-142.7	29.5	1.2924
500.	922.	100.5	-330.1	-304.9	-272.9	-238.7	-201.3	35.7	1.4526

BALLISTIC COEFFICIENT .40, MUZZLE VELOCITY FT/SEC 1300.

RANGE	VELOCITY	HEIGHT	50 YD	100 YD	150 YD	200 YD	250 YD	DRIFT	TIME
0.	1300.	0.0	-1.5	-1.5	-1.5	-1.5	-1.5	0.0	0.0000
50.	1242.	-.1	0.0	2.1	4.8	7.9	11.1	.5	.1183
100.	1190.	2.1	-4.1	0.0	5.5	11.6	18.0	1.9	.2418
150.	1144.	5.8	-14.4	-8.2	0.0	9.1	18.8	4.3	.3704
200.	1107.	11.4	-31.4	-23.1	-12.2	0.0	12.9	7.5	.5039
250.	1075.	19.0	-55.5	-45.1	-31.4	-16.2	0.0	11.3	.6413
300.	1048.	28.9	-87.9	-75.4	-59.0	-40.7	-21.3	16.5	.7859
350.	1024.	40.6	-126.7	-112.2	-93.0	-71.7	-49.0	21.3	.9286
400.	1001.	54.8	-173.9	-157.3	-135.4	-111.0	-85.2	26.9	1.0760
450.	979.	71.6	-229.8	-211.1	-186.5	-159.1	-130.0	33.3	1.2274
500.	958.	91.0	-294.5	-273.8	-246.4	-216.0	-183.6	40.2	1.3822

BALLISTIC COEFFICIENT .40, MUZZLE VELOCITY FT/SEC 1400.

RANGE	VELOCITY	HEIGHT	50 YD	100 YD	150 YD	200 YD	250 YD	DRIFT	TIME
0.	1400.	0.0	-1.5	-1.5	-1.5	-1.5	-1.5	0.0	0.0000
50.	1333.	-.2	0.0	1.7	4.1	6.7	9.6	.5	.1099
100.	1273.	1.7	-3.4	0.0	4.7	10.0	15.8	1.9	.2251
150.	1217.	5.0	-12.2	-7.1	0.0	8.0	16.6	4.3	.3458
200.	1167.	9.9	-26.9	-20.1	-10.6	0.0	11.5	7.6	.4717
250.	1126.	16.7	-48.0	-39.5	-27.6	-14.4	0.0	11.8	.6026
300.	1092.	25.4	-75.9	-65.7	-51.5	-35.6	-18.4	16.8	.7381
350.	1062.	36.3	-111.3	-99.4	-82.9	-64.3	-44.2	22.6	.8783
400.	1037.	49.2	-153.8	-140.2	-121.3	-100.1	-77.1	28.7	1.0203
450.	1013.	65.0	-205.9	-190.6	-169.3	-145.4	-119.6	36.3	1.1703
500.	990.	82.9	-264.7	-247.7	-224.1	-197.5	-168.8	43.7	1.3198

BALLISTIC COEFFICIENT .40, MUZZLE VELOCITY FT/SEC 1500.

RANGE	VELOCITY	HEIGHT	50 YD	100 YD	150 YD	200 YD	250 YD	DRIFT	TIME
0.	1500.	0.0	-1.5	-1.5	-1.5	-1.5	-1.5	0.0	0.0000
50.	1424.	-.2	0.0	1.4	3.4	5.8	8.3	.5	.1028
100.	1355.	1.4	-2.8	0.0	4.1	8.8	13.8	2.0	.2112
150.	1293.	4.3	-10.3	-6.1	0.0	7.0	14.6	4.3	.3242
200.	1235.	8.7	-23.2	-17.5	-9.4	0.0	10.1	7.6	.4434
250.	1183.	14.7	-41.6	-34.5	-24.3	-12.6	0.0	11.8	.5671
300.	1139.	22.5	-66.3	-57.8	-45.6	-31.6	-16.5	17.0	.6967
350.	1103.	32.4	-97.7	-87.8	-73.6	-57.1	-39.5	23.0	.8304
400.	1072.	44.3	-136.2	-124.9	-108.6	-89.8	-69.7	29.7	.9685
450.	1045.	58.6	-182.8	-170.1	-151.8	-130.6	-108.0	37.3	1.1121
500.	1021.	74.9	-235.8	-221.6	-201.3	-177.9	-152.7	45.0	1.2557

BALLISTIC COEFFICIENT .40, MUZZLE VELOCITY FT/SEC 1600.

RANGE	VELOCITY	HEIGHT	50 YD	100 YD	150 YD	200 YD	250 YD	DRIFT	TIME
0.	1600.	0.0	-1.5	-1.5	-1.5	-1.5	-1.5	0.0	0.0000
50.	1520.	-.3	0.0	1.1	2.9	5.0	7.2	.4	.0961
100.	1443.	1.1	-2.3	0.0	3.5	7.6	12.1	1.8	.1976
150.	1372.	3.7	-8.7	-5.3	0.0	6.1	12.9	4.0	.3041
200.	1308.	7.6	-19.8	-15.3	-8.2	0.0	9.0	7.2	.4161
250.	1249.	12.9	-36.0	-30.3	-21.5	-11.3	0.0	11.4	.5336
300.	1196.	20.0	-57.9	-51.1	-40.4	-28.2	-14.7	16.6	.6569
350.	1149.	28.8	-85.7	-77.7	-65.3	-50.9	-35.2	22.6	.7844
400.	1111.	39.6	-120.1	-111.0	-96.8	-80.5	-62.5	29.5	.9174
450.	1079.	52.6	-161.5	-151.2	-135.2	-116.8	-96.6	37.0	1.0542
500.	1051.	67.8	-210.3	-198.8	-181.1	-160.7	-138.2	45.4	1.1952

BALLISTIC COEFFICIENT .40, MUZZLE VELOCITY FT/SEC 1700.

RANGE	VELOCITY	HEIGHT	50 YD	100 YD	150 YD	200 YD	250 YD	DRIFT	TIME
0.	1700.	0.0	-1.5	-1.5	-1.5	-1.5	-1.5	0.0	0.0000
50.	1616.	-.4	0.0	.9	2.5	4.3	6.2	.4	.0906
100.	1535.	.9	-1.8	0.0	3.1	6.7	10.7	1.7	.1859
150.	1457.	3.2	-7.4	-4.6	0.0	5.4	11.4	3.8	.2863
200.	1385.	6.6	-17.0	-13.4	-7.2	0.0	8.0	6.8	.3918
250.	1319.	11.4	-31.2	-26.6	-18.9	-9.9	0.0	10.9	.5029
300.	1260.	17.6	-50.4	-44.9	-35.6	-24.9	-12.9	15.8	.6191
350.	1206.	25.6	-75.2	-68.7	-57.9	-45.3	-31.4	21.7	.7411
400.	1157.	35.5	-106.0	-98.6	-86.3	-71.9	-56.0	28.7	.8690
450.	1118.	47.4	-143.3	-135.0	-121.1	-104.9	-87.0	36.4	1.0011
500.	1085.	61.2	-187.0	-177.8	-162.4	-144.4	-124.5	44.7	1.1365

BALLISTIC COEFFICIENT .40, MUZZLE VELOCITY FT/SEC 1800.

RANGE	VELOCITY	HEIGHT	50 YD	100 YD	150 YD	200 YD	250 YD	DRIFT	TIME
0.	1800.	0.0	-1.5	-1.5	-1.5	-1.5	-1.5	0.0	0.0000
50.	1712.	-.4	0.0	.7	2.1	3.7	5.4	.4	.0853
100.	1627.	.7	-1.5	0.0	2.7	5.9	9.4	1.5	.1754
150.	1545.	2.8	-6.3	-4.0	0.0	4.8	10.0	3.5	.2700
200.	1467.	5.8	-14.7	-11.7	-6.3	0.0	7.0	6.4	.3697
250.	1394.	10.1	-27.2	-23.5	-16.7	-8.8	0.0	10.2	.4745
300.	1328.	15.7	-44.2	-39.7	-31.6	-22.1	-11.6	14.9	.5849
350.	1268.	22.8	-66.1	-60.9	-51.4	-40.3	-28.0	20.6	.7005
400.	1212.	31.7	-93.5	-87.6	-76.8	-64.1	-50.0	27.4	.8222
450.	1163.	42.4	-126.5	-119.9	-107.8	-93.5	-77.6	34.8	.9480
500.	1123.	55.2	-166.2	-158.8	-145.3	-129.5	-111.9	43.3	1.0794

BALLISTIC COEFFICIENT .40, MUZZLE VELOCITY FT/SEC 1900.

RANGE	VELOCITY	HEIGHT	50 YD	100 YD	150 YD	200 YD	250 YD	DRIFT	TIME
0.	1900.	0.0	-1.5	-1.5	-1.5	-1.5	-1.5	0.0	0.0000
50.	1808.	-.4	0.0	.6	1.8	3.2	4.7	.3	.0809
100.	1719.	.6	-1.2	0.0	2.4	5.2	8.3	1.4	.1658
150.	1634.	2.4	-5.3	-3.6	0.0	4.2	8.9	3.3	.2556
200.	1552.	5.1	-12.7	-10.4	-5.6	0.0	6.2	6.0	.3499
250.	1474.	8.9	-23.7	-20.8	-14.9	-7.8	0.0	9.5	.4489
300.	1400.	13.9	-38.7	-35.2	-28.1	-19.7	-10.3	14.0	.5534
350.	1333.	20.4	-58.2	-54.1	-45.8	-36.0	-25.0	19.5	.6633
400.	1273.	28.3	-82.5	-77.9	-68.4	-57.2	-44.7	25.9	.7785
450.	1217.	38.1	-112.3	-107.1	-96.4	-83.7	-69.7	33.2	.8992
500.	1167.	49.7	-147.8	-142.0	-130.2	-116.1	-100.5	41.5	1.0251

BALLISTIC COEFFICIENT .41, MUZZLE VELOCITY FT/SEC 800.

RANGE	VELOCITY	HEIGHT	50 YD	100 YD	150 YD	200 YD	250 YD	DRIFT	TIME
0.	800.	0.0	-1.5	-1.5	-1.5	-1.5	-1.5	0.0	0.0000
50.	785.	1.0	0.0	6.3	13.3	20.6	28.2	.4	.1898
100.	770.	6.3	-12.6	0.0	13.9	28.5	43.7	1.4	.3829
150.	756.	15.4	-39.9	-20.9	0.0	21.8	44.7	3.0	.5793
200.	741.	28.4	-82.2	-57.0	-29.1	0.0	30.4	5.2	.7793
250.	727.	45.8	-140.9	-109.3	-74.4	-38.1	0.0	8.2	.9844
300.	713.	67.5	-215.8	-177.9	-136.1	-92.4	-46.8	11.9	1.1928
350.	699.	94.0	-307.5	-263.3	-214.5	-163.6	-110.3	16.2	1.4046
400.	685.	125.5	-417.7	-367.2	-311.4	-253.2	-192.3	21.5	1.6220
450.	672.	162.4	-546.5	-489.6	-426.9	-361.4	-292.9	27.5	1.8435
500.	658.	204.6	-693.8	-630.6	-560.9	-488.2	-412.1	34.1	2.0685

BALLISTIC COEFFICIENT .41, MUZZLE VELOCITY FT/SEC 900.

RANGE	VELOCITY	HEIGHT	50 YD	100 YD	150 YD	200 YD	250 YD	DRIFT	TIME
0.	900.	0.0	-1.5	-1.5	-1.5	-1.5	-1.5	0.0	0.0000
50.	883.	.6	0.0	4.8	10.3	16.1	22.0	.4	.1688
100.	866.	4.8	-9.7	0.0	11.0	22.5	34.3	1.3	.3407
150.	850.	12.0	-31.0	-16.4	0.0	17.2	35.0	2.8	.5157
200.	834.	22.3	-64.3	-44.9	-23.0	0.0	23.7	4.8	.6937
250.	819.	36.0	-110.0	-85.8	-58.4	-29.7	0.0	7.3	.8748
300.	804.	53.2	-169.0	-139.9	-107.0	-72.5	-37.0	10.6	1.0601
350.	789.	74.1	-241.5	-207.5	-169.1	-128.9	-87.4	14.5	1.2488
400.	774.	98.9	-327.8	-289.0	-245.1	-199.2	-151.7	18.9	1.4407
450.	759.	127.7	-428.3	-384.6	-335.3	-283.6	-230.2	23.9	1.6358
500.	745.	161.0	-544.6	-496.1	-441.3	-383.8	-324.5	29.8	1.8358

BALLISTIC COEFFICIENT .41, MUZZLE VELOCITY FT/SEC 1000.

RANGE	VELOCITY	HEIGHT	50 YD	100 YD	150 YD	200 YD	250 YD	DRIFT	TIME
0.	1000.	0.0	-1.5	-1.5	-1.5	-1.5	-1.5	0.0	0.0000
50.	979.	.4	0.0	3.8	8.2	12.9	17.8	.3	.1517
100.	958.	3.8	-7.6	0.0	8.9	18.2	28.0	1.1	.3065
150.	938.	9.6	-24.6	-13.3	0.0	14.0	28.6	2.6	.4650
200.	919.	18.1	-51.5	-36.4	-18.6	0.0	19.5	4.6	.6261
250.	901.	29.3	-88.8	-69.9	-47.7	-24.4	0.0	7.2	.7909
300.	884.	43.4	-136.9	-114.2	-87.6	-59.6	-30.4	10.5	.9595
350.	867.	60.7	-196.2	-169.7	-138.7	-106.0	-71.9	14.3	1.1311
400.	851.	81.1	-266.9	-236.7	-201.2	-163.9	-124.9	18.6	1.3058
450.	835.	104.9	-349.6	-315.6	-275.7	-233.7	-189.8	23.5	1.4836
500.	820.	132.2	-444.5	-406.7	-362.3	-315.7	-267.0	28.9	1.6644

BALLISTIC COEFFICIENT .41, MUZZLE VELOCITY FT/SEC 1100.

RANGE	VELOCITY	HEIGHT	50 YD	100 YD	150 YD	200 YD	250 YD	DRIFT	TIME
0.	1100.	0.0	-1.5	-1.5	-1.5	-1.5	-1.5	0.0	0.0000
50.	1070.	.2	0.0	3.1	6.7	10.6	14.7	.4	.1384
100.	1044.	3.1	-6.3	0.0	7.2	15.0	23.2	1.6	.2820
150.	1021.	8.0	-20.2	-10.8	0.0	11.7	24.1	2.9	.4258
200.	999.	15.1	-42.6	-30.1	-15.7	0.0	16.4	5.1	.5743
250.	977.	24.6	-73.7	-58.1	-40.1	-20.5	0.0	7.8	.7262
300.	957.	36.5	-113.9	-95.1	-73.6	-50.1	-25.5	11.1	.8812
350.	937.	51.2	-163.8	-141.8	-116.7	-89.3	-60.5	15.0	1.0399
400.	918.	68.5	-223.3	-198.2	-169.4	-138.1	-105.3	19.4	1.2013
450.	900.	88.8	-293.1	-264.9	-232.6	-197.3	-160.4	24.4	1.3662
500.	883.	112.3	-374.0	-342.7	-306.7	-267.6	-226.5	30.1	1.5349

BALLISTIC COEFFICIENT .41, MUZZLE VELOCITY FT/SEC 1200.

RANGE	VELOCITY	HEIGHT	50 YD	100 YD	150 YD	200 YD	250 YD	DRIFT	TIME
0.	1200.	0.0	-1.5	-1.5	-1.5	-1.5	-1.5	0.0	0.0000
50.	1153.	.1	0.0	2.4	5.6	9.0	12.8	.8	.1295
100.	1116.	2.5	-4.9	0.0	6.3	13.2	20.7	1.7	.2599
150.	1084.	6.8	-16.8	-9.5	0.0	10.3	21.6	3.7	.3963
200.	1056.	13.1	-36.1	-26.4	-13.7	0.0	15.1	6.4	.5365
250.	1031.	21.7	-63.9	-51.8	-36.0	-18.8	0.0	10.3	.6833
300.	1009.	32.3	-99.2	-84.7	-65.7	-45.1	-22.5	14.1	.8300
350.	987.	45.4	-142.9	-125.9	-103.8	-79.8	-53.4	18.5	.9801
400.	966.	60.9	-195.4	-176.0	-150.7	-123.3	-93.1	23.5	1.1334
450.	946.	79.2	-257.4	-235.6	-207.1	-176.2	-142.3	29.1	1.2906
500.	927.	100.0	-327.8	-303.6	-271.9	-237.6	-199.9	35.0	1.4487

BALLISTIC COEFFICIENT .41, MUZZLE VELOCITY FT/SEC 1300.

RANGE	VELOCITY	HEIGHT	50 YD	100 YD	150 YD	200 YD	250 YD	DRIFT	TIME
0.	1300.	0.0	-1.5	-1.5	-1.5	-1.5	-1.5	0.0	0.0000
50.	1244.	-.1	0.0	2.1	4.8	7.8	11.0	.5	.1183
100.	1192.	2.1	-4.1	0.0	5.5	11.5	17.9	1.9	.2415
150.	1147.	5.8	-14.4	-8.2	0.0	9.1	18.7	4.2	.3699
200.	1110.	11.4	-31.3	-23.0	-12.1	0.0	12.8	7.3	.5030
250.	1079.	18.9	-55.2	-44.9	-31.2	-16.1	0.0	11.1	.6399
300.	1052.	28.5	-86.6	-74.2	-57.8	-39.6	-20.4	15.6	.7809
350.	1028.	40.5	-126.4	-112.0	-92.8	-71.6	-49.2	21.0	.9273
400.	1005.	54.7	-173.6	-157.0	-135.2	-110.9	-85.2	26.6	1.0744
450.	983.	71.3	-229.1	-210.5	-185.9	-158.6	-129.7	32.8	1.2248
500.	963.	90.5	-293.4	-272.7	-245.4	-215.1	-183.0	39.6	1.3786

BALLISTIC COEFFICIENT .41, MUZZLE VELOCITY FT/SEC 1400.

RANGE	VELOCITY	HEIGHT	50 YD	100 YD	150 YD	200 YD	250 YD	DRIFT	TIME
0.	1400.	0.0	-1.5	-1.5	-1.5	-1.5	-1.5	0.0	0.0000
50.	1335.	-.2	0.0	1.7	4.0	6.7	9.5	.5	.1102
100.	1275.	1.7	-3.4	0.0	4.7	10.0	15.7	1.9	.2251
150.	1221.	5.0	-12.1	-7.0	0.0	7.9	16.5	4.1	.3449
200.	1172.	9.9	-26.7	-19.9	-10.6	0.0	11.4	7.4	.4706
250.	1131.	16.6	-47.6	-39.2	-27.5	-14.3	0.0	11.5	.6011
300.	1096.	25.2	-75.4	-65.2	-51.3	-35.4	-18.2	16.4	.7359
350.	1067.	36.0	-110.3	-98.5	-82.1	-63.6	-43.6	21.9	.8746
400.	1041.	48.9	-152.9	-139.4	-120.7	-99.5	-76.7	28.2	1.0173
450.	1018.	64.1	-203.1	-187.9	-166.9	-143.1	-117.4	34.9	1.1628
500.	996.	81.9	-261.6	-244.6	-221.3	-194.9	-166.3	42.3	1.3119

BALLISTIC COEFFICIENT .41, MUZZLE VELOCITY FT/SEC 1500.

RANGE	VELOCITY	HEIGHT	50 YD	100 YD	150 YD	200 YD	250 YD	DRIFT	TIME
0.	1500.	0.0	-1.5	-1.5	-1.5	-1.5	-1.5	0.0	0.0000
50.	1426.	-.2	0.0	1.4	3.4	5.8	8.3	.5	.1027
100.	1358.	1.4	-2.8	0.0	4.1	8.7	13.7	1.9	.2108
150.	1297.	4.3	-10.3	-6.1	0.0	6.9	14.5	4.2	.3237
200.	1241.	8.6	-23.0	-17.4	-9.3	0.0	10.1	7.4	.4420
250.	1189.	14.6	-41.3	-34.3	-24.2	-12.6	0.0	11.5	.5656
300.	1145.	22.4	-65.8	-57.4	-45.2	-31.3	-16.2	16.6	.6941
350.	1108.	32.1	-97.1	-87.2	-73.0	-56.8	-39.2	22.4	.8275
400.	1077.	43.9	-135.2	-124.0	-107.7	-89.2	-69.0	29.0	.9647
450.	1051.	57.9	-180.7	-168.1	-149.8	-129.0	-106.3	36.2	1.1058
500.	1026.	74.5	-234.7	-220.7	-200.3	-177.2	-152.0	44.4	1.2521

BALLISTIC COEFFICIENT .41, MUZZLE VELOCITY FT/SEC 1600.

RANGE	VELOCITY	HEIGHT	50 YD	100 YD	150 YD	200 YD	250 YD	DRIFT	TIME
0.	1600.	0.0	-1.5	-1.5	-1.5	-1.5	-1.5	0.0	0.0000
50.	1522.	-.3	0.0	1.1	2.9	4.9	7.2	.4	.0960
100.	1446.	1.1	-2.3	0.0	3.5	7.6	12.1	1.7	.1973
150.	1377.	3.7	-8.7	-5.3	0.0	6.1	12.8	3.9	.3036
200.	1314.	7.5	-19.8	-15.2	-8.2	0.0	8.9	7.1	.4154
250.	1256.	12.8	-35.9	-30.2	-21.3	-11.1	0.0	11.2	.5321
300.	1203.	19.8	-57.4	-50.6	-40.0	-27.8	-14.4	16.1	.6540
350.	1156.	28.6	-85.3	-77.3	-65.0	-50.7	-35.1	22.2	.7823
400.	1118.	39.4	-119.4	-110.3	-96.2	-79.8	-62.0	28.9	.9141
450.	1086.	52.1	-160.2	-149.9	-134.0	-115.7	-95.6	36.2	1.0495
500.	1058.	67.2	-208.4	-197.0	-179.3	-158.9	-136.7	44.3	1.1894

BALLISTIC COEFFICIENT .41, MUZZLE VELOCITY FT/SEC 1700.

RANGE	VELOCITY	HEIGHT	50 YD	100 YD	150 YD	200 YD	250 YD	DRIFT	TIME
0.	1700.	0.0	-1.5	-1.5	-1.5	-1.5	-1.5	0.0	0.0000
50.	1618.	-.4	0.0	.9	2.5	4.2	6.2	.4	.0905
100.	1538.	.9	-1.8	0.0	3.1	6.6	10.6	1.6	.1856
150.	1462.	3.2	-7.4	-4.6	0.0	5.4	11.3	3.7	.2857
200.	1391.	6.6	-17.0	-13.3	-7.1	0.0	7.9	6.7	.3909
250.	1327.	11.3	-31.1	-26.5	-18.8	-9.9	0.0	10.6	.5013
300.	1268.	17.5	-50.1	-44.6	-35.4	-24.7	-12.8	15.4	.6169
350.	1214.	25.4	-74.7	-68.2	-57.5	-45.0	-31.2	21.2	.7383
400.	1166.	35.1	-104.9	-97.6	-85.3	-71.0	-55.2	27.8	.8641
450.	1126.	46.8	-141.7	-133.4	-119.6	-103.5	-85.8	35.4	.9951
500.	1092.	60.6	-185.3	-176.1	-160.7	-142.9	-123.1	43.7	1.1305

BALLISTIC COEFFICIENT .41, MUZZLE VELOCITY FT/SEC 1800.

RANGE	VELOCITY	HEIGHT	50 YD	100 YD	150 YD	200 YD	250 YD	DRIFT	TIME
0.	1800.	0.0	-1.5	-1.5	-1.5	-1.5	-1.5	0.0	0.0000
50.	1714.	-.4	0.0	.7	2.1	3.7	5.4	.3	.0853
100.	1631.	.7	-1.5	0.0	2.7	5.8	9.3	1.5	.1752
150.	1551.	2.7	-6.2	-4.0	0.0	4.7	10.0	3.4	.2696
200.	1475.	5.8	-14.6	-11.6	-6.3	0.0	7.0	6.2	.3686
250.	1403.	10.0	-27.0	-23.3	-16.6	-8.7	0.0	9.9	.4730
300.	1337.	15.6	-43.9	-39.4	-31.4	-22.0	-11.5	14.6	.5828
350.	1277.	22.6	-65.6	-60.4	-51.0	-40.0	-27.8	20.1	.6976
400.	1223.	31.3	-92.5	-86.6	-75.9	-63.3	-49.3	26.5	.8174
450.	1173.	42.1	-125.8	-119.2	-107.1	-93.0	-77.3	34.2	.9445
500.	1132.	54.5	-164.5	-157.1	-143.7	-128.0	-110.5	42.2	1.0731

BALLISTIC COEFFICIENT .41, MUZZLE VELOCITY FT/SEC 1900.

RANGE	VELOCITY	HEIGHT	50 YD	100 YD	150 YD	200 YD	250 YD	DRIFT	TIME
0.	1900.	0.0	-1.5	-1.5	-1.5	-1.5	-1.5	0.0	0.0000
50.	1810.	-.4	0.0	.6	1.8	3.2	4.7	.3	.0809
100.	1723.	.6	-1.2	0.0	2.4	5.2	8.3	1.4	.1657
150.	1640.	2.4	-5.3	-3.5	0.0	4.2	8.8	3.2	.2551
200.	1560.	5.1	-12.6	-10.3	-5.6	0.0	6.2	5.8	.3490
250.	1483.	8.9	-23.5	-20.6	-14.7	-7.7	0.0	9.3	.4474
300.	1410.	13.8	-38.4	-34.9	-27.9	-19.5	-10.2	13.6	.5512
350.	1344.	20.2	-57.7	-53.7	-45.4	-35.6	-24.8	19.0	.6603
400.	1284.	28.0	-81.8	-77.2	-67.7	-56.5	-44.2	25.1	.7745
450.	1229.	37.6	-111.1	-105.9	-95.3	-82.7	-68.8	32.3	.8940
500.	1178.	49.1	-146.3	-140.5	-128.7	-114.7	-99.3	40.4	1.0190

BALLISTIC COEFFICIENT .42, MUZZLE VELOCITY FT/SEC 800.

RANGE	VELOCITY	HEIGHT	50 YD	100 YD	150 YD	200 YD	250 YD	DRIFT	TIME
0.	800.	0.0	-1.5	-1.5	-1.5	-1.5	-1.5	0.0	0.0000
50.	785.	1.0	0.0	6.3	13.3	20.5	28.1	.4	.1898
100.	771.	6.3	-12.6	0.0	13.9	28.5	43.6	1.4	.3828
150.	757.	15.3	-39.8	-20.8	0.0	21.9	44.6	2.9	.5788
200.	743.	28.4	-82.2	-56.9	-29.1	0.0	30.3	5.1	.7788
250.	729.	45.7	-140.6	-109.1	-74.3	-37.9	0.0	8.1	.9833
300.	715.	67.3	-215.2	-177.4	-135.7	-92.0	-46.5	11.6	1.1910
350.	701.	93.6	-306.7	-262.5	-213.9	-162.9	-109.8	15.8	1.4024
400.	688.	125.1	-416.6	-366.1	-310.6	-252.3	-191.6	21.0	1.6191
450.	674.	161.6	-544.5	-487.7	-425.2	-359.6	-291.4	26.7	1.8393
500.	661.	203.6	-691.1	-628.0	-558.5	-485.7	-409.9	33.2	2.0634

BALLISTIC COEFFICIENT .42, MUZZLE VELOCITY FT/SEC 900.

RANGE	VELOCITY	HEIGHT	50 YD	100 YD	150 YD	200 YD	250 YD	DRIFT	TIME
0.	900.	0.0	-1.5	-1.5	-1.5	-1.5	-1.5	0.0	0.0000
50.	883.	.6	0.0	4.8	10.3	16.0	22.0	.4	.1688
100.	867.	4.8	-9.7	0.0	10.9	22.4	34.2	1.3	.3406
150.	851.	12.0	-30.9	-16.4	0.0	17.2	34.9	2.7	.5153
200.	836.	22.3	-64.2	-44.8	-22.9	0.0	23.7	4.6	.6930
250.	821.	35.9	-109.8	-85.6	-58.2	-29.6	0.0	7.1	.8738
300.	806.	53.1	-168.7	-139.6	-106.8	-72.4	-36.9	10.4	1.0588
350.	791.	73.9	-240.8	-206.9	-168.6	-128.6	-87.1	14.1	1.2469
400.	777.	98.5	-326.7	-287.9	-244.1	-198.4	-151.0	18.4	1.4379
450.	762.	127.2	-427.0	-383.4	-334.2	-282.7	-229.4	23.3	1.6327
500.	748.	160.3	-542.8	-494.4	-439.6	-382.4	-323.2	29.1	1.8318

BALLISTIC COEFFICIENT .42, MUZZLE VELOCITY FT/SEC 1000.

RANGE	VELOCITY	HEIGHT	50 YD	100 YD	150 YD	200 YD	250 YD	DRIFT	TIME
0.	1000.	0.0	-1.5	-1.5	-1.5	-1.5	-1.5	0.0	0.0000
50.	979.	.4	0.0	3.8	8.2	12.9	17.7	.3	.1517
100.	959.	3.8	-7.5	0.0	8.9	18.2	27.9	1.1	.3062
150.	940.	9.6	-24.6	-13.3	0.0	14.0	28.6	2.6	.4645
200.	921.	18.0	-51.4	-36.4	-18.6	0.0	19.5	4.5	.6255
250.	903.	29.2	-88.6	-69.8	-47.6	-24.3	0.0	7.1	.7901
300.	886.	43.3	-136.6	-114.0	-87.4	-59.5	-30.2	10.2	.9582
350.	870.	60.5	-195.6	-169.3	-138.3	-105.7	-71.6	14.0	1.1293
400.	854.	80.8	-266.1	-235.9	-200.5	-163.2	-124.3	18.2	1.3033
450.	838.	104.4	-348.2	-314.3	-274.4	-232.5	-188.7	22.9	1.4801
500.	823.	131.7	-443.1	-405.4	-361.1	-314.5	-265.8	28.3	1.6608

BALLISTIC COEFFICIENT .42, MUZZLE VELOCITY FT/SEC 1100.

RANGE	VELOCITY	HEIGHT	50 YD	100 YD	150 YD	200 YD	250 YD	DRIFT	TIME
0.	1100.	0.0	-1.5	-1.5	-1.5	-1.5	-1.5	0.0	0.0000
50.	1071.	.2	0.0	3.2	6.7	10.6	14.7	.3	.1383
100.	1045.	3.1	-6.3	0.0	7.2	14.9	23.1	1.7	.2823
150.	1022.	8.0	-20.2	-10.8	0.0	11.7	23.9	3.0	.4260
200.	1001.	15.1	-42.5	-29.9	-15.5	0.0	16.3	5.0	.5737
250.	980.	24.5	-73.5	-57.7	-39.8	-20.4	0.0	7.6	.7251
300.	959.	36.4	-113.5	-94.6	-73.1	-49.8	-25.3	10.8	.8796
350.	940.	51.0	-163.2	-141.1	-116.0	-88.8	-60.3	14.7	1.0378
400.	922.	68.2	-222.5	-197.2	-168.5	-137.5	-104.9	19.0	1.1987
450.	904.	88.5	-292.2	-263.8	-231.5	-196.5	-159.8	23.9	1.3633
500.	887.	111.8	-372.6	-341.0	-305.2	-266.3	-225.6	29.5	1.5312

BALLISTIC COEFFICIENT .42, MUZZLE VELOCITY FT/SEC 1200.

RANGE	VELOCITY	HEIGHT	50 YD	100 YD	150 YD	200 YD	250 YD	DRIFT	TIME
0.	1200.	0.0	-1.5	-1.5	-1.5	-1.5	-1.5	0.0	0.0000
50.	1154.	.1	0.0	2.5	5.6	9.0	12.6	.7	.1291
100.	1117.	2.5	-5.0	0.0	6.2	13.0	20.3	1.9	.2607
150.	1086.	6.8	-16.8	-9.3	0.0	10.2	21.1	3.7	.3959
200.	1058.	13.0	-36.0	-26.1	-13.7	0.0	14.5	6.3	.5357
250.	1034.	21.4	-63.1	-50.7	-35.1	-18.1	0.0	9.6	.6793
300.	1012.	32.3	-99.2	-84.3	-65.7	-45.2	-23.5	14.0	.8294
350.	990.	45.3	-142.9	-125.5	-103.7	-79.9	-54.6	18.3	.9791
400.	970.	60.8	-195.3	-175.5	-150.6	-123.3	-94.4	23.3	1.1321
450.	950.	78.5	-255.5	-233.2	-205.2	-174.5	-142.0	28.2	1.2853
500.	931.	99.4	-326.5	-301.7	-270.6	-236.5	-200.4	34.3	1.4448

BALLISTIC COEFFICIENT .42, MUZZLE VELOCITY FT/SEC 1300.

RANGE	VELOCITY	HEIGHT	50 YD	100 YD	150 YD	200 YD	250 YD	DRIFT	TIME
0.	1300.	0.0	-1.5	-1.5	-1.5	-1.5	-1.5	0.0	0.0000
50.	1245.	-.1	0.0	2.1	4.8	7.8	11.0	.5	.1182
100.	1194.	2.1	-4.2	0.0	5.4	11.4	17.9	2.0	.2419
150.	1150.	5.8	-14.4	-8.1	0.0	9.0	18.8	4.1	.3693
200.	1113.	11.3	-31.2	-22.8	-12.1	0.0	13.0	7.1	.5021
250.	1082.	18.9	-55.2	-44.8	-31.3	-16.2	0.0	11.0	.6396
300.	1056.	28.4	-86.2	-73.7	-57.5	-39.4	-19.9	15.3	.7790
350.	1031.	40.4	-126.2	-111.6	-92.7	-71.6	-48.9	20.8	.9259
400.	1009.	54.5	-173.2	-156.5	-134.9	-110.8	-84.9	26.3	1.0727
450.	988.	71.0	-228.6	-209.8	-185.5	-158.4	-129.2	32.4	1.2228
500.	967.	90.1	-292.7	-271.8	-244.9	-214.7	-182.3	39.1	1.3761

BALLISTIC COEFFICIENT .42, MUZZLE VELOCITY FT/SEC 1400.

RANGE	VELOCITY	HEIGHT	50 YD	100 YD	150 YD	200 YD	250 YD	DRIFT	TIME
0.	1400.	0.0	-1.5	-1.5	-1.5	-1.5	-1.5	0.0	0.0000
50.	1336.	-.2	0.0	1.7	4.0	6.7	9.6	.5	.1100
100.	1278.	1.7	-3.4	0.0	4.7	10.0	15.8	1.8	.2247
150.	1225.	4.9	-12.1	-7.0	0.0	8.0	16.6	4.1	.3445
200.	1176.	9.9	-26.7	-20.0	-10.7	0.0	11.5	7.4	.4706
250.	1135.	16.6	-47.8	-39.4	-27.7	-14.4	0.0	11.5	.6013
300.	1101.	25.1	-75.0	-64.9	-50.9	-34.9	-17.6	16.0	.7338
350.	1072.	35.7	-109.7	-98.0	-81.6	-62.9	-42.8	21.5	.8720
400.	1046.	48.8	-152.8	-139.3	-120.6	-99.3	-76.3	28.0	1.0162
450.	1023.	63.8	-202.2	-187.1	-166.1	-142.1	-116.2	34.4	1.1596
500.	1001.	81.3	-260.0	-243.2	-219.9	-193.2	-164.5	41.5	1.3074

BALLISTIC COEFFICIENT .42, MUZZLE VELOCITY FT/SEC 1500.

RANGE	VELOCITY	HEIGHT	50 YD	100 YD	150 YD	200 YD	250 YD	DRIFT	TIME
0.	1500.	0.0	-1.5	-1.5	-1.5	-1.5	-1.5	0.0	0.0000
50.	1428.	-.2	0.0	1.4	3.4	5.7	8.2	.5	.1027
100.	1361.	1.4	-2.8	0.0	4.0	8.7	13.7	1.8	.2104
150.	1301.	4.3	-10.2	-6.1	0.0	6.9	14.5	4.0	.3229
200.	1246.	8.6	-22.9	-17.3	-9.2	0.0	10.1	7.2	.4410
250.	1195.	14.5	-41.2	-34.3	-24.2	-12.6	0.0	11.4	.5646
300.	1151.	22.2	-65.5	-57.1	-45.0	-31.1	-16.0	16.2	.6920
350.	1114.	31.9	-96.4	-86.7	-72.5	-56.4	-38.7	21.9	.8246
400.	1083.	43.6	-134.2	-123.1	-107.0	-88.5	-68.3	28.4	.9611
450.	1056.	57.5	-179.4	-166.9	-148.7	-127.9	-105.2	35.4	1.1014
500.	1032.	74.1	-233.5	-219.6	-199.4	-176.3	-151.1	43.7	1.2484

BALLISTIC COEFFICIENT .42, MUZZLE VELOCITY FT/SEC 1600.

RANGE	VELOCITY	HEIGHT	50 YD	100 YD	150 YD	200 YD	250 YD	DRIFT	TIME
0.	1600.	0.0	-1.5	-1.5	-1.5	-1.5	-1.5	0.0	0.0000
50.	1523.	-.3	0.0	1.1	2.9	4.9	7.1	.4	.0960
100.	1450.	1.1	-2.3	0.0	3.5	7.6	12.0	1.7	.1971
150.	1382.	3.7	-8.7	-5.3	0.0	6.1	12.7	3.8	.3029
200.	1319.	7.5	-19.6	-15.1	-8.1	0.0	8.8	6.9	.4140
250.	1263.	12.7	-35.6	-29.9	-21.2	-11.1	0.0	10.8	.5302
300.	1211.	19.6	-57.1	-50.3	-39.7	-27.6	-14.3	15.7	.6518
350.	1164.	28.3	-84.4	-76.5	-64.2	-50.1	-34.6	21.5	.7782
400.	1125.	39.0	-118.2	-109.2	-95.2	-79.0	-61.3	28.1	.9095
450.	1092.	51.7	-158.9	-148.7	-133.0	-114.8	-94.9	35.4	1.0449
500.	1064.	66.8	-207.5	-196.1	-178.6	-158.4	-136.3	43.7	1.1860

BALLISTIC COEFFICIENT .42, MUZZLE VELOCITY FT/SEC 1700.

RANGE	VELOCITY	HEIGHT	50 YD	100 YD	150 YD	200 YD	250 YD	DRIFT	TIME
0.	1700.	0.0	-1.5	-1.5	-1.5	-1.5	-1.5	0.0	0.0000
50.	1620.	-.4	0.0	.9	2.4	4.2	6.2	.4	.0904
100.	1542.	.9	-1.8	0.0	3.1	6.6	10.6	1.6	.1854
150.	1468.	3.2	-7.3	-4.6	0.0	5.3	11.2	3.6	.2852
200.	1398.	6.5	-16.9	-13.2	-7.1	0.0	7.9	6.5	.3900
250.	1334.	11.3	-31.0	-26.4	-18.7	-9.9	0.0	10.4	.5002
300.	1276.	17.4	-49.8	-44.4	-35.2	-24.5	-12.7	15.1	.6150
350.	1223.	25.2	-74.0	-67.6	-56.9	-44.5	-30.7	20.6	.7348
400.	1174.	34.9	-104.4	-97.1	-84.9	-70.7	-54.9	27.4	.8615
450.	1134.	46.3	-140.4	-132.2	-118.4	-102.4	-84.7	34.5	.9902
500.	1100.	60.0	-183.6	-174.4	-159.1	-141.4	-121.6	42.7	1.1247

BALLISTIC COEFFICIENT .42, MUZZLE VELOCITY FT/SEC 1800.

RANGE	VELOCITY	HEIGHT	50 YD	100 YD	150 YD	200 YD	250 YD	DRIFT	TIME
0.	1800.	0.0	-1.5	-1.5	-1.5	-1.5	-1.5	0.0	0.0000
50.	1716.	-.4	0.0	.7	2.1	3.6	5.4	.3	.0852
100.	1635.	.7	-1.5	0.0	2.7	5.8	9.3	1.5	.1750
150.	1557.	2.7	-6.2	-4.0	0.0	4.7	9.9	3.4	.2691
200.	1482.	5.7	-14.5	-11.6	-6.2	0.0	6.9	6.0	.3676
250.	1411.	9.9	-26.8	-23.1	-16.5	-8.7	0.0	9.6	.4715
300.	1346.	15.4	-43.6	-39.1	-31.1	-21.8	-11.4	14.2	.5805
350.	1287.	22.4	-65.0	-59.9	-50.5	-39.6	-27.5	19.6	.6945
400.	1233.	31.1	-91.9	-86.0	-75.3	-62.9	-49.0	26.0	.8142
450.	1183.	41.5	-124.2	-117.5	-105.5	-91.5	-75.9	33.1	.9378
500.	1141.	53.9	-162.8	-155.4	-142.1	-126.5	-109.1	41.1	1.0669

BALLISTIC COEFFICIENT .42, MUZZLE VELOCITY FT/SEC 1900.

RANGE	VELOCITY	HEIGHT	50 YD	100 YD	150 YD	200 YD	250 YD	DRIFT	TIME
0.	1900.	0.0	-1.5	-1.5	-1.5	-1.5	-1.5	0.0	0.0000
50.	1812.	-.4	0.0	.6	1.7	3.1	4.7	.3	.0808
100.	1727.	.6	-1.1	0.0	2.3	5.1	8.2	1.3	.1655
150.	1646.	2.4	-5.2	-3.5	0.0	4.2	8.8	3.1	.2547
200.	1567.	5.1	-12.6	-10.3	-5.6	0.0	6.1	5.7	.3482
250.	1492.	8.8	-23.4	-20.5	-14.6	-7.7	0.0	9.0	.4460
300.	1420.	13.7	-38.1	-34.7	-27.6	-19.3	-10.1	13.3	.5490
350.	1355.	20.0	-57.3	-53.3	-45.1	-35.3	-24.6	18.5	.6578
400.	1295.	27.8	-81.1	-76.5	-67.2	-56.0	-43.8	24.5	.7709
450.	1240.	37.2	-110.1	-104.9	-94.4	-81.8	-68.0	31.4	.8892
500.	1190.	48.5	-144.7	-139.0	-127.2	-113.3	-98.0	39.3	1.0127

BALLISTIC COEFFICIENT .43, MUZZLE VELOCITY FT/SEC 800.

RANGE	VELOCITY	HEIGHT	50 YD	100 YD	150 YD	200 YD	250 YD	DRIFT	TIME
0.	800.	0.0	-1.5	-1.5	-1.5	-1.5	-1.5	0.0	0.0000
50.	786.	1.0	0.0	6.3	13.2	20.5	28.1	.4	.1898
100.	772.	6.3	-12.6	0.0	13.9	28.4	43.5	1.3	.3826
150.	758.	15.3	-39.7	-20.8	0.0	21.9	44.5	2.8	.5783
200.	744.	28.3	-82.1	-56.9	-29.2	0.0	30.2	5.0	.7783
250.	730.	45.6	-140.3	-108.8	-74.2	-37.7	0.0	7.9	.9823
300.	717.	67.1	-214.6	-176.8	-135.3	-91.5	-46.2	11.3	1.1892
350.	703.	93.4	-306.1	-262.0	-213.5	-162.5	-109.6	15.5	1.4005
400.	690.	124.6	-415.4	-365.1	-309.6	-251.3	-190.9	20.5	1.6163
450.	677.	160.9	-542.4	-485.7	-423.4	-357.7	-289.8	26.0	1.8351
500.	664.	202.8	-689.1	-626.1	-556.8	-483.9	-408.4	32.4	2.0591

BALLISTIC COEFFICIENT .43, MUZZLE VELOCITY FT/SEC 900.

RANGE	VELOCITY	HEIGHT	50 YD	100 YD	150 YD	200 YD	250 YD	DRIFT	TIME
0.	900.	0.0	-1.5	-1.5	-1.5	-1.5	-1.5	0.0	0.0000
50.	883.	.6	0.0	4.8	10.3	16.0	21.9	.4	.1688
100.	867.	4.8	-9.7	0.0	10.9	22.3	34.2	1.3	.3405
150.	852.	12.0	-30.9	-16.4	0.0	17.1	34.9	2.6	.5150
200.	837.	22.3	-64.0	-44.7	-22.8	0.0	23.7	4.5	.6922
250.	822.	35.8	-109.6	-85.4	-58.1	-29.6	0.0	7.0	.8730
300.	808.	52.9	-168.3	-139.3	-106.5	-72.3	-36.8	10.1	1.0576
350.	794.	73.6	-240.2	-206.3	-168.1	-128.2	-86.7	13.8	1.2449
400.	779.	98.1	-325.6	-286.8	-243.1	-197.5	-150.2	17.9	1.4350
450.	765.	126.7	-425.9	-382.3	-333.2	-281.9	-228.6	22.8	1.6298
500.	751.	159.6	-541.0	-492.6	-438.0	-380.9	-321.7	28.4	1.8278

BALLISTIC COEFFICIENT .43, MUZZLE VELOCITY FT/SEC 1000.

RANGE	VELOCITY	HEIGHT	50 YD	100 YD	150 YD	200 YD	250 YD	DRIFT	TIME
0.	1000.	0.0	-1.5	-1.5	-1.5	-1.5	-1.5	0.0	0.0000
50.	980.	.4	0.0	3.8	8.2	12.8	17.7	.3	.1516
100.	960.	3.7	-7.5	0.0	8.9	18.2	27.9	1.1	.3060
150.	941.	9.6	-24.6	-13.4	0.0	13.9	28.5	2.6	.4646
200.	923.	18.0	-51.4	-36.3	-18.5	0.0	19.4	4.4	.6250
250.	905.	29.2	-88.5	-69.7	-47.4	-24.3	0.0	6.9	.7893
300.	889.	43.2	-136.3	-113.7	-87.0	-59.2	-30.1	10.0	.9569
350.	872.	60.3	-195.1	-168.8	-137.6	-105.2	-71.2	13.6	1.1274
400.	857.	80.5	-265.2	-235.1	-199.5	-162.5	-123.7	17.7	1.3007
450.	842.	104.0	-347.1	-313.3	-273.2	-231.5	-187.8	22.4	1.4771
500.	827.	131.1	-441.7	-404.1	-359.6	-313.2	-264.7	27.7	1.6573

BALLISTIC COEFFICIENT .43, MUZZLE VELOCITY FT/SEC 1100.

RANGE	VELOCITY	HEIGHT	50 YD	100 YD	150 YD	200 YD	250 YD	DRIFT	TIME
0.	1100.	0.0	-1.5	-1.5	-1.5	-1.5	-1.5	0.0	0.0000
50.	1072.	.2	0.0	3.2	6.8	10.6	14.7	.3	.1383
100.	1046.	3.1	-6.4	0.0	7.1	14.9	23.1	1.8	.2827
150.	1024.	8.0	-20.3	-10.7	0.0	11.7	23.9	3.0	.4261
200.	1003.	15.1	-42.6	-29.8	-15.5	0.0	16.3	5.0	.5738
250.	982.	24.5	-73.5	-57.6	-39.8	-20.3	0.0	7.6	.7249
300.	962.	36.3	-113.5	-94.4	-73.0	-49.7	-25.3	10.7	.8791
350.	943.	50.9	-163.2	-140.9	-115.9	-88.7	-60.3	14.6	1.0373
400.	925.	68.0	-221.9	-196.4	-167.8	-136.7	-104.2	18.6	1.1966
450.	908.	88.1	-291.2	-262.6	-230.4	-195.4	-158.8	23.4	1.3604
500.	891.	111.2	-371.1	-339.3	-303.6	-264.7	-224.1	28.8	1.5275

BALLISTIC COEFFICIENT .43, MUZZLE VELOCITY FT/SEC 1200.

RANGE	VELOCITY	HEIGHT	50 YD	100 YD	150 YD	200 YD	250 YD	DRIFT	TIME
0.	1200.	0.0	-1.5	-1.5	-1.5	-1.5	-1.5	0.0	0.0000
50.	1155.	.0	0.0	2.5	5.6	9.0	12.6	.7	.1288
100.	1119.	2.5	-4.9	0.0	6.3	13.1	20.3	1.7	.2598
150.	1088.	6.8	-16.8	-9.4	0.0	10.3	21.0	3.6	.3955
200.	1061.	13.0	-36.0	-26.2	-13.7	0.0	14.3	6.2	.5353
250.	1037.	21.3	-62.9	-50.6	-35.0	-17.9	0.0	9.3	.6781
300.	1015.	31.9	-98.0	-83.3	-64.5	-44.0	-22.5	13.1	.8246
350.	993.	45.2	-142.8	-125.6	-103.7	-79.8	-54.7	18.2	.9782
400.	974.	60.2	-193.5	-173.8	-148.7	-121.4	-92.8	22.3	1.1265
450.	954.	78.1	-254.6	-232.4	-204.2	-173.5	-141.3	27.6	1.2820
500.	936.	98.9	-325.2	-300.6	-269.3	-235.1	-199.3	33.6	1.4409

BALLISTIC COEFFICIENT .43, MUZZLE VELOCITY FT/SEC 1300.

RANGE	VELOCITY	HEIGHT	50 YD	100 YD	150 YD	200 YD	250 YD	DRIFT	TIME
0.	1300.	0.0	-1.5	-1.5	-1.5	-1.5	-1.5	0.0	0.0000
50.	1246.	-.1	0.0	2.1	4.8	7.8	11.0	.5	.1182
100.	1196.	2.0	-4.1	0.0	5.4	11.4	17.8	1.9	.2414
150.	1153.	5.8	-14.3	-8.1	0.0	9.0	18.6	4.0	.3688
200.	1117.	11.3	-31.1	-22.8	-12.0	0.0	12.8	7.0	.5012
250.	1086.	18.8	-54.8	-44.4	-31.0	-16.0	0.0	10.6	.6374
300.	1059.	28.2	-85.8	-73.4	-57.2	-39.2	-20.0	14.9	.7771
350.	1035.	39.9	-124.6	-110.1	-91.3	-70.3	-47.9	19.9	.9206
400.	1013.	54.3	-172.8	-156.3	-134.7	-110.7	-85.2	26.0	1.0711
450.	992.	70.8	-228.1	-209.4	-185.2	-158.2	-129.4	32.1	1.2207
500.	972.	89.3	-290.1	-269.4	-242.5	-212.5	-180.5	38.0	1.3696

BALLISTIC COEFFICIENT .43, MUZZLE VELOCITY FT/SEC 1400.

RANGE	VELOCITY	HEIGHT	50 YD	100 YD	150 YD	200 YD	250 YD	DRIFT	TIME
0.	1400.	0.0	-1.5	-1.5	-1.5	-1.5	-1.5	0.0	0.0000
50.	1338.	-.2	0.0	1.7	4.0	6.6	9.4	.5	.1099
100.	1281.	1.7	-3.3	0.0	4.7	9.9	15.5	1.8	.2243
150.	1228.	4.9	-12.0	-7.0	0.0	7.8	16.3	4.0	.3440
200.	1180.	9.8	-26.5	-19.8	-10.4	0.0	11.3	7.0	.4686
250.	1139.	16.4	-47.2	-38.9	-27.2	-14.2	0.0	11.0	.5982
300.	1105.	25.0	-74.7	-64.6	-50.6	-35.0	-18.0	15.7	.7319
350.	1076.	35.5	-109.1	-97.4	-81.1	-62.8	-43.0	21.0	.8694
400.	1051.	48.3	-151.1	-137.7	-119.0	-98.2	-75.5	27.0	1.0106
450.	1027.	63.5	-201.6	-186.5	-165.5	-142.1	-116.6	34.0	1.1572
500.	1006.	80.9	-259.3	-242.5	-219.2	-193.1	-164.8	41.0	1.3045

BALLISTIC COEFFICIENT .43, MUZZLE VELOCITY FT/SEC 1500.

RANGE	VELOCITY	HEIGHT	50 YD	100 YD	150 YD	200 YD	250 YD	DRIFT	TIME
0.	1500.	0.0	-1.5	-1.5	-1.5	-1.5	-1.5	0.0	0.0000
50.	1429.	-.2	0.0	1.4	3.4	5.7	8.2	.5	.1027
100.	1364.	1.4	-2.8	0.0	4.0	8.6	13.6	1.8	.2102
150.	1305.	4.2	-10.2	-6.0	0.0	6.9	14.3	4.0	.3225
200.	1251.	8.5	-22.8	-17.2	-9.2	0.0	10.0	7.0	.4400
250.	1201.	14.4	-40.9	-34.0	-23.9	-12.4	0.0	11.0	.5625
300.	1156.	22.2	-65.3	-57.0	-44.9	-31.2	-16.3	16.0	.6910
350.	1119.	31.7	-95.8	-86.1	-72.0	-56.0	-38.6	21.5	.8220
400.	1088.	43.3	-133.4	-122.3	-106.2	-87.9	-68.0	27.8	.9578
450.	1061.	57.1	-178.4	-165.9	-147.8	-127.1	-104.7	34.8	1.0978
500.	1037.	73.1	-230.5	-216.7	-196.6	-173.6	-148.7	42.3	1.2403

BALLISTIC COEFFICIENT .43, MUZZLE VELOCITY FT/SEC 1600.

RANGE	VELOCITY	HEIGHT	50 YD	100 YD	150 YD	200 YD	250 YD	DRIFT	TIME
0.	1600.	0.0	-1.5	-1.5	-1.5	-1.5	-1.5	0.0	0.0000
50.	1525.	-.3	0.0	1.1	2.9	4.9	7.1	.4	.0959
100.	1453.	1.1	-2.3	0.0	3.5	7.5	11.9	1.6	.1968
150.	1386.	3.6	-8.6	-5.2	0.0	6.0	12.6	3.7	.3024
200.	1325.	7.4	-19.5	-15.0	-8.0	0.0	8.8	6.7	.4131
250.	1269.	12.7	-35.4	-29.8	-21.1	-11.0	0.0	10.6	.5288
300.	1218.	19.5	-56.7	-50.0	-39.5	-27.4	-14.2	15.3	.6497
350.	1171.	28.1	-83.8	-76.0	-63.7	-49.7	-34.3	21.0	.7754
400.	1132.	38.6	-117.4	-108.4	-94.4	-78.3	-60.7	27.4	.9058
450.	1098.	51.2	-157.6	-147.5	-131.8	-113.7	-93.9	34.6	1.0404
500.	1070.	65.9	-205.0	-193.8	-176.3	-156.2	-134.2	42.5	1.1788

BALLISTIC COEFFICIENT .43, MUZZLE VELOCITY FT/SEC 1700.

RANGE	VELOCITY	HEIGHT	50 YD	100 YD	150 YD	200 YD	250 YD	DRIFT	TIME
0.	1700.	0.0	-1.5	-1.5	-1.5	-1.5	-1.5	0.0	0.0000
50.	1621.	-.4	0.0	.9	2.4	4.2	6.1	.4	.0904
100.	1546.	.9	-1.8	0.0	3.0	6.6	10.5	1.6	.1853
150.	1473.	3.1	-7.3	-4.5	0.0	5.3	11.1	3.5	.2846
200.	1404.	6.5	-16.8	-13.1	-7.1	0.0	7.8	6.3	.3889
250.	1341.	11.2	-30.7	-26.2	-18.6	-9.7	0.0	10.1	.4984
300.	1284.	17.3	-49.5	-44.0	-34.9	-24.3	-12.6	14.7	.6127
350.	1232.	25.0	-73.5	-67.1	-56.5	-44.2	-30.5	20.1	.7321
400.	1183.	34.4	-103.2	-95.9	-83.8	-69.7	-54.1	26.5	.8563
450.	1142.	45.9	-139.2	-131.0	-117.3	-101.4	-83.9	33.7	.9854
500.	1107.	59.4	-181.9	-172.8	-157.6	-140.0	-120.5	41.7	1.1191

BALLISTIC COEFFICIENT .43, MUZZLE VELOCITY FT/SEC 1800.

RANGE	VELOCITY	HEIGHT	50 YD	100 YD	150 YD	200 YD	250 YD	DRIFT	TIME
0.	1800.	0.0	-1.5	-1.5	-1.5	-1.5	-1.5	0.0	0.0000
50.	1718.	-.4	0.0	.7	2.1	3.6	5.3	.3	.0851
100.	1638.	.7	-1.5	0.0	2.7	5.8	9.2	1.4	.1748
150.	1562.	2.7	-6.2	-4.0	0.0	4.6	9.8	3.3	.2687
200.	1489.	5.7	-14.4	-11.5	-6.2	0.0	6.9	5.9	.3668
250.	1419.	9.9	-26.7	-23.0	-16.3	-8.6	0.0	9.4	.4701
300.	1355.	15.3	-43.3	-38.9	-30.9	-21.7	-11.3	13.9	.5788
350.	1296.	22.2	-64.6	-59.5	-50.1	-39.3	-27.3	19.1	.6918
400.	1243.	30.7	-91.1	-85.2	-74.5	-62.2	-48.4	25.2	.8100
450.	1193.	41.1	-123.3	-116.7	-104.8	-90.8	-75.3	32.4	.9340
500.	1150.	53.3	-161.3	-154.0	-140.6	-125.2	-108.0	40.1	1.0613

BALLISTIC COEFFICIENT .43, MUZZLE VELOCITY FT/SEC 1900.

RANGE	VELOCITY	HEIGHT	50 YD	100 YD	150 YD	200 YD	250 YD	DRIFT	TIME
0.	1900.	0.0	-1.5	-1.5	-1.5	-1.5	-1.5	0.0	0.0000
50.	1814.	-.4	0.0	.6	1.7	3.1	4.6	.3	.0807
100.	1731.	.6	-1.1	0.0	2.3	5.1	8.1	1.3	.1654
150.	1651.	2.4	-5.2	-3.5	0.0	4.2	8.7	3.1	.2542
200.	1575.	5.0	-12.5	-10.2	-5.5	0.0	6.1	5.6	.3474
250.	1501.	8.7	-23.2	-20.3	-14.5	-7.6	0.0	8.8	.4446
300.	1430.	13.6	-37.9	-34.5	-27.5	-19.1	-10.0	13.0	.5473
350.	1365.	19.8	-56.8	-52.8	-44.6	-34.9	-24.3	18.0	.6547
400.	1306.	27.5	-80.4	-75.8	-66.5	-55.4	-43.2	23.8	.7670
450.	1251.	36.8	-109.1	-103.9	-93.4	-80.9	-67.3	30.6	.8845
500.	1201.	47.9	-143.3	-137.5	-125.9	-112.0	-96.8	38.3	1.0069

BALLISTIC COEFFICIENT .44, MUZZLE VELOCITY FT/SEC 800.

RANGE	VELOCITY	HEIGHT	50 YD	100 YD	150 YD	200 YD	250 YD	DRIFT	TIME
0.	800.	0.0	-1.5	-1.5	-1.5	-1.5	-1.5	0.0	0.0000
50.	786.	1.0	0.0	6.3	13.2	20.5	28.0	.4	.1898
100.	772.	6.3	-12.6	0.0	13.8	28.4	43.4	1.3	.3824
150.	759.	15.3	-39.6	-20.7	0.0	21.9	44.4	2.7	.5778
200.	745.	28.3	-82.0	-56.8	-29.2	0.0	30.1	4.9	.7778
250.	732.	45.5	-140.1	-108.6	-74.1	-37.6	0.0	7.7	.9812
300.	719.	66.9	-214.0	-176.3	-134.8	-91.0	-45.9	11.0	1.1873
350.	706.	93.2	-305.5	-261.5	-213.1	-162.0	-109.4	15.2	1.3987
400.	693.	124.2	-414.3	-364.0	-308.7	-250.3	-190.2	20.0	1.6134
450.	680.	160.1	-540.3	-483.7	-421.6	-355.9	-288.2	25.2	1.8309
500.	667.	201.9	-687.0	-624.1	-555.0	-482.0	-406.9	31.7	2.0549

BALLISTIC COEFFICIENT .44, MUZZLE VELOCITY FT/SEC 900.

RANGE	VELOCITY	HEIGHT	50 YD	100 YD	150 YD	200 YD	250 YD	DRIFT	TIME
0.	900.	0.0	-1.5	-1.5	-1.5	-1.5	-1.5	0.0	0.0000
50.	884.	.6	0.0	4.8	10.3	16.0	21.9	.4	.1688
100.	868.	4.8	-9.7	0.0	10.9	22.3	34.1	1.2	.3404
150.	853.	12.0	-30.9	-16.4	0.0	17.0	34.8	2.6	.5147
200.	838.	22.2	-63.9	-44.5	-22.7	0.0	23.7	4.4	.6915
250.	824.	35.8	-109.5	-85.3	-58.0	-29.6	0.0	6.9	.8723
300.	810.	52.8	-168.0	-139.0	-106.3	-72.2	-36.6	9.9	1.0563
350.	796.	73.4	-239.6	-205.7	-167.6	-127.8	-86.3	13.4	1.2430
400.	782.	97.8	-324.8	-286.1	-242.5	-197.0	-149.6	17.5	1.4328
450.	768.	126.3	-424.8	-381.2	-332.2	-281.0	-227.7	22.3	1.6269
500.	755.	158.9	-539.1	-490.7	-436.2	-379.4	-320.1	27.7	1.8238

BALLISTIC COEFFICIENT .44, MUZZLE VELOCITY FT/SEC 1000.

RANGE	VELOCITY	HEIGHT	50 YD	100 YD	150 YD	200 YD	250 YD	DRIFT	TIME
0.	1000.	0.0	-1.5	-1.5	-1.5	-1.5	-1.5	0.0	0.0000
50.	980.	.4	0.0	3.8	8.2	12.8	17.7	.3	.1515
100.	961.	3.7	-7.5	0.0	8.9	18.1	27.8	1.1	.3061
150.	942.	9.6	-24.7	-13.4	0.0	13.8	28.3	2.6	.4648
200.	925.	18.0	-51.3	-36.2	-18.4	0.0	19.3	4.3	.6246
250.	907.	29.1	-88.3	-69.5	-47.2	-24.2	0.0	6.8	.7885
300.	891.	43.1	-136.0	-113.3	-86.6	-59.0	-30.0	9.8	.9556
350.	875.	60.1	-194.6	-168.2	-137.0	-104.8	-70.9	13.3	1.1255
400.	859.	80.2	-264.5	-234.3	-198.6	-161.8	-123.1	17.3	1.2984
450.	845.	103.6	-346.2	-312.2	-272.1	-230.7	-187.2	21.9	1.4745
500.	830.	130.5	-440.2	-402.5	-357.9	-311.9	-263.6	27.1	1.6538

BALLISTIC COEFFICIENT .44, MUZZLE VELOCITY FT/SEC 1100.

RANGE	VELOCITY	HEIGHT	50 YD	100 YD	150 YD	200 YD	250 YD	DRIFT	TIME
0.	1100.	0.0	-1.5	-1.5	-1.5	-1.5	-1.5	0.0	0.0000
50.	1072.	.2	0.0	3.2	6.8	10.7	14.7	.3	.1382
100.	1047.	3.1	-6.4	0.0	7.1	14.9	23.0	1.8	.2831
150.	1025.	8.0	-20.3	-10.7	0.0	11.7	23.8	3.0	.4262
200.	1004.	15.1	-42.6	-29.8	-15.5	0.0	16.3	5.0	.5739
250.	984.	24.5	-73.6	-57.6	-39.7	-20.3	0.0	7.6	.7248
300.	965.	36.3	-113.6	-94.3	-73.0	-49.6	-25.3	10.7	.8788
350.	946.	50.8	-163.2	-140.7	-115.8	-88.6	-60.2	14.5	1.0367
400.	928.	67.7	-221.3	-195.6	-167.1	-136.0	-103.5	18.2	1.1945
450.	911.	87.7	-290.2	-261.4	-229.3	-194.3	-157.7	22.9	1.3575
500.	894.	111.1	-371.1	-339.0	-303.4	-264.5	-223.9	28.6	1.5264

BALLISTIC COEFFICIENT .44, MUZZLE VELOCITY FT/SEC 1200.

RANGE	VELOCITY	HEIGHT	50 YD	100 YD	150 YD	200 YD	250 YD	DRIFT	TIME
0.	1200.	0.0	-1.5	-1.5	-1.5	-1.5	-1.5	0.0	0.0000
50.	1156.	.0	0.0	2.4	5.6	9.1	12.6	.6	.1285
100.	1120.	2.5	-4.9	0.0	6.3	13.2	20.2	1.6	.2592
150.	1090.	6.7	-16.8	-9.5	0.0	10.4	20.9	3.5	.3951
200.	1063.	13.0	-36.3	-26.5	-13.9	0.0	13.9	6.3	.5360
250.	1040.	21.2	-62.8	-50.6	-34.8	-17.4	0.0	9.1	.6769
300.	1018.	31.8	-97.8	-83.1	-64.2	-43.3	-22.4	12.9	.8231
350.	997.	44.6	-141.0	-123.9	-101.8	-77.5	-53.1	17.1	.9720
400.	977.	59.9	-192.8	-173.3	-148.1	-120.3	-92.4	21.8	1.1239
450.	958.	77.7	-253.6	-231.6	-203.2	-171.9	-140.6	27.0	1.2787
500.	940.	98.4	-323.9	-299.4	-267.9	-233.2	-198.3	32.9	1.4369

BALLISTIC COEFFICIENT .44, MUZZLE VELOCITY FT/SEC 1300.

RANGE	VELOCITY	HEIGHT	50 YD	100 YD	150 YD	200 YD	250 YD	DRIFT	TIME
0.	1300.	0.0	-1.5	-1.5	-1.5	-1.5	-1.5	0.0	0.0000
50.	1247.	-.1	0.0	2.1	4.8	7.7	10.9	.5	.1181
100.	1199.	2.0	-4.1	0.0	5.5	11.4	17.7	1.8	.2410
150.	1155.	5.8	-14.5	-8.3	0.0	8.8	18.3	4.2	.3698
200.	1119.	11.3	-31.0	-22.8	-11.7	0.0	12.7	6.8	.5005
250.	1089.	18.7	-54.6	-44.3	-30.5	-15.9	0.0	10.4	.6363
300.	1062.	28.2	-85.8	-73.5	-56.9	-39.4	-20.3	14.9	.7768
350.	1039.	39.7	-124.0	-109.6	-90.3	-69.8	-47.6	19.5	.9183
400.	1017.	53.6	-170.7	-154.2	-132.1	-108.7	-83.3	24.9	1.0646
450.	997.	69.9	-225.4	-206.9	-182.0	-155.7	-127.1	30.8	1.2136
500.	977.	88.8	-288.7	-268.1	-240.5	-211.2	-179.5	37.3	1.3656

BALLISTIC COEFFICIENT .44, MUZZLE VELOCITY FT/SEC 1400.

RANGE	VELOCITY	HEIGHT	50 YD	100 YD	150 YD	200 YD	250 YD	DRIFT	TIME
0.	1400.	0.0	-1.5	-1.5	-1.5	-1.5	-1.5	0.0	0.0000
50.	1339.	-.2	0.0	1.7	4.0	6.6	9.4	.5	.1097
100.	1283.	1.7	-3.3	0.0	4.7	9.8	15.5	1.7	.2241
150.	1232.	4.9	-12.0	-7.0	0.0	7.8	16.2	3.9	.3436
200.	1184.	9.8	-26.4	-19.7	-10.4	0.0	11.2	6.9	.4678
250.	1144.	16.3	-47.0	-38.6	-27.0	-14.0	0.0	10.7	.5967
300.	1110.	24.8	-74.3	-64.3	-50.3	-34.8	-17.9	15.4	.7301
350.	1080.	35.3	-108.6	-96.9	-80.6	-62.4	-42.8	20.6	.8670
400.	1055.	48.0	-150.3	-136.9	-118.3	-97.5	-75.1	26.5	1.0076
450.	1032.	62.9	-199.7	-184.7	-163.7	-140.3	-115.1	33.0	1.1515
500.	1011.	80.6	-258.5	-241.7	-218.4	-192.5	-164.4	40.5	1.3016

BALLISTIC COEFFICIENT .44, MUZZLE VELOCITY FT/SEC 1500.

RANGE	VELOCITY	HEIGHT	50 YD	100 YD	150 YD	200 YD	250 YD	DRIFT	TIME
0.	1500.	0.0	-1.5	-1.5	-1.5	-1.5	-1.5	0.0	0.0000
50.	1431.	-.2	0.0	1.4	3.4	5.7	8.1	.5	.1027
100.	1367.	1.4	-2.8	0.0	4.0	8.6	13.5	1.8	.2100
150.	1309.	4.2	-10.2	-6.0	0.0	6.9	14.3	3.9	.3220
200.	1255.	8.5	-22.7	-17.2	-9.2	0.0	9.9	6.9	.4394
250.	1206.	14.4	-40.7	-33.8	-23.8	-12.3	0.0	10.8	.5611
300.	1162.	22.0	-64.7	-56.4	-44.4	-30.6	-15.9	15.5	.6878
350.	1125.	31.5	-95.2	-85.6	-71.5	-55.5	-38.2	21.0	.8192
400.	1094.	43.0	-132.6	-121.5	-105.5	-87.1	-67.4	27.2	.9545
450.	1067.	56.6	-177.0	-164.6	-146.5	-125.9	-103.8	34.0	1.0933
500.	1043.	72.7	-229.4	-215.7	-195.6	-172.7	-148.0	41.7	1.2367

BALLISTIC COEFFICIENT .44, MUZZLE VELOCITY FT/SEC 1600.

RANGE	VELOCITY	HEIGHT	50 YD	100 YD	150 YD	200 YD	250 YD	DRIFT	TIME
0.	1600.	0.0	-1.5	-1.5	-1.5	-1.5	-1.5	0.0	0.0000
50.	1527.	-.3	0.0	1.1	2.9	4.5	7.1	.4	.0959
100.	1456.	1.1	-2.2	0.0	3.5	7.5	11.9	1.6	.1966
150.	1390.	3.6	-8.6	-5.2	0.0	6.0	12.6	3.6	.3019
200.	1330.	7.4	-19.5	-15.0	-8.0	0.0	8.8	6.6	.4123
250.	1275.	12.6	-35.3	-29.7	-21.0	-11.0	0.0	10.4	.5278
300.	1224.	19.4	-56.4	-49.6	-39.2	-27.2	-14.0	15.0	.6475
350.	1178.	27.9	-83.4	-75.6	-63.4	-49.4	-34.0	20.6	.7731
400.	1138.	38.4	-116.7	-107.7	-93.8	-77.8	-60.2	26.9	.9028
450.	1105.	50.8	-156.5	-146.4	-130.7	-112.7	-93.0	33.9	1.0361
500.	1076.	65.4	-203.4	-192.2	-174.8	-154.8	-132.8	41.6	1.1736

BALLISTIC COEFFICIENT .44, MUZZLE VELOCITY FT/SEC 1700.

RANGE	VELOCITY	HEIGHT	50 YD	100 YD	150 YD	200 YD	250 YD	DRIFT	TIME
0.	1700.	0.0	-1.5	-1.5	-1.5	-1.5	-1.5	0.0	0.0000
50.	1623.	-.4	0.0	.9	2.4	4.2	6.1	.4	.0904
100.	1549.	.9	-1.8	0.0	3.0	6.5	10.4	1.5	.1851
150.	1478.	3.1	-7.2	-4.5	0.0	5.3	11.1	3.4	.2841
200.	1410.	6.5	-16.7	-13.1	-7.0	0.0	7.8	6.2	.3881
250.	1348.	11.1	-30.6	-26.0	-18.5	-9.7	0.0	9.8	.4971
300.	1292.	17.1	-49.1	-43.7	-34.7	-24.1	-12.5	14.3	.6106
350.	1240.	24.8	-73.0	-66.7	-56.1	-43.8	-30.2	19.7	.7294
400.	1192.	34.2	-102.5	-95.2	-83.2	-69.1	-53.6	25.9	.8528
450.	1150.	45.4	-138.1	-129.9	-116.4	-100.5	-83.1	32.9	.9810
500.	1115.	58.8	-180.3	-171.2	-156.2	-138.6	-119.2	40.7	1.1137

BALLISTIC COEFFICIENT .44, MUZZLE VELOCITY FT/SEC 1800.

RANGE	VELOCITY	HEIGHT	50 YD	100 YD	150 YD	200 YD	250 YD	DRIFT	TIME
0.	1800.	0.0	-1.5	-1.5	-1.5	-1.5	-1.5	0.0	0.0000
50.	1720.	-.4	0.0	.7	2.1	3.6	5.3	.3	.0850
100.	1642.	.7	-1.5	0.0	2.7	5.7	9.1	1.4	.1746
150.	1567.	2.7	-6.2	-4.0	0.0	4.6	9.7	3.2	.2683
200.	1495.	5.7	-14.4	-11.4	-6.1	0.0	6.9	5.7	.3659
250.	1426.	9.8	-26.5	-22.9	-16.2	-8.6	0.0	9.2	.4688
300.	1363.	15.2	-43.0	-38.6	-30.6	-21.4	-11.2	13.5	.5765
350.	1305.	22.0	-64.1	-59.0	-49.7	-38.9	-26.9	18.6	.6889
400.	1252.	30.5	-90.3	-84.5	-73.9	-61.6	-47.9	24.6	.8063
450.	1203.	40.6	-122.1	-115.5	-103.6	-89.7	-74.3	31.4	.9287
500.	1159.	52.8	-159.9	-152.5	-139.3	-123.9	-106.8	39.2	1.0559

BALLISTIC COEFFICIENT .44, MUZZLE VELOCITY FT/SEC 1900.

RANGE	VELOCITY	HEIGHT	50 YD	100 YD	150 YD	200 YD	250 YD	DRIFT	TIME
0.	1900.	0.0	-1.5	-1.5	-1.5	-1.5	-1.5	0.0	0.0000
50.	1816.	-.4	0.0	.6	1.7	3.1	4.6	.3	.0807
100.	1735.	.6	-1.1	0.0	2.3	5.1	8.1	1.3	.1652
150.	1657.	2.3	-5.2	-3.5	0.0	4.1	8.7	3.0	.2538
200.	1581.	5.0	-12.4	-10.1	-5.5	0.0	6.0	5.4	.3467
250.	1509.	8.7	-23.1	-20.2	-14.4	-7.6	0.0	8.6	.4435
300.	1439.	13.5	-37.7	-34.2	-27.3	-19.0	-10.0	12.6	.5456
350.	1375.	19.7	-56.4	-52.4	-44.3	-34.7	-24.1	17.5	.6523
400.	1316.	27.2	-79.8	-75.2	-65.9	-54.9	-42.8	23.3	.7637
450.	1262.	36.4	-108.1	-102.9	-92.5	-80.1	-66.5	29.8	.8799
500.	1212.	47.4	-141.9	-136.2	-124.6	-110.8	-95.7	37.3	1.0013

BALLISTIC COEFFICIENT .45, MUZZLE VELOCITY FT/SEC 800.

RANGE	VELOCITY	HEIGHT	50 YD	100 YD	150 YD	200 YD	250 YD	DRIFT	TIME
0.	800.	0.0	-1.5	-1.5	-1.5	-1.5	-1.5	0.0	0.0000
50.	786.	1.0	0.0	6.3	13.2	20.5	28.0	.4	.1898
100.	773.	6.3	-12.6	0.0	13.8	28.4	43.4	1.3	.3823
150.	760.	15.2	-39.5	-20.7	0.0	21.9	44.4	2.6	.5773
200.	746.	28.2	-81.9	-56.8	-29.2	0.0	29.9	4.8	.7773
250.	733.	45.4	-139.8	-108.4	-73.9	-37.4	0.0	7.5	.9801
300.	720.	66.7	-213.5	-175.8	-134.5	-90.7	-45.8	10.7	1.1858
350.	708.	92.9	-304.8	-260.9	-212.7	-161.5	-109.2	14.8	1.3968
400.	695.	123.8	-413.1	-362.8	-307.7	-249.3	-189.4	19.5	1.6106
450.	682.	159.6	-539.1	-482.6	-420.6	-354.8	-287.5	24.7	1.8279
500.	670.	201.1	-684.8	-622.0	-553.2	-480.1	-405.3	30.9	2.0507

BALLISTIC COEFFICIENT .45, MUZZLE VELOCITY FT/SEC 900.

RANGE	VELOCITY	HEIGHT	50 YD	100 YD	150 YD	200 YD	250 YD	DRIFT	TIME
0.	900.	0.0	-1.5	-1.5	-1.5	-1.5	-1.5	0.0	0.0000
50.	884.	.6	0.0	4.8	10.3	15.9	21.9	.4	.1688
100.	869.	4.8	-9.7	0.0	10.9	22.2	34.1	1.2	.3403
150.	854.	11.9	-30.8	-16.3	0.0	17.0	34.8	2.5	.5143
200.	840.	22.2	-63.7	-44.4	-22.6	0.0	23.7	4.2	.6908
250.	826.	35.7	-109.3	-85.2	-58.0	-29.7	0.0	6.7	.8716
300.	812.	52.7	-167.6	-138.7	-106.0	-72.1	-36.4	9.7	1.0551
350.	798.	73.2	-238.9	-205.1	-167.0	-127.4	-85.8	13.1	1.2410
400.	784.	97.5	-324.1	-285.5	-242.0	-196.7	-149.2	17.2	1.4308
450.	771.	125.9	-423.6	-380.1	-331.2	-280.2	-226.8	21.8	1.6240
500.	758.	158.2	-537.2	-488.9	-434.5	-377.9	-318.5	27.0	1.8198

BALLISTIC COEFFICIENT .45, MUZZLE VELOCITY FT/SEC 1000.

RANGE	VELOCITY	HEIGHT	50 YD	100 YD	150 YD	200 YD	250 YD	DRIFT	TIME
0.	1000.	0.0	-1.5	-1.5	-1.5	-1.5	-1.5	0.0	0.0000
50.	980.	.4	0.0	3.8	8.2	12.8	17.6	.3	.1517
100.	962.	3.8	-7.6	0.0	8.9	18.1	27.7	1.1	.3064
150.	943.	9.6	-24.7	-13.4	0.0	13.7	28.2	2.6	.4650
200.	926.	18.0	-51.2	-36.1	-18.3	0.0	19.3	4.3	.6242
250.	909.	29.0	-88.1	-69.2	-46.9	-24.1	0.0	6.6	.7877
300.	893.	43.0	-135.6	-112.9	-86.2	-58.8	-29.8	9.6	.9543
350.	877.	59.9	-193.9	-167.5	-136.2	-104.3	-70.6	13.0	1.1237
400.	862.	79.9	-263.6	-233.4	-197.7	-161.2	-122.6	16.9	1.2962
450.	848.	103.2	-345.2	-311.2	-271.0	-229.9	-186.6	21.5	1.4719
500.	833.	130.0	-438.7	-400.8	-356.2	-310.6	-262.4	26.5	1.6503

BALLISTIC COEFFICIENT .45, MUZZLE VELOCITY FT/SEC 1100.

RANGE	VELOCITY	HEIGHT	50 YD	100 YD	150 YD	200 YD	250 YD	DRIFT	TIME
0.	1100.	0.0	-1.5	-1.5	-1.5	-1.5	-1.5	0.0	0.0000
50.	1073.	.2	0.0	3.2	6.8	10.7	14.7	.3	.1382
100.	1048.	3.1	-6.5	0.0	7.1	14.9	23.0	1.9	.2835
150.	1027.	8.0	-20.3	-10.7	0.0	11.7	23.8	3.0	.4263
200.	1006.	15.1	-42.7	-29.8	-15.5	0.0	16.2	5.0	.5739
250.	986.	24.5	-73.7	-57.5	-39.7	-20.3	0.0	7.6	.7247
300.	967.	36.3	-113.6	-94.2	-72.9	-49.6	-25.2	10.6	.8786
350.	949.	50.5	-161.9	-139.3	-114.4	-87.2	-58.8	13.8	1.0328
400.	932.	67.5	-220.6	-194.8	-166.4	-135.3	-102.8	17.9	1.1924
450.	914.	87.3	-289.2	-260.1	-228.2	-193.2	-156.6	22.4	1.3547
500.	898.	110.3	-368.7	-336.4	-300.9	-262.0	-221.4	27.7	1.5210

BALLISTIC COEFFICIENT .45, MUZZLE VELOCITY FT/SEC 1200.

RANGE	VELOCITY	HEIGHT	50 YD	100 YD	150 YD	200 YD	250 YD	DRIFT	TIME
0.	1200.	0.0	-1.5	-1.5	-1.5	-1.5	-1.5	0.0	0.0000
50.	1157.	.0	0.0	2.5	5.6	9.1	12.6	.6	.1281
100.	1122.	2.5	-4.9	0.0	6.3	13.3	20.3	1.6	.2590
150.	1092.	6.7	-16.8	-9.4	0.0	10.6	21.0	3.5	.3947
200.	1065.	13.1	-36.5	-26.7	-14.1	0.0	13.8	6.5	.5367
250.	1042.	21.2	-62.9	-50.7	-34.9	-17.3	0.0	9.1	.6769
300.	1021.	31.7	-97.6	-82.9	-64.0	-42.8	-22.1	12.6	.8219
350.	1000.	44.4	-140.6	-123.4	-101.4	-76.7	-52.5	16.8	.9702
400.	981.	59.7	-192.3	-172.7	-147.6	-119.3	-91.6	21.4	1.1218
450.	962.	77.5	-253.0	-230.9	-202.7	-170.9	-139.8	26.7	1.2765
500.	944.	98.1	-323.5	-299.0	-267.6	-232.3	-197.7	32.6	1.4350

BALLISTIC COEFFICIENT .45, MUZZLE VELOCITY FT/SEC 1300.

RANGE	VELOCITY	HEIGHT	50 YD	100 YD	150 YD	200 YD	250 YD	DRIFT	TIME
0.	1300.	0.0	-1.5	-1.5	-1.5	-1.5	-1.5	0.0	0.0000
50.	1248.	-.1	0.0	2.0	4.8	7.7	10.9	.5	.1180
100.	1201.	2.0	-4.1	0.0	5.4	11.3	17.7	1.7	.2406
150.	1158.	5.8	-14.3	-8.2	0.0	8.8	18.4	3.9	.3684
200.	1123.	11.2	-30.9	-22.7	-11.8	0.0	12.7	6.7	.4995
250.	1092.	18.6	-54.4	-44.2	-30.6	-15.8	0.0	10.2	.6351
300.	1065.	28.2	-86.1	-73.8	-57.5	-39.8	-20.8	14.9	.7773
350.	1042.	39.6	-123.9	-109.6	-90.5	-69.9	-47.7	19.3	.9174
400.	1021.	53.4	-170.0	-153.7	-131.9	-108.3	-82.9	24.5	1.0622
450.	1001.	69.6	-224.4	-206.1	-181.6	-155.0	-126.5	30.3	1.2105
500.	981.	88.3	-287.5	-267.1	-239.8	-210.3	-178.6	36.6	1.3621

BALLISTIC COEFFICIENT .45, MUZZLE VELOCITY FT/SEC 1400.

RANGE	VELOCITY	HEIGHT	50 YD	100 YD	150 YD	200 YD	250 YD	DRIFT	TIME
0.	1400.	0.0	-1.5	-1.5	-1.5	-1.5	-1.5	0.0	0.0000
50.	1340.	-.2	0.0	1.7	4.0	6.6	9.4	.4	.1096
100.	1286.	1.7	-3.3	0.0	4.7	9.8	15.4	1.7	.2239
150.	1235.	4.9	-12.0	-7.0	0.0	7.7	16.1	3.9	.3435
200.	1189.	9.7	-26.3	-19.6	-10.3	0.0	11.2	6.8	.4670
250.	1148.	16.3	-46.8	-38.5	-26.8	-13.9	0.0	10.5	.5954
300.	1114.	24.7	-74.0	-64.0	-49.9	-34.5	-17.8	15.0	.7283
350.	1085.	35.1	-108.1	-96.4	-80.0	-62.0	-42.5	20.2	.8647
400.	1059.	47.7	-149.5	-136.1	-117.4	-96.8	-74.5	25.9	1.0045
450.	1037.	62.5	-198.6	-183.5	-162.5	-139.4	-114.3	32.3	1.1478
500.	1016.	79.7	-255.7	-238.9	-215.6	-189.9	-162.0	39.2	1.2943

BALLISTIC COEFFICIENT .45, MUZZLE VELOCITY FT/SEC 1500.

RANGE	VELOCITY	HEIGHT	50 YD	100 YD	150 YD	200 YD	250 YD	DRIFT	TIME
0.	1500.	0.0	-1.5	-1.5	-1.5	-1.5	-1.5	0.0	0.0000
50.	1432.	-.2	0.0	1.4	3.4	5.6	8.1	.5	.1026
100.	1370.	1.4	-2.7	0.0	4.0	8.5	13.5	1.7	.2097
150.	1313.	4.2	-10.1	-6.0	0.0	6.8	14.2	3.8	.3215
200.	1260.	8.5	-22.5	-17.0	-9.0	0.0	9.9	6.7	.4380
250.	1212.	14.3	-40.5	-33.6	-23.6	-12.3	0.0	10.5	.5597
300.	1168.	21.8	-64.4	-56.1	-44.1	-30.6	-15.8	15.1	.6859
350.	1130.	31.3	-94.7	-85.1	-71.0	-55.2	-38.0	20.5	.8166
400.	1099.	42.7	-131.7	-120.7	-104.7	-86.6	-66.9	26.6	.9512
450.	1072.	56.2	-175.9	-163.6	-145.5	-125.2	-103.0	33.4	1.0895
500.	1047.	72.4	-228.9	-215.2	-195.2	-172.6	-147.9	41.3	1.2345

BALLISTIC COEFFICIENT .45, MUZZLE VELOCITY FT/SEC 1600.

RANGE	VELOCITY	HEIGHT	50 YD	100 YD	150 YD	200 YD	250 YD	DRIFT	TIME
0.	1600.	0.0	-1.5	-1.5	-1.5	-1.5	-1.5	0.0	0.0000
50.	1528.	-.3	0.0	1.1	2.9	4.9	7.0	.4	.0958
100.	1459.	1.1	-2.2	0.0	3.5	7.5	11.8	1.6	.1963
150.	1395.	3.6	-8.6	-5.2	0.0	6.0	12.5	3.6	.3017
200.	1335.	7.4	-19.4	-14.9	-8.0	0.0	8.6	6.5	.4117
250.	1281.	12.5	-35.1	-29.5	-20.8	-10.8	0.0	10.1	.5261
300.	1231.	19.3	-56.1	-49.4	-38.9	-27.0	-14.0	14.6	.6457
350.	1185.	27.7	-82.8	-75.0	-62.8	-48.8	-33.7	20.0	.7699
400.	1145.	38.0	-115.7	-106.8	-92.8	-76.9	-59.6	26.2	.8988
450.	1111.	50.4	-155.4	-145.3	-129.6	-111.7	-92.3	33.1	1.0320
500.	1082.	64.9	-202.3	-191.1	-173.7	-153.7	-132.1	40.9	1.1697

BALLISTIC COEFFICIENT .45, MUZZLE VELOCITY FT/SEC 1700.

RANGE	VELOCITY	HEIGHT	50 YD	100 YD	150 YD	200 YD	250 YD	DRIFT	TIME
0.	1700.	0.0	-1.5	-1.5	-1.5	-1.5	-1.5	0.0	0.0000
50.	1625.	-.4	0.0	.9	2.4	4.2	6.1	.4	.0903
100.	1552.	.9	-1.8	0.0	3.0	6.5	10.4	1.5	.1850
150.	1482.	3.1	-7.2	-4.5	0.0	5.3	11.1	3.3	.2836
200.	1416.	6.5	-16.6	-13.0	-7.0	0.0	7.7	6.1	.3874
250.	1355.	11.1	-30.4	-25.9	-18.4	-9.7	0.0	9.7	.4961
300.	1299.	17.0	-48.9	-43.4	-34.4	-23.9	-12.3	14.0	.6087
350.	1247.	24.6	-72.5	-66.2	-55.7	-43.4	-29.9	19.2	.7267
400.	1200.	33.9	-101.8	-94.5	-82.5	-68.5	-53.0	25.3	.8494
450.	1157.	45.1	-137.3	-129.1	-115.6	-99.9	-82.5	32.3	.9776
500.	1122.	58.2	-178.8	-169.7	-154.7	-137.2	-117.9	39.8	1.1084

BALLISTIC COEFFICIENT .45, MUZZLE VELOCITY FT/SEC 1800.

RANGE	VELOCITY	HEIGHT	50 YD	100 YD	150 YD	200 YD	250 YD	DRIFT	TIME
0.	1800.	0.0	-1.5	-1.5	-1.5	-1.5	-1.5	0.0	0.0000
50.	1721.	-.4	0.0	.7	2.0	3.6	5.3	.3	.0850
100.	1645.	.7	-1.5	0.0	2.6	5.7	9.1	1.4	.1744
150.	1572.	2.7	-6.1	-4.0	0.0	4.6	9.7	3.2	.2679
200.	1501.	5.6	-14.3	-11.4	-6.1	0.0	6.8	5.6	.3651
250.	1434.	9.7	-26.4	-22.8	-16.2	-8.5	0.0	9.0	.4677
300.	1371.	15.1	-42.7	-38.3	-30.4	-21.3	-11.0	13.1	.5746
350.	1314.	21.9	-63.7	-58.6	-49.4	-38.7	-26.8	18.2	.6867
400.	1261.	30.2	-89.6	-83.8	-73.2	-61.0	-47.3	24.0	.8028
450.	1212.	40.3	-121.3	-114.7	-102.8	-89.1	-73.7	30.8	.9249
500.	1168.	52.2	-158.4	-151.1	-137.9	-122.7	-105.6	38.2	1.0505

BALLISTIC COEFFICIENT .45, MUZZLE VELOCITY FT/SEC 1900.

RANGE	VELOCITY	HEIGHT	50 YD	100 YD	150 YD	200 YD	250 YD	DRIFT	TIME
0.	1900.	0.0	-1.5	-1.5	-1.5	-1.5	-1.5	0.0	0.0000
50.	1818.	-.4	0.0	.6	1.7	3.1	4.6	.3	.0806
100.	1738.	.6	-1.1	0.0	2.3	5.0	8.1	1.3	.1651
150.	1662.	2.3	-5.2	-3.5	0.0	4.1	8.6	2.9	.2534
200.	1588.	5.0	-12.4	-10.1	-5.5	0.0	6.0	5.3	.3459
250.	1517.	8.6	-23.0	-20.1	-14.4	-7.5	0.0	8.4	.4424
300.	1448.	13.4	-37.4	-34.0	-27.1	-18.9	-9.9	12.3	.5438
350.	1385.	19.5	-56.0	-52.0	-43.9	-34.3	-23.8	17.0	.6495
400.	1326.	27.0	-79.1	-74.6	-65.3	-54.4	-42.4	22.7	.7603
450.	1273.	36.1	-107.2	-102.1	-91.7	-79.4	-65.8	29.1	.8758
500.	1223.	46.9	-140.6	-134.9	-123.3	-109.6	-94.6	36.3	.9960

BALLISTIC COEFFICIENT .46, MUZZLE VELOCITY FT/SEC 800.

RANGE	VELOCITY	HEIGHT	50 YD	100 YD	150 YD	200 YD	250 YD	DRIFT	TIME
0.	800.	0.0	-1.5	-1.5	-1.5	-1.5	-1.5	0.0	0.0000
50.	787.	1.0	0.0	6.3	13.2	20.4	27.9	.4	.1898
100.	774.	6.3	-12.5	0.0	13.8	28.4	43.3	1.3	.3821
150.	760.	15.2	-39.5	-20.7	0.0	21.9	44.2	2.6	.5770
200.	748.	28.2	-81.8	-56.7	-29.2	0.0	29.8	4.7	.7768
250.	735.	45.3	-139.5	-108.1	-73.7	-37.3	0.0	7.3	.9791
300.	722.	66.6	-213.2	-175.6	-134.3	-90.5	-45.8	10.5	1.1846
350.	710.	92.7	-304.2	-260.3	-212.1	-161.1	-108.9	14.5	1.3950
400.	697.	123.3	-411.8	-361.7	-306.6	-248.3	-188.6	19.0	1.6078
450.	685.	159.1	-537.9	-481.4	-419.5	-353.8	-286.8	24.2	1.8251
500.	673.	200.3	-682.6	-619.9	-551.1	-478.1	-403.6	30.2	2.0465

BALLISTIC COEFFICIENT .46, MUZZLE VELOCITY FT/SEC 900.

RANGE	VELOCITY	HEIGHT	50 YD	100 YD	150 YD	200 YD	250 YD	DRIFT	TIME
0.	900.	0.0	-1.5	-1.5	-1.5	-1.5	-1.5	0.0	0.0000
50.	885.	.6	0.0	4.8	10.3	15.9	21.8	.4	.1688
100.	870.	4.8	-9.7	0.0	10.9	22.2	34.0	1.2	.3402
150.	855.	11.9	-30.8	-16.3	0.0	17.0	34.7	2.5	.5140
200.	841.	22.1	-63.6	-44.3	-22.6	0.0	23.7	4.2	.6904
250.	827.	35.7	-109.2	-85.0	-57.9	-29.6	0.0	6.6	.8709
300.	814.	52.6	-167.3	-138.3	-105.8	-71.8	-36.3	9.5	1.0538
350.	800.	72.9	-238.2	-204.5	-166.5	-126.9	-85.4	12.7	1.2391
400.	787.	97.3	-323.4	-284.8	-241.4	-196.2	-148.8	16.8	1.4289
450.	774.	125.4	-422.4	-378.9	-330.1	-279.2	-225.9	21.3	1.6212
500.	760.	157.6	-535.4	-487.2	-432.9	-376.3	-317.1	26.3	1.8161

BALLISTIC COEFFICIENT .46, MUZZLE VELOCITY FT/SEC 1000.

RANGE	VELOCITY	HEIGHT	50 YD	100 YD	150 YD	200 YD	250 YD	DRIFT	TIME
0.	1000.	0.0	-1.5	-1.5	-1.5	-1.5	-1.5	0.0	0.0000
50.	981.	.4	0.0	3.8	8.3	12.8	17.6	.3	.1518
100.	962.	3.8	-7.6	0.0	8.9	18.0	27.6	1.2	.3066
150.	944.	9.6	-24.8	-13.4	0.0	13.6	28.0	2.7	.4653
200.	928.	17.9	-51.1	-36.0	-18.1	0.0	19.2	4.2	.6237
250.	911.	29.0	-87.9	-69.0	-46.7	-24.0	0.0	6.5	.7869
300.	895.	43.1	-136.1	-113.3	-86.5	-59.4	-30.6	9.7	.9554
350.	880.	59.7	-193.4	-166.9	-135.6	-103.9	-70.3	12.7	1.1220
400.	865.	79.7	-263.0	-232.7	-197.0	-160.7	-122.3	16.6	1.2943
450.	851.	102.9	-344.2	-310.1	-269.9	-229.1	-185.9	21.0	1.4694
500.	837.	129.4	-437.0	-399.2	-354.5	-309.2	-261.2	25.8	1.6469

BALLISTIC COEFFICIENT .46, MUZZLE VELOCITY FT/SEC 1100.

RANGE	VELOCITY	HEIGHT	50 YD	100 YD	150 YD	200 YD	250 YD	DRIFT	TIME
0.	1100.	0.0	-1.5	-1.5	-1.5	-1.5	-1.5	0.0	0.0000
50.	1073.	.2	0.0	3.0	6.8	10.7	14.7	.3	.1382
100.	1050.	3.0	-6.0	0.0	7.5	15.3	23.4	1.2	.2795
150.	1028.	8.0	-20.4	-11.3	0.0	11.7	23.8	3.1	.4264
200.	1008.	15.1	-42.7	-30.6	-15.5	0.0	16.2	5.0	.5740
250.	988.	24.5	-73.7	-58.6	-39.7	-20.3	0.0	7.5	.7247
300.	970.	36.3	-113.7	-95.5	-72.9	-49.6	-25.2	10.6	.8784
350.	952.	50.3	-161.5	-140.3	-113.9	-86.7	-58.3	13.5	1.0311
400.	935.	67.3	-220.0	-195.8	-165.6	-134.5	-102.0	17.5	1.1903
450.	918.	87.0	-288.2	-261.0	-227.0	-192.0	-155.5	21.9	1.3518
500.	902.	109.7	-366.9	-336.6	-298.9	-260.0	-219.4	26.9	1.5167

BALLISTIC COEFFICIENT .46, MUZZLE VELOCITY FT/SEC 1200.

RANGE	VELOCITY	HEIGHT	50 YD	100 YD	150 YD	200 YD	250 YD	DRIFT	TIME
0.	1200.	0.0	-1.5	-1.5	-1.5	-1.5	-1.5	0.0	0.0000
50.	1158.	.0	0.0	2.5	5.6	9.0	12.6	.5	.1278
100.	1123.	2.5	-4.9	0.0	6.3	13.0	20.3	1.6	.2589
150.	1094.	6.7	-16.8	-9.4	0.0	10.1	21.1	3.4	.3943
200.	1068.	12.9	-35.9	-26.0	-13.5	0.0	14.6	5.8	.5331
250.	1044.	21.3	-63.1	-50.8	-35.1	-18.3	0.0	9.2	.6772
300.	1023.	31.6	-97.6	-82.9	-64.1	-43.8	-21.9	12.6	.8214
350.	1003.	44.4	-140.6	-123.4	-101.5	-77.8	-52.3	16.6	.9696
400.	984.	59.6	-192.3	-172.6	-147.5	-120.5	-91.3	21.3	1.1209
450.	965.	77.3	-252.9	-230.8	-202.6	-172.2	-139.3	26.4	1.2752
500.	948.	97.4	-321.8	-297.1	-265.8	-232.0	-195.5	31.7	1.4303

BALLISTIC COEFFICIENT .46, MUZZLE VELOCITY FT/SEC 1300.

RANGE	VELOCITY	HEIGHT	50 YD	100 YD	150 YD	200 YD	250 YD	DRIFT	TIME
0.	1300.	0.0	-1.5	-1.5	-1.5	-1.5	-1.5	0.0	0.0000
50.	1250.	-.1	0.0	2.0	4.7	7.7	10.8	.5	.1180
100.	1203.	2.0	-4.1	0.0	5.4	11.3	17.6	1.7	.2404
150.	1161.	5.7	-14.2	-8.1	0.0	8.9	18.4	3.7	.3673
200.	1125.	11.2	-30.8	-22.6	-11.9	0.0	12.6	6.6	.4988
250.	1095.	18.5	-54.2	-44.0	-30.6	-15.8	0.0	10.0	.6340
300.	1069.	27.9	-84.9	-72.7	-56.6	-38.7	-19.8	14.1	.7726
350.	1046.	39.6	-124.0	-109.7	-90.9	-70.1	-48.1	19.3	.9171
400.	1025.	53.3	-169.8	-153.4	-132.0	-108.2	-83.0	24.3	1.0609
450.	1005.	69.4	-224.1	-205.7	-181.6	-154.8	-126.5	30.0	1.2089
500.	985.	88.0	-286.9	-266.6	-239.7	-210.0	-178.5	36.3	1.3600

BALLISTIC COEFFICIENT .46, MUZZLE VELOCITY FT/SEC 1400.

RANGE	VELOCITY	HEIGHT	50 YD	100 YD	150 YD	200 YD	250 YD	DRIFT	TIME
0.	1400.	0.0	-1.5	-1.5	-1.5	-1.5	-1.5	0.0	0.0000
50.	1342.	-.2	0.0	1.7	4.0	6.6	9.3	.4	.1096
100.	1288.	1.7	-3.3	0.0	4.6	9.9	15.3	1.7	.2237
150.	1238.	4.9	-12.0	-7.0	0.0	7.8	16.0	3.8	.3428
200.	1192.	9.7	-26.4	-19.7	-10.4	0.0	10.9	6.8	.4671
250.	1152.	16.2	-46.6	-38.3	-26.7	-13.7	0.0	10.3	.5942
300.	1118.	24.6	-73.9	-63.9	-50.0	-34.3	-17.9	14.9	.7274
350.	1089.	35.0	-107.6	-95.9	-79.7	-61.4	-42.3	19.8	.8625
400.	1063.	47.6	-149.4	-136.1	-117.5	-96.6	-74.8	25.8	1.0037
450.	1041.	62.1	-197.6	-182.6	-161.7	-138.3	-113.7	31.7	1.1446
500.	1020.	79.1	-254.2	-237.5	-214.3	-188.2	-160.9	38.5	1.2901

BALLISTIC COEFFICIENT .46, MUZZLE VELOCITY FT/SEC 1500.

RANGE	VELOCITY	HEIGHT	50 YD	100 YD	150 YD	200 YD	250 YD	DRIFT	TIME
0.	1500.	0.0	-1.5	-1.5	-1.5	-1.5	-1.5	0.0	0.0000
50.	1434.	-.2	0.0	1.4	3.4	5.6	8.1	.5	.1026
100.	1373.	1.4	-2.7	0.0	4.0	8.5	13.4	1.7	.2095
150.	1316.	4.2	-10.1	-6.0	0.0	6.7	14.1	3.7	.3212
200.	1265.	8.4	-22.5	-17.0	-9.0	0.0	9.8	6.6	.4373
250.	1217.	14.2	-40.3	-33.5	-23.5	-12.3	0.0	10.3	.5586
300.	1172.	21.8	-64.5	-56.3	-44.3	-30.8	-16.1	15.2	.6861
350.	1135.	31.2	-94.6	-85.0	-71.0	-55.3	-38.1	20.4	.8158
400.	1104.	42.4	-130.9	-120.0	-104.0	-86.0	-66.4	26.1	.9482
450.	1077.	55.8	-174.8	-162.5	-144.5	-124.2	-102.1	32.7	1.0857
500.	1053.	71.5	-226.0	-212.4	-192.4	-169.9	-145.3	39.9	1.2267

BALLISTIC COEFFICIENT .46, MUZZLE VELOCITY FT/SEC 1600.

RANGE	VELOCITY	HEIGHT	50 YD	100 YD	150 YD	200 YD	250 YD	DRIFT	TIME
0.	1600.	0.0	-1.5	-1.5	-1.5	-1.5	-1.5	0.0	0.0000
50.	1530.	-.3	0.0	1.1	2.8	4.8	7.0	.4	.0958
100.	1462.	1.1	-2.2	0.0	3.5	7.4	11.7	1.5	.1961
150.	1399.	3.6	-8.5	-5.2	0.0	5.9	12.4	3.5	.3010
200.	1340.	7.3	-19.3	-14.8	-7.9	0.0	8.6	6.3	.4106
250.	1287.	12.5	-34.9	-29.3	-20.7	-10.8	0.0	9.9	.5248
300.	1237.	19.2	-55.8	-49.2	-38.8	-26.9	-14.0	14.4	.6441
350.	1191.	27.5	-82.3	-74.5	-62.4	-48.6	-33.5	19.6	.7675
400.	1151.	37.8	-115.0	-106.1	-92.3	-76.4	-59.2	25.6	.8956
450.	1117.	50.1	-154.7	-144.7	-129.1	-111.3	-91.9	32.6	1.0292
500.	1088.	64.3	-200.5	-189.3	-172.1	-152.2	-130.7	39.9	1.1641

BALLISTIC COEFFICIENT .46, MUZZLE VELOCITY FT/SEC 1700.

RANGE	VELOCITY	HEIGHT	50 YD	100 YD	150 YD	200 YD	250 YD	DRIFT	TIME
0.	1700.	0.0	-1.5	-1.5	-1.5	-1.5	-1.5	0.0	0.0000
50.	1626.	-.4	0.0	.9	2.4	4.1	6.0	.4	.0903
100.	1555.	.9	-1.8	0.0	3.0	6.5	10.3	1.5	.1848
150.	1487.	3.1	-7.2	-4.5	0.0	5.2	11.0	3.2	.2832
200.	1421.	6.4	-16.5	-12.9	-6.9	0.0	7.7	5.9	.3864
250.	1361.	11.0	-30.2	-25.7	-18.3	-9.6	0.0	9.4	.4945
300.	1306.	16.9	-48.6	-43.2	-34.2	-23.8	-12.3	13.6	.6068
350.	1255.	24.4	-72.1	-65.8	-55.4	-43.2	-29.8	18.8	.7245
400.	1208.	33.6	-101.0	-93.8	-81.9	-68.0	-52.7	24.7	.8461
450.	1165.	44.7	-136.0	-127.8	-114.4	-98.8	-81.5	31.4	.9726
500.	1129.	57.7	-177.4	-168.3	-153.4	-136.1	-116.9	38.9	1.1035

BALLISTIC COEFFICIENT .46, MUZZLE VELOCITY FT/SEC 1800.

RANGE	VELOCITY	HEIGHT	50 YD	100 YD	150 YD	200 YD	250 YD	DRIFT	TIME
0.	1800.	0.0	-1.5	-1.5	-1.5	-1.5	-1.5	0.0	0.0000
50.	1723.	-.4	0.0	.7	2.0	3.6	5.3	.3	.0850
100.	1649.	.7	-1.5	0.0	2.6	5.7	9.1	1.3	.1743
150.	1577.	2.7	-6.1	-3.9	0.0	4.6	9.6	3.1	.2675
200.	1508.	5.6	-14.2	-11.3	-6.1	0.0	6.8	5.5	.3645
250.	1441.	9.7	-26.3	-22.7	-16.1	-8.5	0.0	8.8	.4665
300.	1379.	15.0	-42.4	-38.1	-30.2	-21.1	-10.9	12.8	.5728
350.	1322.	21.7	-63.2	-58.2	-48.9	-38.3	-26.4	17.7	.6839
400.	1270.	29.9	-89.0	-83.2	-72.7	-60.5	-46.9	23.4	.7997
450.	1222.	39.9	-120.1	-113.5	-101.7	-88.0	-72.8	29.9	.9200
500.	1177.	51.8	-157.3	-150.0	-136.9	-121.7	-104.7	37.4	1.0460

BALLISTIC COEFFICIENT .46, MUZZLE VELOCITY FT/SEC 1900.

RANGE	VELOCITY	HEIGHT	50 YD	100 YD	150 YD	200 YD	250 YD	DRIFT	TIME
0.	1900.	0.0	-1.5	-1.5	-1.5	-1.5	-1.5	0.0	0.0000
50.	1820.	-.4	0.0	.6	1.7	3.1	4.6	.3	.0805
100.	1742.	.6	-1.1	0.0	2.3	5.0	8.0	1.2	.1650
150.	1667.	2.3	-5.2	-3.4	0.0	4.1	8.6	2.9	.2530
200.	1594.	5.0	-12.3	-10.0	-5.5	0.0	6.0	5.2	.3452
250.	1524.	8.6	-22.9	-20.0	-14.3	-7.5	0.0	8.2	.4413
300.	1457.	13.4	-37.2	-33.8	-26.9	-18.7	-9.8	12.0	.5420
350.	1394.	19.4	-55.6	-51.6	-43.6	-34.1	-23.6	16.6	.6472
400.	1336.	26.8	-78.6	-74.1	-64.9	-54.0	-42.0	22.2	.7575
450.	1283.	35.7	-106.4	-101.2	-90.9	-78.6	-65.2	28.4	.8718
500.	1233.	46.5	-139.6	-133.9	-122.5	-108.8	-93.9	35.6	.9917

BALLISTIC COEFFICIENT .47, MUZZLE VELOCITY FT/SEC 800.

RANGE	VELOCITY	HEIGHT	50 YD	100 YD	150 YD	200 YD	250 YD	DRIFT	TIME
0.	800.	0.0	-1.5	-1.5	-1.5	-1.5	-1.5	0.0	0.0000
50.	787.	1.0	0.0	6.3	13.1	20.4	27.8	.4	.1898
100.	774.	6.3	-12.5	0.0	13.8	28.3	43.2	1.2	.3820
150.	761.	15.2	-39.4	-20.6	0.0	21.8	44.1	2.5	.5768
200.	749.	28.2	-81.7	-56.6	-29.1	0.0	29.7	4.6	.7763
250.	736.	45.2	-139.2	-107.9	-73.5	-37.1	0.0	7.1	.9780
300.	724.	66.5	-212.9	-175.3	-134.0	-90.4	-45.9	10.3	1.1836
350.	711.	92.4	-303.5	-259.7	-211.5	-160.6	-108.7	14.2	1.3932
400.	699.	122.9	-410.6	-360.5	-305.4	-247.2	-187.9	18.5	1.6049
450.	687.	158.6	-536.6	-480.3	-418.3	-352.8	-286.0	23.7	1.8223
500.	675.	199.5	-680.4	-617.8	-548.9	-476.2	-402.0	29.4	2.0423

BALLISTIC COEFFICIENT .47, MUZZLE VELOCITY FT/SEC 900.

RANGE	VELOCITY	HEIGHT	50 YD	100 YD	150 YD	200 YD	250 YD	DRIFT	TIME
0.	900.	0.0	-1.5	-1.5	-1.5	-1.5	-1.5	0.0	0.0000
50.	885.	.6	0.0	4.8	10.2	15.9	21.8	.4	.1688
100.	870.	4.8	-9.6	0.0	10.8	22.2	34.0	1.2	.3401
150.	856.	11.9	-30.7	-16.2	0.0	17.0	34.7	2.4	.5137
200.	842.	22.1	-63.6	-44.3	-22.6	0.0	23.6	4.1	.6900
250.	829.	35.6	-109.0	-84.9	-57.8	-29.5	0.0	6.5	.8701
300.	815.	52.4	-166.9	-138.0	-105.5	-71.5	-36.1	9.2	1.0525
350.	802.	72.8	-237.9	-204.1	-166.2	-126.6	-85.3	12.5	1.2378
400.	789.	97.0	-322.7	-284.2	-240.9	-195.6	-148.3	16.5	1.4269
450.	776.	125.0	-421.1	-377.8	-329.0	-278.1	-224.9	20.8	1.6183
500.	763.	157.0	-534.1	-485.9	-431.8	-375.1	-316.1	25.8	1.8131

BALLISTIC COEFFICIENT .47, MUZZLE VELOCITY FT/SEC 1000.

RANGE	VELOCITY	HEIGHT	50 YD	100 YD	150 YD	200 YD	250 YD	DRIFT	TIME
0.	1000.	0.0	-1.5	-1.5	-1.5	-1.5	-1.5	0.0	0.0000
50.	981.	.4	0.0	3.8	8.3	12.8	17.5	.3	.1519
100.	963.	3.8	-7.6	0.0	8.9	17.9	27.5	1.2	.3068
150.	946.	9.7	-24.8	-13.4	0.0	13.5	27.8	2.7	.4655
200.	929.	17.9	-51.0	-35.8	-18.0	0.0	19.1	4.1	.6233
250.	913.	28.9	-87.7	-68.7	-46.4	-23.9	0.0	6.3	.7860
300.	897.	42.9	-135.3	-112.6	-85.8	-58.8	-30.1	9.4	.9531
350.	882.	59.5	-192.9	-166.3	-135.0	-103.6	-70.1	12.4	1.1204
400.	867.	79.4	-262.3	-231.9	-196.2	-160.2	-122.0	16.3	1.2925
450.	853.	102.5	-343.1	-308.9	-268.7	-228.3	-185.2	20.6	1.4668
500.	839.	128.9	-435.4	-397.5	-352.8	-307.8	-260.0	25.2	1.6434

BALLISTIC COEFFICIENT .47, MUZZLE VELOCITY FT/SEC 1100.

RANGE	VELOCITY	HEIGHT	50 YD	100 YD	150 YD	200 YD	250 YD	DRIFT	TIME
0.	1100.	0.0	-1.5	-1.5	-1.5	-1.5	-1.5	0.0	0.0000
50.	1074.	.2	0.0	3.0	6.8	10.7	14.8	.3	.1381
100.	1051.	3.0	-6.0	0.0	7.6	15.3	23.5	1.2	.2794
150.	1029.	8.0	-20.4	-11.4	0.0	11.7	23.8	3.1	.4266
200.	1009.	15.1	-42.8	-30.7	-15.6	0.0	16.2	5.0	.5741
250.	990.	24.5	-73.8	-58.7	-39.7	-20.3	0.0	7.5	.7246
300.	972.	35.9	-112.3	-94.2	-71.5	-48.2	-23.8	9.8	.8739
350.	955.	50.1	-161.0	-139.9	-113.4	-86.2	-57.8	13.2	1.0295
400.	937.	67.0	-219.3	-195.2	-164.9	-133.8	-101.3	17.1	1.1881
450.	921.	86.6	-287.2	-260.0	-226.0	-191.0	-154.5	21.4	1.3491
500.	905.	109.2	-365.8	-335.6	-297.7	-258.9	-218.3	26.4	1.5138

BALLISTIC COEFFICIENT .47, MUZZLE VELOCITY FT/SEC 1200.

RANGE	VELOCITY	HEIGHT	50 YD	100 YD	150 YD	200 YD	250 YD	DRIFT	TIME
0.	1200.	0.0	-1.5	-1.5	-1.5	-1.5	-1.5	0.0	0.0000
50.	1159.	.0	0.0	2.5	5.6	9.0	12.7	.4	.1275
100.	1125.	2.5	-5.0	0.0	6.2	13.0	20.4	1.5	.2587
150.	1095.	6.7	-16.8	-9.4	0.0	10.1	21.2	3.3	.3939
200.	1070.	12.9	-35.9	-26.0	-13.5	0.0	14.8	5.7	.5325
250.	1047.	21.3	-63.3	-50.9	-35.3	-18.5	0.0	9.2	.6775
300.	1026.	31.6	-97.7	-82.8	-64.1	-43.9	-21.7	12.5	.8210
350.	1006.	44.3	-140.7	-123.3	-101.5	-77.9	-52.0	16.5	.9689
400.	987.	59.5	-192.3	-172.5	-147.5	-120.5	-91.0	21.1	1.1199
450.	969.	77.2	-252.8	-230.6	-202.5	-172.1	-138.8	26.2	1.2739
500.	952.	97.0	-320.7	-295.9	-264.7	-231.0	-194.0	31.1	1.4270

BALLISTIC COEFFICIENT .47, MUZZLE VELOCITY FT/SEC 1300.

RANGE	VELOCITY	HEIGHT	50 YD	100 YD	150 YD	200 YD	250 YD	DRIFT	TIME
0.	1300.	0.0	-1.5	-1.5	-1.5	-1.5	-1.5	0.0	0.0000
50.	1251.	-.1	0.0	2.0	4.7	7.7	10.8	.4	.1179
100.	1205.	2.0	-4.1	0.0	5.4	11.3	17.5	1.7	.2402
150.	1163.	5.7	-14.1	-8.0	0.0	8.9	18.3	3.7	.3669
200.	1128.	11.2	-30.7	-22.5	-11.8	0.0	12.5	6.4	.4980
250.	1098.	18.5	-54.0	-43.9	-30.5	-15.7	0.0	9.8	.6328
300.	1072.	27.8	-84.6	-72.4	-56.3	-38.6	-19.8	13.9	.7711
350.	1049.	39.6	-124.0	-109.8	-91.0	-70.3	-48.4	19.2	.9168
400.	1028.	53.1	-169.5	-153.2	-131.8	-108.1	-83.0	24.0	1.0595
450.	1008.	69.2	-223.7	-205.4	-181.3	-154.7	-126.4	29.7	1.2072
500.	989.	87.8	-286.4	-266.0	-239.3	-209.7	-178.3	35.9	1.3580

BALLISTIC COEFFICIENT .47, MUZZLE VELOCITY FT/SEC 1400.

RANGE	VELOCITY	HEIGHT	50 YD	100 YD	150 YD	200 YD	250 YD	DRIFT	TIME
0.	1400.	0.0	-1.5	-1.5	-1.5	-1.5	-1.5	0.0	0.0000
50.	1343.	-.2	0.0	1.7	4.0	6.6	9.3	.4	.1096
100.	1290.	1.6	-3.3	0.0	4.6	9.8	15.4	1.6	.2235
150.	1241.	4.9	-11.9	-6.9	0.0	7.8	16.1	3.7	.3422
200.	1196.	9.7	-26.2	-19.6	-10.3	0.0	11.2	6.6	.4658
250.	1156.	16.2	-46.7	-38.4	-26.9	-14.0	0.0	10.3	.5944
300.	1122.	24.5	-73.3	-63.3	-49.5	-34.0	-17.2	14.4	.7247
350.	1093.	34.8	-107.1	-95.5	-79.3	-61.2	-41.7	19.4	.8604
400.	1068.	47.2	-148.0	-134.8	-116.3	-95.6	-73.3	25.0	.9992
450.	1045.	62.0	-197.4	-182.5	-161.7	-138.5	-113.3	31.5	1.1435
500.	1024.	78.8	-253.5	-236.9	-213.8	-187.9	-160.0	38.1	1.2877

BALLISTIC COEFFICIENT .47, MUZZLE VELOCITY FT/SEC 1500.

RANGE	VELOCITY	HEIGHT	50 YD	100 YD	150 YD	200 YD	250 YD	DRIFT	TIME
0.	1500.	0.0	-1.5	-1.5	-1.5	-1.5	-1.5	0.0	0.0000
50.	1435.	-.2	0.0	1.4	3.3	5.6	8.0	.5	.1026
100.	1375.	1.4	-2.7	0.0	4.0	8.5	13.3	1.7	.2096
150.	1320.	4.2	-10.0	-5.9	0.0	6.7	14.0	3.6	.3206
200.	1269.	8.4	-22.4	-16.9	-9.0	0.0	9.7	6.4	.4366
250.	1222.	14.1	-40.1	-33.3	-23.4	-12.1	0.0	10.0	.5570
300.	1178.	21.6	-63.8	-55.6	-43.8	-30.3	-15.7	14.6	.6828
350.	1141.	30.9	-93.6	-84.0	-70.1	-54.4	-37.4	19.6	.8116
400.	1109.	42.1	-130.2	-119.3	-103.4	-85.5	-66.1	25.6	.9453
450.	1081.	55.5	-174.0	-161.7	-143.9	-123.6	-101.8	32.2	1.0829
500.	1058.	71.0	-224.6	-210.9	-191.1	-168.6	-144.4	39.1	1.2223

BALLISTIC COEFFICIENT .47, MUZZLE VELOCITY FT/SEC 1600.

RANGE	VELOCITY	HEIGHT	50 YD	100 YD	150 YD	200 YD	250 YD	DRIFT	TIME
0.	1600.	0.0	-1.5	-1.5	-1.5	-1.5	-1.5	0.0	0.0000
50.	1531.	-.3	0.0	1.1	2.8	4.8	6.9	.3	.0957
100.	1465.	1.1	-2.2	0.0	3.4	7.4	11.7	1.5	.1959
150.	1403.	3.6	-8.5	-5.2	0.0	5.9	12.4	3.4	.3005
200.	1345.	7.3	-19.2	-14.8	-7.9	0.0	8.6	6.1	.4099
250.	1292.	12.4	-34.7	-29.2	-20.6	-10.7	0.0	9.6	.5235
300.	1243.	19.0	-55.5	-48.9	-38.5	-26.7	-13.8	14.0	.6421
350.	1198.	27.4	-81.9	-74.2	-62.1	-48.3	-33.3	19.2	.7653
400.	1157.	37.6	-114.5	-105.7	-91.9	-76.1	-59.0	25.2	.8933
450.	1123.	49.6	-153.2	-143.2	-127.8	-110.0	-90.7	31.7	1.0241
500.	1094.	63.8	-199.0	-188.0	-170.8	-151.0	-129.6	39.1	1.1595

BALLISTIC COEFFICIENT .47, MUZZLE VELOCITY FT/SEC 1700.

RANGE	VELOCITY	HEIGHT	50 YD	100 YD	150 YD	200 YD	250 YD	DRIFT	TIME
0.	1700.	0.0	-1.5	-1.5	-1.5	-1.5	-1.5	0.0	0.0000
50.	1628.	-.4	0.0	.9	2.4	4.1	6.0	.4	.0903
100.	1558.	.9	-1.8	0.0	3.0	6.4	10.2	1.4	.1846
150.	1491.	3.1	-7.2	-4.5	0.0	5.2	10.9	3.2	.2828
200.	1427.	6.4	-16.5	-12.9	-6.9	0.0	7.6	5.8	.3857
250.	1367.	10.9	-30.1	-25.6	-18.2	-9.5	0.0	9.2	.4933
300.	1313.	16.8	-48.3	-42.9	-34.0	-23.6	-12.2	13.3	.6051
350.	1262.	24.2	-71.6	-65.3	-54.9	-42.8	-29.5	18.3	.7216
400.	1215.	33.4	-100.5	-93.3	-81.4	-67.5	-52.3	24.2	.8432
450.	1172.	44.5	-135.7	-127.6	-114.2	-98.6	-81.5	31.1	.9707
500.	1136.	57.3	-176.5	-167.5	-152.7	-135.3	-116.3	38.3	1.1001

BALLISTIC COEFFICIENT .47, MUZZLE VELOCITY FT/SEC 1800.

RANGE	VELOCITY	HEIGHT	50 YD	100 YD	150 YD	200 YD	250 YD	DRIFT	TIME
0.	1800.	0.0	-1.5	-1.5	-1.5	-1.5	-1.5	0.0	0.0000
50.	1725.	-.4	0.0	.7	2.0	3.5	5.2	.3	.0850
100.	1652.	.7	-1.4	0.0	2.6	5.6	9.0	1.3	.1741
150.	1581.	2.7	-6.1	-3.9	0.0	4.5	9.6	3.0	.2671
200.	1513.	5.6	-14.2	-11.3	-6.1	0.0	6.7	5.4	.3638
250.	1448.	9.6	-26.1	-22.5	-16.0	-8.4	0.0	8.6	.4654
300.	1387.	14.9	-42.2	-37.9	-30.0	-20.9	-10.8	12.5	.5711
350.	1331.	21.6	-62.9	-57.8	-48.6	-38.0	-26.2	17.3	.6816
400.	1279.	29.7	-88.4	-82.6	-72.2	-60.0	-46.6	22.9	.7967
450.	1231.	39.6	-119.3	-112.7	-101.0	-87.3	-72.2	29.3	.9163
500.	1186.	51.2	-155.8	-148.6	-135.5	-120.3	-103.5	36.5	1.0405

BALLISTIC COEFFICIENT .47, MUZZLE VELOCITY FT/SEC 1900.

RANGE	VELOCITY	HEIGHT	50 YD	100 YD	150 YD	200 YD	250 YD	DRIFT	TIME
0.	1900.	0.0	-1.5	-1.5	-1.5	-1.5	-1.5	0.0	0.0000
50.	1821.	-.4	0.0	.6	1.7	3.1	4.6	.3	.0805
100.	1745.	.6	-1.1	0.0	2.3	5.0	8.0	1.2	.1648
150.	1671.	2.3	-5.1	-3.4	0.0	4.1	8.5	2.8	.2527
200.	1600.	4.9	-12.3	-10.0	-5.4	0.0	6.0	5.1	.3445
250.	1532.	8.6	-22.8	-19.9	-14.2	-7.4	0.0	8.0	.4402
300.	1465.	13.3	-37.0	-33.6	-26.7	-18.6	-9.7	11.7	.5403
350.	1403.	19.2	-55.3	-51.3	-43.3	-33.8	-23.4	16.2	.6449
400.	1345.	26.6	-78.0	-73.5	-64.4	-53.5	-41.6	21.6	.7542
450.	1293.	35.4	-105.5	-100.4	-90.1	-77.9	-64.6	27.7	.8679
500.	1244.	46.0	-138.3	-132.6	-121.2	-107.6	-92.8	34.7	.9864

BALLISTIC COEFFICIENT .48, MUZZLE VELOCITY FT/SEC 800.

RANGE	VELOCITY	HEIGHT	50 YD	100 YD	150 YD	200 YD	250 YD	DRIFT	TIME
0.	800.	0.0	-1.5	-1.5	-1.5	-1.5	-1.5	0.0	0.0000
50.	787.	1.0	0.0	6.2	13.1	20.4	27.8	.4	.1898
100.	775.	6.2	-12.5	0.0	13.8	28.3	43.1	1.2	.3818
150.	762.	15.2	-39.4	-20.7	0.0	21.8	43.9	2.5	.5766
200.	750.	28.1	-81.6	-56.6	-29.0	0.0	29.5	4.5	.7758
250.	737.	45.1	-138.9	-107.7	-73.2	-36.9	0.0	6.9	.9770
300.	725.	66.4	-212.6	-175.1	-133.8	-90.2	-45.9	10.1	1.1825
350.	713.	92.2	-302.9	-259.1	-210.9	-160.1	-108.4	13.9	1.3913
400.	701.	122.5	-409.7	-359.7	-304.6	-246.5	-187.4	18.1	1.6027
450.	689.	158.1	-535.3	-479.1	-417.1	-351.7	-285.3	23.2	1.8194
500.	678.	198.6	-678.1	-615.6	-546.7	-474.1	-400.3	28.7	2.0380

BALLISTIC COEFFICIENT .48, MUZZLE VELOCITY FT/SEC 900.

RANGE	VELOCITY	HEIGHT	50 YD	100 YD	150 YD	200 YD	250 YD	DRIFT	TIME
0.	900.	0.0	-1.5	-1.5	-1.5	-1.5	-1.5	0.0	0.0000
50.	885.	.6	0.0	4.8	10.2	15.9	21.8	.4	.1688
100.	871.	4.8	-9.6	0.0	10.8	22.1	33.9	1.2	.3400
150.	857.	11.9	-30.7	-16.2	0.0	17.0	34.6	2.4	.5134
200.	843.	22.1	-63.5	-44.3	-22.7	0.0	23.5	4.1	.6897
250.	830.	35.5	-108.8	-84.8	-57.7	-29.4	0.0	6.4	.8694
300.	817.	52.3	-166.5	-137.7	-105.2	-71.2	-35.9	9.0	1.0513
350.	804.	72.6	-237.5	-203.8	-166.0	-126.3	-85.1	12.3	1.2366
400.	791.	96.7	-322.0	-283.5	-240.3	-194.9	-147.9	16.1	1.4250
450.	778.	124.5	-419.9	-376.6	-327.9	-276.9	-224.0	20.3	1.6154
500.	766.	156.5	-532.9	-484.7	-430.7	-374.0	-315.2	25.3	1.8102

BALLISTIC COEFFICIENT .48, MUZZLE VELOCITY FT/SEC 1000.

RANGE	VELOCITY	HEIGHT	50 YD	100 YD	150 YD	200 YD	250 YD	DRIFT	TIME
0.	1000.	0.0	-1.5	-1.5	-1.5	-1.5	-1.5	0.0	0.0000
50.	982.	.4	0.0	3.8	8.3	12.7	17.5	.4	.1521
100.	964.	3.8	-7.6	0.0	8.9	17.9	27.4	1.2	.3071
150.	947.	9.7	-24.8	-13.4	0.0	13.4	27.7	2.8	.4657
200.	930.	17.9	-50.9	-35.7	-17.8	0.0	19.1	4.0	.6229
250.	914.	28.8	-87.5	-68.5	-46.1	-23.8	0.0	6.2	.7852
300.	899.	42.7	-134.6	-111.8	-85.0	-58.2	-29.6	9.0	.9509
350.	884.	59.4	-192.4	-165.8	-134.5	-103.3	-70.0	12.2	1.1191
400.	870.	79.2	-261.6	-231.2	-195.4	-159.8	-121.6	15.9	1.2906
450.	856.	102.2	-342.0	-307.8	-267.6	-227.4	-184.6	20.1	1.4643
500.	842.	128.4	-434.2	-396.2	-351.5	-306.9	-259.2	24.7	1.6406

BALLISTIC COEFFICIENT .48, MUZZLE VELOCITY FT/SEC 1100.

RANGE	VELOCITY	HEIGHT	50 YD	100 YD	150 YD	200 YD	250 YD	DRIFT	TIME
0.	1100.	0.0	-1.5	-1.5	-1.5	-1.5	-1.5	0.0	0.0000
50.	1074.	.2	0.0	3.0	6.8	10.7	14.8	.3	.1381
100.	1052.	3.0	-6.0	0.0	7.6	15.4	23.5	1.2	.2793
150.	1031.	8.0	-20.5	-11.4	0.0	11.7	23.8	3.1	.4267
200.	1011.	15.1	-42.8	-30.8	-15.6	0.0	16.2	5.0	.5741
250.	992.	24.4	-73.8	-58.7	-39.7	-20.3	0.0	7.5	.7246
300.	975.	35.8	-112.1	-94.0	-71.2	-47.8	-23.5	9.6	.8728
350.	957.	50.0	-160.6	-139.4	-112.8	-85.6	-57.2	12.9	1.0278
400.	940.	66.8	-218.7	-194.5	-164.1	-133.0	-100.6	16.8	1.1861
450.	924.	86.3	-286.5	-259.4	-225.2	-190.2	-153.7	21.1	1.3470
500.	908.	108.8	-364.7	-334.6	-296.5	-257.6	-217.1	25.9	1.5109

BALLISTIC COEFFICIENT .48, MUZZLE VELOCITY FT/SEC 1200.

RANGE	VELOCITY	HEIGHT	50 YD	100 YD	150 YD	200 YD	250 YD	DRIFT	TIME
0.	1200.	0.0	-1.5	-1.5	-1.5	-1.5	-1.5	0.0	0.0000
50.	1160.	.0	0.0	2.5	5.6	9.0	12.7	.4	.1271
100.	1126.	2.5	-5.0	0.0	6.2	13.0	20.4	1.5	.2586
150.	1097.	6.7	-16.8	-9.3	0.0	10.1	21.3	3.3	.3935
200.	1072.	12.8	-35.9	-25.9	-13.5	0.0	15.0	5.6	.5319
250.	1049.	21.3	-63.5	-51.1	-35.5	-18.7	0.0	9.3	.6777
300.	1028.	31.6	-97.8	-82.8	-64.2	-44.0	-21.5	12.4	.8205
350.	1009.	44.3	-140.7	-123.3	-101.5	-78.0	-51.8	16.4	.9683
400.	990.	59.4	-192.3	-172.4	-147.5	-120.6	-90.6	20.9	1.1190
450.	973.	76.5	-250.7	-228.3	-200.3	-170.0	-136.4	25.2	1.2682
500.	955.	96.5	-319.6	-294.7	-263.6	-230.0	-192.6	30.6	1.4237

BALLISTIC COEFFICIENT .48, MUZZLE VELOCITY FT/SEC 1300.

RANGE	VELOCITY	HEIGHT	50 YD	100 YD	150 YD	200 YD	250 YD	DRIFT	TIME
0.	1300.	0.0	-1.5	-1.5	-1.5	-1.5	-1.5	0.0	0.0000
50.	1252.	-.1	0.0	2.0	4.7	7.6	10.8	.4	.1178
100.	1207.	2.0	-4.1	0.0	5.3	11.2	17.5	1.6	.2400
150.	1166.	5.7	-14.1	-8.0	0.0	8.8	18.2	3.6	.3666
200.	1131.	11.1	-30.6	-22.5	-11.8	0.0	12.5	6.3	.4973
250.	1101.	18.4	-53.9	-43.7	-30.3	-15.6	0.0	9.6	.6317
300.	1075.	27.7	-84.3	-72.1	-56.1	-38.4	-19.7	13.6	.7696
350.	1053.	39.1	-122.2	-108.0	-89.3	-68.7	-46.8	18.1	.9107
400.	1031.	53.0	-169.2	-152.9	-131.6	-108.0	-83.0	23.8	1.0582
450.	1012.	69.0	-223.3	-205.0	-181.0	-154.5	-126.4	29.4	1.2056
500.	993.	87.5	-285.8	-265.5	-238.8	-209.3	-178.1	35.6	1.3559

BALLISTIC COEFFICIENT .48, MUZZLE VELOCITY FT/SEC 1400.

RANGE	VELOCITY	HEIGHT	50 YD	100 YD	150 YD	200 YD	250 YD	DRIFT	TIME
0.	1400.	0.0	-1.5	-1.5	-1.5	-1.5	-1.5	0.0	0.0000
50.	1344.	-.2	0.0	1.7	4.0	6.5	9.3	.4	.1095
100.	1292.	1.6	-3.3	0.0	4.6	9.7	15.2	1.6	.2233
150.	1244.	4.9	-11.9	-6.9	0.0	7.7	15.9	3.6	.3418
200.	1200.	9.6	-26.0	-19.4	-10.2	0.0	11.0	6.3	.4646
250.	1160.	16.1	-46.3	-38.0	-26.5	-13.7	0.0	9.9	.5918
300.	1126.	24.4	-73.0	-63.1	-49.3	-34.0	-17.5	14.1	.7232
350.	1097.	34.6	-106.6	-95.0	-78.9	-61.0	-41.8	19.0	.8582
400.	1072.	46.9	-147.3	-134.1	-115.7	-95.3	-73.3	24.5	.9966
450.	1049.	61.9	-197.2	-182.3	-161.6	-138.7	-114.0	31.3	1.1423
500.	1028.	78.5	-252.7	-236.2	-213.2	-187.6	-160.2	37.6	1.2852

BALLISTIC COEFFICIENT .48, MUZZLE VELOCITY FT/SEC 1500.

RANGE	VELOCITY	HEIGHT	50 YD	100 YD	150 YD	200 YD	250 YD	DRIFT	TIME
0.	1500.	0.0	-1.5	-1.5	-1.5	-1.5	-1.5	0.0	0.0000
50.	1436.	-.2	0.0	1.4	3.3	5.6	8.0	.5	.1027
100.	1377.	1.4	-2.7	0.0	4.0	8.5	13.3	1.6	.2092
150.	1323.	4.2	-10.0	-5.9	0.0	6.8	14.0	3.6	.3202
200.	1273.	8.4	-22.4	-17.0	-9.0	0.0	9.6	6.4	.4364
250.	1226.	14.1	-39.9	-33.2	-23.3	-12.0	0.0	9.8	.5559
300.	1183.	21.5	-63.4	-55.3	-43.4	-29.8	-15.5	14.2	.6805
350.	1146.	30.7	-93.1	-83.6	-69.8	-53.9	-37.2	19.3	.8095
400.	1114.	41.9	-129.4	-118.6	-102.8	-84.7	-65.5	25.1	.9424
450.	1087.	55.1	-172.7	-160.5	-142.7	-122.4	-100.8	31.5	1.0788
500.	1062.	70.6	-223.7	-210.2	-190.4	-167.8	-143.8	38.6	1.2196

BALLISTIC COEFFICIENT .48, MUZZLE VELOCITY FT/SEC 1600.

RANGE	VELOCITY	HEIGHT	50 YD	100 YD	150 YD	200 YD	250 YD	DRIFT	TIME
0.	1600.	0.0	-1.5	-1.5	-1.5	-1.5	-1.5	0.0	0.0000
50.	1533.	-.3	0.0	1.1	2.8	4.8	6.9	.3	.0957
100.	1468.	1.1	-2.2	0.0	3.4	7.4	11.6	1.4	.1957
150.	1406.	3.6	-8.5	-5.1	0.0	5.9	12.3	3.3	.3000
200.	1350.	7.3	-19.2	-14.7	-7.9	0.0	8.5	6.0	.4091
250.	1297.	12.4	-34.6	-29.1	-20.5	-10.7	0.0	9.5	.5225
300.	1249.	18.9	-55.2	-48.6	-38.3	-26.5	-13.7	13.7	.6403
350.	1204.	27.2	-81.4	-73.7	-61.7	-47.9	-33.0	18.7	.7627
400.	1164.	37.2	-113.6	-104.7	-91.0	-75.3	-58.2	24.5	.8894
450.	1129.	49.2	-152.2	-142.3	-126.9	-109.2	-90.0	31.1	1.0204
500.	1100.	63.3	-197.7	-186.6	-169.5	-149.8	-128.5	38.3	1.1551

BALLISTIC COEFFICIENT .48, MUZZLE VELOCITY FT/SEC 1700.

RANGE	VELOCITY	HEIGHT	50 YD	100 YD	150 YD	200 YD	250 YD	DRIFT	TIME
0.	1700.	0.0	-1.5	-1.5	-1.5	-1.5	-1.5	0.0	0.0000
50.	1629.	-.4	0.0	.9	2.4	4.1	6.0	.4	.0902
100.	1561.	.9	-1.8	0.0	3.0	6.4	10.2	1.4	.1845
150.	1495.	3.1	-7.1	-4.4	0.0	5.2	10.8	3.1	.2823
200.	1432.	6.4	-16.4	-12.8	-6.9	0.0	7.5	5.7	.3851
250.	1373.	10.9	-29.9	-25.4	-18.1	-9.4	0.0	9.0	.4920
300.	1319.	16.7	-48.1	-42.7	-33.8	-23.4	-12.1	13.0	.6034
350.	1269.	24.1	-71.2	-64.9	-54.6	-42.5	-29.3	17.9	.7194
400.	1223.	33.1	-99.7	-92.5	-80.7	-66.9	-51.8	23.6	.8397
450.	1180.	43.9	-134.0	-125.9	-112.6	-97.1	-80.1	30.0	.9647
500.	1143.	56.7	-174.6	-165.6	-150.9	-133.6	-114.8	37.2	1.0939

BALLISTIC COEFFICIENT .48, MUZZLE VELOCITY FT/SEC 1800.

RANGE	VELOCITY	HEIGHT	50 YD	100 YD	150 YD	200 YD	250 YD	DRIFT	TIME
0.	1800.	0.0	-1.5	-1.5	-1.5	-1.5	-1.5	0.0	0.0000
50.	1726.	-.4	0.0	.7	2.0	3.5	5.2	.3	.0850
100.	1655.	.7	-1.4	0.0	2.6	5.6	9.0	1.3	.1740
150.	1586.	2.7	-6.1	-3.9	0.0	4.5	9.5	2.9	.2667
200.	1519.	5.6	-14.1	-11.3	-6.0	0.0	6.7	5.3	.3632
250.	1454.	9.6	-26.0	-22.4	-15.9	-8.4	0.0	8.4	.4642
300.	1394.	14.8	-42.0	-37.7	-29.8	-20.8	-10.8	12.2	.5695
350.	1338.	21.4	-62.5	-57.5	-48.3	-37.8	-26.1	16.9	.6795
400.	1287.	29.5	-87.8	-82.1	-71.6	-59.5	-46.2	22.4	.7937
450.	1239.	39.2	-118.4	-112.0	-100.2	-86.6	-71.6	28.6	.9126
500.	1195.	50.8	-154.8	-147.6	-134.6	-119.5	-102.8	35.8	1.0365

BALLISTIC COEFFICIENT .48, MUZZLE VELOCITY FT/SEC 1900.

RANGE	VELOCITY	HEIGHT	50 YD	100 YD	150 YD	200 YD	250 YD	DRIFT	TIME
0.	1900.	0.0	-1.5	-1.5	-1.5	-1.5	-1.5	0.0	0.0000
50.	1823.	-.4	0.0	.6	1.7	3.0	4.5	.3	.0806
100.	1748.	.6	-1.1	0.0	2.3	5.0	7.9	1.2	.1646
150.	1676.	2.3	-5.1	-3.4	0.0	4.0	8.5	2.7	.2523
200.	1606.	4.9	-12.2	-9.9	-5.4	0.0	5.9	4.9	.3438
250.	1539.	8.5	-22.6	-19.8	-14.1	-7.4	0.0	7.8	.4391
300.	1474.	13.2	-36.8	-33.4	-26.6	-18.5	-9.6	11.4	.5387
350.	1412.	19.1	-54.9	-51.0	-43.0	-33.6	-23.2	15.9	.6427
400.	1355.	26.4	-77.6	-73.1	-64.0	-53.2	-41.4	21.1	.7517
450.	1302.	35.1	-104.7	-99.6	-89.4	-77.3	-64.0	27.0	.8641
500.	1253.	45.6	-137.3	-131.7	-120.3	-106.8	-92.0	33.9	.9822

BALLISTIC COEFFICIENT .49, MUZZLE VELOCITY FT/SEC 800.

RANGE	VELOCITY	HEIGHT	50 YD	100 YD	150 YD	200 YD	250 YD	DRIFT	TIME
0.	800.	0.0	-1.5	-1.5	-1.5	-1.5	-1.5	0.0	0.0000
50.	788.	1.0	0.0	6.2	13.1	20.4	27.7	.4	.1898
100.	775.	6.2	-12.5	0.0	13.8	28.3	43.0	1.2	.3816
150.	763.	15.2	-39.4	-20.7	0.0	21.7	43.8	2.5	.5764
200.	751.	28.1	-81.5	-56.5	-29.0	0.0	29.4	4.4	.7753
250.	739.	45.0	-138.6	-107.4	-73.0	-36.8	0.0	6.8	.9759
300.	727.	66.2	-212.3	-174.9	-133.5	-90.1	-46.0	9.9	1.1815
350.	715.	91.9	-302.2	-258.5	-210.3	-159.6	-108.1	13.6	1.3895
400.	703.	122.3	-409.0	-359.1	-304.0	-246.0	-187.2	17.8	1.6009
450.	692.	157.6	-534.0	-477.8	-415.8	-350.7	-284.5	22.7	1.8166
500.	680.	197.8	-675.8	-613.4	-544.5	-472.2	-398.6	28.0	2.0339

BALLISTIC COEFFICIENT .49, MUZZLE VELOCITY FT/SEC 900.

RANGE	VELOCITY	HEIGHT	50 YD	100 YD	150 YD	200 YD	250 YD	DRIFT	TIME
0.	900.	0.0	-1.5	-1.5	-1.5	-1.5	-1.5	0.0	0.0000
50.	885.	.6	0.0	4.8	10.2	15.9	21.7	.4	.1688
100.	871.	4.8	-9.6	0.0	10.8	22.1	33.8	1.1	.3399
150.	858.	11.9	-30.6	-16.2	0.0	17.0	34.6	2.3	.5130
200.	844.	22.1	-63.5	-44.3	-22.7	0.0	23.4	4.0	.6894
250.	831.	35.5	-108.7	-84.6	-57.7	-29.3	0.0	6.2	.8687
300.	818.	52.2	-166.2	-137.3	-105.0	-70.9	-35.8	8.8	1.0500
350.	806.	72.5	-237.1	-203.4	-165.7	-126.0	-85.0	12.1	1.2353
400.	793.	96.5	-321.3	-282.8	-239.7	-194.3	-147.4	15.8	1.4230
450.	781.	124.1	-418.9	-375.6	-327.1	-276.1	-223.3	19.9	1.6131
500.	768.	156.0	-531.6	-483.5	-429.6	-372.9	-314.2	24.8	1.8074

BALLISTIC COEFFICIENT .49, MUZZLE VELOCITY FT/SEC 1000.

RANGE	VELOCITY	HEIGHT	50 YD	100 YD	150 YD	200 YD	250 YD	DRIFT	TIME
0.	1000.	0.0	-1.5	-1.5	-1.5	-1.5	-1.5	0.0	0.0000
50.	982.	.4	0.0	3.8	8.1	12.7	17.5	.4	.1522
100.	964.	3.8	-7.6	0.0	8.6	17.8	27.3	1.3	.3073
150.	948.	9.5	-24.3	-12.9	0.0	13.8	28.1	2.2	.4628
200.	932.	17.8	-50.8	-35.6	-18.4	0.0	19.0	3.9	.6224
250.	916.	28.8	-87.3	-68.2	-46.8	-23.7	0.0	6.1	.7844
300.	901.	42.5	-134.1	-111.2	-85.5	-57.9	-29.4	8.7	.9493
350.	886.	59.2	-192.0	-165.4	-135.3	-103.1	-69.9	11.9	1.1179
400.	872.	79.0	-260.9	-230.4	-196.1	-159.3	-121.3	15.6	1.2887
450.	858.	101.8	-340.9	-306.7	-268.1	-226.6	-183.9	19.7	1.4617
500.	845.	128.0	-433.1	-395.0	-352.1	-306.0	-258.6	24.3	1.6380

BALLISTIC COEFFICIENT .49, MUZZLE VELOCITY FT/SEC 1100.

RANGE	VELOCITY	HEIGHT	50 YD	100 YD	150 YD	200 YD	250 YD	DRIFT	TIME
0.	1100.	0.0	-1.5	-1.5	-1.5	-1.5	-1.5	0.0	0.0000
50.	1075.	.2	0.0	3.0	6.8	10.7	14.5	.3	.1380
100.	1053.	3.0	-6.0	0.0	7.6	15.4	23.0	1.1	.2791
150.	1032.	8.0	-20.5	-11.5	0.0	11.7	23.0	3.1	.4268
200.	1013.	15.1	-42.9	-30.8	-15.6	0.0	15.2	5.1	.5742
250.	995.	24.1	-72.5	-57.5	-38.4	-19.0	0.0	6.6	.7196
300.	977.	35.7	-111.8	-93.8	-70.8	-47.5	-24.8	9.4	.8717
350.	959.	49.8	-160.1	-139.0	-112.3	-85.1	-58.5	12.6	1.0262
400.	943.	66.7	-218.7	-194.6	-164.0	-132.9	-102.6	16.7	1.1856
450.	927.	86.1	-285.9	-258.7	-224.4	-189.4	-155.3	20.7	1.3449
500.	911.	108.4	-363.6	-333.5	-295.3	-256.4	-218.5	25.4	1.5080

BALLISTIC COEFFICIENT .49, MUZZLE VELOCITY FT/SEC 1200.

RANGE	VELOCITY	HEIGHT	50 YD	100 YD	150 YD	200 YD	250 YD	DRIFT	TIME
0.	1200.	0.0	-1.5	-1.5	-1.5	-1.5	-1.5	0.0	0.0000
50.	1161.	.0	0.0	2.5	5.6	8.9	12.5	.4	.1271
100.	1128.	2.5	-5.0	0.0	6.2	12.9	20.0	1.5	.2584
150.	1099.	6.7	-16.8	-9.3	0.0	10.1	20.6	3.2	.3932
200.	1074.	12.8	-35.8	-25.8	-13.4	0.0	14.1	5.5	.5313
250.	1051.	21.0	-62.3	-49.9	-34.4	-17.6	0.0	8.4	.6726
300.	1031.	31.5	-97.7	-82.8	-64.2	-44.1	-22.9	12.3	.8201
350.	1012.	44.2	-140.6	-123.3	-101.5	-78.0	-53.4	16.3	.9676
400.	993.	59.3	-192.1	-172.2	-147.4	-120.6	-92.4	20.8	1.1181
450.	976.	76.1	-249.9	-227.5	-199.6	-169.4	-137.7	24.8	1.2657
500.	959.	96.1	-318.4	-293.5	-262.5	-228.9	-193.7	30.0	1.4204

BALLISTIC COEFFICIENT .49, MUZZLE VELOCITY FT/SEC 1300.

RANGE	VELOCITY	HEIGHT	50 YD	100 YD	150 YD	200 YD	250 YD	DRIFT	TIME
0.	1300.	0.0	-1.5	-1.5	-1.5	-1.5	-1.5	0.0	0.0000
50.	1253.	-.1	0.0	2.0	4.7	7.6	10.7	.4	.1178
100.	1208.	2.0	-4.1	0.0	5.3	11.2	17.4	1.6	.2398
150.	1168.	5.7	-14.1	-8.0	0.0	8.8	18.1	3.5	.3662
200.	1134.	11.1	-30.5	-22.4	-11.7	0.0	12.5	6.2	.4966
250.	1104.	18.4	-53.7	-43.6	-30.2	-15.6	0.0	9.5	.6308
300.	1078.	27.6	-84.0	-71.8	-55.8	-38.2	-19.5	13.3	.7681
350.	1056.	38.9	-121.7	-107.6	-88.9	-68.4	-46.5	17.8	.9088
400.	1035.	52.4	-167.3	-151.1	-129.7	-106.3	-81.3	22.8	1.0525
450.	1016.	68.3	-220.8	-202.6	-178.6	-152.2	-124.1	28.3	1.1990
500.	997.	86.5	-282.5	-262.3	-235.6	-206.3	-175.1	34.2	1.3481

BALLISTIC COEFFICIENT .49, MUZZLE VELOCITY FT/SEC 1400.

RANGE	VELOCITY	HEIGHT	50 YD	100 YD	150 YD	200 YD	250 YD	DRIFT	TIME
0.	1400.	0.0	-1.5	-1.5	-1.5	-1.5	-1.5	0.0	0.0000
50.	1345.	-.2	0.0	1.7	3.9	6.5	9.2	.4	.1095
100.	1294.	1.6	-3.3	0.0	4.5	9.6	15.1	1.6	.2235
150.	1247.	4.8	-11.8	-6.8	0.0	7.6	15.8	3.5	.3414
200.	1203.	9.6	-25.9	-19.3	-10.2	0.0	10.9	6.2	.4638
250.	1164.	16.0	-46.1	-37.7	-26.4	-13.6	0.0	9.7	.5906
300.	1130.	24.2	-72.7	-62.7	-49.1	-33.8	-17.4	13.9	.7216
350.	1101.	34.4	-106.1	-94.4	-78.5	-60.7	-41.6	18.7	.8561
400.	1076.	46.7	-146.6	-133.3	-115.1	-94.7	-72.9	24.1	.9940
450.	1053.	61.1	-194.6	-179.6	-159.1	-136.2	-111.7	30.1	1.1350
500.	1033.	77.8	-250.4	-233.7	-210.9	-185.5	-158.2	36.5	1.2791

BALLISTIC COEFFICIENT .49, MUZZLE VELOCITY FT/SEC 1500.

RANGE	VELOCITY	HEIGHT	50 YD	100 YD	150 YD	200 YD	250 YD	DRIFT	TIME
0.	1500.	0.0	-1.5	-1.5	-1.5	-1.5	-1.5	0.0	0.0000
50.	1438.	-.2	0.0	1.3	3.3	5.6	8.0	.5	.1027
100.	1380.	1.3	-2.7	0.0	4.0	8.4	13.2	1.6	.2088
150.	1326.	4.2	-10.0	-5.9	0.0	6.7	13.9	3.5	.3198
200.	1277.	8.3	-22.2	-16.9	-8.9	0.0	9.6	6.2	.4354
250.	1231.	14.0	-39.8	-33.1	-23.2	-12.0	0.0	9.7	.5549
300.	1188.	21.4	-63.1	-55.1	-43.2	-29.8	-15.4	13.9	.6790
350.	1151.	30.5	-92.6	-83.2	-69.4	-53.7	-36.9	18.9	.8073
400.	1119.	41.7	-128.9	-118.2	-102.3	-84.4	-65.2	24.7	.9401
450.	1091.	54.8	-171.8	-159.7	-141.8	-121.7	-100.1	30.9	1.0755
500.	1067.	70.0	-221.9	-208.5	-188.7	-166.3	-142.3	37.7	1.2144

BALLISTIC COEFFICIENT .49, MUZZLE VELOCITY FT/SEC 1600.

RANGE	VELOCITY	HEIGHT	50 YD	100 YD	150 YD	200 YD	250 YD	DRIFT	TIME
0.	1600.	0.0	-1.5	-1.5	-1.5	-1.5	-1.5	0.0	0.0000
50.	1534.	-.3	0.0	1.1	2.8	4.8	6.9	.3	.0956
100.	1470.	1.1	-2.2	0.0	3.4	7.4	11.6	1.4	.1955
150.	1410.	3.6	-8.4	-5.1	0.0	5.9	12.2	3.2	.2996
200.	1354.	7.3	-19.1	-14.7	-7.9	0.0	8.4	5.9	.4088
250.	1303.	12.3	-34.4	-28.9	-20.4	-10.5	0.0	9.2	.5211
300.	1255.	18.8	-55.0	-48.4	-38.2	-26.3	-13.7	13.5	.6390
350.	1211.	27.0	-81.0	-73.3	-61.3	-47.5	-32.8	18.3	.7604
400.	1170.	37.0	-112.9	-104.1	-90.5	-74.7	-57.9	24.0	.8865
450.	1135.	49.1	-151.9	-142.1	-126.7	-108.9	-90.0	30.8	1.0187
500.	1106.	62.8	-196.4	-185.4	-168.3	-148.6	-127.6	37.5	1.1508

BALLISTIC COEFFICIENT .49, MUZZLE VELOCITY FT/SEC 1700.

RANGE	VELOCITY	HEIGHT	50 YD	100 YD	150 YD	200 YD	250 YD	DRIFT	TIME
0.	1700.	0.0	-1.5	-1.5	-1.5	-1.5	-1.5	0.0	0.0000
50.	1631.	-.4	0.0	.9	2.4	4.1	6.0	.3	.0902
100.	1564.	.9	-1.8	0.0	2.9	6.4	10.1	1.4	.1844
150.	1499.	3.1	-7.1	-4.4	0.0	5.2	10.8	3.0	.2819
200.	1437.	6.4	-16.4	-12.8	-6.9	0.0	7.4	5.6	.3846
250.	1379.	10.8	-29.8	-25.3	-18.0	-9.3	0.0	8.7	.4909
300.	1326.	16.6	-47.8	-42.5	-33.6	-23.3	-12.1	12.8	.6019
350.	1276.	24.0	-70.9	-64.6	-54.3	-42.2	-29.2	17.6	.7175
400.	1230.	32.9	-99.1	-92.0	-80.2	-66.4	-51.5	23.1	.8370
450.	1188.	43.6	-133.2	-125.1	-111.9	-96.3	-79.6	29.4	.9612
500.	1151.	56.2	-173.4	-164.5	-149.8	-132.5	-113.8	36.5	1.0896

BALLISTIC COEFFICIENT .49, MUZZLE VELOCITY FT/SEC 1800.

RANGE	VELOCITY	HEIGHT	50 YD	100 YD	150 YD	200 YD	250 YD	DRIFT	TIME
0.	1800.	0.0	-1.5	-1.5	-1.5	-1.5	-1.5	0.0	0.0000
50.	1728.	-.4	0.0	.7	2.0	3.5	5.2	.3	.0849
100.	1658.	.7	-1.4	0.0	2.6	5.6	8.9	1.3	.1738
150.	1590.	2.7	-6.0	-3.9	0.0	4.5	9.5	2.9	.2664
200.	1524.	5.6	-14.1	-11.2	-6.0	0.0	6.6	5.1	.3626
250.	1461.	9.5	-25.9	-22.3	-15.8	-8.3	0.0	8.2	.4631
300.	1401.	14.7	-41.7	-37.5	-29.7	-20.6	-10.7	11.9	.5679
350.	1346.	21.3	-62.1	-57.1	-48.0	-37.5	-25.9	16.5	.6773
400.	1295.	29.3	-87.3	-81.6	-71.2	-59.2	-45.9	21.9	.7911
450.	1248.	38.9	-117.6	-111.1	-99.4	-85.9	-71.0	28.0	.9089
500.	1204.	50.3	-153.4	-146.3	-133.3	-118.3	-101.7	34.9	1.0314

BALLISTIC COEFFICIENT .49, MUZZLE VELOCITY FT/SEC 1900.

RANGE	VELOCITY	HEIGHT	50 YD	100 YD	150 YD	200 YD	250 YD	DRIFT	TIME
0.	1900.	0.0	-1.5	-1.5	-1.5	-1.5	-1.5	0.0	0.0000
50.	1824.	-.4	0.0	.6	1.7	3.0	4.5	.3	.0806
100.	1751.	.5	-1.1	0.0	2.3	5.0	7.9	1.2	.1645
150.	1680.	2.3	-5.1	-3.4	0.0	4.0	8.4	2.7	.2519
200.	1612.	4.9	-12.1	-9.9	-5.4	0.0	5.9	4.8	.3431
250.	1546.	8.5	-22.5	-19.8	-14.1	-7.4	0.0	7.7	.4382
300.	1481.	13.1	-36.5	-33.2	-26.4	-18.4	-9.5	11.2	.5371
350.	1420.	18.9	-54.6	-50.7	-42.7	-33.3	-23.0	15.5	.6405
400.	1364.	26.2	-77.0	-72.5	-63.4	-52.7	-40.9	20.6	.7487
450.	1312.	34.8	-104.0	-99.0	-88.7	-76.7	-63.4	26.4	.8607
500.	1263.	45.1	-136.0	-130.4	-119.1	-105.7	-90.9	33.0	.9772

BALLISTIC COEFFICIENT .50, MUZZLE VELOCITY FT/SEC 800.

RANGE	VELOCITY	HEIGHT	50 YD	100 YD	150 YD	200 YD	250 YD	DRIFT	TIME
0.	800.	0.0	-1.5	-1.5	-1.5	-1.5	-1.5	0.0	0.0000
50.	788.	1.0	0.0	6.2	13.1	20.3	27.7	.4	.1898
100.	776.	6.2	-12.5	0.0	13.8	28.2	42.9	1.1	.3815
150.	764.	15.2	-39.4	-20.7	0.0	21.7	43.6	2.4	.5763
200.	752.	28.1	-81.4	-56.5	-28.9	0.0	29.3	4.4	.7747
250.	740.	44.9	-138.3	-107.2	-72.7	-36.6	0.0	6.6	.9749
300.	728.	66.1	-212.0	-174.6	-133.2	-89.9	-46.0	9.7	1.1804
350.	717.	91.7	-301.5	-257.9	-209.6	-159.1	-107.9	13.2	1.3877
400.	705.	122.0	-408.3	-358.5	-303.3	-245.6	-187.0	17.4	1.5990
450.	694.	157.1	-532.7	-476.6	-414.5	-349.6	-283.7	22.2	1.8137
500.	682.	197.3	-674.5	-612.2	-543.3	-471.1	-397.9	27.5	2.0311

BALLISTIC COEFFICIENT .50, MUZZLE VELOCITY FT/SEC 900.

RANGE	VELOCITY	HEIGHT	50 YD	100 YD	150 YD	200 YD	250 YD	DRIFT	TIME
0.	900.	0.0	-1.5	-1.5	-1.5	-1.5	-1.5	0.0	0.0000
50.	886.	.6	0.0	4.8	10.2	15.9	21.7	.4	.1688
100.	872.	4.8	-9.6	0.0	10.8	22.1	33.8	1.1	.3397
150.	858.	11.9	-30.6	-16.1	0.0	17.0	34.5	2.2	.5127
200.	845.	22.0	-63.4	-44.2	-22.7	0.0	23.4	3.9	.6891
250.	833.	35.4	-108.5	-84.5	-57.6	-29.2	0.0	6.1	.8680
300.	820.	52.0	-165.8	-137.0	-104.7	-70.6	-35.6	8.6	1.0488
350.	808.	72.3	-236.7	-203.1	-165.4	-125.7	-84.8	11.9	1.2340
400.	795.	96.2	-320.6	-282.1	-239.1	-193.7	-147.0	15.4	1.4211
450.	783.	123.8	-418.1	-374.8	-326.4	-275.3	-222.8	19.5	1.6109
500.	771.	155.5	-530.3	-482.2	-428.4	-371.7	-313.3	24.3	1.8045

BALLISTIC COEFFICIENT .50, MUZZLE VELOCITY FT/SEC 1000.

RANGE	VELOCITY	HEIGHT	50 YD	100 YD	150 YD	200 YD	250 YD	DRIFT	TIME
0.	1000.	0.0	-1.5	-1.5	-1.5	-1.5	-1.5	0.0	0.0000
50.	982.	.4	0.0	3.8	8.1	12.7	17.4	.4	.1524
100.	965.	3.8	-7.6	0.0	8.5	17.7	27.2	1.3	.3075
150.	949.	9.5	-24.2	-12.8	0.0	13.8	28.0	2.2	.4625
200.	933.	17.8	-50.7	-35.5	-18.4	0.0	18.9	3.9	.6220
250.	918.	28.7	-87.0	-68.0	-46.7	-23.6	0.0	5.9	.7836
300.	903.	42.4	-133.9	-111.0	-85.4	-57.8	-29.4	8.5	.9485
350.	888.	59.1	-191.6	-164.9	-135.0	-102.8	-69.7	11.7	1.1166
400.	874.	78.7	-260.2	-229.7	-195.6	-158.8	-120.9	15.3	1.2869
450.	861.	101.5	-339.9	-305.6	-267.2	-225.8	-183.3	19.2	1.4594
500.	848.	127.6	-432.0	-393.8	-351.2	-305.2	-257.9	23.8	1.6355

BALLISTIC COEFFICIENT .50, MUZZLE VELOCITY FT/SEC 1100.

RANGE	VELOCITY	HEIGHT	50 YD	100 YD	150 YD	200 YD	250 YD	DRIFT	TIME
0.	1100.	0.0	-1.5	-1.5	-1.5	-1.5	-1.5	0.0	0.0000
50.	1075.	.2	0.0	3.0	6.6	10.5	14.5	.3	.1380
100.	1053.	3.0	-6.0	0.0	7.2	15.0	22.9	1.1	.2790
150.	1033.	7.8	-19.9	-10.9	0.0	11.6	23.5	2.4	.4229
200.	1015.	14.8	-41.9	-29.9	-15.4	0.0	16.0	4.3	.5697
250.	996.	24.1	-72.4	-57.4	-39.2	-20.0	0.0	6.5	.7188
300.	979.	35.6	-111.6	-93.5	-71.8	-48.7	-24.7	9.2	.8706
350.	962.	49.7	-160.0	-139.0	-113.6	-86.6	-58.7	12.5	1.0256
400.	945.	66.7	-218.7	-194.6	-165.6	-134.8	-102.8	16.6	1.1850
450.	930.	85.8	-285.2	-258.1	-225.5	-190.8	-154.8	20.3	1.3428
500.	914.	108.0	-362.5	-332.4	-296.2	-257.6	-217.7	24.9	1.5052

BALLISTIC COEFFICIENT .50, MUZZLE VELOCITY FT/SEC 1200.

RANGE	VELOCITY	HEIGHT	50 YD	100 YD	150 YD	200 YD	250 YD	DRIFT	TIME
0.	1200.	0.0	-1.5	-1.5	-1.5	-1.5	-1.5	0.0	0.0000
50.	1161.	.0	0.0	2.5	5.6	8.9	12.4	.4	.1270
100.	1129.	2.5	-5.0	0.0	6.2	12.9	19.9	1.5	.2583
150.	1100.	6.7	-16.7	-9.3	0.0	10.0	20.6	3.1	.3928
200.	1076.	12.8	-35.7	-25.8	-13.4	0.0	14.0	5.4	.5307
250.	1054.	20.9	-62.2	-49.8	-34.3	-17.6	0.0	8.2	.6717
300.	1034.	31.2	-96.5	-81.6	-63.0	-43.0	-21.9	11.6	.8156
350.	1015.	43.7	-138.9	-121.6	-99.9	-76.5	-51.9	15.4	.9623
400.	997.	58.5	-189.7	-169.8	-145.1	-118.3	-90.2	19.6	1.1115
450.	979.	75.8	-249.0	-226.7	-198.9	-168.7	-137.1	24.3	1.2632
500.	962.	95.8	-317.7	-292.9	-262.0	-228.5	-193.4	29.6	1.4183

BALLISTIC COEFFICIENT .50, MUZZLE VELOCITY FT/SEC 1300.

RANGE	VELOCITY	HEIGHT	50 YD	100 YD	150 YD	200 YD	250 YD	DRIFT	TIME
0.	1300.	0.0	-1.5	-1.5	-1.5	-1.5	-1.5	0.0	0.0000
50.	1253.	-.1	0.0	2.0	4.7	7.7	10.7	.5	.1183
100.	1210.	2.0	-4.0	0.0	5.3	11.3	17.4	1.6	.2396
150.	1170.	5.7	-14.0	-8.0	0.0	9.0	18.1	3.5	.3658
200.	1136.	11.1	-30.6	-22.6	-12.0	0.0	12.1	6.3	.4976
250.	1107.	18.3	-53.4	-43.5	-30.2	-15.2	0.0	9.3	.6299
300.	1081.	27.5	-83.7	-71.7	-55.8	-37.8	-19.6	13.2	.7674
350.	1058.	38.7	-121.1	-107.1	-88.5	-67.5	-46.3	17.5	.9070
400.	1038.	52.2	-166.4	-150.4	-129.1	-105.2	-80.9	22.4	1.0502
450.	1019.	67.9	-219.7	-201.7	-177.8	-150.8	-123.5	27.8	1.1962
500.	1001.	86.1	-281.2	-261.2	-234.6	-204.7	-174.3	33.6	1.3450

BALLISTIC COEFFICIENT .50, MUZZLE VELOCITY FT/SEC 1400.

RANGE	VELOCITY	HEIGHT	50 YD	100 YD	150 YD	200 YD	250 YD	DRIFT	TIME
0.	1400.	0.0	-1.5	-1.5	-1.5	-1.5	-1.5	0.0	0.0000
50.	1346.	-.2	0.0	1.7	3.9	6.5	9.2	.4	.1095
100.	1296.	1.6	-3.3	0.0	4.5	9.6	15.1	1.6	.2232
150.	1250.	4.8	-11.8	-6.8	0.0	7.6	15.8	3.4	.3409
200.	1207.	9.5	-25.9	-19.2	-10.1	0.0	10.9	6.1	.4631
250.	1167.	15.9	-45.9	-37.6	-26.3	-13.6	0.0	9.5	.5896
300.	1134.	24.1	-72.4	-62.5	-48.8	-33.6	-17.3	13.6	.7201
350.	1105.	34.3	-105.7	-94.1	-78.2	-60.4	-41.4	18.3	.8542
400.	1079.	46.4	-145.9	-132.7	-114.5	-94.2	-72.4	23.6	.9914
450.	1057.	60.8	-193.7	-178.8	-158.3	-135.5	-111.0	29.5	1.1320
500.	1037.	77.3	-249.1	-232.5	-209.8	-184.4	-157.2	35.9	1.2754

BALLISTIC COEFFICIENT .50, MUZZLE VELOCITY FT/SEC 1500.

RANGE	VELOCITY	HEIGHT	50 YD	100 YD	150 YD	200 YD	250 YD	DRIFT	TIME
0.	1500.	0.0	-1.5	-1.5	-1.5	-1.5	-1.5	0.0	0.0000
50.	1439.	-.2	0.0	1.3	3.3	5.5	8.0	.5	.1026
100.	1382.	1.3	-2.7	0.0	4.0	8.4	13.2	1.5	.2087
150.	1329.	4.1	-10.0	-5.9	0.0	6.7	13.9	3.4	.3195
200.	1281.	8.3	-22.1	-16.8	-8.9	0.0	9.7	6.1	.4344
250.	1235.	14.0	-39.8	-33.1	-23.2	-12.1	0.0	9.6	.5543
300.	1193.	21.3	-63.1	-55.0	-43.2	-29.9	-15.4	13.8	.6783
350.	1155.	30.5	-92.6	-83.3	-69.4	-53.9	-37.0	18.8	.8068
400.	1124.	41.4	-128.1	-117.4	-101.5	-83.8	-64.5	24.1	.9369
450.	1096.	54.4	-170.8	-158.8	-141.0	-121.0	-99.3	30.3	1.0722
500.	1072.	69.6	-220.7	-207.3	-187.5	-165.4	-141.2	37.1	1.2106

BALLISTIC COEFFICIENT .50, MUZZLE VELOCITY FT/SEC 1600.

RANGE	VELOCITY	HEIGHT	50 YD	100 YD	150 YD	200 YD	250 YD	DRIFT	TIME
0.	1600.	0.0	-1.5	-1.5	-1.5	-1.5	-1.5	0.0	0.0000
50.	1535.	-.3	0.0	1.1	2.8	4.8	6.9	.3	.0956
100.	1473.	1.1	-2.2	0.0	3.4	7.3	11.5	1.4	.1953
150.	1413.	3.5	-8.4	-5.1	0.0	5.9	12.2	3.2	.2992
200.	1358.	7.2	-19.0	-14.6	-7.8	0.0	8.4	5.8	.4078
250.	1308.	12.2	-34.3	-28.8	-20.3	-10.5	0.0	9.0	.5201
300.	1260.	18.7	-54.6	-48.1	-37.9	-26.1	-13.5	13.1	.6368
350.	1216.	26.9	-80.6	-72.9	-61.0	-47.3	-32.6	18.0	.7584
400.	1176.	36.8	-112.7	-103.9	-90.3	-74.6	-57.8	23.7	.8849
450.	1141.	48.5	-150.3	-140.4	-125.1	-107.5	-88.6	29.8	1.0131
500.	1111.	62.4	-195.2	-184.2	-167.2	-147.6	-126.6	36.8	1.1467

BALLISTIC COEFFICIENT .50, MUZZLE VELOCITY FT/SEC 1700.

RANGE	VELOCITY	HEIGHT	50 YD	100 YD	150 YD	200 YD	250 YD	DRIFT	TIME
0.	1700.	0.0	-1.5	-1.5	-1.5	-1.5	-1.5	0.0	0.0000
50.	1632.	-.4	0.0	.9	2.4	4.1	5.9	.3	.0902
100.	1567.	.9	-1.8	0.0	2.9	6.4	10.1	1.4	.1842
150.	1503.	3.1	-7.1	-4.4	0.0	5.2	10.7	3.0	.2816
200.	1442.	6.3	-16.3	-12.7	-6.9	0.0	7.4	5.4	.3839
250.	1385.	10.8	-29.7	-25.2	-17.9	-9.3	0.0	8.6	.4898
300.	1332.	16.6	-47.6	-42.3	-33.5	-23.2	-12.0	12.5	.6004
350.	1283.	23.8	-70.4	-64.2	-53.9	-41.9	-28.9	17.2	.7151
400.	1237.	32.7	-98.7	-91.5	-79.8	-66.1	-51.2	22.6	.8346
450.	1195.	43.3	-132.5	-124.5	-111.3	-95.9	-79.2	28.9	.9583
500.	1157.	55.9	-172.5	-163.6	-149.0	-131.8	-113.2	35.9	1.0862

BALLISTIC COEFFICIENT .50, MUZZLE VELOCITY FT/SEC 1800.

RANGE	VELOCITY	HEIGHT	50 YD	100 YD	150 YD	200 YD	250 YD	DRIFT	TIME
0.	1800.	0.0	-1.5	-1.5	-1.5	-1.5	-1.5	0.0	0.0000
50.	1729.	-.4	0.0	.7	2.0	3.5	5.2	.3	.0849
100.	1660.	.7	-1.4	0.0	2.6	5.6	8.9	1.2	.1736
150.	1594.	2.6	-6.0	-3.9	0.0	4.5	9.4	2.8	.2660
200.	1529.	5.5	-14.0	-11.2	-6.0	0.0	6.6	5.0	.3619
250.	1467.	9.5	-25.8	-22.2	-15.7	-8.2	0.0	8.0	.4621
300.	1408.	14.6	-41.5	-37.3	-29.5	-20.5	-10.6	11.7	.5664
350.	1354.	21.1	-61.8	-56.8	-47.7	-37.3	-25.7	16.2	.6753
400.	1303.	29.1	-86.7	-81.0	-70.6	-58.6	-45.5	21.4	.7880
450.	1256.	38.6	-116.8	-110.4	-98.8	-85.3	-70.5	27.4	.9056
500.	1212.	49.9	-152.6	-145.4	-132.5	-117.5	-101.0	34.2	1.0277

BALLISTIC COEFFICIENT .50, MUZZLE VELOCITY FT/SEC 1900.

RANGE	VELOCITY	HEIGHT	50 YD	100 YD	150 YD	200 YD	250 YD	DRIFT	TIME
0.	1900.	0.0	-1.5	-1.5	-1.5	-1.5	-1.5	0.0	0.0000
50.	1826.	-.4	0.0	.6	1.7	3.0	4.5	.3	.0805
100.	1754.	.5	-1.1	0.0	2.3	4.9	7.9	1.1	.1643
150.	1685.	2.3	-5.1	-3.4	0.0	4.0	8.4	2.6	.2516
200.	1617.	4.9	-12.1	-9.9	-5.3	0.0	5.9	4.7	.3425
250.	1552.	8.4	-22.5	-19.7	-14.0	-7.4	0.0	7.5	.4373
300.	1489.	13.0	-36.4	-33.1	-26.3	-18.3	-9.4	10.9	.5357
350.	1429.	18.8	-54.3	-50.4	-42.5	-33.2	-22.9	15.2	.6388
400.	1372.	26.0	-76.5	-72.1	-63.0	-52.3	-40.6	20.1	.7460
450.	1321.	34.5	-103.2	-98.3	-88.1	-76.1	-62.8	25.8	.8573
500.	1273.	44.7	-135.1	-129.5	-118.2	-104.9	-90.1	32.3	.9731

BALLISTIC COEFFICIENT .51, MUZZLE VELOCITY FT/SEC 800.

RANGE	VELOCITY	HEIGHT	50 YD	100 YD	150 YD	200 YD	250 YD	DRIFT	TIME
0.	800.	0.0	-1.5	-1.5	-1.5	-1.5	-1.5	0.0	0.0000
50.	788.	1.0	0.0	6.2	13.1	20.3	27.7	.4	.1898
100.	776.	6.2	-12.4	0.0	13.8	28.2	42.9	1.1	.3813
150.	764.	15.2	-39.4	-20.7	0.0	21.6	43.6	2.4	.5761
200.	753.	28.0	-81.3	-56.4	-28.8	0.0	29.3	4.3	.7742
250.	741.	44.8	-138.3	-107.2	-72.7	-36.7	0.0	6.5	.9745
300.	730.	66.0	-211.7	-174.3	-133.0	-89.8	-45.7	9.6	1.1793
350.	718.	91.4	-300.8	-257.2	-208.9	-158.6	-107.2	12.9	1.3858
400.	707.	121.7	-407.6	-357.8	-302.6	-245.1	-186.4	17.1	1.5972
450.	696.	156.7	-531.3	-475.3	-413.2	-348.4	-282.4	21.7	1.8109
500.	685.	196.7	-673.1	-610.9	-542.0	-470.0	-396.6	27.0	2.0282

BALLISTIC COEFFICIENT .51, MUZZLE VELOCITY FT/SEC 900.

RANGE	VELOCITY	HEIGHT	50 YD	100 YD	150 YD	200 YD	250 YD	DRIFT	TIME
0.	900.	0.0	-1.5	-1.5	-1.5	-1.5	-1.5	0.0	0.0000
50.	886.	.6	0.0	4.8	10.2	15.8	21.7	.4	.1688
100.	872.	4.8	-9.6	0.0	10.7	22.1	33.7	1.1	.3396
150.	859.	11.9	-30.5	-16.1	0.0	17.0	34.5	2.2	.5124
200.	846.	22.0	-63.4	-44.2	-22.7	0.0	23.3	3.9	.6887
250.	834.	35.4	-108.3	-84.3	-57.5	-29.1	0.0	6.0	.8673
300.	821.	52.0	-165.6	-136.8	-104.6	-70.5	-35.6	8.5	1.0480
350.	809.	72.2	-236.3	-202.7	-165.1	-125.4	-84.7	11.6	1.2328
400.	797.	95.9	-319.8	-281.4	-238.5	-193.0	-146.5	15.1	1.4191
450.	785.	123.5	-417.3	-374.1	-325.8	-274.7	-222.3	19.2	1.6090
500.	773.	155.1	-529.0	-481.0	-427.3	-370.5	-312.3	23.8	1.8016

BALLISTIC COEFFICIENT .51, MUZZLE VELOCITY FT/SEC 1000.

RANGE	VELOCITY	HEIGHT	50 YD	100 YD	150 YD	200 YD	250 YD	DRIFT	TIME
0.	1000.	0.0	-1.5	-1.5	-1.5	-1.5	-1.5	0.0	0.0000
50.	983.	.4	0.0	3.8	8.1	12.7	17.4	.4	.1525
100.	966.	3.8	-7.6	0.0	8.5	17.7	27.1	1.4	.3077
150.	950.	9.5	-24.2	-12.7	0.0	13.8	27.9	2.2	.4623
200.	934.	17.8	-50.6	-35.3	-18.4	0.0	18.8	3.8	.6216
250.	919.	28.7	-86.8	-67.7	-46.5	-23.6	0.0	5.8	.7828
300.	904.	42.4	-133.6	-110.7	-85.3	-57.7	-29.4	8.4	.9477
350.	890.	59.0	-191.1	-164.4	-134.7	-102.6	-69.6	11.5	1.1153
400.	876.	78.5	-259.5	-228.9	-195.0	-158.3	-120.6	15.0	1.2850
450.	863.	101.2	-339.2	-304.8	-266.7	-225.3	-182.9	18.9	1.4575
500.	850.	127.2	-430.8	-392.6	-350.2	-304.3	-257.2	23.4	1.6329

BALLISTIC COEFFICIENT .51, MUZZLE VELOCITY FT/SEC 1100.

RANGE	VELOCITY	HEIGHT	50 YD	100 YD	150 YD	200 YD	250 YD	DRIFT	TIME
0.	1100.	0.0	-1.5	-1.5	-1.5	-1.5	-1.5	0.0	0.0000
50.	1076.	.2	0.0	3.0	6.6	10.5	14.5	.3	.1380
100.	1054.	3.0	-6.0	0.0	7.2	14.9	22.9	1.1	.2789
150.	1035.	7.8	-19.9	-10.8	0.0	11.5	23.5	2.4	.4227
200.	1016.	14.8	-41.9	-29.9	-15.4	0.0	15.9	4.2	.5692
250.	998.	24.0	-72.3	-57.2	-39.1	-19.9	0.0	6.4	.7180
300.	981.	35.6	-111.5	-93.4	-71.7	-48.6	-24.7	9.1	.8699
350.	964.	49.7	-160.1	-139.1	-113.8	-86.8	-59.0	12.5	1.0254
400.	948.	66.2	-217.2	-193.2	-164.2	-133.5	-101.6	15.9	1.1811
450.	933.	85.5	-284.4	-257.4	-224.8	-190.2	-154.4	20.0	1.3406
500.	917.	107.6	-361.4	-331.3	-295.1	-256.7	-216.8	24.4	1.5023

BALLISTIC COEFFICIENT .51, MUZZLE VELOCITY FT/SEC 1200.

RANGE	VELOCITY	HEIGHT	50 YD	100 YD	150 YD	200 YD	250 YD	DRIFT	TIME
0.	1200.	0.0	-1.5	-1.5	-1.5	-1.5	-1.5	0.0	0.0000
50.	1162.	.0	0.0	2.5	5.6	8.9	12.4	.4	.1270
100.	1130.	2.4	-4.9	0.0	6.2	12.8	19.9	1.4	.2581
150.	1102.	6.6	-16.7	-9.3	0.0	10.0	20.5	3.1	.3925
200.	1077.	12.7	-35.6	-25.7	-13.3	0.0	14.0	5.3	.5301
250.	1056.	20.9	-62.0	-49.6	-34.2	-17.5	0.0	8.1	.6709
300.	1036.	31.1	-96.2	-81.4	-62.8	-42.8	-21.8	11.3	.8144
350.	1017.	43.6	-138.5	-121.2	-99.6	-76.2	-51.7	15.1	.9608
400.	999.	58.3	-189.1	-169.3	-144.5	-117.9	-89.8	19.3	1.1094
450.	982.	75.7	-248.6	-226.4	-198.5	-168.5	-137.0	24.1	1.2617
500.	965.	95.6	-317.4	-292.7	-261.8	-228.4	-193.4	29.4	1.4170

BALLISTIC COEFFICIENT .51, MUZZLE VELOCITY FT/SEC 1300.

RANGE	VELOCITY	HEIGHT	50 YD	100 YD	150 YD	200 YD	250 YD	DRIFT	TIME
0.	1300.	0.0	-1.5	-1.5	-1.5	-1.5	-1.5	0.0	0.0000
50.	1254.	-.1	0.0	2.0	4.8	7.6	10.7	.5	.1182
100.	1212.	2.0	-4.0	0.0	5.5	11.2	17.3	1.5	.2394
150.	1172.	5.7	-14.3	-8.3	0.0	8.5	17.7	3.8	.3678
200.	1138.	11.1	-30.3	-22.4	-11.3	0.0	12.3	6.0	.4958
250.	1110.	18.2	-53.3	-43.3	-29.5	-15.4	0.0	9.2	.6290
300.	1084.	27.4	-83.3	-71.4	-54.8	-37.8	-19.4	12.9	.7656
350.	1061.	38.7	-121.0	-107.0	-87.7	-67.9	-46.4	17.3	.9062
400.	1041.	52.0	-166.0	-150.1	-128.0	-105.3	-80.7	22.1	1.0486
450.	1022.	67.7	-219.3	-201.3	-176.4	-151.0	-123.3	27.5	1.1945
500.	1004.	85.9	-280.8	-260.9	-233.2	-205.0	-174.2	33.4	1.3434

BALLISTIC COEFFICIENT .51, MUZZLE VELOCITY FT/SEC 1400.

RANGE	VELOCITY	HEIGHT	50 YD	100 YD	150 YD	200 YD	250 YD	DRIFT	TIME
0.	1400.	0.0	-1.5	-1.5	-1.5	-1.5	-1.5	0.0	0.0000
50.	1347.	-.2	0.0	1.6	3.9	6.4	9.2	.4	.1095
100.	1298.	1.6	-3.3	0.0	4.5	9.6	15.0	1.5	.2228
150.	1252.	4.8	-11.7	-6.8	0.0	7.6	15.7	3.4	.3405
200.	1210.	9.5	-25.8	-19.2	-10.1	0.0	10.8	6.0	.4624
250.	1171.	15.9	-45.8	-37.6	-26.2	-13.5	0.0	9.3	.5886
300.	1137.	24.1	-72.4	-62.6	-48.9	-33.7	-17.5	13.5	.7197
350.	1108.	34.1	-105.3	-93.8	-77.8	-60.2	-41.2	18.0	.8524
400.	1083.	46.2	-145.3	-132.2	-114.0	-93.8	-72.1	23.2	.9891
450.	1060.	60.5	-192.8	-178.0	-157.6	-134.8	-110.5	29.0	1.1293
500.	1040.	76.9	-247.9	-231.4	-208.7	-183.4	-156.3	35.3	1.2719

BALLISTIC COEFFICIENT .51, MUZZLE VELOCITY FT/SEC 1500.

RANGE	VELOCITY	HEIGHT	50 YD	100 YD	150 YD	200 YD	250 YD	DRIFT	TIME
0.	1500.	0.0	-1.5	-1.5	-1.5	-1.5	-1.5	0.0	0.0000
50.	1440.	-.2	0.0	1.3	3.3	5.5	7.9	.4	.1025
100.	1384.	1.3	-2.7	0.0	3.9	8.4	13.1	1.5	.2085
150.	1332.	4.1	-9.9	-5.9	0.0	6.6	13.8	3.4	.3191
200.	1284.	8.3	-22.1	-16.7	-8.8	0.0	9.6	5.9	.4338
250.	1240.	13.9	-39.5	-32.8	-23.0	-11.9	0.0	9.3	.5528
300.	1198.	21.2	-62.7	-54.6	-42.8	-29.6	-15.2	13.4	.6762
350.	1160.	30.2	-91.7	-82.4	-68.6	-53.1	-36.4	18.1	.8031
400.	1128.	41.2	-127.5	-116.8	-101.0	-83.3	-64.2	23.7	.9344
450.	1101.	54.1	-169.9	-157.8	-140.1	-120.2	-98.7	29.7	1.0689
500.	1076.	69.2	-219.5	-206.1	-186.4	-164.3	-140.4	36.4	1.2068

BALLISTIC COEFFICIENT .51, MUZZLE VELOCITY FT/SEC 1600.

RANGE	VELOCITY	HEIGHT	50 YD	100 YD	150 YD	200 YD	250 YD	DRIFT	TIME
0.	1600.	0.0	-1.5	-1.5	-1.5	-1.5	-1.5	0.0	0.0000
50.	1537.	-.3	0.0	1.1	2.8	4.7	6.8	.3	.0955
100.	1475.	1.1	-2.2	0.0	3.4	7.3	11.5	1.3	.1951
150.	1417.	3.5	-8.4	-5.1	0.0	5.8	12.1	3.1	.2989
200.	1363.	7.2	-18.9	-14.6	-7.8	0.0	8.4	5.6	.4070
250.	1313.	12.2	-34.2	-28.7	-20.2	-10.5	0.0	8.9	.5191
300.	1266.	18.6	-54.4	-47.9	-37.7	-26.0	-13.5	12.8	.6354
350.	1222.	26.7	-80.1	-72.5	-60.6	-47.0	-32.3	17.6	.7560
400.	1182.	36.5	-111.7	-102.9	-89.3	-73.8	-57.0	23.0	.8810
450.	1147.	48.2	-149.5	-139.7	-124.3	-106.9	-88.0	29.2	1.0099
500.	1117.	61.9	-193.9	-183.0	-166.0	-146.5	-125.6	36.1	1.1426

BALLISTIC COEFFICIENT .51, MUZZLE VELOCITY FT/SEC 1700.

RANGE	VELOCITY	HEIGHT	50 YD	100 YD	150 YD	200 YD	250 YD	DRIFT	TIME
0.	1700.	0.0	-1.5	-1.5	-1.5	-1.5	-1.5	0.0	0.0000
50.	1634.	-.4	0.0	.9	2.4	4.1	5.9	.3	.0902
100.	1569.	.9	-1.8	0.0	2.9	6.3	10.0	1.3	.1841
150.	1507.	3.0	-7.1	-4.4	0.0	5.1	10.7	2.9	.2813
200.	1447.	6.3	-16.2	-12.7	-6.8	0.0	7.4	5.3	.3832
250.	1390.	10.7	-29.5	-25.1	-17.8	-9.2	0.0	8.4	.4888
300.	1338.	16.5	-47.4	-42.1	-33.3	-23.1	-12.0	12.3	.5990
350.	1290.	23.7	-70.1	-63.8	-53.6	-41.7	-28.7	16.8	.7131
400.	1244.	32.4	-98.0	-90.9	-79.2	-65.5	-50.8	22.1	.8317
450.	1202.	43.0	-131.5	-123.5	-110.4	-95.0	-78.4	28.2	.9543
500.	1164.	55.4	-171.0	-162.1	-147.5	-130.4	-112.0	35.0	1.0811

BALLISTIC COEFFICIENT .51, MUZZLE VELOCITY FT/SEC 1800.

RANGE	VELOCITY	HEIGHT	50 YD	100 YD	150 YD	200 YD	250 YD	DRIFT	TIME
0.	1800.	0.0	-1.5	-1.5	-1.5	-1.5	-1.5	0.0	0.0000
50.	1730.	-.4	0.0	.7	2.0	3.5	5.1	.3	.0849
100.	1663.	.7	-1.4	0.0	2.6	5.6	8.8	1.2	.1736
150.	1598.	2.6	-6.0	-3.9	0.0	4.5	9.4	2.8	.2657
200.	1534.	5.5	-14.0	-11.1	-6.0	0.0	6.5	4.9	.3613
250.	1473.	9.5	-25.6	-22.1	-15.6	-8.2	0.0	7.8	.4610
300.	1415.	14.6	-41.3	-37.1	-29.3	-20.4	-10.6	11.4	.5649
350.	1361.	21.0	-61.4	-56.5	-47.5	-37.0	-25.6	15.8	.6733
400.	1311.	28.9	-86.2	-80.5	-70.2	-58.3	-45.2	20.9	.7855
450.	1264.	38.3	-116.0	-109.6	-98.0	-84.6	-69.8	26.7	.9020
500.	1221.	49.4	-151.2	-144.1	-131.2	-116.3	-99.9	33.3	1.0227

BALLISTIC COEFFICIENT .51, MUZZLE VELOCITY FT/SEC 1900.

RANGE	VELOCITY	HEIGHT	50 YD	100 YD	150 YD	200 YD	250 YD	DRIFT	TIME
0.	1900.	0.0	-1.5	-1.5	-1.5	-1.5	-1.5	0.0	0.0000
50.	1827.	-.4	0.0	.5	1.7	3.0	4.5	.3	.0805
100.	1757.	.5	-1.1	0.0	2.3	4.9	7.8	1.1	.1641
150.	1689.	2.3	-5.0	-3.4	0.0	4.0	8.4	2.5	.2513
200.	1622.	4.9	-12.0	-9.8	-5.3	0.0	5.9	4.6	.3419
250.	1558.	8.4	-22.4	-19.6	-14.0	-7.3	0.0	7.3	.4365
300.	1496.	13.0	-36.2	-32.9	-26.1	-18.2	-9.4	10.7	.5344
350.	1437.	18.7	-54.1	-50.2	-42.3	-33.0	-22.8	14.9	.6372
400.	1381.	25.8	-76.0	-71.6	-62.6	-51.9	-40.2	19.7	.7433
450.	1329.	34.3	-102.6	-97.7	-87.5	-75.6	-62.4	25.3	.8542
500.	1282.	44.3	-134.1	-128.6	-117.3	-104.0	-89.4	31.6	.9691

BALLISTIC COEFFICIENT .52, MUZZLE VELOCITY FT/SEC 800.

RANGE	VELOCITY	HEIGHT	50 YD	100 YD	150 YD	200 YD	250 YD	DRIFT	TIME
0.	800.	0.0	-1.5	-1.5	-1.5	-1.5	-1.5	0.0	0.0000
50.	788.	1.0	0.0	6.2	13.1	20.3	27.6	.4	.1898
100.	777.	6.2	-12.4	0.0	13.8	28.2	42.8	1.1	.3811
150.	765.	15.2	-39.3	-20.7	0.0	21.5	43.5	2.4	.5759
200.	753.	28.0	-81.2	-56.3	-28.7	0.0	29.3	4.2	.7737
250.	742.	44.8	-138.1	-107.0	-72.5	-36.6	0.0	6.4	.9738
300.	731.	65.9	-211.3	-174.1	-132.7	-89.6	-45.7	9.4	1.1783
350.	720.	91.2	-300.1	-256.6	-208.3	-158.0	-106.8	12.6	1.3840
400.	709.	121.4	-406.9	-357.2	-302.0	-244.6	-186.0	16.8	1.5954
450.	698.	156.2	-529.9	-474.0	-411.9	-347.3	-281.4	21.2	1.8080
500.	687.	196.1	-671.7	-609.6	-540.6	-468.9	-395.6	26.5	2.0254

BALLISTIC COEFFICIENT .52, MUZZLE VELOCITY FT/SEC 900.

RANGE	VELOCITY	HEIGHT	50 YD	100 YD	150 YD	200 YD	250 YD	DRIFT	TIME
0.	900.	0.0	-1.5	-1.5	-1.5	-1.5	-1.5	0.0	0.0000
50.	886.	.6	0.0	4.8	10.2	15.8	21.6	.4	.1688
100.	873.	4.8	-9.6	0.0	10.7	22.1	33.7	1.1	.3395
150.	860.	11.8	-30.5	-16.1	0.0	17.0	34.4	2.1	.5122
200.	847.	22.0	-63.3	-44.2	-22.7	0.0	23.2	3.8	.6884
250.	835.	35.3	-108.2	-84.2	-57.4	-29.0	0.0	5.8	.8666
300.	823.	51.9	-165.4	-136.7	-104.5	-70.4	-35.7	8.3	1.0473
350.	811.	72.0	-235.9	-202.4	-164.8	-125.1	-84.5	11.4	1.2315
400.	799.	95.7	-319.1	-280.7	-237.8	-192.4	-146.0	14.8	1.4172
450.	787.	123.2	-416.6	-373.4	-325.1	-274.1	-221.9	18.8	1.6070
500.	776.	154.6	-527.6	-479.7	-426.0	-369.3	-311.3	23.2	1.7987

BALLISTIC COEFFICIENT .52, MUZZLE VELOCITY FT/SEC 1000.

RANGE	VELOCITY	HEIGHT	50 YD	100 YD	150 YD	200 YD	250 YD	DRIFT	TIME
0.	1000.	0.0	-1.5	-1.5	-1.5	-1.5	-1.5	0.0	0.0000
50.	983.	.4	0.0	3.8	8.0	12.6	17.3	.5	.1526
100.	966.	3.8	-7.7	0.0	8.4	17.6	27.0	1.4	.3080
150.	951.	9.5	-24.1	-12.6	0.0	13.8	27.8	2.1	.4620
200.	935.	17.8	-50.5	-35.2	-18.3	0.0	18.8	3.7	.6211
250.	920.	28.6	-86.6	-67.5	-46.4	-23.5	0.0	5.6	.7821
300.	906.	42.3	-133.4	-110.4	-85.1	-57.6	-29.4	8.2	.9469
350.	892.	58.8	-190.7	-163.9	-134.4	-102.3	-69.4	11.3	1.1140
400.	879.	78.3	-258.8	-228.1	-194.4	-157.8	-120.2	14.6	1.2831
450.	865.	101.0	-338.5	-304.0	-266.1	-224.8	-182.5	18.6	1.4556
500.	853.	126.8	-429.6	-391.3	-349.2	-303.4	-256.4	22.9	1.6304

BALLISTIC COEFFICIENT .52, MUZZLE VELOCITY FT/SEC 1100.

RANGE	VELOCITY	HEIGHT	50 YD	100 YD	150 YD	200 YD	250 YD	DRIFT	TIME
0.	1100.	0.0	-1.5	-1.5	-1.5	-1.5	-1.5	0.0	0.0000
50.	1076.	.2	0.0	3.0	6.6	10.5	14.4	.3	.1379
100.	1055.	3.0	-6.0	0.0	7.2	14.9	22.8	1.1	.2787
150.	1036.	7.8	-19.8	-10.8	0.0	11.5	23.4	2.3	.4224
200.	1017.	14.8	-41.8	-29.8	-15.4	0.0	15.9	4.1	.5687
250.	1000.	23.9	-72.1	-57.1	-39.0	-19.8	0.0	6.2	.7173
300.	983.	35.6	-111.5	-93.5	-71.9	-48.8	-25.0	9.1	.8698
350.	966.	49.7	-160.2	-139.2	-113.9	-87.0	-59.2	12.4	1.0252
400.	951.	66.0	-216.7	-192.7	-163.8	-133.1	-101.3	15.6	1.1794
450.	935.	85.2	-283.7	-256.7	-224.2	-189.6	-153.9	19.6	1.3385
500.	920.	107.2	-360.3	-330.3	-294.2	-255.8	-216.1	23.9	1.4996

BALLISTIC COEFFICIENT .52, MUZZLE VELOCITY FT/SEC 1200.

RANGE	VELOCITY	HEIGHT	50 YD	100 YD	150 YD	200 YD	250 YD	DRIFT	TIME
0.	1200.	0.0	-1.5	-1.5	-1.5	-1.5	-1.5	0.0	0.0000
50.	1163.	.0	0.0	2.5	5.6	8.9	12.4	.3	.1270
100.	1131.	2.4	-4.9	0.0	6.2	12.8	19.8	1.4	.2579
150.	1103.	6.6	-16.7	-9.3	0.0	10.0	20.4	3.0	.3923
200.	1079.	12.7	-35.5	-25.6	-13.3	0.0	14.0	5.2	.5295
250.	1058.	20.8	-61.8	-49.5	-34.1	-17.5	0.0	7.9	.6700
300.	1038.	31.0	-95.9	-81.1	-62.6	-42.7	-21.7	11.1	.8132
350.	1020.	43.4	-138.1	-120.8	-99.2	-76.0	-51.5	14.8	.9592
400.	1002.	58.2	-188.8	-169.1	-144.4	-117.8	-89.9	19.1	1.1084
450.	985.	75.5	-248.5	-226.2	-198.4	-168.6	-137.2	23.9	1.2608
500.	968.	95.5	-317.2	-292.5	-261.6	-228.4	-193.5	29.2	1.4158

BALLISTIC COEFFICIENT .52, MUZZLE VELOCITY FT/SEC 1300.

RANGE	VELOCITY	HEIGHT	50 YD	100 YD	150 YD	200 YD	250 YD	DRIFT	TIME
0.	1300.	0.0	-1.5	-1.5	-1.5	-1.5	-1.5	0.0	0.0000
50.	1255.	-.1	0.0	2.0	4.7	7.5	10.6	.5	.1180
100.	1213.	2.0	-4.1	0.0	5.4	11.0	17.2	1.6	.2399
150.	1174.	5.7	-14.2	-8.1	0.0	8.5	17.7	3.6	.3668
200.	1141.	11.0	-30.2	-22.1	-11.3	0.0	12.4	5.8	.4945
250.	1112.	18.2	-53.2	-43.1	-29.6	-15.5	0.0	9.0	.6281
300.	1087.	27.3	-83.1	-71.0	-54.8	-37.9	-19.3	12.7	.7645
350.	1064.	38.7	-121.3	-107.1	-88.2	-68.5	-46.8	17.4	.9066
400.	1044.	52.0	-166.1	-149.9	-128.3	-105.8	-81.0	22.0	1.0483
450.	1025.	67.6	-219.0	-200.7	-176.5	-151.1	-123.3	27.2	1.1932
500.	1008.	85.7	-280.4	-260.2	-233.2	-205.0	-174.1	33.1	1.3417

BALLISTIC COEFFICIENT .52, MUZZLE VELOCITY FT/SEC 1400.

RANGE	VELOCITY	HEIGHT	50 YD	100 YD	150 YD	200 YD	250 YD	DRIFT	TIME
0.	1400.	0.0	-1.5	-1.5	-1.5	-1.5	-1.5	0.0	0.0000
50.	1348.	-.2	0.0	1.6	3.9	6.5	9.2	.4	.1094
100.	1300.	1.6	-3.3	0.0	4.6	9.6	15.1	1.4	.2225
150.	1255.	4.8	-11.8	-6.9	0.0	7.6	15.8	3.4	.3405
200.	1213.	9.5	-25.8	-19.3	-10.1	0.0	11.0	6.0	.4624
250.	1174.	15.9	-46.0	-37.8	-26.4	-13.7	0.0	9.4	.5893
300.	1141.	23.9	-71.8	-62.0	-48.3	-33.1	-16.6	13.1	.7170
350.	1112.	34.0	-104.8	-93.4	-77.4	-59.7	-40.5	17.7	.8506
400.	1087.	46.0	-144.8	-131.7	-113.4	-93.1	-71.2	22.9	.9870
450.	1064.	60.4	-193.0	-178.3	-157.7	-134.9	-110.2	29.0	1.1291
500.	1044.	76.8	-247.7	-231.4	-208.4	-183.1	-155.7	35.1	1.2708

BALLISTIC COEFFICIENT .52, MUZZLE VELOCITY FT/SEC 1500.

RANGE	VELOCITY	HEIGHT	50 YD	100 YD	150 YD	200 YD	250 YD	DRIFT	TIME
0.	1500.	0.0	-1.5	-1.5	-1.5	-1.5	-1.5	0.0	0.0000
50.	1441.	-.2	0.0	1.3	3.3	5.5	7.9	.4	.1025
100.	1386.	1.3	-2.7	0.0	4.0	8.3	13.1	1.5	.2084
150.	1335.	4.1	-9.9	-5.9	0.0	6.6	13.7	3.4	.3191
200.	1288.	8.3	-22.0	-16.7	-8.7	0.0	9.5	5.8	.4331
250.	1244.	13.9	-39.4	-32.7	-22.8	-11.9	0.0	9.1	.5518
300.	1203.	21.1	-62.4	-54.4	-42.5	-29.4	-15.1	13.1	.6744
350.	1165.	30.1	-91.4	-82.0	-68.2	-52.9	-36.2	17.8	.8012
400.	1133.	40.9	-126.8	-116.1	-100.3	-82.8	-63.8	23.2	.9318
450.	1105.	53.8	-169.1	-157.1	-139.3	-119.6	-98.2	29.2	1.0660
500.	1081.	68.8	-218.4	-205.0	-185.2	-163.3	-139.6	35.8	1.2033

BALLISTIC COEFFICIENT .52, MUZZLE VELOCITY FT/SEC 1600.

RANGE	VELOCITY	HEIGHT	50 YD	100 YD	150 YD	200 YD	250 YD	DRIFT	TIME
0.	1600.	0.0	-1.5	-1.5	-1.5	-1.5	-1.5	0.0	0.0000
50.	1538.	-.3	0.0	1.1	2.8	4.7	6.8	.3	.0955
100.	1478.	1.1	-2.2	0.0	3.4	7.3	11.4	1.3	.1949
150.	1420.	3.5	-8.3	-5.1	0.0	5.8	12.1	3.0	.2984
200.	1367.	7.2	-18.9	-14.5	-7.8	0.0	8.4	5.5	.4064
250.	1317.	12.1	-34.1	-28.6	-20.2	-10.5	0.0	8.7	.5182
300.	1271.	18.5	-54.2	-47.7	-37.6	-25.9	-13.4	12.6	.6340
350.	1228.	26.6	-79.8	-72.2	-60.4	-46.8	-32.1	17.2	.7542
400.	1188.	36.3	-111.2	-102.5	-88.9	-73.4	-56.7	22.6	.8785
450.	1153.	47.9	-148.7	-138.9	-123.6	-106.2	-87.4	28.7	1.0067
500.	1122.	61.5	-192.7	-181.9	-164.9	-145.5	-124.6	35.4	1.1387

BALLISTIC COEFFICIENT .52, MUZZLE VELOCITY FT/SEC 1700.

RANGE	VELOCITY	HEIGHT	50 YD	100 YD	150 YD	200 YD	250 YD	DRIFT	TIME
0.	1700.	0.0	-1.5	-1.5	-1.5	-1.5	-1.5	0.0	0.0000
50.	1635.	-.4	0.0	.9	2.3	4.0	5.9	.3	.0901
100.	1572.	.9	-1.8	0.0	2.9	6.3	10.0	1.3	.1840
150.	1510.	3.0	-7.0	-4.4	0.0	5.1	10.6	2.9	.2810
200.	1451.	6.3	-16.2	-12.6	-6.8	0.0	7.4	5.2	.3826
250.	1395.	10.7	-29.5	-25.0	-17.7	-9.2	0.0	8.3	.4881
300.	1344.	16.4	-47.2	-41.9	-33.1	-22.9	-11.8	12.0	.5975
350.	1296.	23.5	-69.8	-63.6	-53.4	-41.5	-28.5	16.5	.7114
400.	1251.	32.2	-97.5	-90.4	-78.7	-65.1	-50.3	21.7	.8290
450.	1210.	42.7	-130.8	-122.8	-109.6	-94.3	-77.7	27.6	.9510
500.	1171.	54.9	-170.0	-161.1	-146.5	-129.5	-111.0	34.3	1.0771

BALLISTIC COEFFICIENT .52, MUZZLE VELOCITY FT/SEC 1800.

RANGE	VELOCITY	HEIGHT	50 YD	100 YD	150 YD	200 YD	250 YD	DRIFT	TIME
0.	1800.	0.0	-1.5	-1.5	-1.5	-1.5	-1.5	0.0	0.0000
50.	1732.	-.4	0.0	.7	2.0	3.5	5.1	.3	.0849
100.	1665.	.7	-1.4	0.0	2.6	5.5	8.8	1.2	.1734
150.	1601.	2.6	-6.0	-3.9	0.0	4.5	9.3	2.7	.2653
200.	1539.	5.5	-13.9	-11.1	-5.9	0.0	6.5	4.8	.3607
250.	1479.	9.4	-25.5	-22.0	-15.6	-8.1	0.0	7.6	.4600
300.	1421.	14.5	-41.1	-36.9	-29.2	-20.3	-10.5	11.2	.5635
350.	1368.	20.9	-61.1	-56.2	-47.2	-36.8	-25.4	15.5	.6714
400.	1318.	28.7	-85.7	-80.1	-69.8	-57.9	-44.9	20.5	.7830
450.	1272.	38.0	-115.3	-109.0	-97.4	-84.0	-69.4	26.2	.8989
500.	1229.	49.1	-150.3	-143.2	-130.4	-115.5	-99.2	32.7	1.0189

BALLISTIC COEFFICIENT .52, MUZZLE VELOCITY FT/SEC 1900.

RANGE	VELOCITY	HEIGHT	50 YD	100 YD	150 YD	200 YD	250 YD	DRIFT	TIME
0.	1900.	0.0	-1.5	-1.5	-1.5	-1.5	-1.5	0.0	0.0000
50.	1829.	-.4	0.0	.5	1.7	3.0	4.5	.3	.0805
100.	1759.	.5	-1.1	0.0	2.3	4.9	7.8	1.1	.1639
150.	1692.	2.3	-5.0	-3.4	0.0	4.0	8.4	2.5	.2510
200.	1627.	4.8	-12.0	-9.8	-5.3	0.0	5.8	4.5	.3414
250.	1564.	8.4	-22.3	-19.5	-13.9	-7.3	0.0	7.2	.4357
300.	1503.	12.9	-36.0	-32.7	-26.0	-18.1	-9.3	10.5	.5331
350.	1444.	18.6	-53.8	-49.9	-42.1	-32.8	-22.6	14.5	.6353
400.	1389.	25.6	-75.6	-71.2	-62.2	-51.6	-39.9	19.2	.7410
450.	1338.	34.0	-102.0	-97.1	-86.9	-75.0	-61.9	24.8	.8512
500.	1291.	44.0	-133.1	-127.7	-116.4	-103.2	-88.6	30.9	.9652

BALLISTIC COEFFICIENT .53, MUZZLE VELOCITY FT/SEC 800.

RANGE	VELOCITY	HEIGHT	50 YD	100 YD	150 YD	200 YD	250 YD	DRIFT	TIME
0.	800.	0.0	-1.5	-1.5	-1.5	-1.5	-1.5	0.0	0.0000
50.	788.	1.0	0.0	6.2	13.1	20.3	27.6	.4	.1898
100.	777.	6.2	-12.4	0.0	13.8	28.1	42.8	1.1	.3810
150.	766.	15.2	-39.3	-20.7	0.0	21.5	43.5	2.3	.5758
200.	754.	27.9	-81.0	-56.2	-28.6	0.0	29.3	4.1	.7732
250.	743.	44.7	-138.0	-106.9	-72.4	-36.6	0.0	6.3	.9732
300.	732.	65.8	-211.0	-173.8	-132.4	-89.4	-45.5	9.2	1.1772
350.	721.	91.0	-299.6	-256.2	-207.9	-157.8	-106.5	12.4	1.3828
400.	710.	121.1	-406.1	-356.5	-301.3	-244.0	-185.4	16.5	1.5935
450.	699.	155.7	-528.5	-472.7	-410.6	-346.1	-280.2	20.7	1.8052
500.	689.	195.6	-670.3	-608.3	-539.3	-467.7	-394.4	26.0	2.0225

BALLISTIC COEFFICIENT .53, MUZZLE VELOCITY FT/SEC 900.

RANGE	VELOCITY	HEIGHT	50 YD	100 YD	150 YD	200 YD	250 YD	DRIFT	TIME
0.	900.	0.0	-1.5	-1.5	-1.5	-1.5	-1.5	0.0	0.0000
50.	887.	.6	0.0	4.8	10.1	15.8	21.6	.4	.1688
100.	873.	4.8	-9.6	0.0	10.7	22.1	33.6	1.1	.3394
150.	861.	11.8	-30.4	-16.1	0.0	17.0	34.4	2.1	.5119
200.	848.	22.0	-63.3	-44.1	-22.7	0.0	23.1	3.8	.6881
250.	836.	35.2	-108.0	-84.0	-57.3	-28.9	0.0	5.7	.8658
300.	824.	51.8	-165.3	-136.5	-104.4	-70.4	-35.7	8.2	1.0466
350.	813.	71.9	-235.5	-202.0	-164.5	-124.8	-84.3	11.2	1.2303
400.	801.	95.5	-318.5	-280.2	-237.4	-192.0	-145.8	14.5	1.4157
450.	789.	122.9	-415.8	-372.7	-324.5	-273.4	-221.4	18.5	1.6051
500.	778.	154.1	-526.2	-478.3	-424.8	-368.0	-310.3	22.7	1.7959

BALLISTIC COEFFICIENT .53, MUZZLE VELOCITY FT/SEC 1000.

RANGE	VELOCITY	HEIGHT	50 YD	100 YD	150 YD	200 YD	250 YD	DRIFT	TIME
0.	1000.	0.0	-1.5	-1.5	-1.5	-1.5	-1.5	0.0	0.0000
50.	983.	.4	0.0	3.8	8.0	12.6	17.3	.5	.1528
100.	967.	3.8	-7.7	0.0	8.4	17.5	26.9	1.4	.3082
150.	952.	9.5	-24.1	-12.6	0.0	13.7	27.8	2.1	.4618
200.	937.	17.7	-50.4	-35.0	-18.3	0.0	18.8	3.6	.6207
250.	922.	28.6	-86.5	-67.3	-46.4	-23.5	0.0	5.6	.7816
300.	908.	42.2	-133.1	-110.1	-85.0	-57.5	-29.3	8.1	.9460
350.	894.	59.0	-191.7	-164.8	-135.5	-103.5	-70.6	11.6	1.1161
400.	881.	78.1	-258.1	-227.4	-193.9	-157.3	-119.7	14.3	1.2814
450.	868.	100.7	-337.7	-303.2	-265.5	-224.3	-182.0	18.3	1.4538
500.	855.	126.4	-428.4	-390.1	-348.2	-302.5	-255.4	22.5	1.6278

BALLISTIC COEFFICIENT .53, MUZZLE VELOCITY FT/SEC 1100.

RANGE	VELOCITY	HEIGHT	50 YD	100 YD	150 YD	200 YD	250 YD	DRIFT	TIME
0.	1100.	0.0	-1.5	-1.5	-1.5	-1.5	-1.5	0.0	0.0000
50.	1077.	.2	0.0	3.0	6.6	10.4	14.4	.3	.1379
100.	1056.	3.0	-6.0	0.0	7.2	14.9	22.9	1.0	.2786
150.	1037.	7.8	-19.8	-10.8	0.0	11.5	23.5	2.3	.4221
200.	1019.	14.7	-41.7	-29.7	-15.3	0.0	16.0	4.0	.5682
250.	1001.	23.9	-72.1	-57.2	-39.1	-20.0	0.0	6.2	.7172
300.	985.	35.6	-111.6	-93.6	-72.0	-49.0	-25.0	9.1	.8698
350.	968.	49.7	-160.2	-139.3	-114.0	-87.2	-59.2	12.4	1.0249
400.	953.	65.8	-216.2	-192.2	-163.4	-132.7	-100.8	15.3	1.1777
450.	938.	85.0	-283.0	-256.0	-223.6	-189.1	-153.1	19.2	1.3364
500.	923.	106.9	-359.5	-329.5	-293.5	-255.1	-215.2	23.5	1.4974

BALLISTIC COEFFICIENT .53, MUZZLE VELOCITY FT/SEC 1200.

RANGE	VELOCITY	HEIGHT	50 YD	100 YD	150 YD	200 YD	250 YD	DRIFT	TIME
0.	1200.	0.0	-1.5	-1.5	-1.5	-1.5	-1.5	0.0	0.0000
50.	1164.	.0	0.0	2.5	5.6	8.9	12.3	.3	.1270
100.	1132.	2.4	-4.9	0.0	6.2	12.8	19.7	1.4	.2578
150.	1105.	6.6	-16.7	-9.3	0.0	10.0	20.4	3.0	.3920
200.	1081.	12.7	-35.5	-25.7	-13.3	0.0	13.8	5.2	.5295
250.	1059.	20.7	-61.7	-49.4	-33.9	-17.3	0.0	7.8	.6692
300.	1040.	30.9	-95.7	-80.9	-62.4	-42.4	-21.7	10.9	.8121
350.	1022.	43.4	-138.0	-120.8	-99.2	-75.9	-51.7	14.7	.9586
400.	1004.	58.2	-188.8	-169.1	-144.4	-117.7	-90.1	19.0	1.1078
450.	988.	75.4	-248.3	-226.1	-198.4	-168.4	-137.3	23.7	1.2598
500.	971.	95.3	-316.9	-292.2	-261.4	-228.1	-193.5	29.0	1.4145

BALLISTIC COEFFICIENT .53, MUZZLE VELOCITY FT/SEC 1300.

RANGE	VELOCITY	HEIGHT	50 YD	100 YD	150 YD	200 YD	250 YD	DRIFT	TIME
0.	1300.	0.0	-1.5	-1.5	-1.5	-1.5	-1.5	0.0	0.0000
50.	1256.	-.1	0.0	2.0	4.7	7.5	10.6	.4	.1179
100.	1215.	2.0	-4.0	0.0	5.3	11.0	17.2	1.6	.2396
150.	1176.	5.7	-14.1	-8.0	0.0	8.5	17.8	3.5	.3658
200.	1143.	11.0	-30.1	-22.0	-11.4	0.0	12.3	5.7	.4940
250.	1115.	18.1	-53.0	-43.0	-29.6	-15.4	0.0	8.8	.6271
300.	1090.	27.2	-82.9	-70.8	-54.8	-37.8	-19.3	12.5	.7633
350.	1067.	38.3	-120.1	-106.0	-87.3	-67.4	-45.8	16.7	.9023
400.	1047.	52.0	-166.2	-150.1	-128.7	-106.0	-81.3	22.0	1.0480
450.	1028.	67.4	-218.7	-200.5	-176.5	-150.9	-123.2	27.0	1.1918
500.	1011.	85.4	-280.0	-259.8	-233.2	-204.7	-173.9	32.8	1.3401

BALLISTIC COEFFICIENT .53, MUZZLE VELOCITY FT/SEC 1400.

RANGE	VELOCITY	HEIGHT	50 YD	100 YD	150 YD	200 YD	250 YD	DRIFT	TIME
0.	1400.	0.0	-1.5	-1.5	-1.5	-1.5	-1.5	0.0	0.0000
50.	1349.	-.2	0.0	1.6	3.9	6.4	9.1	.4	.1094
100.	1302.	1.6	-3.3	0.0	4.5	9.6	15.0	1.4	.2223
150.	1257.	4.8	-11.7	-6.8	0.0	7.6	15.7	3.2	.3399
200.	1216.	9.5	-25.7	-19.2	-10.1	0.0	10.8	5.8	.4614
250.	1178.	15.8	-45.6	-37.5	-26.1	-13.5	0.0	9.1	.5872
300.	1144.	23.8	-71.6	-61.8	-48.2	-33.1	-16.9	12.8	.7158
350.	1116.	33.8	-104.4	-93.0	-77.1	-59.5	-40.6	17.4	.8487
400.	1090.	45.8	-144.2	-131.2	-113.0	-92.8	-71.2	22.5	.9848
450.	1068.	59.9	-191.2	-176.5	-156.1	-133.4	-109.1	28.1	1.1238
500.	1047.	76.6	-247.5	-231.2	-208.5	-183.3	-156.3	34.9	1.2696

BALLISTIC COEFFICIENT .53, MUZZLE VELOCITY FT/SEC 1500.

RANGE	VELOCITY	HEIGHT	50 YD	100 YD	150 YD	200 YD	250 YD	DRIFT	TIME
0.	1500.	0.0	-1.5	-1.5	-1.5	-1.5	-1.5	0.0	0.0000
50.	1442.	-.2	0.0	1.3	3.3	5.5	7.8	.4	.1024
100.	1388.	1.3	-2.7	0.0	3.9	8.3	13.0	1.4	.2082
150.	1338.	4.1	-9.9	-5.9	0.0	6.6	13.7	3.3	.3185
200.	1291.	8.2	-21.9	-16.6	-8.7	0.0	9.5	5.7	.4325
250.	1248.	13.8	-39.2	-32.6	-22.8	-11.8	0.0	8.9	.5508
300.	1207.	21.0	-62.1	-54.1	-42.4	-29.2	-15.0	12.9	.6730
350.	1170.	29.9	-91.0	-81.7	-67.9	-52.6	-36.0	17.5	.7993
400.	1137.	40.8	-126.6	-115.9	-100.2	-82.7	-63.8	23.0	.9304
450.	1110.	53.5	-168.3	-156.3	-138.6	-118.9	-97.6	28.7	1.0631
500.	1085.	68.3	-217.2	-203.9	-184.2	-162.4	-138.7	35.2	1.1997

BALLISTIC COEFFICIENT .53, MUZZLE VELOCITY FT/SEC 1600.

RANGE	VELOCITY	HEIGHT	50 YD	100 YD	150 YD	200 YD	250 YD	DRIFT	TIME
0.	1600.	0.0	-1.5	-1.5	-1.5	-1.5	-1.5	0.0	0.0000
50.	1539.	-.3	0.0	1.1	2.8	4.7	6.8	.3	.0954
100.	1480.	1.1	-2.2	0.0	3.4	7.2	11.4	1.3	.1947
150.	1423.	3.5	-8.3	-5.1	0.0	5.8	12.0	3.0	.2981
200.	1371.	7.2	-18.8	-14.5	-7.7	0.0	8.3	5.4	.4057
250.	1322.	12.1	-33.9	-28.5	-20.0	-10.4	0.0	8.5	.5171
300.	1276.	18.5	-54.1	-47.6	-37.4	-25.9	-13.4	12.4	.6330
350.	1233.	26.5	-79.6	-72.1	-60.2	-46.7	-32.2	17.0	.7530
400.	1194.	36.2	-110.9	-102.2	-88.7	-73.2	-56.6	22.3	.8769
450.	1158.	47.7	-148.1	-138.3	-123.1	-105.7	-87.1	28.3	1.0043
500.	1128.	61.1	-191.6	-180.8	-163.9	-144.6	-123.8	34.8	1.1350

BALLISTIC COEFFICIENT .53, MUZZLE VELOCITY FT/SEC 1700.

RANGE	VELOCITY	HEIGHT	50 YD	100 YD	150 YD	200 YD	250 YD	DRIFT	TIME
0.	1700.	0.0	-1.5	-1.5	-1.5	-1.5	-1.5	0.0	0.0000
50.	1636.	-.4	0.0	.9	2.3	4.0	5.9	.3	.0901
100.	1574.	.9	-1.8	0.0	2.9	6.3	9.9	1.3	.1839
150.	1514.	3.0	-7.0	-4.4	0.0	5.1	10.6	2.8	.2807
200.	1455.	6.3	-16.1	-12.6	-6.8	0.0	7.3	5.1	.3820
250.	1400.	10.6	-29.3	-24.9	-17.6	-9.1	0.0	8.0	.4868
300.	1349.	16.3	-47.0	-41.7	-32.9	-22.8	-11.8	11.8	.5962
350.	1302.	23.4	-69.4	-63.2	-53.0	-41.1	-28.4	16.1	.7091
400.	1258.	32.0	-97.0	-89.9	-78.2	-64.7	-50.1	21.2	.8266
450.	1216.	42.4	-130.1	-122.1	-109.0	-93.8	-77.4	27.1	.9481
500.	1178.	54.6	-169.1	-160.2	-145.7	-128.8	-110.5	33.7	1.0738

BALLISTIC COEFFICIENT .53, MUZZLE VELOCITY FT/SEC 1800.

RANGE	VELOCITY	HEIGHT	50 YD	100 YD	150 YD	200 YD	250 YD	DRIFT	TIME
0.	1800.	0.0	-1.5	-1.5	-1.5	-1.5	-1.5	0.0	0.0000
50.	1733.	-.4	0.0	.7	2.0	3.5	5.1	.3	.0849
100.	1668.	.7	-1.4	0.0	2.6	5.5	8.8	1.2	.1733
150.	1605.	2.6	-5.9	-3.8	0.0	4.4	9.3	2.6	.2650
200.	1544.	5.5	-13.9	-11.0	-5.9	0.0	6.5	4.7	.3602
250.	1485.	9.4	-25.4	-21.9	-15.5	-8.1	0.0	7.5	.4591
300.	1428.	14.4	-41.0	-36.8	-29.1	-20.2	-10.5	11.0	.5623
350.	1374.	20.8	-60.9	-56.0	-47.0	-36.6	-25.3	15.2	.6698
400.	1325.	28.5	-85.2	-79.6	-69.4	-57.5	-44.6	20.1	.7806
450.	1280.	37.8	-114.6	-108.3	-96.8	-83.4	-68.9	25.7	.8958
500.	1237.	48.8	-149.4	-142.4	-129.6	-114.8	-98.6	32.1	1.0155

BALLISTIC COEFFICIENT .53, MUZZLE VELOCITY FT/SEC 1900.

RANGE	VELOCITY	HEIGHT	50 YD	100 YD	150 YD	200 YD	250 YD	DRIFT	TIME
0.	1900.	0.0	-1.5	-1.5	-1.5	-1.5	-1.5	0.0	0.0000
50.	1830.	-.4	0.0	.5	1.7	3.0	4.4	.3	.0806
100.	1762.	.5	-1.1	0.0	2.2	4.9	7.8	1.1	.1639
150.	1696.	2.3	-5.0	-3.4	0.0	4.0	8.3	2.4	.2507
200.	1632.	4.8	-11.9	-9.8	-5.3	0.0	5.8	4.4	.3409
250.	1570.	8.3	-22.2	-19.5	-13.9	-7.3	0.0	7.1	.4349
300.	1510.	12.8	-35.9	-32.6	-25.9	-18.0	-9.3	10.3	.5320
350.	1452.	18.5	-53.5	-49.7	-41.8	-32.6	-22.4	14.2	.6335
400.	1397.	25.4	-75.2	-70.8	-61.9	-51.3	-39.7	18.9	.7388
450.	1347.	33.8	-101.3	-96.4	-86.4	-74.5	-61.4	24.2	.8482
500.	1299.	43.6	-132.2	-126.8	-115.6	-102.4	-87.9	30.3	.9614

BALLISTIC COEFFICIENT .54, MUZZLE VELOCITY FT/SEC 800.

RANGE	VELOCITY	HEIGHT	50 YD	100 YD	150 YD	200 YD	250 YD	DRIFT	TIME
0.	800.	0.0	-1.5	-1.5	-1.5	-1.5	-1.5	0.0	0.0000
50.	789.	1.0	0.0	6.2	13.1	20.2	27.6	.4	.1898
100.	777.	6.2	-12.4	0.0	13.8	28.1	42.8	1.0	.3808
150.	766.	15.2	-39.3	-20.7	0.0	21.4	43.4	2.3	.5756
200.	755.	27.9	-80.9	-56.2	-28.5	0.0	29.3	4.0	.7727
250.	744.	44.7	-137.8	-106.9	-72.4	-36.7	0.0	6.2	.9727
300.	733.	65.7	-210.7	-173.5	-132.1	-89.3	-45.3	9.0	1.1762
350.	723.	90.9	-299.3	-256.0	-207.6	-157.7	-106.3	12.2	1.3817
400.	712.	120.9	-405.4	-355.8	-300.6	-243.5	-184.8	16.1	1.5917
450.	701.	155.3	-527.5	-471.8	-409.7	-345.4	-279.4	20.3	1.8030
500.	691.	195.0	-668.9	-607.0	-537.9	-466.5	-393.2	25.5	2.0197

BALLISTIC COEFFICIENT .54, MUZZLE VELOCITY FT/SEC 900.

RANGE	VELOCITY	HEIGHT	50 YD	100 YD	150 YD	200 YD	250 YD	DRIFT	TIME
0.	900.	0.0	-1.5	-1.5	-1.5	-1.5	-1.5	0.0	0.0000
50.	887.	.6	0.0	4.8	10.1	15.8	21.6	.4	.1688
100.	874.	4.8	-9.6	0.0	10.7	22.0	33.6	1.1	.3393
150.	861.	11.8	-30.4	-16.1	0.0	17.0	34.3	2.1	.5118
200.	849.	22.0	-63.2	-44.1	-22.6	0.0	23.0	3.7	.6878
250.	837.	35.2	-107.8	-83.9	-57.1	-28.8	0.0	5.6	.8651
300.	826.	51.8	-165.1	-136.4	-104.2	-70.3	-35.7	8.1	1.0459
350.	814.	71.8	-235.1	-201.6	-164.1	-124.5	-84.2	11.0	1.2290
400.	803.	95.3	-318.0	-279.8	-236.9	-191.6	-145.5	14.2	1.4143
450.	791.	122.6	-415.0	-371.9	-323.7	-272.7	-220.9	18.1	1.6031
500.	780.	153.6	-524.8	-477.0	-423.4	-366.8	-309.2	22.2	1.7930

BALLISTIC COEFFICIENT .54, MUZZLE VELOCITY FT/SEC 1000.

RANGE	VELOCITY	HEIGHT	50 YD	100 YD	150 YD	200 YD	250 YD	DRIFT	TIME
0.	1000.	0.0	-1.5	-1.5	-1.5	-1.5	-1.5	0.0	0.0000
50.	984.	.4	0.0	3.8	8.0	12.6	17.3	.5	.1529
100.	968.	3.8	-7.7	0.0	8.3	17.5	26.9	1.5	.3084
150.	953.	9.5	-24.0	-12.5	0.0	13.7	27.8	2.0	.4615
200.	938.	17.7	-50.3	-34.9	-18.3	0.0	18.8	3.6	.6202
250.	923.	28.5	-86.4	-67.2	-46.4	-23.5	0.0	5.5	.7812
300.	909.	42.1	-132.9	-109.8	-84.9	-57.5	-29.2	8.0	.9452
350.	895.	58.8	-190.8	-163.9	-134.8	-102.8	-69.9	11.2	1.1139
400.	883.	77.9	-257.6	-226.9	-193.6	-157.1	-119.4	14.1	1.2801
450.	870.	100.4	-336.9	-302.3	-264.9	-223.8	-181.4	17.9	1.4519
500.	857.	126.0	-427.2	-388.8	-347.2	-301.5	-254.5	22.0	1.6253

BALLISTIC COEFFICIENT .54, MUZZLE VELOCITY FT/SEC 1100.

RANGE	VELOCITY	HEIGHT	50 YD	100 YD	150 YD	200 YD	250 YD	DRIFT	TIME
0.	1100.	0.0	-1.5	-1.5	-1.5	-1.5	-1.5	0.0	0.0000
50.	1077.	.2	0.0	3.0	6.6	10.4	14.4	.3	.1378
100.	1056.	3.0	-6.0	0.0	7.2	14.8	22.9	1.0	.2785
150.	1038.	7.8	-19.8	-10.8	0.0	11.5	23.6	2.2	.4218
200.	1020.	14.7	-41.7	-29.7	-15.3	0.0	16.1	3.9	.5677
250.	1003.	23.9	-72.2	-57.2	-39.3	-20.1	0.0	6.2	.7173
300.	986.	35.6	-111.7	-93.7	-72.1	-49.2	-25.0	9.1	.8697
350.	970.	49.7	-160.3	-139.3	-114.2	-87.4	-59.2	12.4	1.0247
400.	955.	65.6	-215.7	-191.7	-163.0	-132.4	-100.1	15.0	1.1761
450.	940.	84.7	-282.3	-255.3	-222.9	-188.5	-152.3	18.9	1.3344
500.	926.	106.6	-358.7	-328.8	-292.8	-254.6	-214.3	23.2	1.4952

BALLISTIC COEFFICIENT .54, MUZZLE VELOCITY FT/SEC 1200.

RANGE	VELOCITY	HEIGHT	50 YD	100 YD	150 YD	200 YD	250 YD	DRIFT	TIME
0.	1200.	0.0	-1.5	-1.5	-1.5	-1.5	-1.5	0.0	0.0000
50.	1164.	.0	0.0	2.5	5.5	8.9	12.4	.3	.1269
100.	1133.	2.4	-4.9	0.0	6.2	12.9	19.8	1.3	.2576
150.	1106.	6.6	-16.6	-9.2	0.0	10.1	20.4	3.0	.3918
200.	1082.	12.7	-35.7	-25.8	-13.5	0.0	13.7	5.3	.5301
250.	1061.	20.8	-61.8	-49.5	-34.1	-17.2	0.0	7.8	.6693
300.	1042.	30.9	-95.8	-81.1	-62.6	-42.3	-21.7	11.0	.8123
350.	1024.	43.3	-138.0	-120.8	-99.2	-75.6	-51.5	14.6	.9582
400.	1007.	58.1	-188.7	-169.0	-144.3	-117.3	-89.9	18.9	1.1071
450.	990.	75.3	-248.1	-226.0	-198.2	-167.9	-136.9	23.6	1.2589
500.	975.	94.3	-313.7	-289.1	-258.3	-224.5	-190.1	27.7	1.4074

BALLISTIC COEFFICIENT .54, MUZZLE VELOCITY FT/SEC 1300.

RANGE	VELOCITY	HEIGHT	50 YD	100 YD	150 YD	200 YD	250 YD	DRIFT	TIME
0.	1300.	0.0	-1.5	-1.5	-1.5	-1.5	-1.5	0.0	0.0000
50.	1257.	-.1	0.0	2.0	4.7	7.5	10.7	.4	.1178
100.	1216.	2.0	-4.0	0.0	5.3	11.0	17.3	1.5	.2393
150.	1178.	5.6	-14.0	-7.9	0.0	8.6	18.0	3.3	.3648
200.	1146.	10.9	-30.1	-22.0	-11.5	0.0	12.5	5.6	.4935
250.	1117.	18.2	-53.3	-43.2	-30.0	-15.7	0.0	8.9	.6277
300.	1092.	27.1	-82.7	-70.7	-54.8	-37.6	-18.8	12.3	.7621
350.	1070.	38.2	-119.8	-105.7	-87.2	-67.1	-45.2	16.4	.9008
400.	1050.	51.4	-164.4	-148.3	-127.2	-104.2	-79.1	21.0	1.0424
450.	1031.	67.3	-218.4	-200.3	-176.5	-150.7	-122.5	26.8	1.1905
500.	1014.	84.6	-277.3	-257.2	-230.8	-202.1	-170.8	31.6	1.3335

BALLISTIC COEFFICIENT .54, MUZZLE VELOCITY FT/SEC 1400.

RANGE	VELOCITY	HEIGHT	50 YD	100 YD	150 YD	200 YD	250 YD	DRIFT	TIME
0.	1400.	0.0	-1.5	-1.5	-1.5	-1.5	-1.5	0.0	0.0000
50.	1350.	-.2	0.0	1.6	3.9	6.4	9.1	.4	.1094
100.	1303.	1.6	-3.3	0.0	4.5	9.5	14.9	1.4	.2222
150.	1260.	4.8	-11.6	-6.8	0.0	7.5	15.5	3.1	.3392
200.	1219.	9.4	-25.5	-19.0	-10.0	0.0	10.7	5.6	.4604
250.	1181.	15.7	-45.3	-37.2	-25.9	-13.4	0.0	8.8	.5855
300.	1148.	23.8	-71.4	-61.6	-48.1	-33.1	-17.0	12.6	.7145
350.	1119.	33.7	-104.1	-92.8	-77.0	-59.5	-40.7	17.1	.8474
400.	1094.	45.6	-143.6	-130.6	-112.6	-92.5	-71.1	22.1	.9826
450.	1072.	59.6	-190.4	-175.7	-155.5	-132.9	-108.8	27.6	1.1212
500.	1052.	75.8	-244.6	-228.4	-205.9	-180.8	-154.0	33.6	1.2625

BALLISTIC COEFFICIENT .54, MUZZLE VELOCITY FT/SEC 1500.

RANGE	VELOCITY	HEIGHT	50 YD	100 YD	150 YD	200 YD	250 YD	DRIFT	TIME
0.	1500.	0.0	-1.5	-1.5	-1.5	-1.5	-1.5	0.0	0.0000
50.	1443.	-.2	0.0	1.3	3.3	5.5	7.8	.4	.1024
100.	1390.	1.3	-2.7	0.0	3.9	8.3	13.0	1.4	.2081
150.	1341.	4.1	-9.8	-5.9	0.0	6.6	13.6	3.2	.3181
200.	1295.	8.2	-21.9	-16.6	-8.8	0.0	9.4	5.7	.4323
250.	1252.	13.8	-39.1	-32.5	-22.7	-11.7	0.0	8.8	.5498
300.	1212.	20.9	-61.9	-53.9	-42.2	-29.0	-15.0	12.6	.6716
350.	1174.	29.9	-91.1	-81.8	-68.1	-52.7	-36.4	17.5	.7993
400.	1142.	40.5	-125.6	-115.0	-99.4	-81.8	-63.0	22.3	.9269
450.	1114.	53.2	-167.4	-155.5	-137.9	-118.1	-97.1	28.2	1.0602
500.	1090.	68.0	-216.1	-202.8	-183.3	-161.3	-137.9	34.6	1.1964

BALLISTIC COEFFICIENT .54, MUZZLE VELOCITY FT/SEC 1600.

RANGE	VELOCITY	HEIGHT	50 YD	100 YD	150 YD	200 YD	250 YD	DRIFT	TIME
0.	1600.	0.0	-1.5	-1.5	-1.5	-1.5	-1.5	0.0	0.0000
50.	1540.	-.3	0.0	1.1	2.8	4.7	6.8	.3	.0954
100.	1482.	1.1	-2.2	0.0	3.4	7.2	11.4	1.2	.1945
150.	1426.	3.5	-8.3	-5.1	0.0	5.8	12.0	2.9	.2979
200.	1374.	7.1	-18.8	-14.5	-7.7	0.0	8.2	5.4	.4054
250.	1326.	12.0	-33.8	-28.4	-19.9	-10.3	0.0	8.4	.5162
300.	1281.	18.4	-53.8	-47.3	-37.2	-25.6	-13.2	12.1	.6313
350.	1239.	26.3	-79.2	-71.6	-59.8	-46.3	-31.8	16.6	.7507
400.	1200.	35.9	-110.1	-101.5	-87.9	-72.5	-56.0	21.8	.8737
450.	1164.	47.3	-147.1	-137.3	-122.1	-104.8	-86.2	27.6	1.0006
500.	1133.	60.7	-190.6	-179.8	-162.9	-143.5	-123.0	34.1	1.1313

BALLISTIC COEFFICIENT .54, MUZZLE VELOCITY FT/SEC 1700.

RANGE	VELOCITY	HEIGHT	50 YD	100 YD	150 YD	200 YD	250 YD	DRIFT	TIME
0.	1700.	0.0	-1.5	-1.5	-1.5	-1.5	-1.5	0.0	0.0000
50.	1637.	-.4	0.0	.9	2.3	4.0	5.8	.3	.0901
100.	1576.	.9	-1.8	0.0	2.9	6.3	9.9	1.3	.1837
150.	1517.	3.0	-7.0	-4.4	0.0	5.0	10.5	2.8	.2805
200.	1460.	6.2	-16.1	-12.5	-6.7	0.0	7.3	5.0	.3814
250.	1405.	10.6	-29.2	-24.8	-17.5	-9.1	0.0	7.9	.4860
300.	1355.	16.3	-46.9	-41.6	-32.9	-22.8	-11.8	11.6	.5953
350.	1308.	23.3	-69.1	-62.9	-52.7	-41.0	-28.2	15.8	.7074
400.	1264.	31.8	-96.4	-89.4	-77.8	-64.3	-49.7	20.8	.8240
450.	1223.	42.1	-129.2	-121.3	-108.2	-93.1	-76.7	26.5	.9447
500.	1185.	54.2	-167.9	-159.1	-144.6	-127.7	-109.5	32.9	1.0695

BALLISTIC COEFFICIENT .54, MUZZLE VELOCITY FT/SEC 1800.

RANGE	VELOCITY	HEIGHT	50 YD	100 YD	150 YD	200 YD	250 YD	DRIFT	TIME
0.	1800.	0.0	-1.5	-1.5	-1.5	-1.5	-1.5	0.0	0.0000
50.	1734.	-.4	0.0	.7	2.0	3.5	5.1	.3	.0849
100.	1670.	.7	-1.4	0.0	2.5	5.5	8.7	1.1	.1731
150.	1608.	2.6	-5.9	-3.8	0.0	4.4	9.3	2.6	.2647
200.	1548.	5.5	-13.8	-11.0	-5.9	0.0	6.4	4.7	.3598
250.	1490.	9.3	-25.3	-21.8	-15.4	-8.0	0.0	7.3	.4583
300.	1434.	14.4	-40.8	-36.6	-29.0	-20.1	-10.4	10.8	.5612
350.	1381.	20.6	-60.5	-55.6	-46.7	-36.3	-25.0	14.8	.6676
400.	1332.	28.3	-84.8	-79.2	-69.0	-57.2	-44.3	19.7	.7783
450.	1287.	37.5	-114.0	-107.7	-96.2	-82.9	-68.4	25.1	.8929
500.	1245.	48.4	-148.4	-141.4	-128.7	-113.9	-97.8	31.4	1.0116

BALLISTIC COEFFICIENT .54, MUZZLE VELOCITY FT/SEC 1900.

RANGE	VELOCITY	HEIGHT	50 YD	100 YD	150 YD	200 YD	250 YD	DRIFT	TIME
0.	1900.	0.0	-1.5	-1.5	-1.5	-1.5	-1.5	0.0	0.0000
50.	1831.	-.4	0.0	.5	1.7	3.0	4.4	.3	.0806
100.	1764.	.5	-1.1	0.0	2.2	4.9	7.8	1.0	.1639
150.	1700.	2.3	-5.0	-3.3	0.0	4.0	8.3	2.4	.2504
200.	1637.	4.8	-11.9	-9.7	-5.3	0.0	5.8	4.3	.3404
250.	1576.	8.3	-22.1	-19.4	-13.8	-7.2	0.0	6.9	.4341
300.	1517.	12.8	-35.7	-32.5	-25.8	-17.9	-9.2	10.1	.5308
350.	1459.	18.4	-53.2	-49.4	-41.6	-32.4	-22.3	13.9	.6318
400.	1405.	25.3	-74.7	-70.4	-61.5	-50.9	-39.4	18.4	.7364
450.	1355.	33.6	-100.8	-96.0	-86.0	-74.1	-61.1	23.8	.8457
500.	1308.	43.3	-131.4	-126.0	-114.8	-101.7	-87.2	29.6	.9578

BALLISTIC COEFFICIENT .55, MUZZLE VELOCITY FT/SEC 800.

RANGE	VELOCITY	HEIGHT	50 YD	100 YD	150 YD	200 YD	250 YD	DRIFT	TIME
0.	800.	0.0	-1.5	-1.5	-1.5	-1.5	-1.5	0.0	0.0000
50.	789.	1.0	0.0	6.2	13.1	20.2	27.5	.4	.1898
100.	778.	6.2	-12.4	0.0	13.8	28.0	42.7	1.0	.3807
150.	767.	15.1	-39.3	-20.7	0.0	21.3	43.4	2.3	.5755
200.	756.	27.9	-80.8	-56.1	-28.5	0.0	29.4	3.9	.7722
250.	745.	44.6	-137.7	-106.8	-72.3	-36.7	0.0	6.1	.9722
300.	735.	65.5	-210.3	-173.3	-131.8	-89.1	-45.1	8.8	1.1751
350.	724.	90.7	-299.0	-255.7	-207.3	-157.5	-106.1	12.0	1.3807
400.	713.	120.6	-404.6	-355.1	-299.9	-243.0	-184.2	15.8	1.5899
450.	703.	155.0	-526.8	-471.2	-409.0	-344.9	-278.9	20.0	1.8012
500.	693.	194.5	-667.4	-605.6	-536.5	-465.3	-391.9	25.0	2.0168

BALLISTIC COEFFICIENT .55, MUZZLE VELOCITY FT/SEC 900.

RANGE	VELOCITY	HEIGHT	50 YD	100 YD	150 YD	200 YD	250 YD	DRIFT	TIME
0.	900.	0.0	-1.5	-1.5	-1.5	-1.5	-1.5	0.0	0.0000
50.	887.	.6	0.0	4.8	10.1	15.8	21.5	.4	.1688
100.	874.	4.8	-9.6	0.0	10.7	22.0	33.5	1.0	.3392
150.	862.	11.8	-30.4	-16.1	0.0	17.0	34.2	2.1	.5117
200.	850.	21.9	-63.2	-44.0	-22.6	0.0	22.9	3.7	.6874
250.	838.	35.1	-107.6	-83.7	-56.9	-28.7	0.0	5.5	.8644
300.	827.	51.7	-164.9	-136.2	-104.1	-70.2	-35.8	8.0	1.0452
350.	816.	71.6	-234.7	-201.2	-163.7	-124.1	-84.0	10.7	1.2277
400.	804.	95.1	-317.6	-279.4	-236.5	-191.3	-145.4	14.0	1.4130
450.	793.	122.3	-414.2	-371.2	-322.9	-272.1	-220.4	17.8	1.6012
500.	782.	153.2	-524.0	-476.2	-422.6	-366.1	-308.7	21.9	1.7910

BALLISTIC COEFFICIENT .55, MUZZLE VELOCITY FT/SEC 1000.

RANGE	VELOCITY	HEIGHT	50 YD	100 YD	150 YD	200 YD	250 YD	DRIFT	TIME
0.	1000.	0.0	-1.5	-1.5	-1.5	-1.5	-1.5	0.0	0.0000
50.	984.	.4	0.0	3.8	8.0	12.5	17.3	.5	.1530
100.	968.	3.8	-7.7	0.0	8.3	17.4	26.8	1.5	.3086
150.	953.	9.5	-23.9	-12.4	0.0	13.7	27.8	2.0	.4613
200.	939.	17.7	-50.2	-34.8	-18.2	0.0	18.8	3.5	.6198
250.	925.	28.5	-86.3	-67.0	-46.3	-23.6	0.0	5.4	.7808
300.	911.	42.1	-132.6	-109.5	-84.7	-57.4	-29.1	7.8	.9444
350.	897.	58.6	-190.0	-163.0	-134.1	-102.2	-69.2	10.8	1.1116
400.	884.	77.8	-257.1	-226.4	-193.3	-156.8	-119.1	13.9	1.2788
450.	872.	100.2	-336.1	-301.5	-264.3	-223.3	-180.9	17.6	1.4501
500.	859.	125.7	-426.2	-387.7	-346.4	-300.8	-253.7	21.7	1.6231

BALLISTIC COEFFICIENT .55, MUZZLE VELOCITY FT/SEC 1100.

RANGE	VELOCITY	HEIGHT	50 YD	100 YD	150 YD	200 YD	250 YD	DRIFT	TIME
0.	1100.	0.0	-1.5	-1.5	-1.5	-1.5	-1.5	0.0	0.0000
50.	1077.	.2	0.0	3.0	6.6	10.4	14.5	.2	.1378
100.	1057.	3.0	-6.0	0.0	7.2	14.9	22.9	1.0	.2783
150.	1039.	7.8	-19.7	-10.8	0.0	11.5	23.6	2.2	.4215
200.	1021.	14.7	-41.7	-29.8	-15.4	0.0	16.1	3.9	.5678
250.	1004.	23.9	-72.3	-57.3	-39.4	-20.1	0.0	6.3	.7173
300.	988.	35.6	-111.7	-93.8	-72.3	-49.2	-25.0	9.1	.8696
350.	973.	49.1	-158.4	-137.5	-112.4	-85.4	-57.2	11.4	1.0194
400.	957.	65.5	-215.2	-191.2	-162.5	-131.7	-99.5	14.7	1.1744
450.	943.	84.6	-282.3	-255.3	-223.0	-188.4	-152.1	18.8	1.3338
500.	928.	106.3	-357.9	-328.0	-292.2	-253.7	-213.4	22.8	1.4931

BALLISTIC COEFFICIENT .55, MUZZLE VELOCITY FT/SEC 1200.

RANGE	VELOCITY	HEIGHT	50 YD	100 YD	150 YD	200 YD	250 YD	DRIFT	TIME
0.	1200.	0.0	-1.5	-1.5	-1.5	-1.5	-1.5	0.0	0.0000
50.	1165.	.0	0.0	2.5	5.5	8.8	12.4	.3	.1269
100.	1134.	2.4	-4.9	0.0	6.2	12.8	19.9	1.3	.2575
150.	1108.	6.6	-16.6	-9.2	0.0	9.9	20.6	2.9	.3915
200.	1084.	12.6	-35.3	-25.5	-13.2	0.0	14.2	5.0	.5283
250.	1063.	20.8	-62.0	-49.7	-34.3	-17.8	0.0	7.9	.6700
300.	1044.	30.9	-96.0	-81.3	-62.8	-43.0	-21.6	11.0	.8126
350.	1026.	43.3	-138.0	-120.8	-99.2	-76.1	-51.2	14.6	.9577
400.	1009.	58.0	-188.6	-169.0	-144.3	-117.9	-89.4	18.7	1.1065
450.	993.	75.2	-248.0	-225.9	-198.1	-168.4	-136.4	23.4	1.2580
500.	977.	94.0	-312.7	-288.2	-257.4	-224.4	-188.8	27.3	1.4049

BALLISTIC COEFFICIENT .55, MUZZLE VELOCITY FT/SEC 1300.

RANGE	VELOCITY	HEIGHT	50 YD	100 YD	150 YD	200 YD	250 YD	DRIFT	TIME
0.	1300.	0.0	-1.5	-1.5	-1.5	-1.5	-1.5	0.0	0.0000
50.	1257.	-.1	0.0	2.0	4.6	7.5	10.6	.4	.1176
100.	1218.	2.0	-4.0	0.0	5.2	11.0	17.1	1.4	.2389
150.	1180.	5.6	-13.9	-7.9	0.0	8.7	17.8	3.1	.3639
200.	1148.	10.9	-30.0	-22.0	-11.5	0.0	12.2	5.5	.4930
250.	1119.	18.0	-52.8	-42.8	-29.7	-15.3	0.0	8.6	.6256
300.	1095.	27.0	-82.5	-70.5	-54.8	-37.5	-19.1	12.1	.7610
350.	1073.	38.1	-119.4	-105.4	-87.1	-66.9	-45.5	16.1	.8994
400.	1053.	51.2	-163.9	-147.8	-126.9	-103.8	-79.3	20.7	1.0405
450.	1035.	66.6	-216.1	-198.1	-174.5	-148.5	-121.0	25.7	1.1844
500.	1017.	84.2	-276.3	-256.3	-230.1	-201.3	-170.7	31.1	1.3307

BALLISTIC COEFFICIENT .55, MUZZLE VELOCITY FT/SEC 1400.

RANGE	VELOCITY	HEIGHT	50 YD	100 YD	150 YD	200 YD	250 YD	DRIFT	TIME
0.	1400.	0.0	-1.5	-1.5	-1.5	-1.5	-1.5	0.0	0.0000
50.	1351.	-.2	0.0	1.6	3.9	6.4	9.0	.4	.1094
100.	1305.	1.6	-3.2	0.0	4.5	9.5	14.8	1.4	.2221
150.	1262.	4.8	-11.6	-6.7	0.0	7.5	15.5	3.1	.3389
200.	1222.	9.4	-25.5	-19.0	-10.0	0.0	10.7	5.5	.4598
250.	1184.	15.7	-45.2	-37.1	-25.8	-13.4	0.0	8.6	.5847
300.	1151.	23.7	-71.1	-61.4	-47.9	-32.9	-16.9	12.4	.7133
350.	1123.	33.5	-103.6	-92.3	-76.5	-59.1	-40.4	16.8	.8453
400.	1097.	45.4	-143.0	-130.0	-112.1	-92.1	-70.7	21.7	.9805
450.	1075.	59.3	-189.6	-175.0	-154.7	-132.3	-108.2	27.2	1.1186
500.	1055.	75.4	-243.6	-227.3	-204.9	-179.9	-153.2	33.1	1.2594

BALLISTIC COEFFICIENT .55, MUZZLE VELOCITY FT/SEC 1500.

RANGE	VELOCITY	HEIGHT	50 YD	100 YD	150 YD	200 YD	250 YD	DRIFT	TIME
0.	1500.	0.0	-1.5	-1.5	-1.5	-1.5	-1.5	0.0	0.0000
50.	1444.	-.2	0.0	1.3	3.3	5.5	7.8	.4	.1024
100.	1392.	1.3	-2.7	0.0	3.9	8.3	13.0	1.4	.2080
150.	1343.	4.1	-9.8	-5.8	0.0	6.5	13.6	3.1	.3178
200.	1298.	8.2	-21.8	-16.5	-8.7	0.0	9.4	5.5	.4314
250.	1255.	13.7	-39.0	-32.4	-22.7	-11.8	0.0	8.7	.5492
300.	1216.	20.8	-61.7	-53.8	-42.1	-29.0	-14.9	12.4	.6707
350.	1179.	29.7	-90.3	-81.0	-67.4	-52.2	-35.7	16.9	.7960
400.	1147.	40.3	-125.1	-114.5	-98.9	-81.5	-62.6	22.0	.9248
450.	1118.	53.0	-166.9	-155.0	-137.5	-117.9	-96.7	27.9	1.0583
500.	1094.	67.6	-215.1	-201.8	-182.3	-160.5	-137.0	34.0	1.1931

BALLISTIC COEFFICIENT .55, MUZZLE VELOCITY FT/SEC 1600.

RANGE	VELOCITY	HEIGHT	50 YD	100 YD	150 YD	200 YD	250 YD	DRIFT	TIME
0.	1600.	0.0	-1.5	-1.5	-1.5	-1.5	-1.5	0.0	0.0000
50.	1541.	-.3	0.0	1.1	2.8	4.7	6.7	.3	.0954
100.	1484.	1.1	-2.2	0.0	3.4	7.2	11.3	1.2	.1944
150.	1429.	3.5	-8.3	-5.1	0.0	5.7	11.9	2.9	.2976
200.	1378.	7.1	-18.7	-14.4	-7.6	0.0	8.3	5.2	.4046
250.	1330.	12.0	-33.7	-28.3	-19.9	-10.3	0.0	8.2	.5153
300.	1286.	18.3	-53.6	-47.1	-37.0	-25.5	-13.2	11.9	.6300
350.	1244.	26.2	-78.8	-71.3	-59.4	-46.1	-31.6	16.3	.7488
400.	1205.	35.7	-109.6	-101.0	-87.5	-72.2	-55.7	21.3	.8713
450.	1169.	47.0	-146.3	-136.6	-121.4	-104.3	-85.7	27.1	.9977
500.	1138.	60.4	-189.8	-179.0	-162.1	-143.0	-122.4	33.6	1.1286

BALLISTIC COEFFICIENT .55, MUZZLE VELOCITY FT/SEC 1700.

RANGE	VELOCITY	HEIGHT	50 YD	100 YD	150 YD	200 YD	250 YD	DRIFT	TIME
0.	1700.	0.0	-1.5	-1.5	-1.5	-1.5	-1.5	0.0	0.0000
50.	1638.	-.4	0.0	.9	2.3	4.0	5.8	.3	.0900
100.	1578.	.9	-1.8	0.0	2.9	6.2	9.9	1.3	.1836
150.	1520.	3.0	-7.0	-4.3	0.0	5.0	10.5	2.7	.2802
200.	1464.	6.2	-16.0	-12.5	-6.7	0.0	7.3	4.9	.3808
250.	1410.	10.5	-29.1	-24.7	-17.4	-9.1	0.0	7.7	.4851
300.	1360.	16.2	-46.6	-41.3	-32.6	-22.6	-11.7	11.3	.5936
350.	1314.	23.2	-68.9	-62.7	-52.6	-40.8	-28.2	15.6	.7061
400.	1270.	31.7	-96.0	-89.0	-77.4	-64.0	-49.5	20.4	.8219
450.	1230.	41.8	-128.6	-120.7	-107.7	-92.6	-76.3	26.0	.9420
500.	1192.	53.8	-167.0	-158.2	-143.7	-127.0	-108.8	32.3	1.0661

BALLISTIC COEFFICIENT .55, MUZZLE VELOCITY FT/SEC 1800.

RANGE	VELOCITY	HEIGHT	50 YD	100 YD	150 YD	200 YD	250 YD	DRIFT	TIME
0.	1800.	0.0	-1.5	-1.5	-1.5	-1.5	-1.5	0.0	0.0000
50.	1735.	-.4	0.0	.7	2.0	3.4	5.0	.3	.0849
100.	1673.	.7	-1.4	0.0	2.5	5.5	8.7	1.1	.1730
150.	1612.	2.6	-5.9	-3.8	0.0	4.4	9.2	2.5	.2643
200.	1553.	5.4	-13.8	-11.0	-5.9	0.0	6.4	4.6	.3593
250.	1495.	9.3	-25.2	-21.7	-15.4	-8.0	0.0	7.2	.4574
300.	1440.	14.3	-40.7	-36.5	-28.9	-20.0	-10.4	10.6	.5601
350.	1388.	20.5	-60.2	-55.3	-46.4	-36.1	-24.9	14.5	.6660
400.	1339.	28.2	-84.4	-78.8	-68.6	-56.8	-44.0	19.3	.7762
450.	1294.	37.3	-113.5	-107.2	-95.8	-82.4	-68.1	24.7	.8905
500.	1252.	48.0	-147.5	-140.5	-127.8	-113.0	-97.1	30.7	1.0079

BALLISTIC COEFFICIENT .55, MUZZLE VELOCITY FT/SEC 1900.

RANGE	VELOCITY	HEIGHT	50 YD	100 YD	150 YD	200 YD	250 YD	DRIFT	TIME
0.	1900.	0.0	-1.5	-1.5	-1.5	-1.5	-1.5	0.0	0.0000
50.	1833.	-.4	0.0	.5	1.7	3.0	4.4	.3	.0806
100.	1767.	.5	-1.1	0.0	2.2	4.8	7.7	1.0	.1637
150.	1703.	2.3	-5.0	-3.3	0.0	3.9	8.3	2.3	.2501
200.	1641.	4.8	-11.9	-9.7	-5.3	0.0	5.8	4.3	.3400
250.	1581.	8.3	-22.0	-19.3	-13.8	-7.2	0.0	6.8	.4333
300.	1523.	12.7	-35.6	-32.3	-25.7	-17.8	-9.2	9.9	.5297
350.	1467.	18.3	-52.9	-49.2	-41.4	-32.2	-22.1	13.6	.6301
400.	1413.	25.1	-74.3	-70.0	-61.1	-50.6	-39.1	18.1	.7342
450.	1363.	33.3	-100.2	-95.3	-85.3	-73.5	-60.5	23.2	.8426
500.	1316.	43.0	-130.7	-125.3	-114.2	-101.0	-86.6	29.1	.9547

BALLISTIC COEFFICIENT .56, MUZZLE VELOCITY FT/SEC 800.

RANGE	VELOCITY	HEIGHT	50 YD	100 YD	150 YD	200 YD	250 YD	DRIFT	TIME
0.	800.	0.0	-1.5	-1.5	-1.5	-1.5	-1.5	0.0	0.0000
50.	789.	1.0	0.0	6.2	13.1	20.2	27.5	.4	.1898
100.	778.	6.2	-12.3	0.0	13.8	28.0	42.7	1.0	.3805
150.	767.	15.1	-39.2	-20.7	0.0	21.3	43.3	2.3	.5753
200.	757.	27.8	-80.7	-56.0	-28.4	0.0	29.4	3.8	.7717
250.	746.	44.6	-137.6	-106.8	-72.2	-36.7	0.0	6.0	.9717
300.	736.	65.4	-210.0	-173.0	-131.5	-88.9	-44.9	8.6	1.1741
350.	725.	90.6	-298.6	-255.4	-207.0	-157.4	-106.0	11.8	1.3796
400.	715.	120.3	-403.8	-354.4	-299.1	-242.4	-183.6	15.5	1.5880
450.	705.	154.7	-526.0	-470.5	-408.3	-344.4	-278.3	19.7	1.7994
500.	695.	193.9	-665.9	-604.2	-535.1	-464.1	-390.7	24.5	2.0140

BALLISTIC COEFFICIENT .56, MUZZLE VELOCITY FT/SEC 900.

RANGE	VELOCITY	HEIGHT	50 YD	100 YD	150 YD	200 YD	250 YD	DRIFT	TIME
0.	900.	0.0	-1.5	-1.5	-1.5	-1.5	-1.5	0.0	0.0000
50.	887.	.6	0.0	4.8	10.1	15.8	21.5	.4	.1688
100.	875.	4.8	-9.5	0.0	10.7	22.0	33.4	1.0	.3391
150.	863.	11.8	-30.4	-16.1	0.0	16.9	34.1	2.0	.5116
200.	851.	21.9	-63.1	-44.0	-22.6	0.0	22.8	3.6	.6871
250.	839.	35.1	-107.4	-83.6	-56.8	-28.6	0.0	5.3	.8637
300.	828.	51.6	-164.7	-136.1	-103.9	-70.1	-35.8	7.8	1.0445
350.	817.	71.5	-234.2	-200.8	-163.3	-123.8	-83.8	10.5	1.2265
400.	806.	94.9	-317.2	-279.0	-236.1	-191.0	-145.3	13.8	1.4118
450.	795.	122.0	-413.3	-370.4	-322.1	-271.4	-219.9	17.5	1.5992
500.	784.	152.9	-523.1	-475.4	-421.8	-365.4	-308.3	21.5	1.7890

BALLISTIC COEFFICIENT .56, MUZZLE VELOCITY FT/SEC 1000.

RANGE	VELOCITY	HEIGHT	50 YD	100 YD	150 YD	200 YD	250 YD	DRIFT	TIME
0.	1000.	0.0	-1.5	-1.5	-1.5	-1.5	-1.5	0.0	0.0000
50.	984.	.4	0.0	3.9	8.0	12.5	17.2	.6	.1532
100.	969.	3.8	-7.7	0.0	8.2	17.3	26.7	1.6	.3089
150.	954.	9.5	-23.9	-12.3	0.0	13.6	27.8	1.9	.4611
200.	940.	17.7	-50.1	-34.6	-18.2	0.0	18.9	3.4	.6194
250.	926.	28.5	-86.1	-66.9	-46.3	-23.6	0.0	5.3	.7803
300.	912.	42.0	-132.4	-109.2	-84.6	-57.3	-29.0	7.7	.9436
350.	899.	58.3	-189.1	-162.1	-133.4	-101.5	-68.5	10.5	1.1094
400.	886.	77.6	-256.7	-225.8	-192.9	-156.6	-118.8	13.6	1.2775
450.	874.	99.9	-335.3	-300.7	-263.7	-222.7	-180.3	17.3	1.4482
500.	862.	125.3	-425.1	-386.6	-345.5	-300.0	-252.8	21.2	1.6207

BALLISTIC COEFFICIENT .56, MUZZLE VELOCITY FT/SEC 1100.

RANGE	VELOCITY	HEIGHT	50 YD	100 YD	150 YD	200 YD	250 YD	DRIFT	TIME
0.	1100.	0.0	-1.5	-1.5	-1.5	-1.5	-1.5	0.0	0.0000
50.	1078.	.2	0.0	3.0	6.6	10.4	14.5	.2	.1377
100.	1058.	3.0	-6.0	0.0	7.2	14.9	23.0	1.0	.2782
150.	1040.	7.8	-19.7	-10.7	0.0	11.6	23.7	2.1	.4212
200.	1022.	14.7	-41.8	-29.8	-15.5	0.0	16.1	4.0	.5679
250.	1006.	24.0	-72.4	-57.4	-39.5	-20.1	0.0	6.3	.7174
300.	990.	35.5	-111.8	-93.9	-72.4	-49.2	-25.0	9.0	.8696
350.	975.	49.0	-158.1	-137.2	-112.2	-85.0	-56.8	11.2	1.0182
400.	959.	65.3	-214.6	-190.7	-162.1	-131.1	-98.9	14.4	1.1727
450.	945.	84.6	-282.2	-255.4	-223.1	-188.3	-152.0	18.7	1.3333
500.	931.	106.0	-357.2	-327.3	-291.5	-252.7	-212.4	22.4	1.4910

BALLISTIC COEFFICIENT .56, MUZZLE VELOCITY FT/SEC 1200.

RANGE	VELOCITY	HEIGHT	50 YD	100 YD	150 YD	200 YD	250 YD	DRIFT	TIME
0.	1200.	0.0	-1.5	-1.5	-1.5	-1.5	-1.5	0.0	0.0000
50.	1165.	.0	0.0	2.6	5.5	8.8	12.4	.3	.1269
100.	1135.	2.5	-5.1	0.0	5.9	12.5	19.8	1.7	.2596
150.	1109.	6.6	-16.6	-8.9	0.0	9.9	20.7	2.9	.3913
200.	1086.	12.6	-35.3	-25.0	-13.2	0.0	14.5	4.9	.5279
250.	1065.	20.8	-62.2	-49.4	-34.6	-18.1	0.0	8.0	.6707
300.	1046.	31.0	-96.1	-80.8	-63.0	-43.2	-21.5	11.1	.8128
350.	1028.	43.2	-137.9	-120.0	-99.2	-76.2	-50.8	14.5	.9573
400.	1012.	58.0	-188.5	-168.0	-144.3	-117.9	-89.0	18.6	1.1059
450.	996.	74.4	-245.0	-222.0	-195.3	-165.6	-133.1	22.2	1.2509
500.	980.	93.7	-311.8	-286.2	-256.6	-223.6	-187.4	26.8	1.4025

BALLISTIC COEFFICIENT .56, MUZZLE VELOCITY FT/SEC 1300.

RANGE	VELOCITY	HEIGHT	50 YD	100 YD	150 YD	200 YD	250 YD	DRIFT	TIME
0.	1300.	0.0	-1.5	-1.5	-1.5	-1.5	-1.5	0.0	0.0000
50.	1258.	-.1	0.0	2.0	4.6	7.5	10.5	.4	.1175
100.	1219.	2.0	-4.0	0.0	5.3	11.0	17.1	1.4	.2386
150.	1182.	5.6	-13.9	-7.9	0.0	8.6	17.7	3.1	.3637
200.	1150.	10.9	-30.0	-22.0	-11.5	0.0	12.2	5.4	.4925
250.	1122.	18.0	-52.7	-42.7	-29.6	-15.2	0.0	8.4	.6245
300.	1097.	27.0	-82.3	-70.3	-54.6	-37.3	-19.1	11.9	.7598
350.	1075.	37.9	-119.1	-105.1	-86.8	-66.6	-45.3	15.9	.8979
400.	1056.	51.0	-163.4	-147.4	-126.4	-103.4	-79.1	20.3	1.0387
450.	1037.	66.3	-215.4	-197.4	-173.8	-147.9	-120.5	25.3	1.1820
500.	1020.	83.9	-275.4	-255.5	-229.3	-200.5	-170.1	30.7	1.3281

BALLISTIC COEFFICIENT .56, MUZZLE VELOCITY FT/SEC 1400.

RANGE	VELOCITY	HEIGHT	50 YD	100 YD	150 YD	200 YD	250 YD	DRIFT	TIME
0.	1400.	0.0	-1.5	-1.5	-1.5	-1.5	-1.5	0.0	0.0000
50.	1352.	-.2	0.0	1.6	3.9	6.4	9.0	.4	.1093
100.	1307.	1.6	-3.2	0.0	4.5	9.5	14.8	1.4	.2220
150.	1264.	4.8	-11.6	-6.7	0.0	7.5	15.5	3.0	.3387
200.	1225.	9.4	-25.4	-18.9	-10.0	0.0	10.6	5.4	.4593
250.	1188.	15.6	-45.1	-37.0	-25.8	-13.3	0.0	8.5	.5839
300.	1154.	23.7	-71.5	-61.7	-48.3	-33.3	-17.4	12.6	.7143
350.	1126.	33.4	-103.3	-91.9	-76.2	-58.8	-40.2	16.5	.8437
400.	1101.	45.2	-142.5	-129.5	-111.6	-91.6	-70.4	21.3	.9784
450.	1078.	59.0	-188.8	-174.2	-154.0	-131.6	-107.6	26.7	1.1160
500.	1058.	75.0	-242.5	-226.3	-203.9	-179.0	-152.3	32.5	1.2564

BALLISTIC COEFFICIENT .56, MUZZLE VELOCITY FT/SEC 1500.

RANGE	VELOCITY	HEIGHT	50 YD	100 YD	150 YD	200 YD	250 YD	DRIFT	TIME
0.	1500.	0.0	-1.5	-1.5	-1.5	-1.5	-1.5	0.0	0.0000
50.	1445.	-.2	0.0	1.3	3.3	5.4	7.8	.4	.1024
100.	1394.	1.3	-2.6	0.0	3.9	8.2	12.9	1.4	.2078
150.	1346.	4.1	-9.8	-5.8	0.0	6.5	13.5	3.1	.3175
200.	1301.	8.1	-21.7	-16.4	-8.6	0.0	9.3	5.4	.4306
250.	1259.	13.7	-38.8	-32.2	-22.5	-11.7	0.0	8.4	.5479
300.	1220.	20.7	-61.4	-53.4	-41.8	-28.8	-14.8	12.1	.6688
350.	1183.	29.5	-89.9	-80.6	-67.0	-51.9	-35.5	16.5	.7940
400.	1151.	40.1	-124.6	-114.0	-98.4	-81.2	-62.5	21.6	.9227
450.	1123.	52.6	-165.9	-153.9	-136.4	-117.0	-96.0	27.2	1.0547
500.	1098.	67.2	-214.0	-200.8	-181.3	-159.7	-136.4	33.4	1.1898

BALLISTIC COEFFICIENT .56, MUZZLE VELOCITY FT/SEC 1600.

RANGE	VELOCITY	HEIGHT	50 YD	100 YD	150 YD	200 YD	250 YD	DRIFT	TIME
0.	1600.	0.0	-1.5	-1.5	-1.5	-1.5	-1.5	0.0	0.0000
50.	1542.	-.3	0.0	1.1	2.8	4.7	6.7	.3	.0953
100.	1486.	1.1	-2.1	0.0	3.4	7.2	11.3	1.2	.1943
150.	1432.	3.5	-8.3	-5.1	0.0	5.7	11.9	2.8	.2974
200.	1382.	7.1	-18.6	-14.3	-7.6	0.0	8.3	5.1	.4038
250.	1334.	12.0	-33.7	-28.3	-19.9	-10.4	0.0	8.1	.5149
300.	1290.	18.2	-53.4	-46.9	-36.8	-25.5	-13.0	11.7	.6288
350.	1249.	26.0	-78.5	-70.9	-59.1	-45.9	-31.3	16.0	.7470
400.	1211.	35.5	-109.1	-100.5	-87.0	-71.8	-55.2	20.9	.8690
450.	1174.	46.9	-146.2	-136.5	-121.4	-104.3	-85.6	26.9	.9967
500.	1143.	59.9	-188.5	-177.7	-160.9	-141.9	-121.1	32.9	1.1243

BALLISTIC COEFFICIENT .56, MUZZLE VELOCITY FT/SEC 1700.

RANGE	VELOCITY	HEIGHT	50 YD	100 YD	150 YD	200 YD	250 YD	DRIFT	TIME
0.	1700.	0.0	-1.5	-1.5	-1.5	-1.5	-1.5	0.0	0.0000
50.	1639.	-.4	0.0	.9	2.3	4.0	5.8	.3	.0900
100.	1580.	.9	-1.8	0.0	2.9	6.2	9.8	1.2	.1835
150.	1523.	3.0	-7.0	-4.3	0.0	5.0	10.4	2.7	.2799
200.	1468.	6.2	-16.0	-12.4	-6.7	0.0	7.2	4.8	.3802
250.	1415.	10.5	-29.0	-24.6	-17.4	-9.0	0.0	7.6	.4842
300.	1365.	16.1	-46.4	-41.2	-32.5	-22.5	-11.7	11.1	.5924
350.	1319.	23.0	-68.5	-62.3	-52.2	-40.5	-27.9	15.2	.7040
400.	1276.	31.5	-95.6	-88.6	-77.1	-63.7	-49.3	20.1	.8200
450.	1236.	41.6	-128.1	-120.2	-107.2	-92.2	-76.0	25.6	.9397
500.	1198.	53.5	-166.1	-157.4	-142.9	-126.3	-108.2	31.8	1.0628

BALLISTIC COEFFICIENT .56, MUZZLE VELOCITY FT/SEC 1800.

RANGE	VELOCITY	HEIGHT	50 YD	100 YD	150 YD	200 YD	250 YD	DRIFT	TIME
0.	1800.	0.0	-1.5	-1.5	-1.5	-1.5	-1.5	0.0	0.0000
50.	1737.	-.4	0.0	.7	2.0	3.4	5.0	.3	.0849
100.	1675.	.7	-1.4	0.0	2.5	5.5	8.7	1.1	.1729
150.	1615.	2.6	-5.9	-3.8	0.0	4.4	9.2	2.5	.2640
200.	1557.	5.4	-13.7	-11.0	-5.9	0.0	6.3	4.5	.3589
250.	1500.	9.3	-25.1	-21.6	-15.3	-7.9	0.0	7.0	.4566
300.	1445.	14.2	-40.5	-36.3	-28.8	-19.9	-10.4	10.4	.5589
350.	1394.	20.4	-60.0	-55.1	-46.2	-35.9	-24.8	14.3	.6644
400.	1346.	28.0	-84.0	-78.4	-68.3	-56.5	-43.8	18.9	.7741
450.	1301.	37.0	-112.7	-106.4	-95.0	-81.7	-67.5	24.1	.8871
500.	1259.	47.7	-146.6	-139.6	-127.0	-112.2	-96.3	30.1	1.0043

BALLISTIC COEFFICIENT .56, MUZZLE VELOCITY FT/SEC 1900.

RANGE	VELOCITY	HEIGHT	50 YD	100 YD	150 YD	200 YD	250 YD	DRIFT	TIME
0.	1900.	0.0	-1.5	-1.5	-1.5	-1.5	-1.5	0.0	0.0000
50.	1834.	-.4	0.0	.5	1.6	3.0	4.4	.3	.0806
100.	1769.	.5	-1.1	0.0	2.2	4.8	7.7	1.0	.1637
150.	1707.	2.2	-4.9	-3.3	0.0	3.9	8.2	2.3	.2498
200.	1646.	4.8	-11.8	-9.7	-5.2	0.0	5.7	4.2	.3395
250.	1587.	8.2	-21.9	-19.2	-13.7	-7.2	0.0	6.7	.4326
300.	1529.	12.7	-35.4	-32.2	-25.6	-17.7	-9.1	9.7	.5287
350.	1474.	18.2	-52.7	-48.9	-41.2	-32.0	-22.0	13.4	.6285
400.	1420.	25.0	-73.9	-69.6	-60.8	-50.3	-38.8	17.7	.7320
450.	1371.	33.1	-99.6	-94.7	-84.8	-73.0	-60.1	22.8	.8399
500.	1324.	42.7	-129.8	-124.4	-113.4	-100.3	-86.0	28.5	.9511

BALLISTIC COEFFICIENT .57, MUZZLE VELOCITY FT/SEC 800.

RANGE	VELOCITY	HEIGHT	50 YD	100 YD	150 YD	200 YD	250 YD	DRIFT	TIME
0.	800.	0.0	-1.5	-1.5	-1.5	-1.5	-1.5	0.0	0.0000
50.	789.	1.0	0.0	6.2	13.1	20.1	27.5	.4	.1898
100.	779.	6.2	-12.3	0.0	13.8	28.0	42.7	.9	.3803
150.	768.	15.1	-39.2	-20.7	0.0	21.2	43.3	2.2	.5751
200.	757.	27.8	-80.6	-55.9	-28.3	0.0	29.4	3.7	.7712
250.	747.	44.5	-137.5	-106.7	-72.1	-36.8	0.0	5.9	.9712
300.	737.	65.3	-209.6	-172.7	-131.2	-88.8	-44.7	8.4	1.1730
350.	727.	90.5	-298.3	-255.1	-206.7	-157.2	-105.8	11.6	1.3785
400.	716.	120.0	-403.0	-353.7	-298.4	-241.8	-183.0	15.2	1.5862
450.	706.	154.3	-525.2	-469.8	-407.5	-343.9	-277.7	19.4	1.7975
500.	696.	193.4	-664.4	-602.8	-533.6	-462.9	-389.4	24.0	2.0111

BALLISTIC COEFFICIENT .57, MUZZLE VELOCITY FT/SEC 900.

RANGE	VELOCITY	HEIGHT	50 YD	100 YD	150 YD	200 YD	250 YD	DRIFT	TIME
0.	900.	0.0	-1.5	-1.5	-1.5	-1.5	-1.5	0.0	0.0000
50.	888.	.6	0.0	4.8	10.1	15.8	21.5	.4	.1688
100.	875.	4.8	-9.5	0.0	10.7	22.0	33.4	1.0	.3390
150.	863.	11.8	-30.4	-16.1	0.0	16.9	34.0	2.0	.5115
200.	852.	21.9	-63.0	-44.0	-22.5	0.0	22.8	3.5	.6868
250.	840.	35.0	-107.3	-83.5	-56.6	-28.5	0.0	5.2	.8631
300.	829.	51.5	-164.5	-135.9	-103.7	-70.0	-35.8	7.7	1.0437
350.	818.	71.3	-233.8	-200.4	-162.9	-123.5	-83.6	10.3	1.2252
400.	807.	94.7	-316.7	-278.6	-235.7	-190.6	-145.1	13.6	1.4105
450.	797.	121.7	-412.5	-369.6	-321.3	-270.7	-219.4	17.1	1.5973
500.	786.	152.5	-522.3	-474.6	-421.0	-364.7	-307.7	21.2	1.7871

BALLISTIC COEFFICIENT .57, MUZZLE VELOCITY FT/SEC 1000.

RANGE	VELOCITY	HEIGHT	50 YD	100 YD	150 YD	200 YD	250 YD	DRIFT	TIME
0.	1000.	0.0	-1.5	-1.5	-1.5	-1.5	-1.5	0.0	0.0000
50.	984.	.4	0.0	3.9	7.9	12.5	17.2	.6	.1533
100.	969.	3.8	-7.7	0.0	8.2	17.3	26.7	1.6	.3091
150.	955.	9.4	-23.8	-12.3	0.0	13.7	27.8	1.9	.4608
200.	941.	17.7	-50.0	-34.6	-18.3	0.0	18.8	3.4	.6194
250.	927.	28.4	-86.0	-66.7	-46.3	-23.5	0.0	5.3	.7799
300.	914.	41.9	-132.1	-108.9	-84.4	-57.0	-28.9	7.5	.9428
350.	901.	58.1	-188.5	-161.4	-132.9	-100.9	-68.0	10.1	1.1077
400.	888.	77.4	-256.2	-225.3	-192.6	-156.1	-118.5	13.4	1.2762
450.	876.	99.7	-334.5	-299.8	-263.0	-221.9	-179.7	17.0	1.4463
500.	864.	125.0	-424.3	-385.7	-344.8	-299.2	-252.3	20.9	1.6188

BALLISTIC COEFFICIENT .57, MUZZLE VELOCITY FT/SEC 1100.

RANGE	VELOCITY	HEIGHT	50 YD	100 YD	150 YD	200 YD	250 YD	DRIFT	TIME
0.	1100.	0.0	-1.5	-1.5	-1.5	-1.5	-1.5	0.0	0.0000
50.	1078.	.2	0.0	3.0	6.6	10.5	14.5	.2	.1377
100.	1058.	3.0	-6.0	0.0	7.2	14.9	23.0	.9	.2781
150.	1040.	7.8	-19.7	-10.8	0.0	11.6	23.7	2.1	.4212
200.	1024.	14.7	-41.8	-29.9	-15.5	0.0	16.1	4.0	.5681
250.	1007.	24.0	-72.4	-57.5	-39.6	-20.1	0.0	6.3	.7175
300.	991.	35.5	-111.9	-94.0	-72.4	-49.1	-25.0	9.0	.8695
350.	976.	48.9	-157.8	-137.0	-111.8	-84.6	-56.4	11.0	1.0171
400.	962.	65.2	-214.6	-190.7	-162.0	-130.9	-98.7	14.3	1.1722
450.	947.	84.5	-282.2	-255.4	-223.0	-188.1	-151.9	18.6	1.3327
500.	933.	105.7	-356.4	-326.5	-290.6	-251.8	-211.5	22.0	1.4889

BALLISTIC COEFFICIENT .57, MUZZLE VELOCITY FT/SEC 1200.

RANGE	VELOCITY	HEIGHT	50 YD	100 YD	150 YD	200 YD	250 YD	DRIFT	TIME
0.	1200.	0.0	-1.5	-1.5	-1.5	-1.5	-1.5	0.0	0.0000
50.	1166.	.0	0.0	2.5	5.5	8.8	12.3	.3	.1269
100.	1136.	2.5	-5.1	0.0	6.0	12.6	19.4	1.6	.2590
150.	1110.	6.6	-16.6	-9.0	0.0	9.9	20.2	2.8	.3910
200.	1087.	12.6	-35.2	-25.1	-13.2	0.0	13.8	4.8	.5275
250.	1067.	20.6	-61.3	-48.6	-33.7	-17.2	0.0	7.3	.6667
300.	1048.	31.0	-96.3	-81.1	-63.2	-43.4	-22.8	11.1	.8131
350.	1030.	43.2	-137.9	-120.1	-99.2	-76.2	-52.1	14.4	.9569
400.	1014.	57.3	-186.4	-166.1	-142.2	-115.9	-88.3	17.6	1.1000
450.	998.	74.1	-244.3	-221.5	-194.7	-165.0	-134.1	21.8	1.2489
500.	983.	93.5	-311.7	-286.3	-256.5	-223.6	-189.1	26.7	1.4016

BALLISTIC COEFFICIENT .57, MUZZLE VELOCITY FT/SEC 1300.

RANGE	VELOCITY	HEIGHT	50 YD	100 YD	150 YD	200 YD	250 YD	DRIFT	TIME
0.	1300.	0.0	-1.5	-1.5	-1.5	-1.5	-1.5	0.0	0.0000
50.	1259.	-.1	0.0	2.0	4.6	7.5	10.5	.4	.1174
100.	1220.	2.0	-4.0	0.0	5.3	11.0	17.1	1.3	.2383
150.	1184.	5.6	-13.9	-7.9	0.0	8.6	17.7	3.0	.3634
200.	1152.	10.9	-29.9	-22.0	-11.5	0.0	12.1	5.4	.4919
250.	1124.	17.9	-52.6	-42.6	-29.5	-15.2	0.0	8.3	.6238
300.	1100.	26.9	-82.1	-70.1	-54.4	-37.2	-19.0	11.7	.7587
350.	1078.	37.8	-118.7	-104.8	-86.4	-66.4	-45.1	15.6	.8964
400.	1058.	50.8	-162.9	-146.9	-125.9	-103.0	-78.7	20.0	1.0368
450.	1040.	66.1	-214.7	-196.8	-173.1	-147.3	-120.0	24.9	1.1798
500.	1023.	83.7	-275.1	-255.3	-229.0	-200.3	-170.0	30.4	1.3268

BALLISTIC COEFFICIENT .57, MUZZLE VELOCITY FT/SEC 1400.

RANGE	VELOCITY	HEIGHT	50 YD	100 YD	150 YD	200 YD	250 YD	DRIFT	TIME
0.	1400.	0.0	-1.5	-1.5	-1.5	-1.5	-1.5	0.0	0.0000
50.	1353.	-.2	0.0	1.6	3.9	6.3	9.0	.4	.1093
100.	1308.	1.6	-3.2	0.0	4.5	9.4	14.7	1.3	.2218
150.	1267.	4.7	-11.6	-6.7	0.0	7.4	15.4	3.0	.3384
200.	1227.	9.4	-25.4	-18.9	-9.9	0.0	10.6	5.3	.4589
250.	1191.	15.6	-45.0	-36.9	-25.7	-13.3	0.0	8.3	.5831
300.	1158.	23.6	-70.9	-61.2	-47.7	-32.8	-16.9	12.1	.7117
350.	1129.	33.3	-102.9	-91.6	-75.9	-58.6	-40.0	16.2	.8421
400.	1104.	45.0	-142.0	-129.1	-111.1	-91.3	-70.1	21.0	.9765
450.	1082.	58.9	-188.5	-173.9	-153.7	-131.4	-107.5	26.5	1.1147
500.	1062.	74.8	-242.0	-225.8	-203.4	-178.6	-152.1	32.3	1.2547

BALLISTIC COEFFICIENT .57, MUZZLE VELOCITY FT/SEC 1500.

RANGE	VELOCITY	HEIGHT	50 YD	100 YD	150 YD	200 YD	250 YD	DRIFT	TIME
0.	1500.	0.0	-1.5	-1.5	-1.5	-1.5	-1.5	0.0	0.0000
50.	1446.	-.2	0.0	1.3	3.3	5.4	7.7	.4	.1023
100.	1395.	1.3	-2.7	0.0	3.9	8.2	12.8	1.4	.2080
150.	1348.	4.1	-9.8	-5.8	0.0	6.5	13.4	3.0	.3172
200.	1304.	8.1	-21.7	-16.3	-8.6	0.0	9.3	5.3	.4301
250.	1263.	13.6	-38.7	-32.0	-22.4	-11.6	0.0	8.3	.5470
300.	1224.	20.7	-61.2	-53.2	-41.6	-28.7	-14.8	11.9	.6678
350.	1188.	29.4	-89.6	-80.3	-66.7	-51.7	-35.4	16.3	.7924
400.	1155.	40.1	-124.8	-114.1	-98.6	-81.4	-62.8	21.6	.9226
450.	1127.	52.4	-165.2	-153.2	-135.8	-116.4	-95.5	26.8	1.0521
500.	1102.	66.8	-213.0	-199.7	-180.4	-158.8	-135.6	32.9	1.1867

BALLISTIC COEFFICIENT .57, MUZZLE VELOCITY FT/SEC 1600.

RANGE	VELOCITY	HEIGHT	50 YD	100 YD	150 YD	200 YD	250 YD	DRIFT	TIME
0.	1600.	0.0	-1.5	-1.5	-1.5	-1.5	-1.5	0.0	0.0000
50.	1543.	-.3	0.0	1.1	2.8	4.6	6.7	.3	.0953
100.	1488.	1.1	-2.1	0.0	3.4	7.1	11.3	1.2	.1941
150.	1435.	3.5	-8.3	-5.0	0.0	5.7	11.8	2.8	.2971
200.	1385.	7.1	-18.6	-14.3	-7.6	0.0	8.2	5.0	.4033
250.	1338.	11.9	-33.5	-28.1	-19.7	-10.3	0.0	7.9	.5138
300.	1295.	18.2	-53.3	-46.9	-36.8	-25.4	-13.1	11.5	.6279
350.	1254.	26.0	-78.3	-70.8	-59.0	-45.8	-31.4	15.8	.7458
400.	1216.	35.3	-108.7	-100.1	-86.7	-71.5	-55.1	20.6	.8672
450.	1180.	46.5	-144.9	-135.2	-120.1	-103.1	-84.6	26.1	.9920
500.	1149.	59.6	-187.5	-176.8	-160.0	-141.1	-120.6	32.3	1.1211

BALLISTIC COEFFICIENT .57, MUZZLE VELOCITY FT/SEC 1700.

RANGE	VELOCITY	HEIGHT	50 YD	100 YD	150 YD	200 YD	250 YD	DRIFT	TIME
0.	1700.	0.0	-1.5	-1.5	-1.5	-1.5	-1.5	0.0	0.0000
50.	1640.	-.4	0.0	.9	2.3	4.0	5.8	.3	.0900
100.	1583.	.9	-1.8	0.0	2.9	6.2	9.8	1.2	.1834
150.	1526.	3.0	-7.0	-4.3	0.0	5.0	10.4	2.6	.2796
200.	1472.	6.2	-15.9	-12.4	-6.6	0.0	7.2	4.7	.3797
250.	1419.	10.5	-28.9	-24.5	-17.3	-9.0	0.0	7.4	.4834
300.	1370.	16.0	-46.2	-41.0	-32.3	-22.4	-11.6	10.9	.5912
350.	1325.	22.9	-68.2	-62.1	-52.0	-40.4	-27.8	14.9	.7024
400.	1282.	31.3	-95.1	-88.1	-76.6	-63.3	-48.9	19.7	.8176
450.	1242.	41.4	-127.4	-119.5	-106.5	-91.6	-75.4	25.1	.9366
500.	1205.	53.1	-165.2	-156.4	-142.0	-125.4	-107.4	31.1	1.0592

BALLISTIC COEFFICIENT .57, MUZZLE VELOCITY FT/SEC 1800.

RANGE	VELOCITY	HEIGHT	50 YD	100 YD	150 YD	200 YD	250 YD	DRIFT	TIME
0.	1800.	0.0	-1.5	-1.5	-1.5	-1.5	-1.5	0.0	0.0000
50.	1738.	-.4	0.0	.7	2.0	3.4	5.0	.3	.0848
100.	1677.	.7	-1.4	0.0	2.5	5.5	8.6	1.1	.1727
150.	1618.	2.6	-5.9	-3.8	0.0	4.4	9.2	2.4	.2637
200.	1561.	5.4	-13.7	-10.9	-5.9	0.0	6.3	4.4	.3584
250.	1505.	9.2	-25.0	-21.6	-15.3	-7.9	0.0	6.9	.4559
300.	1451.	14.2	-40.4	-36.2	-28.6	-19.8	-10.3	10.2	.5578
350.	1400.	20.3	-59.7	-54.8	-46.0	-35.7	-24.6	14.0	.6627
400.	1353.	27.9	-83.6	-78.1	-68.0	-56.2	-43.5	18.5	.7720
450.	1308.	36.8	-112.1	-105.9	-94.6	-81.3	-67.0	23.7	.8846
500.	1267.	47.4	-145.8	-138.9	-126.3	-111.5	-95.7	29.5	1.0011

BALLISTIC COEFFICIENT .57, MUZZLE VELOCITY FT/SEC 1900.

RANGE	VELOCITY	HEIGHT	50 YD	100 YD	150 YD	200 YD	250 YD	DRIFT	TIME
0.	1900.	0.0	-1.5	-1.5	-1.5	-1.5	-1.5	0.0	0.0000
50.	1835.	-.4	0.0	.5	1.6	2.9	4.4	.3	.0806
100.	1771.	.5	-1.1	0.0	2.2	4.8	7.7	1.0	.1636
150.	1710.	2.2	-4.9	-3.3	0.0	3.9	8.2	2.2	.2495
200.	1650.	4.8	-11.8	-9.6	-5.2	0.0	5.7	4.1	.3391
250.	1592.	8.2	-21.8	-19.2	-13.7	-7.1	0.0	6.5	.4319
300.	1535.	12.6	-35.3	-32.1	-25.5	-17.6	-9.1	9.5	.5276
350.	1480.	18.1	-52.4	-48.7	-41.0	-31.8	-21.8	13.1	.6269
400.	1428.	24.8	-73.6	-69.3	-60.5	-50.1	-38.7	17.4	.7303
450.	1378.	32.9	-99.1	-94.2	-84.3	-72.6	-59.7	22.3	.8373
500.	1332.	42.4	-129.1	-123.7	-112.7	-99.6	-85.4	27.9	.9480

BALLISTIC COEFFICIENT .58, MUZZLE VELOCITY FT/SEC 800.

RANGE	VELOCITY	HEIGHT	50 YD	100 YD	150 YD	200 YD	250 YD	DRIFT	TIME
0.	800.	0.0	-1.5	-1.5	-1.5	-1.5	-1.5	0.0	0.0000
50.	789.	1.0	0.0	6.1	13.1	20.1	27.5	.4	.1898
100.	779.	6.2	-12.3	0.0	13.8	27.9	42.6	.9	.3802
150.	769.	15.1	-39.2	-20.8	0.0	21.2	43.2	2.2	.5750
200.	758.	27.8	-80.5	-55.9	-28.2	0.0	29.4	3.6	.7707
250.	748.	44.5	-137.4	-106.6	-72.0	-36.8	0.0	5.8	.9707
300.	738.	65.2	-209.3	-172.4	-130.9	-88.6	-44.4	8.3	1.1719
350.	728.	90.3	-297.9	-254.9	-206.4	-157.1	-105.6	11.4	1.3775
400.	718.	119.7	-402.2	-353.0	-297.7	-241.3	-182.4	14.8	1.5844
450.	708.	154.0	-524.4	-469.0	-406.8	-343.3	-277.1	19.0	1.7957
500.	698.	192.8	-662.8	-601.3	-532.2	-461.6	-388.1	23.5	2.0083

BALLISTIC COEFFICIENT .58, MUZZLE VELOCITY FT/SEC 900.

RANGE	VELOCITY	HEIGHT	50 YD	100 YD	150 YD	200 YD	250 YD	DRIFT	TIME
0.	900.	0.0	-1.5	-1.5	-1.5	-1.5	-1.5	0.0	0.0000
50.	888.	.6	0.0	4.8	10.1	15.7	21.4	.4	.1688
100.	876.	4.8	-9.5	0.0	10.7	22.0	33.4	1.0	.3389
150.	864.	11.8	-30.4	-16.1	0.0	16.8	34.0	2.0	.5114
200.	853.	21.9	-63.0	-43.9	-22.5	0.0	22.8	3.5	.6864
250.	841.	35.0	-107.2	-83.4	-56.6	-28.5	0.0	5.2	.8628
300.	830.	51.5	-164.3	-135.8	-103.5	-69.9	-35.6	7.6	1.0430
350.	820.	71.2	-233.4	-200.0	-162.5	-123.1	-83.2	10.1	1.2240
400.	809.	94.6	-316.3	-278.2	-235.2	-190.3	-144.7	13.4	1.4093
450.	798.	121.4	-411.6	-368.8	-320.5	-269.9	-218.6	16.8	1.5953
500.	788.	152.2	-521.4	-473.8	-420.2	-364.0	-307.0	20.9	1.7851

BALLISTIC COEFFICIENT .58, MUZZLE VELOCITY FT/SEC 1000.

RANGE	VELOCITY	HEIGHT	50 YD	100 YD	150 YD	200 YD	250 YD	DRIFT	TIME
0.	1000.	0.0	-1.5	-1.5	-1.5	-1.5	-1.5	0.0	0.0000
50.	985.	.4	0.0	3.9	7.9	12.5	17.2	.6	.1535
100.	970.	3.8	-7.7	0.0	8.1	17.3	26.6	1.6	.3093
150.	956.	9.4	-23.8	-12.2	0.0	13.8	27.8	1.9	.4606
200.	942.	17.7	-50.1	-34.6	-18.4	0.0	18.6	3.4	.6196
250.	928.	28.4	-85.9	-66.6	-46.3	-23.3	0.0	5.2	.7795
300.	915.	41.8	-131.8	-108.6	-84.3	-56.7	-28.8	7.4	.9419
350.	902.	58.1	-188.2	-161.1	-132.7	-100.6	-68.0	10.0	1.1068
400.	890.	77.3	-255.7	-224.7	-192.3	-155.5	-118.2	13.2	1.2750
450.	878.	99.4	-333.7	-298.9	-262.4	-221.0	-179.1	16.6	1.4445
500.	866.	124.7	-423.4	-384.8	-344.2	-298.2	-251.7	20.6	1.6170

BALLISTIC COEFFICIENT .58, MUZZLE VELOCITY FT/SEC 1100.

RANGE	VELOCITY	HEIGHT	50 YD	100 YD	150 YD	200 YD	250 YD	DRIFT	TIME
0.	1100.	0.0	-1.5	-1.5	-1.5	-1.5	-1.5	0.0	0.0000
50.	1079.	.2	0.0	3.0	6.6	10.5	14.5	.2	.1377
100.	1059.	3.0	-6.0	0.0	7.2	15.0	23.0	.9	.2779
150.	1041.	7.8	-19.8	-10.9	0.0	11.6	23.7	2.2	.4216
200.	1025.	14.7	-41.9	-30.0	-15.5	0.0	16.1	4.0	.5682
250.	1009.	24.0	-72.5	-57.6	-39.5	-20.1	0.0	6.3	.7175
300.	993.	35.5	-111.9	-94.1	-72.3	-49.1	-24.9	9.0	.8695
350.	978.	48.8	-157.5	-136.7	-111.3	-84.3	-56.1	10.8	1.0160
400.	963.	65.2	-214.6	-190.8	-161.8	-130.9	-98.6	14.3	1.1719
450.	949.	83.9	-280.0	-253.2	-220.6	-185.8	-149.6	17.7	1.3277
500.	936.	105.4	-355.5	-325.8	-289.5	-250.8	-210.6	21.7	1.4868

BALLISTIC COEFFICIENT .58, MUZZLE VELOCITY FT/SEC 1200.

RANGE	VELOCITY	HEIGHT	50 YD	100 YD	150 YD	200 YD	250 YD	DRIFT	TIME
0.	1200.	0.0	-1.5	-1.5	-1.5	-1.5	-1.5	0.0	0.0000
50.	1167.	.0	0.0	2.5	5.5	8.8	12.2	.3	.1268
100.	1137.	2.5	-5.0	0.0	6.0	12.6	19.5	1.5	.2584
150.	1112.	6.6	-16.5	-9.0	0.0	9.9	20.2	2.8	.3907
200.	1089.	12.6	-35.2	-25.2	-13.1	0.0	13.7	4.8	.5271
250.	1069.	20.5	-61.2	-48.6	-33.6	-17.2	0.0	7.2	.6661
300.	1050.	30.6	-94.8	-79.7	-61.7	-42.0	-21.4	10.2	.8078
350.	1033.	42.8	-136.3	-118.7	-97.7	-74.7	-50.7	13.5	.9520
400.	1017.	57.2	-185.9	-165.8	-141.8	-115.5	-88.0	17.3	1.0985
450.	1001.	73.9	-243.8	-221.3	-194.3	-164.7	-133.7	21.5	1.2473
500.	985.	93.4	-311.5	-286.4	-256.4	-223.5	-189.2	26.5	1.4006

BALLISTIC COEFFICIENT .58, MUZZLE VELOCITY FT/SEC 1300.

RANGE	VELOCITY	HEIGHT	50 YD	100 YD	150 YD	200 YD	250 YD	DRIFT	TIME
0.	1300.	0.0	-1.5	-1.5	-1.5	-1.5	-1.5	0.0	0.0000
50.	1260.	-.1	0.0	2.0	4.6	7.6	10.5	.3	.1173
100.	1222.	2.0	-4.0	0.0	5.2	11.2	17.0	1.3	.2382
150.	1186.	5.6	-13.8	-7.9	0.0	8.9	17.6	3.0	.3632
200.	1154.	11.0	-30.3	-22.4	-11.9	0.0	11.7	5.7	.4939
250.	1127.	17.9	-52.5	-42.5	-29.4	-14.6	0.0	8.1	.6231
300.	1102.	26.8	-81.9	-70.0	-54.2	-36.4	-18.9	11.5	.7577
350.	1080.	37.7	-118.5	-104.5	-86.2	-65.4	-45.0	15.4	.8951
400.	1061.	50.7	-162.6	-146.6	-125.6	-101.9	-78.6	19.8	1.0355
450.	1043.	66.0	-214.8	-196.8	-173.2	-146.5	-120.3	24.8	1.1795
500.	1026.	83.6	-274.8	-254.9	-228.7	-199.0	-169.9	30.2	1.3254

BALLISTIC COEFFICIENT .58, MUZZLE VELOCITY FT/SEC 1400.

RANGE	VELOCITY	HEIGHT	50 YD	100 YD	150 YD	200 YD	250 YD	DRIFT	TIME
0.	1400.	0.0	-1.5	-1.5	-1.5	-1.5	-1.5	0.0	0.0000
50.	1353.	-.2	0.0	1.6	3.9	6.3	9.0	.4	.1093
100.	1310.	1.6	-3.2	0.0	4.5	9.4	14.8	1.3	.2217
150.	1269.	4.7	-11.6	-6.7	0.0	7.4	15.5	2.9	.3382
200.	1230.	9.3	-25.3	-18.9	-9.9	0.0	10.7	5.2	.4584
250.	1194.	15.6	-45.0	-37.0	-25.8	-13.4	0.0	8.4	.5832
300.	1161.	23.4	-70.4	-60.7	-47.3	-32.5	-16.4	11.7	.7096
350.	1133.	33.2	-102.6	-91.3	-75.6	-58.3	-39.5	15.9	.8406
400.	1108.	44.9	-141.5	-128.6	-110.7	-90.9	-69.5	20.7	.9747
450.	1085.	58.5	-187.4	-172.9	-152.8	-130.5	-106.4	25.9	1.1114
500.	1065.	74.8	-242.2	-226.1	-203.7	-178.9	-152.1	32.2	1.2545

BALLISTIC COEFFICIENT .58, MUZZLE VELOCITY FT/SEC 1500.

RANGE	VELOCITY	HEIGHT	50 YD	100 YD	150 YD	200 YD	250 YD	DRIFT	TIME
0.	1500.	0.0	-1.5	-1.5	-1.5	-1.5	-1.5	0.0	0.0000
50.	1447.	-.2	0.0	1.3	3.3	5.4	7.7	.4	.1023
100.	1397.	1.3	-2.6	0.0	3.9	8.2	12.8	1.4	.2077
150.	1351.	4.1	-9.8	-5.8	0.0	6.4	13.4	3.0	.3170
200.	1307.	8.1	-21.6	-16.3	-8.6	0.0	9.3	5.2	.4296
250.	1266.	13.6	-38.6	-32.0	-22.3	-11.6	0.0	8.1	.5463
300.	1228.	20.6	-61.1	-53.1	-41.5	-28.6	-14.7	11.7	.6667
350.	1192.	29.3	-89.3	-80.0	-66.5	-51.4	-35.2	16.0	.7909
400.	1159.	39.8	-123.7	-113.1	-97.6	-80.4	-61.9	20.9	.9187
450.	1131.	52.1	-164.5	-152.5	-135.2	-115.8	-95.0	26.3	1.0496
500.	1106.	66.5	-212.1	-198.9	-179.6	-158.1	-134.9	32.4	1.1838

BALLISTIC COEFFICIENT .58, MUZZLE VELOCITY FT/SEC 1600.

RANGE	VELOCITY	HEIGHT	50 YD	100 YD	150 YD	200 YD	250 YD	DRIFT	TIME
0.	1600.	0.0	-1.5	-1.5	-1.5	-1.5	-1.5	0.0	0.0000
50.	1544.	-.3	0.0	1.1	2.8	4.6	6.7	.3	.0953
100.	1490.	1.1	-2.1	0.0	3.4	7.1	11.2	1.1	.1940
150.	1437.	3.5	-8.3	-5.1	0.0	5.6	11.8	2.8	.2970
200.	1388.	7.0	-18.5	-14.2	-7.5	0.0	8.2	4.9	.4028
250.	1342.	11.9	-33.4	-28.0	-19.6	-10.2	0.0	7.8	.5129
300.	1299.	18.1	-53.0	-46.6	-36.5	-25.2	-12.9	11.2	.6263
350.	1259.	25.8	-77.8	-70.3	-58.5	-45.4	-31.1	15.4	.7436
400.	1221.	35.1	-108.0	-99.5	-86.0	-71.0	-54.7	20.2	.8645
450.	1185.	46.3	-144.3	-134.6	-119.5	-102.6	-84.2	25.7	.9895
500.	1153.	59.5	-187.6	-177.0	-160.1	-141.3	-120.9	32.2	1.1206

BALLISTIC COEFFICIENT .58, MUZZLE VELOCITY FT/SEC 1700.

RANGE	VELOCITY	HEIGHT	50 YD	100 YD	150 YD	200 YD	250 YD	DRIFT	TIME
0.	1700.	0.0	-1.5	-1.5	-1.5	-1.5	-1.5	0.0	0.0000
50.	1641.	-.4	0.0	.9	2.3	4.0	5.8	.3	.0900
100.	1584.	.9	-1.7	0.0	2.9	6.2	9.8	1.2	.1833
150.	1529.	3.0	-6.9	-4.3	0.0	5.0	10.3	2.6	.2794
200.	1475.	6.1	-15.8	-12.4	-6.6	0.0	7.2	4.6	.3791
250.	1424.	10.4	-28.8	-24.4	-17.2	-9.0	0.0	7.3	.4826
300.	1375.	16.0	-46.1	-40.9	-32.3	-22.3	-11.6	10.7	.5903
350.	1330.	22.8	-67.9	-61.8	-51.8	-40.2	-27.6	14.7	.7009
400.	1288.	31.2	-94.7	-87.7	-76.2	-63.0	-48.6	19.3	.8155
450.	1248.	41.1	-126.7	-118.9	-105.9	-91.1	-74.9	24.6	.9340
500.	1211.	52.8	-164.3	-155.6	-141.2	-124.7	-106.7	30.6	1.0560

BALLISTIC COEFFICIENT .58, MUZZLE VELOCITY FT/SEC 1800.

RANGE	VELOCITY	HEIGHT	50 YD	100 YD	150 YD	200 YD	250 YD	DRIFT	TIME
0.	1800.	0.0	-1.5	-1.5	-1.5	-1.5	-1.5	0.0	0.0000
50.	1739.	-.4	0.0	.7	1.9	3.4	5.0	.3	.0848
100.	1679.	.7	-1.4	0.0	2.5	5.5	8.6	1.0	.1726
150.	1621.	2.6	-5.8	-3.8	0.0	4.4	9.2	2.4	.2634
200.	1565.	5.4	-13.7	-10.9	-5.9	0.0	6.3	4.3	.3580
250.	1510.	9.2	-25.0	-21.5	-15.3	-7.9	0.0	6.8	.4553
300.	1457.	14.1	-40.2	-36.1	-28.5	-19.7	-10.2	10.0	.5566
350.	1406.	20.2	-59.4	-54.6	-45.8	-35.5	-24.5	13.7	.6613
400.	1359.	27.7	-83.2	-77.7	-67.7	-55.9	-43.3	18.2	.7701
450.	1315.	36.6	-111.7	-105.5	-94.2	-80.9	-66.7	23.3	.8824
500.	1273.	47.1	-145.2	-138.3	-125.8	-111.0	-95.3	29.1	.9987

BALLISTIC COEFFICIENT .58, MUZZLE VELOCITY FT/SEC 1900.

RANGE	VELOCITY	HEIGHT	50 YD	100 YD	150 YD	200 YD	250 YD	DRIFT	TIME
0.	1900.	0.0	-1.5	-1.5	-1.5	-1.5	-1.5	0.0	0.0000
50.	1836.	-.4	0.0	.5	1.6	2.9	4.4	.3	.0806
100.	1774.	.5	-1.1	0.0	2.2	4.8	7.6	1.0	.1635
150.	1713.	2.2	-4.9	-3.3	0.0	3.9	8.2	2.2	.2493
200.	1654.	4.8	-11.7	-9.6	-5.2	0.0	5.7	4.0	.3387
250.	1597.	8.2	-21.8	-19.1	-13.6	-7.1	0.0	6.4	.4311
300.	1541.	12.6	-35.2	-31.9	-25.4	-17.5	-9.0	9.3	.5266
350.	1487.	18.0	-52.2	-48.5	-40.8	-31.7	-21.7	12.8	.6255
400.	1435.	24.7	-73.3	-69.0	-60.3	-49.8	-38.5	17.1	.7285
450.	1386.	32.7	-98.5	-93.7	-83.9	-72.1	-59.3	21.9	.8347
500.	1340.	42.1	-128.4	-123.0	-112.1	-99.0	-84.8	27.4	.9450

BALLISTIC COEFFICIENT .59, MUZZLE VELOCITY FT/SEC 800.

RANGE	VELOCITY	HEIGHT	50 YD	100 YD	150 YD	200 YD	250 YD	DRIFT	TIME
0.	800.	0.0	-1.5	-1.5	-1.5	-1.5	-1.5	0.0	0.0000
50.	790.	1.0	0.0	6.1	13.1	20.1	27.4	.4	.1898
100.	779.	6.2	-12.3	0.0	13.8	27.9	42.6	.9	.3800
150.	769.	15.1	-39.2	-20.8	0.0	21.1	43.2	2.2	.5748
200.	759.	27.7	-80.3	-55.8	-28.1	0.0	29.4	3.6	.7702
250.	749.	44.4	-137.2	-106.5	-71.9	-36.8	0.0	5.8	.9702
300.	739.	65.1	-208.9	-172.1	-130.6	-88.4	-44.2	8.1	1.1709
350.	729.	90.2	-297.5	-254.6	-206.1	-156.9	-105.4	11.3	1.3764
400.	719.	119.5	-401.4	-352.3	-296.9	-240.7	-181.8	14.5	1.5825
450.	709.	153.7	-523.6	-468.3	-406.0	-342.8	-276.5	18.7	1.7939
500.	700.	192.3	-661.2	-599.9	-530.7	-460.4	-386.8	23.0	2.0054

BALLISTIC COEFFICIENT .59, MUZZLE VELOCITY FT/SEC 900.

RANGE	VELOCITY	HEIGHT	50 YD	100 YD	150 YD	200 YD	250 YD	DRIFT	TIME
0.	900.	0.0	-1.5	-1.5	-1.5	-1.5	-1.5	0.0	0.0000
50.	888.	.6	0.0	4.8	10.1	15.7	21.4	.4	.1688
100.	876.	4.8	-9.5	0.0	10.7	21.9	33.4	1.0	.3388
150.	865.	11.8	-30.4	-16.1	0.0	16.8	33.9	2.0	.5113
200.	853.	21.8	-62.9	-43.9	-22.4	0.0	22.8	3.4	.6861
250.	842.	35.0	-107.2	-83.4	-56.6	-28.5	0.0	5.1	.8625
300.	832.	51.4	-164.1	-135.6	-103.4	-69.7	-35.5	7.4	1.0423
350.	821.	71.1	-233.1	-199.8	-162.2	-123.0	-83.0	9.9	1.2231
400.	810.	94.4	-315.8	-277.8	-234.8	-190.0	-144.3	13.1	1.4080
450.	800.	121.1	-410.8	-368.0	-319.6	-269.2	-217.9	16.4	1.5934
500.	789.	151.9	-520.6	-473.0	-419.3	-363.3	-306.2	20.5	1.7832

BALLISTIC COEFFICIENT .59, MUZZLE VELOCITY FT/SEC 1000.

RANGE	VELOCITY	HEIGHT	50 YD	100 YD	150 YD	200 YD	250 YD	DRIFT	TIME
0.	1000.	0.0	-1.5	-1.5	-1.5	-1.5	-1.5	0.0	0.0000
50.	985.	.4	0.0	3.9	7.9	12.5	17.2	.6	.1536
100.	970.	3.8	-7.7	0.0	8.1	17.3	26.6	1.7	.3095
150.	956.	9.4	-23.7	-12.1	0.0	13.9	27.7	1.8	.4603
200.	942.	17.7	-50.1	-34.6	-18.5	0.0	18.5	3.5	.6198
250.	929.	28.4	-85.8	-66.4	-46.2	-23.1	0.0	5.1	.7790
300.	916.	41.8	-131.6	-108.3	-84.1	-56.4	-28.7	7.2	.9411
350.	904.	58.0	-187.9	-160.8	-132.6	-100.2	-67.9	9.9	1.1060
400.	891.	77.1	-255.2	-224.2	-191.9	-154.9	-117.9	13.0	1.2737
450.	879.	99.2	-333.1	-298.3	-262.0	-220.3	-178.7	16.4	1.4430
500.	868.	124.5	-422.6	-383.9	-343.5	-297.3	-251.1	20.3	1.6151

BALLISTIC COEFFICIENT .59, MUZZLE VELOCITY FT/SEC 1100.

RANGE	VELOCITY	HEIGHT	50 YD	100 YD	150 YD	200 YD	250 YD	DRIFT	TIME
0.	1100.	0.0	-1.5	-1.5	-1.5	-1.5	-1.5	0.0	0.0000
50.	1079.	.2	0.0	3.0	6.6	10.5	14.5	.2	.1376
100.	1060.	3.0	-5.9	0.0	7.3	15.0	23.1	.9	.2778
150.	1042.	7.8	-19.9	-11.0	0.0	11.6	23.7	2.3	.4220
200.	1026.	14.8	-41.9	-30.0	-15.4	0.0	16.1	4.0	.5683
250.	1010.	24.0	-72.6	-57.7	-39.4	-20.1	0.0	6.3	.7176
300.	995.	35.0	-110.1	-92.2	-70.3	-47.2	-23.0	8.0	.8634
350.	980.	48.7	-157.2	-136.4	-110.9	-83.9	-55.7	10.6	1.0149
400.	965.	65.2	-214.7	-190.9	-161.7	-130.9	-98.6	14.2	1.1717
450.	952.	83.6	-279.5	-252.7	-219.8	-185.1	-148.9	17.4	1.3260
500.	938.	105.1	-354.7	-325.0	-288.5	-249.9	-209.6	21.3	1.4847

BALLISTIC COEFFICIENT .59, MUZZLE VELOCITY FT/SEC 1200.

RANGE	VELOCITY	HEIGHT	50 YD	100 YD	150 YD	200 YD	250 YD	DRIFT	TIME
0.	1200.	0.0	-1.5	-1.5	-1.5	-1.5	-1.5	0.0	0.0000
50.	1167.	.0	0.0	2.5	5.5	8.8	12.2	.3	.1268
100.	1138.	2.4	-5.0	0.0	6.0	12.6	19.5	1.4	.2578
150.	1113.	6.6	-16.5	-9.1	0.0	9.8	20.1	2.7	.3905
200.	1090.	12.6	-35.1	-25.2	-13.1	0.0	13.7	4.7	.5267
250.	1070.	20.5	-61.1	-48.7	-33.6	-17.1	0.0	7.1	.6655
300.	1052.	30.5	-94.6	-79.7	-61.6	-41.9	-21.3	10.0	.8069
350.	1035.	42.6	-136.0	-118.6	-97.5	-74.5	-50.5	13.3	.9508
400.	1019.	57.0	-185.4	-165.6	-141.4	-115.2	-87.7	17.1	1.0969
450.	1003.	73.8	-243.8	-221.4	-194.2	-164.7	-133.9	21.4	1.2467
500.	988.	93.3	-311.3	-286.5	-256.3	-223.5	-189.2	26.3	1.3997

BALLISTIC COEFFICIENT .59, MUZZLE VELOCITY FT/SEC 1300.

RANGE	VELOCITY	HEIGHT	50 YD	100 YD	150 YD	200 YD	250 YD	DRIFT	TIME
0.	1300.	0.0	-1.5	-1.5	-1.5	-1.5	-1.5	0.0	0.0000
50.	1260.	-.1	0.0	2.0	4.6	7.5	10.5	.3	.1172
100.	1223.	2.0	-4.0	0.0	5.2	11.1	17.0	1.3	.2382
150.	1188.	5.6	-13.8	-7.9	0.0	8.8	17.6	3.0	.3629
200.	1156.	10.9	-30.1	-22.2	-11.7	0.0	11.8	5.5	.4926
250.	1129.	17.8	-52.4	-42.4	-29.3	-14.7	0.0	8.0	.6224
300.	1105.	26.7	-81.8	-69.8	-54.1	-36.6	-18.9	11.3	.7568
350.	1083.	37.6	-118.2	-104.2	-85.9	-65.4	-44.8	15.2	.8938
400.	1063.	50.8	-162.9	-146.9	-126.0	-102.6	-79.1	19.9	1.0359
450.	1045.	66.0	-214.9	-196.9	-173.4	-147.1	-120.6	24.8	1.1792
500.	1029.	83.4	-274.5	-254.6	-228.4	-199.2	-169.7	30.0	1.3241

BALLISTIC COEFFICIENT .59, MUZZLE VELOCITY FT/SEC 1400.

RANGE	VELOCITY	HEIGHT	50 YD	100 YD	150 YD	200 YD	250 YD	DRIFT	TIME
0.	1400.	0.0	-1.5	-1.5	-1.5	-1.5	-1.5	0.0	0.0000
50.	1354.	-.2	0.0	1.6	3.8	6.3	8.9	.4	.1097
100.	1311.	1.6	-3.2	0.0	4.5	9.5	14.7	1.3	.2216
150.	1271.	4.7	-11.5	-6.7	0.0	7.5	15.4	2.9	.3379
200.	1232.	9.4	-25.3	-19.0	-10.0	0.0	10.5	5.3	.4588
250.	1197.	15.5	-44.7	-36.8	-25.6	-13.1	0.0	8.1	.5820
300.	1164.	23.3	-70.1	-60.6	-47.2	-32.1	-16.5	11.6	.7086
350.	1135.	33.2	-102.8	-91.6	-76.0	-58.4	-40.1	16.1	.8412
400.	1111.	44.7	-140.9	-128.2	-110.3	-90.2	-69.3	20.4	.9729
450.	1088.	58.3	-186.6	-172.3	-152.2	-129.6	-106.1	25.5	1.1093
500.	1068.	74.0	-239.6	-223.7	-201.4	-176.3	-150.2	31.1	1.2483

BALLISTIC COEFFICIENT .59, MUZZLE VELOCITY FT/SEC 1500.

RANGE	VELOCITY	HEIGHT	50 YD	100 YD	150 YD	200 YD	250 YD	DRIFT	TIME
0.	1500.	0.0	-1.5	-1.5	-1.5	-1.5	-1.5	0.0	0.0000
50.	1448.	-.2	0.0	1.3	3.2	5.4	7.7	.4	.1023
100.	1399.	1.3	-2.6	0.0	3.9	8.2	12.8	1.3	.2075
150.	1353.	4.1	-9.7	-5.8	0.0	6.4	13.4	2.9	.3167
200.	1310.	8.1	-21.6	-16.3	-8.6	0.0	9.2	5.1	.4292
250.	1270.	13.5	-38.5	-31.9	-22.3	-11.5	0.0	8.0	.5455
300.	1232.	20.5	-60.9	-53.0	-41.4	-28.5	-14.7	11.6	.6657
350.	1196.	29.2	-89.1	-79.9	-66.4	-51.4	-35.2	15.8	.7899
400.	1163.	39.6	-123.1	-112.6	-97.1	-80.0	-61.5	20.5	.9164
450.	1135.	52.1	-164.7	-152.8	-135.4	-116.1	-95.3	26.3	1.0496
500.	1110.	66.2	-211.2	-198.0	-178.7	-157.3	-134.2	31.8	1.1809

BALLISTIC COEFFICIENT .59, MUZZLE VELOCITY FT/SEC 1600.

RANGE	VELOCITY	HEIGHT	50 YD	100 YD	150 YD	200 YD	250 YD	DRIFT	TIME
0.	1600.	0.0	-1.5	-1.5	-1.5	-1.5	-1.5	0.0	0.0000
50.	1545.	-.3	0.0	1.1	2.7	4.6	6.7	.3	.0953
100.	1492.	1.1	-2.1	0.0	3.4	7.1	11.2	1.1	.1939
150.	1440.	3.5	-8.2	-5.0	0.0	5.6	11.7	2.7	.2966
200.	1392.	7.0	-18.5	-14.2	-7.5	0.0	8.2	4.8	.4024
250.	1346.	11.8	-33.3	-28.0	-19.6	-10.2	0.0	7.6	.5121
300.	1304.	18.0	-52.8	-46.4	-36.3	-25.1	-12.9	11.0	.6251
350.	1264.	25.7	-77.5	-70.0	-58.3	-45.2	-30.9	15.1	.7420
400.	1226.	35.0	-107.7	-99.2	-85.7	-70.7	-54.4	19.8	.8627
450.	1191.	46.0	-143.6	-134.1	-119.0	-102.1	-83.8	25.2	.9871
500.	1159.	59.0	-185.9	-175.2	-158.5	-139.7	-119.3	31.3	1.1153

BALLISTIC COEFFICIENT .59, MUZZLE VELOCITY FT/SEC 1700.

RANGE	VELOCITY	HEIGHT	50 YD	100 YD	150 YD	200 YD	250 YD	DRIFT	TIME
0.	1700.	0.0	-1.5	-1.5	-1.5	-1.5	-1.5	0.0	0.0000
50.	1642.	-.4	0.0	.9	2.3	3.9	5.7	.3	.0900
100.	1586.	.9	-1.7	0.0	2.9	6.2	9.7	1.2	.1832
150.	1532.	3.0	-6.9	-4.3	0.0	4.9	10.3	2.5	.2791
200.	1479.	6.1	-15.8	-12.3	-6.6	0.0	7.2	4.5	.3786
250.	1428.	10.4	-28.7	-24.4	-17.2	-9.0	0.0	7.2	.4820
300.	1380.	15.9	-45.9	-40.6	-32.0	-22.2	-11.4	10.4	.5887
350.	1335.	22.8	-67.8	-61.7	-51.6	-40.1	-27.6	14.5	.6998
400.	1293.	31.1	-94.4	-87.5	-76.0	-62.9	-48.5	19.0	.8141
450.	1254.	40.9	-126.3	-118.5	-105.6	-90.8	-74.6	24.3	.9319
500.	1217.	52.5	-163.5	-154.8	-140.5	-124.0	-106.1	30.0	1.0530

BALLISTIC COEFFICIENT .59, MUZZLE VELOCITY FT/SEC 1800.

RANGE	VELOCITY	HEIGHT	50 YD	100 YD	150 YD	200 YD	250 YD	DRIFT	TIME
0.	1800.	0.0	-1.5	-1.5	-1.5	-1.5	-1.5	0.0	0.0000
50.	1740.	-.4	0.0	.7	1.9	3.4	5.0	.3	.0848
100.	1681.	.7	-1.4	0.0	2.5	5.4	8.6	1.0	.1725
150.	1624.	2.6	-5.8	-3.8	0.0	4.4	9.1	2.3	.2632
200.	1568.	5.4	-13.6	-10.9	-5.9	0.0	6.3	4.3	.3576
250.	1514.	9.2	-24.9	-21.5	-15.2	-7.9	0.0	6.7	.4546
300.	1462.	14.1	-40.0	-35.9	-28.4	-19.6	-10.1	9.8	.5555
350.	1412.	20.1	-59.2	-54.4	-45.6	-35.3	-24.3	13.5	.6598
400.	1365.	27.6	-82.8	-77.3	-67.3	-55.6	-43.0	17.9	.7681
450.	1321.	36.4	-111.0	-104.8	-93.5	-80.3	-66.2	22.8	.8796
500.	1280.	46.8	-144.3	-137.4	-124.8	-110.2	-94.4	28.5	.9950

BALLISTIC COEFFICIENT .59, MUZZLE VELOCITY FT/SEC 1900.

RANGE	VELOCITY	HEIGHT	50 YD	100 YD	150 YD	200 YD	250 YD	DRIFT	TIME
0.	1900.	0.0	-1.5	-1.5	-1.5	-1.5	-1.5	0.0	0.0000
50.	1837.	-.4	0.0	.5	1.6	2.9	4.3	.3	.0806
100.	1776.	.5	-1.1	0.0	2.2	4.8	7.6	1.0	.1635
150.	1716.	2.2	-4.9	-3.3	0.0	3.9	8.1	2.1	.2490
200.	1658.	4.7	-11.7	-9.6	-5.2	0.0	5.7	4.0	.3383
250.	1602.	8.1	-21.7	-19.0	-13.6	-7.1	0.0	6.3	.4305
300.	1547.	12.5	-35.0	-31.8	-25.3	-17.5	-9.0	9.2	.5257
350.	1493.	17.9	-52.0	-48.2	-40.6	-31.5	-21.6	12.6	.6241
400.	1442.	24.6	-73.0	-68.7	-60.0	-49.6	-38.3	16.8	.7268
450.	1393.	32.5	-98.0	-93.2	-83.4	-71.7	-59.0	21.4	.8324
500.	1347.	41.9	-127.7	-122.4	-111.5	-98.5	-84.3	26.9	.9421

BALLISTIC COEFFICIENT .60, MUZZLE VELOCITY FT/SEC 800.

RANGE	VELOCITY	HEIGHT	50 YD	100 YD	150 YD	200 YD	250 YD	DRIFT	TIME
0.	800.	0.0	-1.5	-1.5	-1.5	-1.5	-1.5	0.0	0.0000
50.	790.	1.0	0.0	6.1	13.1	20.1	27.4	.4	.1898
100.	780.	6.2	-12.3	0.0	13.8	27.9	42.6	.9	.3798
150.	770.	15.1	-39.2	-20.8	0.0	21.0	43.1	2.1	.5746
200.	760.	27.7	-80.2	-55.7	-28.0	0.0	29.5	3.5	.7697
250.	750.	44.4	-137.1	-106.5	-71.9	-36.8	0.0	5.7	.9697
300.	740.	64.9	-208.6	-171.8	-130.3	-88.2	-44.0	7.9	1.1698
350.	730.	90.0	-297.2	-254.3	-205.8	-156.8	-105.2	11.1	1.3754
400.	720.	119.2	-400.8	-351.7	-296.4	-240.3	-181.4	14.3	1.5810
450.	711.	153.4	-522.7	-467.6	-405.3	-342.2	-275.9	18.4	1.7920
500.	701.	191.9	-660.2	-599.0	-529.7	-459.7	-386.0	22.6	2.0034

BALLISTIC COEFFICIENT .60, MUZZLE VELOCITY FT/SEC 900.

RANGE	VELOCITY	HEIGHT	50 YD	100 YD	150 YD	200 YD	250 YD	DRIFT	TIME
0.	900.	0.0	-1.5	-1.5	-1.5	-1.5	-1.5	0.0	0.0000
50.	888.	.6	0.0	4.7	10.1	15.7	21.4	.4	.1688
100.	876.	4.8	-9.5	0.0	10.7	21.9	33.3	.9	.3387
150.	865.	11.8	-30.4	-16.1	0.0	16.8	33.9	2.0	.5112
200.	854.	21.8	-62.9	-43.9	-22.4	0.0	22.8	3.4	.6858
250.	843.	34.9	-107.1	-83.4	-56.5	-28.6	0.0	5.1	.8621
300.	833.	51.3	-163.9	-135.4	-103.2	-69.6	-35.4	7.3	1.0416
350.	822.	71.0	-232.9	-199.6	-162.0	-122.9	-82.9	9.8	1.2224
400.	812.	94.2	-315.4	-277.4	-234.4	-189.6	-144.0	12.9	1.4067
450.	801.	120.9	-410.3	-367.5	-319.2	-268.9	-217.5	16.2	1.5920
500.	791.	151.5	-519.7	-472.2	-418.5	-362.5	-305.4	20.2	1.7812

BALLISTIC COEFFICIENT .60, MUZZLE VELOCITY FT/SEC 1000.

RANGE	VELOCITY	HEIGHT	50 YD	100 YD	150 YD	200 YD	250 YD	DRIFT	TIME
0.	1000.	0.0	-1.5	-1.5	-1.5	-1.5	-1.5	0.0	0.0000
50.	985.	.4	0.0	3.9	7.9	12.5	17.1	.7	.1537
100.	971.	3.9	-7.8	0.0	8.0	17.3	26.5	1.7	.3098
150.	957.	9.4	-23.7	-12.0	0.0	14.0	27.7	1.8	.4601
200.	943.	17.7	-50.2	-34.6	-18.6	0.0	18.3	3.5	.6200
250.	930.	28.3	-85.6	-66.2	-46.2	-22.9	0.0	5.0	.7786
300.	918.	41.7	-131.3	-108.0	-84.0	-56.1	-28.5	7.1	.9403
350.	905.	57.9	-187.6	-160.5	-132.4	-99.9	-67.8	9.7	1.1052
400.	893.	77.0	-254.6	-223.6	-191.6	-154.3	-117.6	12.7	1.2724
450.	881.	98.9	-332.3	-297.4	-261.3	-219.4	-178.1	16.0	1.4411
500.	870.	124.2	-421.7	-382.9	-342.9	-296.3	-250.5	19.9	1.6132

BALLISTIC COEFFICIENT .60, MUZZLE VELOCITY FT/SEC 1100.

RANGE	VELOCITY	HEIGHT	50 YD	100 YD	150 YD	200 YD	250 YD	DRIFT	TIME
0.	1100.	0.0	-1.5	-1.5	-1.5	-1.5	-1.5	0.0	0.0000
50.	1079.	.2	0.0	3.0	6.7	10.5	14.5	.2	.1376
100.	1060.	3.0	-6.0	0.0	7.3	15.0	23.1	.9	.2779
150.	1043.	7.8	-20.0	-11.0	0.0	11.5	23.6	2.3	.4224
200.	1027.	14.8	-42.0	-30.1	-15.4	0.0	16.1	4.0	.5684
250.	1011.	24.0	-72.6	-57.7	-39.4	-20.1	0.0	6.3	.7177
300.	996.	35.0	-109.9	-92.0	-70.0	-46.9	-22.7	7.8	.8626
350.	982.	48.7	-157.3	-136.4	-110.8	-83.8	-55.6	10.6	1.0148
400.	967.	65.1	-214.8	-190.9	-161.6	-130.8	-98.6	14.2	1.1715
450.	954.	83.4	-278.9	-252.1	-219.0	-184.4	-148.2	17.1	1.3243
500.	940.	104.8	-353.9	-324.1	-287.4	-249.0	-208.7	20.9	1.4826

BALLISTIC COEFFICIENT .60, MUZZLE VELOCITY FT/SEC 1200.

RANGE	VELOCITY	HEIGHT	50 YD	100 YD	150 YD	200 YD	250 YD	DRIFT	TIME
0.	1200.	0.0	-1.5	-1.5	-1.5	-1.5	-1.5	0.0	0.0000
50.	1168.	.0	0.0	2.5	5.5	8.8	12.2	.3	.1268
100.	1139.	2.4	-4.9	0.0	6.1	12.6	19.5	1.3	.2573
150.	1114.	6.6	-16.5	-9.1	0.0	9.8	20.1	2.7	.3902
200.	1092.	12.5	-35.1	-25.3	-13.1	0.0	13.7	4.6	.5263
250.	1072.	20.5	-61.0	-48.7	-33.5	-17.1	0.0	7.0	.6649
300.	1054.	30.4	-94.4	-79.7	-61.4	-41.8	-21.3	9.9	.8061
350.	1037.	42.5	-135.6	-118.5	-97.2	-74.3	-50.3	13.1	.9496
400.	1021.	56.9	-185.1	-165.5	-141.2	-115.0	-87.6	16.9	1.0958
450.	1005.	73.8	-243.7	-221.6	-194.2	-164.7	-134.0	21.3	1.2460
500.	990.	93.2	-311.1	-286.6	-256.2	-223.4	-189.2	26.2	1.3988

BALLISTIC COEFFICIENT .60, MUZZLE VELOCITY FT/SEC 1300.

RANGE	VELOCITY	HEIGHT	50 YD	100 YD	150 YD	200 YD	250 YD	DRIFT	TIME
0.	1300.	0.0	-1.5	-1.5	-1.5	-1.5	-1.5	0.0	0.0000
50.	1261.	-.1	0.0	2.0	4.6	7.5	10.5	.3	.1172
100.	1224.	2.0	-4.0	0.0	5.2	11.0	16.9	1.3	.2381
150.	1190.	5.6	-13.8	-7.8	0.0	8.6	17.5	2.9	.3627
200.	1158.	10.8	-29.9	-21.9	-11.5	0.0	11.9	5.2	.4912
250.	1131.	17.8	-52.3	-42.3	-29.2	-14.9	0.0	7.9	.6217
300.	1107.	26.7	-81.6	-69.6	-53.9	-36.7	-18.9	11.2	.7559
350.	1085.	37.5	-117.9	-104.0	-85.7	-65.6	-44.7	14.9	.8926
400.	1065.	50.8	-163.2	-147.2	-126.3	-103.3	-79.5	19.9	1.0363
450.	1048.	66.0	-214.9	-197.0	-173.5	-147.6	-120.8	24.7	1.1788
500.	1031.	83.2	-274.1	-254.2	-228.1	-199.3	-169.6	29.7	1.3228

BALLISTIC COEFFICIENT .60, MUZZLE VELOCITY FT/SEC 1400.

RANGE	VELOCITY	HEIGHT	50 YD	100 YD	150 YD	200 YD	250 YD	DRIFT	TIME
0.	1400.	0.0	-1.5	-1.5	-1.5	-1.5	-1.5	0.0	0.0000
50.	1355.	-.2	0.0	1.6	3.8	6.3	8.9	.4	.1096
100.	1312.	1.6	-3.2	0.0	4.5	9.4	14.6	1.3	.2215
150.	1273.	4.7	-11.5	-6.7	0.0	7.5	15.3	2.9	.3376
200.	1235.	9.3	-25.2	-18.9	-10.0	0.0	10.4	5.2	.4580
250.	1200.	15.4	-44.5	-36.6	-25.4	-13.0	0.0	7.9	.5808
300.	1167.	23.3	-70.0	-60.4	-47.0	-32.1	-16.5	11.4	.7075
350.	1139.	33.0	-102.0	-90.8	-75.2	-57.8	-39.6	15.5	.8381
400.	1114.	44.5	-140.4	-127.7	-109.9	-90.0	-69.2	20.0	.9710
450.	1092.	58.1	-186.0	-171.7	-151.6	-129.2	-105.9	25.1	1.1071
500.	1072.	73.7	-238.8	-222.8	-200.5	-175.7	-149.7	30.7	1.2457

BALLISTIC COEFFICIENT .60, MUZZLE VELOCITY FT/SEC 1500.

RANGE	VELOCITY	HEIGHT	50 YD	100 YD	150 YD	200 YD	250 YD	DRIFT	TIME
0.	1500.	0.0	-1.5	-1.5	-1.5	-1.5	-1.5	0.0	0.0000
50.	1449.	-.2	0.0	1.3	3.3	5.4	7.7	.4	.1022
100.	1400.	1.3	-2.6	0.0	3.9	8.1	12.8	1.3	.2073
150.	1355.	4.1	-9.8	-5.8	0.0	6.4	13.4	3.0	.3168
200.	1313.	8.1	-21.5	-16.3	-8.5	0.0	9.3	5.0	.4287
250.	1273.	13.5	-38.6	-32.0	-22.3	-11.6	0.0	8.0	.5455
300.	1235.	20.5	-60.8	-53.0	-41.3	-28.5	-14.6	11.5	.6651
350.	1200.	29.0	-88.7	-79.5	-65.9	-51.0	-34.7	15.5	.7878
400.	1168.	39.4	-122.7	-112.2	-96.6	-79.6	-61.0	20.2	.9146
450.	1139.	51.7	-163.2	-151.4	-133.9	-114.8	-93.8	25.5	1.0451
500.	1114.	65.9	-210.3	-197.2	-177.7	-156.5	-133.2	31.3	1.1780

BALLISTIC COEFFICIENT .60, MUZZLE VELOCITY FT/SEC 1600.

RANGE	VELOCITY	HEIGHT	50 YD	100 YD	150 YD	200 YD	250 YD	DRIFT	TIME
0.	1600.	0.0	-1.5	-1.5	-1.5	-1.5	-1.5	0.0	0.0000
50.	1546.	-.3	0.0	1.1	2.7	4.6	6.6	.3	.0953
100.	1494.	1.1	-2.1	0.0	3.3	7.1	11.2	1.1	.1937
150.	1443.	3.5	-8.2	-5.0	0.0	5.6	11.7	2.7	.2964
200.	1395.	7.0	-18.5	-14.2	-7.5	0.0	8.1	4.8	.4022
250.	1350.	11.8	-33.2	-27.9	-19.5	-10.1	0.0	7.5	.5114
300.	1308.	17.9	-52.6	-46.3	-36.2	-24.9	-12.8	10.8	.6241
350.	1268.	25.6	-77.2	-69.8	-58.1	-44.9	-30.8	14.9	.7406
400.	1231.	34.8	-107.3	-98.8	-85.4	-70.3	-54.2	19.5	.8609
450.	1196.	45.8	-143.2	-133.7	-118.6	-101.7	-83.5	24.9	.9853
500.	1164.	58.6	-184.8	-174.2	-157.5	-138.6	-118.4	30.7	1.1117

BALLISTIC COEFFICIENT .60, MUZZLE VELOCITY FT/SEC 1700.

RANGE	VELOCITY	HEIGHT	50 YD	100 YD	150 YD	200 YD	250 YD	DRIFT	TIME
0.	1700.	0.0	-1.5	-1.5	-1.5	-1.5	-1.5	0.0	0.0000
50.	1643.	-.4	0.0	.9	2.3	3.9	5.7	.3	.0900
100.	1588.	.9	-1.7	0.0	2.9	6.1	9.7	1.2	.1830
150.	1535.	3.0	-6.9	-4.3	0.0	4.9	10.3	2.5	.2789
200.	1482.	6.1	-15.7	-12.3	-6.6	0.0	7.2	4.4	.3781
250.	1432.	10.4	-28.6	-24.3	-17.2	-9.0	0.0	7.1	.4814
300.	1385.	15.8	-45.7	-40.5	-31.9	-22.1	-11.4	10.3	.5878
350.	1340.	22.6	-67.4	-61.3	-51.3	-39.8	-27.3	14.1	.6980
400.	1299.	30.9	-93.9	-86.9	-75.5	-62.4	-48.0	18.6	.8116
450.	1260.	40.6	-125.5	-117.7	-104.8	-90.0	-73.9	23.7	.9287
500.	1223.	52.1	-162.6	-153.9	-139.6	-123.2	-105.3	29.4	1.0497

BALLISTIC COEFFICIENT .60, MUZZLE VELOCITY FT/SEC 1800.

RANGE	VELOCITY	HEIGHT	50 YD	100 YD	150 YD	200 YD	250 YD	DRIFT	TIME
0.	1800.	0.0	-1.5	-1.5	-1.5	-1.5	-1.5	0.0	0.0000
50.	1741.	-.4	0.0	.7	1.9	3.4	5.0	.3	.0848
100.	1683.	.7	-1.4	0.0	2.5	5.4	8.6	1.0	.1724
150.	1627.	2.6	-5.8	-3.8	0.0	4.4	9.1	2.3	.2630
200.	1572.	5.4	-13.6	-10.9	-5.9	0.0	6.3	4.2	.3572
250.	1519.	9.1	-24.8	-21.4	-15.1	-7.8	0.0	6.6	.4540
300.	1467.	14.0	-39.9	-35.8	-28.3	-19.5	-10.1	9.6	.5545
350.	1417.	20.1	-59.0	-54.2	-45.4	-35.2	-24.2	13.2	.6585
400.	1371.	27.4	-82.5	-77.0	-67.0	-55.3	-42.7	17.5	.7662
450.	1328.	36.2	-110.5	-104.4	-93.1	-79.9	-65.8	22.4	.8773
500.	1287.	46.5	-143.5	-136.7	-124.2	-109.5	-93.9	27.9	.9921

BALLISTIC COEFFICIENT .60, MUZZLE VELOCITY FT/SEC 1900.

RANGE	VELOCITY	HEIGHT	50 YD	100 YD	150 YD	200 YD	250 YD	DRIFT	TIME
0.	1900.	0.0	-1.5	-1.5	-1.5	-1.5	-1.5	0.0	0.0000
50.	1838.	-.4	0.0	.5	1.6	2.9	4.3	.3	.0806
100.	1778.	.5	-1.1	0.0	2.2	4.8	7.6	1.0	.1634
150.	1719.	2.2	-4.8	-3.2	0.0	3.9	8.1	2.1	.2487
200.	1662.	4.7	-11.7	-9.5	-5.2	0.0	5.6	3.9	.3379
250.	1606.	8.1	-21.6	-18.9	-13.5	-7.0	0.0	6.2	.4298
300.	1552.	12.5	-34.9	-31.7	-25.2	-17.4	-9.0	9.0	.5248
350.	1499.	17.9	-51.8	-48.0	-40.5	-31.4	-21.5	12.3	.6227
400.	1448.	24.5	-72.7	-68.4	-59.7	-49.3	-38.1	16.4	.7250
450.	1400.	32.3	-97.6	-92.8	-83.0	-71.3	-58.7	21.0	.8301
500.	1355.	41.6	-127.2	-121.8	-111.0	-98.0	-84.0	26.4	.9397

BALLISTIC COEFFICIENT .61, MUZZLE VELOCITY FT/SEC 800.

RANGE	VELOCITY	HEIGHT	50 YD	100 YD	150 YD	200 YD	250 YD	DRIFT	TIME
0.	800.	0.0	-1.5	-1.5	-1.5	-1.5	-1.5	0.0	0.0000
50.	790.	1.0	0.0	6.1	13.0	20.0	27.4	.4	.1898
100.	780.	6.2	-12.2	0.0	13.8	27.8	42.6	.8	.3797
150.	770.	15.1	-39.1	-20.8	0.0	21.0	43.1	2.1	.5745
200.	760.	27.7	-80.2	-55.7	-28.0	0.0	29.4	3.4	.7694
250.	750.	44.3	-137.0	-106.4	-71.8	-36.8	0.0	5.6	.9692
300.	741.	64.9	-208.5	-171.8	-130.3	-88.3	-44.1	7.8	1.1695
350.	731.	89.9	-296.8	-253.9	-205.5	-156.5	-105.0	10.9	1.3743
400.	722.	119.1	-400.3	-351.4	-296.0	-240.0	-181.1	14.1	1.5799
450.	712.	153.1	-521.9	-466.8	-404.5	-341.5	-275.3	18.1	1.7902
500.	703.	191.5	-659.4	-598.2	-529.0	-459.0	-385.4	22.3	2.0015

BALLISTIC COEFFICIENT .61, MUZZLE VELOCITY FT/SEC 900.

RANGE	VELOCITY	HEIGHT	50 YD	100 YD	150 YD	200 YD	250 YD	DRIFT	TIME
0.	900.	0.0	-1.5	-1.5	-1.5	-1.5	-1.5	0.0	0.0000
50.	888.	.6	0.0	4.7	10.1	15.7	21.4	.4	.1688
100.	877.	4.8	-9.5	0.0	10.7	21.9	33.3	.9	.3386
150.	866.	11.8	-30.4	-16.1	0.0	16.7	33.9	2.0	.5111
200.	855.	21.8	-62.8	-43.8	-22.3	0.0	22.9	3.3	.6855
250.	844.	34.9	-107.1	-83.3	-56.5	-28.6	0.0	5.0	.8618
300.	834.	51.3	-163.7	-135.2	-103.0	-69.5	-35.2	7.2	1.0409
350.	823.	70.9	-232.7	-199.5	-161.9	-122.8	-82.8	9.7	1.2216
400.	813.	94.1	-314.9	-276.9	-233.9	-189.3	-143.6	12.7	1.4055
450.	803.	120.7	-409.8	-367.1	-318.7	-268.5	-217.1	16.0	1.5908
500.	793.	151.2	-518.8	-471.3	-417.6	-361.8	-304.7	19.8	1.7793

BALLISTIC COEFFICIENT .61, MUZZLE VELOCITY FT/SEC 1000.

RANGE	VELOCITY	HEIGHT	50 YD	100 YD	150 YD	200 YD	250 YD	DRIFT	TIME
0.	1000.	0.0	-1.5	-1.5	-1.5	-1.5	-1.5	0.0	0.0000
50.	985.	.4	0.0	3.9	7.9	12.6	17.1	.7	.1539
100.	971.	3.9	-7.8	0.0	8.0	17.3	26.4	1.8	.3100
150.	958.	9.4	-23.6	-11.9	0.0	14.1	27.7	1.7	.4598
200.	944.	17.7	-50.2	-34.7	-18.7	0.0	18.2	3.6	.6203
250.	931.	28.3	-85.5	-66.1	-46.2	-22.7	0.0	5.0	.7781
300.	919.	41.6	-131.0	-107.7	-83.8	-55.7	-28.4	7.0	.9395
350.	907.	57.8	-187.4	-160.2	-132.3	-99.5	-67.7	9.6	1.1044
400.	894.	77.2	-255.8	-224.7	-192.9	-155.4	-119.0	13.1	1.2746
450.	883.	98.8	-331.7	-296.8	-260.9	-218.8	-177.8	15.8	1.4398
500.	872.	123.9	-420.9	-382.0	-342.2	-295.4	-249.9	19.6	1.6114

BALLISTIC COEFFICIENT .61, MUZZLE VELOCITY FT/SEC 1100.

RANGE	VELOCITY	HEIGHT	50 YD	100 YD	150 YD	200 YD	250 YD	DRIFT	TIME
0.	1100.	0.0	-1.5	-1.5	-1.5	-1.5	-1.5	0.0	0.0000
50.	1079.	.2	0.0	3.0	6.7	10.5	14.5	.2	.1375
100.	1061.	3.0	-6.0	0.0	7.3	15.0	23.1	1.0	.2783
150.	1044.	7.8	-20.0	-11.0	0.0	11.5	23.6	2.4	.4227
200.	1028.	14.8	-42.0	-30.0	-15.3	0.0	16.1	4.1	.5685
250.	1012.	24.0	-72.7	-57.6	-39.3	-20.1	0.0	6.3	.7177
300.	998.	34.9	-109.7	-91.7	-69.7	-46.7	-22.5	7.7	.8618
350.	983.	48.7	-157.4	-136.4	-110.7	-83.8	-55.6	10.6	1.0148
400.	969.	65.1	-214.8	-190.8	-161.4	-130.8	-98.5	14.1	1.1713
450.	956.	83.2	-278.3	-251.2	-218.2	-183.7	-147.5	16.8	1.3227
500.	942.	104.7	-353.9	-323.9	-287.2	-248.8	-208.6	20.8	1.4821

BALLISTIC COEFFICIENT .61, MUZZLE VELOCITY FT/SEC 1200.

RANGE	VELOCITY	HEIGHT	50 YD	100 YD	150 YD	200 YD	250 YD	DRIFT	TIME
0.	1200.	0.0	-1.5	-1.5	-1.5	-1.5	-1.5	0.0	0.0000
50.	1168.	.0	0.0	2.4	5.5	8.8	12.2	.3	.1268
100.	1140.	2.4	-4.9	0.0	6.1	12.7	19.5	1.2	.2567
150.	1115.	6.6	-16.5	-9.2	0.0	9.8	20.1	2.6	.3900
200.	1093.	12.5	-35.0	-25.3	-13.1	0.0	13.7	4.6	.5259
250.	1073.	20.4	-60.8	-48.7	-33.4	-17.1	0.0	6.9	.6643
300.	1055.	30.4	-94.2	-79.6	-61.3	-41.7	-21.2	9.7	.8052
350.	1039.	42.4	-135.3	-118.3	-96.9	-74.0	-50.1	12.9	.9484
400.	1023.	56.8	-185.1	-165.7	-141.2	-115.0	-87.7	16.8	1.0954
450.	1007.	73.7	-243.6	-221.7	-194.2	-164.8	-134.1	21.2	1.2454
500.	993.	93.0	-310.9	-286.6	-256.1	-223.3	-189.2	26.0	1.3978

BALLISTIC COEFFICIENT .61, MUZZLE VELOCITY FT/SEC 1300.

RANGE	VELOCITY	HEIGHT	50 YD	100 YD	150 YD	200 YD	250 YD	DRIFT	TIME
0.	1300.	0.0	-1.5	-1.5	-1.5	-1.5	-1.5	0.0	0.0000
50.	1262.	-.1	0.0	2.0	4.6	7.4	10.4	.3	.1172
100.	1225.	2.0	-4.0	0.0	5.2	10.9	16.9	1.3	.2380
150.	1191.	5.6	-13.8	-7.8	0.0	8.5	17.5	2.9	.3624
200.	1160.	10.8	-29.7	-21.7	-11.3	0.0	12.0	5.0	.4899
250.	1133.	17.8	-52.1	-42.2	-29.2	-15.0	0.0	7.7	.6209
300.	1109.	26.6	-81.4	-69.4	-53.8	-36.9	-18.8	11.0	.7549
350.	1088.	37.4	-117.6	-103.7	-85.4	-65.7	-44.6	14.7	.8915
400.	1068.	50.2	-161.2	-145.3	-124.4	-101.8	-77.8	18.9	1.0306
450.	1051.	65.2	-212.3	-194.4	-171.0	-145.6	-118.5	23.5	1.1722
500.	1034.	82.4	-271.3	-251.4	-225.3	-197.1	-167.0	28.6	1.3162

BALLISTIC COEFFICIENT .61, MUZZLE VELOCITY FT/SEC 1400.

RANGE	VELOCITY	HEIGHT	50 YD	100 YD	150 YD	200 YD	250 YD	DRIFT	TIME
0.	1400.	0.0	-1.5	-1.5	-1.5	-1.5	-1.5	0.0	0.0000
50.	1355.	-.2	0.0	1.6	3.8	6.3	8.9	.4	.1095
100.	1314.	1.6	-3.2	0.0	4.5	9.4	14.5	1.3	.2219
150.	1274.	4.7	-11.5	-6.7	0.0	7.3	15.1	2.9	.3379
200.	1237.	9.3	-25.2	-18.7	-9.8	0.0	10.4	5.1	.4573
250.	1203.	15.4	-44.4	-36.4	-25.2	-13.0	0.0	7.8	.5800
300.	1170.	23.2	-69.8	-60.1	-46.7	-32.0	-16.5	11.2	.7065
350.	1142.	32.8	-101.5	-90.2	-74.6	-57.4	-39.3	15.2	.8361
400.	1117.	44.3	-140.0	-127.1	-109.2	-89.7	-68.9	19.7	.9692
450.	1095.	57.9	-185.4	-170.9	-150.8	-128.8	-105.4	24.8	1.1049
500.	1075.	73.4	-237.9	-221.8	-199.5	-175.0	-149.1	30.2	1.2431

BALLISTIC COEFFICIENT .61, MUZZLE VELOCITY FT/SEC 1500.

RANGE	VELOCITY	HEIGHT	50 YD	100 YD	150 YD	200 YD	250 YD	DRIFT	TIME
0.	1500.	0.0	-1.5	-1.5	-1.5	-1.5	-1.5	0.0	0.0000
50.	1450.	-.2	0.0	1.3	3.2	5.4	7.7	.4	.1022
100.	1402.	1.3	-2.6	0.0	3.9	8.1	12.7	1.3	.2072
150.	1357.	4.1	-9.7	-5.8	0.0	6.4	13.3	2.9	.3164
200.	1315.	8.1	-21.5	-16.3	-8.6	0.0	9.2	5.0	.4286
250.	1276.	13.5	-38.4	-31.9	-22.2	-11.5	0.0	7.8	.5445
300.	1239.	20.4	-60.6	-52.7	-41.1	-28.3	-14.5	11.2	.6637
350.	1204.	28.9	-88.4	-79.2	-65.7	-50.7	-34.6	15.2	.7864
400.	1172.	39.2	-122.2	-111.8	-96.3	-79.2	-60.8	19.8	.9127
450.	1143.	51.4	-162.4	-150.6	-133.2	-114.0	-93.3	25.0	1.0423
500.	1118.	65.7	-209.9	-196.8	-177.4	-156.1	-133.1	31.1	1.1764

BALLISTIC COEFFICIENT .61, MUZZLE VELOCITY FT/SEC 1600.

RANGE	VELOCITY	HEIGHT	50 YD	100 YD	150 YD	200 YD	250 YD	DRIFT	TIME
0.	1600.	0.0	-1.5	-1.5	-1.5	-1.5	-1.5	0.0	0.0000
50.	1547.	-.3	0.0	1.1	2.7	4.6	6.6	.3	.0953
100.	1495.	1.0	-2.1	0.0	3.3	7.1	11.1	1.1	.1936
150.	1445.	3.5	-8.2	-5.0	0.0	5.6	11.7	2.6	.2961
200.	1398.	7.0	-18.4	-14.2	-7.5	0.0	8.1	4.7	.4015
250.	1353.	11.8	-33.1	-27.8	-19.4	-10.1	0.0	7.4	.5107
300.	1312.	17.9	-52.5	-46.1	-36.1	-24.9	-12.8	10.7	.6231
350.	1273.	25.5	-77.0	-69.6	-57.9	-44.8	-30.7	14.6	.7393
400.	1236.	34.7	-107.0	-98.6	-85.2	-70.3	-54.1	19.3	.8596
450.	1201.	45.6	-142.4	-132.9	-117.9	-101.0	-82.9	24.4	.9822
500.	1169.	58.3	-184.0	-173.4	-156.7	-138.0	-117.8	30.2	1.1089

BALLISTIC COEFFICIENT .61, MUZZLE VELOCITY FT/SEC 1700.

RANGE	VELOCITY	HEIGHT	50 YD	100 YD	150 YD	200 YD	250 YD	DRIFT	TIME
0.	1700.	0.0	-1.5	-1.5	-1.5	-1.5	-1.5	0.0	0.0000
50.	1644.	-.4	0.0	.9	2.3	3.9	5.7	.3	.0899
100.	1590.	.9	-1.7	0.0	2.9	6.1	9.7	1.1	.1829
150.	1537.	3.0	-6.9	-4.3	0.0	4.9	10.3	2.4	.2786
200.	1486.	6.1	-15.7	-12.2	-6.5	0.0	7.1	4.4	.3777
250.	1436.	10.3	-28.6	-24.2	-17.1	-8.9	0.0	7.0	.4808
300.	1389.	15.8	-45.6	-40.4	-31.8	-22.0	-11.3	10.1	.5868
350.	1345.	22.5	-67.2	-61.1	-51.1	-39.7	-27.2	13.9	.6967
400.	1304.	30.7	-93.5	-86.6	-75.2	-62.1	-47.8	18.3	.8097
450.	1266.	40.5	-125.0	-117.2	-104.3	-89.6	-73.5	23.3	.9265
500.	1229.	51.9	-161.9	-153.3	-139.0	-122.7	-104.8	29.0	1.0470

BALLISTIC COEFFICIENT .61, MUZZLE VELOCITY FT/SEC 1800.

RANGE	VELOCITY	HEIGHT	50 YD	100 YD	150 YD	200 YD	250 YD	DRIFT	TIME
0.	1800.	0.0	-1.5	-1.5	-1.5	-1.5	-1.5	0.0	0.0000
50.	1742.	-.4	0.0	.7	1.9	3.4	5.0	.3	.0848
100.	1685.	.7	-1.4	0.0	2.5	5.4	8.5	1.0	.1723
150.	1629.	2.6	-5.8	-3.8	0.0	4.4	9.1	2.3	.2628
200.	1576.	5.4	-13.6	-10.8	-5.8	0.0	6.2	4.1	.3568
250.	1523.	9.1	-24.8	-21.4	-15.1	-7.8	0.0	6.5	.4533
300.	1472.	14.0	-39.7	-35.7	-28.1	-19.4	-10.0	9.4	.5535
350.	1423.	20.0	-58.7	-54.0	-45.2	-35.0	-24.1	13.0	.6570
400.	1377.	27.3	-82.1	-76.7	-66.7	-55.0	-42.5	17.2	.7645
450.	1334.	36.1	-110.2	-104.1	-92.8	-79.7	-65.6	22.1	.8756
500.	1293.	46.3	-143.0	-136.2	-123.7	-109.1	-93.5	27.5	.9898

BALLISTIC COEFFICIENT .61, MUZZLE VELOCITY FT/SEC 1900.

RANGE	VELOCITY	HEIGHT	50 YD	100 YD	150 YD	200 YD	250 YD	DRIFT	TIME
0.	1900.	0.0	-1.5	-1.5	-1.5	-1.5	-1.5	0.0	0.0000
50.	1839.	-.4	0.0	.5	1.6	2.9	4.3	.3	.0806
100.	1780.	.5	-1.1	0.0	2.2	4.8	7.5	1.0	.1633
150.	1722.	2.2	-4.8	-3.2	0.0	3.9	8.1	2.1	.2486
200.	1666.	4.7	-11.6	-9.5	-5.2	0.0	5.6	3.8	.3375
250.	1611.	8.1	-21.5	-18.9	-13.5	-7.0	0.0	6.0	.4291
300.	1557.	12.4	-34.8	-31.6	-25.1	-17.4	-9.0	8.8	.5240
350.	1505.	17.8	-51.6	-47.9	-40.3	-31.2	-21.5	12.1	.6216
400.	1455.	24.4	-72.4	-68.1	-59.5	-49.1	-37.9	16.1	.7233
450.	1407.	32.1	-97.1	-92.3	-82.6	-70.9	-58.4	20.7	.8279
500.	1362.	41.3	-126.4	-121.1	-110.3	-97.3	-83.3	25.9	.9365

BALLISTIC COEFFICIENT .62, MUZZLE VELOCITY FT/SEC 800.

RANGE	VELOCITY	HEIGHT	50 YD	100 YD	150 YD	200 YD	250 YD	DRIFT	TIME
0.	800.	0.0	-1.5	-1.5	-1.5	-1.5	-1.5	0.0	0.0000
50.	790.	1.0	0.0	6.1	13.0	20.0	27.4	.4	.1898
100.	780.	6.2	-12.2	0.0	13.8	27.9	42.5	.8	.3797
150.	771.	15.1	-39.1	-20.7	0.0	21.0	43.0	2.1	.5743
200.	761.	27.7	-80.2	-55.7	-28.1	0.0	29.3	3.4	.7694
250.	751.	44.3	-136.9	-106.2	-71.7	-36.6	0.0	5.5	.9687
300.	742.	64.8	-208.3	-171.5	-130.1	-88.0	-44.0	7.7	1.1687
350.	732.	89.8	-296.4	-253.5	-205.2	-156.0	-104.8	10.7	1.3733
400.	723.	118.9	-399.9	-351.0	-295.7	-239.5	-181.0	13.9	1.5788
450.	714.	152.8	-521.0	-465.9	-403.7	-340.5	-274.7	17.8	1.7884
500.	704.	191.2	-658.5	-597.3	-528.2	-458.0	-384.8	21.9	1.9997

BALLISTIC COEFFICIENT .62, MUZZLE VELOCITY FT/SEC 900.

RANGE	VELOCITY	HEIGHT	50 YD	100 YD	150 YD	200 YD	250 YD	DRIFT	TIME
0.	900.	0.0	-1.5	-1.5	-1.5	-1.5	-1.5	0.0	0.0000
50.	889.	.6	0.0	4.7	10.1	15.7	21.4	.4	.1688
100.	877.	4.7	-9.5	0.0	10.8	21.9	33.3	.9	.3385
150.	866.	11.8	-30.3	-16.1	0.0	16.7	33.9	1.9	.5110
200.	855.	21.8	-62.7	-43.8	-22.3	0.0	22.9	3.3	.6851
250.	845.	34.9	-107.0	-83.3	-56.4	-28.6	0.0	5.0	.8615
300.	835.	51.2	-163.5	-135.1	-102.8	-69.4	-35.1	7.1	1.0402
350.	824.	70.8	-232.5	-199.3	-161.7	-122.7	-82.7	9.5	1.2209
400.	814.	93.9	-314.4	-276.5	-233.5	-188.9	-143.2	12.5	1.4042
450.	804.	120.5	-409.3	-366.7	-318.3	-268.2	-216.7	15.8	1.5895
500.	794.	150.9	-517.9	-470.5	-416.7	-361.0	-303.9	19.5	1.7773

BALLISTIC COEFFICIENT .62, MUZZLE VELOCITY FT/SEC 1000.

RANGE	VELOCITY	HEIGHT	50 YD	100 YD	150 YD	200 YD	250 YD	DRIFT	TIME
0.	1000.	0.0	-1.5	-1.5	-1.5	-1.5	-1.5	0.0	0.0000
50.	986.	.4	0.0	3.9	7.8	12.6	17.1	.7	.1540
100.	971.	3.9	-7.8	0.0	7.9	17.3	26.4	1.8	.3102
150.	958.	9.4	-23.5	-11.9	0.0	14.1	27.7	1.7	.4596
200.	945.	17.7	-50.2	-34.7	-18.9	0.0	18.1	3.6	.6205
250.	932.	28.3	-85.4	-65.9	-46.1	-22.6	0.0	4.9	.7777
300.	920.	41.5	-130.8	-107.4	-83.7	-55.4	-28.3	6.8	.9387
350.	908.	57.7	-187.1	-159.8	-132.1	-99.1	-67.5	9.4	1.1036
400.	896.	77.0	-254.8	-223.7	-192.0	-154.3	-118.2	12.7	1.2724
450.	885.	98.6	-331.2	-296.2	-260.6	-218.2	-177.5	15.6	1.4385
500.	873.	123.6	-420.0	-381.1	-341.5	-294.4	-249.3	19.3	1.6095

BALLISTIC COEFFICIENT .62, MUZZLE VELOCITY FT/SEC 1100.

RANGE	VELOCITY	HEIGHT	50 YD	100 YD	150 YD	200 YD	250 YD	DRIFT	TIME
0.	1100.	0.0	-1.5	-1.5	-1.5	-1.5	-1.5	0.0	0.0000
50.	1080.	.2	0.0	3.0	6.7	10.5	14.5	.2	.1375
100.	1061.	3.0	-6.1	0.0	7.3	15.0	23.0	1.1	.2787
150.	1044.	7.8	-20.1	-11.0	0.0	11.5	23.5	2.5	.4231
200.	1029.	14.8	-42.1	-30.0	-15.3	0.0	16.1	4.1	.5687
250.	1013.	24.0	-72.7	-57.6	-39.2	-20.1	0.0	6.3	.7178
300.	999.	34.8	-109.5	-91.3	-69.3	-46.4	-22.2	7.5	.8610
350.	985.	48.7	-157.5	-136.3	-110.6	-83.8	-55.6	10.6	1.0147
400.	971.	65.1	-214.9	-190.6	-161.3	-130.7	-98.5	14.1	1.1711
450.	958.	83.0	-277.7	-250.4	-217.4	-183.0	-146.8	16.5	1.3210
500.	944.	104.6	-353.9	-323.6	-286.9	-248.7	-208.4	20.7	1.4815

BALLISTIC COEFFICIENT .62, MUZZLE VELOCITY FT/SEC 1200.

RANGE	VELOCITY	HEIGHT	50 YD	100 YD	150 YD	200 YD	250 YD	DRIFT	TIME
0.	1200.	0.0	-1.5	-1.5	-1.5	-1.5	-1.5	0.0	0.0000
50.	1169.	.0	0.0	2.4	5.5	8.7	12.1	.3	.1267
100.	1141.	2.4	-4.8	0.0	6.1	12.6	19.5	1.1	.2565
150.	1116.	6.5	-16.4	-9.2	0.0	9.8	20.0	2.6	.3897
200.	1095.	12.5	-35.0	-25.3	-13.1	0.0	13.6	4.5	.5255
250.	1075.	20.4	-60.7	-48.6	-33.4	-17.0	0.0	6.8	.6637
300.	1057.	30.3	-94.0	-79.5	-61.2	-41.6	-21.1	9.6	.8044
350.	1040.	42.3	-135.1	-118.1	-96.8	-73.9	-50.1	12.8	.9475
400.	1025.	56.8	-185.1	-165.7	-141.2	-115.1	-87.9	16.7	1.0949
450.	1010.	73.6	-243.5	-221.7	-194.2	-164.8	-134.2	21.1	1.2447
500.	995.	92.0	-307.4	-283.2	-252.7	-220.0	-186.0	24.7	1.3903

BALLISTIC COEFFICIENT .62, MUZZLE VELOCITY FT/SEC 1300.

RANGE	VELOCITY	HEIGHT	50 YD	100 YD	150 YD	200 YD	250 YD	DRIFT	TIME
0.	1300.	0.0	-1.5	-1.5	-1.5	-1.5	-1.5	0.0	0.0000
50.	1262.	-.1	0.0	2.0	4.6	7.4	10.5	.3	.1172
100.	1226.	2.0	-4.0	0.0	5.3	10.8	17.1	1.3	.2379
150.	1192.	5.6	-13.9	-8.0	0.0	8.3	17.6	3.0	.3634
200.	1162.	10.8	-29.6	-21.7	-11.1	0.0	12.4	4.9	.4895
250.	1135.	17.9	-52.6	-42.7	-29.4	-15.6	0.0	8.1	.6229
300.	1111.	26.5	-81.2	-69.3	-53.3	-36.7	-18.1	10.9	.7540
350.	1090.	37.3	-117.3	-103.4	-84.9	-65.5	-43.7	14.5	.8903
400.	1071.	50.1	-160.8	-144.9	-123.6	-101.5	-76.6	18.7	1.0291
450.	1053.	65.0	-211.7	-193.8	-170.0	-145.1	-117.0	23.2	1.1704
500.	1037.	82.1	-270.4	-250.6	-224.0	-196.3	-165.2	28.2	1.3139

BALLISTIC COEFFICIENT .62, MUZZLE VELOCITY FT/SEC 1400.

RANGE	VELOCITY	HEIGHT	50 YD	100 YD	150 YD	200 YD	250 YD	DRIFT	TIME
0.	1400.	0.0	-1.5	-1.5	-1.5	-1.5	-1.5	0.0	0.0000
50.	1356.	-.2	0.0	1.6	3.8	6.3	8.9	.4	.1095
100.	1315.	1.6	-3.2	0.0	4.4	9.3	14.5	1.3	.2216
150.	1276.	4.7	-11.5	-6.7	0.0	7.3	15.1	2.8	.3375
200.	1240.	9.3	-25.1	-18.6	-9.8	0.0	10.4	4.9	.4566
250.	1205.	15.4	-44.3	-36.3	-25.2	-13.0	0.0	7.7	.5793
300.	1173.	23.3	-70.3	-60.6	-47.3	-32.7	-17.1	11.5	.7080
350.	1145.	32.7	-101.2	-90.0	-74.4	-57.4	-39.2	14.9	.8349
400.	1120.	44.2	-139.5	-126.7	-108.9	-89.4	-68.6	19.4	.9674
450.	1098.	57.6	-184.7	-170.3	-150.3	-128.3	-104.9	24.4	1.1028
500.	1078.	73.1	-237.0	-220.9	-198.8	-174.4	-148.4	29.8	1.2406

BALLISTIC COEFFICIENT .62, MUZZLE VELOCITY FT/SEC 1500.

RANGE	VELOCITY	HEIGHT	50 YD	100 YD	150 YD	200 YD	250 YD	DRIFT	TIME
0.	1500.	0.0	-1.5	-1.5	-1.5	-1.5	-1.5	0.0	0.0000
50.	1450.	-.2	0.0	1.3	3.2	5.4	7.6	.4	.1022
100.	1403.	1.3	-2.6	0.0	3.8	8.1	12.7	1.2	.2071
150.	1359.	4.0	-9.7	-5.8	0.0	6.4	13.2	2.8	.3159
200.	1318.	8.0	-21.4	-16.2	-8.5	0.0	9.1	4.9	.4279
250.	1279.	13.4	-38.2	-31.7	-22.1	-11.4	0.0	7.6	.5434
300.	1242.	20.3	-60.4	-52.5	-41.0	-28.2	-14.5	11.0	.6626
350.	1208.	28.8	-88.1	-79.0	-65.5	-50.6	-34.6	15.0	.7850
400.	1175.	39.2	-122.3	-111.9	-96.5	-79.4	-61.2	19.8	.9126
450.	1147.	51.2	-161.9	-150.1	-132.8	-113.6	-93.0	24.7	1.0401
500.	1122.	65.2	-208.5	-195.4	-176.2	-154.9	-132.0	30.3	1.1724

BALLISTIC COEFFICIENT .62, MUZZLE VELOCITY FT/SEC 1600.

RANGE	VELOCITY	HEIGHT	50 YD	100 YD	150 YD	200 YD	250 YD	DRIFT	TIME
0.	1600.	0.0	-1.5	-1.5	-1.5	-1.5	-1.5	0.0	0.0000
50.	1548.	-.3	0.0	1.1	2.7	4.6	6.6	.3	.0953
100.	1497.	1.0	-2.1	0.0	3.3	7.0	11.1	1.1	.1935
150.	1447.	3.5	-8.2	-5.0	0.0	5.6	11.7	2.6	.2959
200.	1401.	7.0	-18.3	-14.1	-7.4	0.0	8.1	4.6	.4009
250.	1357.	11.7	-33.0	-27.8	-19.4	-10.1	0.0	7.3	.5103
300.	1316.	17.8	-52.4	-46.1	-36.0	-24.9	-12.7	10.5	.6224
350.	1277.	25.4	-76.8	-69.4	-57.7	-44.8	-30.5	14.4	.7382
400.	1240.	34.5	-106.5	-98.1	-84.7	-69.9	-53.6	18.9	.8573
450.	1206.	45.3	-141.9	-132.4	-117.3	-100.6	-82.4	24.0	.9799
500.	1173.	58.2	-184.1	-173.6	-156.9	-138.3	-118.0	30.1	1.1085

BALLISTIC COEFFICIENT .62, MUZZLE VELOCITY FT/SEC 1700.

RANGE	VELOCITY	HEIGHT	50 YD	100 YD	150 YD	200 YD	250 YD	DRIFT	TIME
0.	1700.	0.0	-1.5	-1.5	-1.5	-1.5	-1.5	0.0	0.0000
50.	1645.	-.4	0.0	.9	2.3	3.9	5.7	.3	.0899
100.	1592.	.9	-1.7	0.0	2.8	6.1	9.7	1.1	.1828
150.	1540.	3.0	-6.9	-4.3	0.0	4.9	10.2	2.4	.2783
200.	1489.	6.1	-15.7	-12.2	-6.5	0.0	7.1	4.3	.3773
250.	1440.	10.3	-28.5	-24.2	-17.1	-8.9	0.0	6.9	.4801
300.	1394.	15.7	-45.4	-40.2	-31.7	-21.9	-11.2	9.9	.5858
350.	1350.	22.5	-67.0	-60.9	-51.0	-39.5	-27.1	13.7	.6954
400.	1309.	30.6	-93.1	-86.2	-74.9	-61.8	-47.5	18.0	.8080
450.	1271.	40.3	-124.5	-116.7	-103.9	-89.2	-73.2	22.9	.9243
500.	1235.	51.7	-161.4	-152.8	-138.6	-122.3	-104.4	28.6	1.0449

BALLISTIC COEFFICIENT .62, MUZZLE VELOCITY FT/SEC 1800.

RANGE	VELOCITY	HEIGHT	50 YD	100 YD	150 YD	200 YD	250 YD	DRIFT	TIME
0.	1800.	0.0	-1.5	-1.5	-1.5	-1.5	-1.5	0.0	0.0000
50.	1743.	-.4	0.0	.7	1.9	3.4	4.9	.3	.0848
100.	1687.	.7	-1.4	0.0	2.5	5.4	8.5	1.0	.1722
150.	1632.	2.6	-5.8	-3.7	0.0	4.4	9.0	2.2	.2626
200.	1579.	5.3	-13.5	-10.8	-5.8	0.0	6.2	4.1	.3564
250.	1527.	9.1	-24.7	-21.3	-15.0	-7.8	0.0	6.3	.4527
300.	1477.	13.9	-39.6	-35.5	-28.0	-19.3	-10.0	9.2	.5524
350.	1428.	19.9	-58.6	-53.8	-45.1	-34.9	-24.0	12.8	.6559
400.	1383.	27.2	-81.7	-76.3	-66.3	-54.7	-42.2	16.9	.7625
450.	1340.	35.8	-109.6	-103.4	-92.2	-79.1	-65.1	21.6	.8728
500.	1300.	45.9	-142.1	-135.3	-122.8	-108.3	-92.7	26.9	.9863

BALLISTIC COEFFICIENT .62, MUZZLE VELOCITY FT/SEC 1900.

RANGE	VELOCITY	HEIGHT	50 YD	100 YD	150 YD	200 YD	250 YD	DRIFT	TIME
0.	1900.	0.0	-1.5	-1.5	-1.5	-1.5	-1.5	0.0	0.0000
50.	1840.	-.4	0.0	.5	1.6	2.9	4.3	.3	.0806
100.	1782.	.5	-1.1	0.0	2.2	4.7	7.5	.9	.1633
150.	1725.	2.2	-4.8	-3.2	0.0	3.9	8.0	2.0	.2484
200.	1669.	4.7	-11.6	-9.5	-5.2	0.0	5.6	3.8	.3371
250.	1615.	8.1	-21.5	-18.8	-13.4	-7.0	0.0	5.9	.4284
300.	1562.	12.4	-34.7	-31.5	-25.1	-17.3	-9.0	8.7	.5231
350.	1511.	17.7	-51.4	-47.7	-40.2	-31.1	-21.4	11.9	.6204
400.	1461.	24.2	-72.0	-67.8	-59.2	-48.8	-37.7	15.8	.7215
450.	1414.	32.0	-96.7	-91.9	-82.2	-70.6	-58.0	20.3	.8257
500.	1369.	41.1	-125.8	-120.5	-109.7	-96.8	-82.9	25.4	.9338

BALLISTIC COEFFICIENT .63, MUZZLE VELOCITY FT/SEC 800.

RANGE	VELOCITY	HEIGHT	50 YD	100 YD	150 YD	200 YD	250 YD	DRIFT	TIME
0.	800.	0.0	-1.5	-1.5	-1.5	-1.5	-1.5	0.0	0.0000
50.	790.	1.0	0.0	6.1	13.0	20.0	27.3	.4	.1898
100.	781.	6.2	-12.3	0.0	13.8	27.8	42.4	.9	.3798
150.	771.	15.1	-39.1	-20.7	0.0	21.0	43.0	2.1	.5742
200.	761.	27.6	-80.1	-55.6	-28.0	0.0	29.3	3.3	.7690
250.	752.	44.2	-136.7	-106.1	-71.6	-36.6	0.0	5.4	.9682
300.	743.	64.8	-208.1	-171.3	-130.0	-88.0	-44.1	7.6	1.1682
350.	733.	89.6	-296.0	-253.1	-204.8	-155.8	-104.6	10.5	1.3722
400.	724.	118.7	-399.6	-350.5	-295.4	-239.4	-180.8	13.7	1.5777
450.	715.	152.4	-520.1	-464.9	-402.9	-339.9	-274.0	17.4	1.7865
500.	706.	190.8	-657.7	-596.3	-527.4	-457.4	-384.2	21.6	1.9979

BALLISTIC COEFFICIENT .63, MUZZLE VELOCITY FT/SEC 900.

RANGE	VELOCITY	HEIGHT	50 YD	100 YD	150 YD	200 YD	250 YD	DRIFT	TIME
0.	900.	0.0	-1.5	-1.5	-1.5	-1.5	-1.5	0.0	0.0000
50.	889.	.6	0.0	4.7	10.1	15.7	21.4	.4	.1688
100.	878.	4.7	-9.5	0.0	10.8	21.9	33.3	.9	.3384
150.	867.	11.8	-30.3	-16.1	0.0	16.7	33.8	1.9	.5109
200.	856.	21.8	-62.7	-43.7	-22.2	0.0	22.9	3.2	.6848
250.	846.	34.8	-106.9	-83.3	-56.4	-28.6	0.0	4.9	.8612
300.	836.	51.1	-163.3	-134.9	-102.6	-69.3	-35.0	6.9	1.0394
350.	826.	70.7	-232.3	-199.1	-161.5	-122.6	-82.6	9.4	1.2202
400.	816.	93.7	-313.9	-276.1	-233.0	-188.6	-142.8	12.3	1.4030
450.	806.	120.3	-408.8	-366.2	-317.8	-267.8	-216.4	15.5	1.5882
500.	796.	150.5	-516.9	-469.6	-415.8	-360.3	-303.1	19.1	1.7754

BALLISTIC COEFFICIENT .63, MUZZLE VELOCITY FT/SEC 1000.

RANGE	VELOCITY	HEIGHT	50 YD	100 YD	150 YD	200 YD	250 YD	DRIFT	TIME
0.	1000.	0.0	-1.5	-1.5	-1.5	-1.5	-1.5	0.0	0.0000
50.	986.	.4	0.0	3.6	7.8	12.6	17.0	.7	.1542
100.	972.	3.7	-7.1	0.0	8.5	18.0	27.0	.8	.3047
150.	959.	9.4	-23.5	-12.8	0.0	14.2	27.7	1.6	.4594
200.	946.	17.7	-50.3	-36.0	-19.0	0.0	17.9	3.6	.6207
250.	933.	28.2	-85.2	-67.4	-46.1	-22.4	0.0	4.8	.7773
300.	921.	41.5	-130.6	-109.3	-83.7	-55.2	-28.3	6.7	.9382
350.	909.	57.6	-186.8	-161.8	-132.0	-98.8	-67.4	9.3	1.1028
400.	897.	76.7	-253.8	-225.4	-191.2	-153.3	-117.5	12.3	1.2701
450.	886.	98.4	-330.7	-298.6	-260.2	-217.5	-177.2	15.4	1.4372
500.	875.	123.3	-419.1	-383.5	-340.8	-293.4	-248.6	18.9	1.6077

BALLISTIC COEFFICIENT .63, MUZZLE VELOCITY FT/SEC 1100.

RANGE	VELOCITY	HEIGHT	50 YD	100 YD	150 YD	200 YD	250 YD	DRIFT	TIME
0.	1100.	0.0	-1.5	-1.5	-1.5	-1.5	-1.5	0.0	0.0000
50.	1080.	.2	0.0	3.1	6.7	10.5	14.2	.2	.1375
100.	1062.	3.0	-6.1	0.0	7.3	15.0	22.4	1.1	.2791
150.	1045.	7.9	-20.2	-11.0	0.0	11.4	22.5	2.5	.4235
200.	1030.	14.8	-42.1	-29.9	-15.2	0.0	14.8	4.1	.5688
250.	1015.	23.6	-71.2	-55.9	-37.6	-18.5	0.0	5.3	.7118
300.	1001.	34.8	-109.4	-91.1	-69.1	-46.3	-24.0	7.5	.8606
350.	986.	48.7	-157.6	-136.2	-110.5	-83.8	-57.9	10.6	1.0146
400.	973.	64.4	-212.4	-187.9	-158.6	-128.1	-98.5	13.0	1.1648
450.	959.	82.8	-277.1	-249.6	-216.6	-182.3	-149.0	16.2	1.3193
500.	946.	104.5	-353.9	-323.3	-286.6	-248.6	-211.5	20.6	1.4810

BALLISTIC COEFFICIENT .63, MUZZLE VELOCITY FT/SEC 1200.

RANGE	VELOCITY	HEIGHT	50 YD	100 YD	150 YD	200 YD	250 YD	DRIFT	TIME
0.	1200.	0.0	-1.5	-1.5	-1.5	-1.5	-1.5	0.0	0.0000
50.	1169.	.0	0.0	2.4	5.5	8.7	12.1	.3	.1267
100.	1142.	2.4	-4.8	0.0	6.3	12.6	19.4	1.1	.2564
150.	1117.	6.6	-16.6	-9.4	0.0	9.6	19.7	2.8	.3911
200.	1096.	12.5	-34.9	-25.2	-12.7	0.0	13.6	4.4	.5251
250.	1076.	20.4	-60.6	-48.5	-32.9	-17.0	0.0	6.7	.6631
300.	1058.	30.2	-93.8	-79.3	-60.5	-41.4	-21.1	9.4	.8035
350.	1042.	42.4	-135.3	-118.3	-96.5	-74.2	-50.4	12.8	.9477
400.	1027.	56.8	-185.0	-165.6	-140.6	-115.2	-88.0	16.6	1.0945
450.	1012.	73.5	-243.4	-221.6	-193.5	-164.8	-134.2	21.0	1.2441
500.	998.	91.8	-306.7	-282.5	-251.2	-219.4	-185.4	24.3	1.3883

BALLISTIC COEFFICIENT .63, MUZZLE VELOCITY FT/SEC 1300.

RANGE	VELOCITY	HEIGHT	50 YD	100 YD	150 YD	200 YD	250 YD	DRIFT	TIME
0.	1300.	0.0	-1.5	-1.5	-1.5	-1.5	-1.5	0.0	0.0000
50.	1263.	-.1	0.0	2.0	4.6	7.4	10.4	.3	.1171
100.	1228.	2.0	-4.0	0.0	5.3	10.8	16.9	1.2	.2379
150.	1194.	5.6	-13.9	-7.9	0.0	8.3	17.5	2.9	.3629
200.	1164.	10.7	-29.6	-21.6	-11.1	0.0	12.2	4.9	.4891
250.	1137.	17.8	-52.2	-42.3	-29.1	-15.3	0.0	7.8	.6211
300.	1113.	26.5	-81.0	-69.1	-53.3	-36.6	-18.3	10.7	.7531
350.	1092.	37.2	-117.1	-103.2	-84.7	-65.3	-43.9	14.3	.8892
400.	1073.	49.9	-160.3	-144.4	-123.4	-101.2	-76.7	18.4	1.0276
450.	1056.	64.8	-211.1	-193.2	-169.5	-144.5	-117.1	22.9	1.1685
500.	1039.	81.8	-269.6	-249.7	-223.3	-195.6	-165.1	27.8	1.3115

BALLISTIC COEFFICIENT .63, MUZZLE VELOCITY FT/SEC 1400.

RANGE	VELOCITY	HEIGHT	50 YD	100 YD	150 YD	200 YD	250 YD	DRIFT	TIME
0.	1400.	0.0	-1.5	-1.5	-1.5	-1.5	-1.5	0.0	0.0000
50.	1357.	-.2	0.0	1.6	3.8	6.3	8.8	.4	.1094
100.	1316.	1.6	-3.2	0.0	4.4	9.3	14.5	1.3	.2214
150.	1278.	4.7	-11.4	-6.6	0.0	7.3	15.1	2.8	.3371
200.	1242.	9.2	-25.0	-18.6	-9.8	0.0	10.4	4.9	.4561
250.	1208.	15.3	-44.2	-36.2	-25.2	-13.0	0.0	7.5	.5786
300.	1176.	23.2	-69.8	-60.2	-46.9	-32.3	-16.7	11.1	.7059
350.	1148.	32.6	-101.0	-89.7	-74.3	-57.2	-39.0	14.7	.8336
400.	1123.	44.0	-139.1	-126.3	-108.6	-89.1	-68.3	19.1	.9658
450.	1101.	57.4	-184.1	-169.7	-149.8	-127.8	-104.5	24.0	1.1007
500.	1081.	72.9	-236.4	-220.4	-198.2	-173.8	-147.9	29.4	1.2385

BALLISTIC COEFFICIENT .63, MUZZLE VELOCITY FT/SEC 1500.

RANGE	VELOCITY	HEIGHT	50 YD	100 YD	150 YD	200 YD	250 YD	DRIFT	TIME
0.	1500.	0.0	-1.5	-1.5	-1.5	-1.5	-1.5	0.0	0.0000
50.	1451.	-.2	0.0	1.3	3.2	5.3	7.6	.4	.1022
100.	1405.	1.3	-2.6	0.0	3.8	8.1	12.6	1.2	.2070
150.	1361.	4.0	-9.7	-5.8	0.0	6.4	13.2	2.8	.3157
200.	1321.	8.0	-21.4	-16.2	-8.5	0.0	9.1	4.8	.4273
250.	1282.	13.4	-38.1	-31.6	-22.0	-11.4	0.0	7.5	.5427
300.	1246.	20.3	-60.2	-52.4	-40.9	-28.1	-14.4	10.8	.6616
350.	1212.	28.7	-87.8	-78.7	-65.3	-50.4	-34.4	14.7	.7836
400.	1179.	38.9	-121.4	-111.0	-95.6	-78.7	-60.4	19.2	.9093
450.	1151.	51.0	-161.3	-149.5	-132.3	-113.2	-92.6	24.3	1.0380
500.	1126.	64.9	-207.7	-194.7	-175.5	-154.3	-131.5	29.9	1.1699

BALLISTIC COEFFICIENT .63, MUZZLE VELOCITY FT/SEC 1600.

RANGE	VELOCITY	HEIGHT	50 YD	100 YD	150 YD	200 YD	250 YD	DRIFT	TIME
0.	1600.	0.0	-1.5	-1.5	-1.5	-1.5	-1.5	0.0	0.0000
50.	1549.	-.3	0.0	1.1	2.7	4.6	6.6	.3	.0953
100.	1498.	1.0	-2.1	0.0	3.3	7.0	11.1	1.0	.1933
150.	1450.	3.4	-8.1	-5.0	0.0	5.6	11.6	2.5	.2956
200.	1404.	6.9	-18.3	-14.1	-7.4	0.0	8.0	4.5	.4005
250.	1360.	11.7	-32.9	-27.6	-19.3	-10.1	0.0	7.1	.5093
300.	1319.	17.8	-52.1	-45.8	-35.9	-24.7	-12.7	10.3	.6211
350.	1281.	25.3	-76.5	-69.1	-57.4	-44.5	-30.4	14.1	.7365
400.	1245.	34.4	-106.1	-97.7	-84.4	-69.6	-53.5	18.6	.8555
450.	1211.	45.1	-141.3	-131.8	-116.8	-100.2	-82.1	23.6	.9776
500.	1178.	57.7	-182.6	-172.1	-155.4	-136.9	-116.8	29.3	1.1038

BALLISTIC COEFFICIENT .63, MUZZLE VELOCITY FT/SEC 1700.

RANGE	VELOCITY	HEIGHT	50 YD	100 YD	150 YD	200 YD	250 YD	DRIFT	TIME
0.	1700.	0.0	-1.5	-1.5	-1.5	-1.5	-1.5	0.0	0.0000
50.	1646.	-.4	0.0	.9	2.3	3.9	5.7	.3	.0899
100.	1593.	.9	-1.7	0.0	2.8	6.1	9.6	1.1	.1827
150.	1542.	3.0	-6.8	-4.3	0.0	4.9	10.2	2.4	.2782
200.	1492.	6.1	-15.6	-12.2	-6.5	0.0	7.1	4.2	.3769
250.	1444.	10.3	-28.4	-24.1	-17.0	-8.9	0.0	6.7	.4795
300.	1398.	15.7	-45.3	-40.1	-31.6	-21.9	-11.2	9.8	.5849
350.	1355.	22.4	-66.8	-60.8	-50.9	-39.5	-27.0	13.5	.6945
400.	1314.	30.5	-92.9	-86.0	-74.7	-61.7	-47.4	17.7	.8067
450.	1276.	40.1	-124.1	-116.3	-103.5	-88.9	-72.9	22.6	.9225
500.	1240.	51.3	-160.5	-151.9	-137.7	-121.4	-103.7	28.0	1.0416

BALLISTIC COEFFICIENT .63, MUZZLE VELOCITY FT/SEC 1800.

RANGE	VELOCITY	HEIGHT	50 YD	100 YD	150 YD	200 YD	250 YD	DRIFT	TIME
0.	1800.	0.0	-1.5	-1.5	-1.5	-1.5	-1.5	0.0	0.0000
50.	1743.	-.4	0.0	.7	1.9	3.4	4.9	.3	.0848
100.	1688.	.7	-1.4	0.0	2.5	5.4	8.5	1.0	.1721
150.	1635.	2.6	-5.8	-3.7	0.0	4.4	9.0	2.2	.2624
200.	1582.	5.3	-13.5	-10.8	-5.8	0.0	6.2	4.0	.3561
250.	1531.	9.1	-24.6	-21.2	-15.0	-7.7	0.0	6.2	.4521
300.	1482.	13.8	-39.5	-35.4	-27.9	-19.2	-9.9	9.1	.5515
350.	1434.	19.8	-58.4	-53.6	-44.9	-34.8	-23.9	12.6	.6547
400.	1389.	27.0	-81.4	-76.0	-66.0	-54.4	-42.0	16.6	.7609
450.	1346.	35.6	-109.1	-103.0	-91.8	-78.7	-64.8	21.3	.8708
500.	1306.	45.7	-141.5	-134.7	-122.2	-107.7	-92.2	26.5	.9837

BALLISTIC COEFFICIENT .63, MUZZLE VELOCITY FT/SEC 1900.

RANGE	VELOCITY	HEIGHT	50 YD	100 YD	150 YD	200 YD	250 YD	DRIFT	TIME
0.	1900.	0.0	-1.5	-1.5	-1.5	-1.5	-1.5	0.0	0.0000
50.	1841.	-.4	0.0	.5	1.6	2.9	4.3	.3	.0806
100.	1783.	.5	-1.1	0.0	2.2	4.7	7.5	.9	.1632
150.	1727.	2.2	-4.8	-3.2	0.0	3.9	8.0	2.0	.2483
200.	1673.	4.7	-11.6	-9.4	-5.1	0.0	5.5	3.7	.3368
250.	1619.	8.0	-21.4	-18.7	-13.3	-6.9	0.0	5.8	.4277
300.	1567.	12.3	-34.6	-31.4	-25.0	-17.3	-9.0	8.6	.5223
350.	1517.	17.7	-51.3	-47.6	-40.0	-31.0	-21.3	11.7	.6193
400.	1468.	24.1	-71.8	-67.5	-58.9	-48.6	-37.6	15.5	.7199
450.	1420.	31.8	-96.2	-91.4	-81.7	-70.2	-57.7	19.9	.8235
500.	1376.	40.9	-125.2	-119.9	-109.2	-96.3	-82.5	25.0	.9314

BALLISTIC COEFFICIENT .64, MUZZLE VELOCITY FT/SEC 800.

RANGE	VELOCITY	HEIGHT	50 YD	100 YD	150 YD	200 YD	250 YD	DRIFT	TIME
0.	800.	0.0	-1.5	-1.5	-1.5	-1.5	-1.5	0.0	0.0000
50.	790.	1.0	0.0	6.1	13.0	20.0	27.3	.4	.1898
100.	781.	6.2	-12.3	0.0	13.8	27.8	42.4	.9	.3799
150.	771.	15.1	-39.0	-20.6	0.0	21.0	42.9	2.0	.5740
200.	762.	27.6	-80.1	-55.5	-28.0	0.0	29.2	3.3	.7688
250.	753.	44.2	-136.6	-105.9	-71.5	-36.5	0.0	5.3	.9677
300.	744.	64.7	-208.0	-171.2	-129.9	-87.9	-44.1	7.5	1.1677
350.	734.	89.5	-295.6	-252.6	-204.5	-155.5	-104.4	10.3	1.3711
400.	725.	118.6	-399.2	-350.0	-295.0	-239.0	-180.6	13.5	1.5767
450.	716.	152.1	-519.2	-464.0	-402.1	-339.0	-273.4	17.1	1.7847
500.	707.	190.5	-656.8	-595.4	-526.6	-456.6	-383.6	21.3	1.9960

BALLISTIC COEFFICIENT .64, MUZZLE VELOCITY FT/SEC 900.

RANGE	VELOCITY	HEIGHT	50 YD	100 YD	150 YD	200 YD	250 YD	DRIFT	TIME
0.	900.	0.0	-1.5	-1.5	-1.5	-1.5	-1.5	0.0	0.0000
50.	889.	.6	0.0	4.7	10.1	15.6	21.4	.4	.1688
100.	878.	4.7	-9.5	0.0	10.8	21.8	33.3	.9	.3383
150.	867.	11.8	-30.3	-16.1	0.0	16.6	33.8	1.9	.5108
200.	857.	21.7	-62.6	-43.7	-22.2	0.0	22.9	3.1	.6845
250.	847.	34.8	-106.9	-83.2	-56.3	-28.6	0.0	4.8	.8608
300.	837.	51.0	-163.1	-134.7	-102.4	-69.2	-34.8	6.8	1.0387
350.	827.	70.6	-232.0	-199.0	-161.3	-122.5	-82.4	9.3	1.2195
400.	817.	93.6	-313.4	-275.6	-232.6	-188.2	-142.4	12.0	1.4017
450.	807.	120.1	-408.3	-365.8	-317.4	-267.5	-216.0	15.3	1.5870
500.	798.	150.2	-516.0	-468.7	-414.9	-359.5	-302.3	18.8	1.7734

BALLISTIC COEFFICIENT .64, MUZZLE VELOCITY FT/SEC 1000.

RANGE	VELOCITY	HEIGHT	50 YD	100 YD	150 YD	200 YD	250 YD	DRIFT	TIME
0.	1000.	0.0	-1.5	-1.5	-1.5	-1.5	-1.5	0.0	0.0000
50.	986.	.4	0.0	3.5	7.8	12.6	17.0	.8	.1543
100.	973.	3.7	-7.1	0.0	8.5	18.1	26.9	.8	.3046
150.	960.	9.4	-23.4	-12.8	0.0	14.3	27.6	1.6	.4591
200.	947.	17.8	-50.3	-36.1	-19.1	0.0	17.8	3.7	.6209
250.	934.	28.2	-85.1	-67.4	-46.1	-22.2	0.0	4.7	.7768
300.	922.	41.5	-130.5	-109.2	-83.6	-55.0	-28.4	6.7	.9378
350.	911.	57.5	-186.5	-161.6	-131.8	-98.4	-67.3	9.1	1.1019
400.	899.	76.4	-252.9	-224.5	-190.4	-152.2	-116.7	11.9	1.2679
450.	888.	98.2	-330.1	-298.1	-259.8	-216.9	-176.9	15.1	1.4359
500.	877.	123.0	-418.2	-382.7	-340.1	-292.4	-248.0	18.6	1.6058

BALLISTIC COEFFICIENT .64, MUZZLE VELOCITY FT/SEC 1100.

RANGE	VELOCITY	HEIGHT	50 YD	100 YD	150 YD	200 YD	250 YD	DRIFT	TIME
0.	1100.	0.0	-1.5	-1.5	-1.5	-1.5	-1.5	0.0	0.0000
50.	1080.	.2	0.0	3.1	6.7	10.5	14.2	.2	.1377
100.	1062.	3.0	-6.1	0.0	7.3	14.9	22.3	1.2	.2795
150.	1046.	7.9	-20.2	-11.0	0.0	11.4	22.4	2.6	.4239
200.	1031.	14.8	-42.1	-29.9	-15.2	0.0	14.7	4.1	.5689
250.	1016.	23.5	-71.0	-55.7	-37.4	-18.4	0.0	5.2	.7113
300.	1002.	34.8	-109.5	-91.1	-69.0	-46.3	-24.2	7.5	.8607
350.	988.	48.7	-157.5	-136.1	-110.4	-83.8	-58.1	10.6	1.0146
400.	975.	64.3	-211.9	-187.4	-158.1	-127.7	-98.3	12.8	1.1637
450.	961.	82.7	-276.9	-249.3	-216.3	-182.1	-149.0	16.1	1.3187
500.	949.	103.8	-351.3	-320.7	-284.0	-246.0	-209.3	19.8	1.4759

BALLISTIC COEFFICIENT .64, MUZZLE VELOCITY FT/SEC 1200.

RANGE	VELOCITY	HEIGHT	50 YD	100 YD	150 YD	200 YD	250 YD	DRIFT	TIME
0.	1200.	0.0	-1.5	-1.5	-1.5	-1.5	-1.5	0.0	0.0000
50.	1170.	.0	0.0	2.4	5.5	8.7	12.1	.3	.1267
100.	1142.	2.4	-4.8	0.0	6.2	12.6	19.4	1.1	.2563
150.	1118.	6.6	-16.5	-9.3	0.0	9.6	19.8	2.7	.3902
200.	1097.	12.5	-34.9	-25.2	-12.8	0.0	13.6	4.4	.5247
250.	1078.	20.3	-60.5	-48.4	-33.0	-16.9	0.0	6.6	.6625
300.	1060.	30.2	-93.6	-79.1	-60.6	-41.3	-21.0	9.3	.8027
350.	1044.	42.4	-135.5	-118.5	-96.9	-74.5	-50.7	12.8	.9480
400.	1028.	56.7	-185.0	-165.6	-140.9	-115.2	-88.1	16.6	1.0941
450.	1014.	72.8	-240.7	-218.9	-191.2	-162.3	-131.8	19.8	1.2377
500.	1000.	91.5	-305.9	-281.7	-250.9	-218.8	-184.9	24.0	1.3863

BALLISTIC COEFFICIENT .64, MUZZLE VELOCITY FT/SEC 1300.

RANGE	VELOCITY	HEIGHT	50 YD	100 YD	150 YD	200 YD	250 YD	DRIFT	TIME
0.	1300.	0.0	-1.5	-1.5	-1.5	-1.5	-1.5	0.0	0.0000
50.	1263.	-.1	0.0	2.0	4.6	7.4	10.4	.3	.1171
100.	1229.	2.0	-4.0	0.0	5.2	10.8	16.8	1.2	.2378
150.	1196.	5.6	-13.8	-7.9	0.0	8.3	17.3	2.9	.3624
200.	1166.	10.7	-29.5	-21.6	-11.1	0.0	12.0	4.8	.4887
250.	1139.	17.7	-51.9	-42.0	-28.9	-15.0	0.0	7.5	.6193
300.	1116.	26.4	-80.8	-68.9	-53.2	-36.5	-18.5	10.5	.7522
350.	1094.	37.1	-116.8	-102.9	-84.5	-65.1	-44.1	14.1	.8880
400.	1075.	49.8	-159.9	-144.0	-123.1	-100.8	-76.9	18.1	1.0261
450.	1058.	64.6	-210.5	-192.6	-169.1	-144.0	-117.1	22.6	1.1666
500.	1042.	81.7	-269.4	-249.5	-223.3	-195.5	-165.6	27.6	1.3107

BALLISTIC COEFFICIENT .64, MUZZLE VELOCITY FT/SEC 1400.

RANGE	VELOCITY	HEIGHT	50 YD	100 YD	150 YD	200 YD	250 YD	DRIFT	TIME
0.	1400.	0.0	-1.5	-1.5	-1.5	-1.5	-1.5	0.0	0.0000
50.	1357.	-.2	0.0	1.6	3.8	6.2	8.8	.4	.1093
100.	1317.	1.6	-3.2	0.0	4.4	9.3	14.5	1.2	.2212
150.	1280.	4.7	-11.4	-6.6	0.0	7.3	15.1	2.7	.3366
200.	1244.	9.2	-25.0	-18.6	-9.8	0.0	10.3	4.8	.4557
250.	1211.	15.3	-44.1	-36.2	-25.1	-12.9	0.0	7.4	.5779
300.	1179.	23.0	-69.4	-59.8	-46.5	-31.9	-16.4	10.7	.7038
350.	1151.	32.5	-100.7	-89.5	-74.1	-57.0	-38.9	14.5	.8324
400.	1126.	43.9	-138.7	-126.0	-108.3	-88.8	-68.1	18.8	.9642
450.	1104.	57.2	-183.6	-169.2	-149.4	-127.4	-104.2	23.7	1.0988
500.	1084.	72.6	-235.5	-219.5	-197.5	-173.1	-147.2	28.9	1.2359

BALLISTIC COEFFICIENT .64, MUZZLE VELOCITY FT/SEC 1500.

RANGE	VELOCITY	HEIGHT	50 YD	100 YD	150 YD	200 YD	250 YD	DRIFT	TIME
0.	1500.	0.0	-1.5	-1.5	-1.5	-1.5	-1.5	0.0	0.0000
50.	1452.	-.2	0.0	1.3	3.2	5.3	7.6	.4	.1021
100.	1406.	1.3	-2.6	0.0	3.8	8.1	12.6	1.2	.2068
150.	1363.	4.0	-9.7	-5.7	0.0	6.4	13.2	2.7	.3154
200.	1323.	8.0	-21.3	-16.1	-8.5	0.0	9.1	4.7	.4269
250.	1285.	13.4	-38.0	-31.5	-22.0	-11.4	0.0	7.4	.5420
300.	1249.	20.2	-60.0	-52.2	-40.7	-28.0	-14.4	10.7	.6605
350.	1215.	28.7	-87.7	-78.6	-65.2	-50.3	-34.4	14.6	.7827
400.	1183.	38.8	-121.0	-110.6	-95.2	-78.3	-60.1	18.9	.9074
450.	1154.	51.0	-161.6	-149.9	-132.7	-113.6	-93.1	24.4	1.0384
500.	1129.	64.7	-207.0	-193.9	-174.8	-153.6	-130.9	29.5	1.1673

BALLISTIC COEFFICIENT .64, MUZZLE VELOCITY FT/SEC 1600.

RANGE	VELOCITY	HEIGHT	50 YD	100 YD	150 YD	200 YD	250 YD	DRIFT	TIME
0.	1600.	0.0	-1.5	-1.5	-1.5	-1.5	-1.5	0.0	0.0000
50.	1549.	-.3	0.0	1.0	2.7	4.6	6.6	.3	.0952
100.	1500.	1.0	-2.1	0.0	3.3	7.0	11.0	1.0	.1932
150.	1452.	3.4	-8.1	-5.0	0.0	5.5	11.6	2.5	.2954
200.	1406.	6.9	-18.2	-14.0	-7.4	0.0	8.0	4.4	.4001
250.	1363.	11.7	-32.8	-27.6	-19.3	-10.0	0.0	7.0	.5086
300.	1323.	17.7	-52.0	-45.7	-35.7	-24.7	-12.6	10.1	.6201
350.	1285.	25.2	-76.2	-68.9	-57.3	-44.3	-30.3	13.9	.7352
400.	1249.	34.2	-105.7	-97.3	-84.0	-69.3	-53.2	18.3	.8538
450.	1215.	45.0	-140.9	-131.5	-116.5	-99.9	-81.8	23.3	.9759
500.	1183.	57.4	-181.6	-171.2	-154.5	-136.1	-116.0	28.7	1.1006

BALLISTIC COEFFICIENT .64, MUZZLE VELOCITY FT/SEC 1700.

RANGE	VELOCITY	HEIGHT	50 YD	100 YD	150 YD	200 YD	250 YD	DRIFT	TIME
0.	1700.	0.0	-1.5	-1.5	-1.5	-1.5	-1.5	0.0	0.0000
50.	1647.	-.4	0.0	.9	2.3	3.9	5.7	.3	.0899
100.	1595.	.9	-1.7	0.0	2.8	6.1	9.6	1.1	.1826
150.	1545.	3.0	-6.8	-4.3	0.0	4.9	10.2	2.3	.2780
200.	1495.	6.1	-15.6	-12.1	-6.5	0.0	7.1	4.1	.3764
250.	1447.	10.3	-28.4	-24.0	-17.0	-8.9	0.0	6.6	.4789
300.	1402.	15.6	-45.1	-40.0	-31.5	-21.8	-11.1	9.6	.5839
350.	1359.	22.3	-66.5	-60.5	-50.6	-39.2	-26.8	13.2	.6928
400.	1319.	30.3	-92.5	-85.6	-74.2	-61.3	-47.1	17.4	.8046
450.	1282.	39.9	-123.5	-115.7	-103.0	-88.4	-72.4	22.2	.9200
500.	1246.	51.1	-159.8	-151.2	-137.0	-120.9	-103.1	27.6	1.0389

BALLISTIC COEFFICIENT .64, MUZZLE VELOCITY FT/SEC 1800.

RANGE	VELOCITY	HEIGHT	50 YD	100 YD	150 YD	200 YD	250 YD	DRIFT	TIME
0.	1800.	0.0	-1.5	-1.5	-1.5	-1.5	-1.5	0.0	0.0000
50.	1744.	-.4	0.0	.7	1.9	3.4	4.9	.3	.0848
100.	1690.	.7	-1.4	0.0	2.5	5.4	8.5	.9	.1720
150.	1637.	2.6	-5.8	-3.7	0.0	4.3	9.0	2.2	.2623
200.	1586.	5.3	-13.5	-10.8	-5.8	0.0	6.2	3.9	.3557
250.	1535.	9.0	-24.6	-21.2	-14.9	-7.7	0.0	6.1	.4515
300.	1486.	13.8	-39.3	-35.3	-27.8	-19.1	-9.9	8.9	.5506
350.	1439.	19.8	-58.2	-53.5	-44.8	-34.7	-23.8	12.4	.6537
400.	1394.	26.9	-81.1	-75.7	-65.8	-54.2	-41.9	16.3	.7593
450.	1352.	35.5	-108.7	-102.6	-91.4	-78.4	-64.5	20.9	.8688
500.	1312.	45.5	-140.9	-134.1	-121.6	-107.2	-91.8	26.0	.9812

BALLISTIC COEFFICIENT .64, MUZZLE VELOCITY FT/SEC 1900.

RANGE	VELOCITY	HEIGHT	50 YD	100 YD	150 YD	200 YD	250 YD	DRIFT	TIME
0.	1900.	0.0	-1.5	-1.5	-1.5	-1.5	-1.5	0.0	0.0000
50.	1842.	-.4	0.0	.5	1.6	2.9	4.3	.3	.0805
100.	1785.	.5	-1.1	0.0	2.2	4.7	7.5	.9	.1631
150.	1730.	2.2	-4.8	-3.2	0.0	3.8	8.0	2.0	.2481
200.	1676.	4.7	-11.5	-9.4	-5.1	0.0	5.5	3.6	.3364
250.	1623.	8.0	-21.3	-18.7	-13.3	-6.9	0.0	5.7	.4272
300.	1572.	12.3	-34.5	-31.4	-24.9	-17.2	-8.9	8.4	.5215
350.	1522.	17.6	-51.1	-47.4	-39.9	-30.9	-21.2	11.5	.6182
400.	1474.	24.0	-71.5	-67.3	-58.6	-48.4	-37.3	15.3	.7183
450.	1427.	31.7	-95.9	-91.1	-81.4	-69.9	-57.5	19.6	.8218
500.	1383.	40.6	-124.6	-119.3	-108.5	-95.7	-81.9	24.5	.9285

BALLISTIC COEFFICIENT .65, MUZZLE VELOCITY FT/SEC 800.

RANGE	VELOCITY	HEIGHT	50 YD	100 YD	150 YD	200 YD	250 YD	DRIFT	TIME
0.	800.	0.0	-1.5	-1.5	-1.5	-1.5	-1.5	0.0	0.0000
50.	791.	1.0	0.0	6.1	13.0	20.0	27.3	.4	.1898
100.	781.	6.2	-12.3	0.0	13.7	27.7	42.3	.9	.3800
150.	772.	15.1	-39.0	-20.6	0.0	21.0	42.8	2.0	.5738
200.	763.	27.6	-80.1	-55.5	-28.0	0.0	29.1	3.3	.7686
250.	753.	44.1	-136.4	-105.7	-71.4	-36.4	0.0	5.2	.9672
300.	744.	64.6	-207.9	-171.0	-129.8	-87.8	-44.1	7.4	1.1672
350.	735.	89.4	-295.2	-252.2	-204.1	-155.1	-104.2	10.1	1.3701
400.	726.	118.4	-398.8	-349.6	-294.7	-238.6	-180.4	13.3	1.5756
450.	717.	151.8	-518.3	-463.0	-401.2	-338.2	-272.7	16.8	1.7829
500.	709.	190.1	-655.8	-594.4	-525.8	-455.7	-383.0	21.0	1.9942

BALLISTIC COEFFICIENT .65, MUZZLE VELOCITY FT/SEC 900.

RANGE	VELOCITY	HEIGHT	50 YD	100 YD	150 YD	200 YD	250 YD	DRIFT	TIME
0.	900.	0.0	-1.5	-1.5	-1.5	-1.5	-1.5	0.0	0.0000
50.	889.	.6	0.0	4.7	10.1	15.6	21.4	.4	.1688
100.	878.	4.7	-9.4	0.0	10.8	21.8	33.3	.9	.3382
150.	868.	11.8	-30.3	-16.1	0.0	16.6	33.8	1.9	.5107
200.	857.	21.7	-62.5	-43.6	-22.1	0.0	22.9	3.1	.6842
250.	847.	34.8	-106.8	-83.2	-56.3	-28.6	0.0	4.8	.8605
300.	837.	51.0	-162.9	-134.5	-102.3	-69.1	-34.7	6.7	1.0380
350.	828.	70.5	-231.8	-198.8	-161.1	-122.4	-82.3	9.2	1.2188
400.	818.	93.4	-312.9	-275.2	-232.1	-187.9	-142.1	11.8	1.4004
450.	809.	119.9	-407.8	-365.3	-316.9	-267.1	-215.6	15.1	1.5857
500.	799.	149.9	-515.0	-467.8	-414.0	-358.7	-301.5	18.4	1.7715

BALLISTIC COEFFICIENT .65, MUZZLE VELOCITY FT/SEC 1000.

RANGE	VELOCITY	HEIGHT	50 YD	100 YD	150 YD	200 YD	250 YD	DRIFT	TIME
0.	1000.	0.0	-1.5	-1.5	-1.5	-1.5	-1.5	0.0	0.0000
50.	986.	.4	0.0	3.5	7.8	12.4	17.0	.8	.1544
100.	973.	3.7	-7.1	0.0	8.5	17.6	26.9	.8	.3046
150.	960.	9.4	-23.4	-12.8	0.0	13.7	27.6	1.6	.4590
200.	948.	17.5	-49.4	-35.3	-18.2	0.0	18.6	3.0	.6171
250.	935.	28.2	-85.0	-67.3	-46.0	-23.2	0.0	4.6	.7764
300.	923.	41.4	-130.3	-109.1	-83.6	-56.2	-28.4	6.6	.9374
350.	912.	57.4	-186.2	-161.4	-131.6	-99.7	-67.2	9.0	1.1011
400.	900.	76.2	-252.1	-223.8	-189.7	-153.3	-116.1	11.6	1.2660
450.	889.	98.0	-329.5	-297.7	-259.4	-218.4	-176.6	14.9	1.4346
500.	879.	122.7	-417.3	-381.9	-339.4	-293.8	-247.3	18.3	1.6039

BALLISTIC COEFFICIENT .65, MUZZLE VELOCITY FT/SEC 1100.

RANGE	VELOCITY	HEIGHT	50 YD	100 YD	150 YD	200 YD	250 YD	DRIFT	TIME
0.	1100.	0.0	-1.5	-1.5	-1.5	-1.5	-1.5	0.0	0.0000
50.	1081.	.2	0.0	3.1	6.7	10.5	14.2	.3	.1379
100.	1063.	3.0	-6.2	0.0	7.3	14.9	22.2	1.3	.2800
150.	1047.	7.9	-20.2	-11.0	0.0	11.4	22.3	2.7	.4243
200.	1031.	14.8	-42.1	-29.8	-15.1	0.0	14.6	4.2	.5690
250.	1017.	23.5	-70.9	-55.5	-37.1	-18.2	0.0	5.1	.7109
300.	1003.	34.8	-109.5	-91.0	-69.0	-46.3	-24.4	7.5	.8607
350.	989.	48.7	-157.5	-136.0	-110.3	-83.8	-58.3	10.6	1.0145
400.	976.	64.1	-211.5	-186.9	-157.5	-127.3	-98.1	12.6	1.1626
450.	963.	82.7	-276.9	-249.2	-216.1	-182.1	-149.3	16.1	1.3185
500.	951.	103.6	-350.6	-319.8	-283.1	-245.3	-208.8	19.5	1.4742

BALLISTIC COEFFICIENT .65, MUZZLE VELOCITY FT/SEC 1200.

RANGE	VELOCITY	HEIGHT	50 YD	100 YD	150 YD	200 YD	250 YD	DRIFT	TIME
0.	1200.	0.0	-1.5	-1.5	-1.5	-1.5	-1.5	0.0	0.0000
50.	1170.	.0	0.0	2.4	5.5	8.7	12.1	.3	.1267
100.	1143.	2.4	-4.8	0.0	6.1	12.6	19.3	1.1	.2563
150.	1119.	6.5	-16.4	-9.1	0.0	9.7	19.8	2.5	.3893
200.	1098.	12.4	-34.8	-25.1	-12.9	0.0	13.5	4.3	.5243
250.	1079.	20.3	-60.4	-48.3	-33.1	-16.9	0.0	6.5	.6619
300.	1061.	30.2	-93.9	-79.4	-61.1	-41.7	-21.4	9.4	.8033
350.	1045.	42.4	-135.6	-118.7	-97.4	-74.7	-51.1	12.9	.9482
400.	1030.	56.7	-184.9	-165.6	-141.2	-115.3	-88.2	16.5	1.0936
450.	1016.	72.6	-240.2	-218.4	-191.0	-161.9	-131.5	19.6	1.2362
500.	1002.	91.4	-305.8	-281.6	-251.1	-218.8	-185.0	23.9	1.3855

BALLISTIC COEFFICIENT .65, MUZZLE VELOCITY FT/SEC 1300.

RANGE	VELOCITY	HEIGHT	50 YD	100 YD	150 YD	200 YD	250 YD	DRIFT	TIME
0.	1300.	0.0	-1.5	-1.5	-1.5	-1.5	-1.5	0.0	0.0000
50.	1264.	-.1	0.0	2.0	4.6	7.4	10.3	.3	.1171
100.	1230.	2.0	-4.0	0.0	5.2	10.8	16.7	1.2	.2377
150.	1197.	5.5	-13.8	-7.8	0.0	8.4	17.2	2.8	.3619
200.	1167.	10.7	-29.5	-21.6	-11.2	0.0	11.8	4.7	.4884
250.	1141.	17.6	-51.6	-41.7	-28.7	-14.8	0.0	7.3	.6182
300.	1117.	26.5	-81.0	-69.1	-53.5	-36.8	-19.1	10.7	.7529
350.	1097.	37.0	-116.5	-102.6	-84.4	-64.9	-44.2	13.9	.8868
400.	1078.	49.6	-159.5	-143.6	-122.8	-100.5	-76.8	17.9	1.0247
450.	1060.	64.4	-209.9	-192.1	-168.6	-143.6	-116.9	22.2	1.1649
500.	1044.	81.7	-269.5	-249.6	-223.6	-195.7	-166.2	27.5	1.3104

BALLISTIC COEFFICIENT .65, MUZZLE VELOCITY FT/SEC 1400.

RANGE	VELOCITY	HEIGHT	50 YD	100 YD	150 YD	200 YD	250 YD	DRIFT	TIME
0.	1400.	0.0	-1.5	-1.5	-1.5	-1.5	-1.5	0.0	0.0000
50.	1358.	-.2	0.0	1.6	3.8	6.2	8.8	.4	.1092
100.	1319.	1.6	-3.2	0.0	4.4	9.3	14.5	1.2	.2210
150.	1282.	4.7	-11.4	-6.6	0.0	7.3	15.1	2.6	.3364
200.	1246.	9.2	-24.9	-18.6	-9.7	0.0	10.4	4.7	.4553
250.	1213.	15.3	-44.2	-36.3	-25.2	-13.1	0.0	7.4	.5780
300.	1182.	22.9	-69.1	-59.5	-46.3	-31.7	-16.0	10.5	.7025
350.	1153.	32.7	-101.4	-90.2	-74.8	-57.7	-39.4	14.8	.8342
400.	1129.	43.7	-138.4	-125.6	-108.0	-88.5	-67.6	18.6	.9627
450.	1107.	57.0	-183.1	-168.8	-148.9	-127.0	-103.5	23.4	1.0970
500.	1087.	72.3	-234.8	-218.9	-196.9	-172.5	-146.4	28.6	1.2337

BALLISTIC COEFFICIENT .65, MUZZLE VELOCITY FT/SEC 1500.

RANGE	VELOCITY	HEIGHT	50 YD	100 YD	150 YD	200 YD	250 YD	DRIFT	TIME
0.	1500.	0.0	-1.5	-1.5	-1.5	-1.5	-1.5	0.0	0.0000
50.	1453.	-.2	0.0	1.3	3.2	5.3	7.6	.4	.1021
100.	1408.	1.3	-2.6	0.0	3.8	8.1	12.6	1.2	.2067
150.	1365.	4.0	-9.6	-5.7	0.0	6.3	13.1	2.7	.3152
200.	1326.	8.0	-21.3	-16.1	-8.5	0.0	9.1	4.7	.4266
250.	1288.	13.3	-38.0	-31.5	-21.9	-11.3	0.0	7.3	.5414
300.	1252.	20.1	-59.9	-52.1	-40.6	-27.9	-14.3	10.5	.6595
350.	1219.	28.5	-87.3	-78.2	-64.8	-50.0	-34.1	14.2	.7809
400.	1187.	38.6	-120.6	-110.2	-94.9	-78.0	-59.9	18.6	.9058
450.	1158.	50.6	-160.4	-148.7	-131.5	-112.5	-92.1	23.7	1.0346
500.	1133.	64.4	-206.2	-193.2	-174.1	-152.9	-130.3	29.0	1.1648

BALLISTIC COEFFICIENT .65, MUZZLE VELOCITY FT/SEC 1600.

RANGE	VELOCITY	HEIGHT	50 YD	100 YD	150 YD	200 YD	250 YD	DRIFT	TIME
0.	1600.	0.0	-1.5	-1.5	-1.5	-1.5	-1.5	0.0	0.0000
50.	1550.	-.3	0.0	1.0	2.7	4.5	6.5	.3	.0952
100.	1502.	1.0	-2.1	0.0	3.3	7.0	11.0	1.0	.1932
150.	1454.	3.4	-8.1	-5.0	0.0	5.5	11.5	2.4	.2951
200.	1409.	6.9	-18.2	-14.0	-7.4	0.0	8.0	4.3	.3996
250.	1367.	11.6	-32.7	-27.5	-19.2	-10.0	0.0	6.9	.5080
300.	1327.	17.7	-51.9	-45.6	-35.7	-24.6	-12.6	10.0	.6193
350.	1289.	25.1	-76.0	-68.7	-57.1	-44.2	-30.2	13.7	.7340
400.	1253.	34.2	-105.5	-97.2	-83.9	-69.2	-53.2	18.1	.8527
450.	1220.	44.7	-140.1	-130.7	-115.8	-99.2	-81.2	22.8	.9731
500.	1188.	57.1	-181.0	-170.5	-153.9	-135.5	-115.5	28.3	1.0981

BALLISTIC COEFFICIENT .65, MUZZLE VELOCITY FT/SEC 1700.

RANGE	VELOCITY	HEIGHT	50 YD	100 YD	150 YD	200 YD	250 YD	DRIFT	TIME
0.	1700.	0.0	-1.5	-1.5	-1.5	-1.5	-1.5	0.0	0.0000
50.	1648.	-.4	0.0	.9	2.3	3.9	5.7	.3	.0899
100.	1597.	.8	-1.7	0.0	2.8	6.1	9.6	1.1	.1825
150.	1547.	3.0	-6.8	-4.3	0.0	4.8	10.1	2.3	.2778
200.	1498.	6.0	-15.5	-12.1	-6.4	0.0	7.1	4.1	.3760
250.	1451.	10.2	-28.3	-24.0	-16.9	-8.8	0.0	6.5	.4782
300.	1406.	15.6	-45.0	-39.9	-31.4	-21.7	-11.1	9.4	.5830
350.	1364.	22.2	-66.3	-60.3	-50.4	-39.1	-26.7	13.0	.6915
400.	1324.	30.2	-92.1	-85.3	-73.9	-61.0	-46.9	17.1	.8030
450.	1287.	39.7	-123.0	-115.3	-102.5	-88.0	-72.1	21.8	.9180
500.	1251.	50.8	-159.1	-150.6	-136.4	-120.3	-102.6	27.1	1.0363

BALLISTIC COEFFICIENT .65, MUZZLE VELOCITY FT/SEC 1800.

RANGE	VELOCITY	HEIGHT	50 YD	100 YD	150 YD	200 YD	250 YD	DRIFT	TIME
0.	1800.	0.0	-1.5	-1.5	-1.5	-1.5	-1.5	0.0	0.0000
50.	1745.	-.4	0.0	.7	1.9	3.4	4.9	.2	.0847
100.	1692.	.7	-1.4	0.0	2.5	5.4	8.4	.9	.1719
150.	1640.	2.5	-5.8	-3.7	0.0	4.3	8.9	2.1	.2621
200.	1589.	5.3	-13.4	-10.7	-5.8	0.0	6.2	3.9	.3553
250.	1539.	9.0	-24.5	-21.1	-14.9	-7.7	0.0	6.0	.4509
300.	1491.	13.8	-39.2	-35.2	-27.7	-19.1	-9.9	8.8	.5498
350.	1444.	19.7	-58.0	-53.3	-44.6	-34.5	-23.7	12.2	.6525
400.	1399.	26.8	-80.9	-75.5	-65.5	-54.0	-41.7	16.0	.7577
450.	1357.	35.3	-108.3	-102.3	-91.1	-78.1	-64.2	20.6	.8670
500.	1318.	45.2	-140.3	-133.6	-121.1	-106.7	-91.3	25.6	.9788

BALLISTIC COEFFICIENT .65, MUZZLE VELOCITY FT/SEC 1900.

RANGE	VELOCITY	HEIGHT	50 YD	100 YD	150 YD	200 YD	250 YD	DRIFT	TIME
0.	1900.	0.0	-1.5	-1.5	-1.5	-1.5	-1.5	0.0	0.0000
50.	1843.	-.4	0.0	.5	1.6	2.9	4.3	.3	.0805
100.	1787.	.5	-1.1	0.0	2.2	4.7	7.5	.9	.1630
150.	1732.	2.2	-4.8	-3.2	0.0	3.8	8.0	2.0	.2480
200.	1679.	4.7	-11.5	-9.4	-5.1	0.0	5.5	3.6	.3360
250.	1627.	8.0	-21.3	-18.7	-13.3	-6.9	0.0	5.6	.4267
300.	1577.	12.3	-34.4	-31.3	-24.8	-17.2	-8.9	8.3	.5208
350.	1528.	17.5	-50.9	-47.3	-39.7	-30.8	-21.1	11.4	.6171
400.	1480.	23.9	-71.2	-67.0	-58.3	-48.2	-37.1	15.0	.7167
450.	1433.	31.5	-95.5	-90.8	-81.1	-69.7	-57.2	19.3	.8200
500.	1389.	40.4	-124.0	-118.8	-108.0	-95.3	-81.5	24.1	.9262

BALLISTIC COEFFICIENT .66, MUZZLE VELOCITY FT/SEC 800.

RANGE	VELOCITY	HEIGHT	50 YD	100 YD	150 YD	200 YD	250 YD	DRIFT	TIME
0.	800.	0.0	-1.5	-1.5	-1.5	-1.5	-1.5	0.0	0.0000
50.	791.	1.0	0.0	6.1	13.0	20.0	27.3	.4	.1898
100.	781.	6.2	-12.2	0.0	13.8	27.8	42.3	.8	.3797
150.	772.	15.0	-39.0	-20.6	0.0	21.0	42.8	2.0	.5737
200.	763.	27.6	-80.0	-55.5	-28.0	0.0	29.0	3.3	.7685
250.	754.	44.1	-136.3	-105.7	-71.3	-36.3	0.0	5.1	.9667
300.	745.	64.6	-207.7	-171.0	-129.7	-87.7	-44.2	7.3	1.1667
350.	736.	89.2	-294.8	-251.9	-203.8	-154.7	-104.0	9.9	1.3690
400.	727.	118.3	-398.3	-349.4	-294.3	-238.3	-180.2	13.1	1.5746
450.	719.	151.5	-517.4	-462.3	-400.4	-337.3	-272.0	16.5	1.7810
500.	710.	189.8	-654.9	-593.7	-524.9	-454.8	-382.3	20.7	1.9924

BALLISTIC COEFFICIENT .66, MUZZLE VELOCITY FT/SEC 900.

RANGE	VELOCITY	HEIGHT	50 YD	100 YD	150 YD	200 YD	250 YD	DRIFT	TIME
0.	900.	0.0	-1.5	-1.5	-1.5	-1.5	-1.5	0.0	0.0000
50.	889.	.6	0.0	4.7	10.1	15.6	21.3	.4	.1688
100.	878.	4.7	-9.4	0.0	10.8	21.8	33.3	.8	.3381
150.	868.	11.8	-30.3	-16.1	0.0	16.6	33.7	1.9	.5106
200.	858.	21.7	-62.5	-43.6	-22.1	0.0	22.9	3.0	.6838
250.	848.	34.8	-106.7	-83.1	-56.2	-28.6	0.0	4.7	.8602
300.	838.	50.9	-162.6	-134.4	-102.1	-69.0	-34.6	6.6	1.0373
350.	829.	70.5	-231.6	-198.6	-160.9	-122.3	-82.2	9.0	1.2181
400.	819.	93.2	-312.4	-274.7	-231.6	-187.5	-141.7	11.6	1.3992
450.	810.	119.8	-407.3	-364.9	-316.4	-266.8	-215.2	14.9	1.5845
500.	800.	149.6	-514.3	-467.2	-413.3	-358.2	-300.9	18.2	1.7699

BALLISTIC COEFFICIENT .66, MUZZLE VELOCITY FT/SEC 1000.

RANGE	VELOCITY	HEIGHT	50 YD	100 YD	150 YD	200 YD	250 YD	DRIFT	TIME
0.	1000.	0.0	-1.5	-1.5	-1.5	-1.5	-1.5	0.0	0.0000
50.	986.	.4	0.0	3.5	7.8	12.3	17.0	.8	.1546
100.	974.	3.7	-7.1	0.0	8.5	17.6	26.9	.8	.3045
150.	961.	9.4	-23.4	-12.8	0.0	13.6	27.5	1.6	.4592
200.	948.	17.5	-49.3	-35.2	-18.1	0.0	18.5	3.0	.6169
250.	936.	28.2	-84.8	-67.2	-45.8	-23.2	0.0	4.6	.7760
300.	925.	41.4	-130.2	-109.0	-83.4	-56.2	-28.4	6.5	.9369
350.	913.	57.4	-185.9	-161.2	-131.3	-99.5	-67.1	8.9	1.1003
400.	902.	76.1	-251.8	-223.5	-189.4	-153.1	-116.0	11.5	1.2652
450.	891.	97.9	-329.0	-297.2	-258.8	-218.0	-176.3	14.7	1.4333
500.	880.	122.5	-416.4	-381.1	-338.4	-293.0	-246.7	18.0	1.6021

BALLISTIC COEFFICIENT .66, MUZZLE VELOCITY FT/SEC 1100.

RANGE	VELOCITY	HEIGHT	50 YD	100 YD	150 YD	200 YD	250 YD	DRIFT	TIME
0.	1100.	0.0	-1.5	-1.5	-1.5	-1.5	-1.5	0.0	0.0000
50.	1081.	.2	0.0	3.1	6.8	10.3	14.1	.3	.1381
100.	1063.	3.0	-6.2	0.0	7.3	14.3	22.1	1.3	.2804
150.	1047.	7.9	-20.3	-11.0	0.0	10.5	22.2	2.7	.4247
200.	1033.	14.5	-41.0	-28.7	-14.0	0.0	15.5	3.3	.5641
250.	1018.	23.5	-70.7	-55.3	-36.9	-19.4	0.0	5.0	.7104
300.	1004.	34.8	-109.5	-90.9	-68.9	-47.9	-24.6	7.5	.8608
350.	991.	48.6	-157.5	-135.9	-110.2	-85.7	-58.5	10.5	1.0145
400.	978.	64.0	-211.1	-186.3	-157.0	-129.0	-97.9	12.4	1.1615
450.	965.	82.7	-276.8	-249.0	-216.0	-184.5	-149.5	16.0	1.3183
500.	952.	103.3	-349.8	-318.9	-282.2	-247.2	-208.4	19.2	1.4726

BALLISTIC COEFFICIENT .66, MUZZLE VELOCITY FT/SEC 1200.

RANGE	VELOCITY	HEIGHT	50 YD	100 YD	150 YD	200 YD	250 YD	DRIFT	TIME
0.	1200.	0.0	-1.5	-1.5	-1.5	-1.5	-1.5	0.0	0.0000
50.	1170.	.0	0.0	2.4	5.4	8.7	12.1	.3	.1266
100.	1144.	2.4	-4.8	0.0	6.0	12.5	19.3	1.1	.2562
150.	1120.	6.5	-16.3	-9.1	0.0	9.7	19.9	2.4	.3888
200.	1099.	12.4	-34.8	-25.1	-13.0	0.0	13.6	4.2	.5239
250.	1080.	20.3	-60.4	-48.3	-33.2	-17.0	0.0	6.5	.6617
300.	1063.	30.3	-94.1	-79.6	-61.5	-42.0	-21.7	9.5	.8040
350.	1047.	42.4	-135.8	-118.9	-97.7	-75.0	-51.3	12.9	.9485
400.	1032.	56.1	-182.9	-163.6	-139.4	-113.4	-86.3	15.5	1.0883
450.	1018.	72.4	-239.7	-217.9	-190.7	-161.5	-131.0	19.3	1.2346
500.	1004.	91.3	-305.7	-281.5	-251.3	-218.8	-184.9	23.7	1.3849

BALLISTIC COEFFICIENT .66, MUZZLE VELOCITY FT/SEC 1300.

RANGE	VELOCITY	HEIGHT	50 YD	100 YD	150 YD	200 YD	250 YD	DRIFT	TIME
0.	1300.	0.0	-1.5	-1.5	-1.5	-1.5	-1.5	0.0	0.0000
50.	1264.	-.1	0.0	2.0	4.6	7.4	10.3	.3	.1171
100.	1231.	2.0	-4.0	0.0	5.2	10.8	16.7	1.2	.2377
150.	1199.	5.5	-13.7	-7.8	0.0	8.4	17.2	2.7	.3615
200.	1169.	10.7	-29.4	-21.5	-11.2	0.0	11.8	4.7	.4880
250.	1143.	17.6	-51.6	-41.7	-28.7	-14.8	0.0	7.2	.6176
300.	1119.	26.3	-80.5	-68.6	-53.1	-36.3	-18.6	10.3	.7507
350.	1099.	36.9	-116.2	-102.3	-84.2	-64.7	-44.0	13.7	.8857
400.	1080.	49.5	-159.1	-143.2	-122.5	-100.2	-76.5	17.6	1.0232
450.	1062.	64.4	-210.2	-192.4	-169.1	-144.0	-117.4	22.3	1.1653
500.	1046.	81.6	-269.5	-249.7	-223.8	-195.9	-166.4	27.5	1.3100

BALLISTIC COEFFICIENT .66, MUZZLE VELOCITY FT/SEC 1400.

RANGE	VELOCITY	HEIGHT	50 YD	100 YD	150 YD	200 YD	250 YD	DRIFT	TIME
0.	1400.	0.0	-1.5	-1.5	-1.5	-1.5	-1.5	0.0	0.0000
50.	1359.	-.2	0.0	1.6	3.8	6.2	8.8	.4	.1092
100.	1320.	1.6	-3.2	0.0	4.4	9.3	14.5	1.1	.2208
150.	1283.	4.7	-11.4	-6.6	0.0	7.3	15.1	2.6	.3362
200.	1248.	9.2	-24.9	-18.5	-9.7	0.0	10.4	4.6	.4549
250.	1215.	15.2	-44.1	-36.1	-25.1	-12.9	0.0	7.3	.5770
300.	1184.	22.9	-69.0	-59.4	-46.2	-31.6	-16.1	10.3	.7017
350.	1156.	32.4	-100.7	-89.6	-74.1	-57.1	-39.0	14.4	.8316
400.	1132.	43.6	-138.0	-125.3	-107.6	-88.2	-67.5	18.3	.9611
450.	1110.	56.8	-182.6	-168.3	-148.4	-126.6	-103.3	23.0	1.0952
500.	1090.	72.1	-234.1	-218.3	-196.2	-171.9	-146.0	28.2	1.2316

BALLISTIC COEFFICIENT .66, MUZZLE VELOCITY FT/SEC 1500.

RANGE	VELOCITY	HEIGHT	50 YD	100 YD	150 YD	200 YD	250 YD	DRIFT	TIME
0.	1500.	0.0	-1.5	-1.5	-1.5	-1.5	-1.5	0.0	0.0000
50.	1453.	-.2	0.0	1.3	3.2	5.3	7.6	.4	.1021
100.	1409.	1.3	-2.6	0.0	3.8	8.0	12.6	1.2	.2066
150.	1367.	4.0	-9.6	-5.7	0.0	6.3	13.1	2.6	.3150
200.	1328.	8.0	-21.3	-16.1	-8.4	0.0	9.0	4.6	.4262
250.	1291.	13.3	-37.9	-31.4	-21.8	-11.3	0.0	7.2	.5407
300.	1255.	20.1	-59.8	-52.0	-40.6	-27.9	-14.4	10.4	.6590
350.	1222.	28.4	-87.0	-77.9	-64.6	-49.8	-34.0	14.0	.7796
400.	1191.	38.5	-120.3	-109.9	-94.6	-77.8	-59.7	18.4	.9043
450.	1162.	50.3	-159.6	-147.9	-130.7	-111.7	-91.4	23.2	1.0317
500.	1136.	64.3	-206.1	-193.2	-174.1	-153.0	-130.4	28.9	1.1642

BALLISTIC COEFFICIENT .66, MUZZLE VELOCITY FT/SEC 1600.

RANGE	VELOCITY	HEIGHT	50 YD	100 YD	150 YD	200 YD	250 YD	DRIFT	TIME
0.	1600.	0.0	-1.5	-1.5	-1.5	-1.5	-1.5	0.0	0.0000
50.	1551.	-.3	0.0	1.0	2.7	4.5	6.5	.3	.0952
100.	1503.	1.0	-2.1	0.0	3.3	7.0	11.0	1.0	.1931
150.	1456.	3.4	-8.1	-5.0	0.0	5.5	11.5	2.4	.2949
200.	1412.	6.9	-18.1	-14.0	-7.4	0.0	8.0	4.3	.3992
250.	1370.	11.6	-32.6	-27.4	-19.2	-10.0	0.0	6.8	.5073
300.	1330.	17.6	-51.7	-45.5	-35.6	-24.5	-12.6	9.8	.6184
350.	1293.	25.1	-75.9	-68.6	-57.1	-44.2	-30.2	13.6	.7334
400.	1258.	34.0	-105.0	-96.7	-83.4	-68.7	-52.8	17.7	.8505
450.	1224.	44.5	-139.7	-130.3	-115.4	-98.9	-80.9	22.4	.9712
500.	1192.	57.0	-180.7	-170.3	-153.8	-135.4	-115.5	28.1	1.0969

BALLISTIC COEFFICIENT .66, MUZZLE VELOCITY FT/SEC 1700.

RANGE	VELOCITY	HEIGHT	50 YD	100 YD	150 YD	200 YD	250 YD	DRIFT	TIME
0.	1700.	0.0	-1.5	-1.5	-1.5	-1.5	-1.5	0.0	0.0000
50.	1648.	-.4	0.0	.9	2.3	3.9	5.6	.3	.0899
100.	1598.	.8	-1.7	0.0	2.8	6.0	9.6	1.0	.1824
150.	1549.	3.0	-6.8	-4.3	0.0	4.8	10.1	2.3	.2777
200.	1501.	6.0	-15.5	-12.1	-6.4	0.0	7.0	4.0	.3757
250.	1455.	10.2	-28.2	-23.9	-16.8	-8.8	0.0	6.4	.4776
300.	1410.	15.5	-44.9	-39.8	-31.2	-21.6	-11.1	9.3	.5821
350.	1368.	22.1	-66.1	-60.1	-50.2	-38.9	-26.6	12.8	.6903
400.	1329.	30.1	-91.8	-85.0	-73.7	-60.8	-46.7	16.8	.8015
450.	1292.	39.5	-122.5	-114.8	-102.1	-87.6	-71.8	21.4	.9159
500.	1256.	50.6	-158.6	-150.0	-135.8	-119.8	-102.2	26.7	1.0340

BALLISTIC COEFFICIENT .66, MUZZLE VELOCITY FT/SEC 1800.

RANGE	VELOCITY	HEIGHT	50 YD	100 YD	150 YD	200 YD	250 YD	DRIFT	TIME
0.	1800.	0.0	-1.5	-1.5	-1.5	-1.5	-1.5	0.0	0.0000
50.	1746.	-.4	0.0	.7	1.9	3.4	4.9	.2	.0846
100.	1693.	.7	-1.3	0.0	2.5	5.4	8.4	.9	.1718
150.	1642.	2.5	-5.8	-3.7	0.0	4.3	8.9	2.1	.2619
200.	1592.	5.3	-13.4	-10.7	-5.7	0.0	6.1	3.8	.3549
250.	1543.	9.0	-24.4	-21.1	-14.9	-7.7	0.0	5.9	.4504
300.	1495.	13.7	-39.1	-35.1	-27.6	-19.0	-9.8	8.6	.5489
350.	1449.	19.6	-57.9	-53.1	-44.4	-34.4	-23.6	12.0	.6513
400.	1405.	26.7	-80.6	-75.2	-65.2	-53.7	-41.5	15.8	.7562
450.	1363.	35.2	-107.9	-101.8	-90.6	-77.7	-63.9	20.2	.8648
500.	1324.	45.0	-139.7	-133.0	-120.5	-106.2	-90.8	25.2	.9763

BALLISTIC COEFFICIENT .66, MUZZLE VELOCITY FT/SEC 1900.

RANGE	VELOCITY	HEIGHT	50 YD	100 YD	150 YD	200 YD	250 YD	DRIFT	TIME
0.	1900.	0.0	-1.5	-1.5	-1.5	-1.5	-1.5	0.0	0.0000
50.	1844.	-.4	0.0	.5	1.6	2.9	4.2	.3	.0805
100.	1789.	.5	-1.0	0.0	2.2	4.7	7.4	.9	.1629
150.	1735.	2.2	-4.8	-3.2	0.0	3.8	7.9	1.9	.2479
200.	1682.	4.7	-11.5	-9.4	-5.1	0.0	5.5	3.5	.3357
250.	1631.	8.0	-21.2	-18.6	-13.2	-6.9	0.0	5.5	.4263
300.	1581.	12.2	-34.3	-31.2	-24.7	-17.1	-8.8	8.2	.5200
350.	1533.	17.5	-50.8	-47.1	-39.6	-30.7	-21.1	11.2	.6160
400.	1485.	23.8	-70.9	-66.7	-58.1	-48.0	-37.0	14.7	.7153
450.	1439.	31.4	-95.2	-90.5	-80.8	-69.4	-57.0	19.0	.8183
500.	1396.	40.2	-123.6	-118.4	-107.6	-94.9	-81.1	23.7	.9242

BALLISTIC COEFFICIENT .67, MUZZLE VELOCITY FT/SEC 800.

RANGE	VELOCITY	HEIGHT	50 YD	100 YD	150 YD	200 YD	250 YD	DRIFT	TIME
0.	800.	0.0	-1.5	-1.5	-1.5	-1.5	-1.5	0.0	0.0000
50.	791.	1.0	0.0	6.1	13.0	20.0	27.2	.4	.1898
100.	782.	6.2	-12.2	0.0	13.7	27.8	42.2	.8	.3797
150.	773.	15.0	-39.0	-20.6	0.0	21.0	42.7	1.9	.5735
200.	764.	27.6	-80.0	-55.5	-28.0	0.0	28.9	3.2	.7683
250.	755.	44.1	-136.2	-105.6	-71.2	-36.2	0.0	5.0	.9662
300.	746.	64.5	-207.6	-170.9	-129.7	-87.6	-44.2	7.2	1.1662
350.	737.	89.1	-294.4	-251.5	-203.4	-154.4	-103.7	9.8	1.3680
400.	728.	118.1	-397.9	-348.9	-294.0	-237.9	-180.1	12.9	1.5735
450.	720.	151.2	-516.5	-461.4	-399.6	-336.5	-271.4	16.1	1.7792
500.	711.	189.4	-654.0	-592.7	-524.1	-454.0	-381.6	20.3	1.9905

BALLISTIC COEFFICIENT .67, MUZZLE VELOCITY FT/SEC 900.

RANGE	VELOCITY	HEIGHT	50 YD	100 YD	150 YD	200 YD	250 YD	DRIFT	TIME
0.	900.	0.0	-1.5	-1.5	-1.5	-1.5	-1.5	0.0	0.0000
50.	889.	.6	0.0	4.7	10.1	15.6	21.3	.4	.1688
100.	879.	4.7	-9.4	0.0	10.8	21.8	33.2	.8	.3380
150.	869.	11.8	-30.3	-16.2	0.0	16.5	33.7	1.8	.5105
200.	859.	21.7	-62.4	-43.6	-22.0	0.0	22.9	3.0	.6835
250.	849.	34.7	-106.7	-83.1	-56.2	-28.7	0.0	4.7	.8599
300.	839.	50.8	-162.4	-134.2	-101.9	-68.8	-34.4	6.4	1.0366
350.	830.	70.4	-231.4	-198.4	-160.7	-122.2	-82.1	8.9	1.2173
400.	820.	93.1	-312.1	-274.4	-231.3	-187.3	-141.4	11.4	1.3982
450.	811.	119.6	-406.8	-364.4	-316.0	-266.4	-214.8	14.6	1.5832
500.	802.	149.4	-513.7	-466.6	-412.8	-357.7	-300.4	17.9	1.7685

BALLISTIC COEFFICIENT .67, MUZZLE VELOCITY FT/SEC 1000.

RANGE	VELOCITY	HEIGHT	50 YD	100 YD	150 YD	200 YD	250 YD	DRIFT	TIME
0.	1000.	0.0	-1.5	-1.5	-1.5	-1.5	-1.5	0.0	0.0000
50.	987.	.4	0.0	3.5	7.8	12.3	16.9	.8	.1547
100.	974.	3.7	-7.0	0.0	8.6	17.6	26.8	.8	.3044
150.	961.	9.4	-23.4	-12.9	0.0	13.5	27.4	1.7	.4594
200.	949.	17.5	-49.3	-35.2	-18.0	0.0	18.5	2.9	.6166
250.	937.	28.1	-84.7	-67.1	-45.7	-23.1	0.0	4.5	.7755
300.	926.	41.3	-130.0	-108.9	-83.2	-56.2	-28.4	6.4	.9365
350.	914.	57.3	-185.6	-160.9	-130.9	-99.4	-67.0	8.7	1.0995
400.	903.	76.0	-251.4	-223.3	-189.0	-152.9	-115.9	11.3	1.2644
450.	892.	97.7	-328.4	-296.7	-258.1	-217.6	-176.0	14.4	1.4321
500.	882.	122.3	-415.8	-380.6	-337.7	-292.7	-246.4	17.7	1.6008

BALLISTIC COEFFICIENT .67, MUZZLE VELOCITY FT/SEC 1100.

RANGE	VELOCITY	HEIGHT	50 YD	100 YD	150 YD	200 YD	250 YD	DRIFT	TIME
0.	1100.	0.0	-1.5	-1.5	-1.5	-1.5	-1.5	0.0	0.0000
50.	1081.	.2	0.0	3.1	6.8	10.2	14.1	.3	.1383
100.	1064.	3.0	-6.2	0.0	7.3	14.3	22.0	1.4	.2808
150.	1048.	7.9	-20.3	-11.0	0.0	10.4	22.0	2.8	.4251
200.	1034.	14.5	-41.0	-28.5	-13.9	0.0	15.5	3.2	.5638
250.	1019.	23.4	-70.6	-55.1	-36.7	-19.4	0.0	4.9	.7099
300.	1006.	34.8	-109.5	-90.9	-68.9	-48.0	-24.8	7.5	.8609
350.	992.	48.6	-157.5	-135.8	-110.1	-85.8	-58.7	10.5	1.0144
400.	979.	63.9	-210.6	-185.8	-156.4	-128.7	-97.7	12.2	1.1604
450.	966.	82.6	-276.8	-248.9	-215.9	-184.6	-149.8	16.0	1.3180
500.	954.	103.1	-349.0	-318.0	-281.3	-246.6	-207.9	18.9	1.4709

BALLISTIC COEFFICIENT .67, MUZZLE VELOCITY FT/SEC 1200.

RANGE	VELOCITY	HEIGHT	50 YD	100 YD	150 YD	200 YD	250 YD	DRIFT	TIME
0.	1200.	0.0	-1.5	-1.5	-1.5	-1.5	-1.5	0.0	0.0000
50.	1171.	.0	0.0	2.4	5.4	8.7	12.1	.3	.1266
100.	1145.	2.4	-4.8	0.0	6.0	12.5	19.4	1.1	.2561
150.	1121.	6.5	-16.3	-9.1	0.0	9.7	20.0	2.4	.3886
200.	1101.	12.4	-34.7	-25.0	-13.0	0.0	13.8	4.2	.5236
250.	1082.	20.3	-60.6	-48.5	-33.4	-17.2	0.0	6.6	.6623
300.	1064.	30.3	-94.4	-79.9	-61.8	-42.3	-21.7	9.6	.8047
350.	1048.	42.5	-136.0	-119.1	-97.9	-75.2	-51.2	13.0	.9488
400.	1034.	56.0	-182.6	-163.2	-139.1	-113.2	-85.6	15.3	1.0871
450.	1020.	72.2	-239.1	-217.4	-190.2	-161.0	-130.1	19.0	1.2331
500.	1006.	91.2	-305.6	-281.4	-251.2	-218.8	-184.4	23.6	1.3842

BALLISTIC COEFFICIENT .67, MUZZLE VELOCITY FT/SEC 1300.

RANGE	VELOCITY	HEIGHT	50 YD	100 YD	150 YD	200 YD	250 YD	DRIFT	TIME
0.	1300.	0.0	-1.5	-1.5	-1.5	-1.5	-1.5	0.0	0.0000
50.	1265.	-.1	0.0	2.0	4.5	7.3	10.3	.3	.1171
100.	1232.	2.0	-4.0	0.0	5.1	10.7	16.6	1.2	.2376
150.	1200.	5.5	-13.6	-7.7	0.0	8.4	17.2	2.6	.3610
200.	1171.	10.7	-29.4	-21.5	-11.2	0.0	11.8	4.6	.4876
250.	1145.	17.5	-51.5	-41.6	-28.7	-14.7	0.0	7.1	.6171
300.	1122.	26.2	-80.3	-68.4	-53.0	-36.2	-18.5	10.1	.7496
350.	1101.	36.8	-115.9	-102.1	-84.1	-64.5	-43.8	13.5	.8846
400.	1082.	49.5	-159.2	-143.4	-122.8	-100.4	-76.8	17.6	1.0233
450.	1064.	64.5	-210.6	-192.8	-169.6	-144.4	-117.9	22.4	1.1657
500.	1048.	81.6	-269.6	-249.8	-224.1	-196.1	-166.6	27.4	1.3097

BALLISTIC COEFFICIENT .67, MUZZLE VELOCITY FT/SEC 1400.

RANGE	VELOCITY	HEIGHT	50 YD	100 YD	150 YD	200 YD	250 YD	DRIFT	TIME
0.	1400.	0.0	-1.5	-1.5	-1.5	-1.5	-1.5	0.0	0.0000
50.	1359.	-.2	0.0	1.6	3.8	6.2	8.8	.3	.1091
100.	1321.	1.6	-3.2	0.0	4.4	9.3	14.4	1.1	.2207
150.	1285.	4.7	-11.4	-6.6	0.0	7.3	15.0	2.6	.3360
200.	1251.	9.2	-24.9	-18.5	-9.7	0.0	10.3	4.5	.4544
250.	1218.	15.2	-43.9	-36.0	-24.9	-12.8	0.0	7.1	.5760
300.	1187.	22.8	-68.8	-59.3	-46.1	-31.6	-16.2	10.2	.7009
350.	1159.	32.2	-100.1	-88.9	-73.5	-56.6	-38.6	13.9	.8291
400.	1135.	43.4	-137.6	-124.9	-107.3	-87.9	-67.3	18.0	.9596
450.	1113.	56.6	-182.1	-167.8	-147.9	-126.1	-103.0	22.7	1.0933
500.	1093.	71.8	-233.5	-217.6	-195.5	-171.3	-145.6	27.8	1.2294

BALLISTIC COEFFICIENT .67, MUZZLE VELOCITY FT/SEC 1500.

RANGE	VELOCITY	HEIGHT	50 YD	100 YD	150 YD	200 YD	250 YD	DRIFT	TIME
0.	1500.	0.0	-1.5	-1.5	-1.5	-1.5	-1.5	0.0	0.0000
50.	1454.	-.2	0.0	1.3	3.2	5.3	7.6	.4	.1021
100.	1410.	1.3	-2.6	0.0	3.8	8.0	12.6	1.2	.2065
150.	1369.	4.0	-9.6	-5.7	0.0	6.3	13.1	2.6	.3147
200.	1330.	8.0	-21.2	-16.1	-8.4	0.0	9.1	4.5	.4258
250.	1293.	13.3	-37.9	-31.4	-21.9	-11.4	0.0	7.2	.5408
300.	1259.	20.0	-59.5	-51.8	-40.3	-27.7	-14.0	10.2	.6577
350.	1226.	28.3	-86.8	-77.7	-64.4	-49.7	-33.7	13.8	.7785
400.	1194.	38.5	-120.2	-109.9	-94.6	-77.8	-59.6	18.3	.9037
450.	1166.	50.2	-159.1	-147.4	-130.3	-111.3	-90.8	22.9	1.0298
500.	1140.	63.8	-204.6	-191.6	-172.6	-151.5	-128.8	28.1	1.1597

BALLISTIC COEFFICIENT .67, MUZZLE VELOCITY FT/SEC 1600.

RANGE	VELOCITY	HEIGHT	50 YD	100 YD	150 YD	200 YD	250 YD	DRIFT	TIME
0.	1600.	0.0	-1.5	-1.5	-1.5	-1.5	-1.5	0.0	0.0000
50.	1552.	-.3	0.0	1.0	2.7	4.5	6.5	.3	.0952
100.	1504.	1.0	-2.1	0.0	3.3	7.0	10.9	1.0	.1930
150.	1458.	3.4	-8.1	-4.9	0.0	5.5	11.5	2.4	.2947
200.	1414.	6.9	-18.1	-13.9	-7.3	0.0	7.9	4.2	.3988
250.	1373.	11.6	-32.5	-27.3	-19.1	-9.9	0.0	6.7	.5067
300.	1334.	17.6	-51.7	-45.5	-35.6	-24.6	-12.7	9.8	.6182
350.	1297.	25.0	-75.6	-68.3	-56.7	-43.9	-30.0	13.3	.7318
400.	1262.	33.8	-104.6	-96.2	-83.0	-68.4	-52.5	17.4	.8486
450.	1229.	44.4	-139.2	-129.9	-115.0	-98.5	-80.7	22.1	.9694
500.	1197.	56.7	-179.8	-169.3	-152.8	-134.5	-114.7	27.5	1.0937

BALLISTIC COEFFICIENT .67, MUZZLE VELOCITY FT/SEC 1700.

RANGE	VELOCITY	HEIGHT	50 YD	100 YD	150 YD	200 YD	250 YD	DRIFT	TIME
0.	1700.	0.0	-1.5	-1.5	-1.5	-1.5	-1.5	0.0	0.0000
50.	1649.	-.4	0.0	.9	2.3	3.9	5.6	.3	.0899
100.	1600.	.8	-1.7	0.0	2.8	6.0	9.5	1.0	.1823
150.	1551.	2.9	-6.8	-4.3	0.0	4.8	10.1	2.3	.2775
200.	1504.	6.0	-15.5	-12.1	-6.4	0.0	7.0	3.9	.3754
250.	1458.	10.2	-28.1	-23.9	-16.8	-8.8	0.0	6.3	.4771
300.	1414.	15.5	-44.8	-39.6	-31.1	-21.5	-11.0	9.1	.5812
350.	1373.	22.0	-65.9	-59.9	-50.0	-38.7	-26.5	12.6	.6891
400.	1333.	30.0	-91.7	-84.9	-73.6	-60.7	-46.7	16.7	.8006
450.	1296.	39.4	-122.2	-114.5	-101.7	-87.3	-71.5	21.2	.9143
500.	1262.	50.3	-157.8	-149.2	-135.1	-119.0	-101.5	26.2	1.0311

BALLISTIC COEFFICIENT .67, MUZZLE VELOCITY FT/SEC 1800.

RANGE	VELOCITY	HEIGHT	50 YD	100 YD	150 YD	200 YD	250 YD	DRIFT	TIME
0.	1800.	0.0	-1.5	-1.5	-1.5	-1.5	-1.5	0.0	0.0000
50.	1747.	-.4	0.0	.7	1.9	3.3	4.9	.2	.0846
100.	1695.	.7	-1.3	0.0	2.5	5.3	8.4	.9	.1717
150.	1644.	2.5	-5.7	-3.7	0.0	4.3	8.9	2.1	.2617
200.	1595.	5.3	-13.4	-10.7	-5.7	0.0	6.1	3.7	.3546
250.	1546.	9.0	-24.4	-21.0	-14.8	-7.7	0.0	5.9	.4499
300.	1499.	13.7	-39.0	-35.0	-27.5	-18.9	-9.7	8.5	.5481
350.	1454.	19.5	-57.7	-53.0	-44.3	-34.3	-23.5	11.8	.6502
400.	1410.	26.6	-80.3	-74.9	-65.0	-53.5	-41.3	15.5	.7547
450.	1369.	35.0	-107.5	-101.4	-90.2	-77.3	-63.5	19.9	.8629
500.	1330.	44.8	-139.2	-132.4	-120.0	-105.7	-90.4	24.8	.9741

BALLISTIC COEFFICIENT .67, MUZZLE VELOCITY FT/SEC 1900.

RANGE	VELOCITY	HEIGHT	50 YD	100 YD	150 YD	200 YD	250 YD	DRIFT	TIME
0.	1900.	0.0	-1.5	-1.5	-1.5	-1.5	-1.5	0.0	0.0000
50.	1844.	-.4	0.0	.5	1.6	2.9	4.2	.3	.0804
100.	1790.	.5	-1.0	0.0	2.2	4.7	7.4	.9	.1628
150.	1737.	2.2	-4.8	-3.2	0.0	3.8	7.9	1.9	.2477
200.	1686.	4.6	-11.5	-9.4	-5.1	0.0	5.5	3.4	.3354
250.	1635.	8.0	-21.2	-18.6	-13.2	-6.9	0.0	5.5	.4258
300.	1586.	12.2	-34.2	-31.1	-24.6	-17.0	-8.8	8.0	.5193
350.	1538.	17.4	-50.6	-47.0	-39.4	-30.6	-21.0	11.0	.6150
400.	1491.	23.7	-70.7	-66.5	-57.9	-47.8	-36.8	14.5	.7139
450.	1445.	31.3	-94.9	-90.2	-80.5	-69.1	-56.7	18.7	.8165
500.	1402.	40.0	-123.0	-117.8	-107.0	-94.3	-80.6	23.2	.9216

BALLISTIC COEFFICIENT .68, MUZZLE VELOCITY FT/SEC 800.

RANGE	VELOCITY	HEIGHT	50 YD	100 YD	150 YD	200 YD	250 YD	DRIFT	TIME
0.	800.	0.0	-1.5	-1.5	-1.5	-1.5	-1.5	0.0	0.0000
50.	791.	1.0	0.0	6.1	13.0	20.0	27.2	.4	.1898
100.	782.	6.2	-12.2	0.0	13.7	27.7	42.2	.8	.3797
150.	773.	15.0	-38.9	-20.6	0.0	21.0	42.7	1.9	.5733
200.	764.	27.6	-80.0	-55.5	-28.1	0.0	28.8	3.2	.7681
250.	755.	44.0	-136.0	-105.4	-71.1	-36.1	0.0	5.0	.9656
300.	747.	64.5	-207.4	-170.7	-129.6	-87.5	-44.2	7.2	1.1657
350.	738.	88.9	-294.0	-251.1	-203.1	-154.0	-103.5	9.6	1.3669
400.	730.	117.9	-397.5	-348.5	-293.6	-237.5	-179.9	12.8	1.5725
450.	721.	151.0	-516.0	-460.8	-399.1	-336.0	-271.1	15.9	1.7780
500.	712.	189.1	-653.0	-591.8	-523.2	-453.1	-381.0	20.0	1.9887

BALLISTIC COEFFICIENT .68, MUZZLE VELOCITY FT/SEC 900.

RANGE	VELOCITY	HEIGHT	50 YD	100 YD	150 YD	200 YD	250 YD	DRIFT	TIME
0.	900.	0.0	-1.5	-1.5	-1.5	-1.5	-1.5	0.0	0.0000
50.	890.	.6	0.0	4.7	10.1	15.6	21.3	.4	.1688
100.	879.	4.8	-9.5	0.0	10.7	21.7	33.1	.9	.3386
150.	869.	11.8	-30.3	-16.0	0.0	16.5	33.7	1.8	.5104
200.	859.	21.7	-62.3	-43.3	-22.0	0.0	22.9	2.9	.6832
250.	850.	34.7	-106.6	-82.8	-56.1	-28.7	0.0	4.6	.8595
300.	840.	50.8	-162.2	-133.7	-101.7	-68.7	-34.3	6.3	1.0359
350.	831.	70.3	-231.1	-197.9	-160.5	-122.1	-81.9	8.8	1.2166
400.	821.	93.0	-311.8	-273.8	-231.1	-187.1	-141.3	11.3	1.3974
450.	812.	119.4	-406.3	-363.5	-315.5	-266.0	-214.4	14.4	1.5819
500.	803.	149.2	-513.2	-465.7	-412.3	-357.4	-300.0	17.7	1.7672

BALLISTIC COEFFICIENT .68, MUZZLE VELOCITY FT/SEC 1000.

RANGE	VELOCITY	HEIGHT	50 YD	100 YD	150 YD	200 YD	250 YD	DRIFT	TIME
0.	1000.	0.0	-1.5	-1.5	-1.5	-1.5	-1.5	0.0	0.0000
50.	987.	.4	0.0	3.5	7.8	12.3	16.9	.9	.1548
100.	974.	3.7	-7.0	0.0	8.6	17.6	26.8	.8	.3044
150.	962.	9.4	-23.4	-12.9	0.0	13.4	27.3	1.7	.4597
200.	950.	17.5	-49.2	-35.2	-17.9	0.0	18.5	2.9	.6164
250.	938.	28.1	-84.6	-67.0	-45.5	-23.1	0.0	4.4	.7751
300.	927.	41.3	-129.9	-108.9	-83.0	-56.1	-28.4	6.3	.9360
350.	915.	57.2	-185.3	-160.7	-130.6	-99.2	-66.9	8.6	1.0987
400.	904.	75.9	-251.1	-223.1	-188.6	-152.8	-115.8	11.2	1.2636
450.	893.	98.1	-330.2	-298.7	-259.9	-219.6	-178.0	15.0	1.4353
500.	883.	122.1	-415.2	-380.2	-337.1	-292.2	-246.1	17.5	1.5995

BALLISTIC COEFFICIENT .68, MUZZLE VELOCITY FT/SEC 1100.

RANGE	VELOCITY	HEIGHT	50 YD	100 YD	150 YD	200 YD	250 YD	DRIFT	TIME
0.	1100.	0.0	-1.5	-1.5	-1.5	-1.5	-1.5	0.0	0.0000
50.	1081.	.2	0.0	3.1	6.8	10.2	14.1	.4	.1385
100.	1064.	3.0	-6.2	0.0	7.3	14.2	22.0	1.5	.2812
150.	1049.	7.9	-20.3	-11.0	0.0	10.3	21.9	2.9	.4254
200.	1035.	14.5	-40.9	-28.4	-13.7	0.0	15.5	3.2	.5635
250.	1020.	23.4	-70.5	-54.9	-36.6	-19.4	0.0	4.9	.7097
300.	1007.	34.8	-109.5	-90.8	-68.8	-48.2	-24.9	7.5	.8609
350.	993.	48.6	-157.5	-135.7	-110.0	-86.0	-58.8	10.5	1.0143
400.	981.	63.8	-210.4	-185.5	-156.1	-128.7	-97.6	12.1	1.1598
450.	968.	82.6	-276.7	-248.7	-215.7	-184.8	-149.9	15.9	1.3178
500.	956.	102.9	-348.2	-317.1	-280.4	-246.1	-207.3	18.6	1.4693

BALLISTIC COEFFICIENT .68, MUZZLE VELOCITY FT/SEC 1200.

RANGE	VELOCITY	HEIGHT	50 YD	100 YD	150 YD	200 YD	250 YD	DRIFT	TIME
0.	1200.	0.0	-1.5	-1.5	-1.5	-1.5	-1.5	0.0	0.0000
50.	1171.	.0	0.0	2.4	5.4	8.7	12.0	.3	.1266
100.	1145.	2.4	-4.8	0.0	6.0	12.5	19.2	1.1	.2561
150.	1122.	6.5	-16.3	-9.1	0.0	9.7	19.8	2.4	.3885
200.	1102.	12.4	-34.7	-25.0	-12.9	0.0	13.5	4.1	.5234
250.	1083.	20.2	-60.2	-48.1	-33.0	-16.8	0.0	6.3	.6606
300.	1066.	30.0	-93.2	-78.7	-60.6	-41.1	-20.9	8.8	.8003
350.	1051.	41.9	-133.8	-116.9	-95.7	-73.1	-49.5	11.8	.9421
400.	1036.	55.9	-182.2	-162.9	-138.7	-112.8	-85.9	15.1	1.0859
450.	1022.	72.2	-239.1	-217.3	-190.2	-161.1	-130.7	18.9	1.2326
500.	1008.	91.1	-305.5	-281.3	-251.1	-218.8	-185.1	23.5	1.3836

BALLISTIC COEFFICIENT .68, MUZZLE VELOCITY FT/SEC 1300.

RANGE	VELOCITY	HEIGHT	50 YD	100 YD	150 YD	200 YD	250 YD	DRIFT	TIME
0.	1300.	0.0	-1.5	-1.5	-1.5	-1.5	-1.5	0.0	0.0000
50.	1265.	-.1	0.0	2.0	4.5	7.5	10.3	.3	.1171
100.	1232.	2.0	-4.0	0.0	5.0	10.9	16.5	1.4	.2384
150.	1201.	5.5	-13.6	-7.6	0.0	8.8	17.2	2.6	.3608
200.	1172.	10.8	-29.9	-21.8	-11.7	0.0	11.2	5.1	.4904
250.	1147.	17.5	-51.4	-41.3	-28.7	-14.0	0.0	7.0	.6166
300.	1124.	26.2	-80.1	-68.0	-52.9	-35.2	-18.4	10.0	.7489
350.	1103.	36.7	-115.7	-101.6	-83.9	-63.4	-43.7	13.4	.8837
400.	1084.	49.3	-158.4	-142.3	-122.1	-98.6	-76.2	17.2	1.0208
450.	1067.	63.9	-208.5	-190.3	-167.6	-141.2	-116.0	21.5	1.1603
500.	1051.	80.6	-266.2	-246.0	-220.8	-191.4	-163.4	26.1	1.3021

BALLISTIC COEFFICIENT .68, MUZZLE VELOCITY FT/SEC 1400.

RANGE	VELOCITY	HEIGHT	50 YD	100 YD	150 YD	200 YD	250 YD	DRIFT	TIME
0.	1400.	0.0	-1.5	-1.5	-1.5	-1.5	-1.5	0.0	0.0000
50.	1360.	-.2	0.0	1.6	3.8	6.2	8.8	.3	.1090
100.	1322.	1.6	-3.2	0.0	4.4	9.2	14.3	1.1	.2206
150.	1286.	4.7	-11.4	-6.6	0.0	7.2	14.9	2.5	.3358
200.	1252.	9.1	-24.8	-18.5	-9.7	0.0	10.2	4.5	.4540
250.	1220.	15.1	-43.8	-35.8	-24.8	-12.7	0.0	6.9	.5750
300.	1190.	22.8	-68.7	-59.2	-46.0	-31.5	-16.2	10.1	.7001
350.	1162.	32.1	-99.7	-88.6	-73.2	-56.3	-38.4	13.6	.8275
400.	1137.	43.5	-137.8	-125.0	-107.4	-88.1	-67.7	18.0	.9596
450.	1115.	56.4	-181.5	-167.2	-147.4	-125.7	-102.8	22.4	1.0915
500.	1095.	71.5	-232.7	-216.8	-194.9	-170.7	-145.2	27.4	1.2272

BALLISTIC COEFFICIENT .68, MUZZLE VELOCITY FT/SEC 1500.

RANGE	VELOCITY	HEIGHT	50 YD	100 YD	150 YD	200 YD	250 YD	DRIFT	TIME
0.	1500.	0.0	-1.5	-1.5	-1.5	-1.5	-1.5	0.0	0.0000
50.	1455.	-.3	0.0	1.3	3.2	5.3	7.6	.4	.1020
100.	1412.	1.3	-2.6	0.0	3.8	8.0	12.5	1.1	.2064
150.	1371.	4.0	-9.6	-5.7	0.0	6.3	13.1	2.6	.3145
200.	1332.	7.9	-21.2	-16.0	-8.4	0.0	9.0	4.5	.4255
250.	1296.	13.2	-37.8	-31.3	-21.8	-11.3	0.0	7.0	.5399
300.	1262.	19.9	-59.4	-51.6	-40.2	-27.6	-14.0	10.0	.6567
350.	1229.	28.3	-86.6	-77.6	-64.3	-49.5	-33.7	13.6	.7775
400.	1198.	38.3	-119.7	-109.3	-94.1	-77.3	-59.3	17.9	.9016
450.	1169.	50.0	-158.6	-147.0	-129.8	-110.9	-90.6	22.5	1.0280
500.	1144.	63.6	-204.0	-191.0	-172.0	-151.0	-128.4	27.7	1.1576

BALLISTIC COEFFICIENT .68, MUZZLE VELOCITY FT/SEC 1600.

RANGE	VELOCITY	HEIGHT	50 YD	100 YD	150 YD	200 YD	250 YD	DRIFT	TIME
0.	1600.	0.0	-1.5	-1.5	-1.5	-1.5	-1.5	0.0	0.0000
50.	1552.	-.3	0.0	1.0	2.7	4.5	6.5	.3	.0952
100.	1506.	1.0	-2.1	0.0	3.3	7.0	10.9	1.0	.1930
150.	1460.	3.4	-8.1	-4.9	0.0	5.5	11.5	2.3	.2944
200.	1417.	6.9	-18.1	-13.9	-7.3	0.0	7.9	4.2	.3986
250.	1376.	11.6	-32.5	-27.3	-19.1	-9.9	0.0	6.6	.5064
300.	1337.	17.5	-51.5	-45.3	-35.4	-24.4	-12.5	9.6	.6170
350.	1300.	24.8	-75.3	-68.0	-56.5	-43.6	-29.7	13.0	.7302
400.	1266.	33.7	-104.3	-96.0	-82.8	-68.1	-52.3	17.1	.8472
450.	1233.	44.3	-139.1	-129.7	-115.0	-98.4	-80.6	22.0	.9685
500.	1202.	56.4	-178.9	-168.5	-152.1	-133.7	-113.9	27.0	1.0908

BALLISTIC COEFFICIENT .68, MUZZLE VELOCITY FT/SEC 1700.

RANGE	VELOCITY	HEIGHT	50 YD	100 YD	150 YD	200 YD	250 YD	DRIFT	TIME
0.	1700.	0.0	-1.5	-1.5	-1.5	-1.5	-1.5	0.0	0.0000
50.	1650.	-.4	0.0	.9	2.3	3.9	5.6	.3	.0898
100.	1601.	.8	-1.7	0.0	2.8	6.0	9.5	1.0	.1822
150.	1553.	2.9	-6.8	-4.2	0.0	4.8	10.0	2.2	.2774
200.	1507.	6.0	-15.5	-12.1	-6.4	0.0	7.0	3.9	.3751
250.	1461.	10.1	-28.0	-23.8	-16.7	-8.7	0.0	6.2	.4764
300.	1418.	15.4	-44.7	-39.5	-31.1	-21.5	-11.0	9.0	.5805
350.	1377.	22.0	-65.7	-59.7	-49.8	-38.6	-26.4	12.4	.6882
400.	1338.	29.9	-91.3	-84.5	-73.2	-60.4	-46.4	16.3	.7987
450.	1301.	39.2	-121.6	-113.9	-101.2	-86.8	-71.1	20.7	.9120
500.	1267.	50.1	-157.2	-148.7	-134.5	-118.6	-101.1	25.8	1.0290

BALLISTIC COEFFICIENT .68, MUZZLE VELOCITY FT/SEC 1800.

RANGE	VELOCITY	HEIGHT	50 YD	100 YD	150 YD	200 YD	250 YD	DRIFT	TIME
0.	1800.	0.0	-1.5	-1.5	-1.5	-1.5	-1.5	0.0	0.0000
50.	1748.	-.4	0.0	.7	1.9	3.3	4.9	.2	.0846
100.	1696.	.7	-1.3	0.0	2.5	5.3	8.4	.9	.1716
150.	1646.	2.5	-5.7	-3.7	0.0	4.3	8.9	2.0	.2616
200.	1598.	5.3	-13.3	-10.7	-5.7	0.0	6.1	3.7	.3542
250.	1550.	8.9	-24.4	-21.0	-14.8	-7.7	0.0	5.8	.4495
300.	1504.	13.6	-38.9	-34.9	-27.5	-18.9	-9.7	8.3	.5474
350.	1458.	19.5	-57.5	-52.8	-44.1	-34.1	-23.4	11.6	.6491
400.	1415.	26.5	-80.0	-74.7	-64.7	-53.3	-41.0	15.2	.7532
450.	1374.	34.8	-107.0	-101.0	-89.8	-77.0	-63.2	19.5	.8610
500.	1335.	44.6	-138.8	-132.1	-119.7	-105.4	-90.1	24.5	.9723

BALLISTIC COEFFICIENT .68, MUZZLE VELOCITY FT/SEC 1900.

RANGE	VELOCITY	HEIGHT	50 YD	100 YD	150 YD	200 YD	250 YD	DRIFT	TIME
0.	1900.	0.0	-1.5	-1.5	-1.5	-1.5	-1.5	0.0	0.0000
50.	1845.	-.4	0.0	.5	1.6	2.9	4.2	.2	.0804
100.	1792.	.5	-1.0	0.0	2.2	4.7	7.4	.8	.1627
150.	1740.	2.2	-4.8	-3.2	0.0	3.8	7.9	1.9	.2476
200.	1689.	4.6	-11.4	-9.3	-5.0	0.0	5.5	3.4	.3351
250.	1639.	7.9	-21.1	-18.5	-13.1	-6.8	0.0	5.4	.4253
300.	1590.	12.2	-34.2	-31.0	-24.5	-17.0	-8.8	7.9	.5185
350.	1543.	17.3	-50.5	-46.8	-39.3	-30.5	-20.9	10.8	.6140
400.	1496.	23.6	-70.5	-66.3	-57.7	-47.6	-36.6	14.2	.7125
450.	1451.	31.1	-94.5	-89.8	-80.1	-68.8	-56.5	18.3	.8148
500.	1408.	39.8	-122.5	-117.3	-106.5	-93.9	-80.2	22.9	.9194

BALLISTIC COEFFICIENT .69, MUZZLE VELOCITY FT/SEC 800.

RANGE	VELOCITY	HEIGHT	50 YD	100 YD	150 YD	200 YD	250 YD	DRIFT	TIME
0.	800.	0.0	-1.5	-1.5	-1.5	-1.5	-1.5	0.0	0.0000
50.	791.	1.0	0.0	6.1	13.0	20.0	27.2	.4	.1898
100.	782.	6.2	-12.3	0.0	13.7	27.7	42.1	.8	.3797
150.	774.	15.0	-38.9	-20.5	0.0	21.0	42.6	1.9	.5732
200.	765.	27.6	-80.0	-55.5	-28.1	0.0	28.8	3.2	.7680
250.	756.	44.0	-135.9	-105.3	-71.0	-35.9	0.0	4.9	.9651
300.	748.	64.4	-207.3	-170.6	-129.5	-87.4	-44.2	7.1	1.1652
350.	739.	88.8	-293.5	-250.7	-202.7	-153.6	-103.3	9.4	1.3659
400.	731.	117.8	-397.1	-348.1	-293.3	-237.2	-179.6	12.6	1.5714
450.	722.	150.8	-515.5	-460.4	-398.8	-335.6	-270.9	15.7	1.7769
500.	714.	188.7	-652.1	-590.8	-522.3	-452.2	-380.3	19.7	1.9869

BALLISTIC COEFFICIENT .69, MUZZLE VELOCITY FT/SEC 900.

RANGE	VELOCITY	HEIGHT	50 YD	100 YD	150 YD	200 YD	250 YD	DRIFT	TIME
0.	900.	0.0	-1.5	-1.5	-1.5	-1.5	-1.5	0.0	0.0000
50.	890.	.6	0.0	4.7	10.1	15.6	21.3	.4	.1688
100.	879.	4.7	-9.5	0.0	10.7	21.7	33.1	.9	.3383
150.	870.	11.7	-30.3	-16.1	0.0	16.5	33.7	1.8	.5102
200.	860.	21.7	-62.3	-43.4	-22.0	0.0	22.9	2.9	.6831
250.	850.	34.7	-106.5	-82.9	-56.1	-28.6	0.0	4.6	.8592
300.	841.	50.7	-162.1	-133.8	-101.6	-68.6	-34.3	6.3	1.0355
350.	832.	70.2	-230.9	-197.8	-160.3	-121.8	-81.8	8.7	1.2159
400.	823.	92.9	-311.6	-273.7	-230.9	-186.9	-141.1	11.1	1.3967
450.	814.	119.2	-405.7	-363.2	-315.0	-265.5	-214.0	14.2	1.5807
500.	805.	148.9	-512.6	-465.4	-411.8	-356.8	-299.6	17.5	1.7660

BALLISTIC COEFFICIENT .69, MUZZLE VELOCITY FT/SEC 1000.

RANGE	VELOCITY	HEIGHT	50 YD	100 YD	150 YD	200 YD	250 YD	DRIFT	TIME
0.	1000.	0.0	-1.5	-1.5	-1.5	-1.5	-1.5	0.0	0.0000
50.	987.	.4	0.0	3.5	7.8	12.3	16.9	.9	.1550
100.	975.	3.7	-7.0	0.0	8.7	17.6	26.8	.8	.3043
150.	962.	9.4	-23.5	-13.0	0.0	13.4	27.2	1.7	.4599
200.	951.	17.5	-49.1	-35.1	-17.8	0.0	18.4	2.8	.6161
250.	939.	28.1	-84.4	-66.9	-45.3	-23.0	0.0	4.3	.7746
300.	928.	41.3	-129.7	-108.8	-82.8	-56.1	-28.4	6.3	.9356
350.	916.	57.1	-184.9	-160.5	-130.2	-99.0	-66.8	8.4	1.0979
400.	906.	75.8	-250.8	-222.8	-188.2	-152.6	-115.7	11.0	1.2627
450.	895.	97.8	-329.1	-297.7	-258.7	-218.6	-177.2	14.6	1.4331
500.	885.	121.9	-414.6	-379.7	-336.4	-291.8	-245.8	17.3	1.5982

BALLISTIC COEFFICIENT .69, MUZZLE VELOCITY FT/SEC 1100.

RANGE	VELOCITY	HEIGHT	50 YD	100 YD	150 YD	200 YD	250 YD	DRIFT	TIME
0.	1100.	0.0	-1.5	-1.5	-1.5	-1.5	-1.5	0.0	0.0000
50.	1082.	.2	0.0	3.1	6.4	10.2	14.1	.4	.1388
100.	1065.	3.1	-6.3	0.0	6.6	14.1	21.9	1.6	.2816
150.	1050.	7.7	-19.3	-10.0	0.0	11.2	23.0	1.8	.4193
200.	1035.	14.5	-40.8	-28.3	-15.0	0.0	15.6	3.1	.5632
250.	1021.	23.4	-70.5	-54.9	-38.3	-19.5	0.0	4.9	.7098
300.	1008.	34.8	-109.5	-90.7	-70.8	-48.3	-24.9	7.5	.8610
350.	995.	47.9	-154.8	-132.9	-109.6	-83.4	-56.1	9.3	1.0071
400.	982.	63.8	-210.4	-185.4	-158.8	-128.9	-97.6	12.1	1.1598
450.	970.	82.6	-276.7	-248.6	-218.7	-185.0	-149.8	15.9	1.3176
500.	958.	102.6	-347.4	-316.2	-283.0	-245.5	-206.5	18.3	1.4676

BALLISTIC COEFFICIENT .69, MUZZLE VELOCITY FT/SEC 1200.

RANGE	VELOCITY	HEIGHT	50 YD	100 YD	150 YD	200 YD	250 YD	DRIFT	TIME
0.	1200.	0.0	-1.5	-1.5	-1.5	-1.5	-1.5	0.0	0.0000
50.	1172.	.0	0.0	2.4	5.4	8.7	12.0	.3	.1266
100.	1146.	2.4	-4.8	0.0	6.0	12.5	19.2	1.1	.2560
150.	1123.	6.5	-16.3	-9.0	0.0	9.7	19.8	2.3	.3883
200.	1103.	12.4	-34.7	-25.0	-12.9	0.0	13.5	4.1	.5231
250.	1085.	20.2	-60.1	-48.1	-33.0	-16.8	0.0	6.2	.6602
300.	1068.	29.9	-93.0	-78.6	-60.5	-41.1	-20.9	8.7	.7997
350.	1052.	41.8	-133.6	-116.7	-95.5	-72.9	-49.4	11.7	.9413
400.	1037.	55.7	-181.8	-162.5	-138.4	-112.5	-85.6	14.9	1.0847
450.	1023.	72.1	-239.0	-217.3	-190.2	-161.1	-130.8	18.9	1.2321
500.	1010.	91.1	-305.4	-281.2	-251.1	-218.8	-185.1	23.4	1.3830

BALLISTIC COEFFICIENT .69, MUZZLE VELOCITY FT/SEC 1300.

RANGE	VELOCITY	HEIGHT	50 YD	100 YD	150 YD	200 YD	250 YD	DRIFT	TIME
0.	1300.	0.0	-1.5	-1.5	-1.5	-1.5	-1.5	0.0	0.0000
50.	1266.	-.1	0.0	2.0	4.5	7.4	10.3	.3	.1171
100.	1233.	2.0	-4.0	0.0	5.0	10.9	16.5	1.3	.2383
150.	1203.	5.5	-13.6	-7.6	0.0	8.7	17.2	2.5	.3606
200.	1173.	10.7	-29.8	-21.7	-11.6	0.0	11.3	4.9	.4894
250.	1148.	17.5	-51.3	-41.3	-28.6	-14.1	0.0	6.9	.6161
300.	1125.	26.1	-80.0	-67.9	-52.8	-35.3	-18.4	9.8	.7481
350.	1105.	36.7	-115.5	-101.4	-83.7	-63.4	-43.7	13.2	.8828
400.	1086.	49.2	-158.1	-142.0	-121.8	-98.6	-76.0	17.0	1.0197
450.	1069.	63.7	-208.1	-189.9	-167.2	-141.1	-115.7	21.2	1.1589
500.	1053.	80.4	-265.6	-245.4	-220.2	-191.2	-162.9	25.8	1.3002

BALLISTIC COEFFICIENT .69, MUZZLE VELOCITY FT/SEC 1400.

RANGE	VELOCITY	HEIGHT	50 YD	100 YD	150 YD	200 YD	250 YD	DRIFT	TIME
0.	1400.	0.0	-1.5	-1.5	-1.5	-1.5	-1.5	0.0	0.0000
50.	1361.	-.2	0.0	1.6	3.8	6.2	8.7	.3	.1090
100.	1323.	1.6	-3.2	0.0	4.4	9.3	14.3	1.1	.2206
150.	1288.	4.7	-11.4	-6.6	0.0	7.3	14.9	2.5	.3356
200.	1254.	9.2	-24.9	-18.5	-9.7	0.0	10.1	4.5	.4542
250.	1223.	15.1	-43.7	-35.7	-24.8	-12.6	0.0	6.8	.5746
300.	1192.	22.8	-68.9	-59.4	-46.2	-31.6	-16.5	10.2	.7006
350.	1165.	32.0	-99.5	-88.4	-73.0	-56.0	-38.3	13.5	.8265
400.	1140.	43.2	-136.8	-124.1	-106.5	-87.1	-66.9	17.5	.9565
450.	1118.	56.4	-181.5	-167.2	-147.5	-125.6	-102.9	22.3	1.0911
500.	1098.	71.3	-232.0	-216.1	-194.2	-169.8	-144.6	27.0	1.2251

BALLISTIC COEFFICIENT .69, MUZZLE VELOCITY FT/SEC 1500.

RANGE	VELOCITY	HEIGHT	50 YD	100 YD	150 YD	200 YD	250 YD	DRIFT	TIME
0.	1500.	0.0	-1.5	-1.5	-1.5	-1.5	-1.5	0.0	0.0000
50.	1455.	-.3	0.0	1.3	3.2	5.3	7.5	.4	.1020
100.	1413.	1.3	-2.6	0.0	3.8	8.0	12.5	1.1	.2063
150.	1373.	4.0	-9.6	-5.7	0.0	6.4	13.0	2.5	.3143
200.	1334.	7.9	-21.2	-16.1	-8.5	0.0	8.9	4.5	.4257
250.	1299.	13.2	-37.6	-31.2	-21.7	-11.1	0.0	6.9	.5390
300.	1265.	19.9	-59.3	-51.5	-40.1	-27.4	-14.1	9.8	.6560
350.	1232.	28.2	-86.4	-77.4	-64.1	-49.2	-33.7	13.5	.7764
400.	1202.	38.1	-119.2	-108.9	-93.7	-76.8	-59.0	17.6	.8997
450.	1172.	50.1	-159.2	-147.6	-130.5	-111.4	-91.5	22.7	1.0292
500.	1147.	63.3	-203.3	-190.4	-171.4	-150.2	-128.1	27.4	1.1555

BALLISTIC COEFFICIENT .69, MUZZLE VELOCITY FT/SEC 1600.

RANGE	VELOCITY	HEIGHT	50 YD	100 YD	150 YD	200 YD	250 YD	DRIFT	TIME
0.	1600.	0.0	-1.5	-1.5	-1.5	-1.5	-1.5	0.0	0.0000
50.	1553.	-.3	0.0	1.0	2.7	4.5	6.5	.3	.0952
100.	1507.	1.0	-2.1	0.0	3.3	6.9	10.9	1.0	.1929
150.	1462.	3.4	-8.0	-4.9	0.0	5.5	11.4	2.3	.2942
200.	1419.	6.9	-18.0	-13.8	-7.3	0.0	7.9	4.0	.3980
250.	1379.	11.5	-32.4	-27.2	-19.0	-9.9	0.0	6.5	.5055
300.	1340.	17.5	-51.3	-45.1	-35.3	-24.3	-12.5	9.4	.6159
350.	1304.	24.8	-75.1	-67.8	-56.3	-43.6	-29.7	12.8	.7292
400.	1270.	33.6	-104.0	-95.7	-82.6	-68.0	-52.2	16.9	.8459
450.	1237.	44.1	-138.5	-129.1	-114.4	-98.0	-80.2	21.5	.9662
500.	1206.	56.1	-178.3	-167.9	-151.5	-133.3	-113.5	26.6	1.0886

BALLISTIC COEFFICIENT .69, MUZZLE VELOCITY FT/SEC 1700.

RANGE	VELOCITY	HEIGHT	50 YD	100 YD	150 YD	200 YD	250 YD	DRIFT	TIME
0.	1700.	0.0	-1.5	-1.5	-1.5	-1.5	-1.5	0.0	0.0000
50.	1651.	-.4	0.0	.9	2.3	3.9	5.6	.3	.0898
100.	1602.	.8	-1.7	0.0	2.8	6.0	9.5	1.0	.1822
150.	1555.	2.9	-6.8	-4.2	0.0	4.8	10.0	2.2	.2772
200.	1509.	6.0	-15.4	-12.0	-6.4	0.0	6.9	3.8	.3748
250.	1465.	10.1	-28.0	-23.7	-16.7	-8.7	0.0	6.1	.4758
300.	1421.	15.4	-44.5	-39.4	-30.9	-21.3	-10.9	8.8	.5795
350.	1381.	21.9	-65.4	-59.5	-49.6	-38.4	-26.3	12.2	.6867
400.	1342.	29.8	-91.0	-84.2	-72.9	-60.1	-46.2	16.1	.7971
450.	1306.	39.0	-121.2	-113.6	-100.8	-86.5	-70.9	20.4	.9102
500.	1272.	49.9	-156.7	-148.2	-134.0	-118.1	-100.7	25.4	1.0268

BALLISTIC COEFFICIENT .69, MUZZLE VELOCITY FT/SEC 1800.

RANGE	VELOCITY	HEIGHT	50 YD	100 YD	150 YD	200 YD	250 YD	DRIFT	TIME
0.	1800.	0.0	-1.5	-1.5	-1.5	-1.5	-1.5	0.0	0.0000
50.	1748.	-.4	0.0	.7	1.9	3.3	4.9	.2	.0846
100.	1698.	.7	-1.3	0.0	2.5	5.3	8.4	.9	.1715
150.	1649.	2.5	-5.7	-3.7	0.0	4.3	8.9	2.0	.2614
200.	1600.	5.3	-13.3	-10.6	-5.7	0.0	6.1	3.6	.3539
250.	1553.	8.9	-24.3	-21.0	-14.8	-7.7	0.0	5.7	.4490
300.	1508.	13.6	-38.8	-34.8	-27.4	-18.9	-9.7	8.2	.5467
350.	1463.	19.4	-57.3	-52.6	-44.0	-34.0	-23.3	11.4	.6479
400.	1420.	26.4	-79.7	-74.4	-64.5	-53.1	-40.8	15.0	.7517
450.	1379.	34.7	-106.6	-100.6	-89.5	-76.7	-62.9	19.2	.8592
500.	1341.	44.4	-138.1	-131.4	-119.0	-104.8	-89.5	24.0	.9696

BALLISTIC COEFFICIENT .69, MUZZLE VELOCITY FT/SEC 1900.

RANGE	VELOCITY	HEIGHT	50 YD	100 YD	150 YD	200 YD	250 YD	DRIFT	TIME
0.	1900.	0.0	-1.5	-1.5	-1.5	-1.5	-1.5	0.0	0.0000
50.	1846.	-.4	0.0	.5	1.6	2.9	4.2	.2	.0803
100.	1793.	.5	-1.0	0.0	2.2	4.7	7.4	.8	.1626
150.	1742.	2.2	-4.8	-3.2	0.0	3.8	7.9	1.9	.2475
200.	1691.	4.6	-11.4	-9.3	-5.0	0.0	5.5	3.3	.3348
250.	1642.	7.9	-21.1	-18.5	-13.1	-6.8	0.0	5.3	.4249
300.	1594.	12.1	-34.1	-30.9	-24.5	-16.9	-8.7	7.8	.5178
350.	1547.	17.3	-50.4	-46.7	-39.2	-30.4	-20.8	10.7	.6131
400.	1502.	23.5	-70.3	-66.1	-57.5	-47.4	-36.5	14.0	.7112
450.	1457.	31.0	-94.2	-89.5	-79.8	-68.5	-56.2	18.0	.8130
500.	1414.	39.6	-122.0	-116.8	-106.0	-93.5	-79.8	22.5	.9172

BALLISTIC COEFFICIENT .70, MUZZLE VELOCITY FT/SEC 800.

RANGE	VELOCITY	HEIGHT	50 YD	100 YD	150 YD	200 YD	250 YD	DRIFT	TIME
0.	800.	0.0	-1.5	-1.5	-1.5	-1.5	-1.5	0.0	0.0000
50.	791.	1.0	0.0	6.1	13.0	20.0	27.1	.4	.1898
100.	783.	6.2	-12.3	0.0	13.7	27.7	42.0	.8	.3797
150.	774.	15.0	-38.9	-20.5	0.0	21.1	42.6	1.9	.5730
200.	765.	27.5	-79.9	-55.4	-28.1	0.0	28.7	3.1	.7678
250.	757.	43.9	-135.7	-105.1	-70.9	-35.8	0.0	4.8	.9646
300.	748.	64.4	-207.2	-170.4	-129.4	-87.3	-44.3	7.0	1.1647
350.	740.	88.7	-293.1	-250.2	-202.4	-153.3	-103.1	9.2	1.3648
400.	731.	117.6	-396.6	-347.6	-292.9	-236.8	-179.4	12.4	1.5703
450.	723.	150.6	-515.1	-460.0	-398.4	-335.3	-270.8	15.6	1.7759
500.	715.	188.4	-651.1	-589.8	-521.4	-451.3	-379.6	19.4	1.9850

BALLISTIC COEFFICIENT .70, MUZZLE VELOCITY FT/SEC 900.

RANGE	VELOCITY	HEIGHT	50 YD	100 YD	150 YD	200 YD	250 YD	DRIFT	TIME
0.	900.	0.0	-1.5	-1.5	-1.5	-1.5	-1.5	0.0	0.0000
50.	890.	.6	0.0	4.7	10.1	15.6	21.3	.4	.1688
100.	880.	4.7	-9.4	0.0	10.7	21.7	33.2	.8	.3379
150.	870.	11.7	-30.2	-16.1	0.0	16.4	33.6	1.8	.5101
200.	860.	21.6	-62.2	-43.4	-21.9	0.0	22.9	2.8	.6826
250.	851.	34.7	-106.4	-82.9	-56.0	-28.7	0.0	4.5	.8589
300.	842.	50.7	-162.1	-133.8	-101.6	-68.8	-34.4	6.2	1.0352
350.	833.	70.1	-230.7	-197.7	-160.1	-121.8	-81.7	8.5	1.2152
400.	824.	92.8	-311.3	-273.7	-230.7	-186.9	-141.0	11.0	1.3960
450.	815.	119.0	-405.2	-362.8	-314.5	-265.2	-213.6	14.0	1.5794
500.	806.	148.7	-512.1	-465.0	-411.3	-356.6	-299.2	17.3	1.7647

BALLISTIC COEFFICIENT .70, MUZZLE VELOCITY FT/SEC 1000.

RANGE	VELOCITY	HEIGHT	50 YD	100 YD	150 YD	200 YD	250 YD	DRIFT	TIME
0.	1000.	0.0	-1.5	-1.5	-1.5	-1.5	-1.5	0.0	0.0000
50.	987.	.4	0.0	3.5	7.8	12.3	16.9	.9	.1551
100.	975.	3.7	-7.0	0.0	8.7	17.6	26.7	.7	.3042
150.	963.	9.4	-23.5	-13.0	0.0	13.3	27.1	1.8	.4601
200.	951.	17.5	-49.0	-35.1	-17.7	0.0	18.4	2.8	.6159
250.	940.	28.0	-84.3	-66.9	-45.1	-23.0	0.0	4.3	.7742
300.	929.	41.2	-129.6	-108.7	-82.6	-56.0	-28.5	6.2	.9352
350.	918.	57.0	-184.6	-160.3	-129.8	-98.8	-66.7	8.3	1.0970
400.	907.	75.7	-250.5	-222.6	-187.8	-152.4	-115.6	10.9	1.2619
450.	896.	97.5	-328.0	-296.7	-257.6	-217.7	-176.4	14.2	1.4308
500.	886.	121.7	-414.0	-379.2	-335.7	-291.4	-245.5	17.1	1.5969

BALLISTIC COEFFICIENT .70, MUZZLE VELOCITY FT/SEC 1100.

RANGE	VELOCITY	HEIGHT	50 YD	100 YD	150 YD	200 YD	250 YD	DRIFT	TIME
0.	1100.	0.0	-1.5	-1.5	-1.5	-1.5	-1.5	0.0	0.0000
50.	1082.	.2	0.0	3.1	6.4	10.2	14.1	.5	.1390
100.	1065.	3.1	-6.3	0.0	6.6	14.1	21.9	1.6	.2820
150.	1050.	7.7	-19.3	-9.9	0.0	11.2	23.0	1.8	.4192
200.	1036.	14.5	-40.7	-28.1	-14.9	0.0	15.7	3.1	.5629
250.	1022.	23.4	-70.5	-54.8	-38.3	-19.7	0.0	4.9	.7099
300.	1009.	34.8	-109.5	-90.7	-70.9	-48.5	-24.9	7.5	.8611
350.	996.	47.9	-154.5	-132.5	-109.5	-83.3	-55.8	9.1	1.0063
400.	984.	63.8	-210.4	-185.3	-158.9	-129.0	-97.6	12.1	1.1597
450.	971.	82.6	-276.6	-248.4	-218.8	-185.1	-149.8	15.9	1.3174
500.	959.	102.4	-346.6	-315.3	-282.4	-245.0	-205.7	18.0	1.4659

BALLISTIC COEFFICIENT .70, MUZZLE VELOCITY FT/SEC 1200.

RANGE	VELOCITY	HEIGHT	50 YD	100 YD	150 YD	200 YD	250 YD	DRIFT	TIME
0.	1200.	0.0	-1.5	-1.5	-1.5	-1.5	-1.5	0.0	0.0000
50.	1172.	.0	0.0	2.4	5.4	8.7	12.0	.3	.1265
100.	1147.	2.4	-4.8	0.0	6.0	12.5	19.2	1.0	.2559
150.	1124.	6.5	-16.3	-9.0	0.0	9.7	19.8	2.3	.3881
200.	1104.	12.4	-34.6	-25.0	-12.9	0.0	13.4	4.0	.5229
250.	1086.	20.1	-60.1	-48.0	-32.9	-16.8	0.0	6.1	.6598
300.	1069.	29.9	-92.9	-78.4	-60.4	-41.0	-20.8	8.6	.7991
350.	1054.	41.7	-133.3	-116.5	-95.4	-72.8	-49.2	11.5	.9404
400.	1039.	55.6	-181.5	-162.2	-138.1	-112.2	-85.4	14.7	1.0835
450.	1025.	72.1	-239.0	-217.3	-190.2	-161.1	-130.9	18.8	1.2317
500.	1012.	91.0	-305.3	-281.1	-251.0	-218.7	-185.1	23.3	1.3823

BALLISTIC COEFFICIENT .70, MUZZLE VELOCITY FT/SEC 1300.

RANGE	VELOCITY	HEIGHT	50 YD	100 YD	150 YD	200 YD	250 YD	DRIFT	TIME
0.	1300.	0.0	-1.5	-1.5	-1.5	-1.5	-1.5	0.0	0.0000
50.	1266.	-.1	0.0	2.0	4.5	7.4	10.2	.3	.1171
100.	1234.	2.0	-4.0	0.0	5.1	10.8	16.5	1.3	.2381
150.	1204.	5.5	-13.6	-7.6	0.0	8.6	17.1	2.5	.3604
200.	1175.	10.7	-29.6	-21.6	-11.5	0.0	11.4	4.7	.4884
250.	1150.	17.4	-51.2	-41.2	-28.6	-14.3	0.0	6.8	.6156
300.	1127.	26.1	-79.8	-67.8	-52.6	-35.4	-18.3	9.7	.7474
350.	1107.	36.6	-115.3	-101.2	-83.6	-63.5	-43.6	13.1	.8818
400.	1088.	49.0	-157.8	-141.7	-121.5	-98.6	-75.8	16.8	1.0185
450.	1071.	63.5	-207.6	-189.5	-166.8	-141.0	-115.4	20.9	1.1574
500.	1056.	80.2	-264.9	-244.8	-219.6	-190.9	-162.4	25.4	1.2983

BALLISTIC COEFFICIENT .70, MUZZLE VELOCITY FT/SEC 1400.

RANGE	VELOCITY	HEIGHT	50 YD	100 YD	150 YD	200 YD	250 YD	DRIFT	TIME
0.	1400.	0.0	-1.5	-1.5	-1.5	-1.5	-1.5	0.0	0.0000
50.	1361.	-.2	0.0	1.6	3.8	6.2	8.7	.3	.1090
100.	1324.	1.6	-3.2	0.0	4.4	9.2	14.3	1.1	.2205
150.	1289.	4.6	-11.3	-6.6	0.0	7.3	14.8	2.5	.3354
200.	1256.	9.1	-24.8	-18.4	-9.7	0.0	10.1	4.4	.4536
250.	1225.	15.1	-43.6	-35.7	-24.7	-12.7	0.0	6.8	.5741
300.	1195.	22.7	-68.7	-59.1	-46.0	-31.5	-16.3	9.9	.6994
350.	1167.	32.0	-99.3	-88.2	-72.8	-55.9	-38.2	13.3	.8254
400.	1143.	43.0	-136.5	-123.8	-106.3	-86.9	-66.7	17.3	.9552
450.	1121.	56.1	-180.5	-166.2	-146.5	-124.7	-101.9	21.8	1.0879
500.	1101.	71.0	-231.3	-215.4	-193.5	-169.3	-144.0	26.7	1.2230

BALLISTIC COEFFICIENT .70, MUZZLE VELOCITY FT/SEC 1500.

RANGE	VELOCITY	HEIGHT	50 YD	100 YD	150 YD	200 YD	250 YD	DRIFT	TIME
0.	1500.	0.0	-1.5	-1.5	-1.5	-1.5	-1.5	0.0	0.0000
50.	1456.	-.3	0.0	1.3	3.2	5.3	7.5	.3	.1020
100.	1414.	1.3	-2.6	0.0	3.8	8.0	12.4	1.1	.2062
150.	1374.	4.0	-9.6	-5.7	0.0	6.3	12.9	2.6	.3145
200.	1337.	7.9	-21.2	-16.0	-8.4	0.0	8.8	4.4	.4251
250.	1301.	13.2	-37.5	-31.1	-21.5	-11.0	0.0	6.7	.5382
300.	1267.	19.9	-59.1	-51.4	-39.9	-27.4	-14.1	9.7	.6552
350.	1235.	28.2	-86.4	-77.4	-64.0	-49.3	-33.9	13.4	.7760
400.	1205.	38.0	-118.9	-108.6	-93.3	-76.6	-58.9	17.3	.8983
450.	1176.	49.8	-158.2	-146.6	-129.4	-110.5	-90.6	22.1	1.0258
500.	1151.	63.1	-202.7	-189.8	-170.7	-149.7	-127.6	27.0	1.1533

BALLISTIC COEFFICIENT .70, MUZZLE VELOCITY FT/SEC 1600.

RANGE	VELOCITY	HEIGHT	50 YD	100 YD	150 YD	200 YD	250 YD	DRIFT	TIME
0.	1600.	0.0	-1.5	-1.5	-1.5	-1.5	-1.5	0.0	0.0000
50.	1554.	-.3	0.0	1.0	2.7	4.5	6.5	.2	.0952
100.	1508.	1.0	-2.1	0.0	3.3	6.9	10.8	.9	.1928
150.	1464.	3.4	-8.0	-4.9	0.0	5.5	11.4	2.2	.2939
200.	1422.	6.8	-18.0	-13.8	-7.3	0.0	7.9	4.0	.3977
250.	1382.	11.5	-32.3	-27.1	-18.9	-9.8	0.0	6.3	.5048
300.	1344.	17.4	-51.2	-45.0	-35.2	-24.3	-12.5	9.3	.6151
350.	1308.	24.7	-74.9	-67.6	-56.2	-43.4	-29.7	12.6	.7281
400.	1273.	33.5	-103.9	-95.6	-82.5	-68.0	-52.2	16.8	.8452
450.	1241.	43.9	-137.9	-128.6	-113.9	-97.5	-79.8	21.2	.9640
500.	1211.	55.9	-177.7	-167.3	-150.9	-132.7	-113.1	26.2	1.0863

BALLISTIC COEFFICIENT .70, MUZZLE VELOCITY FT/SEC 1700.

RANGE	VELOCITY	HEIGHT	50 YD	100 YD	150 YD	200 YD	250 YD	DRIFT	TIME
0.	1700.	0.0	-1.5	-1.5	-1.5	-1.5	-1.5	0.0	0.0000
50.	1651.	-.4	0.0	.8	2.3	3.9	5.6	.3	.0898
100.	1604.	.8	-1.7	0.0	2.8	6.0	9.5	1.0	.1820
150.	1557.	2.9	-6.8	-4.2	0.0	4.8	10.0	2.2	.2770
200.	1512.	6.0	-15.4	-12.0	-6.4	0.0	6.9	3.8	.3745
250.	1468.	10.1	-27.9	-23.7	-16.6	-8.6	0.0	6.0	.4753
300.	1425.	15.3	-44.4	-39.3	-30.9	-21.3	-10.9	8.7	.5789
350.	1385.	21.8	-65.3	-59.3	-49.4	-38.3	-26.2	12.0	.6857
400.	1347.	29.7	-90.7	-83.9	-72.6	-59.9	-46.1	15.8	.7958
450.	1311.	38.9	-120.8	-113.2	-100.5	-86.1	-70.6	20.1	.9085
500.	1276.	49.7	-156.2	-147.8	-133.6	-117.7	-100.4	25.1	1.0250

BALLISTIC COEFFICIENT .70, MUZZLE VELOCITY FT/SEC 1800.

RANGE	VELOCITY	HEIGHT	50 YD	100 YD	150 YD	200 YD	250 YD	DRIFT	TIME
0.	1800.	0.0	-1.5	-1.5	-1.5	-1.5	-1.5	0.0	0.0000
50.	1749.	-.4	0.0	.7	1.9	3.3	4.9	.2	.0845
100.	1699.	.7	-1.3	0.0	2.5	5.3	8.4	.8	.1714
150.	1651.	2.5	-5.7	-3.7	0.0	4.3	8.8	2.0	.2613
200.	1603.	5.2	-13.3	-10.6	-5.7	0.0	6.1	3.6	.3535
250.	1557.	8.9	-24.3	-20.9	-14.7	-7.6	0.0	5.6	.4486
300.	1511.	13.6	-38.8	-34.8	-27.3	-18.8	-9.6	8.1	.5461
350.	1467.	19.3	-57.2	-52.5	-43.8	-33.9	-23.2	11.2	.6469
400.	1424.	26.3	-79.5	-74.2	-64.3	-53.0	-40.7	14.8	.7505
450.	1384.	34.5	-106.3	-100.3	-89.1	-76.4	-62.6	18.9	.8574
500.	1346.	44.2	-137.6	-131.0	-118.6	-104.4	-89.1	23.6	.9676

BALLISTIC COEFFICIENT .70, MUZZLE VELOCITY FT/SEC 1900.

RANGE	VELOCITY	HEIGHT	50 YD	100 YD	150 YD	200 YD	250 YD	DRIFT	TIME
0.	1900.	0.0	-1.5	-1.5	-1.5	-1.5	-1.5	0.0	0.0000
50.	1847.	-.4	0.0	.5	1.6	2.8	4.2	.2	.0803
100.	1795.	.5	-1.0	0.0	2.2	4.7	7.4	.8	.1625
150.	1744.	2.2	-4.8	-3.2	0.0	3.7	7.8	1.8	.2473
200.	1694.	4.6	-11.4	-9.3	-5.0	0.0	5.5	3.3	.3344
250.	1646.	7.9	-21.1	-18.5	-13.1	-6.8	0.0	5.2	.4244
300.	1598.	12.1	-34.0	-30.8	-24.4	-16.9	-8.7	7.6	.5171
350.	1552.	17.2	-50.3	-46.6	-39.1	-30.3	-20.8	10.5	.6123
400.	1507.	23.5	-70.1	-65.9	-57.3	-47.3	-36.4	13.8	.7100
450.	1463.	30.8	-93.9	-89.2	-79.5	-68.2	-55.9	17.7	.8113
500.	1420.	39.4	-121.5	-116.3	-105.6	-93.1	-79.4	22.1	.9150

BALLISTIC COEFFICIENT .10, MUZZLE VELOCITY FT/SEC 2000.

RANGE	VELOCITY	HEIGHT	100 YD	150 YD	200 YD	250 YD	300 YD	DRIFT	TIME
0.	2000.	0.0	-1.5	-1.5	-1.5	-1.5	-1.5	0.0	0.0000
100.	1336.	.9	0.0	3.9	9.2	15.4	22.7	6.1	.1848
200.	1013.	8.9	-18.4	-10.7	0.0	12.4	26.9	26.3	.4495
300.	857.	27.8	-68.0	-56.4	-40.4	-21.8	0.0	56.6	.7718
400.	735.	62.7	-160.3	-144.9	-123.5	-98.7	-69.7	96.7	1.1497
500.	623.	121.1	-311.3	-292.0	-265.3	-234.2	-198.0	148.4	1.5932
600.	522.	214.8	-546.0	-522.8	-490.8	-453.5	-410.0	214.6	2.1193
700.	432.	362.7	-904.0	-877.0	-839.7	-796.1	-745.4	299.5	2.7517
800.	352.	594.7	-1448.4	-1417.5	-1374.9	-1325.1	-1267.1	408.7	3.5220
900.	281.	960.5	-2283.1	-2248.4	-2200.4	-2144.4	-2079.1	550.0	4.4750
1000.	220.	1548.1	-3592.5	-3553.9	-3500.6	-3438.4	-3365.8	735.7	5.6804

BALLISTIC COEFFICIENT .10, MUZZLE VELOCITY FT/SEC 2100.

RANGE	VELOCITY	HEIGHT	100 YD	150 YD	200 YD	250 YD	300 YD	DRIFT	TIME
0.	2100.	0.0	-1.5	-1.5	-1.5	-1.5	-1.5	0.0	0.0000
100.	1403.	.7	0.0	3.5	8.3	14.2	21.1	5.7	.1755
200.	1038.	8.1	-16.6	-9.7	0.0	11.8	25.6	25.4	.4301
300.	874.	26.0	-63.4	-53.0	-38.4	-20.7	0.0	56.0	.7470
400.	749.	59.2	-150.8	-136.9	-117.5	-93.8	-66.2	96.1	1.1172
500.	636.	114.8	-294.1	-276.8	-252.5	-222.9	-188.4	147.4	1.5517
600.	534.	204.2	-517.3	-496.5	-467.3	-431.8	-390.4	212.8	2.0665
700.	443.	344.9	-857.0	-832.8	-798.7	-757.2	-709.0	296.3	2.6833
800.	361.	565.1	-1372.8	-1345.2	-1306.2	-1258.9	-1203.7	403.1	3.4333
900.	290.	913.0	-2166.3	-2135.2	-2091.4	-2038.0	-1976.0	541.6	4.3631
1000.	227.	1469.3	-3405.5	-3370.9	-3322.2	-3263.0	-3194.1	722.6	5.5341

BALLISTIC COEFFICIENT .10, MUZZLE VELOCITY FT/SEC 2200.

RANGE	VELOCITY	HEIGHT	100 YD	150 YD	200 YD	250 YD	300 YD	DRIFT	TIME
0.	2200.	0.0	-1.5	-1.5	-1.5	-1.5	-1.5	0.0	0.0000
100.	1476.	.6	0.0	3.1	7.6	13.2	19.6	5.4	.1670
200.	1063.	7.4	-15.2	-9.0	0.0	11.2	24.1	24.6	.4125
300.	892.	24.3	-58.9	-49.6	-36.1	-19.3	0.0	55.0	.7217
400.	764.	55.8	-141.7	-129.4	-111.4	-89.0	-63.2	95.1	1.0856
500.	650.	108.9	-278.1	-262.6	-240.2	-212.2	-180.0	146.1	1.5117
600.	546.	194.2	-490.3	-471.8	-444.9	-411.2	-372.6	210.7	2.0155
700.	454.	328.4	-813.6	-792.0	-760.6	-721.4	-676.3	292.9	2.6185
800.	371.	538.2	-1304.5	-1279.8	-1243.9	-1199.1	-1147.6	397.7	3.3508
900.	298.	867.5	-2054.4	-2026.6	-1986.2	-1935.8	-1877.8	532.5	4.2531
1000.	235.	1394.0	-3226.7	-3195.8	-3150.9	-3094.9	-3030.5	708.7	5.3906

BALLISTIC COEFFICIENT .10, MUZZLE VELOCITY FT/SEC 2300.

RANGE	VELOCITY	HEIGHT	100 YD	150 YD	200 YD	250 YD	300 YD	DRIFT	TIME
0.	2300.	0.0	-1.5	-1.5	-1.5	-1.5	-1.5	0.0	0.0000
100.	1551.	.5	0.0	2.8	6.8	12.1	18.2	5.1	.1593
200.	1091.	6.7	-13.7	-8.2	0.0	10.6	22.8	23.6	.3947
300.	909.	22.6	-54.7	-46.4	-34.2	-18.3	0.0	53.9	.6977
400.	779.	52.7	-133.2	-122.1	-105.8	-84.7	-60.3	93.8	1.0548
500.	663.	103.3	-262.8	-249.1	-228.6	-202.2	-171.7	144.4	1.4726
600.	558.	184.8	-464.9	-448.3	-423.9	-392.1	-355.6	208.3	1.9659
700.	464.	312.8	-772.8	-753.5	-725.0	-688.0	-645.3	289.1	2.5557
800.	380.	512.4	-1239.1	-1217.0	-1184.4	-1142.1	-1093.3	391.8	3.2697
900.	306.	826.2	-1953.3	-1928.5	-1891.8	-1844.2	-1789.4	523.9	4.1508
1000.	242.	1323.5	-3059.4	-3031.9	-2991.1	-2938.2	-2877.2	694.9	5.2524

BALLISTIC COEFFICIENT .10, MUZZLE VELOCITY FT/SEC 2400.

RANGE	VELOCITY	HEIGHT	100 YD	150 YD	200 YD	250 YD	300 YD	DRIFT	TIME
0.	2400.	0.0	-1.5	-1.5	-1.5	-1.5	-1.5	0.0	0.0000
100.	1628.	.4	0.0	2.5	6.2	11.1	16.9	4.8	.1520
200.	1123.	6.1	-12.3	-7.4	0.0	9.9	21.5	22.5	.3779
300.	927.	21.1	-50.7	-43.3	-32.2	-17.3	0.0	52.7	.6743
400.	793.	49.7	-125.1	-115.3	-100.5	-80.7	-57.5	92.4	1.0251
500.	676.	98.1	-248.6	-236.3	-217.7	-193.0	-164.0	142.6	1.4350
600.	570.	176.0	-441.3	-426.6	-404.3	-374.6	-339.9	205.7	1.9187
700.	475.	298.2	-734.9	-717.7	-691.7	-657.0	-616.5	285.2	2.4957
800.	390.	489.0	-1179.9	-1160.2	-1130.5	-1090.9	-1044.7	386.2	3.1941
900.	315.	787.3	-1858.2	-1836.1	-1802.7	-1758.1	-1706.1	515.1	4.0520
1000.	249.	1260.8	-2911.2	-2886.7	-2849.5	-2800.0	-2742.2	682.3	5.1266

BALLISTIC COEFFICIENT .10, MUZZLE VELOCITY FT/SEC 2500.

RANGE	VELOCITY	HEIGHT	100 YD	150 YD	200 YD	250 YD	300 YD	DRIFT	TIME
0.	2500.	0.0	-1.5	-1.5	-1.5	-1.5	-1.5	0.0	0.0000
100.	1706.	.3	0.0	2.2	5.6	10.1	15.7	4.5	.1454
200.	1160.	5.5	-11.1	-6.7	0.0	9.2	20.2	21.4	.3617
300.	945.	19.6	-47.1	-40.5	-30.4	-16.6	0.0	51.4	.6519
400.	807.	46.9	-117.5	-108.8	-95.3	-77.0	-54.8	90.8	.9960
500.	689.	93.1	-235.2	-224.2	-207.3	-184.5	-156.7	140.5	1.3985
600.	582.	167.5	-418.9	-405.8	-385.5	-358.1	-324.8	202.8	1.8725
700.	485.	284.5	-699.2	-683.9	-660.3	-628.3	-589.4	281.2	2.4378
800.	399.	466.4	-1123.0	-1105.4	-1078.5	-1041.9	-997.5	380.1	3.1197
900.	323.	750.7	-1769.0	-1749.2	-1718.9	-1677.7	-1627.8	506.3	3.9568
1000.	256.	1200.5	-2769.0	-2747.1	-2713.4	-2667.6	-2612.1	669.3	5.0027

BALLISTIC COEFFICIENT .10, MUZZLE VELOCITY FT/SEC 2600.

RANGE	VELOCITY	HEIGHT	100 YD	150 YD	200 YD	250 YD	300 YD	DRIFT	TIME
0.	2600.	0.0	-1.5	-1.5	-1.5	-1.5	-1.5	0.0	0.0000
100.	1786.	.2	0.0	2.0	5.0	9.3	14.5	4.2	.1393
200.	1204.	5.0	-10.0	-6.1	0.0	8.5	19.0	20.3	.3463
300.	964.	18.3	-43.5	-37.6	-28.5	-15.7	0.0	49.9	.6295
400.	821.	44.2	-110.3	-102.5	-90.2	-73.3	-52.3	89.1	.9675
500.	702.	88.4	-222.4	-212.6	-197.3	-176.1	-149.9	138.3	1.3629
600.	593.	159.7	-398.1	-386.3	-368.0	-342.6	-311.1	199.9	1.8284
700.	496.	271.6	-665.8	-652.1	-630.7	-601.0	-564.3	277.1	2.3820
800.	408.	445.7	-1071.1	-1055.4	-1030.9	-997.0	-955.0	374.3	3.0498
900.	331.	717.2	-1687.3	-1669.7	-1642.1	-1604.0	-1556.8	497.9	3.8674
1000.	263.	1144.6	-2637.1	-2617.4	-2586.8	-2544.5	-2492.0	656.6	4.8848

BALLISTIC COEFFICIENT .10, MUZZLE VELOCITY FT/SEC 2700.

RANGE	VELOCITY	HEIGHT	100 YD	150 YD	200 YD	250 YD	300 YD	DRIFT	TIME
0.	2700.	0.0	-1.5	-1.5	-1.5	-1.5	-1.5	0.0	0.0000
100.	1866.	.1	0.0	1.8	4.5	8.5	13.4	4.0	.1337
200.	1252.	4.5	-9.1	-5.6	0.0	7.8	17.7	19.3	.3319
300.	982.	17.0	-40.2	-35.0	-26.6	-14.9	0.0	48.3	.6079
400.	835.	41.7	-103.6	-96.5	-85.4	-69.8	-49.9	87.2	.9401
500.	714.	84.0	-210.6	-201.8	-187.9	-168.3	-143.5	136.1	1.3288
600.	605.	152.3	-378.4	-367.9	-351.2	-327.7	-298.0	196.9	1.7855
700.	506.	259.5	-634.6	-622.2	-602.8	-575.4	-540.7	272.9	2.3284
800.	417.	426.0	-1021.5	-1007.4	-985.2	-953.9	-914.2	368.3	2.9816
900.	339.	684.9	-1609.0	-1593.2	-1568.1	-1532.9	-1488.3	489.2	3.7795
1000.	270.	1093.5	-2516.9	-2499.3	-2471.5	-2432.4	-2382.8	644.8	4.7746

BALLISTIC COEFFICIENT .10, MUZZLE VELOCITY FT/SEC 2800.

RANGE	VELOCITY	HEIGHT	100 YD	150 YD	200 YD	250 YD	300 YD	DRIFT	TIME
0.	2800.	0.0	-1.5	-1.5	-1.5	-1.5	-1.5	0.0	0.0000
100.	1948.	.0	0.0	1.6	4.1	7.7	12.4	3.8	.1285
200.	1302.	4.1	-8.2	-5.0	0.0	7.2	16.6	18.3	.3182
300.	1001.	15.8	-37.1	-32.4	-24.8	-14.0	0.0	46.7	.5870
400.	849.	39.3	-97.2	-90.9	-80.8	-66.4	-47.7	85.3	.9135
500.	727.	79.8	-199.4	-191.5	-178.9	-160.8	-137.5	133.8	1.2957
600.	616.	145.3	-360.0	-350.5	-335.3	-313.7	-285.7	193.8	1.7442
700.	516.	248.1	-605.1	-594.1	-576.4	-551.2	-518.4	268.7	2.2767
800.	426.	407.7	-975.8	-963.2	-943.0	-914.2	-876.8	362.5	2.9170
900.	347.	655.7	-1538.1	-1523.9	-1501.2	-1468.8	-1426.7	481.1	3.6981
1000.	277.	1044.6	-2401.7	-2385.9	-2360.7	-2324.6	-2277.9	632.8	4.6666

BALLISTIC COEFFICIENT .10, MUZZLE VELOCITY FT/SEC 2900.

RANGE	VELOCITY	HEIGHT	100 YD	150 YD	200 YD	250 YD	300 YD	DRIFT	TIME
0.	2900.	0.0	-1.5	-1.5	-1.5	-1.5	-1.5	0.0	0.0000
100.	2030.	-.0	0.0	1.4	3.7	7.0	11.4	3.6	.1237
200.	1355.	3.7	-7.4	-4.6	0.0	6.6	15.4	17.4	.3057
300.	1021.	14.7	-34.3	-30.0	-23.1	-13.2	0.0	45.1	.5668
400.	863.	37.1	-91.3	-85.6	-76.4	-63.1	-45.5	83.4	.8877
500.	739.	75.9	-188.8	-181.7	-170.2	-153.7	-131.7	131.4	1.2636
600.	627.	138.7	-342.7	-334.2	-320.4	-300.5	-274.1	190.8	1.7046
700.	526.	237.4	-577.7	-567.8	-551.6	-528.5	-497.7	264.6	2.2274
800.	435.	390.4	-932.7	-921.4	-903.0	-876.5	-841.3	356.8	2.8548
900.	355.	627.9	-1470.9	-1458.1	-1437.4	-1407.6	-1368.0	473.1	3.6191
1000.	284.	999.7	-2296.2	-2282.1	-2259.0	-2225.9	-2181.9	621.4	4.5654

BALLISTIC COEFFICIENT .10, MUZZLE VELOCITY FT/SEC 3000.

RANGE	VELOCITY	HEIGHT	100 YD	150 YD	200 YD	250 YD	300 YD	DRIFT	TIME
0.	3000.	0.0	-1.5	-1.5	-1.5	-1.5	-1.5	0.0	0.0000
100.	2112.	-.1	0.0	1.3	3.4	6.4	10.5	3.4	.1193
200.	1411.	3.4	-6.7	-4.2	0.0	6.1	14.3	16.5	.2938
300.	1040.	13.6	-31.6	-27.8	-21.5	-12.3	0.0	43.5	.5474
400.	876.	35.0	-85.8	-80.7	-72.3	-60.1	-43.6	81.5	.8632
500.	751.	72.2	-179.0	-172.6	-162.1	-146.8	-126.3	129.0	1.2329
600.	638.	132.5	-326.3	-318.7	-306.1	-287.8	-263.1	187.7	1.6663
700.	536.	227.3	-551.8	-542.9	-528.2	-506.8	-478.1	260.4	2.1798
800.	444.	374.3	-892.4	-882.2	-865.4	-841.0	-808.1	351.1	2.7951
900.	363.	601.8	-1407.7	-1396.2	-1377.3	-1349.8	-1312.8	465.2	3.5431
1000.	291.	958.2	-2198.7	-2186.0	-2165.0	-2134.4	-2093.3	610.7	4.4696

BALLISTIC COEFFICIENT .10, MUZZLE VELOCITY FT/SEC 3100.

RANGE	VELOCITY	HEIGHT	100 YD	150 YD	200 YD	250 YD	300 YD	DRIFT	TIME
0.	3100.	0.0	-1.5	-1.5	-1.5	-1.5	-1.5	0.0	0.0000
100.	2193.	-.1	0.0	1.1	3.1	5.9	9.7	3.2	.1151
200.	1471.	3.1	-6.1	-3.8	0.0	5.6	13.3	15.7	.2827
300.	1061.	12.7	-29.1	-25.7	-20.0	-11.6	0.0	41.9	.5286
400.	891.	33.0	-80.4	-75.8	-68.2	-57.0	-41.6	79.4	.8385
500.	763.	68.7	-169.6	-163.9	-154.3	-140.3	-121.0	126.5	1.2028
600.	649.	126.7	-311.1	-304.2	-292.7	-275.9	-252.8	184.6	1.6295
700.	546.	217.8	-527.5	-519.5	-506.1	-486.5	-459.5	256.3	2.1339
800.	453.	359.1	-854.6	-845.5	-830.1	-807.7	-776.9	345.6	2.7379
900.	370.	577.7	-1349.4	-1339.1	-1321.8	-1296.6	-1262.0	457.7	3.4714
1000.	298.	918.2	-2104.9	-2093.4	-2074.2	-2046.2	-2007.7	599.8	4.3755

BALLISTIC COEFFICIENT .10, MUZZLE VELOCITY FT/SEC 3200.

RANGE	VELOCITY	HEIGHT	100 YD	150 YD	200 YD	250 YD	300 YD	DRIFT	TIME
0.	3200.	0.0	-1.5	-1.5	-1.5	-1.5	-1.5	0.0	0.0000
100.	2275.	-.2	0.0	1.0	2.8	5.4	8.9	3.1	.1113
200.	1532.	2.8	-5.6	-3.5	0.0	5.1	12.3	15.0	.2725
300.	1084.	11.8	-26.8	-23.8	-18.5	-10.8	0.0	40.3	.5104
400.	905.	31.1	-75.5	-71.4	-64.3	-54.0	-39.7	77.4	.8150
500.	775.	65.4	-160.8	-155.7	-146.9	-134.0	-116.1	124.1	1.1739
600.	660.	121.1	-296.6	-290.4	-279.8	-264.4	-242.9	181.5	1.5935
700.	555.	208.8	-504.6	-497.4	-485.1	-467.1	-442.0	252.3	2.0896
800.	462.	344.6	-818.6	-810.4	-796.3	-775.7	-747.1	340.1	2.6823
900.	378.	554.5	-1293.4	-1284.2	-1268.3	-1245.2	-1212.9	450.1	3.4011
1000.	304.	881.5	-2018.8	-2008.6	-1990.9	-1965.2	-1929.4	589.5	4.2872

BALLISTIC COEFFICIENT .10, MUZZLE VELOCITY FT/SEC 3300.

RANGE	VELOCITY	HEIGHT	100 YD	150 YD	200 YD	250 YD	300 YD	DRIFT	TIME
0.	3300.	0.0	-1.5	-1.5	-1.5	-1.5	-1.5	0.0	0.0000
100.	2356.	-.2	0.0	.9	2.5	4.9	8.2	3.0	.1078
200.	1594.	2.6	-5.1	-3.2	0.0	4.7	11.4	14.3	.2629
300.	1109.	10.9	-24.7	-22.0	-17.1	-10.0	0.0	38.8	.4930
400.	919.	29.4	-70.8	-67.1	-60.6	-51.2	-37.8	75.4	.7921
500.	787.	62.3	-152.6	-147.9	-139.8	-128.0	-111.3	121.7	1.1458
600.	670.	116.0	-283.0	-277.5	-267.8	-253.6	-233.6	178.4	1.5592
700.	565.	200.4	-483.0	-476.5	-465.2	-448.6	-425.3	248.3	2.0470
800.	470.	331.2	-785.4	-778.0	-765.1	-746.1	-719.5	334.8	2.6297
900.	386.	533.1	-1241.8	-1233.5	-1218.9	-1197.6	-1167.6	442.9	3.3349
1000.	311.	847.1	-1938.3	-1929.1	-1912.9	-1889.2	-1855.9	579.7	4.2029

BALLISTIC COEFFICIENT .10, MUZZLE VELOCITY FT/SEC 3400.

RANGE	VELOCITY	HEIGHT	100 YD	150 YD	200 YD	250 YD	300 YD	DRIFT	TIME
0.	3400.	0.0	-1.5	-1.5	-1.5	-1.5	-1.5	0.0	0.0000
100.	2437.	-.2	0.0	.8	2.3	4.5	7.6	2.8	.1043
200.	1657.	2.3	-4.6	-3.0	0.0	4.4	10.6	13.6	.2539
300.	1136.	10.1	-22.8	-20.3	-15.9	-9.3	0.0	37.3	.4764
400.	934.	27.7	-66.5	-63.2	-57.2	-48.5	-36.0	73.4	.7701
500.	798.	59.3	-144.7	-140.5	-133.1	-122.2	-106.7	119.2	1.1184
600.	681.	110.9	-270.0	-265.0	-256.1	-243.0	-224.4	175.3	1.5255
700.	574.	192.4	-462.7	-456.8	-446.4	-431.1	-409.4	244.3	2.0057
800.	479.	318.2	-753.4	-746.7	-734.8	-717.3	-692.5	329.5	2.5779
900.	393.	512.7	-1192.9	-1185.4	-1172.0	-1152.4	-1124.4	435.9	3.2707
1000.	318.	814.0	-1860.8	-1852.5	-1837.6	-1815.7	-1784.7	569.8	4.1200

BALLISTIC COEFFICIENT .10, MUZZLE VELOCITY FT/SEC 3500.

RANGE	VELOCITY	HEIGHT	100 YD	150 YD	200 YD	250 YD	300 YD	DRIFT	TIME
0.	3500.	0.0	-1.5	-1.5	-1.5	-1.5	-1.5	0.0	0.0000
100.	2518.	-.3	0.0	.7	2.1	4.1	7.0	2.7	.1013
200.	1720.	2.1	-4.2	-2.7	0.0	4.0	9.8	13.0	.2455
300.	1168.	9.4	-21.0	-18.8	-14.7	-8.6	0.0	35.7	.4601
400.	949.	26.2	-62.4	-59.4	-53.9	-45.8	-34.4	71.4	.7487
500.	810.	56.5	-137.3	-133.6	-126.7	-116.6	-102.3	116.8	1.0921
600.	691.	106.3	-258.0	-253.5	-245.2	-233.1	-215.9	172.3	1.4934
700.	584.	184.7	-443.3	-438.1	-428.5	-414.4	-394.3	240.4	1.9658
800.	487.	306.3	-723.8	-717.8	-706.8	-690.6	-667.7	324.4	2.5290
900.	401.	493.2	-1146.0	-1139.2	-1126.9	-1108.7	-1082.9	428.8	3.2080
1000.	324.	783.4	-1789.3	-1781.8	-1768.1	-1747.9	-1719.2	560.5	4.0419

BALLISTIC COEFFICIENT .10, MUZZLE VELOCITY FT/SEC 3600.

RANGE	VELOCITY	HEIGHT	100 YD	150 YD	200 YD	250 YD	300 YD	DRIFT	TIME
0.	3600.	0.0	-1.5	-1.5	-1.5	-1.5	-1.5	0.0	0.0000
100.	2598.	-.3	0.0	.7	1.9	3.8	6.5	2.6	.0982
200.	1784.	2.0	-3.9	-2.5	0.0	3.7	9.1	12.5	.2377
300.	1204.	8.7	-19.4	-17.4	-13.6	-8.0	0.0	34.3	.4448
400.	963.	24.7	-58.6	-55.9	-50.8	-43.3	-32.7	69.5	.7280
500.	821.	53.8	-130.3	-126.9	-120.6	-111.2	-97.9	114.3	1.0662
600.	701.	101.8	-246.3	-242.3	-234.7	-223.5	-207.5	169.3	1.4617
700.	593.	177.5	-425.1	-420.4	-411.5	-398.5	-379.8	236.5	1.9273
800.	495.	294.7	-695.4	-690.0	-679.8	-664.9	-643.6	319.3	2.4811
900.	408.	475.3	-1102.9	-1096.9	-1085.5	-1068.6	-1044.7	422.2	3.1491
1000.	331.	754.6	-1722.0	-1715.3	-1702.6	-1683.9	-1657.2	551.5	3.9670

BALLISTIC COEFFICIENT .10, MUZZLE VELOCITY FT/SEC 3700.

RANGE	VELOCITY	HEIGHT	100 YD	150 YD	200 YD	250 YD	300 YD	DRIFT	TIME
0.	3700.	0.0	-1.5	-1.5	-1.5	-1.5	-1.5	0.0	0.0000
100.	2679.	-.3	0.0	.6	1.8	3.5	6.0	2.5	.0954
200.	1849.	1.8	-3.6	-2.3	0.0	3.5	8.4	12.0	.2303
300.	1242.	8.1	-18.0	-16.1	-12.6	-7.4	0.0	32.9	.4303
400.	979.	23.3	-54.9	-52.5	-47.8	-40.9	-31.0	67.5	.7077
500.	832.	51.3	-123.7	-120.7	-114.8	-106.2	-93.8	111.9	1.0414
600.	712.	97.6	-235.5	-231.8	-224.8	-214.4	-199.6	166.3	1.4315
700.	602.	170.7	-407.8	-403.5	-395.3	-383.2	-365.8	232.7	1.8898
800.	504.	283.8	-668.5	-663.7	-654.3	-640.4	-620.6	314.4	2.4349
900.	415.	458.0	-1061.4	-1056.0	-1045.4	-1029.8	-1007.5	415.6	3.0913
1000.	337.	726.9	-1657.1	-1651.1	-1639.4	-1622.0	-1597.2	542.5	3.8934

BALLISTIC COEFFICIENT .10, MUZZLE VELOCITY FT/SEC 3800.

RANGE	VELOCITY	HEIGHT	100 YD	150 YD	200 YD	250 YD	300 YD	DRIFT	TIME
0.	3800.	0.0	-1.5	-1.5	-1.5	-1.5	-1.5	0.0	0.0000
100.	2760.	-.3	0.0	.5	1.6	3.2	5.5	2.4	.0929
200.	1915.	1.6	-3.3	-2.2	0.0	3.2	7.8	11.5	.2234
300.	1281.	7.6	-16.6	-15.0	-11.7	-6.9	0.0	31.6	.4165
400.	993.	22.0	-51.7	-49.5	-45.2	-38.7	-29.5	65.7	.6891
500.	843.	48.9	-117.4	-114.7	-109.3	-101.2	-89.7	109.5	1.0170
600.	722.	93.6	-225.1	-221.8	-215.3	-205.6	-191.8	163.3	1.4017
700.	611.	164.2	-391.4	-387.6	-380.0	-368.7	-352.6	229.0	1.8537
800.	512.	273.5	-643.2	-638.9	-630.2	-617.3	-598.9	309.6	2.3904
900.	423.	441.5	-1022.0	-1017.2	-1007.4	-992.9	-972.2	409.2	3.0354
1000.	344.	700.9	-1596.6	-1591.2	-1580.4	-1564.2	-1541.2	534.0	3.8234

BALLISTIC COEFFICIENT .10, MUZZLE VELOCITY FT/SEC 3900.

RANGE	VELOCITY	HEIGHT	100 YD	150 YD	200 YD	250 YD	300 YD	DRIFT	TIME
0.	3900.	0.0	-1.5	-1.5	-1.5	-1.5	-1.5	0.0	0.0000
100.	2840.	-.4	0.0	.5	1.5	3.0	5.1	2.3	.0902
200.	1980.	1.5	-3.0	-2.0	0.0	3.0	7.3	11.1	.2167
300.	1323.	7.1	-15.4	-13.9	-10.9	-6.4	0.0	30.4	.4033
400.	1009.	20.8	-48.4	-46.5	-42.5	-36.4	-27.9	63.7	.6696
500.	854.	46.7	-111.7	-109.2	-104.2	-96.7	-86.0	107.3	.9940
600.	732.	89.7	-215.3	-212.3	-206.3	-197.3	-184.5	160.4	1.3730
700.	620.	157.9	-375.7	-372.3	-365.2	-354.7	-339.7	225.2	1.8181
800.	520.	263.5	-618.7	-614.8	-606.8	-594.7	-577.6	304.7	2.3465
900.	430.	426.1	-985.3	-980.9	-971.8	-958.3	-939.0	403.0	2.9821
1000.	350.	676.5	-1539.8	-1535.0	-1524.9	-1509.9	-1488.5	525.7	3.7564

BALLISTIC COEFFICIENT .10, MUZZLE VELOCITY FT/SEC 4000.

RANGE	VELOCITY	HEIGHT	100 YD	150 YD	200 YD	250 YD	300 YD	DRIFT	TIME
0.	4000.	0.0	-1.5	-1.5	-1.5	-1.5	-1.5	0.0	0.0000
100.	2921.	-.4	0.0	.4	1.4	2.8	4.8	2.3	.0880
200.	2047.	1.4	-2.7	-1.9	0.0	2.8	6.8	10.7	.2107
300.	1367.	6.6	-14.3	-13.0	-10.2	-6.0	0.0	29.2	.3910
400.	1025.	19.6	-45.3	-43.6	-39.9	-34.3	-26.3	61.8	.6509
500.	866.	44.4	-105.8	-103.7	-99.0	-92.0	-82.0	104.8	.9703
600.	742.	86.1	-205.8	-203.3	-197.6	-189.2	-177.3	157.5	1.3449
700.	629.	152.1	-361.0	-358.0	-351.4	-341.6	-327.7	221.7	1.7844
800.	528.	254.3	-596.1	-592.6	-585.1	-573.9	-558.0	300.1	2.3051
900.	437.	411.2	-949.7	-945.8	-937.4	-924.8	-906.8	396.8	2.9297
1000.	356.	653.0	-1484.9	-1480.6	-1471.2	-1457.2	-1437.2	517.5	3.6904

BALLISTIC COEFFICIENT .11, MUZZLE VELOCITY FT/SEC 2000.

RANGE	VELOCITY	HEIGHT	100 YD	150 YD	200 YD	250 YD	300 YD	DRIFT	TIME
0.	2000.	0.0	-1.5	-1.5	-1.5	-1.5	-1.5	0.0	0.0000
100.	1384.	.8	0.0	3.6	8.6	14.5	21.3	5.5	.1811
200.	1048.	8.4	-17.2	-9.9	0.0	11.8	25.3	24.0	.4362
300.	895.	26.0	-63.8	-52.9	-38.0	-20.3	0.0	52.2	.7464
400.	778.	57.9	-149.6	-135.1	-115.2	-91.6	-64.5	89.0	1.1058
500.	672.	110.3	-287.4	-269.3	-244.5	-215.0	-181.1	135.7	1.5209
600.	576.	191.9	-496.0	-474.2	-444.4	-409.1	-368.4	194.2	2.0032
700.	488.	316.1	-803.8	-778.4	-743.7	-702.4	-655.0	267.4	2.5694
800.	409.	503.5	-1253.8	-1224.8	-1185.1	-1137.9	-1083.7	359.2	3.2411
900.	338.	785.4	-1911.9	-1879.2	-1834.6	-1781.5	-1720.5	474.7	4.0470
1000.	276.	1214.6	-2889.2	-2852.9	-2803.3	-2744.4	-2676.6	621.6	5.0319

BALLISTIC COEFFICIENT .11, MUZZLE VELOCITY FT/SEC 2100.

RANGE	VELOCITY	HEIGHT	100 YD	150 YD	200 YD	250 YD	300 YD	DRIFT	TIME
0.	2100.	0.0	-1.5	-1.5	-1.5	-1.5	-1.5	0.0	0.0000
100.	1456.	.7	0.0	3.2	7.7	13.3	19.7	5.1	.1721
200.	1076.	7.6	-15.4	-9.0	0.0	11.2	23.9	23.0	.4163
300.	914.	24.1	-59.0	-49.4	-35.9	-19.1	0.0	51.3	.7202
400.	793.	54.5	-140.2	-127.4	-109.4	-87.0	-61.5	88.3	1.0733
500.	686.	104.4	-270.9	-254.9	-232.4	-204.3	-172.6	134.8	1.4800
600.	589.	182.2	-469.1	-449.9	-422.8	-389.2	-351.1	192.7	1.9522
700.	500.	300.4	-761.1	-738.7	-707.1	-667.9	-623.4	264.9	2.5049
800.	420.	478.6	-1188.0	-1162.3	-1126.3	-1081.5	-1030.6	355.0	3.1600
900.	348.	747.1	-1814.1	-1785.2	-1744.7	-1694.3	-1637.1	468.4	3.9472
1000.	284.	1153.1	-2737.4	-2705.3	-2660.3	-2604.3	-2540.7	611.5	4.9029

BALLISTIC COEFFICIENT .11, MUZZLE VELOCITY FT/SEC 2200.

RANGE	VELOCITY	HEIGHT	100 YD	150 YD	200 YD	250 YD	300 YD	DRIFT	TIME
0.	2200.	0.0	-1.5	-1.5	-1.5	-1.5	-1.5	0.0	0.0000
100.	1533.	.5	0.0	2.9	7.0	12.2	18.2	4.8	.1638
200.	1108.	6.9	-13.9	-8.2	0.0	10.5	22.5	22.1	.3981
300.	933.	22.4	-54.6	-46.1	-33.8	-18.0	0.0	50.4	.6952
400.	808.	51.3	-131.3	-119.9	-103.5	-82.5	-58.5	87.3	1.0415
500.	700.	98.8	-255.3	-241.1	-220.6	-194.3	-164.3	133.5	1.4402
600.	601.	173.0	-443.7	-426.6	-402.0	-370.5	-334.5	190.9	1.9026
700.	511.	286.0	-722.1	-702.1	-673.4	-636.7	-594.6	262.2	2.4442
800.	430.	455.9	-1128.4	-1105.6	-1072.8	-1030.8	-982.8	350.9	3.0844
900.	357.	710.8	-1721.7	-1695.9	-1659.1	-1611.8	-1557.8	461.6	3.8502
1000.	292.	1096.8	-2599.0	-2570.4	-2529.4	-2476.9	-2416.9	601.6	4.7819

BALLISTIC COEFFICIENT .11, MUZZLE VELOCITY FT/SEC 2300.

RANGE	VELOCITY	HEIGHT	100 YD	150 YD	200 YD	250 YD	300 YD	DRIFT	TIME
0.	2300.	0.0	-1.5	-1.5	-1.5	-1.5	-1.5	0.0	0.0000
100.	1611.	.4	0.0	2.5	6.3	11.1	16.8	4.5	.1562
200.	1143.	6.2	-12.5	-7.4	0.0	9.8	21.1	21.0	.3804
300.	952.	20.8	-50.5	-42.8	-31.7	-17.0	0.0	49.2	.6707
400.	823.	48.3	-122.9	-112.7	-97.9	-78.3	-55.6	86.0	1.0104
500.	714.	93.6	-240.9	-228.1	-209.6	-185.1	-156.7	132.0	1.4019
600.	614.	164.4	-420.2	-404.9	-382.7	-353.3	-319.3	188.8	1.8552
700.	523.	272.2	-685.0	-667.2	-641.2	-607.0	-567.2	259.0	2.3847
800.	440.	434.1	-1071.4	-1051.0	-1021.4	-982.2	-936.8	346.1	3.0099
900.	366.	677.3	-1636.8	-1613.9	-1580.6	-1536.5	-1485.4	454.9	3.7585
1000.	300.	1042.8	-2466.4	-2440.9	-2403.8	-2354.9	-2298.2	591.1	4.6627

BALLISTIC COEFFICIENT .11, MUZZLE VELOCITY FT/SEC 2400.

RANGE	VELOCITY	HEIGHT	100 YD	150 YD	200 YD	250 YD	300 YD	DRIFT	TIME
0.	2400.	0.0	-1.5	-1.5	-1.5	-1.5	-1.5	0.0	0.0000
100.	1690.	.3	0.0	2.3	5.6	10.1	15.6	4.3	.1492
200.	1186.	5.6	-11.3	-6.7	0.0	9.0	19.9	20.0	.3636
300.	971.	19.4	-46.7	-39.9	-29.8	-16.4	0.0	48.0	.6478
400.	838.	45.4	-115.0	-105.9	-92.5	-74.6	-52.7	84.5	.9802
500.	728.	88.6	-227.2	-215.8	-199.0	-176.6	-149.3	130.2	1.3646
600.	626.	156.3	-398.0	-384.4	-364.3	-337.4	-304.6	186.4	1.8091
700.	534.	259.4	-650.7	-634.8	-611.3	-579.9	-541.6	255.7	2.3281
800.	451.	414.2	-1019.5	-1001.4	-974.5	-938.7	-894.9	341.5	2.9402
900.	375.	645.8	-1557.3	-1536.9	-1506.6	-1466.3	-1417.1	448.0	3.6702
1000.	308.	994.7	-2348.8	-2326.1	-2292.5	-2247.7	-2193.0	581.5	4.5540

BALLISTIC COEFFICIENT .11, MUZZLE VELOCITY FT/SEC 2500.

RANGE	VELOCITY	HEIGHT	100 YD	150 YD	200 YD	250 YD	300 YD	DRIFT	TIME
0.	2500.	0.0	-1.5	-1.5	-1.5	-1.5	-1.5	0.0	0.0000
100.	1770.	.2	0.0	2.0	5.1	9.2	14.3	4.0	.1427
200.	1233.	5.1	-10.2	-6.1	0.0	8.2	18.5	19.0	.3478
300.	991.	18.0	-43.0	-37.0	-27.8	-15.4	0.0	46.6	.6245
400.	853.	42.7	-107.8	-99.7	-87.5	-71.0	-50.4	83.1	.9519
500.	741.	83.9	-214.2	-204.1	-188.8	-168.2	-142.4	128.2	1.3282
600.	639.	148.7	-377.1	-365.0	-346.6	-321.9	-291.0	183.8	1.7643
700.	545.	247.3	-618.3	-604.1	-582.8	-553.9	-517.9	252.2	2.2731
800.	461.	395.0	-969.8	-953.6	-929.2	-896.2	-855.0	336.4	2.8714
900.	384.	616.3	-1483.0	-1464.8	-1437.3	-1400.2	-1353.9	441.0	3.5855
1000.	316.	948.4	-2235.8	-2215.6	-2185.0	-2143.8	-2092.3	571.4	4.4469

BALLISTIC COEFFICIENT .11, MUZZLE VELOCITY FT/SEC 2600.

RANGE	VELOCITY	HEIGHT	100 YD	150 YD	200 YD	250 YD	300 YD	DRIFT	TIME
0.	2600.	0.0	-1.5	-1.5	-1.5	-1.5	-1.5	0.0	0.0000
100.	1852.	.1	0.0	1.8	4.6	8.4	13.2	3.8	.1368
200.	1285.	4.6	-9.1	-5.5	0.0	7.6	17.3	17.9	.3327
300.	1011.	16.6	-39.6	-34.2	-25.9	-14.5	0.0	45.0	.6019
400.	868.	40.1	-100.6	-93.3	-82.3	-67.2	-47.8	81.1	.9222
500.	755.	79.5	-202.1	-193.0	-179.3	-160.3	-136.1	126.1	1.2931
600.	651.	141.5	-357.7	-346.8	-330.3	-307.5	-278.5	181.1	1.7215
700.	556.	235.8	-587.9	-575.2	-555.9	-529.4	-495.5	248.6	2.2200
800.	471.	377.4	-924.2	-909.8	-887.7	-857.4	-818.6	331.5	2.8068
900.	393.	588.8	-1413.9	-1397.6	-1372.8	-1338.7	-1295.1	434.0	3.5046
1000.	324.	905.5	-2131.2	-2113.1	-2085.6	-2047.7	-1999.2	561.7	4.3451

BALLISTIC COEFFICIENT .11, MUZZLE VELOCITY FT/SEC 2700.

RANGE	VELOCITY	HEIGHT	100 YD	150 YD	200 YD	250 YD	300 YD	DRIFT	TIME
0.	2700.	0.0	-1.5	-1.5	-1.5	-1.5	-1.5	0.0	0.0000
100.	1935.	.1	0.0	1.6	4.1	7.6	12.1	3.6	.1313
200.	1339.	4.1	-8.2	-5.0	0.0	6.9	15.9	17.0	.3187
300.	1032.	15.3	-36.2	-31.4	-23.9	-13.5	0.0	43.3	.5791
400.	883.	37.6	-93.9	-87.5	-77.5	-63.6	-45.6	79.2	.8943
500.	768.	75.3	-190.7	-182.6	-170.1	-152.8	-130.3	123.8	1.2591
600.	663.	134.7	-339.2	-329.5	-314.5	-293.7	-266.7	178.3	1.6796
700.	567.	225.1	-559.5	-548.1	-530.6	-506.4	-474.9	244.9	2.1691
800.	481.	360.5	-880.5	-867.6	-847.6	-819.9	-783.9	326.4	2.7433
900.	402.	562.7	-1348.5	-1334.0	-1311.5	-1280.3	-1239.8	427.0	3.4261
1000.	332.	865.6	-2034.3	-2018.2	-1993.2	-1958.5	-1913.6	552.2	4.2484

BALLISTIC COEFFICIENT .11, MUZZLE VELOCITY FT/SEC 2800.

RANGE	VELOCITY	HEIGHT	100 YD	150 YD	200 YD	250 YD	300 YD	DRIFT	TIME
0.	2800.	0.0	-1.5	-1.5	-1.5	-1.5	-1.5	0.0	0.0000
100.	2018.	.0	0.0	1.4	3.7	6.9	11.1	3.4	.1262
200.	1397.	3.7	-7.4	-4.5	0.0	6.3	14.7	16.1	.3056
300.	1054.	14.2	-33.2	-28.9	-22.1	-12.6	0.0	41.6	.5579
400.	899.	35.4	-87.8	-82.0	-72.9	-60.2	-43.5	77.2	.8674
500.	781.	71.4	-179.9	-172.6	-161.3	-145.4	-124.5	121.5	1.2259
600.	675.	128.3	-321.9	-313.2	-299.6	-280.6	-255.4	175.4	1.6394
700.	578.	214.9	-532.6	-522.4	-506.5	-484.4	-455.0	241.1	2.1196
800.	491.	344.9	-840.6	-829.1	-810.9	-785.5	-752.0	321.5	2.6837
900.	411.	538.7	-1288.5	-1275.5	-1255.0	-1226.5	-1188.8	420.3	3.3523
1000.	340.	827.4	-1941.6	-1927.1	-1904.4	-1872.7	-1830.8	542.5	4.1537

BALLISTIC COEFFICIENT .11, MUZZLE VELOCITY FT/SEC 2900.

RANGE	VELOCITY	HEIGHT	100 YD	150 YD	200 YD	250 YD	300 YD	DRIFT	TIME
0.	2900.	0.0	-1.5	-1.5	-1.5	-1.5	-1.5	0.0	0.0000
100.	2102.	-.0	0.0	1.3	3.4	6.3	10.2	3.2	.1216
200.	1458.	3.4	-6.7	-4.1	0.0	5.8	13.6	15.2	.2935
300.	1077.	13.1	-30.5	-26.6	-20.4	-11.7	0.0	40.0	.5375
400.	914.	33.2	-82.0	-76.8	-68.5	-56.9	-41.4	75.2	.8412
500.	794.	67.7	-169.8	-163.4	-153.0	-138.5	-119.0	119.1	1.1942
600.	687.	122.2	-305.7	-297.9	-285.5	-268.1	-244.8	172.5	1.6007
700.	589.	205.5	-507.7	-498.6	-484.1	-463.8	-436.6	237.4	2.0727
800.	500.	330.0	-802.4	-792.0	-775.4	-752.2	-721.1	316.4	2.6252
900.	420.	515.6	-1231.0	-1219.3	-1200.7	-1174.5	-1139.5	413.4	3.2799
1000.	348.	793.1	-1858.5	-1845.5	-1824.8	-1795.7	-1756.9	533.7	4.0667

BALLISTIC COEFFICIENT .11, MUZZLE VELOCITY FT/SEC 3000.

RANGE	VELOCITY	HEIGHT	100 YD	150 YD	200 YD	250 YD	300 YD	DRIFT	TIME
0.	3000.	0.0	-1.5	-1.5	-1.5	-1.5	-1.5	0.0	0.0000
100.	2186.	-.1	0.0	1.2	3.1	5.7	9.3	3.0	.1172
200.	1522.	3.1	-6.1	-3.8	0.0	5.3	12.5	14.5	.2823
300.	1103.	12.1	-28.0	-24.5	-18.8	-10.8	0.0	38.4	.5180
400.	930.	31.2	-76.6	-72.0	-64.4	-53.7	-39.3	73.2	.8161
500.	806.	64.2	-160.4	-154.5	-145.1	-131.8	-113.8	116.7	1.1634
600.	698.	116.5	-290.4	-283.4	-272.1	-256.1	-234.5	169.5	1.5632
700.	600.	196.4	-484.0	-475.9	-462.6	-444.0	-418.8	233.5	2.0269
800.	510.	316.3	-767.3	-758.0	-742.9	-721.6	-692.7	311.6	2.5702
900.	429.	494.6	-1178.8	-1168.3	-1151.3	-1127.3	-1094.9	407.0	3.2126
1000.	356.	760.1	-1778.7	-1767.0	-1748.1	-1721.5	-1685.5	524.7	3.9814

BALLISTIC COEFFICIENT .11, MUZZLE VELOCITY FT/SEC 3100.

RANGE	VELOCITY	HEIGHT	100 YD	150 YD	200 YD	250 YD	300 YD	DRIFT	TIME
0.	3100.	0.0	-1.5	-1.5	-1.5	-1.5	-1.5	0.0	0.0000
100.	2269.	-.1	0.0	1.0	2.8	5.2	8.5	2.9	.1132
200.	1586.	2.8	-5.6	-3.5	0.0	4.9	11.5	13.8	.2717
300.	1132.	11.2	-25.6	-22.5	-17.3	-10.0	0.0	36.7	.4991
400.	945.	29.4	-71.7	-67.5	-60.6	-50.8	-37.5	71.3	.7920
500.	819.	60.9	-151.4	-146.2	-137.5	-125.3	-108.6	114.3	1.1332
600.	710.	111.2	-276.1	-269.9	-259.5	-244.8	-224.8	166.6	1.5271
700.	610.	188.0	-462.0	-454.7	-442.6	-425.5	-402.2	229.8	1.9832
800.	519.	303.1	-733.5	-725.1	-711.3	-691.8	-665.1	306.6	2.5160
900.	437.	474.4	-1128.5	-1119.1	-1103.5	-1081.5	-1051.6	400.4	3.1463
1000.	363.	729.2	-1704.0	-1693.5	-1676.2	-1651.8	-1618.5	516.0	3.8997

BALLISTIC COEFFICIENT .11, MUZZLE VELOCITY FT/SEC 3200.

RANGE	VELOCITY	HEIGHT	100 YD	150 YD	200 YD	250 YD	300 YD	DRIFT	TIME
0.	3200.	0.0	-1.5	-1.5	-1.5	-1.5	-1.5	0.0	0.0000
100.	2352.	-.2	0.0	.9	2.5	4.8	7.8	2.8	.1095
200.	1651.	2.5	-5.1	-3.2	0.0	4.5	10.6	13.1	.2620
300.	1164.	10.4	-23.5	-20.7	-16.0	-9.2	0.0	35.2	.4811
400.	962.	27.5	-66.8	-63.1	-56.7	-47.7	-35.4	69.1	.7678
500.	831.	57.8	-143.0	-138.3	-130.4	-119.2	-103.8	111.9	1.1043
600.	721.	106.1	-262.5	-256.9	-247.4	-233.9	-215.4	163.6	1.4918
700.	620.	180.0	-441.0	-434.4	-423.3	-407.6	-386.0	226.0	1.9404
800.	529.	290.9	-702.4	-694.9	-682.2	-664.2	-639.6	301.8	2.4649
900.	446.	455.6	-1081.8	-1073.4	-1059.1	-1038.9	-1011.2	394.2	3.0833
1000.	371.	700.5	-1634.8	-1625.5	-1609.6	-1587.1	-1556.4	507.7	3.8223

BALLISTIC COEFFICIENT .11, MUZZLE VELOCITY FT/SEC 3300.

RANGE	VELOCITY	HEIGHT	100 YD	150 YD	200 YD	250 YD	300 YD	DRIFT	TIME
0.	3300.	0.0	-1.5	-1.5	-1.5	-1.5	-1.5	0.0	0.0000
100.	2434.	-.2	0.0	.8	2.3	4.4	7.2	2.7	.1060
200.	1717.	2.3	-4.6	-2.9	0.0	4.1	9.8	12.5	.2529
300.	1202.	9.6	-21.7	-19.1	-14.7	-8.5	0.0	33.7	.4641
400.	978.	25.9	-62.4	-59.0	-53.2	-44.9	-33.5	67.1	.7448
500.	843.	54.8	-135.1	-130.9	-123.6	-113.2	-99.0	109.4	1.0761
600.	732.	101.3	-249.9	-244.8	-236.0	-223.6	-206.5	160.6	1.4581
700.	631.	172.5	-421.5	-415.6	-405.4	-390.9	-371.0	222.4	1.8998
800.	538.	279.2	-672.6	-665.8	-654.2	-637.6	-614.8	297.0	2.4148
900.	454.	437.8	-1037.8	-1030.1	-1017.0	-998.4	-972.8	388.0	3.0225
1000.	378.	672.9	-1568.3	-1559.8	-1545.2	-1524.5	-1496.1	499.3	3.7462

BALLISTIC COEFFICIENT .11, MUZZLE VELOCITY FT/SEC 3400.

RANGE	VELOCITY	HEIGHT	100 YD	150 YD	200 YD	250 YD	300 YD	DRIFT	TIME
0.	3400.	0.0	-1.5	-1.5	-1.5	-1.5	-1.5	0.0	0.0000
100.	2517.	-.2	0.0	.8	2.1	4.0	6.6	2.5	.1027
200.	1783.	2.1	-4.2	-2.7	0.0	3.8	9.1	12.0	.2444
300.	1242.	8.9	-19.9	-17.7	-13.6	-7.9	0.0	32.2	.4478
400.	995.	24.3	-58.2	-55.2	-49.8	-42.2	-31.7	65.0	.7225
500.	855.	52.1	-127.8	-124.0	-117.3	-107.8	-94.6	107.0	1.0494
600.	743.	96.7	-237.7	-233.2	-225.1	-213.7	-197.9	157.6	1.4251
700.	641.	165.3	-402.7	-397.4	-388.0	-374.7	-356.2	218.6	1.8597
800.	547.	268.2	-644.8	-638.7	-627.9	-612.7	-591.6	292.4	2.3670
900.	462.	420.8	-995.8	-988.9	-976.9	-959.7	-936.0	381.8	2.9634
1000.	386.	647.4	-1507.0	-1499.4	-1486.0	-1466.9	-1440.6	491.5	3.6747

BALLISTIC COEFFICIENT .11, MUZZLE VELOCITY FT/SEC 3500.

RANGE	VELOCITY	HEIGHT	100 YD	150 YD	200 YD	250 YD	300 YD	DRIFT	TIME
0.	3500.	0.0	-1.5	-1.5	-1.5	-1.5	-1.5	0.0	0.0000
100.	2599.	-.3	0.0	.7	1.9	3.7	6.1	2.5	.0996
200.	1851.	1.9	-3.8	-2.5	0.0	3.5	8.4	11.5	.2365
300.	1284.	8.2	-18.4	-16.3	-12.6	-7.3	0.0	30.9	.4325
400.	1011.	22.9	-54.5	-51.8	-46.8	-39.8	-30.0	63.2	.7018
500.	868.	49.4	-120.6	-117.2	-111.0	-102.2	-90.0	104.5	1.0223
600.	754.	92.4	-226.4	-222.3	-214.9	-204.3	-189.7	154.7	1.3933
700.	651.	158.5	-385.3	-380.6	-371.9	-359.5	-342.5	215.0	1.8217
800.	556.	257.7	-618.2	-612.8	-602.9	-588.7	-569.3	287.7	2.3204
900.	471.	405.0	-956.8	-950.7	-939.6	-923.7	-901.7	375.9	2.9074
1000.	393.	623.2	-1448.8	-1441.9	-1429.6	-1411.9	-1387.5	483.7	3.6054

BALLISTIC COEFFICIENT .11, MUZZLE VELOCITY FT/SEC 3600.

RANGE	VELOCITY	HEIGHT	100 YD	150 YD	200 YD	250 YD	300 YD	DRIFT	TIME
0.	3600.	0.0	-1.5	-1.5	-1.5	-1.5	-1.5	0.0	0.0000
100.	2681.	-.3	0.0	.6	1.8	3.4	5.7	2.3	.0966
200.	1919.	1.8	-3.5	-2.3	0.0	3.3	7.8	11.0	.2290
300.	1328.	7.6	-17.0	-15.1	-11.7	-6.8	0.0	29.6	.4180
400.	1028.	21.5	-50.8	-48.4	-43.8	-37.2	-28.2	61.1	.6805
500.	880.	46.9	-114.0	-110.9	-105.2	-97.0	-85.7	102.0	.9962
600.	765.	88.3	-215.6	-211.9	-205.1	-195.2	-181.7	151.7	1.3622
700.	661.	152.0	-368.6	-364.3	-356.3	-344.8	-329.0	211.3	1.7841
800.	565.	247.8	-593.2	-588.3	-579.2	-566.1	-548.0	283.1	2.2754
900.	479.	389.7	-919.2	-913.7	-903.4	-888.6	-868.3	370.0	2.8520
1000.	401.	599.9	-1392.9	-1386.7	-1375.3	-1358.9	-1336.3	475.9	3.5374

BALLISTIC COEFFICIENT .11, MUZZLE VELOCITY FT/SEC 3700.

RANGE	VELOCITY	HEIGHT	100 YD	150 YD	200 YD	250 YD	300 YD	DRIFT	TIME
0.	3700.	0.0	-1.5	-1.5	-1.5	-1.5	-1.5	0.0	0.0000
100.	2763.	-.3	0.0	.5	1.6	3.1	5.2	2.3	.0940
200.	1987.	1.6	-3.2	-2.1	0.0	3.0	7.2	10.5	.2220
300.	1375.	7.1	-15.7	-14.0	-10.9	-6.3	0.0	28.4	.4044
400.	1045.	20.2	-47.4	-45.2	-41.0	-34.9	-26.5	59.1	.6603
500.	893.	44.5	-107.7	-105.0	-99.7	-92.1	-81.6	99.6	.9713
600.	776.	84.4	-205.4	-202.1	-195.8	-186.6	-174.1	148.8	1.3322
700.	671.	146.0	-352.9	-349.1	-341.7	-331.0	-316.4	207.8	1.7484
800.	574.	238.4	-569.5	-565.2	-556.7	-544.5	-527.8	278.7	2.2320
900.	487.	375.5	-884.3	-879.4	-869.9	-856.2	-837.3	364.3	2.7998
1000.	408.	578.4	-1341.4	-1335.9	-1325.4	-1310.1	-1289.2	468.7	3.4737

BALLISTIC COEFFICIENT .11, MUZZLE VELOCITY FT/SEC 3800.

RANGE	VELOCITY	HEIGHT	100 YD	150 YD	200 YD	250 YD	300 YD	DRIFT	TIME
0.	3800.	0.0	-1.5	-1.5	-1.5	-1.5	-1.5	0.0	0.0000
100.	2845.	-.3	0.0	.5	1.5	2.9	4.8	2.2	.0914
200.	2056.	1.5	-2.9	-2.0	0.0	2.8	6.7	10.1	.2155
300.	1424.	6.6	-14.5	-13.0	-10.1	-5.8	0.0	27.2	.3914
400.	1064.	18.9	-44.2	-42.2	-38.3	-32.7	-24.9	57.2	.6406
500.	906.	42.3	-101.8	-99.4	-94.4	-87.4	-77.6	97.2	.9469
600.	787.	80.7	-195.8	-192.8	-186.9	-178.4	-166.8	146.0	1.3030
700.	680.	140.1	-337.8	-334.4	-327.5	-317.7	-304.0	204.2	1.7130
800.	583.	229.4	-547.0	-543.1	-535.2	-523.9	-508.3	274.2	2.1897
900.	495.	361.9	-850.9	-846.5	-837.7	-825.0	-807.5	358.7	2.7486
1000.	415.	557.7	-1291.8	-1286.9	-1277.1	-1263.0	-1243.5	461.4	3.4110

BALLISTIC COEFFICIENT .11, MUZZLE VELOCITY FT/SEC 3900.

RANGE	VELOCITY	HEIGHT	100 YD	150 YD	200 YD	250 YD	300 YD	DRIFT	TIME
0.	3900.	0.0	-1.5	-1.5	-1.5	-1.5	-1.5	0.0	0.0000
100.	2927.	-.4	0.0	.4	1.3	2.7	4.5	2.1	.0889
200.	2125.	1.3	-2.7	-1.8	0.0	2.6	6.3	9.7	.2092
300.	1475.	6.1	-13.4	-12.1	-9.4	-5.4	0.0	26.1	.3791
400.	1084.	17.8	-41.1	-39.4	-35.8	-30.5	-23.2	55.1	.6209
500.	918.	40.1	-96.2	-94.0	-89.5	-82.9	-73.8	94.7	.9230
600.	797.	77.2	-186.5	-183.9	-178.5	-170.5	-159.7	143.1	1.2744
700.	690.	134.6	-323.7	-320.7	-314.3	-305.1	-292.4	200.8	1.6792
800.	592.	220.9	-525.7	-522.2	-515.0	-504.4	-489.9	269.9	2.1489
900.	503.	348.9	-819.1	-815.1	-807.0	-795.1	-778.8	353.1	2.6988
1000.	422.	538.0	-1244.6	-1240.2	-1231.2	-1218.0	-1199.8	454.2	3.3501

BALLISTIC COEFFICIENT .11, MUZZLE VELOCITY FT/SEC 4000.

RANGE	VELOCITY	HEIGHT	100 YD	150 YD	200 YD	250 YD	300 YD	DRIFT	TIME
0.	4000.	0.0	-1.5	-1.5	-1.5	-1.5	-1.5	0.0	0.0000
100.	3010.	-.4	0.0	.4	1.2	2.5	4.2	2.0	.0866
200.	2194.	1.2	-2.5	-1.7	0.0	2.5	5.9	9.4	.2034
300.	1528.	5.7	-12.5	-11.3	-8.8	-5.1	0.0	25.1	.3677
400.	1106.	16.7	-38.4	-36.8	-33.5	-28.5	-21.8	53.3	.6026
500.	931.	38.1	-91.0	-89.0	-84.8	-78.6	-70.2	92.4	.9001
600.	807.	73.9	-177.9	-175.5	-170.5	-163.1	-152.9	140.2	1.2469
700.	699.	129.3	-310.2	-307.5	-301.6	-292.9	-281.1	197.3	1.6460
800.	601.	212.7	-505.2	-502.1	-495.4	-485.5	-472.0	265.6	2.1089
900.	511.	336.7	-789.2	-785.7	-778.1	-767.0	-751.8	347.8	2.6513
1000.	429.	519.6	-1200.7	-1196.8	-1188.4	-1176.0	-1159.1	447.5	3.2924

BALLISTIC COEFFICIENT .12, MUZZLE VELOCITY FT/SEC 2000.

RANGE	VELOCITY	HEIGHT	100 YD	150 YD	200 YD	250 YD	300 YD	DRIFT	TIME
0.	2000.	0.0	-1.5	-1.5	-1.5	-1.5	-1.5	0.0	0.0000
100.	1427.	.8	0.0	3.4	8.0	13.6	20.0	4.9	.1781
200.	1082.	7.9	-16.0	-9.3	0.0	11.2	24.0	21.8	.4238
300.	929.	24.4	-60.1	-49.9	-36.0	-19.2	0.0	48.2	.7240
400.	815.	54.1	-140.8	-127.2	-108.7	-86.3	-60.7	82.6	1.0694
500.	715.	101.9	-268.6	-251.6	-228.4	-200.4	-168.4	125.4	1.4624
600.	623.	174.7	-458.1	-437.8	-410.0	-376.4	-337.9	178.1	1.9119
700.	538.	282.7	-731.0	-707.3	-674.9	-635.7	-590.8	242.9	2.4300
800.	461.	440.6	-1118.4	-1091.3	-1054.2	-1009.4	-958.2	322.5	3.0322
900.	391.	670.9	-1667.7	-1637.2	-1595.5	-1545.1	-1487.5	420.7	3.7405
1000.	327.	1006.2	-2447.3	-2413.4	-2367.1	-2311.1	-2247.0	542.1	4.5802

BALLISTIC COEFFICIENT .12, MUZZLE VELOCITY FT/SEC 2100.

RANGE	VELOCITY	HEIGHT	100 YD	150 YD	200 YD	250 YD	300 YD	DRIFT	TIME
0.	2100.	0.0	-1.5	-1.5	-1.5	-1.5	-1.5	0.0	0.0000
100.	1503.	.6	0.0	3.0	7.2	12.4	18.5	4.6	.1692
200.	1115.	7.1	-14.4	-8.4	0.0	10.5	22.6	20.9	.4044
300.	949.	22.6	-55.4	-46.4	-33.8	-18.1	0.0	47.4	.6978
400.	831.	50.8	-131.5	-119.4	-102.7	-81.7	-57.6	81.8	1.0364
500.	730.	96.3	-252.6	-237.5	-216.6	-190.3	-160.2	124.5	1.4218
600.	636.	165.7	-432.5	-414.4	-389.3	-357.8	-321.6	176.9	1.8620
700.	551.	268.7	-692.1	-670.9	-641.6	-604.9	-562.7	241.0	2.3693
800.	472.	419.2	-1060.3	-1036.2	-1002.7	-960.7	-912.5	319.4	2.9579
900.	401.	637.7	-1580.2	-1553.0	-1515.3	-1468.1	-1413.8	415.6	3.6469
1000.	337.	956.2	-2319.5	-2289.3	-2247.5	-2195.0	-2134.7	534.4	4.4649

BALLISTIC COEFFICIENT .12, MUZZLE VELOCITY FT/SEC 2200.

RANGE	VELOCITY	HEIGHT	100 YD	150 YD	200 YD	250 YD	300 YD	DRIFT	TIME
0.	2200.	0.0	-1.5	-1.5	-1.5	-1.5	-1.5	0.0	0.0000
100.	1581.	.5	0.0	2.7	6.5	11.4	17.1	4.4	.1613
200.	1153.	6.4	-13.0	-7.6	0.0	9.7	21.2	20.0	.3865
300.	969.	21.0	-51.2	-43.2	-31.7	-17.1	0.0	46.5	.6731
400.	847.	47.7	-122.7	-112.0	-96.7	-77.3	-54.4	80.8	1.0043
500.	744.	91.0	-237.4	-224.1	-205.0	-180.7	-152.1	123.3	1.3823
600.	650.	157.2	-408.6	-392.5	-369.7	-340.5	-306.2	175.3	1.8141
700.	563.	255.4	-655.2	-636.4	-609.7	-575.7	-535.7	238.6	2.3101
800.	483.	398.9	-1005.4	-984.0	-953.5	-914.6	-868.9	315.9	2.8855
900.	411.	607.4	-1500.8	-1476.7	-1442.4	-1398.7	-1347.2	410.5	3.5596
1000.	346.	910.1	-2202.3	-2175.5	-2137.4	-2088.7	-2031.6	526.7	4.3561

BALLISTIC COEFFICIENT .12, MUZZLE VELOCITY FT/SEC 2300.

RANGE	VELOCITY	HEIGHT	100 YD	150 YD	200 YD	250 YD	300 YD	DRIFT	TIME
0.	2300.	0.0	-1.5	-1.5	-1.5	-1.5	-1.5	0.0	0.0000
100.	1661.	.4	0.0	2.4	5.8	10.3	15.7	4.1	.1537
200.	1199.	5.8	-11.6	-6.9	0.0	8.9	19.8	18.9	.3684
300.	990.	19.4	-47.1	-39.9	-29.7	-16.3	0.0	45.2	.6482
400.	862.	44.7	-114.4	-104.9	-91.2	-73.3	-51.6	79.4	.9728
500.	758.	85.9	-223.3	-211.4	-194.3	-171.9	-144.8	121.7	1.3438
600.	663.	149.1	-386.0	-371.8	-351.2	-324.4	-291.9	173.3	1.7672
700.	575.	243.0	-621.0	-604.3	-580.4	-549.1	-511.1	235.9	2.2534
800.	494.	380.0	-954.6	-935.6	-908.2	-872.5	-829.1	312.0	2.8165
900.	421.	578.4	-1425.1	-1403.7	-1372.9	-1332.7	-1283.9	404.8	3.4737
1000.	355.	866.9	-2093.0	-2069.2	-2035.0	-1990.3	-1936.1	518.7	4.2517

BALLISTIC COEFFICIENT .12, MUZZLE VELOCITY FT/SEC 2400.

RANGE	VELOCITY	HEIGHT	100 YD	150 YD	200 YD	250 YD	300 YD	DRIFT	TIME
0.	2400.	0.0	-1.5	-1.5	-1.5	-1.5	-1.5	0.0	0.0000
100.	1742.	.3	0.0	2.1	5.2	9.3	14.4	3.8	.1468
200.	1250.	5.2	-10.4	-6.2	0.0	8.2	18.4	17.9	.3518
300.	1011.	17.9	-43.2	-36.8	-27.6	-15.3	0.0	43.8	.6240
400.	878.	41.9	-106.6	-98.2	-85.8	-69.4	-49.0	77.9	.9424
500.	773.	81.2	-210.0	-199.4	-184.0	-163.5	-138.0	120.0	1.3067
600.	676.	141.6	-364.9	-352.2	-333.6	-309.1	-278.5	171.1	1.7220
700.	587.	231.3	-588.9	-574.1	-552.4	-523.8	-488.1	233.0	2.1986
800.	505.	362.2	-907.0	-890.0	-865.3	-832.5	-791.8	308.0	2.7498
900.	431.	551.9	-1356.2	-1337.2	-1309.3	-1272.5	-1226.6	399.2	3.3933
1000.	364.	826.4	-1990.9	-1969.7	-1938.8	-1897.8	-1846.9	510.6	4.1513

BALLISTIC COEFFICIENT .12, MUZZLE VELOCITY FT/SEC 2500.

RANGE	VELOCITY	HEIGHT	100 YD	150 YD	200 YD	250 YD	300 YD	DRIFT	TIME
0.	2500.	0.0	-1.5	-1.5	-1.5	-1.5	-1.5	0.0	0.0000
100.	1825.	.2	0.0	1.9	4.7	8.4	13.1	3.6	.1405
200.	1303.	4.7	-9.4	-5.6	0.0	7.5	16.9	16.9	.3362
300.	1033.	16.5	-39.4	-33.7	-25.4	-14.1	0.0	42.1	.5994
400.	894.	39.3	-99.5	-91.9	-80.7	-65.8	-46.9	76.2	.9130
500.	787.	76.7	-197.5	-188.0	-174.1	-155.4	-131.8	118.0	1.2705
600.	689.	134.4	-345.1	-333.8	-317.0	-294.6	-266.3	168.7	1.6782
700.	599.	220.2	-558.6	-545.3	-525.8	-499.6	-466.6	229.7	2.1453
800.	516.	345.4	-862.2	-847.1	-824.8	-794.8	-757.1	303.7	2.6853
900.	441.	526.5	-1290.2	-1273.2	-1248.1	-1214.4	-1172.0	393.2	3.3141
1000.	373.	789.0	-1896.7	-1877.8	-1849.8	-1812.4	-1765.3	502.7	4.0562

BALLISTIC COEFFICIENT .12, MUZZLE VELOCITY FT/SEC 2600.

RANGE	VELOCITY	HEIGHT	100 YD	150 YD	200 YD	250 YD	300 YD	DRIFT	TIME
0.	2600.	0.0	-1.5	-1.5	-1.5	-1.5	-1.5	0.0	0.0000
100.	1908.	.1	0.0	1.7	4.2	7.6	12.0	3.4	.1347
200.	1361.	4.2	-8.4	-5.1	0.0	6.8	15.6	16.0	.3217
300.	1056.	15.2	-36.0	-31.0	-23.4	-13.1	0.0	40.5	.5764
400.	911.	36.7	-92.4	-85.7	-75.5	-61.9	-44.4	74.2	.8833
500.	800.	72.5	-185.6	-177.2	-164.5	-147.5	-125.6	115.8	1.2351
600.	702.	127.6	-326.3	-316.2	-301.0	-280.6	-254.3	166.0	1.6355
700.	611.	209.8	-530.4	-518.6	-500.9	-477.0	-446.4	226.4	2.0943
800.	527.	329.6	-820.5	-807.0	-786.8	-759.5	-724.4	299.3	2.6235
900.	451.	503.1	-1229.9	-1214.7	-1192.0	-1161.3	-1121.8	387.4	3.2398
1000.	381.	753.1	-1806.8	-1789.9	-1764.7	-1730.6	-1686.7	494.4	3.9631

BALLISTIC COEFFICIENT .12, MUZZLE VELOCITY FT/SEC 2700.

RANGE	VELOCITY	HEIGHT	100 YD	150 YD	200 YD	250 YD	300 YD	DRIFT	TIME
0.	2700.	0.0	-1.5	-1.5	-1.5	-1.5	-1.5	0.0	0.0000
100.	1993.	.1	0.0	1.5	3.8	6.9	11.0	3.2	.1293
200.	1422.	3.8	-7.6	-4.6	0.0	6.2	14.3	15.1	.3080
300.	1080.	14.0	-32.9	-28.4	-21.5	-12.1	0.0	38.9	.5542
400.	928.	34.3	-86.0	-79.9	-70.8	-58.3	-42.1	72.3	.8551
500.	814.	68.5	-174.6	-167.0	-155.6	-140.0	-119.7	113.6	1.2010
600.	714.	121.3	-308.9	-299.8	-286.1	-267.3	-243.0	163.3	1.5946
700.	622.	199.9	-503.7	-493.1	-477.1	-455.3	-426.9	223.0	2.0446
800.	538.	314.7	-781.1	-769.0	-750.7	-725.7	-693.3	294.7	2.5636
900.	460.	480.6	-1172.0	-1158.4	-1137.8	-1109.7	-1073.3	381.3	3.1666
1000.	390.	720.5	-1725.1	-1710.0	-1687.2	-1655.9	-1615.4	486.7	3.8762

BALLISTIC COEFFICIENT .12, MUZZLE VELOCITY FT/SEC 2800.

RANGE	VELOCITY	HEIGHT	100 YD	150 YD	200 YD	250 YD	300 YD	DRIFT	TIME
0.	2800.	0.0	-1.5	-1.5	-1.5	-1.5	-1.5	0.0	0.0000
100.	2079.	-.0	0.0	1.4	3.4	6.3	10.0	3.0	.1243
200.	1487.	3.4	-6.9	-4.2	0.0	5.7	13.2	14.3	.2954
300.	1108.	12.9	-30.1	-26.0	-19.8	-11.2	0.0	37.2	.5328
400.	945.	32.2	-80.1	-74.7	-66.4	-55.0	-40.0	70.3	.8282
500.	828.	64.7	-164.1	-157.4	-147.0	-132.7	-114.0	111.2	1.1678
600.	727.	115.3	-292.4	-284.3	-271.8	-254.7	-232.3	160.5	1.5548
700.	634.	190.7	-478.9	-469.4	-454.9	-435.0	-408.8	219.5	1.9971
800.	548.	300.8	-744.5	-733.7	-717.1	-694.3	-664.4	290.3	2.5065
900.	470.	460.0	-1119.2	-1107.1	-1088.4	-1062.7	-1029.1	375.6	3.0981
1000.	399.	689.0	-1646.8	-1633.2	-1612.4	-1584.0	-1546.6	478.6	3.7909

BALLISTIC COEFFICIENT .12, MUZZLE VELOCITY FT/SEC 2900.

RANGE	VELOCITY	HEIGHT	100 YD	150 YD	200 YD	250 YD	300 YD	DRIFT	TIME
0.	2900.	0.0	-1.5	-1.5	-1.5	-1.5	-1.5	0.0	0.0000
100.	2163.	-.1	0.0	1.2	3.1	5.7	9.2	2.9	.1198
200.	1553.	3.1	-6.2	-3.8	0.0	5.2	12.1	13.6	.2839
300.	1139.	11.9	-27.5	-23.8	-18.1	-10.3	0.0	35.6	.5125
400.	962.	30.1	-74.4	-69.5	-61.9	-51.6	-37.8	68.2	.8014
500.	841.	61.1	-154.3	-148.2	-138.8	-125.8	-108.5	108.8	1.1356
600.	739.	109.6	-276.9	-269.6	-258.2	-242.6	-221.9	157.6	1.5163
700.	645.	182.0	-455.5	-447.0	-433.7	-415.6	-391.4	216.0	1.9512
800.	558.	287.6	-709.8	-700.1	-684.9	-664.2	-636.6	285.7	2.4510
900.	479.	440.2	-1068.6	-1057.7	-1040.6	-1017.3	-986.2	369.6	3.0309
1000.	407.	660.3	-1575.3	-1563.1	-1544.2	-1518.2	-1483.7	471.1	3.7112

BALLISTIC COEFFICIENT .12, MUZZLE VELOCITY FT/SEC 3000.

RANGE	VELOCITY	HEIGHT	100 YD	150 YD	200 YD	250 YD	300 YD	DRIFT	TIME
0.	3000.	0.0	-1.5	-1.5	-1.5	-1.5	-1.5	0.0	0.0000
100.	2248.	-.1	0.0	1.1	2.8	5.2	8.4	2.8	.1157
200.	1620.	2.8	-5.6	-3.5	0.0	4.8	11.1	12.9	.2731
300.	1174.	10.9	-25.2	-21.9	-16.7	-9.6	0.0	34.0	.4934
400.	979.	28.1	-69.1	-64.8	-57.9	-48.3	-35.6	66.1	.7758
500.	854.	57.9	-145.4	-140.0	-131.4	-119.4	-103.5	106.6	1.1056
600.	751.	104.3	-262.4	-255.9	-245.5	-231.2	-212.1	154.8	1.4795
700.	656.	173.8	-433.6	-426.0	-413.8	-397.2	-374.9	212.4	1.9069
800.	569.	275.4	-677.7	-669.0	-655.2	-636.1	-610.7	281.3	2.3984
900.	489.	422.0	-1022.3	-1012.5	-996.9	-975.5	-946.8	363.9	2.9679
1000.	416.	633.0	-1507.2	-1496.4	-1479.1	-1455.3	-1423.4	463.5	3.6335

BALLISTIC COEFFICIENT .12, MUZZLE VELOCITY FT/SEC 3100.

RANGE	VELOCITY	HEIGHT	100 YD	150 YD	200 YD	250 YD	300 YD	DRIFT	TIME
0.	3100.	0.0	-1.5	-1.5	-1.5	-1.5	-1.5	0.0	0.0000
100.	2332.	-.2	0.0	1.0	2.6	4.7	7.7	2.6	.1116
200.	1687.	2.6	-5.1	-3.2	0.0	4.4	10.2	12.2	.2630
300.	1215.	10.1	-23.0	-20.1	-15.3	-8.8	0.0	32.4	.4746
400.	997.	26.3	-64.3	-60.4	-54.0	-45.3	-33.6	64.0	.7510
500.	868.	54.7	-136.6	-131.7	-123.8	-112.9	-98.2	103.9	1.0745
600.	763.	99.2	-248.7	-242.9	-233.3	-220.2	-202.7	151.8	1.4433
700.	667.	166.0	-412.9	-406.1	-395.0	-379.7	-359.2	208.9	1.8641
800.	579.	263.6	-647.1	-639.3	-626.6	-609.1	-585.7	276.8	2.3467
900.	498.	404.6	-977.8	-969.1	-954.8	-935.1	-908.8	358.1	2.9059
1000.	424.	607.3	-1443.6	-1433.9	-1418.0	-1396.2	-1366.9	456.1	3.5591

BALLISTIC COEFFICIENT .12, MUZZLE VELOCITY FT/SEC 3200.

RANGE	VELOCITY	HEIGHT	100 YD	150 YD	200 YD	250 YD	300 YD	DRIFT	TIME
0.	3200.	0.0	-1.5	-1.5	-1.5	-1.5	-1.5	0.0	0.0000
100.	2416.	-.2	0.0	.9	2.3	4.3	7.0	2.5	.1080
200.	1756.	2.3	-4.7	-2.9	0.0	4.0	9.4	11.7	.2537
300.	1258.	9.3	-21.1	-18.5	-14.1	-8.1	0.0	31.0	.4572
400.	1015.	24.6	-59.7	-56.2	-50.4	-42.4	-31.6	62.0	.7271
500.	881.	51.7	-128.4	-124.1	-116.8	-106.8	-93.3	101.4	1.0451
600.	775.	94.5	-235.9	-230.6	-221.9	-209.8	-193.7	148.9	1.4087
700.	678.	158.7	-393.3	-387.2	-377.0	-363.0	-344.1	205.3	1.8225
800.	589.	252.7	-618.6	-611.7	-600.0	-583.9	-562.4	272.4	2.2978
900.	507.	388.3	-936.6	-928.7	-915.6	-897.5	-873.3	352.6	2.8471
1000.	433.	583.3	-1384.1	-1375.4	-1360.8	-1340.7	-1313.8	448.9	3.4881

BALLISTIC COEFFICIENT .12, MUZZLE VELOCITY FT/SEC 3300.

RANGE	VELOCITY	HEIGHT	100 YD	150 YD	200 YD	250 YD	300 YD	DRIFT	TIME
0.	3300.	0.0	-1.5	-1.5	-1.5	-1.5	-1.5	0.0	0.0000
100.	2500.	-.2	0.0	.8	2.1	4.0	6.5	2.4	.1046
200.	1825.	2.1	-4.2	-2.7	0.0	3.7	8.7	11.1	.2451
300.	1303.	8.6	-19.4	-17.0	-13.0	-7.4	0.0	29.6	.4407
400.	1033.	23.0	-55.4	-52.3	-47.0	-39.5	-29.6	59.9	.7039
500.	894.	48.9	-121.1	-117.1	-110.4	-101.2	-88.8	99.1	1.0175
600.	787.	90.0	-223.7	-219.0	-211.0	-199.9	-185.0	146.0	1.3750
700.	689.	151.8	-375.1	-369.6	-360.2	-347.3	-329.9	201.8	1.7827
800.	599.	242.2	-591.4	-585.2	-574.5	-559.6	-539.8	268.0	2.2497
900.	516.	372.8	-897.3	-890.2	-878.2	-861.5	-839.2	347.0	2.7898
1000.	441.	560.2	-1327.1	-1319.3	-1305.9	-1287.4	-1262.6	441.7	3.4186

BALLISTIC COEFFICIENT .12, MUZZLE VELOCITY FT/SEC 3400.

RANGE	VELOCITY	HEIGHT	100 YD	150 YD	200 YD	250 YD	300 YD	DRIFT	TIME
0.	3400.	0.0	-1.5	-1.5	-1.5	-1.5	-1.5	0.0	0.0000
100.	2584.	-.3	0.0	.7	1.9	3.7	5.9	2.3	.1013
200.	1895.	1.9	-3.9	-2.5	0.0	3.4	8.0	10.6	.2369
300.	1351.	7.9	-17.8	-15.7	-12.0	-6.9	0.0	28.3	.4253
400.	1052.	21.5	-51.5	-48.7	-43.7	-36.9	-27.7	57.8	.6814
500.	908.	46.2	-113.7	-110.2	-104.1	-95.5	-84.0	96.5	.9893
600.	798.	85.7	-212.2	-208.0	-200.6	-190.3	-176.6	143.0	1.3421
700.	700.	145.2	-357.6	-352.7	-344.1	-332.1	-316.0	198.2	1.7436
800.	609.	232.4	-566.1	-560.5	-550.6	-536.9	-518.5	263.7	2.2039
900.	525.	358.2	-860.4	-854.1	-843.0	-827.6	-806.9	341.5	2.7347
1000.	449.	538.9	-1274.8	-1267.7	-1255.4	-1238.3	-1215.3	434.9	3.3532

BALLISTIC COEFFICIENT .12, MUZZLE VELOCITY FT/SEC 3500.

RANGE	VELOCITY	HEIGHT	100 YD	150 YD	200 YD	250 YD	300 YD	DRIFT	TIME
0.	3500.	0.0	-1.5	-1.5	-1.5	-1.5	-1.5	0.0	0.0000
100.	2667.	-.3	0.0	.6	1.8	3.4	5.5	2.2	.0983
200.	1965.	1.8	-3.5	-2.3	0.0	3.2	7.4	10.2	.2294
300.	1401.	7.3	-16.4	-14.5	-11.1	-6.4	0.0	27.0	.4108
400.	1072.	20.1	-47.8	-45.3	-40.7	-34.4	-25.9	55.8	.6598
500.	922.	43.7	-107.0	-103.9	-98.2	-90.2	-79.6	94.0	.9625
600.	810.	81.7	-201.5	-197.7	-190.9	-181.3	-168.6	140.1	1.3105
700.	710.	139.0	-341.4	-336.9	-329.0	-317.9	-303.0	194.7	1.7064
800.	618.	223.0	-541.8	-536.8	-527.7	-515.0	-498.0	259.3	2.1590
900.	534.	344.4	-825.6	-819.9	-809.7	-795.4	-776.3	336.2	2.6816
1000.	457.	518.5	-1224.3	-1218.0	-1206.6	-1190.7	-1169.5	428.0	3.2889

BALLISTIC COEFFICIENT .12, MUZZLE VELOCITY FT/SEC 3600.

RANGE	VELOCITY	HEIGHT	100 YD	150 YD	200 YD	250 YD	300 YD	DRIFT	TIME
0.	3600.	0.0	-1.5	-1.5	-1.5	-1.5	-1.5	0.0	0.0000
100.	2750.	-.3	0.0	.6	1.6	3.1	5.1	2.1	.0954
200.	2036.	1.6	-3.2	-2.1	0.0	2.9	6.9	9.8	.2222
300.	1454.	6.8	-15.2	-13.5	-10.3	-5.9	0.0	25.9	.3971
400.	1094.	18.8	-44.4	-42.1	-37.9	-32.0	-24.1	53.8	.6387
500.	936.	41.4	-100.7	-97.9	-92.6	-85.3	-75.4	91.5	.9366
600.	821.	77.8	-191.2	-187.8	-181.5	-172.7	-160.8	137.2	1.2794
700.	721.	133.1	-325.7	-321.8	-314.4	-304.1	-290.3	191.2	1.6696
800.	628.	214.2	-519.1	-514.6	-506.2	-494.4	-478.7	255.1	2.1160
900.	543.	331.2	-792.4	-787.3	-777.9	-764.6	-746.9	330.9	2.6299
1000.	465.	499.2	-1177.0	-1171.4	-1160.9	-1146.2	-1126.5	421.3	3.2274

BALLISTIC COEFFICIENT .12, MUZZLE VELOCITY FT/SEC 3700.

RANGE	VELOCITY	HEIGHT	100 YD	150 YD	200 YD	250 YD	300 YD	DRIFT	TIME
0.	3700.	0.0	-1.5	-1.5	-1.5	-1.5	-1.5	0.0	0.0000
100.	2834.	-.3	0.0	.5	1.5	2.8	4.7	2.1	.0928
200.	2107.	1.5	-3.0	-1.9	0.0	2.7	6.4	9.4	.2156
300.	1509.	6.3	-14.0	-12.5	-9.6	-5.5	0.0	24.8	.3842
400.	1118.	17.6	-41.3	-39.2	-35.3	-29.9	-22.6	51.8	.6188
500.	951.	39.1	-94.8	-92.3	-87.4	-80.6	-71.4	89.1	.9115
600.	832.	74.2	-181.6	-178.5	-172.7	-164.5	-153.5	134.3	1.2496
700.	731.	127.5	-311.2	-307.6	-300.8	-291.2	-278.4	187.8	1.6345
800.	637.	205.7	-497.4	-493.4	-485.6	-474.7	-460.0	250.9	2.0740
900.	552.	318.9	-761.4	-756.8	-748.1	-735.8	-719.3	325.7	2.5804
1000.	473.	481.0	-1132.3	-1127.3	-1117.6	-1103.9	-1085.6	414.9	3.1680

BALLISTIC COEFFICIENT .12, MUZZLE VELOCITY FT/SEC 3800.

RANGE	VELOCITY	HEIGHT	100 YD	150 YD	200 YD	250 YD	300 YD	DRIFT	TIME
0.	3800.	0.0	-1.5	-1.5	-1.5	-1.5	-1.5	0.0	0.0000
100.	2918.	-.4	0.0	.4	1.3	2.6	4.3	2.0	.0903
200.	2178.	1.4	-2.7	-1.8	0.0	2.6	6.0	9.0	.2092
300.	1564.	5.9	-13.0	-11.6	-8.9	-5.1	0.0	23.8	.3722
400.	1145.	16.5	-38.3	-36.5	-32.9	-27.8	-21.0	49.9	.5992
500.	965.	37.1	-89.3	-87.0	-82.6	-76.2	-67.6	86.7	.8874
600.	843.	70.7	-172.4	-169.7	-164.3	-156.6	-146.4	131.4	1.2204
700.	741.	122.1	-297.1	-294.0	-287.7	-278.8	-266.8	184.3	1.5999
800.	647.	197.7	-477.0	-473.4	-466.2	-456.0	-442.4	246.7	2.0336
900.	560.	306.9	-731.3	-727.3	-719.2	-707.7	-692.4	320.5	2.5316
1000.	481.	463.4	-1089.3	-1084.9	-1075.9	-1063.1	-1046.1	408.4	3.1098

BALLISTIC COEFFICIENT .12, MUZZLE VELOCITY FT/SEC 3900.

RANGE	VELOCITY	HEIGHT	100 YD	150 YD	200 YD	250 YD	300 YD	DRIFT	TIME
0.	3900.	0.0	-1.5	-1.5	-1.5	-1.5	-1.5	0.0	0.0000
100.	3001.	-.4	0.0	.4	1.2	2.4	4.0	1.9	.0877
200.	2249.	1.2	-2.5	-1.7	0.0	2.4	5.6	8.7	.2033
300.	1621.	5.5	-12.0	-10.8	-8.3	-4.8	0.0	22.9	.3606
400.	1175.	15.4	-35.7	-34.1	-30.7	-26.0	-19.6	48.1	.5808
500.	980.	35.0	-83.9	-81.9	-77.7	-71.8	-63.9	84.2	.8631
600.	854.	67.5	-164.0	-161.6	-156.6	-149.4	-139.9	128.7	1.1928
700.	751.	117.1	-284.0	-281.2	-275.3	-267.0	-255.9	181.0	1.5666
800.	656.	190.1	-457.5	-454.3	-447.6	-438.1	-425.4	242.6	1.9939
900.	569.	295.7	-703.5	-699.9	-692.4	-681.7	-667.4	315.6	2.4853
1000.	489.	447.1	-1049.5	-1045.5	-1037.2	-1025.3	-1009.4	402.2	3.0546

BALLISTIC COEFFICIENT .12, MUZZLE VELOCITY FT/SEC 4000.

RANGE	VELOCITY	HEIGHT	100 YD	150 YD	200 YD	250 YD	300 YD	DRIFT	TIME
0.	4000.	0.0	-1.5	-1.5	-1.5	-1.5	-1.5	0.0	0.0000
100.	3085.	-.4	0.0	.3	1.1	2.2	3.7	1.9	.0855
200.	2320.	1.1	-2.2	-1.6	0.0	2.2	5.2	8.4	.1977
300.	1677.	5.1	-11.2	-10.1	-7.8	-4.5	0.0	22.0	.3500
400.	1209.	14.4	-33.2	-31.8	-28.7	-24.2	-18.3	46.2	.5627
500.	995.	33.1	-79.0	-77.2	-73.3	-67.8	-60.4	81.9	.8402
600.	866.	64.3	-155.5	-153.4	-148.7	-142.1	-133.2	125.7	1.1644
700.	761.	112.2	-271.3	-268.9	-263.5	-255.7	-245.3	177.6	1.5340
800.	666.	182.9	-439.0	-436.2	-430.0	-421.1	-409.2	238.6	1.9558
900.	577.	285.0	-676.6	-673.5	-666.5	-656.5	-643.1	310.6	2.4398
1000.	497.	431.4	-1011.0	-1007.5	-999.7	-988.6	-973.8	396.1	3.0004

BALLISTIC COEFFICIENT .13, MUZZLE VELOCITY FT/SEC 2000.

RANGE	VELOCITY	HEIGHT	100 YD	150 YD	200 YD	250 YD	300 YD	DRIFT	TIME
0.	2000.	0.0	-1.5	-1.5	-1.5	-1.5	-1.5	0.0	0.0000
100.	1465.	.7	0.0	3.2	7.6	12.9	19.0	4.5	.1757
200.	1116.	7.5	-15.1	-8.7	0.0	10.8	22.8	19.9	.4133
300.	960.	23.1	-56.9	-47.3	-34.3	-18.1	0.0	44.8	.7047
400.	848.	51.0	-133.3	-120.5	-103.1	-81.5	-57.4	77.1	1.0382
500.	753.	95.2	-253.2	-237.1	-215.4	-188.5	-158.3	116.9	1.4139
600.	665.	161.4	-428.5	-409.2	-383.1	-350.8	-314.6	165.1	1.8382
700.	583.	257.6	-676.0	-653.6	-623.1	-585.4	-543.2	223.6	2.3202
800.	508.	395.2	-1020.1	-994.4	-959.6	-916.6	-868.3	294.3	2.8721
900.	439.	589.7	-1493.3	-1464.4	-1425.3	-1376.8	-1322.5	379.7	3.5074
1000.	376.	865.3	-2146.7	-2114.5	-2071.1	-2017.2	-1956.8	483.6	4.2477

BALLISTIC COEFFICIENT .13, MUZZLE VELOCITY FT/SEC 2100.

RANGE	VELOCITY	HEIGHT	100 YD	150 YD	200 YD	250 YD	300 YD	DRIFT	TIME
0.	2100.	0.0	-1.5	-1.5	-1.5	-1.5	-1.5	0.0	0.0000
100.	1544.	.6	0.0	2.9	6.8	11.7	17.4	4.2	.1670
200.	1155.	6.7	-13.6	-7.9	0.0	9.8	21.3	19.1	.3944
300.	981.	21.3	-52.3	-43.8	-32.0	-17.2	0.0	43.9	.6783
400.	864.	47.7	-124.1	-112.6	-96.9	-77.2	-54.3	76.3	1.0047
500.	768.	89.8	-237.5	-223.3	-203.5	-179.0	-150.3	116.0	1.3733
600.	679.	152.8	-403.7	-386.6	-362.9	-333.4	-299.0	164.0	1.7888
700.	596.	244.6	-638.9	-619.0	-591.4	-557.0	-516.8	221.9	2.2609
800.	520.	375.5	-965.3	-942.5	-910.9	-871.6	-825.7	291.6	2.7998
900.	450.	561.2	-1415.9	-1390.3	-1354.7	-1310.6	-1258.9	375.9	3.4216
1000.	386.	823.0	-2035.3	-2006.8	-1967.4	-1918.3	-1860.9	477.7	4.1426

BALLISTIC COEFFICIENT .13, MUZZLE VELOCITY FT/SEC 2200.

RANGE	VELOCITY	HEIGHT	100 YD	150 YD	200 YD	250 YD	300 YD	DRIFT	TIME
0.	2200.	0.0	-1.5	-1.5	-1.5	-1.5	-1.5	0.0	0.0000
100.	1624.	.5	0.0	2.5	6.1	10.6	16.0	4.0	.1590
200.	1203.	6.0	-12.1	-7.1	0.0	9.1	19.9	18.1	.3757
300.	1003.	19.7	-48.0	-40.4	-29.8	-16.2	0.0	42.8	.6525
400.	881.	44.6	-115.3	-105.2	-91.0	-72.9	-51.3	75.1	.9720
500.	783.	84.6	-222.7	-210.1	-192.4	-169.7	-142.7	114.7	1.3337
600.	692.	144.8	-380.7	-365.5	-344.2	-317.1	-284.7	162.5	1.7415
700.	609.	232.4	-604.3	-586.6	-561.8	-530.1	-492.3	219.9	2.2038
800.	531.	357.4	-915.1	-894.9	-866.6	-830.3	-787.1	288.8	2.7316
900.	460.	533.9	-1342.4	-1319.6	-1287.8	-1247.0	-1198.4	371.4	3.3375
1000.	395.	783.5	-1931.9	-1906.6	-1871.2	-1825.9	-1771.9	471.4	4.0421

BALLISTIC COEFFICIENT .13, MUZZLE VELOCITY FT/SEC 2300.

RANGE	VELOCITY	HEIGHT	100 YD	150 YD	200 YD	250 YD	300 YD	DRIFT	TIME
0.	2300.	0.0	-1.5	-1.5	-1.5	-1.5	-1.5	0.0	0.0000
100.	1705.	.4	0.0	2.3	5.4	9.6	14.6	3.7	.1516
200.	1254.	5.4	-10.9	-6.4	0.0	8.3	18.4	17.2	.3584
300.	1025.	18.1	-43.9	-37.2	-27.6	-15.2	0.0	41.5	.6272
400.	898.	41.7	-107.2	-98.2	-85.4	-68.9	-48.6	73.7	.9404
500.	797.	79.8	-208.8	-197.6	-181.6	-161.0	-135.6	113.2	1.2952
600.	706.	137.2	-358.9	-345.4	-326.2	-301.5	-271.1	160.6	1.6952
700.	621.	220.7	-571.6	-555.8	-533.4	-504.5	-469.1	217.3	2.1480
800.	543.	340.1	-867.6	-849.5	-824.0	-791.0	-750.4	285.4	2.6648
900.	471.	508.9	-1275.4	-1255.2	-1226.4	-1189.3	-1143.7	366.9	3.2586
1000.	405.	746.5	-1835.7	-1813.2	-1781.2	-1739.9	-1689.3	464.9	3.9457

BALLISTIC COEFFICIENT .13, MUZZLE VELOCITY FT/SEC 2400.

RANGE	VELOCITY	HEIGHT	100 YD	150 YD	200 YD	250 YD	300 YD	DRIFT	TIME
0.	2400.	0.0	-1.5	-1.5	-1.5	-1.5	-1.5	0.0	0.0000
100.	1788.	.3	0.0	2.0	4.9	8.6	13.4	3.5	.1449
200.	1310.	4.9	-9.8	-5.7	0.0	7.5	17.0	16.2	.3420
300.	1048.	16.7	-40.2	-34.2	-25.5	-14.2	0.0	40.1	.6031
400.	915.	38.0	-99.5	-91.4	-80.0	-64.9	-45.9	72.0	.9091
500.	812.	75.7	-195.8	-185.8	-171.4	-152.6	-128.8	111.4	1.2578
600.	720.	130.0	-338.4	-326.4	-309.1	-286.5	-258.0	158.4	1.6501
700.	634.	209.9	-541.3	-527.2	-507.1	-480.8	-447.5	214.7	2.0947
800.	555.	324.0	-823.4	-807.4	-784.4	-754.3	-716.3	281.8	2.6009
900.	482.	485.0	-1211.6	-1193.6	-1167.7	-1133.8	-1091.1	361.9	3.1812
1000.	415.	712.0	-1746.0	-1726.0	-1697.2	-1659.6	-1612.1	458.2	3.8533

BALLISTIC COEFFICIENT .13, MUZZLE VELOCITY FT/SEC 2500.

RANGE	VELOCITY	HEIGHT	100 YD	150 YD	200 YD	250 YD	300 YD	DRIFT	TIME
0.	2500.	0.0	-1.5	-1.5	-1.5	-1.5	-1.5	0.0	0.0000
100.	1872.	.2	0.0	1.8	4.4	7.8	12.1	3.3	.1386
200.	1369.	4.4	-8.8	-5.2	0.0	6.9	15.5	15.3	.3268
300.	1073.	15.3	-36.4	-31.1	-23.3	-13.0	0.0	38.4	.5780
400.	933.	36.3	-92.3	-85.1	-74.8	-61.1	-43.7	70.2	.8789
500.	826.	70.8	-183.5	-174.6	-161.6	-144.5	-122.8	109.3	1.2212
600.	733.	123.2	-319.4	-308.6	-293.1	-272.5	-246.5	156.1	1.6069
700.	646.	199.6	-512.7	-500.2	-482.0	-458.0	-427.7	211.7	2.0429
800.	566.	308.7	-781.9	-767.6	-746.9	-719.4	-684.8	277.9	2.5391
900.	492.	462.9	-1152.9	-1136.8	-1113.5	-1082.6	-1043.6	356.9	3.1079
1000.	424.	679.4	-1662.1	-1644.2	-1618.2	-1584.0	-1540.6	451.3	3.7644

BALLISTIC COEFFICIENT .13, MUZZLE VELOCITY FT/SEC 2600.

RANGE	VELOCITY	HEIGHT	100 YD	150 YD	200 YD	250 YD	300 YD	DRIFT	TIME
0.	2600.	0.0	-1.5	-1.5	-1.5	-1.5	-1.5	0.0	0.0000
100.	1957.	.1	0.0	1.6	3.9	7.1	11.0	3.1	.1330
200.	1432.	3.9	-7.9	-4.7	0.0	6.3	14.2	14.4	.3126
300.	1101.	14.0	-33.1	-28.4	-21.3	-12.0	0.0	36.7	.5548
400.	951.	33.9	-85.5	-79.2	-69.8	-57.3	-41.3	68.3	.8494
500.	840.	66.7	-171.9	-163.9	-152.2	-136.5	-116.6	107.1	1.1855
600.	746.	116.8	-301.3	-291.7	-277.6	-258.9	-235.0	153.5	1.5646
700.	659.	189.8	-485.7	-474.5	-458.1	-436.2	-408.3	208.5	1.9925
800.	577.	294.3	-742.8	-730.1	-711.3	-686.3	-654.4	273.9	2.4792
900.	503.	441.8	-1097.1	-1082.7	-1061.6	-1033.5	-997.6	351.6	3.0363
1000.	434.	649.0	-1583.8	-1567.9	-1544.4	-1513.1	-1473.3	444.5	3.6793

BALLISTIC COEFFICIENT .13, MUZZLE VELOCITY FT/SEC 2700.

RANGE	VELOCITY	HEIGHT	100 YD	150 YD	200 YD	250 YD	300 YD	DRIFT	TIME
0.	2700.	0.0	-1.5	-1.5	-1.5	-1.5	-1.5	0.0	0.0000
100.	2044.	.0	0.0	1.4	3.5	6.4	10.1	2.9	.1278
200.	1499.	3.6	-7.1	-4.3	0.0	5.7	13.0	13.6	.2994
300.	1133.	12.9	-30.2	-25.9	-19.5	-11.0	0.0	35.1	.5325
400.	969.	31.7	-79.4	-73.7	-65.2	-53.9	-39.2	66.4	.8218
500.	855.	63.0	-161.3	-154.2	-143.6	-129.4	-111.0	105.0	1.1520
600.	759.	110.7	-284.3	-275.7	-263.0	-245.9	-223.9	150.8	1.5235
700.	671.	180.7	-460.6	-450.7	-435.8	-415.9	-390.2	205.3	1.9443
800.	589.	280.8	-706.5	-695.1	-678.1	-655.3	-626.0	269.8	2.4219
900.	513.	422.1	-1045.3	-1032.6	-1013.4	-987.8	-954.8	346.4	2.9681
1000.	443.	620.3	-1510.1	-1495.9	-1474.7	-1446.2	-1409.6	437.5	3.5971

BALLISTIC COEFFICIENT .13, MUZZLE VELOCITY FT/SEC 2800.

RANGE	VELOCITY	HEIGHT	100 YD	150 YD	200 YD	250 YD	300 YD	DRIFT	TIME
0.	2800.	0.0	-1.5	-1.5	-1.5	-1.5	-1.5	0.0	0.0000
100.	2130.	-.0	0.0	1.3	3.2	5.8	9.2	2.8	.1228
200.	1568.	3.2	-6.4	-3.9	0.0	5.2	11.9	12.9	.2874
300.	1169.	11.8	-27.5	-23.7	-17.8	-10.1	0.0	33.4	.5112
400.	988.	29.5	-73.5	-68.4	-60.7	-50.3	-36.9	64.3	.7939
500.	869.	59.2	-150.9	-144.5	-134.9	-121.9	-105.1	102.4	1.1177
600.	772.	105.0	-268.5	-260.8	-249.2	-233.7	-213.5	148.1	1.4841
700.	683.	172.0	-436.8	-427.9	-414.3	-396.2	-372.7	201.9	1.8972
800.	600.	267.9	-672.0	-661.8	-646.3	-625.6	-598.7	265.6	2.3660
900.	523.	403.5	-996.5	-985.0	-967.5	-944.3	-914.0	341.0	2.9019
1000.	453.	593.6	-1441.9	-1429.2	-1409.8	-1383.9	-1350.3	430.8	3.5189

BALLISTIC COEFFICIENT .13, MUZZLE VELOCITY FT/SEC 2900.

RANGE	VELOCITY	HEIGHT	100 YD	150 YD	200 YD	250 YD	300 YD	DRIFT	TIME
0.	2900.	0.0	-1.5	-1.5	-1.5	-1.5	-1.5	0.0	0.0000
100.	2216.	-.1	0.0	1.1	2.9	5.3	8.4	2.6	.1184
200.	1637.	2.9	-5.8	-3.5	0.0	4.7	10.9	12.2	.2762
300.	1211.	10.8	-25.1	-21.7	-16.4	-9.3	0.0	31.8	.4911
400.	1006.	27.5	-68.0	-63.5	-56.4	-46.9	-34.6	62.2	.7670
500.	884.	55.8	-141.4	-135.7	-126.9	-115.0	-99.6	100.0	1.0852
600.	785.	99.6	-253.6	-246.7	-236.1	-221.9	-203.4	145.2	1.4458
700.	695.	163.9	-414.7	-406.7	-394.4	-377.8	-356.2	198.5	1.8523
800.	611.	256.1	-640.2	-631.0	-616.9	-598.0	-573.3	261.4	2.3131
900.	533.	386.1	-951.1	-940.8	-925.0	-903.7	-875.9	335.8	2.8390
1000.	462.	568.3	-1377.3	-1365.9	-1348.2	-1324.6	-1293.7	423.9	3.4431

BALLISTIC COEFFICIENT .13, MUZZLE VELOCITY FT/SEC 3000.

RANGE	VELOCITY	HEIGHT	100 YD	150 YD	200 YD	250 YD	300 YD	DRIFT	TIME
0.	3000.	0.0	-1.5	-1.5	-1.5	-1.5	-1.5	0.0	0.0000
100.	2302.	-.1	0.0	1.0	2.6	4.8	7.6	2.5	.1143
200.	1706.	2.6	-5.3	-3.2	0.0	4.3	10.0	11.6	.2658
300.	1255.	10.0	-22.9	-19.9	-15.0	-8.5	0.0	30.3	.4724
400.	1026.	25.6	-62.9	-58.8	-52.4	-43.7	-32.3	60.0	.7411
500.	898.	52.6	-132.5	-127.4	-119.4	-108.5	-94.3	97.5	1.0541
600.	798.	94.5	-239.5	-233.4	-223.7	-210.7	-193.7	142.3	1.4088
700.	706.	156.3	-393.9	-386.8	-375.5	-360.3	-340.4	195.1	1.8087
800.	622.	244.7	-609.8	-601.7	-588.8	-571.4	-548.7	257.2	2.2614
900.	543.	369.7	-908.2	-899.0	-884.5	-865.0	-839.4	330.5	2.7780
1000.	471.	544.9	-1317.7	-1307.5	-1291.4	-1269.7	-1241.3	417.4	3.3715

BALLISTIC COEFFICIENT .13, MUZZLE VELOCITY FT/SEC 3100.

RANGE	VELOCITY	HEIGHT	100 YD	150 YD	200 YD	250 YD	300 YD	DRIFT	TIME
0.	3100.	0.0	-1.5	-1.5	-1.5	-1.5	-1.5	0.0	0.0000
100.	2387.	-.2	0.0	.9	2.4	4.4	7.0	2.4	.1103
200.	1777.	2.4	-4.8	-3.0	0.0	4.0	9.2	11.0	.2561
300.	1302.	9.2	-21.0	-18.2	-13.8	-7.8	0.0	28.9	.4544
400.	1045.	23.9	-58.2	-54.5	-48.6	-40.7	-30.2	57.9	.7162
500.	913.	49.5	-124.2	-119.6	-112.2	-102.2	-89.2	95.0	1.0234
600.	810.	89.7	-226.4	-220.9	-212.0	-200.1	-184.5	139.4	1.3729
700.	718.	149.0	-374.2	-367.8	-357.5	-343.5	-325.3	191.6	1.7663
800.	632.	234.1	-581.7	-574.3	-562.5	-546.6	-525.8	253.0	2.2119
900.	553.	354.2	-868.1	-859.9	-846.6	-828.6	-805.2	325.3	2.7195
1000.	480.	522.3	-1260.4	-1251.3	-1236.5	-1216.6	-1190.5	410.7	3.3010

BALLISTIC COEFFICIENT .13, MUZZLE VELOCITY FT/SEC 3200.

RANGE	VELOCITY	HEIGHT	100 YD	150 YD	200 YD	250 YD	300 YD	DRIFT	TIME
0.	3200.	0.0	-1.5	-1.5	-1.5	-1.5	-1.5	0.0	0.0000
100.	2472.	-.2	0.0	.8	2.2	4.0	6.4	2.3	.1067
200.	1848.	2.2	-4.4	-2.7	0.0	3.7	8.5	10.5	.2472
300.	1352.	8.4	-19.2	-16.8	-12.7	-7.2	0.0	27.5	.4377
400.	1065.	22.3	-53.9	-50.6	-45.2	-37.8	-28.2	55.9	.6924
500.	928.	46.7	-116.4	-112.3	-105.5	-96.3	-84.3	92.4	.9940
600.	822.	85.2	-214.0	-209.0	-200.9	-189.9	-175.5	136.5	1.3380
700.	729.	142.2	-355.8	-350.1	-340.6	-327.7	-311.0	188.2	1.7257
800.	643.	224.0	-554.9	-548.3	-537.5	-522.8	-503.6	248.8	2.1638
900.	563.	339.6	-830.1	-822.7	-810.5	-793.9	-772.4	320.1	2.6627
1000.	489.	501.6	-1208.0	-1199.8	-1186.3	-1167.9	-1143.9	404.4	3.2350

BALLISTIC COEFFICIENT .13, MUZZLE VELOCITY FT/SEC 3300.

RANGE	VELOCITY	HEIGHT	100 YD	150 YD	200 YD	250 YD	300 YD	DRIFT	TIME
0.	3300.	0.0	-1.5	-1.5	-1.5	-1.5	-1.5	0.0	0.0000
100.	2557.	-.2	0.0	.7	2.0	3.7	5.9	2.2	.1034
200.	1920.	2.0	-4.0	-2.5	0.0	3.4	7.8	10.0	.2389
300.	1404.	7.8	-17.7	-15.5	-11.7	-6.6	0.0	26.3	.4220
400.	1089.	20.7	-49.7	-46.7	-41.7	-34.9	-26.1	53.6	.6682
500.	943.	44.0	-109.2	-105.5	-99.2	-90.8	-79.7	90.0	.9658
600.	834.	80.9	-202.3	-197.9	-190.4	-180.3	-167.0	133.6	1.3043
700.	740.	135.7	-338.3	-333.2	-324.4	-312.6	-297.1	184.7	1.6859
800.	653.	214.5	-529.8	-524.0	-514.0	-500.4	-482.8	244.7	2.1177
900.	573.	325.8	-794.6	-788.0	-776.7	-761.5	-741.6	315.1	2.6084
1000.	498.	481.6	-1157.6	-1150.2	-1137.7	-1120.8	-1098.7	397.9	3.1701

BALLISTIC COEFFICIENT .13, MUZZLE VELOCITY FT/SEC 3400.

RANGE	VELOCITY	HEIGHT	100 YD	150 YD	200 YD	250 YD	300 YD	DRIFT	TIME
0.	3400.	0.0	-1.5	-1.5	-1.5	-1.5	-1.5	0.0	0.0000
100.	2641.	-.3	0.0	.7	1.8	3.4	5.4	2.1	.1002
200.	1993.	1.8	-3.6	-2.3	0.0	3.1	7.2	9.6	.2309
300.	1459.	7.2	-16.3	-14.3	-10.8	-6.1	0.0	25.1	.4073
400.	1114.	19.3	-45.9	-43.3	-38.7	-32.4	-24.2	51.5	.6457
500.	959.	41.4	-102.2	-98.9	-93.2	-85.3	-75.1	87.4	.9376
600.	846.	76.9	-191.3	-187.4	-180.5	-171.1	-158.8	130.6	1.2715
700.	752.	129.6	-322.0	-317.4	-309.3	-298.4	-284.0	181.3	1.6478
800.	664.	205.4	-506.0	-500.7	-491.6	-479.0	-462.7	240.5	2.0726
900.	582.	312.7	-760.7	-754.7	-744.4	-730.3	-711.9	310.0	2.5553
1000.	507.	463.0	-1110.6	-1104.0	-1092.6	-1076.9	-1056.4	391.7	3.1082

BALLISTIC COEFFICIENT .13, MUZZLE VELOCITY FT/SEC 3500.

RANGE	VELOCITY	HEIGHT	100 YD	150 YD	200 YD	250 YD	300 YD	DRIFT	TIME
0.	3500.	0.0	-1.5	-1.5	-1.5	-1.5	-1.5	0.0	0.0000
100.	2726.	-.3	0.0	.6	1.6	3.1	5.0	2.0	.0973
200.	2066.	1.7	-3.3	-2.1	0.0	2.9	6.7	9.2	.2237
300.	1517.	6.7	-15.0	-13.2	-10.0	-5.7	0.0	24.0	.3935
400.	1141.	17.9	-42.5	-40.1	-35.9	-30.1	-22.5	49.5	.6241
500.	974.	39.1	-95.8	-92.8	-87.6	-80.3	-70.8	84.9	.9109
600.	858.	73.0	-181.0	-177.5	-171.1	-162.4	-151.0	127.7	1.2399
700.	763.	123.8	-306.4	-302.3	-294.9	-284.7	-271.4	177.9	1.6105
800.	674.	196.9	-483.6	-478.9	-470.4	-458.8	-443.6	236.5	2.0293
900.	592.	300.4	-729.1	-723.8	-714.2	-701.1	-684.1	305.1	2.5047
1000.	516.	445.2	-1065.9	-1060.0	-1049.4	-1034.8	-1015.9	385.6	3.0480

BALLISTIC COEFFICIENT .13, MUZZLE VELOCITY FT/SEC 3600.

RANGE	VELOCITY	HEIGHT	100 YD	150 YD	200 YD	250 YD	300 YD	DRIFT	TIME
0.	3600.	0.0	-1.5	-1.5	-1.5	-1.5	-1.5	0.0	0.0000
100.	2810.	-.3	0.0	.5	1.5	2.9	4.6	1.9	.0944
200.	2139.	1.5	-3.0	-2.0	0.0	2.7	6.2	8.8	.2168
300.	1575.	6.2	-13.8	-12.3	-9.3	-5.3	0.0	23.0	.3806
400.	1172.	16.8	-39.5	-37.4	-33.4	-28.0	-21.0	47.6	.6040
500.	989.	36.9	-90.0	-87.3	-82.4	-75.7	-66.9	82.5	.8857
600.	870.	69.5	-171.6	-168.5	-162.6	-154.5	-143.9	125.0	1.2101
700.	774.	118.2	-291.7	-288.1	-281.2	-271.8	-259.4	174.4	1.5745
800.	684.	188.8	-462.3	-458.1	-450.2	-439.4	-425.3	232.4	1.9869
900.	601.	288.5	-698.6	-693.9	-685.1	-672.9	-657.1	300.0	2.4548
1000.	524.	428.3	-1023.6	-1018.3	-1008.5	-995.0	-977.4	379.5	2.9897

BALLISTIC COEFFICIENT .13, MUZZLE VELOCITY FT/SEC 3700.

RANGE	VELOCITY	HEIGHT	100 YD	150 YD	200 YD	250 YD	300 YD	DRIFT	TIME
0.	3700.	0.0	-1.5	-1.5	-1.5	-1.5	-1.5	0.0	0.0000
100.	2895.	-.3	0.0	.5	1.4	2.6	4.3	1.9	.0918
200.	2212.	1.4	-2.7	-1.8	0.0	2.5	5.8	8.5	.2104
300.	1633.	5.8	-12.8	-11.4	-8.7	-4.9	0.0	22.0	.3685
400.	1209.	15.6	-36.5	-34.7	-31.0	-26.0	-19.5	45.7	.5839
500.	1006.	34.8	-84.2	-81.9	-77.3	-71.1	-62.9	80.0	.8600
600.	883.	65.9	-161.9	-159.1	-153.7	-146.2	-136.3	121.8	1.1786
700.	784.	113.0	-277.8	-274.6	-268.2	-259.4	-248.0	171.1	1.5394
800.	694.	181.1	-442.2	-438.5	-431.2	-421.2	-408.1	228.4	1.9462
900.	610.	277.5	-670.3	-666.1	-658.0	-646.7	-631.9	295.3	2.4075
1000.	533.	412.4	-983.8	-979.1	-970.0	-957.5	-941.1	373.7	2.9338

BALLISTIC COEFFICIENT .13, MUZZLE VELOCITY FT/SEC 3800.

RANGE	VELOCITY	HEIGHT	100 YD	150 YD	200 YD	250 YD	300 YD	DRIFT	TIME
0.	3800.	0.0	-1.5	-1.5	-1.5	-1.5	-1.5	0.0	0.0000
100.	2980.	-.4	0.0	.4	1.3	2.4	3.9	1.8	.0893
200.	2285.	1.3	-2.5	-1.7	0.0	2.3	5.4	8.2	.2043
300.	1692.	5.4	-11.8	-10.6	-8.1	-4.6	0.0	21.2	.3571
400.	1246.	14.6	-33.9	-32.3	-28.9	-24.3	-18.1	43.9	.5652
500.	1022.	32.7	-78.8	-76.8	-72.5	-66.7	-59.1	77.5	.8351
600.	895.	62.7	-153.4	-150.9	-145.9	-138.9	-129.7	119.0	1.1500
700.	795.	108.0	-264.7	-261.8	-255.9	-247.7	-237.0	167.7	1.5054
800.	704.	173.7	-423.1	-419.8	-413.0	-403.7	-391.5	224.4	1.9065
900.	619.	266.7	-643.0	-639.3	-631.8	-621.2	-607.5	290.4	2.3607
1000.	541.	397.1	-945.5	-941.4	-933.0	-921.3	-906.0	367.8	2.8790

BALLISTIC COEFFICIENT .13, MUZZLE VELOCITY FT/SEC 3900.

RANGE	VELOCITY	HEIGHT	100 YD	150 YD	200 YD	250 YD	300 YD	DRIFT	TIME
0.	3900.	0.0	-1.5	-1.5	-1.5	-1.5	-1.5	0.0	0.0000
100.	3065.	-.4	0.0	.4	1.1	2.2	3.7	1.7	.0867
200.	2357.	1.1	-2.3	-1.6	0.0	2.2	5.0	7.9	.1985
300.	1752.	5.0	-11.0	-9.9	-7.5	-4.3	0.0	20.3	.3462
400.	1286.	13.6	-31.6	-30.1	-27.0	-22.6	-16.9	42.2	.5473
500.	1038.	30.8	-73.8	-71.9	-68.0	-62.6	-55.5	75.0	.8110
600.	908.	59.5	-145.0	-142.8	-138.1	-131.6	-123.0	116.0	1.1208
700.	806.	103.3	-252.2	-249.6	-244.2	-236.5	-226.5	164.3	1.4721
800.	714.	166.7	-405.0	-402.0	-395.8	-387.1	-375.7	220.4	1.8679
900.	629.	256.7	-617.5	-614.2	-607.2	-597.4	-584.5	285.8	2.3159
1000.	550.	382.8	-909.8	-906.2	-898.4	-887.4	-873.2	362.1	2.8266

BALLISTIC COEFFICIENT .13, MUZZLE VELOCITY FT/SEC 4000.

RANGE	VELOCITY	HEIGHT	100 YD	150 YD	200 YD	250 YD	300 YD	DRIFT	TIME
0.	4000.	0.0	-1.5	-1.5	-1.5	-1.5	-1.5	0.0	0.0000
100.	3150.	-.4	0.0	.3	1.0	2.1	3.4	1.7	.0847
200.	2430.	1.0	-2.1	-1.4	0.0	2.0	4.7	7.6	.1931
300.	1813.	4.7	-10.2	-9.2	-7.1	-4.0	0.0	19.6	.3361
400.	1327.	12.8	-29.4	-28.1	-25.2	-21.1	-15.8	40.6	.5305
500.	1055.	29.1	-69.1	-67.5	-63.9	-58.8	-52.1	72.7	.7880
600.	921.	56.6	-137.2	-135.3	-130.9	-124.8	-116.8	113.2	1.0930
700.	816.	98.8	-240.3	-238.1	-233.0	-225.9	-216.5	161.0	1.4399
800.	724.	160.1	-387.7	-385.2	-379.4	-371.2	-360.5	216.5	1.8304
900.	638.	247.0	-592.9	-590.1	-583.6	-574.4	-562.4	281.1	2.2721
1000.	558.	368.9	-875.3	-872.2	-864.9	-854.7	-841.4	356.4	2.7751

BALLISTIC COEFFICIENT .14, MUZZLE VELOCITY FT/SEC 2000.

RANGE	VELOCITY	HEIGHT	100 YD	150 YD	200 YD	250 YD	300 YD	DRIFT	TIME
0.	2000.	0.0	-1.5	-1.5	-1.5	-1.5	-1.5	0.0	0.0000
100.	1499.	.7	0.0	3.1	7.2	12.3	18.1	4.1	.1736
200.	1150.	7.1	-14.4	-8.2	0.0	10.1	21.9	18.4	.4044
300.	988.	22.0	-54.4	-45.2	-32.9	-17.7	0.0	42.0	.6888
400.	877.	48.3	-127.0	-114.7	-98.3	-78.0	-54.4	72.4	1.0114
500.	786.	89.7	-240.3	-225.0	-204.4	-179.1	-149.6	109.6	1.3726
600.	702.	150.8	-404.2	-385.8	-361.1	-330.7	-295.3	154.3	1.7766
700.	623.	238.1	-632.5	-611.0	-582.2	-546.8	-505.5	207.8	2.2305
800.	550.	360.6	-944.3	-919.7	-886.8	-846.3	-799.1	271.7	2.7436
900.	482.	530.2	-1364.4	-1336.8	-1299.7	-1254.1	-1201.1	347.7	3.3257
1000.	420.	764.2	-1929.4	-1898.7	-1857.6	-1806.9	-1747.9	438.6	3.9919

BALLISTIC COEFFICIENT .14, MUZZLE VELOCITY FT/SEC 2100.

RANGE	VELOCITY	HEIGHT	100 YD	150 YD	200 YD	250 YD	300 YD	DRIFT	TIME
0.	2100.	0.0	-1.5	-1.5	-1.5	-1.5	-1.5	0.0	0.0000
100.	1579.	.6	0.0	2.7	6.4	11.1	16.6	3.9	.1651
200.	1197.	6.4	-12.9	-7.4	0.0	9.3	20.3	17.5	.3852
300.	1010.	20.3	-49.8	-41.6	-30.5	-16.6	0.0	41.1	.6618
400.	894.	45.1	-117.9	-107.0	-92.2	-73.6	-51.5	71.5	.9777
500.	801.	84.4	-224.7	-211.2	-192.6	-169.4	-141.8	108.6	1.3316
600.	716.	142.6	-380.2	-363.9	-341.6	-313.8	-280.6	153.3	1.7279
700.	636.	225.8	-596.9	-577.9	-551.9	-519.4	-480.7	206.3	2.1724
800.	563.	342.4	-892.7	-870.9	-841.3	-804.2	-759.9	269.5	2.6739
900.	494.	504.3	-1292.5	-1268.0	-1234.7	-1192.9	-1143.1	344.6	3.2437
1000.	431.	727.4	-1830.0	-1802.8	-1765.7	-1719.3	-1664.0	434.0	3.8947

BALLISTIC COEFFICIENT .14, MUZZLE VELOCITY FT/SEC 2200.

RANGE	VELOCITY	HEIGHT	100 YD	150 YD	200 YD	250 YD	300 YD	DRIFT	TIME
0.	2200.	0.0	-1.5	-1.5	-1.5	-1.5	-1.5	0.0	0.0000
100.	1660.	.4	0.0	2.4	5.7	10.0	15.1	3.7	.1572
200.	1250.	5.7	-11.5	-6.7	0.0	8.5	18.7	16.6	.3669
300.	1034.	18.6	-45.2	-38.0	-28.0	-15.3	0.0	39.7	.6344
400.	913.	42.0	-109.0	-99.4	-86.0	-69.1	-48.7	70.1	.9439
500.	816.	79.4	-210.2	-198.2	-181.5	-160.3	-134.8	107.4	1.2919
600.	730.	134.8	-357.7	-343.3	-323.3	-297.9	-267.3	151.8	1.6807
700.	650.	214.3	-563.7	-546.8	-523.5	-493.9	-458.2	204.5	2.1164
800.	575.	325.6	-845.1	-825.8	-799.1	-765.3	-724.5	266.9	2.6074
900.	505.	479.9	-1225.2	-1203.4	-1173.4	-1135.4	-1089.5	340.9	3.1643
1000.	441.	692.1	-1735.3	-1711.2	-1677.9	-1635.5	-1584.6	428.7	3.7993

BALLISTIC COEFFICIENT .14, MUZZLE VELOCITY FT/SEC 2300.

RANGE	VELOCITY	HEIGHT	100 YD	150 YD	200 YD	250 YD	300 YD	DRIFT	TIME
0.	2300.	0.0	-1.5	-1.5	-1.5	-1.5	-1.5	0.0	0.0000
100.	1743.	.3	0.0	2.1	5.1	9.0	13.7	3.4	.1499
200.	1306.	5.1	-10.3	-6.0	0.0	7.8	17.1	15.6	.3497
300.	1058.	17.0	-41.1	-34.7	-25.7	-14.1	0.0	38.2	.6084
400.	931.	39.1	-100.9	-92.4	-80.4	-64.9	-46.1	68.6	.9116
500.	832.	74.6	-196.5	-185.8	-170.8	-151.4	-128.0	105.8	1.2531
600.	744.	127.5	-336.5	-323.7	-305.7	-282.4	-254.3	149.9	1.6345
700.	663.	203.3	-532.4	-517.4	-496.4	-469.2	-436.4	202.2	2.0617
800.	587.	309.6	-800.4	-783.3	-759.3	-728.3	-690.8	263.9	2.5430
900.	517.	456.9	-1162.1	-1142.9	-1115.9	-1080.9	-1038.8	336.8	3.0877
1000.	452.	659.8	-1648.9	-1627.5	-1597.5	-1558.7	-1511.8	423.3	3.7095

BALLISTIC COEFFICIENT .14, MUZZLE VELOCITY FT/SEC 2400.

RANGE	VELOCITY	HEIGHT	100 YD	150 YD	200 YD	250 YD	300 YD	DRIFT	TIME
0.	2400.	0.0	-1.5	-1.5	-1.5	-1.5	-1.5	0.0	0.0000
100.	1828.	.2	0.0	1.9	4.6	8.1	12.5	3.2	.1433
200.	1366.	4.6	-9.2	-5.4	0.0	7.0	15.7	14.8	.3338
300.	1085.	15.6	-37.4	-31.6	-23.5	-13.0	0.0	36.7	.5834
400.	950.	36.4	-93.3	-85.7	-74.9	-60.9	-43.5	66.9	.8801
500.	847.	70.1	-183.7	-174.1	-160.6	-143.1	-121.4	103.9	1.2153
600.	758.	120.5	-316.6	-305.1	-288.9	-268.0	-241.9	147.8	1.5897
700.	676.	193.0	-503.2	-489.8	-470.9	-446.4	-416.0	199.6	2.0090
800.	599.	294.5	-758.3	-743.0	-721.4	-693.4	-658.6	260.5	2.4803
900.	528.	435.4	-1103.7	-1086.5	-1062.2	-1030.7	-991.6	332.5	3.0144
1000.	462.	628.8	-1566.8	-1547.7	-1520.7	-1485.7	-1442.2	417.4	3.6217

BALLISTIC COEFFICIENT .14, MUZZLE VELOCITY FT/SEC 2500.

RANGE	VELOCITY	HEIGHT	100 YD	150 YD	200 YD	250 YD	300 YD	DRIFT	TIME
0.	2500.	0.0	-1.5	-1.5	-1.5	-1.5	-1.5	0.0	0.0000
100.	1913.	.2	0.0	1.7	4.1	7.3	11.3	3.0	.1372
200.	1430.	4.1	-8.3	-4.9	0.0	6.4	14.3	13.9	.3189
300.	1115.	14.3	-33.9	-28.8	-21.5	-12.0	0.0	35.1	.5592
400.	968.	34.0	-86.4	-79.7	-69.9	-57.2	-41.2	65.2	.8503
500.	862.	65.9	-171.5	-163.0	-150.8	-135.0	-115.0	101.8	1.1783
600.	772.	114.0	-298.0	-287.8	-273.2	-254.1	-230.2	145.4	1.5464
700.	689.	183.3	-475.7	-463.8	-446.8	-424.5	-396.6	196.8	1.9579
800.	611.	280.5	-719.3	-705.7	-686.2	-660.8	-628.9	257.0	2.4205
900.	539.	415.0	-1048.3	-1033.0	-1011.0	-982.4	-946.5	327.9	2.9430
1000.	472.	600.2	-1491.2	-1474.2	-1449.8	-1418.1	-1378.2	411.6	3.5385

BALLISTIC COEFFICIENT .14, MUZZLE VELOCITY FT/SEC 2600.

RANGE	VELOCITY	HEIGHT	100 YD	150 YD	200 YD	250 YD	300 YD	DRIFT	TIME
0.	2600.	0.0	-1.5	-1.5	-1.5	-1.5	-1.5	0.0	0.0000
100.	2000.	.1	0.0	1.5	3.7	6.6	10.3	2.8	.1316
200.	1499.	3.7	-7.4	-4.4	0.0	5.8	13.1	13.1	.3051
300.	1150.	13.0	-30.8	-26.2	-19.6	-11.0	0.0	33.4	.5359
400.	988.	31.6	-79.8	-73.7	-64.9	-53.4	-38.8	63.2	.8204
500.	877.	61.9	-160.3	-152.8	-141.8	-127.3	-109.1	99.6	1.1429
600.	786.	107.8	-280.4	-271.3	-258.1	-240.8	-218.9	142.9	1.5041
700.	702.	174.0	-449.8	-439.2	-423.7	-403.6	-378.0	193.7	1.9081
800.	623.	267.0	-682.3	-670.2	-652.5	-629.5	-600.2	253.3	2.3620
900.	550.	396.0	-997.1	-983.5	-963.7	-937.7	-904.8	323.2	2.8750
1000.	482.	573.0	-1419.4	-1404.2	-1382.2	-1353.3	-1316.8	405.4	3.4572

BALLISTIC COEFFICIENT .14, MUZZLE VELOCITY FT/SEC 2700.

RANGE	VELOCITY	HEIGHT	100 YD	150 YD	200 YD	250 YD	300 YD	DRIFT	TIME
0.	2700.	0.0	-1.5	-1.5	-1.5	-1.5	-1.5	0.0	0.0000
100.	2088.	.0	0.0	1.4	3.4	6.0	9.3	2.7	.1264
200.	1569.	3.4	-6.7	-4.0	0.0	5.2	11.9	12.4	.2925
300.	1191.	11.9	-28.0	-23.9	-17.9	-10.0	0.0	31.8	.5138
400.	1008.	29.3	-73.6	-68.2	-60.1	-49.6	-36.3	61.0	.7913
500.	892.	58.1	-149.6	-142.8	-132.8	-119.7	-103.0	97.1	1.1075
600.	799.	102.0	-263.9	-255.8	-243.7	-228.0	-207.9	140.2	1.4630
700.	714.	165.4	-425.7	-416.2	-402.1	-383.8	-360.4	190.5	1.8603
800.	635.	254.5	-647.8	-637.0	-621.0	-600.0	-573.3	249.4	2.3060
900.	561.	377.9	-948.4	-936.2	-918.2	-894.5	-864.5	318.3	2.8085
1000.	493.	547.7	-1353.1	-1339.6	-1319.5	-1293.3	-1259.9	399.4	3.3802

BALLISTIC COEFFICIENT .14, MUZZLE VELOCITY FT/SEC 2800.

RANGE	VELOCITY	HEIGHT	100 YD	150 YD	200 YD	250 YD	300 YD	DRIFT	TIME
0.	2800.	0.0	-1.5	-1.5	-1.5	-1.5	-1.5	0.0	0.0000
100.	2175.	-.0	0.0	1.2	3.0	5.4	8.5	2.5	.1215
200.	1640.	3.0	-6.1	-3.6	0.0	4.8	10.9	11.7	.2806
300.	1236.	10.9	-25.5	-21.8	-16.4	-9.2	0.0	30.2	.4930
400.	1028.	27.2	-67.8	-63.0	-55.7	-46.1	-33.8	58.9	.7632
500.	908.	54.6	-139.7	-133.6	-124.5	-112.5	-97.2	94.7	1.0736
600.	813.	96.5	-248.5	-241.2	-230.3	-216.0	-197.5	137.4	1.4234
700.	727.	157.2	-402.9	-394.5	-381.7	-365.0	-343.5	187.3	1.8139
800.	646.	242.6	-615.5	-605.8	-591.2	-572.1	-547.6	245.5	2.2518
900.	572.	361.1	-903.5	-892.6	-876.2	-854.7	-827.1	313.5	2.7456
1000.	503.	523.6	-1290.3	-1278.2	-1259.9	-1236.0	-1205.3	393.1	3.3052

BALLISTIC COEFFICIENT .14, MUZZLE VELOCITY FT/SEC 2900.

RANGE	VELOCITY	HEIGHT	100 YD	150 YD	200 YD	250 YD	300 YD	DRIFT	TIME
0.	2900.	0.0	-1.5	-1.5	-1.5	-1.5	-1.5	0.0	0.0000
100.	2262.	-.1	0.0	1.1	2.7	4.9	7.7	2.4	.1172
200.	1711.	2.7	-5.5	-3.3	0.0	4.4	10.0	11.1	.2698
300.	1284.	10.0	-23.2	-20.0	-15.0	-8.4	0.0	28.7	.4733
400.	1048.	25.3	-62.6	-58.2	-51.6	-42.8	-31.6	56.8	.7366
500.	924.	51.3	-130.4	-125.0	-116.7	-105.7	-91.7	92.2	1.0409
600.	826.	91.3	-234.0	-227.5	-217.5	-204.4	-187.6	134.5	1.3850
700.	739.	149.5	-381.5	-373.9	-362.3	-347.0	-327.4	183.9	1.7690
800.	658.	231.5	-585.0	-576.4	-563.1	-545.6	-523.2	241.5	2.1995
900.	582.	345.1	-860.9	-851.2	-836.2	-816.5	-791.3	308.6	2.6842
1000.	512.	501.3	-1232.1	-1221.3	-1204.7	-1182.8	-1154.8	387.1	3.2341

BALLISTIC COEFFICIENT .14, MUZZLE VELOCITY FT/SEC 3000.

RANGE	VELOCITY	HEIGHT	100 YD	150 YD	200 YD	250 YD	300 YD	DRIFT	TIME
0.	3000.	0.0	-1.5	-1.5	-1.5	-1.5	-1.5	0.0	0.0000
100.	2348.	-.1	0.0	1.0	2.5	4.5	7.1	2.3	.1132
200.	1784.	2.5	-5.0	-3.0	0.0	4.0	9.2	10.5	.2599
300.	1334.	9.2	-21.2	-18.3	-13.7	-7.7	0.0	27.3	.4552
400.	1071.	23.4	-57.4	-53.5	-47.4	-39.4	-29.1	54.5	.7095
500.	940.	48.2	-121.7	-116.9	-109.3	-99.3	-86.4	89.6	1.0093
600.	839.	86.4	-220.3	-214.5	-205.4	-193.4	-177.9	131.6	1.3478
700.	751.	142.3	-361.6	-354.8	-344.2	-330.2	-312.1	180.6	1.7260
800.	669.	221.0	-556.7	-548.9	-536.8	-520.7	-500.1	237.5	2.1494
900.	593.	330.2	-821.3	-812.6	-798.9	-780.9	-757.7	303.7	2.6259
1000.	522.	480.0	-1176.8	-1167.1	-1152.0	-1131.9	-1106.1	381.0	3.1649

BALLISTIC COEFFICIENT .14, MUZZLE VELOCITY FT/SEC 3100.

RANGE	VELOCITY	HEIGHT	100 YD	150 YD	200 YD	250 YD	300 YD	DRIFT	TIME
0.	3100.	0.0	-1.5	-1.5	-1.5	-1.5	-1.5	0.0	0.0000
100.	2434.	-.2	0.0	.9	2.3	4.1	6.5	2.2	.1092
200.	1857.	2.3	-4.5	-2.8	0.0	3.7	8.4	10.0	.2503
300.	1388.	8.5	-19.4	-16.8	-12.6	-7.1	0.0	26.0	.4378
400.	1095.	21.7	-52.9	-49.4	-43.9	-36.5	-27.0	52.3	.6842
500.	956.	45.2	-113.7	-109.3	-102.4	-93.1	-81.3	87.1	.9786
600.	851.	82.0	-208.1	-202.9	-194.6	-183.5	-169.3	128.9	1.3133
700.	763.	135.4	-342.7	-336.6	-326.9	-314.0	-297.5	177.1	1.6839
800.	680.	211.0	-529.6	-522.7	-511.6	-496.8	-477.9	233.4	2.1002
900.	603.	316.0	-783.7	-775.9	-763.4	-746.8	-725.6	298.8	2.5688
1000.	532.	460.2	-1125.6	-1116.9	-1103.1	-1084.6	-1061.0	375.1	3.0990

BALLISTIC COEFFICIENT .14, MUZZLE VELOCITY FT/SEC 3200.

RANGE	VELOCITY	HEIGHT	100 YD	150 YD	200 YD	250 YD	300 YD	DRIFT	TIME
0.	3200.	0.0	-1.5	-1.5	-1.5	-1.5	-1.5	0.0	0.0000
100.	2520.	-.2	0.0	.8	2.1	3.8	5.9	2.1	.1058
200.	1931.	2.1	-4.1	-2.6	0.0	3.4	7.8	9.6	.2418
300.	1444.	7.8	-17.8	-15.5	-11.6	-6.5	0.0	24.7	.4218
400.	1122.	20.2	-48.7	-45.6	-40.5	-33.7	-25.0	50.2	.6601
500.	973.	42.5	-106.1	-102.2	-95.8	-87.4	-76.5	84.5	.9490
600.	865.	77.5	-195.6	-190.9	-183.3	-173.1	-160.0	125.7	1.2768
700.	775.	128.9	-325.0	-319.6	-310.7	-298.8	-283.5	173.7	1.6434
800.	691.	201.7	-504.6	-498.4	-488.1	-474.5	-457.1	229.4	2.0535
900.	614.	302.7	-748.6	-741.6	-730.1	-714.8	-695.2	294.0	2.5142
1000.	541.	441.3	-1076.7	-1068.9	-1056.1	-1039.1	-1017.4	369.1	3.0346

BALLISTIC COEFFICIENT .14, MUZZLE VELOCITY FT/SEC 3300.

RANGE	VELOCITY	HEIGHT	100 YD	150 YD	200 YD	250 YD	300 YD	DRIFT	TIME
0.	3300.	0.0	-1.5	-1.5	-1.5	-1.5	-1.5	0.0	0.0000
100.	2606.	-.2	0.0	.7	1.9	3.4	5.4	2.0	.1024
200.	2005.	1.9	-3.7	-2.3	0.0	3.1	7.2	9.1	.2337
300.	1503.	7.2	-16.3	-14.3	-10.7	-6.0	0.0	23.6	.4068
400.	1153.	18.7	-44.9	-42.2	-37.5	-31.2	-23.2	48.1	.6370
500.	989.	40.0	-99.3	-95.8	-89.9	-82.1	-72.1	82.1	.9212
600.	878.	73.4	-184.4	-180.3	-173.2	-163.8	-151.7	122.8	1.2431
700.	787.	122.8	-308.4	-303.5	-295.3	-284.3	-270.2	170.3	1.6042
800.	702.	192.7	-480.7	-475.1	-465.7	-453.2	-437.1	225.4	2.0078
900.	624.	290.0	-715.3	-709.0	-698.4	-684.3	-666.2	289.2	2.4612
1000.	551.	423.7	-1031.4	-1024.4	-1012.7	-997.0	-976.9	363.4	2.9736

BALLISTIC COEFFICIENT .14, MUZZLE VELOCITY FT/SEC 3400.

RANGE	VELOCITY	HEIGHT	100 YD	150 YD	200 YD	250 YD	300 YD	DRIFT	TIME
0.	3400.	0.0	-1.5	-1.5	-1.5	-1.5	-1.5	0.0	0.0000
100.	2691.	-.3	0.0	.6	1.7	3.2	5.0	1.9	.0992
200.	2080.	1.7	-3.4	-2.2	0.0	2.9	6.6	8.7	.2260
300.	1563.	6.7	-15.1	-13.2	-10.0	-5.6	0.0	22.5	.3928
400.	1187.	17.4	-41.5	-39.0	-34.7	-28.9	-21.4	46.1	.6149
500.	1006.	37.5	-92.6	-89.5	-84.1	-76.8	-67.5	79.5	.8929
600.	891.	69.5	-173.7	-170.0	-163.5	-154.8	-143.6	119.7	1.2098
700.	798.	116.9	-292.6	-288.2	-280.7	-270.5	-257.4	166.9	1.5658
800.	713.	184.3	-458.3	-453.3	-444.7	-433.0	-418.1	221.4	1.9637
900.	634.	278.1	-684.0	-678.4	-668.7	-655.6	-638.8	284.4	2.4101
1000.	560.	406.6	-987.7	-981.4	-970.7	-956.1	-937.5	357.4	2.9133

BALLISTIC COEFFICIENT .14, MUZZLE VELOCITY FT/SEC 3500.

RANGE	VELOCITY	HEIGHT	100 YD	150 YD	200 YD	250 YD	300 YD	DRIFT	TIME
0.	3500.	0.0	-1.5	-1.5	-1.5	-1.5	-1.5	0.0	0.0000
100.	2777.	-.3	0.0	.6	1.6	2.9	4.6	1.9	.0963
200.	2155.	1.6	-3.1	-2.0	0.0	2.7	6.2	8.4	.2190
300.	1623.	6.2	-13.9	-12.2	-9.2	-5.2	0.0	21.6	.3797
400.	1226.	16.2	-38.4	-36.2	-32.2	-26.8	-19.9	44.2	.5940
500.	1023.	35.2	-86.4	-83.6	-78.6	-71.9	-63.2	76.9	.8656
600.	904.	65.8	-163.8	-160.4	-154.4	-146.4	-136.0	116.8	1.1777
700.	810.	111.5	-277.8	-273.9	-266.9	-257.5	-245.4	163.5	1.5290
800.	724.	176.4	-437.0	-432.6	-424.6	-413.8	-400.0	217.4	1.9209
900.	644.	266.7	-654.3	-649.3	-640.3	-628.2	-612.6	279.7	2.3605
1000.	569.	390.9	-947.4	-941.8	-931.8	-918.4	-901.1	351.9	2.8565

BALLISTIC COEFFICIENT .14, MUZZLE VELOCITY FT/SEC 3600.

RANGE	VELOCITY	HEIGHT	100 YD	150 YD	200 YD	250 YD	300 YD	DRIFT	TIME
0.	3600.	0.0	-1.5	-1.5	-1.5	-1.5	-1.5	0.0	0.0000
100.	2862.	-.3	0.0	.5	1.4	2.7	4.3	1.8	.0934
200.	2229.	1.4	-2.8	-1.8	0.0	2.5	5.7	8.0	.2123
300.	1684.	5.7	-12.8	-11.3	-8.6	-4.8	0.0	20.6	.3673
400.	1266.	15.1	-35.6	-33.6	-29.9	-24.9	-18.5	42.4	.5740
500.	1041.	33.0	-80.5	-78.0	-73.4	-67.1	-59.1	74.3	.8390
600.	918.	62.3	-154.4	-151.4	-145.9	-138.3	-128.7	113.8	1.1465
700.	821.	106.2	-263.6	-260.1	-253.7	-244.9	-233.7	160.0	1.4926
800.	734.	168.8	-416.9	-412.9	-405.6	-395.6	-382.7	213.4	1.8793
900.	654.	255.9	-626.3	-621.8	-613.6	-602.3	-587.8	275.0	2.3125
1000.	578.	375.6	-908.4	-903.5	-894.3	-881.7	-865.7	346.2	2.8003

BALLISTIC COEFFICIENT .14, MUZZLE VELOCITY FT/SEC 3700.

RANGE	VELOCITY	HEIGHT	100 YD	150 YD	200 YD	250 YD	300 YD	DRIFT	TIME
0.	3700.	0.0	-1.5	-1.5	-1.5	-1.5	-1.5	0.0	0.0000
100.	2948.	-.4	0.0	.4	1.3	2.5	4.0	1.7	.0909
200.	2304.	1.3	-2.6	-1.7	0.0	2.3	5.3	7.7	.2061
300.	1746.	5.3	-11.9	-10.5	-8.0	-4.5	0.0	19.8	.3558
400.	1308.	14.1	-33.0	-31.2	-27.8	-23.1	-17.2	40.7	.5553
500.	1059.	31.0	-75.1	-72.9	-68.6	-62.8	-55.3	71.9	.8137
600.	932.	59.1	-145.6	-143.0	-137.8	-130.8	-121.9	110.9	1.1166
700.	832.	101.3	-250.4	-247.3	-241.4	-233.2	-222.7	156.7	1.4579
800.	745.	161.6	-397.9	-394.4	-387.6	-378.2	-366.3	209.5	1.8390
900.	663.	245.7	-599.7	-595.8	-588.1	-577.6	-564.1	270.4	2.2660
1000.	588.	361.4	-872.2	-867.8	-859.2	-847.6	-832.6	340.7	2.7469

BALLISTIC COEFFICIENT .14, MUZZLE VELOCITY FT/SEC 3800.

RANGE	VELOCITY	HEIGHT	100 YD	150 YD	200 YD	250 YD	300 YD	DRIFT	TIME
0.	3800.	0.0	-1.5	-1.5	-1.5	-1.5	-1.5	0.0	0.0000
100.	3034.	-.4	0.0	.4	1.2	2.3	3.7	1.7	.0884
200.	2378.	1.2	-2.4	-1.6	0.0	2.2	5.0	7.5	.2002
300.	1809.	5.0	-11.0	-9.8	-7.4	-4.2	0.0	19.0	.3450
400.	1352.	13.1	-30.6	-29.1	-25.9	-21.6	-16.0	39.0	.5375
500.	1079.	29.1	-70.1	-68.1	-64.2	-58.7	-51.8	69.4	.7891
600.	945.	56.1	-137.4	-135.1	-130.4	-123.9	-115.5	108.1	1.0879
700.	843.	96.6	-237.8	-235.1	-229.5	-221.9	-212.2	153.3	1.4238
800.	755.	154.7	-379.9	-376.8	-370.4	-361.7	-350.6	205.6	1.7999
900.	673.	236.0	-574.6	-571.1	-564.0	-554.2	-541.6	265.8	2.2209
1000.	596.	347.7	-837.3	-833.5	-825.6	-814.7	-800.7	335.3	2.6944

BALLISTIC COEFFICIENT .14, MUZZLE VELOCITY FT/SEC 3900.

RANGE	VELOCITY	HEIGHT	100 YD	150 YD	200 YD	250 YD	300 YD	DRIFT	TIME
0.	3900.	0.0	-1.5	-1.5	-1.5	-1.5	-1.5	0.0	0.0000
100.	3120.	-.4	0.0	.3	1.1	2.1	3.4	1.6	.0861
200.	2452.	1.1	-2.1	-1.5	0.0	2.0	4.6	7.2	.1945
300.	1872.	4.6	-10.2	-9.2	-7.0	-3.9	0.0	18.3	.3346
400.	1399.	12.3	-28.5	-27.1	-24.2	-20.1	-14.9	37.5	.5205
500.	1100.	27.4	-65.4	-63.7	-60.0	-54.9	-48.4	67.0	.7653
600.	960.	53.0	-129.4	-127.3	-122.9	-116.8	-109.0	105.1	1.0584
700.	854.	92.2	-226.2	-223.9	-218.7	-211.6	-202.5	150.2	1.3916
800.	765.	148.2	-362.7	-360.0	-354.2	-346.0	-335.6	201.7	1.7617
900.	683.	226.7	-550.5	-547.4	-540.8	-531.7	-520.0	261.3	2.1767
1000.	605.	334.7	-804.4	-801.0	-793.7	-783.5	-770.5	329.9	2.6437

BALLISTIC COEFFICIENT .14, MUZZLE VELOCITY FT/SEC 4000.

RANGE	VELOCITY	HEIGHT	100 YD	150 YD	200 YD	250 YD	300 YD	DRIFT	TIME
0.	4000.	0.0	-1.5	-1.5	-1.5	-1.5	-1.5	0.0	0.0000
100.	3206.	-.4	0.0	.3	1.0	1.9	3.1	1.6	.0839
200.	2526.	1.0	-2.0	-1.4	0.0	1.9	4.3	6.9	.1894
300.	1935.	4.3	-9.4	-8.6	-6.5	-3.7	0.0	17.6	.3251
400.	1448.	11.5	-26.5	-25.4	-22.6	-18.8	-13.9	36.0	.5047
500.	1124.	25.7	-61.1	-59.6	-56.2	-51.4	-45.3	64.7	.7425
600.	974.	50.3	-122.0	-120.3	-116.2	-110.5	-103.1	102.3	1.0310
700.	866.	87.8	-214.6	-212.6	-207.8	-201.1	-192.6	146.7	1.3584
800.	776.	142.0	-346.5	-344.2	-338.7	-331.1	-321.3	197.9	1.7247
900.	692.	217.9	-527.9	-525.3	-519.1	-510.5	-499.5	256.9	2.1344
1000.	614.	322.4	-773.1	-770.2	-763.4	-753.9	-741.7	324.6	2.5946

BALLISTIC COEFFICIENT .15, MUZZLE VELOCITY FT/SEC 2000.

RANGE	VELOCITY	HEIGHT	100 YD	150 YD	200 YD	250 YD	300 YD	DRIFT	TIME
0.	2000.	0.0	-1.5	-1.5	-1.5	-1.5	-1.5	0.0	0.0000
100.	1529.	.7	0.0	3.0	6.9	11.7	17.4	3.8	.1719
200.	1185.	6.8	-13.7	-7.8	0.0	9.7	21.0	17.0	.3965
300.	1013.	21.1	-52.1	-43.3	-31.5	-17.0	0.0	39.5	.6742
400.	904.	46.0	-121.2	-109.4	-93.8	-74.4	-51.7	68.1	.9867
500.	815.	85.0	-229.2	-214.4	-194.9	-170.7	-142.4	103.3	1.3367
600.	735.	142.0	-384.1	-366.3	-342.9	-313.8	-279.8	145.1	1.7246
700.	659.	222.3	-597.1	-576.4	-549.0	-515.1	-475.5	194.6	2.1557
800.	588.	333.3	-884.0	-860.4	-829.1	-790.3	-745.0	253.1	2.6380
900.	522.	484.3	-1264.5	-1237.8	-1202.7	-1159.1	-1108.1	321.9	3.1790
1000.	461.	688.8	-1766.8	-1737.2	-1698.2	-1649.7	-1593.1	403.1	3.7903

BALLISTIC COEFFICIENT .15, MUZZLE VELOCITY FT/SEC 2100.

RANGE	VELOCITY	HEIGHT	100 YD	150 YD	200 YD	250 YD	300 YD	DRIFT	TIME
0.	2100.	0.0	-1.5	-1.5	-1.5	-1.5	-1.5	0.0	0.0000
100.	1610.	.5	0.0	2.6	6.1	10.5	15.7	3.6	.1634
200.	1237.	6.1	-12.3	-7.1	0.0	8.8	19.2	16.1	.3774
300.	1038.	19.2	-47.2	-39.3	-28.7	-15.6	0.0	38.1	.6452
400.	923.	42.8	-112.1	-101.6	-87.5	-70.0	-49.2	67.0	.9523
500.	831.	79.8	-213.9	-200.8	-183.2	-161.3	-135.3	102.3	1.2955
600.	749.	134.1	-360.5	-344.9	-323.7	-297.4	-266.2	144.1	1.6759
700.	673.	210.6	-562.7	-544.5	-519.8	-489.1	-452.7	193.3	2.0985
800.	601.	316.3	-834.7	-813.8	-785.6	-750.6	-708.9	251.2	2.5699
900.	534.	460.5	-1196.7	-1173.3	-1141.5	-1102.1	-1055.3	319.3	3.0997
1000.	472.	655.4	-1674.4	-1648.3	-1613.1	-1569.2	-1517.2	399.3	3.6973

BALLISTIC COEFFICIENT .15, MUZZLE VELOCITY FT/SEC 2200.

RANGE	VELOCITY	HEIGHT	100 YD	150 YD	200 YD	250 YD	300 YD	DRIFT	TIME
0.	2200.	0.0	-1.5	-1.5	-1.5	-1.5	-1.5	0.0	0.0000
100.	1693.	.4	0.0	2.3	5.5	9.5	14.3	3.4	.1556
200.	1294.	5.5	-11.0	-6.3	0.0	8.0	17.6	15.3	.3595
300.	1063.	17.6	-42.9	-36.0	-26.5	-14.4	0.0	36.9	.6188
400.	942.	39.8	-103.6	-94.3	-81.6	-65.6	-46.4	65.8	.9191
500.	847.	74.9	-199.4	-187.9	-172.0	-152.0	-127.9	100.9	1.2554
600.	764.	126.5	-338.4	-324.5	-305.5	-281.4	-252.6	142.6	1.6284
700.	687.	199.6	-530.4	-514.2	-492.0	-463.9	-430.3	191.5	2.0429
800.	614.	300.4	-789.2	-770.7	-745.3	-713.3	-674.8	248.9	2.5050
900.	546.	438.0	-1133.4	-1112.6	-1084.0	-1047.9	-1004.7	316.1	3.0232
1000.	483.	623.7	-1587.3	-1564.2	-1532.4	-1492.3	-1444.3	394.8	3.6069

BALLISTIC COEFFICIENT .15, MUZZLE VELOCITY FT/SEC 2300.

RANGE	VELOCITY	HEIGHT	100 YD	150 YD	200 YD	250 YD	300 YD	DRIFT	TIME
0.	2300.	0.0	-1.5	-1.5	-1.5	-1.5	-1.5	0.0	0.0000
100.	1777.	.3	0.0	2.1	4.9	8.5	12.9	3.2	.1485
200.	1354.	4.9	-9.8	-5.7	0.0	7.2	16.1	14.4	.3427
300.	1091.	16.1	-38.8	-32.6	-24.1	-13.3	0.0	35.3	.5921
400.	962.	36.9	-95.5	-87.3	-75.9	-61.5	-43.8	64.1	.8861
500.	862.	70.2	-185.8	-175.6	-161.3	-143.3	-121.2	99.2	1.2160
600.	779.	119.4	-317.6	-305.2	-288.1	-266.5	-240.0	140.7	1.5822
700.	700.	189.1	-500.0	-485.6	-465.6	-440.4	-409.4	189.3	1.9886
800.	627.	285.4	-746.4	-730.0	-707.2	-678.4	-643.0	246.1	2.4418
900.	558.	416.6	-1073.8	-1055.3	-1029.7	-997.2	-957.4	312.4	2.9488
1000.	494.	594.2	-1506.7	-1486.1	-1457.7	-1421.6	-1377.4	390.1	3.5206

BALLISTIC COEFFICIENT .15, MUZZLE VELOCITY FT/SEC 2400.

RANGE	VELOCITY	HEIGHT	100 YD	150 YD	200 YD	250 YD	300 YD	DRIFT	TIME
0.	2400.	0.0	-1.5	-1.5	-1.5	-1.5	-1.5	0.0	0.0000
100.	1863.	.2	0.0	1.8	4.4	7.7	11.7	3.0	.1418
200.	1419.	4.4	-8.8	-5.1	0.0	6.6	14.6	13.5	.3268
300.	1123.	14.7	-35.1	-29.6	-22.0	-12.1	0.0	33.8	.5668
400.	982.	34.3	-88.0	-80.7	-70.5	-57.3	-41.2	62.3	.8542
500.	878.	65.9	-173.2	-164.1	-151.3	-134.9	-114.7	97.3	1.1780
600.	793.	112.7	-298.2	-287.2	-271.9	-252.2	-227.9	138.6	1.5376
700.	714.	179.3	-471.8	-459.0	-441.1	-418.1	-389.8	186.8	1.9365
800.	639.	271.2	-706.2	-691.5	-671.1	-644.8	-612.5	242.9	2.3804
900.	570.	396.8	-1018.9	-1002.4	-979.4	-949.9	-913.5	308.5	2.8780
1000.	505.	566.3	-1431.2	-1412.9	-1387.3	-1354.5	-1314.1	384.9	3.4372

BALLISTIC COEFFICIENT .15, MUZZLE VELOCITY FT/SEC 2500.

RANGE	VELOCITY	HEIGHT	100 YD	150 YD	200 YD	250 YD	300 YD	DRIFT	TIME
0.	2500.	0.0	-1.5	-1.5	-1.5	-1.5	-1.5	0.0	0.0000
100.	1949.	.1	0.0	1.6	3.9	6.9	10.6	2.8	.1359
200.	1488.	3.9	-7.9	-4.6	0.0	5.9	13.3	12.7	.3123
300.	1160.	13.4	-31.8	-26.9	-20.0	-11.1	0.0	32.1	.5425
400.	1002.	31.8	-81.0	-74.5	-65.2	-53.4	-38.6	60.4	.8232
500.	894.	61.8	-161.5	-153.4	-141.8	-127.0	-108.5	95.3	1.1413
600.	807.	106.4	-279.9	-270.1	-256.2	-238.4	-216.3	136.2	1.4939
700.	727.	169.9	-445.1	-433.7	-417.6	-396.8	-370.9	184.0	1.8857
800.	652.	258.0	-668.9	-655.9	-637.4	-613.6	-584.1	239.7	2.3217
900.	582.	377.9	-966.6	-952.0	-931.2	-904.5	-871.2	304.3	2.8087
1000.	516.	540.1	-1360.3	-1344.0	-1320.9	-1291.2	-1254.3	379.6	3.3567

BALLISTIC COEFFICIENT .15, MUZZLE VELOCITY FT/SEC 2600.

RANGE	VELOCITY	HEIGHT	100 YD	150 YD	200 YD	250 YD	300 YD	DRIFT	TIME
0.	2600.	0.0	-1.5	-1.5	-1.5	-1.5	-1.5	0.0	0.0000
100.	2038.	.1	0.0	1.5	3.5	6.2	9.6	2.6	.1304
200.	1559.	3.5	-7.1	-4.2	0.0	5.4	12.1	12.0	.2990
300.	1204.	12.2	-28.8	-24.5	-18.2	-10.1	0.0	30.5	.5195
400.	1023.	29.4	-74.4	-68.6	-60.3	-49.5	-36.0	58.3	.7930
500.	911.	57.8	-150.1	-142.8	-132.3	-118.9	-102.0	92.8	1.1041
600.	821.	100.3	-262.5	-253.8	-241.3	-225.1	-204.9	133.6	1.4513
700.	740.	161.1	-419.9	-409.8	-395.1	-376.3	-352.6	181.0	1.8361
800.	664.	245.4	-633.5	-621.9	-605.1	-583.6	-556.6	236.1	2.2644
900.	593.	360.3	-918.2	-905.1	-886.3	-862.1	-831.7	299.9	2.7426
1000.	527.	515.5	-1294.2	-1279.7	-1258.7	-1231.8	-1198.1	374.1	3.2794

BALLISTIC COEFFICIENT .15, MUZZLE VELOCITY FT/SEC 2700.

RANGE	VELOCITY	HEIGHT	100 YD	150 YD	200 YD	250 YD	300 YD	DRIFT	TIME
0.	2700.	0.0	-1.5	-1.5	-1.5	-1.5	-1.5	0.0	0.0000
100.	2126.	.0	0.0	1.3	3.2	5.6	8.7	2.5	.1252
200.	1632.	3.2	-6.4	-3.8	0.0	4.9	11.1	11.3	.2865
300.	1252.	11.1	-26.2	-22.3	-16.6	-9.3	0.0	28.9	.4978
400.	1044.	27.3	-68.4	-63.3	-55.6	-45.9	-33.5	56.2	.7640
500.	928.	54.1	-139.7	-133.2	-123.7	-111.5	-96.0	90.3	1.0688
600.	835.	94.7	-246.5	-238.7	-227.3	-212.6	-194.1	130.9	1.4102
700.	753.	152.8	-396.6	-387.5	-374.2	-357.1	-335.5	177.9	1.7886
800.	676.	233.5	-600.4	-590.0	-574.8	-555.3	-530.6	232.3	2.2090
900.	605.	343.6	-872.5	-860.8	-843.7	-821.7	-793.9	295.4	2.6782
1000.	538.	492.2	-1231.9	-1219.0	-1199.9	-1175.5	-1144.6	368.4	3.2045

BALLISTIC COEFFICIENT .15, MUZZLE VELOCITY FT/SEC 2800.

RANGE	VELOCITY	HEIGHT	100 YD	150 YD	200 YD	250 YD	300 YD	DRIFT	TIME
0.	2800.	0.0	-1.5	-1.5	-1.5	-1.5	-1.5	0.0	0.0000
100.	2214.	-.1	0.0	1.2	2.9	5.1	7.9	2.3	.1204
200.	1705.	2.9	-5.8	-3.4	0.0	4.5	10.1	10.7	.2750
300.	1302.	10.2	-23.8	-20.3	-15.2	-8.4	0.0	27.4	.4773
400.	1067.	25.2	-62.7	-58.1	-51.2	-42.2	-31.0	53.9	.7350
500.	945.	50.7	-130.1	-124.3	-115.7	-104.5	-90.4	87.9	1.0352
600.	849.	89.4	-231.4	-224.4	-214.0	-200.6	-183.7	128.0	1.3703
700.	766.	145.0	-374.5	-366.4	-354.4	-338.7	-319.0	174.7	1.7424
800.	689.	222.3	-569.5	-560.2	-546.4	-528.5	-506.0	228.5	2.1557
900.	616.	327.8	-829.7	-819.3	-803.7	-783.6	-758.3	290.8	2.6163
1000.	548.	470.5	-1174.1	-1162.5	-1145.3	-1122.9	-1094.8	362.9	3.1332

BALLISTIC COEFFICIENT .15, MUZZLE VELOCITY FT/SEC 2900.

RANGE	VELOCITY	HEIGHT	100 YD	150 YD	200 YD	250 YD	300 YD	DRIFT	TIME
0.	2900.	0.0	-1.5	-1.5	-1.5	-1.5	-1.5	0.0	0.0000
100.	2302.	-.1	0.0	1.0	2.6	4.7	7.2	2.2	.1162
200.	1778.	2.6	-5.2	-3.2	0.0	4.1	9.3	10.1	.2646
300.	1355.	9.3	-21.7	-18.6	-13.9	-7.7	0.0	26.1	.4585
400.	1092.	23.3	-57.6	-53.4	-47.1	-38.9	-28.6	51.8	.7079
500.	962.	47.4	-120.9	-115.7	-107.9	-97.6	-84.7	85.3	1.0018
600.	863.	84.4	-217.2	-210.9	-201.5	-189.2	-173.7	125.1	1.3315
700.	779.	137.6	-353.8	-346.6	-335.5	-321.2	-303.1	171.3	1.6976
800.	700.	211.7	-540.2	-531.9	-519.3	-502.9	-482.3	224.6	2.1039
900.	627.	313.1	-789.7	-780.4	-766.2	-747.8	-724.6	286.2	2.5570
1000.	558.	449.8	-1119.3	-1109.0	-1093.2	-1072.7	-1046.9	357.2	3.0638

BALLISTIC COEFFICIENT .15, MUZZLE VELOCITY FT/SEC 3000.

RANGE	VELOCITY	HEIGHT	100 YD	150 YD	200 YD	250 YD	300 YD	DRIFT	TIME
0.	3000.	0.0	-1.5	-1.5	-1.5	-1.5	-1.5	0.0	0.0000
100.	2389.	-.1	0.0	.9	2.4	4.2	6.6	2.1	.1122
200.	1853.	2.4	-4.7	-2.9	0.0	3.8	8.5	9.6	.2548
300.	1411.	8.6	-19.8	-17.0	-12.7	-7.1	0.0	24.8	.4407
400.	1119.	21.6	-52.9	-49.2	-43.4	-35.9	-26.4	49.6	.6818
500.	979.	44.4	-112.5	-107.8	-100.6	-91.2	-79.4	82.7	.9698
600.	876.	79.7	-204.1	-198.6	-189.9	-178.7	-164.5	122.3	1.2948
700.	791.	130.7	-334.6	-328.1	-318.0	-304.9	-288.3	168.0	1.6547
800.	712.	201.9	-513.1	-505.7	-494.2	-479.2	-460.3	220.8	2.0545
900.	638.	299.1	-752.0	-743.7	-730.7	-713.8	-692.5	281.5	2.4994
1000.	569.	430.7	-1068.7	-1059.5	-1045.1	-1026.3	-1002.7	351.7	2.9980

BALLISTIC COEFFICIENT .15, MUZZLE VELOCITY FT/SEC 3100.

RANGE	VELOCITY	HEIGHT	100 YD	150 YD	200 YD	250 YD	300 YD	DRIFT	TIME
0.	3100.	0.0	-1.5	-1.5	-1.5	-1.5	-1.5	0.0	0.0000
100.	2476.	-.2	0.0	.8	2.2	3.9	6.1	2.0	.1083
200.	1928.	2.1	-4.3	-2.6	0.0	3.5	7.8	9.2	.2457
300.	1471.	7.9	-18.2	-15.7	-11.7	-6.5	0.0	23.5	.4241
400.	1151.	19.9	-48.6	-45.3	-40.0	-33.1	-24.4	47.4	.6566
500.	997.	41.6	-104.6	-100.5	-93.9	-85.2	-74.4	80.1	.9388
600.	891.	75.2	-191.5	-186.6	-178.6	-168.3	-155.2	119.2	1.2577
700.	804.	124.1	-316.3	-310.5	-301.2	-289.1	-273.9	164.6	1.6126
800.	724.	192.4	-487.4	-480.7	-470.1	-456.3	-438.9	216.8	2.0061
900.	649.	286.0	-716.8	-709.4	-697.5	-681.9	-662.4	276.9	2.4442
1000.	579.	412.3	-1020.3	-1012.0	-998.8	-981.5	-959.8	346.0	2.9334

BALLISTIC COEFFICIENT .15, MUZZLE VELOCITY FT/SEC 3200.

RANGE	VELOCITY	HEIGHT	100 YD	150 YD	200 YD	250 YD	300 YD	DRIFT	TIME
0.	3200.	0.0	-1.5	-1.5	-1.5	-1.5	-1.5	0.0	0.0000
100.	2562.	-.2	0.0	.7	2.0	3.5	5.6	2.0	.1048
200.	2004.	2.0	-3.9	-2.4	0.0	3.2	7.2	8.8	.2372
300.	1532.	7.3	-16.7	-14.4	-10.8	-6.0	0.0	22.4	.4087
400.	1187.	18.5	-44.7	-41.7	-36.9	-30.5	-22.5	45.4	.6329
500.	1015.	38.9	-97.3	-93.6	-87.5	-79.6	-69.5	77.5	.9089
600.	905.	71.0	-179.9	-175.4	-168.1	-158.6	-146.6	116.1	1.2224
700.	816.	117.9	-299.2	-294.0	-285.5	-274.3	-260.3	161.2	1.5721
800.	735.	183.6	-463.2	-457.3	-447.6	-434.8	-418.8	212.9	1.9596
900.	660.	273.5	-683.3	-676.6	-665.7	-651.4	-633.3	272.2	2.3903
1000.	589.	395.2	-975.5	-968.1	-956.0	-940.1	-920.0	340.5	2.8722

BALLISTIC COEFFICIENT .15, MUZZLE VELOCITY FT/SEC 3300.

RANGE	VELOCITY	HEIGHT	100 YD	150 YD	200 YD	250 YD	300 YD	DRIFT	TIME
0.	3300.	0.0	-1.5	-1.5	-1.5	-1.5	-1.5	0.0	0.0000
100.	2649.	-.3	0.0	.7	1.8	3.2	5.1	1.9	.1016
200.	2081.	1.8	-3.6	-2.2	0.0	2.9	6.7	8.4	.2293
300.	1594.	6.7	-15.3	-13.3	-10.0	-5.6	0.0	21.4	.3944
400.	1227.	17.1	-41.2	-38.6	-34.1	-28.2	-20.8	43.4	.6103
500.	1033.	36.4	-90.5	-87.1	-81.6	-74.2	-64.9	74.9	.8799
600.	919.	67.0	-168.9	-164.9	-158.3	-149.4	-138.3	113.1	1.1882
700.	828.	112.0	-283.0	-278.4	-270.6	-260.3	-247.3	157.8	1.5328
800.	747.	175.2	-440.5	-435.1	-426.3	-414.5	-399.6	209.0	1.9145
900.	670.	261.8	-652.2	-646.2	-636.2	-622.9	-606.2	267.6	2.3389
1000.	599.	378.9	-932.6	-926.0	-914.9	-900.1	-881.5	334.9	2.8122

BALLISTIC COEFFICIENT .15, MUZZLE VELOCITY FT/SEC 3400.

RANGE	VELOCITY	HEIGHT	100 YD	150 YD	200 YD	250 YD	300 YD	DRIFT	TIME
0.	3400.	0.0	-1.5	-1.5	-1.5	-1.5	-1.5	0.0	0.0000
100.	2735.	-.3	0.0	.6	1.6	3.0	4.7	1.8	.0984
200.	2157.	1.6	-3.2	-2.1	0.0	2.7	6.2	8.0	.2219
300.	1657.	6.2	-14.1	-12.3	-9.3	-5.2	0.0	20.4	.3808
400.	1269.	15.9	-38.1	-35.7	-31.6	-26.1	-19.3	41.5	.5889
500.	1052.	34.1	-84.1	-81.1	-76.0	-69.2	-60.6	72.3	.8518
600.	934.	63.3	-158.8	-155.2	-149.1	-140.9	-130.6	110.1	1.1552
700.	840.	106.4	-267.7	-263.6	-256.4	-246.9	-234.8	154.3	1.4943
800.	758.	167.2	-418.9	-414.2	-405.9	-395.0	-381.3	205.0	1.8705
900.	681.	250.6	-622.3	-617.0	-607.7	-595.4	-580.0	263.0	2.2882
1000.	609.	363.5	-892.7	-886.8	-876.5	-862.9	-845.7	329.6	2.7549

BALLISTIC COEFFICIENT .15, MUZZLE VELOCITY FT/SEC 3500.

RANGE	VELOCITY	HEIGHT	100 YD	150 YD	200 YD	250 YD	300 YD	DRIFT	TIME
0.	3500.	0.0	-1.5	-1.5	-1.5	-1.5	-1.5	0.0	0.0000
100.	2822.	-.3	0.0	.5	1.5	2.7	4.3	1.7	.0956
200.	2233.	1.5	-3.0	-1.9	0.0	2.5	5.7	7.7	.2151
300.	1720.	5.8	-13.0	-11.4	-8.6	-4.8	0.0	19.6	.3682
400.	1313.	14.8	-35.2	-33.1	-29.3	-24.2	-17.9	39.8	.5689
500.	1072.	31.9	-78.2	-75.5	-70.8	-64.5	-56.5	69.7	.8247
600.	949.	59.8	-149.2	-146.0	-140.3	-132.7	-123.2	107.2	1.1231
700.	852.	101.2	-253.5	-249.8	-243.2	-234.3	-223.2	150.9	1.4575
800.	769.	159.7	-398.7	-394.5	-386.9	-376.8	-364.0	201.1	1.8283
900.	691.	240.1	-594.5	-589.8	-581.2	-569.9	-555.5	258.5	2.2401
1000.	618.	348.8	-854.4	-849.2	-839.7	-827.0	-811.1	324.1	2.6988

BALLISTIC COEFFICIENT .15, MUZZLE VELOCITY FT/SEC 3600.

RANGE	VELOCITY	HEIGHT	100 YD	150 YD	200 YD	250 YD	300 YD	DRIFT	TIME
0.	3600.	0.0	-1.5	-1.5	-1.5	-1.5	-1.5	0.0	0.0000
100.	2908.	-.3	0.0	.5	1.3	2.5	4.0	1.7	.0928
200.	2309.	1.3	-2.7	-1.8	0.0	2.4	5.3	7.4	.2086
300.	1784.	5.3	-12.0	-10.6	-8.0	-4.5	0.0	18.7	.3565
400.	1360.	13.8	-32.6	-30.7	-27.2	-22.5	-16.6	38.1	.5497
500.	1094.	29.8	-72.7	-70.4	-66.0	-60.1	-52.7	67.2	.7984
600.	963.	56.5	-140.2	-137.4	-132.2	-125.1	-116.2	104.2	1.0921
700.	864.	96.2	-239.9	-236.7	-230.5	-222.3	-211.9	147.5	1.4212
800.	780.	152.5	-379.3	-375.6	-368.6	-359.2	-347.3	197.1	1.7867
900.	701.	230.0	-567.8	-563.6	-555.7	-545.1	-531.8	253.9	2.1925
1000.	628.	335.1	-818.7	-814.0	-805.2	-793.5	-778.6	318.8	2.6450

BALLISTIC COEFFICIENT .15, MUZZLE VELOCITY FT/SEC 3700.

RANGE	VELOCITY	HEIGHT	100 YD	150 YD	200 YD	250 YD	300 YD	DRIFT	TIME
0.	3700.	0.0	-1.5	-1.5	-1.5	-1.5	-1.5	0.0	0.0000
100.	2995.	-.4	0.0	.4	1.2	2.3	3.7	1.6	.0902
200.	2384.	1.2	-2.5	-1.6	0.0	2.2	5.0	7.1	.2026
300.	1849.	5.0	-11.1	-9.9	-7.4	-4.1	0.0	18.0	.3455
400.	1409.	12.8	-30.2	-28.6	-25.3	-20.9	-15.4	36.5	.5318
500.	1118.	28.0	-67.7	-65.6	-61.6	-56.1	-49.2	64.8	.7735
600.	979.	53.3	-131.7	-129.2	-124.3	-117.7	-109.5	101.2	1.0615
700.	876.	91.6	-227.4	-224.6	-218.9	-211.2	-201.5	144.2	1.3870
800.	791.	145.8	-361.3	-358.0	-351.5	-342.8	-331.7	193.3	1.7471
900.	712.	220.5	-542.9	-539.2	-531.9	-522.0	-509.6	249.5	2.1472
1000.	637.	321.9	-784.5	-780.4	-772.3	-761.3	-747.5	313.6	2.5925

BALLISTIC COEFFICIENT .15, MUZZLE VELOCITY FT/SEC 3800.

RANGE	VELOCITY	HEIGHT	100 YD	150 YD	200 YD	250 YD	300 YD	DRIFT	TIME
0.	3800.	0.0	-1.5	-1.5	-1.5	-1.5	-1.5	0.0	0.0000
100.	3082.	-.4	0.0	.4	1.1	2.1	3.4	1.6	.0878
200.	2460.	1.1	-2.2	-1.5	0.0	2.0	4.6	6.8	.1968
300.	1915.	4.6	-10.3	-9.2	-7.0	-3.9	0.0	17.3	.3351
400.	1460.	12.0	-28.1	-26.6	-23.6	-19.5	-14.4	35.0	.5149
500.	1145.	26.2	-62.9	-61.1	-57.4	-52.3	-45.8	62.3	.7489
600.	993.	50.5	-124.1	-121.9	-117.4	-111.3	-103.5	98.6	1.0337
700.	888.	87.0	-215.1	-212.6	-207.3	-200.2	-191.1	140.7	1.3521
800.	801.	139.3	-344.0	-341.1	-335.1	-326.9	-316.6	189.4	1.7079
900.	722.	211.4	-519.0	-515.7	-509.0	-499.8	-488.1	245.0	2.1025
1000.	647.	309.4	-752.4	-748.8	-741.3	-731.0	-718.1	308.4	2.5420

BALLISTIC COEFFICIENT .15, MUZZLE VELOCITY FT/SEC 3900.

RANGE	VELOCITY	HEIGHT	100 YD	150 YD	200 YD	250 YD	300 YD	DRIFT	TIME
0.	3900.	0.0	-1.5	-1.5	-1.5	-1.5	-1.5	0.0	0.0000
100.	3169.	-.4	0.0	.3	1.0	2.0	3.2	1.5	.0853
200.	2535.	1.0	-2.0	-1.4	0.0	1.9	4.3	6.6	.1913
300.	1980.	4.3	-9.5	-8.6	-6.5	-3.6	0.0	16.6	.3251
400.	1513.	11.2	-26.1	-24.9	-22.1	-18.3	-13.4	33.6	.4987
500.	1175.	24.5	-58.7	-57.2	-53.7	-48.9	-42.9	60.1	.7260
600.	1009.	47.7	-116.5	-114.6	-110.4	-104.6	-97.4	95.5	1.0044
700.	900.	82.7	-203.6	-201.4	-196.5	-189.8	-181.4	137.3	1.3185
800.	812.	133.1	-327.8	-325.3	-319.7	-312.0	-302.4	185.6	1.6701
900.	732.	202.8	-496.5	-493.7	-487.4	-478.8	-468.0	240.6	2.0594
1000.	656.	297.4	-721.6	-718.5	-711.5	-701.9	-689.9	303.3	2.4924

BALLISTIC COEFFICIENT .15, MUZZLE VELOCITY FT/SEC 4000.

RANGE	VELOCITY	HEIGHT	100 YD	150 YD	200 YD	250 YD	300 YD	DRIFT	TIME
0.	4000.	0.0	-1.5	-1.5	-1.5	-1.5	-1.5	0.0	0.0000
100.	3256.	-.4	0.0	.3	.9	1.8	3.0	1.4	.0832
200.	2611.	.9	-1.8	-1.3	0.0	1.8	4.1	6.4	.1862
300.	2047.	4.0	-8.9	-8.0	-6.1	-3.4	0.0	16.0	.3161
400.	1567.	10.5	-24.4	-23.3	-20.7	-17.1	-12.6	32.4	.4839
500.	1209.	23.0	-54.7	-53.4	-50.1	-45.7	-40.0	57.8	.7034
600.	1025.	45.0	-109.3	-107.6	-103.8	-98.4	-91.6	92.6	.9763
700.	913.	78.7	-193.0	-191.0	-186.5	-180.2	-172.3	134.0	1.2865
800.	823.	127.3	-312.2	-310.0	-304.9	-297.7	-288.6	181.8	1.6332
900.	742.	194.6	-475.0	-472.5	-466.7	-458.7	-448.4	236.2	2.0173
1000.	666.	286.1	-692.6	-689.8	-683.4	-674.4	-663.1	298.3	2.4447

BALLISTIC COEFFICIENT .16, MUZZLE VELOCITY FT/SEC 2000.

RANGE	VELOCITY	HEIGHT	100 YD	150 YD	200 YD	250 YD	300 YD	DRIFT	TIME
0.	2000.	0.0	-1.5	-1.5	-1.5	-1.5	-1.5	0.0	0.0000
100.	1555.	.6	0.0	2.8	6.6	11.2	16.5	3.6	.1704
200.	1219.	6.5	-13.2	-7.5	0.0	9.2	19.9	15.8	.3896
300.	1038.	20.1	-49.6	-41.1	-29.9	-16.1	0.0	36.8	.6591
400.	929.	44.0	-116.3	-104.9	-89.9	-71.5	-50.1	64.3	.9653
500.	841.	81.0	-219.4	-205.2	-186.5	-163.5	-136.7	97.7	1.3049
600.	764.	134.6	-366.8	-349.7	-327.2	-299.6	-267.5	137.2	1.6793
700.	691.	209.4	-567.7	-547.7	-521.5	-489.3	-451.8	183.4	2.0923
800.	623.	311.2	-834.5	-811.8	-781.8	-745.0	-702.1	237.5	2.5492
900.	559.	448.0	-1184.5	-1158.8	-1125.1	-1083.8	-1035.5	300.5	3.0576
1000.	499.	630.4	-1639.7	-1611.2	-1573.7	-1527.8	-1474.2	374.2	3.6260

BALLISTIC COEFFICIENT .16, MUZZLE VELOCITY FT/SEC 2100.

RANGE	VELOCITY	HEIGHT	100 YD	150 YD	200 YD	250 YD	300 YD	DRIFT	TIME
0.	2100.	0.0	-1.5	-1.5	-1.5	-1.5	-1.5	0.0	0.0000
100.	1638.	.5	0.0	2.5	5.9	10.0	15.0	3.4	.1619
200.	1275.	5.8	-11.8	-6.7	0.0	8.3	18.3	15.0	.3707
300.	1063.	18.4	-45.1	-37.6	-27.4	-15.0	0.0	35.8	.6317
400.	949.	40.8	-107.2	-97.1	-83.6	-67.0	-47.0	63.2	.9304
500.	858.	75.9	-204.4	-191.8	-174.9	-154.2	-129.2	96.7	1.2638
600.	779.	126.8	-343.5	-328.3	-308.1	-283.2	-253.2	136.1	1.6302
700.	706.	198.1	-534.0	-516.3	-492.7	-463.7	-428.7	182.2	2.0351
800.	636.	295.1	-787.4	-767.2	-740.2	-707.0	-667.0	235.8	2.4827
900.	571.	425.7	-1120.1	-1097.4	-1067.1	-1029.7	-984.8	298.3	2.9807
1000.	511.	599.5	-1552.7	-1527.5	-1493.8	-1452.3	-1402.3	371.0	3.5364

BALLISTIC COEFFICIENT .16, MUZZLE VELOCITY FT/SEC 2200.

RANGE	VELOCITY	HEIGHT	100 YD	150 YD	200 YD	250 YD	300 YD	DRIFT	TIME
0.	2200.	0.0	-1.5	-1.5	-1.5	-1.5	-1.5	0.0	0.0000
100.	1722.	.4	0.0	2.2	5.3	9.0	13.6	3.1	.1542
200.	1335.	5.2	-10.5	-6.1	0.0	7.6	16.6	14.1	.3530
300.	1093.	16.8	-40.7	-34.0	-24.9	-13.6	0.0	34.3	.6039
400.	969.	37.9	-99.0	-90.0	-77.9	-62.8	-44.7	62.0	.8975
500.	874.	71.1	-190.3	-179.2	-164.0	-145.1	-122.5	95.4	1.2239
600.	794.	119.5	-321.8	-308.4	-290.2	-267.6	-240.4	134.6	1.5830
700.	720.	187.3	-502.3	-486.6	-465.4	-439.0	-407.3	180.4	1.9795
800.	650.	280.1	-743.5	-725.6	-701.3	-671.1	-634.9	233.7	2.4188
900.	584.	404.5	-1059.5	-1039.4	-1012.2	-978.2	-937.3	295.4	2.9058
1000.	522.	570.2	-1470.6	-1448.3	-1418.0	-1380.2	-1334.9	367.0	3.4490

BALLISTIC COEFFICIENT .16, MUZZLE VELOCITY FT/SEC 2300.

RANGE	VELOCITY	HEIGHT	100 YD	150 YD	200 YD	250 YD	300 YD	DRIFT	TIME
0.	2300.	0.0	-1.5	-1.5	-1.5	-1.5	-1.5	0.0	0.0000
100.	1807.	.3	0.0	2.0	4.7	8.1	12.3	2.9	.1472
200.	1399.	4.7	-9.4	-5.4	0.0	6.8	15.1	13.3	.3363
300.	1126.	15.3	-36.8	-30.8	-22.7	-12.5	0.0	32.8	.5774
400.	990.	35.1	-91.0	-83.0	-72.2	-58.5	-41.9	60.3	.8643
500.	891.	66.4	-176.5	-166.6	-153.1	-136.0	-115.2	93.4	1.1830
600.	809.	112.6	-301.4	-289.4	-273.2	-252.7	-227.8	132.7	1.5367
700.	734.	177.3	-472.9	-459.0	-440.0	-416.1	-387.1	178.3	1.9261
800.	663.	265.7	-702.0	-686.1	-664.4	-637.1	-603.9	231.0	2.3562
900.	596.	384.7	-1003.0	-985.1	-960.8	-930.0	-892.7	292.1	2.8336
1000.	534.	543.0	-1394.9	-1375.0	-1347.9	-1313.8	-1272.3	362.8	3.3658

BALLISTIC COEFFICIENT .16, MUZZLE VELOCITY FT/SEC 2400.

RANGE	VELOCITY	HEIGHT	100 YD	150 YD	200 YD	250 YD	300 YD	DRIFT	TIME
0.	2400.	0.0	-1.5	-1.5	-1.5	-1.5	-1.5	0.0	0.0000
100.	1894.	.2	0.0	1.8	4.2	7.3	11.1	2.8	.1407
200.	1468.	4.2	-8.4	-4.9	0.0	6.2	13.7	12.5	.3209
300.	1164.	13.9	-33.2	-27.9	-20.6	-11.3	0.0	31.2	.5521
400.	1011.	32.5	-83.5	-76.4	-66.7	-54.3	-39.2	58.4	.8319
500.	908.	62.1	-164.0	-155.1	-142.9	-127.5	-108.6	91.4	1.1442
600.	824.	106.0	-282.1	-271.5	-256.8	-238.3	-215.6	130.5	1.4915
700.	748.	167.8	-445.2	-432.8	-415.8	-394.1	-367.7	175.8	1.8739
800.	676.	252.3	-663.3	-649.2	-629.6	-604.9	-574.7	228.1	2.2960
900.	608.	366.0	-950.3	-934.4	-912.4	-884.6	-850.6	288.5	2.7641
1000.	545.	517.3	-1323.7	-1306.0	-1281.6	-1250.7	-1212.9	358.2	3.2851

BALLISTIC COEFFICIENT .16, MUZZLE VELOCITY FT/SEC 2500.

RANGE	VELOCITY	HEIGHT	100 YD	150 YD	200 YD	250 YD	300 YD	DRIFT	TIME
0.	2500.	0.0	-1.5	-1.5	-1.5	-1.5	-1.5	0.0	0.0000
100.	1982.	.1	0.0	1.6	3.8	6.6	10.0	2.6	.1347
200.	1540.	3.8	-7.6	-4.4	0.0	5.6	12.5	11.7	.3067
300.	1209.	12.6	-30.1	-25.4	-18.7	-10.3	0.0	29.6	.5280
400.	1033.	29.9	-76.3	-70.0	-61.2	-50.0	-36.2	56.2	.7992
500.	926.	58.0	-152.2	-144.4	-133.3	-119.3	-102.1	89.1	1.1065
600.	838.	99.8	-264.0	-254.6	-241.3	-224.5	-203.9	128.0	1.4474
700.	761.	158.8	-419.1	-408.1	-392.6	-373.0	-348.9	173.0	1.8230
800.	689.	239.6	-627.1	-614.5	-596.9	-574.5	-546.9	224.9	2.2377
900.	621.	348.2	-900.4	-886.2	-866.3	-841.1	-810.1	284.4	2.6961
1000.	557.	493.0	-1257.0	-1241.2	-1219.1	-1191.1	-1156.7	353.3	3.2071

BALLISTIC COEFFICIENT .16, MUZZLE VELOCITY FT/SEC 2600.

RANGE	VELOCITY	HEIGHT	100 YD	150 YD	200 YD	250 YD	300 YD	DRIFT	TIME
0.	2600.	0.0	-1.5	-1.5	-1.5	-1.5	-1.5	0.0	0.0000
100.	2071.	.1	0.0	1.4	3.4	5.9	9.1	2.4	.1293
200.	1614.	3.4	-6.8	-4.0	0.0	5.1	11.4	11.1	.2936
300.	1258.	11.5	-27.2	-23.0	-17.0	-9.4	0.0	28.0	.5053
400.	1056.	27.6	-69.8	-64.2	-56.2	-46.1	-33.5	54.0	.7685
500.	943.	54.2	-141.4	-134.4	-124.4	-111.7	-96.0	86.8	1.0702
600.	853.	94.0	-247.2	-238.7	-226.8	-211.6	-192.7	125.4	1.4048
700.	775.	150.3	-394.7	-384.9	-370.9	-353.2	-331.1	170.1	1.7739
800.	702.	227.5	-592.8	-581.6	-565.6	-545.4	-520.2	221.3	2.1807
900.	633.	331.6	-854.2	-841.6	-823.6	-800.8	-772.5	280.4	2.6314
1000.	568.	470.2	-1194.7	-1180.7	-1160.7	-1135.3	-1103.9	348.2	3.1323

BALLISTIC COEFFICIENT .16, MUZZLE VELOCITY FT/SEC 2700.

RANGE	VELOCITY	HEIGHT	100 YD	150 YD	200 YD	250 YD	300 YD	DRIFT	TIME
0.	2700.	0.0	-1.5	-1.5	-1.5	-1.5	-1.5	0.0	0.0000
100.	2160.	-.0	0.0	1.3	3.1	5.4	8.2	2.3	.1242
200.	1688.	3.1	-6.1	-3.6	0.0	4.6	10.4	10.4	.2815
300.	1311.	10.5	-24.7	-21.0	-15.5	-8.6	0.0	26.5	.4841
400.	1080.	25.5	-63.9	-58.9	-51.6	-42.4	-30.9	51.8	.7389
500.	961.	50.6	-131.0	-124.7	-115.7	-104.1	-89.8	84.2	1.0342
600.	868.	88.5	-231.3	-223.8	-212.9	-199.1	-181.9	122.6	1.3633
700.	788.	142.3	-371.8	-363.1	-350.4	-334.2	-314.1	166.9	1.7262
800.	714.	216.2	-561.0	-551.0	-536.5	-518.0	-495.1	217.7	2.1261
900.	644.	315.9	-810.5	-799.3	-783.0	-762.2	-736.4	276.0	2.5683
1000.	579.	448.5	-1135.8	-1123.3	-1105.1	-1082.0	-1053.4	342.9	3.0594

BALLISTIC COEFFICIENT .16, MUZZLE VELOCITY FT/SEC 2800.

RANGE	VELOCITY	HEIGHT	100 YD	150 YD	200 YD	250 YD	300 YD	DRIFT	TIME
0.	2800.	0.0	-1.5	-1.5	-1.5	-1.5	-1.5	0.0	0.0000
100.	2249.	-.1	0.0	1.1	2.8	4.9	7.5	2.2	.1196
200.	1763.	2.8	-5.5	-3.3	0.0	4.2	9.5	9.9	.2703
300.	1366.	9.6	-22.5	-19.1	-14.2	-7.8	0.0	25.1	.4642
400.	1108.	23.5	-58.4	-54.0	-47.4	-38.9	-28.5	49.6	.7104
500.	979.	47.2	-121.4	-115.9	-107.6	-97.1	-84.0	81.6	.9996
600.	882.	83.2	-216.4	-209.7	-199.8	-187.1	-171.5	119.7	1.3227
700.	802.	134.7	-350.2	-342.4	-330.9	-316.1	-297.8	163.6	1.6797
800.	727.	205.5	-531.0	-522.1	-508.9	-492.0	-471.1	214.0	2.0731
900.	656.	301.1	-769.7	-759.7	-744.8	-725.8	-702.3	271.6	2.5076
1000.	590.	428.4	-1081.4	-1070.2	-1053.7	-1032.6	-1006.5	337.7	2.9902

BALLISTIC COEFFICIENT .16, MUZZLE VELOCITY FT/SEC 2900.

RANGE	VELOCITY	HEIGHT	100 YD	150 YD	200 YD	250 YD	300 YD	DRIFT	TIME
0.	2900.	0.0	-1.5	-1.5	-1.5	-1.5	-1.5	0.0	0.0000
100.	2337.	-.1	0.0	1.0	2.5	4.4	6.8	2.1	.1153
200.	1839.	2.5	-5.0	-3.0	0.0	3.9	8.6	9.4	.2601
300.	1424.	8.8	-20.5	-17.5	-13.0	-7.2	0.0	23.8	.4457
400.	1139.	21.7	-53.5	-49.6	-43.5	-35.8	-26.2	47.4	.6833
500.	998.	44.1	-112.6	-107.6	-100.1	-90.4	-78.5	79.0	.9663
600.	897.	78.4	-202.7	-196.8	-187.7	-176.1	-161.8	116.8	1.2842
700.	815.	127.6	-330.2	-323.2	-312.6	-299.1	-282.4	160.3	1.6352
800.	739.	195.4	-502.8	-494.8	-482.7	-467.3	-448.2	210.2	2.0217
900.	668.	287.2	-731.5	-722.6	-709.0	-691.6	-670.1	267.2	2.4493
1000.	601.	409.2	-1029.6	-1019.7	-1004.6	-985.3	-961.4	332.3	2.9226

BALLISTIC COEFFICIENT .16, MUZZLE VELOCITY FT/SEC 3000.

RANGE	VELOCITY	HEIGHT	100 YD	150 YD	200 YD	250 YD	300 YD	DRIFT	TIME
0.	3000.	0.0	-1.5	-1.5	-1.5	-1.5	-1.5	0.0	0.0000
100.	2425.	-.2	0.0	.9	2.3	4.0	6.2	2.0	.1114
200.	1915.	2.3	-4.5	-2.8	0.0	3.5	7.9	8.9	.2506
300.	1486.	8.1	-18.7	-16.1	-11.9	-6.6	0.0	22.6	.4286
400.	1174.	20.0	-49.2	-45.6	-40.1	-33.0	-24.2	45.4	.6579
500.	1017.	41.1	-104.4	-99.9	-93.0	-84.2	-73.2	76.4	.9342
600.	913.	73.8	-189.7	-184.4	-176.1	-165.4	-152.3	113.7	1.2462
700.	827.	120.9	-311.3	-305.1	-295.4	-283.0	-267.6	157.0	1.5919
800.	751.	186.0	-476.6	-469.5	-458.4	-444.3	-426.7	206.4	1.9726
900.	679.	274.0	-695.4	-687.4	-674.9	-659.0	-639.2	262.7	2.3925
1000.	611.	391.5	-981.9	-973.0	-959.2	-941.5	-919.5	327.1	2.8586

BALLISTIC COEFFICIENT .16, MUZZLE VELOCITY FT/SEC 3100.

RANGE	VELOCITY	HEIGHT	100 YD	150 YD	200 YD	250 YD	300 YD	DRIFT	TIME
0.	3100.	0.0	-1.5	-1.5	-1.5	-1.5	-1.5	0.0	0.0000
100.	2512.	-.2	0.0	.8	2.1	3.7	5.7	1.9	.1076
200.	1993.	2.1	-4.1	-2.5	0.0	3.3	7.3	8.5	.2416
300.	1549.	7.4	-17.1	-14.8	-11.0	-6.1	0.0	21.5	.4126
400.	1215.	18.5	-45.1	-41.9	-36.9	-30.4	-22.2	43.2	.6328
500.	1036.	38.4	-96.7	-92.7	-86.4	-78.3	-68.1	73.7	.9028
600.	928.	69.5	-177.6	-172.8	-165.3	-155.5	-143.3	110.7	1.2095
700.	840.	114.5	-293.4	-287.9	-279.0	-267.6	-253.4	153.5	1.5495
800.	763.	177.0	-451.7	-445.3	-435.2	-422.2	-406.0	202.4	1.9244
900.	690.	261.7	-661.8	-654.6	-643.3	-628.6	-610.3	258.2	2.3381
1000.	622.	374.4	-936.2	-928.3	-915.7	-899.4	-879.1	321.7	2.7956

BALLISTIC COEFFICIENT .16, MUZZLE VELOCITY FT/SEC 3200.

RANGE	VELOCITY	HEIGHT	100 YD	150 YD	200 YD	250 YD	300 YD	DRIFT	TIME
0.	3200.	0.0	-1.5	-1.5	-1.5	-1.5	-1.5	0.0	0.0000
100.	2600.	-.2	0.0	.7	1.9	3.4	5.2	1.8	.1041
200.	2070.	1.9	-3.7	-2.3	0.0	3.0	6.8	8.1	.2334
300.	1613.	6.8	-15.7	-13.6	-10.1	-5.6	0.0	20.5	.3978
400.	1258.	17.1	-41.5	-38.6	-34.0	-28.0	-20.5	41.3	.6095
500.	1056.	35.8	-89.6	-86.0	-80.2	-72.7	-63.3	71.1	.8727
600.	943.	65.5	-166.4	-162.2	-155.2	-146.2	-135.0	107.7	1.1745
700.	853.	108.6	-276.9	-271.9	-263.8	-253.3	-240.2	150.1	1.5091
800.	775.	168.6	-428.4	-422.7	-413.5	-401.4	-386.5	198.6	1.8782
900.	702.	249.9	-629.8	-623.4	-613.0	-599.4	-582.6	253.7	2.2850
1000.	633.	358.5	-893.9	-886.7	-875.2	-860.1	-841.4	316.5	2.7358

BALLISTIC COEFFICIENT .16, MUZZLE VELOCITY FT/SEC 3300.

RANGE	VELOCITY	HEIGHT	100 YD	150 YD	200 YD	250 YD	300 YD	DRIFT	TIME
0.	3300.	0.0	-1.5	-1.5	-1.5	-1.5	-1.5	0.0	0.0000
100.	2687.	-.3	0.0	.6	1.7	3.1	4.8	1.7	.1008
200.	2148.	1.7	-3.4	-2.1	0.0	2.8	6.2	7.7	.2258
300.	1678.	6.3	-14.5	-12.6	-9.4	-5.2	0.0	19.6	.3839
400.	1303.	15.8	-38.2	-35.7	-31.4	-25.9	-18.9	39.4	.5877
500.	1077.	33.4	-83.0	-79.8	-74.5	-67.5	-58.9	68.5	.8436
600.	959.	61.6	-155.7	-151.9	-145.5	-137.1	-126.7	104.6	1.1396
700.	866.	102.9	-261.2	-256.8	-249.3	-239.6	-227.5	146.6	1.4696
800.	787.	160.6	-406.5	-401.4	-392.9	-381.7	-367.9	194.7	1.8333
900.	713.	238.8	-600.0	-594.3	-584.7	-572.2	-556.6	249.2	2.2342
1000.	643.	343.3	-853.5	-847.2	-836.5	-822.6	-805.3	311.2	2.6773

BALLISTIC COEFFICIENT .16, MUZZLE VELOCITY FT/SEC 3400.

RANGE	VELOCITY	HEIGHT	100 YD	150 YD	200 YD	250 YD	300 YD	DRIFT	TIME
0.	3400.	0.0	-1.5	-1.5	-1.5	-1.5	-1.5	0.0	0.0000
100.	2774.	-.3	0.0	.6	1.5	2.8	4.4	1.7	.0977
200.	2225.	1.5	-3.1	-2.0	0.0	2.6	5.8	7.4	.2185
300.	1743.	5.9	-13.3	-11.6	-8.7	-4.8	0.0	18.7	.3709
400.	1351.	14.7	-35.3	-33.0	-29.1	-24.0	-17.5	37.7	.5671
500.	1100.	31.2	-76.9	-74.1	-69.2	-62.7	-54.7	65.9	.8155
600.	975.	58.0	-145.8	-142.4	-136.6	-128.8	-119.2	101.5	1.1063
700.	878.	97.6	-246.6	-242.6	-235.7	-226.7	-215.5	143.2	1.4312
800.	798.	153.0	-385.7	-381.2	-373.3	-363.0	-350.1	190.7	1.7895
900.	723.	228.3	-571.6	-566.5	-557.7	-546.1	-531.6	244.7	2.1844
1000.	653.	329.0	-815.8	-810.1	-800.3	-787.4	-771.3	306.0	2.6212

BALLISTIC COEFFICIENT .16, MUZZLE VELOCITY FT/SEC 3500.

RANGE	VELOCITY	HEIGHT	100 YD	150 YD	200 YD	250 YD	300 YD	DRIFT	TIME
0.	3500.	0.0	-1.5	-1.5	-1.5	-1.5	-1.5	0.0	0.0000
100.	2861.	-.3	0.0	.5	1.4	2.6	4.1	1.6	.0948
200.	2303.	1.4	-2.8	-1.8	0.0	2.4	5.4	7.1	.2118
300.	1809.	5.4	-12.3	-10.8	-8.1	-4.5	0.0	17.9	.3589
400.	1401.	13.6	-32.6	-30.6	-27.0	-22.2	-16.3	36.1	.5477
500.	1127.	29.1	-71.4	-68.8	-64.3	-58.3	-50.9	63.4	.7886
600.	990.	54.8	-136.9	-133.9	-128.5	-121.3	-112.3	98.8	1.0754
700.	892.	92.5	-232.7	-229.1	-222.8	-214.4	-204.0	139.7	1.3938
800.	810.	145.8	-366.2	-362.2	-354.9	-345.3	-333.5	186.9	1.7474
900.	734.	218.4	-545.0	-540.5	-532.3	-521.5	-508.1	240.3	2.1366
1000.	663.	315.4	-779.8	-774.8	-765.7	-753.7	-738.8	300.9	2.5665

BALLISTIC COEFFICIENT .16, MUZZLE VELOCITY FT/SEC 3600.

RANGE	VELOCITY	HEIGHT	100 YD	150 YD	200 YD	250 YD	300 YD	DRIFT	TIME
0.	3600.	0.0	-1.5	-1.5	-1.5	-1.5	-1.5	0.0	0.0000
100.	2949.	-.3	0.0	.4	1.3	2.4	3.8	1.5	.0920
200.	2380.	1.3	-2.6	-1.7	0.0	2.2	5.0	6.8	.2054
300.	1876.	5.0	-11.4	-10.0	-7.5	-4.2	0.0	17.1	.3474
400.	1454.	12.7	-30.3	-28.5	-25.1	-20.7	-15.1	34.5	.5294
500.	1155.	27.2	-66.3	-64.1	-59.9	-54.3	-47.4	61.0	.7631
600.	1007.	51.5	-128.1	-125.4	-120.4	-113.7	-105.4	95.6	1.0434
700.	905.	87.7	-219.6	-216.5	-210.6	-202.8	-193.1	136.2	1.3574
800.	821.	138.9	-347.6	-344.1	-337.3	-328.4	-317.3	182.9	1.7058
900.	745.	208.9	-519.7	-515.6	-508.1	-498.1	-485.6	235.8	2.0898
1000.	673.	302.5	-746.0	-741.5	-733.1	-722.0	-708.1	295.7	2.5136

BALLISTIC COEFFICIENT .16, MUZZLE VELOCITY FT/SEC 3700.

RANGE	VELOCITY	HEIGHT	100 YD	150 YD	200 YD	250 YD	300 YD	DRIFT	TIME
0.	3700.	0.0	-1.5	-1.5	-1.5	-1.5	-1.5	0.0	0.0000
100.	3036.	-.4	0.0	.4	1.2	2.2	3.5	1.5	.0896
200.	2456.	1.2	-2.3	-1.6	0.0	2.1	4.7	6.6	.1995
300.	1943.	4.7	-10.5	-9.3	-7.0	-3.9	0.0	16.5	.3369
400.	1509.	11.8	-28.1	-26.5	-23.4	-19.3	-14.1	33.1	.5123
500.	1189.	25.4	-61.6	-59.6	-55.7	-50.5	-44.1	58.5	.7379
600.	1023.	48.5	-119.8	-117.5	-112.8	-106.6	-98.8	92.6	1.0127
700.	918.	83.2	-207.3	-204.5	-199.1	-191.8	-182.7	132.8	1.3223
800.	832.	132.5	-330.3	-327.1	-320.9	-312.6	-302.2	179.1	1.6662
900.	755.	199.9	-495.8	-492.2	-485.2	-475.9	-464.2	231.4	2.0446
1000.	683.	290.3	-713.7	-709.8	-702.1	-691.7	-678.7	290.7	2.4623

BALLISTIC COEFFICIENT .16, MUZZLE VELOCITY FT/SEC 3800.

RANGE	VELOCITY	HEIGHT	100 YD	150 YD	200 YD	250 YD	300 YD	DRIFT	TIME
0.	3800.	0.0	-1.5	-1.5	-1.5	-1.5	-1.5	0.0	0.0000
100.	3124.	-.4	0.0	.3	1.1	2.0	3.2	1.5	.0873
200.	2533.	1.1	-2.1	-1.4	0.0	1.9	4.4	6.4	.1940
300.	2011.	4.4	-9.7	-8.7	-6.5	-3.6	0.0	15.8	.3269
400.	1564.	11.1	-26.1	-24.7	-21.9	-18.0	-13.1	31.8	.4963
500.	1225.	23.7	-57.3	-55.6	-52.0	-47.1	-41.1	56.2	.7143
600.	1040.	45.6	-112.0	-110.0	-105.7	-99.9	-92.6	89.6	.9828
700.	931.	78.9	-195.7	-193.4	-188.3	-181.5	-173.1	129.5	1.2882
800.	843.	126.3	-313.6	-310.9	-305.1	-297.4	-287.7	175.2	1.6272
900.	766.	191.4	-473.1	-470.0	-463.5	-454.8	-443.9	227.1	2.0007
1000.	693.	278.7	-683.4	-680.0	-672.8	-663.1	-651.0	285.7	2.4126

BALLISTIC COEFFICIENT .16, MUZZLE VELOCITY FT/SEC 3900.

RANGE	VELOCITY	HEIGHT	100 YD	150 YD	200 YD	250 YD	300 YD	DRIFT	TIME
0.	3900.	0.0	-1.5	-1.5	-1.5	-1.5	-1.5	0.0	0.0000
100.	3211.	-.4	0.0	.3	1.0	1.9	3.0	1.4	.0848
200.	2610.	1.0	-1.9	-1.3	0.0	1.8	4.1	6.1	.1884
300.	2079.	4.1	-9.0	-8.1	-6.1	-3.4	0.0	15.2	.3172
400.	1621.	10.3	-24.3	-23.1	-20.4	-16.8	-12.3	30.5	.4808
500.	1263.	22.2	-53.4	-51.9	-48.6	-44.0	-38.4	54.0	.6917
600.	1058.	42.9	-104.9	-103.1	-99.1	-93.7	-86.9	86.7	.9541
700.	945.	74.9	-185.0	-182.9	-178.3	-171.9	-164.0	126.2	1.2555
800.	854.	120.5	-298.0	-295.6	-290.3	-283.0	-274.0	171.4	1.5892
900.	776.	183.2	-451.5	-448.8	-442.9	-434.7	-424.5	222.7	1.9577
1000.	703.	267.5	-654.3	-651.4	-644.7	-635.6	-624.3	280.7	2.3639

BALLISTIC COEFFICIENT .16, MUZZLE VELOCITY FT/SEC 4000.

RANGE	VELOCITY	HEIGHT	100 YD	150 YD	200 YD	250 YD	300 YD	DRIFT	TIME
0.	4000.	0.0	-1.5	-1.5	-1.5	-1.5	-1.5	0.0	0.0000
100.	3299.	-.4	0.0	.3	.9	1.7	2.8	1.3	.0827
200.	2686.	.9	-1.7	-1.2	0.0	1.7	3.8	5.9	.1835
300.	2147.	3.8	-8.4	-7.6	-5.7	-3.2	0.0	14.7	.3085
400.	1677.	9.7	-22.7	-21.6	-19.2	-15.8	-11.5	29.3	.4667
500.	1303.	20.8	-49.8	-48.5	-45.4	-41.2	-35.9	52.0	.6705
600.	1077.	40.5	-98.2	-96.7	-93.0	-87.9	-81.5	83.9	.9265
700.	959.	71.0	-174.5	-172.7	-168.4	-162.4	-155.0	122.8	1.2225
800.	866.	114.9	-283.2	-281.2	-276.2	-269.4	-260.9	167.6	1.5525
900.	787.	175.5	-431.2	-428.9	-423.3	-415.7	-406.1	218.5	1.9163
1000.	712.	257.0	-627.0	-624.4	-618.3	-609.8	-599.1	275.8	2.3172

BALLISTIC COEFFICIENT .17, MUZZLE VELOCITY FT/SEC 2000.

RANGE	VELOCITY	HEIGHT	100 YD	150 YD	200 YD	250 YD	300 YD	DRIFT	TIME
0.	2000.	0.0	-1.5	-1.5	-1.5	-1.5	-1.5	0.0	0.0000
100.	1579.	.6	0.0	2.8	6.4	10.8	15.9	3.4	.1691
200.	1251.	6.3	-12.7	-7.2	0.0	8.8	19.0	14.7	.3836
300.	1060.	19.3	-47.7	-39.4	-28.6	-15.3	0.0	34.6	.6466
400.	953.	42.2	-111.9	-100.8	-86.4	-68.7	-48.3	60.9	.9459
500.	865.	77.5	-210.9	-197.1	-179.1	-157.0	-131.5	92.7	1.2768
600.	791.	128.3	-351.8	-335.3	-313.6	-287.1	-256.5	130.2	1.6396
700.	721.	198.4	-542.2	-522.9	-497.7	-466.7	-431.0	173.7	2.0367
800.	655.	293.0	-793.6	-771.6	-742.7	-707.4	-666.5	224.2	2.4739
900.	592.	418.6	-1119.4	-1094.6	-1062.1	-1022.4	-976.4	282.6	2.9559
1000.	534.	583.8	-1537.8	-1510.2	-1474.1	-1429.9	-1378.9	350.2	3.4897

BALLISTIC COEFFICIENT .17, MUZZLE VELOCITY FT/SEC 2100.

RANGE	VELOCITY	HEIGHT	100 YD	150 YD	200 YD	250 YD	300 YD	DRIFT	TIME
0.	2100.	0.0	-1.5	-1.5	-1.5	-1.5	-1.5	0.0	0.0000
100.	1663.	.5	0.0	2.4	5.7	9.6	14.4	3.1	.1607
200.	1310.	5.6	-11.3	-6.5	0.0	7.9	17.4	13.9	.3647
300.	1089.	17.6	-43.1	-35.8	-26.1	-14.2	0.0	33.4	.6183
400.	974.	39.0	-102.8	-93.0	-80.1	-64.3	-45.3	59.7	.9105
500.	883.	72.4	-195.7	-183.5	-167.4	-147.6	-123.9	91.6	1.2345
600.	806.	120.7	-328.8	-314.2	-294.8	-271.0	-242.6	129.1	1.5905
700.	736.	187.4	-509.3	-492.3	-469.6	-441.9	-408.8	172.5	1.9800
800.	669.	277.6	-747.8	-728.2	-702.4	-670.7	-632.8	222.7	2.4082
900.	605.	397.3	-1056.9	-1035.0	-1005.9	-970.2	-927.7	280.6	2.8798
1000.	546.	554.8	-1454.4	-1430.0	-1397.7	-1358.0	-1310.7	347.3	3.4019

BALLISTIC COEFFICIENT .17, MUZZLE VELOCITY FT/SEC 2200.

RANGE	VELOCITY	HEIGHT	100 YD	150 YD	200 YD	250 YD	300 YD	DRIFT	TIME
0.	2200.	0.0	-1.5	-1.5	-1.5	-1.5	-1.5	0.0	0.0000
100.	1747.	.4	0.0	2.2	5.1	8.7	13.0	3.0	.1531
200.	1374.	5.0	-10.1	-5.8	0.0	7.2	15.8	13.1	.3473
300.	1123.	16.0	-38.9	-32.4	-23.7	-12.9	0.0	32.0	.5908
400.	995.	36.1	-94.3	-85.6	-74.1	-59.7	-42.5	58.2	.8761
500.	900.	67.6	-181.5	-170.6	-156.2	-138.2	-116.7	90.0	1.1932
600.	822.	113.5	-307.2	-294.2	-276.8	-255.3	-229.4	127.5	1.5425
700.	750.	177.1	-478.5	-463.3	-443.0	-417.9	-387.8	170.8	1.9250
800.	682.	263.0	-704.6	-687.3	-664.1	-635.4	-601.0	220.6	2.3442
900.	618.	377.2	-998.5	-979.0	-953.0	-920.7	-881.9	277.9	2.8062
1000.	558.	527.4	-1376.4	-1354.7	-1325.8	-1289.9	-1246.9	343.8	3.3172

BALLISTIC COEFFICIENT .17, MUZZLE VELOCITY FT/SEC 2300.

RANGE	VELOCITY	HEIGHT	100 YD	150 YD	200 YD	250 YD	300 YD	DRIFT	TIME
0.	2300.	0.0	-1.5	-1.5	-1.5	-1.5	-1.5	0.0	0.0000
100.	1834.	.3	0.0	1.9	4.5	7.8	11.7	2.8	.1462
200.	1442.	4.5	-9.0	-5.2	0.0	6.5	14.3	12.3	.3310
300.	1161.	14.5	-35.0	-29.3	-21.5	-11.7	0.0	30.4	.5642
400.	1017.	33.3	-86.3	-78.7	-68.2	-55.3	-39.6	56.4	.8423
500.	918.	63.1	-168.2	-158.6	-145.6	-129.4	-109.8	88.2	1.1531
600.	837.	106.7	-286.9	-275.4	-259.8	-240.3	-216.8	125.5	1.4958
700.	764.	167.3	-449.4	-436.0	-417.7	-395.0	-367.6	168.6	1.8711
800.	696.	249.3	-664.5	-649.2	-628.3	-602.4	-571.1	218.1	2.2826
900.	631.	358.5	-944.4	-927.1	-903.7	-874.5	-839.3	274.9	2.7356
1000.	570.	501.9	-1304.3	-1285.1	-1259.1	-1226.7	-1187.5	340.0	3.2362

BALLISTIC COEFFICIENT .17, MUZZLE VELOCITY FT/SEC 2400.

RANGE	VELOCITY	HEIGHT	100 YD	150 YD	200 YD	250 YD	300 YD	DRIFT	TIME
0.	2400.	0.0	-1.5	-1.5	-1.5	-1.5	-1.5	0.0	0.0000
100.	1921.	.2	0.0	1.7	4.1	7.0	10.5	2.6	.1398
200.	1514.	4.0	-8.1	-4.7	0.0	5.9	13.0	11.6	.3158
300.	1207.	13.2	-31.6	-26.5	-19.5	-10.7	0.0	28.9	.5391
400.	1039.	30.7	-78.9	-72.1	-62.7	-51.0	-36.7	54.4	.8093
500.	936.	58.8	-155.8	-147.3	-135.6	-120.9	-103.1	86.1	1.1141
600.	852.	100.2	-268.0	-257.8	-243.7	-226.1	-204.7	123.3	1.4505
700.	779.	158.0	-422.1	-410.2	-393.8	-373.3	-348.3	166.1	1.8187
800.	709.	236.4	-627.0	-613.4	-594.6	-571.1	-542.6	215.2	2.2229
900.	644.	340.6	-893.3	-878.0	-856.9	-830.5	-798.4	271.4	2.6669
1000.	582.	477.7	-1236.1	-1219.1	-1195.6	-1166.3	-1130.6	335.6	3.1571

BALLISTIC COEFFICIENT .17, MUZZLE VELOCITY FT/SEC 2500.

RANGE	VELOCITY	HEIGHT	100 YD	150 YD	200 YD	250 YD	300 YD	DRIFT	TIME
0.	2500.	0.0	-1.5	-1.5	-1.5	-1.5	-1.5	0.0	0.0000
100.	2011.	.1	0.0	1.5	3.6	6.3	9.5	2.4	.1338
200.	1588.	3.6	-7.3	-4.2	0.0	5.3	11.8	10.9	.3020
300.	1257.	12.0	-28.6	-24.1	-17.7	-9.7	0.0	27.4	.5154
400.	1063.	28.3	-72.2	-66.1	-57.6	-47.0	-34.0	52.5	.7780
500.	955.	54.8	-144.2	-136.6	-126.0	-112.8	-96.5	83.8	1.0759
600.	867.	94.2	-250.2	-241.1	-228.3	-212.5	-193.0	120.8	1.4061
700.	793.	149.3	-396.7	-386.1	-371.2	-352.7	-330.0	163.4	1.7682
800.	722.	224.1	-591.5	-579.4	-562.4	-541.2	-515.2	212.0	2.1646
900.	656.	323.9	-845.7	-832.0	-812.9	-789.1	-759.8	267.6	2.6006
1000.	594.	455.0	-1173.0	-1157.8	-1136.5	-1110.1	-1077.6	331.1	3.0815

BALLISTIC COEFFICIENT .17, MUZZLE VELOCITY FT/SEC 2600.

RANGE	VELOCITY	HEIGHT	100 YD	150 YD	200 YD	250 YD	300 YD	DRIFT	TIME
0.	2600.	0.0	-1.5	-1.5	-1.5	-1.5	-1.5	0.0	0.0000
100.	2100.	.0	0.0	1.3	3.3	5.7	8.6	2.3	.1284
200.	1663.	3.3	-6.5	-3.9	0.0	4.8	10.7	10.3	.2891
300.	1311.	10.9	-25.9	-21.9	-16.1	-8.9	0.0	25.9	.4931
400.	1090.	26.0	-65.8	-60.4	-52.7	-43.1	-31.2	50.2	.7466
500.	974.	51.1	-133.4	-126.7	-117.0	-105.0	-90.2	81.3	1.0388
600.	883.	88.4	-233.5	-225.4	-213.8	-199.4	-181.7	118.0	1.3628
700.	806.	141.0	-372.7	-363.3	-349.8	-333.0	-312.3	160.3	1.7188
800.	736.	212.6	-558.4	-547.6	-532.2	-512.9	-489.3	208.6	2.1083
900.	669.	308.1	-800.9	-788.8	-771.5	-749.8	-723.3	263.6	2.5364
1000.	605.	433.6	-1113.3	-1099.8	-1080.6	-1056.5	-1027.0	326.3	3.0080

BALLISTIC COEFFICIENT .17, MUZZLE VELOCITY FT/SEC 2700.

RANGE	VELOCITY	HEIGHT	100 YD	150 YD	200 YD	250 YD	300 YD	DRIFT	TIME
0.	2700.	0.0	-1.5	-1.5	-1.5	-1.5	-1.5	0.0	0.0000
100.	2190.	-.0	0.0	1.2	3.0	5.1	7.8	2.2	.1233
200.	1739.	2.9	-5.9	-3.5	0.0	4.4	9.8	9.7	.2772
300.	1367.	10.0	-23.5	-19.9	-14.7	-8.1	0.0	24.5	.4724
400.	1119.	23.9	-60.1	-55.2	-48.3	-39.5	-28.7	48.0	.7170
500.	992.	47.7	-123.8	-117.7	-109.0	-98.0	-84.6	79.0	1.0044
600.	898.	83.0	-218.0	-210.7	-200.3	-187.1	-170.9	115.2	1.3210
700.	820.	133.2	-350.1	-341.7	-329.5	-314.1	-295.3	157.1	1.6705
800.	749.	201.7	-527.4	-517.7	-503.8	-486.2	-464.7	205.0	2.0538
900.	681.	293.0	-758.7	-747.8	-732.1	-712.3	-688.1	259.4	2.4738
1000.	617.	413.3	-1057.4	-1045.3	-1027.9	-1005.9	-979.0	321.3	2.9369

BALLISTIC COEFFICIENT .17, MUZZLE VELOCITY FT/SEC 2800.

RANGE	VELOCITY	HEIGHT	100 YD	150 YD	200 YD	250 YD	300 YD	DRIFT	TIME
0.	2800.	0.0	-1.5	-1.5	-1.5	-1.5	-1.5	0.0	0.0000
100.	2279.	-.1	0.0	1.1	2.7	4.7	7.1	2.0	.1187
200.	1816.	2.7	-5.3	-3.2	0.0	4.0	8.9	9.1	.2662
300.	1427.	9.1	-21.4	-18.1	-13.4	-7.4	0.0	23.1	.4529
400.	1153.	22.0	-54.8	-50.5	-44.2	-36.1	-26.3	45.7	.6884
500.	1012.	44.3	-114.3	-108.9	-101.0	-91.0	-78.7	76.3	.9693
600.	914.	77.9	-203.4	-196.9	-187.4	-175.3	-160.6	112.2	1.2801
700.	834.	125.9	-329.2	-321.6	-310.5	-296.5	-279.3	153.9	1.6242
800.	761.	191.4	-498.1	-489.4	-476.8	-460.7	-441.1	201.3	2.0007
900.	693.	279.0	-719.5	-709.8	-695.6	-677.5	-655.4	255.2	2.4141
1000.	628.	394.3	-1005.3	-994.5	-978.7	-958.6	-934.1	316.3	2.8689

BALLISTIC COEFFICIENT .17, MUZZLE VELOCITY FT/SEC 2900.

RANGE	VELOCITY	HEIGHT	100 YD	150 YD	200 YD	250 YD	300 YD	DRIFT	TIME
0.	2900.	0.0	-1.5	-1.5	-1.5	-1.5	-1.5	0.0	0.0000
100.	2368.	-.1	0.0	1.0	2.4	4.2	6.5	2.0	.1145
200.	1893.	2.4	-4.8	-2.9	0.0	3.7	8.2	8.7	.2562
300.	1491.	8.3	-19.5	-16.6	-12.3	-6.7	0.0	21.9	.4350
400.	1192.	20.3	-50.2	-46.3	-40.5	-33.2	-24.2	43.6	.6615
500.	1032.	41.3	-105.6	-100.8	-93.5	-84.3	-73.1	73.6	.9356
600.	931.	73.2	-189.9	-184.1	-175.4	-164.4	-150.9	109.2	1.2410
700.	847.	119.0	-309.5	-302.7	-292.6	-279.7	-264.0	150.5	1.5793
800.	774.	181.8	-470.9	-463.2	-451.6	-436.9	-418.9	197.5	1.9499
900.	705.	265.7	-682.6	-674.0	-661.0	-644.4	-624.2	250.8	2.3562
1000.	640.	376.3	-956.0	-946.4	-931.9	-913.5	-891.0	311.2	2.8027

BALLISTIC COEFFICIENT .17, MUZZLE VELOCITY FT/SEC 3000.

RANGE	VELOCITY	HEIGHT	100 YD	150 YD	200 YD	250 YD	300 YD	DRIFT	TIME
0.	3000.	0.0	-1.5	-1.5	-1.5	-1.5	-1.5	0.0	0.0000
100.	2457.	-.2	0.0	.9	2.2	3.9	5.9	1.9	.1106
200.	1972.	2.2	-4.4	-2.6	0.0	3.4	7.5	8.3	.2469
300.	1556.	7.7	-17.8	-15.3	-11.3	-6.2	0.0	20.9	.4185
400.	1235.	18.7	-46.0	-42.6	-37.3	-30.6	-22.3	41.6	.6363
500.	1053.	38.3	-97.2	-92.9	-86.3	-77.9	-67.5	70.7	.9019
600.	947.	68.8	-177.5	-172.4	-164.4	-154.3	-141.9	106.3	1.2040
700.	861.	112.5	-290.9	-284.9	-275.7	-263.8	-249.3	147.1	1.5357
800.	786.	172.7	-445.3	-438.4	-427.8	-414.3	-397.7	193.7	1.9007
900.	717.	253.2	-648.0	-640.3	-628.3	-613.1	-594.5	246.4	2.3003
1000.	651.	359.6	-910.5	-901.9	-888.7	-871.8	-851.1	306.2	2.7399

BALLISTIC COEFFICIENT .17, MUZZLE VELOCITY FT/SEC 3100.

RANGE	VELOCITY	HEIGHT	100 YD	150 YD	200 YD	250 YD	300 YD	DRIFT	TIME
0.	3100.	0.0	-1.5	-1.5	-1.5	-1.5	-1.5	0.0	0.0000
100.	2545.	-.2	0.0	.8	2.0	3.5	5.4	1.8	.1069
200.	2051.	2.0	-4.0	-2.4	0.0	3.1	6.9	7.9	.2382
300.	1621.	7.0	-16.3	-14.0	-10.4	-5.7	0.0	19.8	.4028
400.	1281.	17.2	-42.2	-39.2	-34.3	-28.1	-20.5	39.6	.6121
500.	1075.	35.6	-89.8	-85.9	-79.9	-72.1	-62.6	68.0	.8703
600.	963.	64.5	-165.5	-160.9	-153.6	-144.3	-132.9	103.1	1.1663
700.	874.	106.5	-274.1	-268.8	-260.3	-249.4	-236.1	143.8	1.4946
800.	799.	164.0	-421.1	-414.9	-405.2	-392.8	-377.6	189.8	1.8525
900.	728.	241.4	-615.5	-608.6	-597.6	-583.7	-566.5	242.0	2.2462
1000.	662.	343.5	-867.0	-859.3	-847.2	-831.7	-812.6	301.0	2.6782

BALLISTIC COEFFICIENT .17, MUZZLE VELOCITY FT/SEC 3200.

RANGE	VELOCITY	HEIGHT	100 YD	150 YD	200 YD	250 YD	300 YD	DRIFT	TIME
0.	3200.	0.0	-1.5	-1.5	-1.5	-1.5	-1.5	0.0	0.0000
100.	2633.	-.2	0.0	.7	1.8	3.2	5.0	1.7	.1034
200.	2130.	1.8	-3.6	-2.2	0.0	2.9	6.4	7.5	.2301
300.	1687.	6.5	-15.0	-12.9	-9.6	-5.3	0.0	18.9	.3885
400.	1329.	15.9	-38.9	-36.1	-31.7	-25.9	-18.9	37.8	.5896
500.	1099.	33.1	-82.9	-79.5	-73.9	-66.8	-58.0	65.4	.8402
600.	980.	60.6	-154.4	-150.3	-143.6	-135.0	-124.4	99.9	1.1303
700.	888.	100.5	-257.5	-252.7	-244.9	-234.8	-222.5	140.1	1.4525
800.	811.	155.9	-398.6	-393.1	-384.2	-372.7	-358.6	185.9	1.8064
900.	740.	230.2	-584.7	-578.5	-568.5	-555.6	-539.7	237.5	2.1935
1000.	673.	328.6	-826.6	-819.7	-808.6	-794.2	-776.6	296.0	2.6193

BALLISTIC COEFFICIENT .17, MUZZLE VELOCITY FT/SEC 3300.

RANGE	VELOCITY	HEIGHT	100 YD	150 YD	200 YD	250 YD	300 YD	DRIFT	TIME
0.	3300.	0.0	-1.5	-1.5	-1.5	-1.5	-1.5	0.0	0.0000
100.	2720.	-.3	0.0	.6	1.6	3.0	4.6	1.6	.1002
200.	2208.	1.6	-3.3	-2.1	0.0	2.7	5.9	7.2	.2226
300.	1754.	6.0	-13.8	-11.9	-8.9	-4.9	0.0	18.0	.3751
400.	1379.	14.8	-35.8	-33.4	-29.3	-24.0	-17.4	36.0	.5685
500.	1126.	30.8	-76.7	-73.6	-68.5	-61.9	-53.7	62.8	.8112
600.	997.	56.9	-144.2	-140.5	-134.3	-126.4	-116.6	96.9	1.0958
700.	902.	95.0	-242.2	-237.9	-230.7	-221.5	-210.1	136.6	1.4126
800.	823.	148.2	-377.2	-372.3	-364.1	-353.5	-340.4	182.0	1.7613
900.	751.	219.7	-556.0	-550.5	-541.3	-529.4	-514.7	233.2	2.1431
1000.	683.	314.3	-788.1	-782.0	-771.7	-758.5	-742.2	290.9	2.5618

BALLISTIC COEFFICIENT .17, MUZZLE VELOCITY FT/SEC 3400.

RANGE	VELOCITY	HEIGHT	100 YD	150 YD	200 YD	250 YD	300 YD	DRIFT	TIME
0.	3400.	0.0	-1.5	-1.5	-1.5	-1.5	-1.5	0.0	0.0000
100.	2808.	-.3	0.0	.5	1.5	2.7	4.2	1.6	.0971
200.	2287.	1.5	-3.0	-1.9	0.0	2.5	5.5	6.9	.2155
300.	1822.	5.6	-12.7	-11.1	-8.2	-4.5	0.0	17.2	.3625
400.	1432.	13.7	-33.1	-30.9	-27.1	-22.2	-16.2	34.4	.5486
500.	1156.	28.7	-71.1	-68.3	-63.6	-57.5	-49.9	60.3	.7838
600.	1013.	53.6	-135.0	-131.7	-126.0	-118.6	-109.6	94.1	1.0638
700.	916.	89.9	-227.9	-224.1	-217.5	-208.9	-198.3	133.1	1.3740
800.	835.	140.9	-357.1	-352.7	-345.2	-335.3	-323.3	178.1	1.7176
900.	762.	209.6	-528.6	-523.8	-515.3	-504.2	-490.6	228.7	2.0936
1000.	694.	300.8	-752.0	-746.5	-737.1	-724.8	-709.7	285.8	2.5064

BALLISTIC COEFFICIENT .17, MUZZLE VELOCITY FT/SEC 3500.

RANGE	VELOCITY	HEIGHT	100 YD	150 YD	200 YD	250 YD	300 YD	DRIFT	TIME
0.	3500.	0.0	-1.5	-1.5	-1.5	-1.5	-1.5	0.0	0.0000
100.	2896.	-.3	0.0	.5	1.4	2.5	3.9	1.5	.0943
200.	2365.	1.3	-2.7	-1.8	0.0	2.3	5.1	6.6	.2090
300.	1890.	5.2	-11.7	-10.3	-7.6	-4.2	0.0	16.5	.3509
400.	1489.	12.7	-30.6	-28.7	-25.2	-20.6	-15.0	32.9	.5299
500.	1191.	26.7	-65.8	-63.4	-59.0	-53.3	-46.3	57.8	.7568
600.	1031.	50.3	-125.9	-123.0	-117.7	-110.9	-102.5	91.0	1.0311
700.	930.	85.0	-214.6	-211.2	-205.1	-197.1	-187.3	129.7	1.3367
800.	847.	134.0	-338.2	-334.3	-327.3	-318.2	-307.0	174.2	1.6753
900.	774.	200.2	-503.1	-498.7	-490.9	-480.6	-467.9	224.3	2.0461
1000.	704.	288.0	-717.8	-713.0	-704.2	-692.8	-678.8	280.8	2.4526

BALLISTIC COEFFICIENT .17, MUZZLE VELOCITY FT/SEC 3600.

RANGE	VELOCITY	HEIGHT	100 YD	150 YD	200 YD	250 YD	300 YD	DRIFT	TIME
0.	3600.	0.0	-1.5	-1.5	-1.5	-1.5	-1.5	0.0	0.0000
100.	2984.	-.3	0.0	.4	1.2	2.3	3.6	1.4	.0915
200.	2443.	1.2	-2.5	-1.6	0.0	2.1	4.8	6.3	.2026
300.	1960.	4.8	-10.8	-9.5	-7.1	-3.9	0.0	15.8	.3398
400.	1546.	11.9	-28.4	-26.7	-23.5	-19.2	-14.0	31.5	.5124
500.	1228.	24.9	-61.0	-58.9	-54.9	-49.6	-43.0	55.4	.7314
600.	1049.	47.2	-117.6	-115.0	-110.2	-103.8	-95.9	88.0	.9998
700.	944.	80.5	-202.2	-199.2	-193.6	-186.1	-176.9	126.3	1.3008
800.	858.	127.4	-320.2	-316.8	-310.4	-301.9	-291.4	170.2	1.6338
900.	785.	191.2	-478.7	-474.9	-467.6	-458.0	-446.2	219.9	1.9996
1000.	715.	275.8	-685.4	-681.2	-673.1	-662.5	-649.4	275.8	2.4003

BALLISTIC COEFFICIENT .17, MUZZLE VELOCITY FT/SEC 3700.

RANGE	VELOCITY	HEIGHT	100 YD	150 YD	200 YD	250 YD	300 YD	DRIFT	TIME
0.	3700.	0.0	-1.5	-1.5	-1.5	-1.5	-1.5	0.0	0.0000
100.	3073.	-.4	0.0	.4	1.1	2.1	3.3	1.4	.0891
200.	2521.	1.1	-2.2	-1.5	0.0	2.0	4.4	6.1	.1970
300.	2029.	4.5	-10.0	-8.9	-6.6	-3.7	0.0	15.2	.3296
400.	1603.	11.1	-26.4	-24.9	-21.9	-17.9	-13.0	30.2	.4962
500.	1268.	23.3	-56.7	-54.9	-51.1	-46.2	-40.0	53.2	.7076
600.	1068.	44.2	-109.4	-107.1	-102.6	-96.7	-89.3	84.7	.9679
700.	959.	76.1	-190.1	-187.5	-182.2	-175.3	-166.7	122.7	1.2649
800.	870.	121.5	-304.2	-301.2	-295.2	-287.3	-277.5	166.7	1.5959
900.	795.	182.7	-455.7	-452.4	-445.6	-436.7	-425.7	215.6	1.9547
1000.	725.	264.3	-654.8	-651.1	-643.6	-633.7	-621.5	270.9	2.3498

BALLISTIC COEFFICIENT .17, MUZZLE VELOCITY FT/SEC 3800.

RANGE	VELOCITY	HEIGHT	100 YD	150 YD	200 YD	250 YD	300 YD	DRIFT	TIME
0.	3800.	0.0	-1.5	-1.5	-1.5	-1.5	-1.5	0.0	0.0000
100.	3161.	-.4	0.0	.3	1.0	1.9	3.1	1.4	.0867
200.	2598.	1.0	-2.0	-1.4	0.0	1.9	4.1	5.9	.1914
300.	2099.	4.2	-9.3	-8.3	-6.2	-3.4	0.0	14.6	.3199
400.	1662.	10.3	-24.5	-23.2	-20.4	-16.7	-12.2	29.0	.4807
500.	1310.	21.8	-52.8	-51.2	-47.7	-43.1	-37.3	51.1	.6849
600.	1089.	41.5	-102.1	-100.2	-96.0	-90.4	-83.6	81.8	.9385
700.	973.	72.0	-179.0	-176.7	-171.9	-165.4	-157.4	119.4	1.2308
800.	883.	115.3	-287.4	-284.9	-279.3	-271.9	-262.7	162.5	1.5549
900.	806.	174.5	-434.0	-431.1	-424.8	-416.5	-406.2	211.3	1.9110
1000.	735.	253.3	-625.8	-622.6	-615.6	-606.4	-594.9	266.0	2.3006

BALLISTIC COEFFICIENT .17, MUZZLE VELOCITY FT/SEC 3900.

RANGE	VELOCITY	HEIGHT	100 YD	150 YD	200 YD	250 YD	300 YD	DRIFT	TIME
0.	3900.	0.0	-1.5	-1.5	-1.5	-1.5	-1.5	0.0	0.0000
100.	3250.	-.4	0.0	.3	.9	1.8	2.9	1.3	.0842
200.	2676.	.9	-1.8	-1.3	0.0	1.7	3.9	5.7	.1860
300.	2168.	3.9	-8.6	-7.7	-5.8	-3.2	0.0	14.0	.3105
400.	1721.	9.7	-22.8	-21.7	-19.1	-15.6	-11.4	27.8	.4659
500.	1353.	20.4	-49.2	-47.8	-44.6	-40.2	-34.9	49.0	.6633
600.	1112.	39.0	-95.3	-93.7	-89.8	-84.6	-78.2	78.9	.9099
700.	988.	68.2	-168.8	-166.9	-162.4	-156.3	-148.8	116.1	1.1984
800.	895.	109.8	-272.7	-270.5	-265.4	-258.4	-249.9	158.8	1.5178
900.	817.	166.8	-413.3	-410.7	-405.0	-397.1	-387.5	206.9	1.8681
1000.	745.	242.8	-598.3	-595.5	-589.1	-580.4	-569.7	261.1	2.2527

BALLISTIC COEFFICIENT .17, MUZZLE VELOCITY FT/SEC 4000.

RANGE	VELOCITY	HEIGHT	100 YD	150 YD	200 YD	250 YD	300 YD	DRIFT	TIME
0.	4000.	0.0	-1.5	-1.5	-1.5	-1.5	-1.5	0.0	0.0000
100.	3338.	-.4	0.0	.2	.8	1.6	2.7	1.3	.0822
200.	2754.	.8	-1.7	-1.2	0.0	1.6	3.6	5.5	.1812
300.	2238.	3.6	-8.0	-7.2	-5.5	-3.0	0.0	13.6	.3021
400.	1780.	9.1	-21.3	-20.3	-17.9	-14.7	-10.7	26.8	.4525
500.	1399.	19.1	-45.9	-44.7	-41.7	-37.7	-32.6	47.2	.6429
600.	1136.	36.7	-89.2	-87.8	-84.2	-79.3	-73.3	76.2	.8831
700.	1003.	64.5	-158.7	-157.1	-152.9	-147.2	-140.2	112.7	1.1655
800.	907.	104.4	-258.2	-256.3	-251.5	-245.0	-236.9	154.9	1.4800
900.	827.	159.4	-393.7	-391.6	-386.2	-378.9	-369.8	202.7	1.8268
1000.	755.	232.9	-572.1	-569.7	-563.8	-555.7	-545.6	256.3	2.2063

BALLISTIC COEFFICIENT .18, MUZZLE VELOCITY FT/SEC 2000.

RANGE	VELOCITY	HEIGHT	100 YD	150 YD	200 YD	250 YD	300 YD	DRIFT	TIME
0.	2000.	0.0	-1.5	-1.5	-1.5	-1.5	-1.5	0.0	0.0000
100.	1600.	.6	0.0	2.7	6.2	10.4	15.3	3.2	.1679
200.	1281.	6.1	-12.3	-7.0	0.0	8.4	18.3	13.8	.3783
300.	1082.	18.7	-46.0	-38.0	-27.5	-14.9	0.0	32.7	.6357
400.	974.	40.6	-107.9	-97.2	-83.2	-66.4	-46.5	57.8	.9282
500.	888.	74.4	-203.3	-189.9	-172.4	-151.4	-126.6	88.2	1.2512
600.	815.	122.8	-338.5	-322.4	-301.5	-276.2	-246.4	123.9	1.6041
700.	748.	189.1	-520.5	-501.8	-477.4	-447.9	-413.2	165.2	1.9885
800.	684.	277.6	-758.5	-737.0	-709.1	-675.5	-635.7	212.6	2.4081
900.	623.	394.0	-1064.2	-1040.1	-1008.7	-970.9	-926.2	267.1	2.8678
1000.	566.	545.5	-1453.3	-1426.5	-1391.7	-1349.6	-1299.9	329.7	3.3736

BALLISTIC COEFFICIENT .18, MUZZLE VELOCITY FT/SEC 2100.

RANGE	VELOCITY	HEIGHT	100 YD	150 YD	200 YD	250 YD	300 YD	DRIFT	TIME
0.	2100.	0.0	-1.5	-1.5	-1.5	-1.5	-1.5	0.0	0.0000
100.	1685.	.5	0.0	2.4	5.5	9.3	13.8	2.9	.1596
200.	1343.	5.5	-11.0	-6.2	0.0	7.6	16.6	13.0	.3596
300.	1115.	16.9	-41.4	-34.3	-24.9	-13.5	0.0	31.3	.6066
400.	996.	37.5	-98.8	-89.3	-76.9	-61.6	-43.6	56.5	.8924
500.	906.	69.3	-188.2	-176.3	-160.7	-141.6	-119.2	87.0	1.2084
600.	831.	115.3	-315.7	-301.5	-282.7	-259.9	-232.9	122.8	1.5546
700.	763.	178.3	-487.9	-471.3	-449.4	-422.8	-391.3	163.9	1.9314
800.	698.	262.7	-713.6	-694.6	-669.6	-639.1	-603.2	211.2	2.3428
900.	636.	373.7	-1003.9	-982.6	-954.4	-920.1	-879.7	265.3	2.7931
1000.	578.	518.0	-1373.0	-1349.3	-1318.0	-1279.9	-1235.0	327.2	3.2875

BALLISTIC COEFFICIENT .18, MUZZLE VELOCITY FT/SEC 2200.

RANGE	VELOCITY	HEIGHT	100 YD	150 YD	200 YD	250 YD	300 YD	DRIFT	TIME
0.	2200.	0.0	-1.5	-1.5	-1.5	-1.5	-1.5	0.0	0.0000
100.	1771.	.4	0.0	2.1	4.9	8.3	12.5	2.8	.1521
200.	1410.	4.9	-9.8	-5.6	0.0	6.9	15.1	12.2	.3423
300.	1153.	15.4	-37.4	-31.1	-22.7	-12.4	0.0	30.0	.5798
400.	1019.	34.5	-90.3	-81.9	-70.8	-57.0	-40.5	54.9	.8574
500.	925.	64.6	-174.1	-163.5	-149.6	-132.4	-111.7	85.4	1.1669
600.	847.	108.2	-294.3	-281.6	-264.9	-244.3	-219.5	121.1	1.5064
700.	777.	168.2	-457.4	-442.7	-423.2	-399.1	-370.1	162.2	1.8762
800.	712.	248.7	-671.8	-654.9	-632.6	-605.1	-572.0	209.2	2.2798
900.	650.	354.7	-947.7	-928.7	-903.7	-872.7	-835.5	262.9	2.7211
1000.	591.	492.4	-1298.9	-1277.8	-1249.9	-1215.6	-1174.2	324.2	3.2054

BALLISTIC COEFFICIENT .18, MUZZLE VELOCITY FT/SEC 2300.

RANGE	VELOCITY	HEIGHT	100 YD	150 YD	200 YD	250 YD	300 YD	DRIFT	TIME
0.	2300.	0.0	-1.5	-1.5	-1.5	-1.5	-1.5	0.0	0.0000
100.	1858.	.3	0.0	1.9	4.4	7.5	11.2	2.6	.1451
200.	1482.	4.4	-8.8	-5.0	0.0	6.2	13.6	11.5	.3261
300.	1199.	13.9	-33.6	-28.0	-20.5	-11.2	0.0	28.4	.5526
400.	1042.	31.8	-82.5	-75.0	-65.0	-52.6	-37.7	53.1	.8234
500.	943.	60.2	-161.1	-151.7	-139.2	-123.7	-105.1	83.5	1.1268
600.	862.	101.5	-274.2	-263.0	-247.9	-229.3	-207.0	119.1	1.4592
700.	792.	158.7	-429.0	-415.9	-398.3	-376.6	-350.6	160.1	1.8225
800.	726.	235.4	-632.4	-617.5	-597.4	-572.6	-542.8	206.8	2.2182
900.	663.	336.5	-894.8	-877.9	-855.4	-827.5	-794.0	259.9	2.6508
1000.	603.	468.1	-1229.0	-1210.2	-1185.2	-1154.2	-1117.0	320.5	3.1252

BALLISTIC COEFFICIENT .18, MUZZLE VELOCITY FT/SEC 2400.

RANGE	VELOCITY	HEIGHT	100 YD	150 YD	200 YD	250 YD	300 YD	DRIFT	TIME
0.	2400.	0.0	-1.5	-1.5	-1.5	-1.5	-1.5	0.0	0.0000
100.	1946.	.2	0.0	1.7	3.9	6.7	10.1	2.4	.1388
200.	1556.	3.9	-7.8	-4.5	0.0	5.6	12.4	10.8	.3114
300.	1250.	12.6	-30.3	-25.3	-18.5	-10.1	0.0	26.9	.5277
400.	1067.	29.2	-75.1	-68.5	-59.4	-48.2	-34.7	51.0	.7899
500.	963.	56.0	-148.7	-140.4	-129.0	-115.1	-98.2	81.3	1.0869
600.	878.	95.2	-255.5	-245.6	-232.0	-215.2	-194.9	116.8	1.4136
700.	807.	149.6	-402.1	-390.5	-374.6	-355.1	-331.4	157.5	1.7700
800.	739.	222.8	-595.4	-582.1	-564.0	-541.6	-514.6	203.9	2.1582
900.	676.	319.5	-845.4	-830.4	-810.0	-784.9	-754.5	256.6	2.5830
1000.	616.	445.2	-1163.8	-1147.2	-1124.5	-1096.6	-1062.8	316.5	3.0481

BALLISTIC COEFFICIENT .18, MUZZLE VELOCITY FT/SEC 2500.

RANGE	VELOCITY	HEIGHT	100 YD	150 YD	200 YD	250 YD	300 YD	DRIFT	TIME
0.	2500.	0.0	-1.5	-1.5	-1.5	-1.5	-1.5	0.0	0.0000
100.	2036.	.1	0.0	1.5	3.5	6.1	9.1	2.3	.1330
200.	1631.	3.5	-7.0	-4.1	0.0	5.1	11.2	10.2	.2977
300.	1303.	11.5	-27.4	-22.9	-16.8	-9.2	0.0	25.4	.5043
400.	1094.	26.8	-68.4	-62.5	-54.3	-44.2	-31.9	48.9	.7578
500.	982.	52.0	-137.2	-129.8	-119.6	-106.9	-91.6	78.9	1.0483
600.	894.	89.3	-238.2	-229.4	-217.1	-201.9	-183.5	114.3	1.3695
700.	821.	141.0	-376.8	-366.5	-352.2	-334.4	-313.0	154.7	1.7187
800.	753.	211.1	-561.0	-549.2	-532.9	-512.6	-488.0	200.8	2.1006
900.	689.	303.4	-799.1	-785.9	-767.5	-744.6	-717.0	253.0	2.5174
1000.	628.	423.7	-1102.8	-1088.1	-1067.7	-1042.3	-1011.6	312.2	2.9736

BALLISTIC COEFFICIENT .18, MUZZLE VELOCITY FT/SEC 2600.

RANGE	VELOCITY	HEIGHT	100 YD	150 YD	200 YD	250 YD	300 YD	DRIFT	TIME
0.	2600.	0.0	-1.5	-1.5	-1.5	-1.5	-1.5	0.0	0.0000
100.	2127.	.0	0.0	1.3	3.2	5.5	8.3	2.1	.1276
200.	1708.	3.2	-6.3	-3.7	0.0	4.6	10.2	9.6	.2851
300.	1361.	10.4	-24.8	-20.8	-15.3	-8.4	0.0	24.0	.4825
400.	1125.	24.6	-62.3	-57.0	-49.6	-40.4	-29.2	46.7	.7267
500.	1002.	48.3	-126.5	-119.9	-110.7	-99.1	-85.2	76.4	1.0108
600.	911.	83.5	-221.4	-213.5	-202.4	-188.6	-171.9	111.3	1.3249
700.	835.	133.0	-353.3	-344.1	-331.1	-315.0	-295.5	151.6	1.6692
800.	767.	199.9	-528.6	-518.0	-503.3	-484.8	-462.5	197.3	2.0443
900.	702.	288.1	-755.4	-743.6	-727.0	-706.2	-681.1	249.0	2.4533
1000.	640.	403.2	-1045.1	-1032.0	-1013.5	-990.4	-962.5	307.5	2.9008

BALLISTIC COEFFICIENT .18, MUZZLE VELOCITY FT/SEC 2700.

RANGE	VELOCITY	HEIGHT	100 YD	150 YD	200 YD	250 YD	300 YD	DRIFT	TIME
0.	2700.	0.0	-1.5	-1.5	-1.5	-1.5	-1.5	0.0	0.0000
100.	2217.	-.0	0.0	1.2	2.9	5.0	7.5	2.0	.1226
200.	1785.	2.8	-5.7	-3.4	0.0	4.2	9.3	9.0	.2736
300.	1422.	9.5	-22.5	-19.0	-13.9	-7.6	0.0	22.7	.4621
400.	1161.	22.6	-56.7	-52.1	-45.3	-36.9	-26.8	44.5	.6970
500.	1023.	44.8	-116.5	-110.7	-102.3	-91.7	-79.1	73.7	.9745
600.	928.	78.2	-206.1	-199.1	-189.0	-176.4	-161.2	108.4	1.2826
700.	849.	125.4	-331.2	-323.0	-311.2	-296.5	-278.8	148.4	1.6211
800.	780.	189.2	-498.0	-488.6	-475.1	-458.3	-438.1	193.7	1.9895
900.	714.	273.9	-714.8	-704.3	-689.1	-670.2	-647.4	245.0	2.3918
1000.	652.	384.1	-992.0	-980.3	-963.4	-942.4	-917.1	302.8	2.8317

BALLISTIC COEFFICIENT .18, MUZZLE VELOCITY FT/SEC 2800.

RANGE	VELOCITY	HEIGHT	100 YD	150 YD	200 YD	250 YD	300 YD	DRIFT	TIME
0.	2800.	0.0	-1.5	-1.5	-1.5	-1.5	-1.5	0.0	0.0000
100.	2307.	-.1	0.0	1.0	2.6	4.5	6.8	1.9	.1180
200.	1864.	2.6	-5.1	-3.1	0.0	3.8	8.5	8.5	.2626
300.	1487.	8.7	-20.4	-17.3	-12.7	-7.0	0.0	21.4	.4431
400.	1203.	20.7	-51.8	-47.6	-41.5	-33.8	-24.5	42.3	.6689
500.	1043.	41.6	-107.4	-102.2	-94.5	-84.9	-73.3	71.1	.9395
600.	945.	73.3	-192.1	-185.8	-176.6	-165.1	-151.2	105.5	1.2423
700.	863.	118.2	-310.5	-303.2	-292.5	-279.0	-262.8	145.0	1.5741
800.	793.	179.3	-469.7	-461.3	-449.1	-433.7	-415.1	190.0	1.9369
900.	727.	260.3	-676.6	-667.3	-653.5	-636.2	-615.3	240.8	2.3322
1000.	664.	366.0	-941.5	-931.1	-915.8	-896.6	-873.4	297.9	2.7642

BALLISTIC COEFFICIENT .18, MUZZLE VELOCITY FT/SEC 2900.

RANGE	VELOCITY	HEIGHT	100 YD	150 YD	200 YD	250 YD	300 YD	DRIFT	TIME
0.	2900.	0.0	-1.5	-1.5	-1.5	-1.5	-1.5	0.0	0.0000
100.	2396.	-.1	0.0	.9	2.3	4.1	6.2	1.8	.1139
200.	1943.	2.3	-4.7	-2.8	0.0	3.5	7.8	8.1	.2529
300.	1553.	8.0	-18.7	-15.9	-11.7	-6.4	0.0	20.3	.4259
400.	1248.	19.1	-47.4	-43.7	-38.1	-31.0	-22.5	40.3	.6425
500.	1065.	38.7	-99.0	-94.4	-87.4	-78.6	-67.9	68.5	.9062
600.	962.	68.6	-178.6	-173.0	-164.6	-154.0	-141.2	102.3	1.2021
700.	878.	111.5	-291.3	-284.8	-275.0	-262.7	-247.8	141.7	1.5293
800.	806.	170.0	-443.0	-435.5	-424.3	-410.3	-393.2	186.3	1.8859
900.	739.	247.6	-640.7	-632.3	-619.7	-603.9	-584.7	236.4	2.2744
1000.	675.	349.0	-894.5	-885.2	-871.2	-853.6	-832.2	293.0	2.6994

BALLISTIC COEFFICIENT .18, MUZZLE VELOCITY FT/SEC 3000.

RANGE	VELOCITY	HEIGHT	100 YD	150 YD	200 YD	250 YD	300 YD	DRIFT	TIME
0.	3000.	0.0	-1.5	-1.5	-1.5	-1.5	-1.5	0.0	0.0000
100.	2485.	-.2	0.0	.8	2.1	3.7	5.7	1.8	.1100
200.	2023.	2.1	-4.2	-2.6	0.0	3.2	7.1	7.7	.2438
300.	1620.	7.3	-17.1	-14.6	-10.7	-5.9	0.0	19.3	.4097
400.	1295.	17.6	-43.5	-40.1	-35.0	-28.5	-20.7	38.3	.6178
500.	1090.	35.8	-91.0	-86.8	-80.4	-72.3	-62.5	65.6	.8725
600.	979.	64.3	-166.2	-161.2	-153.5	-143.8	-132.1	99.2	1.1637
700.	892.	105.1	-273.1	-267.3	-258.4	-247.1	-233.3	138.2	1.4851
800.	819.	161.1	-417.9	-411.2	-401.0	-388.0	-372.4	182.4	1.8363
900.	751.	235.6	-607.3	-599.8	-588.3	-573.8	-556.2	232.2	2.2192
1000.	687.	333.0	-850.4	-842.0	-829.2	-813.1	-793.5	288.1	2.6370

BALLISTIC COEFFICIENT .18, MUZZLE VELOCITY FT/SEC 3100.

RANGE	VELOCITY	HEIGHT	100 YD	150 YD	200 YD	250 YD	300 YD	DRIFT	TIME
0.	3100.	0.0	-1.5	-1.5	-1.5	-1.5	-1.5	0.0	0.0000
100.	2574.	-.2	0.0	.7	1.9	3.4	5.2	1.7	.1062
200.	2103.	1.9	-3.8	-2.3	0.0	3.0	6.6	7.3	.2352
300.	1687.	6.7	-15.6	-13.4	-9.9	-5.4	0.0	18.3	.3946
400.	1345.	16.2	-39.9	-36.9	-32.2	-26.3	-19.0	36.5	.5943
500.	1117.	33.2	-83.8	-80.1	-74.2	-66.8	-57.8	62.8	.8409
600.	997.	60.2	-154.7	-150.2	-143.2	-134.2	-123.4	96.1	1.1265
700.	907.	99.1	-256.2	-251.0	-242.8	-232.4	-219.7	134.6	1.4424
800.	831.	152.8	-394.4	-388.5	-379.1	-367.2	-352.7	178.5	1.7884
900.	763.	224.2	-575.6	-568.9	-558.4	-545.0	-528.7	227.7	2.1650
1000.	698.	317.8	-808.6	-801.1	-789.4	-774.6	-756.5	283.1	2.5760

BALLISTIC COEFFICIENT .18, MUZZLE VELOCITY FT/SEC 3200.

RANGE	VELOCITY	HEIGHT	100 YD	150 YD	200 YD	250 YD	300 YD	DRIFT	TIME
0.	3200.	0.0	-1.5	-1.5	-1.5	-1.5	-1.5	0.0	0.0000
100.	2662.	-.2	0.0	.7	1.7	3.1	4.8	1.6	.1028
200.	2183.	1.7	-3.5	-2.1	0.0	2.8	6.1	7.0	.2272
300.	1756.	6.2	-14.3	-12.4	-9.1	-5.0	0.0	17.5	.3806
400.	1398.	15.0	-36.7	-34.0	-29.7	-24.2	-17.6	34.8	.5725
500.	1147.	30.8	-77.3	-74.0	-68.6	-61.7	-53.4	60.2	.8109
600.	1015.	56.4	-143.9	-140.0	-133.5	-125.2	-115.2	93.0	1.0907
700.	922.	93.5	-240.4	-235.7	-228.2	-218.5	-206.9	131.1	1.4011
800.	844.	144.9	-372.3	-367.0	-358.4	-347.4	-334.0	174.6	1.7418
900.	775.	213.6	-546.0	-540.0	-530.4	-518.0	-503.0	223.4	2.1130
1000.	709.	303.5	-769.7	-763.0	-752.3	-738.5	-721.9	278.1	2.5178

BALLISTIC COEFFICIENT .18, MUZZLE VELOCITY FT/SEC 3300.

RANGE	VELOCITY	HEIGHT	100 YD	150 YD	200 YD	250 YD	300 YD	DRIFT	TIME
0.	3300.	0.0	-1.5	-1.5	-1.5	-1.5	-1.5	0.0	0.0000
100.	2751.	-.3	0.0	.6	1.6	2.9	4.4	1.5	.0997
200.	2262.	1.6	-3.2	-2.0	0.0	2.5	5.6	6.7	.2200
300.	1825.	5.7	-13.2	-11.4	-8.4	-4.6	0.0	16.7	.3676
400.	1454.	13.9	-33.8	-31.5	-27.5	-22.4	-16.2	33.2	.5521
500.	1181.	28.6	-71.4	-68.4	-63.5	-57.1	-49.4	57.7	.7822
600.	1033.	52.8	-133.9	-130.4	-124.4	-116.8	-107.5	89.8	1.0559
700.	936.	88.2	-225.6	-221.5	-214.5	-205.6	-194.8	127.6	1.3613
800.	856.	137.4	-351.5	-346.8	-338.9	-328.7	-316.4	170.6	1.6968
900.	787.	203.4	-518.1	-512.8	-503.9	-492.4	-478.5	219.0	2.0625
1000.	721.	289.9	-732.6	-726.7	-716.8	-704.0	-688.6	273.1	2.4608

BALLISTIC COEFFICIENT .18, MUZZLE VELOCITY FT/SEC 3400.

RANGE	VELOCITY	HEIGHT	100 YD	150 YD	200 YD	250 YD	300 YD	DRIFT	TIME
0.	3400.	0.0	-1.5	-1.5	-1.5	-1.5	-1.5	0.0	0.0000
100.	2839.	-.3	0.0	.5	1.4	2.6	4.1	1.4	.0965
200.	2342.	1.4	-2.9	-1.8	0.0	2.4	5.2	6.4	.2130
300.	1895.	5.3	-12.2	-10.6	-7.8	-4.3	0.0	16.0	.3554
400.	1512.	12.9	-31.3	-29.2	-25.5	-20.8	-15.1	31.7	.5328
500.	1220.	26.6	-66.1	-63.4	-58.8	-53.0	-45.8	55.2	.7548
600.	1052.	49.4	-124.6	-121.4	-115.9	-108.9	-100.3	86.7	1.0221
700.	952.	83.2	-211.8	-208.1	-201.7	-193.4	-183.4	124.1	1.3225
800.	869.	130.4	-332.1	-327.9	-320.5	-311.1	-299.6	166.7	1.6529
900.	798.	193.8	-491.7	-486.9	-478.7	-468.1	-455.2	214.6	2.0132
1000.	732.	277.1	-698.0	-692.8	-683.6	-671.8	-657.5	268.2	2.4061

BALLISTIC COEFFICIENT .18, MUZZLE VELOCITY FT/SEC 3500.

RANGE	VELOCITY	HEIGHT	100 YD	150 YD	200 YD	250 YD	300 YD	DRIFT	TIME
0.	3500.	0.0	-1.5	-1.5	-1.5	-1.5	-1.5	0.0	0.0000
100.	2928.	-.3	0.0	.5	1.3	2.4	3.7	1.4	.0938
200.	2421.	1.3	-2.6	-1.7	0.0	2.2	4.9	6.2	.2066
300.	1965.	4.9	-11.2	-9.8	-7.3	-4.0	0.0	15.3	.3441
400.	1571.	12.0	-29.0	-27.1	-23.7	-19.4	-14.0	30.3	.5151
500.	1261.	24.8	-61.2	-58.9	-54.7	-49.2	-42.5	52.9	.7291
600.	1072.	46.3	-115.9	-113.2	-108.1	-101.5	-93.5	83.7	.9897
700.	967.	78.6	-199.0	-195.8	-189.9	-182.2	-172.9	120.7	1.2859
800.	882.	123.7	-313.6	-309.9	-303.1	-294.4	-283.7	162.7	1.6103
900.	810.	184.7	-466.9	-462.7	-455.1	-445.2	-433.2	210.2	1.9658
1000.	742.	264.9	-665.0	-660.4	-652.0	-641.0	-627.6	263.2	2.3526

BALLISTIC COEFFICIENT .18, MUZZLE VELOCITY FT/SEC 3600.

RANGE	VELOCITY	HEIGHT	100 YD	150 YD	200 YD	250 YD	300 YD	DRIFT	TIME
0.	3600.	0.0	-1.5	-1.5	-1.5	-1.5	-1.5	0.0	0.0000
100.	3017.	-.4	0.0	.4	1.2	2.2	3.5	1.3	.0910
200.	2500.	1.2	-2.4	-1.6	0.0	2.0	4.5	5.9	.2003
300.	2036.	4.6	-10.4	-9.1	-6.8	-3.7	0.0	14.7	.3333
400.	1631.	11.2	-26.9	-25.2	-22.1	-18.0	-13.0	29.0	.4981
500.	1303.	23.1	-56.8	-54.7	-50.8	-45.7	-39.5	50.7	.7046
600.	1094.	43.3	-107.9	-105.5	-100.8	-94.6	-87.2	80.6	.9581
700.	982.	74.1	-186.6	-183.8	-178.3	-171.1	-162.4	117.1	1.2487
800.	894.	117.5	-296.7	-293.4	-287.2	-279.0	-269.1	159.0	1.5699
900.	821.	176.0	-443.2	-439.5	-432.5	-423.3	-412.1	205.8	1.9191
1000.	753.	253.4	-634.1	-630.0	-622.2	-612.0	-599.5	258.3	2.3010

BALLISTIC COEFFICIENT .18, MUZZLE VELOCITY FT/SEC 3700.

RANGE	VELOCITY	HEIGHT	100 YD	150 YD	200 YD	250 YD	300 YD	DRIFT	TIME
0.	3700.	0.0	-1.5	-1.5	-1.5	-1.5	-1.5	0.0	0.0000
100.	3106.	-.4	0.0	.4	1.1	2.0	3.2	1.3	.0887
200.	2579.	1.1	-2.2	-1.4	0.0	1.9	4.2	5.7	.1947
300.	2107.	4.3	-9.6	-8.5	-6.4	-3.5	0.0	14.1	.3234
400.	1691.	10.4	-24.9	-23.5	-20.6	-16.8	-12.2	27.8	.4824
500.	1348.	21.6	-52.8	-51.0	-47.4	-42.6	-36.8	48.6	.6818
600.	1118.	40.6	-100.6	-98.4	-94.1	-88.4	-81.4	77.8	.9283
700.	998.	69.9	-175.1	-172.6	-167.5	-160.9	-152.7	113.6	1.2131
800.	908.	111.4	-280.0	-277.2	-271.4	-263.8	-254.5	154.9	1.5288
900.	832.	167.9	-421.1	-417.9	-411.4	-402.8	-392.3	201.5	1.8745
1000.	764.	242.4	-604.6	-601.1	-593.8	-584.3	-572.7	253.4	2.2508

BALLISTIC COEFFICIENT .18, MUZZLE VELOCITY FT/SEC 3800.

RANGE	VELOCITY	HEIGHT	100 YD	150 YD	200 YD	250 YD	300 YD	DRIFT	TIME
0.	3800.	0.0	-1.5	-1.5	-1.5	-1.5	-1.5	0.0	0.0000
100.	3195.	-.4	0.0	.3	1.0	1.9	3.0	1.3	.0862
200.	2657.	1.0	-2.0	-1.3	0.0	1.8	4.0	5.5	.1892
300.	2178.	4.0	-8.9	-7.9	-5.9	-3.3	0.0	13.6	.3139
400.	1752.	9.7	-23.2	-21.9	-19.3	-15.7	-11.4	26.7	.4676
500.	1395.	20.2	-49.1	-47.6	-44.3	-39.8	-34.4	46.7	.6600
600.	1145.	38.0	-93.7	-91.8	-87.8	-82.4	-75.9	74.8	.8987
700.	1013.	66.1	-164.8	-162.7	-158.0	-151.8	-144.2	110.5	1.1805
800.	921.	105.7	-264.6	-262.1	-256.8	-249.6	-240.9	151.0	1.4895
900.	843.	160.1	-400.0	-397.2	-391.2	-383.2	-373.4	197.1	1.8305
1000.	774.	232.0	-576.8	-573.7	-567.1	-558.1	-547.3	248.6	2.2021

BALLISTIC COEFFICIENT .18, MUZZLE VELOCITY FT/SEC 3900.

RANGE	VELOCITY	HEIGHT	100 YD	150 YD	200 YD	250 YD	300 YD	DRIFT	TIME
0.	3900.	0.0	-1.5	-1.5	-1.5	-1.5	-1.5	0.0	0.0000
100.	3284.	-.4	0.0	.3	.9	1.7	2.7	1.2	.0838
200.	2736.	.9	-1.8	-1.2	0.0	1.7	3.7	5.3	.1839
300.	2249.	3.7	-8.2	-7.4	-5.6	-3.1	0.0	13.1	.3049
400.	1813.	9.1	-21.6	-20.5	-18.1	-14.7	-10.6	25.7	.4534
500.	1445.	18.9	-45.8	-44.5	-41.4	-37.2	-32.1	44.8	.6391
600.	1175.	35.7	-87.5	-85.9	-82.2	-77.2	-71.0	72.1	.8711
700.	1030.	62.3	-154.5	-152.7	-148.4	-142.5	-135.4	107.0	1.1464
800.	934.	100.4	-250.1	-248.0	-243.1	-236.4	-228.2	147.2	1.4516
900.	854.	152.7	-380.1	-377.7	-372.2	-364.7	-355.4	192.8	1.7879
1000.	785.	222.0	-550.3	-547.7	-541.5	-533.2	-523.0	243.8	2.1544

BALLISTIC COEFFICIENT .18, MUZZLE VELOCITY FT/SEC 4000.

RANGE	VELOCITY	HEIGHT	100 YD	150 YD	200 YD	250 YD	300 YD	DRIFT	TIME
0.	4000.	0.0	-1.5	-1.5	-1.5	-1.5	-1.5	0.0	0.0000
100.	3373.	-.4	0.0	.2	.8	1.6	2.5	1.2	.0818
200.	2815.	.8	-1.6	-1.1	0.0	1.6	3.5	5.1	.1792
300.	2320.	3.5	-7.6	-6.9	-5.2	-2.9	0.0	12.6	.2966
400.	1876.	8.6	-20.1	-19.2	-16.9	-13.8	-10.0	24.7	.4404
500.	1496.	17.7	-42.8	-41.6	-38.8	-34.9	-30.1	43.1	.6197
600.	1209.	33.4	-81.6	-80.3	-76.8	-72.2	-66.4	69.4	.8440
700.	1046.	58.8	-145.1	-143.5	-139.5	-134.0	-127.3	103.7	1.1142
800.	947.	95.5	-236.9	-235.1	-230.5	-224.2	-216.6	143.6	1.4159
900.	866.	145.7	-361.3	-359.2	-354.1	-347.1	-338.4	188.6	1.7465
1000.	795.	212.6	-525.4	-523.1	-517.4	-509.6	-500.0	239.1	2.1084

BALLISTIC COEFFICIENT .19, MUZZLE VELOCITY FT/SEC 2000.

RANGE	VELOCITY	HEIGHT	100 YD	150 YD	200 YD	250 YD	300 YD	DRIFT	TIME
0.	2000.	0.0	-1.5	-1.5	-1.5	-1.5	-1.5	0.0	0.0000
100.	1620.	.6	0.0	2.6	6.0	10.1	14.8	3.0	.1668
200.	1309.	5.9	-12.0	-6.7	0.0	8.1	17.6	12.9	.3736
300.	1105.	18.0	-44.4	-36.5	-26.4	-14.2	0.0	30.8	.6249
400.	995.	39.2	-104.3	-93.8	-80.3	-64.1	-45.1	54.9	.9121
500.	909.	71.6	-196.4	-183.3	-166.4	-146.1	-122.4	84.1	1.2279
600.	837.	117.9	-326.6	-310.8	-290.6	-266.2	-237.8	118.3	1.5720
700.	772.	180.9	-501.4	-483.0	-459.4	-431.0	-397.8	157.6	1.9454
800.	710.	264.5	-728.3	-707.4	-680.4	-647.9	-610.0	202.5	2.3507
900.	651.	373.4	-1017.8	-994.2	-963.9	-927.4	-884.7	253.8	2.7920
1000.	596	513.7	-1382.5	-1356.2	-1322.5	-1281.9	-1234.5	312.2	3.2737

BALLISTIC COEFFICIENT .19, MUZZLE VELOCITY FT/SEC 2100.

RANGE	VELOCITY	HEIGHT	100 YD	150 YD	200 YD	250 YD	300 YD	DRIFT	TIME
0.	2100.	0.0	-1.5	-1.5	-1.5	-1.5	-1.5	0.0	0.0000
100.	1705.	.5	0.0	2.3	5.3	9.0	13.3	2.8	.1586
200.	1374.	5.3	-10.7	-6.1	0.0	7.3	15.9	12.2	.3552
300.	1142.	16.3	-39.9	-32.9	-23.9	-12.9	0.0	29.4	.5959
400.	1018.	36.1	-95.2	-85.9	-73.8	-59.2	-42.0	53.6	.8757
500.	928.	66.6	-181.4	-169.8	-154.7	-136.4	-114.9	82.8	1.1847
600.	853.	110.5	-304.0	-290.0	-271.9	-250.0	-224.2	117.1	1.5222
700.	787.	170.4	-469.2	-452.9	-431.7	-406.2	-376.1	156.4	1.8884
800.	725.	250.0	-684.1	-665.6	-641.4	-612.2	-577.8	201.1	2.2855
900.	665.	353.8	-958.8	-937.9	-910.7	-877.8	-839.1	252.1	2.7178
1000.	609.	487.6	-1305.2	-1282.0	-1251.8	-1215.2	-1172.2	309.9	3.1896

BALLISTIC COEFFICIENT .19, MUZZLE VELOCITY FT/SEC 2200.

RANGE	VELOCITY	HEIGHT	100 YD	150 YD	200 YD	250 YD	300 YD	DRIFT	TIME
0.	2200.	0.0	-1.5	-1.5	-1.5	-1.5	-1.5	0.0	0.0000
100.	1791.	.3	0.0	2.1	4.8	8.1	12.0	2.6	.1512
200.	1444.	4.7	-9.5	-5.4	0.0	6.6	14.4	11.5	.3380
300.	1186.	14.8	-35.9	-29.7	-21.6	-11.7	0.0	28.0	.5684
400.	1041.	33.1	-86.8	-78.6	-67.7	-54.6	-38.9	51.9	.8404
500.	948.	61.9	-167.4	-157.1	-143.6	-127.1	-107.6	81.1	1.1428
600.	870.	103.5	-282.7	-270.4	-254.2	-234.4	-211.0	115.4	1.4736
700.	802.	160.5	-438.9	-424.5	-405.6	-382.5	-355.2	154.6	1.8327
800.	739.	236.3	-642.7	-626.3	-604.7	-578.3	-547.1	199.1	2.2222
900.	679.	335.3	-903.5	-885.0	-860.7	-831.0	-795.9	249.7	2.6459
1000.	622.	462.9	-1232.6	-1212.0	-1185.0	-1152.1	-1113.0	307.0	3.1078

BALLISTIC COEFFICIENT .19, MUZZLE VELOCITY FT/SEC 2300.

RANGE	VELOCITY	HEIGHT	100 YD	150 YD	200 YD	250 YD	300 YD	DRIFT	TIME
0.	2300.	0.0	-1.5	-1.5	-1.5	-1.5	-1.5	0.0	0.0000
100.	1879.	.2	0.0	1.8	4.3	7.2	10.8	2.4	.1442
200.	1518.	4.2	-8.5	-4.9	0.0	5.9	13.1	10.8	.3221
300.	1236.	13.4	-32.4	-26.9	-19.6	-10.7	0.0	26.6	.5423
400.	1066.	30.4	-78.9	-71.6	-61.9	-50.0	-35.8	50.0	.8056
500.	967.	57.6	-154.6	-145.5	-133.3	-118.4	-100.7	79.3	1.1026
600.	886.	96.8	-262.8	-251.9	-237.3	-219.4	-198.1	113.2	1.4259
700.	817.	151.1	-410.6	-397.9	-380.8	-360.0	-335.2	152.3	1.7785
800.	753.	223.4	-604.3	-589.8	-570.3	-546.5	-518.1	196.7	2.1611
900.	692.	318.0	-852.4	-835.9	-814.0	-787.3	-755.3	246.9	2.5767
1000.	634.	439.8	-1165.5	-1147.3	-1122.9	-1093.2	-1057.7	303.6	3.0295

BALLISTIC COEFFICIENT .19, MUZZLE VELOCITY FT/SEC 2400.

RANGE	VELOCITY	HEIGHT	100 YD	150 YD	200 YD	250 YD	300 YD	DRIFT	TIME
0.	2400.	0.0	-1.5	-1.5	-1.5	-1.5	-1.5	0.0	0.0000
100.	1969.	.2	0.0	1.6	3.8	6.5	9.7	2.3	.1380
200.	1593.	3.8	-7.6	-4.4	0.0	5.4	11.8	10.1	.3076
300.	1290.	12.1	-29.1	-24.3	-17.7	-9.7	0.0	25.1	.5176
400.	1094.	27.9	-71.7	-65.3	-56.5	-45.7	-32.9	47.9	.7721
500.	988.	53.5	-142.4	-134.3	-123.3	-109.9	-93.8	77.0	1.0626
600.	903.	90.6	-244.2	-234.5	-221.3	-205.2	-185.9	110.8	1.3794
700.	832.	142.2	-384.2	-372.9	-357.5	-338.7	-316.2	149.8	1.7259
800.	767.	211.2	-568.1	-555.2	-537.6	-516.2	-490.4	193.9	2.1015
900.	706.	301.5	-804.1	-789.5	-769.7	-745.6	-716.6	243.6	2.5093
1000.	647.	418.0	-1102.5	-1086.3	-1064.3	-1037.5	-1005.3	299.8	2.9535

BALLISTIC COEFFICIENT .19, MUZZLE VELOCITY FT/SEC 2500.

RANGE	VELOCITY	HEIGHT	100 YD	150 YD	200 YD	250 YD	300 YD	DRIFT	TIME
0.	2500.	0.0	-1.5	-1.5	-1.5	-1.5	-1.5	0.0	0.0000
100.	2060.	.1	0.0	1.4	3.4	5.8	8.8	2.2	.1322
200.	1670.	3.4	-6.8	-4.0	0.0	4.9	10.7	9.5	.2941
300.	1347.	11.0	-26.3	-22.0	-16.1	-8.8	0.0	23.7	.4946
400.	1126.	25.5	-65.2	-59.4	-51.5	-41.8	-30.1	45.7	.7397
500.	1008.	49.5	-131.0	-123.8	-113.9	-101.7	-87.1	74.6	1.0236
600.	921.	84.7	-226.7	-218.2	-206.2	-191.7	-174.1	108.1	1.3341
700.	847.	133.8	-359.3	-349.3	-335.4	-318.4	-297.9	146.9	1.6745
800.	781.	199.7	-534.1	-522.6	-506.7	-487.3	-463.9	190.7	2.0433
900.	719.	285.9	-758.6	-745.7	-727.9	-706.0	-679.7	240.0	2.4436
1000.	660.	397.2	-1043.0	-1028.7	-1008.8	-984.5	-955.3	295.6	2.8794

BALLISTIC COEFFICIENT .19, MUZZLE VELOCITY FT/SEC 2600.

RANGE	VELOCITY	HEIGHT	100 YD	150 YD	200 YD	250 YD	300 YD	DRIFT	TIME
0.	2600.	0.0	-1.5	-1.5	-1.5	-1.5	-1.5	0.0	0.0000
100.	2150.	.0	0.0	1.3	3.1	5.3	7.9	2.0	.1269
200.	1748.	3.1	-6.1	-3.6	0.0	4.4	9.7	9.0	.2817
300.	1409.	10.0	-23.8	-20.0	-14.6	-7.9	0.0	22.3	.4731
400.	1163.	23.4	-59.2	-54.1	-46.9	-38.1	-27.5	43.5	.7087
500.	1029.	45.9	-120.4	-114.0	-105.0	-93.9	-80.7	72.0	.9857
600.	938.	79.2	-210.6	-202.9	-192.2	-178.9	-163.0	105.2	1.2903
700.	862.	125.9	-336.0	-327.0	-314.4	-298.9	-280.4	143.7	1.6243
800.	795.	188.8	-502.3	-492.0	-477.7	-459.9	-438.8	187.3	1.9871
900.	732.	271.3	-716.4	-704.9	-688.8	-668.8	-645.0	236.2	2.3805
1000.	672.	377.8	-987.9	-975.1	-957.2	-935.0	-908.5	291.2	2.8084

BALLISTIC COEFFICIENT .19, MUZZLE VELOCITY FT/SEC 2700.

RANGE	VELOCITY	HEIGHT	100 YD	150 YD	200 YD	250 YD	300 YD	DRIFT	TIME
0.	2700.	0.0	-1.5	-1.5	-1.5	-1.5	-1.5	0.0	0.0000
100.	2241.	-.0	0.0	1.1	2.8	4.8	7.2	1.9	.1220
200.	1828.	2.8	-5.5	-3.3	0.0	4.0	8.9	8.5	.2703
300.	1474.	9.1	-21.6	-18.2	-13.3	-7.2	0.0	21.1	.4532
400.	1206.	21.4	-54.0	-49.4	-42.9	-34.8	-25.1	41.4	.6794
500.	1051.	42.4	-110.3	-104.6	-96.4	-86.3	-74.2	69.0	.9479
600.	956.	74.0	-195.5	-188.7	-178.9	-166.8	-152.3	102.2	1.2476
700.	876.	118.5	-314.5	-306.5	-295.1	-281.0	-264.1	140.6	1.5765
800.	809.	178.5	-472.4	-463.3	-450.3	-434.1	-414.8	183.7	1.9324
900.	745.	257.4	-676.6	-666.3	-651.7	-633.5	-611.7	232.1	2.3190
1000.	684.	359.4	-935.7	-924.3	-908.1	-887.9	-863.7	286.6	2.7392

BALLISTIC COEFFICIENT .19, MUZZLE VELOCITY FT/SEC 2800.

RANGE	VELOCITY	HEIGHT	100 YD	150 YD	200 YD	250 YD	300 YD	DRIFT	TIME
0.	2800.	0.0	-1.5	-1.5	-1.5	-1.5	-1.5	0.0	0.0000
100.	2331.	-.1	0.0	1.0	2.5	4.3	6.6	1.8	.1174
200.	1907.	2.5	-5.0	-3.0	0.0	3.7	8.1	8.0	.2597
300.	1542.	8.3	-19.7	-16.6	-12.2	-6.7	0.0	19.9	.4348
400.	1253.	19.6	-49.3	-45.2	-39.3	-31.9	-23.1	39.3	.6519
500.	1075.	39.2	-101.2	-96.2	-88.7	-79.5	-68.4	66.3	.9123
600.	974.	69.1	-181.5	-175.4	-166.5	-155.4	-142.1	99.2	1.2063
700.	892.	111.4	-293.9	-286.8	-276.4	-263.5	-248.0	137.0	1.5286
800.	822.	168.7	-444.3	-436.2	-424.3	-409.6	-391.9	179.9	1.8791
900.	758.	244.3	-639.2	-630.1	-616.7	-600.2	-580.2	227.9	2.2593
1000.	696.	342.1	-887.1	-876.9	-862.1	-843.6	-821.5	281.8	2.6725

BALLISTIC COEFFICIENT .19, MUZZLE VELOCITY FT/SEC 2900.

RANGE	VELOCITY	HEIGHT	100 YD	150 YD	200 YD	250 YD	300 YD	DRIFT	TIME
0.	2900.	0.0	-1.5	-1.5	-1.5	-1.5	-1.5	0.0	0.0000
100.	2421.	-.1	0.0	.9	2.3	4.0	6.0	1.7	.1133
200.	1988.	2.2	-4.5	-2.7	0.0	3.4	7.5	7.6	.2499
300.	1610.	7.6	-18.0	-15.2	-11.2	-6.1	0.0	18.9	.4178
400.	1302.	18.0	-45.0	-41.4	-36.0	-29.2	-21.1	37.3	.6258
500.	1101.	36.3	-93.0	-88.5	-81.7	-73.2	-63.1	63.5	.8783
600.	992.	64.7	-169.0	-163.5	-155.4	-145.2	-133.0	96.3	1.1681
700.	907.	104.8	-274.9	-268.6	-259.1	-247.2	-233.0	133.6	1.4830
800.	835.	159.6	-418.2	-411.0	-400.2	-386.6	-370.3	176.1	1.8280
900.	770.	232.0	-604.5	-596.4	-584.2	-568.9	-550.6	223.7	2.2023
1000.	708.	325.9	-841.6	-832.5	-819.0	-802.1	-781.7	277.0	2.6085

BALLISTIC COEFFICIENT .19, MUZZLE VELOCITY FT/SEC 3000.

RANGE	VELOCITY	HEIGHT	100 YD	150 YD	200 YD	250 YD	300 YD	DRIFT	TIME
0.	3000.	0.0	-1.5	-1.5	-1.5	-1.5	-1.5	0.0	0.0000
100.	2511.	-.2	0.0	.8	2.0	3.6	5.5	1.7	.1095
200.	2069.	2.0	-4.1	-2.5	0.0	3.1	6.8	7.2	.2411
300.	1679.	7.0	-16.4	-14.0	-10.3	-5.6	0.0	18.0	.4022
400.	1354.	16.6	-41.3	-38.1	-33.1	-26.9	-19.4	35.5	.6019
500.	1130.	33.6	-85.5	-81.5	-75.3	-67.5	-58.1	60.9	.8458
600.	1010.	60.5	-156.7	-151.9	-144.4	-135.1	-123.9	93.1	1.1292
700.	922.	98.6	-257.2	-251.5	-242.9	-232.0	-218.9	130.0	1.4388
800.	848.	151.1	-393.7	-387.3	-377.4	-365.0	-350.0	172.2	1.7785
900.	783.	220.4	-571.7	-564.4	-553.3	-539.3	-522.4	219.4	2.1466
1000.	720.	310.4	-798.5	-790.4	-778.1	-762.5	-743.8	272.1	2.5461

BALLISTIC COEFFICIENT .19, MUZZLE VELOCITY FT/SEC 3100.

RANGE	VELOCITY	HEIGHT	100 YD	150 YD	200 YD	250 YD	300 YD	DRIFT	TIME
0.	3100.	0.0	-1.5	-1.5	-1.5	-1.5	-1.5	0.0	0.0000
100.	2600.	-.2	0.0	.7	1.9	3.3	5.0	1.6	.1057
200.	2150.	1.8	-3.7	-2.3	0.0	2.9	6.3	6.9	.2326
300.	1748.	6.5	-15.0	-12.9	-9.5	-5.2	0.0	17.1	.3874
400.	1409.	15.3	-37.9	-35.0	-30.5	-24.7	-17.8	33.7	.5788
500.	1163.	31.1	-78.7	-75.1	-69.4	-62.2	-53.6	58.2	.8144
600.	1029.	56.4	-145.3	-141.0	-134.2	-125.6	-115.3	89.9	1.0914
700.	938.	92.8	-240.7	-235.7	-227.7	-217.7	-205.6	126.5	1.3960
800.	862.	142.9	-370.7	-364.9	-355.8	-344.3	-330.6	168.2	1.7300
900.	795.	209.5	-541.0	-534.5	-524.3	-511.4	-495.9	215.0	2.0928
1000.	732.	296.0	-758.5	-751.3	-739.9	-725.6	-708.4	267.3	2.4862

BALLISTIC COEFFICIENT .19, MUZZLE VELOCITY FT/SEC 3200.

RANGE	VELOCITY	HEIGHT	100 YD	150 YD	200 YD	250 YD	300 YD	DRIFT	TIME
0.	3200.	0.0	-1.5	-1.5	-1.5	-1.5	-1.5	0.0	0.0000
100.	2689.	-.2	0.0	.6	1.7	3.0	4.6	1.5	.1023
200.	2231.	1.7	-3.4	-2.1	0.0	2.6	5.8	6.6	.2248
300.	1819.	6.0	-13.8	-11.9	-8.7	-4.8	0.0	16.3	.3738
400.	1467.	14.2	-34.9	-32.3	-28.1	-22.9	-16.5	32.2	.5577
500.	1201.	28.8	-72.6	-69.4	-64.1	-57.5	-49.6	55.6	.7849
600.	1048.	52.8	-135.0	-131.1	-124.9	-116.9	-107.4	86.8	1.0558
700.	954.	87.3	-225.3	-220.8	-213.5	-204.2	-193.1	122.9	1.3546
800.	875.	135.5	-349.7	-344.5	-336.1	-325.6	-312.8	164.5	1.6846
900.	807.	199.2	-512.1	-506.3	-496.9	-485.0	-470.7	210.7	2.0407
1000.	743.	282.2	-720.6	-714.2	-703.7	-690.5	-674.6	262.3	2.4279

BALLISTIC COEFFICIENT .19, MUZZLE VELOCITY FT/SEC 3300.

RANGE	VELOCITY	HEIGHT	100 YD	150 YD	200 YD	250 YD	300 YD	DRIFT	TIME
0.	3300.	0.0	-1.5	-1.5	-1.5	-1.5	-1.5	0.0	0.0000
100.	2778.	-.3	0.0	.6	1.5	2.8	4.2	1.4	.0991
200.	2312.	1.5	-3.1	-1.9	0.0	2.5	5.4	6.3	.2176
300.	1890.	5.5	-12.7	-11.0	-8.1	-4.4	0.0	15.6	.3611
400.	1527.	13.1	-32.2	-29.9	-26.1	-21.2	-15.3	30.7	.5380
500.	1242.	26.7	-67.0	-64.2	-59.4	-53.2	-45.9	53.2	.7569
600.	1069.	49.1	-124.8	-121.4	-115.6	-108.3	-99.4	83.4	1.0193
700.	970.	82.4	-211.4	-207.3	-200.6	-192.0	-181.7	119.6	1.3159
800.	888.	128.1	-329.0	-324.3	-316.7	-306.9	-295.1	160.3	1.6381
900.	819.	189.3	-484.8	-479.6	-471.0	-459.9	-446.6	206.2	1.9899
1000.	755.	269.2	-685.0	-679.2	-669.6	-657.4	-642.6	257.4	2.3717

BALLISTIC COEFFICIENT .19, MUZZLE VELOCITY FT/SEC 3400.

RANGE	VELOCITY	HEIGHT	100 YD	150 YD	200 YD	250 YD	300 YD	DRIFT	TIME
0.	3400.	0.0	-1.5	-1.5	-1.5	-1.5	-1.5	0.0	0.0000
100.	2867.	-.3	0.0	.5	1.4	2.5	3.9	1.4	.0960
200.	2392.	1.4	-2.8	-1.8	0.0	2.3	5.0	6.0	.2107
300.	1961.	5.1	-11.7	-10.2	-7.5	-4.1	0.0	14.9	.3492
400.	1587.	12.2	-29.8	-27.8	-24.2	-19.7	-14.2	29.3	.5195
500.	1286.	24.8	-62.0	-59.5	-55.0	-49.4	-42.5	50.9	.7303
600.	1092.	45.9	-115.9	-112.8	-107.5	-100.7	-92.4	80.3	.9856
700.	986.	77.4	-197.7	-194.1	-187.9	-180.0	-170.3	115.9	1.2764
800.	902.	121.2	-309.8	-305.8	-298.7	-289.6	-278.6	156.3	1.5938
900.	831.	180.1	-459.2	-454.7	-446.7	-436.5	-424.1	201.8	1.9409
1000.	766.	256.9	-651.4	-646.3	-637.4	-626.1	-612.3	252.5	2.3170

BALLISTIC COEFFICIENT .19, MUZZLE VELOCITY FT/SEC 3500.

RANGE	VELOCITY	HEIGHT	100 YD	150 YD	200 YD	250 YD	300 YD	DRIFT	TIME
0.	3500.	0.0	-1.5	-1.5	-1.5	-1.5	-1.5	0.0	0.0000
100.	2956.	-.3	0.0	.5	1.3	2.3	3.6	1.3	.0933
200.	2472.	1.3	-2.5	-1.6	0.0	2.1	4.7	5.8	.2043
300.	2034.	4.7	-10.8	-9.5	-7.0	-3.8	0.0	14.3	.3383
400.	1648.	11.4	-27.6	-25.8	-22.5	-18.3	-13.2	28.0	.5022
500.	1331.	23.1	-57.5	-55.2	-51.1	-45.8	-39.4	48.7	.7054
600.	1117.	42.9	-107.6	-104.9	-100.0	-93.6	-86.0	77.2	.9532
700.	1002.	72.8	-184.9	-181.7	-176.0	-168.6	-159.6	112.3	1.2382
800.	916.	114.8	-292.0	-288.4	-281.9	-273.4	-263.2	152.3	1.5513
900.	843.	171.3	-435.0	-430.9	-423.6	-414.0	-402.5	197.4	1.8932
1000.	777.	245.3	-619.6	-615.1	-606.9	-596.3	-583.5	247.6	2.2639

BALLISTIC COEFFICIENT .19, MUZZLE VELOCITY FT/SEC 3600.

RANGE	VELOCITY	HEIGHT	100 YD	150 YD	200 YD	250 YD	300 YD	DRIFT	TIME
0.	3600.	0.0	-1.5	-1.5	-1.5	-1.5	-1.5	0.0	0.0000
100.	3046.	-.4	0.0	.4	1.2	2.1	3.3	1.3	.0905
200.	2552.	1.1	-2.3	-1.5	0.0	2.0	4.4	5.6	.1982
300.	2106.	4.4	-10.0	-8.8	-6.5	-3.6	0.0	13.7	.3277
400.	1710.	10.6	-25.6	-24.0	-21.0	-17.0	-12.3	26.8	.4858
500.	1378.	21.6	-53.3	-51.3	-47.6	-42.6	-36.7	46.6	.6816
600.	1145.	40.0	-100.0	-97.6	-93.1	-87.2	-80.0	74.2	.9218
700.	1019.	68.5	-172.9	-170.1	-164.8	-157.9	-149.6	108.7	1.2011
800.	930.	108.7	-275.3	-272.1	-266.1	-258.2	-248.7	148.4	1.5098
900.	854.	163.0	-412.2	-408.6	-401.8	-393.0	-382.2	193.0	1.8468
1000.	788.	234.2	-589.6	-585.7	-578.1	-568.3	-556.3	242.7	2.2125

BALLISTIC COEFFICIENT .19, MUZZLE VELOCITY FT/SEC 3700.

RANGE	VELOCITY	HEIGHT	100 YD	150 YD	200 YD	250 YD	300 YD	DRIFT	TIME
0.	3700.	0.0	-1.5	-1.5	-1.5	-1.5	-1.5	0.0	0.0000
100.	3135.	-.4	0.0	.3	1.0	2.0	3.1	1.3	.0883
200.	2631.	1.0	-2.1	-1.4	0.0	1.8	4.1	5.4	.1927
300.	2179.	4.1	-9.2	-8.2	-6.1	-3.3	0.0	13.1	.3179
400.	1773.	9.9	-23.7	-22.4	-19.6	-15.9	-11.5	25.8	.4708
500.	1429.	20.1	-49.6	-47.9	-44.4	-39.8	-34.2	44.7	.6595
600.	1176.	37.5	-93.1	-91.1	-86.9	-81.4	-74.7	71.5	.8926
700.	1036.	64.5	-161.8	-159.4	-154.5	-148.1	-140.3	105.3	1.1657
800.	943.	103.0	-259.7	-257.0	-251.4	-244.0	-235.1	144.6	1.4702
900.	866.	155.1	-390.6	-387.6	-381.3	-373.0	-363.0	188.7	1.8019
1000.	799.	223.7	-561.1	-557.7	-550.7	-541.5	-530.4	237.9	2.1623

BALLISTIC COEFFICIENT .19, MUZZLE VELOCITY FT/SEC 3800.

RANGE	VELOCITY	HEIGHT	100 YD	150 YD	200 YD	250 YD	300 YD	DRIFT	TIME
0.	3800.	0.0	-1.5	-1.5	-1.5	-1.5	-1.5	0.0	0.0000
100.	3225.	-.4	0.0	.3	.9	1.8	2.8	1.2	.0858
200.	2711.	.9	-1.9	-1.3	0.0	1.7	3.8	5.2	.1874
300.	2251.	3.8	-8.5	-7.7	-5.7	-3.1	0.0	12.7	.3089
400.	1836.	9.3	-22.1	-20.9	-18.3	-14.9	-10.7	24.8	.4565
500.	1481.	18.8	-46.2	-44.7	-41.4	-37.2	-31.9	42.9	.6385
600.	1211.	35.1	-86.7	-85.0	-81.1	-75.9	-69.6	68.6	.8636
700.	1054.	60.7	-151.4	-149.4	-144.8	-138.8	-131.5	101.8	1.1313
800.	958.	97.4	-244.6	-242.2	-237.0	-230.2	-221.8	140.6	1.4303
900.	878.	147.6	-370.4	-367.8	-361.9	-354.2	-344.8	184.4	1.7583
1000.	810.	213.8	-534.3	-531.4	-524.9	-516.3	-505.9	233.1	2.1140

BALLISTIC COEFFICIENT .19, MUZZLE VELOCITY FT/SEC 3900.

RANGE	VELOCITY	HEIGHT	100 YD	150 YD	200 YD	250 YD	300 YD	DRIFT	TIME
0.	3900.	0.0	-1.5	-1.5	-1.5	-1.5	-1.5	0.0	0.0000
100.	3314.	-.4	0.0	.3	.9	1.7	2.6	1.1	.0834
200.	2791.	.8	-1.7	-1.2	0.0	1.6	3.6	5.0	.1821
300.	2323.	3.6	-7.9	-7.1	-5.3	-2.9	0.0	12.2	.2999
400.	1900.	8.7	-20.6	-19.6	-17.2	-14.0	-10.0	23.8	.4427
500.	1535.	17.6	-43.1	-41.8	-38.8	-34.8	-29.9	41.2	.6186
600.	1249.	32.8	-80.9	-79.4	-75.8	-71.0	-65.1	66.0	.8364
700.	1073.	57.1	-141.7	-140.0	-135.8	-130.1	-123.3	98.4	1.0978
800.	972.	92.3	-230.6	-228.6	-223.8	-217.4	-209.6	136.7	1.3923
900.	890.	140.4	-351.1	-348.8	-343.4	-336.1	-327.3	180.0	1.7151
1000.	821.	204.2	-508.7	-506.1	-500.1	-492.1	-482.3	228.3	2.0661

BALLISTIC COEFFICIENT .19, MUZZLE VELOCITY FT/SEC 4000.

RANGE	VELOCITY	HEIGHT	100 YD	150 YD	200 YD	250 YD	300 YD	DRIFT	TIME
0.	4000.	0.0	-1.5	-1.5	-1.5	-1.5	-1.5	0.0	0.0000
100.	3404.	-.4	0.0	.2	.8	1.5	2.4	1.1	.0814
200.	2871.	.8	-1.5	-1.1	0.0	1.5	3.4	4.8	.1773
300.	2395.	3.3	-7.3	-6.7	-5.0	-2.8	0.0	11.8	.2919
400.	1964.	8.1	-19.2	-18.3	-16.1	-13.1	-9.4	22.9	.4302
500.	1590.	16.5	-40.3	-39.2	-36.4	-32.7	-28.0	39.6	.6002
600.	1287.	30.8	-75.6	-74.3	-71.0	-66.5	-60.9	63.5	.8107
700.	1093.	53.8	-132.7	-131.2	-127.4	-122.1	-115.6	95.2	1.0657
800.	987.	87.6	-217.8	-216.0	-211.6	-205.6	-198.2	133.1	1.3564
900.	902.	133.7	-332.8	-330.8	-325.9	-319.1	-310.8	175.7	1.6736
1000.	831.	195.2	-484.7	-482.5	-477.0	-469.5	-460.2	223.6	2.0205

BALLISTIC COEFFICIENT .20, MUZZLE VELOCITY FT/SEC 2000.

RANGE	VELOCITY	HEIGHT	100 YD	150 YD	200 YD	250 YD	300 YD	DRIFT	TIME
0.	2000.	0.0	-1.5	-1.5	-1.5	-1.5	-1.5	0.0	0.0000
100.	1637.	.6	0.0	2.6	5.9	9.8	14.3	2.8	.1659
200.	1336.	5.8	-11.7	-6.6	0.0	7.8	17.0	12.2	.3695
300.	1128.	17.4	-43.0	-35.3	-25.4	-13.7	0.0	29.1	.6153
400.	1013.	38.0	-101.5	-91.2	-78.1	-62.4	-44.2	52.6	.8990
500.	929.	69.1	-190.1	-177.4	-160.9	-141.4	-118.5	80.4	1.2066
600.	858.	113.5	-315.8	-300.5	-280.7	-257.3	-229.9	113.2	1.5430
700.	795.	173.7	-484.2	-466.3	-443.2	-415.8	-383.9	150.8	1.9066
800.	735.	253.0	-701.6	-681.1	-654.8	-623.5	-587.0	193.5	2.2994
900.	677.	355.6	-977.0	-954.0	-924.4	-889.2	-848.1	241.9	2.7246
1000.	623.	486.6	-1321.7	-1296.1	-1263.2	-1224.1	-1178.4	296.8	3.1865

BALLISTIC COEFFICIENT .20, MUZZLE VELOCITY FT/SEC 2100.

RANGE	VELOCITY	HEIGHT	100 YD	150 YD	200 YD	250 YD	300 YD	DRIFT	TIME
0.	2100.	0.0	-1.5	-1.5	-1.5	-1.5	-1.5	0.0	0.0000
100.	1723.	.4	0.0	2.3	5.2	8.7	12.9	2.6	.1577
200.	1403.	5.2	-10.4	-5.9	0.0	7.1	15.3	11.5	.3510
300.	1169.	15.7	-38.6	-31.8	-23.0	-12.4	0.0	27.8	.5863
400.	1038.	34.8	-91.9	-82.8	-71.1	-56.9	-40.4	50.8	.8602
500.	949.	64.2	-175.2	-163.9	-149.2	-131.5	-110.9	79.0	1.1630
600.	874.	106.4	-293.9	-280.3	-262.7	-241.4	-216.7	112.1	1.4939
700.	810.	163.4	-452.3	-436.4	-415.9	-391.1	-362.2	149.5	1.8493
800.	749.	238.9	-658.2	-640.0	-616.6	-588.3	-555.3	192.1	2.2345
900.	692.	336.7	-919.6	-899.1	-872.8	-840.9	-803.8	240.4	2.6514
1000.	636.	461.5	-1246.5	-1223.8	-1194.6	-1159.1	-1117.9	294.8	3.1034

BALLISTIC COEFFICIENT .20, MUZZLE VELOCITY FT/SEC 2200.

RANGE	VELOCITY	HEIGHT	100 YD	150 YD	200 YD	250 YD	300 YD	DRIFT	TIME
0.	2200.	0.0	-1.5	-1.5	-1.5	-1.5	-1.5	0.0	0.0000
100.	1810.	.3	0.0	2.0	4.6	7.8	11.6	2.5	.1504
200.	1476.	4.6	-9.3	-5.3	0.0	6.4	13.9	10.8	.3341
300.	1219.	14.2	-34.7	-28.7	-20.8	-11.2	0.0	26.4	.5588
400.	1063.	31.9	-83.6	-75.6	-65.1	-52.4	-37.4	49.2	.8251
500.	969.	59.7	-161.7	-151.7	-138.5	-122.6	-103.9	77.5	1.1219
600.	892.	99.3	-272.2	-260.2	-244.4	-225.3	-202.9	110.1	1.4435
700.	825.	153.6	-422.3	-408.3	-389.9	-367.6	-341.4	147.6	1.7934
800.	764.	225.5	-617.5	-601.4	-580.4	-554.9	-525.0	190.1	2.1712
900.	706.	318.8	-865.6	-847.5	-823.9	-795.2	-761.6	238.1	2.5801
1000.	650.	438.0	-1176.6	-1156.6	-1130.3	-1098.4	-1061.0	292.1	3.0234

BALLISTIC COEFFICIENT .20, MUZZLE VELOCITY FT/SEC 2300.

RANGE	VELOCITY	HEIGHT	100 YD	150 YD	200 YD	250 YD	300 YD	DRIFT	TIME
0.	2300.	0.0	-1.5	-1.5	-1.5	-1.5	-1.5	0.0	0.0000
100.	1899.	.2	0.0	1.8	4.1	7.0	10.4	2.3	.1435
200.	1551.	4.1	-8.3	-4.7	0.0	5.7	12.5	10.1	.3185
300.	1272.	12.9	-31.2	-25.9	-18.8	-10.2	0.0	24.9	.5329
400.	1091.	29.2	-75.8	-68.7	-59.2	-47.7	-34.1	47.1	.7895
500.	990.	55.3	-148.7	-139.8	-127.9	-113.6	-96.6	75.4	1.0804
600.	909.	92.7	-252.5	-241.9	-227.6	-210.5	-190.1	107.8	1.3953
700.	841.	144.3	-394.2	-381.7	-365.1	-345.1	-321.3	145.3	1.7386
800.	779.	212.9	-579.3	-565.1	-546.1	-523.2	-496.1	187.6	2.1096
900.	719.	301.8	-814.9	-798.9	-777.6	-751.8	-721.2	235.2	2.5105
1000.	663.	415.6	-1110.8	-1093.0	-1069.3	-1040.7	-1006.7	288.8	2.9453

BALLISTIC COEFFICIENT .20, MUZZLE VELOCITY FT/SEC 2400.

RANGE	VELOCITY	HEIGHT	100 YD	150 YD	200 YD	250 YD	300 YD	DRIFT	TIME
0.	2400.	0.0	-1.5	-1.5	-1.5	-1.5	-1.5	0.0	0.0000
100.	1989.	.2	0.0	1.6	3.7	6.3	9.4	2.2	.1372
200.	1628.	3.7	-7.4	-4.3	0.0	5.2	11.3	9.5	.3041
300.	1329.	11.7	-28.1	-23.4	-17.0	-9.2	0.0	23.5	.5087
400.	1123.	26.7	-68.7	-62.4	-53.9	-43.5	-31.2	45.0	.7557
500.	1011.	51.2	-136.5	-128.6	-118.0	-105.0	-89.6	73.0	1.0399
600.	927.	86.5	-234.1	-224.6	-211.8	-196.3	-177.8	105.3	1.3485
700.	856.	135.6	-368.1	-357.0	-342.1	-323.9	-302.4	142.7	1.6857
800.	793.	201.0	-543.8	-531.1	-514.1	-493.4	-468.8	184.8	2.0501
900.	733.	285.9	-767.9	-753.7	-734.5	-711.2	-683.5	232.1	2.4437
1000.	676.	394.6	-1049.4	-1033.6	-1012.3	-986.4	-955.6	285.1	2.8700

BALLISTIC COEFFICIENT .20, MUZZLE VELOCITY FT/SEC 2500.

RANGE	VELOCITY	HEIGHT	100 YD	150 YD	200 YD	250 YD	300 YD	DRIFT	TIME
0.	2500.	0.0	-1.5	-1.5	-1.5	-1.5	-1.5	0.0	0.0000
100.	2081.	.1	0.0	1.4	3.3	5.7	8.5	2.0	.1314
200.	1706.	3.3	-6.7	-3.9	0.0	4.7	10.3	8.9	.2908
300.	1390.	10.6	-25.4	-21.2	-15.4	-8.4	0.0	22.2	.4860
400.	1160.	24.4	-62.4	-56.7	-49.0	-39.6	-28.5	42.8	.7233
500.	1033.	47.2	-124.8	-117.8	-108.2	-96.4	-82.5	70.2	.9990
600.	945.	80.8	-217.3	-208.8	-197.3	-183.2	-166.4	102.8	1.3038
700.	871.	127.7	-344.5	-334.6	-321.1	-304.7	-285.1	140.1	1.6360
800.	807.	189.7	-510.3	-499.1	-483.7	-464.9	-442.6	181.6	1.9919
900.	747.	270.8	-723.6	-710.9	-693.6	-672.4	-647.4	228.5	2.3785
1000.	689.	374.8	-992.0	-977.9	-958.7	-935.1	-907.3	281.1	2.7971

BALLISTIC COEFFICIENT .20, MUZZLE VELOCITY FT/SEC 2600.

RANGE	VELOCITY	HEIGHT	100 YD	150 YD	200 YD	250 YD	300 YD	DRIFT	TIME
0.	2600.	0.0	-1.5	-1.5	-1.5	-1.5	-1.5	0.0	0.0000
100.	2172.	.0	0.0	1.3	3.0	5.1	7.7	1.9	.1262
200.	1786.	3.0	-6.0	-3.5	0.0	4.3	9.3	8.4	.2787
300.	1455.	9.6	-23.0	-19.3	-14.0	-7.6	0.0	20.9	.4650
400.	1204.	22.3	-56.7	-51.7	-44.7	-36.1	-26.0	40.7	.6927
500.	1056.	43.5	-114.3	-108.1	-99.3	-88.7	-76.0	67.5	.9606
600.	964.	75.3	-201.1	-193.6	-183.1	-170.3	-155.0	99.7	1.2590
700.	887.	119.6	-320.5	-311.8	-299.5	-284.6	-266.8	136.5	1.5834
800.	821.	179.0	-478.7	-468.7	-454.7	-437.7	-417.4	178.1	1.9350
900.	760.	256.4	-681.8	-670.5	-654.8	-635.5	-612.7	224.6	2.3148
1000.	702.	355.9	-937.7	-925.2	-907.7	-886.3	-861.0	276.7	2.7259

BALLISTIC COEFFICIENT .20, MUZZLE VELOCITY FT/SEC 2700.

RANGE	VELOCITY	HEIGHT	100 YD	150 YD	200 YD	250 YD	300 YD	DRIFT	TIME
0.	2700.	0.0	-1.5	-1.5	-1.5	-1.5	-1.5	0.0	0.0000
100.	2263.	-.0	0.0	1.1	2.7	4.6	7.0	1.8	.1214
200.	1866.	2.7	-5.4	-3.2	0.0	3.9	8.5	7.9	.2673
300.	1523.	8.8	-20.9	-17.5	-12.8	-6.9	0.0	19.7	.4455
400.	1252.	20.4	-51.6	-47.1	-40.8	-33.0	-23.7	38.6	.6637
500.	1080.	40.2	-104.7	-99.1	-91.3	-81.5	-69.9	64.8	.9236
600.	982.	70.2	-186.1	-179.4	-169.9	-158.2	-144.3	96.7	1.2159
700.	903.	112.2	-298.9	-291.1	-280.1	-266.4	-250.2	133.1	1.5342
800.	835.	168.9	-449.4	-440.5	-427.8	-412.2	-393.7	174.5	1.8803
900.	773.	243.1	-643.0	-633.0	-618.8	-601.2	-580.4	220.7	2.2538
1000.	714.	338.3	-887.2	-876.1	-860.3	-840.8	-817.6	272.2	2.6576

BALLISTIC COEFFICIENT .20, MUZZLE VELOCITY FT/SEC 2800.

RANGE	VELOCITY	HEIGHT	100 YD	150 YD	200 YD	250 YD	300 YD	DRIFT	TIME
0.	2800.	0.0	-1.5	-1.5	-1.5	-1.5	-1.5	0.0	0.0000
100.	2354.	-.1	0.0	1.0	2.4	4.2	6.3	1.7	.1169
200.	1948.	2.4	-4.9	-2.9	0.0	3.6	7.8	7.5	.2570
300.	1593.	8.0	-19.0	-16.0	-11.7	-6.4	0.0	18.7	.4275
400.	1302.	18.7	-47.0	-43.1	-37.3	-30.2	-21.7	36.6	.6364
500.	1108.	37.1	-96.0	-91.0	-83.8	-74.9	-64.3	62.0	.8880
600.	1001.	65.4	-172.2	-166.2	-157.6	-146.9	-134.2	93.5	1.1739
700.	919.	105.3	-278.9	-272.0	-261.9	-249.4	-234.5	129.6	1.4866
800.	849.	159.5	-421.9	-414.0	-402.4	-388.2	-371.2	170.7	1.8270
900.	786.	230.4	-606.5	-597.6	-584.6	-568.6	-549.5	216.5	2.1943
1000.	727.	321.6	-839.8	-829.9	-815.5	-797.7	-776.4	267.5	2.5913

BALLISTIC COEFFICIENT .20, MUZZLE VELOCITY FT/SEC 2900.

RANGE	VELOCITY	HEIGHT	100 YD	150 YD	200 YD	250 YD	300 YD	DRIFT	TIME
0.	2900.	0.0	-1.5	-1.5	-1.5	-1.5	-1.5	0.0	0.0000
100.	2444.	-.1	0.0	.9	2.2	3.8	5.8	1.6	.1126
200.	2030.	2.2	-4.4	-2.6	0.0	3.3	7.2	7.1	.2475
300.	1662.	7.4	-17.4	-14.7	-10.7	-5.8	0.0	17.7	.4109
400.	1355.	17.2	-43.1	-39.5	-34.3	-27.7	-19.9	34.8	.6114
500.	1139.	34.3	-88.0	-83.6	-77.0	-68.8	-59.1	59.3	.8541
600.	1021.	60.9	-159.3	-154.0	-146.1	-136.3	-124.6	90.3	1.1337
700.	935.	98.9	-260.4	-254.2	-245.0	-233.5	-219.9	126.2	1.4410
800.	863.	150.5	-396.2	-389.0	-378.5	-365.4	-349.9	166.8	1.7754
900.	799.	218.4	-572.2	-564.3	-552.4	-537.7	-520.2	212.2	2.1367
1000.	739.	305.8	-795.2	-786.4	-773.2	-756.8	-737.4	262.7	2.5271

BALLISTIC COEFFICIENT .20, MUZZLE VELOCITY FT/SEC 3000.

RANGE	VELOCITY	HEIGHT	100 YD	150 YD	200 YD	250 YD	300 YD	DRIFT	TIME
0.	3000.	0.0	-1.5	-1.5	-1.5	-1.5	-1.5	0.0	0.0000
100.	2534.	-.2	0.0	.8	2.0	3.5	5.3	1.6	.1090
200.	2112.	2.0	-4.0	-2.4	0.0	3.0	6.6	6.8	.2386
300.	1733.	6.8	-15.9	-13.5	-9.9	-5.4	0.0	16.8	.3955
400.	1411.	15.8	-39.4	-36.3	-31.5	-25.4	-18.3	33.0	.5876
500.	1174.	31.7	-81.0	-77.0	-71.0	-63.5	-54.6	56.7	.8224
600.	1040.	56.8	-147.4	-142.7	-135.4	-126.4	-115.7	87.1	1.0947
700.	952.	92.9	-243.1	-237.6	-229.1	-218.6	-206.1	122.6	1.3967
800.	876.	142.3	-372.5	-366.2	-356.5	-344.5	-330.2	163.0	1.7263
900.	812.	207.2	-540.3	-533.3	-522.4	-508.9	-492.8	207.9	2.0815
1000.	751.	291.1	-753.8	-746.0	-733.9	-718.8	-701.0	258.0	2.4658

BALLISTIC COEFFICIENT .20, MUZZLE VELOCITY FT/SEC 3100.

RANGE	VELOCITY	HEIGHT	100 YD	150 YD	200 YD	250 YD	300 YD	DRIFT	TIME
0.	3100.	0.0	-1.5	-1.5	-1.5	-1.5	-1.5	0.0	0.0000
100.	2624.	-.2	0.0	.7	1.8	3.2	4.8	1.5	.1051
200.	2193.	1.8	-3.6	-2.2	0.0	2.8	6.1	6.5	.2302
300.	1805.	6.2	-14.5	-12.4	-9.1	-5.0	0.0	16.0	.3811
400.	1471.	14.6	-36.3	-33.4	-29.0	-23.5	-16.9	31.4	.5654
500.	1215.	29.3	-74.4	-70.9	-65.4	-58.5	-50.2	54.1	.7910
600.	1061.	52.9	-136.4	-132.2	-125.6	-117.3	-107.4	83.9	1.0572
700.	968.	87.3	-227.3	-222.4	-214.7	-205.0	-193.4	119.2	1.3547
800.	891.	134.2	-349.6	-344.0	-335.2	-324.1	-310.9	158.9	1.6769
900.	824.	196.5	-510.2	-503.8	-493.9	-481.5	-466.6	203.5	2.0273
1000.	763.	277.0	-714.5	-707.5	-696.5	-682.6	-666.1	253.1	2.4056

BALLISTIC COEFFICIENT .20, MUZZLE VELOCITY FT/SEC 3200.

RANGE	VELOCITY	HEIGHT	100 YD	150 YD	200 YD	250 YD	300 YD	DRIFT	TIME
0.	3200.	0.0	-1.5	-1.5	-1.5	-1.5	-1.5	0.0	0.0000
100.	2713.	-.3	0.0	.6	1.6	2.9	4.4	1.4	.1019
200.	2275.	1.6	-3.3	-2.0	0.0	2.6	5.6	6.2	.2226
300.	1877.	5.7	-13.3	-11.5	-8.4	-4.6	0.0	15.2	.3678
400.	1532.	13.5	-33.4	-30.9	-26.8	-21.7	-15.6	29.9	.5449
500.	1258.	27.1	-68.6	-65.5	-60.4	-54.0	-46.4	51.6	.7619
600.	1084.	49.3	-126.2	-122.5	-116.4	-108.7	-99.5	80.7	1.0208
700.	985.	82.0	-212.0	-207.7	-200.5	-191.6	-180.9	115.5	1.3126
800.	905.	126.8	-328.5	-323.5	-315.4	-305.1	-292.9	154.9	1.6299
900.	837.	186.5	-481.9	-476.3	-467.1	-455.6	-441.9	199.1	1.9750
1000.	775.	263.8	-677.8	-671.5	-661.4	-648.6	-633.3	248.2	2.3478

BALLISTIC COEFFICIENT .20, MUZZLE VELOCITY FT/SEC 3300.

RANGE	VELOCITY	HEIGHT	100 YD	150 YD	200 YD	250 YD	300 YD	DRIFT	TIME
0.	3300.	0.0	-1.5	-1.5	-1.5	-1.5	-1.5	0.0	0.0000
100.	2803.	-.3	0.0	.6	1.5	2.7	4.1	1.4	.0988
200.	2356.	1.5	-3.0	-1.9	0.0	2.4	5.2	5.9	.2156
300.	1950.	5.3	-12.3	-10.6	-7.8	-4.3	0.0	14.6	.3555
400.	1594.	12.5	-30.8	-28.6	-24.9	-20.1	-14.5	28.6	.5259
500.	1303.	25.2	-63.3	-60.6	-55.9	-49.9	-42.9	49.3	.7346
600.	1109.	45.9	-116.9	-113.6	-107.9	-100.8	-92.3	77.5	.9860
700.	1002.	76.9	-197.8	-193.9	-187.3	-179.0	-169.1	111.8	1.2718
800.	919.	119.7	-308.7	-304.2	-296.7	-287.3	-275.9	150.8	1.5842
900.	849.	177.0	-455.4	-450.4	-442.0	-431.3	-418.6	194.7	1.9245
1000.	787.	251.3	-643.1	-637.6	-628.2	-616.4	-602.2	243.3	2.2916

BALLISTIC COEFFICIENT .20, MUZZLE VELOCITY FT/SEC 3400.

RANGE	VELOCITY	HEIGHT	100 YD	150 YD	200 YD	250 YD	300 YD	DRIFT	TIME
0.	3400.	0.0	-1.5	-1.5	-1.5	-1.5	-1.5	0.0	0.0000
100.	2892.	-.3	0.0	.5	1.4	2.5	3.8	1.3	.0956
200.	2437.	1.3	-2.7	-1.7	0.0	2.2	4.8	5.7	.2086
300.	2023.	4.9	-11.3	-9.8	-7.3	-4.0	0.0	13.9	.3438
400.	1657.	11.6	-28.5	-26.5	-23.1	-18.7	-13.4	27.3	.5078
500.	1351.	23.4	-58.6	-56.1	-51.8	-46.3	-39.7	47.1	.7088
600.	1136.	42.8	-108.5	-105.5	-100.3	-93.7	-85.8	74.5	.9529
700.	1019.	72.1	-184.5	-181.1	-175.0	-167.3	-158.1	108.2	1.2323
800.	934.	113.1	-290.3	-286.3	-279.5	-270.6	-260.1	146.8	1.5402
900.	861.	168.0	-430.3	-425.9	-418.1	-408.2	-396.3	190.2	1.8749
1000.	798.	239.4	-610.4	-605.5	-596.9	-585.8	-572.6	238.4	2.2368

BALLISTIC COEFFICIENT .20, MUZZLE VELOCITY FT/SEC 3500.

RANGE	VELOCITY	HEIGHT	100 YD	150 YD	200 YD	250 YD	300 YD	DRIFT	TIME
0.	3500.	0.0	-1.5	-1.5	-1.5	-1.5	-1.5	0.0	0.0000
100.	2982.	-.3	0.0	.4	1.2	2.3	3.5	1.3	.0930
200.	2518.	1.2	-2.5	-1.6	0.0	2.0	4.5	5.5	.2026
300.	2097.	4.6	-10.4	-9.1	-6.7	-3.7	0.0	13.4	.3331
400.	1720.	10.8	-26.4	-24.6	-21.5	-17.4	-12.5	26.1	.4910
500.	1401.	21.7	-54.3	-52.1	-48.1	-43.0	-36.9	45.1	.6846
600.	1168.	39.9	-100.5	-97.9	-93.1	-87.0	-79.6	71.5	.9203
700.	1037.	67.7	-172.2	-169.2	-163.6	-156.4	-147.8	104.6	1.1945
800.	949.	106.9	-272.9	-269.5	-263.1	-254.9	-245.1	142.9	1.4975
900.	874.	159.8	-407.5	-403.6	-396.4	-387.2	-376.1	186.1	1.8287
1000.	810.	228.3	-579.7	-575.3	-567.4	-557.1	-544.9	233.6	2.1842

BALLISTIC COEFFICIENT .20, MUZZLE VELOCITY FT/SEC 3600.

RANGE	VELOCITY	HEIGHT	100 YD	150 YD	200 YD	250 YD	300 YD	DRIFT	TIME
0.	3600.	0.0	-1.5	-1.5	-1.5	-1.5	-1.5	0.0	0.0000
100.	3072.	-.4	0.0	.4	1.1	2.1	3.2	1.2	.0901
200.	2598.	1.1	-2.2	-1.5	0.0	1.9	4.2	5.2	.1964
300.	2170.	4.2	-9.6	-8.5	-6.3	-3.4	0.0	12.8	.3227
400.	1784.	10.1	-24.5	-23.0	-20.0	-16.2	-11.7	25.0	.4753
500.	1454.	20.3	-50.4	-48.5	-44.9	-40.1	-34.4	43.1	.6618
600.	1204.	37.2	-93.4	-91.1	-86.7	-81.0	-74.1	68.6	.8896
700.	1055.	63.6	-160.7	-158.1	-152.9	-146.2	-138.2	101.1	1.1576
800.	963.	101.0	-256.7	-253.7	-247.8	-240.2	-231.0	138.9	1.4561
900.	887.	151.4	-384.6	-381.2	-374.6	-366.0	-355.7	181.4	1.7806
1000.	821.	217.5	-550.3	-546.5	-539.2	-529.6	-518.2	228.6	2.1323

BALLISTIC COEFFICIENT .20, MUZZLE VELOCITY FT/SEC 3700.

RANGE	VELOCITY	HEIGHT	100 YD	150 YD	200 YD	250 YD	300 YD	DRIFT	TIME
0.	3700.	0.0	-1.5	-1.5	-1.5	-1.5	-1.5	0.0	0.0000
100.	3162.	-.4	0.0	.3	1.0	1.9	3.0	1.2	.0878
200.	2679.	1.0	-2.0	-1.3	0.0	1.8	3.9	5.1	.1909
300.	2244.	4.0	-8.9	-7.9	-5.9	-3.2	0.0	12.3	.3134
400.	1849.	9.4	-22.8	-21.4	-18.7	-15.2	-10.9	24.0	.4607
500.	1509.	18.9	-46.9	-45.3	-41.9	-37.4	-32.1	41.4	.6404
600.	1242.	34.8	-86.9	-84.9	-80.9	-75.5	-69.1	65.9	.8607
700.	1075.	59.7	-150.1	-147.7	-143.0	-136.8	-129.2	97.6	1.1223
800.	979.	95.4	-241.3	-238.6	-233.2	-226.1	-217.5	134.9	1.4154
900.	899.	143.8	-363.7	-360.7	-354.6	-346.6	-336.9	177.0	1.7355
1000.	832.	207.5	-522.9	-519.6	-512.9	-504.0	-493.2	223.9	2.0827

BALLISTIC COEFFICIENT .20, MUZZLE VELOCITY FT/SEC 3800.

RANGE	VELOCITY	HEIGHT	100 YD	150 YD	200 YD	250 YD	300 YD	DRIFT	TIME
0.	3800.	0.0	-1.5	-1.5	-1.5	-1.5	-1.5	0.0	0.0000
100.	3252.	-.4	0.0	.3	.9	1.7	2.8	1.1	.0854
200.	2760.	.9	-1.8	-1.3	0.0	1.7	3.7	4.9	.1857
300.	2317.	3.7	-8.3	-7.4	-5.5	-3.0	0.0	11.9	.3043
400.	1915.	8.8	-21.2	-20.1	-17.5	-14.2	-10.2	23.1	.4468
500.	1564.	17.7	-43.8	-42.3	-39.2	-35.0	-30.0	39.7	.6204
600.	1281.	32.6	-81.0	-79.3	-75.5	-70.5	-64.5	63.2	.8330
700.	1096.	56.1	-140.2	-138.2	-133.8	-127.9	-120.9	94.2	1.0880
800.	993.	90.4	-227.6	-225.3	-220.3	-213.6	-205.6	131.4	1.3782
900.	912.	136.6	-344.0	-341.5	-335.8	-328.3	-319.3	172.7	1.6916
1000.	843.	197.8	-496.8	-494.0	-487.7	-479.3	-469.3	219.0	2.0339

BALLISTIC COEFFICIENT .20, MUZZLE VELOCITY FT/SEC 3900.

RANGE	VELOCITY	HEIGHT	100 YD	150 YD	200 YD	250 YD	300 YD	DRIFT	TIME
0.	3900.	0.0	-1.5	-1.5	-1.5	-1.5	-1.5	0.0	0.0000
100.	3342.	-.4	0.0	.2	.8	1.6	2.5	1.1	.0831
200.	2840.	.8	-1.6	-1.2	0.0	1.6	3.5	4.7	.1803
300.	2390.	3.4	-7.6	-6.9	-5.2	-2.8	0.0	11.4	.2956
400.	1980.	8.3	-19.7	-18.7	-16.4	-13.3	-9.5	22.1	.4334
500.	1621.	16.6	-40.8	-39.6	-36.7	-32.8	-28.1	38.1	.6011
600.	1323.	30.5	-75.5	-74.1	-70.6	-65.9	-60.3	60.7	.8066
700.	1120.	52.6	-130.9	-129.2	-125.2	-119.7	-113.1	90.8	1.0546
800.	1009.	85.3	-213.8	-211.8	-207.2	-201.0	-193.4	127.4	1.3392
900.	925.	129.7	-325.4	-323.2	-318.0	-311.0	-302.5	168.3	1.6487
1000.	854.	188.7	-472.2	-469.7	-464.0	-456.2	-446.7	214.2	1.9866

BALLISTIC COEFFICIENT .20, MUZZLE VELOCITY FT/SEC 4000.

RANGE	VELOCITY	HEIGHT	100 YD	150 YD	200 YD	250 YD	300 YD	DRIFT	TIME
0.	4000.	0.0	-1.5	-1.5	-1.5	-1.5	-1.5	0.0	0.0000
100.	3432.	-.4	0.0	.2	.7	1.5	2.4	1.1	.0810
200.	2921.	.7	-1.5	-1.1	0.0	1.5	3.2	4.6	.1759
300.	2463.	3.2	-7.1	-6.5	-4.8	-2.7	0.0	11.0	.2877
400.	2047.	7.8	-18.4	-17.6	-15.4	-12.5	-9.0	21.4	.4214
500.	1677.	15.6	-38.1	-37.1	-34.4	-30.8	-26.4	36.7	.5834
600.	1367.	28.6	-70.6	-69.4	-66.1	-61.8	-56.5	58.4	.7820
700.	1146.	49.5	-122.5	-121.0	-117.2	-112.2	-106.0	87.6	1.0229
800.	1025.	80.6	-200.8	-199.2	-194.9	-189.1	-182.0	123.5	1.3017
900.	939.	123.3	-307.9	-306.0	-301.2	-294.6	-286.7	164.1	1.6073
1000.	866.	180.0	-448.8	-446.8	-441.4	-434.1	-425.3	209.5	1.9406

BALLISTIC COEFFICIENT .21, MUZZLE VELOCITY FT/SEC 2000.

RANGE	VELOCITY	HEIGHT	100 YD	150 YD	200 YD	250 YD	300 YD	DRIFT	TIME
0.	2000.	0.0	-1.5	-1.5	-1.5	-1.5	-1.5	0.0	0.0000
100.	1653.	.6	0.0	2.5	5.7	9.5	13.9	2.7	.1651
200.	1360.	5.7	-11.4	-6.4	0.0	7.6	16.4	11.5	.3656
300.	1150.	16.9	-41.7	-34.2	-24.6	-13.2	0.0	27.6	.6066
400.	1031.	36.8	-98.3	-88.3	-75.5	-60.3	-42.7	50.1	.8848
500.	948.	66.9	-184.5	-171.9	-155.9	-136.9	-114.9	76.9	1.1870
600.	877.	109.7	-306.4	-291.4	-272.2	-249.4	-223.0	108.6	1.5171
700.	815.	167.4	-468.7	-451.1	-428.7	-402.1	-371.3	144.6	1.8714
800.	757.	243.0	-677.8	-657.8	-632.1	-601.7	-566.6	185.4	2.2533
900.	702.	340.1	-941.4	-918.9	-890.0	-855.8	-816.3	231.4	2.6649
1000.	649.	463.6	-1269.6	-1244.6	-1212.5	-1174.5	-1130.5	283.4	3.1102

BALLISTIC COEFFICIENT .21, MUZZLE VELOCITY FT/SEC 2100.

RANGE	VELOCITY	HEIGHT	100 YD	150 YD	200 YD	250 YD	300 YD	DRIFT	TIME
0.	2100.	0.0	-1.5	-1.5	-1.5	-1.5	-1.5	0.0	0.0000
100.	1740.	.4	0.0	2.2	5.1	8.5	12.5	2.5	.1570
200.	1431.	5.0	-10.2	-5.7	0.0	6.8	14.8	10.9	.3474
300.	1197.	15.3	-37.5	-30.8	-22.2	-12.0	0.0	26.3	.5777
400.	1057.	33.6	-88.8	-79.9	-68.5	-54.9	-38.9	48.3	.8459
500.	969.	62.1	-170.0	-158.9	-144.6	-127.5	-107.6	75.7	1.1443
600.	894.	102.5	-284.2	-270.8	-253.7	-233.2	-209.2	107.3	1.4666
700.	831.	157.2	-437.0	-421.4	-401.4	-377.5	-349.6	143.2	1.8138
800.	772.	229.1	-635.0	-617.2	-594.4	-567.1	-535.1	184.0	2.1884
900.	716.	321.7	-884.8	-864.8	-839.2	-808.4	-772.5	229.9	2.5919
1000.	662.	439.2	-1195.8	-1173.5	-1145.0	-1110.9	-1070.9	281.4	3.0275

BALLISTIC COEFFICIENT .21, MUZZLE VELOCITY FT/SEC 2200.

RANGE	VELOCITY	HEIGHT	100 YD	150 YD	200 YD	250 YD	300 YD	DRIFT	TIME
0.	2200.	0.0	-1.5	-1.5	-1.5	-1.5	-1.5	0.0	0.0000
100.	1828.	.3	0.0	2.0	4.5	7.6	11.2	2.3	.1497
200.	1506.	4.5	-9.0	-5.1	0.0	6.2	13.4	10.2	.3307
300.	1250.	13.8	-33.6	-27.7	-20.1	-10.8	0.0	24.9	.5503
400.	1085.	30.7	-80.5	-72.7	-62.4	-50.1	-35.7	46.5	.8098
500.	990.	57.5	-156.1	-146.3	-133.5	-118.1	-100.1	73.9	1.1015
600.	913.	95.5	-262.7	-250.9	-235.6	-217.1	-195.5	105.2	1.4158
700.	847.	147.5	-407.3	-393.5	-375.6	-354.0	-328.8	141.3	1.7575
800.	787.	216.0	-594.8	-579.1	-558.6	-534.0	-505.1	182.0	2.1252
900.	730.	304.3	-832.0	-814.3	-791.2	-763.5	-731.1	227.7	2.5210
1000.	676.	416.4	-1127.3	-1107.6	-1082.0	-1051.2	-1015.2	278.8	2.9480

BALLISTIC COEFFICIENT .21, MUZZLE VELOCITY FT/SEC 2300.

RANGE	VELOCITY	HEIGHT	100 YD	150 YD	200 YD	250 YD	300 YD	DRIFT	TIME
0.	2300.	0.0	-1.5	-1.5	-1.5	-1.5	-1.5	0.0	0.0000
100.	1917.	.2	0.0	1.7	4.0	6.8	10.1	2.2	.1429
200.	1582.	4.0	-8.1	-4.6	0.0	5.5	12.1	9.6	.3154
300.	1306.	12.5	-30.2	-25.0	-18.1	-9.8	0.0	23.5	.5246
400.	1117.	28.0	-72.9	-65.9	-56.7	-45.6	-32.6	44.5	.7746
500.	1011.	53.1	-143.2	-134.5	-122.9	-109.1	-92.8	71.7	1.0595
600.	931.	89.0	-243.2	-232.7	-218.9	-202.3	-182.7	102.9	1.3673
700.	862.	138.4	-379.4	-367.2	-351.0	-331.7	-308.9	138.9	1.7024
800.	802.	203.6	-557.0	-543.1	-524.6	-502.5	-476.4	179.5	2.0632
900.	744.	287.8	-782.2	-766.5	-745.7	-720.8	-691.5	224.9	2.4517
1000.	689.	394.9	-1063.2	-1045.8	-1022.7	-995.0	-962.4	275.7	2.8709

BALLISTIC COEFFICIENT .21, MUZZLE VELOCITY FT/SEC 2400.

RANGE	VELOCITY	HEIGHT	100 YD	150 YD	200 YD	250 YD	300 YD	DRIFT	TIME
0.	2400.	0.0	-1.5	-1.5	-1.5	-1.5	-1.5	0.0	0.0000
100.	2008.	.1	0.0	1.5	3.6	6.1	9.1	2.0	.1366
200.	1660.	3.6	-7.3	-4.2	0.0	5.0	10.9	9.0	.3011
300.	1366.	11.3	-27.3	-22.6	-16.4	-8.9	0.0	22.1	.5007
400.	1153.	25.7	-66.3	-60.1	-51.8	-41.7	-29.9	42.6	.7418
500.	1034.	48.9	-130.6	-122.9	-112.5	-100.0	-85.2	69.0	1.0169
600.	950.	82.9	-224.9	-215.7	-203.2	-188.2	-170.4	100.3	1.3201
700.	878.	129.8	-353.6	-342.8	-328.2	-310.7	-290.0	136.3	1.6492
800.	816.	191.9	-521.8	-509.4	-492.8	-472.8	-449.1	176.6	2.0033
900.	758.	272.2	-735.6	-721.7	-703.0	-680.5	-653.9	221.7	2.3845
1000.	703.	374.4	-1003.0	-987.5	-966.7	-941.7	-912.1	272.0	2.7957

BALLISTIC COEFFICIENT .21, MUZZLE VELOCITY FT/SEC 2500.

RANGE	VELOCITY	HEIGHT	100 YD	150 YD	200 YD	250 YD	300 YD	DRIFT	TIME
0.	2500.	0.0	-1.5	-1.5	-1.5	-1.5	-1.5	0.0	0.0000
100.	2100.	.1	0.0	1.4	3.2	5.5	8.2	1.9	.1310
200.	1739.	3.2	-6.5	-3.7	0.0	4.6	9.9	8.4	.2880
300.	1430.	10.2	-24.6	-20.5	-14.9	-8.0	0.0	20.8	.4784
400.	1197.	23.4	-59.9	-54.4	-46.9	-37.8	-27.1	40.3	.7088
500.	1057.	45.1	-119.4	-112.5	-103.2	-91.8	-78.4	66.4	.9770
600.	968.	77.3	-208.4	-200.1	-188.9	-175.2	-159.2	97.7	1.2754
700.	894.	121.8	-329.7	-320.0	-306.9	-290.9	-272.2	133.4	1.5978
800.	831.	180.8	-488.8	-477.8	-462.8	-444.5	-423.1	173.3	1.9449
900.	772.	257.5	-692.2	-679.9	-663.0	-642.5	-618.4	218.2	2.3196
1000.	716.	355.2	-946.7	-933.0	-914.2	-891.4	-864.7	268.1	2.7231

BALLISTIC COEFFICIENT .21, MUZZLE VELOCITY FT/SEC 2600.

RANGE	VELOCITY	HEIGHT	100 YD	150 YD	200 YD	250 YD	300 YD	DRIFT	TIME
0.	2600.	0.0	-1.5	-1.5	-1.5	-1.5	-1.5	0.0	0.0000
100.	2191.	.0	0.0	1.2	2.9	5.0	7.4	1.8	.1256
200.	1820.	2.9	-5.8	-3.4	0.0	4.2	9.0	7.9	.2759
300.	1499.	9.3	-22.3	-18.6	-13.5	-7.3	0.0	19.6	.4577
400.	1245.	21.3	-54.4	-49.5	-42.8	-34.5	-24.7	38.2	.6783
500.	1082.	41.6	-109.3	-103.2	-94.7	-84.3	-72.2	63.7	.9390
600.	988.	71.9	-192.5	-185.2	-175.0	-162.5	-147.9	94.7	1.2305
700.	911.	113.9	-306.5	-297.9	-286.0	-271.5	-254.5	129.9	1.5457
800.	845.	170.3	-457.7	-447.9	-434.3	-417.7	-398.3	169.8	1.8879
900.	786.	243.6	-651.4	-640.3	-625.1	-606.4	-584.5	214.3	2.2562
1000.	729.	337.0	-894.0	-881.8	-864.8	-844.1	-819.7	263.8	2.6527

BALLISTIC COEFFICIENT .21, MUZZLE VELOCITY FT/SEC 2700.

RANGE	VELOCITY	HEIGHT	100 YD	150 YD	200 YD	250 YD	300 YD	DRIFT	TIME
0.	2700.	0.0	-1.5	-1.5	-1.5	-1.5	-1.5	0.0	0.0000
100.	2283.	-.0	0.0	1.1	2.6	4.5	6.8	1.7	.1208
200.	1902.	2.6	-5.3	-3.1	0.0	3.8	8.2	7.5	.2648
300.	1569.	8.5	-20.3	-17.0	-12.4	-6.7	0.0	18.6	.4388
400.	1296.	19.5	-49.6	-45.2	-39.0	-31.5	-22.5	36.2	.6499
500.	1111.	38.3	-99.8	-94.4	-86.7	-77.2	-66.1	60.9	.9014
600.	1008.	66.9	-177.7	-171.2	-161.9	-150.6	-137.2	91.6	1.1870
700.	928.	106.7	-285.3	-277.7	-266.9	-253.7	-238.0	126.5	1.4964
800.	860.	160.4	-428.7	-420.0	-407.6	-392.5	-374.6	166.1	1.8326
900.	799.	230.4	-612.9	-603.1	-589.2	-572.2	-552.1	210.2	2.1945
1000.	742.	319.8	-844.3	-833.4	-817.9	-799.1	-776.7	259.2	2.5841

BALLISTIC COEFFICIENT .21, MUZZLE VELOCITY FT/SEC 2800.

RANGE	VELOCITY	HEIGHT	100 YD	150 YD	200 YD	250 YD	300 YD	DRIFT	TIME
0.	2800.	0.0	-1.5	-1.5	-1.5	-1.5	-1.5	0.0	0.0000
100.	2374.	-.1	0.0	1.0	2.4	4.1	6.1	1.6	.1163
200.	1984.	2.4	-4.7	-2.8	0.0	3.5	7.5	7.1	.2545
300.	1640.	7.8	-18.4	-15.5	-11.3	-6.1	0.0	17.5	.4209
400.	1350.	17.9	-45.2	-41.3	-35.7	-28.8	-20.6	34.2	.6231
500.	1144.	35.2	-91.3	-86.5	-79.4	-70.8	-60.6	58.1	.8657
600.	1028.	62.2	-164.0	-158.2	-149.8	-139.4	-127.1	88.3	1.1448
700.	945.	100.1	-265.9	-259.2	-249.3	-237.2	-222.9	123.1	1.4493
800.	874.	151.4	-402.3	-394.6	-383.3	-369.5	-353.1	162.5	1.7804
900.	813.	218.1	-577.2	-568.5	-555.8	-540.3	-521.9	206.1	2.1351
1000.	755.	303.6	-798.1	-788.4	-774.3	-757.1	-736.6	254.6	2.5182

BALLISTIC COEFFICIENT .21, MUZZLE VELOCITY FT/SEC 2900.

RANGE	VELOCITY	HEIGHT	100 YD	150 YD	200 YD	250 YD	300 YD	DRIFT	TIME
0.	2900.	0.0	-1.5	-1.5	-1.5	-1.5	-1.5	0.0	0.0000
100.	2465.	-.1	0.0	.9	2.2	3.7	5.6	1.5	.1122
200.	2067.	2.1	-4.3	-2.6	0.0	3.2	6.9	6.7	.2451
300.	1711.	7.1	-16.8	-14.2	-10.4	-5.6	0.0	16.6	.4046
400.	1407.	16.4	-41.3	-37.8	-32.7	-26.4	-18.9	32.5	.5983
500.	1181.	32.5	-83.7	-79.3	-72.9	-65.0	-55.7	55.4	.8320
600.	1048.	57.9	-151.6	-146.4	-138.7	-129.2	-118.0	85.2	1.1050
700.	962.	93.7	-247.4	-241.3	-232.3	-221.2	-208.1	119.4	1.4025
800.	889.	142.4	-376.4	-369.4	-359.1	-346.5	-331.5	158.3	1.7271
900.	826.	206.4	-543.6	-535.8	-524.3	-510.0	-493.2	201.8	2.0775
1000.	767.	288.4	-754.8	-746.1	-733.3	-717.4	-698.7	249.9	2.4546

BALLISTIC COEFFICIENT .21, MUZZLE VELOCITY FT/SEC 3000.

RANGE	VELOCITY	HEIGHT	100 YD	150 YD	200 YD	250 YD	300 YD	DRIFT	TIME
0.	3000.	0.0	-1.5	-1.5	-1.5	-1.5	-1.5	0.0	0.0000
100.	2555.	-.2	0.0	.8	1.9	3.4	5.1	1.5	.1085
200.	2150.	1.9	-3.9	-2.3	0.0	2.9	6.4	6.4	.2365
300.	1784.	6.5	-15.4	-13.1	-9.5	-5.2	0.0	15.8	.3898
400.	1468.	15.1	-37.9	-34.8	-30.1	-24.3	-17.4	30.9	.5753
500.	1223.	30.0	-76.9	-73.0	-67.1	-59.8	-51.2	52.8	.8001
600.	1071.	53.6	-139.4	-134.8	-127.7	-119.0	-108.6	81.7	1.0642
700.	979.	87.7	-230.2	-224.8	-216.6	-206.4	-194.3	115.7	1.3577
800.	904.	134.2	-352.7	-346.5	-337.1	-325.4	-311.6	154.3	1.6768
900.	839.	195.4	-512.0	-505.1	-494.5	-481.4	-465.9	197.4	2.0216
1000.	780.	274.0	-713.8	-706.1	-694.4	-679.8	-662.5	245.1	2.3926

BALLISTIC COEFFICIENT .21, MUZZLE VELOCITY FT/SEC 3100.

RANGE	VELOCITY	HEIGHT	100 YD	150 YD	200 YD	250 YD	300 YD	DRIFT	TIME
0.	3100.	0.0	-1.5	-1.5	-1.5	-1.5	-1.5	0.0	0.0000
100.	2645.	-.2	0.0	.7	1.8	3.1	4.7	1.4	.1047
200.	2233.	1.8	-3.5	-2.2	0.0	2.7	5.9	6.1	.2282
300.	1857.	6.0	-14.1	-12.0	-8.8	-4.8	0.0	15.0	.3755
400.	1530.	14.0	-34.8	-32.1	-27.8	-22.4	-16.1	29.3	.5537
500.	1268.	27.7	-70.7	-67.3	-61.9	-55.2	-47.3	50.3	.7699
600.	1095.	49.8	-128.7	-124.6	-118.1	-110.0	-100.5	78.4	1.0263
700.	997.	82.2	-214.4	-209.6	-202.0	-192.6	-181.5	112.1	1.3143
800.	919.	126.4	-330.6	-325.1	-316.4	-305.7	-293.0	150.2	1.6278
900.	852.	185.1	-482.7	-476.5	-466.8	-454.7	-440.4	193.0	1.9676
1000.	792.	260.5	-675.8	-669.0	-658.2	-644.7	-628.9	240.3	2.3331

BALLISTIC COEFFICIENT .21, MUZZLE VELOCITY FT/SEC 3200.

RANGE	VELOCITY	HEIGHT	100 YD	150 YD	200 YD	250 YD	300 YD	DRIFT	TIME
0.	3200.	0.0	-1.5	-1.5	-1.5	-1.5	-1.5	0.0	0.0000
100.	2735.	-.3	0.0	.6	1.6	2.8	4.3	1.4	.1014
200.	2315.	1.6	-3.2	-2.0	0.0	2.5	5.4	5.8	.2207
300.	1931.	5.6	-13.0	-11.1	-8.2	-4.4	0.0	14.3	.3627
400.	1594.	12.9	-32.1	-29.7	-25.7	-20.8	-14.8	28.0	.5339
500.	1315.	25.7	-65.2	-62.2	-57.2	-51.0	-43.6	48.1	.7418
600.	1122.	46.3	-118.9	-115.2	-109.3	-101.9	-93.0	75.3	.9902
700.	1015.	77.0	-199.6	-195.3	-188.4	-179.7	-169.4	108.5	1.2725
800.	934.	119.2	-310.1	-305.1	-297.3	-287.3	-275.5	146.2	1.5810
900.	865.	175.3	-455.0	-449.5	-440.6	-429.4	-416.1	188.6	1.9151
1000.	804.	247.7	-639.8	-633.7	-623.8	-611.4	-596.6	235.4	2.2750

BALLISTIC COEFFICIENT .21, MUZZLE VELOCITY FT/SEC 3300.

RANGE	VELOCITY	HEIGHT	100 YD	150 YD	200 YD	250 YD	300 YD	DRIFT	TIME
0.	3300.	0.0	-1.5	-1.5	-1.5	-1.5	-1.5	0.0	0.0000
100.	2825.	-.3	0.0	.5	1.5	2.6	4.0	1.3	.0983
200.	2397.	1.4	-2.9	-1.8	0.0	2.3	5.0	5.6	.2137
300.	2005.	5.1	-11.9	-10.3	-7.5	-4.1	0.0	13.7	.3505
400.	1658.	12.0	-29.6	-27.5	-23.8	-19.2	-13.8	26.7	.5152
500.	1364.	23.8	-60.2	-57.5	-53.0	-47.2	-40.4	45.9	.7151
600.	1153.	43.1	-110.0	-106.8	-101.3	-94.4	-86.2	72.2	.9555
700.	1033.	72.1	-185.8	-182.0	-175.6	-167.6	-158.0	104.8	1.2319
800.	949.	112.4	-290.8	-286.4	-279.1	-269.9	-259.0	142.2	1.5352
900.	878.	166.1	-429.2	-424.3	-416.1	-405.8	-393.5	184.2	1.8647
1000.	816.	235.6	-606.0	-600.6	-591.4	-579.9	-566.3	230.5	2.2188

BALLISTIC COEFFICIENT .21, MUZZLE VELOCITY FT/SEC 3400.

RANGE	VELOCITY	HEIGHT	100 YD	150 YD	200 YD	250 YD	300 YD	DRIFT	TIME
0.	3400.	0.0	-1.5	-1.5	-1.5	-1.5	-1.5	0.0	0.0000
100.	2915.	-.3	0.0	.5	1.3	2.4	3.7	1.3	.0953
200.	2479.	1.3	-2.6	-1.7	0.0	2.1	4.7	5.3	.2069
300.	2080.	4.8	-11.0	-9.5	-7.0	-3.8	0.0	13.1	.3389
400.	1722.	11.1	-27.4	-25.5	-22.1	-17.8	-12.8	25.5	.4975
500.	1416.	22.1	-55.7	-53.3	-49.1	-43.8	-37.5	43.8	.6900
600.	1187.	40.1	-101.9	-99.0	-94.0	-87.6	-80.0	69.2	.9223
700.	1052.	67.5	-173.0	-169.6	-163.8	-156.3	-147.4	101.2	1.1925
800.	965.	106.0	-272.9	-269.1	-262.4	-253.8	-243.7	138.2	1.4914
900.	891.	157.3	-404.6	-400.3	-392.7	-383.1	-371.7	179.6	1.8147
1000.	828.	224.1	-574.1	-569.3	-561.0	-550.2	-537.6	225.6	2.1643

BALLISTIC COEFFICIENT .21, MUZZLE VELOCITY FT/SEC 3500.

RANGE	VELOCITY	HEIGHT	100 YD	150 YD	200 YD	250 YD	300 YD	DRIFT	TIME
0.	3500.	0.0	-1.5	-1.5	-1.5	-1.5	-1.5	0.0	0.0000
100.	3005.	-.3	0.0	.4	1.2	2.2	3.4	1.2	.0925
200.	2560.	1.2	-2.4	-1.6	0.0	2.0	4.4	5.2	.2008
300.	2155.	4.4	-10.1	-8.9	-6.5	-3.5	0.0	12.6	.3285
400.	1788.	10.4	-25.4	-23.7	-20.6	-16.6	-11.9	24.4	.4815
500.	1471.	20.6	-51.7	-49.6	-45.7	-40.7	-34.8	41.9	.6666
600.	1226.	37.4	-94.6	-92.0	-87.4	-81.4	-74.3	66.3	.8910
700.	1072.	63.2	-161.1	-158.1	-152.7	-145.7	-137.4	97.6	1.1546
800.	980.	99.9	-255.8	-252.4	-246.2	-238.2	-228.7	134.1	1.4478
900.	904.	149.1	-381.6	-377.8	-370.9	-361.9	-351.2	175.2	1.7666
1000.	839.	213.2	-543.9	-539.7	-531.9	-522.0	-510.1	220.7	2.1111

BALLISTIC COEFFICIENT .21, MUZZLE VELOCITY FT/SEC 3600.

RANGE	VELOCITY	HEIGHT	100 YD	150 YD	200 YD	250 YD	300 YD	DRIFT	TIME
0.	3600.	0.0	-1.5	-1.5	-1.5	-1.5	-1.5	0.0	0.0000
100.	3096.	-.4	0.0	.4	1.1	2.0	3.1	1.2	.0899
200.	2641.	1.1	-2.2	-1.4	0.0	1.9	4.1	4.9	.1948
300.	2229.	4.1	-9.3	-8.3	-6.1	-3.3	0.0	12.0	.3184
400.	1854.	9.7	-23.6	-22.1	-19.2	-15.5	-11.1	23.3	.4660
500.	1528.	19.2	-48.0	-46.2	-42.6	-38.0	-32.4	40.1	.6445
600.	1266.	34.8	-87.8	-85.6	-81.3	-75.8	-69.1	63.5	.8610
700.	1094.	59.2	-150.0	-147.5	-142.4	-136.0	-128.2	94.1	1.1178
800.	996.	94.1	-239.9	-237.0	-231.2	-223.8	-214.9	130.1	1.4060
900.	918.	141.2	-360.0	-356.7	-350.2	-341.9	-331.9	170.7	1.7197
1000.	851.	202.9	-515.7	-512.0	-504.8	-495.6	-484.5	215.9	2.0598

BALLISTIC COEFFICIENT .21, MUZZLE VELOCITY FT/SEC 3700.

RANGE	VELOCITY	HEIGHT	100 YD	150 YD	200 YD	250 YD	300 YD	DRIFT	TIME
0.	3700.	0.0	-1.5	-1.5	-1.5	-1.5	-1.5	0.0	0.0000
100.	3186.	-.4	0.0	.3	1.0	1.9	2.9	1.1	.0874
200.	2723.	1.0	-2.0	-1.3	0.0	1.7	3.8	4.8	.1894
300.	2304.	3.8	-8.7	-7.7	-5.7	-3.1	0.0	11.6	.3092
400.	1920.	9.1	-21.9	-20.7	-18.0	-14.5	-10.4	22.5	.4519
500.	1585.	17.9	-44.7	-43.1	-39.8	-35.5	-30.3	38.5	.6240
600.	1308.	32.6	-81.8	-79.8	-75.9	-70.6	-64.4	61.0	.8330
700.	1118.	55.5	-140.0	-137.7	-133.1	-127.0	-119.8	90.7	1.0830
800.	1012.	89.0	-225.6	-223.1	-217.8	-210.8	-202.6	126.5	1.3674
900.	932.	133.9	-339.8	-337.0	-331.0	-323.2	-313.9	166.3	1.6749
1000.	863.	193.1	-488.9	-485.7	-479.0	-470.4	-460.0	211.0	2.0097

BALLISTIC COEFFICIENT .21, MUZZLE VELOCITY FT/SEC 3800.

RANGE	VELOCITY	HEIGHT	100 YD	150 YD	200 YD	250 YD	300 YD	DRIFT	TIME
0.	3800.	0.0	-1.5	-1.5	-1.5	-1.5	-1.5	0.0	0.0000
100.	3277.	-.4	0.0	.3	.9	1.7	2.7	1.1	.0852
200.	2804.	.9	-1.8	-1.2	0.0	1.6	3.6	4.6	.1842
300.	2378.	3.6	-8.0	-7.2	-5.3	-2.9	0.0	11.2	.3004
400.	1988.	8.5	-20.4	-19.3	-16.8	-13.6	-9.7	21.6	.4383
500.	1643.	16.8	-41.7	-40.3	-37.2	-33.2	-28.3	36.9	.6045
600.	1352.	30.5	-76.2	-74.6	-70.9	-66.0	-60.2	58.5	.8063
700.	1145.	52.0	-130.4	-128.5	-124.2	-118.6	-111.8	87.3	1.0485
800.	1029.	83.8	-211.4	-209.2	-204.3	-197.9	-190.1	122.5	1.3275
900.	945.	127.1	-321.1	-318.6	-313.1	-305.8	-297.1	162.1	1.6318
1000.	875.	184.1	-464.2	-461.5	-455.3	-447.3	-437.6	206.4	1.9624

BALLISTIC COEFFICIENT .21, MUZZLE VELOCITY FT/SEC 3900.

RANGE	VELOCITY	HEIGHT	100 YD	150 YD	200 YD	250 YD	300 YD	DRIFT	TIME
0.	3900.	0.0	-1.5	-1.5	-1.5	-1.5	-1.5	0.0	0.0000
100.	3367.	-.4	0.0	.2	.8	1.6	2.5	1.0	.0827
200.	2886.	.8	-1.6	-1.1	0.0	1.5	3.3	4.4	.1789
300.	2452.	3.3	-7.4	-6.7	-5.0	-2.7	0.0	10.7	.2918
400.	2056.	7.9	-19.0	-18.1	-15.8	-12.8	-9.1	20.7	.4255
500.	1701.	15.7	-38.9	-37.7	-34.9	-31.1	-26.6	35.4	.5859
600.	1399.	28.5	-71.1	-69.7	-66.3	-61.8	-56.3	56.2	.7808
700.	1175.	48.8	-122.0	-120.3	-116.4	-111.1	-104.7	84.1	1.0163
800.	1045.	79.0	-198.3	-196.4	-191.9	-185.8	-178.6	118.6	1.2891
900.	960.	120.2	-302.6	-300.4	-295.4	-288.6	-280.4	157.6	1.5876
1000.	887.	175.0	-439.7	-437.3	-431.7	-424.1	-415.0	201.4	1.9134

BALLISTIC COEFFICIENT .21, MUZZLE VELOCITY FT/SEC 4000.

RANGE	VELOCITY	HEIGHT	100 YD	150 YD	200 YD	250 YD	300 YD	DRIFT	TIME
0.	4000.	0.0	-1.5	-1.5	-1.5	-1.5	-1.5	0.0	0.0000
100.	3458.	-.4	0.0	.2	.7	1.4	2.3	1.0	.0808
200.	2968.	.7	-1.4	-1.0	0.0	1.4	3.1	4.3	.1744
300.	2526.	3.1	-6.9	-6.3	-4.7	-2.6	0.0	10.4	.2842
400.	2123.	7.5	-17.7	-16.9	-14.8	-12.0	-8.6	20.0	.4136
500.	1760.	14.8	-36.4	-35.4	-32.8	-29.2	-24.9	34.1	.5688
600.	1448.	26.8	-66.5	-65.3	-62.2	-57.9	-52.8	54.0	.7571
700.	1209.	45.8	-113.9	-112.5	-108.9	-103.9	-97.9	80.9	.9847
800.	1063.	74.5	-185.9	-184.3	-180.2	-174.5	-167.6	114.7	1.2519
900.	974.	114.0	-285.7	-283.9	-279.3	-272.9	-265.1	153.4	1.5465
1000.	899.	166.6	-417.1	-415.1	-409.9	-402.8	-394.2	196.7	1.8675

BALLISTIC COEFFICIENT .22, MUZZLE VELOCITY FT/SEC 2000.

RANGE	VELOCITY	HEIGHT	100 YD	150 YD	200 YD	250 YD	300 YD	DRIFT	TIME
0.	2000.	0.0	-1.5	-1.5	-1.5	-1.5	-1.5	0.0	0.0000
100.	1668.	.5	0.0	2.5	5.6	9.3	13.6	2.5	.1644
200.	1384.	5.5	-11.2	-6.2	0.0	7.4	16.0	10.9	.3621
300.	1173.	16.5	-40.8	-33.4	-24.0	-12.9	0.0	26.3	.5994
400.	1048.	35.8	-95.6	-85.7	-73.2	-58.5	-41.2	47.9	.8723
500.	966.	64.9	-179.5	-167.2	-151.6	-133.1	-111.6	73.9	1.1698
600.	895.	106.2	-297.7	-282.9	-264.2	-242.0	-216.2	104.3	1.4928
700.	834.	161.6	-454.6	-437.3	-415.5	-389.6	-359.5	138.9	1.8392
800.	778.	234.0	-656.5	-636.8	-611.8	-582.3	-547.8	178.0	2.2116
900.	724.	326.6	-910.0	-887.8	-859.8	-826.5	-787.7	222.0	2.6116
1000.	672.	443.4	-1223.6	-1198.9	-1167.7	-1130.8	-1087.6	271.4	3.0419

BALLISTIC COEFFICIENT .22, MUZZLE VELOCITY FT/SEC 2100.

RANGE	VELOCITY	HEIGHT	100 YD	150 YD	200 YD	250 YD	300 YD	DRIFT	TIME
0.	2100.	0.0	-1.5	-1.5	-1.5	-1.5	-1.5	0.0	0.0000
100.	1755.	.4	0.0	2.2	5.0	8.3	12.1	2.4	.1562
200.	1456.	4.9	-9.9	-5.6	0.0	6.6	14.3	10.3	.3441
300.	1224.	14.8	-36.4	-29.9	-21.5	-11.5	0.0	24.8	.5696
400.	1076.	32.5	-86.1	-77.3	-66.2	-52.9	-37.5	46.0	.8326
500.	987.	60.1	-164.8	-153.8	-139.9	-123.3	-104.1	72.4	1.1255
600.	914.	98.8	-274.9	-261.8	-245.1	-225.2	-202.1	102.6	1.4403
700.	850.	151.5	-423.1	-407.8	-388.3	-365.1	-338.1	137.5	1.7812
800.	793.	220.4	-614.1	-596.7	-574.3	-547.8	-517.0	176.7	2.1466
900.	739.	308.5	-853.9	-834.3	-809.2	-779.3	-744.7	220.4	2.5382
1000.	686.	419.8	-1151.4	-1129.5	-1101.6	-1068.5	-1030.0	269.5	2.9600

BALLISTIC COEFFICIENT .22, MUZZLE VELOCITY FT/SEC 2200.

RANGE	VELOCITY	HEIGHT	100 YD	150 YD	200 YD	250 YD	300 YD	DRIFT	TIME
0.	2200.	0.0	-1.5	-1.5	-1.5	-1.5	-1.5	0.0	0.0000
100.	1844.	.3	0.0	1.9	4.4	7.4	10.9	2.2	.1491
200.	1533.	4.4	-8.9	-5.0	0.0	6.0	12.9	9.7	.3276
300.	1280.	13.4	-32.7	-26.9	-19.4	-10.4	0.0	23.5	.5426
400.	1108.	29.7	-77.8	-70.2	-60.1	-48.2	-34.3	44.1	.7962
500.	1009.	55.5	-150.9	-141.3	-128.8	-113.8	-96.4	70.5	1.0823
600.	933.	92.0	-254.1	-242.5	-227.5	-209.6	-188.7	100.7	1.3904
700.	867.	142.0	-393.6	-380.1	-362.6	-341.7	-317.3	135.5	1.7245
800.	808.	207.5	-574.3	-558.9	-538.8	-515.0	-487.1	174.6	2.0831
900.	753.	291.5	-801.8	-784.5	-762.0	-735.1	-703.8	218.3	2.4676
1000.	700.	397.5	-1083.7	-1064.4	-1039.4	-1009.5	-974.7	267.0	2.8804

BALLISTIC COEFFICIENT .22, MUZZLE VELOCITY FT/SEC 2300.

RANGE	VELOCITY	HEIGHT	100 YD	150 YD	200 YD	250 YD	300 YD	DRIFT	TIME
0.	2300.	0.0	-1.5	-1.5	-1.5	-1.5	-1.5	0.0	0.0000
100.	1934.	.2	0.0	1.7	4.0	6.6	9.8	2.1	.1423
200.	1611.	3.9	-7.9	-4.5	0.0	5.4	11.7	9.1	.3124
300.	1338.	12.1	-29.4	-24.3	-17.5	-9.4	0.0	22.2	.5173
400.	1143.	27.0	-70.4	-63.5	-54.5	-43.8	-31.2	42.1	.7608
500.	1032.	51.2	-138.0	-129.5	-118.2	-104.8	-89.0	68.2	1.0399
600.	952.	85.6	-234.6	-224.4	-210.9	-194.7	-175.8	98.3	1.3414
700.	883.	132.9	-365.9	-354.0	-338.2	-319.4	-297.3	133.0	1.6688
800.	823.	195.3	-536.8	-523.2	-505.2	-483.6	-458.5	172.0	2.0209
900.	768.	275.4	-752.9	-737.6	-717.3	-693.1	-664.7	215.5	2.3985
1000.	714.	376.6	-1021.0	-1004.0	-981.4	-954.5	-923.0	263.9	2.8038

BALLISTIC COEFFICIENT .22, MUZZLE VELOCITY FT/SEC 2400.

RANGE	VELOCITY	HEIGHT	100 YD	150 YD	200 YD	250 YD	300 YD	DRIFT	TIME
0.	2400.	0.0	-1.5	-1.5	-1.5	-1.5	-1.5	0.0	0.0000
100.	2025.	.1	0.0	1.5	3.5	6.0	8.8	1.9	.1361
200.	1690.	3.5	-7.1	-4.1	0.0	4.9	10.6	8.5	.2983
300.	1402.	10.9	-26.5	-21.9	-15.8	-8.5	0.0	20.9	.4935
400.	1186.	24.6	-63.7	-57.7	-49.5	-39.8	-28.4	40.0	.7272
500.	1056.	46.9	-125.6	-118.0	-107.9	-95.7	-81.5	65.4	.9969
600.	971.	79.8	-217.2	-208.1	-195.9	-181.3	-164.3	96.0	1.2955
700.	900.	124.4	-340.1	-329.5	-315.3	-298.2	-278.3	130.2	1.6146
800.	838.	183.7	-501.9	-489.7	-473.5	-454.0	-431.3	169.0	1.9604
900.	782.	260.1	-706.9	-693.3	-675.0	-653.1	-627.5	212.3	2.3311
1000.	728.	356.8	-962.1	-946.9	-926.6	-902.3	-873.9	260.3	2.7292

BALLISTIC COEFFICIENT .22, MUZZLE VELOCITY FT/SEC 2500.

RANGE	VELOCITY	HEIGHT	100 YD	150 YD	200 YD	250 YD	300 YD	DRIFT	TIME
0.	2500.	0.0	-1.5	-1.5	-1.5	-1.5	-1.5	0.0	0.0000
100.	2117.	.1	0.0	1.4	3.2	5.4	8.0	1.8	.1303
200.	1770.	3.2	-5.4	-3.7	0.0	4.4	9.6	8.0	.2854
300.	1469.	9.9	-23.9	-19.9	-14.4	-7.7	0.0	19.6	.4716
400.	1233.	22.5	-57.9	-52.4	-45.1	-36.3	-25.9	37.9	.6956
500.	1081.	43.2	-114.7	-107.9	-98.8	-87.7	-74.8	62.8	.9570
600.	991.	74.1	-200.3	-192.2	-181.2	-167.9	-152.4	93.1	1.2490
700.	917.	116.4	-316.1	-306.6	-293.8	-278.3	-260.3	127.1	1.5620
800.	853.	172.9	-469.3	-458.5	-443.8	-426.1	-405.5	165.8	1.9018
900.	796.	245.7	-664.1	-651.9	-635.4	-615.5	-592.3	208.7	2.2659
1000.	741.	337.9	-906.6	-893.1	-874.7	-852.6	-826.8	256.3	2.6563

BALLISTIC COEFFICIENT .22, MUZZLE VELOCITY FT/SEC 2600.

RANGE	VELOCITY	HEIGHT	100 YD	150 YD	200 YD	250 YD	300 YD	DRIFT	TIME
0.	2600.	0.0	-1.5	-1.5	-1.5	-1.5	-1.5	0.0	0.0000
100.	2209.	.0	0.0	1.2	2.9	4.9	7.2	1.7	.1251
200.	1852.	2.8	-5.7	-3.3	0.0	4.0	8.7	7.5	.2735
300.	1540.	9.0	-21.7	-18.1	-13.1	-7.1	0.0	18.5	.4513
400.	1285.	20.5	-52.5	-47.7	-41.0	-33.0	-23.6	35.9	.6654
500.	1111.	39.7	-104.6	-98.6	-90.3	-80.2	-68.5	60.0	.9181
600.	1011.	68.8	-184.5	-177.3	-167.4	-155.3	-141.2	90.0	1.2037
700.	934.	108.8	-293.8	-285.3	-273.7	-259.6	-243.2	123.8	1.5110
800.	868.	162.5	-438.6	-429.0	-415.7	-399.7	-380.8	162.2	1.8445
900.	810.	232.1	-623.8	-613.0	-598.1	-580.0	-558.8	204.9	2.2025
1000.	755.	320.3	-854.9	-842.9	-826.3	-806.2	-782.7	252.1	2.5863

BALLISTIC COEFFICIENT .22, MUZZLE VELOCITY FT/SEC 2700.

RANGE	VELOCITY	HEIGHT	100 YD	150 YD	200 YD	250 YD	300 YD	DRIFT	TIME
0.	2700.	0.0	-1.5	-1.5	-1.5	-1.5	-1.5	0.0	0.0000
100.	2301.	-.1	0.0	1.1	2.6	4.4	6.6	1.6	.1204
200.	1935.	2.6	-5.2	-3.0	0.0	3.7	8.0	7.1	.2626
300.	1611.	8.2	-19.7	-16.5	-11.9	-6.5	0.0	17.5	.4326
400.	1339.	18.7	-47.7	-43.5	-37.4	-30.1	-21.5	34.0	.6374
500.	1144.	36.5	-95.4	-90.1	-82.5	-73.4	-62.6	57.2	.8808
600.	1032.	63.6	-169.4	-163.0	-153.9	-142.9	-130.0	86.5	1.1582
700.	952.	101.7	-272.8	-265.4	-254.8	-242.0	-226.9	120.3	1.4612
800.	883.	152.8	-409.9	-401.4	-389.3	-374.6	-357.4	158.4	1.7886
900.	824.	219.2	-585.9	-576.3	-562.7	-546.2	-526.9	200.7	2.1406
1000.	768.	303.6	-806.4	-795.8	-780.6	-762.3	-740.8	247.6	2.5181

BALLISTIC COEFFICIENT .22, MUZZLE VELOCITY FT/SEC 2800.

RANGE	VELOCITY	HEIGHT	100 YD	150 YD	200 YD	250 YD	300 YD	DRIFT	TIME
0.	2800.	0.0	-1.5	-1.5	-1.5	-1.5	-1.5	0.0	0.0000
100.	2393.	-.1	0.0	.9	2.3	4.0	6.0	1.5	.1159
200.	2018.	2.3	-4.6	-2.8	0.0	3.4	7.3	6.7	.2524
300.	1684.	7.5	-17.9	-15.1	-10.9	-5.9	0.0	16.5	.4152
400.	1397.	17.2	-43.6	-39.8	-34.3	-27.5	-19.7	32.1	.6112
500.	1182.	33.6	-87.3	-82.5	-75.6	-67.2	-57.4	54.5	.8454
600.	1054.	59.0	-155.9	-150.2	-142.0	-131.9	-120.1	83.2	1.1158
700.	969.	95.3	-254.0	-247.4	-237.7	-226.0	-212.2	117.0	1.4146
800.	899.	143.7	-383.3	-375.7	-364.7	-351.2	-335.5	154.5	1.7347
900.	837.	207.1	-550.6	-542.0	-529.7	-514.5	-496.8	196.5	2.0807
1000.	781.	287.8	-760.7	-751.2	-737.5	-720.7	-701.0	242.9	2.4518

BALLISTIC COEFFICIENT .22, MUZZLE VELOCITY FT/SEC 2900.

RANGE	VELOCITY	HEIGHT	100 YD	150 YD	200 YD	250 YD	300 YD	DRIFT	TIME
0.	2900.	0.0	-1.5	-1.5	-1.5	-1.5	-1.5	0.0	0.0000
100.	2484.	-.2	0.0	.8	2.1	3.6	5.5	1.5	.1118
200.	2102.	2.1	-4.2	-2.5	0.0	3.1	6.7	6.4	.2432
300.	1757.	6.9	-16.4	-13.8	-10.0	-5.4	0.0	15.6	.3992
400.	1458.	15.8	-39.9	-36.5	-31.4	-25.3	-18.1	30.5	.5870
500.	1226.	30.9	-80.0	-75.8	-69.5	-61.8	-52.8	51.9	.8122
600.	1077.	54.7	-143.6	-138.5	-130.9	-121.7	-110.9	80.0	1.0750
700.	988.	89.1	-235.9	-229.9	-221.1	-210.3	-197.7	113.3	1.3678
800.	914.	135.1	-358.4	-351.6	-341.5	-329.2	-314.8	150.4	1.6824
900.	851.	195.7	-517.7	-510.1	-498.7	-484.9	-468.7	192.2	2.0231
1000.	794.	273.1	-718.4	-709.9	-697.3	-681.9	-663.9	238.3	2.3884

BALLISTIC COEFFICIENT .22, MUZZLE VELOCITY FT/SEC 3000.

RANGE	VELOCITY	HEIGHT	100 YD	150 YD	200 YD	250 YD	300 YD	DRIFT	TIME
0.	3000.	0.0	-1.5	-1.5	-1.5	-1.5	-1.5	0.0	0.0000
100.	2574.	-.2	0.0	.8	1.9	3.3	5.0	1.4	.1081
200.	2186.	1.9	-3.8	-2.3	0.0	2.8	6.2	6.1	.2345
300.	1831.	6.4	-15.0	-12.7	-9.3	-5.0	0.0	14.9	.3846
400.	1522.	14.5	-36.6	-33.5	-29.0	-23.3	-16.6	29.0	.5645
500.	1271.	28.5	-73.5	-69.7	-64.0	-56.9	-48.5	49.5	.7810
600.	1103.	50.8	-132.3	-127.7	-120.9	-112.3	-102.3	76.7	1.0360
700.	1006.	83.2	-219.0	-213.7	-205.7	-195.7	-184.0	109.6	1.3226
800.	930.	127.1	-335.2	-329.2	-320.0	-308.7	-295.3	146.5	1.6321
900.	864.	185.0	-486.8	-480.0	-469.7	-456.9	-441.9	187.8	1.9671
1000.	806.	259.1	-678.5	-671.0	-659.5	-645.3	-628.6	233.5	2.3267

BALLISTIC COEFFICIENT .22, MUZZLE VELOCITY FT/SEC 3100.

RANGE	VELOCITY	HEIGHT	100 YD	150 YD	200 YD	250 YD	300 YD	DRIFT	TIME
0.	3100.	0.0	-1.5	-1.5	-1.5	-1.5	-1.5	0.0	0.0000
100.	2665.	-.2	0.0	.7	1.7	3.0	4.6	1.3	.1043
200.	2269.	1.7	-3.5	-2.1	0.0	2.6	5.7	5.8	.2264
300.	1905.	5.8	-13.7	-11.7	-8.5	-4.6	0.0	14.1	.3707
400.	1586.	13.4	-33.6	-31.0	-26.7	-21.5	-15.4	27.5	.5435
500.	1320.	26.3	-67.6	-64.2	-58.9	-52.4	-44.7	47.1	.7513
600.	1132.	47.1	-122.0	-117.9	-111.6	-103.8	-94.5	73.5	.9982
700.	1025.	77.7	-203.3	-198.6	-191.2	-182.0	-171.3	105.8	1.2785
800.	945.	119.7	-314.0	-308.6	-300.2	-289.7	-277.4	142.5	1.5840
900.	878.	174.9	-458.0	-452.0	-442.5	-430.7	-416.9	183.4	1.9130
1000.	819.	245.8	-640.8	-634.1	-623.6	-610.5	-595.1	228.6	2.2664

BALLISTIC COEFFICIENT .22, MUZZLE VELOCITY FT/SEC 3200.

RANGE	VELOCITY	HEIGHT	100 YD	150 YD	200 YD	250 YD	300 YD	DRIFT	TIME
0.	3200.	0.0	-1.5	-1.5	-1.5	-1.5	-1.5	0.0	0.0000
100.	2755.	-.3	0.0	.6	1.6	2.8	4.2	1.3	.1011
200.	2352.	1.6	-3.1	-1.9	0.0	2.4	5.3	5.5	.2190
300.	1981.	5.4	-12.6	-10.8	-7.9	-4.3	0.0	13.5	.3579
400.	1651.	12.4	-31.0	-28.6	-24.7	-19.9	-14.2	26.2	.5240
500.	1370.	24.4	-62.3	-59.3	-54.5	-48.5	-41.3	44.9	.7238
600.	1164.	43.7	-112.6	-109.0	-103.2	-95.9	-87.4	70.4	.9622
700.	1044.	72.6	-188.8	-184.6	-177.8	-169.4	-159.4	102.1	1.2366
800.	962.	112.4	-293.4	-288.6	-280.8	-271.2	-259.8	138.3	1.5357
900.	892.	165.3	-430.7	-425.3	-416.6	-405.8	-393.0	178.8	1.8598
1000.	831.	233.4	-605.8	-599.8	-590.1	-578.1	-563.9	223.7	2.2087

BALLISTIC COEFFICIENT .22, MUZZLE VELOCITY FT/SEC 3300.

RANGE	VELOCITY	HEIGHT	100 YD	150 YD	200 YD	250 YD	300 YD	DRIFT	TIME
0.	3300.	0.0	-1.5	-1.5	-1.5	-1.5	-1.5	0.0	0.0000
100.	2846.	-.3	0.0	.5	1.4	2.5	3.9	1.2	.0979
200.	2434.	1.4	-2.8	-1.8	0.0	2.2	4.9	5.3	.2119
300.	2057.	5.0	-11.6	-10.0	-7.3	-4.0	0.0	12.9	.3461
400.	1717.	11.5	-28.6	-26.5	-22.9	-18.4	-13.1	25.0	.5057
500.	1424.	22.6	-57.6	-54.9	-50.5	-44.8	-38.2	42.8	.6978
600.	1202.	40.6	-104.2	-101.0	-95.6	-88.9	-80.9	67.3	.9281
700.	1064.	67.9	-175.4	-171.7	-165.5	-157.6	-148.4	98.5	1.1961
800.	978.	105.8	-274.5	-270.2	-263.1	-254.1	-243.5	134.2	1.4896
900.	906.	156.3	-405.3	-400.5	-392.5	-382.4	-370.5	174.3	1.8086
1000.	843.	221.6	-572.6	-567.3	-558.4	-547.1	-533.9	218.8	2.1522

BALLISTIC COEFFICIENT .22, MUZZLE VELOCITY FT/SEC 3400.

RANGE	VELOCITY	HEIGHT	100 YD	150 YD	200 YD	250 YD	300 YD	DRIFT	TIME
0.	3400.	0.0	-1.5	-1.5	-1.5	-1.5	-1.5	0.0	0.0000
100.	2936.	-.3	0.0	.5	1.3	2.3	3.6	1.2	.0949
200.	2517.	1.3	-2.6	-1.6	0.0	2.1	4.5	5.1	.2054
300.	2132.	4.6	-10.7	-9.3	-6.8	-3.7	0.0	12.3	.3349
400.	1783.	10.7	-26.5	-24.6	-21.3	-17.2	-12.2	23.9	.4888
500.	1481.	21.0	-53.3	-51.0	-46.8	-41.7	-35.5	40.9	.6734
600.	1242.	37.8	-96.5	-93.7	-88.7	-82.5	-75.1	64.5	.8956
700.	1086.	63.4	-162.7	-159.4	-153.6	-146.4	-137.8	94.7	1.1556
800.	995.	99.5	-256.8	-253.1	-246.5	-238.2	-228.3	130.1	1.4451
900.	920.	147.7	-381.3	-377.1	-369.7	-360.3	-349.3	169.8	1.7587
1000.	856.	210.4	-541.5	-536.8	-528.6	-518.2	-505.9	213.9	2.0975

BALLISTIC COEFFICIENT .22, MUZZLE VELOCITY FT/SEC 3500.

RANGE	VELOCITY	HEIGHT	100 YD	150 YD	200 YD	250 YD	300 YD	DRIFT	TIME
0.	3500.	0.0	-1.5	-1.5	-1.5	-1.5	-1.5	0.0	0.0000
100.	3027.	-.3	0.0	.4	1.2	2.1	3.3	1.1	.0922
200.	2599.	1.2	-2.3	-1.5	0.0	1.9	4.2	4.9	.1993
300.	2208.	4.3	-9.9	-8.6	-6.3	-3.4	0.0	11.9	.3245
400.	1851.	10.0	-24.5	-22.9	-19.9	-16.0	-11.4	22.9	.4730
500.	1539.	19.6	-49.5	-47.4	-43.6	-38.8	-33.1	39.1	.6509
600.	1284.	35.2	-89.5	-87.0	-82.5	-76.7	-69.8	61.7	.8651
700.	1110.	59.2	-151.3	-148.4	-143.1	-136.3	-128.3	91.2	1.1179
800.	1011.	93.8	-240.9	-237.6	-231.6	-223.8	-214.6	126.3	1.4036
900.	934.	139.8	-359.1	-355.4	-348.6	-339.9	-329.6	165.4	1.7110
1000.	868.	199.9	-512.2	-508.1	-500.5	-490.8	-479.4	209.0	2.0446

BALLISTIC COEFFICIENT .22, MUZZLE VELOCITY FT/SEC 3600.

RANGE	VELOCITY	HEIGHT	100 YD	150 YD	200 YD	250 YD	300 YD	DRIFT	TIME
0.	3600.	0.0	-1.5	-1.5	-1.5	-1.5	-1.5	0.0	0.0000
100.	3118.	-.4	0.0	.4	1.1	2.0	3.0	1.1	.0896
200.	2681.	1.0	-2.1	-1.4	0.0	1.8	4.0	4.7	.1932
300.	2283.	4.0	-9.1	-8.0	-5.9	-3.2	0.0	11.4	.3146
400.	1919.	9.3	-22.8	-21.4	-18.6	-14.9	-10.7	21.9	.4580
500.	1597.	18.3	-46.0	-44.2	-40.7	-36.2	-30.8	37.5	.6295
600.	1328.	32.8	-83.1	-81.0	-76.8	-71.4	-65.0	59.1	.8359
700.	1137.	55.4	-140.8	-138.3	-133.4	-127.1	-119.6	87.7	1.0818
800.	1028.	88.2	-225.2	-222.4	-216.8	-209.5	-201.0	122.2	1.3610
900.	949.	132.2	-338.0	-334.9	-328.6	-320.5	-310.8	160.9	1.6641
1000.	880.	189.8	-484.3	-480.8	-473.8	-464.8	-454.0	204.0	1.9924

BALLISTIC COEFFICIENT .22, MUZZLE VELOCITY FT/SEC 3700.

RANGE	VELOCITY	HEIGHT	100 YD	150 YD	200 YD	250 YD	300 YD	DRIFT	TIME
0.	3700.	0.0	-1.5	-1.5	-1.5	-1.5	-1.5	0.0	0.0000
100.	3208.	-.4	0.0	.3	1.0	1.8	2.8	1.1	.0872
200.	2763.	.9	-1.9	-1.3	0.0	1.7	3.7	4.6	.1880
300.	2359.	3.7	-8.4	-7.5	-5.5	-3.0	0.0	11.0	.3056
400.	1987.	8.7	-21.2	-19.9	-17.3	-14.0	-9.9	21.1	.4441
500.	1657.	17.1	-42.8	-41.3	-38.0	-33.8	-28.8	35.9	.6096
600.	1375.	30.7	-77.4	-75.6	-71.7	-66.6	-60.6	56.7	.8088
700.	1167.	51.8	-131.0	-128.9	-124.3	-118.4	-111.4	84.3	1.0465
800.	1045.	83.0	-210.7	-208.3	-203.1	-196.4	-188.3	118.3	1.3206
900.	963.	125.1	-318.5	-315.7	-309.9	-302.3	-293.2	156.5	1.6192
1000.	893.	180.4	-458.4	-455.3	-448.9	-440.4	-430.4	199.2	1.9426

BALLISTIC COEFFICIENT .22, MUZZLE VELOCITY FT/SEC 3800.

RANGE	VELOCITY	HEIGHT	100 YD	150 YD	200 YD	250 YD	300 YD	DRIFT	TIME
0.	3800.	0.0	-1.5	-1.5	-1.5	-1.5	-1.5	0.0	0.0000
100.	3299.	-.4	0.0	.3	.9	1.7	2.6	1.0	.0848
200.	2845.	.9	-1.7	-1.2	0.0	1.6	3.5	4.4	.1827
300.	2434.	3.5	-7.8	-7.0	-5.2	-2.8	0.0	10.6	.2968
400.	2056.	8.2	-19.7	-18.7	-16.3	-13.1	-9.4	20.3	.4310
500.	1717.	16.0	-39.9	-38.6	-35.6	-31.6	-26.9	34.5	.5906
600.	1424.	28.7	-72.2	-70.6	-67.0	-62.2	-56.6	54.4	.7828
700.	1201.	48.5	-122.2	-120.4	-116.2	-110.7	-104.1	81.0	1.0131
800.	1064.	78.0	-197.2	-195.1	-190.3	-184.0	-176.5	114.3	1.2811
900.	978.	118.3	-299.8	-297.4	-292.0	-284.9	-276.4	152.1	1.5747
1000.	906.	171.4	-433.8	-431.2	-425.2	-417.3	-407.9	194.3	1.8937

BALLISTIC COEFFICIENT .22, MUZZLE VELOCITY FT/SEC 3900.

RANGE	VELOCITY	HEIGHT	100 YD	150 YD	200 YD	250 YD	300 YD	DRIFT	TIME
0.	3900.	0.0	-1.5	-1.5	-1.5	-1.5	-1.5	0.0	0.0000
100.	3390.	-.4	0.0	.2	.8	1.5	2.4	1.0	.0824
200.	2927.	.8	-1.6	-1.1	0.0	1.5	3.2	4.2	.1777
300.	2509.	3.2	-7.2	-6.5	-4.9	-2.7	0.0	10.2	.2885
400.	2125.	7.6	-18.4	-17.4	-15.2	-12.3	-8.8	19.5	.4183
500.	1777.	15.0	-37.3	-36.1	-33.4	-29.7	-25.3	33.1	.5729
600.	1475.	26.8	-67.4	-66.0	-62.7	-58.3	-53.0	52.2	.7582
700.	1238.	45.5	-114.2	-112.6	-108.7	-103.6	-97.4	77.9	.9813
800.	1084.	73.3	-184.2	-182.4	-178.0	-172.1	-165.1	110.3	1.2418
900.	992.	112.2	-283.1	-281.0	-276.1	-269.4	-261.5	148.1	1.5338
1000.	918.	162.8	-410.5	-408.3	-402.7	-395.4	-386.5	189.5	1.8459

BALLISTIC COEFFICIENT .22, MUZZLE VELOCITY FT/SEC 4000.

RANGE	VELOCITY	HEIGHT	100 YD	150 YD	200 YD	250 YD	300 YD	DRIFT	TIME
0.	4000.	0.0	-1.5	-1.5	-1.5	-1.5	-1.5	0.0	0.0000
100.	3481.	-.4	0.0	.2	.7	1.4	2.2	1.0	.0805
200.	3010.	.7	-1.4	-1.0	0.0	1.4	3.0	4.1	.1731
300.	2583.	3.0	-6.7	-6.1	-4.6	-2.5	0.0	9.8	.2808
400.	2194.	7.2	-17.1	-16.4	-14.3	-11.6	-8.2	18.8	.4068
500.	1838.	14.1	-34.9	-33.9	-31.4	-27.9	-23.8	31.9	.5564
600.	1528.	25.2	-63.1	-61.9	-58.9	-54.7	-49.7	50.2	.7354
700.	1276.	42.7	-106.8	-105.4	-101.9	-97.0	-91.2	75.0	.9512
800.	1106.	69.0	-172.6	-171.1	-167.0	-161.5	-154.8	106.5	1.2052
900.	1008.	106.0	-266.3	-264.6	-260.1	-253.8	-246.3	143.7	1.4915
1000.	931.	154.8	-388.8	-386.9	-381.8	-374.9	-366.6	184.8	1.8002

BALLISTIC COEFFICIENT .23, MUZZLE VELOCITY FT/SEC 2000.

RANGE	VELOCITY	HEIGHT	100 YD	150 YD	200 YD	250 YD	300 YD	DRIFT	TIME
0.	2000.	0.0	-1.5	-1.5	-1.5	-1.5	-1.5	0.0	0.0000
100.	1681.	.5	0.0	2.4	5.5	9.1	13.2	2.4	.1637
200.	1406.	5.4	-11.0	-6.1	0.0	7.2	15.5	10.4	.3590
300.	1196.	16.0	-39.7	-32.4	-23.2	-12.4	0.0	24.9	.5914
400.	1065.	34.7	-92.9	-83.2	-71.0	-56.6	-40.0	45.7	.8599
500.	983.	63.0	-174.5	-162.4	-147.2	-129.2	-108.4	70.8	1.1524
600.	913.	102.8	-289.1	-274.6	-256.2	-234.6	-209.8	100.1	1.4690
700.	852.	156.4	-441.8	-424.8	-403.4	-378.2	-349.2	133.7	1.8097
800.	797.	226.0	-637.3	-617.9	-593.5	-564.7	-531.5	171.4	2.1737
900.	745.	314.6	-881.8	-859.9	-832.5	-800.1	-762.8	213.5	2.5631
1000.	695.	425.6	-1182.7	-1158.4	-1127.9	-1091.9	-1050.4	260.6	2.9805

BALLISTIC COEFFICIENT .23, MUZZLE VELOCITY FT/SEC 2100.

RANGE	VELOCITY	HEIGHT	100 YD	150 YD	200 YD	250 YD	300 YD	DRIFT	TIME
0.	2100.	0.0	-1.5	-1.5	-1.5	-1.5	-1.5	0.0	0.0000
100.	1769.	.4	0.0	2.1	4.9	8.1	11.8	2.3	.1556
200.	1481.	4.8	-9.7	-5.5	0.0	6.5	13.9	9.7	.3411
300.	1250.	14.4	-35.5	-29.1	-20.9	-11.2	0.0	23.6	.5625
400.	1096.	31.5	-83.6	-75.0	-64.1	-51.1	-36.2	43.8	.8203
500.	1005.	58.1	-159.8	-149.1	-135.5	-119.2	-100.6	69.2	1.1077
600.	932.	95.7	-266.9	-254.1	-237.7	-218.2	-195.9	98.6	1.4172
700.	869.	146.4	-410.4	-395.4	-376.3	-353.6	-327.5	132.2	1.7510
800.	813.	212.6	-595.1	-578.0	-556.2	-530.2	-500.4	169.9	2.1082
900.	760.	296.8	-826.2	-806.9	-782.4	-753.2	-719.7	211.9	2.4896
1000.	709.	402.7	-1111.8	-1090.4	-1063.1	-1030.6	-993.4	258.8	2.8991

BALLISTIC COEFFICIENT .23, MUZZLE VELOCITY FT/SEC 2200.

RANGE	VELOCITY	HEIGHT	100 YD	150 YD	200 YD	250 YD	300 YD	DRIFT	TIME
0.	2200.	0.0	-1.5	-1.5	-1.5	-1.5	-1.5	0.0	0.0000
100.	1858.	.3	0.0	1.9	4.4	7.3	10.6	2.1	.1484
200.	1558.	4.3	-8.7	-4.9	0.0	5.8	12.5	9.2	.3250
300.	1308.	13.0	-31.9	-26.2	-18.8	-10.1	0.0	22.3	.5356
400.	1130.	28.7	-75.5	-67.9	-58.0	-46.5	-33.0	41.9	.7836
500.	1028.	53.6	-146.1	-136.6	-124.3	-109.8	-93.0	67.3	1.0641
600.	952.	88.9	-246.2	-234.8	-220.0	-202.7	-182.4	96.5	1.3667
700.	886.	137.0	-381.1	-367.8	-350.6	-330.3	-306.7	130.1	1.6938
800.	828.	199.9	-555.6	-540.4	-520.8	-497.6	-470.6	167.8	2.0444
900.	775.	280.1	-774.8	-757.8	-735.6	-709.6	-679.2	209.7	2.4190
1000.	723.	380.9	-1045.3	-1026.3	-1001.7	-972.8	-939.1	256.3	2.8199

BALLISTIC COEFFICIENT .23, MUZZLE VELOCITY FT/SEC 2300.

RANGE	VELOCITY	HEIGHT	100 YD	150 YD	200 YD	250 YD	300 YD	DRIFT	TIME
0.	2300.	0.0	-1.5	-1.5	-1.5	-1.5	-1.5	0.0	0.0000
100.	1949.	.2	0.0	1.7	3.9	6.5	9.5	2.0	.1417
200.	1637.	3.9	-7.8	-4.4	0.0	5.2	11.3	8.6	.3098
300.	1370.	11.8	-28.6	-23.6	-17.0	-9.1	0.0	21.0	.5105
400.	1171.	26.1	-68.1	-61.4	-52.6	-42.2	-30.0	39.9	.7483
500.	1052.	49.2	-132.8	-124.4	-113.4	-100.4	-85.1	64.7	1.0200
600.	971.	82.8	-227.5	-217.4	-204.2	-188.5	-170.3	94.4	1.3192
700.	903.	128.0	-353.5	-341.8	-326.4	-308.1	-286.7	127.5	1.6376
800.	843.	187.8	-518.4	-505.0	-487.4	-466.4	-442.1	165.1	1.9818
900.	789.	264.3	-726.4	-711.3	-691.5	-668.0	-640.5	207.0	2.3498
1000.	737.	360.4	-983.2	-966.4	-944.4	-918.3	-887.8	253.2	2.7430

BALLISTIC COEFFICIENT .23, MUZZLE VELOCITY FT/SEC 2400.

RANGE	VELOCITY	HEIGHT	100 YD	150 YD	200 YD	250 YD	300 YD	DRIFT	TIME
0.	2400.	0.0	-1.5	-1.5	-1.5	-1.5	-1.5	0.0	0.0000
100.	2041.	.1	0.0	1.5	3.5	5.9	8.6	1.9	.1356
200.	1717.	3.4	-6.9	-4.0	0.0	4.8	10.3	8.1	.2958
300.	1436.	10.6	-25.8	-21.4	-15.4	-8.2	0.0	19.7	.4872
400.	1218.	23.8	-61.7	-55.7	-47.8	-38.2	-27.3	37.8	.7148
500.	1077.	45.2	-121.0	-113.6	-103.7	-91.7	-78.0	62.2	.9782
600.	991.	76.8	-209.5	-200.6	-188.7	-174.4	-157.9	91.7	1.2710
700.	920.	119.5	-328.0	-317.6	-303.7	-287.0	-267.8	124.6	1.5831
800.	859.	176.4	-483.7	-471.9	-456.0	-436.9	-414.9	162.1	1.9209
900.	804.	249.3	-681.0	-667.7	-649.8	-628.3	-603.6	203.7	2.2823
1000.	751.	341.1	-925.4	-910.5	-890.7	-866.8	-839.4	249.7	2.6688

BALLISTIC COEFFICIENT .23, MUZZLE VELOCITY FT/SEC 2500.

RANGE	VELOCITY	HEIGHT	100 YD	150 YD	200 YD	250 YD	300 YD	DRIFT	TIME
0.	2500.	0.0	-1.5	-1.5	-1.5	-1.5	-1.5	0.0	0.0000
100.	2133.	.1	0.0	1.3	3.1	5.3	7.8	1.7	.1299
200.	1798.	3.1	-6.2	-3.6	0.0	4.3	9.3	7.6	.2831
300.	1506.	9.7	-23.3	-19.3	-14.0	-7.5	0.0	18.6	.4655
400.	1269.	21.7	-55.9	-50.6	-43.4	-34.8	-24.8	35.8	.6833
500.	1107.	41.5	-110.3	-103.6	-94.7	-83.9	-71.4	59.5	.9380
600.	1012.	71.2	-192.7	-184.8	-174.0	-161.1	-146.1	88.7	1.2241
700.	938.	111.6	-304.3	-295.0	-282.4	-267.3	-249.9	121.5	1.5302
800.	874.	165.9	-452.2	-441.6	-427.3	-410.1	-390.1	159.0	1.8634
900.	818.	235.1	-638.5	-626.5	-610.4	-591.0	-568.5	200.0	2.2166
1000.	765.	322.8	-870.9	-857.6	-839.7	-818.2	-793.2	245.7	2.5962

BALLISTIC COEFFICIENT .23, MUZZLE VELOCITY FT/SEC 2600.

RANGE	VELOCITY	HEIGHT	100 YD	150 YD	200 YD	250 YD	300 YD	DRIFT	TIME
0.	2600.	0.0	-1.5	-1.5	-1.5	-1.5	-1.5	0.0	0.0000
100.	2226.	-.0	0.0	1.2	2.8	4.8	7.0	1.6	.1247
200.	1881.	2.8	-5.6	-3.2	0.0	3.9	8.5	7.1	.2713
300.	1578.	8.8	-21.1	-17.6	-12.7	-6.9	0.0	17.5	.4457
400.	1323.	19.8	-50.7	-46.0	-39.5	-31.7	-22.5	33.8	.6537
500.	1140.	38.0	-100.4	-94.5	-86.4	-76.6	-65.2	56.7	.8991
600.	1034.	65.7	-176.5	-169.4	-159.7	-147.9	-134.2	85.2	1.1766
700.	957.	104.2	-282.0	-273.8	-262.4	-248.7	-232.7	118.1	1.4785
800.	890.	155.5	-421.2	-411.7	-398.7	-383.1	-364.8	155.1	1.8041
900.	832.	221.7	-598.8	-588.1	-573.5	-555.9	-535.3	196.2	2.1530
1000.	778.	305.4	-819.8	-807.9	-791.7	-772.2	-749.3	241.4	2.5256

BALLISTIC COEFFICIENT .23, MUZZLE VELOCITY FT/SEC 2700.

RANGE	VELOCITY	HEIGHT	100 YD	150 YD	200 YD	250 YD	300 YD	DRIFT	TIME
0.	2700.	0.0	-1.5	-1.5	-1.5	-1.5	-1.5	0.0	0.0000
100.	2318.	-.1	0.0	1.1	2.5	4.3	6.4	1.6	.1199
200.	1965.	2.5	-5.1	-3.0	0.0	3.6	7.7	6.7	.2605
300.	1651.	8.0	-19.2	-16.0	-11.6	-6.3	0.0	16.5	.4272
400.	1381.	18.1	-46.2	-42.0	-36.1	-28.9	-20.6	32.0	.6261
500.	1179.	34.9	-91.7	-86.4	-79.0	-70.1	-59.7	54.0	.8623
600.	1056.	60.8	-162.1	-155.8	-146.9	-136.2	-123.7	82.0	1.1324
700.	975.	97.2	-261.4	-254.1	-243.7	-231.2	-216.7	114.5	1.4285
800.	906.	145.9	-392.8	-384.4	-372.6	-358.3	-341.7	151.2	1.7479
900.	846.	209.1	-561.4	-552.0	-538.7	-522.6	-503.9	192.0	2.0909
1000.	792.	289.2	-772.2	-761.7	-746.9	-729.1	-708.2	237.0	2.4577

BALLISTIC COEFFICIENT .23, MUZZLE VELOCITY FT/SEC 2800.

RANGE	VELOCITY	HEIGHT	100 YD	150 YD	200 YD	250 YD	300 YD	DRIFT	TIME
0.	2800.	0.0	-1.5	-1.5	-1.5	-1.5	-1.5	0.0	0.0000
100.	2410.	-.1	0.0	.9	2.3	3.9	5.8	1.5	.1155
200.	2050.	2.3	-4.6	-2.7	0.0	3.3	7.1	6.4	.2505
300.	1725.	7.3	-17.5	-14.7	-10.6	-5.7	0.0	15.6	.4100
400.	1442.	16.6	-42.2	-38.4	-33.0	-26.5	-18.9	30.3	.6005
500.	1223.	32.1	-83.8	-79.1	-72.4	-64.2	-54.7	51.3	.8271
600.	1080.	56.2	-148.9	-143.3	-135.2	-125.4	-114.0	78.6	1.0897
700.	993.	91.0	-243.1	-236.6	-227.1	-215.7	-202.4	111.3	1.3822
800.	922.	136.9	-366.4	-359.0	-348.2	-335.2	-319.9	147.2	1.6934
900.	860.	197.2	-526.4	-518.1	-505.9	-491.3	-474.1	187.7	2.0305
1000.	805.	273.8	-727.4	-718.0	-704.5	-688.3	-669.2	232.3	2.3914

BALLISTIC COEFFICIENT .23, MUZZLE VELOCITY FT/SEC 2900.

RANGE	VELOCITY	HEIGHT	100 YD	150 YD	200 YD	250 YD	300 YD	DRIFT	TIME
0.	2900.	0.0	-1.5	-1.5	-1.5	-1.5	-1.5	0.0	0.0000
100.	2501.	-.2	0.0	.8	2.1	3.6	5.3	1.4	.1115
200.	2134.	2.0	-4.1	-2.5	0.0	3.0	6.5	6.1	.2413
300.	1799.	6.7	-15.9	-13.4	-9.8	-5.3	0.0	14.8	.3944
400.	1507.	15.2	-38.6	-35.2	-30.3	-24.3	-17.3	28.7	.5767
500.	1270.	29.5	-76.8	-72.6	-66.5	-59.0	-50.2	48.8	.7944
600.	1107.	52.1	-136.9	-131.9	-124.5	-115.5	-105.0	75.4	1.0490
700.	1012.	84.8	-225.2	-219.4	-210.7	-200.2	-188.0	107.5	1.3351
800.	938.	128.5	-342.0	-335.4	-325.5	-313.5	-299.5	143.2	1.6411
900.	874.	186.3	-494.9	-487.4	-476.3	-462.8	-447.1	183.6	1.9742
1000.	818.	259.2	-685.4	-677.1	-664.8	-649.8	-632.3	227.5	2.3274

BALLISTIC COEFFICIENT .23, MUZZLE VELOCITY FT/SEC 3000.

RANGE	VELOCITY	HEIGHT	100 YD	150 YD	200 YD	250 YD	300 YD	DRIFT	TIME
0.	3000.	0.0	-1.5	-1.5	-1.5	-1.5	-1.5	0.0	0.0000
100.	2592.	-.2	0.0	.7	1.9	3.3	4.9	1.4	.1077
200.	2218.	1.9	-3.7	-2.2	0.0	2.8	6.0	5.8	.2328
300.	1874.	6.2	-14.6	-12.4	-9.0	-4.8	0.0	14.1	.3799
400.	1572.	14.0	-35.4	-32.5	-28.0	-22.4	-16.0	27.3	.5550
500.	1319.	27.3	-70.5	-66.8	-61.2	-54.2	-46.2	46.4	.7638
600.	1137.	48.3	-126.1	-121.6	-114.9	-106.6	-96.9	72.2	1.0105
700.	1032.	79.1	-208.6	-203.4	-195.5	-185.8	-174.6	103.8	1.2897
800.	955.	120.7	-319.3	-313.3	-304.3	-293.2	-280.4	139.1	1.5904
900.	889.	175.5	-463.7	-457.1	-447.0	-434.5	-420.0	178.9	1.9164
1000.	831.	245.7	-646.4	-638.9	-627.7	-613.8	-597.8	222.8	2.2658

BALLISTIC COEFFICIENT .23, MUZZLE VELOCITY FT/SEC 3100.

RANGE	VELOCITY	HEIGHT	100 YD	150 YD	200 YD	250 YD	300 YD	DRIFT	TIME
0.	3100.	0.0	-1.5	-1.5	-1.5	-1.5	-1.5	0.0	0.0000
100.	2683.	-.2	0.0	.7	1.7	3.0	4.5	1.3	.1039
200.	2302.	1.7	-3.4	-2.1	0.0	2.5	5.5	5.5	.2248
300.	1951.	5.7	-13.4	-11.4	-8.3	-4.5	0.0	13.4	.3664
400.	1638.	13.0	-32.6	-29.9	-25.8	-20.7	-14.7	25.9	.5342
500.	1371.	25.2	-64.9	-61.6	-56.4	-50.0	-42.6	44.2	.7348
600.	1172.	44.6	-116.1	-112.1	-105.9	-98.3	-89.3	68.9	.9724
700.	1052.	73.5	-192.7	-188.1	-180.8	-171.9	-161.5	99.7	1.2440
800.	971.	113.5	-298.9	-293.6	-285.3	-275.1	-263.2	135.3	1.5431
900.	903.	165.6	-435.2	-429.3	-420.0	-408.5	-395.1	174.3	1.8614
1000.	844.	232.7	-609.4	-602.8	-592.5	-579.8	-564.8	217.8	2.2054

BALLISTIC COEFFICIENT .23, MUZZLE VELOCITY FT/SEC 3200.

RANGE	VELOCITY	HEIGHT	100 YD	150 YD	200 YD	250 YD	300 YD	DRIFT	TIME
0.	3200.	0.0	-1.5	-1.5	-1.5	-1.5	-1.5	0.0	0.0000
100.	2774.	-.3	0.0	.6	1.5	2.7	4.1	1.2	.1007
200.	2385.	1.5	-3.1	-1.9	0.0	2.4	5.1	5.3	.2174
300.	2027.	5.3	-12.3	-10.5	-7.7	-4.2	0.0	12.8	.3539
400.	1705.	12.0	-30.0	-27.7	-23.9	-19.2	-13.6	24.7	.5152
500.	1426.	23.3	-59.8	-56.9	-52.1	-46.3	-39.3	42.1	.7079
600.	1211.	41.4	-107.2	-103.6	-97.9	-90.9	-82.6	65.9	.9370
700.	1074.	68.6	-178.6	-174.5	-167.8	-159.6	-149.9	96.0	1.2017
800.	988.	106.5	-278.8	-274.1	-266.5	-257.1	-246.0	131.1	1.4950
900.	918.	156.2	-408.6	-403.4	-394.8	-384.2	-371.7	169.8	1.8084
1000.	856.	220.6	-574.9	-569.1	-559.6	-547.8	-534.0	212.9	2.1473

BALLISTIC COEFFICIENT .23, MUZZLE VELOCITY FT/SEC 3300.

RANGE	VELOCITY	HEIGHT	100 YD	150 YD	200 YD	250 YD	300 YD	DRIFT	TIME
0.	3300.	0.0	-1.5	-1.5	-1.5	-1.5	-1.5	0.0	0.0000
100.	2864.	-.3	0.0	.5	1.4	2.5	3.8	1.2	.0976
200.	2469.	1.4	-2.8	-1.7	0.0	2.2	4.8	5.0	.2105
300.	2104.	4.9	-11.3	-9.8	-7.1	-3.9	0.0	12.2	.3422
400.	1773.	11.1	-27.7	-25.7	-22.2	-17.8	-12.6	23.6	.4976
500.	1484.	21.6	-55.3	-52.7	-48.3	-42.9	-36.4	40.1	.6826
600.	1253.	38.4	-99.1	-96.0	-90.7	-84.2	-76.5	63.0	.9035
700.	1097.	64.0	-165.6	-161.9	-155.8	-148.1	-139.1	92.3	1.1610
800.	1006.	99.9	-260.1	-255.9	-248.9	-240.2	-229.9	126.9	1.4483
900.	933.	147.5	-383.9	-379.2	-371.4	-361.5	-350.0	165.3	1.7574
1000.	869.	209.1	-542.6	-537.4	-528.7	-517.7	-504.9	208.0	2.0909

BALLISTIC COEFFICIENT .23, MUZZLE VELOCITY FT/SEC 3400.

RANGE	VELOCITY	HEIGHT	100 YD	150 YD	200 YD	250 YD	300 YD	DRIFT	TIME
0.	3400.	0.0	-1.5	-1.5	-1.5	-1.5	-1.5	0.0	0.0000
100.	2955.	-.3	0.0	.5	1.3	2.3	3.5	1.1	.0945
200.	2551.	1.2	-2.5	-1.6	0.0	2.0	4.4	4.8	.2039
300.	2181.	4.5	-10.4	-9.0	-6.6	-3.6	0.0	11.7	.3310
400.	1841.	10.4	-25.7	-23.8	-20.6	-16.5	-11.8	22.5	.4810
500.	1543.	20.1	-51.3	-49.0	-44.9	-39.8	-33.9	38.3	.6590
600.	1296.	35.7	-91.8	-89.0	-84.2	-78.1	-71.0	60.3	.8719
700.	1123.	59.6	-153.6	-150.4	-144.7	-137.6	-129.3	88.7	1.1217
800.	1023.	93.7	-242.6	-238.9	-232.5	-224.3	-214.8	122.7	1.4030
900.	948.	139.2	-360.6	-356.5	-349.2	-340.0	-329.4	160.8	1.7075
1000.	882.	198.2	-512.0	-507.4	-499.3	-489.2	-477.3	203.0	2.0357

BALLISTIC COEFFICIENT .23, MUZZLE VELOCITY FT/SEC 3500.

RANGE	VELOCITY	HEIGHT	100 YD	150 YD	200 YD	250 YD	300 YD	DRIFT	TIME
0.	3500.	0.0	-1.5	-1.5	-1.5	-1.5	-1.5	0.0	0.0000
100.	3046.	-.3	0.0	.4	1.1	2.1	3.2	1.1	.0919
200.	2634.	1.1	-2.3	-1.5	0.0	1.9	4.1	4.7	.1979
300.	2257.	4.2	-9.6	-8.4	-6.2	-3.4	0.0	11.2	.3210
400.	1910.	9.7	-23.8	-22.2	-19.2	-15.4	-11.0	21.6	.4655
500.	1603.	18.7	-47.6	-45.6	-41.9	-37.1	-31.5	36.7	.6371
600.	1343.	33.3	-85.2	-82.8	-78.3	-72.6	-65.9	57.7	.8421
700.	1153.	55.7	-142.6	-139.8	-134.6	-128.0	-120.2	85.2	1.0842
800.	1041.	88.0	-226.3	-223.1	-217.2	-209.6	-200.7	118.6	1.3594
900.	963.	131.5	-338.9	-335.3	-328.6	-320.1	-310.0	156.3	1.6596
1000.	895.	188.1	-483.9	-479.9	-472.5	-463.0	-451.9	198.3	1.9837

BALLISTIC COEFFICIENT .23, MUZZLE VELOCITY FT/SEC 3600.

RANGE	VELOCITY	HEIGHT	100 YD	150 YD	200 YD	250 YD	300 YD	DRIFT	TIME
0.	3600.	0.0	-1.5	-1.5	-1.5	-1.5	-1.5	0.0	0.0000
100.	3138.	-.4	0.0	.3	1.0	1.9	3.0	1.0	.0893
200.	2717.	1.0	-2.1	-1.4	0.0	1.8	3.8	4.5	.1920
300.	2333.	3.9	-8.9	-7.8	-5.8	-3.1	0.0	10.8	.3112
400.	1979.	9.0	-22.0	-20.7	-17.9	-14.4	-10.2	20.6	.4506
500.	1663.	17.5	-44.2	-42.5	-39.1	-34.6	-29.4	35.1	.6162
600.	1391.	31.0	-79.1	-77.0	-72.9	-67.6	-61.4	55.2	.8137
700.	1186.	52.0	-132.6	-130.2	-125.4	-119.2	-111.9	81.8	1.0481
800.	1060.	82.5	-211.1	-208.3	-202.9	-195.8	-187.5	114.5	1.3170
900.	978.	124.0	-318.1	-315.0	-308.9	-300.9	-291.5	151.7	1.6122
1000.	909.	178.2	-456.3	-452.9	-446.0	-437.2	-426.8	193.1	1.9306

BALLISTIC COEFFICIENT .23, MUZZLE VELOCITY FT/SEC 3700.

RANGE	VELOCITY	HEIGHT	100 YD	150 YD	200 YD	250 YD	300 YD	DRIFT	TIME
0.	3700.	0.0	-1.5	-1.5	-1.5	-1.5	-1.5	0.0	0.0000
100.	3229.	-.4	0.0	.3	.9	1.8	2.7	1.0	.0869
200.	2800.	.9	-1.9	-1.3	0.0	1.6	3.6	4.3	.1868
300.	2409.	3.6	-8.2	-7.3	-5.4	-2.9	0.0	10.4	.3023
400.	2049.	8.4	-20.5	-19.3	-16.8	-13.5	-9.6	19.9	.4374
500.	1724.	16.3	-41.2	-39.7	-36.5	-32.4	-27.5	33.7	.5969
600.	1442.	29.0	-73.7	-71.9	-68.1	-63.2	-57.3	53.0	.7874
700.	1223.	48.6	-123.5	-121.4	-117.0	-111.2	-104.4	78.6	1.0140
800.	1080.	77.5	-197.2	-194.8	-189.7	-183.2	-175.3	110.5	1.2766
900.	993.	117.4	-299.9	-297.2	-291.4	-284.1	-275.3	147.7	1.5692
1000.	922.	169.0	-430.9	-428.0	-421.6	-413.4	-403.6	188.2	1.8804

BALLISTIC COEFFICIENT .23, MUZZLE VELOCITY FT/SEC 3800.

RANGE	VELOCITY	HEIGHT	100 YD	150 YD	200 YD	250 YD	300 YD	DRIFT	TIME
0.	3800.	0.0	-1.5	-1.5	-1.5	-1.5	-1.5	0.0	0.0000
100.	3320.	-.4	0.0	.3	.8	1.6	2.5	1.0	.0846
200.	2883.	.8	-1.7	-1.2	0.0	1.5	3.4	4.2	.1815
300.	2485.	3.4	-7.6	-6.8	-5.1	-2.7	0.0	10.0	.2937
400.	2119.	7.9	-19.1	-18.1	-15.7	-12.7	-9.0	19.1	.4243
500.	1786.	15.3	-38.4	-37.1	-34.2	-30.4	-25.8	32.4	.5787
600.	1496.	27.1	-68.8	-67.2	-63.7	-59.1	-53.6	50.8	.7623
700.	1261.	45.5	-115.2	-113.4	-109.3	-104.0	-97.5	75.5	.9815
800.	1102.	72.8	-184.3	-182.2	-177.6	-171.4	-164.1	106.6	1.2375
900.	1009.	110.8	-281.5	-279.1	-273.9	-267.0	-258.8	143.2	1.5242
1000.	936.	160.3	-407.2	-404.6	-398.8	-391.1	-381.9	183.4	1.8317

BALLISTIC COEFFICIENT .23, MUZZLE VELOCITY FT/SEC 3900.

RANGE	VELOCITY	HEIGHT	100 YD	150 YD	200 YD	250 YD	300 YD	DRIFT	TIME
0.	3900.	0.0	-1.5	-1.5	-1.5	-1.5	-1.5	0.0	0.0000
100.	3412.	-.4	0.0	.2	.8	1.5	2.3	.9	.0821
200.	2966.	.7	-1.5	-1.1	0.0	1.4	3.2	4.0	.1764
300.	2561.	3.2	-7.0	-6.4	-4.7	-2.6	0.0	9.6	.2854
400.	2189.	7.4	-17.8	-16.9	-14.8	-11.9	-8.5	18.4	.4120
500.	1849.	14.4	-35.9	-34.8	-32.1	-28.5	-24.2	31.1	.5613
600.	1550.	25.4	-64.3	-63.0	-59.7	-55.4	-50.3	48.8	.7387
700.	1302.	42.6	-107.6	-106.1	-102.3	-97.2	-91.2	72.5	.9504
800.	1127.	68.3	-172.3	-170.6	-166.2	-160.5	-153.6	102.8	1.1995
900.	1025.	104.5	-264.2	-262.3	-257.4	-250.9	-243.2	138.7	1.4804
1000.	949.	152.0	-384.6	-382.5	-377.1	-369.8	-361.3	178.6	1.7840

BALLISTIC COEFFICIENT .23, MUZZLE VELOCITY FT/SEC 4000.

RANGE	VELOCITY	HEIGHT	100 YD	150 YD	200 YD	250 YD	300 YD	DRIFT	TIME
0.	4000.	0.0	-1.5	-1.5	-1.5	-1.5	-1.5	0.0	0.0000
100.	3503.	-.4	0.0	.2	.7	1.4	2.2	.9	.0802
200.	3049.	.7	-1.4	-1.0	0.0	1.3	3.0	3.9	.1720
300.	2637.	3.0	-6.5	-5.9	-4.4	-2.4	0.0	9.3	.2779
400.	2259.	7.0	-16.6	-15.9	-13.9	-11.2	-8.0	17.8	.4009
500.	1912.	13.5	-33.6	-32.7	-30.2	-26.8	-22.8	30.0	.5453
600.	1605.	23.9	-60.2	-59.1	-56.1	-52.1	-47.2	46.9	.7166
700.	1344.	40.0	-100.7	-99.4	-95.9	-91.2	-85.5	69.8	.9214
800.	1153.	64.3	-161.5	-160.1	-156.1	-150.7	-144.2	99.3	1.1644
900.	1042.	98.6	-248.1	-246.5	-242.0	-235.9	-228.6	134.4	1.4384
1000.	963.	144.3	-363.6	-361.8	-356.8	-350.1	-342.0	174.0	1.7384

BALLISTIC COEFFICIENT .24, MUZZLE VELOCITY FT/SEC 2000.

RANGE	VELOCITY	HEIGHT	100 YD	150 YD	200 YD	250 YD	300 YD	DRIFT	TIME
0.	2000.	0.0	-1.5	-1.5	-1.5	-1.5	-1.5	0.0	0.0000
100.	1694.	.5	0.0	2.4	5.4	8.9	12.9	2.3	.1630
200.	1427.	5.3	-10.8	-6.0	0.0	7.0	15.0	9.9	.3562
300.	1219.	15.6	-38.7	-31.5	-22.6	-12.0	0.0	23.7	.5844
400.	1082.	33.7	-90.2	-80.7	-68.7	-54.6	-38.6	43.6	.8477
500.	999.	61.2	-169.9	-157.9	-143.0	-125.4	-105.4	68.0	1.1362
600.	929.	99.9	-281.7	-267.3	-249.4	-228.2	-204.2	96.4	1.4480
700.	869.	151.7	-430.0	-413.2	-392.3	-367.7	-339.6	128.9	1.7822
800.	815.	218.8	-619.7	-600.6	-576.7	-548.5	-516.5	165.2	2.1388
900.	764.	303.8	-856.2	-834.6	-807.7	-776.0	-740.0	205.7	2.5189
1000.	715.	409.9	-1146.1	-1122.1	-1092.2	-1057.0	-1017.0	250.8	2.9249

BALLISTIC COEFFICIENT .24, MUZZLE VELOCITY FT/SEC 2100.

RANGE	VELOCITY	HEIGHT	100 YD	150 YD	200 YD	250 YD	300 YD	DRIFT	TIME
0.	2100.	0.0	-1.5	-1.5	-1.5	-1.5	-1.5	0.0	0.0000
100.	1782.	.4	0.0	2.1	4.8	7.9	11.6	2.2	.1551
200.	1503.	4.7	-9.6	-5.3	0.0	6.3	13.6	9.3	.3384
300.	1275.	14.1	-34.7	-28.4	-20.4	-10.9	0.0	22.4	.5561
400.	1115.	30.6	-81.2	-72.8	-62.1	-49.5	-35.0	41.8	.8088
500.	1022.	56.4	-155.1	-144.6	-131.2	-115.5	-97.3	66.3	1.0909
600.	949.	92.7	-259.4	-246.8	-230.8	-211.9	-190.0	94.8	1.3956
700.	886.	141.7	-398.6	-383.9	-365.2	-343.1	-317.7	127.2	1.7229
800.	831.	205.5	-577.7	-560.8	-539.5	-514.3	-485.1	163.7	2.0729
900.	779.	286.3	-801.1	-782.2	-758.1	-729.8	-697.0	204.1	2.4454
1000.	730.	387.4	-1076.1	-1055.0	-1028.3	-996.8	-960.4	249.1	2.8437

BALLISTIC COEFFICIENT .24, MUZZLE VELOCITY FT/SEC 2200.

RANGE	VELOCITY	HEIGHT	100 YD	150 YD	200 YD	250 YD	300 YD	DRIFT	TIME
0.	2200.	0.0	-1.5	-1.5	-1.5	-1.5	-1.5	0.0	0.0000
100.	1872.	.3	0.0	1.9	4.3	7.1	10.4	2.0	.1478
200.	1581.	4.2	-8.6	-4.8	0.0	5.7	12.2	8.8	.3225
300.	1335.	12.7	-31.2	-25.6	-18.3	-9.8	0.0	21.2	.5295
400.	1153.	27.9	-73.5	-66.1	-56.4	-45.1	-32.0	40.1	.7731
500.	1045.	51.9	-141.7	-132.4	-120.3	-106.1	-89.7	64.4	1.0475
600.	969.	86.3	-239.5	-228.3	-213.8	-196.8	-177.2	92.9	1.3463
700.	904.	132.3	-369.5	-356.4	-339.5	-319.7	-296.8	125.1	1.6652
800.	847.	192.9	-538.4	-523.5	-504.2	-481.6	-455.3	161.5	2.0086
900.	794.	269.9	-750.3	-733.5	-711.7	-686.3	-656.8	201.9	2.3745
1000.	744.	366.1	-1010.6	-992.0	-967.8	-939.5	-906.8	246.6	2.7647

BALLISTIC COEFFICIENT .24, MUZZLE VELOCITY FT/SEC 2300.

RANGE	VELOCITY	HEIGHT	100 YD	150 YD	200 YD	250 YD	300 YD	DRIFT	TIME
0.	2300.	0.0	-1.5	-1.5	-1.5	-1.5	-1.5	0.0	0.0000
100.	1963.	.2	0.0	1.7	3.8	6.4	9.3	1.9	.1412
200.	1661.	3.8	-7.6	-4.3	0.0	5.1	11.0	8.2	.3074
300.	1399.	11.5	-28.0	-23.0	-16.5	-8.8	0.0	19.9	.5044
400.	1199.	25.3	-66.2	-59.6	-50.9	-40.7	-28.9	37.9	.7368
500.	1071.	47.5	-128.5	-120.3	-109.5	-96.7	-81.9	61.7	1.0029
600.	990.	79.9	-220.2	-210.3	-197.3	-182.0	-164.3	90.4	1.2965
700.	922.	123.5	-342.2	-330.7	-315.5	-297.6	-277.0	122.4	1.6087
800.	862.	180.9	-501.5	-488.3	-471.0	-450.6	-427.0	158.8	1.9456
900.	809.	254.3	-702.3	-687.4	-668.0	-645.0	-618.4	199.1	2.3051
1000.	758.	346.0	-949.2	-932.7	-911.1	-885.5	-856.0	243.5	2.6877

BALLISTIC COEFFICIENT .24, MUZZLE VELOCITY FT/SEC 2400.

RANGE	VELOCITY	HEIGHT	100 YD	150 YD	200 YD	250 YD	300 YD	DRIFT	TIME
0.	2400.	0.0	-1.5	-1.5	-1.5	-1.5	-1.5	0.0	0.0000
100.	2055.	.1	0.0	1.5	3.4	5.7	8.4	1.8	.1351
200.	1742.	3.4	-6.8	-3.9	0.0	4.6	10.0	7.7	.2937
300.	1468.	10.4	-25.2	-20.8	-15.0	-8.0	0.0	18.7	.4814
400.	1250.	23.0	-59.8	-54.0	-46.2	-36.9	-26.3	35.8	.7036
500.	1100.	43.6	-116.9	-109.6	-99.8	-88.2	-74.9	59.1	.9608
600.	1011.	74.0	-202.3	-193.5	-181.8	-167.9	-151.9	87.6	1.2479
700.	940.	115.2	-316.9	-306.7	-293.0	-276.8	-258.1	119.5	1.5539
800.	878.	169.8	-467.4	-455.7	-440.1	-421.5	-400.2	155.7	1.8848
900.	824.	239.5	-657.3	-644.1	-626.5	-605.7	-581.7	195.7	2.2372
1000.	773.	327.1	-892.2	-877.6	-858.1	-834.9	-808.2	240.0	2.6135

BALLISTIC COEFFICIENT .24, MUZZLE VELOCITY FT/SEC 2500.

RANGE	VELOCITY	HEIGHT	100 YD	150 YD	200 YD	250 YD	300 YD	DRIFT	TIME
0.	2500.	0.0	-1.5	-1.5	-1.5	-1.5	-1.5	0.0	0.0000
100.	2148.	.1	0.0	1.3	3.1	5.2	7.6	1.7	.1295
200.	1825.	3.0	-6.1	-3.5	0.0	4.2	9.1	7.2	.2810
300.	1540.	9.4	-22.8	-18.9	-13.6	-7.3	0.0	17.6	.4601
400.	1303.	20.9	-54.2	-49.0	-42.0	-33.6	-23.8	33.9	.6724
500.	1133.	39.9	-106.3	-99.8	-91.0	-80.5	-68.3	56.4	.9204
600.	1033.	68.2	-185.0	-177.3	-166.7	-154.0	-139.5	84.3	1.1988
700.	959.	107.3	-293.3	-284.2	-271.8	-257.1	-240.1	116.2	1.5003
800.	894.	159.3	-435.8	-425.4	-411.3	-394.5	-375.1	152.4	1.8260
900.	838.	225.5	-615.1	-603.4	-587.5	-568.6	-546.8	192.0	2.1711
1000.	787.	309.2	-838.5	-825.6	-807.9	-786.9	-762.6	236.0	2.5409

BALLISTIC COEFFICIENT .24, MUZZLE VELOCITY FT/SEC 2600.

RANGE	VELOCITY	HEIGHT	100 YD	150 YD	200 YD	250 YD	300 YD	DRIFT	TIME
0.	2600.	0.0	-1.5	-1.5	-1.5	-1.5	-1.5	0.0	0.0000
100.	2241.	-.0	0.0	1.2	2.8	4.7	6.9	1.6	.1244
200.	1908.	2.7	-5.5	-3.2	0.0	3.8	8.2	6.8	.2694
300.	1614.	8.6	-20.6	-17.2	-12.4	-6.6	0.0	16.6	.4404
400.	1361.	19.1	-49.2	-44.6	-38.2	-30.6	-21.7	32.0	.6434
500.	1171.	36.6	-96.8	-91.0	-83.0	-73.5	-62.4	53.6	.8818
600.	1056.	63.0	-169.6	-162.7	-153.1	-141.6	-128.3	81.0	1.1527
700.	978.	100.0	-271.3	-263.3	-252.1	-238.7	-223.2	112.8	1.4484
800.	911.	149.0	-405.1	-395.9	-383.1	-367.8	-350.1	148.4	1.7665
900.	853.	212.4	-575.9	-565.5	-551.1	-533.9	-514.0	188.1	2.1072
1000.	800.	292.1	-787.9	-776.4	-760.4	-741.3	-719.2	231.7	2.4701

BALLISTIC COEFFICIENT .24, MUZZLE VELOCITY FT/SEC 2700.

RANGE	VELOCITY	HEIGHT	100 YD	150 YD	200 YD	250 YD	300 YD	DRIFT	TIME
0.	2700.	0.0	-1.5	-1.5	-1.5	-1.5	-1.5	0.0	0.0000
100.	2333.	-.1	0.0	1.0	2.5	4.2	6.2	1.5	.1196
200.	1993.	2.5	-5.0	-2.9	0.0	3.5	7.5	6.4	.2586
300.	1688.	7.8	-18.7	-15.7	-11.3	-6.1	0.0	15.7	.4223
400.	1422.	17.5	-44.8	-40.7	-34.9	-27.9	-19.8	30.2	.6161
500.	1215.	33.5	-88.3	-83.2	-75.9	-67.2	-57.1	51.0	.8453
600.	1080.	58.2	-155.5	-149.3	-140.6	-130.1	-118.0	77.7	1.1083
700.	997.	93.1	-250.9	-243.8	-233.6	-221.3	-207.2	109.2	1.3980
800.	928.	139.6	-377.3	-369.0	-357.4	-343.4	-327.3	144.5	1.7101
900.	868.	200.0	-539.0	-529.8	-516.7	-501.0	-482.8	183.9	2.0449
1000.	814.	276.2	-741.0	-730.7	-716.2	-698.7	-678.5	227.2	2.4020

BALLISTIC COEFFICIENT .24, MUZZLE VELOCITY FT/SEC 2800.

RANGE	VELOCITY	HEIGHT	100 YD	150 YD	200 YD	250 YD	300 YD	DRIFT	TIME
0.	2800.	0.0	-1.5	-1.5	-1.5	-1.5	-1.5	0.0	0.0000
100.	2425.	-.1	0.0	.9	2.2	3.8	5.7	1.4	.1150
200.	2079.	2.2	-4.5	-2.6	0.0	3.2	6.9	6.0	.2486
300.	1763.	7.1	-17.1	-14.3	-10.4	-5.5	0.0	14.8	.4054
400.	1487.	16.0	-40.9	-37.2	-32.0	-25.5	-18.1	28.6	.5908
500.	1263.	30.8	-80.7	-76.1	-69.5	-61.5	-52.3	48.4	.8105
600.	1108.	53.8	-142.6	-137.1	-129.2	-119.6	-108.5	74.4	1.0656
700.	1017.	86.6	-232.0	-225.6	-216.4	-205.1	-192.2	105.4	1.3491
800.	945.	130.9	-351.7	-344.3	-333.8	-321.0	-306.2	140.7	1.6564
900.	882.	188.2	-504.6	-496.3	-484.5	-470.0	-453.4	179.5	1.9841
1000.	828.	261.1	-696.9	-687.7	-674.6	-658.5	-640.0	222.5	2.3356

BALLISTIC COEFFICIENT .24, MUZZLE VELOCITY FT/SEC 2900.

RANGE	VELOCITY	HEIGHT	100 YD	150 YD	200 YD	250 YD	300 YD	DRIFT	TIME
0.	2900.	0.0	-1.5	-1.5	-1.5	-1.5	-1.5	0.0	0.0000
100.	2517.	-.2	0.0	.8	2.0	3.5	5.2	1.4	.1111
200.	2163.	2.0	-4.0	-2.4	0.0	2.9	6.4	5.7	.2395
300.	1839.	6.6	-15.6	-13.1	-9.5	-5.1	0.0	14.1	.3902
400.	1553.	14.7	-37.5	-34.2	-29.4	-23.5	-16.7	27.1	.5678
500.	1313.	28.3	-73.9	-69.8	-63.8	-56.5	-47.9	46.0	.7784
600.	1139.	49.7	-130.9	-126.1	-118.8	-110.0	-99.8	71.2	1.0250
700.	1037.	80.6	-214.4	-208.7	-200.3	-190.0	-178.0	101.7	1.3019
800.	962.	122.6	-327.2	-320.7	-311.1	-299.3	-285.6	136.4	1.6028
900.	897.	177.4	-472.8	-465.5	-454.7	-441.4	-426.1	175.2	1.9263
1000.	841.	246.8	-655.4	-647.3	-635.2	-620.5	-603.5	217.7	2.2712

BALLISTIC COEFFICIENT .24, MUZZLE VELOCITY FT/SEC 3000.

RANGE	VELOCITY	HEIGHT	100 YD	150 YD	200 YD	250 YD	300 YD	DRIFT	TIME
0.	3000.	0.0	-1.5	-1.5	-1.5	-1.5	-1.5	0.0	0.0000
100.	2608.	-.2	0.0	.7	1.8	3.2	4.8	1.3	.1074
200.	2248.	1.8	-3.7	-2.2	0.0	2.7	5.8	5.5	.2314
300.	1915.	6.0	-14.3	-12.1	-8.8	-4.7	0.0	13.4	.3760
400.	1620.	13.6	-34.4	-31.5	-27.0	-21.7	-15.4	25.7	.5463
500.	1365.	26.1	-67.9	-64.3	-58.7	-52.0	-44.1	43.7	.7485
600.	1174.	46.0	-120.6	-116.3	-109.6	-101.5	-92.1	68.1	.9869
700.	1058.	75.0	-198.2	-193.1	-185.4	-176.0	-164.9	97.9	1.2564
800.	979.	114.8	-304.7	-298.9	-290.0	-279.2	-266.6	132.3	1.5516
900.	913.	167.0	-442.7	-436.2	-426.2	-414.1	-399.9	170.6	1.8692
1000.	854.	233.6	-617.1	-609.8	-598.7	-585.3	-569.5	212.9	2.2095

BALLISTIC COEFFICIENT .24, MUZZLE VELOCITY FT/SEC 3100.

RANGE	VELOCITY	HEIGHT	100 YD	150 YD	200 YD	250 YD	300 YD	DRIFT	TIME
0.	3100.	0.0	-1.5	-1.5	-1.5	-1.5	-1.5	0.0	0.0000
100.	2699.	-.2	0.0	.6	1.7	2.9	4.4	1.2	.1037
200.	2332.	1.6	-3.3	-2.0	0.0	2.5	5.4	5.2	.2233
300.	1993.	5.6	-13.1	-11.1	-8.1	-4.4	0.0	12.7	.3624
400.	1687.	12.5	-31.6	-29.0	-25.0	-20.0	-14.2	24.5	.5261
500.	1421.	24.1	-62.5	-59.2	-54.2	-48.0	-40.7	41.5	.7199
600.	1215.	42.5	-111.0	-107.1	-101.0	-93.6	-84.9	64.9	.9492
700.	1080.	69.8	-183.3	-178.7	-171.6	-162.9	-152.8	94.1	1.2123
800.	997.	107.5	-283.8	-278.6	-270.5	-260.6	-248.9	128.1	1.5020
900.	928.	157.2	-414.8	-409.0	-399.9	-388.7	-375.6	166.0	1.8142
1000.	868.	220.9	-580.9	-574.4	-564.3	-551.9	-537.3	207.9	2.1490

BALLISTIC COEFFICIENT .24, MUZZLE VELOCITY FT/SEC 3200.

RANGE	VELOCITY	HEIGHT	100 YD	150 YD	200 YD	250 YD	300 YD	DRIFT	TIME
0.	3200.	0.0	-1.5	-1.5	-1.5	-1.5	-1.5	0.0	0.0000
100.	2790.	-.3	0.0	.6	1.5	2.7	4.0	1.2	.1004
200.	2416.	1.5	-3.0	-1.9	0.0	2.3	5.0	5.0	.2160
300.	2070.	5.1	-12.0	-10.3	-7.5	-4.0	0.0	12.1	.3501
400.	1756.	11.6	-29.1	-26.9	-23.1	-18.5	-13.1	23.3	.5075
500.	1480.	22.3	-57.7	-54.8	-50.1	-44.4	-37.6	39.6	.6937
600.	1258.	39.4	-102.4	-99.0	-93.4	-86.4	-78.4	61.9	.9143
700.	1105.	65.0	-169.6	-165.6	-159.1	-151.0	-141.6	90.4	1.1702
800.	1015.	100.8	-264.3	-259.8	-252.3	-243.1	-232.3	123.9	1.4543
900.	943.	148.2	-389.0	-383.9	-375.5	-365.1	-353.0	161.6	1.7617
1000.	881.	209.0	-546.8	-541.1	-531.8	-520.2	-506.8	202.9	2.0903

BALLISTIC COEFFICIENT .24, MUZZLE VELOCITY FT/SEC 3300.

RANGE	VELOCITY	HEIGHT	100 YD	150 YD	200 YD	250 YD	300 YD	DRIFT	TIME
0.	3300.	0.0	-1.5	-1.5	-1.5	-1.5	-1.5	0.0	0.0000
100.	2882.	-.3	0.0	.5	1.4	2.4	3.7	1.1	.0972
200.	2500.	1.4	-2.7	-1.7	0.0	2.1	4.6	4.8	.2092
300.	2148.	4.8	-11.1	-9.5	-7.0	-3.8	0.0	11.6	.3386
400.	1825.	10.8	-27.0	-24.9	-21.5	-17.2	-12.2	22.3	.4902
500.	1541.	20.7	-53.4	-50.8	-46.5	-41.2	-34.9	37.8	.6692
600.	1303.	36.5	-94.7	-91.7	-86.5	-80.1	-72.6	59.1	.8815
700.	1133.	60.5	-157.1	-153.5	-147.5	-140.0	-131.3	86.8	1.1295
800.	1033.	94.4	-246.2	-242.1	-235.3	-226.7	-216.7	119.8	1.4079
900.	959.	139.5	-364.3	-359.7	-352.0	-342.3	-331.1	156.8	1.7093
1000.	894.	198.0	-515.9	-510.8	-502.2	-491.5	-479.0	198.2	2.0351

BALLISTIC COEFFICIENT .24, MUZZLE VELOCITY FT/SEC 3400.

RANGE	VELOCITY	HEIGHT	100 YD	150 YD	200 YD	250 YD	300 YD	DRIFT	TIME
0.	3400.	0.0	-1.5	-1.5	-1.5	-1.5	-1.5	0.0	0.0000
100.	2973.	-.3	0.0	.4	1.2	2.2	3.4	1.1	.0943
200.	2584.	1.2	-2.5	-1.6	0.0	2.0	4.3	4.6	.2026
300.	2225.	4.4	-10.2	-8.8	-6.5	-3.5	0.0	11.1	.3277
400.	1895.	10.0	-24.9	-23.1	-20.0	-16.0	-11.4	21.3	.4738
500.	1602.	19.3	-49.4	-47.2	-43.3	-38.3	-32.5	36.1	.6462
600.	1351.	34.0	-87.7	-85.1	-80.3	-74.3	-67.4	56.5	.8506
700.	1165.	56.3	-145.6	-142.5	-136.9	-130.0	-121.8	83.2	1.0905
800.	1052.	88.4	-229.3	-225.7	-219.4	-211.4	-202.1	115.6	1.3628
900.	975.	131.4	-341.5	-337.5	-330.4	-321.4	-311.0	152.3	1.6595
1000.	908.	187.1	-485.3	-480.9	-472.9	-463.0	-451.4	192.9	1.9785

BALLISTIC COEFFICIENT .24, MUZZLE VELOCITY FT/SEC 3500.

RANGE	VELOCITY	HEIGHT	100 YD	150 YD	200 YD	250 YD	300 YD	DRIFT	TIME
0.	3500.	0.0	-1.5	-1.5	-1.5	-1.5	-1.5	0.0	0.0000
100.	3065.	-.3	0.0	.4	1.1	2.0	3.1	1.0	.0916
200.	2667.	1.1	-2.2	-1.5	0.0	1.8	4.0	4.4	.1966
300.	2303.	4.1	-9.4	-8.2	-6.0	-3.3	0.0	10.7	.3178
400.	1965.	9.4	-23.1	-21.6	-18.6	-14.9	-10.6	20.4	.4588
500.	1663.	18.0	-45.9	-44.0	-40.3	-35.7	-30.2	34.5	.6249
600.	1401.	31.6	-81.4	-79.1	-74.7	-69.2	-62.6	54.1	.8215
700.	1200.	52.5	-135.2	-132.5	-127.4	-120.9	-113.3	79.8	1.0537
800.	1072.	82.8	-213.6	-210.5	-204.7	-197.3	-188.6	111.6	1.3195
900.	990.	124.1	-321.0	-317.5	-310.9	-302.6	-292.8	148.1	1.6131
1000.	922.	177.1	-457.3	-453.4	-446.1	-436.8	-425.9	188.0	1.9250

BALLISTIC COEFFICIENT .24, MUZZLE VELOCITY FT/SEC 3600.

RANGE	VELOCITY	HEIGHT	100 YD	150 YD	200 YD	250 YD	300 YD	DRIFT	TIME
0.	3600.	0.0	-1.5	-1.5	-1.5	-1.5	-1.5	0.0	0.0000
100.	3156.	-.4	0.0	.3	1.0	1.9	2.9	1.0	.0890
200.	2750.	1.0	-2.0	-1.4	0.0	1.7	3.8	4.2	.1908
300.	2380.	3.8	-8.7	-7.7	-5.6	-3.0	0.0	10.2	.3081
400.	2036.	8.7	-21.5	-20.1	-17.4	-14.0	-9.9	19.6	.4444
500.	1726.	16.8	-42.7	-41.0	-37.6	-33.3	-28.2	33.0	.6044
600.	1454.	29.5	-75.7	-73.7	-69.7	-64.5	-58.4	51.8	.7941
700.	1239.	49.0	-125.8	-123.4	-118.7	-112.7	-105.6	76.6	1.0184
800.	1094.	77.6	-199.1	-196.4	-191.0	-184.1	-176.0	107.5	1.2775
900.	1007.	116.8	-300.6	-297.6	-291.5	-283.7	-274.6	143.5	1.5651
1000.	936.	167.7	-431.0	-427.6	-420.9	-412.3	-402.1	183.0	1.8732

BALLISTIC COEFFICIENT .24, MUZZLE VELOCITY FT/SEC 3700.

RANGE	VELOCITY	HEIGHT	100 YD	150 YD	200 YD	250 YD	300 YD	DRIFT	TIME
0.	3700.	0.0	-1.5	-1.5	-1.5	-1.5	-1.5	0.0	0.0000
100.	3248.	-.4	0.0	.3	.9	1.7	2.7	1.0	.0866
200.	2834.	.9	-1.8	-1.2	0.0	1.6	3.5	4.1	.1855
300.	2456.	3.5	-8.0	-7.1	-5.3	-2.8	0.0	9.9	.2993
400.	2107.	8.2	-20.0	-18.8	-16.3	-13.1	-9.3	18.8	.4312
500.	1789.	15.7	-39.8	-38.3	-35.2	-31.2	-26.4	31.8	.5858
600.	1509.	27.6	-70.6	-68.8	-65.1	-60.3	-54.6	49.6	.7685
700.	1279.	45.8	-117.1	-115.1	-110.8	-105.1	-98.5	73.5	.9851
800.	1118.	72.8	-185.8	-183.5	-178.5	-172.0	-164.5	103.7	1.2377
900.	1023.	110.0	-281.6	-279.0	-273.4	-266.1	-257.6	138.9	1.5191
1000.	951.	158.8	-406.3	-403.4	-397.2	-389.1	-379.6	178.1	1.8230

BALLISTIC COEFFICIENT .24, MUZZLE VELOCITY FT/SEC 3800.

RANGE	VELOCITY	HEIGHT	100 YD	150 YD	200 YD	250 YD	300 YD	DRIFT	TIME
0.	3800.	0.0	-1.5	-1.5	-1.5	-1.5	-1.5	0.0	0.0000
100.	3339.	-.4	0.0	.3	.8	1.6	2.5	1.0	.0844
200.	2918.	.8	-1.7	-1.1	0.0	1.5	3.3	4.0	.1806
300.	2533.	3.3	-7.4	-6.6	-4.9	-2.7	0.0	9.5	.2910
400.	2178.	7.7	-18.6	-17.5	-15.3	-12.3	-8.7	18.1	.4185
500.	1852.	14.7	-37.1	-35.8	-33.0	-29.2	-24.7	30.5	.5680
600.	1564.	25.9	-65.9	-64.4	-60.9	-56.4	-51.1	47.7	.7444
700.	1322.	42.9	-109.2	-107.5	-103.5	-98.2	-92.0	70.5	.9535
800.	1145.	68.2	-173.3	-171.2	-166.7	-160.7	-153.5	99.7	1.1983
900.	1040.	103.6	-263.8	-261.5	-256.3	-249.6	-241.5	134.4	1.4742
1000.	965.	150.5	-383.4	-380.8	-375.1	-367.6	-358.7	173.4	1.7749

BALLISTIC COEFFICIENT .24, MUZZLE VELOCITY FT/SEC 3900.

RANGE	VELOCITY	HEIGHT	100 YD	150 YD	200 YD	250 YD	300 YD	DRIFT	TIME
0.	3900.	0.0	-1.5	-1.5	-1.5	-1.5	-1.5	0.0	0.0000
100.	3431.	-.4	0.0	.2	.7	1.4	2.3	.9	.0819
200.	3001.	.7	-1.5	-1.0	0.0	1.4	3.1	3.8	.1753
300.	2610.	3.1	-6.8	-6.2	-4.6	-2.5	0.0	9.1	.2826
400.	2249.	7.2	-17.3	-16.5	-14.4	-11.5	-8.2	17.4	.4066
500.	1916.	13.8	-34.7	-33.6	-31.0	-27.4	-23.3	29.3	.5511
600.	1621.	24.2	-61.6	-60.3	-57.1	-52.9	-47.9	45.7	.7213
700.	1366.	40.2	-102.1	-100.6	-96.9	-91.9	-86.1	67.7	.9234
800.	1175.	64.0	-162.1	-160.4	-156.2	-150.5	-143.9	96.1	1.1615
900.	1058.	97.6	-247.3	-245.4	-240.7	-234.3	-226.8	130.0	1.4311
1000.	980.	142.3	-361.0	-358.9	-353.7	-346.6	-338.2	168.4	1.7262

BALLISTIC COEFFICIENT .24, MUZZLE VELOCITY FT/SEC 4000.

RANGE	VELOCITY	HEIGHT	100 YD	150 YD	200 YD	250 YD	300 YD	DRIFT	TIME
0.	4000.	0.0	-1.5	-1.5	-1.5	-1.5	-1.5	0.0	0.0000
100.	3523.	-.4	0.0	.2	.7	1.3	2.1	.9	.0799
200.	3085.	.7	-1.3	-1.0	0.0	1.3	2.9	3.7	.1711
300.	2686.	2.9	-6.3	-5.8	-4.3	-2.4	0.0	8.8	.2752
400.	2320.	6.8	-16.2	-15.4	-13.5	-10.9	-7.7	16.8	.3954
500.	1981.	13.0	-32.4	-31.5	-29.1	-25.8	-21.9	28.2	.5353
600.	1677.	22.8	-57.7	-56.6	-53.7	-49.7	-45.0	44.0	.7000
700.	1413.	37.7	-95.5	-94.3	-90.8	-86.3	-80.7	65.1	.8951
800.	1209.	60.0	-151.6	-150.1	-146.2	-141.0	-134.7	92.5	1.1254
900.	1077.	92.0	-232.0	-230.4	-226.0	-220.1	-213.0	125.8	1.3897
1000.	995.	134.8	-340.5	-338.8	-333.9	-327.3	-319.4	163.7	1.6803

BALLISTIC COEFFICIENT .25, MUZZLE VELOCITY FT/SEC 2000.

RANGE	VELOCITY	HEIGHT	100 YD	150 YD	200 YD	250 YD	300 YD	DRIFT	TIME
0.	2000.	0.0	-1.5	-1.5	-1.5	-1.5	-1.5	0.0	0.0000
100.	1705.	.5	0.0	2.4	5.3	8.7	12.7	2.2	.1624
200.	1446.	5.3	-10.6	-5.9	0.0	6.9	14.7	9.5	.3538
300.	1241.	15.3	-38.0	-30.9	-22.0	-11.7	0.0	22.6	.5784
400.	1099.	32.8	-87.9	-78.4	-66.7	-52.9	-37.3	41.6	.8365
500.	1013.	59.9	-166.5	-154.7	-140.0	-122.8	-103.3	65.8	1.1237
600.	945.	97.3	-275.2	-261.0	-243.3	-222.7	-199.2	93.1	1.4292
700.	885.	147.3	-419.0	-402.5	-381.9	-357.8	-330.5	124.3	1.7565
800.	832.	212.2	-603.6	-584.6	-561.1	-533.6	-502.4	159.5	2.1065
900.	782.	294.0	-832.8	-811.5	-785.0	-754.1	-718.9	198.6	2.4782
1000.	735.	395.8	-1113.1	-1089.4	-1060.0	-1025.6	-986.6	241.9	2.8743

BALLISTIC COEFFICIENT .25, MUZZLE VELOCITY FT/SEC 2100.

RANGE	VELOCITY	HEIGHT	100 YD	150 YD	200 YD	250 YD	300 YD	DRIFT	TIME
0.	2100.	0.0	-1.5	-1.5	-1.5	-1.5	-1.5	0.0	0.0000
100.	1794.	.4	0.0	2.1	4.7	7.8	11.3	2.1	.1546
200.	1524.	4.7	-9.4	-5.3	0.0	6.1	13.2	8.9	.3361
300.	1299.	13.8	-33.9	-27.7	-19.8	-10.6	0.0	21.4	.5499
400.	1135.	29.9	-79.4	-71.1	-60.5	-48.2	-34.1	40.0	.7990
500.	1038.	54.7	-150.9	-140.5	-127.3	-111.9	-94.3	63.5	1.0753
600.	965.	90.2	-253.0	-240.5	-224.7	-206.2	-185.1	91.4	1.3763
700.	903.	137.4	-387.8	-373.2	-354.8	-333.3	-308.6	122.6	1.6967
800.	848.	199.0	-561.7	-545.0	-524.0	-499.4	-471.2	157.9	2.0401
900.	798.	276.8	-778.4	-759.7	-736.0	-708.3	-676.6	197.0	2.4048
1000.	749.	373.7	-1043.9	-1023.0	-996.7	-966.0	-930.7	240.2	2.7931

BALLISTIC COEFFICIENT .25, MUZZLE VELOCITY FT/SEC 2200.

RANGE	VELOCITY	HEIGHT	100 YD	150 YD	200 YD	250 YD	300 YD	DRIFT	TIME
0.	2200.	0.0	-1.5	-1.5	-1.5	-1.5	-1.5	0.0	0.0000
100.	1884.	.3	0.0	1.8	4.2	7.0	10.2	1.9	.1474
200.	1603.	4.2	-8.4	-4.7	0.0	5.5	11.9	8.4	.3202
300.	1361.	12.4	-30.5	-24.9	-17.8	-9.6	0.0	20.2	.5236
400.	1178.	27.1	-71.4	-64.0	-54.6	-43.5	-30.8	38.0	.7615
500.	1063.	50.3	-137.4	-128.2	-116.4	-102.6	-86.7	61.5	1.0314
600.	986.	83.5	-232.5	-221.4	-207.2	-190.7	-171.5	89.2	1.3251
700.	921.	128.1	-358.8	-345.9	-329.3	-310.0	-287.7	120.4	1.6385
800.	864.	186.5	-522.6	-507.9	-488.9	-466.8	-441.4	155.7	1.9753
900.	813.	260.6	-727.8	-711.2	-689.9	-665.0	-636.4	194.7	2.3335
1000.	764.	352.8	-979.1	-960.7	-937.0	-909.4	-877.5	237.7	2.7141

BALLISTIC COEFFICIENT .25, MUZZLE VELOCITY FT/SEC 2300.

RANGE	VELOCITY	HEIGHT	100 YD	150 YD	200 YD	250 YD	300 YD	DRIFT	TIME
0.	2300.	0.0	-1.5	-1.5	-1.5	-1.5	-1.5	0.0	0.0000
100.	1976.	.2	0.0	1.6	3.8	6.3	9.1	1.8	.1406
200.	1684.	3.7	-7.5	-4.3	0.0	5.0	10.7	7.8	.3053
300.	1428.	11.2	-27.4	-22.5	-16.1	-8.6	0.0	18.9	.4989
400.	1227.	24.6	-64.4	-57.9	-49.4	-39.4	-27.9	36.0	.7262
500.	1091.	46.0	-124.6	-116.4	-105.8	-93.3	-79.0	58.9	.9869
600.	1008.	77.3	-213.4	-203.6	-190.8	-175.8	-158.6	86.6	1.2749
700.	940.	119.4	-331.9	-320.4	-305.5	-288.0	-268.0	117.7	1.5818
800.	880.	174.7	-486.0	-472.9	-455.9	-435.9	-413.0	152.8	1.9117
900.	828.	245.2	-680.1	-665.4	-646.3	-623.8	-598.0	191.8	2.2636
1000.	779.	333.0	-918.4	-902.1	-880.9	-855.8	-827.2	234.5	2.6370

BALLISTIC COEFFICIENT .25, MUZZLE VELOCITY FT/SEC 2400.

RANGE	VELOCITY	HEIGHT	100 YD	150 YD	200 YD	250 YD	300 YD	DRIFT	TIME
0.	2400.	0.0	-1.5	-1.5	-1.5	-1.5	-1.5	0.0	0.0000
100.	2068.	.1	0.0	1.4	3.4	5.6	8.2	1.7	.1346
200.	1766.	3.3	-6.7	-3.8	0.0	4.5	9.7	7.3	.2916
300.	1499.	10.1	-24.7	-20.3	-14.6	-7.8	0.0	17.8	.4760
400.	1280.	22.3	-58.2	-52.4	-44.8	-35.7	-25.3	34.0	.6933
500.	1123.	42.1	-113.1	-105.9	-96.3	-85.0	-72.0	56.3	.9447
600.	1030.	71.4	-195.6	-186.9	-175.4	-161.8	-146.3	83.8	1.2260
700.	959.	111.1	-306.6	-296.5	-283.1	-267.2	-249.1	114.6	1.5263
800.	897.	163.7	-452.3	-440.7	-425.4	-407.3	-386.5	149.7	1.8508
900.	843.	230.6	-635.4	-622.4	-605.2	-584.8	-561.5	188.4	2.1953
1000.	793.	314.5	-862.0	-847.5	-828.4	-805.8	-779.8	231.0	2.5626

BALLISTIC COEFFICIENT .25, MUZZLE VELOCITY FT/SEC 2500.

RANGE	VELOCITY	HEIGHT	100 YD	150 YD	200 YD	250 YD	300 YD	DRIFT	TIME
0.	2500.	0.0	−1.5	−1.5	−1.5	−1.5	−1.5	0.0	0.0000
100.	2161.	.0	0.0	1.3	3.0	5.1	7.4	1.6	.1289
200.	1849.	3.0	−6.0	−3.5	0.0	4.1	8.9	6.9	.2791
300.	1573.	9.2	−22.3	−18.5	−13.3	−7.1	0.0	16.8	.4553
400.	1336.	20.3	−52.8	−47.6	−40.7	−32.5	−23.0	32.2	.6627
500.	1160.	38.5	−102.8	−96.4	−87.7	−77.4	−65.6	53.5	.9042
600.	1053.	65.7	−178.6	−170.8	−160.4	−148.1	−133.9	80.4	1.1767
700.	979.	103.3	−283.3	−274.3	−262.1	−247.8	−231.2	111.3	1.4726
800.	914.	153.0	−420.2	−409.9	−396.0	−379.6	−360.6	146.1	1.7900
900.	858.	216.8	−593.8	−582.2	−566.6	−548.2	−526.8	184.6	2.1290
1000.	807.	296.8	−809.0	−796.1	−778.8	−758.3	−734.5	227.0	2.4899

BALLISTIC COEFFICIENT .25, MUZZLE VELOCITY FT/SEC 2600.

RANGE	VELOCITY	HEIGHT	100 YD	150 YD	200 YD	250 YD	300 YD	DRIFT	TIME
0.	2600.	0.0	−1.5	−1.5	−1.5	−1.5	−1.5	0.0	0.0000
100.	2254.	−.0	0.0	1.1	2.7	4.6	6.7	1.5	.1240
200.	1934.	2.7	−5.4	−3.1	0.0	3.7	8.0	6.5	.2677
300.	1647.	8.4	−20.2	−16.8	−12.1	−6.5	0.0	15.8	.4358
400.	1397.	18.5	−47.9	−43.3	−37.0	−29.6	−20.9	30.3	.6339
500.	1204.	35.2	−93.6	−87.9	−80.0	−70.7	−59.9	50.9	.8659
600.	1078.	60.6	−163.3	−156.5	−147.1	−135.8	−122.9	77.1	1.1304
700.	998.	96.1	−261.4	−253.5	−242.5	−229.4	−214.3	107.8	1.4202
800.	932.	143.2	−390.5	−381.4	−368.8	−353.9	−336.6	142.3	1.7316
900.	873.	204.3	−556.0	−545.7	−531.6	−514.8	−495.4	181.0	2.0667
1000.	821.	280.1	−758.9	−747.5	−731.8	−713.1	−691.5	222.6	2.4188

BALLISTIC COEFFICIENT .25, MUZZLE VELOCITY FT/SEC 2700.

RANGE	VELOCITY	HEIGHT	100 YD	150 YD	200 YD	250 YD	300 YD	DRIFT	TIME
0.	2700.	0.0	−1.5	−1.5	−1.5	−1.5	−1.5	0.0	0.0000
100.	2347.	−.1	0.0	1.0	2.4	4.1	6.1	1.4	.1193
200.	2020.	2.4	−4.9	−2.9	0.0	3.4	7.3	6.1	.2570
300.	1722.	7.6	−18.3	−15.3	−11.0	−5.9	0.0	14.9	.4178
400.	1461.	16.9	−43.6	−39.6	−33.8	−27.0	−19.2	28.6	.6072
500.	1252.	32.3	−85.3	−80.3	−73.1	−64.6	−54.8	48.2	.8296
600.	1106.	55.9	−149.5	−143.5	−134.9	−124.7	−112.9	73.8	1.0860
700.	1018.	89.3	−241.2	−234.2	−224.2	−212.2	−198.5	104.1	1.3694
800.	949.	133.9	−362.9	−354.9	−343.4	−329.8	−314.0	138.3	1.6749
900.	888.	191.6	−518.4	−509.4	−496.5	−481.2	−463.4	176.3	2.0018
1000.	835.	264.4	−712.4	−702.4	−688.1	−671.0	−651.3	218.1	2.3503

BALLISTIC COEFFICIENT .25, MUZZLE VELOCITY FT/SEC 2800.

RANGE	VELOCITY	HEIGHT	100 YD	150 YD	200 YD	250 YD	300 YD	DRIFT	TIME
0.	2800.	0.0	−1.5	−1.5	−1.5	−1.5	−1.5	0.0	0.0000
100.	2440.	−.1	0.0	.9	2.2	3.8	5.6	1.3	.1146
200.	2105.	2.2	−4.4	−2.6	0.0	3.1	6.7	5.8	.2471
300.	1799.	7.0	−16.7	−14.0	−10.1	−5.4	0.0	14.1	.4013
400.	1528.	15.5	−39.8	−36.2	−31.0	−24.8	−17.5	27.1	.5824
500.	1302.	29.6	−78.0	−73.5	−67.0	−59.1	−50.1	45.7	.7955
600.	1137.	51.5	−137.1	−131.7	−123.9	−114.5	−103.7	70.5	1.0436
700.	1039.	82.9	−222.5	−216.1	−207.0	−196.1	−183.5	100.3	1.3199
800.	967.	125.4	−337.8	−330.5	−320.2	−307.6	−293.2	134.4	1.6210
900.	904.	180.0	−484.5	−476.3	−464.6	−450.5	−434.3	171.9	1.9407
1000.	849.	249.6	−669.0	−659.9	−647.0	−631.3	−613.3	213.4	2.2838

BALLISTIC COEFFICIENT .25, MUZZLE VELOCITY FT/SEC 2900.

RANGE	VELOCITY	HEIGHT	100 YD	150 YD	200 YD	250 YD	300 YD	DRIFT	TIME
0.	2900.	0.0	−1.5	−1.5	−1.5	−1.5	−1.5	0.0	0.0000
100.	2532.	−.2	0.0	.8	2.0	3.4	5.1	1.3	.1108
200.	2191.	2.0	−4.0	−2.4	0.0	2.9	6.2	5.5	.2380
300.	1876.	6.4	−15.2	−12.8	−9.3	−5.0	0.0	13.3	.3860
400.	1596.	14.3	−36.5	−33.3	−28.5	−22.8	−16.2	25.7	.5597
500.	1355.	27.3	−71.5	−67.5	−61.6	−54.4	−46.1	43.5	.7642
600.	1173.	47.6	−125.9	−121.1	−114.0	−105.4	−95.5	67.4	1.0036
700.	1060.	77.0	−205.2	−199.6	−191.3	−181.2	−169.7	96.5	1.2727
800.	985.	117.2	−313.8	−307.3	−297.9	−286.4	−273.1	130.2	1.5675
900.	920.	169.2	−452.7	−445.5	−434.9	−421.9	−407.0	167.3	1.8818
1000.	863.	235.6	−628.2	−620.2	−608.4	−593.9	−577.4	208.5	2.2192

BALLISTIC COEFFICIENT .25, MUZZLE VELOCITY FT/SEC 3000.

RANGE	VELOCITY	HEIGHT	100 YD	150 YD	200 YD	250 YD	300 YD	DRIFT	TIME
0.	3000.	0.0	-1.5	-1.5	-1.5	-1.5	-1.5	0.0	0.0000
100.	2623.	-.2	0.0	.7	1.8	3.1	4.7	1.2	.1070
200.	2276.	1.8	-3.6	-2.2	0.0	2.6	5.7	5.3	.2299
300.	1954.	5.9	-14.0	-11.8	-8.6	-4.6	0.0	12.7	.3722
400.	1664.	13.2	-33.5	-30.6	-26.3	-21.0	-14.9	24.4	.5386
500.	1411.	25.1	-65.6	-62.1	-56.6	-50.0	-42.3	41.3	.7345
600.	1215.	43.9	-115.6	-111.4	-104.8	-96.9	-87.7	64.2	.9645
700.	1084.	71.5	-189.4	-184.4	-176.8	-167.5	-156.8	92.8	1.2271
800.	1003.	109.6	-291.5	-285.8	-277.1	-266.5	-254.3	126.0	1.5159
900.	936.	159.2	-423.6	-417.1	-407.3	-395.4	-381.6	162.9	1.8254
1000.	876.	222.8	-590.9	-583.7	-572.8	-559.6	-544.3	203.8	2.1579

BALLISTIC COEFFICIENT .25, MUZZLE VELOCITY FT/SEC 3100.

RANGE	VELOCITY	HEIGHT	100 YD	150 YD	200 YD	250 YD	300 YD	DRIFT	TIME
0.	3100.	0.0	-1.5	-1.5	-1.5	-1.5	-1.5	0.0	0.0000
100.	2715.	-.2	0.0	.6	1.6	2.8	4.3	1.2	.1034
200.	2361.	1.6	-3.3	-2.0	0.0	2.4	5.3	5.0	.2220
300.	2032.	5.4	-12.8	-10.9	-7.9	-4.3	0.0	12.1	.3589
400.	1733.	12.2	-30.8	-28.2	-24.3	-19.4	-13.7	23.2	.5187
500.	1471.	23.2	-60.4	-57.2	-52.3	-46.2	-39.1	39.2	.7068
600.	1259.	40.6	-106.5	-102.6	-96.7	-89.4	-80.8	61.1	.9279
700.	1110.	66.4	-174.9	-170.4	-163.5	-154.9	-145.0	89.0	1.1830
800.	1021.	102.4	-270.8	-265.7	-257.7	-248.0	-236.6	121.7	1.4656
900.	952.	149.6	-396.1	-390.3	-381.4	-370.4	-357.6	158.2	1.7700
1000.	891.	210.2	-554.7	-548.3	-538.4	-526.2	-512.0	198.6	2.0962

BALLISTIC COEFFICIENT .25, MUZZLE VELOCITY FT/SEC 3200.

RANGE	VELOCITY	HEIGHT	100 YD	150 YD	200 YD	250 YD	300 YD	DRIFT	TIME
0.	3200.	0.0	-1.5	-1.5	-1.5	-1.5	-1.5	0.0	0.0000
100.	2806.	-.3	0.0	.6	1.5	2.6	3.9	1.1	.1002
200.	2445.	1.5	-2.9	-1.8	0.0	2.3	4.9	4.8	.2146
300.	2111.	5.0	-11.8	-10.1	-7.3	-3.9	0.0	11.5	.3467
400.	1803.	11.3	-28.4	-26.2	-22.5	-18.0	-12.7	22.1	.5006
500.	1532.	21.5	-55.8	-53.0	-48.4	-42.8	-36.2	37.4	.6812
600.	1305.	37.6	-98.2	-94.9	-89.4	-82.6	-74.7	58.3	.8938
700.	1139.	61.7	-161.6	-157.7	-151.3	-143.4	-134.2	85.3	1.1409
800.	1040.	95.7	-251.5	-247.0	-239.7	-230.7	-220.2	117.4	1.4173
900.	968.	140.9	-371.1	-366.1	-357.8	-347.6	-335.8	153.9	1.7181
1000.	905.	198.5	-521.3	-515.8	-506.6	-495.3	-482.2	193.6	2.0374

BALLISTIC COEFFICIENT .25, MUZZLE VELOCITY FT/SEC 3300.

RANGE	VELOCITY	HEIGHT	100 YD	150 YD	200 YD	250 YD	300 YD	DRIFT	TIME
0.	3300.	0.0	-1.5	-1.5	-1.5	-1.5	-1.5	0.0	0.0000
100.	2898.	-.3	0.0	.5	1.3	2.4	3.6	1.1	.0971
200.	2530.	1.3	-2.7	-1.7	0.0	2.1	4.5	4.6	.2080
300.	2189.	4.6	-10.8	-9.3	-6.8	-3.6	0.0	11.0	.3353
400.	1874.	10.5	-26.2	-24.2	-20.9	-16.7	-11.8	21.1	.4835
500.	1594.	20.0	-51.6	-49.2	-44.9	-39.7	-33.6	35.7	.6573
600.	1354.	34.9	-90.9	-88.0	-82.9	-76.6	-69.3	55.7	.8621
700.	1172.	57.5	-149.9	-146.4	-140.5	-133.2	-124.7	81.9	1.1017
800.	1060.	89.4	-233.8	-229.8	-223.0	-214.6	-204.9	113.2	1.3707
900.	984.	132.4	-346.9	-342.4	-334.8	-325.3	-314.4	149.2	1.6658
1000.	919.	187.5	-490.1	-485.1	-476.7	-466.2	-454.0	188.5	1.9803

BALLISTIC COEFFICIENT .25, MUZZLE VELOCITY FT/SEC 3400.

RANGE	VELOCITY	HEIGHT	100 YD	150 YD	200 YD	250 YD	300 YD	DRIFT	TIME
0.	3400.	0.0	-1.5	-1.5	-1.5	-1.5	-1.5	0.0	0.0000
100.	2989.	-.3	0.0	.4	1.2	2.2	3.3	1.0	.0940
200.	2613.	1.2	-2.4	-1.5	0.0	1.9	4.2	4.4	.2014
300.	2267.	4.3	-10.0	-8.7	-6.3	-3.4	0.0	10.6	.3247
400.	1945.	9.7	-24.3	-22.5	-19.5	-15.6	-11.0	20.2	.4676
500.	1657.	18.6	-47.9	-45.7	-41.8	-36.9	-31.2	34.1	.6347
600.	1405.	32.4	-84.2	-81.6	-76.9	-71.1	-64.3	53.2	.8315
700.	1211.	53.4	-138.7	-135.6	-130.2	-123.4	-115.4	78.2	1.0622
800.	1081.	83.7	-217.5	-214.0	-207.8	-200.0	-190.9	109.2	1.3261
900.	1001.	124.4	-324.2	-320.2	-313.3	-304.5	-294.3	144.4	1.6148
1000.	934.	177.2	-461.1	-456.6	-448.9	-439.2	-427.8	183.6	1.9253

BALLISTIC COEFFICIENT .25, MUZZLE VELOCITY FT/SEC 3500.

RANGE	VELOCITY	HEIGHT	100 YD	150 YD	200 YD	250 YD	300 YD	DRIFT	TIME
0.	3500.	0.0	-1.5	-1.5	-1.5	-1.5	-1.5	0.0	0.0000
100.	3081.	-.3	0.0	.4	1.1	2.0	3.1	1.0	.0915
200.	2697.	1.1	-2.2	-1.4	0.0	1.8	3.9	4.2	.1955
300.	2345.	4.0	-9.2	-8.1	-5.9	-3.2	0.0	10.2	.3149
400.	2017.	9.1	-22.5	-21.0	-18.1	-14.5	-10.3	19.3	.4528
500.	1720.	17.3	-44.4	-42.5	-38.9	-34.4	-29.1	32.6	.6137
600.	1460.	30.2	-78.2	-75.9	-71.6	-66.2	-59.8	50.9	.8034
700.	1251.	49.8	-128.8	-126.2	-121.1	-114.8	-107.3	75.0	1.0261
800.	1105.	78.2	-202.3	-199.2	-193.5	-186.2	-177.7	105.1	1.2826
900.	1018.	117.0	-303.2	-299.8	-293.4	-285.2	-275.6	139.9	1.5662
1000.	949.	167.4	-433.7	-429.9	-422.7	-413.6	-403.0	178.6	1.8718

BALLISTIC COEFFICIENT .25, MUZZLE VELOCITY FT/SEC 3600.

RANGE	VELOCITY	HEIGHT	100 YD	150 YD	200 YD	250 YD	300 YD	DRIFT	TIME
0.	3600.	0.0	-1.5	-1.5	-1.5	-1.5	-1.5	0.0	0.0000
100.	3173.	-.4	0.0	.3	1.0	1.8	2.8	.9	.0886
200.	2781.	1.0	-2.0	-1.3	0.0	1.7	3.7	4.0	.1896
300.	2423.	3.7	-8.5	-7.5	-5.5	-3.0	0.0	9.7	.3053
400.	2089.	8.5	-20.9	-19.6	-17.0	-13.6	-9.6	18.5	.4386
500.	1784.	16.2	-41.4	-39.7	-36.4	-32.2	-27.2	31.2	.5941
600.	1516.	28.2	-72.8	-70.8	-66.9	-61.8	-55.8	48.7	.7766
700.	1293.	46.4	-119.8	-117.5	-112.9	-106.9	-100.0	71.8	.9915
800.	1131.	73.1	-188.3	-185.6	-180.4	-173.6	-165.6	101.0	1.2407
900.	1035.	110.0	-283.7	-280.7	-274.7	-267.1	-258.2	135.3	1.5189
1000.	963.	158.3	-408.1	-404.8	-398.2	-389.7	-379.8	173.7	1.8201

BALLISTIC COEFFICIENT .25, MUZZLE VELOCITY FT/SEC 3700.

RANGE	VELOCITY	HEIGHT	100 YD	150 YD	200 YD	250 YD	300 YD	DRIFT	TIME
0.	3700.	0.0	-1.5	-1.5	-1.5	-1.5	-1.5	0.0	0.0000
100.	3265.	-.4	0.0	.3	.9	1.7	2.6	.9	.0864
200.	2866.	.9	-1.8	-1.2	0.0	1.6	3.4	3.9	.1844
300.	2500.	3.5	-7.8	-7.0	-5.2	-2.8	0.0	9.4	.2967
400.	2161.	7.9	-19.4	-18.3	-15.9	-12.7	-9.0	17.8	.4256
500.	1849.	15.2	-38.5	-37.1	-34.1	-30.1	-25.4	30.0	.5758
600.	1573.	26.4	-67.9	-66.2	-62.5	-57.8	-52.2	46.7	.7520
700.	1337.	43.4	-111.6	-109.7	-105.4	-99.9	-93.3	69.0	.9594
800.	1161.	68.5	-175.5	-173.2	-168.3	-162.0	-154.5	97.2	1.2009
900.	1053.	103.4	-265.4	-262.8	-257.4	-250.3	-241.9	130.9	1.4734
1000.	979.	149.5	-383.8	-380.9	-374.8	-366.9	-357.6	168.7	1.7692

BALLISTIC COEFFICIENT .25, MUZZLE VELOCITY FT/SEC 3800.

RANGE	VELOCITY	HEIGHT	100 YD	150 YD	200 YD	250 YD	300 YD	DRIFT	TIME
0.	3800.	0.0	-1.5	-1.5	-1.5	-1.5	-1.5	0.0	0.0000
100.	3357.	-.4	0.0	.2	.8	1.5	2.4	.9	.0841
200.	2950.	.8	-1.6	-1.1	0.0	1.5	3.2	3.8	.1794
300.	2577.	3.2	-7.2	-6.5	-4.8	-2.6	0.0	9.1	.2883
400.	2233.	7.5	-18.1	-17.1	-14.9	-11.9	-8.5	17.2	.4134
500.	1915.	14.2	-36.0	-34.7	-31.9	-28.2	-23.9	28.8	.5586
600.	1630.	24.7	-63.3	-61.9	-58.5	-54.1	-48.9	44.8	.7284
700.	1383.	40.6	-104.1	-102.4	-98.5	-93.3	-87.2	66.1	.9284
800.	1194.	64.2	-163.9	-161.9	-157.5	-151.5	-144.6	93.5	1.1631
900.	1072.	97.3	-248.4	-246.1	-241.1	-234.4	-226.6	126.5	1.4291
1000.	993.	141.7	-362.1	-359.7	-354.1	-346.7	-338.0	164.3	1.7228

BALLISTIC COEFFICIENT .25, MUZZLE VELOCITY FT/SEC 3900.

RANGE	VELOCITY	HEIGHT	100 YD	150 YD	200 YD	250 YD	300 YD	DRIFT	TIME
0.	3900.	0.0	-1.5	-1.5	-1.5	-1.5	-1.5	0.0	0.0000
100.	3449.	-.4	0.0	.2	.7	1.4	2.2	.8	.0817
200.	3034.	.7	-1.4	-1.0	0.0	1.4	3.0	3.6	.1744
300.	2655.	3.0	-6.7	-6.1	-4.5	-2.5	0.0	8.7	.2802
400.	2305.	7.0	-16.9	-16.1	-14.0	-11.2	-8.0	16.5	.4015
500.	1980.	13.3	-33.6	-32.5	-30.0	-26.5	-22.4	27.7	.5418
600.	1688.	23.2	-59.2	-58.0	-54.9	-50.8	-45.8	43.0	.7060
700.	1432.	38.1	-97.3	-95.9	-92.3	-87.5	-81.7	63.5	.8992
800.	1230.	60.1	-153.1	-151.4	-147.3	-141.8	-135.2	89.9	1.1260
900.	1093.	91.5	-232.5	-230.6	-226.0	-219.8	-212.4	122.1	1.3862
1000.	1009.	133.8	-340.4	-338.3	-333.2	-326.3	-318.1	159.2	1.6740

BALLISTIC COEFFICIENT .25, MUZZLE VELOCITY FT/SEC 4000.

RANGE	VELOCITY	HEIGHT	100 YD	150 YD	200 YD	250 YD	300 YD	DRIFT	TIME
0.	4000.	0.0	-1.5	-1.5	-1.5	-1.5	-1.5	0.0	0.0000
100.	3541.	-.4	0.0	.2	.7	1.3	2.1	.8	.0798
200.	3119.	.6	-1.3	-1.0	0.0	1.3	2.8	3.6	.1702
300.	2732.	2.8	-6.2	-5.7	-4.2	-2.3	0.0	8.4	.2729
400.	2377.	6.6	-15.7	-15.1	-13.1	-10.6	-7.5	16.0	.3907
500.	2047.	12.6	-31.5	-30.6	-28.2	-25.0	-21.1	26.7	.5268
600.	1747.	21.8	-55.5	-54.4	-51.6	-47.7	-43.1	41.4	.6854
700.	1483.	35.7	-91.1	-89.9	-86.5	-82.0	-76.7	61.0	.8719
800.	1268.	56.4	-143.2	-141.9	-138.0	-132.8	-126.7	86.5	1.0914
900.	1116.	86.1	-217.8	-216.3	-211.9	-206.1	-199.2	117.9	1.3450
1000.	1025.	126.3	-320.0	-318.3	-313.5	-307.0	-299.4	154.4	1.6272

BALLISTIC COEFFICIENT .26, MUZZLE VELOCITY FT/SEC 2000.

RANGE	VELOCITY	HEIGHT	100 YD	150 YD	200 YD	250 YD	300 YD	DRIFT	TIME
0.	2000.	0.0	-1.5	-1.5	-1.5	-1.5	-1.5	0.0	0.0000
100.	1716.	.5	0.0	2.3	5.2	8.6	12.4	2.1	.1619
200.	1465.	5.2	-10.5	-5.8	0.0	6.7	14.4	9.0	.3514
300.	1262.	15.0	-37.2	-30.2	-21.5	-11.4	0.0	21.6	.5726
400.	1116.	32.1	-85.9	-76.5	-65.0	-51.5	-36.3	39.9	.8267
500.	1028.	58.3	-162.3	-150.6	-136.2	-119.3	-100.3	63.2	1.1089
600.	960.	94.6	-268.1	-254.1	-236.8	-216.5	-193.7	89.7	1.4094
700.	900.	143.3	-408.8	-392.4	-372.2	-348.6	-322.0	120.1	1.7323
800.	848.	206.2	-588.6	-569.9	-546.8	-519.8	-489.4	154.2	2.0764
900.	799.	285.2	-811.4	-790.3	-764.3	-734.0	-699.7	192.0	2.4407
1000.	753.	383.1	-1083.1	-1059.7	-1030.8	-997.1	-959.1	233.7	2.8279

BALLISTIC COEFFICIENT .26, MUZZLE VELOCITY FT/SEC 2100.

RANGE	VELOCITY	HEIGHT	100 YD	150 YD	200 YD	250 YD	300 YD	DRIFT	TIME
0.	2100.	0.0	-1.5	-1.5	-1.5	-1.5	-1.5	0.0	0.0000
100.	1805.	.4	0.0	2.1	4.6	7.7	11.1	2.0	.1540
200.	1544.	4.6	-9.3	-5.2	0.0	6.0	12.9	8.5	.3339
300.	1322.	13.5	-33.3	-27.1	-19.4	-10.3	0.0	20.4	.5444
400.	1155.	29.1	-77.4	-69.2	-58.9	-46.8	-33.1	38.3	.7888
500.	1053.	53.3	-147.0	-136.7	-123.8	-108.7	-91.5	61.0	1.0609
600.	981.	87.6	-246.2	-233.9	-218.4	-200.3	-179.7	87.9	1.3566
700.	919.	133.5	-377.8	-363.4	-345.3	-324.2	-300.1	118.3	1.6722
800.	864.	193.1	-546.8	-530.4	-509.7	-485.6	-458.1	152.5	2.0094
900.	815.	268.2	-757.4	-738.9	-715.6	-688.6	-657.6	190.3	2.3671
1000.	768.	361.3	-1014.5	-993.9	-968.0	-937.9	-903.5	232.0	2.7466

BALLISTIC COEFFICIENT .26, MUZZLE VELOCITY FT/SEC 2200.

RANGE	VELOCITY	HEIGHT	100 YD	150 YD	200 YD	250 YD	300 YD	DRIFT	TIME
0.	2200.	0.0	-1.5	-1.5	-1.5	-1.5	-1.5	0.0	0.0000
100.	1896.	.3	0.0	1.8	4.1	6.8	9.9	1.9	.1469
200.	1624.	4.1	-8.3	-4.6	0.0	5.4	11.6	8.0	.3180
300.	1386.	12.1	-29.8	-24.4	-17.4	-9.3	0.0	19.2	.5183
400.	1203.	26.3	-69.6	-62.3	-53.1	-42.2	-29.8	36.2	.7513
500.	1080.	48.8	-133.4	-124.3	-112.7	-99.2	-83.7	58.8	1.0158
600.	1003.	81.0	-225.8	-214.9	-201.0	-184.8	-166.1	85.7	1.3050
700.	938.	124.3	-349.1	-336.4	-320.1	-301.2	-279.4	116.0	1.6139
800.	881.	180.7	-507.9	-493.4	-474.8	-453.1	-428.3	150.1	1.9440
900.	830.	252.1	-707.0	-690.6	-669.7	-645.4	-617.4	188.0	2.2953
1000.	783.	340.8	-950.2	-932.1	-908.8	-881.8	-850.7	229.4	2.6673

BALLISTIC COEFFICIENT .26, MUZZLE VELOCITY FT/SEC 2300.

RANGE	VELOCITY	HEIGHT	100 YD	150 YD	200 YD	250 YD	300 YD	DRIFT	TIME
0.	2300.	0.0	-1.5	-1.5	-1.5	-1.5	-1.5	0.0	0.0000
100.	1988.	.2	0.0	1.6	3.7	6.2	8.9	1.7	.1402
200.	1705.	3.7	-7.4	-4.2	0.0	4.9	10.5	7.5	.3032
300.	1455.	11.0	-26.8	-22.0	-15.8	-8.4	0.0	18.1	.4939
400.	1254.	23.9	-62.9	-56.4	-48.1	-38.2	-27.1	34.3	.7168
500.	1112.	44.6	-121.0	-112.9	-102.5	-90.2	-76.2	56.3	.9719
600.	1025.	74.8	-206.9	-197.2	-184.7	-169.9	-153.2	83.0	1.2545
700.	957.	115.5	-322.1	-310.8	-296.3	-279.0	-259.5	113.2	1.5565
800.	898.	169.0	-471.9	-458.9	-442.3	-422.6	-400.3	147.4	1.8807
900.	846.	236.9	-659.6	-645.1	-626.4	-604.2	-579.1	185.0	2.2249
1000.	797.	321.3	-890.3	-874.1	-853.3	-828.7	-800.8	226.3	2.5903

BALLISTIC COEFFICIENT .26, MUZZLE VELOCITY FT/SEC 2400.

RANGE	VELOCITY	HEIGHT	100 YD	150 YD	200 YD	250 YD	300 YD	DRIFT	TIME
0.	2400.	0.0	-1.5	-1.5	-1.5	-1.5	-1.5	0.0	0.0000
100.	2081.	.1	0.0	1.4	3.3	5.5	8.1	1.6	.1341
200.	1788.	3.3	-6.6	-3.8	0.0	4.4	9.5	7.0	.2899
300.	1528.	9.9	-24.2	-19.9	-14.3	-7.6	0.0	17.0	.4714
400.	1310.	21.7	-56.8	-51.0	-43.5	-34.6	-24.5	32.4	.6840
500.	1148.	40.7	-109.7	-102.6	-93.2	-82.1	-69.4	53.6	.9296
600.	1048.	69.1	-189.6	-181.0	-169.7	-156.4	-141.2	80.3	1.2062
700.	977.	107.4	-297.2	-287.2	-274.0	-258.5	-240.7	110.1	1.5008
800.	915.	157.9	-437.9	-426.4	-411.3	-393.6	-373.3	144.0	1.8181
900.	861.	222.4	-615.3	-602.4	-585.5	-565.5	-542.7	181.5	2.1563
1000.	812.	303.0	-834.4	-820.0	-801.2	-779.0	-753.7	222.7	2.5155

BALLISTIC COEFFICIENT .26, MUZZLE VELOCITY FT/SEC 2500.

RANGE	VELOCITY	HEIGHT	100 YD	150 YD	200 YD	250 YD	300 YD	DRIFT	TIME
0.	2500.	0.0	-1.5	-1.5	-1.5	-1.5	-1.5	0.0	0.0000
100.	2174.	.0	0.0	1.3	3.0	5.0	7.3	1.5	.1285
200.	1872.	2.9	-5.9	-3.4	0.0	4.0	8.7	6.6	.2773
300.	1603.	9.0	-21.9	-18.1	-13.0	-6.9	0.0	16.0	.4508
400.	1369.	19.8	-51.4	-46.4	-39.6	-31.5	-22.2	30.6	.6536
500.	1190.	37.2	-99.7	-93.3	-84.9	-74.8	-63.2	50.9	.8894
600.	1073.	63.4	-172.6	-164.9	-154.8	-142.7	-128.8	76.7	1.1560
700.	997.	99.7	-273.9	-265.0	-253.2	-239.0	-222.9	106.7	1.4465
800.	933.	147.6	-406.4	-396.3	-382.7	-366.5	-348.0	140.4	1.7578
900.	876.	209.1	-574.7	-563.3	-548.0	-529.8	-509.0	177.9	2.0907
1000.	826.	285.6	-781.7	-769.0	-752.0	-731.8	-708.7	218.7	2.4424

BALLISTIC COEFFICIENT .26, MUZZLE VELOCITY FT/SEC 2600.

RANGE	VELOCITY	HEIGHT	100 YD	150 YD	200 YD	250 YD	300 YD	DRIFT	TIME
0.	2600.	0.0	-1.5	-1.5	-1.5	-1.5	-1.5	0.0	0.0000
100.	2267.	-.0	0.0	1.1	2.7	4.5	6.6	1.4	.1236
200.	1957.	2.6	-5.3	-3.1	0.0	3.7	7.9	6.2	.2660
300.	1678.	8.2	-19.8	-16.4	-11.8	-6.3	0.0	15.0	.4316
400.	1432.	18.0	-46.7	-42.2	-36.0	-28.7	-20.3	28.8	.6253
500.	1237.	34.1	-90.7	-85.1	-77.4	-68.3	-57.7	48.3	.8515
600.	1101.	58.3	-157.6	-150.9	-141.6	-130.6	-118.0	73.4	1.1096
700.	1018.	92.5	-252.3	-244.4	-233.6	-220.8	-206.1	103.1	1.3937
800.	951.	137.8	-377.0	-368.0	-355.6	-341.0	-324.1	136.5	1.6989
900.	892.	196.0	-535.5	-525.4	-511.5	-495.0	-476.1	173.6	2.0247
1000.	840.	269.1	-732.1	-720.9	-705.4	-687.1	-666.1	214.2	2.3710

BALLISTIC COEFFICIENT .26, MUZZLE VELOCITY FT/SEC 2700.

RANGE	VELOCITY	HEIGHT	100 YD	150 YD	200 YD	250 YD	300 YD	DRIFT	TIME
0.	2700.	0.0	-1.5	-1.5	-1.5	-1.5	-1.5	0.0	0.0000
100.	2360.	-.1	0.0	1.0	2.4	4.1	6.0	1.4	.1189
200.	2044.	2.4	-4.8	-2.8	0.0	3.3	7.2	5.9	.2556
300.	1755.	7.5	-18.0	-15.0	-10.8	-5.8	0.0	14.2	.4139
400.	1499.	16.5	-42.5	-38.5	-32.9	-26.2	-18.5	27.2	.5989
500.	1287.	31.2	-82.7	-77.7	-70.7	-62.3	-52.7	45.8	.8155
600.	1133.	53.7	-144.1	-138.1	-129.7	-119.6	-108.1	70.1	1.0650
700.	1039.	85.7	-232.2	-225.2	-215.3	-203.7	-190.2	99.4	1.3423
800.	969.	128.9	-350.4	-342.5	-331.2	-317.8	-302.5	132.8	1.6435
900.	909.	183.9	-499.4	-490.5	-477.8	-462.8	-445.5	169.2	1.9615
1000.	855.	253.7	-686.3	-676.3	-662.2	-645.6	-626.3	209.7	2.3023

BALLISTIC COEFFICIENT .26, MUZZLE VELOCITY FT/SEC 2800.

RANGE	VELOCITY	HEIGHT	100 YD	150 YD	200 YD	250 YD	300 YD	DRIFT	TIME
0.	2800.	0.0	-1.5	-1.5	-1.5	-1.5	-1.5	0.0	0.0000
100.	2453.	-.1	0.0	.9	2.2	3.7	5.5	1.3	.1143
200.	2130.	2.1	-4.3	-2.6	0.0	3.1	6.6	5.5	.2456
300.	1832.	6.8	-16.4	-13.7	-9.9	-5.3	0.0	13.4	.3976
400.	1568.	15.1	-38.9	-35.3	-30.2	-24.1	-17.0	25.7	.5748
500.	1341.	28.6	-75.6	-71.1	-64.7	-57.1	-48.3	43.4	.7820
600.	1169.	49.4	-131.9	-126.6	-118.9	-109.7	-99.1	66.8	1.0224
700.	1061.	79.5	-213.8	-207.6	-198.7	-188.0	-175.6	95.6	1.2930
800.	988.	120.3	-324.9	-317.8	-307.6	-295.3	-281.2	128.6	1.5878
900.	925.	172.6	-466.0	-457.9	-446.4	-432.7	-416.8	164.7	1.9003
1000.	869.	239.1	-643.4	-634.4	-621.7	-606.4	-588.7	204.9	2.2354

BALLISTIC COEFFICIENT .26, MUZZLE VELOCITY FT/SEC 2900.

RANGE	VELOCITY	HEIGHT	100 YD	150 YD	200 YD	250 YD	300 YD	DRIFT	TIME
0.	2900.	0.0	-1.5	-1.5	-1.5	-1.5	-1.5	0.0	0.0000
100.	2545.	-.2	0.0	.8	2.0	3.4	5.0	1.2	.1105
200.	2216.	1.9	-3.9	-2.3	0.0	2.8	6.1	5.2	.2367
300.	1910.	6.3	-15.0	-12.6	-9.1	-4.9	0.0	12.7	.3826
400.	1637.	13.9	-35.6	-32.4	-27.8	-22.1	-15.6	24.4	.5523
500.	1397.	26.3	-69.3	-65.3	-59.5	-52.4	-44.3	41.1	.7510
600.	1211.	45.6	-121.0	-116.3	-109.3	-100.8	-91.1	63.6	.9823
700.	1085.	73.7	-196.8	-191.3	-183.2	-173.3	-161.9	91.7	1.2453
800.	1006.	112.2	-301.2	-294.9	-285.6	-274.3	-261.3	124.3	1.5341
900.	941.	162.0	-434.9	-427.8	-417.3	-404.6	-390.0	160.2	1.8415
1000.	884.	225.4	-603.1	-595.2	-583.6	-569.4	-553.2	199.9	2.1705

BALLISTIC COEFFICIENT .26, MUZZLE VELOCITY FT/SEC 3000.

RANGE	VELOCITY	HEIGHT	100 YD	150 YD	200 YD	250 YD	300 YD	DRIFT	TIME
0.	3000.	0.0	-1.5	-1.5	-1.5	-1.5	-1.5	0.0	0.0000
100.	2637.	-.2	0.0	.7	1.8	3.1	4.6	1.2	.1068
200.	2302.	1.8	-3.5	-2.1	0.0	2.6	5.6	5.0	.2286
300.	1989.	5.8	-13.7	-11.6	-8.4	-4.5	0.0	12.1	.3687
400.	1706.	12.8	-32.7	-29.8	-25.6	-20.4	-14.4	23.2	.5316
500.	1457.	24.3	-63.7	-60.1	-54.8	-48.3	-40.8	39.1	.7222
600.	1255.	42.1	-111.4	-107.1	-100.7	-92.9	-84.0	60.7	.9447
700.	1112.	68.3	-181.4	-176.5	-169.0	-159.9	-149.5	88.0	1.1997
800.	1026.	104.7	-279.3	-273.7	-265.1	-254.7	-242.8	120.1	1.4823
900.	958.	152.0	-405.8	-399.5	-389.8	-378.1	-364.7	155.6	1.7841
1000.	898.	212.6	-565.9	-558.8	-548.1	-535.1	-520.2	195.0	2.1082

BALLISTIC COEFFICIENT .26, MUZZLE VELOCITY FT/SEC 3100.

RANGE	VELOCITY	HEIGHT	100 YD	150 YD	200 YD	250 YD	300 YD	DRIFT	TIME
0.	3100.	0.0	-1.5	-1.5	-1.5	-1.5	-1.5	0.0	0.0000
100.	2729.	-.2	0.0	.6	1.6	2.8	4.2	1.1	.1031
200.	2387.	1.6	-3.2	-2.0	0.0	2.4	5.2	4.8	.2207
300.	2069.	5.3	-12.6	-10.7	-7.7	-4.1	0.0	11.5	.3556
400.	1777.	11.8	-30.1	-27.6	-23.6	-18.9	-13.3	22.0	.5122
500.	1519.	22.4	-58.6	-55.5	-50.6	-44.6	-37.7	37.1	.6949
600.	1302.	38.9	-102.5	-98.8	-92.9	-85.7	-77.4	57.7	.9087
700.	1143.	63.3	-167.3	-162.9	-156.1	-147.7	-138.0	84.2	1.1556
800.	1045.	97.7	-259.1	-254.1	-246.3	-236.7	-225.7	115.8	1.4324
900.	975.	142.7	-378.8	-373.2	-364.4	-353.6	-341.2	151.0	1.7287
1000.	913.	200.3	-530.7	-524.4	-514.6	-502.7	-488.8	189.9	2.0468

BALLISTIC COEFFICIENT .26, MUZZLE VELOCITY FT/SEC 3200.

RANGE	VELOCITY	HEIGHT	100 YD	150 YD	200 YD	250 YD	300 YD	DRIFT	TIME
0.	3200.	0.0	-1.5	-1.5	-1.5	-1.5	-1.5	0.0	0.0000
100.	2821.	-.3	0.0	.6	1.5	2.6	3.8	1.1	.0999
200.	2472.	1.4	-2.9	-1.8	0.0	2.2	4.8	4.6	.2135
300.	2148.	4.9	-11.5	-9.9	-7.2	-3.8	0.0	11.0	.3438
400.	1848.	11.0	-27.7	-25.5	-21.9	-17.5	-12.3	21.0	.4944
500.	1582.	20.8	-54.2	-51.4	-46.9	-41.4	-34.9	35.4	.6700
600.	1352.	36.0	-94.7	-91.3	-85.9	-79.3	-71.6	55.1	.8755
700.	1177.	58.9	-154.7	-150.9	-144.6	-136.8	-127.8	80.6	1.1143
800.	1065.	91.3	-240.6	-236.2	-229.0	-220.1	-209.8	111.7	1.3847
900.	991.	134.3	-354.6	-349.6	-341.5	-331.5	-320.0	146.7	1.6773
1000.	928.	189.0	-498.0	-492.5	-483.5	-472.4	-459.6	184.9	1.9880

BALLISTIC COEFFICIENT .26, MUZZLE VELOCITY FT/SEC 3300.

RANGE	VELOCITY	HEIGHT	100 YD	150 YD	200 YD	250 YD	300 YD	DRIFT	TIME
0.	3300.	0.0	-1.5	-1.5	-1.5	-1.5	-1.5	0.0	0.0000
100.	2913.	-.3	0.0	.5	1.3	2.3	3.5	1.1	.0969
200.	2557.	1.3	-2.6	-1.7	0.0	2.1	4.4	4.4	.2068
300.	2227.	4.6	-10.6	-9.2	-6.7	-3.6	0.0	10.5	.3325
400.	1920.	10.2	-25.6	-23.7	-20.4	-16.3	-11.5	20.1	.4777
500.	1645.	19.3	-50.1	-47.7	-43.5	-38.4	-32.4	33.8	.6465
600.	1404.	33.4	-87.5	-84.6	-79.6	-73.5	-66.3	52.5	.8440
700.	1216.	54.7	-143.1	-139.7	-133.9	-126.7	-118.4	77.1	1.0744
800.	1089.	85.0	-222.6	-218.8	-212.1	-203.9	-194.3	107.2	1.3363
900.	1009.	126.0	-330.8	-326.4	-319.0	-309.7	-299.0	142.0	1.6248
1000.	943.	178.3	-467.7	-462.9	-454.6	-444.3	-432.4	180.0	1.9316

BALLISTIC COEFFICIENT .26, MUZZLE VELOCITY FT/SEC 3400.

RANGE	VELOCITY	HEIGHT	100 YD	150 YD	200 YD	250 YD	300 YD	DRIFT	TIME
0.	3400.	0.0	-1.5	-1.5	-1.5	-1.5	-1.5	0.0	0.0000
100.	3005.	-.3	0.0	.4	1.2	2.2	3.3	1.0	.0937
200.	2641.	1.2	-2.4	-1.5	0.0	1.9	4.1	4.2	.2003
300.	2306.	4.2	-9.8	-8.5	-6.2	-3.3	0.0	10.1	.3219
400.	1993.	9.5	-23.7	-22.0	-19.0	-15.1	-10.7	19.2	.4618
500.	1710.	18.0	-46.4	-44.3	-40.5	-35.7	-30.1	32.2	.6244
600.	1459.	31.1	-81.2	-78.6	-74.0	-68.3	-61.6	50.2	.8146
700.	1257.	50.8	-132.7	-129.7	-124.3	-117.7	-109.9	73.7	1.0366
800.	1114.	79.3	-206.8	-203.3	-197.2	-189.6	-180.7	103.0	1.2914
900.	1026.	118.1	-308.6	-304.7	-297.9	-289.3	-279.3	137.2	1.5738
1000.	959.	168.0	-438.7	-434.3	-426.7	-417.2	-406.0	174.7	1.8752

BALLISTIC COEFFICIENT .26, MUZZLE VELOCITY FT/SEC 3500.

RANGE	VELOCITY	HEIGHT	100 YD	150 YD	200 YD	250 YD	300 YD	DRIFT	TIME
0.	3500.	0.0	-1.5	-1.5	-1.5	-1.5	-1.5	0.0	0.0000
100.	3097.	-.3	0.0	.4	1.1	2.0	3.0	1.0	.0913
200.	2726.	1.1	-2.2	-1.4	0.0	1.8	3.9	4.1	.1945
300.	2384.	3.9	-9.0	-7.9	-5.8	-3.1	0.0	9.7	.3122
400.	2066.	8.9	-22.0	-20.5	-17.7	-14.1	-10.0	18.4	.4474
500.	1775.	16.8	-43.1	-41.3	-37.8	-33.3	-28.1	30.9	.6042
600.	1517.	29.0	-75.4	-73.2	-69.0	-63.6	-57.4	48.0	.7871
700.	1301.	47.4	-123.2	-120.5	-115.6	-109.4	-102.1	70.6	1.0012
800.	1141.	74.0	-192.1	-189.1	-183.5	-176.3	-168.1	99.0	1.2483
900.	1044.	110.9	-288.1	-284.7	-278.4	-270.4	-261.1	132.7	1.5251
1000.	974.	158.6	-411.9	-408.2	-401.2	-392.3	-381.9	169.8	1.8218

BALLISTIC COEFFICIENT .26, MUZZLE VELOCITY FT/SEC 3600.

RANGE	VELOCITY	HEIGHT	100 YD	150 YD	200 YD	250 YD	300 YD	DRIFT	TIME
0.	3600.	0.0	-1.5	-1.5	-1.5	-1.5	-1.5	0.0	0.0000
100.	3189.	-.4	0.0	.3	1.0	1.8	2.8	.9	.0884
200.	2810.	1.0	-2.0	-1.3	0.0	1.7	3.6	3.9	.1888
300.	2463.	3.6	-8.3	-7.3	-5.4	-2.9	0.0	9.3	.3027
400.	2139.	8.3	-20.4	-19.2	-16.5	-13.2	-9.4	17.6	.4336
500.	1840.	15.7	-40.2	-38.6	-35.3	-31.2	-26.3	29.6	.5848
600.	1575.	27.1	-70.3	-68.3	-64.4	-59.4	-53.6	46.0	.7612
700.	1346.	44.2	-114.6	-112.4	-107.8	-102.0	-95.2	67.6	.9676
800.	1172.	69.3	-179.2	-176.6	-171.3	-164.7	-157.0	95.3	1.2081
900.	1063.	104.1	-269.0	-266.1	-260.2	-252.8	-244.0	128.1	1.4777
1000.	989.	149.9	-387.6	-384.4	-377.8	-369.6	-359.9	165.1	1.7713

BALLISTIC COEFFICIENT .26, MUZZLE VELOCITY FT/SEC 3700.

RANGE	VELOCITY	HEIGHT	100 YD	150 YD	200 YD	250 YD	300 YD	DRIFT	TIME
0.	3700.	0.0	-1.5	-1.5	-1.5	-1.5	-1.5	0.0	0.0000
100.	3281.	-.4	0.0	.3	.9	1.6	2.6	.9	.0863
200.	2895.	.9	-1.7	-1.2	0.0	1.6	3.4	3.8	.1836
300.	2541.	3.4	-7.7	-6.9	-5.1	-2.7	0.0	9.0	.2943
400.	2212.	7.7	-19.0	-17.9	-15.5	-12.4	-8.7	17.0	.4208
500.	1907.	14.7	-37.4	-36.0	-33.0	-29.2	-24.6	28.4	.5669
600.	1633.	25.3	-65.5	-63.8	-60.2	-55.6	-50.1	44.1	.7370
700.	1394.	41.3	-106.8	-104.8	-100.6	-95.2	-88.8	64.8	.9360
800.	1209.	64.7	-166.6	-164.4	-159.6	-153.4	-146.2	91.4	1.1678
900.	1084.	97.6	-250.9	-248.4	-243.0	-236.0	-227.8	123.5	1.4312
1000.	1006.	141.3	-363.6	-360.8	-354.8	-347.1	-338.0	160.0	1.7201

BALLISTIC COEFFICIENT .26, MUZZLE VELOCITY FT/SEC 3800.

RANGE	VELOCITY	HEIGHT	100 YD	150 YD	200 YD	250 YD	300 YD	DRIFT	TIME
0.	3800.	0.0	-1.5	-1.5	-1.5	-1.5	-1.5	0.0	0.0000
100.	3373.	-.4	0.0	.2	.8	1.5	2.4	.9	.0839
200.	2980.	.8	-1.6	-1.1	0.0	1.4	3.1	3.6	.1786
300.	2619.	3.2	-7.1	-6.4	-4.7	-2.5	0.0	8.6	.2859
400.	2285.	7.3	-17.7	-16.7	-14.5	-11.6	-8.2	16.3	.4086
500.	1974.	13.8	-34.9	-33.7	-31.0	-27.3	-23.1	27.3	.5499
600.	1692.	23.7	-61.2	-59.7	-56.4	-52.0	-47.0	42.3	.7141
700.	1444.	38.7	-99.7	-98.1	-94.2	-89.1	-83.2	62.2	.9063
800.	1246.	60.6	-155.6	-153.7	-149.2	-143.4	-136.7	87.8	1.1305
900.	1107.	91.6	-234.5	-232.4	-227.4	-220.9	-213.3	119.1	1.3872
1000.	1022.	133.2	-341.1	-338.8	-333.2	-326.0	-317.5	155.0	1.6702

BALLISTIC COEFFICIENT .26, MUZZLE VELOCITY FT/SEC 3900.

RANGE	VELOCITY	HEIGHT	100 YD	150 YD	200 YD	250 YD	300 YD	DRIFT	TIME
0.	3900.	0.0	-1.5	-1.5	-1.5	-1.5	-1.5	0.0	0.0000
100.	3466.	-.4	0.0	.2	.7	1.4	2.2	.8	.0815
200.	3065.	.7	-1.4	-1.0	0.0	1.4	3.0	3.5	.1735
300.	2697.	3.0	-6.6	-6.0	-4.4	-2.4	0.0	8.3	.2779
400.	2357.	6.8	-16.5	-15.7	-13.7	-11.0	-7.7	15.7	.3970
500.	2041.	12.9	-32.7	-31.7	-29.1	-25.8	-21.7	26.3	.5338
600.	1752.	22.3	-57.2	-56.0	-52.9	-48.9	-44.1	40.6	.6924
700.	1497.	36.2	-93.2	-91.8	-88.3	-83.5	-77.9	59.7	.8777
800.	1286.	56.8	-145.4	-143.8	-139.7	-134.3	-127.9	84.3	1.0946
900.	1132.	86.0	-219.3	-217.5	-212.9	-206.9	-199.6	114.8	1.3444
1000.	1038.	125.5	-320.1	-318.2	-313.1	-306.3	-298.3	150.1	1.6219

BALLISTIC COEFFICIENT .26, MUZZLE VELOCITY FT/SEC 4000.

RANGE	VELOCITY	HEIGHT	100 YD	150 YD	200 YD	250 YD	300 YD	DRIFT	TIME
0.	4000.	0.0	-1.5	-1.5	-1.5	-1.5	-1.5	0.0	0.0000
100.	3558.	-.4	0.0	.2	.6	1.3	2.0	.8	.0797
200.	3150.	.6	-1.3	-.9	0.0	1.3	2.8	3.4	.1693
300.	2775.	2.8	-6.1	-5.6	-4.1	-2.3	0.0	8.0	.2707
400.	2430.	6.4	-15.4	-14.7	-12.8	-10.3	-7.3	15.2	.3863
500.	2109.	12.2	-30.5	-29.7	-27.4	-24.2	-20.5	25.3	.5189
600.	1813.	20.9	-53.6	-52.6	-49.7	-46.0	-41.5	39.1	.6723
700.	1550.	34.1	-87.4	-86.2	-82.9	-78.5	-73.2	57.5	.8515
800.	1327.	53.3	-136.0	-134.7	-131.0	-125.9	-119.9	81.1	1.0611
900.	1159.	80.8	-205.3	-203.8	-199.6	-194.0	-187.2	110.7	1.3038
1000.	1055.	118.5	-300.7	-299.1	-294.3	-288.1	-280.5	145.4	1.5759

BALLISTIC COEFFICIENT .27, MUZZLE VELOCITY FT/SEC 2000.

RANGE	VELOCITY	HEIGHT	100 YD	150 YD	200 YD	250 YD	300 YD	DRIFT	TIME
0.	2000.	0.0	-1.5	-1.5	-1.5	-1.5	-1.5	0.0	0.0000
100.	1726.	.5	0.0	2.3	5.1	8.5	12.2	2.0	.1614
200.	1482.	5.1	-10.3	-5.7	0.0	6.6	14.1	8.6	.3491
300.	1281.	14.7	-36.5	-29.6	-21.1	-11.2	0.0	20.7	.5674
400.	1133.	31.3	-84.0	-74.8	-63.4	-50.2	-35.3	38.3	.8174
500.	1042.	56.8	-158.3	-146.8	-132.6	-116.0	-97.4	60.7	1.0948
600.	974.	92.3	-262.2	-248.4	-231.4	-211.5	-189.1	86.7	1.3924
700.	915.	139.6	-399.5	-383.4	-363.5	-340.3	-314.2	116.2	1.7101
800.	862.	200.6	-574.6	-556.1	-533.4	-506.9	-477.1	149.3	2.0481
900.	815.	277.1	-791.7	-771.0	-745.4	-715.6	-682.1	185.9	2.4061
1000.	770.	371.6	-1055.6	-1032.5	-1004.1	-971.0	-933.8	226.2	2.7851

BALLISTIC COEFFICIENT .27, MUZZLE VELOCITY FT/SEC 2100.

RANGE	VELOCITY	HEIGHT	100 YD	150 YD	200 YD	250 YD	300 YD	DRIFT	TIME
0.	2100.	0.0	-1.5	-1.5	-1.5	-1.5	-1.5	0.0	0.0000
100.	1815.	.4	0.0	2.0	4.6	7.5	10.9	1.9	.1535
200.	1562.	4.5	-9.2	-5.1	0.0	5.9	12.6	8.1	.3320
300.	1343.	13.2	-32.7	-26.6	-18.9	-10.1	0.0	19.5	.5394
400.	1176.	28.4	-75.7	-67.5	-57.3	-45.5	-32.1	36.6	.7793
500.	1069.	51.9	-143.4	-133.2	-120.5	-105.7	-88.9	58.6	1.0473
600.	996.	85.3	-240.2	-228.0	-212.7	-194.9	-174.8	84.7	1.3386
700.	934.	129.9	-368.7	-354.5	-336.6	-315.9	-292.4	114.3	1.6497
800.	880.	187.6	-533.0	-516.8	-496.3	-472.7	-445.9	147.5	1.9807
900.	831.	260.3	-738.0	-719.7	-696.7	-670.1	-639.9	184.1	2.3320
1000.	785.	350.1	-987.4	-967.1	-941.5	-912.0	-878.4	224.4	2.7034

BALLISTIC COEFFICIENT .27, MUZZLE VELOCITY FT/SEC 2200.

RANGE	VELOCITY	HEIGHT	100 YD	150 YD	200 YD	250 YD	300 YD	DRIFT	TIME
0.	2200.	0.0	-1.5	-1.5	-1.5	-1.5	-1.5	0.0	0.0000
100.	1906.	.3	0.0	1.8	4.1	6.7	9.8	1.8	.1466
200.	1643.	4.0	-8.2	-4.6	0.0	5.3	11.4	7.6	.3162
300.	1410.	11.9	-29.3	-23.9	-17.0	-9.1	0.0	18.4	.5134
400.	1227.	25.7	-68.0	-60.8	-51.7	-41.0	-29.0	34.6	.7422
500.	1098.	47.4	-129.9	-120.9	-109.5	-96.2	-81.1	56.3	1.0018
600.	1019.	78.6	-219.6	-208.9	-195.2	-179.2	-161.1	82.4	1.2861
700.	954.	120.7	-339.9	-327.3	-311.3	-292.7	-271.6	111.9	1.5906
800.	897.	175.4	-494.8	-480.4	-462.1	-440.9	-416.7	145.2	1.9158
900.	847.	244.3	-687.7	-671.5	-650.9	-627.0	-599.8	181.7	2.2597
1000.	800.	329.7	-923.5	-905.6	-882.7	-856.1	-825.9	221.8	2.6239

BALLISTIC COEFFICIENT .27, MUZZLE VELOCITY FT/SEC 2300.

RANGE	VELOCITY	HEIGHT	100 YD	150 YD	200 YD	250 YD	300 YD	DRIFT	TIME
0.	2300.	0.0	-1.5	-1.5	-1.5	-1.5	-1.5	0.0	0.0000
100.	1999.	.2	0.0	1.6	3.6	6.1	8.8	1.7	.1399
200.	1725.	3.6	-7.3	-4.1	0.0	4.8	10.3	7.1	.3014
300.	1482.	10.7	-26.3	-21.6	-15.4	-8.2	0.0	17.2	.4892
400.	1281.	23.3	-61.3	-55.0	-46.8	-37.1	-26.2	32.7	.7077
500.	1133.	43.3	-117.6	-109.7	-99.4	-87.4	-73.8	53.8	.9578
600.	1042.	72.5	-200.8	-191.3	-178.9	-164.4	-148.1	79.6	1.2351
700.	974.	112.0	-313.2	-302.1	-287.7	-270.8	-251.7	109.1	1.5329
800.	915.	163.7	-458.3	-445.7	-429.2	-409.9	-388.1	142.1	1.8507
900.	862.	229.2	-640.6	-626.3	-607.8	-586.1	-561.6	178.6	2.1888
1000.	815.	310.6	-864.3	-848.5	-827.9	-803.8	-776.5	218.7	2.5469

BALLISTIC COEFFICIENT .27, MUZZLE VELOCITY FT/SEC 2400.

RANGE	VELOCITY	HEIGHT	100 YD	150 YD	200 YD	250 YD	300 YD	DRIFT	TIME
0.	2400.	0.0	-1.5	-1.5	-1.5	-1.5	-1.5	0.0	0.0000
100.	2092.	.1	0.0	1.4	3.3	5.4	7.9	1.6	.1338
200.	1808.	3.2	-6.5	-3.7	0.0	4.4	9.3	6.7	.2881
300.	1556.	9.7	-23.8	-19.5	-14.0	-7.5	0.0	16.2	.4672
400.	1338.	21.2	-55.5	-49.8	-42.4	-33.7	-23.8	30.9	.6755
500.	1172.	39.6	-107.0	-99.9	-90.7	-79.8	-67.4	51.4	.9168
600.	1067.	66.6	-183.1	-174.6	-163.5	-150.5	-135.6	76.5	1.1849
700.	995.	103.9	-288.3	-278.4	-265.5	-250.2	-232.9	105.9	1.4767
800.	933.	152.8	-424.8	-413.6	-398.8	-381.3	-361.5	138.7	1.7883
900.	878.	215.1	-597.0	-584.3	-567.7	-548.1	-525.7	175.2	2.1203
1000.	830.	292.5	-808.7	-794.6	-776.1	-754.3	-729.4	215.0	2.4716

BALLISTIC COEFFICIENT .27, MUZZLE VELOCITY FT/SEC 2500.

RANGE	VELOCITY	HEIGHT	100 YD	150 YD	200 YD	250 YD	300 YD	DRIFT	TIME
0.	2500.	0.0	-1.5	-1.5	-1.5	-1.5	-1.5	0.0	0.0000
100.	2186.	.0	0.0	1.3	2.9	4.9	7.2	1.4	.1282
200.	1893.	2.9	-5.9	-3.3	0.0	4.0	8.5	6.3	.2758
300.	1631.	8.8	-21.5	-17.7	-12.7	-6.8	0.0	15.2	.4466
400.	1400.	19.2	-50.2	-45.2	-38.5	-30.6	-21.6	29.1	.6453
500.	1219.	36.0	-96.9	-90.6	-82.2	-72.3	-61.1	48.5	.8755
600.	1094.	61.3	-167.1	-159.5	-149.5	-137.7	-124.1	73.3	1.1366
700.	1016.	96.3	-265.2	-256.4	-244.7	-230.9	-215.1	102.4	1.4219
800.	951.	142.5	-393.6	-383.5	-370.1	-354.3	-336.3	135.1	1.7274
900.	894.	201.8	-556.7	-545.4	-530.3	-512.6	-492.2	171.5	2.0543
1000.	844.	275.3	-756.5	-743.9	-727.2	-707.4	-684.8	210.9	2.3981

BALLISTIC COEFFICIENT .27, MUZZLE VELOCITY FT/SEC 2600.

RANGE	VELOCITY	HEIGHT	100 YD	150 YD	200 YD	250 YD	300 YD	DRIFT	TIME
0.	2600.	0.0	-1.5	-1.5	-1.5	-1.5	-1.5	0.0	0.0000
100.	2279.	-.0	0.0	1.1	2.6	4.4	6.5	1.4	.1232
200.	1979.	2.6	-5.3	-3.0	0.0	3.6	7.7	5.9	.2644
300.	1708.	8.0	-19.4	-16.1	-11.6	-6.1	0.0	14.3	.4276
400.	1466.	17.5	-45.6	-41.2	-35.1	-27.9	-19.7	27.4	.6175
500.	1269.	33.0	-88.1	-82.6	-75.0	-65.9	-55.7	45.9	.8379
600.	1125.	56.3	-152.4	-145.8	-136.7	-125.8	-113.5	70.0	1.0901
700.	1037.	89.1	-243.7	-235.9	-225.3	-212.6	-198.3	98.7	1.3685
800.	970.	133.1	-365.3	-356.4	-344.3	-329.8	-313.5	131.5	1.6701
900.	911.	188.8	-517.6	-507.6	-494.0	-477.6	-459.2	167.0	1.9873
1000.	859.	259.1	-707.5	-696.4	-681.3	-663.2	-642.7	206.4	2.3266

BALLISTIC COEFFICIENT .27, MUZZLE VELOCITY FT/SEC 2700.

RANGE	VELOCITY	HEIGHT	100 YD	150 YD	200 YD	250 YD	300 YD	DRIFT	TIME
0.	2700.	0.0	-1.5	-1.5	-1.5	-1.5	-1.5	0.0	0.0000
100.	2373.	-.1	0.0	1.0	2.4	4.0	5.9	1.3	.1186
200.	2067.	2.3	-4.7	-2.8	0.0	3.3	7.1	5.6	.2541
300.	1785.	7.3	-17.7	-14.7	-10.6	-5.6	0.0	13.6	.4103
400.	1535.	16.1	-41.6	-37.6	-32.1	-25.5	-18.0	25.9	.5916
500.	1322.	30.2	-80.3	-75.4	-68.5	-60.2	-50.9	43.5	.8026
600.	1161.	51.7	-139.2	-133.3	-125.0	-115.1	-103.8	66.7	1.0455
700.	1059.	82.5	-223.8	-216.9	-207.3	-195.7	-182.6	94.9	1.3169
800.	989.	124.0	-338.1	-330.2	-319.2	-306.0	-290.9	127.3	1.6123
900.	928.	176.9	-482.0	-473.1	-460.7	-445.8	-428.9	162.6	1.9239
1000.	874.	244.3	-663.3	-653.4	-639.6	-623.1	-604.3	202.1	2.2593

BALLISTIC COEFFICIENT .27, MUZZLE VELOCITY FT/SEC 2800.

RANGE	VELOCITY	HEIGHT	100 YD	150 YD	200 YD	250 YD	300 YD	DRIFT	TIME
0.	2800.	0.0	-1.5	-1.5	-1.5	-1.5	-1.5	0.0	0.0000
100.	2465.	-.1	0.0	.9	2.1	3.6	5.4	1.2	.1140
200.	2153.	2.1	-4.3	-2.5	0.0	3.0	6.5	5.3	.2443
300.	1864.	6.7	-16.1	-13.4	-9.7	-5.2	0.0	12.8	.3940
400.	1605.	14.7	-38.0	-34.5	-29.4	-23.4	-16.5	24.5	.5677
500.	1378.	27.7	-73.4	-69.0	-62.8	-55.2	-46.6	41.2	.7697
600.	1203.	47.6	-127.5	-122.2	-114.6	-105.6	-95.3	63.4	1.0033
700.	1084.	76.3	-205.8	-199.6	-190.8	-180.2	-168.2	91.0	1.2672
800.	1008.	115.5	-312.9	-305.8	-295.8	-283.7	-270.0	123.0	1.5563
900.	945.	165.9	-449.4	-441.5	-430.2	-416.6	-401.2	158.3	1.8634
1000.	889.	229.5	-619.6	-610.8	-598.3	-583.2	-566.0	196.9	2.1899

BALLISTIC COEFFICIENT .27, MUZZLE VELOCITY FT/SEC 2900.

RANGE	VELOCITY	HEIGHT	100 YD	150 YD	200 YD	250 YD	300 YD	DRIFT	TIME
0.	2900.	0.0	-1.5	-1.5	-1.5	-1.5	-1.5	0.0	0.0000
100.	2558.	-.2	0.0	.8	1.9	3.3	4.9	1.2	.1101
200.	2240.	1.9	-3.9	-2.3	0.0	2.8	5.9	5.0	.2356
300.	1943.	6.2	-14.7	-12.4	-8.9	-4.8	0.0	12.2	.3794
400.	1675.	13.5	-34.8	-31.7	-27.1	-21.5	-15.2	23.2	.5457
500.	1438.	25.5	-67.3	-63.4	-57.7	-50.8	-42.8	39.1	.7393
600.	1248.	43.8	-117.0	-112.3	-105.3	-97.1	-87.5	60.4	.9637
700.	1112.	70.7	-189.3	-183.8	-175.7	-166.1	-154.9	87.2	1.2197
800.	1028.	107.6	-289.5	-283.3	-274.1	-263.0	-250.3	118.8	1.5024
900.	962.	155.3	-418.2	-411.2	-400.8	-388.4	-374.1	153.5	1.8032
1000.	904.	215.9	-579.9	-572.1	-560.6	-546.8	-530.9	191.9	2.1247

BALLISTIC COEFFICIENT .27, MUZZLE VELOCITY FT/SEC 3000.

RANGE	VELOCITY	HEIGHT	100 YD	150 YD	200 YD	250 YD	300 YD	DRIFT	TIME
0.	3000.	0.0	-1.5	-1.5	-1.5	-1.5	-1.5	0.0	0.0000
100.	2650.	-.2	0.0	.7	1.7	3.0	4.5	1.2	.1065
200.	2326.	1.7	-3.5	-2.1	0.0	2.5	5.5	4.8	.2274
300.	2023.	5.7	-13.5	-11.4	-8.2	-4.4	0.0	11.6	.3658
400.	1746.	12.5	-31.9	-29.2	-25.0	-19.9	-14.0	22.1	.5254
500.	1501.	23.5	-61.9	-58.4	-53.1	-46.8	-39.4	37.1	.7107
600.	1295.	40.5	-107.6	-103.4	-97.1	-89.4	-80.6	57.5	.9268
700.	1142.	65.4	-174.2	-169.3	-162.0	-153.1	-142.8	83.4	1.1741
800.	1048.	100.3	-268.3	-262.7	-254.3	-244.1	-232.4	114.6	1.4513
900.	979.	145.5	-389.5	-383.3	-373.8	-362.3	-349.1	148.8	1.7456
1000.	919.	203.3	-543.0	-536.0	-525.5	-512.8	-498.1	186.9	2.0617

BALLISTIC COEFFICIENT .27, MUZZLE VELOCITY FT/SEC 3100.

RANGE	VELOCITY	HEIGHT	100 YD	150 YD	200 YD	250 YD	300 YD	DRIFT	TIME
0.	3100.	0.0	-1.5	-1.5	-1.5	-1.5	-1.5	0.0	0.0000
100.	2742.	-.2	0.0	.6	1.6	2.8	4.1	1.1	.1028
200.	2412.	1.6	-3.2	-1.9	0.0	2.4	5.1	4.6	.2196
300.	2103.	5.2	-12.4	-10.5	-7.6	-4.1	0.0	11.0	.3528
400.	1818.	11.5	-29.4	-26.9	-23.1	-18.3	-12.9	20.9	.5061
500.	1564.	21.7	-57.1	-54.0	-49.1	-43.2	-36.5	35.3	.6843
600.	1345.	37.4	-99.1	-95.4	-89.5	-82.5	-74.4	54.7	.8915
700.	1177.	60.6	-160.8	-156.4	-149.7	-141.4	-131.9	79.8	1.1308
800.	1070.	93.2	-247.6	-242.7	-234.9	-225.5	-214.7	109.9	1.3988
900.	997.	136.3	-362.9	-357.3	-348.6	-338.0	-325.8	144.1	1.6898
1000.	935.	191.4	-508.7	-502.5	-492.9	-481.1	-467.5	181.8	2.0007

BALLISTIC COEFFICIENT .27, MUZZLE VELOCITY FT/SEC 3200.

RANGE	VELOCITY	HEIGHT	100 YD	150 YD	200 YD	250 YD	300 YD	DRIFT	TIME
0.	3200.	0.0	-1.5	-1.5	-1.5	-1.5	-1.5	0.0	0.0000
100.	2834.	-.3	0.0	.5	1.4	2.5	3.8	1.0	.0996
200.	2497.	1.4	-2.9	-1.8	0.0	2.2	4.7	4.4	.2125
300.	2183.	4.8	-11.3	-9.7	-7.0	-3.7	0.0	10.5	.3408
400.	1891.	10.7	-27.1	-24.9	-21.4	-17.0	-12.0	20.0	.4886
500.	1629.	20.1	-52.7	-50.0	-45.5	-40.0	-33.8	33.6	.6597
600.	1398.	34.6	-91.5	-88.2	-82.8	-76.3	-68.8	52.1	.8587
700.	1218.	56.2	-148.5	-144.6	-138.4	-130.8	-122.0	76.2	1.0893
800.	1093.	86.8	-229.4	-225.0	-217.9	-209.1	-199.1	105.7	1.3505
900.	1015.	127.7	-338.2	-333.2	-325.2	-315.4	-304.2	139.4	1.6360
1000.	951.	180.2	-476.6	-471.1	-462.2	-451.3	-438.8	176.7	1.9416

BALLISTIC COEFFICIENT .27, MUZZLE VELOCITY FT/SEC 3300.

RANGE	VELOCITY	HEIGHT	100 YD	150 YD	200 YD	250 YD	300 YD	DRIFT	TIME
0.	3300.	0.0	-1.5	-1.5	-1.5	-1.5	-1.5	0.0	0.0000
100.	2926.	-.3	0.0	.5	1.3	2.3	3.5	1.0	.0967
200.	2582.	1.3	-2.6	-1.6	0.0	2.0	4.4	4.2	.2058
300.	2262.	4.5	-10.4	-9.0	-6.5	-3.5	0.0	10.1	.3300
400.	1964.	10.0	-25.1	-23.2	-19.9	-15.8	-11.2	19.1	.4723
500.	1694.	18.7	-48.7	-46.3	-42.3	-37.2	-31.3	32.1	.6367
600.	1454.	32.2	-84.6	-81.8	-76.9	-70.8	-63.8	49.8	.8281
700.	1260.	52.2	-137.3	-134.0	-128.3	-121.2	-113.0	72.8	1.0502
800.	1119.	80.9	-212.6	-208.8	-202.2	-194.1	-184.8	101.5	1.3042
900.	1033.	119.7	-315.0	-310.7	-303.3	-294.2	-283.7	134.8	1.5838
1000.	966.	169.9	-447.0	-442.2	-434.1	-424.0	-412.3	171.9	1.8856

BALLISTIC COEFFICIENT .27, MUZZLE VELOCITY FT/SEC 3400.

RANGE	VELOCITY	HEIGHT	100 YD	150 YD	200 YD	250 YD	300 YD	DRIFT	TIME
0.	3400.	0.0	-1.5	-1.5	-1.5	-1.5	-1.5	0.0	0.0000
100.	3019.	-.3	0.0	.4	1.2	2.1	3.2	.9	.0935
200.	2667.	1.2	-2.3	-1.5	0.0	1.9	4.1	4.0	.1993
300.	2342.	4.1	-9.6	-8.4	-6.1	-3.3	0.0	9.6	.3195
400.	2038.	9.3	-23.2	-21.6	-18.5	-14.8	-10.4	18.3	.4568
500.	1759.	17.4	-45.2	-43.1	-39.3	-34.6	-29.2	30.6	.6151
600.	1512.	29.9	-78.5	-76.0	-71.5	-65.8	-59.3	47.5	.7992
700.	1304.	48.5	-127.4	-124.4	-119.1	-112.5	-104.9	69.6	1.0133
800.	1149.	75.4	-197.2	-193.9	-187.8	-180.3	-171.6	97.4	1.2593
900.	1052.	112.1	-293.5	-289.7	-282.9	-274.4	-264.6	130.1	1.5332
1000.	983.	159.8	-418.5	-414.4	-406.8	-397.4	-386.5	166.6	1.8291

BALLISTIC COEFFICIENT .27, MUZZLE VELOCITY FT/SEC 3500.

RANGE	VELOCITY	HEIGHT	100 YD	150 YD	200 YD	250 YD	300 YD	DRIFT	TIME
0.	3500.	0.0	-1.5	-1.5	-1.5	-1.5	-1.5	0.0	0.0000
100.	3111.	-.4	0.0	.4	1.1	1.9	2.9	1.0	.0911
200.	2752.	1.0	-2.1	-1.4	0.0	1.7	3.8	3.9	.1936
300.	2421.	3.9	-8.8	-7.8	-5.7	-3.1	0.0	9.3	.3098
400.	2112.	8.6	-21.5	-20.0	-17.3	-13.8	-9.7	17.5	.4425
500.	1826.	16.3	-42.0	-40.2	-36.7	-32.3	-27.2	29.3	.5953
600.	1571.	27.9	-73.0	-70.8	-66.6	-61.4	-55.3	45.5	.7726
700.	1351.	45.2	-118.3	-115.7	-110.9	-104.8	-97.6	66.7	.9789
800.	1182.	70.3	-183.2	-180.3	-174.7	-167.7	-159.6	93.5	1.2168
900.	1072.	105.0	-273.5	-270.2	-264.0	-256.1	-246.9	125.5	1.4845
1000.	999.	150.4	-391.9	-388.3	-381.3	-372.6	-362.4	161.5	1.7748

BALLISTIC COEFFICIENT .27, MUZZLE VELOCITY FT/SEC 3600.

RANGE	VELOCITY	HEIGHT	100 YD	150 YD	200 YD	250 YD	300 YD	DRIFT	TIME
0.	3600.	0.0	-1.5	-1.5	-1.5	-1.5	-1.5	0.0	0.0000
100.	3203.	-.4	0.0	.3	1.0	1.8	2.7	.9	.0883
200.	2837.	.9	-1.9	-1.3	0.0	1.6	3.5	3.7	.1877
300.	2500.	3.6	-8.2	-7.2	-5.3	-2.9	0.0	8.9	.3005
400.	2186.	8.1	-20.0	-18.7	-16.2	-12.9	-9.1	16.8	.4287
500.	1893.	15.2	-39.1	-37.5	-34.3	-30.2	-25.5	28.1	.5763
600.	1631.	26.0	-67.9	-66.0	-62.2	-57.3	-51.6	43.5	.7471
700.	1400.	42.2	-110.1	-107.8	-103.4	-97.7	-91.0	63.8	.9458
800.	1219.	65.6	-170.5	-168.0	-162.9	-156.3	-148.7	89.6	1.1760
900.	1094.	98.4	-255.0	-252.2	-246.4	-239.1	-230.5	120.9	1.4372
1000.	1016.	141.7	-367.3	-364.2	-357.8	-349.6	-340.1	156.5	1.7224

BALLISTIC COEFFICIENT .27, MUZZLE VELOCITY FT/SEC 3700.

RANGE	VELOCITY	HEIGHT	100 YD	150 YD	200 YD	250 YD	300 YD	DRIFT	TIME
0.	3700.	0.0	-1.5	-1.5	-1.5	-1.5	-1.5	0.0	0.0000
100.	3296.	-.4	0.0	.3	.9	1.6	2.5	.9	.0860
200.	2923.	.9	-1.7	-1.2	0.0	1.5	3.3	3.6	.1828
300.	2579.	3.3	-7.5	-6.7	-4.9	-2.7	0.0	8.6	.2921
400.	2259.	7.6	-18.6	-17.5	-15.1	-12.1	-8.5	16.2	.4164
500.	1961.	14.2	-36.5	-35.1	-32.1	-28.3	-23.9	27.0	.5589
600.	1691.	24.4	-63.4	-61.7	-58.2	-53.6	-48.3	41.7	.7236
700.	1452.	39.5	-102.7	-100.7	-96.6	-91.3	-85.0	61.2	.9153
800.	1258.	61.4	-159.0	-156.8	-152.0	-146.0	-138.8	86.1	1.1378
900.	1118.	92.3	-238.1	-235.6	-230.3	-223.5	-215.5	116.6	1.3924
1000.	1032.	133.4	-344.3	-341.5	-335.6	-328.0	-319.1	151.6	1.6719

BALLISTIC COEFFICIENT .27, MUZZLE VELOCITY FT/SEC 3800.

RANGE	VELOCITY	HEIGHT	100 YD	150 YD	200 YD	250 YD	300 YD	DRIFT	TIME
0.	3800.	0.0	-1.5	-1.5	-1.5	-1.5	-1.5	0.0	0.0000
100.	3389.	-.4	0.0	.2	.8	1.5	2.3	.8	.0837
200.	3008.	.8	-1.5	-1.1	0.0	1.4	3.1	3.5	.1776
300.	2657.	3.1	-7.0	-6.3	-4.6	-2.5	0.0	8.3	.2839
400.	2333.	7.1	-17.3	-16.4	-14.2	-11.4	-8.0	15.6	.4044
500.	2029.	13.4	-34.0	-32.9	-30.2	-26.6	-22.4	26.0	.5423
600.	1752.	22.9	-59.2	-57.8	-54.6	-50.3	-45.3	40.1	.7014
700.	1506.	36.9	-95.9	-94.2	-90.5	-85.5	-79.6	58.7	.8862
800.	1299.	57.5	-148.4	-146.5	-142.2	-136.5	-129.8	82.7	1.1013
900.	1145.	86.5	-222.2	-220.1	-215.2	-208.8	-201.3	112.2	1.3481
1000.	1050.	125.6	-322.6	-320.3	-314.9	-307.8	-299.4	146.6	1.6227

BALLISTIC COEFFICIENT .27, MUZZLE VELOCITY FT/SEC 3900.

RANGE	VELOCITY	HEIGHT	100 YD	150 YD	200 YD	250 YD	300 YD	DRIFT	TIME
0.	3900.	0.0	-1.5	-1.5	-1.5	-1.5	-1.5	0.0	0.0000
100.	3481.	-.4	0.0	.2	.7	1.4	2.1	.8	.0813
200.	3094.	.7	-1.4	-1.0	0.0	1.3	2.9	3.3	.1728
300.	2736.	2.9	-6.4	-5.8	-4.3	-2.3	0.0	7.9	.2759
400.	2406.	6.7	-16.1	-15.3	-13.3	-10.7	-7.6	15.0	.3929
500.	2098.	12.6	-31.8	-30.8	-28.3	-25.0	-21.1	25.0	.5264
600.	1813.	21.5	-55.4	-54.2	-51.2	-47.2	-42.5	38.5	.6802
700.	1560.	34.6	-89.7	-88.3	-84.8	-80.2	-74.7	56.4	.8588
800.	1342.	53.8	-138.7	-137.1	-133.1	-127.8	-121.5	79.4	1.0664
900.	1175.	81.2	-208.0	-206.2	-201.7	-195.7	-188.7	108.1	1.3067
1000.	1068.	118.3	-302.4	-300.5	-295.5	-288.8	-281.0	141.8	1.5748

BALLISTIC COEFFICIENT .27, MUZZLE VELOCITY FT/SEC 4000.

RANGE	VELOCITY	HEIGHT	100 YD	150 YD	200 YD	250 YD	300 YD	DRIFT	TIME
0.	4000.	0.0	-1.5	-1.5	-1.5	-1.5	-1.5	0.0	0.0000
100.	3574.	-.4	0.0	.2	.6	1.2	2.0	.8	.0795
200.	3179.	.6	-1.2	-.9	0.0	1.2	2.7	3.2	.1684
300.	2815.	2.7	-5.9	-5.5	-4.1	-2.2	0.0	7.7	.2688
400.	2479.	6.3	-15.0	-14.4	-12.6	-10.1	-7.1	14.5	.3823
500.	2166.	11.8	-29.7	-29.0	-26.7	-23.5	-19.8	24.1	.5117
600.	1876.	20.2	-51.9	-50.9	-48.2	-44.4	-40.0	37.1	.6606
700.	1615.	32.6	-84.0	-82.9	-79.7	-75.4	-70.2	54.2	.8331
800.	1387.	50.5	-129.8	-128.6	-124.9	-119.9	-114.0	76.3	1.0338
900.	1209.	76.2	-194.5	-193.1	-188.9	-183.3	-176.7	104.0	1.2661
1000.	1088.	111.4	-283.6	-282.1	-277.5	-271.2	-263.8	137.1	1.5289

BALLISTIC COEFFICIENT .28, MUZZLE VELOCITY FT/SEC 2000.

RANGE	VELOCITY	HEIGHT	100 YD	150 YD	200 YD	250 YD	300 YD	DRIFT	TIME
0.	2000.	0.0	-1.5	-1.5	-1.5	-1.5	-1.5	0.0	0.0000
100.	1735.	.5	0.0	2.3	5.1	8.3	12.0	1.9	.1611
200.	1499.	5.0	-10.1	-5.6	0.0	6.5	13.8	8.3	.3471
300.	1300.	14.4	-35.9	-29.1	-20.7	-11.0	0.0	19.8	.5625
400.	1150.	30.6	-82.3	-73.2	-62.0	-49.0	-34.4	36.7	.8087
500.	1056.	55.4	-154.7	-143.3	-129.3	-113.1	-94.8	58.4	1.0818
600.	988.	90.4	-257.2	-243.6	-226.8	-207.3	-185.4	84.1	1.3777
700.	929.	136.2	-390.8	-374.9	-355.3	-332.6	-307.0	112.5	1.6893
800.	877.	195.6	-562.2	-544.0	-521.6	-495.6	-466.4	144.8	2.0228
900.	830.	269.7	-773.4	-752.9	-727.7	-698.5	-665.7	180.2	2.3737
1000.	786.	361.0	-1030.1	-1007.4	-979.3	-946.9	-910.4	219.2	2.7452

BALLISTIC COEFFICIENT .28, MUZZLE VELOCITY FT/SEC 2100.

RANGE	VELOCITY	HEIGHT	100 YD	150 YD	200 YD	250 YD	300 YD	DRIFT	TIME
0.	2100.	0.0	-1.5	-1.5	-1.5	-1.5	-1.5	0.0	0.0000
100.	1825.	.4	0.0	2.0	4.5	7.4	10.7	1.8	.1532
200.	1579.	4.5	-9.1	-5.1	0.0	5.8	12.4	7.8	.3301
300.	1364.	13.0	-32.1	-26.1	-18.6	-9.9	0.0	18.7	.5348
400.	1197.	27.7	-74.0	-66.0	-55.9	-44.3	-31.2	35.0	.7703
500.	1084.	50.6	-140.0	-130.0	-117.4	-102.9	-86.5	56.4	1.0345
600.	1010.	83.4	-235.3	-223.3	-208.2	-190.8	-171.0	82.1	1.3237
700.	949.	126.5	-360.0	-346.0	-328.3	-308.1	-285.1	110.6	1.6282
800.	894.	182.8	-521.2	-505.2	-485.0	-461.8	-435.5	143.0	1.9555
900.	846.	252.9	-719.8	-701.7	-679.0	-653.0	-623.3	178.3	2.2990
1000.	801.	339.7	-962.2	-942.2	-917.0	-888.0	-855.1	217.3	2.6632

BALLISTIC COEFFICIENT .28, MUZZLE VELOCITY FT/SEC 2200.

RANGE	VELOCITY	HEIGHT	100 YD	150 YD	200 YD	250 YD	300 YD	DRIFT	TIME
0.	2200.	0.0	-1.5	-1.5	-1.5	-1.5	-1.5	0.0	0.0000
100.	1916.	.3	0.0	1.8	4.0	6.6	9.6	1.7	.1463
200.	1660.	4.0	-8.0	-4.5	0.0	5.2	11.1	7.3	.3144
300.	1433.	11.7	-28.8	-23.5	-16.7	-8.9	0.0	17.6	.5091
400.	1250.	25.1	-66.6	-59.5	-50.5	-40.0	-28.2	33.1	.7338
500.	1117.	46.2	-126.6	-117.8	-106.5	-93.5	-78.7	54.0	.9888
600.	1034.	76.5	-214.0	-203.4	-189.9	-174.2	-156.5	79.3	1.2688
700.	969.	117.7	-332.2	-319.8	-304.1	-285.8	-265.0	108.4	1.5707
800.	913.	170.3	-481.8	-467.6	-449.6	-428.7	-405.0	140.2	1.8877
900.	862.	237.1	-669.6	-653.7	-633.4	-609.9	-583.2	175.8	2.2261
1000.	816.	319.7	-899.1	-881.4	-858.9	-832.7	-803.1	214.8	2.5838

BALLISTIC COEFFICIENT .28, MUZZLE VELOCITY FT/SEC 2300.

RANGE	VELOCITY	HEIGHT	100 YD	150 YD	200 YD	250 YD	300 YD	DRIFT	TIME
0.	2300.	0.0	-1.5	-1.5	-1.5	-1.5	-1.5	0.0	0.0000
100.	2009.	.2	0.0	1.6	3.6	6.0	8.6	1.6	.1395
200.	1743.	3.6	-7.2	-4.1	0.0	4.7	10.1	6.9	.2999
300.	1506.	10.5	-25.9	-21.2	-15.1	-8.0	0.0	16.5	.4851
400.	1306.	22.7	-60.0	-53.8	-45.6	-36.2	-25.5	31.3	.6995
500.	1154.	42.2	-115.0	-107.1	-97.0	-85.2	-71.8	51.7	.9458
600.	1058.	70.3	-195.1	-185.7	-173.5	-159.4	-143.3	76.4	1.2169
700.	990.	109.1	-305.6	-294.7	-280.4	-263.9	-245.2	105.5	1.5125
800.	931.	158.8	-446.0	-433.5	-417.2	-398.4	-377.0	137.2	1.8231
900.	878.	222.2	-623.2	-609.2	-590.9	-569.6	-545.6	172.7	2.1553
1000.	832.	300.8	-840.2	-824.5	-804.2	-780.6	-753.9	211.5	2.5063

BALLISTIC COEFFICIENT .28, MUZZLE VELOCITY FT/SEC 2400.

RANGE	VELOCITY	HEIGHT	100 YD	150 YD	200 YD	250 YD	300 YD	DRIFT	TIME
0.	2400.	0.0	-1.5	-1.5	-1.5	-1.5	-1.5	0.0	0.0000
100.	2103.	.1	0.0	1.4	3.2	5.4	7.8	1.5	.1336
200.	1828.	3.2	-6.4	-3.7	0.0	4.3	9.1	6.4	.2866
300.	1581.	9.6	-23.4	-19.2	-13.7	-7.3	0.0	15.5	.4633
400.	1366.	20.6	-54.3	-48.7	-41.4	-32.8	-23.1	29.5	.6676
500.	1199.	38.4	-103.9	-96.9	-87.8	-77.1	-64.9	48.9	.9027
600.	1085.	64.6	-177.8	-169.4	-158.5	-145.7	-131.0	73.4	1.1668
700.	1011.	101.0	-280.8	-271.1	-258.3	-243.4	-226.3	102.2	1.4559
800.	950.	148.0	-412.6	-401.5	-386.9	-369.8	-350.3	133.8	1.7601
900.	895.	208.3	-580.3	-567.7	-551.3	-532.1	-510.1	169.3	2.0870
1000.	847.	282.8	-784.8	-770.9	-752.7	-731.3	-706.9	207.8	2.4305

BALLISTIC COEFFICIENT .28, MUZZLE VELOCITY FT/SEC 2500.

RANGE	VELOCITY	HEIGHT	100 YD	150 YD	200 YD	250 YD	300 YD	DRIFT	TIME
0.	2500.	0.0	-1.5	-1.5	-1.5	-1.5	-1.5	0.0	0.0000
100.	2196.	.0	0.0	1.2	2.9	4.8	7.0	1.4	.1279
200.	1913.	2.9	-5.8	-3.3	0.0	3.9	8.3	6.1	.2744
300.	1657.	8.7	-21.1	-17.4	-12.4	-6.6	0.0	14.6	.4429
400.	1430.	18.8	-49.2	-44.2	-37.6	-29.8	-21.0	27.8	.6379
500.	1248.	35.0	-94.4	-88.1	-79.9	-70.2	-59.1	46.3	.8630
600.	1115.	59.3	-162.0	-154.6	-144.6	-133.0	-119.8	70.1	1.1184
700.	1033.	93.1	-257.0	-248.3	-236.8	-223.2	-207.7	98.3	1.3986
800.	968.	138.1	-382.5	-372.5	-359.3	-343.7	-326.1	130.3	1.7005
900.	912.	194.7	-538.9	-527.7	-512.8	-495.3	-475.5	165.1	2.0180
1000.	862.	265.8	-733.1	-720.6	-704.1	-684.7	-662.6	203.6	2.3566

BALLISTIC COEFFICIENT .28, MUZZLE VELOCITY FT/SEC 2600.

RANGE	VELOCITY	HEIGHT	100 YD	150 YD	200 YD	250 YD	300 YD	DRIFT	TIME
0.	2600.	0.0	-1.5	-1.5	-1.5	-1.5	-1.5	0.0	0.0000
100.	2290.	-.0	0.0	1.1	2.6	4.4	6.4	1.3	.1229
200.	2000.	2.6	-5.2	-3.0	0.0	3.5	7.6	5.7	.2631
300.	1735.	7.9	-19.1	-15.8	-11.3	-6.1	0.0	13.7	.4242
400.	1499.	17.1	-44.6	-40.2	-34.3	-27.2	-19.1	26.2	.6102
500.	1300.	32.0	-85.8	-80.3	-72.8	-64.0	-53.9	43.8	.8256
600.	1150.	54.4	-147.7	-141.1	-132.1	-121.5	-109.4	66.8	1.0718
700.	1056.	86.1	-235.7	-228.0	-217.5	-205.2	-191.1	94.5	1.3449
800.	988.	128.5	-353.4	-344.6	-332.7	-318.5	-302.4	126.3	1.6407
900.	929.	182.2	-501.0	-491.1	-477.7	-461.8	-443.6	160.8	1.9523
1000.	877.	250.0	-685.4	-674.4	-659.4	-641.8	-621.6	199.2	2.2858

BALLISTIC COEFFICIENT .28, MUZZLE VELOCITY FT/SEC 2700.

RANGE	VELOCITY	HEIGHT	100 YD	150 YD	200 YD	250 YD	300 YD	DRIFT	TIME
0.	2700.	0.0	-1.5	-1.5	-1.5	-1.5	-1.5	0.0	0.0000
100.	2384.	-.1	0.0	1.0	2.3	4.0	5.8	1.3	.1183
200.	2088.	2.3	-4.7	-2.7	0.0	3.2	6.9	5.4	.2527
300.	1814.	7.2	-17.4	-14.4	-10.4	-5.5	0.0	12.9	.4069
400.	1569.	15.7	-40.7	-36.8	-31.4	-24.9	-17.6	24.7	.5850
500.	1356.	29.3	-78.3	-73.4	-66.6	-58.5	-49.3	41.5	.7911
600.	1191.	49.9	-134.9	-129.0	-120.9	-111.1	-100.1	63.5	1.0276
700.	1080.	79.5	-216.2	-209.4	-199.8	-188.5	-175.6	90.7	1.2931
800.	1008.	119.5	-326.5	-318.7	-307.8	-294.8	-280.2	122.1	1.5826
900.	946.	170.7	-466.6	-457.8	-445.5	-430.9	-414.4	156.6	1.8900
1000.	892.	234.8	-639.7	-629.9	-616.3	-600.1	-581.8	194.3	2.2150

BALLISTIC COEFFICIENT .28, MUZZLE VELOCITY FT/SEC 2800.

RANGE	VELOCITY	HEIGHT	100 YD	150 YD	200 YD	250 YD	300 YD	DRIFT	TIME
0.	2800.	0.0	-1.5	-1.5	-1.5	-1.5	-1.5	0.0	0.0000
100.	2477.	-.1	0.0	.9	2.1	3.6	5.3	1.2	.1137
200.	2175.	2.1	-4.2	-2.5	0.0	3.0	6.3	5.0	.2430
300.	1893.	6.6	-15.8	-13.2	-9.5	-5.1	0.0	12.2	.3909
400.	1640.	14.4	-37.2	-33.7	-28.8	-22.8	-16.1	23.4	.5613
500.	1415.	26.9	-71.5	-67.2	-61.0	-53.5	-45.1	39.2	.7583
600.	1236.	45.9	-123.5	-118.3	-110.9	-102.0	-91.9	60.4	.9860
700.	1108.	73.4	-198.5	-192.4	-183.7	-173.3	-161.5	86.8	1.2432
800.	1028.	111.1	-301.6	-294.6	-284.7	-272.8	-259.4	117.8	1.5264
900.	964.	159.5	-433.4	-425.6	-414.4	-401.0	-385.9	151.9	1.8273
1000.	908.	220.6	-597.7	-589.0	-576.6	-561.7	-544.9	189.3	2.1472

BALLISTIC COEFFICIENT .28, MUZZLE VELOCITY FT/SEC 2900.

RANGE	VELOCITY	HEIGHT	100 YD	150 YD	200 YD	250 YD	300 YD	DRIFT	TIME
0.	2900.	0.0	-1.5	-1.5	-1.5	-1.5	-1.5	0.0	0.0000
100.	2570.	-.2	0.0	.8	1.9	3.3	4.8	1.1	.1099
200.	2262.	1.9	-3.8	-2.3	0.0	2.7	5.8	4.8	.2344
300.	1974.	6.0	-14.5	-12.1	-8.7	-4.7	0.0	11.6	.3763
400.	1711.	13.2	-34.1	-31.0	-26.4	-21.0	-14.8	22.1	.5395
500.	1478.	24.7	-65.6	-61.7	-56.0	-49.3	-41.5	37.1	.7283
600.	1284.	42.3	-113.3	-108.6	-101.8	-93.7	-84.3	57.4	.9467
700.	1139.	67.9	-182.4	-176.9	-169.0	-159.5	-148.6	83.0	1.1958
800.	1048.	103.4	-279.0	-272.8	-263.7	-252.9	-240.4	113.6	1.4733
900.	982.	149.2	-402.8	-395.8	-385.6	-373.4	-359.4	147.2	1.7673
1000.	924.	207.3	-558.5	-550.8	-539.5	-525.9	-510.4	184.3	2.0818

BALLISTIC COEFFICIENT .28, MUZZLE VELOCITY FT/SEC 3000.

RANGE	VELOCITY	HEIGHT	100 YD	150 YD	200 YD	250 YD	300 YD	DRIFT	TIME
0.	3000.	0.0	-1.5	-1.5	-1.5	-1.5	-1.5	0.0	0.0000
100.	2662.	-.2	0.0	.7	1.7	3.0	4.4	1.1	.1063
200.	2348.	1.7	-3.5	-2.1	0.0	2.5	5.4	4.6	.2264
300.	2054.	5.6	-13.3	-11.2	-8.1	-4.3	0.0	11.1	.3630
400.	1784.	12.2	-31.3	-28.6	-24.4	-19.4	-13.6	21.1	.5198
500.	1542.	22.8	-60.3	-56.9	-51.7	-45.4	-38.2	35.3	.7006
600.	1334.	39.0	-104.2	-100.1	-93.8	-86.3	-77.7	54.6	.9104
700.	1174.	62.9	-168.1	-163.3	-156.0	-147.2	-137.2	79.4	1.1513
800.	1071.	95.9	-256.9	-251.4	-243.1	-233.1	-221.6	108.9	1.4189
900.	1000.	139.5	-374.4	-368.2	-358.8	-347.6	-334.6	142.4	1.7092
1000.	940.	194.9	-522.2	-515.3	-504.9	-492.4	-478.0	179.3	2.0187

BALLISTIC COEFFICIENT .28, MUZZLE VELOCITY FT/SEC 3100.

RANGE	VELOCITY	HEIGHT	100 YD	150 YD	200 YD	250 YD	300 YD	DRIFT	TIME
0.	3100.	0.0	-1.5	-1.5	-1.5	-1.5	-1.5	0.0	0.0000
100.	2754.	-.2	0.0	.6	1.6	2.7	4.0	1.0	.1026
200.	2434.	1.5	-3.1	-1.9	0.0	2.3	5.0	4.4	.2184
300.	2135.	5.1	-12.1	-10.3	-7.5	-4.0	0.0	10.5	.3501
400.	1857.	11.3	-28.8	-26.4	-22.6	-17.9	-12.6	20.0	.5007
500.	1607.	21.1	-55.6	-52.6	-47.8	-42.0	-35.4	33.6	.6745
600.	1388.	36.1	-95.9	-92.3	-86.6	-79.6	-71.6	51.9	.8756
700.	1215.	58.1	-154.8	-150.5	-143.9	-135.8	-126.5	75.7	1.1074
800.	1095.	89.1	-237.5	-232.6	-225.0	-215.7	-205.1	104.6	1.3685
900.	1019.	130.4	-348.1	-342.6	-334.1	-323.6	-311.6	137.6	1.6531
1000.	956.	183.1	-488.2	-482.1	-472.6	-461.0	-447.7	174.1	1.9571

BALLISTIC COEFFICIENT .28, MUZZLE VELOCITY FT/SEC 3200.

RANGE	VELOCITY	HEIGHT	100 YD	150 YD	200 YD	250 YD	300 YD	DRIFT	TIME
0.	3200.	0.0	-1.5	-1.5	-1.5	-1.5	-1.5	0.0	0.0000
100.	2847.	-.3	0.0	.5	1.4	2.5	3.7	1.0	.0993
200.	2520.	1.4	-2.8	-1.8	0.0	2.1	4.6	4.2	.2116
300.	2215.	4.7	-11.2	-9.5	-6.9	-3.7	0.0	10.1	.3384
400.	1931.	10.5	-26.6	-24.5	-20.9	-16.6	-11.7	19.1	.4836
500.	1673.	19.6	-51.4	-48.7	-44.3	-38.9	-32.8	32.0	.6505
600.	1444.	33.4	-88.7	-85.5	-80.1	-73.7	-66.4	49.5	.8437
700.	1258.	53.9	-143.0	-139.2	-133.0	-125.5	-116.9	72.2	1.0667
800.	1122.	82.9	-219.7	-215.4	-208.4	-199.8	-190.0	100.4	1.3202
900.	1037.	122.0	-323.7	-318.8	-310.9	-301.2	-290.2	132.9	1.5989
1000.	973.	172.2	-456.7	-451.3	-442.5	-431.7	-419.5	169.0	1.8980

BALLISTIC COEFFICIENT .28, MUZZLE VELOCITY FT/SEC 3300.

RANGE	VELOCITY	HEIGHT	100 YD	150 YD	200 YD	250 YD	300 YD	DRIFT	TIME
0.	3300.	0.0	-1.5	-1.5	-1.5	-1.5	-1.5	0.0	0.0000
100.	2939.	-.3	0.0	.5	1.3	2.3	3.4	.9	.0963
200.	2606.	1.3	-2.6	-1.6	0.0	2.0	4.3	4.0	.2048
300.	2296.	4.4	-10.3	-8.8	-6.4	-3.4	0.0	9.6	.3275
400.	2005.	9.7	-24.6	-22.7	-19.5	-15.5	-10.9	18.3	.4673
500.	1740.	18.2	-47.6	-45.2	-41.2	-36.2	-30.4	30.5	.6280
600.	1503.	31.0	-82.1	-79.2	-74.4	-68.4	-61.5	47.2	.8135
700.	1303.	50.0	-132.3	-129.0	-123.3	-116.4	-108.3	69.0	1.0284
800.	1153.	77.2	-203.5	-199.7	-193.3	-185.3	-176.1	96.2	1.2740
900.	1057.	114.1	-301.0	-296.7	-289.5	-280.5	-270.2	128.2	1.5466
1000.	989.	162.2	-428.0	-423.2	-415.2	-405.2	-393.7	164.3	1.8424

BALLISTIC COEFFICIENT .28, MUZZLE VELOCITY FT/SEC 3400.

RANGE	VELOCITY	HEIGHT	100 YD	150 YD	200 YD	250 YD	300 YD	DRIFT	TIME
0.	3400.	0.0	-1.5	-1.5	-1.5	-1.5	-1.5	0.0	0.0000
100.	3032.	-.3	0.0	.4	1.2	2.1	3.2	.9	.0933
200.	2691.	1.1	-2.3	-1.5	0.0	1.9	4.0	3.9	.1984
300.	2376.	4.1	-9.5	-8.2	-6.0	-3.2	0.0	9.2	.3171
400.	2080.	9.1	-22.7	-21.1	-18.1	-14.4	-10.1	17.4	.4519
500.	1807.	16.9	-44.1	-42.0	-38.3	-33.7	-28.3	29.2	.6068
600.	1563.	28.9	-76.2	-73.7	-69.3	-63.7	-57.3	45.1	.7856
700.	1351.	46.5	-122.7	-119.8	-114.6	-108.1	-100.7	66.0	.9924
800.	1187.	71.8	-188.8	-185.5	-179.6	-172.1	-163.6	92.2	1.2298
900.	1078.	106.7	-280.0	-276.3	-269.6	-261.2	-251.6	123.5	1.4959
1000.	1006.	152.3	-400.0	-395.8	-388.4	-379.1	-368.4	159.0	1.7857

BALLISTIC COEFFICIENT .28, MUZZLE VELOCITY FT/SEC 3500.

RANGE	VELOCITY	HEIGHT	100 YD	150 YD	200 YD	250 YD	300 YD	DRIFT	TIME
0.	3500.	0.0	-1.5	-1.5	-1.5	-1.5	-1.5	0.0	0.0000
100.	3124.	-.4	0.0	.4	1.0	1.9	2.9	.9	.0910
200.	2777.	1.0	-2.1	-1.4	0.0	1.7	3.7	3.7	.1926
300.	2455.	3.8	-8.7	-7.6	-5.6	-3.0	0.0	8.9	.3076
400.	2155.	8.5	-21.1	-19.7	-16.9	-13.5	-9.5	16.7	.4380
500.	1875.	15.8	-40.9	-39.2	-35.8	-31.4	-26.5	27.9	.5872
600.	1623.	26.9	-70.8	-68.7	-64.6	-59.4	-53.4	43.1	.7594
700.	1401.	43.3	-113.9	-111.5	-106.7	-100.6	-93.7	63.1	.9585
800.	1226.	67.0	-175.4	-172.6	-167.1	-160.2	-152.2	88.4	1.1880
900.	1101.	99.8	-260.6	-257.4	-251.3	-243.5	-234.5	119.0	1.4474
1000.	1023.	143.1	-373.8	-370.3	-363.4	-354.8	-344.8	153.8	1.7312

BALLISTIC COEFFICIENT .28, MUZZLE VELOCITY FT/SEC 3600.

RANGE	VELOCITY	HEIGHT	100 YD	150 YD	200 YD	250 YD	300 YD	DRIFT	TIME
0.	3600.	0.0	-1.5	-1.5	-1.5	-1.5	-1.5	0.0	0.0000
100.	3217.	-.4	0.0	.3	.9	1.7	2.7	.8	.0881
200.	2862.	.9	-1.9	-1.3	0.0	1.6	3.5	3.5	.1868
300.	2535.	3.5	-8.0	-7.1	-5.2	-2.8	0.0	8.5	.2984
400.	2229.	7.9	-19.6	-18.4	-15.8	-12.6	-8.9	16.1	.4246
500.	1943.	14.8	-38.2	-36.6	-33.5	-29.4	-24.7	26.8	.5688
600.	1684.	25.1	-65.9	-64.1	-60.3	-55.5	-49.8	41.3	.7346
700.	1454.	40.4	-106.1	-104.0	-99.6	-93.9	-87.4	60.4	.9265
800.	1266.	62.5	-163.3	-160.8	-155.8	-149.3	-141.8	84.7	1.1480
900.	1127.	93.4	-242.8	-240.0	-234.3	-227.1	-218.6	114.4	1.4002
1000.	1041.	134.4	-349.4	-346.3	-340.0	-331.9	-322.6	148.7	1.6780

BALLISTIC COEFFICIENT .28, MUZZLE VELOCITY FT/SEC 3700.

RANGE	VELOCITY	HEIGHT	100 YD	150 YD	200 YD	250 YD	300 YD	DRIFT	TIME
0.	3700.	0.0	-1.5	-1.5	-1.5	-1.5	-1.5	0.0	0.0000
100.	3310.	-.4	0.0	.3	.8	1.6	2.5	.8	.0858
200.	2948.	.8	-1.7	-1.1	0.0	1.5	3.3	3.5	.1818
300.	2614.	3.3	-7.4	-6.6	-4.9	-2.6	0.0	8.2	.2899
400.	2304.	7.4	-18.2	-17.1	-14.9	-11.9	-8.3	15.5	.4123
500.	2013.	13.9	-35.6	-34.2	-31.3	-27.6	-23.2	25.7	.5516
600.	1746.	23.6	-61.5	-59.9	-56.5	-52.0	-46.7	39.6	.7117
700.	1509.	37.8	-99.0	-97.1	-93.1	-87.8	-81.7	57.9	.8966
800.	1308.	58.5	-152.2	-150.0	-145.4	-139.4	-132.4	81.3	1.1106
900.	1155.	87.5	-226.7	-224.3	-219.1	-212.4	-204.5	110.3	1.3563
1000.	1059.	126.4	-326.8	-324.1	-318.4	-310.9	-302.1	143.7	1.6273

BALLISTIC COEFFICIENT .28, MUZZLE VELOCITY FT/SEC 3800.

RANGE	VELOCITY	HEIGHT	100 YD	150 YD	200 YD	250 YD	300 YD	DRIFT	TIME
0.	3800.	0.0	-1.5	-1.5	-1.5	-1.5	-1.5	0.0	0.0000
100.	3403.	-.4	0.0	.2	.8	1.5	2.3	.8	.0836
200.	3034.	.8	-1.5	-1.1	0.0	1.4	3.0	3.3	.1769
300.	2694.	3.1	-6.8	-6.2	-4.6	-2.5	0.0	7.9	.2819
400.	2378.	6.9	-17.0	-16.1	-13.9	-11.1	-7.8	14.9	.4005
500.	2082.	13.0	-33.2	-32.1	-29.4	-25.9	-21.8	24.7	.5352
600.	1809.	22.1	-57.5	-56.1	-52.9	-48.7	-43.8	38.1	.6899
700.	1564.	35.5	-92.5	-91.0	-87.2	-82.3	-76.6	55.6	.8685
800.	1352.	54.7	-142.1	-140.3	-136.0	-130.5	-123.9	78.1	1.0751
900.	1188.	81.9	-211.4	-209.4	-204.6	-198.3	-190.9	105.9	1.3123
1000.	1079.	118.8	-305.8	-303.6	-298.3	-291.3	-283.0	138.8	1.5782

BALLISTIC COEFFICIENT .28, MUZZLE VELOCITY FT/SEC 3900.

RANGE	VELOCITY	HEIGHT	100 YD	150 YD	200 YD	250 YD	300 YD	DRIFT	TIME
0.	3900.	0.0	-1.5	-1.5	-1.5	-1.5	-1.5	0.0	0.0000
100.	3496.	-.4	0.0	.2	.7	1.3	2.1	.7	.0811
200.	3120.	.7	-1.4	-1.0	0.0	1.3	2.8	3.2	.1721
300.	2773.	2.9	-6.3	-5.8	-4.3	-2.3	0.0	7.6	.2739
400.	2452.	6.5	-15.8	-15.0	-13.0	-10.5	-7.4	14.3	.3890
500.	2151.	12.2	-31.1	-30.1	-27.6	-24.4	-20.5	23.8	.5197
600.	1872.	20.7	-53.8	-52.6	-49.7	-45.8	-41.1	36.5	.6691
700.	1621.	33.2	-86.6	-85.2	-81.7	-77.2	-71.8	53.3	.8415
800.	1399.	51.3	-132.9	-131.4	-127.4	-122.2	-116.0	74.9	1.0410
900.	1224.	76.8	-197.7	-195.9	-191.5	-185.6	-178.7	101.8	1.2708
1000.	1100.	111.7	-286.4	-284.5	-279.5	-273.0	-265.3	134.0	1.5305

BALLISTIC COEFFICIENT .28, MUZZLE VELOCITY FT/SEC 4000.

RANGE	VELOCITY	HEIGHT	100 YD	150 YD	200 YD	250 YD	300 YD	DRIFT	TIME
0.	4000.	0.0	-1.5	-1.5	-1.5	-1.5	-1.5	0.0	0.0000
100.	3589.	-.4	0.0	.2	.6	1.2	1.9	.8	.0794
200.	3206.	.6	-1.2	-.9	0.0	1.2	2.7	3.1	.1678
300.	2853.	2.7	-5.8	-5.4	-4.0	-2.2	0.0	7.4	.2669
400.	2526.	6.1	-14.7	-14.1	-12.3	-9.9	-7.0	13.9	.3789
500.	2221.	11.5	-29.0	-28.3	-26.0	-23.0	-19.3	22.9	.5054
600.	1935.	19.5	-50.4	-49.5	-46.8	-43.1	-38.8	35.2	.6502
700.	1677.	31.3	-81.1	-80.1	-76.9	-72.6	-67.6	51.3	.8167
800.	1448.	48.2	-124.5	-123.2	-119.6	-114.7	-108.9	72.1	1.0094
900.	1261.	72.1	-185.0	-183.6	-179.5	-174.1	-167.6	98.0	1.2318
1000.	1124.	105.1	-268.3	-266.8	-262.2	-256.1	-248.9	129.3	1.4849

BALLISTIC COEFFICIENT .29, MUZZLE VELOCITY FT/SEC 2000.

RANGE	VELOCITY	HEIGHT	100 YD	150 YD	200 YD	250 YD	300 YD	DRIFT	TIME
0.	2000.	0.0	-1.5	-1.5	-1.5	-1.5	-1.5	0.0	0.0000
100.	1744.	.5	0.0	2.2	5.0	8.2	11.8	1.9	.1607
200.	1514.	5.0	-10.0	-5.5	0.0	6.4	13.5	8.0	.3454
300.	1318.	14.2	-35.4	-28.6	-20.3	-10.8	0.0	19.0	.5582
400.	1168.	30.0	-80.7	-71.7	-60.6	-47.9	-33.6	35.3	.8006
500.	1069.	54.2	-151.5	-140.2	-126.4	-110.5	-92.5	56.3	1.0701
600.	1001.	88.1	-251.3	-237.8	-221.2	-202.1	-180.5	81.1	1.3606
700.	943.	133.1	-382.9	-367.2	-347.8	-325.5	-300.4	109.1	1.6701
800.	891.	190.7	-549.6	-531.6	-509.4	-484.0	-455.3	140.3	1.9970
900.	844.	262.8	-756.1	-735.9	-710.9	-682.3	-650.0	174.8	2.3431
1000.	801.	351.2	-1006.3	-983.9	-956.1	-924.3	-888.5	212.	2.7080

BALLISTIC COEFFICIENT .29, MUZZLE VELOCITY FT/SEC 2100.

RANGE	VELOCITY	HEIGHT	100 YD	150 YD	200 YD	250 YD	300 YD	DRIFT	TIME
0.	2100.	0.0	-1.5	-1.5	-1.5	-1.5	-1.5	0.0	0.0000
100.	1834.	.4	0.0	2.0	4.5	7.3	10.5	1.8	.1529
200.	1595.	4.4	-8.9	-5.0	0.0	5.7	12.1	7.5	.3284
300.	1384.	12.8	-31.6	-25.7	-18.2	-9.7	0.0	17.9	.5304
400.	1218.	27.1	-72.5	-64.6	-54.6	-43.3	-30.4	33.6	.7621
500.	1100.	49.4	-136.9	-127.0	-114.5	-100.3	-84.2	54.2	1.0223
600.	1024.	81.2	-229.5	-217.7	-202.7	-185.6	-166.3	79.1	1.3065
700.	963.	123.4	-352.1	-338.3	-320.8	-300.9	-278.4	107.1	1.6084
800.	909.	177.8	-508.3	-492.5	-472.6	-449.8	-424.1	138.3	1.9286
900.	861.	246.1	-702.6	-684.8	-662.4	-636.8	-607.8	172.9	2.2678
1000.	816.	330.2	-939.1	-919.4	-894.4	-866.0	-833.8	210.7	2.6260

BALLISTIC COEFFICIENT .29, MUZZLE VELOCITY FT/SEC 2200.

RANGE	VELOCITY	HEIGHT	100 YD	150 YD	200 YD	250 YD	300 YD	DRIFT	TIME
0.	2200.	0.0	-1.5	-1.5	-1.5	-1.5	-1.5	0.0	0.0000
100.	1926.	.3	0.0	1.7	4.0	6.6	9.4	1.7	.1459
200.	1677.	3.9	-7.9	-4.5	0.0	5.2	10.9	7.1	.3128
300.	1455.	11.5	-28.3	-23.1	-16.4	-8.7	0.0	16.9	.5050
400.	1272.	24.5	-65.2	-58.3	-49.3	-39.0	-27.4	31.8	.7260
500.	1135.	45.1	-124.0	-115.3	-104.2	-91.3	-76.8	52.1	.9777
600.	1048.	74.8	-209.7	-199.2	-185.8	-170.3	-153.0	76.8	1.2547
700.	984.	114.4	-323.7	-311.5	-295.9	-277.8	-257.6	104.6	1.5488
800.	928.	165.7	-470.1	-456.1	-438.3	-417.6	-394.5	135.7	1.8620
900.	877.	230.7	-653.5	-637.8	-617.8	-594.6	-568.5	170.4	2.1957
1000.	832.	310.5	-876.4	-858.9	-836.6	-810.9	-781.9	208.1	2.5462

BALLISTIC COEFFICIENT .29, MUZZLE VELOCITY FT/SEC 2300.

RANGE	VELOCITY	HEIGHT	100 YD	150 YD	200 YD	250 YD	300 YD	DRIFT	TIME
0.	2300.	0.0	-1.5	-1.5	-1.5	-1.5	-1.5	0.0	0.0000
100.	2019.	.2	0.0	1.5	3.5	5.9	8.5	1.5	.1392
200.	1761.	3.5	-7.1	-4.0	0.0	4.7	9.9	6.6	.2983
300.	1529.	10.4	-25.5	-20.9	-14.9	-7.9	0.0	15.8	.4813
400.	1331.	22.2	-58.8	-52.6	-44.6	-35.3	-24.8	30.0	.6920
500.	1176.	41.0	-112.0	-104.3	-94.3	-82.6	-69.5	49.4	.9330
600.	1075.	68.4	-190.0	-180.7	-168.7	-154.8	-139.0	73.5	1.2001
700.	1005.	105.9	-297.3	-286.4	-272.4	-256.1	-237.8	101.6	1.4903
800.	946.	154.5	-435.2	-422.9	-406.9	-388.2	-367.2	132.9	1.7985
900.	894.	216.0	-607.7	-593.8	-575.7	-554.8	-531.2	167.4	2.1249
1000.	847.	291.6	-817.7	-802.2	-782.2	-758.9	-732.7	204.8	2.4681

BALLISTIC COEFFICIENT .29, MUZZLE VELOCITY FT/SEC 2400.

RANGE	VELOCITY	HEIGHT	100 YD	150 YD	200 YD	250 YD	300 YD	DRIFT	TIME
0.	2400.	0.0	-1.5	-1.5	-1.5	-1.5	-1.5	0.0	0.0000
100.	2113.	.1	0.0	1.4	3.2	5.3	7.7	1.4	.1331
200.	1846.	3.2	-6.4	-3.6	0.0	4.2	9.0	6.2	.2852
300.	1605.	9.4	-23.0	-18.9	-13.5	-7.2	0.0	14.9	.4596
400.	1393.	20.2	-53.2	-47.7	-40.4	-32.0	-22.5	28.2	.6603
500.	1225.	37.3	-101.4	-94.5	-85.4	-75.0	-63.0	46.7	.8905
600.	1104.	62.7	-172.9	-164.7	-153.8	-141.2	-126.9	70.4	1.1497
700.	1027.	97.9	-272.7	-263.1	-250.4	-235.8	-219.0	98.3	1.4334
800.	966.	143.7	-401.9	-390.9	-376.4	-359.6	-340.5	129.3	1.7346
900.	911.	201.6	-563.4	-551.0	-534.7	-515.9	-494.4	163.4	2.0533
1000.	863.	273.8	-762.7	-749.0	-730.8	-709.9	-686.0	201.0	2.3918

BALLISTIC COEFFICIENT .29, MUZZLE VELOCITY FT/SEC 2500.

RANGE	VELOCITY	HEIGHT	100 YD	150 YD	200 YD	250 YD	300 YD	DRIFT	TIME
0.	2500.	0.0	-1.5	-1.5	-1.5	-1.5	-1.5	0.0	0.0000
100.	2207.	.0	0.0	1.2	2.9	4.8	6.9	1.3	.1276
200.	1932.	2.8	-5.7	-3.3	0.0	3.8	8.1	5.8	.2731
300.	1683.	8.5	-20.8	-17.1	-12.2	-6.5	0.0	14.0	.4394
400.	1460.	18.4	-48.2	-43.3	-36.8	-29.1	-20.5	26.6	.6310
500.	1276.	34.1	-92.1	-86.0	-77.8	-68.3	-57.4	44.3	.8515
600.	1137.	57.5	-157.6	-150.2	-140.4	-129.0	-116.0	67.2	1.1017
700.	1050.	90.2	-249.4	-240.8	-229.3	-216.0	-200.8	94.4	1.3764
800.	985.	133.6	-371.0	-361.2	-348.1	-332.8	-315.5	125.5	1.6728
900.	929.	188.5	-523.1	-512.1	-497.4	-480.2	-460.8	159.3	1.9854
1000.	878.	257.2	-711.7	-699.5	-683.1	-664.0	-642.4	196.8	2.3181

BALLISTIC COEFFICIENT .29, MUZZLE VELOCITY FT/SEC 2600.

RANGE	VELOCITY	HEIGHT	100 YD	150 YD	200 YD	250 YD	300 YD	DRIFT	TIME
0.	2600.	0.0	-1.5	-1.5	-1.5	-1.5	-1.5	0.0	0.0000
100.	2301.	-.0	0.0	1.1	2.6	4.3	6.3	1.3	.1227
200.	2020.	2.5	-5.1	-3.0	0.0	3.5	7.4	5.5	.2618
300.	1761.	7.8	-18.8	-15.6	-11.1	-5.9	0.0	13.2	.4209
400.	1530.	16.8	-43.8	-39.5	-33.5	-26.6	-18.7	25.0	.6038
500.	1331.	31.1	-83.7	-78.3	-70.9	-62.2	-52.3	41.8	.8145
600.	1176.	52.7	-143.6	-137.1	-128.2	-117.8	-105.9	63.9	1.0554
700.	1075.	83.2	-228.3	-220.7	-210.4	-198.2	-184.4	90.6	1.3225
800.	1006.	124.1	-342.2	-333.5	-321.7	-307.8	-292.0	121.4	1.6127
900.	947.	176.4	-486.4	-476.7	-463.3	-447.7	-429.9	155.3	1.9208
1000.	894.	241.6	-664.5	-653.7	-638.9	-621.5	-601.8	192.4	2.2471

BALLISTIC COEFFICIENT .29, MUZZLE VELOCITY FT/SEC 2700.

RANGE	VELOCITY	HEIGHT	100 YD	150 YD	200 YD	250 YD	300 YD	DRIFT	TIME
0.	2700.	0.0	-1.5	-1.5	-1.5	-1.5	-1.5	0.0	0.0000
100.	2394.	-.1	0.0	1.0	2.3	3.9	5.7	1.2	.1181
200.	2107.	2.3	-4.6	-2.7	0.0	3.2	6.8	5.2	.2516
300.	1841.	7.1	-17.1	-14.2	-10.2	-5.4	0.0	12.4	.4040
400.	1601.	15.3	-39.9	-36.1	-30.7	-24.3	-17.1	23.6	.5788
500.	1389.	28.5	-76.3	-71.5	-64.8	-56.8	-47.8	39.5	.7801
600.	1222.	48.3	-130.9	-125.1	-117.1	-107.5	-96.7	60.6	1.0108
700.	1102.	76.7	-209.1	-202.4	-193.0	-181.9	-169.2	86.7	1.2705
800.	1026.	115.2	-315.6	-307.8	-297.1	-284.4	-269.9	117.1	1.5544
900.	965.	164.6	-451.1	-442.4	-430.3	-416.0	-399.7	150.6	1.8559
1000.	910.	226.4	-618.8	-609.2	-595.8	-579.9	-561.7	187.3	2.1752

BALLISTIC COEFFICIENT .29, MUZZLE VELOCITY FT/SEC 2800.

RANGE	VELOCITY	HEIGHT	100 YD	150 YD	200 YD	250 YD	300 YD	DRIFT	TIME
0.	2800.	0.0	-1.5	-1.5	-1.5	-1.5	-1.5	0.0	0.0000
100.	2488.	-.1	0.0	.9	2.1	3.5	5.2	1.1	.1135
200.	2195.	2.1	-4.2	-2.4	0.0	2.9	6.2	4.9	.2419
300.	1921.	6.5	-15.6	-13.0	-9.4	-5.0	0.0	11.7	.3882
400.	1673.	14.1	-36.5	-33.0	-28.2	-22.3	-15.7	22.3	.5554
500.	1451.	26.1	-69.8	-65.5	-59.4	-52.1	-43.8	37.4	.7482
600.	1270.	44.4	-119.9	-114.7	-107.4	-98.6	-88.6	57.5	.9696
700.	1133.	70.8	-191.8	-185.7	-177.2	-166.9	-155.3	82.8	1.2206
800.	1047.	107.1	-291.3	-284.4	-274.7	-263.0	-249.7	112.9	1.4988
900.	983.	153.6	-418.5	-410.8	-399.8	-386.6	-371.7	145.9	1.7934
1000.	927.	212.4	-577.3	-568.7	-556.5	-541.9	-525.3	182.3	2.1071

BALLISTIC COEFFICIENT .29, MUZZLE VELOCITY FT/SEC 2900.

RANGE	VELOCITY	HEIGHT	100 YD	150 YD	200 YD	250 YD	300 YD	DRIFT	TIME
0.	2900.	0.0	-1.5	-1.5	-1.5	-1.5	-1.5	0.0	0.0000
100.	2581.	-.2	0.0	.8	1.9	3.2	4.7	1.1	.1097
200.	2282.	1.9	-3.8	-2.2	0.0	2.7	5.7	4.6	.2333
300.	2002.	6.0	-14.2	-12.0	-8.6	-4.6	0.0	11.1	.3736
400.	1746.	12.9	-33.4	-30.4	-25.9	-20.5	-14.4	21.2	.5341
500.	1516.	24.0	-64.0	-60.2	-54.6	-47.9	-40.3	35.4	.7186
600.	1320.	40.9	-109.9	-105.4	-98.6	-90.6	-81.4	54.6	.9311
700.	1169.	65.3	-176.1	-170.8	-162.9	-153.5	-142.9	79.1	1.1733
800.	1070.	99.1	-267.9	-261.8	-252.9	-242.1	-229.9	108.2	1.4425
900.	1002.	143.4	-388.3	-381.4	-371.4	-359.2	-345.5	141.2	1.7330
1000.	943.	199.5	-539.1	-531.6	-520.4	-506.9	-491.7	177.4	2.0423

BALLISTIC COEFFICIENT .29, MUZZLE VELOCITY FT/SEC 3000.

RANGE	VELOCITY	HEIGHT	100 YD	150 YD	200 YD	250 YD	300 YD	DRIFT	TIME
0.	3000.	0.0	-1.5	-1.5	-1.5	-1.5	-1.5	0.0	0.0000
100.	2673.	-.2	0.0	.7	1.7	2.9	4.3	1.1	.1060
200.	2369.	1.7	-3.4	-2.0	0.0	2.5	5.3	4.5	.2254
300.	2084.	5.5	-13.0	-11.0	-7.9	-4.2	0.0	10.6	.3603
400.	1819.	11.9	-30.7	-28.0	-23.9	-18.9	-13.3	20.1	.5144
500.	1582.	22.2	-59.0	-55.5	-50.4	-44.3	-37.2	33.7	.6915
600.	1373.	37.7	-101.1	-97.0	-90.9	-83.5	-75.0	52.0	.8952
700.	1209.	60.4	-162.1	-157.3	-150.2	-141.5	-131.7	75.4	1.1286
800.	1095.	92.0	-247.3	-241.8	-233.6	-223.7	-212.5	103.9	1.3904
900.	1021.	133.9	-360.3	-354.2	-345.0	-333.9	-321.2	136.4	1.6749
1000.	960.	187.0	-502.7	-495.8	-485.6	-473.3	-459.2	172.1	1.9778

BALLISTIC COEFFICIENT .29, MUZZLE VELOCITY FT/SEC 3100.

RANGE	VELOCITY	HEIGHT	100 YD	150 YD	200 YD	250 YD	300 YD	DRIFT	TIME
0.	3100.	0.0	-1.5	-1.5	-1.5	-1.5	-1.5	0.0	0.0000
100.	2766.	-.2	0.0	.6	1.5	2.7	4.0	1.0	.1024
200.	2456.	1.5	-3.1	-1.9	0.0	2.3	4.9	4.2	.2175
300.	2165.	5.0	-11.9	-10.2	-7.3	-3.9	0.0	10.1	.3475
400.	1894.	11.0	-28.3	-25.9	-22.1	-17.5	-12.3	19.1	.4958
500.	1648.	20.5	-54.3	-51.3	-46.6	-40.9	-34.4	32.0	.6657
600.	1430.	34.9	-93.2	-89.6	-84.0	-77.1	-69.3	49.4	.8613
700.	1253.	55.9	-149.5	-145.4	-138.8	-130.8	-121.7	72.0	1.0863
800.	1122.	85.4	-228.3	-223.5	-216.0	-206.8	-196.4	99.6	1.3399
900.	1040.	124.9	-334.2	-328.8	-320.4	-310.1	-298.4	131.5	1.6182
1000.	977.	175.5	-469.3	-463.3	-453.9	-442.5	-429.5	166.9	1.9163

BALLISTIC COEFFICIENT .29, MUZZLE VELOCITY FT/SEC 3200.

RANGE	VELOCITY	HEIGHT	100 YD	150 YD	200 YD	250 YD	300 YD	DRIFT	TIME
0.	3200.	0.0	-1.5	-1.5	-1.5	-1.5	-1.5	0.0	0.0000
100.	2858.	-.3	0.0	.5	1.4	2.5	3.7	.9	.0991
200.	2542.	1.4	-2.8	-1.7	0.0	2.1	4.5	4.1	.2106
300.	2246.	4.7	-11.0	-9.4	-6.8	-3.6	0.0	9.7	.3362
400.	1969.	10.3	-26.1	-24.0	-20.5	-16.3	-11.4	18.3	.4788
500.	1715.	19.0	-50.2	-47.5	-43.2	-37.9	-31.8	30.5	.6419
600.	1489.	32.3	-86.1	-82.9	-77.7	-71.4	-64.1	47.0	.8298
700.	1299.	51.8	-138.1	-134.4	-128.3	-120.9	-112.4	68.6	1.0461
800.	1153.	79.6	-211.6	-207.3	-200.4	-191.9	-182.3	95.6	1.2935
900.	1060.	116.7	-310.3	-305.5	-297.7	-288.2	-277.3	126.8	1.5640
1000.	994.	165.2	-439.6	-434.2	-425.6	-415.0	-402.9	162.3	1.8597

BALLISTIC COEFFICIENT .29, MUZZLE VELOCITY FT/SEC 3300.

RANGE	VELOCITY	HEIGHT	100 YD	150 YD	200 YD	250 YD	300 YD	DRIFT	TIME
0.	3300.	0.0	-1.5	-1.5	-1.5	-1.5	-1.5	0.0	0.0000
100.	2951.	-.3	0.0	.5	1.3	2.2	3.4	.9	.0961
200.	2628.	1.2	-2.5	-1.6	0.0	2.0	4.2	3.9	.2039
300.	2327.	4.3	-10.1	-8.7	-6.3	-3.4	0.0	9.3	.3253
400.	2044.	9.5	-24.2	-22.3	-19.1	-15.2	-10.7	17.5	.4630
500.	1783.	17.7	-46.5	-44.2	-40.2	-35.3	-29.6	29.1	.6201
600.	1550.	30.0	-79.8	-77.0	-72.2	-66.4	-59.6	44.9	.8006
700.	1347.	48.1	-127.9	-124.6	-119.0	-112.1	-104.2	65.5	1.0087
800.	1189.	73.8	-195.5	-191.8	-185.4	-177.6	-168.5	91.3	1.2463
900.	1082.	109.0	-288.4	-284.2	-277.1	-268.2	-258.1	122.2	1.5123
1000.	1011.	155.0	-410.1	-405.4	-397.5	-387.7	-376.4	157.0	1.8013

BALLISTIC COEFFICIENT .29, MUZZLE VELOCITY FT/SEC 3400.

RANGE	VELOCITY	HEIGHT	100 YD	150 YD	200 YD	250 YD	300 YD	DRIFT	TIME
0.	3400.	0.0	-1.5	-1.5	-1.5	-1.5	-1.5	0.0	0.0000
100.	3044.	-.3	0.0	.4	1.1	2.1	3.1	.9	.0931
200.	2714.	1.1	-2.3	-1.5	0.0	1.8	3.9	3.7	.1976
300.	2407.	4.0	-9.3	-8.1	-5.9	-3.2	0.0	8.9	.3150
400.	2120.	8.9	-22.3	-20.7	-17.8	-14.1	-9.9	16.7	.4477
500.	1852.	16.5	-43.1	-41.1	-37.4	-32.8	-27.6	27.8	.5992
600.	1611.	27.9	-74.1	-71.6	-67.2	-61.7	-55.4	42.9	.7730
700.	1397.	44.7	-118.6	-115.7	-110.6	-104.2	-96.8	62.6	.9732
800.	1228.	68.7	-181.4	-178.1	-172.2	-165.0	-156.5	87.4	1.2027
900.	1106.	101.7	-267.8	-264.2	-257.5	-249.3	-239.9	117.4	1.4612
1000.	1029.	145.3	-382.6	-378.6	-371.2	-362.1	-351.6	151.8	1.7446

BALLISTIC COEFFICIENT .29, MUZZLE VELOCITY FT/SEC 3500.

RANGE	VELOCITY	HEIGHT	100 YD	150 YD	200 YD	250 YD	300 YD	DRIFT	TIME
0.	3500.	0.0	-1.5	-1.5	-1.5	-1.5	-1.5	0.0	0.0000
100.	3137.	-.4	0.0	.3	1.0	1.9	2.9	.9	.0908
200.	2800.	1.0	-2.1	-1.4	0.0	1.7	3.6	3.6	.1920
300.	2488.	3.7	-8.6	-7.5	-5.5	-2.9	0.0	8.5	.3056
400.	2195.	8.3	-20.7	-19.3	-16.6	-13.2	-9.3	16.0	.4339
500.	1921.	15.4	-40.1	-38.3	-34.9	-30.7	-25.8	26.7	.5802
600.	1673.	26.1	-68.8	-66.8	-62.7	-57.6	-51.7	41.0	.7475
700.	1451.	41.7	-110.2	-107.8	-103.0	-97.1	-90.2	59.9	.9402
800.	1269.	64.0	-168.5	-165.7	-160.3	-153.5	-145.7	83.8	1.1617
900.	1133.	95.0	-249.0	-245.9	-239.7	-232.1	-223.3	112.9	1.4127
1000.	1047.	136.5	-357.4	-353.9	-347.1	-338.6	-328.8	146.7	1.6909

BALLISTIC COEFFICIENT .29, MUZZLE VELOCITY FT/SEC 3600.

RANGE	VELOCITY	HEIGHT	100 YD	150 YD	200 YD	250 YD	300 YD	DRIFT	TIME
0.	3600.	0.0	-1.5	-1.5	-1.5	-1.5	-1.5	0.0	0.0000
100.	3230.	-.4	0.0	.3	.9	1.7	2.6	.8	.0879
200.	2886.	.9	-1.8	-1.2	0.0	1.6	3.4	3.4	.1861
300.	2568.	3.5	-7.9	-7.0	-5.1	-2.8	0.0	8.2	.2964
400.	2270.	7.7	-19.2	-18.0	-15.5	-12.4	-8.7	15.4	.4207
500.	1991.	14.4	-37.3	-35.8	-32.7	-28.7	-24.1	25.5	.5617
600.	1736.	24.4	-64.1	-62.3	-58.6	-53.8	-48.3	39.3	.7232
700.	1507.	38.9	-102.6	-100.5	-96.2	-90.6	-84.2	57.3	.9087
800.	1313.	59.7	-156.8	-154.4	-149.4	-143.1	-135.8	80.2	1.1225
900.	1163.	88.8	-231.8	-229.1	-223.5	-216.4	-208.1	108.4	1.3659
1000.	1067.	127.7	-332.9	-329.9	-323.7	-315.7	-306.5	141.3	1.6362

BALLISTIC COEFFICIENT .29, MUZZLE VELOCITY FT/SEC 3700.

RANGE	VELOCITY	HEIGHT	100 YD	150 YD	200 YD	250 YD	300 YD	DRIFT	TIME
0.	3700.	0.0	-1.5	-1.5	-1.5	-1.5	-1.5	0.0	0.0000
100.	3323.	-.4	0.0	.3	.8	1.6	2.4	.8	.0857
200.	2972.	.8	-1.7	-1.1	0.0	1.5	3.2	3.3	.1811
300.	2648.	3.2	-7.3	-6.5	-4.8	-2.6	0.0	7.9	.2881
400.	2345.	7.3	-17.9	-16.8	-14.6	-11.6	-8.2	14.8	.4086
500.	2061.	13.5	-34.8	-33.5	-30.6	-26.9	-22.6	24.6	.5451
600.	1799.	22.8	-59.9	-58.3	-54.9	-50.4	-45.3	37.7	.7008
700.	1564.	36.4	-95.8	-94.0	-90.0	-84.8	-78.8	55.0	.8799
800.	1358.	55.9	-146.3	-144.2	-139.6	-133.7	-125.8	77.0	1.0861
900.	1198.	83.1	-216.3	-213.9	-208.8	-202.1	-194.4	104.2	1.3220
1000.	1087.	119.9	-311.1	-308.4	-302.7	-295.3	-286.7	136.4	1.5857

BALLISTIC COEFFICIENT .29, MUZZLE VELOCITY FT/SEC 3800.

RANGE	VELOCITY	HEIGHT	100 YD	150 YD	200 YD	250 YD	300 YD	DRIFT	TIME
0.	3800.	0.0	-1.5	-1.5	-1.5	-1.5	-1.5	0.0	0.0000
100.	3416.	-.4	0.0	.2	.7	1.4	2.2	.8	.0834
200.	3059.	.7	-1.5	-1.0	0.0	1.4	3.0	3.2	.1761
300.	2728.	3.0	-6.7	-6.1	-4.5	-2.4	0.0	7.6	.2801
400.	2420.	6.8	-16.7	-15.8	-13.7	-10.9	-7.7	14.3	.3970
500.	2132.	12.7	-32.5	-31.4	-28.8	-25.3	-21.3	23.6	.5290
600.	1863.	21.4	-55.9	-54.6	-51.5	-47.3	-42.5	36.2	.6795
700.	1621.	34.1	-89.5	-88.0	-84.3	-79.5	-73.8	52.7	.8522
800.	1406.	52.3	-136.6	-134.8	-130.6	-125.1	-118.6	73.9	1.0512
900.	1234.	77.9	-202.1	-200.0	-195.3	-189.1	-181.8	100.2	1.2798
1000.	1110.	112.6	-290.8	-288.6	-283.4	-276.4	-268.4	131.5	1.5368

BALLISTIC COEFFICIENT .29, MUZZLE VELOCITY FT/SEC 3900.

RANGE	VELOCITY	HEIGHT	100 YD	150 YD	200 YD	250 YD	300 YD	DRIFT	TIME
0.	3900.	0.0	-1.5	-1.5	-1.5	-1.5	-1.5	0.0	0.0000
100.	3509.	-.4	0.0	.2	.7	1.3	2.1	.7	.0808
200.	3145.	.7	-1.4	-1.0	0.0	1.3	2.8	3.1	.1714
300.	2808.	2.8	-6.2	-5.7	-4.2	-2.3	0.0	7.3	.2724
400.	2495.	6.4	-15.5	-14.8	-12.8	-10.3	-7.2	13.7	.3857
500.	2202.	11.9	-30.4	-29.4	-27.0	-23.8	-20.0	22.7	.5136
600.	1927.	20.1	-52.4	-51.3	-48.4	-44.5	-39.9	34.8	.6594
700.	1679.	32.0	-83.8	-82.5	-79.1	-74.6	-69.3	50.6	.8261
800.	1456.	49.0	-127.8	-126.3	-122.4	-117.3	-111.2	70.9	1.0182
900.	1273.	73.0	-188.9	-187.2	-182.8	-177.1	-170.2	96.3	1.2393
1000.	1135.	105.9	-272.4	-270.6	-265.7	-259.2	-251.6	126.9	1.4904

BALLISTIC COEFFICIENT .29, MUZZLE VELOCITY FT/SEC 4000.

RANGE	VELOCITY	HEIGHT	100 YD	150 YD	200 YD	250 YD	300 YD	DRIFT	TIME
0.	4000.	0.0	-1.5	-1.5	-1.5	-1.5	-1.5	0.0	0.0000
100.	3602.	-.4	0.0	.1	.6	1.2	1.9	.8	.0793
200.	3232.	.6	-1.2	-.9	0.0	1.2	2.6	3.0	.1671
300.	2888.	2.6	-5.7	-5.3	-3.9	-2.1	0.0	7.1	.2652
400.	2569.	6.0	-14.4	-13.9	-12.1	-9.7	-6.8	13.3	.3755
500.	2272.	11.2	-28.4	-27.7	-25.4	-22.4	-18.9	21.9	.4997
600.	1993.	18.9	-49.1	-48.2	-45.5	-41.9	-37.6	33.5	.6406
700.	1737.	30.1	-78.6	-77.5	-74.4	-70.2	-65.2	48.7	.8019
800.	1508.	46.0	-119.7	-118.6	-115.0	-110.2	-104.5	68.2	.9873
900.	1314.	68.5	-176.8	-175.5	-171.5	-166.1	-159.7	92.6	1.2012
1000.	1164.	99.4	-254.7	-253.2	-248.7	-242.8	-235.6	122.2	1.4442

BALLISTIC COEFFICIENT .30, MUZZLE VELOCITY FT/SEC 2000.

RANGE	VELOCITY	HEIGHT	100 YD	150 YD	200 YD	250 YD	300 YD	DRIFT	TIME
0.	2000.	0.0	-1.5	-1.5	-1.5	-1.5	-1.5	0.0	0.0000
100.	1752.	.5	0.0	2.2	5.0	8.1	11.6	1.8	.1603
200.	1529.	4.9	-9.9	-5.5	0.0	6.3	13.3	7.7	.3437
300.	1336.	14.0	-34.9	-28.2	-20.0	-10.6	0.0	18.3	.5543
400.	1185.	29.4	-79.3	-70.4	-59.4	-46.9	-32.8	34.0	.7930
500.	1082.	53.1	-148.7	-137.5	-123.8	-108.2	-90.5	54.5	1.0596
600.	1013.	86.5	-247.3	-233.9	-217.5	-198.7	-177.5	78.9	1.3485
700.	956.	130.0	-374.9	-359.3	-340.1	-318.2	-293.5	105.7	1.6506
800.	904.	186.2	-538.1	-520.3	-498.4	-473.4	-445.1	136.1	1.9734
900.	858.	256.4	-740.1	-720.0	-695.4	-667.2	-635.4	169.7	2.3144
1000.	815.	342.3	-984.6	-962.3	-934.9	-903.6	-868.3	206.5	2.6735

BALLISTIC COEFFICIENT .30, MUZZLE VELOCITY FT/SEC 2100.

RANGE	VELOCITY	HEIGHT	100 YD	150 YD	200 YD	250 YD	300 YD	DRIFT	TIME
0.	2100.	0.0	-1.5	-1.5	-1.5	-1.5	-1.5	0.0	0.0000
100.	1842.	.4	0.0	1.9	4.4	7.2	10.4	1.7	.1526
200.	1610.	4.4	-8.8	-4.9	0.0	5.6	11.9	7.2	.3267
300.	1403.	12.6	-31.1	-25.3	-17.9	-9.5	0.0	17.2	.5264
400.	1237.	26.6	-71.2	-63.4	-53.6	-42.4	-29.7	32.3	.7548
500.	1115.	48.3	-134.0	-124.2	-111.9	-97.9	-82.1	52.2	1.0110
600.	1038.	79.2	-224.1	-212.4	-197.6	-180.8	-161.8	76.2	1.2903
700.	977.	120.4	-344.2	-330.5	-313.3	-293.7	-271.6	103.6	1.5886
800.	923.	173.4	-497.0	-481.4	-461.8	-439.3	-414.0	134.1	1.9045
900.	875.	239.9	-687.0	-669.4	-647.3	-622.0	-593.6	167.8	2.2390
1000.	831.	321.5	-917.6	-898.1	-873.5	-845.4	-813.8	204.6	2.5911

BALLISTIC COEFFICIENT .30, MUZZLE VELOCITY FT/SEC 2200.

RANGE	VELOCITY	HEIGHT	100 YD	150 YD	200 YD	250 YD	300 YD	DRIFT	TIME
0.	2200.	0.0	-1.5	-1.5	-1.5	-1.5	-1.5	0.0	0.0000
100.	1935.	.3	0.0	1.7	3.9	6.5	9.3	1.6	.1456
200.	1693.	3.9	-7.9	-4.4	0.0	5.1	10.8	6.8	.3113
300.	1476.	11.3	-27.9	-22.7	-16.1	-8.5	0.0	16.2	.5011
400.	1294.	24.1	-64.1	-57.2	-48.4	-38.2	-26.9	30.5	.7190
500.	1153.	44.1	-121.4	-112.7	-101.7	-89.0	-74.8	50.1	.9664
600.	1063.	72.8	-204.2	-193.8	-180.6	-165.4	-148.4	73.8	1.2376
700.	998.	111.3	-315.7	-303.6	-288.2	-270.4	-250.6	101.0	1.5282
800.	942.	161.4	-459.3	-445.4	-427.9	-407.5	-384.8	131.5	1.8382
900.	892.	224.3	-637.5	-621.9	-602.1	-579.2	-553.7	165.1	2.1652
1000.	847.	301.8	-855.0	-837.7	-815.8	-790.3	-762.0	201.9	2.5107

BALLISTIC COEFFICIENT .30, MUZZLE VELOCITY FT/SEC 2300.

RANGE	VELOCITY	HEIGHT	100 YD	150 YD	200 YD	250 YD	300 YD	DRIFT	TIME
0.	2300.	0.0	-1.5	-1.5	-1.5	-1.5	-1.5	0.0	0.0000
100.	2028.	.2	0.0	1.5	3.5	5.8	8.4	1.5	.1389
200.	1777.	3.5	-7.0	-4.0	0.0	4.6	9.7	6.4	.2970
300.	1551.	10.2	-25.1	-20.5	-14.6	-7.7	0.0	15.2	.4778
400.	1354.	21.8	-57.8	-51.7	-43.7	-34.6	-24.3	28.8	.6853
500.	1199.	40.0	-109.4	-101.7	-91.8	-80.4	-67.5	47.3	.9210
600.	1091.	66.6	-185.3	-176.1	-164.2	-150.5	-135.0	70.7	1.1842
700.	1020.	102.8	-289.3	-278.6	-264.7	-248.7	-230.6	97.9	1.4690
800.	962.	150.0	-423.6	-411.3	-395.5	-377.2	-356.6	128.3	1.7722
900.	909.	209.5	-591.3	-577.5	-559.7	-539.1	-515.9	161.8	2.0930
1000.	862.	283.1	-796.5	-781.2	-761.4	-738.5	-712.7	198.5	2.4320

BALLISTIC COEFFICIENT .30, MUZZLE VELOCITY FT/SEC 2400.

RANGE	VELOCITY	HEIGHT	100 YD	150 YD	200 YD	250 YD	300 YD	DRIFT	TIME
0.	2400.	0.0	-1.5	-1.5	-1.5	-1.5	-1.5	0.0	0.0000
100.	2122.	.1	0.0	1.4	3.1	5.2	7.6	1.4	.1328
200.	1863.	3.1	-6.3	-3.6	0.0	4.2	8.8	5.9	.2837
300.	1628.	9.2	-22.7	-18.6	-13.2	-7.0	0.0	14.3	.4561
400.	1419.	19.8	-52.2	-46.7	-39.6	-31.3	-22.0	27.0	.6536
500.	1250.	36.4	-99.1	-92.3	-83.4	-73.0	-61.3	44.8	.8795
600.	1123.	60.9	-168.4	-160.2	-149.6	-137.1	-123.1	67.5	1.1336
700.	1043.	95.0	-265.1	-255.6	-243.1	-228.6	-212.2	94.5	1.4122
800.	982.	139.4	-390.7	-379.8	-365.6	-348.9	-330.3	124.7	1.7085
900.	927.	195.7	-548.4	-536.1	-520.1	-501.4	-480.4	158.0	2.0228
1000.	878.	265.7	-742.4	-728.8	-711.0	-690.2	-666.8	194.6	2.3559

BALLISTIC COEFFICIENT .30, MUZZLE VELOCITY FT/SEC 2500.

RANGE	VELOCITY	HEIGHT	100 YD	150 YD	200 YD	250 YD	300 YD	DRIFT	TIME
0.	2500.	0.0	-1.5	-1.5	-1.5	-1.5	-1.5	0.0	0.0000
100.	2216.	.0	0.0	1.2	2.8	4.7	6.8	1.3	.1273
200.	1949.	2.8	-5.7	-3.2	0.0	3.8	8.0	5.6	.2718
300.	1706.	8.4	-20.5	-16.9	-12.0	-6.4	0.0	13.4	.4362
400.	1488.	18.0	-47.3	-42.5	-36.0	-28.5	-20.0	25.4	.6246
500.	1303.	33.2	-89.9	-83.9	-75.8	-66.4	-55.8	42.3	.8405
600.	1160.	55.8	-153.1	-145.9	-136.2	-124.9	-112.2	64.2	1.0850
700.	1067.	87.4	-242.2	-233.8	-222.4	-209.3	-194.4	90.7	1.3555
800.	1002.	129.3	-360.1	-350.4	-337.4	-322.4	-305.4	120.8	1.6463
900.	945.	182.8	-509.1	-498.2	-483.6	-466.7	-447.6	154.1	1.9557
1000.	894.	249.3	-692.3	-680.2	-664.0	-645.2	-624.0	190.5	2.2825

BALLISTIC COEFFICIENT .30, MUZZLE VELOCITY FT/SEC 2600.

RANGE	VELOCITY	HEIGHT	100 YD	150 YD	200 YD	250 YD	300 YD	DRIFT	TIME
0.	2600.	0.0	-1.5	-1.5	-1.5	-1.5	-1.5	0.0	0.0000
100.	2310.	-.0	0.0	1.1	2.5	4.3	6.2	1.2	.1224
200.	2038.	2.5	-5.1	-2.9	0.0	3.4	7.3	5.3	.2608
300.	1786.	7.6	-18.6	-15.4	-11.0	-5.8	0.0	12.7	.4180
400.	1559.	16.4	-43.0	-38.8	-32.9	-26.0	-18.3	24.0	.5980
500.	1361.	30.3	-81.8	-76.5	-69.1	-60.6	-50.9	40.0	.8042
600.	1204.	51.1	-139.6	-133.2	-124.3	-114.1	-102.4	61.0	1.0390
700.	1094.	80.5	-221.5	-214.0	-203.7	-191.7	-178.1	86.9	1.3014
800.	1023.	120.0	-331.6	-323.1	-311.3	-297.6	-282.1	116.7	1.5860
900.	964.	170.4	-471.4	-461.8	-448.6	-433.1	-415.7	149.6	1.8885
1000.	911.	233.3	-643.7	-633.0	-618.3	-601.1	-581.7	185.6	2.2081

BALLISTIC COEFFICIENT .30, MUZZLE VELOCITY FT/SEC 2700.

RANGE	VELOCITY	HEIGHT	100 YD	150 YD	200 YD	250 YD	300 YD	DRIFT	TIME
0.	2700.	0.0	-1.5	-1.5	-1.5	-1.5	-1.5	0.0	0.0000
100.	2404.	-.1	0.0	1.0	2.3	3.8	5.6	1.2	.1178
200.	2126.	2.3	-4.6	-2.6	0.0	3.1	6.7	5.0	.2504
300.	1866.	7.0	-16.8	-14.0	-10.0	-5.3	0.0	11.9	.4010
400.	1632.	15.0	-39.2	-35.4	-30.1	-23.8	-16.7	22.6	.5731
500.	1422.	27.7	-74.6	-69.8	-63.2	-55.3	-46.5	37.8	.7701
600.	1252.	46.8	-127.4	-121.7	-113.8	-104.3	-93.7	57.9	.9956
700.	1125.	74.2	-202.7	-196.0	-186.7	-175.7	-163.3	83.0	1.2493
800.	1044.	111.3	-305.5	-297.8	-287.2	-274.7	-260.5	112.5	1.5279
900.	982.	158.9	-436.7	-428.1	-416.2	-402.0	-386.1	145.0	1.8238
1000.	928.	218.6	-599.5	-589.9	-576.7	-561.0	-543.3	180.7	2.1377

BALLISTIC COEFFICIENT .30, MUZZLE VELOCITY FT/SEC 2800.

RANGE	VELOCITY	HEIGHT	100 YD	150 YD	200 YD	250 YD	300 YD	DRIFT	TIME
0.	2800.	0.0	-1.5	-1.5	-1.5	-1.5	-1.5	0.0	0.0000
100.	2498.	-.1	0.0	.8	2.1	3.5	5.1	1.1	.1134
200.	2214.	2.0	-4.1	-2.4	0.0	2.9	6.1	4.7	.2409
300.	1948.	6.4	-15.4	-12.8	-9.2	-4.9	0.0	11.3	.3855
400.	1705.	13.8	-35.8	-32.4	-27.6	-21.8	-15.3	21.4	.5500
500.	1487.	25.4	-68.2	-64.0	-58.0	-50.8	-42.6	35.7	.7385
600.	1302.	43.0	-116.6	-111.5	-104.2	-95.6	-85.8	54.9	.9546
700.	1160.	68.3	-185.7	-179.7	-171.3	-161.3	-149.8	79.1	1.1995
800.	1067.	103.0	-280.7	-273.9	-264.3	-252.8	-239.7	107.9	1.4700
900.	1001.	148.1	-404.4	-396.8	-385.9	-373.0	-358.3	140.2	1.7609
1000.	945.	205.0	-559.0	-550.5	-538.5	-524.2	-507.8	175.8	2.0705

BALLISTIC COEFFICIENT .30, MUZZLE VELOCITY FT/SEC 2900.

RANGE	VELOCITY	HEIGHT	100 YD	150 YD	200 YD	250 YD	300 YD	DRIFT	TIME
0.	2900.	0.0	-1.5	-1.5	-1.5	-1.5	-1.5	0.0	0.0000
100.	2591.	-.2	0.0	.7	1.9	3.2	4.7	1.1	.1095
200.	2302.	1.8	-3.7	-2.2	0.0	2.6	5.6	4.5	.2323
300.	2030.	5.9	-14.0	-11.8	-8.5	-4.5	0.0	10.7	.3712
400.	1778.	12.7	-32.8	-29.8	-25.4	-20.1	-14.1	20.3	.5291
500.	1553.	23.4	-62.7	-58.9	-53.3	-46.8	-39.2	33.9	.7098
600.	1355.	39.6	-107.0	-102.5	-95.9	-88.0	-78.9	52.2	.9171
700.	1200.	63.0	-170.5	-165.3	-157.5	-148.3	-137.8	75.4	1.1526
800.	1092.	95.5	-258.6	-252.6	-243.7	-233.2	-221.1	103.5	1.4157
900.	1021.	138.0	-374.6	-367.9	-357.9	-346.1	-332.5	135.4	1.7005
1000.	962.	191.9	-520.3	-512.8	-501.7	-488.6	-473.5	170.6	2.0035

BALLISTIC COEFFICIENT .30, MUZZLE VELOCITY FT/SEC 3000.

RANGE	VELOCITY	HEIGHT	100 YD	150 YD	200 YD	250 YD	300 YD	DRIFT	TIME
0.	3000.	0.0	-1.5	-1.5	-1.5	-1.5	-1.5	0.0	0.0000
100.	2684.	-.2	0.0	.7	1.7	2.9	4.3	1.0	.1058
200.	2389.	1.7	-3.4	-2.0	0.0	2.4	5.2	4.3	.2244
300.	2112.	5.4	-12.9	-10.8	-7.8	-4.1	0.0	10.2	.3579
400.	1853.	11.7	-30.2	-27.5	-23.4	-18.5	-13.0	19.3	.5096
500.	1620.	21.6	-57.6	-54.2	-49.2	-43.1	-36.2	32.2	.6828
600.	1411.	36.5	-98.4	-94.3	-88.3	-81.0	-72.7	49.5	.8814
700.	1244.	58.2	-157.0	-152.3	-145.2	-136.7	-127.0	71.9	1.1084
800.	1119.	88.5	-238.4	-233.0	-224.9	-215.2	-204.1	99.2	1.3637
900.	1040.	128.7	-347.1	-341.0	-331.9	-320.9	-308.5	130.6	1.6421
1000.	979.	179.8	-484.7	-478.0	-467.9	-455.7	-441.8	165.4	1.9395

BALLISTIC COEFFICIENT .30, MUZZLE VELOCITY FT/SEC 3100.

RANGE	VELOCITY	HEIGHT	100 YD	150 YD	200 YD	250 YD	300 YD	DRIFT	TIME
0.	3100.	0.0	-1.5	-1.5	-1.5	-1.5	-1.5	0.0	0.0000
100.	2776.	-.2	0.0	.6	1.5	2.7	3.9	.9	.1021
200.	2476.	1.5	-3.0	-1.9	0.0	2.3	4.8	4.0	.2165
300.	2193.	5.0	-11.8	-10.0	-7.2	-3.8	0.0	9.7	.3453
400.	1928.	10.8	-27.8	-25.4	-21.7	-17.2	-12.1	18.4	.4914
500.	1687.	20.0	-53.1	-50.2	-45.5	-39.9	-33.5	30.6	.6576
600.	1471.	33.8	-90.8	-87.2	-81.6	-74.8	-67.2	47.1	.8481
700.	1290.	53.8	-144.7	-140.6	-134.1	-126.2	-117.2	68.5	1.0664
800.	1151.	82.0	-220.0	-215.2	-207.8	-198.8	-188.5	94.9	1.3133
900.	1061.	120.0	-321.7	-316.4	-308.0	-297.8	-286.3	125.8	1.5858
1000.	997.	168.5	-451.7	-445.8	-436.5	-425.2	-412.4	160.1	1.8775

BALLISTIC COEFFICIENT .30, MUZZLE VELOCITY FT/SEC 3200.

RANGE	VELOCITY	HEIGHT	100 YD	150 YD	200 YD	250 YD	300 YD	DRIFT	TIME
0.	3200.	0.0	-1.5	-1.5	-1.5	-1.5	-1.5	0.0	0.0000
100.	2869.	-.3	0.0	.5	1.4	2.4	3.6	.9	.0989
200.	2562.	1.4	-2.8	-1.7	0.0	2.1	4.5	3.9	.2097
300.	2275.	4.6	-10.8	-9.3	-6.7	-3.6	0.0	9.3	.3340
400.	2004.	10.1	-25.7	-23.5	-20.1	-16.0	-11.2	17.5	.4745
500.	1756.	18.6	-49.1	-46.5	-42.2	-37.0	-31.0	29.1	.6343
600.	1532.	31.3	-83.9	-80.8	-75.6	-69.4	-62.2	44.9	.8174
700.	1339.	49.9	-133.8	-130.1	-124.1	-116.8	-108.4	65.3	1.0273
800.	1187.	76.2	-203.4	-199.2	-192.3	-184.0	-174.5	90.8	1.2658
900.	1084.	111.8	-298.1	-293.4	-285.7	-276.3	-265.6	121.0	1.5312
1000.	1015.	157.9	-421.0	-415.7	-407.2	-396.8	-384.9	154.9	1.8178

BALLISTIC COEFFICIENT .30, MUZZLE VELOCITY FT/SEC 3300.

RANGE	VELOCITY	HEIGHT	100 YD	150 YD	200 YD	250 YD	300 YD	DRIFT	TIME
0.	3300.	0.0	-1.5	-1.5	-1.5	-1.5	-1.5	0.0	0.0000
100.	2962.	-.3	0.0	.5	1.2	2.2	3.3	.9	.0959
200.	2649.	1.2	-2.5	-1.6	0.0	1.9	4.2	3.8	.2031
300.	2356.	4.3	-10.0	-8.6	-6.2	-3.3	0.0	8.9	.3233
400.	2081.	9.3	-23.7	-21.8	-18.7	-14.8	-10.4	16.7	.4586
500.	1825.	17.3	-45.5	-43.1	-39.2	-34.4	-28.8	27.8	.6127
600.	1594.	29.1	-77.8	-75.0	-70.3	-64.5	-57.8	42.8	.7888
700.	1390.	46.3	-123.8	-120.5	-115.1	-108.3	-100.5	62.3	.9904
800.	1227.	70.8	-188.4	-184.6	-178.4	-170.6	-161.8	86.8	1.2206
900.	1109.	104.2	-276.6	-272.4	-265.3	-256.6	-246.6	116.3	1.4789
1000.	1033.	147.9	-392.3	-387.6	-379.8	-370.1	-359.0	149.7	1.7598

BALLISTIC COEFFICIENT .30, MUZZLE VELOCITY FT/SEC 3400.

RANGE	VELOCITY	HEIGHT	100 YD	150 YD	200 YD	250 YD	300 YD	DRIFT	TIME
0.	3400.	0.0	-1.5	-1.5	-1.5	-1.5	-1.5	0.0	0.0000
100.	3055.	-.3	0.0	.4	1.1	2.0	3.1	.8	.0929
200.	2735.	1.1	-2.3	-1.4	0.0	1.8	3.9	3.6	.1968
300.	2437.	4.0	-9.2	-8.0	-5.8	-3.1	0.0	8.5	.3129
400.	2157.	8.7	-22.0	-20.3	-17.4	-13.8	-9.7	16.0	.4438
500.	1895.	16.1	-42.2	-40.2	-36.6	-32.1	-26.9	26.6	.5923
600.	1657.	27.1	-72.2	-69.8	-65.4	-60.0	-53.8	40.9	.7617
700.	1444.	43.1	-114.9	-112.1	-107.0	-100.7	-93.5	59.5	.9558
800.	1269.	65.8	-174.8	-171.5	-165.7	-158.5	-150.3	83.1	1.1778
900.	1136.	97.3	-257.1	-253.4	-246.9	-238.8	-229.5	111.8	1.4293
1000.	1052.	138.5	-365.6	-361.6	-354.3	-345.3	-335.0	144.5	1.7035

BALLISTIC COEFFICIENT .30, MUZZLE VELOCITY FT/SEC 3500.

RANGE	VELOCITY	HEIGHT	100 YD	150 YD	200 YD	250 YD	300 YD	DRIFT	TIME
0.	3500.	0.0	-1.5	-1.5	-1.5	-1.5	-1.5	0.0	0.0000
100.	3149.	-.4	0.0	.3	1.0	1.8	2.8	.9	.0906
200.	2822.	1.0	-2.0	-1.3	0.0	1.7	3.6	3.5	.1912
300.	2518.	3.7	-8.5	-7.4	-5.4	-2.9	0.0	8.2	.3039
400.	2233.	8.1	-20.4	-19.0	-16.3	-13.0	-9.1	15.4	.4303
500.	1965.	15.0	-39.2	-37.5	-34.2	-30.0	-25.1	25.5	.5735
600.	1720.	25.3	-67.1	-65.0	-61.0	-56.0	-50.1	39.1	.7365
700.	1501.	40.2	-106.8	-104.4	-99.7	-93.9	-87.1	56.9	.9233
800.	1313.	61.4	-162.4	-159.7	-154.3	-147.7	-139.9	79.6	1.1378
900.	1168.	90.7	-238.6	-235.5	-229.5	-222.0	-213.3	107.2	1.3804
1000.	1072.	129.8	-340.8	-337.3	-330.7	-322.3	-312.6	139.4	1.6494

BALLISTIC COEFFICIENT .30, MUZZLE VELOCITY FT/SEC 3600.

RANGE	VELOCITY	HEIGHT	100 YD	150 YD	200 YD	250 YD	300 YD	DRIFT	TIME
0.	3600.	0.0	-1.5	-1.5	-1.5	-1.5	-1.5	0.0	0.0000
100.	3242.	-.4	0.0	.3	.9	1.7	2.6	.8	.0877
200.	2908.	.9	-1.8	-1.2	0.0	1.6	3.4	3.3	.1855
300.	2598.	3.4	-7.8	-6.9	-5.0	-2.7	0.0	7.9	.2946
400.	2309.	7.6	-18.9	-17.7	-15.2	-12.1	-8.5	14.7	.4171
500.	2036.	14.1	-36.5	-35.0	-32.0	-28.1	-23.6	24.4	.5556
600.	1784.	23.6	-62.6	-60.8	-57.1	-52.4	-47.0	37.5	.7130
700.	1558.	37.5	-99.6	-97.5	-93.2	-87.7	-81.4	54.5	.8930
800.	1360.	57.3	-151.2	-148.8	-143.8	-137.6	-130.4	76.2	1.0994
900.	1204.	84.7	-222.2	-219.5	-214.0	-207.0	-198.8	102.8	1.3343
1000.	1094.	121.6	-317.9	-314.9	-308.7	-300.9	-291.9	134.4	1.5968

BALLISTIC COEFFICIENT .30, MUZZLE VELOCITY FT/SEC 3700.

RANGE	VELOCITY	HEIGHT	100 YD	150 YD	200 YD	250 YD	300 YD	DRIFT	TIME
0.	3700.	0.0	-1.5	-1.5	-1.5	-1.5	-1.5	0.0	0.0000
100.	3335.	-.4	0.0	.3	.8	1.5	2.4	.8	.0856
200.	2995.	.8	-1.6	-1.1	0.0	1.5	3.1	3.2	.1804
300.	2679.	3.2	-7.2	-6.4	-4.7	-2.5	0.0	7.6	.2863
400.	2384.	7.1	-17.6	-16.5	-14.3	-11.4	-8.0	14.2	.4051
500.	2107.	13.2	-34.1	-32.8	-30.0	-26.3	-22.1	23.5	.5390
600.	1849.	22.2	-58.4	-56.8	-53.5	-49.1	-44.0	36.0	.6910
700.	1616.	35.1	-92.9	-91.1	-87.2	-82.1	-76.2	52.3	.8645
800.	1409.	53.5	-141.0	-138.9	-134.4	-128.6	-121.8	73.0	1.0635
900.	1242.	79.2	-207.3	-204.9	-199.9	-193.3	-185.7	98.8	1.2910
1000.	1118.	114.1	-296.9	-294.3	-288.7	-281.4	-273.0	129.6	1.5471

BALLISTIC COEFFICIENT .30, MUZZLE VELOCITY FT/SEC 3800.

RANGE	VELOCITY	HEIGHT	100 YD	150 YD	200 YD	250 YD	300 YD	DRIFT	TIME
0.	3800.	0.0	-1.5	-1.5	-1.5	-1.5	-1.5	0.0	0.0000
100.	3428.	-.4	0.0	.2	.7	1.4	2.2	.8	.0832
200.	3082.	.7	-1.5	-1.0	0.0	1.4	3.0	3.1	.1756
300.	2760.	3.0	-6.7	-6.0	-4.4	-2.4	0.0	7.3	.2786
400.	2460.	6.7	-16.4	-15.5	-13.4	-10.7	-7.5	13.7	.3936
500.	2178.	12.4	-31.8	-30.7	-28.1	-24.7	-20.7	22.6	.5231
600.	1915.	20.8	-54.6	-53.3	-50.2	-46.1	-41.3	34.6	.6703
700.	1675.	32.9	-86.9	-85.4	-81.7	-76.9	-71.4	50.2	.8378
800.	1460.	50.1	-131.8	-130.0	-125.9	-120.4	-114.0	70.1	1.0297
900.	1281.	74.2	-193.6	-191.6	-186.9	-180.7	-173.6	94.9	1.2495
1000.	1145.	106.9	-277.2	-275.0	-269.8	-262.9	-255.0	124.7	1.4979

BALLISTIC COEFFICIENT .30, MUZZLE VELOCITY FT/SEC 3900.

RANGE	VELOCITY	HEIGHT	100 YD	150 YD	200 YD	250 YD	300 YD	DRIFT	TIME
0.	3900.	0.0	-1.5	-1.5	-1.5	-1.5	-1.5	0.0	0.0000
100.	3522.	-.4	0.0	.2	.7	1.3	2.0	.7	.0806
200.	3169.	.6	-1.3	-1.0	0.0	1.3	2.8	3.0	.1706
300.	2840.	2.8	-6.1	-5.6	-4.1	-2.2	0.0	7.0	.2705
400.	2535.	6.3	-15.3	-14.5	-12.6	-10.1	-7.1	13.2	.3826
500.	2249.	11.6	-29.8	-28.9	-26.5	-23.3	-19.6	21.8	.5082
600.	1980.	19.5	-51.1	-50.0	-47.1	-43.3	-38.9	33.2	.6501
700.	1734.	30.9	-81.4	-80.1	-76.7	-72.3	-67.1	48.2	.8121
800.	1513.	47.0	-123.4	-121.9	-118.0	-113.0	-107.0	67.2	.9974
900.	1323.	69.5	-181.0	-179.3	-175.0	-169.3	-162.6	91.1	1.2099
1000.	1175.	100.4	-259.6	-257.7	-252.9	-246.6	-239.2	120.2	1.4519

BALLISTIC COEFFICIENT .30, MUZZLE VELOCITY FT/SEC 4000.

RANGE	VELOCITY	HEIGHT	100 YD	150 YD	200 YD	250 YD	300 YD	DRIFT	TIME
0.	4000.	0.0	-1.5	-1.5	-1.5	-1.5	-1.5	0.0	0.0000
100.	3615.	-.4	0.0	.1	.6	1.2	1.9	.7	.0791
200.	3256.	.6	-1.2	-.9	0.0	1.2	2.6	2.9	.1664
300.	2921.	2.6	-5.7	-5.2	-3.9	-2.1	0.0	6.8	.2639
400.	2611.	5.9	-14.2	-13.6	-11.9	-9.5	-6.7	12.7	.3724
500.	2320.	11.0	-27.8	-27.1	-24.9	-21.9	-18.4	21.0	.4943
600.	2047.	18.4	-47.9	-47.1	-44.4	-40.8	-36.6	32.1	.6321
700.	1794.	29.1	-76.3	-75.3	-72.2	-68.0	-63.1	46.4	.7886
800.	1567.	44.2	-115.7	-114.5	-111.0	-106.2	-100.6	64.7	.9677
900.	1367.	65.3	-169.5	-168.2	-164.3	-158.9	-152.5	87.6	1.1729
1000.	1209.	94.2	-242.8	-241.4	-237.0	-231.0	-224.0	115.6	1.4067

BALLISTIC COEFFICIENT .31, MUZZLE VELOCITY FT/SEC 2000.

RANGE	VELOCITY	HEIGHT	100 YD	150 YD	200 YD	250 YD	300 YD	DRIFT	TIME
0.	2000.	0.0	-1.5	-1.5	-1.5	-1.5	-1.5	0.0	0.0000
100.	1759.	.5	0.0	2.2	4.9	8.0	11.5	1.7	.1598
200.	1542.	4.9	-9.8	-5.4	0.0	6.2	13.1	7.4	.3421
300.	1352.	13.8	-34.4	-27.8	-19.6	-10.4	0.0	17.6	.5503
400.	1202.	28.9	-78.0	-69.1	-58.3	-45.9	-32.1	32.7	.7860
500.	1096.	52.0	-145.7	-134.6	-121.1	-105.6	-88.3	52.5	1.0483
600.	1026.	84.5	-241.9	-228.6	-212.4	-193.8	-173.1	76.2	1.3329
700.	968.	127.5	-368.6	-353.1	-334.2	-312.5	-288.4	102.9	1.6349
800.	917.	182.0	-527.5	-509.8	-488.2	-463.4	-435.8	132.2	1.9513
900.	871.	250.5	-725.3	-705.4	-681.0	-653.2	-622.1	165.0	2.2877
1000.	829.	334.0	-964.1	-942.0	-914.9	-884.0	-849.5	200.8	2.6410

BALLISTIC COEFFICIENT .31, MUZZLE VELOCITY FT/SEC 2100.

RANGE	VELOCITY	HEIGHT	100 YD	150 YD	200 YD	250 YD	300 YD	DRIFT	TIME
0.	2100.	0.0	-1.5	-1.5	-1.5	-1.5	-1.5	0.0	0.0000
100.	1850.	.4	0.0	1.9	4.4	7.1	10.2	1.6	.1521
200.	1624.	4.3	-8.7	-4.8	0.0	5.5	11.7	6.9	.3252
300.	1422.	12.4	-30.7	-24.9	-17.6	-9.3	0.0	16.6	.5227
400.	1257.	26.1	-70.0	-62.2	-52.6	-41.5	-29.1	31.0	.7478
500.	1131.	47.3	-131.3	-121.6	-109.5	-95.6	-80.1	50.3	1.0001
600.	1051.	77.4	-219.4	-207.8	-193.2	-176.6	-158.0	73.7	1.2759
700.	990.	118.0	-338.1	-324.5	-307.5	-288.1	-266.5	100.8	1.5727
800.	936.	169.3	-486.7	-471.2	-451.8	-429.6	-404.8	130.1	1.8822
900.	888.	234.0	-672.2	-654.7	-632.9	-608.0	-580.1	162.9	2.2114
1000.	845.	313.3	-897.2	-877.8	-853.6	-825.8	-794.9	198.8	2.5579

BALLISTIC COEFFICIENT .31, MUZZLE VELOCITY FT/SEC 2200.

RANGE	VELOCITY	HEIGHT	100 YD	150 YD	200 YD	250 YD	300 YD	DRIFT	TIME
0.	2200.	0.0	-1.5	-1.5	-1.5	-1.5	-1.5	0.0	0.0000
100.	1943.	.3	0.0	1.7	3.9	6.4	9.2	1.6	.1452
200.	1708.	3.9	-7.8	-4.3	0.0	5.0	10.6	6.5	.3098
300.	1496.	11.1	-27.5	-22.4	-15.9	-8.4	0.0	15.6	.4976
400.	1315.	23.6	-63.0	-56.1	-47.4	-37.4	-26.3	29.3	.7122
500.	1173.	43.1	-118.9	-110.3	-99.4	-86.9	-73.0	48.2	.9555
600.	1078.	70.9	-199.2	-188.9	-175.9	-160.9	-144.2	71.0	1.2217
700.	1012.	109.0	-309.7	-297.7	-282.5	-265.0	-245.5	98.1	1.5120
800.	956.	157.3	-448.6	-434.9	-417.5	-397.5	-375.2	127.4	1.8145
900.	906.	218.5	-622.8	-607.4	-587.9	-565.3	-540.2	160.1	2.1371
1000.	861.	293.7	-834.8	-817.7	-796.0	-770.9	-743.1	195.9	2.4770

BALLISTIC COEFFICIENT .31, MUZZLE VELOCITY FT/SEC 2300.

RANGE	VELOCITY	HEIGHT	100 YD	150 YD	200 YD	250 YD	300 YD	DRIFT	TIME
0.	2300.	0.0	-1.5	-1.5	-1.5	-1.5	-1.5	0.0	0.0000
100.	2037.	.2	0.0	1.5	3.5	5.7	8.3	1.5	.1387
200.	1792.	3.4	-6.9	-3.9	0.0	4.5	9.6	6.1	.2957
300.	1572.	10.1	-24.8	-20.2	-14.4	-7.6	0.0	14.7	.4746
400.	1377.	21.4	-56.7	-50.7	-42.8	-33.8	-23.7	27.6	.6786
500.	1222.	39.0	-107.1	-99.5	-89.7	-78.4	-65.7	45.4	.9102
600.	1108.	64.9	-180.9	-171.7	-160.0	-146.5	-131.3	68.0	1.1692
700.	1035.	100.1	-282.2	-271.6	-257.9	-242.1	-224.3	94.5	1.4498
800.	976.	146.0	-413.3	-401.1	-385.5	-367.5	-347.2	124.1	1.7485
900.	924.	203.8	-576.9	-563.2	-545.6	-525.3	-502.4	156.7	2.0645
1000.	877.	275.3	-776.9	-761.7	-742.1	-719.6	-694.2	192.5	2.3980

BALLISTIC COEFFICIENT .31, MUZZLE VELOCITY FT/SEC 2400.

RANGE	VELOCITY	HEIGHT	100 YD	150 YD	200 YD	250 YD	300 YD	DRIFT	TIME
0.	2400.	0.0	-1.5	-1.5	-1.5	-1.5	-1.5	0.0	0.0000
100.	2131.	.1	0.0	1.4	3.1	5.2	7.5	1.3	.1326
200.	1879.	3.1	-6.2	-3.5	0.0	4.1	8.7	5.7	.2825
300.	1650.	9.1	-22.4	-18.3	-13.0	-6.9	0.0	13.7	.4531
400.	1444.	19.4	-51.3	-45.9	-38.9	-30.7	-21.5	26.0	.6476
500.	1274.	35.5	-97.1	-90.3	-81.6	-71.3	-59.8	43.0	.8694
600.	1143.	59.3	-164.2	-156.1	-145.6	-133.3	-119.5	64.8	1.1184
700.	1059.	92.2	-257.8	-248.4	-236.1	-221.8	-205.6	90.9	1.3917
800.	997.	135.4	-380.5	-369.7	-355.7	-339.3	-320.9	120.4	1.6842
900.	942.	190.2	-534.6	-522.4	-506.6	-488.2	-467.5	153.0	1.9945
1000.	894.	257.8	-722.8	-709.3	-691.7	-671.3	-648.2	188.5	2.3211

BALLISTIC COEFFICIENT .31, MUZZLE VELOCITY FT/SEC 2500.

RANGE	VELOCITY	HEIGHT	100 YD	150 YD	200 YD	250 YD	300 YD	DRIFT	TIME
0.	2500.	0.0	-1.5	-1.5	-1.5	-1.5	-1.5	0.0	0.0000
100.	2225.	.0	0.0	1.2	2.8	4.7	6.7	1.3	.1271
200.	1966.	2.8	-5.6	-3.2	0.0	3.7	7.9	5.4	.2706
300.	1729.	8.3	-20.2	-16.6	-11.8	-6.2	0.0	12.9	.4333
400.	1515.	17.6	-46.5	-41.7	-35.3	-27.9	-19.6	24.4	.6188
500.	1330.	32.4	-88.1	-82.1	-74.1	-64.8	-54.4	40.6	.8306
600.	1185.	54.2	-149.3	-142.1	-132.6	-121.4	-108.9	61.6	1.0701
700.	1085.	84.9	-235.6	-227.2	-216.0	-203.0	-188.4	87.2	1.3356
800.	1018.	125.5	-350.0	-340.4	-327.7	-312.8	-296.1	116.4	1.6216
900.	961.	177.1	-494.5	-483.7	-469.3	-452.6	-433.8	148.7	1.9250
1000.	910.	241.4	-672.3	-660.3	-644.3	-625.7	-604.9	184.1	2.2460

BALLISTIC COEFFICIENT .31, MUZZLE VELOCITY FT/SEC 2600.

RANGE	VELOCITY	HEIGHT	100 YD	150 YD	200 YD	250 YD	300 YD	DRIFT	TIME
0.	2600.	0.0	-1.5	-1.5	-1.5	-1.5	-1.5	0.0	0.0000
100.	2319.	-.0	0.0	1.1	2.5	4.2	6.1	1.2	.1221
200.	2055.	2.5	-5.0	-2.9	0.0	3.4	7.2	5.1	.2597
300.	1809.	7.5	-18.3	-15.1	-10.8	-5.7	0.0	12.1	.4152
400.	1587.	16.1	-42.3	-38.1	-32.3	-25.5	-17.9	23.0	.5925
500.	1390.	29.6	-80.1	-74.8	-67.5	-59.0	-49.5	38.3	.7945
600.	1232.	49.6	-136.1	-129.7	-121.0	-110.8	-99.5	58.4	1.0244
700.	1115.	78.1	-215.2	-207.8	-197.6	-185.7	-172.4	83.4	1.2814
800.	1039.	116.1	-321.6	-313.1	-301.5	-288.0	-272.8	112.2	1.5605
900.	980.	165.0	-457.4	-447.9	-434.8	-419.6	-402.5	144.2	1.8581
1000.	928.	225.9	-625.2	-614.6	-600.1	-583.1	-564.1	179.4	2.1730

BALLISTIC COEFFICIENT .31, MUZZLE VELOCITY FT/SEC 2700.

RANGE	VELOCITY	HEIGHT	100 YD	150 YD	200 YD	250 YD	300 YD	DRIFT	TIME
0.	2700.	0.0	-1.5	-1.5	-1.5	-1.5	-1.5	0.0	0.0000
100.	2414.	-.1	0.0	.9	2.3	3.8	5.5	1.1	.1176
200.	2143.	2.2	-4.5	-2.6	0.0	3.1	6.6	4.8	.2495
300.	1890.	6.9	-16.6	-13.8	-9.9	-5.2	0.0	11.5	.3984
400.	1660.	14.7	-38.5	-34.8	-29.5	-23.3	-16.4	21.7	.5679
500.	1453.	27.1	-73.1	-68.4	-61.8	-54.1	-45.3	36.2	.7612
600.	1281.	45.5	-124.2	-118.6	-110.6	-101.4	-90.9	55.4	.9814
700.	1149.	71.8	-196.7	-190.1	-180.9	-170.1	-157.9	79.5	1.2293
800.	1062.	107.6	-295.8	-288.3	-277.7	-265.4	-251.5	108.0	1.5023
900.	1000.	153.6	-423.1	-414.6	-402.7	-388.9	-373.2	139.6	1.7931
1000.	945.	211.6	-582.0	-572.6	-559.4	-544.0	-526.6	174.6	2.1034

BALLISTIC COEFFICIENT .31, MUZZLE VELOCITY FT/SEC 2800.

RANGE	VELOCITY	HEIGHT	100 YD	150 YD	200 YD	250 YD	300 YD	DRIFT	TIME
0.	2800.	0.0	-1.5	-1.5	-1.5	-1.5	-1.5	0.0	0.0000
100.	2507.	-.1	0.0	.8	2.0	3.5	5.0	1.1	.1132
200.	2232.	2.0	-4.1	-2.4	0.0	2.8	6.0	4.5	.2400
300.	1972.	6.3	-15.1	-12.6	-9.0	-4.8	0.0	10.8	.3829
400.	1735.	13.5	-35.2	-31.9	-27.1	-21.4	-15.0	20.5	.5451
500.	1520.	24.8	-66.8	-62.7	-56.7	-49.6	-41.6	34.2	.7300
600.	1334.	41.8	-113.7	-108.7	-101.5	-93.0	-83.4	52.5	.9413
700.	1188.	66.1	-180.2	-174.4	-166.0	-156.0	-144.9	75.7	1.1798
800.	1087.	99.4	-271.7	-265.0	-255.4	-244.1	-231.3	103.4	1.4448
900.	1019.	142.9	-391.3	-383.7	-373.0	-360.2	-345.8	134.8	1.7302
1000.	962.	197.8	-540.8	-532.4	-520.4	-506.2	-490.3	169.4	2.0337

BALLISTIC COEFFICIENT .31, MUZZLE VELOCITY FT/SEC 2900.

RANGE	VELOCITY	HEIGHT	100 YD	150 YD	200 YD	250 YD	300 YD	DRIFT	TIME
0.	2900.	0.0	-1.5	-1.5	-1.5	-1.5	-1.5	0.0	0.0000
100.	2600.	-.2	0.0	.7	1.8	3.1	4.6	1.0	.1092
200.	2320.	1.8	-3.7	-2.2	0.0	2.6	5.6	4.3	.2314
300.	2055.	5.8	-13.9	-11.6	-8.4	-4.5	0.0	10.3	.3689
400.	1809.	12.4	-32.3	-29.3	-24.9	-19.7	-13.8	19.5	.5243
500.	1587.	22.9	-61.4	-57.7	-52.2	-45.7	-38.3	32.5	.7016
600.	1390.	38.4	-104.3	-99.8	-93.3	-85.5	-76.5	49.8	.9036
700.	1232.	60.9	-165.5	-160.3	-152.7	-143.6	-133.2	72.0	1.1334
800.	1115.	92.0	-250.0	-244.1	-235.3	-224.9	-213.0	99.0	1.3904
900.	1039.	133.0	-361.8	-355.1	-345.3	-333.6	-320.2	130.0	1.6694
1000.	980.	185.0	-502.8	-495.4	-484.5	-471.5	-456.6	164.1	1.9670

BALLISTIC COEFFICIENT .31, MUZZLE VELOCITY FT/SEC 3000.

RANGE	VELOCITY	HEIGHT	100 YD	150 YD	200 YD	250 YD	300 YD	DRIFT	TIME
0.	3000.	0.0	-1.5	-1.5	-1.5	-1.5	-1.5	0.0	0.0000
100.	2693.	-.2	0.0	.7	1.7	2.9	4.2	1.0	.1057
200.	2407.	1.6	-3.3	-2.0	0.0	2.4	5.1	4.2	.2236
300.	2138.	5.3	-12.7	-10.7	-7.7	-4.1	0.0	9.8	.3559
400.	1885.	11.5	-29.7	-27.0	-23.0	-18.2	-12.7	18.5	.5052
500.	1656.	21.1	-56.5	-53.2	-48.1	-42.1	-35.3	30.8	.6752
600.	1449.	35.5	-96.0	-92.0	-86.0	-78.8	-70.6	47.3	.8690
700.	1278.	56.3	-152.4	-147.7	-140.7	-132.3	-122.7	68.6	1.0899
800.	1146.	85.2	-230.3	-225.0	-217.0	-207.3	-196.4	94.7	1.3383
900.	1060.	123.8	-334.7	-328.8	-319.7	-308.9	-296.6	125.2	1.6111
1000.	998.	173.1	-467.8	-461.2	-451.2	-439.1	-425.5	158.9	1.9031

BALLISTIC COEFFICIENT .31, MUZZLE VELOCITY FT/SEC 3100.

RANGE	VELOCITY	HEIGHT	100 YD	150 YD	200 YD	250 YD	300 YD	DRIFT	TIME
0.	3100.	0.0	-1.5	-1.5	-1.5	-1.5	-1.5	0.0	0.0000
100.	2787.	-.3	0.0	.6	1.5	2.6	3.9	.9	.1019
200.	2495.	1.5	-3.0	-1.8	0.0	2.2	4.7	3.9	.2158
300.	2220.	4.9	-11.6	-9.9	-7.1	-3.8	0.0	9.3	.3432
400.	1961.	10.6	-27.3	-25.0	-21.3	-16.9	-11.8	17.6	.4871
500.	1724.	19.5	-52.1	-49.1	-44.5	-38.9	-32.7	29.3	.6501
600.	1511.	32.8	-88.5	-85.0	-79.5	-72.8	-65.3	45.0	.8361
700.	1327.	52.0	-140.5	-136.4	-129.9	-122.1	-113.3	65.3	1.0484
800.	1182.	78.9	-212.5	-207.8	-200.4	-191.5	-181.4	90.5	1.2884
900.	1083.	115.2	-309.8	-304.5	-296.2	-286.2	-274.9	120.3	1.5544
1000.	1017.	161.9	-435.2	-429.4	-420.1	-409.0	-396.4	153.7	1.8408

BALLISTIC COEFFICIENT .31, MUZZLE VELOCITY FT/SEC 3200.

RANGE	VELOCITY	HEIGHT	100 YD	150 YD	200 YD	250 YD	300 YD	DRIFT	TIME
0.	3200.	0.0	-1.5	-1.5	-1.5	-1.5	-1.5	0.0	0.0000
100.	2880.	-.3	0.0	.5	1.4	2.4	3.6	.9	.0987
200.	2582.	1.3	-2.7	-1.7	0.0	2.0	4.4	3.8	.2089
300.	2302.	4.5	-10.7	-9.1	-6.6	-3.5	0.0	8.9	.3320
400.	2038.	9.9	-25.3	-23.2	-19.8	-15.7	-11.0	16.8	.4706
500.	1794.	18.1	-48.2	-45.6	-41.3	-36.2	-30.3	27.9	.6274
600.	1574.	30.4	-81.9	-78.8	-73.7	-67.6	-60.5	42.9	.8062
700.	1378.	48.2	-129.8	-126.2	-120.2	-113.1	-104.8	62.3	1.0100
800.	1223.	73.2	-196.5	-192.3	-185.5	-177.3	-167.9	86.5	1.2414
900.	1109.	107.3	-286.9	-282.2	-274.6	-265.4	-254.8	115.5	1.5002
1000.	1035.	151.4	-404.9	-399.7	-391.2	-381.0	-369.2	148.4	1.7807

BALLISTIC COEFFICIENT .31, MUZZLE VELOCITY FT/SEC 3300.

RANGE	VELOCITY	HEIGHT	100 YD	150 YD	200 YD	250 YD	300 YD	DRIFT	TIME
0.	3300.	0.0	-1.5	-1.5	-1.5	-1.5	-1.5	0.0	0.0000
100.	2973.	-.3	0.0	.5	1.2	2.2	3.3	.9	.0958
200.	2668.	1.2	-2.5	-1.5	0.0	1.9	4.1	3.6	.2023
300.	2384.	4.2	-9.8	-8.5	-6.1	-3.3	0.0	8.6	.3214
400.	2115.	9.2	-23.3	-21.5	-18.4	-14.6	-10.2	16.1	.4549
500.	1864.	16.9	-44.6	-42.3	-38.4	-33.6	-28.2	26.6	.6059
600.	1637.	28.3	-75.9	-73.1	-68.5	-62.7	-56.2	40.9	.7778
700.	1433.	44.8	-120.2	-117.0	-111.6	-104.9	-97.3	59.4	.9739
800.	1265.	68.0	-181.9	-178.3	-172.1	-164.4	-155.7	82.7	1.1970
900.	1137.	100.0	-266.1	-262.0	-255.1	-246.4	-236.6	111.0	1.4486
1000.	1055.	141.7	-376.6	-372.1	-364.4	-354.7	-343.9	143.2	1.7225

BALLISTIC COEFFICIENT .31, MUZZLE VELOCITY FT/SEC 3400.

RANGE	VELOCITY	HEIGHT	100 YD	150 YD	200 YD	250 YD	300 YD	DRIFT	TIME
0.	3400.	0.0	-1.5	-1.5	-1.5	-1.5	-1.5	0.0	0.0000
100.	3066.	-.3	0.0	.4	1.1	2.0	3.0	.8	.0927
200.	2755.	1.1	-2.2	-1.4	0.0	1.8	3.8	3.5	.1961
300.	2465.	3.9	-9.1	-7.8	-5.7	-3.0	0.0	8.2	.3112
400.	2192.	8.6	-21.6	-20.0	-17.1	-13.6	-9.5	15.4	.4402
500.	1935.	15.7	-41.4	-39.4	-35.8	-31.4	-26.3	25.5	.5860
600.	1701.	26.3	-70.5	-68.0	-63.8	-58.4	-52.3	39.0	.7512
700.	1490.	41.6	-111.6	-108.8	-103.8	-97.6	-90.5	56.7	.9398
800.	1310.	63.3	-168.8	-165.6	-159.9	-152.8	-144.7	79.0	1.1550
900.	1170.	93.1	-246.8	-243.2	-236.7	-228.8	-219.6	106.3	1.3979
1000.	1075.	132.5	-350.6	-346.5	-339.4	-330.5	-320.4	138.0	1.6662

BALLISTIC COEFFICIENT .31, MUZZLE VELOCITY FT/SEC 3500.

RANGE	VELOCITY	HEIGHT	100 YD	150 YD	200 YD	250 YD	300 YD	DRIFT	TIME
0.	3500.	0.0	-1.5	-1.5	-1.5	-1.5	-1.5	0.0	0.0000
100.	3159.	-.4	0.0	.3	1.0	1.8	2.8	.8	.0904
200.	2842.	1.0	-2.0	-1.3	0.0	1.7	3.6	3.3	.1903
300.	2546.	3.6	-8.3	-7.3	-5.4	-2.9	0.0	7.9	.3021
400.	2269.	8.0	-20.0	-18.7	-16.1	-12.7	-8.9	14.8	.4269
500.	2007.	14.7	-38.5	-36.8	-33.5	-29.3	-24.6	24.4	.5674
600.	1766.	24.6	-65.5	-63.5	-59.6	-54.6	-48.8	37.4	.7268
700.	1548.	38.9	-103.9	-101.5	-96.9	-91.1	-84.4	54.3	.9084
800.	1357.	59.0	-157.0	-154.2	-149.0	-142.3	-134.7	75.7	1.1158
900.	1206.	86.8	-229.4	-226.4	-220.5	-213.0	-204.4	101.9	1.3506
1000.	1098.	124.0	-326.4	-323.0	-316.5	-308.2	-298.6	132.9	1.6122

BALLISTIC COEFFICIENT .31, MUZZLE VELOCITY FT/SEC 3600.

RANGE	VELOCITY	HEIGHT	100 YD	150 YD	200 YD	250 YD	300 YD	DRIFT	TIME
0.	3600.	0.0	-1.5	-1.5	-1.5	-1.5	-1.5	0.0	0.0000
100.	3253.	-.4	0.0	.3	.9	1.7	2.6	.7	.0875
200.	2929.	.9	-1.8	-1.2	0.0	1.5	3.3	3.2	.1848
300.	2627.	3.4	-7.7	-6.8	-5.0	-2.7	0.0	7.5	.2929
400.	2345.	7.5	-18.6	-17.4	-15.0	-11.9	-8.4	14.2	.4139
500.	2079.	13.7	-35.8	-34.3	-31.3	-27.5	-23.0	23.4	.5495
600.	1831.	23.0	-61.1	-59.3	-55.7	-51.1	-45.7	35.8	.7035
700.	1607.	36.3	-96.8	-94.7	-90.5	-85.1	-78.9	51.9	.8784
800.	1407.	55.0	-146.1	-143.7	-138.9	-132.7	-125.6	72.4	1.0780
900.	1245.	81.0	-213.7	-211.0	-205.5	-198.6	-190.6	97.7	1.3053
1000.	1124.	116.0	-304.2	-301.2	-295.1	-287.5	-278.6	127.9	1.5598

BALLISTIC COEFFICIENT .31, MUZZLE VELOCITY FT/SEC 3700.

RANGE	VELOCITY	HEIGHT	100 YD	150 YD	200 YD	250 YD	300 YD	DRIFT	TIME
0.	3700.	0.0	-1.5	-1.5	-1.5	-1.5	-1.5	0.0	0.0000
100.	3346.	-.4	0.0	.2	.8	1.5	2.4	.8	.0854
200.	3016.	.8	-1.6	-1.1	0.0	1.4	3.1	3.1	.1798
300.	2708.	3.1	-7.1	-6.3	-4.7	-2.5	0.0	7.3	.2848
400.	2421.	7.0	-17.3	-16.3	-14.1	-11.2	-7.8	13.7	.4020
500.	2151.	12.9	-33.4	-32.2	-29.4	-25.8	-21.6	22.5	.5335
600.	1897.	21.6	-57.0	-55.5	-52.2	-47.9	-42.8	34.4	.6819
700.	1667.	34.0	-90.4	-88.6	-84.7	-79.7	-73.8	49.8	.8508
800.	1459.	51.5	-136.4	-134.4	-129.9	-124.2	-117.5	69.5	1.0433
900.	1286.	75.8	-199.2	-197.0	-192.0	-185.5	-178.0	93.8	1.2627
1000.	1152.	108.7	-283.8	-281.3	-275.7	-268.5	-260.1	123.0	1.5099

BALLISTIC COEFFICIENT .31, MUZZLE VELOCITY FT/SEC 3800.

RANGE	VELOCITY	HEIGHT	100 YD	150 YD	200 YD	250 YD	300 YD	DRIFT	TIME
0.	3800.	0.0	-1.5	-1.5	-1.5	-1.5	-1.5	0.0	0.0000
100.	3440.	-.4	0.0	.2	.7	1.4	2.2	.7	.0831
200.	3103.	.7	-1.5	-1.0	0.0	1.3	2.9	3.0	.1751
300.	2790.	2.9	-6.5	-5.9	-4.3	-2.4	0.0	7.1	.2769
400.	2498.	6.6	-16.1	-15.3	-13.2	-10.5	-7.4	13.2	.3907
500.	2223.	12.1	-31.2	-30.1	-27.6	-24.2	-20.3	21.7	.5179
600.	1964.	20.3	-53.4	-52.1	-49.0	-45.0	-40.3	33.1	.6616
700.	1727.	31.9	-84.5	-83.0	-79.4	-74.7	-69.2	47.8	.8244
800.	1513.	48.2	-127.5	-125.8	-121.6	-116.3	-110.0	66.6	1.0102
900.	1329.	70.9	-186.1	-184.2	-179.5	-173.5	-166.5	90.1	1.2222
1000.	1184.	101.8	-265.1	-263.0	-257.8	-251.1	-243.3	118.3	1.4619

BALLISTIC COEFFICIENT .31, MUZZLE VELOCITY FT/SEC 3900.

RANGE	VELOCITY	HEIGHT	100 YD	150 YD	200 YD	250 YD	300 YD	DRIFT	TIME
0.	3900.	0.0	-1.5	-1.5	-1.5	-1.5	-1.5	0.0	0.0000
100.	3533.	-.4	0.0	.2	.7	1.3	2.0	.6	.0806
200.	3191.	.6	-1.3	-.9	0.0	1.3	2.7	2.8	.1700
300.	2871.	2.7	-6.0	-5.5	-4.1	-2.2	0.0	6.7	.2690
400.	2573.	6.2	-15.0	-14.3	-12.4	-9.9	-7.0	12.7	.3796
500.	2294.	11.4	-29.2	-28.3	-25.9	-22.8	-19.1	20.8	.5031
600.	2031.	19.0	-50.0	-48.9	-46.1	-42.3	-37.9	31.8	.6421
700.	1788.	29.9	-79.2	-77.9	-74.6	-70.2	-65.1	46.0	.7996
800.	1568.	45.2	-119.4	-118.0	-114.2	-109.2	-103.4	64.0	.9789
900.	1373.	66.5	-174.1	-172.5	-168.2	-162.6	-156.0	86.4	1.1835
1000.	1219.	95.5	-248.0	-246.2	-241.5	-235.2	-227.9	113.8	1.4157

BALLISTIC COEFFICIENT .31, MUZZLE VELOCITY FT/SEC 4000.

RANGE	VELOCITY	HEIGHT	100 YD	150 YD	200 YD	250 YD	300 YD	DRIFT	TIME
0.	4000.	0.0	-1.5	-1.5	-1.5	-1.5	-1.5	0.0	0.0000
100.	3627.	-.5	0.0	.1	.6	1.2	1.9	.7	.0790
200.	3278.	.6	-1.2	-.9	0.0	1.2	2.5	2.8	.1660
300.	2953.	2.6	-5.6	-5.1	-3.8	-2.0	0.0	6.5	.2622
400.	2649.	5.8	-14.0	-13.4	-11.6	-9.3	-6.6	12.3	.3696
500.	2366.	10.8	-27.3	-26.6	-24.4	-21.5	-18.1	20.2	.4896
600.	2099.	18.0	-46.8	-46.0	-43.4	-39.8	-35.7	30.7	.6242
700.	1849.	28.2	-74.3	-73.3	-70.2	-66.1	-61.3	44.3	.7765
800.	1623.	42.5	-111.9	-110.8	-107.3	-102.6	-97.1	61.5	.9496
900.	1421.	62.4	-163.1	-161.8	-157.8	-152.5	-146.4	83.1	1.1473
1000.	1256.	89.7	-232.3	-230.9	-226.5	-220.6	-213.8	109.6	1.3726

BALLISTIC COEFFICIENT .32, MUZZLE VELOCITY FT/SEC 2000.

RANGE	VELOCITY	HEIGHT	100 YD	150 YD	200 YD	250 YD	300 YD	DRIFT	TIME
0.	2000.	0.0	-1.5	-1.5	-1.5	-1.5	-1.5	0.0	0.0000
100.	1767.	.5	0.0	2.2	4.9	7.9	11.3	1.7	.1596
200.	1555.	4.8	-9.7	-5.4	0.0	6.1	12.9	7.2	.3408
300.	1368.	13.6	-33.9	-27.4	-19.3	-10.2	0.0	17.0	.5466
400.	1219.	28.4	-76.7	-68.0	-57.2	-45.1	-31.5	31.5	.7792
500.	1109.	51.0	-143.0	-132.1	-118.6	-103.4	-86.4	50.7	1.0383
600.	1038.	82.7	-236.8	-223.7	-207.6	-189.3	-168.9	73.6	1.3183
700.	980.	124.6	-360.8	-345.5	-326.7	-305.4	-281.6	99.6	1.6160
800.	929.	178.2	-517.7	-500.2	-478.7	-454.3	-427.2	128.6	1.9306
900.	883.	244.8	-711.0	-691.2	-667.1	-639.7	-609.1	160.5	2.2618
1000.	841.	326.2	-944.5	-922.6	-895.8	-865.3	-831.3	195.3	2.6098

BALLISTIC COEFFICIENT .32, MUZZLE VELOCITY FT/SEC 2100.

RANGE	VELOCITY	HEIGHT	100 YD	150 YD	200 YD	250 YD	300 YD	DRIFT	TIME
0.	2100.	0.0	-1.5	-1.5	-1.5	-1.5	-1.5	0.0	0.0000
100.	1858.	.4	0.0	1.9	4.3	7.1	10.1	1.6	.1517
200.	1638.	4.3	-8.7	-4.8	0.0	5.5	11.6	6.7	.3239
300.	1439.	12.2	-30.4	-24.6	-17.4	-9.2	0.0	16.0	.5195
400.	1275.	25.6	-69.0	-61.2	-51.6	-40.7	-28.5	29.9	.7414
500.	1148.	46.3	-128.8	-119.2	-107.2	-93.5	-78.2	48.5	.9899
600.	1063.	75.9	-215.5	-203.9	-189.5	-173.1	-154.7	71.5	1.2634
700.	1003.	115.1	-330.5	-317.0	-300.2	-281.0	-259.6	97.4	1.5536
800.	949.	165.5	-476.9	-461.5	-442.3	-420.4	-395.9	126.4	1.8608
900.	901.	228.4	-658.1	-640.7	-619.1	-594.5	-567.0	158.3	2.1849
1000.	858.	305.7	-878.2	-858.9	-834.9	-807.6	-777.0	193.3	2.5266

BALLISTIC COEFFICIENT .32, MUZZLE VELOCITY FT/SEC 2200.

RANGE	VELOCITY	HEIGHT	100 YD	150 YD	200 YD	250 YD	300 YD	DRIFT	TIME
0.	2200.	0.0	-1.5	-1.5	-1.5	-1.5	-1.5	0.0	0.0000
100.	1950.	.3	0.0	1.7	3.8	6.3	9.1	1.5	.1450
200.	1722.	3.8	-7.7	-4.3	0.0	4.9	10.5	6.3	.3085
300.	1515.	11.0	-27.2	-22.1	-15.7	-8.3	0.0	15.0	.4945
400.	1335.	23.2	-62.0	-55.2	-46.6	-36.7	-25.7	28.3	.7060
500.	1192.	42.1	-116.4	-107.9	-97.2	-84.8	-71.0	46.3	.9446
600.	1093.	69.3	-195.0	-184.7	-171.9	-157.1	-140.6	68.6	1.2077
700.	1025.	106.2	-302.3	-290.4	-275.4	-258.1	-238.9	94.7	1.4927
800.	969.	153.9	-440.1	-426.5	-409.4	-389.6	-367.6	123.9	1.7951
900.	919.	213.0	-608.9	-593.6	-574.3	-552.1	-527.3	155.4	2.1103
1000.	874.	286.3	-816.3	-799.3	-777.9	-753.2	-725.7	190.4	2.4456

BALLISTIC COEFFICIENT .32, MUZZLE VELOCITY FT/SEC 2300.

RANGE	VELOCITY	HEIGHT	100 YD	150 YD	200 YD	250 YD	300 YD	DRIFT	TIME
0.	2300.	0.0	-1.5	-1.5	-1.5	-1.5	-1.5	0.0	0.0000
100.	2045.	.2	0.0	1.5	3.4	5.7	8.2	1.4	.1385
200.	1807.	3.4	-6.9	-3.8	0.0	4.5	9.5	5.9	.2944
300.	1592.	9.9	-24.5	-20.0	-14.2	-7.5	0.0	14.1	.4715
400.	1399.	21.0	-55.8	-49.8	-42.1	-33.2	-23.2	26.5	.6725
500.	1244.	38.2	-105.0	-97.5	-87.9	-76.7	-64.2	43.7	.9005
600.	1126.	63.3	-176.7	-167.7	-156.1	-142.8	-127.8	65.5	1.1548
700.	1048.	98.0	-276.7	-266.1	-252.6	-237.1	-219.6	91.7	1.4343
800.	990.	142.7	-404.8	-392.8	-377.4	-359.6	-339.6	120.6	1.7286
900.	938.	198.5	-563.4	-549.8	-532.5	-512.5	-490.0	152.0	2.0376
1000.	891.	267.9	-758.5	-743.4	-724.2	-701.9	-676.9	186.8	2.3660

BALLISTIC COEFFICIENT .32, MUZZLE VELOCITY FT/SEC 2400.

RANGE	VELOCITY	HEIGHT	100 YD	150 YD	200 YD	250 YD	300 YD	DRIFT	TIME
0.	2400.	0.0	-1.5	-1.5	-1.5	-1.5	-1.5	0.0	0.0000
100.	2139.	.1	0.0	1.3	3.1	5.1	7.4	1.3	.1325
200.	1894.	3.1	-6.1	-3.5	0.0	4.0	8.6	5.5	.2814
300.	1670.	9.0	-22.1	-18.1	-12.8	-6.8	0.0	13.2	.4502
400.	1468.	19.0	-50.5	-45.1	-38.2	-30.1	-21.0	25.0	.6418
500.	1298.	34.7	-95.1	-88.4	-79.7	-69.6	-58.3	41.3	.8596
600.	1164.	57.8	-160.4	-152.4	-141.9	-129.8	-116.3	62.3	1.1041
700.	1074.	89.8	-251.4	-242.1	-229.9	-215.7	-199.9	87.7	1.3732
800.	1011.	132.1	-372.1	-361.5	-347.5	-331.3	-313.3	116.8	1.6639
900.	957.	184.8	-520.9	-508.9	-493.2	-475.0	-454.6	148.1	1.9663
1000.	908.	250.6	-704.6	-691.3	-673.9	-653.6	-631.1	182.8	2.2884

BALLISTIC COEFFICIENT .32, MUZZLE VELOCITY FT/SEC 2500.

RANGE	VELOCITY	HEIGHT	100 YD	150 YD	200 YD	250 YD	300 YD	DRIFT	TIME
0.	2500.	0.0	-1.5	-1.5	-1.5	-1.5	-1.5	0.0	0.0000
100.	2233.	.0	0.0	1.2	2.8	4.6	6.7	1.2	.1269
200.	1982.	2.7	-5.5	-3.1	0.0	3.7	7.8	5.2	.2694
300.	1750.	8.2	-20.0	-16.4	-11.7	-6.2	0.0	12.4	.4306
400.	1540.	17.3	-45.8	-41.0	-34.7	-27.4	-19.2	23.5	.6134
500.	1356.	31.6	-86.4	-80.4	-72.6	-63.4	-53.1	39.0	.8216
600.	1209.	52.8	-145.9	-138.7	-129.3	-118.3	-106.0	59.1	1.0561
700.	1103.	82.5	-229.4	-221.1	-210.1	-197.2	-182.8	83.9	1.3167
800.	1033.	121.9	-340.7	-331.2	-318.6	-304.0	-287.5	112.4	1.5984
900.	976.	172.0	-481.5	-470.8	-456.7	-440.2	-421.6	143.8	1.8973
1000.	926.	234.3	-654.5	-642.6	-626.9	-608.6	-588.0	178.3	2.2129

BALLISTIC COEFFICIENT .32, MUZZLE VELOCITY FT/SEC 2600.

RANGE	VELOCITY	HEIGHT	100 YD	150 YD	200 YD	250 YD	300 YD	DRIFT	TIME
0.	2600.	0.0	-1.5	-1.5	-1.5	-1.5	-1.5	0.0	0.0000
100.	2328.	-.0	0.0	1.0	2.5	4.2	6.0	1.2	.1220
200.	2071.	2.5	-5.0	-2.9	0.0	3.4	7.1	4.9	.2585
300.	1831.	7.4	-18.1	-14.9	-10.7	-5.6	0.0	11.7	.4127
400.	1614.	15.8	-41.6	-37.4	-31.7	-25.0	-17.5	22.1	.5873
500.	1418.	28.9	-78.5	-73.2	-66.1	-57.7	-48.3	36.7	.7857
600.	1258.	48.3	-132.9	-126.6	-118.0	-107.9	-96.7	56.0	1.0107
700.	1135.	75.9	-209.7	-202.4	-192.4	-180.6	-167.5	80.2	1.2636
800.	1056.	112.6	-312.6	-304.2	-292.7	-279.3	-264.3	108.0	1.5370
900.	996.	160.0	-444.7	-435.3	-422.4	-407.3	-390.5	139.3	1.8300
1000.	943.	219.2	-608.3	-597.8	-583.5	-566.7	-548.0	173.6	2.1405

BALLISTIC COEFFICIENT .32, MUZZLE VELOCITY FT/SEC 2700.

RANGE	VELOCITY	HEIGHT	100 YD	150 YD	200 YD	250 YD	300 YD	DRIFT	TIME
0.	2700.	0.0	-1.5	-1.5	-1.5	-1.5	-1.5	0.0	0.0000
100.	2422.	-.1	0.0	.9	2.2	3.8	5.5	1.1	.1174
200.	2160.	2.2	-4.5	-2.6	0.0	3.1	6.5	4.6	.2484
300.	1913.	6.8	-16.5	-13.6	-9.8	-5.2	0.0	11.1	.3962
400.	1688.	14.5	-37.9	-34.2	-29.0	-22.9	-16.0	20.9	.5631
500.	1484.	26.4	-71.6	-67.0	-60.5	-52.8	-44.2	34.7	.7526
600.	1311.	44.2	-121.3	-115.7	-107.9	-98.7	-88.4	53.1	.9682
700.	1173.	69.7	-191.8	-185.2	-176.2	-165.4	-153.4	76.4	1.2119
800.	1080.	104.1	-286.8	-279.3	-269.0	-256.7	-242.9	103.6	1.4778
900.	1016.	148.7	-410.7	-402.3	-390.6	-376.8	-361.3	134.6	1.7647
1000.	961.	204.6	-564.3	-555.0	-542.0	-526.7	-509.5	168.5	2.0683

BALLISTIC COEFFICIENT .32, MUZZLE VELOCITY FT/SEC 2800.

RANGE	VELOCITY	HEIGHT	100 YD	150 YD	200 YD	250 YD	300 YD	DRIFT	TIME
0.	2800.	0.0	-1.5	-1.5	-1.5	-1.5	-1.5	0.0	0.0000
100.	2516.	-.1	0.0	.8	2.0	3.4	5.0	1.0	.1130
200.	2249.	2.0	-4.0	-2.4	0.0	2.8	5.9	4.4	.2391
300.	1996.	6.2	-15.0	-12.5	-8.9	-4.7	0.0	10.4	.3806
400.	1763.	13.3	-34.7	-31.4	-26.6	-21.0	-14.7	19.7	.5405
500.	1552.	24.3	-65.6	-61.5	-55.5	-48.6	-40.6	32.8	.7221
600.	1366.	40.6	-111.0	-106.0	-98.9	-90.6	-81.1	50.3	.9284
700.	1217.	64.0	-175.3	-169.5	-161.2	-151.5	-140.4	72.4	1.1616
800.	1108.	96.2	-263.4	-256.8	-247.3	-236.1	-223.5	99.2	1.4208
900.	1037.	138.2	-379.0	-371.6	-360.9	-348.4	-334.1	129.7	1.7012
1000.	979.	191.1	-523.9	-515.6	-503.7	-489.8	-474.0	163.3	1.9991

BALLISTIC COEFFICIENT .32, MUZZLE VELOCITY FT/SEC 2900.

RANGE	VELOCITY	HEIGHT	100 YD	150 YD	200 YD	250 YD	300 YD	DRIFT	TIME
0.	2900.	0.0	-1.5	-1.5	-1.5	-1.5	-1.5	0.0	0.0000
100.	2610.	-.2	0.0	.7	1.8	3.1	4.6	1.0	.1090
200.	2337.	1.8	-3.7	-2.2	0.0	2.6	5.5	4.2	.2306
300.	2079.	5.7	-13.7	-11.5	-8.2	-4.3	0.0	9.9	.3665
400.	1839.	12.2	-31.8	-28.9	-24.5	-19.4	-13.6	18.7	.5202
500.	1621.	22.4	-60.2	-56.5	-51.1	-44.7	-37.4	31.1	.6939
600.	1424.	37.4	-101.9	-97.4	-90.9	-83.2	-74.5	47.7	.8915
700.	1263.	59.0	-161.0	-155.8	-148.2	-139.2	-129.1	68.9	1.1155
800.	1139.	88.9	-242.2	-236.3	-227.6	-217.3	-205.7	94.9	1.3666
900.	1058.	128.4	-350.0	-343.4	-333.6	-322.0	-309.0	124.8	1.6403
1000.	998.	178.5	-486.6	-479.3	-468.4	-455.6	-441.1	158.1	1.9327

BALLISTIC COEFFICIENT .32, MUZZLE VELOCITY FT/SEC 3000.

RANGE	VELOCITY	HEIGHT	100 YD	150 YD	200 YD	250 YD	300 YD	DRIFT	TIME
0.	3000.	0.0	-1.5	-1.5	-1.5	-1.5	-1.5	0.0	0.0000
100.	2703.	-.2	0.0	.7	1.6	2.8	4.2	1.0	.1055
200.	2425.	1.6	-3.3	-2.0	0.0	2.4	5.1	4.0	.2227
300.	2162.	5.3	-12.5	-10.6	-7.6	-4.0	0.0	9.4	.3537
400.	1915.	11.3	-29.2	-26.6	-22.7	-17.9	-12.5	17.8	.5013
500.	1690.	20.7	-55.4	-52.1	-47.2	-41.2	-34.5	29.6	.6679
600.	1486.	34.5	-93.8	-89.8	-83.9	-76.7	-68.7	45.3	.8573
700.	1312.	54.5	-148.1	-143.5	-136.6	-128.3	-118.9	65.6	1.0726
800.	1174.	82.4	-223.4	-218.1	-210.2	-200.7	-190.0	90.8	1.3158
900.	1081.	119.4	-323.5	-317.6	-308.6	-298.0	-285.9	120.0	1.5820
1000.	1017.	166.8	-452.0	-445.5	-435.5	-423.7	-410.3	152.8	1.8684

BALLISTIC COEFFICIENT .32, MUZZLE VELOCITY FT/SEC 3100.

RANGE	VELOCITY	HEIGHT	100 YD	150 YD	200 YD	250 YD	300 YD	DRIFT	TIME
0.	3100.	0.0	-1.5	-1.5	-1.5	-1.5	-1.5	0.0	0.0000
100.	2796.	-.3	0.0	.6	1.5	2.6	3.8	.9	.1019
200.	2512.	1.5	-3.0	-1.8	0.0	2.2	4.7	3.8	.2151
300.	2245.	4.8	-11.5	-9.8	-7.0	-3.7	0.0	9.0	.3414
400.	1993.	10.5	-26.9	-24.6	-20.9	-16.6	-11.6	16.9	.4832
500.	1760.	19.1	-51.1	-48.2	-43.6	-38.1	-31.9	28.1	.6433
600.	1549.	31.9	-86.5	-83.1	-77.6	-71.0	-63.5	43.0	.8252
700.	1363.	50.4	-136.6	-132.6	-126.2	-118.5	-109.8	62.4	1.0319
800.	1215.	76.1	-205.8	-201.2	-193.9	-185.1	-175.2	86.5	1.2656
900.	1107.	110.9	-298.9	-293.7	-285.5	-275.6	-264.4	115.1	1.5250
1000.	1036.	155.8	-419.6	-413.9	-404.8	-393.8	-381.3	147.5	1.8057

BALLISTIC COEFFICIENT .32, MUZZLE VELOCITY FT/SEC 3200.

RANGE	VELOCITY	HEIGHT	100 YD	150 YD	200 YD	250 YD	300 YD	DRIFT	TIME
0.	3200.	0.0	-1.5	-1.5	-1.5	-1.5	-1.5	0.0	0.0000
100.	2889.	-.3	0.0	.5	1.3	2.4	3.5	.9	.0986
200.	2600.	1.3	-2.7	-1.7	0.0	2.0	4.3	3.6	.2082
300.	2327.	4.5	-10.6	-9.0	-6.5	-3.5	0.0	8.6	.3302
400.	2070.	9.7	-24.9	-22.8	-19.5	-15.4	-10.8	16.2	.4668
500.	1831.	17.8	-47.3	-44.7	-40.5	-35.4	-29.7	26.8	.6210
600.	1613.	29.6	-80.1	-77.0	-72.0	-65.9	-58.9	41.0	.7956
700.	1418.	46.7	-126.3	-122.7	-116.8	-109.7	-101.6	59.4	.9940
800.	1258.	70.6	-190.2	-186.1	-179.4	-171.3	-162.1	82.6	1.2[illegible]91
900.	1135.	103.3	-277.0	-272.4	-264.9	-255.8	-245.3	110.6	1.4720
1000.	1056.	145.5	-389.8	-384.7	-376.3	-366.2	-354.6	142.2	1.7455

BALLISTIC COEFFICIENT .32, MUZZLE VELOCITY FT/SEC 3300.

RANGE	VELOCITY	HEIGHT	100 YD	150 YD	200 YD	250 YD	300 YD	DRIFT	TIME
0.	3300.	0.0	-1.5	-1.5	-1.5	-1.5	-1.5	0.0	0.0000
100.	2983.	-.3	0.0	.4	1.2	2.2	3.2	.8	.0957
200.	2687.	1.2	-2.4	-1.5	0.0	1.9	4.0	3.5	.2016
300.	2410.	4.2	-9.7	-8.4	-6.1	-3.2	0.0	8.3	.3197
400.	2148.	9.0	-23.0	-21.2	-18.1	-14.3	-10.0	15.5	.4515
500.	1902.	16.5	-43.8	-41.5	-37.7	-33.0	-27.6	25.6	.5999
600.	1678.	27.6	-74.2	-71.5	-66.9	-61.2	-54.8	39.1	.7679
700.	1475.	43.4	-117.0	-113.9	-108.5	-101.9	-94.3	56.7	.9586
800.	1303.	65.6	-176.2	-172.6	-166.4	-158.9	-150.3	78.9	1.1753
900.	1168.	95.9	-256.3	-252.3	-245.3	-236.8	-227.1	105.8	1.4191
1000.	1077.	135.9	-362.2	-357.7	-350.0	-340.6	-329.8	137.0	1.6872

BALLISTIC COEFFICIENT .32, MUZZLE VELOCITY FT/SEC 3400.

RANGE	VELOCITY	HEIGHT	100 YD	150 YD	200 YD	250 YD	300 YD	DRIFT	TIME
0.	3400.	0.0	-1.5	-1.5	-1.5	-1.5	-1.5	0.0	0.0000
100.	3076.	-.3	0.0	.4	1.1	2.0	3.0	.8	.0927
200.	2774.	1.1	-2.2	-1.4	0.0	1.8	3.8	3.3	.1953
300.	2492.	3.8	-8.9	-7.8	-5.6	-3.0	0.0	7.9	.3096
400.	2225.	8.4	-21.3	-19.7	-16.9	-13.4	-9.4	14.8	.4369
500.	1974.	15.4	-40.6	-38.6	-35.1	-30.7	-25.7	24.4	.5800
600.	1743.	25.7	-68.9	-66.5	-62.3	-57.1	-51.0	37.4	.7419
700.	1534.	40.4	-108.7	-105.9	-101.0	-94.9	-87.8	54.2	.9254
800.	1351.	61.0	-163.5	-160.3	-154.7	-147.7	-139.7	75.4	1.1341
900.	1205.	89.3	-237.9	-234.3	-228.0	-220.1	-211.1	101.3	1.3696
1000.	1100.	126.9	-336.7	-332.7	-325.7	-317.0	-306.9	131.8	1.6310

BALLISTIC COEFFICIENT .32, MUZZLE VELOCITY FT/SEC 3500.

RANGE	VELOCITY	HEIGHT	100 YD	150 YD	200 YD	250 YD	300 YD	DRIFT	TIME
0.	3500.	0.0	-1.5	-1.5	-1.5	-1.5	-1.5	0.0	0.0000
100.	3170.	-.4	0.0	.3	1.0	1.8	2.7	.8	.0901
200.	2861.	1.0	-2.0	-1.3	0.0	1.6	3.5	3.2	.1897
300.	2573.	3.6	-8.2	-7.2	-5.3	-2.8	0.0	7.6	.3004
400.	2303.	7.9	-19.8	-18.4	-15.8	-12.5	-8.8	14.2	.4237
500.	2047.	14.4	-37.8	-36.1	-32.9	-28.8	-24.1	23.5	.5620
600.	1809.	24.0	-64.1	-62.1	-58.2	-53.3	-47.6	35.8	.7177
700.	1594.	37.7	-101.2	-98.8	-94.3	-88.5	-82.0	51.9	.8946
800.	1401.	56.8	-152.0	-149.3	-144.1	-137.5	-130.0	72.1	1.0954
900.	1245.	83.3	-221.2	-218.2	-212.3	-204.9	-196.5	97.1	1.3231
1000.	1127.	118.7	-313.4	-310.0	-303.5	-295.3	-285.9	126.7	1.5772

BALLISTIC COEFFICIENT .32, MUZZLE VELOCITY FT/SEC 3600.

RANGE	VELOCITY	HEIGHT	100 YD	150 YD	200 YD	250 YD	300 YD	DRIFT	TIME
0.	3600.	0.0	-1.5	-1.5	-1.5	-1.5	-1.5	0.0	0.0000
100.	3264.	-.4	0.0	.3	.9	1.7	2.5	.7	.0873
200.	2949.	.9	-1.8	-1.2	0.0	1.5	3.3	3.1	.1840
300.	2655.	3.3	-7.6	-6.7	-4.9	-2.6	0.0	7.3	.2914
400.	2380.	7.3	-18.3	-17.1	-14.8	-11.7	-8.2	13.6	.4108
500.	2120.	13.5	-35.2	-33.7	-30.7	-26.9	-22.5	22.4	.5442
600.	1876.	22.4	-59.8	-58.0	-54.4	-49.8	-44.6	34.3	.6947
700.	1654.	35.2	-94.3	-92.2	-88.1	-82.7	-76.6	49.6	.8652
800.	1454.	53.1	-141.7	-139.3	-134.6	-128.4	-121.4	69.0	1.0588
900.	1287.	77.7	-206.0	-203.3	-198.0	-191.0	-183.2	93.0	1.2785
1000.	1155.	111.0	-292.3	-289.3	-283.4	-275.7	-266.9	121.9	1.5261

BALLISTIC COEFFICIENT .32, MUZZLE VELOCITY FT/SEC 3700.

RANGE	VELOCITY	HEIGHT	100 YD	150 YD	200 YD	250 YD	300 YD	DRIFT	TIME
0.	3700.	0.0	-1.5	-1.5	-1.5	-1.5	-1.5	0.0	0.0000
100.	3357.	-.4	0.0	.2	.8	1.5	2.3	.7	.0853
200.	3036.	.8	-1.6	-1.1	0.0	1.4	3.1	3.0	.1792
300.	2736.	3.1	-7.0	-6.3	-4.6	-2.5	0.0	7.1	.2833
400.	2456.	6.9	-17.0	-16.0	-13.8	-11.0	-7.7	13.1	.3990
500.	2192.	12.6	-32.8	-31.6	-28.8	-25.3	-21.2	21.6	.5283
600.	1943.	21.0	-55.8	-54.4	-51.0	-46.8	-41.8	33.0	.6738
700.	1715.	33.0	-88.0	-86.3	-82.4	-77.5	-71.7	47.6	.8380
800.	1509.	49.6	-132.2	-130.2	-125.8	-120.2	-113.6	66.2	1.0246
900.	1330.	72.7	-192.1	-189.9	-184.9	-178.5	-171.1	89.3	1.2368
1000.	1189.	103.8	-272.3	-269.8	-264.3	-257.2	-249.0	117.0	1.4759

BALLISTIC COEFFICIENT .32, MUZZLE VELOCITY FT/SEC 3800.

RANGE	VELOCITY	HEIGHT	100 YD	150 YD	200 YD	250 YD	300 YD	DRIFT	TIME
0.	3800.	0.0	-1.5	-1.5	-1.5	-1.5	-1.5	0.0	0.0000
100.	3451.	-.4	0.0	.2	.7	1.4	2.2	.7	.0830
200.	3124.	.7	-1.4	-1.0	0.0	1.3	2.9	2.9	.1746
300.	2818.	2.9	-6.5	-5.8	-4.3	-2.3	0.0	6.8	.2756
400.	2533.	6.5	-15.9	-15.0	-13.0	-10.4	-7.3	12.7	.3880
500.	2265.	11.9	-30.7	-29.6	-27.1	-23.8	-19.9	20.8	.5132
600.	2011.	19.8	-52.2	-50.9	-47.9	-43.9	-39.3	31.7	.6537
700.	1777.	30.9	-82.4	-80.9	-77.3	-72.7	-67.3	45.7	.8125
800.	1564.	46.5	-123.7	-122.0	-117.9	-112.7	-106.5	63.5	.9926
900.	1376.	68.1	-179.5	-177.6	-173.0	-167.1	-160.1	85.7	1.1973
1000.	1225.	97.2	-254.4	-252.2	-247.1	-240.6	-232.8	112.5	1.4286

BALLISTIC COEFFICIENT .32, MUZZLE VELOCITY FT/SEC 3900.

RANGE	VELOCITY	HEIGHT	100 YD	150 YD	200 YD	250 YD	300 YD	DRIFT	TIME
0.	3900.	0.0	-1.5	-1.5	-1.5	-1.5	-1.5	0.0	0.0000
100.	3545.	-.4	0.0	.2	.6	1.3	2.0	.6	.0805
200.	3211.	.6	-1.3	-.9	0.0	1.2	2.7	2.8	.1695
300.	2900.	2.7	-6.0	-5.4	-4.0	-2.2	0.0	6.5	.2678
400.	2610.	6.1	-14.8	-14.1	-12.2	-9.7	-6.8	12.2	.3768
500.	2337.	11.2	-28.7	-27.8	-25.5	-22.4	-18.8	20.0	.4985
600.	2079.	18.6	-48.9	-47.9	-45.0	-41.3	-37.0	30.4	.6343
700.	1839.	29.1	-77.2	-76.0	-72.7	-68.4	-63.3	43.9	.7881
800.	1621.	43.6	-115.8	-114.4	-110.7	-105.8	-99.9	61.0	.9617
900.	1424.	63.8	-167.9	-166.4	-162.1	-156.6	-150.1	82.2	1.1593
1000.	1263.	91.1	-237.9	-236.2	-231.5	-225.3	-218.0	108.1	1.3834

BALLISTIC COEFFICIENT .32, MUZZLE VELOCITY FT/SEC 4000.

RANGE	VELOCITY	HEIGHT	100 YD	150 YD	200 YD	250 YD	300 YD	DRIFT	TIME
0.	4000.	0.0	-1.5	-1.5	-1.5	-1.5	-1.5	0.0	0.0000
100.	3638.	-.5	0.0	.1	.6	1.2	1.8	.7	.0788
200.	3299.	.6	-1.1	-.9	0.0	1.2	2.5	2.7	.1653
300.	2982.	2.5	-5.5	-5.1	-3.8	-2.0	0.0	6.3	.2611
400.	2686.	5.7	-13.8	-13.2	-11.5	-9.1	-6.4	11.8	.3670
500.	2409.	10.5	-26.8	-26.2	-24.0	-21.1	-17.7	19.4	.4851
600.	2147.	17.5	-45.9	-45.0	-42.5	-38.9	-34.9	29.4	.6169
700.	1902.	27.4	-72.4	-71.4	-68.4	-64.3	-59.6	42.3	.7654
800.	1677.	41.1	-108.7	-107.6	-104.1	-99.4	-94.0	58.7	.9334
900.	1475.	59.9	-157.5	-156.2	-152.3	-147.1	-141.0	79.1	1.1242
1000.	1303.	85.6	-222.9	-221.5	-217.2	-211.3	-204.6	104.0	1.3409

BALLISTIC COEFFICIENT .33, MUZZLE VELOCITY FT/SEC 2000.

RANGE	VELOCITY	HEIGHT	100 YD	150 YD	200 YD	250 YD	300 YD	DRIFT	TIME
0.	2000.	0.0	-1.5	-1.5	-1.5	-1.5	-1.5	0.0	0.0000
100.	1773.	.5	0.0	2.2	4.8	7.8	11.2	1.6	.1594
200.	1568.	4.8	-9.7	-5.3	0.0	6.0	12.7	6.9	.3395
300.	1384.	13.4	-33.5	-27.0	-19.0	-10.0	0.0	16.4	.5432
400.	1235.	28.0	-75.7	-67.0	-56.4	-44.4	-31.0	30.5	.7734
500.	1123.	50.0	-140.5	-129.6	-116.3	-101.4	-84.6	49.0	1.0236
600.	1048.	81.4	-233.7	-220.7	-204.7	-186.7	-166.7	71.9	1.3085
700.	992.	122.6	-355.7	-340.5	-321.9	-300.9	-277.5	97.3	1.6029
800.	941.	174.5	-508.3	-490.9	-469.7	-445.7	-418.9	125.1	1.9109
900.	895.	239.9	-698.7	-679.2	-655.3	-628.3	-598.2	156.5	2.2393
1000.	854.	319.0	-926.6	-905.0	-878.4	-848.4	-815.0	190.3	2.5812

BALLISTIC COEFFICIENT .33, MUZZLE VELOCITY FT/SEC 2100.

RANGE	VELOCITY	HEIGHT	100 YD	150 YD	200 YD	250 YD	300 YD	DRIFT	TIME
0.	2100.	0.0	-1.5	-1.5	-1.5	-1.5	-1.5	0.0	0.0000
100.	1865.	.4	0.0	1.9	4.3	7.0	10.0	1.5	.1514
200.	1651.	4.2	-8.6	-4.8	0.0	5.4	11.4	6.5	.3226
300.	1456.	12.0	-30.0	-24.3	-17.1	-9.0	0.0	15.4	.5162
400.	1293.	25.2	-68.0	-60.3	-50.8	-40.0	-27.9	28.9	.7355
500.	1164.	45.4	-126.5	-117.0	105.1	-91.5	-76.5	46.8	.9802
600.	1076.	74.1	-210.8	-199.3	·185.0	-168.8	-150.8	69.0	1.2489
700.	1015.	112.6	-323.9	-310.5	-293.8	-274.9	-253.9	94.4	1.5366
800.	962.	161.8	-467.6	-452.3	-433.2	-411.6	-387.6	122.8	1.8405
900.	914.	223.3	-645.2	-627.9	-606.5	-582.2	-555.1	154.0	2.1605
1000.	870.	298.6	-860.6	-841.4	-817.6	-790.6	-760.6	188.1	2.4973

BALLISTIC COEFFICIENT .33, MUZZLE VELOCITY FT/SEC 2200.

RANGE	VELOCITY	HEIGHT	100 YD	150 YD	200 YD	250 YD	300 YD	DRIFT	TIME
0.	2200.	0.0	-1.5	-1.5	-1.5	-1.5	-1.5	0.0	0.0000
100.	1958.	.3	0.0	1.7	3.8	6.2	9.0	1.5	.1447
200.	1735.	3.8	-7.6	-4.2	0.0	4.9	10.3	6.1	.3074
300.	1533.	10.8	-26.9	-21.8	-15.4	-8.1	0.0	14.5	.4915
400.	1355.	22.8	-61.1	-54.3	-45.8	-36.1	-25.2	27.2	.7003
500.	1212.	41.2	-114.1	-105.6	-95.0	-82.8	-69.3	44.5	.9345
600.	1108.	67.7	-191.0	-180.8	-168.1	-153.5	-137.2	66.2	1.1943
700.	1038.	103.6	-295.4	-283.5	-268.7	-251.6	-232.6	91.5	1.4745
800.	982.	150.0	-430.0	-416.4	-399.4	-380.0	-358.3	120.0	1.7725
900.	933.	208.0	-596.3	-581.1	-562.0	-540.1	-515.7	151.1	2.0856
1000.	888.	279.3	-798.7	-781.8	-760.6	-736.2	-709.1	185.1	2.4155

BALLISTIC COEFFICIENT .33, MUZZLE VELOCITY FT/SEC 2300.

RANGE	VELOCITY	HEIGHT	100 YD	150 YD	200 YD	250 YD	300 YD	DRIFT	TIME
0.	2300.	0.0	-1.5	-1.5	-1.5	-1.5	-1.5	0.0	0.0000
100.	2052.	.2	0.0	1.5	3.4	5.6	8.1	1.4	.1382
200.	1821.	3.4	-6.8	-3.8	0.0	4.4	9.3	5.7	.2932
300.	1611.	9.8	-24.2	-19.7	-14.0	-7.4	0.0	13.6	.4686
400.	1421.	20.6	-54.9	-49.0	-41.3	-32.6	-22.7	25.5	.6669
500.	1265.	37.4	-103.1	-95.6	-86.1	-75.1	-62.8	42.0	.8911
600.	1143.	61.8	-172.9	-164.0	-152.5	-139.3	-124.6	63.1	1.1412
700.	1063.	95.4	-269.7	-259.2	-245.9	-230.5	-213.3	88.4	1.4151
800.	1004.	138.9	-395.0	-383.0	-367.8	-350.2	-330.5	116.6	1.7057
900.	952.	193.6	-550.8	-537.3	-520.2	-500.4	-478.2	147.5	2.0120
1000.	905.	261.0	-741.1	-726.1	-707.1	-685.1	-660.5	181.4	2.3353

BALLISTIC COEFFICIENT .33, MUZZLE VELOCITY FT/SEC 2400.

RANGE	VELOCITY	HEIGHT	100 YD	150 YD	200 YD	250 YD	300 YD	DRIFT	TIME
0.	2400.	0.0	-1.5	-1.5	-1.5	-1.5	-1.5	0.0	0.0000
100.	2147.	.1	0.0	1.3	3.1	5.0	7.3	1.3	.1322
200.	1908.	3.0	-6.1	-3.5	0.0	4.0	8.4	5.4	.2804
300.	1690.	8.9	-21.8	-17.8	-12.6	-6.7	0.0	12.8	.4475
400.	1492.	18.7	-49.7	-44.4	-37.5	-29.5	-20.6	24.0	.6364
500.	1321.	34.0	-93.3	-86.7	-78.1	-68.1	-57.0	39.7	.8505
600.	1186.	56.4	-156.9	-149.0	-138.6	-126.7	-113.3	60.0	1.0908
700.	1090.	87.4	-245.4	-236.1	-224.1	-210.1	-194.5	84.6	1.3555
800.	1025.	128.5	-362.5	-351.9	-338.1	-322.2	-304.4	112.8	1.6408
900.	971.	180.5	-510.0	-498.1	-482.5	-464.6	-444.6	144.0	1.9433
1000.	922.	243.8	-687.5	-674.2	-657.0	-637.1	-614.8	177.3	2.2573

BALLISTIC COEFFICIENT .33, MUZZLE VELOCITY FT/SEC 2500.

RANGE	VELOCITY	HEIGHT	100 YD	150 YD	200 YD	250 YD	300 YD	DRIFT	TIME
0.	2500.	0.0	-1.5	-1.5	-1.5	-1.5	-1.5	0.0	0.0000
100.	2241.	.0	0.0	1.2	2.7	4.5	6.6	1.2	.1268
200.	1997.	2.7	-5.5	-3.1	0.0	3.6	7.7	5.0	.2685
300.	1770.	8.0	-19.7	-16.2	-11.5	-6.1	0.0	12.0	.4281
400.	1565.	17.0	-45.1	-40.4	-34.2	-27.0	-18.8	22.6	.6086
500.	1381.	31.0	-84.7	-78.8	-71.0	-62.0	-51.8	37.4	.8127
600.	1233.	51.5	-142.8	-135.8	-126.4	-115.6	-103.4	56.9	1.0434
700.	1122.	80.2	-223.6	-215.4	-204.5	-191.8	-177.6	80.8	1.2988
800.	1048.	118.9	-333.0	-323.6	-311.1	-296.6	-280.4	108.9	1.5787
900.	991.	167.7	-470.6	-460.1	-446.0	-429.7	-411.4	139.7	1.8735
1000.	940.	227.7	-637.8	-626.1	-610.5	-592.4	-572.1	172.8	2.1816

BALLISTIC COEFFICIENT .33, MUZZLE VELOCITY FT/SEC 2600.

RANGE	VELOCITY	HEIGHT	100 YD	150 YD	200 YD	250 YD	300 YD	DRIFT	TIME
0.	2600.	0.0	-1.5	-1.5	-1.5	-1.5	-1.5	0.0	0.0000
100.	2336.	-.0	0.0	1.0	2.5	4.1	6.0	1.1	.1218
200.	2086.	2.4	-4.9	-2.8	0.0	3.3	7.0	4.7	.2576
300.	1852.	7.3	-17.9	-14.8	-10.5	-5.5	0.0	11.3	.4103
400.	1639.	15.5	-41.0	-36.8	-31.2	-24.6	-17.2	21.3	.5826
500.	1446.	28.3	-77.1	-71.9	-64.8	-56.6	-47.3	35.3	.7777
600.	1285.	47.1	-129.9	-123.7	-115.2	-105.3	-94.2	53.8	.9980
700.	1157.	73.7	-204.2	-196.9	-187.0	-175.4	-162.5	77.0	1.2452
800.	1072.	109.4	-304.1	-295.8	-284.4	-271.2	-256.4	104.1	1.5146
900.	1011.	155.7	-433.9	-424.6	-411.9	-397.0	-380.3	135.0	1.8056
1000.	959.	212.5	-591.2	-580.8	-566.7	-550.1	-531.7	167.9	2.1076

BALLISTIC COEFFICIENT .33, MUZZLE VELOCITY FT/SEC 2700.

RANGE	VELOCITY	HEIGHT	100 YD	150 YD	200 YD	250 YD	300 YD	DRIFT	TIME
0.	2700.	0.0	−1.5	−1.5	−1.5	−1.5	−1.5	0.0	0.0000
100.	2430.	−.1	0.0	.9	2.2	3.7	5.4	1.1	.1171
200.	2175.	2.2	−4.4	−2.6	0.0	3.0	6.4	4.4	.2475
300.	1935.	6.7	−16.3	−13.5	−9.6	−5.1	0.0	10.7	.3939
400.	1714.	14.2	−37.4	−33.6	−28.5	−22.4	−15.7	20.1	.5585
500.	1514.	25.9	−70.4	−65.7	−59.3	−51.7	−43.3	33.3	.7449
600.	1339.	43.1	−118.6	−113.1	−105.4	−96.3	−86.1	50.9	.9561
700.	1200.	67.6	−186.5	−180.0	−171.0	−160.4	−148.5	73.1	1.1931
800.	1099.	100.9	−278.7	−271.3	−261.0	−248.9	−235.3	99.7	1.4551
900.	1032.	144.1	−398.9	−390.5	−379.0	−365.3	−350.1	129.8	1.7374
1000.	977.	198.3	−548.3	−539.0	−526.2	−511.0	−494.1	162.8	2.0362

BALLISTIC COEFFICIENT .33, MUZZLE VELOCITY FT/SEC 2800.

RANGE	VELOCITY	HEIGHT	100 YD	150 YD	200 YD	250 YD	300 YD	DRIFT	TIME
0.	2800.	0.0	−1.5	−1.5	−1.5	−1.5	−1.5	0.0	0.0000
100.	2524.	−.1	0.0	.8	2.0	3.4	4.9	1.0	.1128
200.	2264.	2.0	−4.0	−2.3	0.0	2.8	5.9	4.2	.2383
300.	2018.	6.1	−14.8	−12.3	−8.8	−4.7	0.0	10.1	.3785
400.	1790.	13.1	−34.2	−30.9	−26.2	−20.6	−14.4	19.0	.5364
500.	1583.	23.8	−64.4	−60.3	−54.5	−47.5	−39.8	31.5	.7148
600.	1397.	39.6	−108.6	−103.6	−96.6	−88.3	−79.0	48.2	.9167
700.	1246.	62.1	−170.8	−165.1	−156.8	−147.1	−136.3	69.4	1.1445
800.	1130.	93.1	−255.7	−249.1	−239.7	−228.6	−216.2	95.2	1.3981
900.	1054.	133.7	−367.6	−360.2	−349.6	−337.2	−323.2	124.8	1.6736
1000.	996.	184.9	−508.3	−500.1	−488.4	−474.5	−459.0	157.6	1.9668

BALLISTIC COEFFICIENT .33, MUZZLE VELOCITY FT/SEC 2900.

RANGE	VELOCITY	HEIGHT	100 YD	150 YD	200 YD	250 YD	300 YD	DRIFT	TIME
0.	2900.	0.0	−1.5	−1.5	−1.5	−1.5	−1.5	0.0	0.0000
100.	2618.	−.2	0.0	.7	1.8	3.1	4.5	.9	.1088
200.	2353.	1.8	−3.6	−2.2	0.0	2.5	5.4	4.0	.2299
300.	2102.	5.6	−13.6	−11.4	−8.1	−4.3	0.0	9.6	.3647
400.	1867.	12.0	−31.3	−28.4	−24.1	−19.0	−13.3	18.0	.5160
500.	1652.	21.9	−59.2	−55.5	−50.1	−43.8	−36.6	29.9	.6870
600.	1458.	36.5	−99.7	−95.3	−88.9	−81.3	−72.6	45.7	.8804
700.	1294.	57.3	−156.9	−151.8	−144.2	−135.4	−125.3	66.0	1.0994
800.	1165.	86.0	−235.0	−229.1	−220.5	−210.4	−198.9	90.9	1.3440
900.	1077.	124.1	−339.0	−332.4	−322.7	−311.3	−298.4	119.9	1.6125
1000.	1015.	172.5	−471.4	−464.1	−453.3	−440.7	−426.3	152.3	1.9000

BALLISTIC COEFFICIENT .33, MUZZLE VELOCITY FT/SEC 3000.

RANGE	VELOCITY	HEIGHT	100 YD	150 YD	200 YD	250 YD	300 YD	DRIFT	TIME
0.	3000.	0.0	−1.5	−1.5	−1.5	−1.5	−1.5	0.0	0.0000
100.	2711.	−.2	0.0	.6	1.6	2.8	4.1	.9	.1054
200.	2441.	1.6	−3.3	−2.0	0.0	2.4	5.0	3.8	.2219
300.	2186.	5.2	−12.4	−10.4	−7.5	−4.0	0.0	9.1	.3517
400.	1944.	11.1	−28.8	−26.2	−22.3	−17.6	−12.3	17.2	.4975
500.	1723.	20.2	−54.4	−51.2	−46.3	−40.4	−33.8	28.4	.6612
600.	1522.	33.7	−91.8	−87.9	−82.0	−75.0	−67.1	43.4	.8468
700.	1346.	52.9	−144.3	−139.8	−133.0	−124.7	−115.5	62.8	1.0568
800.	1205.	79.5	−216.4	−211.2	−203.4	−194.0	−183.4	86.7	1.2928
900.	1103.	115.2	−312.8	−307.0	−298.2	−287.6	−275.7	115.1	1.5539
1000.	1035.	160.9	−437.0	−430.6	−420.8	−409.0	−395.8	147.0	1.8353

BALLISTIC COEFFICIENT .33, MUZZLE VELOCITY FT/SEC 3100.

RANGE	VELOCITY	HEIGHT	100 YD	150 YD	200 YD	250 YD	300 YD	DRIFT	TIME
0.	3100.	0.0	−1.5	−1.5	−1.5	−1.5	−1.5	0.0	0.0000
100.	2805.	−.3	0.0	.6	1.5	2.6	3.8	.9	.1018
200.	2529.	1.5	−2.9	−1.8	0.0	2.2	4.6	3.7	.2144
300.	2269.	4.8	−11.4	−9.6	−6.9	−3.7	0.0	8.7	.3396
400.	2022.	10.3	−26.5	−24.3	−20.6	−16.3	−11.4	16.3	.4796
500.	1794.	18.7	−50.2	−47.4	−42.8	−37.4	−31.3	27.0	.6372
600.	1586.	31.2	−84.7	−81.3	−75.9	−69.4	−62.0	41.3	.8152
700.	1400.	48.9	−133.1	−129.1	−122.7	−115.2	−106.6	59.7	1.0166
800.	1248.	73.5	−199.7	−195.1	−187.9	−179.2	−169.4	82.7	1.2440
900.	1132.	106.9	−288.9	−283.8	−275.6	−265.9	−254.8	110.2	1.4972
1000.	1055.	150.0	−405.2	−399.5	−390.4	−379.6	−367.3	141.6	1.7724

BALLISTIC COEFFICIENT .33, MUZZLE VELOCITY FT/SEC 3200.

RANGE	VELOCITY	HEIGHT	100 YD	150 YD	200 YD	250 YD	300 YD	DRIFT	TIME
0.	3200.	0.0	-1.5	-1.5	-1.5	-1.5	-1.5	0.0	0.0000
100.	2898.	-.3	0.0	.5	1.3	2.3	3.5	.8	.0986
200.	2617.	1.3	-2.7	-1.6	0.0	2.0	4.3	3.5	.2074
300.	2352.	4.4	-10.4	-8.9	-6.5	-3.4	0.0	8.3	.3286
400.	2101.	9.6	-24.5	-22.5	-19.2	-15.2	-10.6	15.6	.4635
500.	1866.	17.4	-46.4	-43.9	-39.8	-34.7	-29.0	25.7	.6148
600.	1651.	28.9	-78.4	-75.3	-70.4	-64.4	-57.5	39.3	.7860
700.	1457.	45.3	-123.1	-119.6	-113.8	-106.8	-98.8	56.9	.9795
800.	1293.	68.2	-184.7	-180.6	-174.1	-166.0	-156.8	79.0	1.1987
900.	1164.	99.2	-267.2	-262.7	-255.3	-246.2	-235.9	105.5	1.4434
1000.	1077.	139.9	-375.9	-370.8	-362.6	-352.5	-341.0	136.3	1.7120

BALLISTIC COEFFICIENT .33, MUZZLE VELOCITY FT/SEC 3300.

RANGE	VELOCITY	HEIGHT	100 YD	150 YD	200 YD	250 YD	300 YD	DRIFT	TIME
0.	3300.	0.0	-1.5	-1.5	-1.5	-1.5	-1.5	0.0	0.0000
100.	2992.	-.3	0.0	.4	1.2	2.1	3.2	.8	.0955
200.	2704.	1.2	-2.4	-1.5	0.0	1.9	4.0	3.4	.2010
300.	2434.	4.1	-9.6	-8.3	-6.0	-3.2	0.0	8.0	.3179
400.	2179.	8.9	-22.6	-20.9	-17.8	-14.1	-9.8	14.9	.4481
500.	1938.	16.2	-43.1	-40.9	-37.0	-32.4	-27.1	24.6	.5943
600.	1717.	26.9	-72.6	-70.0	-65.4	-59.8	-53.4	37.5	.7586
700.	1517.	42.1	-114.2	-111.1	-105.7	-99.2	-91.8	54.3	.9447
800.	1341.	63.3	-171.0	-167.5	-161.4	-153.9	-145.4	75.4	1.1554
900.	1202.	92.3	-247.7	-243.7	-236.8	-228.4	-218.9	101.0	1.3922
1000.	1100.	130.5	-348.9	-344.5	-336.8	-327.5	-316.9	131.1	1.6539

BALLISTIC COEFFICIENT .33, MUZZLE VELOCITY FT/SEC 3400.

RANGE	VELOCITY	HEIGHT	100 YD	150 YD	200 YD	250 YD	300 YD	DRIFT	TIME
0.	3400.	0.0	-1.5	-1.5	-1.5	-1.5	-1.5	0.0	0.0000
100.	3086.	-.3	0.0	.4	1.1	2.0	2.9	.8	.0926
200.	2792.	1.1	-2.2	-1.4	0.0	1.7	3.7	3.2	.1948
300.	2517.	3.8	-8.8	-7.7	-5.6	-3.0	0.0	7.6	.3081
400.	2257.	8.3	-21.0	-19.5	-16.6	-13.2	-9.2	14.3	.4340
500.	2011.	15.1	-39.9	-38.0	-34.5	-30.2	-25.2	23.5	.5747
600.	1783.	25.1	-67.5	-65.2	-61.0	-55.8	-49.8	35.9	.7332
700.	1577.	39.2	-106.1	-103.4	-98.5	-92.4	-85.4	51.8	.9122
800.	1392.	58.9	-158.7	-155.6	-150.0	-143.1	-135.1	72.0	1.1148
900.	1242.	85.9	-229.9	-226.4	-220.1	-212.3	-203.4	96.7	1.3435
1000.	1127.	121.8	-324.1	-320.3	-313.2	-304.6	-294.6	125.9	1.5978

BALLISTIC COEFFICIENT .33, MUZZLE VELOCITY FT/SEC 3500.

RANGE	VELOCITY	HEIGHT	100 YD	150 YD	200 YD	250 YD	300 YD	DRIFT	TIME
0.	3500.	0.0	-1.5	-1.5	-1.5	-1.5	-1.5	0.0	0.0000
100.	3179.	-.4	0.0	.3	1.0	1.8	2.7	.7	.0899
200.	2879.	1.0	-2.0	-1.3	0.0	1.6	3.5	3.1	.1890
300.	2599.	3.5	-8.2	-7.1	-5.2	-2.8	0.0	7.4	.2989
400.	2335.	7.7	-19.5	-18.1	-15.6	-12.3	-8.6	13.7	.4208
500.	2085.	14.1	-37.2	-35.5	-32.3	-28.2	-23.6	22.5	.5566
600.	1851.	23.4	-62.9	-60.8	-57.0	-52.1	-46.6	34.4	.7095
700.	1638.	36.6	-98.7	-96.4	-91.9	-86.2	-79.7	49.6	.8818
800.	1445.	54.9	-147.7	-145.0	-139.9	-133.4	-126.0	68.9	1.0771
900.	1284.	80.1	-213.8	-210.7	-205.0	-197.7	-189.3	92.6	1.2976
1000.	1157.	113.8	-301.8	-298.4	-292.0	-283.9	-274.6	121.1	1.5450

BALLISTIC COEFFICIENT .33, MUZZLE VELOCITY FT/SEC 3600.

RANGE	VELOCITY	HEIGHT	100 YD	150 YD	200 YD	250 YD	300 YD	DRIFT	TIME
0.	3600.	0.0	-1.5	-1.5	-1.5	-1.5	-1.5	0.0	0.0000
100.	3273.	-.4	0.0	.3	.9	1.6	2.5	.7	.0873
200.	2967.	.9	-1.8	-1.2	0.0	1.5	3.2	3.0	.1835
300.	2681.	3.3	-7.5	-6.6	-4.8	-2.6	0.0	7.0	.2898
400.	2412.	7.2	-18.1	-16.9	-14.6	-11.5	-8.1	13.1	.4080
500.	2158.	13.2	-34.6	-33.1	-30.2	-26.4	-22.1	21.6	.5394
600.	1919.	21.9	-58.6	-56.8	-53.3	-48.8	-43.6	32.9	.6870
700.	1699.	34.2	-92.0	-90.0	-85.9	-80.6	-74.5	47.5	.8530
800.	1501.	51.3	-137.6	-135.2	-130.5	-124.5	-117.6	65.9	1.0409
900.	1328.	74.7	-199.0	-196.4	-191.1	-184.3	-176.6	88.7	1.2539
1000.	1191.	106.2	-280.8	-277.9	-272.0	-264.5	-255.9	116.1	1.4929

BALLISTIC COEFFICIENT .33, MUZZLE VELOCITY FT/SEC 3700.

RANGE	VELOCITY	HEIGHT	100 YD	150 YD	200 YD	250 YD	300 YD	DRIFT	TIME
0.	3700.	0.0	-1.5	-1.5	-1.5	-1.5	-1.5	0.0	0.0000
100.	3367.	-.4	0.0	.2	.8	1.5	2.3	.7	.0851
200.	3055.	.8	-1.6	-1.1	0.0	1.4	3.0	2.9	.1786
300.	2763.	3.1	-6.9	-6.2	-4.6	-2.4	0.0	6.8	.2821
400.	2489.	6.8	-16.8	-15.8	-13.7	-10.8	-7.6	12.7	.3964
500.	2231.	12.4	-32.3	-31.1	-28.4	-24.8	-20.8	20.8	.5237
600.	1987.	20.5	-54.7	-53.2	-50.0	-45.7	-40.9	31.6	.6661
700.	1762.	32.0	-86.0	-84.2	-80.4	-75.5	-69.8	45.6	.8265
800.	1557.	48.0	-128.6	-126.6	-122.3	-116.6	-110.1	63.2	1.0079
900.	1375.	69.9	-185.8	-183.6	-178.7	-172.3	-165.0	85.1	1.2133
1000.	1228.	99.4	-262.0	-259.5	-254.1	-247.0	-238.9	111.5	1.4443

BALLISTIC COEFFICIENT .33, MUZZLE VELOCITY FT/SEC 3800.

RANGE	VELOCITY	HEIGHT	100 YD	150 YD	200 YD	250 YD	300 YD	DRIFT	TIME
0.	3800.	0.0	-1.5	-1.5	-1.5	-1.5	-1.5	0.0	0.0000
100.	3461.	-.4	0.0	.2	.7	1.4	2.1	.7	.0829
200.	3143.	.7	-1.4	-1.0	0.0	1.3	2.8	2.8	.1740
300.	2845.	2.9	-6.4	-5.7	-4.2	-2.3	0.0	6.5	.2741
400.	2567.	6.4	-15.6	-14.8	-12.8	-10.2	-7.2	12.2	.3852
500.	2304.	11.7	-30.2	-29.1	-26.6	-23.3	-19.6	20.0	.5086
600.	2056.	19.3	-51.2	-50.0	-46.9	-43.0	-38.5	30.4	.6465
700.	1824.	30.1	-80.4	-79.0	-75.5	-70.9	-65.6	43.8	.8014
800.	1614.	45.0	-120.2	-118.6	-114.5	-109.3	-103.3	60.7	.9762
900.	1424.	65.4	-173.6	-171.7	-167.1	-161.3	-154.5	81.6	1.1742
1000.	1267.	93.1	-244.8	-242.7	-237.6	-231.1	-223.5	107.1	1.3980

BALLISTIC COEFFICIENT .33, MUZZLE VELOCITY FT/SEC 3900.

RANGE	VELOCITY	HEIGHT	100 YD	150 YD	200 YD	250 YD	300 YD	DRIFT	TIME
0.	3900.	0.0	-1.5	-1.5	-1.5	-1.5	-1.5	0.0	0.0000
100.	3555.	-.4	0.0	.2	.6	1.2	2.0	.6	.0804
200.	3231.	.6	-1.3	-.9	0.0	1.2	2.7	2.7	.1690
300.	2927.	2.7	-5.9	-5.4	-4.0	-2.2	0.0	6.3	.2666
400.	2644.	6.0	-14.6	-13.9	-12.0	-9.6	-6.7	11.7	.3744
500.	2377.	11.0	-28.2	-27.4	-25.0	-22.0	-18.4	19.3	.4941
600.	2125.	18.1	-47.9	-46.9	-44.1	-40.5	-36.2	29.2	.6275
700.	1888.	28.2	-75.4	-74.2	-70.9	-66.7	-61.6	42.0	.7773
800.	1672.	42.2	-112.7	-111.3	-107.6	-102.7	-97.0	58.2	.9463
900.	1475.	61.3	-162.5	-161.0	-156.8	-151.3	-144.9	78.3	1.1373
1000.	1308.	87.2	-229.0	-227.3	-222.6	-216.5	-209.3	102.9	1.3537

BALLISTIC COEFFICIENT .33, MUZZLE VELOCITY FT/SEC 4000.

RANGE	VELOCITY	HEIGHT	100 YD	150 YD	200 YD	250 YD	300 YD	DRIFT	TIME
0.	4000.	0.0	-1.5	-1.5	-1.5	-1.5	-1.5	0.0	0.0000
100.	3649.	-.5	0.0	.1	.6	1.1	1.8	.6	.0786
200.	3319.	.6	-1.1	-.9	0.0	1.1	2.5	2.6	.1649
300.	3010.	2.5	-5.4	-5.0	-3.7	-2.0	0.0	6.1	.2597
400.	2721.	5.6	-13.6	-13.0	-11.3	-9.0	-6.4	11.4	.3647
500.	2450.	10.3	-26.4	-25.7	-23.6	-20.7	-17.4	18.6	.4808
600.	2194.	17.1	-45.0	-44.2	-41.6	-38.1	-34.1	28.2	.6102
700.	1952.	26.6	-70.8	-69.8	-66.8	-62.8	-58.1	40.5	.7553
800.	1730.	39.7	-105.7	-104.6	-101.2	-96.6	-91.3	56.0	.9184
900.	1528.	57.7	-152.5	-151.3	-147.4	-142.2	-136.2	75.4	1.1031
1000.	1351.	81.9	-214.6	-213.2	-209.0	-203.2	-196.5	99.0	1.3123

BALLISTIC COEFFICIENT .34, MUZZLE VELOCITY FT/SEC 2000.

RANGE	VELOCITY	HEIGHT	100 YD	150 YD	200 YD	250 YD	300 YD	DRIFT	TIME
0.	2000.	0.0	-1.5	-1.5	-1.5	-1.5	-1.5	0.0	0.0000
100.	1780.	.5	0.0	2.1	4.8	7.7	11.0	1.6	.1591
200.	1579.	4.7	-9.6	-5.3	0.0	5.9	12.5	6.7	.3382
300.	1399.	13.3	-33.1	-26.7	-18.8	-9.9	0.0	15.9	.5401
400.	1251.	27.5	-74.6	-66.0	-55.4	-43.6	-30.4	29.4	.7673
500.	1136.	49.2	-138.5	-127.7	-114.6	-99.8	-83.3	47.6	1.0205
600.	1060.	79.5	-228.4	-215.5	-199.7	-181.9	-162.1	69.2	1.2932
700.	1003.	119.9	-348.4	-333.4	-314.9	-294.2	-271.1	94.2	1.5854
800.	953.	171.0	-499.2	-482.1	-461.0	-437.3	-410.9	121.7	1.8917
900.	907.	234.7	-685.1	-665.8	-642.1	-615.4	-585.7	152.2	2.2146
1000.	865.	312.2	-909.5	-888.0	-861.7	-832.1	-799.1	185.4	2.5535

BALLISTIC COEFFICIENT .34, MUZZLE VELOCITY FT/SEC 2100.

RANGE	VELOCITY	HEIGHT	100 YD	150 YD	200 YD	250 YD	300 YD	DRIFT	TIME
0.	2100.	0.0	-1.5	-1.5	-1.5	-1.5	-1.5	0.0	0.0000
100.	1871.	.3	0.0	1.9	4.3	6.9	9.9	1.5	.1511
200.	1663.	4.2	-8.5	-4.7	0.0	5.4	11.3	6.3	.3214
300.	1473.	11.9	-29.7	-23.9	-16.9	-8.9	0.0	14.9	.5131
400.	1310.	24.8	-66.9	-59.3	-49.9	-39.2	-27.4	27.8	.7295
500.	1180.	44.5	-124.4	-114.9	-103.1	-89.7	-75.0	45.2	.9711
600.	1089.	72.6	-206.9	-195.5	-181.4	-165.3	-147.6	66.8	1.2365
700.	1026.	110.4	-318.3	-305.0	-288.5	-269.8	-249.1	91.9	1.5220
800.	974.	158.4	-458.7	-443.5	-424.7	-403.3	-379.7	119.4	1.8210
900.	926.	218.5	-632.8	-615.7	-594.5	-570.4	-543.9	149.8	2.1370
1000.	883.	291.8	-843.5	-824.4	-800.9	-774.1	-744.6	183.1	2.4690

BALLISTIC COEFFICIENT .34, MUZZLE VELOCITY FT/SEC 2200.

RANGE	VELOCITY	HEIGHT	100 YD	150 YD	200 YD	250 YD	300 YD	DRIFT	TIME
0.	2200.	0.0	-1.5	-1.5	-1.5	-1.5	-1.5	0.0	0.0000
100.	1965.	.3	0.0	1.7	3.8	6.2	8.9	1.4	.1444
200.	1747.	3.8	-7.6	-4.2	0.0	4.8	10.2	5.9	.3063
300.	1550.	10.7	-26.6	-21.6	-15.3	-8.0	0.0	14.0	.4887
400.	1374.	22.4	-60.1	-53.4	-45.0	-35.4	-24.7	26.2	.6945
500.	1231.	40.4	-112.2	-103.8	-93.3	-81.2	-67.8	42.9	.9256
600.	1123.	66.3	-187.2	-177.1	-164.5	-150.0	-134.0	63.9	1.1815
700.	1051.	101.3	-289.4	-277.7	-262.9	-246.0	-227.3	88.7	1.4583
800.	995.	146.6	-421.0	-407.6	-390.7	-371.4	-350.0	116.4	1.7521
900.	945.	203.5	-584.9	-569.9	-550.9	-529.2	-505.1	147.1	2.0631
1000.	900.	272.6	-781.7	-765.0	-743.9	-719.8	-693.1	180.0	2.3865

BALLISTIC COEFFICIENT .34, MUZZLE VELOCITY FT/SEC 2300.

RANGE	VELOCITY	HEIGHT	100 YD	150 YD	200 YD	250 YD	300 YD	DRIFT	TIME
0.	2300.	0.0	-1.5	-1.5	-1.5	-1.5	-1.5	0.0	0.0000
100.	2059.	.2	0.0	1.5	3.4	5.5	8.0	1.3	.1379
200.	1834.	3.4	-6.8	-3.8	0.0	4.3	9.2	5.5	.2923
300.	1628.	9.7	-23.9	-19.5	-13.8	-7.3	0.0	13.1	.4659
400.	1442.	20.3	-54.2	-48.3	-40.7	-32.0	-22.3	24.7	.6619
500.	1286.	36.6	-101.3	-93.9	-84.4	-73.6	-61.5	40.5	.8825
600.	1161.	60.4	-169.4	-160.5	-149.1	-136.2	-121.6	60.9	1.1285
700.	1077.	93.0	-263.4	-253.0	-239.7	-224.6	-207.6	85.2	1.3974
800.	1017.	135.5	-385.9	-374.0	-358.9	-341.6	-322.2	112.8	1.6846
900.	965.	189.1	-539.4	-526.0	-508.9	-489.5	-467.6	143.4	1.9887
1000.	918.	254.5	-724.6	-709.8	-690.8	-669.2	-644.9	176.3	2.3061

BALLISTIC COEFFICIENT .34, MUZZLE VELOCITY FT/SEC 2400.

RANGE	VELOCITY	HEIGHT	100 YD	150 YD	200 YD	250 YD	300 YD	DRIFT	TIME
0.	2400.	0.0	-1.5	-1.5	-1.5	-1.5	-1.5	0.0	0.0000
100.	2154.	.1	0.0	1.3	3.0	5.0	7.2	1.2	.1319
200.	1921.	3.0	-6.1	-3.4	0.0	3.9	8.3	5.2	.2795
300.	1708.	8.8	-21.6	-17.6	-12.5	-6.6	0.0	12.3	.4449
400.	1514.	18.4	-49.0	-43.8	-36.9	-29.1	-20.3	23.2	.6316
500.	1344.	33.3	-91.7	-85.1	-76.6	-66.8	-55.8	38.2	.8423
600.	1207.	55.1	-153.8	-145.8	-135.6	-123.8	-110.7	57.8	1.0782
700.	1107.	85.3	-239.8	-230.6	-218.6	-204.9	-189.5	81.6	1.3386
800.	1039.	125.0	-353.4	-342.8	-329.1	-313.4	-295.9	108.9	1.6186
900.	985.	175.6	-497.4	-485.5	-470.0	-452.4	-432.7	139.4	1.9169
1000.	936.	237.5	-671.7	-658.5	-641.3	-621.7	-599.8	172.2	2.2281

BALLISTIC COEFFICIENT .34, MUZZLE VELOCITY FT/SEC 2500.

RANGE	VELOCITY	HEIGHT	100 YD	150 YD	200 YD	250 YD	300 YD	DRIFT	TIME
0.	2500.	0.0	-1.5	-1.5	-1.5	-1.5	-1.5	0.0	0.0000
100.	2249.	.0	0.0	1.2	2.7	4.5	6.5	1.2	.1266
200.	2011.	2.7	-5.4	-3.1	0.0	3.6	7.6	4.9	.2676
300.	1789.	8.0	-19.5	-16.0	-11.4	-6.0	0.0	11.6	.4258
400.	1588.	16.8	-44.5	-39.8	-33.7	-26.5	-18.5	21.8	.6040
500.	1406.	30.3	-83.2	-77.4	-69.7	-60.7	-50.7	36.0	.8047
600.	1257.	50.3	-139.7	-132.8	-123.5	-112.8	-100.7	54.7	1.0308
700.	1141.	78.1	-218.3	-210.1	-199.3	-186.8	-172.7	77.8	1.2818
800.	1063.	115.5	-324.0	-314.7	-302.3	-288.0	-272.0	104.9	1.5561
900.	1005.	163.0	-458.2	-447.8	-433.8	-417.7	-399.7	135.0	1.8468
1000.	955.	221.5	-622.0	-610.4	-594.9	-577.1	-557.0	167.5	2.1517

BALLISTIC COEFFICIENT .34, MUZZLE VELOCITY FT/SEC 2600.

RANGE	VELOCITY	HEIGHT	100 YD	150 YD	200 YD	250 YD	300 YD	DRIFT	TIME
0.	2600.	0.0	-1.5	-1.5	-1.5	-1.5	-1.5	0.0	0.0000
100.	2343.	-.0	0.0	1.0	2.4	4.1	5.9	1.1	.1216
200.	2100.	2.4	-4.9	-2.8	0.0	3.2	6.9	4.6	.2569
300.	1872.	7.2	-17.7	-14.6	-10.3	-5.5	0.0	10.9	.4080
400.	1663.	15.3	-40.4	-36.3	-30.7	-24.2	-16.9	20.5	.5782
500.	1473.	27.7	-75.7	-70.6	-63.6	-55.4	-46.3	34.0	.7699
600.	1311.	45.9	-127.2	-121.1	-112.6	-102.9	-91.9	51.7	.9862
700.	1180.	71.6	-199.0	-191.8	-182.0	-170.6	-157.9	73.9	1.2278
800.	1090.	106.3	-296.1	-287.9	-276.6	-263.6	-249.1	100.3	1.4932
900.	1026.	151.1	-421.9	-412.7	-400.0	-385.4	-369.0	130.3	1.7787
1000.	974.	206.5	-576.0	-565.7	-551.6	-535.4	-517.2	162.6	2.0777

BALLISTIC COEFFICIENT .34, MUZZLE VELOCITY FT/SEC 2700.

RANGE	VELOCITY	HEIGHT	100 YD	150 YD	200 YD	250 YD	300 YD	DRIFT	TIME
0.	2700.	0.0	-1.5	-1.5	-1.5	-1.5	-1.5	0.0	0.0000
100.	2438.	-.1	0.0	.9	2.2	3.7	5.4	1.0	.1168
200.	2190.	2.2	-4.4	-2.5	0.0	3.0	6.3	4.3	.2467
300.	1955.	6.6	-16.1	-13.3	-9.5	-5.0	0.0	10.3	.3919
400.	1739.	14.0	-36.9	-33.2	-28.1	-22.1	-15.4	19.4	.5545
500.	1542.	25.4	-69.2	-64.6	-58.2	-50.8	-42.4	32.1	.7378
600.	1367.	42.1	-116.2	-110.7	-103.0	-94.1	-84.0	48.9	.9447
700.	1226.	65.7	-182.0	-175.5	-166.6	-156.2	-144.5	70.2	1.1767
800.	1119.	98.0	-271.3	-263.9	-253.8	-241.8	-228.4	95.9	1.4340
900.	1048.	140.2	-389.0	-380.7	-369.3	-355.8	-340.7	125.6	1.7139
1000.	992.	192.9	-535.0	-525.8	-513.1	-498.1	-481.4	158.0	2.0088

BALLISTIC COEFFICIENT .34, MUZZLE VELOCITY FT/SEC 2800.

RANGE	VELOCITY	HEIGHT	100 YD	150 YD	200 YD	250 YD	300 YD	DRIFT	TIME
0.	2800.	0.0	-1.5	-1.5	-1.5	-1.5	-1.5	0.0	0.0000
100.	2532.	-.1	0.0	.8	2.0	3.3	4.9	1.0	.1127
200.	2279.	2.0	-3.9	-2.3	0.0	2.7	5.8	4.1	.2374
300.	2040.	6.1	-14.6	-12.2	-8.7	-4.6	0.0	9.7	.3767
400.	1816.	12.9	-33.7	-30.4	-25.8	-20.3	-14.1	18.3	.5324
500.	1612.	23.3	-63.3	-59.3	-53.5	-46.6	-38.9	30.3	.7079
600.	1427.	38.6	-106.3	-101.4	-94.5	-86.3	-77.0	46.3	.9058
700.	1274.	60.4	-166.8	-161.1	-152.9	-143.4	-132.6	66.7	1.1289
800.	1153.	90.2	-248.6	-242.1	-232.8	-221.9	-209.5	91.4	1.3767
900.	1071.	129.5	-356.9	-349.5	-339.1	-326.8	-312.9	120.2	1.6473
1000.	1012.	179.6	-495.0	-486.9	-475.3	-461.6	-446.2	152.6	1.9386

BALLISTIC COEFFICIENT .34, MUZZLE VELOCITY FT/SEC 2900.

RANGE	VELOCITY	HEIGHT	100 YD	150 YD	200 YD	250 YD	300 YD	DRIFT	TIME
0.	2900.	0.0	-1.5	-1.5	-1.5	-1.5	-1.5	0.0	0.0000
100.	2626.	-.2	0.0	.7	1.8	3.1	4.5	.9	.1086
200.	2368.	1.8	-3.6	-2.1	0.0	2.5	5.3	3.9	.2291
300.	2124.	5.6	-13.4	-11.2	-8.0	-4.2	0.0	9.2	.3627
400.	1893.	11.9	-30.9	-28.0	-23.7	-18.7	-13.1	17.3	.5124
500.	1682.	21.5	-58.2	-54.5	-49.2	-42.9	-35.9	28.7	.6805
600.	1491.	35.6	-97.7	-93.3	-86.9	-79.4	-70.9	43.9	.8699
700.	1325.	55.6	-153.1	-148.0	-140.5	-131.7	-121.8	63.3	1.0838
800.	1192.	83.3	-228.5	-222.7	-214.2	-204.1	-192.8	87.2	1.3230
900.	1097.	120.0	-328.8	-322.2	-312.6	-301.3	-288.6	115.3	1.5862
1000.	1032.	167.3	-458.2	-450.9	-440.3	-427.7	-413.6	147.2	1.8711

BALLISTIC COEFFICIENT .34, MUZZLE VELOCITY FT/SEC 3000.

RANGE	VELOCITY	HEIGHT	100 YD	150 YD	200 YD	250 YD	300 YD	DRIFT	TIME
0.	3000.	0.0	-1.5	-1.5	-1.5	-1.5	-1.5	0.0	0.0000
100.	2720.	-.2	0.0	.6	1.6	2.8	4.1	.9	.1052
200.	2457.	1.6	-3.2	-1.9	0.0	2.3	4.9	3.7	.2212
300.	2208.	5.1	-12.2	-10.3	-7.4	-3.9	0.0	8.8	.3501
400.	1972.	11.0	-28.4	-25.8	-22.0	-17.3	-12.1	16.5	.4939
500.	1754.	19.9	-53.5	-50.3	-45.5	-39.6	-33.1	27.3	.6552
600.	1556.	32.9	-90.0	-86.1	-80.3	-73.3	-65.5	41.7	.8371
700.	1379.	51.4	-140.9	-136.4	-129.6	-121.4	-112.3	60.2	1.0420
800.	1235.	77.0	-210.5	-205.4	-197.6	-188.3	-177.9	83.2	1.2727
900.	1125.	111.3	-303.1	-297.3	-288.6	-278.1	-266.4	110.5	1.5276
1000.	1053.	155.4	-423.0	-416.6	-406.9	-395.2	-382.2	141.5	1.8037

BALLISTIC COEFFICIENT .34, MUZZLE VELOCITY FT/SEC 3100.

RANGE	VELOCITY	HEIGHT	100 YD	150 YD	200 YD	250 YD	300 YD	DRIFT	TIME
0.	3100.	0.0	-1.5	-1.5	-1.5	-1.5	-1.5	0.0	0.0000
100.	2813.	-.3	0.0	.6	1.5	2.5	3.7	.8	.1016
200.	2545.	1.4	-2.9	-1.8	0.0	2.1	4.6	3.5	.2137
300.	2291.	4.7	-11.2	-9.5	-6.8	-3.6	0.0	8.4	.3379
400.	2051.	10.1	-26.2	-24.0	-20.4	-16.1	-11.2	15.7	.4764
500.	1826.	18.4	-49.4	-46.6	-42.1	-36.7	-30.7	26.0	.6313
600.	1621.	30.4	-83.0	-79.7	-74.3	-67.8	-60.6	39.6	.8057
700.	1436.	47.5	-130.0	-126.1	-119.8	-112.3	-103.8	57.2	1.0026
800.	1281.	71.2	-194.1	-189.6	-182.4	-173.8	-164.2	79.2	1.2241
900.	1158.	103.2	-280.0	-274.9	-266.8	-257.2	-246.3	105.7	1.4716
1000.	1075.	144.7	-391.7	-386.1	-377.1	-366.4	-354.3	136.0	1.7407

BALLISTIC COEFFICIENT .34, MUZZLE VELOCITY FT/SEC 3200.

RANGE	VELOCITY	HEIGHT	100 YD	150 YD	200 YD	250 YD	300 YD	DRIFT	TIME
0.	3200.	0.0	-1.5	-1.5	-1.5	-1.5	-1.5	0.0	0.0000
100.	2907.	-.3	0.0	.5	1.3	2.3	3.4	.8	.0985
200.	2633.	1.3	-2.6	-1.6	0.0	2.0	4.2	3.4	.2068
300.	2374.	4.4	-10.3	-8.8	-6.4	-3.4	0.0	8.0	.3269
400.	2130.	9.4	-24.2	-22.2	-18.9	-14.9	-10.4	15.0	.4603
500.	1899.	17.1	-45.7	-43.2	-39.1	-34.1	-28.5	24.8	.6094
600.	1687.	28.2	-76.8	-73.8	-68.9	-62.9	-56.2	37.8	.7771
700.	1495.	44.0	-120.2	-116.7	-111.0	-104.0	-96.1	54.5	.9659
800.	1329.	66.0	-179.5	-175.5	-168.9	-161.0	-152.0	75.5	1.1792
900.	1195.	95.8	-259.0	-254.5	-247.1	-238.2	-228.0	101.1	1.4182
1000.	1099.	134.8	-362.9	-358.0	-349.8	-339.8	-328.6	130.7	1.6803

BALLISTIC COEFFICIENT .34, MUZZLE VELOCITY FT/SEC 3300.

RANGE	VELOCITY	HEIGHT	100 YD	150 YD	200 YD	250 YD	300 YD	DRIFT	TIME
0.	3300.	0.0	-1.5	-1.5	-1.5	-1.5	-1.5	0.0	0.0000
100.	3001.	-.3	0.0	.4	1.2	2.1	3.2	.8	.0953
200.	2720.	1.2	-2.4	-1.5	0.0	1.8	3.9	3.3	.2004
300.	2457.	4.1	-9.5	-8.2	-5.9	-3.1	0.0	7.7	.3164
400.	2208.	8.8	-22.4	-20.6	-17.6	-13.9	-9.7	14.4	.4452
500.	1972.	15.9	-42.4	-40.2	-36.4	-31.8	-26.5	23.7	.5890
600.	1754.	26.3	-71.3	-68.6	-64.1	-58.6	-52.3	36.0	.7502
700.	1556.	40.9	-111.6	-108.5	-103.2	-96.7	-89.4	52.0	.9320
800.	1379.	61.3	-166.3	-162.8	-156.7	-149.3	-141.0	72.1	1.1369
900.	1236.	89.0	-240.0	-236.1	-229.3	-220.9	-211.5	96.7	1.3675
1000.	1126.	125.6	-336.7	-332.3	-324.7	-315.5	-305.1	125.5	1.6223

BALLISTIC COEFFICIENT .34, MUZZLE VELOCITY FT/SEC 3400.

RANGE	VELOCITY	HEIGHT	100 YD	150 YD	200 YD	250 YD	300 YD	DRIFT	TIME
0.	3400.	0.0	-1.5	-1.5	-1.5	-1.5	-1.5	0.0	0.0000
100.	3095.	-.3	0.0	.4	1.1	1.9	2.9	.7	.0925
200.	2808.	1.1	-2.2	-1.4	0.0	1.7	3.7	3.1	.1943
300.	2540.	3.8	-8.7	-7.6	-5.5	-3.0	0.0	7.4	.3066
400.	2287.	8.2	-20.7	-19.2	-16.4	-13.0	-9.0	13.7	.4310
500.	2046.	14.8	-39.3	-37.5	-33.9	-29.7	-24.8	22.6	.5698
600.	1822.	24.5	-66.2	-63.9	-59.7	-54.6	-48.7	34.4	.7250
700.	1618.	38.1	-103.6	-101.0	-96.1	-90.1	-83.2	49.7	.8998
800.	1432.	57.0	-154.4	-151.4	-145.8	-139.0	-131.1	68.9	1.0971
900.	1278.	82.8	-222.6	-219.3	-212.9	-205.3	-196.4	92.4	1.3192
1000.	1156.	117.2	-313.0	-309.2	-302.2	-293.7	-283.9	120.6	1.5675

BALLISTIC COEFFICIENT .34, MUZZLE VELOCITY FT/SEC 3500.

RANGE	VELOCITY	HEIGHT	100 YD	150 YD	200 YD	250 YD	300 YD	DRIFT	TIME
0.	3500.	0.0	-1.5	-1.5	-1.5	-1.5	-1.5	0.0	0.0000
100.	3189.	-.4	0.0	.3	1.0	1.8	2.7	.7	.0898
200.	2896.	1.0	-1.9	-1.3	0.0	1.6	3.4	3.0	.1887
300.	2623.	3.5	-8.0	-7.1	-5.1	-2.7	0.0	7.1	.2974
400.	2365.	7.6	-19.2	-17.9	-15.3	-12.1	-8.5	13.2	.4181
500.	2121.	13.9	-36.6	-34.9	-31.7	-27.7	-23.2	21.7	.5518
600.	1890.	22.9	-61.7	-59.7	-55.8	-51.0	-45.6	33.0	.7018
700.	1680.	35.6	-96.5	-94.2	-89.7	-84.1	-77.7	47.6	.8702
800.	1489.	53.2	-143.7	-141.1	-135.9	-129.5	-122.3	65.9	1.0599
900.	1323.	77.2	-207.0	-204.1	-198.3	-191.1	-182.9	88.5	1.2741
1000.	1191.	109.2	-290.8	-287.5	-281.1	-273.1	-264.0	115.5	1.5136

BALLISTIC COEFFICIENT .34, MUZZLE VELOCITY FT/SEC 3600.

RANGE	VELOCITY	HEIGHT	100 YD	150 YD	200 YD	250 YD	300 YD	DRIFT	TIME
0.	3600.	0.0	-1.5	-1.5	-1.5	-1.5	-1.5	0.0	0.0000
100.	3283.	-.4	0.0	.3	.9	1.6	2.5	.7	.0873
200.	2984.	.9	-1.7	-1.2	0.0	1.5	3.2	2.9	.1830
300.	2705.	3.2	-7.4	-6.5	-4.8	-2.6	0.0	6.8	.2886
400.	2443.	7.1	-17.8	-16.7	-14.3	-11.4	-7.9	12.7	.4052
500.	2195.	13.0	-34.0	-32.6	-29.7	-26.0	-21.7	20.8	.5348
600.	1960.	21.4	-57.5	-55.8	-52.3	-47.8	-42.7	31.6	.6797
700.	1743.	33.3	-90.0	-88.0	-83.9	-78.7	-72.7	45.5	.8420
800.	1546.	49.7	-134.0	-131.7	-127.0	-121.1	-114.2	63.0	1.0248
900.	1370.	72.0	-192.8	-190.3	-185.0	-178.4	-170.6	84.7	1.2313
1000.	1228.	102.0	-270.8	-268.0	-262.1	-254.8	-246.2	110.8	1.4629

BALLISTIC COEFFICIENT .34, MUZZLE VELOCITY FT/SEC 3700.

RANGE	VELOCITY	HEIGHT	100 YD	150 YD	200 YD	250 YD	300 YD	DRIFT	TIME
0.	3700.	0.0	-1.5	-1.5	-1.5	-1.5	-1.5	0.0	0.0000
100.	3377.	-.4	0.0	.2	.8	1.5	2.3	.7	.0850
200.	3073.	.8	-1.6	-1.1	0.0	1.4	3.0	2.8	.1781
300.	2788.	3.0	-6.8	-6.1	-4.5	-2.4	0.0	6.6	.2807
400.	2521.	6.7	-16.6	-15.6	-13.5	-10.7	-7.5	12.3	.3940
500.	2268.	12.2	-31.8	-30.6	-27.9	-24.4	-20.4	20.1	.5194
600.	2029.	20.1	-53.7	-52.3	-49.0	-44.9	-40.1	30.4	.6593
700.	1806.	31.2	-84.1	-82.4	-78.6	-73.7	-68.1	43.7	.8159
800.	1603.	46.5	-125.2	-123.3	-119.0	-113.4	-107.0	60.5	.9923
900.	1420.	67.4	-180.0	-177.8	-172.9	-166.6	-159.5	81.2	1.1911
1000.	1268.	95.4	-252.7	-250.3	-244.9	-237.9	-229.9	106.4	1.4152

BALLISTIC COEFFICIENT .34, MUZZLE VELOCITY FT/SEC 3800.

RANGE	VELOCITY	HEIGHT	100 YD	150 YD	200 YD	250 YD	300 YD	DRIFT	TIME
0.	3800.	0.0	-1.5	-1.5	-1.5	-1.5	-1.5	0.0	0.0000
100.	3471.	-.4	0.0	.2	.7	1.3	2.1	.7	.0828
200.	3161.	.7	-1.4	-1.0	0.0	1.3	2.8	2.7	.1734
300.	2871.	2.8	-6.3	-5.7	-4.2	-2.3	0.0	6.3	.2728
400.	2598.	6.3	-15.4	-14.6	-12.6	-10.0	-7.0	11.8	.3829
500.	2342.	11.5	-29.7	-28.7	-26.2	-23.0	-19.2	19.3	.5046
600.	2099.	18.9	-50.3	-49.1	-46.1	-42.2	-37.7	29.2	.6399
700.	1870.	29.3	-78.7	-77.2	-73.7	-69.2	-64.0	42.0	.7911
800.	1662.	43.6	-117.2	-115.5	-111.5	-106.4	-100.4	58.1	.9615
900.	1472.	63.1	-168.3	-166.5	-162.0	-156.2	-149.5	77.9	1.1533
1000.	1310.	89.3	-236.1	-234.1	-229.1	-222.7	-215.2	102.1	1.3697

BALLISTIC COEFFICIENT .34, MUZZLE VELOCITY FT/SEC 3900.

RANGE	VELOCITY	HEIGHT	100 YD	150 YD	200 YD	250 YD	300 YD	DRIFT	TIME
0.	3900.	0.0	-1.5	-1.5	-1.5	-1.5	-1.5	0.0	0.0000
100.	3565.	-.4	0.0	.2	.6	1.2	1.9	.6	.0804
200.	3250.	.6	-1.2	-.9	0.0	1.2	2.6	2.6	.1684
300.	2953.	2.6	-5.8	-5.3	-3.9	-2.1	0.0	6.1	.2652
400.	2676.	5.9	-14.3	-13.7	-11.9	-9.4	-6.6	11.3	.3720
500.	2416.	10.8	-27.8	-26.9	-24.7	-21.6	-18.1	18.6	.4901
600.	2168.	17.8	-47.0	-46.0	-43.3	-39.6	-35.5	28.1	.6210
700.	1935.	27.5	-73.8	-72.6	-69.4	-65.1	-60.3	40.3	.7677
800.	1721.	40.9	-109.7	-108.4	-104.7	-99.8	-94.3	55.7	.9318
900.	1525.	59.2	-157.7	-156.2	-152.1	-146.6	-140.3	74.8	1.1173
1000.	1353.	83.7	-221.0	-219.3	-214.8	-208.6	-201.7	98.1	1.3265

BALLISTIC COEFFICIENT .34, MUZZLE VELOCITY FT/SEC 4000.

RANGE	VELOCITY	HEIGHT	100 YD	150 YD	200 YD	250 YD	300 YD	DRIFT	TIME
0.	4000.	0.0	-1.5	-1.5	-1.5	-1.5	-1.5	0.0	0.0000
100.	3659.	-.5	0.0	.1	.6	1.1	1.8	.6	.0785
200.	3338.	.5	-1.1	-.9	0.0	1.1	2.5	2.5	.1644
300.	3037.	2.5	-5.4	-5.0	-3.7	-2.0	0.0	5.9	.2586
400.	2754.	5.6	-13.4	-12.9	-11.2	-8.9	-6.3	11.0	.3625
500.	2489.	10.2	-26.0	-25.3	-23.2	-20.4	-17.1	18.0	.4770
600.	2238.	16.8	-44.2	-43.4	-40.8	-37.5	-33.5	27.1	.6042
700.	2001.	26.0	-69.3	-68.3	-65.4	-61.4	-56.8	38.9	.7459
800.	1780.	38.6	-103.1	-102.0	-98.6	-94.1	-88.8	53.7	.9050
900.	1580.	55.7	-148.0	-146.8	-143.0	-137.9	-132.0	72.0	1.0841
1000.	1399.	78.6	-207.1	-205.8	-201.5	-195.9	-189.3	94.3	1.2859

BALLISTIC COEFFICIENT .35, MUZZLE VELOCITY FT/SEC 2000.

RANGE	VELOCITY	HEIGHT	100 YD	150 YD	200 YD	250 YD	300 YD	DRIFT	TIME
0.	2000.	0.0	-1.5	-1.5	-1.5	-1.5	-1.5	0.0	0.0000
100.	1786.	.5	0.0	2.1	4.7	7.7	10.9	1.6	.1588
200.	1590.	4.7	-9.5	-5.2	0.0	5.8	12.4	6.5	.3370
300.	1413.	13.1	-32.8	-26.4	-18.5	-9.8	0.0	15.3	.5371
400.	1267.	27.1	-73.6	-65.0	-54.6	-42.9	-29.9	28.5	.7617
500.	1150.	48.3	-136.0	-125.4	-112.3	-97.7	-81.4	45.9	1.0109
600.	1071.	78.1	-224.7	-211.9	-196.2	-178.7	-159.1	67.2	1.2818
700.	1013.	118.1	-343.7	-328.8	-310.5	-290.1	-267.3	92.1	1.5732
800.	964.	167.9	-491.2	-474.1	-453.2	-429.9	-403.8	118.7	1.8745
900.	918.	230.0	-673.1	-653.9	-630.4	-604.1	-574.8	148.3	2.1924
1000.	877.	305.8	-893.3	-871.9	-845.8	-816.6	-784.0	180.8	2.5272

BALLISTIC COEFFICIENT .35, MUZZLE VELOCITY FT/SEC 2100.

RANGE	VELOCITY	HEIGHT	100 YD	150 YD	200 YD	250 YD	300 YD	DRIFT	TIME
0.	2100.	0.0	-1.5	-1.5	-1.5	-1.5	-1.5	0.0	0.0000
100.	1878.	.3	0.0	1.9	4.2	6.9	9.8	1.4	.1509
200.	1674.	4.2	-8.4	-4.7	0.0	5.3	11.1	6.1	.3203
300.	1488.	11.7	-29.3	-23.6	-16.7	-8.7	0.0	14.4	.5103
400.	1327.	24.4	-66.0	-58.5	-49.1	-38.5	-26.9	26.9	.7242
500.	1197.	43.8	-122.5	-113.1	-101.4	-88.2	-73.7	43.8	.9629
600.	1102.	71.2	-203.2	-191.8	-177.8	-162.0	-144.5	64.7	1.2245
700.	1038.	108.0	-311.7	-298.5	-282.2	-263.6	-243.3	88.9	1.5054
800.	985.	155.4	-451.0	-435.9	-417.2	-396.0	-372.8	116.3	1.8038
900.	938.	213.9	-621.1	-604.1	-583.1	-559.3	-533.1	145.9	2.1147
1000.	894.	286.0	-829.0	-810.0	-786.8	-760.3	-731.2	178.8	2.4443

BALLISTIC COEFFICIENT .35, MUZZLE VELOCITY FT/SEC 2200.

RANGE	VELOCITY	HEIGHT	100 YD	150 YD	200 YD	250 YD	300 YD	DRIFT	TIME
0.	2200.	0.0	-1.5	-1.5	-1.5	-1.5	-1.5	0.0	0.0000
100.	1971.	.2	0.0	1.7	3.8	6.1	8.8	1.4	.1441
200.	1759.	3.7	-7.5	-4.2	0.0	4.8	10.1	5.7	.3051
300.	1566.	10.6	-26.4	-21.4	-15.1	-7.9	0.0	13.6	.4862
400.	1392.	22.1	-59.3	-52.7	-44.3	-34.8	-24.2	25.3	.6894
500.	1250.	39.6	-110.4	-102.1	-91.6	-79.7	-66.5	41.4	.9172
600.	1138.	65.0	-183.9	-173.9	-161.4	-147.0	-131.2	61.9	1.1699
700.	1063.	99.3	-284.1	-272.5	-257.9	-241.2	-222.7	86.1	1.4439
800.	1007.	143.7	-413.5	-400.2	-383.5	-364.4	-343.3	113.3	1.7347
900.	958.	198.7	-572.4	-557.4	-538.6	-517.1	-493.3	142.8	2.0384
1000.	913.	266.5	-766.4	-749.8	-728.9	-705.0	-678.6	175.3	2.3596

BALLISTIC COEFFICIENT .35, MUZZLE VELOCITY FT/SEC 2300.

RANGE	VELOCITY	HEIGHT	100 YD	150 YD	200 YD	250 YD	300 YD	DRIFT	TIME
0.	2300.	0.0	-1.5	-1.5	-1.5	-1.5	-1.5	0.0	0.0000
100.	2066.	.2	0.0	1.5	3.4	5.5	7.9	1.3	.1376
200.	1846.	3.3	-6.7	-3.8	0.0	4.3	9.1	5.4	.2913
300.	1645.	9.6	-23.7	-19.2	-13.6	-7.2	0.0	12.7	.4635
400.	1462.	20.0	-53.5	-47.6	-40.1	-31.5	-21.9	23.8	.6570
500.	1306.	35.9	-99.7	-92.3	-82.9	-72.1	-60.2	39.1	.8743
600.	1180.	59.1	-166.2	-157.3	-146.1	-133.2	-118.9	58.7	1.1164
700.	1091.	90.9	-257.9	-247.6	-234.4	-219.4	-202.7	82.5	1.3816
800.	1029.	132.7	-378.7	-366.9	-351.8	-334.7	-315.6	109.8	1.6671
900.	978.	184.4	-527.3	-514.0	-497.1	-477.8	-456.3	139.1	1.9642
1000.	931.	248.5	-709.5	-694.7	-675.9	-654.5	-630.6	171.5	2.2789

BALLISTIC COEFFICIENT .35, MUZZLE VELOCITY FT/SEC 2400.

RANGE	VELOCITY	HEIGHT	100 YD	150 YD	200 YD	250 YD	300 YD	DRIFT	TIME
0.	2400.	0.0	-1.5	-1.5	-1.5	-1.5	-1.5	0.0	0.0000
100.	2161.	.1	0.0	1.3	3.0	5.0	7.1	1.2	.1316
200.	1934.	3.0	-6.0	-3.4	0.0	3.9	8.2	5.0	.2785
300.	1726.	8.7	-21.3	-17.4	-12.3	-6.5	0.0	11.9	.4425
400.	1535.	18.1	-48.4	-43.1	-36.4	-28.6	-20.0	22.4	.6271
500.	1366.	32.7	-90.3	-83.7	-75.2	-65.5	-54.7	36.9	.8345
600.	1229.	53.8	-150.9	-142.9	-132.8	-121.1	-108.2	55.7	1.0664
700.	1123.	83.2	-234.6	-225.3	-213.5	-199.9	-184.7	78.8	1.3225
800.	1053.	122.0	-345.5	-334.9	-321.4	-305.8	-288.6	105.4	1.5989
900.	998.	171.0	-485.5	-473.6	-458.4	-440.9	-421.4	135.0	1.8918
1000.	950.	231.6	-656.6	-643.4	-626.5	-607.1	-585.5	167.2	2.2001

BALLISTIC COEFFICIENT .35, MUZZLE VELOCITY FT/SEC 2500.

RANGE	VELOCITY	HEIGHT	100 YD	150 YD	200 YD	250 YD	300 YD	DRIFT	TIME
0.	2500.	0.0	-1.5	-1.5	-1.5	-1.5	-1.5	0.0	0.0000
100.	2256.	.0	0.0	1.2	2.7	4.5	6.4	1.1	.1264
200.	2024.	2.7	-5.4	-3.1	0.0	3.5	7.5	4.7	.2667
300.	1807.	7.9	-19.3	-15.8	-11.2	-5.9	0.0	11.2	.4235
400.	1610.	16.5	-43.9	-39.3	-33.1	-26.0	-18.2	21.0	.5995
500.	1430.	29.8	-81.9	-76.1	-68.4	-59.6	-49.7	34.7	.7973
600.	1281.	49.1	-137.0	-130.1	-120.9	-110.2	-98.4	52.7	1.0193
700.	1160.	76.2	-213.4	-205.3	-194.6	-182.1	-168.4	75.0	1.2659
800.	1078.	112.3	-315.7	-306.4	-294.2	-280.0	-264.3	101.1	1.5346
900.	1019.	158.5	-446.5	-436.0	-422.3	-406.3	-388.6	130.5	1.8212
1000.	968.	216.1	-608.6	-597.0	-581.7	-563.9	-544.3	162.9	2.1257

BALLISTIC COEFFICIENT .35, MUZZLE VELOCITY FT/SEC 2600.

RANGE	VELOCITY	HEIGHT	100 YD	150 YD	200 YD	250 YD	300 YD	DRIFT	TIME
0.	2600.	0.0	-1.5	-1.5	-1.5	-1.5	-1.5	0.0	0.0000
100.	2351.	-.0	0.0	1.0	2.4	4.0	5.8	1.1	.1215
200.	2114.	2.4	-4.8	-2.8	0.0	3.2	6.8	4.4	.2559
300.	1891.	7.2	-17.5	-14.4	-10.2	-5.4	0.0	10.5	.4060
400.	1686.	15.1	-39.9	-35.8	-30.3	-23.8	-16.6	19.8	.5741
500.	1499.	27.2	-74.5	-69.4	-62.5	-54.4	-45.4	32.7	.7628
600.	1336.	44.9	-124.8	-118.7	-110.4	-100.7	-89.9	49.8	.9755
700.	1204.	69.8	-194.6	-187.5	-177.8	-166.4	-153.8	71.2	1.2122
800.	1107.	103.4	-288.7	-280.5	-269.4	-256.5	-242.1	96.8	1.4729
900.	1041.	146.7	-410.5	-401.3	-388.8	-374.2	-358.0	125.7	1.7529
1000.	988.	201.1	-562.6	-552.4	-538.5	-522.3	-504.3	157.9	2.0509

BALLISTIC COEFFICIENT .35, MUZZLE VELOCITY FT/SEC 2700.

RANGE	VELOCITY	HEIGHT	100 YD	150 YD	200 YD	250 YD	300 YD	DRIFT	TIME
0.	2700.	0.0	-1.5	-1.5	-1.5	-1.5	-1.5	0.0	0.0000
100.	2445.	-.1	0.0	.9	2.2	3.7	5.3	1.0	.1167
200.	2204.	2.2	-4.4	-2.5	0.0	3.0	6.2	4.2	.2460
300.	1975.	6.5	-15.9	-13.2	-9.4	-4.9	0.0	9.9	.3897
400.	1763.	13.8	-36.4	-32.7	-27.7	-21.8	-15.2	18.7	.5506
500.	1569.	24.9	-68.2	-63.6	-57.3	-49.9	-41.6	30.9	.7313
600.	1395.	41.1	-114.0	-108.5	-100.9	-92.1	-82.2	47.1	.9343
700.	1252.	64.0	-177.9	-171.5	-162.7	-152.3	-140.8	67.5	1.1615
800.	1139.	95.1	-264.2	-256.9	-246.8	-235.0	-221.8	92.3	1.4134
900.	1064.	136.0	-378.0	-369.8	-358.4	-345.1	-330.3	121.1	1.6880
1000.	1008.	187.1	-520.0	-510.9	-498.3	-483.5	-467.0	152.6	1.9783

BALLISTIC COEFFICIENT .35, MUZZLE VELOCITY FT/SEC 2800.

RANGE	VELOCITY	HEIGHT	100 YD	150 YD	200 YD	250 YD	300 YD	DRIFT	TIME
0.	2800.	0.0	-1.5	-1.5	-1.5	-1.5	-1.5	0.0	0.0000
100.	2540.	-.1	0.0	.8	2.0	3.3	4.8	.9	.1125
200.	2293.	1.9	-3.9	-2.3	0.0	2.7	5.7	3.9	.2367
300.	2060.	6.0	-14.5	-12.1	-8.6	-4.6	0.0	9.4	.3748
400.	1840.	12.7	-33.3	-30.1	-25.5	-20.1	-14.0	17.7	.5290
500.	1640.	22.9	-62.3	-58.3	-52.5	-45.8	-38.2	29.2	.7016
600.	1457.	37.8	-104.4	-99.5	-92.6	-84.5	-75.4	44.5	.8958
700.	1302.	58.8	-162.8	-157.2	-149.1	-139.7	-129.0	64.0	1.1138
800.	1177.	87.7	-242.4	-235.9	-226.7	-215.9	-203.7	88.0	1.3572
900.	1089.	125.6	-346.9	-339.6	-329.2	-317.1	-303.4	115.8	1.6223
1000.	1028.	174.0	-480.5	-472.4	-460.9	-447.4	-432.2	147.2	1.9080

BALLISTIC COEFFICIENT .35, MUZZLE VELOCITY FT/SEC 2900.

RANGE	VELOCITY	HEIGHT	100 YD	150 YD	200 YD	250 YD	300 YD	DRIFT	TIME
0.	2900.	0.0	-1.5	-1.5	-1.5	-1.5	-1.5	0.0	0.0000
100.	2634.	-.2	0.0	.7	1.8	3.0	4.4	.9	.1085
200.	2383.	1.8	-3.5	-2.1	0.0	2.5	5.3	3.8	.2283
300.	2144.	5.5	-13.3	-11.1	-7.9	-4.2	0.0	8.9	.3611
400.	1919.	11.7	-30.6	-27.7	-23.5	-18.5	-12.9	16.8	.5091
500.	1711.	21.1	-57.2	-53.6	-48.4	-42.1	-35.1	27.7	.6744
600.	1522.	34.8	-95.9	-91.5	-85.2	-77.7	-69.4	42.2	.8605
700.	1355.	54.2	-149.7	-144.6	-137.3	-128.5	-118.7	60.9	1.0699
800.	1220.	80.8	-222.5	-216.7	-208.3	-198.3	-187.1	83.7	1.3^31
900.	1117.	116.4	-319.6	-313.1	-303.6	-292.4	-279.8	111.0	1.5619
1000.	1048.	162.0	-444.8	-437.6	-427.1	-414.6	-400.6	142.1	1.8416

BALLISTIC COEFFICIENT .35, MUZZLE VELOCITY FT/SEC 3000.

RANGE	VELOCITY	HEIGHT	100 YD	150 YD	200 YD	250 YD	300 YD	DRIFT	TIME
0.	3000.	0.0	-1.5	-1.5	-1.5	-1.5	-1.5	0.0	0.0000
100.	2727.	-.2	0.0	.6	1.6	2.8	4.0	.9	.1051
200.	2471.	1.6	-3.2	-1.9	0.0	2.3	4.9	3.6	.2206
300.	2228.	5.1	-12.1	-10.2	-7.3	-3.9	0.0	8.5	.3485
400.	1998.	10.8	-28.1	-25.5	-21.7	-17.0	-11.9	16.0	.4907
500.	1784.	19.5	-52.7	-49.6	-44.8	-39.0	-32.5	26.3	.6497
600.	1588.	32.2	-88.4	-84.5	-78.8	-71.8	-64.1	40.1	.8281
700.	1411.	50.0	-137.7	-133.2	-126.5	-118.4	-109.4	57.8	1.0283
800.	1265.	74.6	-204.9	-199.8	-192.1	-182.9	-172.6	79.8	1.2532
900.	1149.	107.6	-294.2	-288.4	-279.8	-269.4	-257.8	106.1	1.5026
1000.	1071.	150.3	-409.9	-403.5	-393.9	-382.4	-369.5	136.2	1.7737

BALLISTIC COEFFICIENT .35, MUZZLE VELOCITY FT/SEC 3100.

RANGE	VELOCITY	HEIGHT	100 YD	150 YD	200 YD	250 YD	300 YD	DRIFT	TIME
0.	3100.	0.0	-1.5	-1.5	-1.5	-1.5	-1.5	0.0	0.0000
100.	2821.	-.3	0.0	.6	1.4	2.5	3.7	.8	.1014
200.	2560.	1.4	-2.9	-1.8	0.0	2.1	4.5	3.4	.2130
300.	2312.	4.7	-11.1	-9.4	-6.8	-3.6	0.0	8.1	.3363
400.	2078.	10.0	-25.8	-23.6	-20.1	-15.8	-11.0	15.1	.4731
500.	1857.	18.1	-48.6	-45.8	-41.4	-36.1	-30.1	25.0	.6258
600.	1655.	29.8	-81.5	-78.2	-72.9	-66.5	-59.3	38.1	.7972
700.	1471.	46.2	-127.1	-123.2	-117.0	-109.6	-101.2	54.9	.9895
800.	1313.	69.1	-189.2	-184.7	-177.6	-169.1	-159.5	76.0	1.2060
900.	1186.	99.7	-271.5	-266.5	-258.5	-249.0	-238.2	101.3	1.4465
1000.	1095.	139.7	-379.2	-373.6	-364.8	-354.1	-342.1	130.7	1.7106

BALLISTIC COEFFICIENT .35, MUZZLE VELOCITY FT/SEC 3200.

RANGE	VELOCITY	HEIGHT	100 YD	150 YD	200 YD	250 YD	300 YD	DRIFT	TIME
0.	3200.	0.0	-1.5	-1.5	-1.5	-1.5	-1.5	0.0	0.0000
100.	2915.	-.3	0.0	.5	1.3	2.3	3.4	.8	.0984
200.	2648.	1.3	-2.6	-1.6	0.0	2.0	4.2	3.3	.2062
300.	2396.	4.3	-10.2	-8.8	-6.3	-3.3	0.0	7.8	.3254
400.	2157.	9.3	-23.8	-21.9	-18.6	-14.7	-10.2	14.5	.4573
500.	1931.	16.8	-45.0	-42.6	-38.5	-33.5	-28.0	23.9	.6045
600.	1722.	27.6	-75.4	-72.5	-67.6	-61.6	-55.0	36.3	.7687
700.	1532.	42.9	-117.6	-114.2	-108.5	-101.6	-93.8	52.3	.9536
800.	1364.	64.0	-174.8	-171.0	-164.4	-156.5	-147.6	72.4	1.1615
900.	1227.	92.5	-251.1	-246.8	-239.4	-230.5	-220.5	96.8	1.3938
1000.	1122.	130.0	-351.0	-346.2	-338.0	-328.1	-317.0	125.4	1.6503

BALLISTIC COEFFICIENT .35, MUZZLE VELOCITY FT/SEC 3300.

RANGE	VELOCITY	HEIGHT	100 YD	150 YD	200 YD	250 YD	300 YD	DRIFT	TIME
0.	3300.	0.0	-1.5	-1.5	-1.5	-1.5	-1.5	0.0	0.0000
100.	3009.	-.3	0.0	.4	1.2	2.1	3.1	.7	.0952
200.	2736.	1.2	-2.4	-1.5	0.0	1.8	3.9	3.2	.1998
300.	2479.	4.0	-9.4	-8.1	-5.8	-3.1	0.0	7.4	.3150
400.	2236.	8.7	-22.1	-20.4	-17.4	-13.7	-9.6	13.9	.4425
500.	2005.	15.6	-41.7	-39.6	-35.8	-31.3	-26.1	22.8	.5842
600.	1790.	25.7	-70.0	-67.4	-62.9	-57.4	-51.2	34.7	.7426
700.	1594.	39.9	-109.2	-106.2	-100.9	-94.5	-87.3	50.0	.9203
800.	1417.	59.4	-162.1	-158.6	-152.6	-145.3	-137.0	69.1	1.1199
900.	1270.	85.9	-232.8	-228.9	-222.1	-213.9	-204.6	92.5	1.3439
1000.	1153.	121.0	-325.5	-321.2	-313.7	-304.5	-294.2	120.3	1.5925

BALLISTIC COEFFICIENT .35, MUZZLE VELOCITY FT/SEC 3400.

RANGE	VELOCITY	HEIGHT	100 YD	150 YD	200 YD	250 YD	300 YD	DRIFT	TIME
0.	3400.	0.0	-1.5	-1.5	-1.5	-1.5	-1.5	0.0	0.0000
100.	3103.	-.3	0.0	.4	1.1	1.9	2.9	.7	.0924
200.	2824.	1.0	-2.1	-1.4	0.0	1.7	3.6	3.0	.1936
300.	2562.	3.7	-8.6	-7.5	-5.4	-2.9	0.0	7.1	.3051
400.	2315.	8.1	-20.4	-19.0	-16.2	-12.8	-8.9	13.3	.4283
500.	2080.	14.6	-38.7	-36.8	-33.4	-29.1	-24.3	21.8	.5649
600.	1859.	24.0	-65.0	-62.8	-58.6	-53.5	-47.7	33.1	.7175
700.	1657.	37.2	-101.4	-98.8	-94.0	-88.1	-81.3	47.7	.8886
800.	1473.	55.3	-150.5	-147.5	-142.0	-135.2	-127.5	66.0	1.0807
900.	1315.	80.0	-216.1	-212.8	-206.6	-198.9	-190.2	88.5	1.2969
1000.	1187.	112.7	-302.2	-298.5	-291.6	-283.1	-273.4	115.3	1.5372

BALLISTIC COEFFICIENT .35, MUZZLE VELOCITY FT/SEC 3500.

RANGE	VELOCITY	HEIGHT	100 YD	150 YD	200 YD	250 YD	300 YD	DRIFT	TIME
0.	3500.	0.0	-1.5	-1.5	-1.5	-1.5	-1.5	0.0	0.0000
100.	3197.	-.4	0.0	.3	1.0	1.7	2.7	.7	.0898
200.	2913.	1.0	-1.9	-1.3	0.0	1.6	3.4	3.0	.1882
300.	2645.	3.5	-8.0	-7.0	-5.1	-2.7	0.0	6.9	.2962
400.	2394.	7.5	-19.0	-17.7	-15.1	-12.0	-8.4	12.8	.4155
500.	2155.	13.6	-36.0	-34.4	-31.2	-27.3	-22.8	20.9	.5475
600.	1929.	22.4	-60.6	-58.7	-54.8	-50.1	-44.7	31.8	.6949
700.	1720.	34.7	-94.4	-92.1	-87.7	-82.2	-75.8	45.6	.8592
800.	1531.	51.6	-140.2	-137.6	-132.5	-126.2	-118.9	63.1	1.0444
900.	1362.	74.6	-201.0	-198.1	-192.3	-185.3	-177.1	84.7	1.2525
1000.	1226.	105.1	-281.1	-277.9	-271.5	-263.7	-254.6	110.5	1.4850

BALLISTIC COEFFICIENT .35, MUZZLE VELOCITY FT/SEC 3600.

RANGE	VELOCITY	HEIGHT	100 YD	150 YD	200 YD	250 YD	300 YD	DRIFT	TIME
0.	3600.	0.0	-1.5	-1.5	-1.5	-1.5	-1.5	0.0	0.0000
100.	3291.	-.4	0.0	.3	.9	1.6	2.4	.7	.0871
200.	3001.	.8	-1.7	-1.1	0.0	1.5	3.2	2.8	.1823
300.	2728.	3.2	-7.3	-6.5	-4.8	-2.5	0.0	6.6	.2874
400.	2472.	7.0	-17.6	-16.5	-14.2	-11.2	-7.8	12.2	.4028
500.	2229.	12.8	-33.6	-32.2	-29.3	-25.6	-21.4	20.1	.5307
600.	1999.	21.0	-56.5	-54.8	-51.3	-46.9	-41.8	30.4	.6728
700.	1784.	32.5	-88.1	-86.1	-82.1	-77.0	-71.0	43.7	.8318
800.	1589.	48.2	-130.8	-128.5	-123.9	-118.0	-111.2	60.4	1.0101
900.	1412.	69.6	-187.2	-184.7	-179.5	-172.9	-165.2	81.0	1.2103
1000.	1266.	98.1	-261.9	-259.0	-253.3	-245.9	-237.4	105.9	1.4350

BALLISTIC COEFFICIENT .35, MUZZLE VELOCITY FT/SEC 3700.

RANGE	VELOCITY	HEIGHT	100 YD	150 YD	200 YD	250 YD	300 YD	DRIFT	TIME
0.	3700.	0.0	-1.5	-1.5	-1.5	-1.5	-1.5	0.0	0.0000
100.	3386.	-.4	0.0	.2	.8	1.5	2.3	.7	.0849
200.	3090.	.8	-1.6	-1.1	0.0	1.4	3.0	2.7	.1778
300.	2811.	3.0	-6.8	-6.1	-4.4	-2.4	0.0	6.4	.2796
400.	2550.	6.6	-16.4	-15.4	-13.3	-10.6	-7.4	11.8	.3916
500.	2304.	12.0	-31.3	-30.1	-27.4	-24.0	-20.0	19.4	.5154
600.	2069.	19.7	-52.8	-51.3	-48.1	-44.0	-39.2	29.3	.6527
700.	1849.	30.4	-82.4	-80.7	-76.9	-72.2	-66.6	42.0	.8062
800.	1648.	45.2	-122.2	-120.3	-116.0	-110.5	-104.1	58.0	.9781
900.	1465.	65.1	-174.9	-172.7	-167.9	-161.8	-154.6	77.7	1.1712
1000.	1308.	91.8	-244.4	-242.0	-236.6	-229.8	-221.8	101.6	1.3883

BALLISTIC COEFFICIENT .35, MUZZLE VELOCITY FT/SEC 3800.

RANGE	VELOCITY	HEIGHT	100 YD	150 YD	200 YD	250 YD	300 YD	DRIFT	TIME
0.	3800.	0.0	-1.5	-1.5	-1.5	-1.5	-1.5	0.0	0.0000
100.	3480.	-.4	0.0	.2	.7	1.3	2.1	.7	.0827
200.	3178.	.7	-1.4	-1.0	0.0	1.3	2.8	2.6	.1728
300.	2895.	2.8	-6.2	-5.6	-4.2	-2.2	0.0	6.2	.2718
400.	2629.	6.2	-15.2	-14.4	-12.5	-9.9	-6.9	11.4	.3805
500.	2378.	11.3	-29.3	-28.3	-25.8	-22.6	-18.9	18.6	.5006
600.	2140.	18.5	-49.4	-48.2	-45.3	-41.4	-37.0	28.2	.6337
700.	1915.	28.6	-77.1	-75.7	-72.3	-67.8	-62.6	40.4	.7820
800.	1708.	42.4	-114.3	-112.7	-108.8	-103.6	-97.7	55.6	.9477
900.	1519.	61.0	-163.6	-161.8	-157.4	-151.6	-144.9	74.6	1.1341
1000.	1352.	85.9	-228.4	-226.4	-221.5	-215.1	-207.6	97.6	1.3439

BALLISTIC COEFFICIENT .35, MUZZLE VELOCITY FT/SEC 3900.

RANGE	VELOCITY	HEIGHT	100 YD	150 YD	200 YD	250 YD	300 YD	DRIFT	TIME
0.	3900.	0.0	-1.5	-1.5	-1.5	-1.5	-1.5	0.0	0.0000
100.	3574.	-.4	0.0	.2	.6	1.2	1.9	.6	.0803
200.	3267.	.6	-1.2	-.9	0.0	1.2	2.6	2.5	.1680
300.	2978.	2.6	-5.7	-5.3	-3.9	-2.1	0.0	5.9	.2642
400.	2707.	5.8	-14.2	-13.5	-11.7	-9.3	-6.5	10.9	.3699
500.	2452.	10.6	-27.3	-26.5	-24.3	-21.2	-17.8	17.9	.4863
600.	2210.	17.4	-46.3	-45.3	-42.6	-38.9	-34.8	27.0	.6152
700.	1980.	26.9	-72.2	-71.1	-67.9	-63.7	-58.8	38.7	.7585
800.	1768.	39.8	-107.1	-105.9	-102.2	-97.4	-91.8	53.4	.9190
900.	1574.	57.2	-153.3	-151.9	-147.8	-142.4	-136.1	71.6	1.0991
1000.	1399.	80.5	-213.7	-212.2	-207.6	-201.5	-194.6	93.6	1.3013

BALLISTIC COEFFICIENT .35, MUZZLE VELOCITY FT/SEC 4000.

RANGE	VELOCITY	HEIGHT	100 YD	150 YD	200 YD	250 YD	300 YD	DRIFT	TIME
0.	4000.	0.0	-1.5	-1.5	-1.5	-1.5	-1.5	0.0	0.0000
100.	3668.	-.5	0.0	.1	.6	1.1	1.8	.6	.0784
200.	3356.	.5	-1.1	-.8	0.0	1.1	2.4	2.5	.1640
300.	3062.	2.4	-5.3	-4.9	-3.6	-2.0	0.0	5.7	.2574
400.	2785.	5.5	-13.2	-12.7	-11.0	-8.8	-6.2	10.6	.3602
500.	2526.	10.0	-25.6	-25.0	-22.9	-20.1	-16.8	17.4	.4736
600.	2280.	16.4	-43.4	-42.7	-40.1	-36.7	-32.8	26.1	.5984
700.	2047.	25.4	-67.9	-67.0	-64.1	-60.1	-55.6	37.4	.7375
800.	1829.	37.5	-100.6	-99.6	-96.2	-91.7	-86.5	51.5	.8924
900.	1629.	53.8	-143.9	-142.8	-138.9	-133.9	-128.0	68.9	1.0662
1000.	1448.	75.7	-200.5	-199.3	-195.0	-189.4	-182.9	90.1	1.2618

BALLISTIC COEFFICIENT .36, MUZZLE VELOCITY FT/SEC 2000.

RANGE	VELOCITY	HEIGHT	100 YD	150 YD	200 YD	250 YD	300 YD	DRIFT	TIME
0.	2000.	0.0	-1.5	-1.5	-1.5	-1.5	-1.5	0.0	0.0000
100.	1791.	.5	0.0	2.1	4.7	7.6	10.8	1.5	.1585
200.	1600.	4.7	-9.4	-5.2	0.0	5.8	12.2	6.3	.3358
300.	1427.	13.0	-32.5	-26.1	-18.3	-9.6	0.0	14.8	.5344
400.	1281.	26.7	-72.7	-64.2	-53.8	-42.2	-29.4	27.6	.7566
500.	1164.	47.5	-134.0	-123.4	-110.5	-96.0	-79.9	44.5	1.0027
600.	1082.	76.9	-221.3	-208.6	-193.1	-175.7	-156.4	65.4	1.2715
700.	1024.	115.6	-337.1	-322.3	-304.1	-283.8	-261.4	89.2	1.5569
800.	974.	164.7	-482.8	-465.8	-445.1	-421.9	-396.2	115.5	1.8565
900.	929.	225.7	-662.2	-643.1	-619.8	-593.7	-564.8	144.7	2.1719
1000.	888.	299.8	-878.1	-856.9	-831.0	-802.0	-769.9	176.4	2.5024

BALLISTIC COEFFICIENT .36, MUZZLE VELOCITY FT/SEC 2100.

RANGE	VELOCITY	HEIGHT	100 YD	150 YD	200 YD	250 YD	300 YD	DRIFT	TIME
0.	2100.	0.0	-1.5	-1.5	-1.5	-1.5	-1.5	0.0	0.0000
100.	1884.	.3	0.0	1.9	4.2	6.8	9.7	1.4	.1507
200.	1685.	4.1	-8.4	-4.6	0.0	5.2	11.0	5.9	.3192
300.	1503.	11.6	-29.0	-23.4	-16.5	-8.6	0.0	13.9	.5076
400.	1343.	24.1	-65.2	-57.7	-48.5	-38.0	-26.5	26.0	.7192
500.	1214.	43.0	-120.7	-111.3	-99.8	-86.7	-72.4	42.3	.9549
600.	1115.	69.9	-199.7	-188.4	-174.6	-158.9	-141.6	62.7	1.2132
700.	1048.	106.4	-307.6	-294.4	-278.3	-259.9	-239.8	86.9	1.4940
800.	996.	152.1	-442.3	-427.2	-408.8	-387.8	-364.9	113.0	1.7847
900.	949.	209.6	-610.1	-593.1	-572.4	-548.8	-523.0	142.2	2.0934
1000.	906.	279.6	-812.5	-793.7	-770.6	-744.4	-715.7	173.9	2.4168

BALLISTIC COEFFICIENT .36, MUZZLE VELOCITY FT/SEC 2200.

RANGE	VELOCITY	HEIGHT	100 YD	150 YD	200 YD	250 YD	300 YD	DRIFT	TIME
0.	2200.	0.0	-1.5	-1.5	-1.5	-1.5	-1.5	0.0	0.0000
100.	1977.	.2	0.0	1.7	3.7	6.1	8.7	1.3	.1438
200.	1771.	3.7	-7.5	-4.2	0.0	4.7	9.9	5.6	.3043
300.	1581.	10.5	-26.1	-21.1	-14.9	-7.8	0.0	13.1	.4838
400.	1410.	21.7	-58.6	-52.0	-43.7	-34.2	-23.8	24.5	.6846
500.	1268.	38.9	-108.7	-100.4	-90.0	-78.3	-65.2	40.0	.9093
600.	1153.	63.8	-181.0	-171.1	-158.6	-144.5	-128.8	60.1	1.1596
700.	1076.	97.1	-278.3	-266.7	-252.1	-235.6	-217.3	83.3	1.4278
800.	1019.	140.4	-404.7	-391.5	-374.9	-356.0	-335.1	109.8	1.7148
900.	969.	195.0	-563.1	-548.2	-529.5	-508.3	-484.8	139.4	2.0194
1000.	925.	260.7	-751.6	-735.0	-714.2	-690.7	-664.5	170.7	2.3337

BALLISTIC COEFFICIENT .36, MUZZLE VELOCITY FT/SEC 2300.

RANGE	VELOCITY	HEIGHT	100 YD	150 YD	200 YD	250 YD	300 YD	DRIFT	TIME
0.	2300.	0.0	-1.5	-1.5	-1.5	-1.5	-1.5	0.0	0.0000
100.	2072.	.2	0.0	1.5	3.3	5.5	7.8	1.2	.1373
200.	1858.	3.3	-6.6	-3.7	0.0	4.3	9.0	5.2	.2902
300.	1661.	9.5	-23.5	-19.0	-13.5	-7.1	0.0	12.3	.4611
400.	1482.	19.7	-52.8	-46.9	-39.5	-31.0	-21.6	23.0	.6523
500.	1326.	35.3	-98.2	-90.8	-81.5	-70.9	-59.1	37.8	.8668
600.	1199.	57.9	-163.3	-154.5	-143.4	-130.6	-116.4	56.8	1.1052
700.	1106.	88.9	-252.8	-242.5	-229.5	-214.6	-198.0	79.8	1.3665
800.	1042.	129.4	-370.1	-358.3	-343.5	-326.5	-307.6	106.2	1.6468
900.	990.	180.8	-518.1	-504.8	-488.2	-469.0	-447.7	135.7	1.9447
1000.	943.	243.0	-695.5	-680.8	-662.3	-641.0	-617.4	167.1	2.2535

BALLISTIC COEFFICIENT .36, MUZZLE VELOCITY FT/SEC 2400.

RANGE	VELOCITY	HEIGHT	100 YD	150 YD	200 YD	250 YD	300 YD	DRIFT	TIME
0.	2400.	0.0	-1.5	-1.5	-1.5	-1.5	-1.5	0.0	0.0000
100.	2167.	.1	0.0	1.3	3.0	4.9	7.1	1.1	.1314
200.	1946.	3.0	-6.0	-3.4	0.0	3.9	8.1	4.9	.2776
300.	1742.	8.6	-21.2	-17.2	-12.2	-6.4	0.0	11.5	.4405
400.	1556.	17.9	-47.9	-42.6	-35.9	-28.2	-19.6	21.6	.6229
500.	1388.	32.1	-88.9	-82.3	-73.9	-64.3	-53.6	35.6	.8272
600.	1250.	52.7	-148.1	-140.3	-130.2	-118.6	-105.8	53.7	1.0554
700.	1140.	81.3	-229.6	-220.4	-208.7	-195.2	-180.2	76.1	1.3072
800.	1067.	119.1	-337.9	-327.4	-314.0	-298.5	-281.4	102.1	1.5798
900.	1011.	167.4	-476.3	-464.5	-449.4	-432.0	-412.8	131.4	1.8719
1000.	963.	226.1	-642.5	-629.4	-612.6	-593.3	-572.0	162.6	2.1738

BALLISTIC COEFFICIENT .36, MUZZLE VELOCITY FT/SEC 2500.

RANGE	VELOCITY	HEIGHT	100 YD	150 YD	200 YD	250 YD	300 YD	DRIFT	TIME
0.	2500.	0.0	-1.5	-1.5	-1.5	-1.5	-1.5	0.0	0.0000
100.	2262.	.0	0.0	1.1	2.7	4.4	6.4	1.1	.1262
200.	2036.	2.6	-5.3	-3.0	0.0	3.5	7.4	4.6	.2660
300.	1825.	7.8	-19.1	-15.7	-11.1	-5.8	0.0	10.8	.4215
400.	1631.	16.3	-43.3	-38.8	-32.7	-25.6	-17.9	20.3	.5955
500.	1454.	29.2	-80.7	-74.9	-67.3	-58.5	-48.8	33.5	.7905
600.	1303.	48.1	-134.5	-127.6	-118.5	-107.9	-96.3	50.8	1.0085
700.	1181.	74.3	-208.9	-200.8	-190.2	-177.9	-164.3	72.3	1.2508
800.	1094.	109.5	-308.5	-299.3	-287.1	-273.1	-257.5	97.8	1.5155
900.	1033.	154.5	-436.0	-425.7	-412.0	-396.2	-378.7	126.4	1.7981
1000.	982.	210.3	-593.4	-581.9	-566.7	-549.1	-529.7	157.8	2.0967

BALLISTIC COEFFICIENT .36, MUZZLE VELOCITY FT/SEC 2600.

RANGE	VELOCITY	HEIGHT	100 YD	150 YD	200 YD	250 YD	300 YD	DRIFT	TIME
0.	2600.	0.0	-1.5	-1.5	-1.5	-1.5	-1.5	0.0	0.0000
100.	2357.	-.0	0.0	1.0	2.4	4.0	5.8	1.0	.1213
200.	2127.	2.4	-4.8	-2.7	0.0	3.2	6.8	4.3	.2551
300.	1908.	7.1	-17.3	-14.3	-10.1	-5.3	0.0	10.2	.4042
400.	1708.	14.9	-39.4	-35.3	-29.8	-23.4	-16.3	19.1	.5702
500.	1524.	26.7	-73.4	-68.3	-61.5	-53.5	-44.6	31.6	.7564
600.	1361.	44.0	-122.5	-116.4	-108.2	-98.5	-87.9	48.0	.9650
700.	1228.	68.1	-190.5	-183.4	-173.8	-162.5	-150.1	68.6	1.1974
800.	1125.	100.6	-281.8	-273.6	-262.6	-249.8	-235.6	93.3	1.4534
900.	1056.	142.8	-400.2	-391.0	-378.6	-364.2	-348.2	121.6	1.7291
1000.	1002.	195.4	-547.8	-537.6	-523.9	-507.9	-490.1	152.7	2.0216

BALLISTIC COEFFICIENT .36, MUZZLE VELOCITY FT/SEC 2700.

RANGE	VELOCITY	HEIGHT	100 YD	150 YD	200 YD	250 YD	300 YD	DRIFT	TIME
0.	2700.	0.0	-1.5	-1.5	-1.5	-1.5	-1.5	0.0	0.0000
100.	2452.	-.1	0.0	.9	2.2	3.6	5.3	1.0	.1165
200.	2217.	2.1	-4.3	-2.5	0.0	2.9	6.2	4.0	.2452
300.	1993.	6.5	-15.8	-13.0	-9.3	-4.9	0.0	9.6	.3880
400.	1785.	13.6	-36.0	-32.3	-27.4	-21.5	-15.0	18.1	.5471
500.	1595.	24.5	-67.1	-62.6	-56.3	-49.0	-40.9	29.8	.7250
600.	1422.	40.2	-111.9	-106.4	-98.9	-90.1	-80.3	45.3	.9241
700.	1277.	62.4	-174.2	-167.8	-159.0	-148.8	-137.4	65.0	1.1474
800.	1161.	92.5	-257.7	-250.4	-240.5	-228.7	-215.7	88.9	1.3940
900.	1080.	131.9	-367.4	-359.2	-348.0	-334.8	-320.1	116.6	1.6625
1000.	1023.	181.6	-505.7	-496.6	-484.1	-469.5	-453.2	147.5	1.9490

BALLISTIC COEFFICIENT .36, MUZZLE VELOCITY FT/SEC 2800.

RANGE	VELOCITY	HEIGHT	100 YD	150 YD	200 YD	250 YD	300 YD	DRIFT	TIME
0.	2800.	0.0	-1.5	-1.5	-1.5	-1.5	-1.5	0.0	0.0000
100.	2547.	-.1	0.0	.8	1.9	3.3	4.8	.9	.1123
200.	2307.	1.9	-3.9	-2.3	0.0	2.7	5.7	3.8	.2360
300.	2079.	5.9	-14.3	-11.9	-8.5	-4.5	0.0	9.0	.3728
400.	1864.	12.5	-32.9	-29.6	-25.1	-19.7	-13.7	17.0	.5253
500.	1667.	22.5	-61.4	-57.4	-51.7	-45.0	-37.6	28.2	.6957
600.	1487.	37.0	-102.5	-97.6	-90.8	-82.8	-73.8	42.8	.8863
700.	1330.	57.3	-159.4	-153.8	-145.8	-136.5	-126.0	61.6	1.1001
800.	1203.	85.1	-236.3	-229.9	-220.7	-210.0	-198.1	84.6	1.3377
900.	1108.	121.9	-337.5	-330.3	-320.0	-308.0	-294.5	111.6	1.5985
1000.	1043.	168.7	-466.9	-458.9	-447.5	-434.1	-419.2	142.1	1.8789

BALLISTIC COEFFICIENT .36, MUZZLE VELOCITY FT/SEC 2900.

RANGE	VELOCITY	HEIGHT	100 YD	150 YD	200 YD	250 YD	300 YD	DRIFT	TIME
0.	2900.	0.0	-1.5	-1.5	-1.5	-1.5	-1.5	0.0	0.0000
100.	2641.	-.2	0.0	.7	1.8	3.0	4.4	.9	.1084
200.	2396.	1.7	-3.5	-2.1	0.0	2.5	5.2	3.7	.2277
300.	2163.	5.4	-13.1	-10.9	-7.8	-4.1	0.0	8.6	.3593
400.	1943.	11.5	-30.2	-27.3	-23.1	-18.2	-12.7	16.2	.5058
500.	1739.	20.7	-56.4	-52.8	-47.6	-41.5	-34.6	26.7	.6690
600.	1553.	34.1	-94.2	-89.9	-83.7	-76.3	-68.0	40.7	.8517
700.	1385.	52.8	-146.4	-141.4	-134.1	-125.5	-115.8	58.5	1.0563
800.	1248.	78.5	-217.2	-211.4	-203.1	-193.2	-182.2	80.5	1.2850
900.	1139.	112.7	-310.5	-304.0	-294.6	-283.6	-271.2	106.7	1.5374
1000.	1065.	156.9	-431.7	-424.5	-414.1	-401.8	-388.0	136.9	1.8124

BALLISTIC COEFFICIENT .36, MUZZLE VELOCITY FT/SEC 3000.

RANGE	VELOCITY	HEIGHT	100 YD	150 YD	200 YD	250 YD	300 YD	DRIFT	TIME
0.	3000.	0.0	-1.5	-1.5	-1.5	-1.5	-1.5	0.0	0.0000
100.	2735.	-.2	0.0	.6	1.6	2.7	4.0	.9	.1049
200.	2485.	1.6	-3.2	-1.9	0.0	2.3	4.8	3.5	.2200
300.	2248.	5.0	-12.0	-10.1	-7.3	-3.8	0.0	8.3	.3470
400.	2023.	10.7	-27.8	-25.2	-21.4	-16.8	-11.7	15.4	.4877
500.	1812.	19.2	-52.0	-48.8	-44.0	-38.3	-31.9	25.4	.6443
600.	1620.	31.5	-86.8	-83.0	-77.2	-70.4	-62.7	38.6	.8194
700.	1444.	48.8	-134.9	-130.5	-123.8	-115.8	-106.8	55.6	1.0159
800.	1295.	72.5	-200.0	-195.0	-187.3	-178.2	-167.9	76.7	1.2357
900.	1174.	104.4	-286.5	-280.8	-272.2	-261.9	-250.4	102.1	1.4803
1000.	1090.	145.4	-397.7	-391.4	-381.8	-370.4	-357.6	131.1	1.7450

BALLISTIC COEFFICIENT .36, MUZZLE VELOCITY FT/SEC 3100.

RANGE	VELOCITY	HEIGHT	100 YD	150 YD	200 YD	250 YD	300 YD	DRIFT	TIME
0.	3100.	0.0	-1.5	-1.5	-1.5	-1.5	-1.5	0.0	0.0000
100.	2829.	-.3	0.0	.6	1.4	2.5	3.7	.8	.1011
200.	2574.	1.4	-2.9	-1.8	0.0	2.1	4.5	3.3	.2124
300.	2332.	4.6	-11.0	-9.4	-6.7	-3.6	0.0	7.8	.3349
400.	2103.	9.9	-25.6	-23.3	-19.8	-15.6	-10.9	14.7	.4704
500.	1886.	17.8	-48.0	-45.2	-40.8	-35.5	-29.6	24.1	.6209
600.	1687.	29.1	-80.1	-76.8	-71.5	-65.2	-58.1	36.7	.7891
700.	1505.	45.1	-124.5	-120.6	-114.5	-107.1	-98.8	52.8	.9773
800.	1345.	67.1	-184.5	-180.0	-173.0	-164.6	-155.1	72.9	1.1886
900.	1215.	96.6	-264.1	-259.1	-251.2	-241.8	-231.1	97.3	1.4238
1000.	1117.	135.0	-367.6	-362.0	-353.2	-342.7	-330.8	125.7	1.6819

BALLISTIC COEFFICIENT .36, MUZZLE VELOCITY FT/SEC 3200.

RANGE	VELOCITY	HEIGHT	100 YD	150 YD	200 YD	250 YD	300 YD	DRIFT	TIME
0.	3200.	0.0	-1.5	-1.5	-1.5	-1.5	-1.5	0.0	0.0000
100.	2923.	-.3	0.0	.5	1.3	2.3	3.4	.8	.0982
200.	2662.	1.3	-2.6	-1.6	0.0	2.0	4.2	3.2	.2057
300.	2416.	4.3	-10.1	-8.6	-6.2	-3.3	0.0	7.5	.3240
400.	2183.	9.2	-23.6	-21.6	-18.4	-14.5	-10.1	14.0	.4545
500.	1961.	16.5	-44.4	-41.9	-37.9	-33.0	-27.5	23.0	.5997
600.	1756.	27.1	-74.1	-71.2	-66.4	-60.5	-53.9	35.0	.7612
700.	1568.	41.9	-115.3	-111.9	-106.3	-99.4	-91.7	50.4	.9423
800.	1398.	62.2	-170.6	-166.7	-160.3	-152.4	-143.6	69.5	1.1450
900.	1258.	89.5	-244.2	-239.8	-232.6	-223.7	-213.8	92.9	1.3715
1000.	1147.	125.5	-340.1	-335.2	-327.2	-317.4	-306.4	120.4	1.6218

BALLISTIC COEFFICIENT .36, MUZZLE VELOCITY FT/SEC 3300.

RANGE	VELOCITY	HEIGHT	100 YD	150 YD	200 YD	250 YD	300 YD	DRIFT	TIME
0.	3300.	0.0	-1.5	-1.5	-1.5	-1.5	-1.5	0.0	0.0000
100.	3017.	-.3	0.0	.4	1.2	2.1	3.1	.7	.0951
200.	2751.	1.2	-2.4	-1.5	0.0	1.8	3.9	3.1	.1994
300.	2500.	4.0	-9.3	-8.0	-5.8	-3.1	0.0	7.2	.3138
400.	2262.	8.5	-21.9	-20.1	-17.2	-13.6	-9.4	13.4	.4400
500.	2037.	15.4	-41.2	-39.0	-35.3	-30.8	-25.6	22.0	.5798
600.	1825.	25.2	-68.8	-66.3	-61.8	-56.4	-50.2	33.4	.7353
700.	1631.	38.9	-107.0	-104.0	-98.7	-92.4	-85.2	48.0	.9092
800.	1454.	57.8	-158.3	-154.8	-148.9	-141.7	-133.4	66.3	1.1042
900.	1303.	83.2	-226.3	-222.5	-215.7	-207.6	-198.4	88.7	1.3222
1000.	1181.	116.7	-315.3	-311.0	-303.5	-294.5	-284.2	115.3	1.5644

BALLISTIC COEFFICIENT .36, MUZZLE VELOCITY FT/SEC 3400.

RANGE	VELOCITY	HEIGHT	100 YD	150 YD	200 YD	250 YD	300 YD	DRIFT	TIME
0.	3400.	0.0	-1.5	-1.5	-1.5	-1.5	-1.5	0.0	0.0000
100.	3111.	-.3	0.0	.4	1.0	1.9	2.8	.7	.0923
200.	2839.	1.0	-2.1	-1.4	0.0	1.7	3.6	2.9	.1929
300.	2584.	3.7	-8.5	-7.4	-5.4	-2.9	0.0	6.9	.3039
400.	2342.	8.0	-20.2	-18.7	-16.0	-12.6	-8.8	12.9	.4260
500.	2112.	14.3	-38.2	-36.3	-33.0	-28.7	-23.9	21.0	.5607
600.	1895.	23.5	-63.9	-61.7	-57.7	-52.6	-46.8	31.9	.7107
700.	1695.	36.3	-99.4	-96.8	-92.1	-86.1	-79.4	45.8	.8781
800.	1512.	53.8	-147.0	-144.0	-138.6	-131.8	-124.2	63.3	1.0656
900.	1351.	77.4	-210.1	-206.8	-200.7	-193.1	-184.5	84.8	1.2759
1000.	1220.	108.6	-292.7	-289.0	-282.2	-273.8	-264.2	110.4	1.5097

BALLISTIC COEFFICIENT .36, MUZZLE VELOCITY FT/SEC 3500.

RANGE	VELOCITY	HEIGHT	100 YD	150 YD	200 YD	250 YD	300 YD	DRIFT	TIME
0.	3500.	0.0	-1.5	-1.5	-1.5	-1.5	-1.5	0.0	0.0000
100.	3205.	-.4	0.0	.3	1.0	1.7	2.6	.7	.0897
200.	2928.	.9	-1.9	-1.3	0.0	1.6	3.3	2.9	.1877
300.	2667.	3.4	-7.9	-6.9	-5.0	-2.7	0.0	6.7	.2950
400.	2421.	7.4	-18.8	-17.5	-15.0	-11.8	-8.3	12.4	.4131
500.	2187.	13.4	-35.5	-33.9	-30.8	-26.8	-22.4	20.2	.5433
600.	1965.	22.0	-59.6	-57.7	-53.9	-49.2	-43.8	30.6	.6882
700.	1759.	33.9	-92.6	-90.3	-85.9	-80.4	-74.2	43.9	.8494
800.	1571.	50.2	-137.0	-134.4	-129.4	-123.1	-116.0	60.6	1.0302
900.	1401.	72.1	-195.5	-192.6	-186.9	-179.8	-171.8	81.1	1.2323
1000.	1261.	101.3	-272.3	-269.1	-262.8	-254.9	-246.1	105.8	1.4583

BALLISTIC COEFFICIENT .36, MUZZLE VELOCITY FT/SEC 3600.

RANGE	VELOCITY	HEIGHT	100 YD	150 YD	200 YD	250 YD	300 YD	DRIFT	TIME
0.	3600.	0.0	-1.5	-1.5	-1.5	-1.5	-1.5	0.0	0.0000
100.	3300.	-.4	0.0	.3	.9	1.6	2.4	.6	.0869
200.	3017.	.8	-1.7	-1.1	0.0	1.5	3.1	2.7	.1819
300.	2750.	3.2	-7.3	-6.4	-4.7	-2.5	0.0	6.4	.2862
400.	2500.	7.0	-17.4	-16.3	-14.0	-11.1	-7.7	11.9	.4007
500.	2262.	12.6	-33.1	-31.7	-28.9	-25.2	-21.0	19.4	.5269
600.	2036.	20.6	-55.6	-53.9	-50.5	-46.1	-41.1	29.3	.6667
700.	1825.	31.7	-86.4	-84.4	-80.4	-75.4	-69.5	42.0	.8222
800.	1631.	46.9	-127.8	-125.5	-120.9	-115.1	-108.4	58.0	.9961
900.	1454.	67.4	-182.3	-179.8	-174.6	-168.1	-160.5	77.6	1.1912
1000.	1303.	94.6	-253.7	-250.9	-245.2	-237.9	-229.5	101.4	1.4092

BALLISTIC COEFFICIENT .36, MUZZLE VELOCITY FT/SEC 3700.

RANGE	VELOCITY	HEIGHT	100 YD	150 YD	200 YD	250 YD	300 YD	DRIFT	TIME
0.	3700.	0.0	-1.5	-1.5	-1.5	-1.5	-1.5	0.0	0.0000
100.	3394.	-.4	0.0	.2	.8	1.4	2.2	.7	.0849
200.	3106.	.8	-1.5	-1.1	0.0	1.3	2.9	2.7	.1774
300.	2834.	3.0	-6.7	-6.0	-4.4	-2.3	0.0	6.2	.2783
400.	2579.	6.5	-16.2	-15.3	-13.1	-10.4	-7.3	11.5	.3894
500.	2337.	11.8	-30.9	-29.7	-27.1	-23.7	-19.8	18.7	.5117
600.	2107.	19.3	-51.9	-50.5	-47.3	-43.3	-38.6	28.2	.6468
700.	1891.	29.7	-80.8	-79.1	-75.4	-70.7	-65.2	40.4	.7970
800.	1691.	43.9	-119.4	-117.5	-113.2	-107.8	-101.6	55.7	.9649
900.	1509.	63.0	-170.2	-168.2	-163.3	-157.3	-150.2	74.4	1.1527
1000.	1348.	88.5	-236.8	-234.5	-229.1	-222.4	-214.6	97.3	1.3635

BALLISTIC COEFFICIENT .36, MUZZLE VELOCITY FT/SEC 3800.

RANGE	VELOCITY	HEIGHT	100 YD	150 YD	200 YD	250 YD	300 YD	DRIFT	TIME
0.	3800.	0.0	-1.5	-1.5	-1.5	-1.5	-1.5	0.0	0.0000
100.	3488.	-.4	0.0	.2	.7	1.3	2.1	.6	.0825
200.	3195.	.7	-1.4	-1.0	0.0	1.3	2.8	2.6	.1724
300.	2918.	2.8	-6.2	-5.6	-4.1	-2.2	0.0	6.0	.2709
400.	2657.	6.1	-15.1	-14.3	-12.3	-9.8	-6.8	11.0	.3785
500.	2412.	11.1	-28.9	-27.9	-25.4	-22.3	-18.6	18.0	.4971
600.	2178.	18.2	-48.6	-47.4	-44.5	-40.7	-36.2	27.1	.6277
700.	1957.	28.0	-75.7	-74.3	-70.9	-66.4	-61.2	38.8	.7733
800.	1752.	41.2	-111.8	-110.2	-106.3	-101.2	-95.3	53.4	.9352
900.	1564.	59.1	-159.4	-157.6	-153.2	-147.6	-140.9	71.5	1.1167
1000.	1395.	82.9	-221.5	-219.5	-214.6	-208.3	-200.8	93.4	1.3199

BALLISTIC COEFFICIENT .36, MUZZLE VELOCITY FT/SEC 3900.

RANGE	VELOCITY	HEIGHT	100 YD	150 YD	200 YD	250 YD	300 YD	DRIFT	TIME
0.	3900.	0.0	-1.5	-1.5	-1.5	-1.5	-1.5	0.0	0.0000
100.	3583.	-.4	0.0	.2	.6	1.2	1.9	.6	.0802
200.	3284.	.6	-1.2	-.9	0.0	1.2	2.6	2.4	.1677
300.	3001.	2.6	-5.7	-5.2	-3.8	-2.0	0.0	5.7	.2630
400.	2736.	5.7	-14.0	-13.4	-11.6	-9.2	-6.5	10.6	.3678
500.	2486.	10.4	-27.0	-26.2	-23.9	-20.9	-17.5	17.3	.4828
600.	2249.	17.1	-45.5	-44.6	-41.9	-38.3	-34.2	26.1	.6098
700.	2024.	26.3	-70.9	-69.8	-66.6	-62.5	-57.7	37.3	.7504
800.	1813.	38.7	-104.7	-103.5	-99.9	-95.1	-89.7	51.3	.9069
900.	1621.	55.4	-149.3	-147.9	-143.8	-138.4	-132.3	68.6	1.0819
1000.	1445.	77.7	-207.3	-205.8	-201.2	-195.3	-188.5	89.6	1.2783

BALLISTIC COEFFICIENT .36, MUZZLE VELOCITY FT/SEC 4000.

RANGE	VELOCITY	HEIGHT	100 YD	150 YD	200 YD	250 YD	300 YD	DRIFT	TIME
0.	4000.	0.0	-1.5	-1.5	-1.5	-1.5	-1.5	0.0	0.0000
100.	3677.	-.5	0.0	.1	.5	1.1	1.7	.6	.0784
200.	3373.	.5	-1.1	-.8	0.0	1.1	2.4	2.4	.1635
300.	3085.	2.4	-5.2	-4.9	-3.6	-1.9	0.0	5.6	.2566
400.	2815.	5.4	-13.1	-12.6	-10.9	-8.7	-6.1	10.3	.3584
500.	2561.	9.9	-25.2	-24.6	-22.5	-19.8	-16.5	16.7	.4701
600.	2320.	16.1	-42.7	-42.0	-39.5	-36.1	-32.2	25.2	.5932
700.	2091.	24.8	-66.6	-65.8	-62.8	-58.9	-54.4	36.0	.7294
800.	1876.	36.5	-98.4	-97.4	-94.1	-89.6	-84.4	49.4	.8808
900.	1677.	52.2	-140.2	-139.1	-135.3	-130.3	-124.5	66.0	1.0501
1000.	1496.	73.0	-194.5	-193.3	-189.1	-183.5	-177.0	86.1	1.2394

BALLISTIC COEFFICIENT .37, MUZZLE VELOCITY FT/SEC 2000.

RANGE	VELOCITY	HEIGHT	100 YD	150 YD	200 YD	250 YD	300 YD	DRIFT	TIME
0.	2000.	0.0	-1.5	-1.5	-1.5	-1.5	-1.5	0.0	0.0000
100.	1797.	.5	0.0	2.1	4.7	7.6	10.7	1.5	.1583
200.	1610.	4.6	-9.3	-5.1	0.0	5.8	12.1	6.1	.3347
300.	1440.	12.8	-32.2	-25.9	-18.2	-9.5	0.0	14.4	.5319
400.	1295.	26.4	-71.9	-63.5	-53.2	-41.7	-29.0	26.7	.7519
500.	1178.	46.8	-132.3	-121.8	-108.9	-94.5	-78.7	43.2	.9954
600.	1094.	75.5	-217.6	-205.0	-189.6	-172.3	-153.3	63.4	1.2600
700.	1034.	113.5	-331.3	-316.5	-298.6	-278.4	-256.2	86.7	1.5425
800.	985.	162.0	-475.8	-459.0	-438.4	-415.4	-390.0	112.9	1.8413
900.	940.	221.5	-651.4	-632.4	-609.3	-583.4	-554.8	141.1	2.1518
1000.	899.	294.1	-863.7	-842.6	-816.9	-788.2	-756.4	172.2	2.4786

BALLISTIC COEFFICIENT .37, MUZZLE VELOCITY FT/SEC 2100.

RANGE	VELOCITY	HEIGHT	100 YD	150 YD	200 YD	250 YD	300 YD	DRIFT	TIME
0.	2100.	0.0	-1.5	-1.5	-1.5	-1.5	-1.5	0.0	0.0000
100.	1889.	.3	0.0	1.9	4.2	6.7	9.6	1.3	.1504
200.	1695.	4.1	-8.3	-4.6	0.0	5.2	10.9	5.7	.3181
300.	1517.	11.5	-28.8	-23.2	-16.3	-8.5	0.0	13.5	.5053
400.	1359.	23.8	-64.4	-57.0	-47.8	-37.5	-26.1	25.2	.7146
500.	1230.	42.3	-118.9	-109.6	-98.2	-85.2	-71.0	40.9	.9469
600.	1129.	68.6	-196.4	-185.2	-171.5	-155.9	-138.9	60.7	1.2023
700.	1060.	103.9	-300.8	-287.7	-271.7	-253.6	-233.7	83.9	1.4768
800.	1007.	149.6	-435.6	-420.7	-402.4	-381.6	-358.9	110.3	1.7695
900.	960.	205.4	-599.2	-582.4	-561.8	-538.5	-512.9	138.5	2.0725
1000.	917.	274.0	-798.1	-779.4	-756.5	-730.6	-702.1	169.6	2.3924

BALLISTIC COEFFICIENT .37, MUZZLE VELOCITY FT/SEC 2200.

RANGE	VELOCITY	HEIGHT	100 YD	150 YD	200 YD	250 YD	300 YD	DRIFT	TIME
0.	2200.	0.0	-1.5	-1.5	-1.5	-1.5	-1.5	0.0	0.0000
100.	1983.	.2	0.0	1.6	3.7	6.0	8.6	1.3	.1436
200.	1781.	3.7	-7.4	-4.1	0.0	4.6	9.8	5.4	.3034
300.	1596.	10.4	-25.9	-20.9	-14.7	-7.7	0.0	12.7	.4814
400.	1427.	21.5	-57.9	-51.3	-43.1	-33.8	-23.4	23.7	.6801
500.	1285.	38.3	-107.2	-98.9	-88.6	-77.0	-64.1	38.7	.9020
600.	1170.	62.4	-177.4	-167.6	-155.2	-141.2	-125.7	57.9	1.1471
700.	1088.	95.2	-273.2	-261.7	-247.2	-230.9	-212.9	80.8	1.4136
800.	1030.	137.9	-398.3	-385.2	-368.7	-350.1	-329.4	107.1	1.6997
900.	981.	190.5	-551.2	-536.5	-517.8	-496.9	-473.7	135.3	1.9961
1000.	936.	255.2	-737.6	-721.1	-700.5	-677.2	-651.4	166.4	2.3091

BALLISTIC COEFFICIENT .37, MUZZLE VELOCITY FT/SEC 2300.

RANGE	VELOCITY	HEIGHT	100 YD	150 YD	200 YD	250 YD	300 YD	DRIFT	TIME
0.	2300.	0.0	-1.5	-1.5	-1.5	-1.5	-1.5	0.0	0.0000
100.	2078.	.2	0.0	1.5	3.3	5.4	7.7	1.2	.1370
200.	1869.	3.3	-6.6	-3.7	0.0	4.2	8.9	5.0	.2893
300.	1676.	9.4	-23.2	-18.9	-13.3	-7.0	0.0	11.9	.4589
400.	1500.	19.4	-52.2	-46.4	-39.0	-30.5	-21.2	22.2	.6480
500.	1345.	34.7	-96.8	-89.5	-80.3	-69.7	-58.0	36.5	.8597
600.	1218.	56.7	-160.6	-151.8	-140.7	-128.1	-114.1	54.9	1.0944
700.	1120.	87.0	-247.9	-237.6	-224.7	-209.9	-193.6	77.2	1.3519
800.	1054.	126.5	-362.5	-350.8	-336.1	-319.1	-300.5	102.9	1.6284
900.	1002.	176.4	-506.6	-493.4	-476.8	-457.8	-436.8	131.5	1.9211
1000.	956.	237.3	-681.0	-666.4	-647.9	-626.8	-603.5	162.4	2.2271

BALLISTIC COEFFICIENT .37, MUZZLE VELOCITY FT/SEC 2400.

RANGE	VELOCITY	HEIGHT	100 YD	150 YD	200 YD	250 YD	300 YD	DRIFT	TIME
0.	2400.	0.0	-1.5	-1.5	-1.5	-1.5	-1.5	0.0	0.0000
100.	2173.	.1	0.0	1.3	3.0	4.9	7.0	1.1	.1311
200.	1958.	2.9	-5.9	-3.3	0.0	3.8	8.0	4.7	.2768
300.	1758.	8.5	-20.9	-17.1	-12.0	-6.3	0.0	11.1	.4383
400.	1575.	17.6	-47.3	-42.1	-35.4	-27.8	-19.4	20.9	.6190
500.	1409.	31.5	-87.6	-81.1	-72.8	-63.2	-52.7	34.4	.8203
600.	1270.	51.7	-145.6	-137.8	-127.8	-116.4	-103.7	51.9	1.0449
700.	1158.	79.5	-225.3	-216.2	-204.5	-191.2	-176.4	73.6	1.2933
800.	1080.	116.3	-330.6	-320.2	-306.9	-291.7	-274.8	98.8	1.5615
900.	1024.	163.2	-465.2	-453.5	-438.5	-421.4	-402.3	127.3	1.8482
1000.	975.	220.6	-628.6	-615.6	-598.9	-579.9	-558.8	158.0	2.1476

BALLISTIC COEFFICIENT .37, MUZZLE VELOCITY FT/SEC 2500.

RANGE	VELOCITY	HEIGHT	100 YD	150 YD	200 YD	250 YD	300 YD	DRIFT	TIME
0.	2500.	0.0	-1.5	-1.5	-1.5	-1.5	-1.5	0.0	0.0000
100.	2268.	.0	0.0	1.1	2.7	4.4	6.3	1.0	.1260
200.	2048.	2.6	-5.3	-3.0	0.0	3.5	7.3	4.4	.2652
300.	1841.	7.7	-19.0	-15.5	-11.0	-5.8	0.0	10.5	.4198
400.	1651.	16.1	-42.9	-38.3	-32.2	-25.3	-17.6	19.7	.5917
500.	1477.	28.7	-79.5	-73.8	-66.2	-57.5	-47.9	32.3	.7838
600.	1326.	47.1	-132.2	-125.4	-116.3	-105.9	-94.3	49.0	.9986
700.	1202.	72.7	-204.8	-196.9	-186.2	-174.1	-160.6	69.8	1.2366
800.	1110.	106.8	-301.7	-292.6	-280.4	-266.6	-251.1	94.5	1.4971
900.	1046.	151.0	-427.2	-417.0	-403.3	-387.7	-370.3	122.9	1.7781
1000.	996.	205.0	-579.8	-568.4	-553.2	-535.9	-516.6	153.1	2.0701

BALLISTIC COEFFICIENT .37, MUZZLE VELOCITY FT/SEC 2600.

RANGE	VELOCITY	HEIGHT	100 YD	150 YD	200 YD	250 YD	300 YD	DRIFT	TIME
0.	2600.	0.0	-1.5	-1.5	-1.5	-1.5	-1.5	0.0	0.0000
100.	2364.	-.0	0.0	1.0	2.4	4.0	5.7	1.0	.1211
200.	2139.	2.4	-4.8	-2.7	0.0	3.2	6.7	4.2	.2546
300.	1926.	7.0	-17.2	-14.1	-10.0	-5.3	0.0	9.9	.4024
400.	1728.	14.7	-38.9	-34.9	-29.4	-23.1	-16.0	18.5	.5666
500.	1548.	26.3	-72.4	-67.4	-60.5	-52.6	-43.8	30.5	.7503
600.	1385.	43.1	-120.4	-114.3	-106.1	-96.6	-86.0	46.3	.9553
700.	1251.	66.5	-186.7	-179.6	-170.0	-158.9	-146.6	66.2	1.1836
800.	1144.	98.1	-275.3	-267.2	-256.2	-243.6	-229.5	90.1	1.4349
900.	1071.	139.1	-390.6	-381.5	-369.2	-354.9	-339.1	117.6	1.7067
1000.	1016.	190.2	-534.5	-524.3	-510.6	-494.8	-477.2	148.0	1.9947

BALLISTIC COEFFICIENT .37, MUZZLE VELOCITY FT/SEC 2700.

RANGE	VELOCITY	HEIGHT	100 YD	150 YD	200 YD	250 YD	300 YD	DRIFT	TIME
0.	2700.	0.0	-1.5	-1.5	-1.5	-1.5	-1.5	0.0	0.0000
100.	2459.	-.1	0.0	.9	2.1	3.6	5.2	.9	.1164
200.	2229.	2.1	-4.3	-2.5	0.0	2.9	6.1	3.9	.2446
300.	2011.	6.4	-15.6	-12.9	-9.2	-4.9	0.0	9.3	.3863
400.	1807.	13.4	-35.6	-31.9	-27.0	-21.2	-14.7	17.5	.5436
500.	1620.	24.1	-66.1	-61.6	-55.4	-48.2	-40.1	28.8	.7190
600.	1448.	39.5	-110.0	-104.6	-97.2	-88.5	-78.8	43.7	.9152
700.	1302.	60.9	-170.6	-164.2	-155.5	-145.5	-134.1	62.6	1.1337
800.	1183.	90.1	-251.8	-244.6	-234.7	-223.1	-210.2	85.7	1.3759
900.	1097.	128.3	-358.2	-350.1	-338.9	-326.0	-311.4	112.6	1.6399
1000.	1037.	176.5	-492.5	-483.4	-471.0	-456.6	-440.4	142.6	1.9214

BALLISTIC COEFFICIENT .37, MUZZLE VELOCITY FT/SEC 2800.

RANGE	VELOCITY	HEIGHT	100 YD	150 YD	200 YD	250 YD	300 YD	DRIFT	TIME
0.	2800.	0.0	-1.5	-1.5	-1.5	-1.5	-1.5	0.0	0.0000
100.	2553.	-.1	0.0	.8	1.9	3.3	4.7	.9	.1121
200.	2319.	1.9	-3.9	-2.3	0.0	2.7	5.6	3.7	.2353
300.	2097.	5.9	-14.2	-11.9	-8.4	-4.5	0.0	8.8	.3714
400.	1886.	12.3	-32.5	-29.3	-24.8	-19.5	-13.5	16.5	.5222
500.	1692.	22.1	-60.6	-56.6	-50.9	-44.3	-36.9	27.2	.6902
600.	1515.	36.2	-100.8	-96.0	-89.2	-81.2	-72.3	41.3	.8776
700.	1357.	56.0	-156.3	-150.8	-142.8	-133.5	-123.1	59.4	1.0875
800.	1228.	82.9	-230.9	-224.5	-215.4	-204.8	-192.9	81.5	1.3200
900.	1128.	118.4	-328.8	-321.7	-311.4	-299.5	-286.1	107.6	1.5757
1000.	1059.	163.6	-453.8	-445.9	-434.5	-421.2	-406.4	137.1	1.8505

BALLISTIC COEFFICIENT .37, MUZZLE VELOCITY FT/SEC 2900.

RANGE	VELOCITY	HEIGHT	100 YD	150 YD	200 YD	250 YD	300 YD	DRIFT	TIME
0.	2900.	0.0	-1.5	-1.5	-1.5	-1.5	-1.5	0.0	0.0000
100.	2648.	-.2	0.0	.7	1.7	3.0	4.3	.8	.1082
200.	2409.	1.7	-3.5	-2.1	0.0	2.4	5.2	3.6	.2271
300.	2182.	5.4	-13.0	-10.8	-7.7	-4.1	0.0	8.3	.3577
400.	1966.	11.4	-29.8	-27.0	-22.8	-18.0	-12.5	15.7	.5028
500.	1766.	20.4	-55.6	-52.1	-46.9	-40.8	-34.0	25.8	.6637
600.	1582.	33.4	-92.7	-88.4	-82.2	-74.9	-66.7	39.2	.8435
700.	1415.	51.6	-143.5	-138.5	-131.3	-122.7	-113.2	56.3	1.0439
800.	1275.	76.4	-212.2	-206.5	-198.2	-188.5	-177.6	77.5	1.2680
900.	1162.	109.4	-302.3	-295.8	-286.5	-275.5	-263.3	102.7	1.5147
1000.	1083.	151.8	-418.5	-411.4	-401.0	-388.8	-375.3	131.7	1.7827

BALLISTIC COEFFICIENT .37, MUZZLE VELOCITY FT/SEC 3000.

RANGE	VELOCITY	HEIGHT	100 YD	150 YD	200 YD	250 YD	300 YD	DRIFT	TIME
0.	3000.	0.0	-1.5	-1.5	-1.5	-1.5	-1.5	0.0	0.0000
100.	2742.	-.2	0.0	.6	1.6	2.7	4.0	.8	.1048
200.	2498.	1.6	-3.2	-1.9	0.0	2.3	4.8	3.4	.2195
300.	2267.	5.0	-11.9	-10.0	-7.2	-3.8	0.0	8.0	.3456
400.	2047.	10.5	-27.5	-25.0	-21.1	-16.6	-11.6	14.9	.4849
500.	1840.	18.9	-51.3	-48.2	-43.4	-37.8	-31.4	24.6	.6396
600.	1650.	30.9	-85.4	-81.6	-75.9	-69.2	-61.6	37.3	.8117
700.	1476.	47.6	-132.2	-127.9	-121.2	-113.3	-104.4	53.5	1.0039
800.	1325.	70.6	-195.3	-190.3	-182.7	-173.7	-163.6	73.7	1.2189
900.	1202.	101.2	-278.6	-273.0	-264.3	-254.2	-242.8	98.0	1.4571
1000.	1109.	140.9	-386.3	-380.1	-370.5	-359.3	-346.6	126.3	1.7177

BALLISTIC COEFFICIENT .37, MUZZLE VELOCITY FT/SEC 3100.

RANGE	VELOCITY	HEIGHT	100 YD	150 YD	200 YD	250 YD	300 YD	DRIFT	TIME
0.	3100.	0.0	-1.5	-1.5	-1.5	-1.5	-1.5	0.0	0.0000
100.	2836.	-.3	0.0	.6	1.4	2.5	3.6	.7	.1009
200.	2587.	1.4	-2.9	-1.7	0.0	2.1	4.4	3.2	.2119
300.	2352.	4.6	-10.9	-9.3	-6.7	-3.5	0.0	7.6	.3336
400.	2127.	9.7	-25.3	-23.1	-19.6	-15.4	-10.7	14.2	.4676
500.	1915.	17.5	-47.4	-44.6	-40.2	-35.0	-29.1	23.3	.6164
600.	1718.	28.6	-78.8	-75.5	-70.2	-64.0	-56.9	35.3	.7815
700.	1539.	44.1	-122.2	-118.3	-112.2	-104.9	-96.6	50.8	.9662
800.	1377.	65.3	-180.3	-175.9	-168.9	-160.6	-151.1	70.1	1.1727
900.	1245.	93.6	-257.2	-252.2	-244.3	-235.0	-224.4	93.5	1.4022
1000.	1139.	130.7	-356.9	-351.3	-342.6	-332.2	-320.4	120.9	1.6548

BALLISTIC COEFFICIENT .37, MUZZLE VELOCITY FT/SEC 3200.

RANGE	VELOCITY	HEIGHT	100 YD	150 YD	200 YD	250 YD	300 YD	DRIFT	TIME
0.	3200.	0.0	-1.5	-1.5	-1.5	-1.5	-1.5	0.0	0.0000
100.	2930.	-.3	0.0	.5	1.3	2.3	3.3	.7	.0980
200.	2676.	1.3	-2.6	-1.6	0.0	1.9	4.1	3.1	.2050
300.	2436.	4.2	-10.0	-8.5	-6.2	-3.3	0.0	7.3	.3226
400.	2208.	9.1	-23.3	-21.4	-18.2	-14.3	-10.0	13.6	.4520
500.	1990.	16.2	-43.8	-41.3	-37.4	-32.5	-27.1	22.2	.5951
600.	1788.	26.6	-73.0	-70.1	-65.3	-59.5	-53.0	33.8	.7543
700.	1602.	40.9	-113.1	-109.7	-104.2	-97.4	-89.8	48.5	.9316
800.	1433.	60.5	-166.8	-162.9	-156.6	-148.8	-140.1	66.8	1.1297
900.	1290.	86.8	-237.8	-233.4	-226.3	-217.6	-207.8	89.2	1.3507
1000.	1173.	121.6	-330.8	-325.9	-318.0	-308.3	-297.4	116.0	1.5966

BALLISTIC COEFFICIENT .37, MUZZLE VELOCITY FT/SEC 3300.

RANGE	VELOCITY	HEIGHT	100 YD	150 YD	200 YD	250 YD	300 YD	DRIFT	TIME
0.	3300.	0.0	-1.5	-1.5	-1.5	-1.5	-1.5	0.0	0.0000
100.	3024.	-.3	0.0	.4	1.2	2.1	3.1	.7	.0949
200.	2765.	1.1	-2.3	-1.5	0.0	1.8	3.8	3.0	.1989
300.	2520.	3.9	-9.2	-8.0	-5.7	-3.1	0.0	7.0	.3126
400.	2288.	8.4	-21.6	-19.9	-16.9	-13.4	-9.3	13.0	.4374
500.	2066.	15.1	-40.6	-38.5	-34.8	-30.3	-25.2	21.3	.5754
600.	1858.	24.7	-67.7	-65.1	-60.7	-55.3	-49.2	32.2	.7284
700.	1666.	38.1	-105.0	-102.0	-96.8	-90.6	-83.4	46.2	.8991
800.	1491.	56.2	-154.7	-151.3	-145.4	-138.2	-130.1	63.7	1.0894
900.	1337.	80.7	-220.5	-216.7	-210.0	-202.0	-192.8	85.2	1.3024
1000.	1212.	112.8	-306.1	-301.8	-294.4	-285.5	-275.3	110.7	1.5383

BALLISTIC COEFFICIENT .37, MUZZLE VELOCITY FT/SEC 3400.

RANGE	VELOCITY	HEIGHT	100 YD	150 YD	200 YD	250 YD	300 YD	DRIFT	TIME
0.	3400.	0.0	-1.5	-1.5	-1.5	-1.5	-1.5	0.0	0.0000
100.	3119.	-.3	0.0	.4	1.0	1.9	2.8	.7	.0922
200.	2853.	1.0	-2.1	-1.3	0.0	1.7	3.6	2.8	.1925
300.	2604.	3.6	-8.5	-7.4	-5.4	-2.8	0.0	6.7	.3027
400.	2367.	7.9	-20.0	-18.5	-15.8	-12.5	-8.7	12.4	.4236
500.	2142.	14.1	-37.7	-35.9	-32.5	-28.3	-23.6	20.4	.5569
600.	1929.	23.1	-63.0	-60.8	-56.7	-51.7	-46.0	30.8	.7045
700.	1731.	35.5	-97.5	-95.0	-90.3	-84.4	-77.8	44.1	.8685
800.	1550.	52.4	-143.8	-140.9	-135.5	-128.8	-121.2	60.9	1.0518
900.	1387.	75.0	-204.6	-201.3	-195.3	-187.8	-179.3	81.4	1.2564
1000.	1253.	105.1	-284.3	-280.7	-274.0	-265.6	-256.1	106.0	1.4848

BALLISTIC COEFFICIENT .37, MUZZLE VELOCITY FT/SEC 3500.

RANGE	VELOCITY	HEIGHT	100 YD	150 YD	200 YD	250 YD	300 YD	DRIFT	TIME
0.	3500.	0.0	-1.5	-1.5	-1.5	-1.5	-1.5	0.0	0.0000
100.	3213.	-.4	0.0	.3	.9	1.7	2.6	.7	.0896
200.	2942.	.9	-1.9	-1.2	0.0	1.6	3.3	2.7	.1870
300.	2688.	3.4	-7.8	-6.8	-5.0	-2.6	0.0	6.4	.2938
400.	2447.	7.3	-18.5	-17.3	-14.8	-11.7	-8.1	12.0	.4108
500.	2218.	13.2	-35.1	-33.5	-30.4	-26.5	-22.1	19.5	.5396
600.	2000.	21.6	-58.7	-56.7	-53.0	-48.3	-43.1	29.5	.6821
700.	1797.	33.1	-90.9	-88.7	-84.3	-78.9	-72.7	42.3	.8403
800.	1611.	48.9	-134.0	-131.4	-126.5	-120.2	-113.2	58.3	1.0167
900.	1440.	70.0	-190.6	-187.7	-182.1	-175.1	-167.2	77.9	1.2139
1000.	1296.	97.9	-264.5	-261.3	-255.1	-247.3	-238.5	101.5	1.4339

BALLISTIC COEFFICIENT .37, MUZZLE VELOCITY FT/SEC 3600.

RANGE	VELOCITY	HEIGHT	100 YD	150 YD	200 YD	250 YD	300 YD	DRIFT	TIME
0.	3600.	0.0	-1.5	-1.5	-1.5	-1.5	-1.5	0.0	0.0000
100.	3308.	-.4	0.0	.3	.8	1.6	2.4	.6	.0868
200.	3032.	.8	-1.7	-1.1	0.0	1.5	3.1	2.6	.1815
300.	2771.	3.1	-7.2	-6.4	-4.7	-2.5	0.0	6.2	.2850
400.	2526.	6.9	-17.2	-16.1	-13.9	-11.0	-7.7	11.5	.3986
500.	2294.	12.4	-32.7	-31.3	-28.5	-24.8	-20.7	18.7	.5231
600.	2072.	20.2	-54.7	-53.1	-49.7	-45.3	-40.3	28.3	.6607
700.	1863.	31.0	-84.8	-82.9	-78.9	-73.8	-68.0	40.5	.8133
800.	1671.	45.7	-125.0	-122.8	-118.3	-112.4	-105.9	55.8	.9835
900.	1495.	65.3	-177.7	-175.2	-170.1	-163.5	-156.1	74.5	1.1731
1000.	1341.	91.4	-246.4	-243.6	-237.9	-230.6	-222.4	97.2	1.3854

BALLISTIC COEFFICIENT .37, MUZZLE VELOCITY FT/SEC 3700.

RANGE	VELOCITY	HEIGHT	100 YD	150 YD	200 YD	250 YD	300 YD	DRIFT	TIME
0.	3700.	0.0	-1.5	-1.5	-1.5	-1.5	-1.5	0.0	0.0000
100.	3402.	-.4	0.0	.2	.8	1.4	2.2	.7	.0848
200.	3121.	.8	-1.5	-1.1	0.0	1.3	2.9	2.6	.1770
300.	2855.	2.9	-6.6	-5.9	-4.3	-2.3	0.0	6.0	.2772
400.	2606.	6.5	-16.0	-15.1	-12.9	-10.3	-7.2	11.1	.3873
500.	2369.	11.6	-30.5	-29.4	-26.7	-23.3	-19.5	18.1	.5082
600.	2144.	19.0	-51.2	-49.8	-46.6	-42.6	-37.9	27.2	.6413
700.	1930.	29.1	-79.3	-77.8	-74.0	-69.3	-63.9	38.9	.7888
800.	1733.	42.8	-116.8	-115.0	-110.7	-105.4	-99.2	53.5	.9527
900.	1552.	61.2	-166.1	-164.1	-159.2	-153.2	-146.3	71.5	1.1359
1000.	1388.	85.5	-230.0	-227.7	-222.3	-215.6	-207.9	93.2	1.3403

BALLISTIC COEFFICIENT .37, MUZZLE VELOCITY FT/SEC 3800.

RANGE	VELOCITY	HEIGHT	100 YD	150 YD	200 YD	250 YD	300 YD	DRIFT	TIME
0.	3800.	0.0	-1.5	-1.5	-1.5	-1.5	-1.5	0.0	0.0000
100.	3497.	-.4	0.0	.2	.7	1.3	2.0	.6	.0824
200.	3210.	.7	-1.4	-1.0	0.0	1.3	2.7	2.5	.1721
300.	2939.	2.7	-6.1	-5.5	-4.1	-2.2	0.0	5.8	.2696
400.	2685.	6.1	-14.9	-14.1	-12.2	-9.6	-6.8	10.7	.3764
500.	2444.	10.9	-28.5	-27.5	-25.1	-21.9	-18.3	17.4	.4935
600.	2215.	17.9	-47.9	-46.7	-43.8	-40.0	-35.7	26.2	.6225
700.	1998.	27.4	-74.3	-72.9	-69.5	-65.1	-60.1	37.4	.7652
800.	1795.	40.2	-109.5	-107.9	-104.0	-99.0	-93.2	51.4	.9236
900.	1609.	57.4	-155.5	-153.7	-149.4	-143.7	-137.2	68.6	1.1003
1000.	1438.	80.1	-215.2	-213.2	-208.4	-202.1	-194.9	89.5	1.2977

BALLISTIC COEFFICIENT .37, MUZZLE VELOCITY FT/SEC 3900.

RANGE	VELOCITY	HEIGHT	100 YD	150 YD	200 YD	250 YD	300 YD	DRIFT	TIME
0.	3900.	0.0	-1.5	-1.5	-1.5	-1.5	-1.5	0.0	0.0000
100.	3591.	-.4	0.0	.2	.6	1.2	1.9	.6	.0801
200.	3299.	.6	-1.2	-.9	0.0	1.2	2.5	2.3	.1671
300.	3024.	2.5	-5.6	-5.1	-3.8	-2.0	0.0	5.5	.2621
400.	2764.	5.7	-13.8	-13.2	-11.5	-9.1	-6.4	10.3	.3660
500.	2519.	10.3	-26.6	-25.9	-23.6	-20.7	-17.3	16.8	.4798
600.	2287.	16.8	-44.8	-43.9	-41.2	-37.7	-33.6	25.2	.6046
700.	2066.	25.7	-69.6	-68.6	-65.5	-61.3	-56.6	35.9	.7427
800.	1857.	37.8	-102.6	-101.3	-97.8	-93.0	-87.6	49.3	.8957
900.	1666.	53.8	-145.7	-144.3	-140.3	-135.0	-128.9	65.8	1.0664
1000.	1491.	75.1	-201.5	-199.9	-195.5	-189.6	-182.8	85.8	1.2567

BALLISTIC COEFFICIENT .37, MUZZLE VELOCITY FT/SEC 4000.

RANGE	VELOCITY	HEIGHT	100 YD	150 YD	200 YD	250 YD	300 YD	DRIFT	TIME
0.	4000.	0.0	-1.5	-1.5	-1.5	-1.5	-1.5	0.0	0.0000
100.	3686.	-.5	0.0	.1	.5	1.1	1.7	.6	.0783
200.	3389.	.5	-1.1	-.8	0.0	1.1	2.4	2.3	.1632
300.	3108.	2.4	-5.2	-4.8	-3.6	-1.9	0.0	5.4	.2558
400.	2843.	5.3	-12.9	-12.4	-10.7	-8.5	-6.0	9.9	.3564
500.	2594.	9.7	-24.9	-24.3	-22.3	-19.5	-16.3	16.2	.4670
600.	2358.	15.9	-42.1	-41.4	-38.9	-35.6	-31.7	24.4	.5884
700.	2134.	24.3	-65.5	-64.6	-61.7	-57.9	-53.3	34.7	.7221
800.	1921.	35.6	-96.4	-95.5	-92.2	-87.8	-82.6	47.6	.8704
900.	1724.	50.7	-136.8	-135.7	-132.0	-127.0	-121.2	63.4	1.0350
1000.	1544.	70.6	-189.2	-188.0	-183.8	-178.3	-171.8	82.6	1.2191

BALLISTIC COEFFICIENT .38, MUZZLE VELOCITY FT/SEC 2000.

RANGE	VELOCITY	HEIGHT	100 YD	150 YD	200 YD	250 YD	300 YD	DRIFT	TIME
0.	2000.	0.0	-1.5	-1.5	-1.5	-1.5	-1.5	0.0	0.0000
100.	1802.	.4	0.0	2.1	4.6	7.5	10.6	1.4	.1580
200.	1620.	4.6	-9.3	-5.1	0.0	5.7	12.0	5.9	.3336
300.	1453.	12.7	-31.9	-25.6	-18.0	-9.4	0.0	14.0	.5294
400.	1309.	26.0	-71.0	-62.7	-52.5	-41.0	-28.5	25.9	.7471
500.	1192.	46.1	-130.5	-120.0	-107.3	-93.0	-77.3	41.8	.9877
600.	1105.	74.2	-214.4	-201.9	-186.6	-169.4	-150.6	61.6	1.2498
700.	1044.	111.7	-326.8	-312.2	-294.4	-274.3	-252.4	84.6	1.5309
800.	995.	159.0	-467.8	-451.1	-430.8	-407.9	-382.8	109.9	1.8242
900.	950.	217.6	-641.4	-622.6	-599.7	-574.0	-545.7	137.8	2.1329
1000.	909.	288.7	-849.9	-829.0	-803.5	-774.9	-743.5	168.2	2.4557

BALLISTIC COEFFICIENT .38, MUZZLE VELOCITY FT/SEC 2100.

RANGE	VELOCITY	HEIGHT	100 YD	150 YD	200 YD	250 YD	300 YD	DRIFT	TIME
0.	2100.	0.0	-1.5	-1.5	-1.5	-1.5	-1.5	0.0	0.0000
100.	1895.	.3	0.0	1.9	4.1	6.7	9.5	1.3	.1502
200.	1705.	4.1	-8.2	-4.5	0.0	5.1	10.8	5.5	.3171
300.	1531.	11.4	-28.5	-23.0	-16.2	-8.5	0.0	13.1	.5030
400.	1374.	23.5	-63.7	-56.3	-47.3	-37.0	-25.7	24.4	.7103
500.	1245.	41.6	-117.3	-108.1	-96.7	-83.9	-69.8	39.7	.9398
600.	1142.	67.4	-193.3	-182.2	-168.6	-153.2	-136.2	58.9	1.1918
700.	1071.	102.1	-295.9	-282.9	-267.1	-249.1	-229.3	81.6	1.4637
800.	1018.	146.5	-427.4	-412.5	-394.4	-373.9	-351.3	107.1	1.7514
900.	971.	202.3	-591.3	-574.6	-554.2	-531.1	-505.7	135.7	2.0565
1000.	928.	268.7	-784.7	-766.2	-743.5	-717.9	-689.6	165.6	2.3695

BALLISTIC COEFFICIENT .38, MUZZLE VELOCITY FT/SEC 2200.

RANGE	VELOCITY	HEIGHT	100 YD	150 YD	200 YD	250 YD	300 YD	DRIFT	TIME
0.	2200.	0.0	-1.5	-1.5	-1.5	-1.5	-1.5	0.0	0.0000
100.	1989.	.2	0.0	1.6	3.7	6.0	8.5	1.2	.1434
200.	1791.	3.6	-7.4	-4.1	0.0	4.6	9.7	5.2	.3025
300.	1610.	10.3	-25.6	-20.7	-14.6	-7.7	0.0	12.3	.4791
400.	1444.	21.2	-57.3	-50.8	-42.6	-33.4	-23.2	23.0	.6761
500.	1302.	37.7	-105.7	-97.5	-87.3	-75.7	-63.0	37.5	.8949
600.	1186.	61.3	-174.6	-164.8	-152.5	-138.7	-123.4	56.1	1.1368
700.	1101.	93.3	-268.3	-256.8	-242.5	-226.4	-208.5	78.4	1.3999
800.	1041.	134.8	-390.0	-377.0	-360.6	-342.2	-321.7	103.8	1.6807
900.	992.	187.3	-543.1	-528.4	-510.0	-489.3	-466.3	132.4	1.9793
1000.	948.	250.0	-724.2	-707.9	-687.4	-664.4	-638.9	162.2	2.2855

BALLISTIC COEFFICIENT .38, MUZZLE VELOCITY FT/SEC 2300.

RANGE	VELOCITY	HEIGHT	100 YD	150 YD	200 YD	250 YD	300 YD	DRIFT	TIME
0.	2300.	0.0	-1.5	-1.5	-1.5	-1.5	-1.5	0.0	0.0000
100.	2084.	.1	0.0	1.4	3.3	5.4	7.7	1.1	.1369
200.	1879.	3.2	-6.6	-3.7	0.0	4.2	8.8	4.9	.2885
300.	1691.	9.3	-23.0	-18.7	-13.2	-6.9	0.0	11.5	.4569
400.	1518.	19.2	-51.7	-45.9	-38.6	-30.2	-21.0	21.6	.6442
500.	1363.	34.2	-95.5	-88.2	-79.1	-68.6	-57.1	35.4	.8531
600.	1236.	55.7	-158.1	-149.4	-138.4	-125.8	-112.0	53.2	1.0847
700.	1135.	85.4	-243.8	-233.8	-220.9	-206.2	-190.1	75.1	1.3395
800.	1066.	123.9	-355.4	-343.9	-329.2	-312.4	-294.0	99.9	1.6112
900.	1013.	173.2	-498.4	-485.4	-468.9	-449.9	-429.3	128.4	1.9036
1000.	967.	232.7	-669.2	-654.8	-636.4	-615.4	-592.4	158.6	2.2053

BALLISTIC COEFFICIENT .38, MUZZLE VELOCITY FT/SEC 2400.

RANGE	VELOCITY	HEIGHT	100 YD	150 YD	200 YD	250 YD	300 YD	DRIFT	TIME
0.	2400.	0.0	-1.5	-1.5	-1.5	-1.5	-1.5	0.0	0.0000
100.	2179.	.1	0.0	1.3	2.9	4.8	6.9	1.0	.1309
200.	1969.	2.9	-5.9	-3.3	0.0	3.8	8.0	4.6	.2759
300.	1773.	8.4	-20.8	-16.9	-12.0	-6.3	0.0	10.8	.4366
400.	1593.	17.4	-46.8	-41.7	-35.0	-27.5	-19.1	20.3	.6152
500.	1429.	31.1	-86.4	-80.0	-71.7	-62.3	-51.8	33.3	.8140
600.	1290.	50.7	-143.2	-135.5	-125.6	-114.3	-101.6	50.2	1.0352
700.	1176.	77.9	-221.2	-212.2	-200.6	-187.4	-172.7	71.3	1.2800
800.	1094.	113.7	-324.0	-313.7	-300.4	-285.3	-268.5	95.8	1.5442
900.	1036.	159.3	-455.0	-443.4	-428.5	-411.5	-392.6	123.4	1.8262
1000.	988.	216.1	-616.9	-604.1	-587.5	-568.6	-547.6	154.0	2.1252

BALLISTIC COEFFICIENT .38, MUZZLE VELOCITY FT/SEC 2500.

RANGE	VELOCITY	HEIGHT	100 YD	150 YD	200 YD	250 YD	300 YD	DRIFT	TIME
0.	2500.	0.0	-1.5	-1.5	-1.5	-1.5	-1.5	0.0	0.0000
100.	2274.	.0	0.0	1.1	2.6	4.4	6.3	1.0	.1257
200.	2060.	2.6	-5.3	-3.0	0.0	3.4	7.2	4.3	.2644
300.	1857.	7.6	-18.8	-15.4	-10.8	-5.7	0.0	10.1	.4177
400.	1670.	15.9	-42.4	-37.9	-31.9	-25.0	-17.4	19.0	.5882
500.	1499.	28.3	-78.4	-72.8	-65.2	-56.6	-47.2	31.3	.7777
600.	1347.	46.2	-130.1	-123.3	-114.3	-104.0	-92.6	47.4	.9893
700.	1223.	71.1	-200.9	-193.0	-182.5	-170.4	-157.2	67.4	1.2231
800.	1126.	104.3	-295.2	-286.2	-274.2	-260.4	-245.2	91.4	1.4795
900.	1060.	147.0	-416.5	-406.4	-392.8	-377.3	-360.3	118.7	1.7543
1000.	1008.	200.4	-568.2	-557.0	-541.9	-524.7	-505.7	149.1	2.0473

BALLISTIC COEFFICIENT .38, MUZZLE VELOCITY FT/SEC 2600.

RANGE	VELOCITY	HEIGHT	100 YD	150 YD	200 YD	250 YD	300 YD	DRIFT	TIME
0.	2600.	0.0	-1.5	-1.5	-1.5	-1.5	-1.5	0.0	0.0000
100.	2370.	-.0	0.0	1.0	2.4	3.9	5.7	1.0	.1209
200.	2150.	2.3	-4.7	-2.7	0.0	3.2	6.6	4.0	.2538
300.	1942.	7.0	-17.0	-14.0	-9.9	-5.2	0.0	9.6	.4007
400.	1748.	14.5	-38.5	-34.5	-29.1	-22.8	-15.8	17.9	.5634
500.	1571.	25.9	-71.5	-66.5	-59.7	-51.8	-43.1	29.5	.7447
600.	1409.	42.2	-118.4	-112.4	-104.3	-94.8	-84.4	44.7	.9463
700.	1273.	65.1	-183.3	-176.2	-166.8	-155.7	-143.6	63.9	1.1710
800.	1163.	95.7	-269.4	-261.3	-250.5	-237.9	-224.0	87.0	1.4175
900.	1086.	135.5	-381.5	-372.5	-360.3	-346.1	-330.5	113.8	1.6851
1000.	1029.	185.8	-523.1	-513.1	-499.5	-483.8	-466.4	143.9	1.9715

BALLISTIC COEFFICIENT .38, MUZZLE VELOCITY FT/SEC 2700.

RANGE	VELOCITY	HEIGHT	100 YD	150 YD	200 YD	250 YD	300 YD	DRIFT	TIME
0.	2700.	0.0	-1.5	-1.5	-1.5	-1.5	-1.5	0.0	0.0000
100.	2465.	-.1	0.0	.9	2.1	3.6	5.2	.9	.1162
200.	2241.	2.1	-4.3	-2.5	0.0	2.9	6.1	3.8	.2441
300.	2028.	6.4	-15.5	-12.8	-9.1	-4.8	0.0	9.1	.3848
400.	1828.	13.3	-35.2	-31.6	-26.6	-20.9	-14.5	16.9	.5405
500.	1643.	23.7	-65.3	-60.8	-54.6	-47.5	-39.5	27.8	.7138
600.	1474.	38.7	-108.3	-102.8	-95.4	-86.8	-77.3	42.2	.9065
700.	1327.	59.6	-167.4	-161.1	-152.4	-142.4	-131.2	60.5	1.1213
800.	1206.	87.9	-246.4	-239.2	-229.4	-217.9	-205.1	82.7	1.3588
900.	1115.	124.9	-349.7	-341.6	-330.4	-317.5	-303.2	108.8	1.6182
1000.	1051.	171.8	-480.4	-471.4	-459.0	-444.7	-428.7	138.1	1.8958

BALLISTIC COEFFICIENT .38, MUZZLE VELOCITY FT/SEC 2800.

RANGE	VELOCITY	HEIGHT	100 YD	150 YD	200 YD	250 YD	300 YD	DRIFT	TIME
0.	2800.	0.0	-1.5	-1.5	-1.5	-1.5	-1.5	0.0	0.0000
100.	2560.	-.1	0.0	.8	1.9	3.2	4.7	.8	.1119
200.	2331.	1.9	-3.8	-2.3	0.0	2.6	5.6	3.6	.2348
300.	2114.	5.8	-14.1	-11.7	-8.3	-4.4	0.0	8.5	.3698
400.	1907.	12.2	-32.2	-29.0	-24.5	-19.3	-13.4	16.0	.5193
500.	1717.	21.8	-59.8	-55.8	-50.2	-43.6	-36.3	26.3	.6849
600.	1542.	35.5	-99.2	-94.5	-87.7	-79.8	-71.0	39.9	.8695
700.	1384.	54.7	-153.3	-147.8	-139.8	-130.7	-120.4	57.2	1.0751
800.	1253.	80.8	-226.0	-219.7	-210.7	-200.2	-188.4	78.6	1.3038
900.	1148.	115.2	-320.7	-313.6	-303.4	-291.6	-278.4	103.8	1.5540
1000.	1075.	159.1	-442.2	-434.3	-423.0	-409.8	-395.2	132.6	1.8247

BALLISTIC COEFFICIENT .38, MUZZLE VELOCITY FT/SEC 2900.

RANGE	VELOCITY	HEIGHT	100 YD	150 YD	200 YD	250 YD	300 YD	DRIFT	TIME
0.	2900.	0.0	-1.5	-1.5	-1.5	-1.5	-1.5	0.0	0.0000
100.	2654.	-.2	0.0	.7	1.7	2.9	4.3	.8	.1081
200.	2421.	1.7	-3.5	-2.1	0.0	2.4	5.1	3.5	.2265
300.	2199.	5.3	-12.9	-10.8	-7.7	-4.0	0.0	8.1	.3564
400.	1988.	11.2	-29.5	-26.7	-22.6	-17.7	-12.3	15.1	.4998
500.	1791.	20.1	-55.0	-51.4	-46.3	-40.2	-33.5	24.9	.6590
600.	1610.	32.8	-91.2	-87.0	-80.8	-73.5	-65.5	37.8	.8357
700.	1444.	50.4	-140.9	-136.0	-128.8	-120.3	-110.9	54.3	1.0327
800.	1302.	74.4	-207.5	-201.8	-193.6	-183.9	-173.1	74.6	1.2515
900.	1186.	106.3	-294.9	-288.5	-279.3	-268.4	-256.2	99.0	1.4935
1000.	1101.	147.4	-407.3	-400.2	-390.0	-377.9	-364.4	127.1	1.7566

BALLISTIC COEFFICIENT .38, MUZZLE VELOCITY FT/SEC 3000.

RANGE	VELOCITY	HEIGHT	100 YD	150 YD	200 YD	250 YD	300 YD	DRIFT	TIME
0.	3000.	0.0	-1.5	-1.5	-1.5	-1.5	-1.5	0.0	0.0000
100.	2748.	-.2	0.0	.6	1.6	2.7	3.9	.8	.1047
200.	2511.	1.6	-3.1	-1.9	0.0	2.2	4.7	3.3	.2190
300.	2285.	4.9	-11.8	-9.9	-7.1	-3.7	0.0	7.8	.3441
400.	2069.	10.4	-27.1	-24.7	-20.9	-16.4	-11.4	14.5	.4821
500.	1866.	18.6	-50.6	-47.5	-42.7	-37.1	-30.9	23.7	.6347
600.	1679.	30.3	-84.1	-80.3	-74.6	-67.9	-60.5	36.0	.8043
700.	1507.	46.6	-129.8	-125.5	-118.8	-111.0	-102.3	51.6	.9930
800.	1354.	68.8	-191.2	-186.3	-178.7	-169.8	-159.8	71.1	1.2038
900.	1229.	98.3	-271.7	-266.1	-257.6	-247.5	-236.3	94.4	1.4363
1000.	1130.	136.6	-375.7	-369.5	-360.0	-348.8	-336.4	121.7	1.6916

BALLISTIC COEFFICIENT .38, MUZZLE VELOCITY FT/SEC 3100.

RANGE	VELOCITY	HEIGHT	100 YD	150 YD	200 YD	250 YD	300 YD	DRIFT	TIME
0.	3100.	0.0	-1.5	-1.5	-1.5	-1.5	-1.5	0.0	0.0000
100.	2843.	-.3	0.0	.6	1.4	2.5	3.6	.7	.1008
200.	2600.	1.4	-2.8	-1.7	0.0	2.1	4.4	3.1	.2113
300.	2370.	4.5	-10.8	-9.2	-6.6	-3.5	0.0	7.4	.3323
400.	2150.	9.6	-25.0	-22.8	-19.4	-15.2	-10.6	13.7	.4651
500.	1942.	17.2	-46.8	-44.0	-39.7	-34.5	-28.7	22.6	.6120
600.	1748.	28.1	-77.7	-74.3	-69.2	-63.0	-56.0	34.2	.7747
700.	1571.	43.1	-120.1	-116.2	-110.1	-102.9	-94.7	49.0	.9560
800.	1409.	63.6	-176.4	-172.0	-165.1	-156.8	-147.5	67.5	1.1576
900.	1273.	91.0	-251.0	-246.0	-238.2	-228.9	-218.5	90.0	1.3824
1000.	1163.	126.6	-347.0	-341.4	-332.8	-322.4	-310.8	116.4	1.6289

BALLISTIC COEFFICIENT .38, MUZZLE VELOCITY FT/SEC 3200.

RANGE	VELOCITY	HEIGHT	100 YD	150 YD	200 YD	250 YD	300 YD	DRIFT	TIME
0.	3200.	0.0	-1.5	-1.5	-1.5	-1.5	-1.5	0.0	0.0000
100.	2937.	-.3	0.0	.5	1.3	2.2	3.3	.7	.0978
200.	2689.	1.3	-2.6	-1.6	0.0	1.9	4.1	3.0	.2046
300.	2454.	4.2	-9.9	-8.5	-6.1	-3.2	0.0	7.1	.3214
400.	2231.	9.0	-23.1	-21.2	-18.0	-14.2	-9.9	13.1	.4497
500.	2018.	16.0	-43.3	-40.8	-36.9	-32.1	-26.7	21.5	.5910
600.	1819.	26.1	-71.9	-69.0	-64.2	-58.5	-52.0	32.6	.7475
700.	1635.	40.0	-111.1	-107.7	-102.2	-95.4	-87.9	46.7	.9216
800.	1467.	59.0	-163.3	-159.4	-153.1	-145.4	-136.8	64.3	1.1154
900.	1321.	84.3	-232.0	-227.6	-220.5	-211.9	-202.2	85.8	1.3312
1000.	1201.	117.5	-321.1	-316.2	-308.3	-298.7	-287.9	111.3	1.5698

BALLISTIC COEFFICIENT .38, MUZZLE VELOCITY FT/SEC 3300.

RANGE	VELOCITY	HEIGHT	100 YD	150 YD	200 YD	250 YD	300 YD	DRIFT	TIME
0.	3300.	0.0	-1.5	-1.5	-1.5	-1.5	-1.5	0.0	0.0000
100.	3031.	-.3	0.0	.4	1.2	2.0	3.1	.7	.0948
200.	2778.	1.1	-2.3	-1.4	0.0	1.8	3.8	2.9	.1982
300.	2539.	3.9	-9.2	-7.9	-5.7	-3.0	0.0	6.8	.3114
400.	2312.	8.3	-21.4	-19.7	-16.8	-13.2	-9.2	12.6	.4352
500.	2095.	14.9	-40.1	-38.0	-34.4	-29.9	-24.9	20.6	.5715
600.	1890.	24.3	-66.7	-64.2	-59.8	-54.5	-48.4	31.1	.7222
700.	1700.	37.2	-103.1	-100.1	-95.0	-88.8	-81.7	44.6	.8896
800.	1527.	54.8	-151.6	-148.1	-142.4	-135.2	-127.1	61.4	1.0759
900.	1371.	78.3	-215.1	-211.2	-204.8	-196.7	-187.6	81.9	1.2836
1000.	1242.	109.2	-297.7	-293.4	-286.2	-277.3	-267.2	106.4	1.5138

BALLISTIC COEFFICIENT .38, MUZZLE VELOCITY FT/SEC 3400.

RANGE	VELOCITY	HEIGHT	100 YD	150 YD	200 YD	250 YD	300 YD	DRIFT	TIME
0.	3400.	0.0	-1.5	-1.5	-1.5	-1.5	-1.5	0.0	0.0000
100.	3126.	-.3	0.0	.4	1.0	1.9	2.8	.7	.0921
200.	2867.	1.0	-2.1	-1.3	0.0	1.7	3.5	2.7	.1920
300.	2623.	3.6	-8.4	-7.3	-5.3	-2.8	0.0	6.5	.3015
400.	2392.	7.8	-19.8	-18.3	-15.7	-12.3	-8.6	12.0	.4214
500.	2171.	13.9	-37.2	-35.4	-32.1	-27.9	-23.2	19.7	.5529
600.	1961.	22.7	-62.0	-59.9	-55.9	-50.9	-45.3	29.8	.6985
700.	1766.	34.7	-95.8	-93.3	-88.6	-82.8	-76.3	42.6	.8595
800.	1587.	51.1	-140.9	-138.0	-132.7	-126.0	-118.6	58.6	1.0390
900.	1423.	72.9	-199.7	-196.5	-190.5	-183.0	-174.6	78.2	1.2384
1000.	1286.	101.6	-276.3	-272.7	-266.1	-257.7	-248.4	101.8	1.4606

BALLISTIC COEFFICIENT .38, MUZZLE VELOCITY FT/SEC 3500.

RANGE	VELOCITY	HEIGHT	100 YD	150 YD	200 YD	250 YD	300 YD	DRIFT	TIME
0.	3500.	0.0	-1.5	-1.5	-1.5	-1.5	-1.5	0.0	0.0000
100.	3220.	-.4	0.0	.3	.9	1.7	2.6	.7	.0895
200.	2956.	.9	-1.9	-1.2	0.0	1.5	3.3	2.7	.1865
300.	2707.	3.4	-7.7	-6.8	-5.0	-2.6	0.0	6.3	.2928
400.	2472.	7.3	-18.3	-17.1	-14.6	-11.5	-8.0	11.6	.4087
500.	2248.	13.0	-34.7	-33.1	-30.0	-26.2	-21.8	18.9	.5362
600.	2034.	21.2	-57.8	-55.9	-52.3	-47.6	-42.4	28.6	.6765
700.	1833.	32.5	-89.4	-87.1	-82.9	-77.5	-71.3	40.8	.8319
800.	1648.	47.7	-131.3	-128.7	-123.8	-117.7	-110.6	56.1	1.0044
900.	1479.	68.0	-186.0	-183.1	-177.7	-170.7	-162.8	74.8	1.1965
1000.	1331.	94.8	-257.1	-254.0	-247.9	-240.1	-231.4	97.4	1.4108

BALLISTIC COEFFICIENT .38, MUZZLE VELOCITY FT/SEC 3600.

RANGE	VELOCITY	HEIGHT	100 YD	150 YD	200 YD	250 YD	300 YD	DRIFT	TIME
0.	3600.	0.0	-1.5	-1.5	-1.5	-1.5	-1.5	0.0	0.0000
100.	3315.	-.4	0.0	.3	.8	1.6	2.4	.6	.0867
200.	3046.	.8	-1.7	-1.1	0.0	1.5	3.1	2.5	.1810
300.	2791.	3.1	-7.1	-6.3	-4.6	-2.4	0.0	6.0	.2841
400.	2552.	6.8	-17.0	-16.0	-13.7	-10.8	-7.5	11.1	.3965
500.	2324.	12.2	-32.3	-30.9	-28.1	-24.5	-20.4	18.1	.5197
600.	2106.	19.9	-53.9	-52.3	-48.9	-44.6	-39.7	27.3	.6553
700.	1900.	30.4	-83.4	-81.5	-77.5	-72.5	-66.8	39.1	.8052
800.	1710.	44.6	-122.5	-120.3	-115.8	-110.0	-103.4	53.7	.9715
900.	1536.	63.5	-173.6	-171.2	-166.0	-159.5	-152.2	71.6	1.1568
1000.	1378.	88.5	-239.6	-237.0	-231.3	-224.0	-215.9	93.3	1.3632

BALLISTIC COEFFICIENT .38, MUZZLE VELOCITY FT/SEC 3700.

RANGE	VELOCITY	HEIGHT	100 YD	150 YD	200 YD	250 YD	300 YD	DRIFT	TIME
0.	3700.	0.0	-1.5	-1.5	-1.5	-1.5	-1.5	0.0	0.0000
100.	3410.	-.4	0.0	.2	.8	1.4	2.2	.6	.0846
200.	3135.	.7	-1.5	-1.1	0.0	1.3	2.9	2.5	.1766
300.	2876.	2.9	-6.5	-5.9	-4.3	-2.3	0.0	5.8	.2762
400.	2631.	6.4	-15.8	-14.9	-12.8	-10.2	-7.1	10.7	.3854
500.	2400.	11.5	-30.1	-29.0	-26.4	-23.1	-19.2	17.5	.5049
600.	2179.	18.7	-50.4	-49.0	-45.8	-41.9	-37.3	26.3	.6358
700.	1969.	28.5	-78.0	-76.4	-72.7	-68.1	-62.7	37.5	.7809
800.	1773.	41.8	-114.6	-112.7	-108.5	-103.2	-97.1	51.5	.9415
900.	1593.	59.5	-162.3	-160.2	-155.5	-149.6	-142.7	68.7	1.1202
1000.	1429.	82.8	-223.8	-221.5	-216.2	-209.7	-202.0	89.4	1.3190

BALLISTIC COEFFICIENT .38, MUZZLE VELOCITY FT/SEC 3800.

RANGE	VELOCITY	HEIGHT	100 YD	150 YD	200 YD	250 YD	300 YD	DRIFT	TIME
0.	3800.	0.0	-1.5	-1.5	-1.5	-1.5	-1.5	0.0	0.0000
100.	3504.	-.4	0.0	.2	.7	1.3	2.0	.6	.0823
200.	3225.	.7	-1.4	-1.0	0.0	1.3	2.7	2.4	.1717
300.	2960.	2.7	-6.0	-5.5	-4.0	-2.1	0.0	5.6	.2686
400.	2711.	6.0	-14.8	-14.0	-12.0	-9.5	-6.7	10.4	.3747
500.	2475.	10.8	-28.1	-27.2	-24.8	-21.6	-18.1	16.8	.4904
600.	2251.	17.6	-47.2	-46.1	-43.2	-39.4	-35.1	25.4	.6177
700.	2037.	26.8	-73.1	-71.8	-68.4	-64.0	-59.0	36.1	.7579
800.	1836.	39.3	-107.4	-105.8	-101.9	-96.9	-91.2	49.5	.9130
900.	1651.	55.8	-152.0	-150.2	-145.9	-140.2	-133.8	65.9	1.0852
1000.	1481.	77.5	-209.4	-207.5	-202.6	-196.4	-189.3	85.8	1.2769

BALLISTIC COEFFICIENT .38, MUZZLE VELOCITY FT/SEC 3900.

RANGE	VELOCITY	HEIGHT	100 YD	150 YD	200 YD	250 YD	300 YD	DRIFT	TIME
0.	3900.	0.0	-1.5	-1.5	-1.5	-1.5	-1.5	0.0	0.0000
100.	3599.	-.4	0.0	.2	.6	1.2	1.8	.6	.0801
200.	3314.	.6	-1.2	-.9	0.0	1.2	2.5	2.3	.1668
300.	3045.	2.5	-5.5	-5.1	-3.8	-2.0	0.0	5.3	.2611
400.	2791.	5.6	-13.7	-13.1	-11.3	-9.0	-6.3	9.9	.3642
500.	2551.	10.2	-26.3	-25.5	-23.3	-20.4	-17.0	16.2	.4767
600.	2323.	16.5	-44.2	-43.3	-40.6	-37.1	-33.1	24.3	.5999
700.	2106.	25.2	-68.5	-67.4	-64.3	-60.3	-55.6	34.7	.7356
800.	1900.	36.9	-100.6	-99.4	-95.8	-91.2	-85.8	47.5	.8855
900.	1710.	52.4	-142.4	-141.0	-137.0	-131.8	-125.7	63.3	1.0518
1000.	1535.	72.7	-196.2	-194.7	-190.3	-184.5	-177.7	82.4	1.2372

BALLISTIC COEFFICIENT .38, MUZZLE VELOCITY FT/SEC 4000.

RANGE	VELOCITY	HEIGHT	100 YD	150 YD	200 YD	250 YD	300 YD	DRIFT	TIME
0.	4000.	0.0	-1.5	-1.5	-1.5	-1.5	-1.5	0.0	0.0000
100.	3694.	-.5	0.0	.1	.5	1.1	1.7	.5	.0781
200.	3404.	.5	-1.1	-.8	0.0	1.1	2.4	2.3	.1629
300.	3130.	2.4	-5.2	-4.8	-3.6	-1.9	0.0	5.3	.2549
400.	2871.	5.3	-12.8	-12.3	-10.6	-8.5	-5.9	9.6	.3547
500.	2626.	9.6	-24.6	-24.0	-21.9	-19.2	-16.0	15.7	.4640
600.	2395.	15.6	-41.5	-40.8	-38.3	-35.1	-31.2	23.5	.5838
700.	2174.	23.8	-64.3	-63.5	-60.6	-56.8	-52.3	33.5	.7151
800.	1964.	34.8	-94.6	-93.6	-90.3	-86.0	-80.8	45.8	.8604
900.	1769.	49.3	-133.8	-132.7	-129.0	-124.1	-118.3	61.0	1.0213
1000.	1590.	68.4	-184.4	-183.1	-179.0	-173.6	-167.2	79.3	1.2004

BALLISTIC COEFFICIENT .39, MUZZLE VELOCITY FT/SEC 2000.

RANGE	VELOCITY	HEIGHT	100 YD	150 YD	200 YD	250 YD	300 YD	DRIFT	TIME
0.	2000.	0.0	-1.5	-1.5	-1.5	-1.5	-1.5	0.0	0.0000
100.	1807.	.4	0.0	2.1	4.6	7.4	10.5	1.4	.1578
200.	1629.	4.6	-9.2	-5.1	0.0	5.7	11.9	5.8	.3327
300.	1465.	12.6	-31.6	-25.4	-17.8	-9.3	0.0	13.6	.5271
400.	1323.	25.7	-70.3	-62.0	-51.8	-40.5	-28.1	25.1	.7428
500.	1206.	45.4	-128.8	-118.5	-105.8	-91.6	-76.1	40.6	.9807
600.	1116.	73.1	-211.3	-198.9	-183.7	-166.7	-148.1	59.8	1.2400
700.	1054.	109.7	-321.3	-306.7	-289.0	-269.2	-247.5	82.2	1.5170
800.	1004.	156.5	-461.4	-444.8	-424.5	-401.9	-377.1	107.4	1.8101
900.	960.	213.8	-631.4	-612.7	-589.9	-564.4	-536.5	134.5	2.1141
1000.	919.	283.5	-836.5	-815.7	-790.4	-762.1	-731.1	164.3	2.4335

BALLISTIC COEFFICIENT .39, MUZZLE VELOCITY FT/SEC 2100.

RANGE	VELOCITY	HEIGHT	100 YD	150 YD	200 YD	250 YD	300 YD	DRIFT	TIME
0.	2100.	0.0	-1.5	-1.5	-1.5	-1.5	-1.5	0.0	0.0000
100.	1900.	.3	0.0	1.8	4.1	6.6	9.4	1.3	.1501
200.	1714.	4.0	-8.2	-4.5	0.0	5.1	10.7	5.4	.3162
300.	1544.	11.3	-28.3	-22.8	-16.0	-8.4	0.0	12.7	.5009
400.	1389.	23.2	-63.0	-55.6	-46.6	-36.4	-25.3	23.7	.7059
500.	1260.	41.0	-115.8	-106.6	-95.3	-82.6	-68.6	38.5	.9328
600.	1155.	66.5	-190.9	-179.9	-166.4	-151.1	-134.3	57.4	1.1833
700.	1082.	100.4	-291.4	-278.5	-262.8	-244.9	-225.4	79.5	1.4515
800.	1028.	144.2	-421.4	-406.7	-388.7	-368.3	-345.9	104.7	1.7376
900.	981.	198.0	-579.9	-563.4	-543.1	-520.2	-495.1	131.8	2.0349
1000.	939.	263.7	-771.7	-753.3	-730.8	-705.3	-677.4	161.7	2.3471

BALLISTIC COEFFICIENT .39, MUZZLE VELOCITY FT/SEC 2200.

RANGE	VELOCITY	HEIGHT	100 YD	150 YD	200 YD	250 YD	300 YD	DRIFT	TIME
0.	2200.	0.0	-1.5	-1.5	-1.5	-1.5	-1.5	0.0	0.0000
100.	1994.	.2	0.0	1.6	3.6	5.9	8.5	1.2	.1433
200.	1801.	3.6	-7.3	-4.1	0.0	4.6	9.6	5.1	.3016
300.	1624.	10.2	-25.4	-20.5	-14.4	-7.6	0.0	12.0	.4770
400.	1460.	20.9	-56.7	-50.2	-42.1	-32.9	-22.8	22.3	.6720
500.	1319.	37.1	-104.3	-96.2	-86.1	-74.6	-62.0	36.4	.8884
600.	1203.	60.2	-172.0	-162.2	-150.1	-136.4	-121.2	54.4	1.1270
700.	1114.	91.6	-263.8	-252.4	-238.2	-222.2	-204.6	76.1	1.3869
800.	1052.	132.2	-383.0	-370.0	-353.8	-335.6	-315.4	100.9	1.6644
900.	1003.	183.2	-532.1	-517.5	-499.3	-478.7	-456.0	128.5	1.9575
1000.	959.	244.9	-711.1	-694.9	-674.6	-651.8	-626.6	158.2	2.2624

BALLISTIC COEFFICIENT .39, MUZZLE VELOCITY FT/SEC 2300.

RANGE	VELOCITY	HEIGHT	100 YD	150 YD	200 YD	250 YD	300 YD	DRIFT	TIME
0.	2300.	0.0	-1.5	-1.5	-1.5	-1.5	-1.5	0.0	0.0000
100.	2089.	.1	0.0	1.4	3.3	5.3	7.6	1.1	.1368
200.	1890.	3.2	-6.5	-3.6	0.0	4.1	8.7	4.7	.2877
300.	1705.	9.2	-22.8	-18.5	-13.0	-6.8	0.0	11.2	.4548
400.	1535.	18.9	-51.1	-45.4	-38.1	-29.8	-20.7	20.9	.6405
500.	1382.	33.7	-94.2	-87.0	-77.9	-67.5	-56.1	34.2	.8466
600.	1254.	54.7	-155.7	-147.1	-136.1	-123.7	-110.0	51.5	1.0752
700.	1151.	83.5	-239.0	-229.0	-216.2	-201.7	-185.8	72.5	1.3250
800.	1079.	121.3	-348.7	-337.2	-322.6	-306.0	-287.8	97.0	1.5946
900.	1025.	169.2	-487.7	-474.8	-458.4	-439.8	-419.3	124.6	1.8817
1000.	979.	227.3	-655.1	-640.8	-622.6	-601.8	-579.1	154.1	2.1799

BALLISTIC COEFFICIENT .39, MUZZLE VELOCITY FT/SEC 2400.

RANGE	VELOCITY	HEIGHT	100 YD	150 YD	200 YD	250 YD	300 YD	DRIFT	TIME
0.	2400.	0.0	-1.5	-1.5	-1.5	-1.5	-1.5	0.0	0.0000
100.	2185.	.1	0.0	1.3	2.9	4.8	6.9	1.0	.1308
200.	1979.	2.9	-5.8	-3.3	0.0	3.8	7.9	4.4	.2750
300.	1788.	8.3	-20.6	-16.8	-11.9	-6.2	0.0	10.5	.4348
400.	1611.	17.2	-46.3	-41.2	-34.6	-27.1	-18.8	19.6	.6115
500.	1449.	30.6	-85.3	-79.0	-70.8	-61.4	-51.0	32.2	.8081
600.	1310.	49.8	-141.0	-133.4	-123.5	-112.2	-99.8	48.6	1.0259
700.	1195.	76.3	-217.2	-208.3	-196.8	-183.7	-169.1	68.9	1.2667
800.	1109.	111.3	-317.6	-307.4	-294.3	-279.3	-262.6	92.8	1.5276
900.	1048.	156.4	-447.4	-435.9	-421.2	-404.2	-385.5	120.4	1.8093
1000.	1000.	210.8	-603.1	-590.4	-574.0	-555.2	-534.4	149.5	2.0993

BALLISTIC COEFFICIENT .39, MUZZLE VELOCITY FT/SEC 2500.

RANGE	VELOCITY	HEIGHT	100 YD	150 YD	200 YD	250 YD	300 YD	DRIFT	TIME
0.	2500.	0.0	-1.5	-1.5	-1.5	-1.5	-1.5	0.0	0.0000
100.	2280.	.0	0.0	1.1	2.6	4.3	6.2	1.0	.1255
200.	2070.	2.6	-5.2	-3.0	0.0	3.4	7.2	4.2	.2636
300.	1872.	7.6	-18.6	-15.2	-10.8	-5.6	0.0	9.8	.4159
400.	1689.	15.7	-42.0	-37.5	-31.5	-24.7	-17.2	18.4	.5848
500.	1520.	27.9	-77.5	-71.9	-64.4	-55.9	-46.5	30.3	.7721
600.	1369.	45.4	-128.1	-121.4	-112.4	-102.2	-90.9	45.8	.9804
700.	1244.	69.6	-197.4	-189.6	-179.1	-167.2	-154.0	65.2	1.2107
800.	1143.	101.9	-289.2	-280.2	-268.3	-254.7	-239.6	88.5	1.4626
900.	1073.	143.6	-407.8	-397.7	-384.2	-368.9	-352.0	115.1	1.7341
1000.	1021.	195.3	-554.9	-543.7	-528.7	-511.7	-492.9	144.5	2.0212

BALLISTIC COEFFICIENT .39, MUZZLE VELOCITY FT/SEC 2600.

RANGE	VELOCITY	HEIGHT	100 YD	150 YD	200 YD	250 YD	300 YD	DRIFT	TIME
0.	2600.	0.0	-1.5	-1.5	-1.5	-1.5	-1.5	0.0	0.0000
100.	2376.	-.1	0.0	1.0	2.3	3.9	5.6	.9	.1207
200.	2161.	2.3	-4.7	-2.7	0.0	3.1	6.6	3.9	.2530
300.	1957.	6.9	-16.9	-13.9	-9.9	-5.2	0.0	9.3	.3991
400.	1767.	14.3	-38.2	-34.1	-28.8	-22.5	-15.6	17.4	.5602
500.	1593.	25.5	-70.6	-65.6	-58.9	-51.1	-42.5	28.6	.7393
600.	1432.	41.5	-116.7	-110.7	-102.6	-93.2	-82.9	43.2	.9379
700.	1296.	63.7	-180.0	-172.9	-163.6	-152.6	-140.5	61.8	1.1586
800.	1183.	93.5	-263.9	-255.9	-245.1	-232.6	-218.8	84.1	1.4009
900.	1101.	132.2	-373.0	-364.0	-351.9	-337.8	-322.3	110.2	1.6644
1000.	1042.	180.9	-510.4	-500.4	-487.0	-471.3	-454.1	139.4	1.9456

BALLISTIC COEFFICIENT .39, MUZZLE VELOCITY FT/SEC 2700.

RANGE	VELOCITY	HEIGHT	100 YD	150 YD	200 YD	250 YD	300 YD	DRIFT	TIME
0.	2700.	0.0	-1.5	-1.5	-1.5	-1.5	-1.5	0.0	0.0000
100.	2471.	-.1	0.0	.9	2.1	3.5	5.1	.9	.1161
200.	2252.	2.1	-4.2	-2.5	0.0	2.8	6.0	3.7	.2435
300.	2044.	6.3	-15.4	-12.7	-9.0	-4.7	0.0	8.8	.3833
400.	1847.	13.1	-34.8	-31.3	-26.3	-20.7	-14.3	16.4	.5377
500.	1666.	23.4	-64.5	-60.0	-53.9	-46.8	-38.9	26.9	.7087
600.	1499.	38.0	-106.6	-101.2	-93.9	-85.4	-75.9	40.8	.8983
700.	1351.	58.4	-164.4	-158.2	-149.6	-139.6	-128.5	58.4	1.1097
800.	1229.	85.8	-241.4	-234.2	-224.4	-213.1	-200.4	79.9	1.3426
900.	1133.	121.7	-341.6	-333.5	-322.5	-309.7	-295.5	105.2	1.5975
1000.	1065.	167.8	-470.2	-461.2	-449.0	-434.8	-418.9	134.2	1.8737

BALLISTIC COEFFICIENT .39, MUZZLE VELOCITY FT/SEC 2800.

RANGE	VELOCITY	HEIGHT	100 YD	150 YD	200 YD	250 YD	300 YD	DRIFT	TIME
0.	2800.	0.0	-1.5	-1.5	-1.5	-1.5	-1.5	0.0	0.0000
100.	2566.	-.2	0.0	.8	1.9	3.2	4.7	.8	.1117
200.	2343.	1.9	-3.8	-2.3	0.0	2.6	5.5	3.5	.2343
300.	2130.	5.8	-14.0	-11.6	-8.3	-4.3	0.0	8.3	.3684
400.	1928.	12.1	-31.9	-28.7	-24.2	-19.0	-13.2	15.5	.5166
500.	1740.	21.5	-59.1	-55.2	-49.5	-43.0	-35.8	25.4	.6803
600.	1568.	34.9	-97.8	-93.1	-86.3	-78.5	-69.8	38.6	.8621
700.	1410.	53.6	-150.6	-145.1	-137.2	-128.1	-117.9	55.2	1.0638
800.	1278.	78.9	-221.3	-215.0	-206.0	-195.5	-184.0	75.8	1.2878
900.	1169.	112.1	-313.3	-306.2	-296.1	-284.3	-271.3	100.2	1.5336
1000.	1091.	154.7	-431.2	-423.3	-412.1	-399.0	-384.5	128.2	1.7999

BALLISTIC COEFFICIENT .39, MUZZLE VELOCITY FT/SEC 2900.

RANGE	VELOCITY	HEIGHT	100 YD	150 YD	200 YD	250 YD	300 YD	DRIFT	TIME
0.	2900.	0.0	-1.5	-1.5	-1.5	-1.5	-1.5	0.0	0.0000
100.	2660.	-.2	0.0	.7	1.7	2.9	4.3	.8	.1080
200.	2433.	1.7	-3.4	-2.0	0.0	2.4	5.1	3.3	.2259
300.	2216.	5.3	-12.8	-10.7	-7.6	-4.0	0.0	7.9	.3551
400.	2009.	11.1	-29.2	-26.4	-22.4	-17.5	-12.2	14.7	.4973
500.	1815.	19.8	-54.3	-50.7	-45.7	-39.6	-33.0	24.1	.6543
600.	1637.	32.2	-89.9	-85.7	-79.6	-72.3	-64.3	36.6	.8285
700.	1472.	49.4	-138.4	-133.5	-126.4	-118.0	-108.6	52.4	1.0218
800.	1329.	72.6	-203.3	-197.6	-189.5	-179.9	-169.2	72.0	1.2366
900.	1211.	103.5	-288.0	-281.7	-272.6	-261.7	-249.7	95.5	1.4734
1000.	1119.	143.2	-397.0	-389.9	-379.8	-367.7	-354.4	122.8	1.7319

BALLISTIC COEFFICIENT .39, MUZZLE VELOCITY FT/SEC 3000.

RANGE	VELOCITY	HEIGHT	100 YD	150 YD	200 YD	250 YD	300 YD	DRIFT	TIME
0.	3000.	0.0	-1.5	-1.5	-1.5	-1.5	-1.5	0.0	0.0000
100.	2755.	-.2	0.0	.6	1.6	2.7	3.9	.8	.1046
200.	2523.	1.5	-3.1	-1.9	0.0	2.2	4.7	3.3	.2185
300.	2302.	4.9	-11.7	-9.9	-7.0	-3.7	0.0	7.6	.3429
400.	2091.	10.3	-26.9	-24.4	-20.6	-16.2	-11.3	14.0	.4796
500.	1891.	18.3	-50.0	-46.9	-42.2	-36.7	-30.5	23.0	.6304
600.	1706.	29.8	-82.8	-79.1	-73.4	-66.8	-59.4	34.7	.7974
700.	1536.	45.6	-127.6	-123.3	-116.7	-109.0	-100.3	49.8	.9829
800.	1383.	67.1	-187.2	-182.3	-174.7	-165.9	-156.0	68.4	1.1888
900.	1255.	95.6	-265.4	-259.9	-251.4	-241.5	-230.3	91.0	1.4171
1000.	1152.	132.6	-365.8	-359.7	-350.3	-339.2	-326.8	117.4	1.6669

BALLISTIC COEFFICIENT .39, MUZZLE VELOCITY FT/SEC 3100.

RANGE	VELOCITY	HEIGHT	100 YD	150 YD	200 YD	250 YD	300 YD	DRIFT	TIME
0.	3100.	0.0	-1.5	-1.5	-1.5	-1.5	-1.5	0.0	0.0000
100.	2849.	-.3	0.0	.6	1.4	2.4	3.6	.7	.1007
200.	2612.	1.4	-2.8	-1.7	0.0	2.1	4.4	3.0	.2108
300.	2387.	4.5	-10.8	-9.1	-6.5	-3.4	0.0	7.2	.3310
400.	2172.	9.5	-24.8	-22.6	-19.1	-15.0	-10.4	13.3	.4626
500.	1968.	17.0	-46.2	-43.4	-39.2	-34.0	-28.3	21.8	.6078
600.	1777.	27.6	-76.6	-73.3	-68.2	-62.0	-55.1	33.0	.7683
700.	1601.	42.2	-118.0	-114.1	-108.1	-100.9	-92.9	47.3	.9463
800.	1440.	62.1	-173.0	-168.6	-161.7	-153.5	-144.3	65.1	1.1439
900.	1302.	88.4	-245.0	-240.0	-232.3	-223.0	-212.7	86.6	1.3631
1000.	1189.	122.8	-337.9	-332.4	-323.9	-313.6	-302.1	112.1	1.6047

BALLISTIC COEFFICIENT .39, MUZZLE VELOCITY FT/SEC 3200.

RANGE	VELOCITY	HEIGHT	100 YD	150 YD	200 YD	250 YD	300 YD	DRIFT	TIME
0.	3200.	0.0	-1.5	-1.5	-1.5	-1.5	-1.5	0.0	0.0000
100.	2944.	-.3	0.0	.5	1.3	2.2	3.3	.7	.0976
200.	2701.	1.3	-2.5	-1.6	0.0	1.9	4.0	2.9	.2042
300.	2472.	4.2	-9.9	-8.4	-6.0	-3.2	0.0	6.9	.3202
400.	2253.	8.9	-22.9	-21.0	-17.8	-14.0	-9.8	12.8	.4475
500.	2045.	15.8	-42.8	-40.4	-36.4	-31.7	-26.4	20.9	.5873
600.	1848.	25.6	-70.9	-68.0	-63.3	-57.6	-51.2	31.5	.7415
700.	1667.	39.2	-109.3	-105.9	-100.4	-93.7	-86.3	45.1	.9125
800.	1500.	57.5	-160.1	-156.2	-149.9	-142.3	-133.8	62.0	1.1020
900.	1352.	82.0	-226.7	-222.4	-215.3	-206.7	-197.1	82.6	1.3132
1000.	1230.	114.0	-312.7	-307.9	-300.0	-290.5	-279.9	107.1	1.5461

BALLISTIC COEFFICIENT .39, MUZZLE VELOCITY FT/SEC 3300.

RANGE	VELOCITY	HEIGHT	100 YD	150 YD	200 YD	250 YD	300 YD	DRIFT	TIME
0.	3300.	0.0	-1.5	-1.5	-1.5	-1.5	-1.5	0.0	0.0000
100.	3038.	-.3	0.0	.4	1.1	2.0	3.0	.7	.0947
200.	2791.	1.1	-2.3	-1.4	0.0	1.8	3.8	2.8	.1979
300.	2557.	3.9	-9.1	-7.8	-5.6	-3.0	0.0	6.6	.3102
400.	2334.	8.3	-21.2	-19.5	-16.6	-13.1	-9.1	12.2	.4331
500.	2122.	14.7	-39.6	-37.5	-33.9	-29.5	-24.5	19.9	.5677
600.	1920.	23.9	-65.9	-63.3	-59.0	-53.7	-47.7	30.1	.7166
700.	1733.	36.5	-101.4	-98.4	-93.4	-87.2	-80.2	43.0	.8808
800.	1561.	53.5	-148.7	-145.2	-139.5	-132.5	-124.5	59.2	1.0635
900.	1404.	76.2	-210.2	-206.3	-199.8	-191.9	-182.9	78.8	1.2660
1000.	1273.	106.0	-290.1	-285.8	-278.6	-269.8	-259.8	102.5	1.4912

BALLISTIC COEFFICIENT .39, MUZZLE VELOCITY FT/SEC 3400.

RANGE	VELOCITY	HEIGHT	100 YD	150 YD	200 YD	250 YD	300 YD	DRIFT	TIME
0.	3400.	0.0	-1.5	-1.5	-1.5	-1.5	-1.5	0.0	0.0000
100.	3133.	-.3	0.0	.4	1.0	1.8	2.8	.7	.0920
200.	2880.	1.0	-2.0	-1.3	0.0	1.7	3.5	2.6	.1915
300.	2641.	3.6	-8.3	-7.3	-5.3	-2.8	0.0	6.3	.3005
400.	2415.	7.7	-19.6	-18.2	-15.5	-12.2	-8.5	11.7	.4194
500.	2199.	13.7	-36.8	-35.0	-31.7	-27.5	-22.9	19.1	.5494
600.	1993.	22.3	-61.1	-59.0	-55.0	-50.1	-44.5	28.8	.6928
700.	1800.	34.0	-94.2	-91.8	-87.1	-81.3	-74.8	41.1	.8512
800.	1623.	49.8	-138.1	-135.3	-130.0	-123.3	-115.9	56.5	1.0267
900.	1459.	70.9	-195.3	-192.1	-186.1	-178.6	-170.3	75.3	1.2218
1000.	1318.	98.6	-269.1	-265.6	-259.0	-250.7	-241.4	97.9	1.4383

BALLISTIC COEFFICIENT .39, MUZZLE VELOCITY FT/SEC 3500.

RANGE	VELOCITY	HEIGHT	100 YD	150 YD	200 YD	250 YD	300 YD	DRIFT	TIME
0.	3500.	0.0	-1.5	-1.5	-1.5	-1.5	-1.5	0.0	0.0000
100.	3227.	-.4	0.0	.3	.9	1.7	2.6	.7	.0894
200.	2969.	.9	-1.8	-1.2	0.0	1.5	3.3	2.6	.1862
300.	2726.	3.3	-7.7	-6.7	-4.9	-2.6	0.0	6.1	.2918
400.	2495.	7.2	-18.2	-16.9	-14.5	-11.4	-8.0	11.3	.4069
500.	2276.	12.9	-34.2	-32.7	-29.6	-25.8	-21.5	18.3	.5327
600.	2066.	20.9	-57.0	-55.1	-51.5	-46.9	-41.7	27.6	.6711
700.	1868.	31.8	-87.8	-85.6	-81.4	-76.1	-69.9	39.4	.8236
800.	1685.	46.6	-128.7	-126.2	-121.4	-115.3	-108.3	54.1	.9929
900.	1517.	66.2	-181.9	-179.1	-173.6	-166.8	-158.9	72.0	1.1806
1000.	1366.	91.9	-250.5	-247.3	-241.3	-233.7	-225.0	93.7	1.3894

BALLISTIC COEFFICIENT .39, MUZZLE VELOCITY FT/SEC 3600.

RANGE	VELOCITY	HEIGHT	100 YD	150 YD	200 YD	250 YD	300 YD	DRIFT	TIME
0.	3600.	0.0	-1.5	-1.5	-1.5	-1.5	-1.5	0.0	0.0000
100.	3322.	-.4	0.0	.3	.8	1.5	2.4	.6	.0866
200.	3059.	.8	-1.7	-1.1	0.0	1.4	3.1	2.4	.1805
300.	2810.	3.1	-7.1	-6.3	-4.6	-2.4	0.0	5.8	.2832
400.	2576.	6.7	-16.9	-15.8	-13.6	-10.7	-7.4	10.8	.3946
500.	2352.	12.1	-31.9	-30.6	-27.8	-24.2	-20.1	17.6	.5166
600.	2139.	19.6	-53.2	-51.6	-48.3	-44.0	-39.1	26.5	.6503
700.	1936.	29.8	-82.1	-80.2	-76.3	-71.3	-65.6	37.7	.7977
800.	1748.	43.5	-120.2	-118.1	-113.6	-107.8	-101.3	51.7	.9607
900.	1575.	61.8	-169.8	-167.4	-162.4	-156.0	-148.6	69.0	1.1418
1000.	1416.	85.8	-233.5	-230.9	-225.3	-218.1	-210.0	89.6	1.3426

BALLISTIC COEFFICIENT .39, MUZZLE VELOCITY FT/SEC 3700.

RANGE	VELOCITY	HEIGHT	100 YD	150 YD	200 YD	250 YD	300 YD	DRIFT	TIME
0.	3700.	0.0	-1.5	-1.5	-1.5	-1.5	-1.5	0.0	0.0000
100.	3417.	-.4	0.0	.2	.8	1.4	2.2	.6	.0845
200.	3149.	.7	-1.5	-1.0	0.0	1.3	2.8	2.5	.1761
300.	2895.	2.9	-6.5	-5.8	-4.3	-2.3	0.0	5.7	.2754
400.	2656.	6.3	-15.7	-14.8	-12.7	-10.1	-7.0	10.4	.3836
500.	2429.	11.3	-29.8	-28.6	-26.0	-22.7	-18.9	17.0	.5017
600.	2212.	18.4	-49.7	-48.3	-45.2	-41.3	-36.7	25.5	.6311
700.	2005.	28.0	-76.8	-75.1	-71.5	-66.9	-61.6	36.3	.7737
800.	1812.	40.9	-112.4	-110.5	-106.4	-101.2	-95.0	49.7	.9310
900.	1633.	57.9	-158.8	-156.7	-152.0	-146.1	-139.2	66.1	1.1055
1000.	1469.	80.3	-218.2	-215.9	-210.7	-204.2	-196.5	86.0	1.2992

BALLISTIC COEFFICIENT .39, MUZZLE VELOCITY FT/SEC 3800.

RANGE	VELOCITY	HEIGHT	100 YD	150 YD	200 YD	250 YD	300 YD	DRIFT	TIME
0.	3800.	0.0	-1.5	-1.5	-1.5	-1.5	-1.5	0.0	0.0000
100.	3512.	-.4	0.0	.2	.7	1.3	2.0	.6	.0821
200.	3239.	.7	-1.3	-1.0	0.0	1.2	2.7	2.4	.1713
300.	2980.	2.7	-6.0	-5.4	-4.0	-2.1	0.0	5.5	.2679
400.	2736.	5.9	-14.6	-13.8	-11.9	-9.4	-6.6	10.1	.3729
500.	2505.	10.7	-27.9	-26.9	-24.5	-21.4	-17.8	16.4	.4877
600.	2285.	17.3	-46.6	-45.4	-42.5	-38.8	-34.5	24.5	.6130
700.	2075.	26.3	-71.9	-70.6	-67.2	-62.9	-57.9	34.9	.7507
800.	1876.	38.4	-105.3	-103.8	-99.9	-95.0	-89.3	47.7	.9027
900.	1692.	54.3	-148.7	-147.0	-142.6	-137.0	-130.6	63.5	1.0712
1000.	1524.	75.2	-204.3	-202.4	-197.6	-191.4	-184.3	82.5	1.2581

BALLISTIC COEFFICIENT .39, MUZZLE VELOCITY FT/SEC 3900.

RANGE	VELOCITY	HEIGHT	100 YD	150 YD	200 YD	250 YD	300 YD	DRIFT	TIME
0.	3900.	0.0	-1.5	-1.5	-1.5	-1.5	-1.5	0.0	0.0000
100.	3607.	-.4	0.0	.1	.6	1.2	1.8	.5	.0800
200.	3328.	.6	-1.2	-.9	0.0	1.2	2.5	2.2	.1664
300.	3065.	2.5	-5.5	-5.1	-3.7	-2.0	0.0	5.2	.2602
400.	2816.	5.6	-13.6	-13.0	-11.2	-8.9	-6.2	9.7	.3626
500.	2581.	10.0	-26.0	-25.2	-23.0	-20.2	-16.8	15.7	.4738
600.	2357.	16.3	-43.6	-42.7	-40.1	-36.6	-32.6	23.6	.5955
700.	2144.	24.8	-67.4	-66.4	-63.3	-59.3	-54.6	33.5	.7290
800.	1941.	36.1	-98.8	-97.6	-94.1	-89.5	-84.1	45.9	.8760
900.	1752.	51.0	-139.4	-138.0	-134.1	-128.9	-122.9	60.9	1.0386
1000.	1579.	70.6	-191.5	-190.0	-185.6	-179.9	-173.2	79.2	1.2192

BALLISTIC COEFFICIENT .39, MUZZLE VELOCITY FT/SEC 4000.

RANGE	VELOCITY	HEIGHT	100 YD	150 YD	200 YD	250 YD	300 YD	DRIFT	TIME
0.	4000.	0.0	-1.5	-1.5	-1.5	-1.5	-1.5	0.0	0.0000
100.	3702.	-.5	0.0	.1	.5	1.1	1.7	.5	.0780
200.	3418.	.5	-1.1	-.8	0.0	1.1	2.4	2.2	.1624
300.	3150.	2.3	-5.1	-4.7	-3.5	-1.9	0.0	5.1	.2540
400.	2896.	5.2	-12.7	-12.2	-10.6	-8.4	-5.8	9.4	.3532
500.	2657.	9.5	-24.4	-23.7	-21.7	-19.0	-15.8	15.2	.4614
600.	2430.	15.4	-40.9	-40.2	-37.8	-34.5	-30.7	22.8	.5794
700.	2213.	23.4	-63.4	-62.5	-59.7	-55.9	-51.5	32.3	.7088
800.	2006.	34.0	-92.9	-91.9	-88.7	-84.3	-79.3	44.2	.8512
900.	1813.	48.1	-131.0	-129.9	-126.3	-121.3	-115.7	58.7	1.0084
1000.	1634.	66.4	-179.9	-178.6	-174.6	-169.1	-162.8	76.2	1.1829

BALLISTIC COEFFICIENT .40, MUZZLE VELOCITY FT/SEC 2000.

RANGE	VELOCITY	HEIGHT	100 YD	150 YD	200 YD	250 YD	300 YD	DRIFT	TIME
0.	2000.	0.0	-1.5	-1.5	-1.5	-1.5	-1.5	0.0	0.0000
100.	1811.	.4	0.0	2.1	4.6	7.4	10.4	1.3	.1575
200.	1637.	4.5	-9.2	-5.0	0.0	5.6	11.7	5.6	.3319
300.	1477.	12.5	-31.3	-25.2	-17.6	-9.2	0.0	13.2	.5248
400.	1336.	25.5	-69.6	-61.4	-51.3	-40.1	-27.8	24.5	.7390
500.	1219.	44.8	-127.3	-116.9	-104.3	-90.3	-75.0	39.4	.9740
600.	1128.	71.9	-208.4	-196.0	-180.9	-164.1	-145.7	58.2	1.2307
700.	1063.	108.2	-317.3	-302.8	-285.2	-265.5	-244.1	80.3	1.5065
800.	1013.	154.4	-456.0	-439.5	-419.4	-396.9	-372.4	105.2	1.7980
900.	969.	210.9	-624.2	-605.6	-583.0	-557.7	-530.2	132.0	2.1000
1000.	929.	278.8	-824.4	-803.8	-778.6	-750.5	-719.9	160.7	2.4133

BALLISTIC COEFFICIENT .40, MUZZLE VELOCITY FT/SEC 2100.

RANGE	VELOCITY	HEIGHT	100 YD	150 YD	200 YD	250 YD	300 YD	DRIFT	TIME
0.	2100.	0.0	-1.5	-1.5	-1.5	-1.5	-1.5	0.0	0.0000
100.	1904.	.3	0.0	1.8	4.1	6.6	9.4	1.2	.1499
200.	1723.	4.0	-8.1	-4.5	0.0	5.1	10.6	5.2	.3153
300.	1556.	11.2	-28.1	-22.6	-15.9	-8.3	0.0	12.4	.4989
400.	1403.	22.9	-62.3	-55.0	-46.1	-36.0	-24.9	23.0	.7019
500.	1275.	40.5	-114.5	-105.4	-94.2	-81.5	-67.7	37.4	.9268
600.	1169.	65.2	-187.7	-176.8	-163.4	-148.2	-131.6	55.5	1.1726
700.	1093.	98.6	-286.6	-273.9	-258.2	-240.5	-221.1	77.2	1.4386
800.	1038.	141.3	-413.5	-399.0	-381.1	-360.8	-338.7	101.6	1.7204
900.	991.	195.1	-572.5	-556.1	-536.0	-513.2	-488.3	129.2	2.0199
1000.	949.	258.9	-759.5	-741.2	-718.9	-693.6	-665.9	157.9	2.3260

BALLISTIC COEFFICIENT .40, MUZZLE VELOCITY FT/SEC 2200.

RANGE	VELOCITY	HEIGHT	100 YD	150 YD	200 YD	250 YD	300 YD	DRIFT	TIME
0.	2200.	0.0	-1.5	-1.5	-1.5	-1.5	-1.5	0.0	0.0000
100.	1999.	.2	0.0	1.6	3.6	5.9	8.4	1.2	.1431
200.	1810.	3.6	-7.2	-4.0	0.0	4.5	9.5	4.9	.3008
300.	1636.	10.1	-25.2	-20.3	-14.3	-7.5	0.0	11.6	.4752
400.	1476.	20.7	-56.1	-49.6	-41.6	-32.5	-22.5	21.6	.6682
500.	1335.	36.6	-103.1	-95.0	-85.0	-73.7	-61.2	35.3	.8825
600.	1219.	59.2	-169.5	-159.8	-147.7	-134.2	-119.1	52.7	1.1177
700.	1127.	89.9	-259.4	-248.1	-234.1	-218.2	-200.7	73.9	1.3744
800.	1063.	130.0	-377.1	-364.2	-348.1	-330.0	-310.0	98.4	1.6502
900.	1013.	180.2	-524.6	-510.0	-492.0	-471.6	-449.1	125.8	1.9418
1000.	969.	240.9	-701.0	-684.9	-664.8	-642.2	-617.1	154.9	2.2438

BALLISTIC COEFFICIENT .40, MUZZLE VELOCITY FT/SEC 2300.

RANGE	VELOCITY	HEIGHT	100 YD	150 YD	200 YD	250 YD	300 YD	DRIFT	TIME
0.	2300.	0.0	-1.5	-1.5	-1.5	-1.5	-1.5	0.0	0.0000
100.	2094.	.1	0.0	1.4	3.2	5.3	7.5	1.1	.1367
200.	1899.	3.2	-6.5	-3.6	0.0	4.1	8.6	4.6	.2870
300.	1718.	9.1	-22.6	-18.3	-12.9	-6.7	0.0	10.8	.4529
400.	1551.	18.7	-50.6	-44.9	-37.7	-29.5	-20.5	20.3	.6370
500.	1399.	33.2	-93.0	-85.9	-76.8	-66.6	-55.3	33.2	.8407
600.	1272.	53.8	-153.3	-144.7	-133.9	-121.6	-108.1	49.8	1.0658
700.	1167.	81.9	-235.0	-225.0	-212.4	-198.0	-182.3	70.3	1.3125
800.	1091.	118.9	-342.4	-331.0	-316.6	-300.2	-282.2	94.2	1.5790
900.	1036.	165.5	-477.9	-465.1	-448.8	-430.4	-410.1	121.0	1.8613
1000.	990.	223.4	-645.0	-630.8	-612.7	-592.2	-569.7	150.7	2.1608

BALLISTIC COEFFICIENT .40, MUZZLE VELOCITY FT/SEC 2400.

RANGE	VELOCITY	HEIGHT	100 YD	150 YD	200 YD	250 YD	300 YD	DRIFT	TIME
0.	2400.	0.0	-1.5	-1.5	-1.5	-1.5	-1.5	0.0	0.0000
100.	2190.	.1	0.0	1.3	2.9	4.8	6.8	1.0	.1307
200.	1989.	2.9	-5.8	-3.3	0.0	3.7	7.8	4.3	.2744
300.	1802.	8.2	-20.4	-16.6	-11.7	-6.2	0.0	10.2	.4330
400.	1628.	17.0	-45.8	-40.7	-34.2	-26.8	-18.6	19.0	.6082
500.	1468.	30.1	-84.3	-77.9	-69.8	-60.5	-50.2	31.2	.8023
600.	1329.	48.9	-139.0	-131.4	-121.6	-110.4	-98.1	47.1	1.0173
700.	1213.	74.8	-213.6	-204.7	-193.3	-180.2	-165.9	66.8	1.2543
800.	1123.	108.9	-311.6	-301.4	-288.4	-273.5	-257.1	90.0	1.5115
900.	1060.	152.5	-436.9	-425.5	-410.8	-394.0	-375.6	116.4	1.7866
1000.	1011.	206.9	-593.2	-580.5	-564.2	-545.6	-525.1	146.1	2.0798

BALLISTIC COEFFICIENT .40, MUZZLE VELOCITY FT/SEC 2500.

RANGE	VELOCITY	HEIGHT	100 YD	150 YD	200 YD	250 YD	300 YD	DRIFT	TIME
0.	2500.	0.0	-1.5	-1.5	-1.5	-1.5	-1.5	0.0	0.0000
100.	2285.	.0	0.0	1.1	2.6	4.3	6.2	1.0	.1254
200.	2081.	2.6	-5.2	-3.0	0.0	3.4	7.1	4.0	.2628
300.	1886.	7.5	-18.5	-15.1	-10.7	-5.6	0.0	9.6	.4144
400.	1706.	15.5	-41.5	-37.1	-31.2	-24.4	-16.9	17.9	.5816
500.	1540.	27.5	-76.5	-71.0	-63.6	-55.1	-45.8	29.4	.7668
600.	1390.	44.6	-126.2	-119.5	-110.7	-100.5	-89.3	44.3	.9720
700.	1264.	68.2	-194.0	-186.2	-175.9	-164.0	-151.0	63.1	1.1985
800.	1160.	99.7	-283.6	-274.7	-262.9	-249.4	-234.4	85.7	1.4467
900.	1087.	140.3	-399.3	-389.3	-376.0	-360.8	-344.0	111.7	1.7144
1000.	1033.	190.9	-543.2	-532.1	-517.3	-500.4	-481.7	140.4	1.9979

BALLISTIC COEFFICIENT .40, MUZZLE VELOCITY FT/SEC 2600.

RANGE	VELOCITY	HEIGHT	100 YD	150 YD	200 YD	250 YD	300 YD	DRIFT	TIME
0.	2600.	0.0	-1.5	-1.5	-1.5	-1.5	-1.5	0.0	0.0000
100.	2381.	-.1	0.0	1.0	2.3	3.9	5.6	.9	.1206
200.	2172.	2.3	-4.7	-2.7	0.0	3.1	6.5	3.8	.2523
300.	1972.	6.8	-16.7	-13.8	-9.8	-5.1	0.0	9.0	.3974
400.	1786.	14.2	-37.8	-33.8	-28.5	-22.3	-15.5	16.9	.5574
500.	1614.	25.1	-69.7	-64.8	-58.1	-50.4	-41.8	27.7	.7341
600.	1455.	40.8	-115.0	-109.1	-101.1	-91.8	-81.6	41.8	.9300
700.	1318.	62.4	-176.9	-169.9	-160.6	-149.7	-137.8	59.7	1.1469
800.	1204.	91.4	-258.9	-250.9	-240.2	-227.9	-214.2	81.4	1.3854
900.	1117.	129.1	-365.0	-356.1	-344.1	-330.2	-314.8	106.7	1.6447
1000.	1056.	176.4	-498.7	-488.7	-475.4	-459.9	-442.9	135.1	1.9212

BALLISTIC COEFFICIENT .40, MUZZLE VELOCITY FT/SEC 2700.

RANGE	VELOCITY	HEIGHT	100 YD	150 YD	200 YD	250 YD	300 YD	DRIFT	TIME
0.	2700.	0.0	-1.5	-1.5	-1.5	-1.5	-1.5	0.0	0.0000
100.	2477.	-.1	0.0	.9	2.1	3.5	5.1	.8	.1159
200.	2263.	2.1	-4.2	-2.4	0.0	2.8	6.0	3.6	.2429
300.	2059.	6.2	-15.3	-12.6	-8.9	-4.7	0.0	8.5	.3818
400.	1866.	13.0	-34.5	-30.9	-26.0	-20.4	-14.1	15.9	.5346
500.	1688.	23.0	-63.7	-59.3	-53.2	-46.2	-38.3	26.1	.7038
600.	1523.	37.4	-105.2	-99.8	-92.5	-84.1	-74.7	39.5	.8910
700.	1375.	57.2	-161.7	-155.5	-146.9	-137.1	-126.1	56.5	1.0987
800.	1252.	83.8	-236.7	-229.6	-219.8	-208.6	-196.0	77.2	1.3274
900.	1151.	118.7	-334.1	-326.1	-315.1	-302.5	-288.4	101.7	1.5778
1000.	1080.	163.0	-457.9	-449.0	-436.8	-422.8	-407.1	129.6	1.8472

BALLISTIC COEFFICIENT .40, MUZZLE VELOCITY FT/SEC 2800.

RANGE	VELOCITY	HEIGHT	100 YD	150 YD	200 YD	250 YD	300 YD	DRIFT	TIME
0.	2800.	0.0	-1.5	-1.5	-1.5	-1.5	-1.5	0.0	0.0000
100.	2571.	-.2	0.0	.8	1.9	3.2	4.6	.8	.1116
200.	2354.	1.9	-3.8	-2.2	0.0	2.6	5.5	3.4	.2337
300.	2146.	5.7	-13.9	-11.6	-8.2	-4.3	0.0	8.0	.3672
400.	1948.	11.9	-31.6	-28.4	-24.0	-18.8	-13.0	15.0	.5139
500.	1763.	21.2	-58.4	-54.5	-48.9	-42.4	-35.2	24.6	.6757
600.	1593.	34.3	-96.4	-91.8	-85.1	-77.3	-68.7	37.3	.8550
700.	1436.	52.5	-148.1	-142.7	-134.8	-125.8	-115.7	53.4	1.0534
800.	1302.	77.0	-216.9	-210.6	-201.7	-191.3	-179.8	73.2	1.2729
900.	1191.	109.3	-306.4	-299.4	-289.3	-277.7	-264.8	96.8	1.5143
1000.	1108.	150.7	-420.8	-413.0	-401.8	-388.9	-374.5	124.0	1.7761

BALLISTIC COEFFICIENT .40, MUZZLE VELOCITY FT/SEC 2900.

RANGE	VELOCITY	HEIGHT	100 YD	150 YD	200 YD	250 YD	300 YD	DRIFT	TIME
0.	2900.	0.0	-1.5	-1.5	-1.5	-1.5	-1.5	0.0	0.0000
100.	2666.	-.2	0.0	.7	1.7	2.9	4.2	.8	.1078
200.	2444.	1.7	-3.4	-2.0	0.0	2.4	5.1	3.2	.2253
300.	2232.	5.3	-12.7	-10.6	-7.6	-4.0	0.0	7.7	.3539
400.	2030.	11.0	-29.0	-26.2	-22.1	-17.3	-12.0	14.3	.4949
500.	1839.	19.5	-53.7	-50.2	-45.2	-39.2	-32.5	23.4	.6503
600.	1662.	31.7	-88.7	-84.5	-78.4	-71.2	-63.3	35.4	.8218
700.	1500.	48.4	-136.1	-131.3	-124.2	-115.8	-106.5	50.6	1.0116
800.	1355.	71.0	-199.4	-193.9	-185.8	-176.2	-165.6	69.6	1.2228
900.	1235.	100.9	-281.9	-275.6	-266.5	-255.7	-243.8	92.2	1.4550
1000.	1139.	139.3	-387.2	-380.3	-370.2	-358.2	-344.9	118.6	1.7083

BALLISTIC COEFFICIENT .40, MUZZLE VELOCITY FT/SEC 3000.

RANGE	VELOCITY	HEIGHT	100 YD	150 YD	200 YD	250 YD	300 YD	DRIFT	TIME
0.	3000.	0.0	-1.5	-1.5	-1.5	-1.5	-1.5	0.0	0.0000
100.	2760.	-.2	0.0	.6	1.5	2.6	3.9	.8	.1046
200.	2534.	1.5	-3.1	-1.9	0.0	2.2	4.6	3.2	.2180
300.	2318.	4.9	-11.6	-9.8	-7.0	-3.7	0.0	7.3	.3417
400.	2112.	10.2	-26.6	-24.2	-20.4	-16.0	-11.1	13.6	.4773
500.	1915.	18.1	-49.4	-46.4	-41.7	-36.2	-30.1	22.3	.6266
600.	1733.	29.3	-81.7	-78.0	-72.4	-65.8	-58.5	33.6	.7911
700.	1565.	44.7	-125.6	-121.3	-114.8	-107.1	-98.5	48.1	.9735
800.	1411.	65.5	-183.6	-178.7	-171.2	-162.4	-152.6	66.0	1.1752
900.	1282.	93.1	-259.5	-254.0	-245.6	-235.7	-224.7	87.8	1.3986
1000.	1174.	129.1	-357.3	-351.3	-341.9	-330.9	-318.7	113.5	1.6448

BALLISTIC COEFFICIENT .40, MUZZLE VELOCITY FT/SEC 3100.

RANGE	VELOCITY	HEIGHT	100 YD	150 YD	200 YD	250 YD	300 YD	DRIFT	TIME
0.	3100.	0.0	-1.5	-1.5	-1.5	-1.5	-1.5	0.0	0.0000
100.	2855.	-.3	0.0	.5	1.4	2.4	3.6	.7	.1006
200.	2624.	1.4	-2.8	-1.7	0.0	2.1	4.3	2.9	.2102
300.	2404.	4.5	-10.7	-9.0	-6.5	-3.4	0.0	7.0	.3299
400.	2193.	9.4	-24.6	-22.4	-19.0	-14.9	-10.3	12.9	.4604
500.	1993.	16.8	-45.7	-42.9	-38.7	-33.6	-27.9	21.1	.6039
600.	1805.	27.1	-75.5	-72.3	-67.2	-61.0	-54.2	32.0	.7622
700.	1631.	41.4	-116.1	-112.3	-106.3	-99.1	-91.2	45.7	.9371
800.	1471.	60.6	-169.7	-165.3	-158.5	-150.3	-141.2	62.8	1.1308
900.	1331.	86.2	-239.7	-234.7	-227.1	-217.8	-207.6	83.5	1.3456
1000.	1215.	119.4	-329.7	-324.3	-315.8	-305.5	-294.1	108.1	1.5820

BALLISTIC COEFFICIENT .40, MUZZLE VELOCITY FT/SEC 3200.

RANGE	VELOCITY	HEIGHT	100 YD	150 YD	200 YD	250 YD	300 YD	DRIFT	TIME
0.	3200.	0.0	-1.5	-1.5	-1.5	-1.5	-1.5	0.0	0.0000
100.	2950.	-.3	0.0	.5	1.3	2.2	3.3	.7	.0975
200.	2713.	1.2	-2.5	-1.6	0.0	1.9	4.0	2.9	.2038
300.	2489.	4.1	-9.8	-8.4	-6.0	-3.2	0.0	6.7	.3192
400.	2275.	8.8	-22.7	-20.8	-17.6	-13.9	-9.6	12.4	.4453
500.	2070.	15.6	-42.3	-39.9	-36.0	-31.2	-26.0	20.2	.5835
600.	1877.	25.2	-69.9	-67.1	-62.4	-56.7	-50.4	30.5	.7356
700.	1698.	38.5	-107.5	-104.2	-98.7	-92.1	-84.7	43.6	.9037
800.	1532.	56.3	-157.2	-153.3	-147.1	-139.5	-131.1	59.8	1.0899
900.	1383.	79.9	-221.7	-217.4	-210.3	-201.9	-192.3	79.6	1.2961
1000.	1258.	110.7	-305.0	-300.2	-292.3	-282.9	-272.3	103.2	1.5239

BALLISTIC COEFFICIENT .40, MUZZLE VELOCITY FT/SEC 3300.

RANGE	VELOCITY	HEIGHT	100 YD	150 YD	200 YD	250 YD	300 YD	DRIFT	TIME
0.	3300.	0.0	-1.5	-1.5	-1.5	-1.5	-1.5	0.0	0.0000
100.	3045.	-.3	0.0	.4	1.1	2.0	3.0	.6	.0946
200.	2803.	1.1	-2.3	-1.4	0.0	1.7	3.7	2.8	.1976
300.	2574.	3.8	-9.0	-7.7	-5.6	-3.0	0.0	6.4	.3092
400.	2356.	8.2	-21.0	-19.3	-16.4	-13.0	-9.0	11.9	.4311
500.	2148.	14.5	-39.2	-37.1	-33.5	-29.2	-24.2	19.3	.5644
600.	1950.	23.5	-65.0	-62.4	-58.1	-52.9	-47.0	29.1	.7111
700.	1765.	35.8	-99.8	-96.8	-91.8	-85.7	-78.8	41.6	.8726
800.	1594.	52.3	-145.9	-142.5	-136.8	-129.9	-121.9	57.1	1.0517
900.	1437.	74.3	-205.8	-201.9	-195.5	-187.7	-178.7	76.0	1.2500
1000.	1303.	102.9	-282.7	-278.5	-271.3	-262.6	-252.7	98.6	1.4692

BALLISTIC COEFFICIENT .40, MUZZLE VELOCITY FT/SEC 3400.

RANGE	VELOCITY	HEIGHT	100 YD	150 YD	200 YD	250 YD	300 YD	DRIFT	TIME
0.	3400.	0.0	-1.5	-1.5	-1.5	-1.5	-1.5	0.0	0.0000
100.	3139.	-.3	0.0	.4	1.0	1.8	2.7	.6	.0919
200.	2892.	1.0	-2.0	-1.3	0.0	1.6	3.5	2.6	.1913
300.	2659.	3.6	-8.2	-7.2	-5.2	-2.8	0.0	6.1	.2995
400.	2437.	7.6	-19.4	-18.0	-15.3	-12.1	-8.4	11.3	.4173
500.	2225.	13.6	-36.4	-34.6	-31.3	-27.2	-22.6	18.5	.5462
600.	2023.	21.9	-60.4	-58.3	-54.3	-49.4	-43.9	27.8	.6876
700.	1833.	33.4	-92.8	-90.4	-85.7	-80.0	-73.6	39.7	.8435
800.	1657.	48.8	-135.6	-132.8	-127.5	-121.0	-113.6	54.5	1.0156
900.	1495.	69.1	-191.1	-187.9	-182.0	-174.7	-166.4	72.5	1.2061
1000.	1351.	95.7	-262.6	-259.1	-252.4	-244.3	-235.1	94.2	1.4177

BALLISTIC COEFFICIENT .40, MUZZLE VELOCITY FT/SEC 3500.

RANGE	VELOCITY	HEIGHT	100 YD	150 YD	200 YD	250 YD	300 YD	DRIFT	TIME
0.	3500.	0.0	-1.5	-1.5	-1.5	-1.5	-1.5	0.0	0.0000
100.	3234.	-.4	0.0	.3	.9	1.7	2.5	.6	.0893
200.	2982.	.9	-1.8	-1.2	0.0	1.5	3.2	2.6	.1859
300.	2744.	3.3	-7.6	-6.7	-4.8	-2.6	0.0	5.9	.2908
400.	2518.	7.1	-18.0	-16.8	-14.3	-11.3	-7.9	11.0	.4051
500.	2303.	12.7	-33.9	-32.3	-29.3	-25.5	-21.2	17.8	.5296
600.	2097.	20.6	-56.3	-54.4	-50.8	-46.3	-41.1	26.7	.6662
700.	1902.	31.2	-86.5	-84.3	-80.1	-74.8	-68.8	38.1	.8163
800.	1720.	45.5	-126.4	-123.8	-119.0	-113.0	-106.1	52.1	.9820
900.	1553.	64.5	-178.2	-175.3	-169.9	-163.2	-155.4	69.4	1.1659
1000.	1401.	89.2	-244.4	-241.2	-235.2	-227.7	-219.0	90.1	1.3692

BALLISTIC COEFFICIENT .40, MUZZLE VELOCITY FT/SEC 3600.

RANGE	VELOCITY	HEIGHT	100 YD	150 YD	200 YD	250 YD	300 YD	DRIFT	TIME
0.	3600.	0.0	-1.5	-1.5	-1.5	-1.5	-1.5	0.0	0.0000
100.	3329.	-.4	0.0	.3	.8	1.5	2.3	.6	.0865
200.	3072.	.8	-1.6	-1.1	0.0	1.4	3.0	2.4	.1803
300.	2829.	3.1	-7.0	-6.2	-4.5	-2.4	0.0	5.6	.2821
400.	2598.	6.7	-16.7	-15.7	-13.4	-10.6	-7.4	10.5	.3928
500.	2380.	11.9	-31.6	-30.2	-27.4	-23.9	-19.9	17.0	.5134
600.	2170.	19.2	-52.5	-50.9	-47.6	-43.3	-38.5	25.6	.6453
700.	1971.	29.2	-80.8	-78.9	-75.0	-70.1	-64.5	36.5	.7905
800.	1784.	42.6	-118.1	-116.0	-111.5	-105.9	-99.4	50.0	.9506
900.	1612.	60.3	-166.3	-163.9	-158.9	-152.6	-145.3	66.4	1.1274
1000.	1454.	83.3	-228.0	-225.4	-219.8	-212.8	-204.7	86.3	1.3235

BALLISTIC COEFFICIENT .40, MUZZLE VELOCITY FT/SEC 3700.

RANGE	VELOCITY	HEIGHT	100 YD	150 YD	200 YD	250 YD	300 YD	DRIFT	TIME
0.	3700.	0.0	-1.5	-1.5	-1.5	-1.5	-1.5	0.0	0.0000
100.	3424.	-.4	0.0	.2	.7	1.4	2.2	.6	.0844
200.	3162.	.7	-1.5	-1.0	0.0	1.3	2.8	2.4	.1757
300.	2914.	2.9	-6.5	-5.8	-4.2	-2.3	0.0	5.5	.2747
400.	2679.	6.2	-15.5	-14.6	-12.6	-9.9	-6.9	10.1	.3818
500.	2456.	11.2	-29.5	-28.3	-25.7	-22.5	-18.7	16.4	.4988
600.	2244.	18.1	-49.1	-47.8	-44.7	-40.8	-36.2	24.7	.6268
700.	2041.	27.5	-75.7	-74.1	-70.5	-65.9	-60.6	35.1	.7670
800.	1849.	40.0	-110.5	-108.7	-104.5	-99.3	-93.2	48.0	.9213
900.	1672.	56.5	-155.6	-153.5	-148.9	-143.0	-136.1	63.8	1.0920
1000.	1509.	78.0	-213.1	-210.8	-205.7	-199.1	-191.5	82.7	1.2808

BALLISTIC COEFFICIENT .40, MUZZLE VELOCITY FT/SEC 3800.

RANGE	VELOCITY	HEIGHT	100 YD	150 YD	200 YD	250 YD	300 YD	DRIFT	TIME
0.	3800.	0.0	-1.5	-1.5	-1.5	-1.5	-1.5	0.0	0.0000
100.	3519.	-.4	0.0	.2	.7	1.3	2.0	.5	.0820
200.	3252.	.7	-1.3	-.9	0.0	1.2	2.6	2.3	.1708
300.	2999.	2.7	-6.0	-5.4	-4.0	-2.1	0.0	5.3	.2668
400.	2760.	5.9	-14.5	-13.7	-11.8	-9.3	-6.6	9.8	.3714
500.	2533.	10.5	-27.6	-26.6	-24.2	-21.1	-17.7	15.9	.4849
600.	2317.	17.0	-46.0	-44.9	-42.0	-38.3	-34.1	23.7	.6086
700.	2111.	25.8	-70.9	-69.5	-66.2	-61.9	-57.0	33.7	.7443
800.	1915.	37.6	-103.6	-102.1	-98.3	-93.3	-87.7	46.1	.8937
900.	1732.	53.0	-145.7	-144.0	-139.7	-134.1	-127.9	61.2	1.0582
1000.	1564.	73.1	-199.7	-197.8	-193.0	-186.8	-179.8	79.4	1.2407

BALLISTIC COEFFICIENT .40, MUZZLE VELOCITY FT/SEC 3900.

RANGE	VELOCITY	HEIGHT	100 YD	150 YD	200 YD	250 YD	300 YD	DRIFT	TIME
0.	3900.	0.0	-1.5	-1.5	-1.5	-1.5	-1.5	0.0	0.0000
100.	3614.	-.4	0.0	.1	.6	1.2	1.8	.5	.0799
200.	3342.	.6	-1.2	-.9	0.0	1.1	2.5	2.2	.1661
300.	3084.	2.5	-5.5	-5.0	-3.7	-2.0	0.0	5.1	.2595
400.	2840.	5.5	-13.4	-12.8	-11.1	-8.8	-6.1	9.3	.3607
500.	2610.	9.9	-25.7	-25.0	-22.8	-19.9	-16.6	15.2	.4711
600.	2390.	16.0	-43.1	-42.2	-39.5	-36.1	-32.1	22.8	.5913
700.	2181.	24.3	-66.4	-65.4	-62.3	-58.3	-53.6	32.4	.7225
800.	1980.	35.3	-97.0	-95.9	-92.4	-87.8	-82.5	44.3	.8668
900.	1793.	49.8	-136.6	-135.3	-131.4	-126.2	-120.2	58.8	1.0262
1000.	1621.	68.6	-187.0	-185.6	-181.2	-175.5	-168.8	76.2	1.2021

BALLISTIC COEFFICIENT .40, MUZZLE VELOCITY FT/SEC 4000.

RANGE	VELOCITY	HEIGHT	100 YD	150 YD	200 YD	250 YD	300 YD	DRIFT	TIME
0.	4000.	0.0	-1.5	-1.5	-1.5	-1.5	-1.5	0.0	0.0000
100.	3709.	-.5	0.0	.1	.5	1.1	1.7	.5	.0779
200.	3432.	.5	-1.0	-.8	0.0	1.1	2.3	2.1	.1621
300.	3170.	2.3	-5.1	-4.7	-3.5	-1.9	0.0	4.9	.2531
400.	2921.	5.2	-12.6	-12.1	-10.5	-8.3	-5.8	9.1	.3519
500.	2686.	9.4	-24.1	-23.4	-21.5	-18.8	-15.6	14.7	.4587
600.	2463.	15.1	-40.4	-39.7	-37.3	-34.1	-30.3	22.1	.5754
700.	2250.	23.0	-62.5	-61.6	-58.8	-55.1	-50.7	31.3	.7030
800.	2047.	33.3	-91.4	-90.4	-87.2	-82.9	-77.9	42.7	.8428
900.	1855.	46.9	-128.4	-127.3	-123.7	-118.9	-113.3	56.6	.9966
1000.	1677.	64.6	-175.8	-174.6	-170.6	-165.2	-159.0	73.3	1.1667

BALLISTIC COEFFICIENT .41, MUZZLE VELOCITY FT/SEC 2000.

RANGE	VELOCITY	HEIGHT	100 YD	150 YD	200 YD	250 YD	300 YD	DRIFT	TIME
0.	2000.	0.0	-1.5	-1.5	-1.5	-1.5	-1.5	0.0	0.0000
100.	1816.	.4	0.0	2.1	4.6	7.4	10.4	1.3	.1573
200.	1645.	4.5	-9.1	-5.0	0.0	5.6	11.6	5.5	.3310
300.	1488.	12.4	-31.1	-24.9	-17.4	-9.1	0.0	12.8	.5227
400.	1348.	25.2	-69.0	-60.7	-50.7	-39.5	-27.5	23.7	.7349
500.	1232.	44.2	-125.9	-115.6	-103.1	-89.1	-74.0	38.4	.9679
600.	1139.	70.9	-205.8	-193.4	-178.4	-161.7	-143.6	56.7	1.2220
700.	1073.	106.3	-312.2	-297.8	-280.3	-260.7	-239.6	78.1	1.4935
800.	1023.	151.5	-448.0	-431.5	-411.5	-389.2	-365.0	102.3	1.7810
900.	979.	206.9	-613.5	-594.9	-572.4	-547.3	-520.2	128.5	2.0800
1000.	939.	274.1	-812.4	-791.8	-766.8	-738.9	-708.7	157.2	2.3931

BALLISTIC COEFFICIENT .41, MUZZLE VELOCITY FT/SEC 2100.

RANGE	VELOCITY	HEIGHT	100 YD	150 YD	200 YD	250 YD	300 YD	DRIFT	TIME
0.	2100.	0.0	-1.5	-1.5	-1.5	-1.5	-1.5	0.0	0.0000
100.	1909.	.3	0.0	1.8	4.0	6.5	9.3	1.2	.1498
200.	1732.	4.0	-8.1	-4.5	0.0	5.0	10.5	5.1	.3146
300.	1568.	11.1	-27.9	-22.4	-15.7	-8.2	0.0	12.1	.4970
400.	1417.	22.7	-61.7	-54.5	-45.6	-35.6	-24.6	22.3	.6982
500.	1290.	39.9	-113.0	-104.0	-92.9	-80.3	-66.6	36.3	.9204
600.	1183.	64.2	-185.2	-174.3	-161.0	-145.9	-129.5	53.9	1.1636
700.	1104.	97.0	-282.3	-269.7	-254.0	-236.5	-217.3	75.1	1.4267
800.	1047.	139.4	-408.7	-394.2	-376.4	-356.4	-334.4	99.6	1.7089
900.	1001.	191.1	-561.8	-545.6	-525.5	-503.0	-478.3	125.6	1.9994
1000.	959.	254.3	-747.3	-729.2	-706.9	-681.9	-654.4	154.3	2.3050

BALLISTIC COEFFICIENT .41, MUZZLE VELOCITY FT/SEC 2200.

RANGE	VELOCITY	HEIGHT	100 YD	150 YD	200 YD	250 YD	300 YD	DRIFT	TIME
0.	2200.	0.0	-1.5	-1.5	-1.5	-1.5	-1.5	0.0	0.0000
100.	2004.	.2	0.0	1.6	3.6	5.9	8.3	1.2	.1430
200.	1819.	3.6	-7.2	-4.0	0.0	4.5	9.5	4.8	.2999
300.	1649.	10.0	-25.0	-20.2	-14.2	-7.4	0.0	11.3	.4734
400.	1491.	20.5	-55.5	-49.1	-41.1	-32.1	-22.2	21.0	.6646
500.	1351.	36.1	-101.8	-93.8	-83.9	-72.6	-60.2	34.3	.8765
600.	1234.	58.3	-167.3	-157.7	-145.8	-132.2	-117.4	51.3	1.1095
700.	1140.	88.3	-255.3	-244.0	-230.1	-214.3	-197.0	71.8	1.3622
800.	1074.	127.4	-370.2	-357.3	-341.4	-323.4	-303.6	95.6	1.6339
900.	1024.	176.4	-514.3	-499.9	-482.0	-461.7	-439.4	122.2	1.9213
1000.	980.	235.8	-687.4	-671.3	-651.4	-628.9	-604.1	150.7	2.2198

BALLISTIC COEFFICIENT .41, MUZZLE VELOCITY FT/SEC 2300.

RANGE	VELOCITY	HEIGHT	100 YD	150 YD	200 YD	250 YD	300 YD	DRIFT	TIME
0.	2300.	0.0	-1.5	-1.5	-1.5	-1.5	-1.5	0.0	0.0000
100.	2099.	.1	0.0	1.4	3.2	5.2	7.5	1.1	.1366
200.	1908.	3.2	-6.4	-3.6	0.0	4.1	8.5	4.5	.2864
300.	1731.	9.0	-22.4	-18.2	-12.8	-6.7	0.0	10.6	.4514
400.	1567.	18.5	-50.1	-44.5	-37.3	-29.2	-20.2	19.7	.6338
500.	1417.	32.7	-91.9	-84.8	-75.8	-65.7	-54.5	32.2	.8351
600.	1289.	52.9	-151.1	-142.6	-131.8	-119.7	-106.3	48.4	1.0573
700.	1183.	80.4	-231.2	-221.3	-208.7	-194.5	-178.9	68.2	1.3006
800.	1104.	116.6	-336.4	-325.1	-310.7	-294.5	-276.6	91.6	1.5638
900.	1047.	162.8	-470.9	-458.1	-441.9	-423.7	-403.6	118.3	1.8460
1000.	1001.	218.4	-631.7	-617.6	-599.6	-579.3	-557.0	146.5	2.1365

BALLISTIC COEFFICIENT .41, MUZZLE VELOCITY FT/SEC 2400.

RANGE	VELOCITY	HEIGHT	100 YD	150 YD	200 YD	250 YD	300 YD	DRIFT	TIME
0.	2400.	0.0	-1.5	-1.5	-1.5	-1.5	-1.5	0.0	0.0000
100.	2195.	.1	0.0	1.3	2.9	4.7	6.8	1.0	.1305
200.	1999.	2.9	-5.8	-3.2	0.0	3.7	7.7	4.2	.2739
300.	1815.	8.2	-20.3	-16.5	-11.6	-6.1	0.0	9.9	.4313
400.	1645.	16.8	-45.4	-40.3	-33.8	-26.5	-18.4	18.5	.6051
500.	1487.	29.7	-83.3	-77.0	-68.9	-59.6	-49.5	30.2	.7968
600.	1348.	48.1	-137.1	-129.5	-119.7	-108.7	-96.5	45.6	1.0092
700.	1232.	73.3	-210.0	-201.1	-189.8	-176.9	-162.7	64.6	1.2423
800.	1138.	106.8	-306.1	-296.0	-283.0	-268.3	-252.1	87.4	1.4967
900.	1073.	149.3	-428.6	-417.3	-402.7	-386.1	-367.8	113.2	1.7681
1000.	1023.	202.1	-580.4	-567.7	-551.6	-533.1	-512.9	141.8	2.0556

BALLISTIC COEFFICIENT .41, MUZZLE VELOCITY FT/SEC 2500.

RANGE	VELOCITY	HEIGHT	100 YD	150 YD	200 YD	250 YD	300 YD	DRIFT	TIME
0.	2500.	0.0	-1.5	-1.5	-1.5	-1.5	-1.5	0.0	0.0000
100.	2291.	.0	0.0	1.1	2.6	4.3	6.1	.9	.1253
200.	2090.	2.6	-5.2	-2.9	0.0	3.4	7.1	3.9	.2624
300.	1900.	7.4	-18.3	-15.0	-10.6	-5.5	0.0	9.3	.4129
400.	1723.	15.3	-41.1	-36.7	-30.8	-24.1	-16.7	17.3	.5786
500.	1560.	27.1	-75.7	-70.2	-62.8	-54.4	-45.2	28.5	.7619
600.	1410.	43.9	-124.5	-117.8	-109.0	-98.9	-87.8	43.0	.9641
700.	1284.	66.9	-190.9	-183.2	-172.9	-161.1	-148.2	61.1	1.1874
800.	1178.	97.7	-278.6	-269.7	-257.9	-244.5	-229.7	83.1	1.4319
900.	1101.	137.2	-391.3	-381.3	-368.1	-353.0	-336.3	108.3	1.6955
1000.	1045.	187.0	-533.1	-522.0	-507.3	-490.5	-472.0	136.8	1.9775

BALLISTIC COEFFICIENT .41, MUZZLE VELOCITY FT/SEC 2600.

RANGE	VELOCITY	HEIGHT	100 YD	150 YD	200 YD	250 YD	300 YD	DRIFT	TIME
0.	2600.	0.0	-1.5	-1.5	-1.5	-1.5	-1.5	0.0	0.0000
100.	2386.	-.1	0.0	1.0	2.3	3.8	5.5	.9	.1205
200.	2182.	2.3	-4.6	-2.6	0.0	3.1	6.4	3.7	.2517
300.	1987.	6.8	-16.6	-13.6	-9.7	-5.1	0.0	8.8	.3959
400.	1803.	14.0	-37.4	-33.5	-28.2	-22.1	-15.3	16.4	.5545
500.	1634.	24.8	-69.0	-64.0	-57.4	-49.7	-41.3	26.8	.7293
600.	1477.	40.1	-113.4	-107.5	-99.6	-90.4	-80.2	40.5	.9224
700.	1339.	61.2	-174.0	-167.1	-157.8	-147.1	-135.3	57.8	1.1361
800.	1225.	89.4	-254.1	-246.2	-235.6	-223.3	-209.8	78.7	1.3705
900.	1133.	126.1	-357.5	-348.6	-336.7	-322.9	-307.7	103.4	1.6257
1000.	1069.	172.3	-488.0	-478.1	-464.8	-449.5	-432.6	131.1	1.8987

BALLISTIC COEFFICIENT .41, MUZZLE VELOCITY FT/SEC 2700.

RANGE	VELOCITY	HEIGHT	100 YD	150 YD	200 YD	250 YD	300 YD	DRIFT	TIME
0.	2700.	0.0	-1.5	-1.5	-1.5	-1.5	-1.5	0.0	0.0000
100.	2482.	-.1	0.0	.9	2.1	3.5	5.0	.8	.1158
200.	2273.	2.1	-4.2	-2.4	0.0	2.8	5.9	3.5	.2422
300.	2074.	6.2	-15.1	-12.5	-8.8	-4.7	0.0	8.3	.3803
400.	1884.	12.8	-34.2	-30.6	-25.8	-20.2	-14.0	15.4	.5321
500.	1709.	22.7	-63.0	-58.6	-52.5	-45.6	-37.8	25.3	.6992
600.	1547.	36.8	-103.8	-98.5	-91.2	-82.9	-73.6	38.3	.8840
700.	1398.	56.1	-159.0	-152.9	-144.4	-134.6	-123.8	54.6	1.0881
800.	1274.	82.1	-232.5	-225.4	-215.8	-204.6	-192.2	74.7	1.3135
900.	1171.	115.9	-327.2	-319.2	-308.3	-295.8	-281.8	98.4	1.5591
1000.	1096.	159.0	-447.7	-438.9	-426.8	-412.9	-397.3	125.6	1.8246

BALLISTIC COEFFICIENT .41, MUZZLE VELOCITY FT/SEC 2800.

RANGE	VELOCITY	HEIGHT	100 YD	150 YD	200 YD	250 YD	300 YD	DRIFT	TIME
0.	2800.	0.0	-1.5	-1.5	-1.5	-1.5	-1.5	0.0	0.0000
100.	2577.	-.2	0.0	.8	1.9	3.2	4.6	.8	.1115
200.	2364.	1.9	-3.8	-2.2	0.0	2.6	5.4	3.3	.2332
300.	2161.	5.7	-13.8	-11.4	-8.1	-4.2	0.0	7.8	.3657
400.	1966.	11.8	-31.3	-28.2	-23.7	-18.6	-12.9	14.6	.5114
500.	1785.	20.9	-57.8	-53.9	-48.3	-41.9	-34.8	23.9	.6716
600.	1617.	33.8	-95.1	-90.5	-83.8	-76.1	-67.6	36.1	.8481
700.	1462.	51.5	-145.8	-140.4	-132.6	-123.6	-113.7	51.6	1.0434
800.	1326.	75.4	-212.9	-206.7	-197.8	-187.5	-176.2	70.8	1.2591
900.	1214.	106.7	-300.2	-293.3	-283.2	-271.7	-258.9	93.6	1.4963
1000.	1126.	146.8	-411.1	-403.3	-392.2	-379.3	-365.2	120.0	1.7532

BALLISTIC COEFFICIENT .41, MUZZLE VELOCITY FT/SEC 2900.

RANGE	VELOCITY	HEIGHT	100 YD	150 YD	200 YD	250 YD	300 YD	DRIFT	TIME
0.	2900.	0.0	-1.5	-1.5	-1.5	-1.5	-1.5	0.0	0.0000
100.	2672.	-.2	0.0	.7	1.7	2.9	4.2	.7	.1077
200.	2455.	1.7	-3.4	-2.0	0.0	2.4	5.0	3.2	.2248
300.	2247.	5.2	-12.6	-10.5	-7.5	-3.9	0.0	7.5	.3528
400.	2049.	10.9	-28.7	-26.0	-21.9	-17.2	-11.9	13.9	.4926
500.	1861.	19.3	-53.0	-49.6	-44.5	-38.6	-32.0	22.6	.6459
600.	1687.	31.2	-87.5	-83.3	-77.3	-70.1	-62.3	34.3	.8154
700.	1527.	47.5	-134.1	-129.3	-122.2	-113.9	-104.7	49.0	1.0024
800.	1381.	69.4	-195.7	-190.1	-182.1	-172.5	-162.0	67.1	1.2090
900.	1260.	98.3	-275.8	-269.6	-260.5	-249.8	-237.9	89.0	1.4366
1000.	1159.	135.6	-378.1	-371.2	-361.1	-349.2	-336.1	114.6	1.6856

BALLISTIC COEFFICIENT .41, MUZZLE VELOCITY FT/SEC 3000.

RANGE	VELOCITY	HEIGHT	100 YD	150 YD	200 YD	250 YD	300 YD	DRIFT	TIME
0.	3000.	0.0	-1.5	-1.5	-1.5	-1.5	-1.5	0.0	0.0000
100.	2766.	-.2	0.0	.6	1.5	2.6	3.8	.8	.1044
200.	2545.	1.5	-3.1	-1.9	0.0	2.2	4.6	3.1	.2175
300.	2333.	4.8	-11.5	-9.7	-6.9	-3.7	0.0	7.2	.3406
400.	2131.	10.1	-26.4	-24.0	-20.2	-15.9	-11.0	13.2	.4751
500.	1939.	17.9	-48.9	-45.9	-41.2	-35.8	-29.7	21.6	.6228
600.	1759.	28.8	-80.6	-77.0	-71.4	-64.9	-57.5	32.6	.7850
700.	1593.	43.9	-123.7	-119.4	-112.9	-105.3	-96.8	46.6	.9646
800.	1440.	64.1	-180.4	-175.5	-168.1	-159.4	-149.6	63.8	1.1627
900.	1308.	90.8	-254.1	-248.6	-240.3	-230.5	-219.5	84.7	1.3814
1000.	1199.	125.5	-348.5	-342.5	-333.2	-322.3	-310.1	109.4	1.6215

BALLISTIC COEFFICIENT .41, MUZZLE VELOCITY FT/SEC 3100.

RANGE	VELOCITY	HEIGHT	100 YD	150 YD	200 YD	250 YD	300 YD	DRIFT	TIME
0.	3100.	0.0	-1.5	-1.5	-1.5	-1.5	-1.5	0.0	0.0000
100.	2861.	-.3	0.0	.5	1.4	2.4	3.5	.6	.1004
200.	2635.	1.4	-2.8	-1.7	0.0	2.0	4.3	2.9	.2099
300.	2419.	4.4	-10.6	-9.0	-6.4	-3.4	0.0	6.8	.3288
400.	2213.	9.3	-24.3	-22.2	-18.8	-14.7	-10.2	12.5	.4583
500.	2017.	16.6	-45.2	-42.5	-38.3	-33.2	-27.5	20.5	.6004
600.	1831.	26.7	-74.6	-71.3	-66.3	-60.1	-53.4	31.0	.7566
700.	1660.	40.6	-114.3	-110.5	-104.6	-97.5	-89.6	44.2	.9286
800.	1501.	59.3	-166.6	-162.3	-155.5	-147.4	-138.4	60.6	1.1186
900.	1359.	84.0	-234.7	-229.8	-222.2	-213.1	-202.9	80.6	1.3291
1000.	1242.	116.1	-321.9	-316.5	-308.1	-297.9	-286.6	104.3	1.5603

BALLISTIC COEFFICIENT .41, MUZZLE VELOCITY FT/SEC 3200.

RANGE	VELOCITY	HEIGHT	100 YD	150 YD	200 YD	250 YD	300 YD	DRIFT	TIME
0.	3200.	0.0	-1.5	-1.5	-1.5	-1.5	-1.5	0.0	0.0000
100.	2956.	-.3	0.0	.5	1.3	2.2	3.2	.6	.0974
200.	2724.	1.2	-2.5	-1.6	0.0	1.9	4.0	2.8	.2033
300.	2505.	4.1	-9.7	-8.3	-6.0	-3.2	0.0	6.5	.3183
400.	2295.	8.7	-22.5	-20.6	-17.5	-13.7	-9.5	12.0	.4433
500.	2095.	15.4	-41.9	-39.5	-35.6	-30.9	-25.6	19.6	.5802
600.	1904.	24.9	-69.1	-66.3	-61.6	-56.0	-49.6	29.5	.7304
700.	1727.	37.8	-105.9	-102.6	-97.1	-90.6	-83.2	42.1	.8956
800.	1564.	55.1	-154.5	-150.7	-144.4	-137.0	-128.5	57.8	1.0785
900.	1413.	77.9	-217.2	-212.9	-205.9	-197.5	-188.0	76.8	1.2802
1000.	1287.	107.7	-297.9	-293.2	-285.3	-276.0	-265.4	99.5	1.5030

BALLISTIC COEFFICIENT .41, MUZZLE VELOCITY FT/SEC 3300.

RANGE	VELOCITY	HEIGHT	100 YD	150 YD	200 YD	250 YD	300 YD	DRIFT	TIME
0.	3300.	0.0	-1.5	-1.5	-1.5	-1.5	-1.5	0.0	0.0000
100.	3051.	-.3	0.0	.4	1.1	2.0	3.0	.6	.0945
200.	2814.	1.1	-2.3	-1.4	0.0	1.7	3.7	2.7	.1971
300.	2590.	3.8	-8.9	-7.7	-5.5	-2.9	0.0	6.2	.3082
400.	2377.	8.1	-20.8	-19.1	-16.3	-12.8	-8.9	11.5	.4291
500.	2173.	14.4	-38.8	-36.7	-33.1	-28.8	-23.9	18.7	.5609
600.	1978.	23.2	-64.1	-61.6	-57.3	-52.1	-46.2	28.2	.7057
700.	1795.	35.2	-98.4	-95.4	-90.4	-84.4	-77.5	40.2	.8650
800.	1627.	51.2	-143.4	-140.0	-134.3	-127.4	-119.5	55.1	1.0406
900.	1471.	72.4	-201.6	-197.8	-191.3	-183.6	-174.7	73.3	1.2346
1000.	1334.	100.1	-276.4	-272.2	-265.0	-256.3	-246.5	95.1	1.4495

BALLISTIC COEFFICIENT .41, MUZZLE VELOCITY FT/SEC 3400.

RANGE	VELOCITY	HEIGHT	100 YD	150 YD	200 YD	250 YD	300 YD	DRIFT	TIME
0.	3400.	0.0	-1.5	-1.5	-1.5	-1.5	-1.5	0.0	0.0000
100.	3145.	-.3	0.0	.4	1.0	1.8	2.7	.6	.0918
200.	2904.	1.0	-2.0	-1.3	0.0	1.6	3.4	2.6	.1910
300.	2675.	3.5	-8.2	-7.1	-5.1	-2.7	0.0	5.9	.2984
400.	2458.	7.5	-19.2	-17.8	-15.2	-11.9	-8.3	11.0	.4155
500.	2251.	13.4	-36.0	-34.3	-31.0	-26.9	-22.4	18.0	.5432
600.	2052.	21.6	-59.7	-57.5	-53.6	-48.7	-43.3	27.0	.6828
700.	1864.	32.8	-91.4	-89.0	-84.3	-78.7	-72.3	38.4	.8359
800.	1690.	47.7	-133.3	-130.5	-125.2	-118.8	-111.5	52.7	1.0051
900.	1529.	67.4	-187.4	-184.2	-178.3	-171.0	-162.9	70.0	1.1918
1000.	1384.	93.1	-256.5	-253.0	-246.3	-238.3	-229.2	90.8	1.3981

BALLISTIC COEFFICIENT .41, MUZZLE VELOCITY FT/SEC 3500.

RANGE	VELOCITY	HEIGHT	100 YD	150 YD	200 YD	250 YD	300 YD	DRIFT	TIME
0.	3500.	0.0	-1.5	-1.5	-1.5	-1.5	-1.5	0.0	0.0000
100.	3240.	-.4	0.0	.3	.9	1.7	2.5	.6	.0892
200.	2994.	.9	-1.8	-1.2	0.0	1.5	3.2	2.5	.1854
300.	2761.	3.3	-7.6	-6.6	-4.8	-2.6	0.0	5.8	.2900
400.	2539.	7.1	-17.9	-16.6	-14.2	-11.2	-7.8	10.6	.4034
500.	2328.	12.6	-33.5	-32.0	-29.0	-25.2	-20.9	17.3	.5267
600.	2126.	20.2	-55.6	-53.7	-50.1	-45.6	-40.5	25.9	.6614
700.	1934.	30.7	-85.3	-83.1	-79.0	-73.7	-67.7	36.9	.8095
800.	1755.	44.6	-124.3	-121.8	-117.0	-111.0	-104.1	50.4	.9722
900.	1589.	63.0	-174.7	-171.9	-166.6	-159.8	-152.0	67.0	1.1522
1000.	1436.	86.8	-239.0	-235.8	-229.9	-222.4	-213.8	86.9	1.3509

BALLISTIC COEFFICIENT .41, MUZZLE VELOCITY FT/SEC 3600.

RANGE	VELOCITY	HEIGHT	100 YD	150 YD	200 YD	250 YD	300 YD	DRIFT	TIME
0.	3600.	0.0	-1.5	-1.5	-1.5	-1.5	-1.5	0.0	0.0000
100.	3335.	-.4	0.0	.3	.8	1.5	2.3	.6	.0865
200.	3084.	.8	-1.6	-1.1	0.0	1.4	3.0	2.3	.1800
300.	2846.	3.0	-6.9	-6.1	-4.5	-2.4	0.0	5.5	.2811
400.	2620.	6.6	-16.5	-15.5	-13.3	-10.5	-7.3	10.1	.3910
500.	2406.	11.8	-31.2	-29.9	-27.1	-23.7	-19.7	16.5	.5107
600.	2200.	19.0	-51.9	-50.3	-46.9	-42.8	-38.0	24.8	.6410
700.	2004.	28.7	-79.7	-77.8	-73.9	-69.1	-63.5	35.3	.7839
800.	1820.	41.7	-116.0	-113.9	-109.5	-103.9	-97.6	48.3	.9408
900.	1649.	58.8	-163.1	-160.7	-155.7	-149.5	-142.3	64.1	1.1142
1000.	1491.	81.0	-222.9	-220.2	-214.7	-207.7	-199.8	83.1	1.3054

BALLISTIC COEFFICIENT .41, MUZZLE VELOCITY FT/SEC 3700.

RANGE	VELOCITY	HEIGHT	100 YD	150 YD	200 YD	250 YD	300 YD	DRIFT	TIME
0.	3700.	0.0	-1.5	-1.5	-1.5	-1.5	-1.5	0.0	0.0000
100.	3430.	-.4	0.0	.2	.7	1.4	2.1	.6	.0843
200.	3174.	.7	-1.5	-1.0	0.0	1.3	2.8	2.3	.1752
300.	2931.	2.8	-6.4	-5.7	-4.2	-2.3	0.0	5.4	.2737
400.	2701.	6.2	-15.4	-14.5	-12.5	-9.9	-6.9	9.9	.3803
500.	2483.	11.1	-29.1	-28.0	-25.5	-22.2	-18.5	16.0	.4961
600.	2274.	17.8	-48.5	-47.2	-44.1	-40.2	-35.7	23.9	.6224
700.	2075.	27.0	-74.5	-73.0	-69.4	-64.9	-59.6	33.9	.7604
800.	1885.	39.2	-108.6	-106.8	-102.8	-97.5	-91.5	46.4	.9121
900.	1710.	55.1	-152.5	-150.5	-145.9	-140.0	-133.3	61.5	1.0791
1000.	1547.	75.9	-208.5	-206.2	-201.1	-194.6	-187.1	79.7	1.2639

BALLISTIC COEFFICIENT .41, MUZZLE VELOCITY FT/SEC 3800.

RANGE	VELOCITY	HEIGHT	100 YD	150 YD	200 YD	250 YD	300 YD	DRIFT	TIME
0.	3800.	0.0	-1.5	-1.5	-1.5	-1.5	-1.5	0.0	0.0000
100.	3525.	-.4	0.0	.2	.7	1.3	2.0	.5	.0820
200.	3265.	.6	-1.3	-.9	0.0	1.2	2.6	2.2	.1705
300.	3017.	2.6	-5.9	-5.3	-3.9	-2.1	0.0	5.1	.2661
400.	2782.	5.8	-14.3	-13.6	-11.7	-9.2	-6.4	9.5	.3696
500.	2560.	10.4	-27.2	-26.3	-24.0	-20.9	-17.4	15.4	.4821
600.	2348.	16.8	-45.5	-44.3	-41.5	-37.8	-33.7	23.1	.6047
700.	2145.	25.4	-69.9	-68.6	-65.3	-61.0	-56.1	32.7	.7383
800.	1952.	36.8	-101.9	-100.4	-96.6	-91.7	-86.1	44.6	.8850
900.	1771.	51.8	-143.0	-141.3	-137.1	-131.5	-125.3	59.1	1.0462
1000.	1604.	71.2	-195.3	-193.4	-188.7	-182.5	-175.6	76.5	1.2243

BALLISTIC COEFFICIENT .41, MUZZLE VELOCITY FT/SEC 3900.

RANGE	VELOCITY	HEIGHT	100 YD	150 YD	200 YD	250 YD	300 YD	DRIFT	TIME
0.	3900.	0.0	-1.5	-1.5	-1.5	-1.5	-1.5	0.0	0.0000
100.	3621.	-.4	0.0	.1	.6	1.1	1.8	.5	.0798
200.	3355.	.6	-1.2	-.9	0.0	1.1	2.5	2.1	.1658
300.	3103.	2.5	-5.4	-5.0	-3.7	-2.0	0.0	4.9	.2589
400.	2863.	5.4	-13.3	-12.7	-11.0	-8.7	-6.0	9.1	.3592
500.	2637.	9.8	-25.4	-24.7	-22.5	-19.7	-16.4	14.8	.4686
600.	2422.	15.8	-42.5	-41.7	-39.1	-35.6	-31.7	22.1	.5873
700.	2216.	23.9	-65.5	-64.5	-61.4	-57.4	-52.8	31.4	.7168
800.	2019.	34.6	-95.5	-94.4	-90.9	-86.3	-81.0	42.8	.8587
900.	1833.	48.7	-134.1	-132.8	-128.9	-123.8	-117.8	56.7	1.0147
1000.	1661.	66.8	-183.1	-181.6	-177.3	-171.6	-164.9	73.5	1.1866

BALLISTIC COEFFICIENT .41, MUZZLE VELOCITY FT/SEC 4000.

RANGE	VELOCITY	HEIGHT	100 YD	150 YD	200 YD	250 YD	300 YD	DRIFT	TIME
0.	4000.	0.0	-1.5	-1.5	-1.5	-1.5	-1.5	0.0	0.0000
100.	3716.	-.5	0.0	.1	.5	1.1	1.7	.5	.0779
200.	3445.	.5	-1.0	-.8	0.0	1.1	2.3	2.1	.1618
300.	3189.	2.3	-5.0	-4.6	-3.5	-1.9	0.0	4.8	.2523
400.	2945.	5.1	-12.4	-11.9	-10.3	-8.2	-5.7	8.8	.3501
500.	2714.	9.2	-23.8	-23.2	-21.2	-18.6	-15.5	14.3	.4564
600.	2495.	14.9	-39.9	-39.2	-36.8	-33.6	-29.9	21.4	.5717
700.	2286.	22.6	-61.6	-60.7	-58.0	-54.2	-49.9	30.3	.6973
800.	2086.	32.7	-89.8	-88.8	-85.7	-81.4	-76.5	41.3	.8346
900.	1896.	45.9	-126.1	-125.0	-121.4	-116.6	-111.1	54.7	.9855
1000.	1719.	62.9	-172.0	-170.7	-166.8	-161.5	-155.3	70.7	1.1515

BALLISTIC COEFFICIENT .42, MUZZLE VELOCITY FT/SEC 2000.

RANGE	VELOCITY	HEIGHT	100 YD	150 YD	200 YD	250 YD	300 YD	DRIFT	TIME
0.	2000.	0.0	-1.5	-1.5	-1.5	-1.5	-1.5	0.0	0.0000
100.	1820.	.4	0.0	2.1	4.5	7.3	10.3	1.2	.1571
200.	1653.	4.5	-9.1	-5.0	0.0	5.5	11.5	5.3	.3303
300.	1499.	12.3	-30.9	-24.7	-17.3	-9.0	0.0	12.4	.5207
400.	1360.	24.9	-68.3	-60.1	-50.2	-39.1	-27.1	23.1	.7312
500.	1245.	43.7	-124.5	-114.3	-101.9	-88.0	-73.1	37.3	.9620
600.	1150.	69.9	-203.1	-190.8	-175.9	-159.3	-141.3	55.1	1.2131
700.	1082.	104.9	-308.4	-294.0	-276.6	-257.2	-236.3	76.3	1.4834
800.	1031.	149.6	-443.1	-426.6	-406.7	-384.6	-360.7	100.3	1.7697
900.	988.	204.2	-606.7	-588.2	-565.8	-540.9	-514.0	126.1	2.0665
1000.	948.	269.8	-801.2	-780.6	-755.8	-728.1	-698.3	153.8	2.3741

BALLISTIC COEFFICIENT .42, MUZZLE VELOCITY FT/SEC 2100.

RANGE	VELOCITY	HEIGHT	100 YD	150 YD	200 YD	250 YD	300 YD	DRIFT	TIME
0.	2100.	0.0	-1.5	-1.5	-1.5	-1.5	-1.5	0.0	0.0000
100.	1913.	.3	0.0	1.8	4.0	6.5	9.2	1.2	.1497
200.	1740.	4.0	-8.0	-4.4	0.0	5.0	10.4	5.0	.3140
300.	1579.	11.0	-27.6	-22.3	-15.6	-8.2	0.0	11.7	.4952
400.	1431.	22.4	-61.2	-54.0	-45.1	-35.2	-24.3	21.7	.6948
500.	1304.	39.4	-111.8	-102.8	-91.7	-79.3	-65.7	35.2	.9145
600.	1197.	63.3	-182.9	-172.2	-158.9	-144.0	-127.7	52.5	1.1555
700.	1115.	95.4	-278.2	-265.6	-250.1	-232.7	-213.7	73.1	1.4154
800.	1057.	136.6	-401.0	-386.6	-368.9	-349.0	-327.2	96.6	1.6918
900.	1010.	188.5	-555.0	-538.9	-518.9	-496.6	-472.1	123.2	1.9855
1000.	969.	250.7	-738.1	-720.1	-698.0	-673.2	-646.0	151.4	2.2886

BALLISTIC COEFFICIENT .42, MUZZLE VELOCITY FT/SEC 2200.

RANGE	VELOCITY	HEIGHT	100 YD	150 YD	200 YD	250 YD	300 YD	DRIFT	TIME
0.	2200.	0.0	-1.5	-1.5	-1.5	-1.5	-1.5	0.0	0.0000
100.	2008.	.2	0.0	1.6	3.6	5.8	8.3	1.1	.1428
200.	1828.	3.6	-7.2	-4.0	0.0	4.5	9.4	4.7	.2994
300.	1660.	9.9	-24.8	-20.0	-14.0	-7.3	0.0	11.0	.4716
400.	1506.	20.2	-55.0	-48.6	-40.7	-31.8	-22.0	20.4	.6613
500.	1366.	35.7	-100.7	-92.7	-82.8	-71.6	-59.4	33.3	.8709
600.	1250.	57.4	-165.0	-155.4	-143.5	-130.1	-115.4	49.7	1.1007
700.	1153.	87.1	-252.3	-241.1	-227.3	-211.6	-194.5	70.1	1.3529
800.	1085.	125.1	-364.3	-351.4	-335.6	-317.7	-298.1	93.0	1.6195
900.	1034.	173.1	-505.4	-491.0	-473.2	-453.0	-431.0	119.0	1.9031
1000.	990.	232.2	-678.3	-662.3	-642.5	-620.1	-595.7	147.7	2.2030

BALLISTIC COEFFICIENT .42, MUZZLE VELOCITY FT/SEC 2300.

RANGE	VELOCITY	HEIGHT	100 YD	150 YD	200 YD	250 YD	300 YD	DRIFT	TIME
0.	2300.	0.0	-1.5	-1.5	-1.5	-1.5	-1.5	0.0	0.0000
100.	2104.	.1	0.0	1.4	3.2	5.2	7.4	1.1	.1364
200.	1917.	3.2	-6.4	-3.6	0.0	4.0	8.5	4.4	.2859
300.	1743.	9.0	-22.3	-18.1	-12.7	-6.6	0.0	10.3	.4498
400.	1582.	18.3	-49.7	-44.1	-36.9	-28.8	-20.0	19.2	.6307
500.	1434.	32.3	-90.9	-83.9	-74.9	-64.8	-53.8	31.3	.8299
600.	1306.	52.1	-149.1	-140.7	-129.9	-117.8	-104.5	46.9	1.0492
700.	1199.	79.1	-227.8	-218.0	-205.4	-191.3	-175.8	66.2	1.2894
800.	1117.	114.4	-330.8	-319.5	-305.1	-289.0	-271.3	89.0	1.5491
900.	1058.	159.2	-461.0	-448.3	-432.2	-414.1	-394.1	114.6	1.8253
1000.	1011.	214.8	-622.6	-608.5	-590.6	-570.5	-548.3	143.4	2.1190

BALLISTIC COEFFICIENT .42, MUZZLE VELOCITY FT/SEC 2400.

RANGE	VELOCITY	HEIGHT	100 YD	150 YD	200 YD	250 YD	300 YD	DRIFT	TIME
0.	2400.	0.0	-1.5	-1.5	-1.5	-1.5	-1.5	0.0	0.0000
100.	2200.	.1	0.0	1.3	2.9	4.7	6.7	1.0	.1304
200.	2008.	2.8	-5.7	-3.2	0.0	3.7	7.7	4.1	.2733
300.	1828.	8.1	-20.1	-16.4	-11.5	-6.0	0.0	9.7	.4299
400.	1660.	16.7	-45.0	-39.9	-33.5	-26.1	-18.1	18.0	.6021
500.	1505.	29.3	-82.4	-76.1	-68.0	-58.8	-48.8	29.4	.7918
600.	1366.	47.4	-135.2	-127.7	-118.0	-107.0	-95.0	44.3	1.0015
700.	1250.	72.0	-206.8	-198.0	-186.7	-173.9	-159.8	62.7	1.2313
800.	1153.	104.9	-301.5	-291.4	-278.5	-263.8	-247.8	85.1	1.4836
900.	1085.	146.3	-420.7	-409.4	-394.9	-378.4	-360.3	110.0	1.7502
1000.	1034.	197.8	-569.2	-556.6	-540.5	-522.1	-502.0	138.0	2.0338

BALLISTIC COEFFICIENT .42, MUZZLE VELOCITY FT/SEC 2500.

RANGE	VELOCITY	HEIGHT	100 YD	150 YD	200 YD	250 YD	300 YD	DRIFT	TIME
0.	2500.	0.0	-1.5	-1.5	-1.5	-1.5	-1.5	0.0	0.0000
100.	2295.	.0	0.0	1.1	2.6	4.2	6.1	.9	.1252
200.	2100.	2.5	-5.1	-2.9	0.0	3.3	7.0	3.9	.2619
300.	1913.	7.4	-18.2	-14.9	-10.5	-5.5	0.0	9.1	.4116
400.	1739.	15.2	-40.8	-36.4	-30.5	-23.8	-16.5	16.9	.5760
500.	1579.	26.8	-74.9	-69.4	-62.1	-53.7	-44.6	27.7	.7572
600.	1430.	43.2	-122.9	-116.3	-107.4	-97.5	-86.4	41.7	.9568
700.	1303.	65.7	-188.0	-180.2	-170.0	-158.3	-145.5	59.2	1.1766
800.	1197.	95.7	-273.8	-265.0	-253.2	-239.9	-225.2	80.5	1.4176
900.	1115.	134.3	-383.8	-373.9	-360.7	-345.7	-329.2	105.2	1.6775
1000.	1057.	182.5	-521.4	-510.4	-495.7	-479.0	-460.7	132.7	1.9540

BALLISTIC COEFFICIENT .42, MUZZLE VELOCITY FT/SEC 2600.

RANGE	VELOCITY	HEIGHT	100 YD	150 YD	200 YD	250 YD	300 YD	DRIFT	TIME
0.	2600.	0.0	-1.5	-1.5	-1.5	-1.5	-1.5	0.0	0.0000
100.	2391.	-.1	0.0	1.0	2.3	3.8	5.5	.9	.1204
200.	2191.	2.3	-4.6	-2.6	0.0	3.1	6.4	3.6	.2513
300.	2000.	6.7	-16.5	-13.6	-9.6	-5.0	0.0	8.5	.3947
400.	1820.	13.9	-37.1	-33.2	-27.9	-21.8	-15.1	15.9	.5518
500.	1653.	24.5	-68.2	-63.3	-56.7	-49.1	-40.7	26.0	.7249
600.	1499.	39.5	-112.0	-106.1	-98.2	-89.0	-79.0	39.2	.9153
700.	1361.	60.1	-171.4	-164.5	-155.3	-144.6	-132.9	56.0	1.1259
800.	1245.	87.6	-249.7	-241.8	-231.3	-219.1	-205.7	76.3	1.3566
900.	1150.	123.3	-350.5	-341.6	-329.8	-316.0	-301.0	100.2	1.6077
1000.	1082.	168.5	-478.3	-468.5	-455.3	-440.0	-423.3	127.5	1.8780

BALLISTIC COEFFICIENT .42, MUZZLE VELOCITY FT/SEC 2700.

RANGE	VELOCITY	HEIGHT	100 YD	150 YD	200 YD	250 YD	300 YD	DRIFT	TIME
0.	2700.	0.0	-1.5	-1.5	-1.5	-1.5	-1.5	0.0	0.0000
100.	2487.	-.1	0.0	.9	2.1	3.5	5.0	.8	.1158
200.	2283.	2.1	-4.2	-2.4	0.0	2.8	5.9	3.4	.2417
300.	2088.	6.1	-15.0	-12.4	-8.8	-4.6	0.0	8.0	.3791
400.	1902.	12.7	-33.9	-30.4	-25.6	-20.0	-13.9	15.0	.5297
500.	1729.	22.4	-62.3	-58.0	-51.9	-45.0	-37.3	24.5	.6950
600.	1569.	36.2	-102.5	-97.3	-90.0	-81.7	-72.5	37.1	.8775
700.	1422.	55.0	-156.6	-150.5	-142.1	-132.4	-121.6	52.9	1.0782
800.	1296.	80.3	-228.3	-221.3	-211.7	-200.6	-188.3	72.3	1.2997
900.	1191.	113.3	-320.7	-312.9	-302.0	-289.6	-275.7	95.3	1.5414
1000.	1111.	155.3	-438.1	-429.3	-417.3	-403.4	-388.0	121.7	1.8028

BALLISTIC COEFFICIENT .42, MUZZLE VELOCITY FT/SEC 2800.

RANGE	VELOCITY	HEIGHT	100 YD	150 YD	200 YD	250 YD	300 YD	DRIFT	TIME
0.	2800.	0.0	-1.5	-1.5	-1.5	-1.5	-1.5	0.0	0.0000
100.	2582.	-.2	0.0	.8	1.9	3.1	4.6	.8	.1114
200.	2374.	1.8	-3.7	-2.2	0.0	2.6	5.4	3.2	.2326
300.	2175.	5.6	-13.7	-11.4	-8.0	-4.2	0.0	7.6	.3644
400.	1984.	11.7	-31.0	-27.9	-23.5	-18.4	-12.8	14.1	.5089
500.	1806.	20.6	-57.1	-53.3	-47.8	-41.4	-34.4	23.2	.6675
600.	1640.	33.3	-93.9	-89.3	-82.7	-75.0	-66.6	35.0	.8419
700.	1487.	50.6	-143.6	-138.2	-130.5	-121.6	-111.7	50.0	1.0340
800.	1350.	73.8	-209.2	-203.1	-194.3	-184.1	-172.8	68.5	1.2462
900.	1236.	104.2	-294.3	-287.4	-277.4	-265.9	-253.3	90.6	1.4790
1000.	1144.	143.1	-401.9	-394.2	-383.2	-370.4	-356.4	116.1	1.7313

BALLISTIC COEFFICIENT .42, MUZZLE VELOCITY FT/SEC 2900.

RANGE	VELOCITY	HEIGHT	100 YD	150 YD	200 YD	250 YD	300 YD	DRIFT	TIME
0.	2900.	0.0	-1.5	-1.5	-1.5	-1.5	-1.5	0.0	0.0000
100.	2677.	-.2	0.0	.7	1.7	2.9	4.2	.7	.1075
200.	2465.	1.7	-3.4	-2.0	0.0	2.4	5.0	3.1	.2244
300.	2262.	5.2	-12.5	-10.5	-7.5	-3.9	0.0	7.3	.3516
400.	2067.	10.8	-28.5	-25.7	-21.7	-17.0	-11.8	13.5	.4903
500.	1883.	19.0	-52.5	-49.1	-44.1	-38.1	-31.6	22.0	.6422
600.	1711.	30.7	-86.4	-82.2	-76.2	-69.1	-61.3	33.2	.8093
700.	1553.	46.6	-132.2	-127.4	-120.4	-112.1	-103.0	47.4	.9937
800.	1407.	68.0	-192.3	-186.8	-178.8	-169.3	-158.9	64.9	1.1966
900.	1284.	96.0	-270.4	-264.2	-255.2	-244.5	-232.8	86.1	1.4200
1000.	1181.	132.1	-369.7	-362.8	-352.8	-340.9	-327.9	110.8	1.6639

BALLISTIC COEFFICIENT .42, MUZZLE VELOCITY FT/SEC 3000.

RANGE	VELOCITY	HEIGHT	100 YD	150 YD	200 YD	250 YD	300 YD	DRIFT	TIME
0.	3000.	0.0	-1.5	-1.5	-1.5	-1.5	-1.5	0.0	0.0000
100.	2772.	-.2	0.0	.6	1.5	2.6	3.8	.7	.1042
200.	2555.	1.5	-3.1	-1.8	0.0	2.2	4.6	3.0	.2170
300.	2348.	4.8	-11.5	-9.7	-6.9	-3.6	0.0	7.0	.3396
400.	2150.	10.0	-26.2	-23.8	-20.1	-15.7	-10.9	12.9	.4730
500.	1961.	17.7	-48.4	-45.4	-40.8	-35.4	-29.3	21.0	.6192
600.	1784.	28.4	-79.7	-76.1	-70.6	-64.1	-56.8	31.6	.7797
700.	1620.	43.1	-121.8	-117.6	-111.2	-103.6	-95.1	45.0	.9559
800.	1468.	62.8	-177.3	-172.5	-165.1	-156.4	-146.7	61.7	1.1507
900.	1334.	88.8	-249.3	-243.8	-235.5	-225.8	-214.8	82.0	1.3657
1000.	1223.	122.2	-340.8	-334.7	-325.5	-314.7	-302.5	105.6	1.6003

BALLISTIC COEFFICIENT .42, MUZZLE VELOCITY FT/SEC 3100.

RANGE	VELOCITY	HEIGHT	100 YD	150 YD	200 YD	250 YD	300 YD	DRIFT	TIME
0.	3100.	0.0	-1.5	-1.5	-1.5	-1.5	-1.5	0.0	0.0000
100.	2866.	-.3	0.0	.5	1.4	2.4	3.5	.6	.1003
200.	2645.	1.4	-2.8	-1.7	0.0	2.0	4.2	2.8	.2095
300.	2434.	4.4	-10.5	-8.9	-6.4	-3.3	0.0	6.6	.3276
400.	2233.	9.3	-24.2	-22.0	-18.6	-14.6	-10.1	12.2	.4564
500.	2040.	16.4	-44.8	-42.1	-37.9	-32.8	-27.3	19.9	.5972
600.	1857.	26.3	-73.6	-70.4	-65.4	-59.3	-52.6	30.0	.7510
700.	1687.	39.9	-112.7	-108.9	-103.0	-96.0	-88.2	42.8	.9206
800.	1530.	58.1	-163.9	-159.6	-152.9	-144.8	-135.9	58.6	1.1074
900.	1388.	82.1	-230.1	-225.2	-217.6	-208.6	-198.6	77.9	1.3134
1000.	1268.	113.1	-314.8	-309.3	-300.9	-290.8	-279.7	100.7	1.5398

BALLISTIC COEFFICIENT .42, MUZZLE VELOCITY FT/SEC 3200.

RANGE	VELOCITY	HEIGHT	100 YD	150 YD	200 YD	250 YD	300 YD	DRIFT	TIME
0.	3200.	0.0	-1.5	-1.5	-1.5	-1.5	-1.5	0.0	0.0000
100.	2961.	-.3	0.0	.5	1.2	2.2	3.2	.6	.0973
200.	2735.	1.2	-2.5	-1.6	0.0	1.9	4.0	2.7	.2029
300.	2520.	4.1	-9.7	-8.3	-5.9	-3.2	0.0	6.4	.3174
400.	2315.	8.6	-22.3	-20.4	-17.3	-13.6	-9.4	11.7	.4414
500.	2118.	15.2	-41.4	-39.1	-35.2	-30.5	-25.3	19.0	.5767
600.	1931.	24.5	-68.3	-65.5	-60.8	-55.3	-49.0	28.7	.7254
700.	1756.	37.1	-104.4	-101.1	-95.7	-89.2	-81.8	40.8	.8881
800.	1594.	54.0	-152.0	-148.2	-142.0	-134.6	-126.1	55.9	1.0677
900.	1444.	76.1	-213.1	-208.9	-201.9	-193.5	-184.1	74.2	1.2655
1000.	1315.	104.9	-291.4	-286.7	-278.9	-269.6	-259.1	96.1	1.4836

BALLISTIC COEFFICIENT .42, MUZZLE VELOCITY FT/SEC 3300.

RANGE	VELOCITY	HEIGHT	100 YD	150 YD	200 YD	250 YD	300 YD	DRIFT	TIME
0.	3300.	0.0	-1.5	-1.5	-1.5	-1.5	-1.5	0.0	0.0000
100.	3056.	-.3	0.0	.4	1.1	2.0	3.0	.6	.0944
200.	2825.	1.1	-2.3	-1.4	0.0	1.7	3.7	2.6	.1966
300.	2606.	3.8	-8.9	-7.6	-5.5	-2.9	0.0	6.1	.3072
400.	2397.	8.0	-20.6	-19.0	-16.1	-12.7	-8.8	11.2	.4274
500.	2197.	14.2	-38.4	-36.4	-32.8	-28.5	-23.6	18.2	.5580
600.	2005.	22.8	-63.4	-60.9	-56.7	-51.5	-45.7	27.4	.7010
700.	1825.	34.6	-97.0	-94.1	-89.1	-83.1	-76.3	39.0	.8578
800.	1658.	50.2	-141.1	-137.8	-132.1	-125.1	-117.4	53.3	1.0304
900.	1503.	70.7	-197.7	-194.0	-187.6	-179.8	-171.1	70.8	1.2203
1000.	1364.	97.4	-270.2	-266.1	-258.9	-250.3	-240.6	91.7	1.4303

BALLISTIC COEFFICIENT .42, MUZZLE VELOCITY FT/SEC 3400.

RANGE	VELOCITY	HEIGHT	100 YD	150 YD	200 YD	250 YD	300 YD	DRIFT	TIME
0.	3400.	0.0	-1.5	-1.5	-1.5	-1.5	-1.5	0.0	0.0000
100.	3151.	-.3	0.0	.3	1.0	1.8	2.7	.6	.0916
200.	2915.	1.0	-2.0	-1.3	0.0	1.6	3.4	2.5	.1907
300.	2691.	3.5	-8.1	-7.1	-5.1	-2.7	0.0	5.8	.2976
400.	2479.	7.5	-19.1	-17.7	-15.0	-11.8	-8.2	10.7	.4137
500.	2275.	13.3	-35.7	-33.9	-30.6	-26.6	-22.1	17.4	.5402
600.	2080.	21.3	-58.9	-56.8	-52.8	-48.0	-42.6	26.1	.6779
700.	1895.	32.3	-90.2	-87.8	-83.1	-77.5	-71.2	37.2	.8292
800.	1722.	46.8	-131.1	-128.3	-123.0	-116.6	-109.4	50.9	.9951
900.	1563.	65.9	-184.0	-180.8	-174.9	-167.6	-159.6	67.6	1.1784
1000.	1416.	90.7	-251.0	-247.5	-240.9	-232.8	-223.9	87.6	1.3800

BALLISTIC COEFFICIENT .42, MUZZLE VELOCITY FT/SEC 3500.

RANGE	VELOCITY	HEIGHT	100 YD	150 YD	200 YD	250 YD	300 YD	DRIFT	TIME
0.	3500.	0.0	-1.5	-1.5	-1.5	-1.5	-1.5	0.0	0.0000
100.	3247.	-.4	0.0	.3	.9	1.7	2.5	.6	.0891
200.	3005.	.9	-1.8	-1.2	0.0	1.5	3.2	2.4	.1850
300.	2777.	3.3	-7.5	-6.5	-4.8	-2.5	0.0	5.6	.2889
400.	2560.	7.0	-17.7	-16.4	-14.1	-11.1	-7.7	10.3	.4015
500.	2353.	12.4	-33.2	-31.7	-28.7	-25.0	-20.8	16.8	.5240
600.	2155.	20.0	-54.9	-53.1	-49.5	-45.0	-40.0	25.1	.6570
700.	1965.	30.2	-84.2	-82.0	-77.8	-72.6	-66.7	35.7	.8029
800.	1788.	43.8	-122.4	-119.9	-115.2	-109.2	-102.4	48.8	.9630
900.	1623.	61.5	-171.4	-168.6	-163.3	-156.6	-149.0	64.7	1.1390
1000.	1471.	84.6	-233.8	-230.7	-224.8	-217.3	-208.9	83.8	1.3332

BALLISTIC COEFFICIENT .42, MUZZLE VELOCITY FT/SEC 3600.

RANGE	VELOCITY	HEIGHT	100 YD	150 YD	200 YD	250 YD	300 YD	DRIFT	TIME
0.	3600.	0.0	-1.5	-1.5	-1.5	-1.5	-1.5	0.0	0.0000
100.	3342.	-.4	0.0	.3	.8	1.5	2.3	.5	.0864
200.	3096.	.8	-1.6	-1.1	0.0	1.4	3.0	2.3	.1797
300.	2862.	3.0	-6.9	-6.1	-4.4	-2.4	0.0	5.3	.2802
400.	2641.	6.5	-16.4	-15.4	-13.2	-10.4	-7.2	9.9	.3895
500.	2431.	11.6	-30.9	-29.6	-26.8	-23.4	-19.5	16.1	.5079
600.	2229.	18.7	-51.3	-49.7	-46.4	-42.2	-37.5	24.1	.6368
700.	2036.	28.3	-78.6	-76.8	-72.9	-68.1	-62.6	34.2	.7778
800.	1854.	40.9	-114.2	-112.1	-107.7	-102.2	-95.9	46.7	.9320
900.	1684.	57.5	-160.1	-157.7	-152.7	-146.5	-139.4	61.9	1.1018
1000.	1528.	79.0	-218.2	-215.6	-210.1	-203.2	-195.3	80.2	1.2889

BALLISTIC COEFFICIENT .42, MUZZLE VELOCITY FT/SEC 3700.

RANGE	VELOCITY	HEIGHT	100 YD	150 YD	200 YD	250 YD	300 YD	DRIFT	TIME
0.	3700.	0.0	-1.5	-1.5	-1.5	-1.5	-1.5	0.0	0.0000
100.	3437.	-.4	0.0	.2	.7	1.4	2.1	.6	.0843
200.	3186.	.7	-1.5	-1.0	0.0	1.3	2.8	2.2	.1749
300.	2948.	2.8	-6.3	-5.7	-4.2	-2.2	0.0	5.2	.2727
400.	2723.	6.1	-15.3	-14.4	-12.4	-9.8	-6.8	9.6	.3788
500.	2508.	11.0	-28.9	-27.8	-25.3	-22.0	-18.3	15.5	.4937
600.	2304.	17.6	-48.0	-46.6	-43.6	-39.7	-35.3	23.2	.6184
700.	2107.	26.6	-73.6	-72.0	-68.5	-64.0	-58.8	32.9	.7546
800.	1920.	38.5	-107.0	-105.2	-101.2	-96.0	-90.1	44.9	.9039
900.	1746.	53.9	-149.8	-147.8	-143.3	-137.4	-130.8	59.4	1.0675
1000.	1585.	74.0	-204.2	-202.0	-196.9	-190.5	-183.1	77.0	1.2481

BALLISTIC COEFFICIENT .42, MUZZLE VELOCITY FT/SEC 3800.

RANGE	VELOCITY	HEIGHT	100 YD	150 YD	200 YD	250 YD	300 YD	DRIFT	TIME
0.	3800.	0.0	-1.5	-1.5	-1.5	-1.5	-1.5	0.0	0.0000
100.	3532.	-.4	0.0	.2	.7	1.3	2.0	.5	.0819
200.	3277.	.6	-1.3	-.9	0.0	1.2	2.6	2.2	.1704
300.	3034.	2.6	-5.9	-5.3	-3.9	-2.1	0.0	5.0	.2653
400.	2804.	5.8	-14.2	-13.5	-11.6	-9.2	-6.4	9.3	.3685
500.	2586.	10.3	-27.0	-26.0	-23.7	-20.7	-17.2	15.0	.4798
600.	2378.	16.6	-44.9	-43.8	-41.0	-37.3	-33.2	22.4	.6007
700.	2178.	25.0	-68.9	-67.6	-64.3	-60.1	-55.3	31.6	.7323
800.	1988.	36.1	-100.3	-98.7	-95.0	-90.2	-84.6	43.1	.8766
900.	1809.	50.7	-140.4	-138.7	-134.5	-129.0	-122.8	57.1	1.0349
1000.	1643.	69.4	-191.3	-189.4	-184.7	-178.6	-171.7	73.8	1.2091

BALLISTIC COEFFICIENT .42, MUZZLE VELOCITY FT/SEC 3900.

RANGE	VELOCITY	HEIGHT	100 YD	150 YD	200 YD	250 YD	300 YD	DRIFT	TIME
0.	3900.	0.0	-1.5	-1.5	-1.5	-1.5	-1.5	0.0	0.0000
100.	3627.	-.4	0.0	.1	.6	1.1	1.8	.5	.0797
200.	3367.	.6	-1.1	-.9	0.0	1.1	2.5	2.0	.1654
300.	3120.	2.4	-5.4	-5.0	-3.7	-2.0	0.0	4.8	.2582
400.	2886.	5.4	-13.2	-12.6	-10.9	-8.6	-6.0	8.8	.3578
500.	2663.	9.7	-25.2	-24.5	-22.3	-19.5	-16.2	14.4	.4662
600.	2452.	15.6	-42.0	-41.2	-38.6	-35.2	-31.2	21.5	.5835
700.	2249.	23.5	-64.7	-63.7	-60.7	-56.7	-52.1	30.5	.7115
800.	2056.	34.0	-94.1	-93.0	-89.5	-85.0	-79.7	41.5	.8510
900.	1872.	47.6	-131.7	-130.4	-126.5	-121.4	-115.5	54.8	1.0037
1000.	1701.	65.2	-179.3	-177.9	-173.6	-167.9	-161.3	70.9	1.1719

BALLISTIC COEFFICIENT .42, MUZZLE VELOCITY FT/SEC 4000.

RANGE	VELOCITY	HEIGHT	100 YD	150 YD	200 YD	250 YD	300 YD	DRIFT	TIME
0.	4000.	0.0	-1.5	-1.5	-1.5	-1.5	-1.5	0.0	0.0000
100.	3722.	-.5	0.0	.1	.5	1.0	1.7	.5	.0778
200.	3458.	.5	-1.0	-.8	0.0	1.1	2.3	2.0	.1615
300.	3206.	2.3	-5.0	-4.6	-3.4	-1.8	0.0	4.7	.2517
400.	2968.	5.1	-12.3	-11.8	-10.2	-8.1	-5.7	8.6	.3488
500.	2741.	9.1	-23.6	-23.0	-21.0	-18.4	-15.3	13.9	.4541
600.	2526.	14.8	-39.5	-38.8	-36.4	-33.3	-29.6	20.8	.5683
700.	2320.	22.2	-60.8	-59.9	-57.2	-53.5	-49.2	29.4	.6920
800.	2123.	32.1	-88.5	-87.5	-84.4	-80.1	-75.2	40.0	.8271
900.	1935.	44.9	-123.9	-122.8	-119.3	-114.5	-109.0	52.9	.9754
1000.	1760.	61.4	-168.5	-167.3	-163.4	-158.1	-152.0	68.2	1.1376

BALLISTIC COEFFICIENT .43, MUZZLE VELOCITY FT/SEC 2000.

RANGE	VELOCITY	HEIGHT	100 YD	150 YD	200 YD	250 YD	300 YD	DRIFT	TIME
0.	2000.	0.0	-1.5	-1.5	-1.5	-1.5	-1.5	0.0	0.0000
100.	1824.	.4	0.0	2.0	4.5	7.3	10.2	1.2	.1570
200.	1661.	4.5	-9.0	-4.9	0.0	5.5	11.4	5.2	.3295
300.	1509.	12.2	-30.7	-24.5	-17.2	-8.9	0.0	12.1	.5189
400.	1372.	24.7	-67.7	-59.5	-49.7	-38.7	-26.8	22.5	.7277
500.	1257.	43.1	-123.2	-113.0	-100.7	-86.9	-72.1	36.3	.9563
600.	1162.	68.9	-200.6	-188.3	-173.5	-157.0	-139.2	53.6	1.2048
700.	1092.	103.2	-303.9	-289.6	-272.4	-253.1	-232.3	74.2	1.4718
800.	1040.	146.8	-435.3	-418.9	-399.3	-377.2	-353.5	97.4	1.7534
900.	997.	200.6	-596.9	-578.5	-556.3	-531.6	-504.9	122.9	2.0483
1000.	957.	265.5	-790.0	-769.5	-744.9	-717.4	-687.7	150.5	2.3553

BALLISTIC COEFFICIENT .43, MUZZLE VELOCITY FT/SEC 2100.

RANGE	VELOCITY	HEIGHT	100 YD	150 YD	200 YD	250 YD	300 YD	DRIFT	TIME
0.	2100.	0.0	-1.5	-1.5	-1.5	-1.5	-1.5	0.0	0.0000
100.	1918.	.3	0.0	1.8	4.0	6.5	9.2	1.2	.1495
200.	1748.	4.0	-8.0	-4.4	0.0	4.9	10.3	4.8	.3132
300.	1590.	10.9	-27.5	-22.1	-15.5	-8.1	0.0	11.4	.4935
400.	1444.	22.2	-60.7	-53.5	-44.7	-34.8	-24.1	21.1	.6915
500.	1317.	38.9	-110.6	-101.6	-90.7	-78.3	-64.9	34.3	.9092
600.	1211.	62.4	-180.6	-169.8	-156.6	-141.8	-125.7	51.0	1.1470
700.	1127.	93.9	-274.3	-261.8	-246.4	-229.1	-210.3	71.2	1.4044
800.	1067.	134.5	-395.2	-380.8	-363.3	-343.5	-322.0	94.3	1.6784
900.	1020.	184.8	-545.0	-528.9	-509.1	-486.9	-462.7	119.8	1.9662
1000.	978.	245.9	-725.4	-707.4	-685.5	-660.8	-633.9	147.5	2.2667

BALLISTIC COEFFICIENT .43, MUZZLE VELOCITY FT/SEC 2200.

RANGE	VELOCITY	HEIGHT	100 YD	150 YD	200 YD	250 YD	300 YD	DRIFT	TIME
0.	2200.	0.0	-1.5	-1.5	-1.5	-1.5	-1.5	0.0	0.0000
100.	2013.	.2	0.0	1.6	3.6	5.8	8.2	1.1	.1427
200.	1836.	3.5	-7.1	-3.9	0.0	4.4	9.3	4.6	.2989
300.	1672.	9.9	-24.6	-19.8	-13.9	-7.3	0.0	10.7	.4700
400.	1519.	20.0	-54.6	-48.2	-40.3	-31.5	-21.7	19.9	.6583
500.	1381.	35.2	-99.6	-91.6	-81.7	-70.7	-58.5	32.3	.8654
600.	1265.	56.6	-162.9	-153.3	-141.5	-128.3	-113.7	48.3	1.0927
700.	1168.	85.4	-248.0	-236.8	-223.0	-207.5	-190.5	67.8	1.3400
800.	1096.	123.0	-358.7	-345.8	-330.1	-312.4	-293.0	90.6	1.6057
900.	1043.	170.3	-498.1	-483.7	-466.0	-446.1	-424.2	116.3	1.8879
1000.	1000.	227.4	-665.5	-649.5	-629.8	-607.7	-583.4	143.7	2.1803

BALLISTIC COEFFICIENT .43, MUZZLE VELOCITY FT/SEC 2300.

RANGE	VELOCITY	HEIGHT	100 YD	150 YD	200 YD	250 YD	300 YD	DRIFT	TIME
0.	2300.	0.0	-1.5	-1.5	-1.5	-1.5	-1.5	0.0	0.0000
100.	2108.	.1	0.0	1.4	3.2	5.2	7.4	1.0	.1362
200.	1926.	3.2	-6.4	-3.6	0.0	4.0	8.4	4.3	.2853
300.	1755.	8.9	-22.1	-17.9	-12.6	-6.5	0.0	10.0	.4482
400.	1597.	18.2	-49.3	-43.7	-36.5	-28.5	-19.8	18.6	.6277
500.	1450.	31.9	-90.0	-83.0	-74.1	-64.0	-53.1	30.4	.8249
600.	1322.	51.3	-147.3	-138.9	-128.2	-116.1	-103.0	45.6	1.0416
700.	1215.	77.8	-224.7	-214.8	-202.3	-188.3	-173.0	64.4	1.2789
800.	1130.	112.4	-325.5	-314.2	-299.9	-283.9	-266.4	86.5	1.5352
900.	1069.	156.2	-453.4	-440.7	-424.6	-406.6	-387.0	111.7	1.8085
1000.	1022.	210.2	-610.4	-596.3	-578.5	-558.4	-536.6	139.4	2.0963

BALLISTIC COEFFICIENT .43, MUZZLE VELOCITY FT/SEC 2400.

RANGE	VELOCITY	HEIGHT	100 YD	150 YD	200 YD	250 YD	300 YD	DRIFT	TIME
0.	2400.	0.0	-1.5	-1.5	-1.5	-1.5	-1.5	0.0	0.0000
100.	2204.	.1	0.0	1.2	2.9	4.7	6.7	.9	.1303
200.	2017.	2.8	-5.7	-3.2	0.0	3.7	7.6	4.0	.2727
300.	1840.	8.1	-20.0	-16.3	-11.5	-6.0	0.0	9.4	.4286
400.	1675.	16.5	-44.6	-39.6	-33.2	-25.9	-17.9	17.5	.5993
500.	1523.	29.0	-81.6	-75.3	-67.3	-58.2	-48.2	28.5	.7872
600.	1384.	46.7	-133.5	-126.0	-116.4	-105.4	-93.5	42.9	.9940
700.	1267.	70.8	-203.8	-195.0	-183.8	-171.0	-157.1	60.9	1.2208
800.	1169.	102.6	-295.7	-285.7	-272.9	-258.3	-242.3	82.3	1.4676
900.	1098.	143.4	-413.2	-402.0	-387.6	-371.1	-353.1	107.0	1.7329
1000.	1044.	194.2	-559.8	-547.3	-531.3	-513.0	-493.1	134.7	2.0153

BALLISTIC COEFFICIENT .43, MUZZLE VELOCITY FT/SEC 2500.

RANGE	VELOCITY	HEIGHT	100 YD	150 YD	200 YD	250 YD	300 YD	DRIFT	TIME
0.	2500.	0.0	-1.5	-1.5	-1.5	-1.5	-1.5	0.0	0.0000
100.	2300.	.0	0.0	1.1	2.6	4.2	6.0	.9	.1251
200.	2109.	2.5	-5.1	-2.9	0.0	3.3	7.0	3.7	.2613
300.	1926.	7.3	-18.1	-14.8	-10.4	-5.5	0.0	8.9	.4103
400.	1755.	15.0	-40.4	-36.0	-30.2	-23.6	-16.3	16.4	.5732
500.	1597.	26.4	-74.1	-68.6	-61.4	-53.1	-44.0	26.9	.7527
600.	1450.	42.6	-121.4	-114.8	-106.1	-96.2	-85.2	40.5	.9499
700.	1322.	64.6	-185.2	-177.6	-167.4	-155.8	-143.0	57.5	1.1666
800.	1215.	93.9	-269.3	-260.5	-248.9	-235.6	-221.0	78.1	1.4039
900.	1130.	131.5	-376.7	-366.9	-353.8	-338.9	-322.5	102.1	1.6602
1000.	1069.	178.7	-511.4	-500.4	-485.9	-469.3	-451.1	129.1	1.9335

BALLISTIC COEFFICIENT .43, MUZZLE VELOCITY FT/SEC 2600.

RANGE	VELOCITY	HEIGHT	100 YD	150 YD	200 YD	250 YD	300 YD	DRIFT	TIME
0.	2600.	0.0	-1.5	-1.5	-1.5	-1.5	-1.5	0.0	0.0000
100.	2396.	-.1	0.0	1.0	2.3	3.8	5.5	.9	.1203
200.	2200.	2.3	-4.6	-2.6	0.0	3.0	6.4	3.5	.2508
300.	2013.	6.7	-16.4	-13.5	-9.5	-5.0	0.0	8.3	.3934
400.	1836.	13.7	-36.8	-32.9	-27.7	-21.6	-15.0	15.5	.5496
500.	1672.	24.2	-67.5	-62.6	-56.1	-48.5	-40.2	25.3	.7207
600.	1520.	38.9	-110.7	-104.8	-96.9	-87.8	-77.9	38.1	.9089
700.	1381.	59.0	-168.8	-162.0	-152.8	-142.2	-130.6	54.3	1.1160
800.	1265.	85.9	-245.5	-237.6	-227.2	-215.0	-201.8	73.9	1.3432
900.	1168.	120.7	-343.9	-335.1	-323.3	-309.7	-294.7	97.2	1.5905
1000.	1096.	164.6	-468.1	-458.4	-445.3	-430.1	-413.5	123.6	1.8561

BALLISTIC COEFFICIENT .43, MUZZLE VELOCITY FT/SEC 2700.

RANGE	VELOCITY	HEIGHT	100 YD	150 YD	200 YD	250 YD	300 YD	DRIFT	TIME
0.	2700.	0.0	-1.5	-1.5	-1.5	-1.5	-1.5	0.0	0.0000
100.	2492.	-.1	0.0	.9	2.1	3.4	5.0	.8	.1157
200.	2292.	2.0	-4.1	-2.4	0.0	2.8	5.8	3.3	.2412
300.	2101.	6.1	-14.9	-12.3	-8.7	-4.6	0.0	7.9	.3780
400.	1919.	12.6	-33.6	-30.1	-25.4	-19.9	-13.7	14.6	.5275
500.	1748.	22.2	-61.7	-57.4	-51.4	-44.5	-36.8	23.9	.6911
600.	1591.	35.7	-101.3	-96.0	-88.9	-80.6	-71.4	36.0	.8713
700.	1444.	54.1	-154.5	-148.4	-140.0	-130.4	-119.6	51.3	1.0692
800.	1318.	78.7	-224.5	-217.5	-208.0	-196.9	-184.7	70.0	1.2868
900.	1212.	110.8	-314.7	-306.9	-296.1	-283.7	-269.9	92.3	1.5245
1000.	1127.	151.6	-428.9	-420.2	-408.3	-394.5	-379.2	118.0	1.7818

BALLISTIC COEFFICIENT .43, MUZZLE VELOCITY FT/SEC 2800.

RANGE	VELOCITY	HEIGHT	100 YD	150 YD	200 YD	250 YD	300 YD	DRIFT	TIME
0.	2800.	0.0	-1.5	-1.5	-1.5	-1.5	-1.5	0.0	0.0000
100.	2587.	-.2	0.0	.8	1.9	3.1	4.5	.7	.1113
200.	2384.	1.8	-3.7	-2.2	0.0	2.5	5.3	3.1	.2321
300.	2189.	5.6	-13.6	-11.3	-8.0	-4.2	0.0	7.4	.3633
400.	2002.	11.6	-30.7	-27.7	-23.3	-18.2	-12.6	13.8	.5068
500.	1826.	20.4	-56.6	-52.7	-47.3	-40.9	-33.9	22.5	.6637
600.	1662.	32.8	-92.8	-88.2	-81.7	-74.1	-65.7	34.0	.8360
700.	1511.	49.7	-141.6	-136.3	-128.6	-119.7	-109.9	48.5	1.0253
800.	1374.	72.3	-205.7	-199.6	-190.9	-180.7	-169.6	66.3	1.2338
900.	1258.	101.9	-288.6	-281.7	-271.8	-260.4	-247.8	87.6	1.4622
1000.	1162.	139.7	-393.5	-385.8	-374.9	-362.2	-348.2	112.5	1.7106

BALLISTIC COEFFICIENT .43, MUZZLE VELOCITY FT/SEC 2900.

RANGE	VELOCITY	HEIGHT	100 YD	150 YD	200 YD	250 YD	300 YD	DRIFT	TIME
0.	2900.	0.0	-1.5	-1.5	-1.5	-1.5	-1.5	0.0	0.0000
100.	2682.	-.2	0.0	.7	1.7	2.9	4.1	.7	.1074
200.	2474.	1.7	-3.4	-2.0	0.0	2.3	4.9	3.0	.2239
300.	2276.	5.1	-12.4	-10.4	-7.4	-3.9	0.0	7.1	.3504
400.	2085.	10.7	-28.2	-25.5	-21.5	-16.8	-11.6	13.1	.4881
500.	1904.	18.8	-52.0	-48.6	-43.6	-37.8	-31.3	21.4	.6388
600.	1735.	30.3	-85.4	-81.3	-75.3	-68.3	-60.5	32.2	.8038
700.	1578.	45.9	-130.4	-125.6	-118.6	-110.4	-101.4	46.0	.9855
800.	1433.	66.6	-189.2	-183.7	-175.8	-166.4	-156.1	62.9	1.1849
900.	1308.	93.9	-265.3	-259.1	-250.2	-239.6	-228.0	83.3	1.4042
1000.	1203.	128.9	-362.0	-355.1	-345.2	-333.5	-320.6	107.2	1.6437

BALLISTIC COEFFICIENT .43, MUZZLE VELOCITY FT/SEC 3000.

RANGE	VELOCITY	HEIGHT	100 YD	150 YD	200 YD	250 YD	300 YD	DRIFT	TIME
0.	3000.	0.0	-1.5	-1.5	-1.5	-1.5	-1.5	0.0	0.0000
100.	2777.	-.2	0.0	.6	1.5	2.6	3.8	.7	.1040
200.	2565.	1.5	-3.0	-1.8	0.0	2.2	4.6	2.9	.2166
300.	2362.	4.8	-11.4	-9.6	-6.8	-3.6	0.0	6.8	.3386
400.	2168.	9.9	-26.0	-23.5	-19.9	-15.6	-10.8	12.5	.4709
500.	1982.	17.4	-47.9	-44.9	-40.3	-34.9	-28.9	20.3	.6156
600.	1808.	28.0	-78.8	-75.1	-69.6	-63.2	-56.0	30.7	.7742
700.	1645.	42.4	-120.2	-116.0	-109.6	-102.1	-93.6	43.7	.9482
800.	1495.	61.6	-174.5	-169.6	-162.3	-153.7	-144.0	59.7	1.1394
900.	1360.	86.8	-244.5	-239.1	-230.9	-221.2	-210.3	79.2	1.3502
1000.	1247.	119.2	-333.7	-327.7	-318.5	-307.7	-295.7	102.2	1.5807

BALLISTIC COEFFICIENT .43, MUZZLE VELOCITY FT/SEC 3100.

RANGE	VELOCITY	HEIGHT	100 YD	150 YD	200 YD	250 YD	300 YD	DRIFT	TIME
0.	3100.	0.0	-1.5	-1.5	-1.5	-1.5	-1.5	0.0	0.0000
100.	2872.	-.3	0.0	.5	1.4	2.4	3.5	.6	.1002
200.	2655.	1.3	-2.7	-1.7	0.0	2.0	4.2	2.7	.2091
300.	2449.	4.4	-10.4	-8.8	-6.3	-3.3	0.0	6.4	.3266
400.	2251.	9.2	-24.0	-21.8	-18.5	-14.5	-10.1	11.9	.4546
500.	2062.	16.2	-44.3	-41.6	-37.5	-32.5	-26.9	19.4	.5939
600.	1882.	26.0	-72.8	-69.6	-64.6	-58.6	-51.9	29.1	.7460
700.	1714.	39.3	-111.1	-107.3	-101.5	-94.5	-86.8	41.5	.9130
800.	1559.	57.0	-161.4	-157.1	-150.4	-142.4	-133.6	56.8	1.0969
900.	1416.	80.2	-225.9	-221.0	-213.5	-204.5	-194.6	75.3	1.2989
1000.	1294.	110.3	-308.3	-302.9	-294.6	-284.6	-273.5	97.4	1.5211

BALLISTIC COEFFICIENT .43, MUZZLE VELOCITY FT/SEC 3200.

RANGE	VELOCITY	HEIGHT	100 YD	150 YD	200 YD	250 YD	300 YD	DRIFT	TIME
0.	3200.	0.0	-1.5	-1.5	-1.5	-1.5	-1.5	0.0	0.0000
100.	2967.	-.3	0.0	.5	1.2	2.2	3.2	.6	.0973
200.	2745.	1.2	-2.5	-1.5	0.0	1.8	3.9	2.6	.2025
300.	2535.	4.1	-9.6	-8.2	-5.9	-3.1	0.0	6.2	.3164
400.	2334.	8.5	-22.1	-20.3	-17.2	-13.5	-9.3	11.4	.4397
500.	2141.	15.1	-41.1	-38.7	-34.9	-30.2	-25.0	18.5	.5740
600.	1956.	24.2	-67.5	-64.7	-60.1	-54.5	-48.3	27.8	.7206
700.	1783.	36.5	-103.1	-99.8	-94.4	-87.9	-80.7	39.6	.8812
800.	1623.	52.9	-149.5	-145.8	-139.6	-132.2	-123.9	54.1	1.0574
900.	1474.	74.4	-209.2	-205.0	-198.0	-189.7	-180.4	71.7	1.2514
1000.	1343.	102.3	-285.2	-280.6	-272.8	-263.6	-253.2	92.8	1.4650

BALLISTIC COEFFICIENT .43, MUZZLE VELOCITY FT/SEC 3300.

RANGE	VELOCITY	HEIGHT	100 YD	150 YD	200 YD	250 YD	300 YD	DRIFT	TIME
0.	3300.	0.0	-1.5	-1.5	-1.5	-1.5	-1.5	0.0	0.0000
100.	3062.	-.3	0.0	.4	1.1	2.0	2.9	.6	.0943
200.	2836.	1.1	-2.2	-1.4	0.0	1.7	3.6	2.5	.1962
300.	2621.	3.8	-8.8	-7.6	-5.5	-2.9	0.0	5.9	.3062
400.	2416.	7.9	-20.5	-18.8	-16.0	-12.6	-8.7	10.9	.4257
500.	2219.	14.0	-38.1	-36.0	-32.5	-28.2	-23.4	17.7	.5551
600.	2031.	22.5	-62.7	-60.3	-56.0	-50.9	-45.1	26.6	.6966
700.	1853.	34.0	-95.7	-92.9	-87.9	-81.9	-75.2	37.8	.8510
800.	1688.	49.3	-138.9	-135.6	-130.0	-123.1	-115.4	51.6	1.0207
900.	1534.	69.2	-194.3	-190.6	-184.2	-176.5	-167.9	68.5	1.2072
1000.	1394.	95.0	-264.5	-260.5	-253.4	-244.8	-235.2	88.6	1.4124

BALLISTIC COEFFICIENT .43, MUZZLE VELOCITY FT/SEC 3400.

RANGE	VELOCITY	HEIGHT	100 YD	150 YD	200 YD	250 YD	300 YD	DRIFT	TIME
0.	3400.	0.0	-1.5	-1.5	-1.5	-1.5	-1.5	0.0	0.0000
100.	3157.	-.3	0.0	.3	1.0	1.8	2.7	.6	.0915
200.	2926.	1.0	-2.0	-1.3	0.0	1.6	3.4	2.4	.1903
300.	2706.	3.5	-8.1	-7.0	-5.1	-2.7	0.0	5.7	.2968
400.	2498.	7.4	-18.9	-17.5	-14.9	-11.7	-8.2	10.4	.4123
500.	2298.	13.1	-35.3	-33.6	-30.3	-26.3	-21.9	16.9	.5375
600.	2107.	21.0	-58.3	-56.2	-52.3	-47.5	-42.1	25.4	.6738
700.	1924.	31.8	-89.1	-86.7	-82.1	-76.5	-70.2	36.1	.8230
800.	1753.	45.9	-129.2	-126.4	-121.1	-114.8	-107.6	49.3	.9860
900.	1595.	64.5	-180.7	-177.6	-171.7	-164.5	-156.5	65.4	1.1657
1000.	1449.	88.4	-245.9	-242.5	-235.9	-227.9	-219.0	84.6	1.3631

BALLISTIC COEFFICIENT .43, MUZZLE VELOCITY FT/SEC 3500.

RANGE	VELOCITY	HEIGHT	100 YD	150 YD	200 YD	250 YD	300 YD	DRIFT	TIME
0.	3500.	0.0	-1.5	-1.5	-1.5	-1.5	-1.5	0.0	0.0000
100.	3252.	-.4	0.0	.3	.9	1.6	2.5	.6	.0889
200.	3016.	.9	-1.8	-1.2	0.0	1.5	3.2	2.3	.1847
300.	2792.	3.2	-7.5	-6.5	-4.8	-2.5	0.0	5.5	.2883
400.	2580.	6.9	-17.6	-16.3	-14.0	-11.0	-7.6	10.1	.4001
500.	2377.	12.3	-32.9	-31.3	-28.4	-24.7	-20.5	16.3	.5212
600.	2182.	19.7	-54.3	-52.4	-48.9	-44.4	-39.4	24.4	.6528
700.	1995.	29.7	-83.1	-80.9	-76.8	-71.5	-65.7	34.6	.7967
800.	1820.	42.9	-120.5	-118.0	-113.3	-107.3	-100.6	47.2	.9540
900.	1657.	60.2	-168.5	-165.7	-160.4	-153.7	-146.1	62.6	1.1270
1000.	1506.	82.5	-229.2	-226.0	-220.1	-212.7	-204.3	80.9	1.3169

BALLISTIC COEFFICIENT .43, MUZZLE VELOCITY FT/SEC 3600.

RANGE	VELOCITY	HEIGHT	100 YD	150 YD	200 YD	250 YD	300 YD	DRIFT	TIME
0.	3600.	0.0	-1.5	-1.5	-1.5	-1.5	-1.5	0.0	0.0000
100.	3347.	-.4	0.0	.3	.8	1.5	2.3	.5	.0863
200.	3107.	.8	-1.6	-1.1	0.0	1.4	2.9	2.3	.1795
300.	2878.	3.0	-6.8	-6.0	-4.4	-2.3	0.0	5.2	.2794
400.	2661.	6.5	-16.3	-15.2	-13.0	-10.3	-7.2	9.6	.3880
500.	2455.	11.5	-30.6	-29.3	-26.6	-23.1	-19.2	15.6	.5053
600.	2257.	18.5	-50.7	-49.2	-45.9	-41.8	-37.1	23.4	.6330
700.	2067.	27.8	-77.6	-75.8	-71.9	-67.1	-61.7	33.2	.7718
800.	1887.	40.2	-112.5	-110.5	-106.0	-100.6	-94.3	45.2	.9236
900.	1719.	56.3	-157.2	-154.9	-149.9	-143.8	-136.8	59.8	1.0900
1000.	1563.	77.1	-214.0	-211.4	-205.9	-199.0	-191.2	77.5	1.2736

BALLISTIC COEFFICIENT .43, MUZZLE VELOCITY FT/SEC 3700.

RANGE	VELOCITY	HEIGHT	100 YD	150 YD	200 YD	250 YD	300 YD	DRIFT	TIME
0.	3700.	0.0	-1.5	-1.5	-1.5	-1.5	-1.5	0.0	0.0000
100.	3443.	-.4	0.0	.2	.7	1.4	2.1	.6	.0842
200.	3198.	.7	-1.4	-1.0	0.0	1.3	2.8	2.2	.1747
300.	2965.	2.8	-6.3	-5.6	-4.1	-2.2	0.0	5.1	.2720
400.	2743.	6.1	-15.1	-14.3	-12.2	-9.7	-6.7	9.3	.3773
500.	2533.	10.8	-28.6	-27.5	-25.0	-21.8	-18.1	15.1	.4913
600.	2332.	17.4	-47.4	-46.1	-43.1	-39.2	-34.9	22.6	.6147
700.	2139.	26.2	-72.7	-71.1	-67.6	-63.1	-58.0	32.0	.7491
800.	1954.	37.8	-105.4	-103.7	-99.6	-94.5	-88.6	43.5	.8958
900.	1782.	52.8	-147.3	-145.3	-140.8	-135.0	-128.4	57.5	1.0566
1000.	1621.	72.2	-200.1	-197.9	-192.9	-186.4	-179.1	74.3	1.2330

BALLISTIC COEFFICIENT .43, MUZZLE VELOCITY FT/SEC 3800.

RANGE	VELOCITY	HEIGHT	100 YD	150 YD	200 YD	250 YD	300 YD	DRIFT	TIME
0.	3800.	0.0	-1.5	-1.5	-1.5	-1.5	-1.5	0.0	0.0000
100.	3538.	-.4	0.0	.2	.7	1.2	1.9	.5	.0819
200.	3288.	.6	-1.3	-.9	0.0	1.2	2.6	2.1	.1701
300.	3051.	2.6	-5.8	-5.3	-3.9	-2.1	0.0	4.9	.2645
400.	2825.	5.7	-14.1	-13.4	-11.5	-9.1	-6.3	9.0	.3669
500.	2611.	10.2	-26.7	-25.8	-23.5	-20.5	-17.0	14.5	.4773
600.	2406.	16.4	-44.4	-43.3	-40.5	-36.9	-32.8	21.7	.5972
700.	2210.	24.6	-68.1	-66.8	-63.5	-59.3	-54.5	30.7	.7271
800.	2023.	35.5	-98.8	-97.4	-93.6	-88.8	-83.4	41.8	.8691
900.	1845.	49.6	-138.1	-136.4	-132.2	-126.8	-120.7	55.3	1.0245
1000.	1680.	67.8	-187.5	-185.7	-181.0	-175.0	-168.2	71.3	1.1948

BALLISTIC COEFFICIENT .43, MUZZLE VELOCITY FT/SEC 3900.

RANGE	VELOCITY	HEIGHT	100 YD	150 YD	200 YD	250 YD	300 YD	DRIFT	TIME
0.	3900.	0.0	-1.5	-1.5	-1.5	-1.5	-1.5	0.0	0.0000
100.	3633.	-.4	0.0	.1	.6	1.1	1.8	.5	.0796
200.	3379.	.6	-1.1	-.9	0.0	1.1	2.4	2.0	.1650
300.	3137.	2.4	-5.4	-4.9	-3.7	-2.0	0.0	4.7	.2575
400.	2907.	5.4	-13.1	-12.5	-10.8	-8.6	-6.0	8.6	.3567
500.	2689.	9.6	-24.9	-24.2	-22.1	-19.3	-16.0	13.9	.4639
600.	2481.	15.4	-41.6	-40.7	-38.2	-34.8	-30.9	20.9	.5800
700.	2282.	23.2	-63.8	-62.9	-59.9	-56.0	-51.3	29.5	.7062
800.	2091.	33.4	-92.7	-91.6	-88.2	-83.7	-78.4	40.2	.8436
900.	1909.	46.7	-129.6	-128.3	-124.5	-119.4	-113.5	53.1	.9938
1000.	1740.	63.7	-175.9	-174.5	-170.3	-164.6	-158.1	68.5	1.1584

BALLISTIC COEFFICIENT .43, MUZZLE VELOCITY FT/SEC 4000.

RANGE	VELOCITY	HEIGHT	100 YD	150 YD	200 YD	250 YD	300 YD	DRIFT	TIME
0.	4000.	0.0	-1.5	-1.5	-1.5	-1.5	-1.5	0.0	0.0000
100.	3729.	-.5	0.0	.1	.5	1.0	1.6	.5	.0778
200.	3470.	.5	-1.0	-.8	0.0	1.1	2.3	2.0	.1612
300.	3224.	2.3	-4.9	-4.6	-3.4	-1.8	0.0	4.6	.2510
400.	2989.	5.0	-12.2	-11.7	-10.2	-8.1	-5.6	8.4	.3476
500.	2767.	9.1	-23.4	-22.8	-20.8	-18.2	-15.1	13.5	.4520
600.	2555.	14.6	-39.0	-38.3	-36.0	-32.8	-29.2	20.2	.5648
700.	2353.	21.9	-60.1	-59.2	-56.5	-52.8	-48.5	28.6	.6873
800.	2159.	31.5	-87.2	-86.2	-83.1	-78.9	-74.0	38.7	.8201
900.	1974.	44.0	-121.8	-120.7	-117.2	-112.5	-107.0	51.1	.9655
1000.	1800.	60.0	-165.4	-164.2	-160.3	-155.0	-148.9	65.9	1.1247

BALLISTIC COEFFICIENT .44, MUZZLE VELOCITY FT/SEC 2000.

RANGE	VELOCITY	HEIGHT	100 YD	150 YD	200 YD	250 YD	300 YD	DRIFT	TIME
0.	2000.	0.0	-1.5	-1.5	-1.5	-1.5	-1.5	0.0	0.0000
100.	1828.	.4	0.0	2.0	4.5	7.2	10.2	1.2	.1569
200.	1668.	4.4	-9.0	-4.9	0.0	5.5	11.4	5.1	.3288
300.	1519.	12.1	-30.5	-24.4	-17.0	-8.8	0.0	11.8	.5172
400.	1384.	24.4	-67.1	-59.0	-49.2	-38.2	-26.5	21.9	.7243
500.	1270.	42.6	-122.0	-111.8	-99.5	-85.9	-71.2	35.4	.9509
600.	1173.	68.2	-199.0	-186.8	-172.1	-155.7	-138.0	52.6	1.1989
700.	1102.	101.8	-300.0	-285.8	-268.6	-249.5	-228.9	72.4	1.4614
800.	1048.	145.3	-431.7	-415.4	-395.8	-374.0	-350.4	95.9	1.7446
900.	1005.	198.0	-590.0	-571.8	-549.7	-525.1	-498.6	120.5	2.0349
1000.	966.	262.0	-781.0	-760.6	-736.1	-708.8	-679.4	147.8	2.3396

BALLISTIC COEFFICIENT .44, MUZZLE VELOCITY FT/SEC 2100.

RANGE	VELOCITY	HEIGHT	100 YD	150 YD	200 YD	250 YD	300 YD	DRIFT	TIME
0.	2100.	0.0	-1.5	-1.5	-1.5	-1.5	-1.5	0.0	0.0000
100.	1922.	.3	0.0	1.8	4.0	6.4	9.1	1.2	.1494
200.	1755.	3.9	-7.9	-4.3	0.0	4.9	10.2	4.7	.3125
300.	1600.	10.9	-27.3	-21.9	-15.4	-8.0	0.0	11.1	.4917
400.	1456.	22.0	-60.1	-53.0	-44.3	-34.5	-23.8	20.6	.6883
500.	1330.	38.5	-109.5	-100.5	-89.7	-77.4	-64.1	33.4	.9040
600.	1224.	61.5	-178.4	-167.7	-154.7	-140.0	-123.9	49.6	1.1392
700.	1138.	92.6	-270.9	-258.4	-243.1	-226.0	-207.3	69.4	1.3945
800.	1076.	132.4	-389.5	-375.2	-357.8	-338.2	-316.9	91.9	1.6653
900.	1029.	182.4	-538.7	-522.6	-503.0	-481.0	-456.9	117.5	1.9532
1000.	987.	242.5	-716.7	-698.8	-677.0	-652.5	-625.8	144.7	2.2510

BALLISTIC COEFFICIENT .44, MUZZLE VELOCITY FT/SEC 2200.

RANGE	VELOCITY	HEIGHT	100 YD	150 YD	200 YD	250 YD	300 YD	DRIFT	TIME
0.	2200.	0.0	-1.5	-1.5	-1.5	-1.5	-1.5	0.0	0.0000
100.	2017.	.2	0.0	1.6	3.6	5.7	8.2	1.1	.1425
200.	1844.	3.5	-7.1	-3.9	0.0	4.4	9.2	4.5	.2983
300.	1682.	9.8	-24.5	-19.7	-13.8	-7.2	0.0	10.4	.4684
400.	1533.	19.9	-54.1	-47.8	-39.9	-31.2	-21.5	19.3	.6553
500.	1396.	34.8	-98.7	-90.7	-80.9	-70.0	-57.9	31.5	.8607
600.	1280.	55.8	-161.0	-151.5	-139.7	-126.6	-112.1	47.0	1.0853
700.	1181.	84.1	-244.6	-233.4	-219.7	-204.4	-187.5	66.0	1.3295
800.	1108.	121.0	-353.3	-340.6	-324.9	-307.4	-288.1	88.3	1.5925
900.	1053.	167.2	-489.7	-475.3	-457.7	-438.0	-416.3	113.2	1.8704
1000.	1009.	224.1	-657.2	-641.3	-621.6	-599.7	-575.6	141.0	2.1645

BALLISTIC COEFFICIENT .44, MUZZLE VELOCITY FT/SEC 2300.

RANGE	VELOCITY	HEIGHT	100 YD	150 YD	200 YD	250 YD	300 YD	DRIFT	TIME
0.	2300.	0.0	-1.5	-1.5	-1.5	-1.5	-1.5	0.0	0.0000
100.	2113.	.1	0.0	1.4	3.2	5.2	7.3	1.0	.1360
200.	1934.	3.1	-6.4	-3.5	0.0	4.0	8.3	4.2	.2846
300.	1766.	8.8	-22.0	-17.8	-12.5	-6.5	0.0	9.8	.4468
400.	1611.	18.0	-48.9	-43.3	-36.2	-28.2	-19.5	18.1	.6247
500.	1466.	31.5	-89.1	-82.1	-73.2	-63.3	-52.4	29.5	.8200
600.	1338.	50.6	-145.6	-137.2	-126.6	-114.6	-101.6	44.3	1.0345
700.	1231.	76.5	-221.6	-211.7	-199.3	-185.4	-170.2	62.5	1.2684
800.	1143.	110.4	-320.4	-309.1	-295.0	-279.1	-261.7	84.2	1.5216
900.	1080.	153.4	-445.8	-433.2	-417.2	-399.3	-379.8	108.8	1.7919
1000.	1032.	206.9	-601.9	-587.9	-570.1	-550.3	-528.5	136.5	2.0798

BALLISTIC COEFFICIENT .44, MUZZLE VELOCITY FT/SEC 2400.

RANGE	VELOCITY	HEIGHT	100 YD	150 YD	200 YD	250 YD	300 YD	DRIFT	TIME
0.	2400.	0.0	-1.5	-1.5	-1.5	-1.5	-1.5	0.0	0.0000
100.	2209.	.1	0.0	1.2	2.8	4.6	6.6	.9	.1302
200.	2025.	2.8	-5.7	-3.2	0.0	3.6	7.6	3.9	.2721
300.	1851.	8.0	-19.9	-16.2	-11.3	-5.9	0.0	9.2	.4270
400.	1690.	16.3	-44.2	-39.3	-32.9	-25.6	-17.7	17.0	.5966
500.	1539.	28.7	-80.8	-74.6	-66.6	-57.5	-47.6	27.8	.7827
600.	1402.	46.0	-131.9	-124.5	-114.9	-104.0	-92.2	41.7	.9870
700.	1284.	69.6	-201.0	-192.3	-181.1	-168.4	-154.6	59.1	1.2109
800.	1186.	100.8	-291.1	-281.2	-268.4	-253.9	-238.1	80.0	1.4543
900.	1110.	140.7	-406.2	-395.1	-380.6	-364.4	-346.6	104.1	1.7165
1000.	1056.	190.0	-548.8	-536.4	-520.4	-502.3	-482.5	130.9	1.9937

BALLISTIC COEFFICIENT .44, MUZZLE VELOCITY FT/SEC 2500.

RANGE	VELOCITY	HEIGHT	100 YD	150 YD	200 YD	250 YD	300 YD	DRIFT	TIME
0.	2500.	0.0	-1.5	-1.5	-1.5	-1.5	-1.5	0.0	0.0000
100.	2305.	-.0	0.0	1.1	2.5	4.2	6.0	.9	.1249
200.	2117.	2.5	-5.1	-2.9	0.0	3.3	6.9	3.6	.2606
300.	1938.	7.3	-18.0	-14.7	-10.4	-5.4	0.0	8.6	.4089
400.	1770.	14.9	-40.1	-35.7	-30.0	-23.4	-16.2	16.0	.5708
500.	1614.	26.1	-73.4	-67.9	-60.7	-52.4	-43.4	26.1	.7483
600.	1469.	42.0	-119.9	-113.4	-104.8	-94.8	-84.0	39.3	.9432
700.	1341.	63.5	-182.7	-175.1	-165.0	-153.4	-140.8	55.8	1.1572
800.	1233.	92.2	-265.2	-256.4	-244.9	-231.7	-217.3	75.9	1.3912
900.	1145.	128.9	-370.0	-360.2	-347.2	-332.4	-316.1	99.2	1.6435
1000.	1081.	175.1	-502.1	-491.1	-476.7	-460.2	-442.2	125.7	1.9140

BALLISTIC COEFFICIENT .44, MUZZLE VELOCITY FT/SEC 2600.

RANGE	VELOCITY	HEIGHT	100 YD	150 YD	200 YD	250 YD	300 YD	DRIFT	TIME
0.	2600.	0.0	-1.5	-1.5	-1.5	-1.5	-1.5	0.0	0.0000
100.	2401.	-.1	0.0	1.0	2.3	3.8	5.4	.8	.1202
200.	2209.	2.3	-4.6	-2.6	0.0	3.0	6.3	3.4	.2503
300.	2026.	6.6	-16.3	-13.4	-9.5	-5.0	0.0	8.1	.3922
400.	1852.	13.6	-36.5	-32.6	-27.4	-21.4	-14.8	15.0	.5470
500.	1690.	23.9	-66.9	-62.0	-55.5	-48.0	-39.7	24.6	.7166
600.	1540.	38.4	-109.4	-103.5	-95.7	-86.7	-76.8	37.0	.9026
700.	1402.	58.1	-166.5	-159.7	-150.6	-140.1	-128.5	52.7	1.1069
800.	1285.	84.3	-241.6	-233.8	-223.4	-211.4	-198.2	71.7	1.3307
900.	1186.	118.2	-337.8	-329.0	-317.3	-303.8	-288.9	94.3	1.5741
1000.	1111.	161.1	-459.1	-449.4	-436.3	-421.4	-404.8	120.1	1.8362

BALLISTIC COEFFICIENT .44, MUZZLE VELOCITY FT/SEC 2700.

RANGE	VELOCITY	HEIGHT	100 YD	150 YD	200 YD	250 YD	300 YD	DRIFT	TIME
0.	2700.	0.0	-1.5	-1.5	-1.5	-1.5	-1.5	0.0	0.0000
100.	2496.	-.1	0.0	.9	2.1	3.4	4.9	.8	.1156
200.	2301.	2.0	-4.1	-2.4	0.0	2.7	5.8	3.3	.2408
300.	2114.	6.1	-14.8	-12.2	-8.6	-4.5	0.0	7.6	.3767
400.	1935.	12.5	-33.3	-29.9	-25.1	-19.7	-13.6	14.2	.5253
500.	1767.	21.9	-61.1	-56.8	-50.8	-44.0	-36.4	23.2	.6873
600.	1611.	35.2	-100.1	-94.9	-87.7	-79.5	-70.5	34.9	.8652
700.	1467.	53.2	-152.3	-146.3	-137.9	-128.4	-117.8	49.7	1.0604
800.	1339.	77.2	-220.9	-214.0	-204.5	-193.5	-181.5	67.9	1.2747
900.	1232.	108.5	-309.1	-301.3	-290.6	-278.3	-264.7	89.5	1.5086
1000.	1144.	148.2	-420.3	-411.6	-399.7	-386.1	-371.0	114.5	1.7617

BALLISTIC COEFFICIENT .44, MUZZLE VELOCITY FT/SEC 2800.

RANGE	VELOCITY	HEIGHT	100 YD	150 YD	200 YD	250 YD	300 YD	DRIFT	TIME
0.	2800.	0.0	-1.5	-1.5	-1.5	-1.5	-1.5	0.0	0.0000
100.	2592.	-.2	0.0	.8	1.9	3.1	4.5	.7	.1112
200.	2393.	1.8	-3.7	-2.2	0.0	2.5	5.3	3.1	.2317
300.	2202.	5.6	-13.5	-11.2	-7.9	-4.2	0.0	7.2	.3623
400.	2018.	11.5	-30.5	-27.4	-23.1	-18.1	-12.5	13.4	.5047
500.	1845.	20.2	-56.1	-52.2	-46.8	-40.5	-33.6	21.9	.6603
600.	1684.	32.3	-91.8	-87.2	-80.7	-73.1	-64.8	33.0	.8304
700.	1534.	48.9	-139.7	-134.4	-126.8	-118.0	-108.3	47.0	1.0171
800.	1397.	71.0	-202.6	-196.4	-187.7	-177.7	-166.6	64.3	1.2223
900.	1280.	99.7	-283.4	-276.5	-266.7	-255.4	-242.9	84.9	1.4467
1000.	1182.	136.5	-385.6	-377.9	-367.1	-354.5	-340.6	109.0	1.6908

BALLISTIC COEFFICIENT .44, MUZZLE VELOCITY FT/SEC 2900.

RANGE	VELOCITY	HEIGHT	100 YD	150 YD	200 YD	250 YD	300 YD	DRIFT	TIME
0.	2900.	0.0	-1.5	-1.5	-1.5	-1.5	-1.5	0.0	0.0000
100.	2687.	-.2	0.0	.7	1.7	2.8	4.1	.7	.1073
200.	2484.	1.6	-3.3	-2.0	0.0	2.3	4.9	2.9	.2235
300.	2289.	5.1	-12.4	-10.3	-7.3	-3.9	0.0	6.9	.3494
400.	2102.	10.6	-28.0	-25.3	-21.4	-16.7	-11.6	12.8	.4863
500.	1924.	18.6	-51.6	-48.2	-43.2	-37.4	-31.0	20.8	.6356
600.	1757.	29.9	-84.4	-80.3	-74.4	-67.4	-59.7	31.3	.7985
700.	1602.	45.1	-128.7	-123.9	-117.0	-108.8	-99.8	44.6	.9776
800.	1458.	65.4	-186.3	-180.9	-173.0	-163.7	-153.4	61.0	1.1739
900.	1332.	91.9	-260.6	-254.4	-245.6	-235.1	-223.5	80.7	1.3894
1000.	1226.	125.9	-354.7	-347.9	-338.0	-326.4	-313.5	103.8	1.6244

BALLISTIC COEFFICIENT .44, MUZZLE VELOCITY FT/SEC 3000.

RANGE	VELOCITY	HEIGHT	100 YD	150 YD	200 YD	250 YD	300 YD	DRIFT	TIME
0.	3000.	0.0	-1.5	-1.5	-1.5	-1.5	-1.5	0.0	0.0000
100.	2782.	-.2	0.0	.6	1.5	2.6	3.8	.7	.1039
200.	2574.	1.5	-3.0	-1.8	0.0	2.1	4.5	2.9	.2162
300.	2376.	4.7	-11.3	-9.5	-6.8	-3.6	0.0	6.6	.3375
400.	2186.	9.8	-25.8	-23.4	-19.7	-15.4	-10.7	12.1	.4690
500.	2003.	17.3	-47.5	-44.5	-40.0	-34.6	-28.7	19.8	.6126
600.	1831.	27.7	-77.9	-74.3	-68.9	-62.5	-55.3	29.8	.7693
700.	1670.	41.7	-118.7	-114.5	-108.1	-100.6	-92.3	42.4	.9408
800.	1522.	60.4	-171.9	-167.1	-159.8	-151.2	-141.7	57.9	1.1290
900.	1386.	84.9	-240.2	-234.8	-226.5	-216.9	-206.2	76.7	1.3357
1000.	1271.	116.4	-327.0	-321.0	-311.9	-301.2	-289.3	98.9	1.5620

BALLISTIC COEFFICIENT .44, MUZZLE VELOCITY FT/SEC 3100.

RANGE	VELOCITY	HEIGHT	100 YD	150 YD	200 YD	250 YD	300 YD	DRIFT	TIME
0.	3100.	0.0	-1.5	-1.5	-1.5	-1.5	-1.5	0.0	0.0000
100.	2877.	-.3	0.0	.5	1.4	2.4	3.5	.6	.1001
200.	2665.	1.3	-2.7	-1.7	0.0	2.0	4.2	2.7	.2086
300.	2463.	4.3	-10.4	-8.8	-6.3	-3.3	0.0	6.2	.3257
400.	2269.	9.1	-23.8	-21.6	-18.3	-14.4	-10.0	11.6	.4528
500.	2083.	16.0	-43.9	-41.2	-37.1	-32.1	-26.6	18.8	.5906
600.	1905.	25.6	-72.1	-68.9	-63.9	-57.9	-51.3	28.3	.7414
700.	1740.	38.7	-109.8	-106.0	-100.2	-93.3	-85.6	40.3	.9061
800.	1586.	56.0	-159.0	-154.7	-148.1	-140.2	-131.3	55.1	1.0870
900.	1443.	78.5	-222.0	-217.2	-209.7	-200.8	-190.9	72.9	1.2853
1000.	1320.	107.6	-302.0	-296.6	-288.3	-278.4	-267.4	94.1	1.5026

BALLISTIC COEFFICIENT .44, MUZZLE VELOCITY FT/SEC 3200.

RANGE	VELOCITY	HEIGHT	100 YD	150 YD	200 YD	250 YD	300 YD	DRIFT	TIME
0.	3200.	0.0	-1.5	-1.5	-1.5	-1.5	-1.5	0.0	0.0000
100.	2972.	-.3	0.0	.5	1.2	2.1	3.2	.6	.0972
200.	2755.	1.2	-2.5	-1.5	0.0	1.8	3.9	2.6	.2022
300.	2549.	4.0	-9.5	-8.1	-5.8	-3.1	0.0	6.0	.3154
400.	2352.	8.5	-22.0	-20.1	-17.0	-13.4	-9.3	11.1	.4381
500.	2162.	14.9	-40.6	-38.3	-34.5	-29.9	-24.7	18.0	.5708
600.	1981.	23.8	-66.7	-64.0	-59.3	-53.8	-47.7	27.0	.7157
700.	1810.	35.9	-101.7	-98.5	-93.1	-86.7	-79.5	38.4	.8743
800.	1651.	52.0	-147.3	-143.6	-137.4	-130.1	-121.9	52.4	1.0480
900.	1504.	72.8	-205.6	-201.4	-194.5	-186.2	-177.0	69.4	1.2383
1000.	1370.	99.8	-279.5	-274.9	-267.2	-258.1	-247.8	89.8	1.4476

BALLISTIC COEFFICIENT .44, MUZZLE VELOCITY FT/SEC 3300.

RANGE	VELOCITY	HEIGHT	100 YD	150 YD	200 YD	250 YD	300 YD	DRIFT	TIME
0.	3300.	0.0	-1.5	-1.5	-1.5	-1.5	-1.5	0.0	0.0000
100.	3067.	-.3	0.0	.4	1.1	2.0	2.9	.6	.0942
200.	2846.	1.1	-2.2	-1.4	0.0	1.7	3.6	2.5	.1957
300.	2635.	3.7	-8.7	-7.5	-5.4	-2.9	0.0	5.8	.3055
400.	2434.	7.9	-20.3	-18.7	-15.9	-12.4	-8.6	10.6	.4239
500.	2241.	13.9	-37.7	-35.7	-32.2	-27.9	-23.2	17.2	.5525
600.	2057.	22.3	-62.1	-59.6	-55.4	-50.3	-44.6	25.8	.6923
700.	1881.	33.5	-94.5	-91.7	-86.7	-80.7	-74.1	36.6	.8446
800.	1717.	48.4	-136.8	-133.6	-127.9	-121.1	-113.4	50.0	1.0114
900.	1565.	67.8	-191.0	-187.4	-181.1	-173.4	-164.8	66.3	1.1949
1000.	1424.	92.8	-259.3	-255.3	-248.3	-239.7	-230.2	85.6	1.3957

BALLISTIC COEFFICIENT .44, MUZZLE VELOCITY FT/SEC 3400.

RANGE	VELOCITY	HEIGHT	100 YD	150 YD	200 YD	250 YD	300 YD	DRIFT	TIME
0.	3400.	0.0	-1.5	-1.5	-1.5	-1.5	-1.5	0.0	0.0000
100.	3162.	-.3	0.0	.3	1.0	1.8	2.7	.6	.0914
200.	2936.	1.0	-2.0	-1.3	0.0	1.6	3.4	2.3	.1897
300.	2721.	3.5	-8.0	-7.0	-5.1	-2.7	0.0	5.5	.2960
400.	2517.	7.4	-18.8	-17.4	-14.8	-11.7	-8.1	10.2	.4108
500.	2320.	13.0	-35.0	-33.3	-30.1	-26.1	-21.6	16.5	.5348
600.	2132.	20.8	-57.7	-55.6	-51.7	-47.0	-41.6	24.7	.6697
700.	1952.	31.3	-88.0	-85.6	-81.0	-75.5	-69.2	35.1	.8168
800.	1783.	45.1	-127.4	-124.6	-119.4	-113.1	-105.9	47.8	.9776
900.	1627.	63.1	-177.6	-174.6	-168.7	-161.6	-153.5	63.3	1.1536
1000.	1481.	86.3	-241.1	-237.7	-231.2	-223.3	-214.3	81.8	1.3469

BALLISTIC COEFFICIENT .44, MUZZLE VELOCITY FT/SEC 3500.

RANGE	VELOCITY	HEIGHT	100 YD	150 YD	200 YD	250 YD	300 YD	DRIFT	TIME
0.	3500.	0.0	-1.5	-1.5	-1.5	-1.5	-1.5	0.0	0.0000
100.	3258.	-.4	0.0	.3	.9	1.6	2.5	.5	.0888
200.	3027.	.9	-1.8	-1.2	0.0	1.5	3.2	2.3	.1844
300.	2807.	3.2	-7.4	-6.5	-4.7	-2.5	0.0	5.4	.2876
400.	2599.	6.9	-17.4	-16.2	-13.9	-10.9	-7.5	9.8	.3986
500.	2399.	12.2	-32.6	-31.1	-28.2	-24.4	-20.3	15.9	.5188
600.	2208.	19.5	-53.8	-51.9	-48.4	-43.9	-38.9	23.7	.6490
700.	2024.	29.3	-82.1	-79.9	-75.8	-70.6	-64.8	33.6	.7910
800.	1851.	42.2	-118.8	-116.3	-111.7	-105.7	-99.0	45.8	.9460
900.	1689.	59.0	-165.7	-162.9	-157.7	-150.9	-143.4	60.6	1.1156
1000.	1539.	80.6	-224.9	-221.7	-215.9	-208.4	-200.1	78.3	1.3018

BALLISTIC COEFFICIENT .44, MUZZLE VELOCITY FT/SEC 3600.

RANGE	VELOCITY	HEIGHT	100 YD	150 YD	200 YD	250 YD	300 YD	DRIFT	TIME
0.	3600.	0.0	-1.5	-1.5	-1.5	-1.5	-1.5	0.0	0.0000
100.	3353.	-.4	0.0	.3	.8	1.5	2.3	.5	.0862
200.	3118.	.8	-1.6	-1.1	0.0	1.3	2.9	2.2	.1792
300.	2894.	3.0	-6.8	-6.0	-4.4	-2.4	0.0	5.1	.2788
400.	2681.	6.4	-16.1	-15.1	-12.9	-10.2	-7.1	9.3	.3864
500.	2478.	11.4	-30.3	-29.1	-26.3	-22.9	-19.0	15.2	.5029
600.	2283.	18.2	-50.2	-48.6	-45.3	-41.3	-36.6	22.7	.6291
700.	2097.	27.4	-76.7	-74.9	-71.0	-66.3	-60.8	32.2	.7664
800.	1919.	39.5	-111.0	-109.0	-104.6	-99.2	-92.9	43.9	.9160
900.	1752.	55.2	-154.7	-152.4	-147.5	-141.4	-134.3	58.0	1.0794
1000.	1597.	75.3	-210.0	-207.4	-201.9	-195.2	-187.3	74.9	1.2590

BALLISTIC COEFFICIENT .44, MUZZLE VELOCITY FT/SEC 3700.

RANGE	VELOCITY	HEIGHT	100 YD	150 YD	200 YD	250 YD	300 YD	DRIFT	TIME
0.	3700.	0.0	-1.5	-1.5	-1.5	-1.5	-1.5	0.0	0.0000
100.	3448.	-.4	0.0	.2	.7	1.4	2.1	.5	.0842
200.	3208.	.7	-1.4	-1.0	0.0	1.3	2.7	2.2	.1744
300.	2980.	2.8	-6.3	-5.6	-4.1	-2.2	0.0	5.0	.2715
400.	2763.	6.0	-15.0	-14.2	-12.2	-9.6	-6.7	9.1	.3761
500.	2556.	10.7	-28.3	-27.2	-24.7	-21.5	-17.9	14.7	.4889
600.	2359.	17.2	-47.0	-45.7	-42.7	-38.8	-34.4	22.0	.6112
700.	2169.	25.8	-71.7	-70.2	-66.7	-62.2	-57.1	31.0	.7436
800.	1987.	37.1	-103.9	-102.1	-98.1	-93.0	-87.2	42.1	.8881
900.	1816.	51.8	-144.8	-142.9	-138.4	-132.6	-126.0	55.7	1.0460
1000.	1657.	70.6	-196.4	-194.3	-189.3	-182.8	-175.5	71.9	1.2192

BALLISTIC COEFFICIENT .44, MUZZLE VELOCITY FT/SEC 3800.

RANGE	VELOCITY	HEIGHT	100 YD	150 YD	200 YD	250 YD	300 YD	DRIFT	TIME
0.	3800.	0.0	-1.5	-1.5	-1.5	-1.5	-1.5	0.0	0.0000
100.	3544.	-.4	0.0	.2	.6	1.2	1.9	.5	.0819
200.	3299.	.6	-1.3	-.9	0.0	1.2	2.6	2.1	.1696
300.	3067.	2.6	-5.8	-5.2	-3.8	-2.1	0.0	4.8	.2639
400.	2845.	5.7	-13.9	-13.2	-11.4	-9.0	-6.3	8.7	.3654
500.	2634.	10.1	-26.5	-25.6	-23.3	-20.3	-16.9	14.2	.4752
600.	2434.	16.2	-43.9	-42.9	-40.1	-36.5	-32.4	21.1	.5936
700.	2241.	24.3	-67.3	-66.1	-62.8	-58.6	-53.9	29.9	.7223
800.	2056.	34.9	-97.5	-96.1	-92.4	-87.6	-82.1	40.6	.8621
900.	1880.	48.6	-135.8	-134.2	-130.0	-124.7	-118.5	53.5	1.0144
1000.	1717.	66.2	-184.0	-182.3	-177.6	-171.7	-164.8	69.0	1.1813

BALLISTIC COEFFICIENT .44, MUZZLE VELOCITY FT/SEC 3900.

RANGE	VELOCITY	HEIGHT	100 YD	150 YD	200 YD	250 YD	300 YD	DRIFT	TIME
0.	3900.	0.0	-1.5	-1.5	-1.5	-1.5	-1.5	0.0	0.0000
100.	3639.	-.4	0.0	.1	.6	1.1	1.8	.4	.0795
200.	3390.	.6	-1.1	-.9	0.0	1.1	2.4	1.9	.1649
300.	3153.	2.4	-5.3	-4.9	-3.6	-1.9	0.0	4.6	.2567
400.	2927.	5.3	-13.0	-12.5	-10.7	-8.5	-5.9	8.4	.3554
500.	2713.	9.5	-24.7	-24.1	-21.9	-19.1	-15.9	13.6	.4619
600.	2509.	15.2	-41.2	-40.4	-37.8	-34.4	-30.6	20.3	.5770
700.	2313.	22.9	-63.1	-62.2	-59.1	-55.2	-50.7	28.7	.7014
800.	2125.	32.8	-91.5	-90.4	-86.9	-82.5	-77.3	38.9	.8366
900.	1945.	45.8	-127.5	-126.3	-122.4	-117.4	-111.6	51.4	.9844
1000.	1777.	62.3	-172.8	-171.4	-167.1	-161.5	-155.1	66.3	1.1457

BALLISTIC COEFFICIENT .44, MUZZLE VELOCITY FT/SEC 4000.

RANGE	VELOCITY	HEIGHT	100 YD	150 YD	200 YD	250 YD	300 YD	DRIFT	TIME
0.	4000.	0.0	-1.5	-1.5	-1.5	-1.5	-1.5	0.0	0.0000
100.	3735.	-.5	0.0	.1	.5	1.0	1.6	.5	.0777
200.	3481.	.5	-1.0	-.8	0.0	1.0	2.3	1.9	.1610
300.	3240.	2.3	-4.9	-4.5	-3.4	-1.8	0.0	4.5	.2504
400.	3010.	5.0	-12.1	-11.6	-10.1	-8.0	-5.6	8.1	.3462
500.	2791.	9.0	-23.1	-22.6	-20.6	-18.0	-15.0	13.2	.4499
600.	2583.	14.4	-38.6	-37.9	-35.6	-32.5	-28.9	19.7	.5617
700.	2385.	21.6	-59.3	-58.5	-55.8	-52.1	-47.9	27.7	.6825
800.	2194.	31.0	-86.0	-85.1	-82.0	-77.8	-73.0	37.6	.8136
900.	2011.	43.2	-120.0	-118.9	-115.4	-110.7	-105.3	49.6	.9565
1000.	1838.	58.7	-162.5	-161.3	-157.5	-152.3	-146.2	63.8	1.1127

BALLISTIC COEFFICIENT .45, MUZZLE VELOCITY FT/SEC 2000.

RANGE	VELOCITY	HEIGHT	100 YD	150 YD	200 YD	250 YD	300 YD	DRIFT	TIME
0.	2000.	0.0	-1.5	-1.5	-1.5	-1.5	-1.5	0.0	0.0000
100.	1832.	.4	0.0	2.0	4.5	7.2	10.1	1.2	.1568
200.	1675.	4.4	-8.9	-4.9	0.0	5.4	11.3	4.9	.3281
300.	1529.	12.0	-30.3	-24.2	-16.9	-8.8	0.0	11.5	.5156
400.	1395.	24.2	-66.6	-58.6	-48.8	-38.0	-26.2	21.4	.7214
500.	1281.	42.2	-120.8	-110.7	-98.5	-85.0	-70.3	34.5	.9457
600.	1185.	67.2	-196.1	-184.0	-169.4	-153.1	-135.6	51.0	1.1895
700.	1111.	100.4	-296.4	-282.2	-265.2	-246.2	-225.7	70.7	1.4516
800.	1057.	142.6	-424.0	-407.9	-388.4	-366.7	-343.3	93.0	1.7284
900.	1013.	195.6	-583.9	-565.8	-543.8	-519.4	-493.1	118.4	2.0227
1000.	974.	257.7	-769.6	-749.5	-725.1	-698.0	-668.7	144.4	2.3206

BALLISTIC COEFFICIENT .45, MUZZLE VELOCITY FT/SEC 2100.

RANGE	VELOCITY	HEIGHT	100 YD	150 YD	200 YD	250 YD	300 YD	DRIFT	TIME
0.	2100.	0.0	-1.5	-1.5	-1.5	-1.5	-1.5	0.0	0.0000
100.	1926.	.3	0.0	1.8	3.9	6.4	9.0	1.1	.1492
200.	1762.	3.9	-7.9	-4.3	0.0	4.9	10.2	4.6	.3118
300.	1610.	10.8	-27.1	-21.7	-15.2	-7.9	0.0	10.8	.4901
400.	1469.	21.8	-59.7	-52.5	-43.9	-34.1	-23.6	20.0	.6852
500.	1343.	38.0	-108.5	-99.5	-88.7	-76.6	-63.3	32.5	.8991
600.	1237.	60.8	-176.6	-165.9	-152.9	-138.3	-122.5	48.4	1.1322
700.	1150.	91.2	-267.2	-254.7	-239.6	-222.6	-204.1	67.6	1.3839
800.	1086.	130.4	-384.3	-370.0	-352.7	-333.2	-312.1	89.7	1.6527
900.	1038.	179.1	-529.6	-513.6	-494.1	-472.2	-448.4	114.4	1.9355
1000.	996.	238.1	-705.1	-687.3	-665.7	-641.3	-614.9	141.2	2.2309

BALLISTIC COEFFICIENT .45, MUZZLE VELOCITY FT/SEC 2200.

RANGE	VELOCITY	HEIGHT	100 YD	150 YD	200 YD	250 YD	300 YD	DRIFT	TIME
0.	2200.	0.0	-1.5	-1.5	-1.5	-1.5	-1.5	0.0	0.0000
100.	2021.	.2	0.0	1.6	3.5	5.7	8.1	1.1	.1424
200.	1851.	3.5	-7.1	-3.9	0.0	4.4	9.1	4.4	.2975
300.	1693.	9.7	-24.3	-19.6	-13.7	-7.1	0.0	10.2	.4669
400.	1546.	19.7	-53.7	-47.4	-39.6	-30.9	-21.3	18.8	.6525
500.	1410.	34.4	-97.7	-89.8	-80.0	-69.1	-57.2	30.6	.8557
600.	1294.	55.1	-159.4	-149.9	-138.2	-125.0	-110.8	45.8	1.0785
700.	1195.	82.9	-241.7	-230.6	-217.0	-201.7	-185.0	64.4	1.3202
800.	1119.	119.1	-348.4	-335.8	-320.2	-302.7	-283.6	86.1	1.5801
900.	1063.	164.7	-483.1	-468.9	-451.3	-431.6	-410.2	110.7	1.8564
1000.	1019.	219.8	-645.4	-629.6	-610.2	-588.2	-564.4	137.3	2.1435

BALLISTIC COEFFICIENT .45, MUZZLE VELOCITY FT/SEC 2300.

RANGE	VELOCITY	HEIGHT	100 YD	150 YD	200 YD	250 YD	300 YD	DRIFT	TIME
0.	2300.	0.0	-1.5	-1.5	-1.5	-1.5	-1.5	0.0	0.0000
100.	2117.	.1	0.0	1.4	3.2	5.1	7.3	.9	.1358
200.	1941.	3.1	-6.3	-3.5	0.0	3.9	8.3	4.1	.2840
300.	1777.	8.8	-21.9	-17.7	-12.4	-6.5	0.0	9.5	.4455
400.	1624.	17.8	-48.5	-42.9	-35.8	-28.0	-19.3	17.6	.6220
500.	1482.	31.2	-88.3	-81.2	-72.4	-62.6	-51.7	28.7	.8154
600.	1354.	50.0	-144.1	-135.7	-125.1	-113.3	-100.3	43.2	1.0280
700.	1247.	75.3	-218.7	-208.9	-196.6	-182.8	-167.6	60.8	1.2587
800.	1157.	108.7	-316.2	-304.9	-290.9	-275.1	-257.7	82.1	1.5099
900.	1091.	150.7	-438.9	-426.3	-410.5	-392.8	-373.2	106.0	1.7764
1000.	1042.	202.6	-590.6	-576.5	-558.9	-539.3	-517.5	132.7	2.0585

BALLISTIC COEFFICIENT .45, MUZZLE VELOCITY FT/SEC 2400.

RANGE	VELOCITY	HEIGHT	100 YD	150 YD	200 YD	250 YD	300 YD	DRIFT	TIME
0.	2400.	0.0	-1.5	-1.5	-1.5	-1.5	-1.5	0.0	0.0000
100.	2213.	.1	0.0	1.2	2.8	4.6	6.6	.9	.1300
200.	2033.	2.8	-5.7	-3.2	0.0	3.6	7.5	3.8	.2717
300.	1863.	7.9	-19.7	-16.0	-11.2	-5.8	0.0	8.9	.4255
400.	1703.	16.2	-43.9	-39.0	-32.6	-25.4	-17.6	16.6	.5941
500.	1556.	28.4	-80.1	-73.9	-65.9	-56.9	-47.2	27.0	.7786
600.	1419.	45.4	-130.4	-123.1	-113.5	-102.7	-91.0	40.6	.9804
700.	1301.	68.5	-198.3	-189.7	-178.5	-165.9	-152.3	57.4	1.2014
800.	1202.	99.0	-286.8	-276.9	-264.1	-249.8	-234.2	77.7	1.4417
900.	1123.	138.0	-399.4	-388.3	-373.9	-357.8	-340.3	101.3	1.7004
1000.	1067.	186.4	-539.3	-527.0	-511.0	-493.1	-473.6	127.6	1.9748

BALLISTIC COEFFICIENT .45, MUZZLE VELOCITY FT/SEC 2500.

RANGE	VELOCITY	HEIGHT	100 YD	150 YD	200 YD	250 YD	300 YD	DRIFT	TIME
0.	2500.	0.0	-1.5	-1.5	-1.5	-1.5	-1.5	0.0	0.0000
100.	2309.	-.0	0.0	1.1	2.5	4.2	6.0	.8	.1248
200.	2125.	2.5	-5.0	-2.9	0.0	3.3	6.9	3.5	.2601
300.	1949.	7.2	-17.9	-14.6	-10.3	-5.4	0.0	8.4	.4077
400.	1784.	14.8	-39.8	-35.5	-29.8	-23.2	-16.0	15.6	.5686
500.	1631.	25.8	-72.7	-67.3	-60.1	-51.9	-42.9	25.4	.7443
600.	1488.	41.4	-118.6	-112.1	-103.5	-93.6	-82.9	38.2	.9368
700.	1360.	62.5	-180.4	-172.7	-162.7	-151.2	-138.7	54.2	1.1482
800.	1251.	90.4	-261.0	-252.3	-240.8	-227.6	-213.3	73.6	1.3783
900.	1160.	126.4	-363.7	-354.0	-341.0	-326.2	-310.2	96.4	1.6275
1000.	1094.	171.5	-492.8	-481.9	-467.6	-451.1	-433.3	122.2	1.8944

BALLISTIC COEFFICIENT .45, MUZZLE VELOCITY FT/SEC 2600.

RANGE	VELOCITY	HEIGHT	100 YD	150 YD	200 YD	250 YD	300 YD	DRIFT	TIME
0.	2600.	0.0	-1.5	-1.5	-1.5	-1.5	-1.5	0.0	0.0000
100.	2405.	-.1	0.0	1.0	2.3	3.8	5.4	.8	.1201
200.	2218.	2.2	-4.5	-2.6	0.0	3.0	6.3	3.3	.2498
300.	2038.	6.6	-16.2	-13.3	-9.4	-4.9	0.0	7.9	.3912
400.	1867.	13.5	-36.2	-32.3	-27.1	-21.2	-14.6	14.6	.5446
500.	1708.	23.6	-66.2	-61.4	-54.9	-47.4	-39.2	23.9	.7127
600.	1559.	37.9	-108.2	-102.4	-94.7	-85.7	-75.8	36.0	.8969
700.	1422.	57.1	-164.3	-157.6	-148.5	-138.0	-126.5	51.1	1.0982
800.	1304.	82.7	-237.9	-230.2	-219.8	-207.9	-194.7	69.6	1.3187
900.	1204.	115.8	-332.1	-323.4	-311.7	-298.3	-283.5	91.5	1.5585
1000.	1125.	157.7	-450.4	-440.8	-427.8	-412.9	-396.4	116.7	1.8168

BALLISTIC COEFFICIENT .45, MUZZLE VELOCITY FT/SEC 2700.

RANGE	VELOCITY	HEIGHT	100 YD	150 YD	200 YD	250 YD	300 YD	DRIFT	TIME
0.	2700.	0.0	-1.5	-1.5	-1.5	-1.5	-1.5	0.0	0.0000
100.	2501.	-.1	0.0	.9	2.0	3.4	4.9	.8	.1156
200.	2310.	2.0	-4.1	-2.4	0.0	2.7	5.7	3.2	.2403
300.	2126.	6.0	-14.7	-12.1	-8.6	-4.5	0.0	7.4	.3756
400.	1950.	12.4	-33.1	-29.6	-24.9	-19.5	-13.5	13.9	.5231
500.	1785.	21.7	-60.6	-56.3	-50.4	-43.6	-36.1	22.6	.6839
600.	1632.	34.7	-98.9	-93.8	-86.7	-78.6	-69.5	34.0	.8596
700.	1489.	52.4	-150.3	-144.3	-136.0	-126.5	-116.0	48.3	1.0520
800.	1360.	75.9	-217.6	-210.7	-201.3	-190.4	-178.4	65.9	1.2633
900.	1252.	106.3	-303.8	-296.1	-285.4	-273.2	-259.7	86.8	1.4934
1000.	1161.	145.0	-412.2	-403.7	-391.8	-378.2	-363.2	111.1	1.7425

BALLISTIC COEFFICIENT .45, MUZZLE VELOCITY FT/SEC 2800.

RANGE	VELOCITY	HEIGHT	100 YD	150 YD	200 YD	250 YD	300 YD	DRIFT	TIME
0.	2800.	0.0	-1.5	-1.5	-1.5	-1.5	-1.5	0.0	0.0000
100.	2596.	-.2	0.0	.8	1.8	3.1	4.5	.7	.1111
200.	2401.	1.8	-3.7	-2.2	0.0	2.5	5.3	3.0	.2313
300.	2214.	5.5	-13.4	-11.1	-7.9	-4.1	0.0	7.0	.3613
400.	2034.	11.4	-30.3	-27.2	-22.9	-17.9	-12.4	13.1	.5029
500.	1864.	19.9	-55.5	-51.7	-46.3	-40.0	-33.1	21.3	.6566
600.	1705.	31.9	-90.8	-86.2	-79.7	-72.2	-63.9	32.1	.8250
700.	1557.	48.2	-138.0	-132.7	-125.1	-116.4	-106.7	45.7	1.0095
800.	1420.	69.7	-199.5	-193.3	-184.7	-174.7	-163.7	62.3	1.2111
900.	1302.	97.7	-278.5	-271.7	-261.9	-250.7	-238.3	82.3	1.4320
1000.	1203.	133.5	-378.3	-370.7	-359.9	-347.4	-333.6	105.7	1.6722

BALLISTIC COEFFICIENT .45, MUZZLE VELOCITY FT/SEC 2900.

RANGE	VELOCITY	HEIGHT	100 YD	150 YD	200 YD	250 YD	300 YD	DRIFT	TIME
0.	2900.	0.0	-1.5	-1.5	-1.5	-1.5	-1.5	0.0	0.0000
100.	2691.	-.2	0.0	.7	1.7	2.8	4.1	.7	.1073
200.	2493.	1.6	-3.3	-2.0	0.0	2.3	4.9	2.9	.2232
300.	2302.	5.1	-12.3	-10.2	-7.3	-3.8	0.0	6.7	.3485
400.	2118.	10.5	-27.8	-25.1	-21.1	-16.5	-11.4	12.4	.4841
500.	1943.	18.4	-51.1	-47.7	-42.8	-37.0	-30.6	20.3	.6323
600.	1778.	29.5	-83.6	-79.5	-73.6	-66.7	-59.0	30.4	.7937
700.	1625.	44.4	-127.0	-122.2	-115.3	-107.3	-98.3	43.3	.9700
800.	1483.	64.2	-183.5	-178.1	-170.2	-161.0	-150.8	59.1	1.1632
900.	1355.	90.1	-256.3	-250.2	-241.3	-230.9	-219.5	78.2	1.3756
1000.	1248.	123.1	-348.0	-341.2	-331.4	-319.8	-307.1	100.6	1.6062

BALLISTIC COEFFICIENT .45, MUZZLE VELOCITY FT/SEC 3000.

RANGE	VELOCITY	HEIGHT	100 YD	150 YD	200 YD	250 YD	300 YD	DRIFT	TIME
0.	3000.	0.0	-1.5	-1.5	-1.5	-1.5	-1.5	0.0	0.0000
100.	2786.	-.2	0.0	.6	1.5	2.6	3.7	.7	.1039
200.	2583.	1.5	-3.0	-1.8	0.0	2.1	4.5	2.8	.2159
300.	2389.	4.7	-11.2	-9.5	-6.7	-3.6	0.0	6.4	.3366
400.	2202.	9.7	-25.6	-23.2	-19.6	-15.3	-10.6	11.8	.4673
500.	2023.	17.1	-47.1	-44.1	-39.6	-34.3	-28.4	19.3	.6096
600.	1853.	27.3	-77.1	-73.5	-68.1	-61.7	-54.6	28.9	.7644
700.	1695.	41.1	-117.2	-113.0	-106.6	-99.2	-90.9	41.1	.9337
800.	1547.	59.4	-169.4	-164.7	-157.4	-148.9	-139.4	56.2	1.1192
900.	1411.	83.2	-236.1	-230.8	-222.6	-213.0	-202.4	74.3	1.3221
1000.	1295.	113.8	-320.8	-314.9	-305.8	-295.2	-283.4	95.9	1.5446

BALLISTIC COEFFICIENT .45, MUZZLE VELOCITY FT/SEC 3100.

RANGE	VELOCITY	HEIGHT	100 YD	150 YD	200 YD	250 YD	300 YD	DRIFT	TIME
0.	3100.	0.0	-1.5	-1.5	-1.5	-1.5	-1.5	0.0	0.0000
100.	2882.	-.3	0.0	.5	1.4	2.3	3.4	.6	.1000
200.	2674.	1.3	-2.7	-1.7	0.0	2.0	4.2	2.6	.2082
300.	2476.	4.3	-10.3	-8.7	-6.2	-3.3	0.0	6.1	.3248
400.	2286.	9.0	-23.6	-21.5	-18.2	-14.2	-9.9	11.3	.4510
500.	2103.	15.8	-43.6	-40.9	-36.8	-31.8	-26.4	18.3	.5880
600.	1928.	25.3	-71.4	-68.2	-63.2	-57.3	-50.7	27.5	.7371
700.	1765.	38.1	-108.4	-104.7	-98.9	-92.0	-84.4	39.1	.8994
800.	1613.	55.0	-156.7	-152.5	-145.8	-137.9	-129.2	53.4	1.0774
900.	1471.	76.9	-218.3	-213.6	-206.1	-197.2	-187.4	70.6	1.2722
1000.	1345.	105.2	-296.3	-291.1	-282.8	-272.9	-262.0	91.2	1.4858

BALLISTIC COEFFICIENT .45, MUZZLE VELOCITY FT/SEC 3200.

RANGE	VELOCITY	HEIGHT	100 YD	150 YD	200 YD	250 YD	300 YD	DRIFT	TIME
0.	3200.	0.0	-1.5	-1.5	-1.5	-1.5	-1.5	0.0	0.0000
100.	2977.	-.3	0.0	.5	1.2	2.1	3.1	.6	.0972
200.	2764.	1.2	-2.4	-1.5	0.0	1.8	3.9	2.5	.2019
300.	2562.	4.0	-9.4	-8.1	-5.8	-3.1	0.0	5.9	.3145
400.	2369.	8.4	-21.8	-20.0	-16.9	-13.3	-9.2	10.8	.4364
500.	2183.	14.7	-40.2	-38.0	-34.1	-29.6	-24.5	17.5	.5681
600.	2004.	23.6	-66.1	-63.4	-58.7	-53.3	-47.2	26.3	.7117
700.	1836.	35.4	-100.5	-97.3	-92.0	-85.6	-78.5	37.3	.8682
800.	1679.	51.1	-145.2	-141.6	-135.4	-128.1	-120.0	50.9	1.0390
900.	1532.	71.4	-202.3	-198.2	-191.3	-183.1	-173.9	67.3	1.2261
1000.	1398.	97.6	-274.2	-269.7	-262.0	-252.9	-242.8	86.9	1.4312

BALLISTIC COEFFICIENT .45, MUZZLE VELOCITY FT/SEC 3300.

RANGE	VELOCITY	HEIGHT	100 YD	150 YD	200 YD	250 YD	300 YD	DRIFT	TIME
0.	3300.	0.0	-1.5	-1.5	-1.5	-1.5	-1.5	0.0	0.0000
100.	3072.	-.3	0.0	.4	1.1	2.0	2.9	.6	.0942
200.	2855.	1.1	-2.2	-1.4	0.0	1.7	3.6	2.4	.1954
300.	2649.	3.7	-8.7	-7.5	-5.4	-2.8	0.0	5.6	.3047
400.	2452.	7.8	-20.1	-18.5	-15.8	-12.3	-8.5	10.3	.4224
500.	2262.	13.8	-37.4	-35.4	-31.9	-27.6	-22.9	16.8	.5499
600.	2081.	22.0	-61.3	-59.0	-54.8	-49.6	-44.0	25.1	.6879
700.	1907.	33.0	-93.4	-90.6	-85.7	-79.7	-73.1	35.6	.8388
800.	1745.	47.6	-134.9	-131.7	-126.2	-119.3	-111.7	48.6	1.0031
900.	1594.	66.4	-188.0	-184.4	-178.1	-170.4	-161.9	64.2	1.1832
1000.	1454.	90.7	-254.6	-250.7	-243.7	-235.1	-225.7	82.9	1.3802

BALLISTIC COEFFICIENT .45, MUZZLE VELOCITY FT/SEC 3400.

RANGE	VELOCITY	HEIGHT	100 YD	150 YD	200 YD	250 YD	300 YD	DRIFT	TIME
0.	3400.	0.0	-1.5	-1.5	-1.5	-1.5	-1.5	0.0	0.0000
100.	3168.	-.4	0.0	.3	1.0	1.8	2.7	.5	.0913
200.	2946.	1.0	-2.0	-1.3	0.0	1.6	3.4	2.3	.1893
300.	2735.	3.4	-8.0	-7.0	-5.0	-2.7	0.0	5.4	.2952
400.	2534.	7.3	-18.7	-17.3	-14.7	-11.6	-8.0	9.9	.4093
500.	2342.	12.9	-34.8	-33.0	-29.8	-25.9	-21.4	16.1	.5325
600.	2157.	20.5	-57.1	-55.0	-51.2	-46.5	-41.1	24.0	.6657
700.	1980.	30.8	-86.9	-84.5	-80.0	-74.5	-68.2	34.0	.8108
800.	1813.	44.3	-125.6	-122.8	-117.7	-111.4	-104.3	46.4	.9693
900.	1657.	61.9	-174.9	-171.8	-166.0	-159.0	-150.9	61.3	1.1425
1000.	1512.	84.4	-236.8	-233.4	-227.0	-219.1	-210.2	79.1	1.3320

BALLISTIC COEFFICIENT .45, MUZZLE VELOCITY FT/SEC 3500.

RANGE	VELOCITY	HEIGHT	100 YD	150 YD	200 YD	250 YD	300 YD	DRIFT	TIME
0.	3500.	0.0	-1.5	-1.5	-1.5	-1.5	-1.5	0.0	0.0000
100.	3263.	-.4	0.0	.3	.9	1.6	2.5	.5	.0888
200.	3037.	.9	-1.8	-1.2	0.0	1.5	3.1	2.2	.1841
300.	2822.	3.2	-7.4	-6.4	-4.7	-2.5	0.0	5.2	.2868
400.	2617.	6.8	-17.3	-16.1	-13.7	-10.8	-7.5	9.5	.3970
500.	2421.	12.0	-32.3	-30.8	-27.9	-24.2	-20.1	15.5	.5164
600.	2233.	19.2	-53.3	-51.4	-47.9	-43.5	-38.6	23.1	.6454
700.	2052.	28.9	-81.2	-79.0	-74.9	-69.8	-64.0	32.7	.7857
800.	1881.	41.5	-117.2	-114.7	-110.0	-104.1	-97.5	44.4	.9381
900.	1720.	57.8	-163.0	-160.3	-155.0	-148.4	-141.0	58.7	1.1047
1000.	1571.	78.8	-221.0	-217.9	-212.0	-204.7	-196.4	75.8	1.2877

BALLISTIC COEFFICIENT .45, MUZZLE VELOCITY FT/SEC 3600.

RANGE	VELOCITY	HEIGHT	100 YD	150 YD	200 YD	250 YD	300 YD	DRIFT	TIME
0.	3600.	0.0	-1.5	-1.5	-1.5	-1.5	-1.5	0.0	0.0000
100.	3358.	-.4	0.0	.3	.8	1.5	2.3	.5	.0861
200.	3128.	.8	-1.6	-1.1	0.0	1.3	2.9	2.2	.1789
300.	2908.	3.0	-6.8	-6.0	-4.4	-2.3	0.0	5.0	.2783
400.	2699.	6.4	-16.0	-15.0	-12.8	-10.1	-7.0	9.1	.3852
500.	2500.	11.3	-30.1	-28.8	-26.1	-22.7	-18.8	14.8	.5009
600.	2309.	18.0	-49.7	-48.2	-44.9	-40.9	-36.2	22.1	.6257
700.	2125.	27.0	-75.8	-74.0	-70.1	-65.4	-60.0	31.3	.7610
800.	1949.	38.9	-109.5	-107.5	-103.1	-97.7	-91.5	42.6	.9086
900.	1784.	54.1	-152.4	-150.1	-145.2	-139.2	-132.1	56.2	1.0695
1000.	1631.	73.7	-206.2	-203.7	-198.2	-191.5	-183.6	72.5	1.2452

BALLISTIC COEFFICIENT .45, MUZZLE VELOCITY FT/SEC 3700.

RANGE	VELOCITY	HEIGHT	100 YD	150 YD	200 YD	250 YD	300 YD	DRIFT	TIME
0.	3700.	0.0	-1.5	-1.5	-1.5	-1.5	-1.5	0.0	0.0000
100.	3454.	-.4	0.0	.2	.7	1.4	2.1	.5	.0841
200.	3219.	.7	-1.4	-1.0	0.0	1.3	2.7	2.1	.1742
300.	2995.	2.8	-6.2	-5.6	-4.1	-2.1	0.0	4.8	.2706
400.	2782.	6.0	-14.9	-14.0	-12.0	-9.5	-6.6	8.8	.3745
500.	2579.	10.6	-28.1	-27.0	-24.5	-21.3	-17.7	14.3	.4868
600.	2384.	17.0	-46.5	-45.2	-42.2	-38.3	-34.1	21.3	.6077
700.	2198.	25.4	-70.9	-69.4	-65.9	-61.4	-56.4	30.1	.7387
800.	2019.	36.5	-102.5	-100.8	-96.8	-91.7	-86.0	40.9	.8812
900.	1849.	50.8	-142.7	-140.7	-136.2	-130.5	-124.0	54.0	1.0365
1000.	1691.	69.1	-193.0	-190.8	-185.8	-179.4	-172.3	69.6	1.2061

BALLISTIC COEFFICIENT .45, MUZZLE VELOCITY FT/SEC 3800.

RANGE	VELOCITY	HEIGHT	100 YD	150 YD	200 YD	250 YD	300 YD	DRIFT	TIME
0.	3800.	0.0	-1.5	-1.5	-1.5	-1.5	-1.5	0.0	0.0000
100.	3549.	-.4	0.0	.2	.6	1.2	1.9	.5	.0819
200.	3310.	.6	-1.3	-.9	0.0	1.2	2.6	2.0	.1694
300.	3082.	2.6	-5.7	-5.2	-3.8	-2.0	0.0	4.7	.2634
400.	2864.	5.6	-13.8	-13.1	-11.3	-8.9	-6.2	8.5	.3642
500.	2657.	10.0	-26.2	-25.4	-23.1	-20.1	-16.7	13.8	.4731
600.	2460.	16.0	-43.5	-42.5	-39.7	-36.1	-32.0	20.5	.5904
700.	2270.	24.0	-66.5	-65.3	-62.1	-57.9	-53.1	29.0	.7175
800.	2088.	34.4	-96.2	-94.8	-91.1	-86.3	-80.9	39.3	.8551
900.	1915.	47.8	-133.8	-132.3	-128.1	-122.8	-116.6	51.9	1.0054
1000.	1752.	64.8	-180.9	-179.1	-174.5	-168.6	-161.8	66.8	1.1690

BALLISTIC COEFFICIENT .45, MUZZLE VELOCITY FT/SEC 3900.

RANGE	VELOCITY	HEIGHT	100 YD	150 YD	200 YD	250 YD	300 YD	DRIFT	TIME
0.	3900.	0.0	-1.5	-1.5	-1.5	-1.5	-1.5	0.0	0.0000
100.	3645.	-.4	0.0	.1	.6	1.1	1.8	.4	.0794
200.	3401.	.6	-1.1	-.9	0.0	1.1	2.4	1.9	.1647
300.	3169.	2.4	-5.3	-4.9	-3.6	-1.9	0.0	4.4	.2559
400.	2947.	5.3	-12.9	-12.3	-10.6	-8.4	-5.8	8.1	.3540
500.	2736.	9.4	-24.5	-23.8	-21.7	-18.9	-15.7	13.2	.4598
600.	2535.	15.1	-40.8	-40.0	-37.4	-34.1	-30.3	19.8	.5739
700.	2343.	22.6	-62.4	-61.5	-58.5	-54.6	-50.1	27.9	.6970
800.	2158.	32.3	-90.3	-89.2	-85.7	-81.3	-76.2	37.8	.8302
900.	1980.	44.9	-125.6	-124.4	-120.5	-115.5	-109.7	49.8	.9752
1000.	1813.	60.9	-169.8	-168.5	-164.1	-158.6	-152.2	64.1	1.1336

BALLISTIC COEFFICIENT .45, MUZZLE VELOCITY FT/SEC 4000.

RANGE	VELOCITY	HEIGHT	100 YD	150 YD	200 YD	250 YD	300 YD	DRIFT	TIME
0.	4000.	0.0	-1.5	-1.5	-1.5	-1.5	-1.5	0.0	0.0000
100.	3740.	-.5	0.0	.1	.5	1.0	1.6	.5	.0777
200.	3492.	.5	-1.0	-.8	0.0	1.0	2.2	1.9	.1607
300.	3256.	2.2	-4.8	-4.5	-3.4	-1.8	0.0	4.3	.2496
400.	3030.	5.0	-12.0	-11.5	-10.0	-7.9	-5.5	7.9	.3451
500.	2815.	8.9	-23.0	-22.4	-20.5	-17.9	-14.9	12.9	.4481
600.	2611.	14.2	-38.2	-37.6	-35.3	-32.2	-28.6	19.1	.5586
700.	2415.	21.3	-58.7	-57.9	-55.2	-51.5	-47.4	27.0	.6782
800.	2227.	30.5	-84.9	-84.0	-81.0	-76.8	-72.0	36.5	.8075
900.	2047.	42.4	-118.2	-117.2	-113.8	-109.1	-103.7	48.1	.9482
1000.	1876.	57.4	-159.6	-158.5	-154.7	-149.5	-143.5	61.8	1.1010

BALLISTIC COEFFICIENT .46, MUZZLE VELOCITY FT/SEC 2000.

RANGE	VELOCITY	HEIGHT	100 YD	150 YD	200 YD	250 YD	300 YD	DRIFT	TIME
0.	2000.	0.0	-1.5	-1.5	-1.5	-1.5	-1.5	0.0	0.0000
100.	1835.	.4	0.0	2.0	4.4	7.1	10.0	1.2	.1567
200.	1681.	4.4	-8.9	-4.9	0.0	5.4	11.2	4.8	.3274
300.	1538.	11.9	-30.1	-24.1	-16.8	-8.7	0.0	11.3	.5140
400.	1406.	24.0	-66.0	-58.0	-48.3	-37.6	-25.9	20.8	.7181
500.	1293.	41.7	-119.7	-109.7	-97.5	-84.1	-69.5	33.6	.9408
600.	1196.	66.4	-194.2	-182.2	-167.7	-151.5	-134.1	49.8	1.1828
700.	1121.	99.0	-292.7	-278.7	-261.7	-242.9	-222.6	69.0	1.4419
800.	1065.	141.2	-420.5	-404.5	-385.0	-363.5	-340.3	91.5	1.7199
900.	1022.	192.2	-574.4	-556.4	-534.5	-510.3	-484.2	115.3	2.0050
1000.	983.	254.2	-760.5	-740.5	-716.2	-689.3	-660.2	141.7	2.3049

BALLISTIC COEFFICIENT .46, MUZZLE VELOCITY FT/SEC 2100.

RANGE	VELOCITY	HEIGHT	100 YD	150 YD	200 YD	250 YD	300 YD	DRIFT	TIME
0.	2100.	0.0	-1.5	-1.5	-1.5	-1.5	-1.5	0.0	0.0000
100.	1929.	.3	0.0	1.8	3.9	6.3	9.0	1.1	.1491
200.	1769.	3.9	-7.9	-4.3	0.0	4.8	10.1	4.5	.3113
300.	1620.	10.7	-26.9	-21.6	-15.1	-7.9	0.0	10.5	.4884
400.	1481.	21.6	-59.2	-52.1	-43.5	-33.8	-23.3	19.5	.6822
500.	1356.	37.7	-107.5	-98.7	-87.9	-75.8	-62.7	31.7	.8946
600.	1250.	60.0	-174.6	-164.0	-151.0	-136.6	-120.9	47.1	1.1249
700.	1161.	89.9	-263.9	-251.5	-236.4	-219.5	-201.2	65.8	1.3741
800.	1096.	128.4	-379.2	-365.0	-347.7	-328.4	-307.5	87.6	1.6406
900.	1046.	176.9	-524.1	-508.2	-488.7	-467.0	-443.5	112.3	1.9238
1000.	1005.	234.8	-696.5	-678.8	-657.2	-633.1	-606.9	138.5	2.2154

BALLISTIC COEFFICIENT .46, MUZZLE VELOCITY FT/SEC 2200.

RANGE	VELOCITY	HEIGHT	100 YD	150 YD	200 YD	250 YD	300 YD	DRIFT	TIME
0.	2200.	0.0	-1.5	-1.5	-1.5	-1.5	-1.5	0.0	0.0000
100.	2025.	.2	0.0	1.6	3.5	5.7	8.0	1.0	.1423
200.	1858.	3.5	-7.0	-3.9	0.0	4.4	9.1	4.2	.2967
300.	1703.	9.6	-24.1	-19.4	-13.6	-7.1	0.0	9.9	.4654
400.	1558.	19.5	-53.3	-47.1	-39.4	-30.6	-21.2	18.4	.6499
500.	1424.	34.0	-96.8	-88.9	-79.3	-68.3	-56.5	29.8	.8512
600.	1308.	54.3	-157.5	-148.0	-136.5	-123.3	-109.2	44.5	1.0713
700.	1210.	81.6	-238.5	-227.5	-214.0	-198.7	-182.2	62.6	1.3101
800.	1130.	117.1	-343.4	-330.8	-315.4	-297.8	-279.0	83.8	1.5672
900.	1073.	161.8	-475.2	-461.1	-443.7	-424.0	-402.8	107.8	1.8400
1000.	1028.	216.7	-637.3	-621.6	-602.4	-580.4	-556.9	134.6	2.1283

BALLISTIC COEFFICIENT .46, MUZZLE VELOCITY FT/SEC 2300.

RANGE	VELOCITY	HEIGHT	100 YD	150 YD	200 YD	250 YD	300 YD	DRIFT	TIME
0.	2300.	0.0	-1.5	-1.5	-1.5	-1.5	-1.5	0.0	0.0000
100.	2121.	.1	0.0	1.4	3.2	5.1	7.3	.9	.1356
200.	1949.	3.1	-6.3	-3.5	0.0	3.9	8.2	4.0	.2835
300.	1787.	8.7	-21.8	-17.6	-12.3	-6.5	0.0	9.3	.4442
400.	1637.	17.7	-48.2	-42.5	-35.6	-27.8	-19.1	17.2	.6195
500.	1497.	30.8	-87.5	-80.4	-71.7	-62.0	-51.2	28.0	.8111
600.	1370.	49.3	-142.4	-134.0	-123.5	-111.8	-98.9	42.0	1.0211
700.	1262.	74.2	-215.9	-206.1	-193.8	-180.2	-165.1	59.2	1.2492
800.	1171.	106.8	-311.3	-300.0	-286.1	-270.5	-253.2	79.7	1.4966
900.	1103.	148.1	-432.1	-419.5	-403.8	-386.2	-366.8	103.3	1.7610
1000.	1052.	199.0	-580.9	-566.9	-549.4	-529.9	-508.3	129.5	2.0400

BALLISTIC COEFFICIENT .46, MUZZLE VELOCITY FT/SEC 2400.

RANGE	VELOCITY	HEIGHT	100 YD	150 YD	200 YD	250 YD	300 YD	DRIFT	TIME
0.	2400.	0.0	-1.5	-1.5	-1.5	-1.5	-1.5	0.0	0.0000
100.	2217.	.1	0.0	1.2	2.8	4.6	6.5	.9	.1299
200.	2041.	2.8	-5.6	-3.2	0.0	3.6	7.4	3.7	.2712
300.	1873.	7.9	-19.6	-15.9	-11.1	-5.8	0.0	8.7	.4243
400.	1717.	16.0	-43.5	-38.6	-32.2	-25.1	-17.4	16.1	.5915
500.	1571.	28.1	-79.4	-73.2	-65.3	-56.4	-46.7	26.3	.7747
600.	1436.	44.8	-129.1	-121.7	-112.2	-101.5	-89.9	39.5	.9744
700.	1318.	67.5	-195.9	-187.3	-176.1	-163.7	-150.1	55.9	1.1926
800.	1218.	97.4	-282.7	-272.8	-260.1	-245.9	-230.4	75.6	1.4296
900.	1136.	135.7	-393.6	-382.6	-368.3	-352.2	-334.8	98.8	1.6863
1000.	1077.	183.0	-530.2	-517.9	-502.0	-484.2	-464.9	124.3	1.9564

BALLISTIC COEFFICIENT .46, MUZZLE VELOCITY FT/SEC 2500.

RANGE	VELOCITY	HEIGHT	100 YD	150 YD	200 YD	250 YD	300 YD	DRIFT	TIME
0.	2500.	0.0	-1.5	-1.5	-1.5	-1.5	-1.5	0.0	0.0000
100.	2313.	-.0	0.0	1.1	2.5	4.1	5.9	.8	.1247
200.	2133.	2.5	-5.0	-2.9	0.0	3.3	6.8	3.5	.2598
300.	1961.	7.2	-17.8	-14.5	-10.2	-5.3	0.0	8.2	.4065
400.	1798.	14.6	-39.5	-35.2	-29.5	-22.9	-15.8	15.2	.5662
500.	1647.	25.6	-72.1	-66.7	-59.5	-51.3	-42.5	24.7	.7406
600.	1506.	40.9	-117.4	-110.8	-102.3	-92.5	-81.8	37.1	.9310
700.	1378.	61.6	-178.1	-170.5	-160.5	-149.0	-136.6	52.7	1.1396
800.	1269.	88.9	-257.2	-248.5	-237.1	-224.0	-209.9	71.6	1.3666
900.	1176.	124.2	-358.4	-348.6	-335.7	-321.1	-305.1	93.9	1.6134
1000.	1107.	168.2	-484.2	-473.3	-459.0	-442.7	-425.0	119.0	1.8759

BALLISTIC COEFFICIENT .46, MUZZLE VELOCITY FT/SEC 2600.

RANGE	VELOCITY	HEIGHT	100 YD	150 YD	200 YD	250 YD	300 YD	DRIFT	TIME
0.	2600.	0.0	-1.5	-1.5	-1.5	-1.5	-1.5	0.0	0.0000
100.	2409.	-.1	0.0	1.0	2.3	3.7	5.4	.8	.1199
200.	2226.	2.2	-4.5	-2.6	0.0	3.0	6.2	3.3	.2494
300.	2049.	6.6	-16.1	-13.2	-9.3	-4.9	0.0	7.7	.3901
400.	1881.	13.4	-35.9	-32.1	-26.9	-21.0	-14.5	14.3	.5426
500.	1724.	23.4	-65.6	-60.8	-54.4	-46.9	-38.8	23.3	.7091
600.	1578.	37.4	-107.1	-101.4	-93.6	-84.7	-74.9	35.0	.8914
700.	1442.	56.3	-162.4	-155.7	-146.6	-136.2	-124.8	49.7	1.0903
800.	1323.	81.3	-234.5	-226.8	-216.5	-204.6	-191.6	67.7	1.3075
900.	1223.	113.6	-326.7	-318.0	-306.4	-293.0	-278.3	88.9	1.5435
1000.	1140.	154.4	-442.2	-432.6	-419.7	-404.8	-388.5	113.4	1.7981

BALLISTIC COEFFICIENT .46, MUZZLE VELOCITY FT/SEC 2700.

RANGE	VELOCITY	HEIGHT	100 YD	150 YD	200 YD	250 YD	300 YD	DRIFT	TIME
0.	2700.	0.0	-1.5	-1.5	-1.5	-1.5	-1.5	0.0	0.0000
100.	2505.	-.1	0.0	.9	2.0	3.4	4.9	.8	.1155
200.	2318.	2.0	-4.1	-2.3	0.0	2.7	5.7	3.1	.2399
300.	2138.	6.0	-14.6	-12.1	-8.6	-4.5	0.0	7.3	.3748
400.	1965.	12.3	-32.8	-29.4	-24.7	-19.3	-13.3	13.5	.5211
500.	1803.	21.5	-60.0	-55.8	-49.9	-43.1	-35.6	22.0	.6804
600.	1651.	34.3	-97.9	-92.8	-85.8	-77.6	-68.7	33.0	.8544
700.	1510.	51.6	-148.5	-142.5	-134.3	-124.8	-114.4	46.9	1.0444
800.	1381.	74.5	-214.4	-207.6	-198.2	-187.3	-175.4	64.0	1.2523
900.	1272.	104.2	-298.8	-291.2	-280.6	-268.4	-254.9	84.3	1.4789
1000.	1179.	142.0	-404.9	-396.4	-384.7	-371.1	-356.1	108.0	1.7246

BALLISTIC COEFFICIENT .46, MUZZLE VELOCITY FT/SEC 2800.

RANGE	VELOCITY	HEIGHT	100 YD	150 YD	200 YD	250 YD	300 YD	DRIFT	TIME
0.	2800.	0.0	-1.5	-1.5	-1.5	-1.5	-1.5	0.0	0.0000
100.	2601.	-.2	0.0	.8	1.8	3.1	4.4	.7	.1110
200.	2410.	1.8	-3.7	-2.1	0.0	2.5	5.2	2.9	.2309
300.	2226.	5.5	-13.3	-11.1	-7.8	-4.1	0.0	6.9	.3604
400.	2050.	11.3	-30.1	-27.0	-22.7	-17.8	-12.3	12.7	.5010
500.	1882.	19.7	-55.0	-51.2	-45.8	-39.6	-32.8	20.7	.6535
600.	1725.	31.5	-89.8	-85.2	-78.8	-71.4	-63.1	31.2	.8199
700.	1578.	47.5	-136.4	-131.1	-123.6	-114.9	-105.3	44.4	1.0022
800.	1442.	68.5	-196.8	-190.7	-182.1	-172.2	-161.2	60.5	1.2010
900.	1324.	95.8	-274.1	-267.2	-257.5	-246.4	-234.0	79.9	1.4182
1000.	1223.	130.6	-371.4	-363.8	-353.1	-340.7	-326.9	102.6	1.6541

BALLISTIC COEFFICIENT .46, MUZZLE VELOCITY FT/SEC 2900.

RANGE	VELOCITY	HEIGHT	100 YD	150 YD	200 YD	250 YD	300 YD	DRIFT	TIME
0.	2900.	0.0	-1.5	-1.5	-1.5	-1.5	-1.5	0.0	0.0000
100.	2696.	-.2	0.0	.7	1.7	2.8	4.1	.7	.1072
200.	2501.	1.6	-3.3	-2.0	0.0	2.3	4.8	2.8	.2229
300.	2314.	5.0	-12.2	-10.2	-7.2	-3.8	0.0	6.5	.3475
400.	2134.	10.4	-27.6	-24.9	-21.0	-16.4	-11.3	12.1	.4826
500.	1962.	18.3	-50.7	-47.3	-42.4	-36.6	-30.3	19.7	.6293
600.	1799.	29.1	-82.7	-78.7	-72.7	-65.9	-58.3	29.6	.7888
700.	1648.	43.8	-125.5	-120.8	-113.9	-105.9	-97.0	42.1	.9632
800.	1507.	63.1	-181.0	-175.6	-167.8	-158.6	-148.5	57.4	1.1534
900.	1378.	88.3	-252.0	-246.0	-237.1	-226.8	-215.4	75.8	1.3619
1000.	1270.	120.4	-341.6	-334.9	-325.0	-313.6	-301.0	97.6	1.5888

BALLISTIC COEFFICIENT .46, MUZZLE VELOCITY FT/SEC 3000.

RANGE	VELOCITY	HEIGHT	100 YD	150 YD	200 YD	250 YD	300 YD	DRIFT	TIME
0.	3000.	0.0	-1.5	-1.5	-1.5	-1.5	-1.5	0.0	0.0000
100.	2791.	-.2	0.0	.6	1.5	2.6	3.7	.7	.1039
200.	2592.	1.5	-3.0	-1.8	0.0	2.1	4.5	2.7	.2155
300.	2401.	4.7	-11.2	-9.4	-6.7	-3.5	0.0	6.3	.3358
400.	2218.	9.7	-25.4	-23.0	-19.4	-15.2	-10.5	11.5	.4656
500.	2042.	16.9	-46.7	-43.8	-39.3	-34.0	-28.1	18.8	.6069
600.	1874.	27.0	-76.3	-72.7	-67.3	-60.9	-53.9	28.1	.7599
700.	1718.	40.5	-115.7	-111.6	-105.3	-97.9	-89.6	39.9	.9269
800.	1572.	58.4	-167.1	-162.4	-155.2	-146.7	-137.3	54.6	1.1100
900.	1437.	81.6	-232.5	-227.2	-219.1	-209.5	-199.0	72.1	1.3098
1000.	1319.	111.3	-314.9	-309.0	-299.9	-289.4	-277.6	92.9	1.5276

BALLISTIC COEFFICIENT .46, MUZZLE VELOCITY FT/SEC 3100.

RANGE	VELOCITY	HEIGHT	100 YD	150 YD	200 YD	250 YD	300 YD	DRIFT	TIME
0.	3100.	0.0	-1.5	-1.5	-1.5	-1.5	-1.5	0.0	0.0000
100.	2886.	-.3	0.0	.5	1.3	2.3	3.4	.6	.1000
200.	2683.	1.3	-2.7	-1.6	0.0	2.0	4.1	2.5	.2078
300.	2489.	4.3	-10.2	-8.7	-6.2	-3.3	0.0	5.9	.3241
400.	2302.	8.9	-23.4	-21.4	-18.1	-14.1	-9.8	11.0	.4495
500.	2122.	15.7	-43.1	-40.5	-36.4	-31.5	-26.1	17.8	.5850
600.	1951.	25.0	-70.6	-67.5	-62.6	-56.6	-50.1	26.8	.7327
700.	1789.	37.6	-107.2	-103.5	-97.8	-90.9	-83.3	38.0	.8933
800.	1638.	54.1	-154.6	-150.4	-143.8	-135.9	-127.2	51.8	1.0685
900.	1498.	75.4	-214.8	-210.1	-202.7	-193.9	-184.1	68.4	1.2599
1000.	1371.	102.9	-291.0	-285.7	-277.5	-267.6	-256.8	88.3	1.4697

BALLISTIC COEFFICIENT .46, MUZZLE VELOCITY FT/SEC 3200.

RANGE	VELOCITY	HEIGHT	100 YD	150 YD	200 YD	250 YD	300 YD	DRIFT	TIME
0.	3200.	0.0	-1.5	-1.5	-1.5	-1.5	-1.5	0.0	0.0000
100.	2982.	-.3	0.0	.4	1.2	2.1	3.1	.6	.0972
200.	2774.	1.2	-2.4	-1.5	0.0	1.8	3.8	2.4	.2014
300.	2575.	4.0	-9.4	-8.0	-5.8	-3.0	0.0	5.7	.3138
400.	2385.	8.3	-21.6	-19.8	-16.8	-13.2	-9.1	10.5	.4348
500.	2203.	14.6	-39.9	-37.7	-33.9	-29.4	-24.3	17.0	.5656
600.	2027.	23.3	-65.4	-62.7	-58.2	-52.7	-46.6	25.6	.7077
700.	1861.	34.9	-99.3	-96.1	-90.8	-84.4	-77.3	36.2	.8619
800.	1705.	50.2	-143.2	-139.6	-133.5	-126.3	-118.2	49.4	1.0304
900.	1560.	70.1	-199.2	-195.2	-188.3	-180.2	-171.0	65.3	1.2147
1000.	1426.	95.5	-269.4	-264.9	-257.3	-248.2	-238.0	84.2	1.4157

BALLISTIC COEFFICIENT .46, MUZZLE VELOCITY FT/SEC 3300.

RANGE	VELOCITY	HEIGHT	100 YD	150 YD	200 YD	250 YD	300 YD	DRIFT	TIME
0.	3300.	0.0	-1.5	-1.5	-1.5	-1.5	-1.5	0.0	0.0000
100.	3077.	-.3	0.0	.4	1.1	1.9	2.9	.6	.0942
200.	2864.	1.1	-2.2	-1.4	0.0	1.7	3.6	2.3	.1951
300.	2662.	3.7	-8.6	-7.4	-5.4	-2.8	0.0	5.5	.3039
400.	2469.	7.8	-20.0	-18.4	-15.6	-12.2	-8.5	10.1	.4209
500.	2283.	13.6	-37.1	-35.1	-31.6	-27.4	-22.7	16.3	.5474
600.	2104.	21.7	-60.8	-58.4	-54.3	-49.2	-43.5	24.5	.6844
700.	1933.	32.6	-92.4	-89.6	-84.7	-78.8	-72.2	34.6	.8332
800.	1773.	46.8	-133.2	-130.0	-124.5	-117.6	-110.1	47.1	.9952
900.	1623.	65.2	-185.0	-181.4	-175.2	-167.6	-159.1	62.3	1.1720
1000.	1484.	88.7	-250.1	-246.1	-239.2	-230.7	-221.3	80.3	1.3652

BALLISTIC COEFFICIENT .46, MUZZLE VELOCITY FT/SEC 3400.

RANGE	VELOCITY	HEIGHT	100 YD	150 YD	200 YD	250 YD	300 YD	DRIFT	TIME
0.	3400.	0.0	-1.5	-1.5	-1.5	-1.5	-1.5	0.0	0.0000
100.	3172.	-.4	0.0	.3	1.0	1.8	2.6	.5	.0911
200.	2955.	1.0	-2.0	-1.3	0.0	1.6	3.3	2.2	.1890
300.	2748.	3.4	-7.9	-6.9	-5.0	-2.7	0.0	5.2	.2945
400.	2551.	7.2	-18.5	-17.2	-14.6	-11.5	-7.9	9.7	.4078
500.	2362.	12.7	-34.5	-32.8	-29.6	-25.7	-21.2	15.7	.5301
600.	2181.	20.3	-56.5	-54.5	-50.7	-46.0	-40.6	23.3	.6620
700.	2006.	30.4	-86.0	-83.6	-79.1	-73.7	-67.4	33.1	.8056
800.	1841.	43.7	-124.1	-121.3	-116.2	-110.0	-102.9	45.1	.9619
900.	1687.	60.8	-172.3	-169.2	-163.5	-156.4	-148.4	59.5	1.1320
1000.	1543.	82.6	-232.8	-229.4	-223.0	-215.2	-206.3	76.7	1.3180

BALLISTIC COEFFICIENT .46, MUZZLE VELOCITY FT/SEC 3500.

RANGE	VELOCITY	HEIGHT	100 YD	150 YD	200 YD	250 YD	300 YD	DRIFT	TIME
0.	3500.	0.0	-1.5	-1.5	-1.5	-1.5	-1.5	0.0	0.0000
100.	3268.	-.4	0.0	.3	.9	1.6	2.4	.5	.0888
200.	3046.	.9	-1.8	-1.2	0.0	1.5	3.1	2.2	.1838
300.	2835.	3.2	-7.3	-6.4	-4.6	-2.5	0.0	5.1	.2859
400.	2634.	6.8	-17.2	-16.0	-13.6	-10.7	-7.5	9.3	.3958
500.	2442.	11.9	-32.0	-30.5	-27.6	-24.0	-19.9	15.0	.5140
600.	2257.	19.0	-52.8	-51.0	-47.5	-43.1	-38.2	22.5	.6420
700.	2079.	28.5	-80.2	-78.0	-74.0	-68.9	-63.2	31.7	.7802
800.	1910.	40.9	-115.7	-113.3	-108.7	-102.9	-96.3	43.2	.9310
900.	1751.	56.8	-160.7	-158.0	-152.8	-146.2	-138.8	56.9	1.0949
1000.	1603.	77.2	-217.1	-214.1	-208.3	-201.1	-192.8	73.4	1.2743

BALLISTIC COEFFICIENT .46, MUZZLE VELOCITY FT/SEC 3600.

RANGE	VELOCITY	HEIGHT	100 YD	150 YD	200 YD	250 YD	300 YD	DRIFT	TIME
0.	3600.	0.0	-1.5	-1.5	-1.5	-1.5	-1.5	0.0	0.0000
100.	3364.	-.4	0.0	.3	.8	1.5	2.2	.5	.0860
200.	3138.	.8	-1.6	-1.1	0.0	1.3	2.9	2.1	.1786
300.	2922.	3.0	-6.7	-6.0	-4.3	-2.3	0.0	4.9	.2777
400.	2717.	6.3	-15.9	-14.9	-12.7	-10.1	-7.0	8.9	.3840
500.	2521.	11.2	-29.9	-28.6	-25.9	-22.6	-18.7	14.5	.4988
600.	2333.	17.8	-49.3	-47.8	-44.5	-40.5	-35.8	21.5	.6224
700.	2153.	26.7	-75.0	-73.2	-69.4	-64.7	-59.2	30.4	.7562
800.	1979.	38.2	-108.1	-106.1	-101.7	-96.3	-90.1	41.3	.9013
900.	1816.	53.1	-150.1	-147.9	-142.9	-136.9	-129.9	54.5	1.0596
1000.	1663.	72.2	-202.8	-200.3	-194.8	-188.1	-180.4	70.2	1.2325

BALLISTIC COEFFICIENT .46, MUZZLE VELOCITY FT/SEC 3700.

RANGE	VELOCITY	HEIGHT	100 YD	150 YD	200 YD	250 YD	300 YD	DRIFT	TIME
0.	3700.	0.0	-1.5	-1.5	-1.5	-1.5	-1.5	0.0	0.0000
100.	3459.	-.4	0.0	.2	.7	1.3	2.1	.5	.0841
200.	3229.	.7	-1.4	-1.0	0.0	1.3	2.7	2.1	.1739
300.	3009.	2.7	-6.2	-5.5	-4.0	-2.1	0.0	4.7	.2699
400.	2800.	6.0	-14.8	-14.0	-12.0	-9.4	-6.6	8.7	.3737
500.	2600.	10.5	-27.9	-26.8	-24.3	-21.1	-17.6	14.0	.4847
600.	2409.	16.8	-46.1	-44.8	-41.8	-38.0	-33.7	20.8	.6046
700.	2226.	25.1	-70.2	-68.7	-65.2	-60.7	-55.8	29.3	.7341
800.	2049.	36.0	-101.3	-99.6	-95.6	-90.5	-84.9	39.8	.8747
900.	1881.	49.9	-140.6	-138.6	-134.2	-128.4	-122.0	52.4	1.0272
1000.	1724.	67.6	-189.7	-187.6	-182.6	-176.3	-169.2	67.4	1.1937

BALLISTIC COEFFICIENT .46, MUZZLE VELOCITY FT/SEC 3800.

RANGE	VELOCITY	HEIGHT	100 YD	150 YD	200 YD	250 YD	300 YD	DRIFT	TIME
0.	3800.	0.0	-1.5	-1.5	-1.5	-1.5	-1.5	0.0	0.0000
100.	3555.	-.4	0.0	.2	.6	1.2	1.9	.5	.0819
200.	3320.	.6	-1.3	-.9	0.0	1.2	2.5	2.0	.1692
300.	3096.	2.6	-5.7	-5.2	-3.8	-2.0	0.0	4.6	.2629
400.	2883.	5.6	-13.7	-13.0	-11.2	-8.8	-6.1	8.3	.3630
500.	2679.	9.9	-26.0	-25.1	-22.8	-19.9	-16.5	13.4	.4710
600.	2485.	15.8	-43.1	-42.0	-39.3	-35.8	-31.7	20.0	.5873
700.	2299.	23.7	-65.8	-64.6	-61.4	-57.2	-52.5	28.2	.7130
800.	2119.	33.8	-94.9	-93.5	-89.9	-85.1	-79.7	38.2	.8487
900.	1948.	46.9	-131.9	-130.3	-126.2	-120.9	-114.8	50.4	.9966
1000.	1786.	63.6	-178.0	-176.2	-171.7	-165.7	-159.0	64.8	1.1574

BALLISTIC COEFFICIENT .46, MUZZLE VELOCITY FT/SEC 3900.

RANGE	VELOCITY	HEIGHT	100 YD	150 YD	200 YD	250 YD	300 YD	DRIFT	TIME
0.	3900.	0.0	-1.5	-1.5	-1.5	-1.5	-1.5	0.0	0.0000
100.	3650.	-.4	0.0	.1	.6	1.1	1.7	.4	.0792
200.	3412.	.5	-1.1	-.8	0.0	1.1	2.4	1.8	.1643
300.	3184.	2.4	-5.2	-4.8	-3.6	-1.9	0.0	4.3	.2552
400.	2966.	5.2	-12.8	-12.2	-10.5	-8.3	-5.8	7.9	.3529
500.	2759.	9.3	-24.3	-23.7	-21.5	-18.8	-15.6	12.9	.4580
600.	2561.	14.9	-40.4	-39.6	-37.0	-33.7	-29.9	19.2	.5707
700.	2372.	22.3	-61.7	-60.8	-57.8	-54.0	-49.5	27.1	.6926
800.	2189.	31.8	-89.2	-88.1	-84.7	-80.3	-75.2	36.7	.8240
900.	2015.	44.1	-123.9	-122.7	-118.8	-113.9	-108.2	48.4	.9671
1000.	1849.	59.7	-167.1	-165.8	-161.5	-156.0	-149.7	62.2	1.1225

BALLISTIC COEFFICIENT .46, MUZZLE VELOCITY FT/SEC 4000.

RANGE	VELOCITY	HEIGHT	100 YD	150 YD	200 YD	250 YD	300 YD	DRIFT	TIME
0.	4000.	0.0	-1.5	-1.5	-1.5	-1.5	-1.5	0.0	0.0000
100.	3746.	-.5	0.0	.1	.5	1.0	1.6	.5	.0776
200.	3503.	.5	-1.0	-.8	0.0	1.0	2.2	1.8	.1603
300.	3271.	2.2	-4.8	-4.5	-3.4	-1.8	0.0	4.2	.2491
400.	3049.	4.9	-11.9	-11.4	-9.9	-7.9	-5.5	7.7	.3440
500.	2838.	8.8	-22.7	-22.2	-20.3	-17.7	-14.7	12.5	.4459
600.	2637.	14.1	-37.9	-37.2	-34.9	-31.8	-28.2	18.6	.5558
700.	2444.	21.0	-58.0	-57.2	-54.6	-50.9	-46.7	26.2	.6739
800.	2259.	30.1	-83.9	-83.0	-80.0	-75.9	-71.1	35.5	.8018
900.	2082.	41.6	-116.5	-115.5	-112.1	-107.4	-102.0	46.6	.9398
1000.	1912.	56.3	-157.2	-156.1	-152.3	-147.1	-141.1	59.9	1.0905

BALLISTIC COEFFICIENT .47, MUZZLE VELOCITY FT/SEC 2000.

RANGE	VELOCITY	HEIGHT	100 YD	150 YD	200 YD	250 YD	300 YD	DRIFT	TIME
0.	2000.	0.0	-1.5	-1.5	-1.5	-1.5	-1.5	0.0	0.0000
100.	1839.	.4	0.0	2.0	4.4	7.1	10.0	1.2	.1566
200.	1688.	4.4	-8.8	-4.8	0.0	5.3	11.1	4.7	.3267
300.	1547.	11.9	-29.9	-23.9	-16.7	-8.7	0.0	11.0	.5125
400.	1416.	23.8	-65.6	-57.6	-47.9	-37.3	-25.7	20.3	.7153
500.	1304.	41.3	-118.6	-108.6	-96.6	-83.2	-68.7	32.8	.9361
600.	1208.	65.6	-192.1	-180.2	-165.7	-149.7	-132.3	48.5	1.1755
700.	1131.	97.8	-289.4	-275.5	-258.6	-239.9	-219.6	67.4	1.4327
800.	1074.	138.8	-413.7	-397.8	-378.5	-357.2	-334.0	88.9	1.7053
900.	1030.	190.0	-568.8	-550.9	-529.1	-505.1	-479.0	113.3	1.9937
1000.	990.	251.3	-753.1	-733.2	-709.0	-682.4	-653.4	139.3	2.2917

BALLISTIC COEFFICIENT .47, MUZZLE VELOCITY FT/SEC 2100.

RANGE	VELOCITY	HEIGHT	100 YD	150 YD	200 YD	250 YD	300 YD	DRIFT	TIME
0.	2100.	0.0	-1.5	-1.5	-1.5	-1.5	-1.5	0.0	0.0000
100.	1933.	.3	0.0	1.8	3.9	6.3	8.9	1.1	.1489
200.	1776.	3.9	-7.8	-4.3	0.0	4.8	10.0	4.4	.3108
300.	1629.	10.6	-26.7	-21.5	-15.0	-7.8	0.0	10.3	.4871
400.	1492.	21.4	-58.8	-51.7	-43.1	-33.5	-23.1	19.0	.6795
500.	1368.	37.3	-106.5	-97.7	-86.9	-75.0	-62.0	30.9	.8898
600.	1263.	59.3	-172.8	-162.2	-149.3	-134.9	-119.3	45.9	1.1181
700.	1173.	88.9	-261.7	-249.4	-234.2	-217.5	-199.3	64.6	1.3669
800.	1106.	126.6	-374.4	-360.3	-343.0	-323.9	-303.0	85.5	1.6289
900.	1055.	173.8	-515.5	-499.7	-480.2	-458.7	-435.3	109.3	1.9068
1000.	1013.	231.9	-689.0	-671.4	-649.7	-625.9	-599.8	136.0	2.2015

BALLISTIC COEFFICIENT .47, MUZZLE VELOCITY FT/SEC 2200.

RANGE	VELOCITY	HEIGHT	100 YD	150 YD	200 YD	250 YD	300 YD	DRIFT	TIME
0.	2200.	0.0	-1.5	-1.5	-1.5	-1.5	-1.5	0.0	0.0000
100.	2028.	.2	0.0	1.6	3.5	5.7	8.0	1.0	.1422
200.	1865.	3.5	-7.0	-3.8	0.0	4.4	9.0	4.1	.2961
300.	1712.	9.6	-24.0	-19.3	-13.5	-7.0	0.0	9.7	.4640
400.	1570.	19.4	-53.0	-46.7	-39.1	-30.3	-21.0	18.0	.6475
500.	1437.	33.7	-96.0	-88.2	-78.6	-67.7	-56.0	29.1	.8473
600.	1322.	53.7	-155.8	-146.4	-134.9	-121.9	-107.9	43.4	1.0648
700.	1223.	80.5	-235.6	-224.7	-211.3	-196.0	-179.7	61.0	1.3009
800.	1142.	115.3	-338.7	-326.2	-310.8	-293.4	-274.7	81.7	1.5552
900.	1082.	159.4	-469.1	-455.0	-437.8	-418.2	-397.1	105.5	1.8267
1000.	1037.	212.7	-626.6	-611.0	-591.8	-570.0	-546.7	131.2	2.1089

BALLISTIC COEFFICIENT .47, MUZZLE VELOCITY FT/SEC 2300.

RANGE	VELOCITY	HEIGHT	100 YD	150 YD	200 YD	250 YD	300 YD	DRIFT	TIME
0.	2300.	0.0	-1.5	-1.5	-1.5	-1.5	-1.5	0.0	0.0000
100.	2124.	.1	0.0	1.4	3.1	5.1	7.2	.9	.1356
200.	1956.	3.1	-6.3	-3.5	0.0	3.9	8.2	3.9	.2830
300.	1797.	8.7	-21.6	-17.4	-12.2	-6.4	0.0	9.1	.4428
400.	1649.	17.5	-47.8	-42.2	-35.3	-27.5	-19.0	16.8	.6172
500.	1511.	30.5	-86.7	-79.8	-71.1	-61.4	-50.7	27.3	.8071
600.	1385.	48.7	-140.9	-132.5	-122.1	-110.5	-97.6	40.8	1.0147
700.	1277.	73.2	-213.4	-203.7	-191.5	-177.9	-163.0	57.7	1.2408
800.	1185.	105.1	-307.0	-295.9	-282.0	-266.4	-249.3	77.7	1.4848
900.	1114.	145.7	-425.7	-413.2	-397.5	-380.1	-360.8	100.8	1.7464
1000.	1061.	195.7	-572.1	-558.2	-540.8	-521.4	-500.0	126.5	2.0231

BALLISTIC COEFFICIENT .47, MUZZLE VELOCITY FT/SEC 2400.

RANGE	VELOCITY	HEIGHT	100 YD	150 YD	200 YD	250 YD	300 YD	DRIFT	TIME
0.	2400.	0.0	-1.5	-1.5	-1.5	-1.5	-1.5	0.0	0.0000
100.	2221.	.1	0.0	1.2	2.8	4.6	6.5	.8	.1298
200.	2048.	2.8	-5.6	-3.2	0.0	3.5	7.4	3.6	.2707
300.	1884.	7.8	-19.5	-15.8	-11.1	-5.8	0.0	8.5	.4232
400.	1730.	15.9	-43.2	-38.3	-32.0	-24.9	-17.2	15.7	.5894
500.	1586.	27.8	-78.7	-72.5	-64.7	-55.8	-46.2	25.7	.7709
600.	1452.	44.3	-127.8	-120.4	-111.0	-100.4	-88.8	38.5	.9685
700.	1334.	66.6	-193.7	-185.0	-174.0	-161.6	-148.2	54.5	1.1845
800.	1234.	95.9	-279.1	-269.3	-256.7	-242.5	-227.1	73.7	1.4187
900.	1150.	133.2	-387.0	-376.0	-361.8	-345.9	-328.5	96.0	1.6704
1000.	1089.	179.7	-521.7	-509.4	-493.6	-476.0	-456.7	121.2	1.9389

BALLISTIC COEFFICIENT .47, MUZZLE VELOCITY FT/SEC 2500.

RANGE	VELOCITY	HEIGHT	100 YD	150 YD	200 YD	250 YD	300 YD	DRIFT	TIME
0.	2500.	0.0	-1.5	-1.5	-1.5	-1.5	-1.5	0.0	0.0000
100.	2317.	-.0	0.0	1.1	2.5	4.1	5.9	.8	.1245
200.	2141.	2.5	-5.0	-2.9	0.0	3.2	6.7	3.4	.2594
300.	1971.	7.1	-17.6	-14.4	-10.1	-5.3	0.0	8.0	.4053
400.	1812.	14.5	-39.2	-34.9	-29.2	-22.7	-15.7	14.8	.5640
500.	1663.	25.3	-71.5	-66.1	-58.9	-50.9	-42.1	24.1	.7370
600.	1524.	40.4	-116.2	-109.7	-101.1	-91.5	-81.0	36.2	.9255
700.	1396.	60.7	-176.0	-168.4	-158.4	-147.1	-134.9	51.3	1.1315
800.	1286.	87.4	-253.7	-245.0	-233.6	-220.7	-206.7	69.6	1.3555
900.	1193.	122.0	-352.9	-343.1	-330.3	-315.8	-300.0	91.3	1.5989
1000.	1120.	165.0	-476.0	-465.2	-450.9	-434.7	-417.2	115.8	1.8581

BALLISTIC COEFFICIENT .47, MUZZLE VELOCITY FT/SEC 2600.

RANGE	VELOCITY	HEIGHT	100 YD	150 YD	200 YD	250 YD	300 YD	DRIFT	TIME
0.	2600.	0.0	-1.5	-1.5	-1.5	-1.5	-1.5	0.0	0.0000
100.	2413.	-.1	0.0	1.0	2.2	3.7	5.3	.8	.1198
200.	2233.	2.2	-4.5	-2.6	0.0	2.9	6.2	3.2	.2491
300.	2060.	6.5	-16.0	-13.1	-9.3	-4.8	0.0	7.5	.3890
400.	1895.	13.3	-35.7	-31.9	-26.7	-20.8	-14.4	13.9	.5406
500.	1740.	23.2	-65.1	-60.3	-53.9	-46.5	-38.4	22.7	.7059
600.	1596.	36.9	-106.0	-100.3	-92.6	-83.7	-74.0	34.1	.8861
700.	1461.	55.5	-160.5	-153.8	-144.7	-134.4	-123.1	48.4	1.0824
800.	1342.	80.0	-231.3	-223.7	-213.4	-201.6	-188.7	65.8	1.2969
900.	1241.	111.6	-321.7	-313.1	-301.5	-288.2	-273.7	86.4	1.5296
1000.	1155.	151.7	-435.5	-425.9	-413.0	-398.2	-382.1	110.6	1.7821

BALLISTIC COEFFICIENT .47, MUZZLE VELOCITY FT/SEC 2700.

RANGE	VELOCITY	HEIGHT	100 YD	150 YD	200 YD	250 YD	300 YD	DRIFT	TIME
0.	2700.	0.0	-1.5	-1.5	-1.5	-1.5	-1.5	0.0	0.0000
100.	2509.	-.1	0.0	.8	2.0	3.4	4.8	.8	.1154
200.	2326.	2.0	-4.0	-2.3	0.0	2.7	5.7	3.0	.2395
300.	2149.	6.0	-14.5	-12.0	-8.5	-4.4	0.0	7.1	.3737
400.	1979.	12.2	-32.5	-29.2	-24.5	-19.1	-13.2	13.1	.5189
500.	1819.	21.3	-59.5	-55.2	-49.4	-42.6	-35.2	21.4	.6771
600.	1670.	33.9	-97.0	-91.9	-84.9	-76.7	-67.9	32.2	.8494
700.	1530.	50.9	-146.8	-140.9	-132.7	-123.2	-112.9	45.6	1.0371
800.	1401.	73.3	-211.5	-204.7	-195.4	-184.5	-172.7	62.1	1.2420
900.	1291.	102.3	-294.2	-286.6	-276.0	-263.9	-250.6	81.9	1.4653
1000.	1197.	139.2	-398.0	-389.5	-377.8	-364.3	-349.6	104.9	1.7073

BALLISTIC COEFFICIENT .47, MUZZLE VELOCITY FT/SEC 2800.

RANGE	VELOCITY	HEIGHT	100 YD	150 YD	200 YD	250 YD	300 YD	DRIFT	TIME
0.	2800.	0.0	-1.5	-1.5	-1.5	-1.5	-1.5	0.0	0.0000
100.	2605.	-.2	0.0	.8	1.8	3.1	4.4	.7	.1109
200.	2418.	1.8	-3.7	-2.1	0.0	2.5	5.2	2.9	.2305
300.	2238.	5.5	-13.3	-11.0	-7.8	-4.1	0.0	6.7	.3595
400.	2064.	11.2	-29.9	-26.8	-22.6	-17.6	-12.2	12.4	.4991
500.	1899.	19.6	-54.6	-50.8	-45.5	-39.3	-32.5	20.2	.6505
600.	1744.	31.2	-89.0	-84.4	-78.0	-70.6	-62.4	30.4	.8154
700.	1600.	46.8	-134.8	-129.5	-122.0	-113.4	-103.8	43.1	.9952
800.	1465.	67.4	-194.1	-188.0	-179.5	-169.6	-158.7	58.8	1.1911
900.	1345.	94.0	-269.9	-263.0	-253.4	-242.3	-230.0	77.6	1.4051
1000.	1243.	127.9	-365.1	-357.5	-346.8	-334.5	-320.8	99.6	1.6374

BALLISTIC COEFFICIENT .47, MUZZLE VELOCITY FT/SEC 2900.

RANGE	VELOCITY	HEIGHT	100 YD	150 YD	200 YD	250 YD	300 YD	DRIFT	TIME
0.	2900.	0.0	-1.5	-1.5	-1.5	-1.5	-1.5	0.0	0.0000
100.	2700.	-.2	0.0	.7	1.7	2.8	4.0	.7	.1072
200.	2509.	1.6	-3.3	-2.0	0.0	2.3	4.8	2.8	.2226
300.	2326.	5.0	-12.1	-10.1	-7.2	-3.8	0.0	6.4	.3467
400.	2149.	10.3	-27.4	-24.8	-20.8	-16.3	-11.2	11.8	.4809
500.	1979.	18.1	-50.2	-46.8	-41.9	-36.2	-30.0	19.2	.6261
600.	1819.	28.8	-81.9	-77.9	-71.9	-65.1	-57.6	28.8	.7842
700.	1670.	43.2	-124.1	-119.4	-112.5	-104.6	-95.8	40.9	.9566
800.	1530.	62.1	-178.7	-173.3	-165.5	-156.4	-146.3	55.7	1.1443
900.	1401.	86.6	-248.1	-242.1	-233.3	-223.0	-211.7	73.6	1.3492
1000.	1291.	117.9	-335.7	-329.0	-319.2	-307.8	-295.3	94.7	1.5724

BALLISTIC COEFFICIENT .47, MUZZLE VELOCITY FT/SEC 3000.

RANGE	VELOCITY	HEIGHT	100 YD	150 YD	200 YD	250 YD	300 YD	DRIFT	TIME
0.	3000.	0.0	-1.5	-1.5	-1.5	-1.5	-1.5	0.0	0.0000
100.	2795.	-.2	0.0	.6	1.5	2.5	3.7	.7	.1039
200.	2600.	1.5	-3.0	-1.8	0.0	2.1	4.4	2.7	.2151
300.	2413.	4.6	-11.1	-9.3	-6.7	-3.5	0.0	6.2	.3350
400.	2233.	9.6	-25.2	-22.9	-19.3	-15.1	-10.4	11.3	.4642
500.	2060.	16.8	-46.3	-43.4	-38.9	-33.6	-27.8	18.3	.6041
600.	1895.	26.7	-75.5	-72.0	-66.7	-60.3	-53.3	27.4	.7557
700.	1741.	40.0	-114.5	-110.4	-104.1	-96.7	-88.6	38.9	.9210
800.	1596.	57.5	-164.9	-160.2	-153.1	-144.6	-135.3	53.0	1.1011
900.	1462.	80.1	-228.9	-223.6	-215.6	-206.0	-195.6	70.0	1.2975
1000.	1342.	109.0	-309.4	-303.6	-294.6	-284.0	-272.4	90.1	1.5119

BALLISTIC COEFFICIENT .47, MUZZLE VELOCITY FT/SEC 3100.

RANGE	VELOCITY	HEIGHT	100 YD	150 YD	200 YD	250 YD	300 YD	DRIFT	TIME
0.	3100.	0.0	-1.5	-1.5	-1.5	-1.5	-1.5	0.0	0.0000
100.	2891.	-.3	0.0	.5	1.3	2.3	3.4	.6	.1000
200.	2691.	1.3	-2.7	-1.6	0.0	2.0	4.1	2.5	.2076
300.	2501.	4.3	-10.2	-8.6	-6.2	-3.2	0.0	5.8	.3234
400.	2317.	8.9	-23.3	-21.2	-17.9	-14.0	-9.7	10.7	.4479
500.	2141.	15.6	-42.8	-40.2	-36.1	-31.3	-25.8	17.4	.5828
600.	1972.	24.7	-69.9	-66.8	-61.9	-56.0	-49.5	26.0	.7285
700.	1812.	37.0	-105.9	-102.3	-96.6	-89.7	-82.1	36.9	.8872
800.	1663.	53.2	-152.6	-148.4	-141.9	-134.1	-125.4	50.3	1.0602
900.	1524.	74.1	-211.7	-207.0	-199.7	-190.9	-181.1	66.5	1.2486
1000.	1396.	100.8	-286.0	-280.8	-272.6	-262.8	-252.0	85.7	1.4546

BALLISTIC COEFFICIENT .47, MUZZLE VELOCITY FT/SEC 3200.

RANGE	VELOCITY	HEIGHT	100 YD	150 YD	200 YD	250 YD	300 YD	DRIFT	TIME
0.	3200.	0.0	-1.5	-1.5	-1.5	-1.5	-1.5	0.0	0.0000
100.	2986.	-.3	0.0	.4	1.2	2.1	3.1	.6	.0971
200.	2782.	1.2	-2.4	-1.5	0.0	1.8	3.8	2.4	.2010
300.	2588.	4.0	-9.3	-8.0	-5.7	-3.0	0.0	5.6	.3130
400.	2401.	8.3	-21.5	-19.7	-16.7	-13.1	-9.0	10.3	.4334
500.	2222.	14.5	-39.6	-37.4	-33.6	-29.1	-24.0	16.6	.5632
600.	2049.	23.0	-64.9	-62.2	-57.7	-52.2	-46.2	24.9	.7040
700.	1885.	34.5	-98.2	-95.1	-89.8	-83.5	-76.4	35.2	.8564
800.	1731.	49.4	-141.4	-137.8	-131.8	-124.6	-116.5	48.0	1.0225
900.	1587.	68.8	-196.4	-192.3	-185.5	-177.4	-168.3	63.4	1.2039
1000.	1453.	93.5	-265.0	-260.5	-252.9	-243.9	-233.8	81.6	1.4014

BALLISTIC COEFFICIENT .47, MUZZLE VELOCITY FT/SEC 3300.

RANGE	VELOCITY	HEIGHT	100 YD	150 YD	200 YD	250 YD	300 YD	DRIFT	TIME
0.	3300.	0.0	-1.5	-1.5	-1.5	-1.5	-1.5	0.0	0.0000
100.	3082.	-.3	0.0	.4	1.1	1.9	2.9	.6	.0942
200.	2873.	1.1	-2.2	-1.4	0.0	1.7	3.6	2.3	.1948
300.	2674.	3.7	-8.6	-7.4	-5.3	-2.8	0.0	5.3	.3031
400.	2485.	7.7	-19.9	-18.3	-15.5	-12.2	-8.4	9.8	.4195
500.	2302.	13.5	-36.8	-34.8	-31.4	-27.2	-22.5	15.9	.5451
600.	2126.	21.5	-60.2	-57.8	-53.7	-48.7	-43.1	23.8	.6806
700.	1958.	32.1	-91.3	-88.6	-83.8	-77.9	-71.4	33.7	.8278
800.	1799.	46.1	-131.5	-128.3	-122.8	-116.1	-108.6	45.8	.9875
900.	1651.	64.0	-182.4	-178.8	-172.7	-165.1	-156.7	60.5	1.1617
1000.	1513.	86.9	-246.0	-242.0	-235.2	-226.8	-217.5	77.9	1.3515

BALLISTIC COEFFICIENT .47, MUZZLE VELOCITY FT/SEC 3400.

RANGE	VELOCITY	HEIGHT	100 YD	150 YD	200 YD	250 YD	300 YD	DRIFT	TIME
0.	3400.	0.0	-1.5	-1.5	-1.5	-1.5	-1.5	0.0	0.0000
100.	3177.	-.4	0.0	.3	1.0	1.8	2.6	.5	.0910
200.	2964.	1.0	-2.0	-1.3	0.0	1.6	3.3	2.2	.1887
300.	2761.	3.4	-7.9	-6.9	-5.0	-2.7	0.0	5.1	.2939
400.	2568.	7.2	-18.4	-17.0	-14.5	-11.4	-7.9	9.4	.4064
500.	2382.	12.6	-34.2	-32.5	-29.3	-25.4	-21.0	15.2	.5278
600.	2203.	20.1	-56.1	-54.0	-50.2	-45.5	-40.2	22.7	.6587
700.	2032.	30.0	-85.1	-82.7	-78.3	-72.8	-66.6	32.2	.8006
800.	1868.	43.0	-122.4	-119.7	-114.6	-108.4	-101.3	43.7	.9543
900.	1715.	59.7	-169.8	-166.7	-161.0	-154.0	-146.0	57.7	1.1219
1000.	1573.	81.0	-229.1	-225.7	-219.4	-211.6	-202.8	74.4	1.3050

BALLISTIC COEFFICIENT .47, MUZZLE VELOCITY FT/SEC 3500.

RANGE	VELOCITY	HEIGHT	100 YD	150 YD	200 YD	250 YD	300 YD	DRIFT	TIME
0.	3500.	0.0	-1.5	-1.5	-1.5	-1.5	-1.5	0.0	0.0000
100.	3273.	-.4	0.0	.3	.9	1.6	2.4	.5	.0888
200.	3056.	.9	-1.7	-1.2	0.0	1.4	3.1	2.1	.1834
300.	2848.	3.2	-7.2	-6.3	-4.6	-2.4	0.0	4.9	.2852
400.	2651.	6.7	-17.0	-15.9	-13.6	-10.7	-7.4	9.1	.3945
500.	2462.	11.8	-31.8	-30.3	-27.4	-23.8	-19.7	14.7	.5119
600.	2280.	18.8	-52.2	-50.5	-47.0	-42.7	-37.8	21.9	.6386
700.	2105.	28.1	-79.4	-77.3	-73.3	-68.2	-62.5	30.9	.7756
800.	1938.	40.2	-114.3	-111.9	-107.3	-101.5	-95.0	42.0	.9242
900.	1780.	55.8	-158.5	-155.8	-150.6	-144.1	-136.8	55.3	1.0857
1000.	1633.	75.6	-213.6	-210.6	-204.9	-197.6	-189.5	71.2	1.2615

BALLISTIC COEFFICIENT .47, MUZZLE VELOCITY FT/SEC 3600.

RANGE	VELOCITY	HEIGHT	100 YD	150 YD	200 YD	250 YD	300 YD	DRIFT	TIME
0.	3600.	0.0	-1.5	-1.5	-1.5	-1.5	-1.5	0.0	0.0000
100.	3368.	-.4	0.0	.2	.8	1.5	2.2	.5	.0859
200.	3147.	.8	-1.6	-1.1	0.0	1.3	2.9	2.0	.1783
300.	2936.	2.9	-6.7	-5.9	-4.3	-2.3	0.0	4.7	.2767
400.	2734.	6.3	-15.8	-14.8	-12.7	-10.0	-6.9	8.7	.3827
500.	2542.	11.1	-29.7	-28.4	-25.7	-22.4	-18.5	14.1	.4967
600.	2357.	17.7	-48.8	-47.4	-44.1	-40.1	-35.5	21.0	.6193
700.	2179.	26.3	-74.1	-72.4	-68.6	-63.9	-58.6	29.6	.7513
800.	2008.	37.7	-106.8	-104.9	-100.5	-95.2	-89.1	40.2	.8950
900.	1846.	52.3	-148.1	-145.9	-141.0	-135.0	-128.1	52.9	1.0508
1000.	1695.	70.7	-199.5	-197.1	-191.6	-184.9	-177.3	68.1	1.2203

BALLISTIC COEFFICIENT .47, MUZZLE VELOCITY FT/SEC 3700.

RANGE	VELOCITY	HEIGHT	100 YD	150 YD	200 YD	250 YD	300 YD	DRIFT	TIME
0.	3700.	0.0	-1.5	-1.5	-1.5	-1.5	-1.5	0.0	0.0000
100.	3464.	-.4	0.0	.2	.7	1.3	2.0	.5	.0840
200.	3238.	.7	-1.4	-1.0	0.0	1.3	2.7	2.0	.1736
300.	3023.	2.7	-6.1	-5.5	-4.0	-2.1	0.0	4.6	.2694
400.	2817.	5.9	-14.7	-13.9	-11.9	-9.4	-6.5	8.5	.3724
500.	2621.	10.4	-27.6	-26.6	-24.1	-20.9	-17.4	13.6	.4826
600.	2433.	16.6	-45.6	-44.3	-41.4	-37.6	-33.3	20.2	.6014
700.	2253.	24.8	-69.5	-68.0	-64.5	-60.1	-55.2	28.5	.7298
800.	2079.	35.4	-100.0	-98.3	-94.4	-89.3	-83.6	38.6	.8681
900.	1913.	49.1	-138.7	-136.8	-132.4	-126.7	-120.3	50.9	1.0189
1000.	1757.	66.3	-186.8	-184.7	-179.7	-173.4	-166.3	65.4	1.1822

BALLISTIC COEFFICIENT .47, MUZZLE VELOCITY FT/SEC 3800.

RANGE	VELOCITY	HEIGHT	100 YD	150 YD	200 YD	250 YD	300 YD	DRIFT	TIME
0.	3800.	0.0	-1.5	-1.5	-1.5	-1.5	-1.5	0.0	0.0000
100.	3560.	-.4	0.0	.2	.6	1.2	1.9	.5	.0819
200.	3330.	.6	-1.3	-.9	0.0	1.2	2.5	2.0	.1690
300.	3110.	2.6	-5.7	-5.2	-3.8	-2.0	0.0	4.5	.2624
400.	2901.	5.5	-13.7	-13.0	-11.2	-8.8	-6.1	8.2	.3621
500.	2700.	9.8	-25.8	-24.9	-22.7	-19.7	-16.3	13.1	.4693
600.	2510.	15.7	-42.7	-41.7	-39.0	-35.4	-31.4	19.5	.5847
700.	2326.	23.4	-65.1	-63.9	-60.7	-56.6	-51.9	27.5	.7087
800.	2149.	33.4	-93.8	-92.5	-88.8	-84.1	-78.7	37.2	.8429
900.	1980.	46.1	-130.0	-128.5	-124.4	-119.1	-113.0	48.8	.9881
1000.	1820.	62.3	-175.1	-173.4	-168.8	-163.0	-156.2	62.8	1.1462

BALLISTIC COEFFICIENT .47, MUZZLE VELOCITY FT/SEC 3900.

RANGE	VELOCITY	HEIGHT	100 YD	150 YD	200 YD	250 YD	300 YD	DRIFT	TIME
0.	3900.	0.0	-1.5	-1.5	-1.5	-1.5	-1.5	0.0	0.0000
100.	3656.	-.4	0.0	.1	.6	1.1	1.7	.4	.0791
200.	3422.	.5	-1.1	-.8	0.0	1.1	2.4	1.8	.1639
300.	3198.	2.4	-5.2	-4.8	-3.5	-1.9	0.0	4.2	.2548
400.	2984.	5.2	-12.7	-12.2	-10.5	-8.3	-5.8	7.8	.3519
500.	2780.	9.2	-24.1	-23.4	-21.3	-18.6	-15.4	12.5	.4558
600.	2586.	14.7	-40.0	-39.2	-36.7	-33.4	-29.6	18.7	.5680
700.	2399.	22.0	-61.1	-60.2	-57.2	-53.4	-49.0	26.4	.6885
800.	2220.	31.4	-88.1	-87.0	-83.7	-79.3	-74.2	35.7	.8183
900.	2048.	43.4	-122.3	-121.0	-117.3	-112.3	-106.6	47.0	.9593
1000.	1883.	58.6	-164.5	-163.1	-159.0	-153.4	-147.1	60.3	1.1118

BALLISTIC COEFFICIENT .47, MUZZLE VELOCITY FT/SEC 4000.

RANGE	VELOCITY	HEIGHT	100 YD	150 YD	200 YD	250 YD	300 YD	DRIFT	TIME
0.	4000.	0.0	-1.5	-1.5	-1.5	-1.5	-1.5	0.0	0.0000
100.	3751.	-.5	0.0	.1	.5	1.0	1.6	.5	.0776
200.	3513.	.5	-1.0	-.7	0.0	1.0	2.2	1.8	.1600
300.	3285.	2.2	-4.8	-4.5	-3.4	-1.8	0.0	4.2	.2487
400.	3068.	4.9	-11.8	-11.4	-9.9	-7.8	-5.4	7.6	.3429
500.	2860.	8.7	-22.5	-22.0	-20.1	-17.5	-14.5	12.2	.4442
600.	2662.	13.9	-37.5	-36.9	-34.6	-31.5	-27.9	18.1	.5531
700.	2473.	20.8	-57.4	-56.6	-54.0	-50.3	-46.2	25.5	.6700
800.	2290.	29.7	-82.9	-82.0	-79.1	-74.9	-70.1	34.5	.7962
900.	2115.	41.0	-115.0	-114.0	-110.6	-105.9	-100.5	45.3	.9324
1000.	1947.	55.3	-154.8	-153.7	-150.0	-144.7	-138.8	58.2	1.0804

BALLISTIC COEFFICIENT .48, MUZZLE VELOCITY FT/SEC 2000.

RANGE	VELOCITY	HEIGHT	100 YD	150 YD	200 YD	250 YD	300 YD	DRIFT	TIME
0.	2000.	0.0	-1.5	-1.5	-1.5	-1.5	-1.5	0.0	0.0000
100.	1842.	.4	0.0	2.0	4.4	7.0	9.9	1.1	.1565
200.	1694.	4.4	-8.8	-4.8	0.0	5.3	11.1	4.6	.3261
300.	1555.	11.8	-29.8	-23.8	-16.6	-8.7	0.0	10.8	.5112
400.	1427.	23.6	-65.1	-57.2	-47.6	-37.0	-25.4	19.8	.7125
500.	1315.	40.9	-117.7	-107.8	-95.8	-82.5	-68.1	32.0	.9320
600.	1219.	64.8	-190.2	-178.3	-163.9	-148.0	-130.7	47.3	1.1688
700.	1140.	96.5	-286.1	-272.2	-255.5	-236.9	-216.7	65.8	1.4236
800.	1082.	137.2	-409.5	-393.6	-374.4	-353.2	-330.1	87.2	1.6953
900.	1038.	186.9	-560.1	-542.2	-520.7	-496.8	-470.8	110.4	1.9774
1000.	999.	247.1	-741.5	-721.6	-697.6	-671.1	-642.3	135.9	2.2724

BALLISTIC COEFFICIENT .48, MUZZLE VELOCITY FT/SEC 2100.

RANGE	VELOCITY	HEIGHT	100 YD	150 YD	200 YD	250 YD	300 YD	DRIFT	TIME
0.	2100.	0.0	-1.5	-1.5	-1.5	-1.5	-1.5	0.0	0.0000
100.	1936.	.3	0.0	1.7	3.9	6.3	8.9	1.0	.1488
200.	1782.	3.9	-7.8	-4.3	0.0	4.7	9.9	4.3	.3103
300.	1638.	10.6	-26.6	-21.4	-14.9	-7.8	0.0	10.1	.4858
400.	1503.	21.2	-58.4	-51.4	-42.7	-33.3	-22.9	18.6	.6769
500.	1380.	36.9	-105.6	-96.9	-86.0	-74.2	-61.2	30.1	.8854
600.	1275.	58.6	-171.3	-160.8	-147.8	-133.6	-118.1	44.9	1.1122
700.	1185.	87.5	-258.0	-245.8	-230.6	-214.0	-195.9	62.7	1.3561
800.	1115.	124.8	-369.7	-355.8	-338.4	-319.5	-298.8	83.5	1.6175
900.	1063.	171.6	-509.9	-494.2	-474.8	-453.5	-430.1	107.3	1.8951
1000.	1022.	227.7	-677.6	-660.1	-638.5	-614.8	-588.9	132.6	2.1818

BALLISTIC COEFFICIENT .48, MUZZLE VELOCITY FT/SEC 2200.

RANGE	VELOCITY	HEIGHT	100 YD	150 YD	200 YD	250 YD	300 YD	DRIFT	TIME
0.	2200.	0.0	-1.5	-1.5	-1.5	-1.5	-1.5	0.0	0.0000
100.	2032.	.2	0.0	1.6	3.5	5.6	7.9	1.0	.1421
200.	1872.	3.4	-6.9	-3.8	0.0	4.3	9.0	4.0	.2956
300.	1722.	9.5	-23.8	-19.2	-13.4	-7.0	0.0	9.4	.4627
400.	1581.	19.2	-52.6	-46.4	-38.8	-30.1	-20.8	17.5	.6450
500.	1451.	33.4	-95.2	-87.4	-77.9	-67.1	-55.5	28.4	.8431
600.	1335.	53.1	-154.4	-145.1	-133.6	-120.7	-106.8	42.4	1.0591
700.	1237.	79.5	-233.2	-222.3	-208.9	-193.8	-177.6	59.5	1.2928
800.	1153.	114.0	-335.4	-323.0	-307.7	-290.4	-271.9	80.1	1.5462
900.	1093.	156.8	-462.1	-448.1	-430.9	-411.4	-390.6	102.8	1.8116
1000.	1045.	209.9	-619.4	-603.8	-584.7	-563.1	-539.9	128.7	2.0950

BALLISTIC COEFFICIENT .48, MUZZLE VELOCITY FT/SEC 2300.

RANGE	VELOCITY	HEIGHT	100 YD	150 YD	200 YD	250 YD	300 YD	DRIFT	TIME
0.	2300.	0.0	-1.5	-1.5	-1.5	-1.5	-1.5	0.0	0.0000
100.	2128.	.1	0.0	1.4	3.1	5.0	7.2	.9	.1355
200.	1963.	3.1	-6.2	-3.5	0.0	3.9	8.1	3.8	.2824
300.	1807.	8.6	-21.5	-17.3	-12.2	-6.3	0.0	8.8	.4416
400.	1661.	17.4	-47.5	-41.9	-35.0	-27.3	-18.8	16.4	.6149
500.	1525.	30.2	-86.0	-79.1	-70.5	-60.8	-50.2	26.6	.8034
600.	1399.	48.1	-139.5	-131.2	-120.8	-109.2	-96.5	39.8	1.0088
700.	1292.	72.1	-210.9	-201.2	-189.1	-175.5	-160.7	56.2	1.2322
800.	1199.	103.5	-303.1	-292.0	-278.2	-262.7	-245.8	75.7	1.4736
900.	1126.	143.3	-419.6	-407.1	-391.5	-374.1	-355.1	98.3	1.7322
1000.	1071.	192.4	-563.3	-549.4	-532.1	-512.8	-491.6	123.4	2.0057

BALLISTIC COEFFICIENT .48, MUZZLE VELOCITY FT/SEC 2400.

RANGE	VELOCITY	HEIGHT	100 YD	150 YD	200 YD	250 YD	300 YD	DRIFT	TIME
0.	2400.	0.0	-1.5	-1.5	-1.5	-1.5	-1.5	0.0	0.0000
100.	2224.	.1	0.0	1.2	2.8	4.5	6.5	.8	.1297
200.	2055.	2.8	-5.6	-3.1	0.0	3.5	7.3	3.6	.2702
300.	1894.	7.8	-19.4	-15.7	-11.0	-5.8	0.0	8.3	.4221
400.	1742.	15.8	-43.0	-38.1	-31.8	-24.8	-17.1	15.4	.5873
500.	1601.	27.5	-78.0	-71.9	-64.1	-55.3	-45.7	25.0	.7671
600.	1468.	43.7	-126.5	-119.2	-109.8	-99.3	-87.7	37.4	.9627
700.	1350.	65.6	-191.3	-182.7	-171.8	-159.5	-146.1	53.0	1.1761
800.	1250.	94.3	-275.2	-265.4	-252.9	-238.9	-223.5	71.7	1.4072
900.	1164.	130.9	-381.3	-370.3	-356.2	-340.4	-323.1	93.5	1.6562
1000.	1100.	176.5	-513.3	-501.1	-485.4	-467.9	-448.7	118.2	1.9217

BALLISTIC COEFFICIENT .48, MUZZLE VELOCITY FT/SEC 2500.

RANGE	VELOCITY	HEIGHT	100 YD	150 YD	200 YD	250 YD	300 YD	DRIFT	TIME
0.	2500.	0.0	-1.5	-1.5	-1.5	-1.5	-1.5	0.0	0.0000
100.	2321.	-.0	0.0	1.1	2.5	4.1	5.8	.8	.1244
200.	2148.	2.5	-5.0	-2.8	0.0	3.2	6.7	3.3	.2589
300.	1982.	7.1	-17.5	-14.3	-10.0	-5.2	0.0	7.8	.4041
400.	1825.	14.4	-39.0	-34.7	-29.0	-22.6	-15.6	14.4	.5620
500.	1678.	25.1	-70.9	-65.5	-58.4	-50.4	-41.7	23.5	.7335
600.	1540.	39.9	-115.1	-108.6	-100.1	-90.5	-80.1	35.2	.9201
700.	1413.	59.8	-173.9	-166.3	-156.4	-145.2	-133.0	49.9	1.1235
800.	1303.	86.0	-250.3	-241.7	-230.3	-217.5	-203.6	67.7	1.3447
900.	1209.	119.7	-347.3	-337.6	-324.8	-310.4	-294.8	88.7	1.5841
1000.	1133.	161.9	-468.2	-457.4	-443.2	-427.2	-409.8	112.8	1.8408

BALLISTIC COEFFICIENT .48, MUZZLE VELOCITY FT/SEC 2600.

RANGE	VELOCITY	HEIGHT	100 YD	150 YD	200 YD	250 YD	300 YD	DRIFT	TIME
0.	2600.	0.0	-1.5	-1.5	-1.5	-1.5	-1.5	0.0	0.0000
100.	2417.	-.1	0.0	1.0	2.2	3.7	5.3	.8	.1197
200.	2241.	2.2	-4.5	-2.6	0.0	2.9	6.1	3.2	.2487
300.	2071.	6.5	-15.9	-13.0	-9.2	-4.8	0.0	7.3	.3878
400.	1908.	13.2	-35.5	-31.7	-26.5	-20.7	-14.3	13.6	.5389
500.	1756.	22.9	-64.6	-59.8	-53.4	-46.1	-38.1	22.1	.7025
600.	1614.	36.5	-105.0	-99.3	-91.5	-82.8	-73.2	33.2	.8809
700.	1481.	54.7	-158.6	-151.9	-142.9	-132.7	-121.5	47.0	1.0749
800.	1361.	78.7	-228.4	-220.7	-210.4	-198.8	-185.9	64.0	1.2867
900.	1258.	109.6	-316.9	-308.3	-296.7	-283.6	-269.2	84.1	1.5160
1000.	1171.	148.5	-427.5	-418.0	-405.1	-390.5	-374.5	107.3	1.7635

BALLISTIC COEFFICIENT .48, MUZZLE VELOCITY FT/SEC 2700.

RANGE	VELOCITY	HEIGHT	100 YD	150 YD	200 YD	250 YD	300 YD	DRIFT	TIME
0.	2700.	0.0	-1.5	-1.5	-1.5	-1.5	-1.5	0.0	0.0000
100.	2513.	-.1	0.0	.8	2.0	3.4	4.8	.8	.1154
200.	2333.	2.0	-4.0	-2.3	0.0	2.7	5.6	3.0	.2392
300.	2160.	5.9	-14.4	-11.9	-8.4	-4.4	0.0	6.9	.3726
400.	1993.	12.1	-32.4	-29.0	-24.3	-18.9	-13.1	12.8	.5173
500.	1836.	21.1	-59.1	-54.9	-49.0	-42.3	-35.0	20.9	.6743
600.	1688.	33.5	-96.0	-91.0	-84.0	-75.9	-67.2	31.3	.8446
700.	1550.	50.2	-145.2	-139.3	-131.2	-121.7	-111.5	44.4	1.0302
800.	1422.	72.1	-208.7	-202.0	-192.6	-181.9	-170.2	60.4	1.2322
900.	1311.	100.5	-289.8	-282.2	-271.7	-259.6	-246.5	79.6	1.4523
1000.	1215.	136.4	-391.4	-383.0	-371.3	-357.9	-343.3	102.0	1.6905

BALLISTIC COEFFICIENT .48, MUZZLE VELOCITY FT/SEC 2800.

RANGE	VELOCITY	HEIGHT	100 YD	150 YD	200 YD	250 YD	300 YD	DRIFT	TIME
0.	2800.	0.0	-1.5	-1.5	-1.5	-1.5	-1.5	0.0	0.0000
100.	2609.	-.2	0.0	.8	1.8	3.1	4.4	.6	.1108
200.	2425.	1.8	-3.6	-2.1	0.0	2.5	5.2	2.8	.2301
300.	2249.	5.4	-13.2	-10.9	-7.8	-4.1	0.0	6.6	.3587
400.	2079.	11.1	-29.6	-26.6	-22.4	-17.4	-12.0	12.1	.4971
500.	1916.	19.4	-54.2	-50.4	-45.1	-39.0	-32.2	19.7	.6478
600.	1763.	30.8	-88.2	-83.6	-77.3	-69.8	-61.7	29.6	.8108
700.	1620.	46.1	-133.3	-128.0	-120.6	-111.9	-102.4	41.9	.9883
800.	1487.	66.3	-191.6	-185.5	-177.1	-167.2	-156.4	57.1	1.1817
900.	1366.	92.3	-265.9	-259.1	-249.6	-238.5	-226.3	75.4	1.3927
1000.	1263.	125.4	-359.0	-351.4	-340.8	-328.4	-314.9	96.7	1.6210

BALLISTIC COEFFICIENT .48, MUZZLE VELOCITY FT/SEC 2900.

RANGE	VELOCITY	HEIGHT	100 YD	150 YD	200 YD	250 YD	300 YD	DRIFT	TIME
0.	2900.	0.0	-1.5	-1.5	-1.5	-1.5	-1.5	0.0	0.0000
100.	2704.	-.2	0.0	.7	1.6	2.8	4.0	.6	.1071
200.	2517.	1.6	-3.3	-2.0	0.0	2.3	4.8	2.7	.2223
300.	2337.	5.0	-12.1	-10.1	-7.1	-3.8	0.0	6.3	.3459
400.	2163.	10.3	-27.2	-24.6	-20.6	-16.1	-11.1	11.5	.4791
500.	1997.	17.9	-49.8	-46.5	-41.6	-36.0	-29.7	18.7	.6236
600.	1839.	28.5	-81.2	-77.2	-71.3	-64.6	-57.1	28.1	.7804
700.	1691.	42.6	-122.7	-118.1	-111.2	-103.3	-94.5	39.8	.9502
800.	1553.	61.2	-176.5	-171.2	-163.3	-154.3	-144.3	54.2	1.1356
900.	1424.	85.1	-244.5	-238.6	-229.7	-219.6	-208.3	71.5	1.3372
1000.	1313.	115.6	-330.1	-323.5	-313.7	-302.4	-289.9	91.9	1.5568

BALLISTIC COEFFICIENT .48, MUZZLE VELOCITY FT/SEC 3000.

RANGE	VELOCITY	HEIGHT	100 YD	150 YD	200 YD	250 YD	300 YD	DRIFT	TIME
0.	3000.	0.0	-1.5	-1.5	-1.5	-1.5	-1.5	0.0	0.0000
100.	2799.	-.2	0.0	.6	1.5	2.5	3.7	.7	.1040
200.	2608.	1.5	-2.9	-1.8	0.0	2.1	4.4	2.6	.2147
300.	2425.	4.6	-11.0	-9.3	-6.6	-3.4	0.0	6.0	.3341
400.	2248.	9.5	-25.1	-22.7	-19.2	-15.0	-10.4	11.0	.4627
500.	2078.	16.6	-45.9	-43.0	-38.5	-33.3	-27.5	17.8	.6012
600.	1915.	26.4	-74.9	-71.4	-66.1	-59.7	-52.8	26.7	.7519
700.	1762.	39.4	-113.2	-109.2	-102.9	-95.5	-87.5	37.8	.9149
800.	1620.	56.5	-162.8	-158.1	-151.0	-142.6	-133.4	51.5	1.0925
900.	1486.	78.6	-225.5	-220.3	-212.3	-202.8	-192.5	67.9	1.2859
1000.	1365.	106.8	-304.3	-298.6	-289.7	-279.1	-267.6	87.5	1.4969

BALLISTIC COEFFICIENT .48, MUZZLE VELOCITY FT/SEC 3100.

RANGE	VELOCITY	HEIGHT	100 YD	150 YD	200 YD	250 YD	300 YD	DRIFT	TIME
0.	3100.	0.0	-1.5	-1.5	-1.5	-1.5	-1.5	0.0	0.0000
100.	2895.	-.3	0.0	.5	1.3	2.3	3.4	.6	.1000
200.	2699.	1.3	-2.7	-1.6	0.0	1.9	4.1	2.4	.2073
300.	2512.	4.2	-10.1	-8.6	-6.1	-3.2	0.0	5.7	.3227
400.	2332.	8.8	-23.1	-21.0	-17.8	-13.9	-9.6	10.5	.4466
500.	2159.	15.4	-42.5	-39.8	-35.8	-30.9	-25.5	16.9	.5800
600.	1993.	24.5	-69.3	-66.1	-61.3	-55.5	-49.0	25.4	.7247
700.	1835.	36.6	-104.9	-101.2	-95.5	-88.8	-81.2	36.0	.8818
800.	1687.	52.4	-150.7	-146.5	-140.0	-132.3	-123.6	48.9	1.0522
900.	1549.	72.8	-208.8	-204.1	-196.8	-188.1	-178.3	64.6	1.2379
1000.	1421.	98.8	-281.2	-276.0	-267.9	-258.2	-247.4	83.1	1.4399

BALLISTIC COEFFICIENT .48, MUZZLE VELOCITY FT/SEC 3200.

RANGE	VELOCITY	HEIGHT	100 YD	150 YD	200 YD	250 YD	300 YD	DRIFT	TIME
0.	3200.	0.0	-1.5	-1.5	-1.5	-1.5	-1.5	0.0	0.0000
100.	2991.	-.3	0.0	.5	1.2	2.1	3.1	.6	.0969
200.	2790.	1.2	-2.4	-1.5	0.0	1.8	3.8	2.4	.2009
300.	2600.	3.9	-9.3	-7.9	-5.7	-3.0	0.0	5.5	.3123
400.	2416.	8.2	-21.4	-19.6	-16.5	-13.0	-9.0	10.0	.4320
500.	2240.	14.4	-39.4	-37.1	-33.4	-28.9	-23.9	16.2	.5610
600.	2070.	22.8	-64.3	-61.5	-57.0	-51.7	-45.6	24.2	.7002
700.	1908.	34.0	-97.3	-94.1	-88.8	-82.6	-75.6	34.3	.8512
800.	1756.	48.7	-139.7	-136.1	-130.1	-122.9	-114.9	46.6	1.0150
900.	1613.	67.6	-193.6	-189.5	-182.7	-174.7	-165.7	61.5	1.1933
1000.	1480.	91.6	-260.7	-256.1	-248.6	-239.7	-229.7	79.2	1.3874

BALLISTIC COEFFICIENT .48, MUZZLE VELOCITY FT/SEC 3300.

RANGE	VELOCITY	HEIGHT	100 YD	150 YD	200 YD	250 YD	300 YD	DRIFT	TIME
0.	3300.	0.0	-1.5	-1.5	-1.5	-1.5	-1.5	0.0	0.0000
100.	3086.	-.3	0.0	.4	1.1	1.9	2.8	.6	.0942
200.	2882.	1.1	-2.1	-1.4	0.0	1.7	3.5	2.2	.1945
300.	2687.	3.6	-8.5	-7.3	-5.3	-2.8	0.0	5.2	.3024
400.	2500.	7.7	-19.7	-18.2	-15.5	-12.1	-8.4	9.6	.4184
500.	2321.	13.4	-36.5	-34.5	-31.1	-26.9	-22.3	15.5	.5428
600.	2148.	21.3	-59.7	-57.3	-53.3	-48.3	-42.7	23.2	.6773
700.	1982.	31.7	-90.3	-87.6	-82.8	-77.0	-70.4	32.7	.8224
800.	1825.	45.4	-129.9	-126.7	-121.3	-114.6	-107.2	44.5	.9804
900.	1678.	62.9	-179.8	-176.3	-170.2	-162.7	-154.3	58.7	1.1518
1000.	1541.	85.2	-242.2	-238.2	-231.4	-223.1	-213.8	75.6	1.3384

BALLISTIC COEFFICIENT .48, MUZZLE VELOCITY FT/SEC 3400.

RANGE	VELOCITY	HEIGHT	100 YD	150 YD	200 YD	250 YD	300 YD	DRIFT	TIME
0.	3400.	0.0	-1.5	-1.5	-1.5	-1.5	-1.5	0.0	0.0000
100.	3182.	-.4	0.0	.3	1.0	1.7	2.6	.5	.0909
200.	2973.	1.0	-2.0	-1.3	0.0	1.5	3.3	2.1	.1886
300.	2774.	3.4	-7.9	-6.8	-4.9	-2.6	0.0	5.0	.2930
400.	2584.	7.1	-18.3	-16.9	-14.4	-11.3	-7.8	9.2	.4052
500.	2401.	12.5	-34.0	-32.2	-29.1	-25.2	-20.9	14.9	.5257
600.	2225.	19.9	-55.6	-53.5	-49.7	-45.1	-39.9	22.2	.6554
700.	2056.	29.6	-84.3	-81.8	-77.4	-72.0	-65.9	31.4	.7958
800.	1895.	42.4	-121.0	-118.3	-113.2	-107.1	-100.1	42.5	.9476
900.	1743.	58.7	-167.6	-164.4	-158.8	-151.8	-144.0	56.1	1.1128
1000.	1602.	79.4	-225.6	-222.1	-215.8	-208.1	-199.4	72.2	1.2925

BALLISTIC COEFFICIENT .48, MUZZLE VELOCITY FT/SEC 3500.

RANGE	VELOCITY	HEIGHT	100 YD	150 YD	200 YD	250 YD	300 YD	DRIFT	TIME
0.	3500.	0.0	-1.5	-1.5	-1.5	-1.5	-1.5	0.0	0.0000
100.	3277.	-.4	0.0	.3	.9	1.6	2.4	.6	.0889
200.	3065.	.9	-1.7	-1.2	0.0	1.4	3.1	2.1	.1832
300.	2861.	3.1	-7.2	-6.3	-4.6	-2.4	0.0	4.8	.2845
400.	2667.	6.7	-16.9	-15.8	-13.5	-10.6	-7.3	8.9	.3933
500.	2481.	11.7	-31.5	-30.1	-27.2	-23.6	-19.5	14.3	.5098
600.	2303.	18.6	-51.8	-50.1	-46.6	-42.3	-37.4	21.3	.6355
700.	2130.	27.8	-78.5	-76.5	-72.5	-67.5	-61.8	30.1	.7710
800.	1965.	39.7	-112.9	-110.6	-106.0	-100.3	-93.8	40.8	.9176
900.	1809.	54.9	-156.2	-153.7	-148.5	-142.0	-134.7	53.7	1.0766
1000.	1663.	74.2	-210.3	-207.4	-201.7	-194.5	-186.4	69.1	1.2497

BALLISTIC COEFFICIENT .48, MUZZLE VELOCITY FT/SEC 3600.

RANGE	VELOCITY	HEIGHT	100 YD	150 YD	200 YD	250 YD	300 YD	DRIFT	TIME
0.	3600.	0.0	-1.5	-1.5	-1.5	-1.5	-1.5	0.0	0.0000
100.	3373.	-.4	0.0	.2	.8	1.5	2.2	.4	.0858
200.	3156.	.8	-1.6	-1.1	0.0	1.3	2.8	2.0	.1780
300.	2949.	2.9	-6.6	-5.9	-4.3	-2.3	0.0	4.6	.2760
400.	2750.	6.2	-15.7	-14.8	-12.6	-9.9	-6.9	8.5	.3816
500.	2561.	11.0	-29.4	-28.2	-25.5	-22.2	-18.4	13.7	.4946
600.	2380.	17.5	-48.4	-46.9	-43.6	-39.7	-35.1	20.4	.6161
700.	2205.	26.0	-73.4	-71.7	-67.9	-63.3	-58.0	28.8	.7471
800.	2036.	37.2	-105.7	-103.7	-99.4	-94.1	-88.0	39.1	.8889
900.	1876.	51.4	-146.1	-143.9	-139.0	-133.0	-126.2	51.4	1.0421
1000.	1726.	69.4	-196.5	-194.0	-188.6	-182.0	-174.4	66.1	1.2088

BALLISTIC COEFFICIENT .48, MUZZLE VELOCITY FT/SEC 3700.

RANGE	VELOCITY	HEIGHT	100 YD	150 YD	200 YD	250 YD	300 YD	DRIFT	TIME
0.	3700.	0.0	-1.5	-1.5	-1.5	-1.5	-1.5	0.0	0.0000
100.	3469.	-.4	0.0	.2	.7	1.3	2.0	.5	.0839
200.	3248.	.7	-1.4	-1.0	0.0	1.3	2.7	2.0	.1733
300.	3036.	2.7	-6.1	-5.5	-4.0	-2.1	0.0	4.5	.2688
400.	2834.	5.9	-14.6	-13.8	-11.8	-9.3	-6.5	8.2	.3710
500.	2641.	10.3	-27.4	-26.4	-23.9	-20.8	-17.3	13.3	.4808
600.	2456.	16.4	-45.2	-44.0	-41.0	-37.2	-33.0	19.7	.5986
700.	2279.	24.5	-68.7	-67.3	-63.9	-59.4	-54.5	27.8	.7254
800.	2107.	34.9	-98.9	-97.3	-93.3	-88.3	-82.7	37.6	.8624
900.	1943.	48.3	-136.9	-135.0	-130.6	-124.9	-118.6	49.5	1.0107
1000.	1789.	65.1	-184.1	-182.0	-177.1	-170.7	-163.7	63.5	1.1716

BALLISTIC COEFFICIENT .48, MUZZLE VELOCITY FT/SEC 3800.

RANGE	VELOCITY	HEIGHT	100 YD	150 YD	200 YD	250 YD	300 YD	DRIFT	TIME
0.	3800.	0.0	-1.5	-1.5	-1.5	-1.5	-1.5	0.0	0.0000
100.	3565.	-.4	0.0	.2	.6	1.2	1.9	.5	.0818
200.	3339.	.6	-1.2	-.9	0.0	1.2	2.5	1.9	.1688
300.	3124.	2.5	-5.6	-5.1	-3.8	-2.0	0.0	4.4	.2618
400.	2918.	5.5	-13.6	-12.9	-11.1	-8.8	-6.1	8.0	.3612
500.	2721.	9.7	-25.6	-24.8	-22.5	-19.6	-16.2	12.8	.4675
600.	2533.	15.5	-42.4	-41.4	-38.6	-35.1	-31.1	19.1	.5819
700.	2352.	23.1	-64.5	-63.3	-60.1	-56.0	-51.3	26.8	.7048
800.	2178.	32.9	-92.7	-91.4	-87.7	-83.1	-77.7	36.1	.8370
900.	2011.	45.4	-128.4	-126.9	-122.8	-117.5	-111.5	47.5	.9806
1000.	1852.	61.2	-172.6	-170.9	-166.3	-160.5	-153.8	61.0	1.1359

BALLISTIC COEFFICIENT .48, MUZZLE VELOCITY FT/SEC 3900.

RANGE	VELOCITY	HEIGHT	100 YD	150 YD	200 YD	250 YD	300 YD	DRIFT	TIME
0.	3900.	0.0	-1.5	-1.5	-1.5	-1.5	-1.5	0.0	0.0000
100.	3660.	-.5	0.0	.1	.6	1.1	1.7	.4	.0790
200.	3431.	.5	-1.1	-.8	0.0	1.1	2.4	1.7	.1638
300.	3211.	2.4	-5.2	-4.8	-3.5	-1.9	0.0	4.1	.2543
400.	3001.	5.2	-12.6	-12.1	-10.4	-8.2	-5.7	7.6	.3506
500.	2801.	9.2	-24.0	-23.3	-21.2	-18.5	-15.3	12.3	.4546
600.	2610.	14.6	-39.7	-38.9	-36.4	-33.1	-29.3	18.3	.5653
700.	2426.	21.7	-60.5	-59.5	-56.6	-52.8	-48.4	25.7	.6846
800.	2249.	31.0	-87.2	-86.1	-82.8	-78.4	-73.4	34.8	.8131
900.	2079.	42.7	-120.6	-119.4	-115.6	-110.7	-105.0	45.6	.9515
1000.	1916.	57.6	-162.2	-160.8	-156.7	-151.2	-144.9	58.6	1.1022

BALLISTIC COEFFICIENT .48, MUZZLE VELOCITY FT/SEC 4000.

RANGE	VELOCITY	HEIGHT	100 YD	150 YD	200 YD	250 YD	300 YD	DRIFT	TIME
0.	4000.	0.0	-1.5	-1.5	-1.5	-1.5	-1.5	0.0	0.0000
100.	3756.	-.5	0.0	.1	.5	1.0	1.6	.4	.0775
200.	3523.	.5	-1.0	-.7	0.0	1.0	2.2	1.7	.1598
300.	3299.	2.2	-4.8	-4.4	-3.3	-1.8	0.0	4.0	.2480
400.	3085.	4.9	-11.8	-11.3	-9.8	-7.8	-5.4	7.4	.3421
500.	2881.	8.6	-22.4	-21.8	-20.0	-17.4	-14.4	11.9	.4425
600.	2686.	13.8	-37.2	-36.6	-34.3	-31.2	-27.7	17.7	.5505
700.	2500.	20.6	-56.9	-56.1	-53.5	-49.9	-45.8	24.9	.6665
800.	2320.	29.3	-82.0	-81.1	-78.1	-74.0	-69.3	33.6	.7909
900.	2147.	40.4	-113.6	-112.6	-109.2	-104.6	-99.3	44.1	.9254
1000.	1981.	54.3	-152.5	-151.4	-147.7	-142.5	-136.6	56.4	1.0706

BALLISTIC COEFFICIENT .49, MUZZLE VELOCITY FT/SEC 2000.

RANGE	VELOCITY	HEIGHT	100 YD	150 YD	200 YD	250 YD	300 YD	DRIFT	TIME
0.	2000.	0.0	-1.5	-1.5	-1.5	-1.5	-1.5	0.0	0.0000
100.	1845.	.4	0.0	2.0	4.4	7.0	9.9	1.1	.1563
200.	1700.	4.3	-8.7	-4.8	0.0	5.3	11.0	4.5	.3254
300.	1564.	11.7	-29.6	-23.7	-16.5	-8.6	0.0	10.5	.5099
400.	1437.	23.5	-64.8	-56.8	-47.3	-36.7	-25.3	19.4	.7102
500.	1325.	40.5	-116.7	-106.8	-94.9	-81.7	-67.3	31.2	.9275
600.	1230.	64.1	-188.5	-176.6	-162.3	-146.5	-129.3	46.2	1.1627
700.	1150.	95.4	-283.3	-269.3	-252.7	-234.2	-214.1	64.3	1.4153
800.	1091.	135.3	-404.4	-388.5	-369.4	-348.3	-325.3	85.1	1.6836
900.	1045.	185.0	-555.1	-537.2	-515.8	-492.0	-466.2	108.6	1.9671
1000.	1006.	244.3	-734.4	-714.5	-690.7	-664.3	-635.6	133.7	2.2596

BALLISTIC COEFFICIENT .49, MUZZLE VELOCITY FT/SEC 2100.

RANGE	VELOCITY	HEIGHT	100 YD	150 YD	200 YD	250 YD	300 YD	DRIFT	TIME
0.	2100.	0.0	-1.5	-1.5	-1.5	-1.5	-1.5	0.0	0.0000
100.	1939.	.3	0.0	1.7	3.9	6.2	8.8	1.0	.1486
200.	1788.	3.9	-7.8	-4.3	0.0	4.7	9.9	4.2	.3097
300.	1646.	10.5	-26.5	-21.3	-14.8	-7.8	0.0	9.8	.4845
400.	1514.	21.1	-58.0	-51.1	-42.5	-33.1	-22.7	18.1	.6745
500.	1392.	36.5	-104.8	-96.1	-85.3	-73.6	-60.6	29.4	.8813
600.	1287.	57.9	-169.6	-159.1	-146.2	-132.1	-116.6	43.7	1.1057
700.	1197.	86.5	-255.4	-243.2	-228.2	-211.7	-193.6	61.3	1.3480
800.	1125.	123.1	-365.3	-351.4	-334.2	-315.3	-294.7	81.6	1.6066
900.	1072.	168.9	-502.5	-486.9	-467.5	-446.3	-423.1	104.6	1.8800
1000.	1030.	225.0	-670.6	-653.2	-631.7	-608.1	-582.3	130.3	2.1688

BALLISTIC COEFFICIENT .49, MUZZLE VELOCITY FT/SEC 2200.

RANGE	VELOCITY	HEIGHT	100 YD	150 YD	200 YD	250 YD	300 YD	DRIFT	TIME
0.	2200.	0.0	-1.5	-1.5	-1.5	-1.5	-1.5	0.0	0.0000
100.	2035.	.2	0.0	1.5	3.4	5.6	7.9	1.0	.1420
200.	1878.	3.4	-6.9	-3.8	0.0	4.3	8.9	3.9	.2951
300.	1731.	9.5	-23.7	-19.1	-13.4	-6.9	0.0	9.2	.4616
400.	1593.	19.1	-52.3	-46.1	-38.5	-29.9	-20.6	17.1	.6427
500.	1463.	33.0	-94.4	-86.6	-77.1	-66.4	-54.8	27.7	.8391
600.	1348.	52.5	-152.9	-143.6	-132.2	-119.3	-105.4	41.3	1.0529
700.	1250.	78.4	-230.5	-219.6	-206.3	-191.3	-175.1	58.0	1.2841
800.	1166.	112.0	-330.3	-317.9	-302.7	-285.5	-267.0	77.8	1.5330
900.	1102.	154.4	-455.8	-441.9	-424.8	-405.5	-384.7	100.4	1.7980
1000.	1054.	206.2	-609.3	-593.8	-574.8	-553.3	-530.2	125.5	2.0765

BALLISTIC COEFFICIENT .49, MUZZLE VELOCITY FT/SEC 2300.

RANGE	VELOCITY	HEIGHT	100 YD	150 YD	200 YD	250 YD	300 YD	DRIFT	TIME
0.	2300.	0.0	-1.5	-1.5	-1.5	-1.5	-1.5	0.0	0.0000
100.	2131.	.1	0.0	1.4	3.1	5.0	7.1	.9	.1355
200.	1969.	3.1	-6.2	-3.4	0.0	3.9	8.1	3.7	.2818
300.	1816.	8.6	-21.4	-17.2	-12.1	-6.3	0.0	8.6	.4404
400.	1673.	17.3	-47.2	-41.6	-34.8	-27.1	-18.7	16.0	.6127
500.	1538.	29.9	-85.3	-78.4	-69.9	-60.2	-49.7	26.0	.7996
600.	1414.	47.5	-138.1	-129.9	-119.6	-108.0	-95.4	38.8	1.0031
700.	1306.	71.2	-208.5	-198.9	-186.9	-173.4	-158.7	54.7	1.2241
800.	1213.	102.0	-299.5	-288.5	-274.8	-259.4	-242.6	73.9	1.4633
900.	1137.	141.2	-414.2	-401.8	-386.4	-369.0	-350.1	96.0	1.7196
1000.	1081.	189.3	-555.2	-541.4	-524.3	-505.0	-484.0	120.6	1.9898

BALLISTIC COEFFICIENT .49, MUZZLE VELOCITY FT/SEC 2400.

RANGE	VELOCITY	HEIGHT	100 YD	150 YD	200 YD	250 YD	300 YD	DRIFT	TIME
0.	2400.	0.0	-1.5	-1.5	-1.5	-1.5	-1.5	0.0	0.0000
100.	2228.	.1	0.0	1.2	2.8	4.5	6.4	.8	.1297
200.	2062.	2.7	-5.5	-3.1	0.0	3.5	7.3	3.5	.2697
300.	1903.	7.8	-19.3	-15.6	-11.0	-5.8	0.0	8.1	.4211
400.	1754.	15.7	-42.7	-37.8	-31.6	-24.6	-16.9	15.0	.5852
500.	1615.	27.2	-77.4	-71.3	-63.5	-54.8	-45.2	24.4	.7635
600.	1484.	43.2	-125.3	-118.0	-108.7	-98.2	-86.7	36.5	.9572
700.	1366.	64.8	-189.2	-180.7	-169.8	-157.6	-144.2	51.6	1.1684
800.	1265.	92.9	-271.7	-261.9	-249.5	-235.6	-220.2	69.8	1.3966
900.	1178.	128.8	-376.2	-365.2	-351.2	-335.5	-318.3	91.2	1.6431
1000.	1112.	173.5	-505.6	-493.4	-477.9	-460.5	-441.3	115.3	1.9054

BALLISTIC COEFFICIENT .49, MUZZLE VELOCITY FT/SEC 2500.

RANGE	VELOCITY	HEIGHT	100 YD	150 YD	200 YD	250 YD	300 YD	DRIFT	TIME
0.	2500.	0.0	-1.5	-1.5	-1.5	-1.5	-1.5	0.0	0.0000
100.	2324.	-.0	0.0	1.1	2.5	4.1	5.8	.8	.1244
200.	2155.	2.5	-5.0	-2.8	0.0	3.2	6.7	3.2	.2584
300.	1992.	7.1	-17.4	-14.2	-10.0	-5.2	0.0	7.6	.4032
400.	1837.	14.3	-38.8	-34.5	-28.9	-22.5	-15.5	14.1	.5603
500.	1692.	24.8	-70.3	-64.9	-57.9	-49.9	-41.2	22.9	.7302
600.	1557.	39.5	-114.1	-107.6	-99.2	-89.7	-79.2	34.4	.9153
700.	1430.	59.1	-172.1	-164.5	-154.7	-143.6	-131.4	48.6	1.1163
800.	1320.	84.8	-247.2	-238.5	-227.3	-214.6	-200.6	65.9	1.3346
900.	1226.	117.7	-342.4	-332.7	-320.1	-305.7	-290.0	86.4	1.5708
1000.	1147.	159.0	-460.8	-450.0	-436.0	-420.1	-402.7	109.9	1.8243

BALLISTIC COEFFICIENT .49, MUZZLE VELOCITY FT/SEC 2600.

RANGE	VELOCITY	HEIGHT	100 YD	150 YD	200 YD	250 YD	300 YD	DRIFT	TIME
0.	2600.	0.0	-1.5	-1.5	-1.5	-1.5	-1.5	0.0	0.0000
100.	2421.	-.1	0.0	1.0	2.2	3.7	5.3	.7	.1196
200.	2248.	2.2	-4.5	-2.6	0.0	2.9	6.1	3.1	.2483
300.	2081.	6.4	-15.8	-12.9	-9.1	-4.8	0.0	7.1	.3867
400.	1921.	13.1	-35.3	-31.5	-26.4	-20.6	-14.2	13.3	.5372
500.	1771.	22.7	-64.1	-59.4	-53.0	-45.7	-37.8	21.6	.6996
600.	1631.	36.1	-104.0	-98.3	-90.6	-82.0	-72.4	32.3	.8761
700.	1499.	54.0	-156.9	-150.3	-141.3	-131.2	-120.1	45.8	1.0679
800.	1379.	77.5	-225.5	-217.8	-207.6	-196.0	-183.3	62.3	1.2769
900.	1276.	107.8	-312.6	-304.0	-292.5	-279.5	-265.2	81.9	1.5036
1000.	1188.	145.8	-420.8	-411.3	-398.5	-384.0	-368.1	104.4	1.7473

BALLISTIC COEFFICIENT .49, MUZZLE VELOCITY FT/SEC 2700.

RANGE	VELOCITY	HEIGHT	100 YD	150 YD	200 YD	250 YD	300 YD	DRIFT	TIME
0.	2700.	0.0	-1.5	-1.5	-1.5	-1.5	-1.5	0.0	0.0000
100.	2517.	-.1	0.0	.8	2.0	3.3	4.8	.7	.1153
200.	2340.	2.0	-4.0	-2.3	0.0	2.7	5.6	2.9	.2389
300.	2170.	5.9	-14.4	-11.9	-8.3	-4.3	0.0	6.7	.3717
400.	2007.	12.0	-32.2	-28.8	-24.2	-18.8	-13.0	12.5	.5156
500.	1851.	20.9	-58.6	-54.4	-48.6	-41.9	-34.7	20.4	.6713
600.	1705.	33.1	-95.1	-90.1	-83.1	-75.1	-66.4	30.5	.8400
700.	1569.	49.6	-143.7	-137.9	-129.7	-120.4	-110.2	43.3	1.0238
800.	1442.	71.1	-206.3	-199.6	-190.3	-179.6	-168.0	58.8	1.2233
900.	1330.	98.8	-285.7	-278.2	-267.7	-255.7	-242.7	77.4	1.4400
1000.	1234.	133.9	-385.4	-377.1	-365.4	-352.0	-337.6	99.3	1.6750

BALLISTIC COEFFICIENT .49, MUZZLE VELOCITY FT/SEC 2800.

RANGE	VELOCITY	HEIGHT	100 YD	150 YD	200 YD	250 YD	300 YD	DRIFT	TIME
0.	2800.	0.0	-1.5	-1.5	-1.5	-1.5	-1.5	0.0	0.0000
100.	2612.	-.2	0.0	.8	1.8	3.0	4.4	.6	.1106
200.	2433.	1.8	-3.6	-2.1	0.0	2.5	5.2	2.7	.2296
300.	2259.	5.4	-13.2	-10.9	-7.7	-4.0	0.0	6.4	.3578
400.	2092.	11.0	-29.5	-26.5	-22.3	-17.3	-12.0	11.8	.4957
500.	1932.	19.2	-53.8	-50.0	-44.8	-38.6	-31.9	19.2	.6451
600.	1781.	30.5	-87.4	-82.9	-76.6	-69.2	-61.1	28.8	.8068
700.	1640.	45.6	-132.0	-126.6	-119.3	-110.7	-101.3	40.9	.9822
800.	1508.	65.3	-189.4	-183.3	-174.9	-165.0	-154.3	55.6	1.1729
900.	1387.	90.7	-262.2	-255.3	-245.9	-234.8	-222.7	73.3	1.3806
1000.	1283.	123.0	-353.4	-345.8	-335.3	-322.9	-309.5	94.1	1.6058

BALLISTIC COEFFICIENT .49, MUZZLE VELOCITY FT/SEC 2900.

RANGE	VELOCITY	HEIGHT	100 YD	150 YD	200 YD	250 YD	300 YD	DRIFT	TIME
0.	2900.	0.0	-1.5	-1.5	-1.5	-1.5	-1.5	0.0	0.0000
100.	2708.	-.2	0.0	.7	1.6	2.8	4.0	.6	.1070
200.	2524.	1.6	-3.3	-2.0	0.0	2.2	4.7	2.7	.2220
300.	2348.	5.0	-12.0	-10.1	-7.1	-3.8	0.0	6.1	.3452
400.	2177.	10.2	-27.0	-24.4	-20.5	-16.0	-11.0	11.2	.4775
500.	2013.	17.8	-49.5	-46.2	-41.3	-35.7	-29.4	18.3	.6210
600.	1858.	28.1	-80.4	-76.4	-70.6	-63.9	-56.3	27.3	.7759
700.	1711.	42.0	-121.4	-116.8	-109.9	-102.1	-93.4	38.7	.9442
800.	1575.	60.3	-174.4	-169.2	-161.3	-152.4	-142.4	52.8	1.1274
900.	1447.	83.7	-241.3	-235.4	-226.5	-216.5	-205.2	69.5	1.3261
1000.	1334.	113.5	-325.1	-318.6	-308.8	-297.6	-285.1	89.4	1.5425

BALLISTIC COEFFICIENT .49, MUZZLE VELOCITY FT/SEC 3000.

RANGE	VELOCITY	HEIGHT	100 YD	150 YD	200 YD	250 YD	300 YD	DRIFT	TIME
0.	3000.	0.0	-1.5	-1.5	-1.5	-1.5	-1.5	0.0	0.0000
100.	2803.	-.2	0.0	.6	1.5	2.5	3.7	.7	.1039
200.	2616.	1.5	-2.9	-1.8	0.0	2.1	4.4	2.5	.2144
300.	2436.	4.6	-11.0	-9.2	-6.6	-3.4	0.0	5.8	.3332
400.	2262.	9.5	-24.9	-22.6	-19.1	-14.8	-10.3	10.8	.4613
500.	2095.	16.5	-45.6	-42.7	-38.3	-33.0	-27.4	17.4	.5990
600.	1935.	26.1	-74.2	-70.8	-65.5	-59.2	-52.3	26.1	.7481
700.	1784.	39.0	-112.1	-108.1	-101.9	-94.5	-86.6	36.9	.9096
800.	1642.	55.7	-160.9	-156.3	-149.2	-140.8	-131.7	50.1	1.0848
900.	1510.	77.3	-222.5	-217.3	-209.4	-199.9	-189.7	66.0	1.2752
1000.	1389.	104.8	-299.5	-293.8	-284.9	-274.4	-263.0	84.9	1.4826

BALLISTIC COEFFICIENT .49, MUZZLE VELOCITY FT/SEC 3100.

RANGE	VELOCITY	HEIGHT	100 YD	150 YD	200 YD	250 YD	300 YD	DRIFT	TIME
0.	3100.	0.0	-1.5	-1.5	-1.5	-1.5	-1.5	0.0	0.0000
100.	2899.	-.3	0.0	.5	1.3	2.3	3.4	.6	.1000
200.	2707.	1.3	-2.7	-1.6	0.0	1.9	4.1	2.4	.2070
300.	2524.	4.2	-10.1	-8.5	-6.1	-3.2	0.0	5.6	.3220
400.	2347.	8.8	-23.0	-20.9	-17.7	-13.8	-9.5	10.2	.4452
500.	2177.	15.3	-42.1	-39.5	-35.5	-30.7	-25.3	16.5	.5776
600.	2013.	24.2	-68.7	-65.6	-60.7	-55.0	-48.5	24.7	.7212
700.	1857.	36.1	-103.7	-100.1	-94.4	-87.7	-80.2	35.0	.8762
800.	1711.	51.6	-148.8	-144.7	-138.2	-130.6	-121.9	47.6	1.0445
900.	1574.	71.6	-206.0	-201.3	-194.1	-185.5	-175.8	62.8	1.2278
1000.	1446.	96.9	-277.0	-271.8	-263.8	-254.2	-243.4	80.8	1.4266

BALLISTIC COEFFICIENT .49, MUZZLE VELOCITY FT/SEC 3200.

RANGE	VELOCITY	HEIGHT	100 YD	150 YD	200 YD	250 YD	300 YD	DRIFT	TIME
0.	3200.	0.0	-1.5	-1.5	-1.5	-1.5	-1.5	0.0	0.0000
100.	2995.	-.3	0.0	.5	1.2	2.1	3.1	.5	.0968
200.	2798.	1.2	-2.4	-1.5	0.0	1.8	3.8	2.3	.2008
300.	2611.	3.9	-9.3	-7.9	-5.6	-3.0	0.0	5.3	.3115
400.	2431.	8.1	-21.2	-19.4	-16.4	-12.8	-8.9	9.8	.4306
500.	2258.	14.2	-39.1	-36.8	-33.1	-28.6	-23.7	15.9	.5588
600.	2091.	22.6	-63.7	-61.0	-56.5	-51.2	-45.2	23.6	.6968
700.	1931.	33.6	-96.4	-93.2	-87.9	-81.7	-74.8	33.4	.8463
800.	1780.	48.0	-138.2	-134.6	-128.6	-121.5	-113.6	45.4	1.0081
900.	1639.	66.5	-191.0	-186.9	-180.2	-172.2	-163.3	59.8	1.1836
1000.	1507.	89.9	-256.8	-252.2	-244.7	-235.9	-226.0	76.9	1.3745

BALLISTIC COEFFICIENT .49, MUZZLE VELOCITY FT/SEC 3300.

RANGE	VELOCITY	HEIGHT	100 YD	150 YD	200 YD	250 YD	300 YD	DRIFT	TIME
0.	3300.	0.0	-1.5	-1.5	-1.5	-1.5	-1.5	0.0	0.0000
100.	3090.	-.3	0.0	.4	1.1	1.9	2.8	.6	.0941
200.	2890.	1.1	-2.1	-1.4	0.0	1.7	3.5	2.2	.1943
300.	2698.	3.6	-8.5	-7.3	-5.3	-2.8	0.0	5.1	.3019
400.	2515.	7.6	-19.6	-18.1	-15.4	-12.0	-8.3	9.4	.4172
500.	2339.	13.3	-36.2	-34.3	-30.9	-26.7	-22.1	15.2	.5409
600.	2169.	21.0	-59.1	-56.8	-52.7	-47.7	-42 2	22.6	.6738
700.	2005.	31.4	-89.5	-86.8	-82.0	-76.2	-69.7	31.9	.8179
800.	1850.	44.8	-128.4	-125.3	-119.9	-113.2	-105.8	43.4	.9736
900.	1704.	61.9	-177.5	-174.0	-167.8	-160.3	-152.0	57.1	1.1425
1000.	1568.	83.7	-238.7	-234.8	-228.0	-219.6	-210.4	73.4	1.3264

BALLISTIC COEFFICIENT .49, MUZZLE VELOCITY FT/SEC 3400.

RANGE	VELOCITY	HEIGHT	100 YD	150 YD	200 YD	250 YD	300 YD	DRIFT	TIME
0.	3400.	0.0	-1.5	-1.5	-1.5	-1.5	-1.5	0.0	0.0000
100.	3186.	-.4	0.0	.3	1.0	1.7	2.6	.5	.0909
200.	2981.	1.0	-1.9	-1.3	0.0	1.5	3.3	2.1	.1884
300.	2786.	3.4	-7.8	-6.8	-4.9	-2.6	0.0	4.9	.2924
400.	2599.	7.1	-18.2	-16.8	-14.3	-11.2	-7.8	9.0	.4040
500.	2419.	12.4	-33.7	-32.0	-28.8	-25.0	-20.7	14.5	.5237
600.	2247.	19.7	-55.2	-53.1	-49.3	-44.7	-39.5	21.7	.6525
700.	2080.	29.3	-83.3	-80.9	-76.5	-71.2	-65.1	30.5	.7909
800.	1920.	41.8	-119.7	-117.0	-112.0	-105.9	-98.9	41.5	.9414
900.	1770.	57.7	-165.4	-162.3	-156.6	-149.8	-142.0	54.5	1.1039
1000.	1630.	78.0	-222.2	-218.7	-212.4	-204.8	-196.2	70.1	1.2805

BALLISTIC COEFFICIENT .49, MUZZLE VELOCITY FT/SEC 3500.

RANGE	VELOCITY	HEIGHT	100 YD	150 YD	200 YD	250 YD	300 YD	DRIFT	TIME
0.	3500.	0.0	-1.5	-1.5	-1.5	-1.5	-1.5	0.0	0.0000
100.	3282.	-.4	0.0	.3	.9	1.6	2.4	.6	.0888
200.	3073.	.9	-1.7	-1.2	0.0	1.4	3.0	2.0	.1831
300.	2873.	3.1	-7.1	-6.3	-4.6	-2.4	0.0	4.7	.2839
400.	2682.	6.6	-16.8	-15.7	-13.4	-10.5	-7.3	8.7	.3920
500.	2500.	11.6	-31.3	-29.9	-27.0	-23.4	-19.4	14.0	.5082
600.	2324.	18.5	-51.3	-49.7	-46.2	-41.9	-37.1	20.8	.6325
700.	2155.	27.5	-77.8	-75.8	-71.7	-66.7	-61.1	29.3	.7665
800.	1992.	39.1	-111.6	-109.4	-104.7	-99.0	-92.6	39.7	.9113
900.	1837.	54.0	-154.3	-151.8	-146.6	-140.1	-132.9	52.3	1.0684
1000.	1692.	72.9	-207.2	-204.4	-198.6	-191.4	-183.4	67.1	1.2383

BALLISTIC COEFFICIENT .49, MUZZLE VELOCITY FT/SEC 3600.

RANGE	VELOCITY	HEIGHT	100 YD	150 YD	200 YD	250 YD	300 YD	DRIFT	TIME
0.	3600.	0.0	-1.5	-1.5	-1.5	-1.5	-1.5	0.0	0.0000
100.	3378.	-.4	0.0	.3	.8	1.4	2.2	.4	.0857
200.	3165.	.8	-1.6	-1.1	0.0	1.3	2.8	1.9	.1776
300.	2961.	2.9	-6.6	-5.8	-4.2	-2.3	0.0	4.5	.2754
400.	2766.	6.2	-15.6	-14.6	-12.5	-9.9	-6.9	8.3	.3805
500.	2580.	10.9	-29.2	-28.0	-25.3	-22.0	-18.2	13.4	.4928
600.	2402.	17.3	-48.0	-46.5	-43.3	-39.4	-34.8	20.0	.6134
700.	2229.	25.7	-72.8	-71.0	-67.3	-62.7	-57.4	28.1	.7430
800.	2063.	36.7	-104.5	-102.5	-98.2	-93.0	-86.9	38.1	.8829
900.	1905.	50.6	-144.3	-142.1	-137.3	-131.4	-124.6	50.0	1.0342
1000.	1755.	68.2	-193.7	-191.2	-185.9	-179.3	-171.7	64.2	1.1981

BALLISTIC COEFFICIENT .49, MUZZLE VELOCITY FT/SEC 3700.

RANGE	VELOCITY	HEIGHT	100 YD	150 YD	200 YD	250 YD	300 YD	DRIFT	TIME
0.	3700.	0.0	-1.5	-1.5	-1.5	-1.5	-1.5	0.0	0.0000
100.	3474.	-.4	0.0	.2	.7	1.3	2.0	.5	.0839
200.	3257.	.7	-1.4	-1.0	0.0	1.3	2.7	1.9	.1729
300.	3049.	2.7	-6.1	-5.4	-4.0	-2.1	0.0	4.4	.2681
400.	2850.	5.8	-14.5	-13.7	-11.7	-9.2	-6.4	8.0	.3699
500.	2660.	10.3	-27.2	-26.2	-23.8	-20.6	-17.1	13.0	.4790
600.	2479.	16.3	-44.8	-43.6	-40.7	-36.9	-32.7	19.2	.5958
700.	2304.	24.2	-68.1	-66.7	-63.3	-58.9	-54.0	27.1	.7215
800.	2135.	34.5	-97.9	-96.2	-92.4	-87.3	-81.7	36.6	.8568
900.	1973.	47.5	-135.1	-133.3	-128.9	-123.2	-117.0	48.1	1.0028
1000.	1819.	64.0	-181.4	-179.3	-174.5	-168.1	-161.2	61.7	1.1611

BALLISTIC COEFFICIENT .49, MUZZLE VELOCITY FT/SEC 3800.

RANGE	VELOCITY	HEIGHT	100 YD	150 YD	200 YD	250 YD	300 YD	DRIFT	TIME
0.	3800.	0.0	-1.5	-1.5	-1.5	-1.5	-1.5	0.0	0.0000
100.	3569.	-.4	0.0	.2	.6	1.2	1.9	.5	.0818
200.	3348.	.6	-1.2	-.9	0.0	1.2	2.5	1.9	.1685
300.	3137.	2.5	-5.6	-5.1	-3.8	-2.0	0.0	4.3	.2613
400.	2934.	5.5	-13.5	-12.8	-11.0	-8.7	-6.0	7.8	.3599
500.	2741.	9.7	-25.4	-24.6	-22.3	-19.4	-16.1	12.5	.4657
600.	2556.	15.3	-42.0	-41.0	-38.3	-34.8	-30.8	18.6	.5791
700.	2378.	22.8	-63.8	-62.7	-59.5	-55.4	-50.8	26.1	.7008
800.	2206.	32.5	-91.8	-90.4	-86.8	-82.2	-76.8	35.2	.8318
900.	2041.	44.7	-126.9	-125.4	-121.3	-116.1	-110.1	46.3	.9734
1000.	1884.	60.1	-170.1	-168.5	-163.9	-158.1	-151.4	59.2	1.1261

BALLISTIC COEFFICIENT .49, MUZZLE VELOCITY FT/SEC 3900.

RANGE	VELOCITY	HEIGHT	100 YD	150 YD	200 YD	250 YD	300 YD	DRIFT	TIME
0.	3900.	0.0	-1.5	-1.5	-1.5	-1.5	-1.5	0.0	0.0000
100.	3665.	-.5	0.0	.1	.5	1.1	1.7	.4	.0790
200.	3440.	.5	-1.1	-.8	0.0	1.1	2.3	1.7	.1636
300.	3225.	2.3	-5.2	-4.7	-3.5	-1.9	0.0	4.0	.2537
400.	3018.	5.1	-12.5	-12.0	-10.3	-8.2	-5.7	7.4	.3497
500.	2821.	9.1	-23.8	-23.1	-21.0	-18.3	-15.2	12.0	.4527
600.	2632.	14.4	-39.4	-38.5	-36.1	-32.8	-29.0	17.8	.5627
700.	2452.	21.5	-59.9	-59.0	-56.1	-52.3	-47.9	25.0	.6808
800.	2278.	30.6	-86.2	-85.1	-81.8	-77.5	-72.5	33.9	.8078
900.	2110.	42.1	-119.2	-118.0	-114.3	-109.3	-103.7	44.4	.9447
1000.	1949.	56.6	-160.0	-158.6	-154.5	-149.0	-142.8	56.9	1.0927

BALLISTIC COEFFICIENT .49, MUZZLE VELOCITY FT/SEC 4000.

RANGE	VELOCITY	HEIGHT	100 YD	150 YD	200 YD	250 YD	300 YD	DRIFT	TIME
0.	4000.	0.0	-1.5	-1.5	-1.5	-1.5	-1.5	0.0	0.0000
100.	3761.	-.5	0.0	.1	.5	1.0	1.6	.4	.0775
200.	3532.	.5	-1.0	-.8	0.0	1.0	2.2	1.7	.1597
300.	3313.	2.2	-4.7	-4.4	-3.3	-1.8	0.0	4.0	.2475
400.	3103.	4.8	-11.7	-11.3	-9.8	-7.8	-5.4	7.3	.3413
500.	2902.	8.6	-22.3	-21.7	-19.9	-17.3	-14.4	11.7	.4412
600.	2709.	13.7	-36.9	-36.3	-34.1	-31.0	-27.5	17.3	.5482
700.	2526.	20.4	-56.4	-55.6	-53.0	-49.5	-45.3	24.3	.6630
800.	2349.	28.9	-81.2	-80.3	-77.3	-73.3	-68.5	32.8	.7862
900.	2179.	39.7	-112.1	-111.2	-107.8	-103.2	-97.9	42.8	.9184
1000.	2015.	53.4	-150.5	-149.5	-145.7	-140.6	-134.7	54.9	1.0618

BALLISTIC COEFFICIENT .50, MUZZLE VELOCITY FT/SEC 2000.

RANGE	VELOCITY	HEIGHT	100 YD	150 YD	200 YD	250 YD	300 YD	DRIFT	TIME
0.	2000.	0.0	-1.5	-1.5	-1.5	-1.5	-1.5	0.0	0.0000
100.	1848.	.4	0.0	2.0	4.3	7.0	9.8	1.1	.1561
200.	1705.	4.3	-8.7	-4.7	0.0	5.3	11.0	4.4	.3249
300.	1571.	11.7	-29.5	-23.5	-16.5	-8.5	0.0	10.3	.5086
400.	1446.	23.3	-64.3	-56.4	-46.9	-36.4	-25.0	18.9	.7076
500.	1336.	40.2	-115.9	-106.0	-94.2	-81.0	-66.8	30.6	.9238
600.	1241.	63.5	-186.9	-175.0	-160.8	-145.0	-127.9	45.2	1.1567
700.	1160.	94.3	-280.3	-266.4	-249.9	-231.5	-211.5	62.8	1.4070
800.	1099.	133.6	-399.8	-383.9	-365.0	-344.0	-321.2	83.3	1.6730
900.	1053.	182.2	-547.4	-529.5	-508.3	-484.5	-458.9	106.0	1.9522
1000.	1013.	241.7	-727.6	-707.8	-684.1	-657.8	-629.3	131.6	2.2475

BALLISTIC COEFFICIENT .50, MUZZLE VELOCITY FT/SEC 2100.

RANGE	VELOCITY	HEIGHT	100 YD	150 YD	200 YD	250 YD	300 YD	DRIFT	TIME
0.	2100.	0.0	-1.5	-1.5	-1.5	-1.5	-1.5	0.0	0.0000
100.	1942.	.3	0.0	1.7	3.9	6.2	8.8	1.0	.1485
200.	1794.	3.8	-7.7	-4.3	0.0	4.7	9.8	4.1	.3091
300.	1655.	10.5	-26.3	-21.2	-14.7	-7.7	0.0	9.6	.4833
400.	1524.	20.9	-57.7	-50.7	-42.2	-32.8	-22.5	17.7	.6722
500.	1403.	36.2	-104.0	-95.3	-84.6	-72.9	-60.0	28.7	.8774
600.	1299.	57.3	-168.0	-157.6	-144.8	-130.7	-115.3	42.7	1.0998
700.	1209.	85.4	-252.6	-240.5	-225.5	-209.1	-191.1	59.8	1.3395
800.	1135.	121.8	-362.0	-348.1	-331.0	-312.2	-291.7	80.1	1.5979
900.	1081.	166.6	-496.4	-480.8	-461.5	-440.4	-417.3	102.3	1.8672
1000.	1038.	221.2	-660.1	-642.8	-621.4	-598.0	-572.3	127.1	2.1505

BALLISTIC COEFFICIENT .50, MUZZLE VELOCITY FT/SEC 2200.

RANGE	VELOCITY	HEIGHT	100 YD	150 YD	200 YD	250 YD	300 YD	DRIFT	TIME
0.	2200.	0.0	-1.5	-1.5	-1.5	-1.5	-1.5	0.0	0.0000
100.	2038.	.2	0.0	1.5	3.4	5.6	7.9	1.0	.1419
200.	1884.	3.4	-6.9	-3.8	0.0	4.3	8.9	3.9	.2947
300.	1739.	9.4	-23.6	-19.0	-13.3	-6.9	0.0	9.1	.4606
400.	1603.	18.9	-51.9	-45.7	-38.2	-29.6	-20.4	16.7	.6404
500.	1476.	32.7	-93.6	-85.9	-76.4	-65.8	-54.3	27.0	.8352
600.	1361.	51.9	-151.5	-142.2	-130.8	-118.1	-104.2	40.3	1.0473
700.	1263.	77.4	-228.0	-217.2	-203.9	-189.0	-172.9	56.6	1.2760
800.	1178.	110.6	-326.6	-314.3	-299.1	-282.1	-263.6	76.1	1.5230
900.	1113.	152.2	-450.0	-436.1	-419.1	-399.9	-379.2	98.2	1.7850
1000.	1063.	203.5	-602.1	-586.7	-567.7	-546.4	-523.3	123.0	2.0627

BALLISTIC COEFFICIENT .50, MUZZLE VELOCITY FT/SEC 2300.

RANGE	VELOCITY	HEIGHT	100 YD	150 YD	200 YD	250 YD	300 YD	DRIFT	TIME
0.	2300.	0.0	-1.5	-1.5	-1.5	-1.5	-1.5	0.0	0.0000
100.	2135.	.1	0.0	1.4	3.1	5.0	7.1	.9	.1354
200.	1976.	3.0	-6.1	-3.4	0.0	3.9	8.0	3.6	.2813
300.	1825.	8.5	-21.3	-17.1	-12.0	-6.3	0.0	8.5	.4393
400.	1684.	17.1	-46.8	-41.4	-34.6	-26.9	-18.5	15.6	.6106
500.	1551.	29.7	-84.7	-77.9	-69.4	-59.7	-49.3	25.4	.7963
600.	1428.	47.0	-136.9	-128.7	-118.5	-106.9	-94.4	37.9	.9978
700.	1320.	70.3	-206.3	-196.7	-184.8	-171.3	-156.7	53.4	1.2164
800.	1227.	100.5	-295.7	-284.7	-271.1	-255.7	-239.0	72.0	1.4524
900.	1149.	138.9	-408.2	-395.8	-380.5	-363.2	-344.4	93.5	1.7054
1000.	1091.	186.2	-547.2	-533.5	-516.5	-497.2	-476.3	117.8	1.9737

BALLISTIC COEFFICIENT .50, MUZZLE VELOCITY FT/SEC 2400.

RANGE	VELOCITY	HEIGHT	100 YD	150 YD	200 YD	250 YD	300 YD	DRIFT	TIME
0.	2400.	0.0	-1.5	-1.5	-1.5	-1.5	-1.5	0.0	0.0000
100.	2231.	.1	0.0	1.2	2.8	4.5	6.4	.8	.1296
200.	2068.	2.7	-5.5	-3.1	0.0	3.5	7.3	3.4	.2692
300.	1912.	7.7	-19.2	-15.6	-10.9	-5.7	0.0	8.0	.4202
400.	1766.	15.6	-42.4	-37.6	-31.4	-24.4	-16.8	14.6	.5832
500.	1628.	27.0	-76.8	-70.7	-63.0	-54.3	-44.7	23.8	.7602
600.	1499.	42.8	-124.2	-116.9	-107.6	-97.2	-85.7	35.6	.9521
700.	1381.	63.9	-187.1	-178.7	-167.8	-155.7	-142.3	50.3	1.1608
800.	1280.	91.5	-268.4	-258.8	-246.4	-232.5	-217.2	68.1	1.3867
900.	1193.	126.9	-371.4	-360.5	-346.6	-330.9	-313.8	89.0	1.6307
1000.	1123.	170.6	-498.1	-486.0	-470.5	-453.1	-434.0	112.5	1.8893

BALLISTIC COEFFICIENT .50, MUZZLE VELOCITY FT/SEC 2500.

RANGE	VELOCITY	HEIGHT	100 YD	150 YD	200 YD	250 YD	300 YD	DRIFT	TIME
0.	2500.	0.0	-1.5	-1.5	-1.5	-1.5	-1.5	0.0	0.0000
100.	2328.	-.0	0.0	1.1	2.5	4.0	5.8	.8	.1243
200.	2161.	2.4	-4.9	-2.8	0.0	3.2	6.7	3.1	.2579
300.	2001.	7.0	-17.4	-14.1	-10.0	-5.2	0.0	7.4	.4023
400.	1849.	14.2	-38.5	-34.2	-28.7	-22.3	-15.3	13.8	.5583
500.	1706.	24.6	-69.8	-64.4	-57.5	-49.5	-40.8	22.3	.7270
600.	1573.	39.0	-113.1	-106.7	-98.3	-88.8	-78.4	33.5	.9106
700.	1447.	58.3	-170.3	-162.8	-153.1	-142.0	-129.8	47.4	1.1094
800.	1336.	83.6	-244.4	-235.7	-224.6	-212.0	-198.0	64.3	1.3254
900.	1242.	115.8	-337.8	-328.2	-315.7	-301.4	-285.7	84.2	1.5583
1000.	1160.	156.2	-453.8	-443.0	-429.2	-413.3	-395.9	107.1	1.8084

BALLISTIC COEFFICIENT .50, MUZZLE VELOCITY FT/SEC 2600.

RANGE	VELOCITY	HEIGHT	100 YD	150 YD	200 YD	250 YD	300 YD	DRIFT	TIME
0.	2600.	0.0	-1.5	-1.5	-1.5	-1.5	-1.5	0.0	0.0000
100.	2424.	-.1	0.0	1.0	2.2	3.7	5.3	.7	.1195
200.	2254.	2.2	-4.5	-2.5	0.0	2.9	6.1	3.0	.2480
300.	2091.	6.4	-15.8	-12.9	-9.1	-4.8	0.0	7.0	.3860
400.	1934.	13.0	-35.1	-31.3	-26.2	-20.4	-14.1	13.0	.5354
500.	1786.	22.6	-63.7	-58.9	-52.5	-45.4	-37.4	21.1	.6967
600.	1647.	35.7	-103.2	-97.4	-89.8	-81.2	-71.7	31.6	.8716
700.	1517.	53.3	-155.4	-148.7	-139.8	-129.8	-118.7	44.7	1.0614
800.	1397.	76.4	-222.8	-215.2	-205.0	-193.5	-180.8	60.7	1.2678
900.	1293.	106.0	-308.4	-299.8	-288.4	-275.5	-261.2	79.7	1.4915
1000.	1204.	143.2	-414.5	-405.0	-392.2	-377.9	-362.0	101.7	1.7317

BALLISTIC COEFFICIENT .50, MUZZLE VELOCITY FT/SEC 2700.

RANGE	VELOCITY	HEIGHT	100 YD	150 YD	200 YD	250 YD	300 YD	DRIFT	TIME
0.	2700.	0.0	-1.5	-1.5	-1.5	-1.5	-1.5	0.0	0.0000
100.	2520.	-.1	0.0	.8	2.0	3.3	4.8	.7	.1152
200.	2347.	2.0	-4.0	-2.3	0.0	2.7	5.5	2.9	.2385
300.	2180.	5.8	-14.3	-11.8	-8.3	-4.3	0.0	6.6	.3708
400.	2020.	11.9	-32.0	-28.7	-24.0	-18.7	-12.9	12.2	.5140
500.	1866.	20.7	-58.1	-54.0	-48.1	-41.5	-34.3	19.8	.6683
600.	1722.	32.8	-94.3	-89.3	-82.3	-74.3	-65.7	29.7	.8356
700.	1588.	48.9	-142.3	-136.5	-128.3	-119.0	-109.0	42.2	1.0175
800.	1461.	70.0	-203.8	-197.2	-187.9	-177.3	-165.8	57.3	1.2143
900.	1349.	97.2	-281.9	-274.5	-264.0	-252.0	-239.1	75.4	1.4283
1000.	1252.	131.4	-379.3	-371.0	-359.3	-346.0	-331.7	96.5	1.6593

BALLISTIC COEFFICIENT .50, MUZZLE VELOCITY FT/SEC 2800.

RANGE	VELOCITY	HEIGHT	100 YD	150 YD	200 YD	250 YD	300 YD	DRIFT	TIME
0.	2800.	0.0	-1.5	-1.5	-1.5	-1.5	-1.5	0.0	0.0000
100.	2616.	-.2	0.0	.8	1.8	3.0	4.4	.6	.1105
200.	2440.	1.8	-3.6	-2.1	0.0	2.5	5.1	2.6	.2292
300.	2269.	5.4	-13.1	-10.8	-7.7	-4.0	0.0	6.3	.3569
400.	2105.	11.0	-29.3	-26.3	-22.2	-17.2	-11.9	11.6	.4942
500.	1948.	19.1	-53.4	-49.7	-44.5	-38.3	-31.6	18.8	.6424
600.	1799.	30.2	-86.7	-82.1	-75.9	-68.5	-60.5	28.1	.8025
700.	1659.	45.0	-130.7	-125.4	-118.1	-109.5	-100.1	39.8	.9763
800.	1528.	64.4	-187.2	-181.2	-172.9	-163.0	-152.3	54.1	1.1647
900.	1407.	89.3	-258.7	-251.8	-242.5	-231.4	-219.4	71.3	1.3693
1000.	1302.	120.8	-348.0	-340.4	-330.0	-317.6	-304.3	91.5	1.5911

BALLISTIC COEFFICIENT .50, MUZZLE VELOCITY FT/SEC 2900.

RANGE	VELOCITY	HEIGHT	100 YD	150 YD	200 YD	250 YD	300 YD	DRIFT	TIME
0.	2900.	0.0	-1.5	-1.5	-1.5	-1.5	-1.5	0.0	0.0000
100.	2712.	-.2	0.0	.7	1.6	2.7	4.0	.6	.1070
200.	2532.	1.6	-3.3	-2.0	0.0	2.2	4.7	2.6	.2216
300.	2358.	4.9	-12.0	-10.0	-7.1	-3.7	0.0	6.0	.3444
400.	2191.	10.1	-26.9	-24.3	-20.4	-15.9	-10.9	11.0	.4761
500.	2030.	17.6	-49.1	-45.9	-41.0	-35.4	-29.2	17.8	.6186
600.	1876.	27.9	-79.7	-75.8	-69.9	-63.3	-55.8	26.6	.7721
700.	1731.	41.5	-120.2	-115.7	-108.8	-101.1	-92.3	37.8	.9386
800.	1596.	59.4	-172.4	-167.2	-159.4	-150.5	-140.5	51.4	1.1194
900.	1469.	82.3	-238.0	-232.2	-223.4	-213.4	-202.2	67.6	1.3152
1000.	1355.	111.4	-320.2	-313.7	-303.9	-292.8	-280.4	86.9	1.5285

BALLISTIC COEFFICIENT .50, MUZZLE VELOCITY FT/SEC 3000.

RANGE	VELOCITY	HEIGHT	100 YD	150 YD	200 YD	250 YD	300 YD	DRIFT	TIME
0.	3000.	0.0	-1.5	-1.5	-1.5	-1.5	-1.5	0.0	0.0000
100.	2807.	-.2	0.0	.6	1.5	2.5	3.6	.7	.1038
200.	2623.	1.4	-2.9	-1.8	0.0	2.1	4.4	2.5	.2140
300.	2446.	4.6	-10.9	-9.2	-6.5	-3.4	0.0	5.7	.3325
400.	2276.	9.4	-24.7	-22.4	-18.9	-14.7	-10.2	10.5	.4598
500.	2112.	16.3	-45.3	-42.4	-38.0	-32.8	-27.1	17.0	.5966
600.	1954.	25.9	-73.6	-70.2	-64.9	-58.6	-51.8	25.4	.7445
700.	1804.	38.5	-111.0	-107.0	-100.8	-93.5	-85.6	35.9	.9041
800.	1664.	55.0	-159.1	-154.5	-147.5	-139.1	-130.0	48.8	1.0773
900.	1533.	76.1	-219.7	-214.5	-206.6	-197.2	-187.0	64.3	1.2651
1000.	1411.	102.8	-295.1	-289.3	-280.5	-270.0	-258.7	82.6	1.4690

BALLISTIC COEFFICIENT .50, MUZZLE VELOCITY FT/SEC 3100.

RANGE	VELOCITY	HEIGHT	100 YD	150 YD	200 YD	250 YD	300 YD	DRIFT	TIME
0.	3100.	0.0	-1.5	-1.5	-1.5	-1.5	-1.5	0.0	0.0000
100.	2903.	-.3	0.0	.5	1.3	2.3	3.3	.6	.0999
200.	2715.	1.3	-2.6	-1.6	0.0	1.9	4.1	2.3	.2068
300.	2534.	4.2	-10.0	-8.5	-6.1	-3.2	0.0	5.5	.3213
400.	2361.	8.7	-22.8	-20.8	-17.6	-13.8	-9.5	10.0	.4439
500.	2193.	15.1	-41.8	-39.3	-35.2	-30.5	-25.1	16.1	.5755
600.	2032.	24.0	-68.1	-65.1	-60.2	-54.5	-48.1	24.1	.7179
700.	1878.	35.7	-102.7	-99.1	-93.5	-86.8	-79.3	34.1	.8711
800.	1733.	50.9	-147.2	-143.1	-136.7	-129.1	-120.5	46.3	1.0375
900.	1598.	70.5	-203.4	-198.7	-191.5	-182.9	-173.3	61.1	1.2180
1000.	1471.	95.2	-272.9	-267.8	-259.7	-250.2	-239.5	78.5	1.4135

BALLISTIC COEFFICIENT .50, MUZZLE VELOCITY FT/SEC 3200.

RANGE	VELOCITY	HEIGHT	100 YD	150 YD	200 YD	250 YD	300 YD	DRIFT	TIME
0.	3200.	0.0	-1.5	-1.5	-1.5	-1.5	-1.5	0.0	0.0000
100.	2999.	-.3	0.0	.5	1.2	2.1	3.1	.5	.0967
200.	2806.	1.2	-2.4	-1.5	0.0	1.8	3.7	2.3	.2005
300.	2622.	3.9	-9.2	-7.8	-5.6	-2.9	0.0	5.2	.3108
400.	2445.	8.1	-21.1	-19.3	-16.3	-12.7	-8.8	9.5	.4293
500.	2275.	14.1	-38.8	-36.5	-32.8	-28.4	-23.5	15.5	.5566
600.	2111.	22.3	-63.2	-60.5	-56.0	-50.7	-44.8	23.1	.6935
700.	1953.	33.2	-95.4	-92.2	-87.0	-80.8	-74.0	32.6	.8414
800.	1803.	47.4	-136.7	-133.0	-127.0	-120.0	-112.1	44.2	1.0011
900.	1663.	65.5	-188.7	-184.6	-177.8	-169.9	-161.1	58.2	1.1745
1000.	1532.	88.3	-253.2	-248.6	-241.1	-232.3	-222.5	74.8	1.3624

BALLISTIC COEFFICIENT .50, MUZZLE VELOCITY FT/SEC 3300.

RANGE	VELOCITY	HEIGHT	100 YD	150 YD	200 YD	250 YD	300 YD	DRIFT	TIME
0.	3300.	0.0	-1.5	-1.5	-1.5	-1.5	-1.5	0.0	0.0000
100.	3095.	-.3	0.0	.4	1.1	1.9	2.8	.6	.0941
200.	2898.	1.1	-2.1	-1.4	0.0	1.7	3.5	2.2	.1942
300.	2710.	3.6	-8.4	-7.3	-5.2	-2.7	0.0	5.0	.3013
400.	2530.	7.6	-19.5	-18.0	-15.2	-11.9	-8.3	9.2	.4160
500.	2356.	13.2	-36.0	-34.1	-30.7	-26.5	-21.9	14.8	.5389
600.	2189.	20.8	-58.7	-56.4	-52.2	-47.3	-41.8	22.0	.6707
700.	2028.	31.0	-88.7	-86.0	-81.2	-75.4	-69.0	31.2	.8134
800.	1874.	44.1	-126.9	-123.9	-118.4	-111.8	-104.5	42.2	.9670
900.	1730.	60.9	-175.2	-171.8	-165.6	-158.1	-149.9	55.5	1.1337
1000.	1594.	82.2	-235.2	-231.4	-224.6	-216.3	-207.1	71.4	1.3147

BALLISTIC COEFFICIENT .50, MUZZLE VELOCITY FT/SEC 3400.

RANGE	VELOCITY	HEIGHT	100 YD	150 YD	200 YD	250 YD	300 YD	DRIFT	TIME
0.	3400.	0.0	-1.5	-1.5	-1.5	-1.5	-1.5	0.0	0.0000
100.	3190.	-.4	0.0	.3	1.0	1.7	2.6	.5	.0909
200.	2989.	.9	-1.9	-1.2	0.0	1.5	3.3	2.0	.1880
300.	2797.	3.3	-7.8	-6.8	-4.9	-2.6	0.0	4.8	.2920
400.	2613.	7.0	-18.1	-16.7	-14.2	-11.1	-7.7	8.8	.4027
500.	2437.	12.3	-33.4	-31.7	-28.6	-24.8	-20.4	14.1	.5216
600.	2267.	19.5	-54.7	-52.6	-48.9	-44.3	-39.1	21.1	.6494
700.	2103.	29.0	-82.7	-80.3	-75.9	-70.5	-64.5	29.8	.7869
800.	1945.	41.2	-118.4	-115.7	-110.7	-104.6	-97.7	40.4	.9352
900.	1797.	56.9	-163.3	-160.2	-154.7	-147.8	-140.0	53.1	1.0956
1000.	1657.	76.6	-219.1	-215.6	-209.4	-201.8	-193.1	68.1	1.2695

BALLISTIC COEFFICIENT .50, MUZZLE VELOCITY FT/SEC 3500.

RANGE	VELOCITY	HEIGHT	100 YD	150 YD	200 YD	250 YD	300 YD	DRIFT	TIME
0.	3500.	0.0	-1.5	-1.5	-1.5	-1.5	-1.5	0.0	0.0000
100.	3286.	-.4	0.0	.3	.9	1.6	2.4	.5	.0887
200.	3081.	.9	-1.7	-1.2	0.0	1.4	3.0	2.0	.1829
300.	2885.	3.1	-7.1	-6.3	-4.5	-2.4	0.0	4.6	.2834
400.	2697.	6.6	-16.7	-15.6	-13.3	-10.4	-7.2	8.5	.3911
500.	2518.	11.6	-31.1	-29.7	-26.8	-23.3	-19.3	13.7	.5064
600.	2345.	18.3	-51.0	-49.3	-45.8	-41.5	-36.8	20.3	.6298
700.	2178.	27.1	-77.0	-75.1	-71.0	-66.0	-60.4	28.5	.7622
800.	2017.	38.6	-110.5	-108.3	-103.6	-97.9	-91.5	38.7	.9056
900.	1864.	53.2	-152.4	-149.8	-144.6	-138.2	-131.0	50.8	1.0601
1000.	1720.	71.6	-204.2	-201.4	-195.6	-188.5	-180.5	65.2	1.2275

BALLISTIC COEFFICIENT .50, MUZZLE VELOCITY FT/SEC 3600.

RANGE	VELOCITY	HEIGHT	100 YD	150 YD	200 YD	250 YD	300 YD	DRIFT	TIME
0.	3600.	0.0	-1.5	-1.5	-1.5	-1.5	-1.5	0.0	0.0000
100.	3382.	-.4	0.0	.3	.8	1.4	2.2	.4	.0857
200.	3173.	.8	-1.6	-1.0	0.0	1.3	2.8	1.9	.1773
300.	2973.	2.9	-6.6	-5.8	-4.2	-2.3	0.0	4.4	.2750
400.	2781.	6.2	-15.5	-14.5	-12.4	-9.8	-6.8	8.1	.3792
500.	2598.	10.8	-29.0	-27.7	-25.1	-21.8	-18.1	13.1	.4910
600.	2423.	17.1	-47.6	-46.1	-43.0	-39.0	-34.5	19.5	.6106
700.	2253.	25.5	-72.1	-70.4	-66.7	-62.1	-56.8	27.4	.7391
800.	2089.	36.2	-103.4	-101.4	-97.2	-91.9	-85.8	37.1	.8772
900.	1932.	49.9	-142.6	-140.4	-135.7	-129.7	-122.9	48.7	1.0267
1000.	1784.	67.0	-191.2	-188.6	-183.4	-176.8	-169.3	62.5	1.1883

BALLISTIC COEFFICIENT .50, MUZZLE VELOCITY FT/SEC 3700.

RANGE	VELOCITY	HEIGHT	100 YD	150 YD	200 YD	250 YD	300 YD	DRIFT	TIME
0.	3700.	0.0	-1.5	-1.5	-1.5	-1.5	-1.5	0.0	0.0000
100.	3478.	-.4	0.0	.2	.7	1.3	2.0	.5	.0838
200.	3265.	.7	-1.4	-1.0	0.0	1.2	2.6	1.9	.1728
300.	3061.	2.7	-6.0	-5.4	-4.0	-2.1	0.0	4.3	.2675
400.	2866.	5.8	-14.4	-13.6	-11.6	-9.1	-6.4	7.8	.3689
500.	2679.	10.2	-27.0	-26.0	-23.6	-20.5	-17.0	12.6	.4772
600.	2500.	16.2	-44.5	-43.3	-40.4	-36.7	-32.5	18.8	.5934
700.	2328.	24.0	-67.5	-66.1	-62.7	-58.3	-53.4	26.4	.7177
800.	2161.	34.0	-96.8	-95.2	-91.3	-86.3	-80.7	35.7	.8513
900.	2001.	46.8	-133.6	-131.8	-127.4	-121.8	-115.5	46.8	.9957
1000.	1849.	62.9	-179.0	-177.0	-172.1	-165.9	-158.9	60.0	1.1517

BALLISTIC COEFFICIENT .50, MUZZLE VELOCITY FT/SEC 3800.

RANGE	VELOCITY	HEIGHT	100 YD	150 YD	200 YD	250 YD	300 YD	DRIFT	TIME
0.	3800.	0.0	-1.5	-1.5	-1.5	-1.5	-1.5	0.0	0.0000
100.	3574.	-.4	0.0	.2	.6	1.2	1.9	.5	.0818
200.	3357.	.6	-1.2	-.9	0.0	1.1	2.5	1.8	.1683
300.	3149.	2.5	-5.6	-5.1	-3.7	-2.0	0.0	4.2	.2607
400.	2950.	5.4	-13.4	-12.7	-10.9	-8.6	-5.9	7.6	.3587
500.	2760.	9.6	-25.3	-24.4	-22.2	-19.3	-16.0	12.2	.4643
600.	2577.	15.2	-41.6	-40.7	-38.0	-34.5	-30.5	18.1	.5767
700.	2402.	22.6	-63.3	-62.1	-59.0	-55.0	-50.3	25.5	.6973
800.	2233.	32.1	-90.8	-89.5	-85.9	-81.4	-76.0	34.4	.8268
900.	2071.	44.1	-125.3	-123.8	-119.8	-114.6	-108.6	45.0	.9662
1000.	1915.	59.2	-167.9	-166.3	-161.8	-156.1	-149.4	57.7	1.1171

BALLISTIC COEFFICIENT .50, MUZZLE VELOCITY FT/SEC 3900.

RANGE	VELOCITY	HEIGHT	100 YD	150 YD	200 YD	250 YD	300 YD	DRIFT	TIME
0.	3900.	0.0	-1.5	-1.5	-1.5	-1.5	-1.5	0.0	0.0000
100.	3670.	-.5	0.0	.1	.5	1.1	1.7	.4	.0790
200.	3449.	.5	-1.1	-.8	0.0	1.1	2.3	1.7	.1634
300.	3237.	2.3	-5.1	-4.7	-3.5	-1.9	0.0	4.0	.2532
400.	3034.	5.1	-12.5	-11.9	-10.3	-8.1	-5.6	7.2	.3488
500.	2840.	9.0	-23.6	-22.9	-20.8	-18.1	-15.0	11.7	.4509
600.	2655.	14.3	-39.0	-38.2	-35.8	-32.5	-28.8	17.4	.5604
700.	2477.	21.3	-59.4	-58.4	-55.5	-51.7	-47.4	24.4	.6773
800.	2305.	30.2	-85.4	-84.3	-81.0	-76.7	-71.7	33.0	.8030
900.	2140.	41.5	-117.9	-116.6	-113.0	-108.1	-102.5	43.3	.9382
1000.	1980.	55.6	-157.7	-156.4	-152.3	-146.8	-140.6	55.3	1.0835

BALLISTIC COEFFICIENT .50, MUZZLE VELOCITY FT/SEC 4000.

RANGE	VELOCITY	HEIGHT	100 YD	150 YD	200 YD	250 YD	300 YD	DRIFT	TIME
0.	4000.	0.0	-1.5	-1.5	-1.5	-1.5	-1.5	0.0	0.0000
100.	3766.	-.5	0.0	.1	.5	1.0	1.6	.4	.0774
200.	3541.	.5	-1.0	-.8	0.0	1.0	2.2	1.7	.1596
300.	3326.	2.2	-4.7	-4.4	-3.3	-1.8	0.0	3.9	.2471
400.	3119.	4.8	-11.6	-11.2	-9.7	-7.7	-5.4	7.1	.3405
500.	2921.	8.5	-22.1	-21.6	-19.7	-17.2	-14.3	11.4	.4398
600.	2732.	13.5	-36.6	-36.0	-33.8	-30.7	-27.2	16.9	.5458
700.	2551.	20.1	-55.8	-55.1	-52.5	-48.9	-44.8	23.7	.6595
800.	2377.	28.6	-80.3	-79.5	-76.5	-72.4	-67.7	31.9	.7813
900.	2209.	39.2	-110.9	-110.0	-106.6	-102.0	-96.7	41.7	.9122
1000.	2047.	52.5	-148.6	-147.6	-143.8	-138.8	-132.9	53.4	1.0535

BALLISTIC COEFFICIENT .51, MUZZLE VELOCITY FT/SEC 2000.

RANGE	VELOCITY	HEIGHT	100 YD	150 YD	200 YD	250 YD	300 YD	DRIFT	TIME
0.	2000.	0.0	-1.5	-1.5	-1.5	-1.5	-1.5	0.0	0.0000
100.	1851.	.4	0.0	2.0	4.3	7.0	9.8	1.0	.1559
200.	1711.	4.3	-8.7	-4.7	0.0	5.3	10.9	4.3	.3243
300.	1579.	11.6	-29.4	-23.4	-16.4	-8.5	0.0	10.1	.5073
400.	1456.	23.1	-64.0	-56.0	-46.6	-36.1	-24.8	18.5	.7051
500.	1346.	39.8	-115.0	-105.1	-93.4	-80.2	-66.1	29.9	.9196
600.	1251.	62.8	-185.3	-173.4	-159.3	-143.5	-126.6	44.2	1.1509
700.	1170.	93.2	-277.7	-263.8	-247.4	-229.0	-209.2	61.4	1.3991
800.	1108.	132.0	-395.7	-379.8	-361.0	-340.0	-317.4	81.5	1.6632
900.	1060.	179.9	-541.1	-523.2	-502.1	-478.5	-453.0	103.8	1.9398
1000.	1021.	237.7	-716.6	-696.8	-673.3	-647.0	-618.7	128.3	2.2290

BALLISTIC COEFFICIENT .51, MUZZLE VELOCITY FT/SEC 2100.

RANGE	VELOCITY	HEIGHT	100 YD	150 YD	200 YD	250 YD	300 YD	DRIFT	TIME
0.	2100.	0.0	-1.5	-1.5	-1.5	-1.5	-1.5	0.0	0.0000
100.	1946.	.3	0.0	1.7	3.9	6.2	8.7	1.0	.1484
200.	1800.	3.8	-7.7	-4.3	0.0	4.7	9.8	4.0	.3085
300.	1663.	10.4	-26.2	-21.0	-14.7	-7.6	0.0	9.4	.4821
400.	1534.	20.8	-57.3	-50.4	-41.9	-32.5	-22.4	17.3	.6699
500.	1414.	35.9	-103.2	-94.5	-83.9	-72.2	-59.5	28.0	.8736
600.	1310.	56.7	-166.5	-156.2	-143.4	-129.4	-114.1	41.7	1.0942
700.	1221.	84.3	-250.0	-237.9	-223.1	-206.7	-188.8	58.3	1.3314
800.	1146.	120.0	-357.0	-343.2	-326.2	-307.4	-287.1	78.0	1.5858
900.	1089.	164.4	-490.5	-474.9	-455.8	-434.7	-411.8	100.2	1.8548
1000.	1045.	218.8	-653.9	-636.6	-615.4	-591.9	-566.5	125.0	2.1387

BALLISTIC COEFFICIENT .51, MUZZLE VELOCITY FT/SEC 2200.

RANGE	VELOCITY	HEIGHT	100 YD	150 YD	200 YD	250 YD	300 YD	DRIFT	TIME
0.	2200.	0.0	-1.5	-1.5	-1.5	-1.5	-1.5	0.0	0.0000
100.	2042.	.2	0.0	1.5	3.4	5.5	7.8	1.0	.1419
200.	1890.	3.4	-6.8	-3.8	0.0	4.2	8.8	3.8	.2943
300.	1747.	9.4	-23.5	-18.9	-13.2	-6.9	0.0	8.9	.4594
400.	1614.	18.8	-51.6	-45.4	-37.9	-29.4	-20.2	16.3	.6381
500.	1488.	32.4	-92.9	-85.3	-75.8	-65.3	-53.8	26.4	.8316
600.	1374.	51.3	-150.1	-140.9	-129.6	-116.9	-103.1	39.4	1.0418
700.	1275.	76.5	-225.9	-215.2	-201.9	-187.1	-171.1	55.3	1.2690
800.	1190.	109.0	-322.7	-310.4	-295.3	-278.4	-260.1	74.2	1.5124
900.	1123.	150.0	-444.3	-430.5	-413.5	-394.5	-373.8	95.9	1.7723
1000.	1072.	200.2	-593.3	-577.9	-559.0	-537.9	-514.9	120.1	2.0461

BALLISTIC COEFFICIENT .51, MUZZLE VELOCITY FT/SEC 2300.

RANGE	VELOCITY	HEIGHT	100 YD	150 YD	200 YD	250 YD	300 YD	DRIFT	TIME
0.	2300.	0.0	-1.5	-1.5	-1.5	-1.5	-1.5	0.0	0.0000
100.	2138.	.1	0.0	1.4	3.1	5.0	7.1	.9	.1354
200.	1982.	3.0	-6.1	-3.4	0.0	3.9	8.0	3.5	.2808
300.	1834.	8.5	-21.2	-17.1	-12.0	-6.2	0.0	8.3	.4385
400.	1695.	17.0	-46.5	-41.1	-34.3	-26.6	-18.3	15.3	.6085
500.	1564.	29.4	-84.1	-77.3	-68.9	-59.2	-48.8	24.8	.7931
600.	1442.	46.6	-135.8	-127.6	-117.5	-105.9	-93.4	37.0	.9929
700.	1334.	69.5	-204.4	-194.9	-183.0	-169.6	-155.0	52.2	1.2097
800.	1241.	99.2	-292.4	-281.5	-267.9	-252.5	-235.9	70.3	1.4427
900.	1161.	136.8	-402.8	-390.6	-375.4	-358.0	-339.3	91.3	1.6927
1000.	1101.	183.3	-539.5	-525.9	-509.0	-489.7	-468.9	115.1	1.9583

BALLISTIC COEFFICIENT .51, MUZZLE VELOCITY FT/SEC 2400.

RANGE	VELOCITY	HEIGHT	100 YD	150 YD	200 YD	250 YD	300 YD	DRIFT	TIME
0.	2400.	0.0	-1.5	-1.5	-1.5	-1.5	-1.5	0.0	0.0000
100.	2234.	.1	0.0	1.2	2.7	4.5	6.4	.8	.1295
200.	2075.	2.7	-5.5	-3.1	0.0	3.5	7.3	3.3	.2687
300.	1921.	7.7	-19.1	-15.5	-10.9	-5.7	0.0	7.8	.4193
400.	1777.	15.5	-42.2	-37.4	-31.2	-24.3	-16.7	14.3	.5815
500.	1641.	26.8	-76.2	-70.2	-62.5	-53.8	-44.3	23.3	.7571
600.	1514.	42.3	-123.2	-116.0	-106.7	-96.3	-84.9	34.7	.9474
700.	1396.	63.2	-185.3	-176.9	-166.2	-154.0	-140.7	49.1	1.1540
800.	1295.	90.3	-265.5	-255.9	-243.6	-229.6	-214.5	66.4	1.3775
900.	1207.	124.8	-366.2	-355.4	-341.5	-325.9	-308.8	86.6	1.6173
1000.	1135.	168.2	-492.0	-480.0	-464.7	-447.2	-428.3	110.2	1.8760

BALLISTIC COEFFICIENT .51, MUZZLE VELOCITY FT/SEC 2500.

RANGE	VELOCITY	HEIGHT	100 YD	150 YD	200 YD	250 YD	300 YD	DRIFT	TIME
0.	2500.	0.0	-1.5	-1.5	-1.5	-1.5	-1.5	0.0	0.0000
100.	2331.	-.0	0.0	1.1	2.5	4.0	5.8	.7	.1242
200.	2168.	2.4	-4.9	-2.8	0.0	3.2	6.6	3.1	.2575
300.	2011.	7.0	-17.3	-14.1	-9.9	-5.2	0.0	7.3	.4014
400.	1861.	14.1	-38.2	-33.9	-28.4	-22.1	-15.2	13.4	.5562
500.	1720.	24.4	-69.2	-63.9	-57.0	-49.0	-40.4	21.8	.7238
600.	1588.	38.6	-112.2	-105.7	-97.5	-87.9	-77.6	32.7	.9060
700.	1464.	57.6	-168.6	-161.1	-151.4	-140.3	-128.3	46.2	1.1026
800.	1353.	82.4	-241.5	-232.9	-221.9	-209.2	-195.4	62.7	1.3161
900.	1257.	114.0	-333.4	-323.8	-311.4	-297.1	-281.6	82.1	1.5462
1000.	1174.	153.9	-448.1	-437.4	-423.6	-407.8	-390.5	104.7	1.7948

BALLISTIC COEFFICIENT .51, MUZZLE VELOCITY FT/SEC 2600.

RANGE	VELOCITY	HEIGHT	100 YD	150 YD	200 YD	250 YD	300 YD	DRIFT	TIME
0.	2600.	0.0	-1.5	-1.5	-1.5	-1.5	-1.5	0.0	0.0000
100.	2428.	-.1	0.0	1.0	2.2	3.6	5.2	.7	.1194
200.	2261.	2.2	-4.4	-2.5	0.0	2.9	6.0	3.0	.2476
300.	2100.	6.4	-15.7	-12.8	-9.1	-4.8	0.0	6.9	.3853
400.	1946.	12.9	-34.9	-31.1	-26.0	-20.3	-13.9	12.7	.5337
500.	1800.	22.4	-63.2	-58.4	-52.1	-45.0	-37.0	20.6	.6938
600.	1663.	35.4	-102.3	-96.6	-89.0	-80.5	-70.9	30.8	.8674
700.	1534.	52.7	-153.9	-147.3	-138.4	-128.4	-117.3	43.6	1.0551
800.	1415.	75.3	-220.2	-212.6	-202.5	-191.1	-178.3	59.1	1.2587
900.	1311.	104.3	-304.2	-295.6	-284.2	-271.4	-257.1	77.6	1.4793
1000.	1221.	140.7	-408.4	-398.9	-386.2	-372.0	-356.0	99.0	1.7165

BALLISTIC COEFFICIENT .51, MUZZLE VELOCITY FT/SEC 2700.

RANGE	VELOCITY	HEIGHT	100 YD	150 YD	200 YD	250 YD	300 YD	DRIFT	TIME
0.	2700.	0.0	-1.5	-1.5	-1.5	-1.5	-1.5	0.0	0.0000
100.	2524.	-.1	0.0	.8	2.0	3.3	4.7	.7	.1152
200.	2354.	2.0	-4.0	-2.3	0.0	2.6	5.5	2.8	.2382
300.	2190.	5.8	-14.2	-11.8	-8.3	-4.3	0.0	6.5	.3700
400.	2032.	11.9	-31.8	-28.5	-23.9	-18.6	-12.9	12.0	.5125
500.	1881.	20.5	-57.7	-53.6	-47.8	-41.2	-34.0	19.4	.6657
600.	1739.	32.5	-93.6	-88.6	-81.6	-73.7	-65.1	29.1	.8317
700.	1606.	48.4	-140.9	-135.1	-127.0	-117.8	-107.7	41.1	1.0114
800.	1481.	69.0	-201.5	-194.9	-185.6	-175.1	-163.6	55.8	1.2058
900.	1367.	95.6	-278.3	-270.9	-260.4	-248.5	-235.6	73.4	1.4171
1000.	1270.	129.1	-373.7	-365.5	-353.8	-340.7	-326.3	93.9	1.6448

BALLISTIC COEFFICIENT .51, MUZZLE VELOCITY FT/SEC 2800.

RANGE	VELOCITY	HEIGHT	100 YD	150 YD	200 YD	250 YD	300 YD	DRIFT	TIME
0.	2800.	0.0	-1.5	-1.5	-1.5	-1.5	-1.5	0.0	0.0000
100.	2620.	-.2	0.0	.8	1 8	3.0	4.3	.6	.1104
200.	2447.	1.8	-3.6	-2.1	0.0	2.5	5.1	2.6	.2289
300.	2279.	5.3	-13.0	-10.8	-7.6	-4.0	0.0	6.1	.3561
400.	2118.	10.9	-29.1	-26.1	-22.0	-17.1	-11.8	11.2	.4925
500.	1963.	18.9	-53.1	-49.3	-44.1	-38.0	-31.4	18.3	.6399
600.	1816.	29.9	-85.9	-81.4	-75.2	-67.8	-59.9	27.4	.7986
700.	1678.	44.5	-129.4	-124.1	-116.9	-108.3	-99.0	38.8	.9706
800.	1548.	63.5	-185.2	-179.2	-170.9	-161.1	-150.5	52.8	1.1569
900.	1427.	87.9	-255.4	-248.6	-239.3	-228.2	-216.3	69.4	1.3587
1000.	1321.	118.7	-343.0	-335.4	-325.0	-312.8	-299.6	89.0	1.5772

BALLISTIC COEFFICIENT .51, MUZZLE VELOCITY FT/SEC 2900.

RANGE	VELOCITY	HEIGHT	100 YD	150 YD	200 YD	250 YD	300 YD	DRIFT	TIME
0.	2900.	0.0	-1.5	-1.5	-1.5	-1.5	-1.5	0.0	0.0000
100.	2715.	-.2	0.0	.7	1.6	2.7	4.0	.6	.1069
200.	2539.	1.6	-3.2	-1.9	0.0	2.2	4.7	2.5	.2213
300.	2368.	4.9	-11.9	-9.9	-7.0	-3.7	0.0	5.9	.3436
400.	2204.	10.1	-26.7	-24.1	-20.3	-15.8	-10.9	10.7	.4748
500.	2045.	17.5	-48.8	-45.6	-40.7	-35.2	-29.0	17.4	.6164
600.	1893.	27.6	-79.1	-75.2	-69.4	-62.7	-55.3	26.0	.7685
700.	1751.	41.1	-119.1	-114.6	-107.8	-100.0	-91.4	36.8	.9333
800.	1617.	58.6	-170.5	-165.3	-157.5	-148.6	-138.7	50.0	1.1117
900.	1491.	81.0	-235.0	-229.2	-220.4	-210.5	-199.3	65.8	1.3049
1000.	1376.	109.4	-315.5	-309.0	-299.3	-288.2	-275.9	84.5	1.5148

BALLISTIC COEFFICIENT .51, MUZZLE VELOCITY FT/SEC 3000.

RANGE	VELOCITY	HEIGHT	100 YD	150 YD	200 YD	250 YD	300 YD	DRIFT	TIME
0.	3000.	0.0	-1.5	-1.5	-1.5	-1.5	-1.5	0.0	0.0000
100.	2811.	-.2	0.0	.6	1.4	2.5	3.6	.6	.1037
200.	2630.	1.4	-2.9	-1.7	0.0	2.1	4.3	2.4	.2138
300.	2457.	4.5	-10.9	-9.1	-6.5	-3.4	0.0	5.6	.3318
400.	2289.	9.3	-24.6	-22.3	-18.8	-14.6	-10.1	10.3	.4585
500.	2127.	16.2	-45.0	-42.1	-37.7	-32.5	-26.9	16.6	.5943
600.	1972.	25.6	-73.0	-69.5	-64.3	-58.1	-51.3	24.8	.7408
700.	1824.	38.0	-110.0	-105.9	-99.8	-92.5	-84.6	35.0	.8990
800.	1686.	54.2	-157.4	-152.8	-145.8	-137.5	-128.4	47.5	1.0701
900.	1556.	74.9	-217.1	-211.9	-204.0	-194.6	-184.5	62.6	1.2556
1000.	1434.	101.1	-291.0	-285.2	-276.5	-266.1	-254.8	80.3	1.4565

BALLISTIC COEFFICIENT .51, MUZZLE VELOCITY FT/SEC 3100.

RANGE	VELOCITY	HEIGHT	100 YD	150 YD	200 YD	250 YD	300 YD	DRIFT	TIME
0.	3100.	0.0	-1.5	-1.5	-1.5	-1.5	-1.5	0.0	0.0000
100.	2907.	-.3	0.0	.5	1.3	2.3	3.3	.5	.0999
200.	2722.	1.3	-2.6	-1.6	0.0	1.9	4.0	2.3	.2065
300.	2545.	4.2	-10.0	-8.4	-6.0	-3.2	0.0	5.3	.3206
400.	2374.	8.7	-22.7	-20.7	-17.4	-13.6	-9.4	9.8	.4425
500.	2209.	15.0	-41.6	-39.0	-35.0	-30.3	-24.9	15.8	.5734
600.	2051.	23.8	-67.6	-64.6	-59.7	-54.0	-47.7	23.6	.7146
700.	1899.	35.3	-101.8	-98.2	-92.6	-86.0	-78.5	33.3	.8664
800.	1756.	50.2	-145.6	-141.6	-135.1	-127.5	-119.0	45.1	1.0307
900.	1621.	69.4	-200.8	-196.2	-189.0	-180.4	-170.8	59.4	1.2085
1000.	1495.	93.5	-269.1	-264.0	-256.0	-246.4	-235.8	76.3	1.4012

BALLISTIC COEFFICIENT .51, MUZZLE VELOCITY FT/SEC 3200.

RANGE	VELOCITY	HEIGHT	100 YD	150 YD	200 YD	250 YD	300 YD	DRIFT	TIME
0.	3200.	0.0	-1.5	-1.5	-1.5	-1.5	-1.5	0.0	0.0000
100.	3003.	-.3	0.0	.5	1.2	2.1	3.1	.5	.0966
200.	2814.	1.2	-2.4	-1.5	0.0	1.8	3.7	2.2	.2001
300.	2633.	3.9	-9.2	-7.8	-5.6	-2.9	0.0	5.1	.3102
400.	2459.	8.0	-21.0	-19.2	-16.2	-12.7	-8.8	9.3	.4281
500.	2291.	14.0	-38.6	-36.3	-32.6	-28.2	-23.3	15.1	.5546
600.	2130.	22.1	-62.7	-60.0	-55.6	-50.3	-44.4	22.5	.6904
700.	1974.	32.8	-94.5	-91.3	-86.1	-80.0	-73.1	31.7	.8366
800.	1826.	46.7	-135.3	-131.6	-125.7	-118.6	-110.8	43.1	.9947
900.	1687.	64.5	-186.4	-182.2	-175.6	-167.7	-158.9	56.6	1.1656
1000.	1557.	86.8	-249.8	-245.2	-237.8	-229.0	-219.2	72.8	1.3509

BALLISTIC COEFFICIENT .51, MUZZLE VELOCITY FT/SEC 3300.

RANGE	VELOCITY	HEIGHT	100 YD	150 YD	200 YD	250 YD	300 YD	DRIFT	TIME
0.	3300.	0.0	-1.5	-1.5	-1.5	-1.5	-1.5	0.0	0.0000
100.	3098.	-.3	0.0	.4	1.1	1.9	2.8	.6	.0941
200.	2905.	1.1	-2.1	-1.4	0.0	1.6	3.5	2.1	.1940
300.	2720.	3.6	-8.4	-7.3	-5.2	-2.7	0.0	4.9	.3007
400.	2543.	7.5	-19.4	-17.9	-15.1	-11.8	-8.2	9.0	.4148
500.	2373.	13.1	-35.7	-33.9	-30.4	-26.3	-21.7	14.5	.5369
600.	2208.	20.7	-58.2	-56.0	-51.8	-46.9	-41.4	21.5	.6678
700.	2049.	30.7	-87.9	-85.3	-80.4	-74.7	-68.3	30.4	.8091
800.	1897.	43.6	-125.7	-122.7	-117.1	-110.6	-103.3	41.1	.9610
900.	1754.	60.0	-173.1	-169.8	-163.6	-156.2	-148.0	54.1	1.1254
1000.	1620.	80.8	-231.9	-228.2	-221.3	-213.1	-203.9	69.4	1.3033

BALLISTIC COEFFICIENT .51, MUZZLE VELOCITY FT/SEC 3400.

RANGE	VELOCITY	HEIGHT	100 YD	150 YD	200 YD	250 YD	300 YD	DRIFT	TIME
0.	3400.	0.0	-1.5	-1.5	-1.5	-1.5	-1.5	0.0	0.0000
100.	3194.	-.4	0.0	.3	1.0	1.7	2.6	.5	.0908
200.	2997.	.9	-1.9	-1.2	0.0	1.5	3.3	2.0	.1877
300.	2808.	3.3	-7.7	-6.7	-4.9	-2.6	0.0	4.7	.2914
400.	2628.	7.0	-17.9	-16.6	-14.1	-11.0	-7.6	8.6	.4016
500.	2454.	12.2	-33.2	-31.5	-28.4	-24.6	-20.3	13.8	.5198
600.	2287.	19.3	-54.3	-52.2	-48.5	-43.9	-38.8	20.6	.6465
700.	2125.	28.6	-81.9	-79.5	-75.2	-69.8	-63.8	29.0	.7825
800.	1969.	40.7	-117.2	-114.4	-109.5	-103.4	-96.5	39.3	.9292
900.	1822.	56.0	-161.4	-158.3	-152.8	-145.8	-138.1	51.6	1.0875
1000.	1684.	75.3	-216.1	-212.7	-206.6	-198.9	-190.3	66.3	1.2589

BALLISTIC COEFFICIENT .51, MUZZLE VELOCITY FT/SEC 3500.

RANGE	VELOCITY	HEIGHT	100 YD	150 YD	200 YD	250 YD	300 YD	DRIFT	TIME
0.	3500.	0.0	-1.5	-1.5	-1.5	-1.5	-1.5	0.0	0.0000
100.	3290.	-.4	0.0	.3	.9	1.6	2.4	.5	.0886
200.	3089.	.9	-1.7	-1.2	0.0	1.4	3.0	2.0	.1828
300.	2896.	3.1	-7.1	-6.3	-4.5	-2.4	0.0	4.5	.2830
400.	2712.	6.6	-16.6	-15.5	-13.2	-10.4	-7.2	8.3	.3901
500.	2535.	11.5	-30.9	-29.5	-26.6	-23.1	-19.1	13.4	.5046
600.	2365.	18.1	-50.6	-48.9	-45.4	-41.2	-36.4	19.9	.6271
700.	2201.	26.9	-76.4	-74.4	-70.4	-65.4	-59.8	27.9	.7584
800.	2042.	38.2	-109.5	-107.2	-102.6	-96.9	-90.5	37.8	.9002
900.	1890.	52.4	-150.6	-148.1	-142.9	-136.6	-129.3	49.5	1.0526
1000.	1748.	70.4	-201.7	-198.8	-193.0	-186.0	-178.0	63.5	1.2177

BALLISTIC COEFFICIENT .51, MUZZLE VELOCITY FT/SEC 3600.

RANGE	VELOCITY	HEIGHT	100 YD	150 YD	200 YD	250 YD	300 YD	DRIFT	TIME
0.	3600.	0.0	-1.5	-1.5	-1.5	-1.5	-1.5	0.0	0.0000
100.	3386.	-.4	0.0	.3	.8	1.4	2.2	.4	.0857
200.	3181.	.8	-1.5	-1.0	0.0	1.3	2.8	1.8	.1770
300.	2984.	2.9	-6.5	-5.8	-4.2	-2.2	0.0	4.3	.2745
400.	2796.	6.1	-15.5	-14.5	-12.4	-9.7	-6.8	8.0	.3785
500.	2616.	10.7	-28.8	-27.5	-25.0	-21.6	-17.9	12.8	.4892
600.	2443.	17.0	-47.2	-45.7	-42.6	-38.6	-34.1	19.0	.6078
700.	2276.	25.2	-71.5	-69.7	-66.1	-61.4	-56.2	26.7	.7352
800.	2115.	35.7	-102.3	-100.3	-96.2	-90.8	-84.9	36.1	.8719
900.	1960.	49.1	-141.0	-138.7	-134.1	-128.1	-121.4	47.4	1.0195
1000.	1813.	65.9	-188.6	-186.0	-180.9	-174.2	-166.8	60.7	1.1784

BALLISTIC COEFFICIENT .51, MUZZLE VELOCITY FT/SEC 3700.

RANGE	VELOCITY	HEIGHT	100 YD	150 YD	200 YD	250 YD	300 YD	DRIFT	TIME
0.	3700.	0.0	-1.5	-1.5	-1.5	-1.5	-1.5	0.0	0.0000
100.	3482.	-.4	0.0	.2	.7	1.3	2.0	.5	.0838
200.	3273.	.7	-1.4	-1.0	0.0	1.2	2.6	1.9	.1727
300.	3073.	2.7	-6.0	-5.4	-3.9	-2.1	0.0	4.2	.2672
400.	2881.	5.7	-14.3	-13.5	-11.5	-9.1	-6.3	7.7	.3679
500.	2697.	10.1	-26.9	-25.9	-23.4	-20.4	-16.9	12.4	.4757
600.	2521.	16.0	-44.2	-43.0	-40.1	-36.4	-32.2	18.4	.5911
700.	2351.	23.7	-67.0	-65.5	-62.1	-57.8	-52.9	25.8	.7143
800.	2187.	33.6	-95.8	-94.2	-90.3	-85.4	-79.8	34.8	.8462
900.	2029.	46.2	-132.1	-130.3	-125.9	-120.4	-114.1	45.6	.9889
1000.	1878.	61.9	-176.6	-174.6	-169.7	-163.6	-156.6	58.3	1.1423

BALLISTIC COEFFICIENT .51, MUZZLE VELOCITY FT/SEC 3800.

RANGE	VELOCITY	HEIGHT	100 YD	150 YD	200 YD	250 YD	300 YD	DRIFT	TIME
0.	3800.	0.0	-1.5	-1.5	-1.5	-1.5	-1.5	0.0	0.0000
100.	3578.	-.4	0.0	.2	.6	1.2	1.8	.5	.0818
200.	3365.	.6	-1.2	-.9	0.0	1.1	2.5	1.8	.1680
300.	3161.	2.5	-5.5	-5.0	-3.7	-2.0	0.0	4.1	.2601
400.	2965.	5.4	-13.3	-12.6	-10.9	-8.6	-5.9	7.4	.3579
500.	2778.	9.5	-25.0	-24.2	-22.0	-19.1	-15.8	11.9	.4624
600.	2598.	15.1	-41.3	-40.4	-37.7	-34.3	-30.3	17.7	.5743
700.	2426.	22.4	-62.7	-61.6	-58.4	-54.5	-49.8	24.8	.6938
800.	2259.	31.7	-90.0	-88.7	-85.1	-80.5	-75.2	33.5	.8220
900.	2099.	43.5	-124.0	-122.5	-118.5	-113.4	-107.4	43.9	.9598
1000.	1944.	58.2	-165.8	-164.2	-159.7	-154.0	-147.4	56.1	1.1083

BALLISTIC COEFFICIENT .51, MUZZLE VELOCITY FT/SEC 3900.

RANGE	VELOCITY	HEIGHT	100 YD	150 YD	200 YD	250 YD	300 YD	DRIFT	TIME
0.	3900.	0.0	-1.5	-1.5	-1.5	-1.5	-1.5	0.0	0.0000
100.	3674.	-.5	0.0	.1	.5	1.1	1.7	.4	.0790
200.	3458.	.5	-1.1	-.8	0.0	1.1	2.3	1.6	.1632
300.	3250.	2.3	-5.1	-4.7	-3.5	-1.8	0.0	3.8	.2526
400.	3050.	5.1	-12.4	-11.8	-10.2	-8.1	-5.6	7.1	.3478
500.	2859.	8.9	-23.4	-22.7	-20.7	-18.0	-14.9	11.4	.4494
600.	2676.	14.2	-38.7	-37.9	-35.5	-32.2	-28.6	17.0	.5579
700.	2501.	21.1	-58.9	-58.0	-55.1	-51.3	-47.0	23.9	.6742
800.	2332.	29.8	-84.5	-83.5	-80.2	-75.9	-71.0	32.2	.7984
900.	2168.	40.9	-116.4	-115.2	-111.6	-106.7	-101.2	42.1	.9315
1000.	2011.	54.8	-155.8	-154.5	-150.4	-145.0	-138.9	53.9	1.0754

BALLISTIC COEFFICIENT .51, MUZZLE VELOCITY FT/SEC 4000.

RANGE	VELOCITY	HEIGHT	100 YD	150 YD	200 YD	250 YD	300 YD	DRIFT	TIME
0.	4000.	0.0	-1.5	-1.5	-1.5	-1.5	-1.5	0.0	0.0000
100.	3770.	-.5	0.0	.1	.5	1.0	1.6	.4	.0774
200.	3550.	.5	-1.0	-.8	0.0	1.0	2.2	1.7	.1594
300.	3338.	2.2	-4.7	-4.4	-3.3	-1.7	0.0	3.8	.2467
400.	3135.	4.8	-11.6	-11.2	-9.7	-7.6	-5.3	7.0	.3396
500.	2940.	8.5	-21.9	-21.4	-19.5	-17.0	-14.1	11.1	.4380
600.	2754.	13.4	-36.4	-35.8	-33.5	-30.5	-27.0	16.5	.5437
700.	2575.	19.9	-55.3	-54.7	-52.0	-48.5	-44.4	23.1	.6563
800.	2404.	28.2	-79.6	-78.8	-75.7	-71.7	-67.0	31.1	.7770
900.	2238.	38.7	-109.7	-108.8	-105.4	-100.9	-95.6	40.7	.9063
1000.	2078.	51.7	-146.6	-145.6	-141.8	-136.8	-131.0	51.9	1.0451

BALLISTIC COEFFICIENT .52, MUZZLE VELOCITY FT/SEC 2000.

RANGE	VELOCITY	HEIGHT	100 YD	150 YD	200 YD	250 YD	300 YD	DRIFT	TIME
0.	2000.	0.0	-1.5	-1.5	-1.5	-1.5	-1.5	0.0	0.0000
100.	1854.	.4	0.0	2.0	4.3	6.9	9.7	1.0	.1557
200.	1716.	4.3	-8.6	-4.7	0.0	5.2	10.9	4.2	.3237
300.	1586.	11.5	-29.2	-23.3	-16.3	-8.4	0.0	9.9	.5061
400.	1465.	23.0	-63.6	-55.6	-46.3	-35.8	-24.6	18.1	.7027
500.	1356.	39.5	-114.3	-104.4	-92.7	-79.7	-65.6	29.2	.9162
600.	1262.	62.2	-183.7	-171.8	-157.8	-142.1	-125.2	43.1	1.1451
700.	1180.	92.2	-275.1	-261.2	-244.9	-226.5	-206.8	60.1	1.3914
800.	1116.	130.5	-391.6	-375.7	-357.0	-336.1	-313.6	79.8	1.6534
900.	1068.	177.7	-535.5	-517.6	-496.6	-473.1	-447.8	101.8	1.9284
1000.	1028.	235.3	-710.4	-690.6	-667.2	-641.1	-613.0	126.3	2.2177

BALLISTIC COEFFICIENT .52, MUZZLE VELOCITY FT/SEC 2100.

RANGE	VELOCITY	HEIGHT	100 YD	150 YD	200 YD	250 YD	300 YD	DRIFT	TIME
0.	2100.	0.0	-1.5	-1.5	-1.5	-1.5	-1.5	0.0	0.0000
100.	1948.	.3	0.0	1.7	3.8	6.2	8.7	1.0	.1483
200.	1805.	3.8	-7.7	-4.2	0.0	4.7	9.7	3.9	.3081
300.	1670.	10.4	-26.1	-20.9	-14.6	-7.6	0.0	9.2	.4810
400.	1544.	20.7	-57.0	-50.1	-41.6	-32.3	-22.2	17.0	.6678
500.	1425.	35.6	-102.5	-93.9	-83.3	-71.6	-59.0	27.4	.8701
600.	1322.	56.2	-165.1	-154.8	-142.1	-128.1	-113.0	40.8	1.0888
700.	1232.	83.4	-247.8	-235.7	-220.9	-204.6	-186.9	57.1	1.3243
800.	1155.	118.7	-354.0	-340.2	-323.3	-304.6	-284.4	76.5	1.5777
900.	1098.	162.2	-484.7	-469.3	-450.2	-429.2	-406.5	98.0	1.8426
1000.	1053.	215.4	-644.4	-627.2	-606.0	-582.6	-557.4	122.0	2.1218

BALLISTIC COEFFICIENT .52, MUZZLE VELOCITY FT/SEC 2200.

RANGE	VELOCITY	HEIGHT	100 YD	150 YD	200 YD	250 YD	300 YD	DRIFT	TIME
0.	2200.	0.0	-1.5	-1.5	-1.5	-1.5	-1.5	0.0	0.0000
100.	2045.	.2	0.0	1.5	3.4	5.5	7.8	1.0	.1418
200.	1896.	3.4	-6.8	-3.8	0.0	4.2	8.8	3.7	.2939
300.	1755.	9.3	-23.4	-18.8	-13.1	-6.8	0.0	8.7	.4583
400.	1624.	18.7	-51.2	-45.2	-37.6	-29.2	-20.1	15.9	.6360
500.	1500.	32.2	-92.3	-84.7	-75.2	-64.7	-53.3	25.8	.8282
600.	1386.	50.8	-148.8	-139.7	-128.4	-115.7	-102.1	38.4	1.0366
700.	1288.	75.6	-223.6	-212.9	-199.7	-184.9	-169.0	54.0	1.2613
800.	1203.	107.6	-319.2	-307.0	-291.9	-275.1	-256.9	72.5	1.5027
900.	1133.	148.0	-438.9	-425.2	-408.2	-389.3	-368.8	93.8	1.7601
1000.	1080.	197.4	-585.7	-570.5	-551.6	-530.5	-507.8	117.6	2.0316

BALLISTIC COEFFICIENT .52, MUZZLE VELOCITY FT/SEC 2300.

RANGE	VELOCITY	HEIGHT	100 YD	150 YD	200 YD	250 YD	300 YD	DRIFT	TIME
0.	2300.	0.0	-1.5	-1.5	-1.5	-1.5	-1.5	0.0	0.0000
100.	2141.	.1	0.0	1.4	3.0	5.0	7.0	.9	.1353
200.	1988.	3.0	-6.1	-3.4	0.0	3.8	8.0	3.4	.2804
300.	1842.	8.4	-21.1	-17.0	-12.0	-6.2	0.0	8.1	.4376
400.	1705.	16.9	-46.2	-40.8	-34.1	-26.4	-18.1	14.9	.6065
500.	1576.	29.2	-83.5	-76.8	-68.3	-58.7	-48.4	24.2	.7899
600.	1455.	46.1	-134.7	-126.5	-116.4	-104.9	-92.5	36.1	.9879
700.	1348.	68.6	-202.3	-192.8	-181.0	-167.6	-153.1	50.9	1.2023
800.	1254.	97.9	-289.3	-278.5	-264.9	-249.6	-233.0	68.7	1.4335
900.	1173.	135.1	-398.9	-386.7	-371.5	-354.2	-335.6	89.5	1.6826
1000.	1112.	180.6	-532.4	-518.9	-502.0	-482.8	-462.1	112.5	1.9437

BALLISTIC COEFFICIENT .52, MUZZLE VELOCITY FT/SEC 2400.

RANGE	VELOCITY	HEIGHT	100 YD	150 YD	200 YD	250 YD	300 YD	DRIFT	TIME
0.	2400.	0.0	-1.5	-1.5	-1.5	-1.5	-1.5	0.0	0.0000
100.	2238.	.1	0.0	1.2	2.7	4.5	6.3	.8	.1295
200.	2081.	2.7	-5.4	-3.1	0.0	3.5	7.2	3.2	.2682
300.	1930.	7.6	-19.0	-15.5	-10.9	-5.6	0.0	7.6	.4183
400.	1788.	15.4	-41.9	-37.2	-31.0	-24.1	-16.5	14.0	.5797
500.	1654.	26.5	-75.7	-69.8	-62.1	-53.4	-44.0	22.7	.7542
600.	1528.	41.9	-122.2	-115.0	-105.9	-95.4	-84.1	33.9	.9428
700.	1411.	62.4	-183.5	-175.1	-164.4	-152.2	-139.0	47.9	1.1471
800.	1310.	89.1	-262.4	-252.9	-240.6	-226.7	-211.6	64.8	1.3679
900.	1222.	122.9	-361.5	-350.8	-337.0	-321.4	-304.4	84.5	1.6052
1000.	1148.	165.2	-484.2	-472.3	-457.0	-439.5	-420.7	107.2	1.8592

BALLISTIC COEFFICIENT .52, MUZZLE VELOCITY FT/SEC 2500.

RANGE	VELOCITY	HEIGHT	100 YD	150 YD	200 YD	250 YD	300 YD	DRIFT	TIME
0.	2500.	0.0	-1.5	-1.5	-1.5	-1.5	-1.5	0.0	0.0000
100.	2334.	-.0	0.0	1.1	2.4	4.0	5.7	.7	.1242
200.	2174.	2.4	-4.9	-2.7	0.0	3.2	6.6	3.0	.2571
300.	2019.	6.9	-17.2	-14.0	-9.9	-5.1	0.0	7.1	.4005
400.	1872.	14.0	-38.0	-33.7	-28.2	-21.9	-15.1	13.1	.5546
500.	1733.	24.2	-68.8	-63.5	-56.6	-48.7	-40.2	21.3	.7212
600.	1603.	38.3	-111.2	-104.8	-96.6	-87.1	-76.8	31.9	.9015
700.	1480.	56.9	-166.9	-159.5	-149.9	-138.7	-126.8	45.1	1.0960
800.	1369.	81.3	-238.8	-230.3	-219.3	-206.6	-192.9	61.1	1.3072
900.	1273.	112.4	-329.5	-319.9	-307.5	-293.3	-277.9	80.1	1.5352
1000.	1190.	151.1	-441.2	-430.5	-416.8	-400.9	-383.9	101.9	1.7787

BALLISTIC COEFFICIENT .52, MUZZLE VELOCITY FT/SEC 2600.

RANGE	VELOCITY	HEIGHT	100 YD	150 YD	200 YD	250 YD	300 YD	DRIFT	TIME
0.	2600.	0.0	-1.5	-1.5	-1.5	-1.5	-1.5	0.0	0.0000
100.	2431.	-.1	0.0	1.0	2.2	3.6	5.2	.7	.1193
200.	2267.	2.2	-4.4	-2.5	0.0	2.8	6.0	2.9	.2472
300.	2109.	6.3	-15.6	-12.8	-9.0	-4.7	0.0	6.7	.3843
400.	1957.	12.8	-34.7	-30.9	-25.8	-20.2	-13.9	12.4	.5321
500.	1813.	22.2	-62.8	-58.0	-51.7	-44.6	-36.7	20.1	.6911
600.	1678.	35.0	-101.5	-95.8	-88.2	-79.7	-70.3	30.1	.8631
700.	1551.	52.1	-152.6	-145.9	-137.1	-127.1	-116.1	42.5	1.0493
800.	1432.	74.3	-217.9	-210.3	-200.3	-188.9	-176.3	57.6	1.2506
900.	1328.	102.7	-300.5	-291.9	-280.6	-267.8	-253.6	75.6	1.4683
1000.	1237.	138.5	-403.1	-393.6	-381.0	-366.7	-351.0	96.6	1.7029

BALLISTIC COEFFICIENT .52, MUZZLE VELOCITY FT/SEC 2700.

RANGE	VELOCITY	HEIGHT	100 YD	150 YD	200 YD	250 YD	300 YD	DRIFT	TIME
0.	2700.	0.0	-1.5	-1.5	-1.5	-1.5	-1.5	0.0	0.0000
100.	2527.	-.1	0.0	.8	2.0	3.3	4.7	.7	.1151
200.	2360.	2.0	-4.0	-2.3	0.0	2.6	5.5	2.8	.2379
300.	2199.	5.8	-14.2	-11.7	-8.2	-4.3	0.0	6.3	.3693
400.	2044.	11.8	-31.7	-28.4	-23.7	-18.5	-12.8	11.7	.5111
500.	1895.	20.4	-57.4	-53.3	-47.5	-40.9	-33.8	19.0	.6633
600.	1755.	32.1	-92.8	-87.9	-80.9	-73.1	-64.5	28.3	.8277
700.	1623.	47.8	-139.5	-133.8	-125.7	-116.5	-106.5	40.1	1.0056
800.	1499.	68.1	-199.3	-192.8	-183.5	-173.0	-161.6	54.4	1.1978
900.	1386.	94.2	-274.8	-267.4	-256.9	-245.2	-232.3	71.5	1.4062
1000.	1287.	127.0	-368.5	-360.4	-348.7	-335.7	-321.3	91.5	1.6311

BALLISTIC COEFFICIENT .52, MUZZLE VELOCITY FT/SEC 2800.

RANGE	VELOCITY	HEIGHT	100 YD	150 YD	200 YD	250 YD	300 YD	DRIFT	TIME
0.	2800.	0.0	-1.5	-1.5	-1.5	-1.5	-1.5	0.0	0.0000
100.	2623.	-.2	0.0	.8	1.8	3.0	4.3	.6	.1103
200.	2453.	1.8	-3.6	-2.1	0.0	2.4	5.1	2.5	.2286
300.	2289.	5.3	-13.0	-10.7	-7.6	-4.0	0.0	6.0	.3554
400.	2130.	10.8	-29.0	-26.0	-21.9	-17.0	-11.7	11.0	.4912
500.	1977.	18.7	-52.7	-48.9	-43.7	-37.7	-31.1	17.9	.6372
600.	1832.	29.6	-85.3	-80.8	-74.6	-67.3	-59.4	26.8	.7951
700.	1696.	44.0	-128.2	-122.9	-115.7	-107.2	-98.0	37.9	.9651
800.	1568.	62.7	-183.3	-177.3	-169.0	-159.3	-148.7	51.5	1.1495
900.	1447.	86.6	-252.4	-245.6	-236.3	-225.4	-213.5	67.7	1.3487
1000.	1341.	116.7	-338.2	-330.7	-320.4	-308.2	-295.0	86.7	1.5641

BALLISTIC COEFFICIENT .52, MUZZLE VELOCITY FT/SEC 2900.

RANGE	VELOCITY	HEIGHT	100 YD	150 YD	200 YD	250 YD	300 YD	DRIFT	TIME
0.	2900.	0.0	-1.5	-1.5	-1.5	-1.5	-1.5	0.0	0.0000
100.	2719.	-.2	0.0	.6	1.6	2.7	3.9	.6	.1068
200.	2545.	1.6	-3.2	-1.9	0.0	2.2	4.7	2.5	.2210
300.	2378.	4.9	-11.8	-9.9	-7.0	-3.7	0.0	5.7	.3428
400.	2216.	10.0	-26.6	-24.0	-20.1	-15.7	-10.8	10.5	.4734
500.	2060.	17.3	-48.5	-45.2	-40.4	-34.9	-28.8	17.0	.6140
600.	1910.	27.4	-78.5	-74.6	-68.9	-62.2	-54.9	25.4	.7653
700.	1769.	40.6	-118.1	-113.5	-106.8	-99.0	-90.5	35.9	.9283
800.	1637.	57.8	-168.7	-163.5	-155.8	-147.0	-137.2	48.8	1.1046
900.	1512.	79.8	-232.3	-226.4	-217.7	-207.8	-196.8	64.1	1.2953
1000.	1397.	107.5	-311.2	-304.7	-295.0	-284.0	-271.7	82.3	1.5020

BALLISTIC COEFFICIENT .52, MUZZLE VELOCITY FT/SEC 3000.

RANGE	VELOCITY	HEIGHT	100 YD	150 YD	200 YD	250 YD	300 YD	DRIFT	TIME
0.	3000.	0.0	-1.5	-1.5	-1.5	-1.5	-1.5	0.0	0.0000
100.	2815.	-.2	0.0	.6	1.4	2.5	3.6	.6	.1036
200.	2637.	1.4	-2.9	-1.7	0.0	2.1	4.3	2.4	.2136
300.	2467.	4.5	-10.8	-9.1	-6.5	-3.4	0.0	5.5	.3312
400.	2302.	9.3	-24.5	-22.2	-18.7	-14.6	-10.1	10.1	.4572
500.	2143.	16.1	-44.7	-41.9	-37.5	-32.3	-26.7	16.3	.5924
600.	1989.	25.4	-72.5	-69.0	-63.8	-57.6	-50.8	24.2	.7375
700.	1844.	37.7	-109.1	-105.1	-99.0	-91.7	-83.9	34.2	.8944
800.	1706.	53.5	-155.7	-151.1	-144.1	-135.9	-126.9	46.3	1.0632
900.	1578.	73.8	-214.6	-209.4	-201.5	-192.3	-182.1	61.0	1.2465
1000.	1457.	99.4	-287.1	-281.4	-272.6	-262.3	-251.1	78.2	1.4443

BALLISTIC COEFFICIENT .52, MUZZLE VELOCITY FT/SEC 3100.

RANGE	VELOCITY	HEIGHT	100 YD	150 YD	200 YD	250 YD	300 YD	DRIFT	TIME
0.	3100.	0.0	-1.5	-1.5	-1.5	-1.5	-1.5	0.0	0.0000
100.	2910.	-.3	0.0	.5	1.3	2.3	3.3	.5	.0999
200.	2729.	1.3	-2.6	-1.6	0.0	1.9	4.0	2.2	.2062
300.	2555.	4.2	-9.9	-8.4	-6.0	-3.2	0.0	5.2	.3198
400.	2387.	8.6	-22.6	-20.5	-17.3	-13.5	-9.3	9.5	.4413
500.	2225.	14.9	-41.3	-38.8	-34.8	-30.0	-24.8	15.4	.5714
600.	2069.	23.5	-67.1	-64.0	-59.2	-53.5	-47.2	23.0	.7113
700.	1919.	34.9	-101.0	-97.4	-91.8	-85.2	-77.8	32.5	.8621
800.	1777.	49.6	-144.2	-140.2	-133.8	-126.2	-117.8	44.0	1.0245
900.	1644.	68.4	-198.5	-193.9	-186.7	-178.2	-168.7	57.9	1.1999
1000.	1519.	92.0	-265.6	-260.5	-252.5	-243.0	-232.5	74.3	1.3898

BALLISTIC COEFFICIENT .52, MUZZLE VELOCITY FT/SEC 3200.

RANGE	VELOCITY	HEIGHT	100 YD	150 YD	200 YD	250 YD	300 YD	DRIFT	TIME
0.	3200.	0.0	-1.5	-1.5	-1.5	-1.5	-1.5	0.0	0.0000
100.	3006.	-.3	0.0	.5	1.2	2.1	3.0	.5	.0965
200.	2821.	1.2	-2.4	-1.5	0.0	1.8	3.7	2.2	.1998
300.	2643.	3.9	-9.1	-7.8	-5.6	-2.9	0.0	5.0	.3096
400.	2472.	8.0	-20.9	-19.0	-16.1	-12.6	-8.7	9.1	.4269
500.	2307.	13.9	-38.3	-36.0	-32.4	-28.0	-23.1	14.8	.5527
600.	2148.	21.9	-62.3	-59.5	-55.2	-49.9	-44.0	22.0	.6875
700.	1994.	32.5	-93.7	-90.5	-85.4	-79.3	-72.4	31.0	.8324
800.	1848.	46.2	-133.9	-130.3	-124.4	-117.4	-109.6	42.0	.9887
900.	1711.	63.5	-184.2	-180.0	-173.5	-165.6	-156.8	55.2	1.1571
1000.	1582.	85.4	-246.6	-242.0	-234.7	-225.9	-216.1	70.8	1.3400

BALLISTIC COEFFICIENT .52, MUZZLE VELOCITY FT/SEC 3300.

RANGE	VELOCITY	HEIGHT	100 YD	150 YD	200 YD	250 YD	300 YD	DRIFT	TIME
0.	3300.	0.0	-1.5	-1.5	-1.5	-1.5	-1.5	0.0	0.0000
100.	3102.	-.3	0.0	.4	1.1	1.9	2.8	.6	.0941
200.	2913.	1.1	-2.1	-1.4	0.0	1.6	3.4	2.1	.1939
300.	2731.	3.6	-8.3	-7.2	-5.2	-2.7	0.0	4.8	.3000
400.	2557.	7.5	-19.3	-17.8	-15.0	-11.8	-8.1	8.8	.4136
500.	2389.	13.0	-35.5	-33.6	-30.2	-26.1	-21.6	14.2	.5350
600.	2227.	20.5	-57.8	-55.6	-51.4	-46.6	-41.1	21.1	.6651
700.	2071.	30.3	-87.1	-84.5	-79.6	-74.0	-67.6	29.6	.8047
800.	1920.	43.1	-124.5	-121.5	-116.0	-109.5	-102.3	40.2	.9555
900.	1779.	59.2	-171.2	-167.9	-161.7	-154.4	-146.2	52.7	1.1177
1000.	1645.	79.5	-229.0	-225.3	-218.4	-210.2	-201.2	67.6	1.2930

BALLISTIC COEFFICIENT .52, MUZZLE VELOCITY FT/SEC 3400.

RANGE	VELOCITY	HEIGHT	100 YD	150 YD	200 YD	250 YD	300 YD	DRIFT	TIME
0.	3400.	0.0	-1.5	-1.5	-1.5	-1.5	-1.5	0.0	0.0000
100.	3198.	-.4	0.0	.3	.9	1.7	2.6	.5	.0908
200.	3005.	.9	-1.9	-1.2	0.0	1.5	3.2	1.9	.1874
300.	2819.	3.3	-7.7	-6.7	-4.9	-2.5	0.0	4.6	.2908
400.	2641.	7.0	-17.9	-16.5	-14.1	-11.0	-7.6	8.4	.4007
500.	2471.	12.1	-33.0	-31.3	-28.3	-24.4	-20.2	13.5	.5180
600.	2306.	19.1	-53.9	-51.8	-48.2	-43.5	-38.5	20.1	.6439
700.	2146.	28.4	-81.2	-78.8	-74.6	-69.2	-63.3	28.4	.7788
800.	1993.	40.2	-116.0	-113.3	-108.4	-102.3	-95.5	38.3	.9237
900.	1847.	55.3	-159.6	-156.6	-151.1	-144.1	-136.5	50.4	1.0802
1000.	1710.	74.1	-213.3	-209.9	-203.8	-196.1	-187.6	64.5	1.2487

BALLISTIC COEFFICIENT .52, MUZZLE VELOCITY FT/SEC 3500.

RANGE	VELOCITY	HEIGHT	100 YD	150 YD	200 YD	250 YD	300 YD	DRIFT	TIME
0.	3500.	0.0	-1.5	-1.5	-1.5	-1.5	-1.5	0.0	0.0000
100.	3294.	-.4	0.0	.3	.9	1.6	2.4	.5	.0885
200.	3097.	.9	-1.7	-1.2	0.0	1.4	3.0	2.0	.1826
300.	2907.	3.1	-7.1	-6.2	-4.5	-2.4	0.0	4.5	.2826
400.	2726.	6.5	-16.6	-15.4	-13.1	-10.3	-7.1	8.1	.3890
500.	2552.	11.4	-30.7	-29.3	-26.4	-23.0	-18.9	13.1	.5028
600.	2384.	18.0	-50.2	-48.5	-45.0	-40.9	-36.0	19.4	.6244
700.	2222.	26.6	-75.8	-73.8	-69.7	-64.9	-59.2	27.2	.7547
800.	2066.	37.7	-108.4	-106.1	-101.5	-95.9	-89.5	36.8	.8948
900.	1916.	51.7	-149.1	-146.5	-141.3	-135.1	-127.8	48.3	1.0458
1000.	1775.	69.3	-199.2	-196.3	-190.6	-183.6	-175.6	61.8	1.2083

BALLISTIC COEFFICIENT .52, MUZZLE VELOCITY FT/SEC 3600.

RANGE	VELOCITY	HEIGHT	100 YD	150 YD	200 YD	250 YD	300 YD	DRIFT	TIME
0.	3600.	0.0	-1.5	-1.5	-1.5	-1.5	-1.5	0.0	0.0000
100.	3390.	-.4	0.0	.2	.8	1.4	2.2	.4	.0857
200.	3189.	.8	-1.5	-1.0	0.0	1.3	2.8	1.8	.1768
300.	2996.	2.8	-6.5	-5.7	-4.2	-2.2	0.0	4.2	.2738
400.	2810.	6.1	-15.4	-14.4	-12.3	-9.7	-6.7	7.8	.3776
500.	2633.	10.7	-28.6	-27.4	-24.8	-21.5	-17.8	12.5	.4876
600.	2463.	16.8	-46.9	-45.4	-42.3	-38.3	-33.9	18.6	.6054
700.	2298.	24.9	-70.9	-69.1	-65.5	-60.9	-55.7	26.1	.7317
800.	2139.	35.3	-101.4	-99.4	-95.3	-90.0	-84.1	35.3	.8671
900.	1986.	48.4	-139.4	-137.1	-132.5	-126.5	-119.9	46.2	1.0124
1000.	1840.	64.9	-186.3	-183.9	-178.7	-172.1	-164.7	59.2	1.1697

BALLISTIC COEFFICIENT .52, MUZZLE VELOCITY FT/SEC 3700.

RANGE	VELOCITY	HEIGHT	100 YD	150 YD	200 YD	250 YD	300 YD	DRIFT	TIME
0.	3700.	0.0	-1.5	-1.5	-1.5	-1.5	-1.5	0.0	0.0000
100.	3486.	-.4	0.0	.2	.7	1.3	2.0	.5	.0837
200.	3281.	.7	-1.4	-1.0	0.0	1.2	2.6	1.8	.1727
300.	3084.	2.7	-6.0	-5.4	-3.9	-2.1	0.0	4.1	.2668
400.	2895.	5.7	-14.3	-13.5	-11.5	-9.1	-6.3	7.5	.3672
500.	2714.	10.0	-26.7	-25.7	-23.2	-20.2	-16.7	12.1	.4743
600.	2541.	15.9	-43.9	-42.7	-39.7	-36.1	-31.9	18.0	.5886
700.	2374.	23.5	-66.4	-65.0	-61.5	-57.3	-52.4	25.2	.7107
800.	2212.	33.2	-95.0	-93.4	-89.4	-84.6	-79.0	33.9	.8415
900.	2056.	45.6	-130.7	-128.9	-124.5	-119.1	-112.7	44.5	.9824
1000.	1907.	61.0	-174.6	-172.5	-167.6	-161.6	-154.6	56.9	1.1339

BALLISTIC COEFFICIENT .52, MUZZLE VELOCITY FT/SEC 3800.

RANGE	VELOCITY	HEIGHT	100 YD	150 YD	200 YD	250 YD	300 YD	DRIFT	TIME
0.	3800.	0.0	-1.5	-1.5	-1.5	-1.5	-1.5	0.0	0.0000
100.	3582.	-.4	0.0	.2	.6	1.2	1.8	.5	.0817
200.	3373.	.6	-1.2	-.9	0.0	1.2	2.5	1.7	.1678
300.	3173.	2.5	-5.5	-5.0	-3.7	-1.9	0.0	4.0	.2595
400.	2980.	5.4	-13.2	-12.6	-10.8	-8.5	-5.9	7.3	.3572
500.	2795.	9.5	-24.9	-24.1	-21.9	-19.0	-15.8	11.7	.4612
600.	2619.	14.9	-41.0	-40.0	-37.4	-33.9	-30.0	17.3	.5718
700.	2449.	22.1	-62.1	-61.0	-57.9	-53.9	-49.4	24.2	.6903
800.	2285.	31.3	-89.1	-87.8	-84.3	-79.7	-74.5	32.7	.8173
900.	2126.	42.9	-122.6	-121.1	-117.2	-112.0	-106.1	42.7	.9533
1000.	1974.	57.3	-163.7	-162.1	-157.7	-151.9	-145.4	54.6	1.0998

BALLISTIC COEFFICIENT .52, MUZZLE VELOCITY FT/SEC 3900.

RANGE	VELOCITY	HEIGHT	100 YD	150 YD	200 YD	250 YD	300 YD	DRIFT	TIME
0.	3900.	0.0	-1.5	-1.5	-1.5	-1.5	-1.5	0.0	0.0000
100.	3679.	-.5	0.0	.1	.5	1.1	1.7	.4	.0790
200.	3466.	.5	-1.1	-.8	0.0	1.1	2.3	1.6	.1630
300.	3262.	2.3	-5.0	-4.6	-3.4	-1.8	0.0	3.7	.2520
400.	3065.	5.0	-12.3	-11.8	-10.1	-8.0	-5.6	6.9	.3469
500.	2877.	8.9	-23.2	-22.6	-20.6	-17.9	-14.9	11.1	.4479
600.	2697.	14.1	-38.4	-37.6	-35.2	-32.0	-28.4	16.6	.5559
700.	2524.	20.9	-58.4	-57.5	-54.7	-50.9	-46.7	23.3	.6711
800.	2357.	29.5	-83.8	-82.7	-79.5	-75.2	-70.3	31.4	.7941
900.	2196.	40.4	-115.2	-114.0	-110.4	-105.6	-100.1	41.1	.9256
1000.	2041.	54.0	-154.0	-152.7	-148.6	-143.3	-137.2	52.5	1.0676

BALLISTIC COEFFICIENT .52, MUZZLE VELOCITY FT/SEC 4000.

RANGE	VELOCITY	HEIGHT	100 YD	150 YD	200 YD	250 YD	300 YD	DRIFT	TIME
0.	4000.	0.0	-1.5	-1.5	-1.5	-1.5	-1.5	0.0	0.0000
100.	3775.	-.5	0.0	.1	.5	1.0	1.6	.4	.0773
200.	3558.	.5	-1.0	-.8	0.0	1.0	2.2	1.6	.1593
300.	3350.	2.2	-4.7	-4.4	-3.2	-1.7	0.0	3.7	.2462
400.	3150.	4.8	-11.5	-11.1	-9.6	-7.6	-5.3	6.8	.3387
500.	2959.	8.4	-21.8	-21.3	-19.4	-16.9	-14.0	10.8	.4365
600.	2775.	13.3	-36.1	-35.5	-33.2	-30.2	-26.7	16.1	.5414
700.	2599.	19.7	-54.9	-54.2	-51.5	-48.1	-44.0	22.6	.6533
800.	2430.	27.9	-78.8	-78.0	-75.0	-71.0	-66.3	30.4	.7726
900.	2266.	38.2	-108.6	-107.7	-104.2	-99.8	-94.6	39.7	.9006
1000.	2109.	50.9	-145.0	-144.0	-140.2	-135.2	-129.4	50.6	1.0377

BALLISTIC COEFFICIENT .53, MUZZLE VELOCITY FT/SEC 2000.

RANGE	VELOCITY	HEIGHT	100 YD	150 YD	200 YD	250 YD	300 YD	DRIFT	TIME
0.	2000.	0.0	-1.5	-1.5	-1.5	-1.5	-1.5	0.0	0.0000
100.	1856.	.4	0.0	2.0	4.3	6.9	9.7	1.0	.1555
200.	1721.	4.3	-8.6	-4.6	0.0	5.2	10.8	4.1	.3232
300.	1594.	11.5	-29.1	-23.2	-16.2	-8.4	0.0	9.7	.5049
400.	1474.	22.8	-63.2	-55.3	-46.0	-35.6	-24.4	17.7	.7004
500.	1365.	39.2	-113.5	-103.6	-92.0	-78.9	-65.0	28.6	.9123
600.	1272.	61.6	-182.4	-170.5	-156.5	-140.9	-124.1	42.2	1.1400
700.	1190.	91.2	-272.7	-258.9	-242.6	-224.3	-204.8	58.8	1.3842
800.	1125.	129.0	-387.7	-371.8	-353.2	-332.4	-310.0	78.1	1.6439
900.	1075.	175.6	-529.9	-512.0	-491.1	-467.6	-442.5	99.8	1.9170
1000.	1035.	231.9	-701.0	-681.2	-657.9	-631.9	-603.9	123.5	2.2016

BALLISTIC COEFFICIENT .53, MUZZLE VELOCITY FT/SEC 2100.

RANGE	VELOCITY	HEIGHT	100 YD	150 YD	200 YD	250 YD	300 YD	DRIFT	TIME
0.	2100.	0.0	-1.5	-1.5	-1.5	-1.5	-1.5	0.0	0.0000
100.	1951.	.3	0.0	1.7	3.8	6.2	8.7	.9	.1482
200.	1810.	3.8	-7.6	-4.2	0.0	4.7	9.7	3.8	.3076
300.	1678.	10.3	-26.0	-20.8	-14.5	-7.5	0.0	9.0	.4798
400.	1553.	20.5	-56.7	-49.8	-41.4	-32.1	-22.1	16.6	.6659
500.	1436.	35.3	-101.8	-93.3	-82.7	-71.1	-58.5	26.8	.8668
600.	1333.	55.6	-163.9	-153.6	-140.9	-126.9	-111.9	39.9	1.0838
700.	1243.	82.5	-245.5	-233.5	-218.8	-202.5	-184.9	55.8	1.3171
800.	1166.	117.0	-349.5	-335.8	-318.9	-300.3	-280.3	74.6	1.5665
900.	1107.	160.2	-479.4	-464.0	-445.0	-424.0	-401.5	96.0	1.8311
1000.	1061.	212.6	-637.0	-619.8	-598.8	-575.5	-550.4	119.6	2.1083

BALLISTIC COEFFICIENT .53, MUZZLE VELOCITY FT/SEC 2200.

RANGE	VELOCITY	HEIGHT	100 YD	150 YD	200 YD	250 YD	300 YD	DRIFT	TIME
0.	2200.	0.0	-1.5	-1.5	-1.5	-1.5	-1.5	0.0	0.0000
100.	2047.	.2	0.0	1.5	3.4	5.5	7.8	.9	.1416
200.	1901.	3.4	-6.8	-3.8	0.0	4.2	8.7	3.7	.2935
300.	1763.	9.3	-23.3	-18.7	-13.1	-6.8	0.0	8.5	.4573
400.	1633.	18.6	-51.0	-44.9	-37.4	-29.0	-20.0	15.6	.6342
500.	1511.	31.9	-91.7	-84.1	-74.7	-64.2	-52.9	25.2	.8251
600.	1398.	50.3	-147.7	-138.6	-127.3	-114.7	-101.1	37.6	1.0317
700.	1300.	74.8	-221.4	-210.8	-197.6	-182.9	-167.1	52.7	1.2542
800.	1215.	106.4	-316.1	-304.0	-288.9	-272.0	-254.0	70.9	1.4938
900.	1143.	145.9	-433.8	-420.1	-403.1	-384.2	-363.9	91.7	1.7482
1000.	1090.	194.6	-578.5	-563.4	-544.5	-523.5	-500.9	115.1	2.0175

BALLISTIC COEFFICIENT .53, MUZZLE VELOCITY FT/SEC 2300.

RANGE	VELOCITY	HEIGHT	100 YD	150 YD	200 YD	250 YD	300 YD	DRIFT	TIME
0.	2300.	0.0	-1.5	-1.5	-1.5	-1.5	-1.5	0.0	0.0000
100.	2144.	.1	0.0	1.3	3.0	4.9	7.0	.8	.1352
200.	1993.	3.0	-6.1	-3.4	0.0	3.8	7.9	3.4	.2801
300.	1850.	8.4	-21.0	-16.9	-11.8	-6.1	0.0	7.9	.4364
400.	1715.	16.8	-46.0	-40.6	-33.8	-26.2	-18.0	14.6	.6045
500.	1588.	29.0	-83.0	-76.3	-67.8	-58.2	-48.1	23.7	.7869
600.	1469.	45.6	-133.6	-125.5	-115.3	-103.9	-91.6	35.3	.9831
700.	1361.	67.9	-200.4	-191.0	-179.2	-165.8	-151.5	49.7	1.1956
800.	1268.	96.6	-286.0	-275.3	-261.7	-246.4	-230.1	67.0	1.4240
900.	1187.	132.9	-393.3	-381.2	-365.9	-348.7	-330.4	87.1	1.6689
1000.	1122.	177.9	-525.4	-512.0	-495.0	-475.9	-455.5	110.0	1.9293

BALLISTIC COEFFICIENT .53, MUZZLE VELOCITY FT/SEC 2400.

RANGE	VELOCITY	HEIGHT	100 YD	150 YD	200 YD	250 YD	300 YD	DRIFT	TIME
0.	2400.	0.0	-1.5	-1.5	-1.5	-1.5	-1.5	0.0	0.0000
100.	2241.	.1	0.0	1.2	2.7	4.4	6.3	.8	.1294
200.	2087.	2.7	-5.4	-3.1	0.0	3.5	7.2	3.2	.2679
300.	1938.	7.6	-18.9	-15.4	-10.8	-5.6	0.0	7.4	.4173
400.	1798.	15.3	-41.7	-36.9	-30.8	-23.9	-16.4	13.7	.5779
500.	1666.	26.3	-75.2	-69.3	-61.6	-53.0	-43.7	22.2	.7513
600.	1542.	41.5	-121.2	-114.1	-104.9	-94.5	-83.3	33.2	.9384
700.	1426.	61.7	-181.8	-173.5	-162.8	-150.7	-137.6	46.8	1.1407
800.	1324.	87.9	-259.6	-250.1	-237.9	-224.1	-209.1	63.2	1.3592
900.	1236.	121.3	-357.6	-346.9	-333.1	-317.6	-300.8	82.6	1.5945
1000.	1160.	162.6	-477.7	-465.9	-450.6	-433.3	-414.6	104.7	1.8449

BALLISTIC COEFFICIENT .53, MUZZLE VELOCITY FT/SEC 2500.

RANGE	VELOCITY	HEIGHT	100 YD	150 YD	200 YD	250 YD	300 YD	DRIFT	TIME
0.	2500.	0.0	-1.5	-1.5	-1.5	-1.5	-1.5	0.0	0.0000
100.	2337.	-.0	0.0	1.1	2.4	4.0	5.7	.7	.1241
200.	2180.	2.4	-4.9	-2.7	0.0	3.2	6.6	2.9	.2567
300.	2028.	6.9	-17.1	-14.0	-9.9	-5.1	0.0	7.0	.3997
400.	1883.	13.9	-37.8	-33.6	-28.1	-21.8	-15.0	12.8	.5530
500.	1746.	24.0	-68.4	-63.1	-56.2	-48.3	-39.8	20.9	.7186
600.	1617.	37.9	-110.3	-103.9	-95.7	-86.2	-76.0	31.2	.8971
700.	1496.	56.3	-165.4	-158.0	-148.4	-137.3	-125.4	44.0	1.0899
800.	1385.	80.2	-236.2	-227.7	-216.8	-204.1	-190.5	59.6	1.2987
900.	1288.	110.7	-325.3	-315.7	-303.4	-289.2	-273.9	78.1	1.5235
1000.	1204.	148.7	-435.3	-424.7	-411.1	-395.2	-378.2	99.4	1.7647

BALLISTIC COEFFICIENT .53, MUZZLE VELOCITY FT/SEC 2600.

RANGE	VELOCITY	HEIGHT	100 YD	150 YD	200 YD	250 YD	300 YD	DRIFT	TIME
0.	2600.	0.0	-1.5	-1.5	-1.5	-1.5	-1.5	0.0	0.0000
100.	2434.	-.1	0.0	1.0	2.2	3.6	5.2	.7	.1191
200.	2273.	2.2	-4.4	-2.5	0.0	2.8	6.0	2.8	.2468
300.	2118.	6.3	-15.5	-12.7	-8.9	-4.7	0.0	6.5	.3833
400.	1969.	12.8	-34.5	-30.7	-25.7	-20.0	-13.8	12.1	.5304
500.	1827.	22.0	-62.4	-57.6	-51.4	-44.3	-36.5	19.7	.6886
600.	1693.	34.7	-100.7	-95.0	-87.5	-79.0	-69.6	29.4	.8591
700.	1567.	51.5	-151.2	-144.6	-135.8	-125.9	-115.0	41.5	1.0437
800.	1449.	73.4	-215.7	-208.1	-198.1	-186.7	-174.3	56.3	1.2427
900.	1344.	101.2	-296.9	-288.4	-277.1	-264.3	-250.3	73.8	1.4577
1000.	1253.	136.3	-397.8	-388.3	-375.8	-361.6	-346.0	94.3	1.6895

BALLISTIC COEFFICIENT .53, MUZZLE VELOCITY FT/SEC 2700.

RANGE	VELOCITY	HEIGHT	100 YD	150 YD	200 YD	250 YD	300 YD	DRIFT	TIME
0.	2700.	0.0	-1.5	-1.5	-1.5	-1.5	-1.5	0.0	0.0000
100.	2530.	-.1	0.0	.8	2.0	3.3	4.7	.7	.1150
200.	2367.	2.0	-3.9	-2.3	0.0	2.6	5.5	2.7	.2375
300.	2208.	5.8	-14.1	-11.7	-8.2	-4.3	0.0	6.2	.3686
400.	2055.	11.7	-31.5	-28.2	-23.6	-18.4	-12.7	11.5	.5096
500.	1909.	20.2	-57.0	-53.0	-47.2	-40.7	-33.5	18.6	.6610
600.	1770.	31.9	-92.1	-87.2	-80.3	-72.5	-63.9	27.7	.8241
700.	1640.	47.3	-138.3	-132.6	-124.5	-115.4	-105.4	39.2	1.0002
800.	1517.	67.3	-197.4	-190.8	-181.6	-171.2	-159.8	53.1	1.1904
900.	1404.	92.8	-271.5	-264.2	-253.8	-242.0	-229.2	69.7	1.3960
1000.	1305.	124.9	-363.6	-355.4	-343.9	-330.8	-316.6	89.2	1.6178

BALLISTIC COEFFICIENT .53, MUZZLE VELOCITY FT/SEC 2800.

RANGE	VELOCITY	HEIGHT	100 YD	150 YD	200 YD	250 YD	300 YD	DRIFT	TIME
0.	2800.	0.0	-1.5	-1.5	-1.5	-1.5	-1.5	0.0	0.0000
100.	2626.	-.2	0.0	.7	1.8	3.0	4.3	.6	.1103
200.	2459.	1.8	-3.6	-2.1	0.0	2.4	5.0	2.5	.2284
300.	2298.	5.3	-12.9	-10.7	-7.6	-4.0	0.0	5.9	.3547
400.	2142.	10.8	-28.9	-25.9	-21.8	-16.9	-11.7	10.8	.4901
500.	1991.	18.6	-52.4	-48.6	-43.5	-37.4	-30.8	17.5	.6351
600.	1848.	29.3	-84.7	-80.2	-74.0	-66.8	-58.9	26.2	.7916
700.	1713.	43.5	-127.0	-121.8	-114.6	-106.1	-96.9	36.9	.9599
800.	1586.	61.9	-181.5	-175.5	-167.2	-157.6	-147.0	50.2	1.1424
900.	1467.	85.3	-249.4	-242.7	-233.4	-222.5	-210.7	65.9	1.3388
1000.	1359.	114.8	-333.8	-326.3	-316.0	-303.9	-290.8	84.5	1.5515

BALLISTIC COEFFICIENT .53, MUZZLE VELOCITY FT/SEC 2900.

RANGE	VELOCITY	HEIGHT	100 YD	150 YD	200 YD	250 YD	300 YD	DRIFT	TIME
0.	2900.	0.0	-1.5	-1.5	-1.5	-1.5	-1.5	0.0	0.0000
100.	2722.	-.2	0.0	.6	1.6	2.7	3.9	.6	.1068
200.	2552.	1.6	-3.2	-1.9	0.0	2.2	4.6	2.4	.2206
300.	2387.	4.9	-11.8	-9.8	-7.0	-3.7	0.0	5.6	.3422
400.	2228.	10.0	-26.5	-23.9	-20.1	-15.6	-10.8	10.3	.4722
500.	2075.	17.2	-48.1	-44.9	-40.1	-34.6	-28.5	16.6	.6116
600.	1927.	27.1	-78.0	-74.1	-68.4	-61.8	-54.4	24.9	.7620
700.	1787.	40.2	-117.1	-112.5	-105.8	-98.1	-89.6	35.1	.9235
800.	1656.	57.1	-167.0	-161.9	-154.2	-145.4	-135.6	47.6	1.0979
900.	1533.	78.7	-229.6	-223.8	-215.2	-205.3	-194.3	62.5	1.2862
1000.	1417.	105.8	-307.0	-300.5	-291.0	-280.0	-267.7	80.1	1.4898

BALLISTIC COEFFICIENT .53, MUZZLE VELOCITY FT/SEC 3000.

RANGE	VELOCITY	HEIGHT	100 YD	150 YD	200 YD	250 YD	300 YD	DRIFT	TIME
0.	3000.	0.0	-1.5	-1.5	-1.5	-1.5	-1.5	0.0	0.0000
100.	2818.	-.2	0.0	.6	1.4	2.5	3.6	.6	.1035
200.	2644.	1.4	-2.9	-1.7	0.0	2.0	4.3	2.3	.2133
300.	2476.	4.5	-10.8	-9.0	-6.4	-3.4	0.0	5.4	.3305
400.	2314.	9.2	-24.4	-22.1	-18.6	-14.5	-10.0	9.8	.4560
500.	2157.	16.0	-44.4	-41.5	-37.2	-32.1	-26.5	15.9	.5901
600.	2006.	25.1	-72.0	-68.5	-63.3	-57.2	-50.4	23.7	.7345
700.	1862.	37.2	-108.0	-104.0	-97.9	-90.8	-82.9	33.3	.8894
800.	1727.	52.9	-154.2	-149.6	-142.6	-134.5	-125.5	45.2	1.0567
900.	1599.	72.8	-212.1	-207.0	-199.1	-189.9	-179.8	59.4	1.2377
1000.	1479.	97.8	-283.3	-277.6	-268.9	-258.7	-247.5	76.1	1.4325

BALLISTIC COEFFICIENT .53, MUZZLE VELOCITY FT/SEC 3100.

RANGE	VELOCITY	HEIGHT	100 YD	150 YD	200 YD	250 YD	300 YD	DRIFT	TIME
0.	3100.	0.0	-1.5	-1.5	-1.5	-1.5	-1.5	0.0	0.0000
100.	2914.	-.3	0.0	.5	1.3	2.2	3.3	.5	.0998
200.	2735.	1.3	-2.6	-1.6	0.0	1.9	4.0	2.2	.2059
300.	2564.	4.1	-9.9	-8.4	-6.0	-3.1	0.0	5.1	.3192
400.	2400.	8.6	-22.4	-20.5	-17.3	-13.5	-9.3	9.4	.4402
500.	2240.	14.8	-41.1	-38.6	-34.6	-29.9	-24.6	15.1	.5697
600.	2086.	23.3	-66.5	-63.6	-58.8	-53.1	-46.8	22.5	.7082
700.	1938.	34.6	-100.1	-96.6	-91.0	-84.4	-77.1	31.7	.8576
800.	1798.	49.0	-142.8	-138.8	-132.4	-124.8	-116.5	42.9	1.0182
900.	1666.	67.4	-196.3	-191.8	-184.6	-176.1	-166.7	56.4	1.1917
1000.	1542.	90.5	-262.2	-257.2	-249.2	-239.8	-229.3	72.3	1.3788

BALLISTIC COEFFICIENT .53, MUZZLE VELOCITY FT/SEC 3200.

RANGE	VELOCITY	HEIGHT	100 YD	150 YD	200 YD	250 YD	300 YD	DRIFT	TIME
0.	3200.	0.0	-1.5	-1.5	-1.5	-1.5	-1.5	0.0	0.0000
100.	3010.	-.3	0.0	.5	1.2	2.1	3.0	.5	.0965
200.	2828.	1.2	-2.4	-1.4	0.0	1.8	3.7	2.1	.1995
300.	2653.	3.8	-9.1	-7.7	-5.5	-2.9	0.0	4.9	.3091
400.	2485.	8.0	-20.8	-18.9	-16.0	-12.5	-8.6	9.0	.4259
500.	2322.	13.8	-38.1	-35.8	-32.2	-27.8	-22.9	14.4	.5508
600.	2166.	21.7	-61.8	-59.0	-54.7	-49.5	-43.6	21.5	.6844
700.	2014.	32.2	-93.0	-89.8	-84.7	-78.6	-71.8	30.3	.8283
800.	1870.	45.6	-132.6	-128.9	-123.1	-116.1	-108.3	40.9	.9826
900.	1734.	62.7	-182.2	-178.1	-171.6	-163.7	-154.9	53.8	1.1494
1000.	1606.	84.1	-243.4	-238.8	-231.6	-222.9	-213.1	69.0	1.3294

BALLISTIC COEFFICIENT .53, MUZZLE VELOCITY FT/SEC 3300.

RANGE	VELOCITY	HEIGHT	100 YD	150 YD	200 YD	250 YD	300 YD	DRIFT	TIME
0.	3300.	0.0	-1.5	-1.5	-1.5	-1.5	-1.5	0.0	0.0000
100.	3106.	-.3	0.0	.4	1.1	1.9	2.8	.5	.0940
200.	2920.	1.1	-2.1	-1.4	0.0	1.6	3.4	2.1	.1937
300.	2741.	3.6	-8.3	-7.2	-5.1	-2.7	0.0	4.7	.2994
400.	2570.	7.4	-19.2	-17.7	-14.9	-11.7	-8.1	8.6	.4126
500.	2405.	12.9	-35.3	-33.4	-30.0	-26.0	-21.5	13.9	.5334
600.	2245.	20.3	-57.4	-55.2	-51.1	-46.3	-40.8	20.6	.6625
700.	2091.	30.0	-86.4	-83.8	-79.0	-73.3	-67.0	28.9	.8008
800.	1942.	42.6	-123.3	-120.3	-114.8	-108.4	-101.2	39.2	.9498
900.	1802.	58.4	-169.3	-166.0	-159.8	-152.6	-144.4	51.4	1.1100
1000.	1670.	78.3	-226.2	-222.4	-215.6	-207.6	-198.5	65.8	1.2831

BALLISTIC COEFFICIENT .53, MUZZLE VELOCITY FT/SEC 3400.

RANGE	VELOCITY	HEIGHT	100 YD	150 YD	200 YD	250 YD	300 YD	DRIFT	TIME
0.	3400.	0.0	-1.5	-1.5	-1.5	-1.5	-1.5	0.0	0.0000
100.	3202.	-.4	0.0	.3	.9	1.7	2.5	.4	.0908
200.	3012.	.9	-1.9	-1.2	0.0	1.5	3.2	1.9	.1872
300.	2829.	3.3	-7.6	-6.6	-4.8	-2.5	0.0	4.5	.2901
400.	2654.	6.9	-17.8	-16.4	-14.0	-10.9	-7.6	8.2	.3996
500.	2486.	12.0	-32.8	-31.1	-28.1	-24.2	-20.1	13.2	.5164
600.	2324.	19.0	-53.5	-51.4	-47.8	-43.2	-38.2	19.7	.6413
700.	2167.	28.1	-80.5	-78.1	-73.9	-68.5	-62.6	27.7	.7748
800.	2016.	39.7	-115.0	-112.2	-107.4	-101.2	-94.6	37.4	.9185
900.	1871.	54.5	-157.8	-154.8	-149.3	-142.4	-134.9	49.0	1.0727
1000.	1735.	73.0	-210.7	-207.3	-201.3	-193.6	-185.2	62.8	1.2394

BALLISTIC COEFFICIENT .53, MUZZLE VELOCITY FT/SEC 3500.

RANGE	VELOCITY	HEIGHT	100 YD	150 YD	200 YD	250 YD	300 YD	DRIFT	TIME
0.	3500.	0.0	-1.5	-1.5	-1.5	-1.5	-1.5	0.0	0.0000
100.	3298.	-.4	0.0	.3	.9	1.6	2.4	.5	.0884
200.	3104.	.8	-1.7	-1.2	0.0	1.4	3.0	1.9	.1824
300.	2918.	3.1	-7.1	-6.2	-4.5	-2.4	0.0	4.4	.2822
400.	2739.	6.5	-16.5	-15.3	-13.0	-10.3	-7.0	7.9	.3880
500.	2568.	11.3	-30.5	-29.1	-26.2	-22.8	-18.8	12.8	.5012
600.	2403.	17.8	-49.9	-48.2	-44.7	-40.6	-35.8	19.0	.6221
700.	2243.	26.3	-75.2	-73.2	-69.2	-64.4	-58.7	26.6	.7513
800.	2089.	37.2	-107.4	-105.1	-100.5	-94.9	-88.5	35.9	.8897
900.	1941.	51.1	-147.5	-144.9	-139.7	-133.5	-126.3	47.1	1.0388
1000.	1801.	68.3	-196.8	-193.9	-188.1	-181.2	-173.2	60.2	1.1992

BALLISTIC COEFFICIENT .53, MUZZLE VELOCITY FT/SEC 3600.

RANGE	VELOCITY	HEIGHT	100 YD	150 YD	200 YD	250 YD	300 YD	DRIFT	TIME
0.	3600.	0.0	-1.5	-1.5	-1.5	-1.5	-1.5	0.0	0.0000
100.	3394.	-.4	0.0	.2	.8	1.4	2.1	.4	.0857
200.	3196.	.7	-1.5	-1.0	0.0	1.3	2.8	1.8	.1767
300.	3006.	2.8	-6.4	-5.7	-4.2	-2.2	0.0	4.1	.2733
400.	2824.	6.1	-15.3	-14.3	-12.2	-9.6	-6.7	7.6	.3765
500.	2649.	10.6	-28.4	-27.2	-24.6	-21.3	-17.7	12.2	.4861
600.	2482.	16.7	-46.5	-45.1	-42.0	-38.0	-33.6	18.1	.6030
700.	2319.	24.7	-70.3	-68.6	-64.9	-60.3	-55.2	25.5	.7282
800.	2163.	34.9	-100.4	-98.5	-94.3	-89.0	-83.2	34.4	.8620
900.	2011.	47.8	-137.9	-135.8	-131.1	-125.1	-118.6	45.1	1.0060
1000.	1867.	63.9	-183.9	-181.5	-176.3	-169.7	-162.4	57.6	1.1606

BALLISTIC COEFFICIENT .53, MUZZLE VELOCITY FT/SEC 3700.

RANGE	VELOCITY	HEIGHT	100 YD	150 YD	200 YD	250 YD	300 YD	DRIFT	TIME
0.	3700.	0.0	-1.5	-1.5	-1.5	-1.5	-1.5	0.0	0.0000
100.	3490.	-.4	0.0	.2	.7	1.3	2.0	.4	.0836
200.	3289.	.7	-1.4	-1.0	0.0	1.2	2.6	1.8	.1724
300.	3095.	2.7	-6.0	-5.4	-3.9	-2.1	0.0	4.1	.2665
400.	2909.	5.7	-14.2	-13.4	-11.5	-9.1	-6.3	7.4	.3664
500.	2731.	10.0	-26.5	-25.5	-23.1	-20.1	-16.6	11.8	.4727
600.	2560.	15.7	-43.5	-42.3	-39.4	-35.8	-31.6	17.5	.5861
700.	2395.	23.3	-65.9	-64.4	-61.1	-56.8	-51.9	24.6	.7075
800.	2236.	32.9	-94.2	-92.5	-88.7	-83.8	-78.2	33.2	.8371
900.	2082.	45.0	-129.3	-127.4	-123.1	-117.6	-111.3	43.3	.9759
1000.	1934.	60.1	-172.6	-170.5	-165.7	-159.6	-152.6	55.4	1.1257

BALLISTIC COEFFICIENT .53, MUZZLE VELOCITY FT/SEC 3800.

RANGE	VELOCITY	HEIGHT	100 YD	150 YD	200 YD	250 YD	300 YD	DRIFT	TIME
0.	3800.	0.0	-1.5	-1.5	-1.5	-1.5	-1.5	0.0	0.0000
100.	3586.	-.4	0.0	.2	.6	1.2	1.8	.5	.0817
200.	3381.	.6	-1.2	-.9	0.0	1.2	2.4	1.7	.1676
300.	3184.	2.5	-5.4	-5.0	-3.7	-1.9	0.0	3.9	.2590
400.	2994.	5.3	-13.1	-12.5	-10.8	-8.4	-5.9	7.1	.3561
500.	2813.	9.4	-24.8	-24.0	-21.8	-18.9	-15.7	11.5	.4598
600.	2638.	14.8	-40.7	-39.8	-37.1	-33.7	-29.8	16.9	.5698
700.	2471.	21.9	-61.6	-60.5	-57.5	-53.4	-48.9	23.7	.6872
800.	2309.	31.0	-88.3	-87.0	-83.5	-78.9	-73.8	31.9	.8130
900.	2153.	42.3	-121.3	-119.9	-116.0	-110.7	-105.0	41.7	.9475
1000.	2002.	56.5	-161.9	-160.3	-155.9	-150.1	-143.7	53.3	1.0920

BALLISTIC COEFFICIENT .53, MUZZLE VELOCITY FT/SEC 3900.

RANGE	VELOCITY	HEIGHT	100 YD	150 YD	200 YD	250 YD	300 YD	DRIFT	TIME
0.	3900.	0.0	-1.5	-1.5	-1.5	-1.5	-1.5	0.0	0.0000
100.	3683.	-.5	0.0	.1	.5	1.1	1.7	.4	.0790
200.	3474.	.5	-1.1	-.8	0.0	1.1	2.3	1.6	.1628
300.	3273.	2.3	-5.0	-4.6	-3.4	-1.8	0.0	3.7	.2518
400.	3080.	5.0	-12.3	-11.7	-10.1	-8.0	-5.5	6.8	.3463
500.	2894.	8.8	-23.1	-22.5	-20.5	-17.8	-14.7	10.9	.4468
600.	2717.	14.0	-38.2	-37.4	-35.0	-31.8	-28.1	16.2	.5538
700.	2546.	20.7	-57.9	-57.0	-54.2	-50.5	-46.2	22.8	.6680
800.	2382.	29.2	-83.0	-81.9	-78.7	-74.4	-69.5	30.7	.7897
900.	2223.	39.9	-114.1	-112.9	-109.3	-104.5	-99.0	40.1	.9200
1000.	2070.	53.2	-152.2	-150.8	-146.8	-141.5	-135.4	51.1	1.0598

BALLISTIC COEFFICIENT .53, MUZZLE VELOCITY FT/SEC 4000.

RANGE	VELOCITY	HEIGHT	100 YD	150 YD	200 YD	250 YD	300 YD	DRIFT	TIME
0.	4000.	0.0	-1.5	-1.5	-1.5	-1.5	-1.5	0.0	0.0000
100.	3779.	-.5	0.0	.1	.5	1.0	1.5	.4	.0773
200.	3566.	.5	-1.0	-.8	0.0	1.0	2.1	1.6	.1592
300.	3362.	2.1	-4.6	-4.3	-3.2	-1.7	0.0	3.6	.2457
400.	3165.	4.7	-11.4	-11.0	-9.5	-7.5	-5.2	6.6	.3377
500.	2976.	8.4	-21.7	-21.2	-19.3	-16.8	-13.9	10.6	.4355
600.	2795.	13.2	-35.9	-35.3	-33.0	-30.0	-26.6	15.8	.5396
700.	2622.	19.5	-54.4	-53.7	-51.1	-47.6	-43.6	22.0	.6502
800.	2455.	27.6	-78.1	-77.3	-74.2	-70.3	-65.7	29.7	.7685
900.	2294.	37.7	-107.5	-106.5	-103.2	-98.7	-93.5	38.7	.8950
1000.	2138.	50.2	-143.4	-142.4	-138.6	-133.7	-127.9	49.4	1.0306

BALLISTIC COEFFICIENT .54, MUZZLE VELOCITY FT/SEC 2000.

RANGE	VELOCITY	HEIGHT	100 YD	150 YD	200 YD	250 YD	300 YD	DRIFT	TIME
0.	2000.	0.0	-1.5	-1.5	-1.5	-1.5	-1.5	0.0	0.0000
100.	1859.	.4	0.0	2.0	4.3	6.9	9.7	.9	.1553
200.	1726.	4.3	-8.6	-4.6	0.0	5.2	10.7	4.0	.3228
300.	1600.	11.4	-29.0	-23.1	-16.1	-8.3	0.0	9.5	.5037
400.	1482.	22.7	-62.9	-55.0	-45.7	-35.3	-24.2	17.3	.6982
500.	1375.	38.9	-112.8	-102.9	-91.3	-78.4	-64.5	28.0	.9091
600.	1281.	61.1	-181.0	-169.1	-155.2	-139.6	-123.0	41.3	1.1349
700.	1200.	90.3	-270.4	-256.6	-240.3	-222.1	-202.7	57.6	1.3772
800.	1133.	127.5	-384.0	-368.1	-349.5	-328.8	-306.6	76.5	1.6348
900.	1082.	173.9	-525.2	-507.4	-486.5	-463.2	-438.2	98.1	1.9072
1000.	1042.	229.4	-694.2	-674.4	-651.2	-625.3	-597.6	121.4	2.1895

BALLISTIC COEFFICIENT .54, MUZZLE VELOCITY FT/SEC 2100.

RANGE	VELOCITY	HEIGHT	100 YD	150 YD	200 YD	250 YD	300 YD	DRIFT	TIME
0.	2100.	0.0	-1.5	-1.5	-1.5	-1.5	-1.5	0.0	0.0000
100.	1954.	.3	0.0	1.7	3.8	6.1	8.6	.9	.1481
200.	1815.	3.8	-7.6	-4.2	0.0	4.6	9.6	3.8	.3071
300.	1685.	10.3	-25.9	-20.7	-14.4	-7.5	0.0	8.8	.4787
400.	1562.	20.4	-56.4	-49.6	-41.2	-31.9	-21.9	16.3	.6640
500.	1446.	35.0	-101.2	-92.6	-82.2	-70.6	-58.1	26.3	.8636
600.	1343.	55.1	-162.6	-152.3	-139.8	-125.9	-110.9	39.0	1.0789
700.	1254.	81.7	-243.5	-231.5	-216.9	-200.7	-183.2	54.7	1.3106
800.	1176.	115.9	-346.6	-332.9	-316.2	-297.6	-277.7	73.2	1.5586
900.	1115.	158.2	-474.1	-458.7	-439.9	-419.0	-396.6	94.0	1.8197
1000.	1069.	209.9	-629.7	-612.5	-591.6	-568.4	-543.5	117.2	2.0947

BALLISTIC COEFFICIENT .54, MUZZLE VELOCITY FT/SEC 2200.

RANGE	VELOCITY	HEIGHT	100 YD	150 YD	200 YD	250 YD	300 YD	DRIFT	TIME
0.	2200.	0.0	-1.5	-1.5	-1.5	-1.5	-1.5	0.0	0.0000
100.	2050.	.2	0.0	1.5	3.4	5.5	7.7	.9	.1415
200.	1906.	3.4	-6.8	-3.8	0.0	4.2	8.7	3.6	.2932
300.	1771.	9.2	-23.2	-18.7	-13.0	-6.7	0.0	8.3	.4564
400.	1643.	18.4	-50.7	-44.7	-37.1	-28.8	-19.8	15.3	.6324
500.	1522.	31.7	-91.1	-83.6	-74.2	-63.7	-52.5	24.7	.8221
600.	1410.	49.9	-146.5	-137.5	-126.1	-113.6	-100.1	36.7	1.0269
700.	1312.	74.0	-219.5	-209.0	-195.7	-181.1	-165.4	51.6	1.2476
800.	1227.	105.0	-312.7	-300.7	-285.5	-268.8	-250.8	69.2	1.4843
900.	1153.	144.5	-430.2	-416.7	-399.7	-380.8	-360.7	90.1	1.7394
1000.	1098.	192.0	-571.5	-556.4	-537.5	-516.5	-494.1	112.7	2.0037

BALLISTIC COEFFICIENT .54, MUZZLE VELOCITY FT/SEC 2300.

RANGE	VELOCITY	HEIGHT	100 YD	150 YD	200 YD	250 YD	300 YD	DRIFT	TIME
0.	2300.	0.0	-1.5	-1.5	-1.5	-1.5	-1.5	0.0	0.0000
100.	2147.	.1	0.0	1.3	3.0	4.9	6.9	.8	.1350
200.	1999.	3.0	-6.1	-3.4	0.0	3.8	7.8	3.3	.2798
300.	1858.	8.3	-20.8	-16.8	-11.7	-6.1	0.0	7.7	.4352
400.	1725.	16.7	-45.7	-40.4	-33.6	-26.0	-18.0	14.3	.6028
500.	1599.	28.7	-82.5	-75.8	-67.3	-57.8	-47.7	23.2	.7838
600.	1482.	45.2	-132.5	-124.5	-114.3	-102.9	-90.8	34.5	.9784
700.	1374.	67.1	-198.6	-189.2	-177.4	-164.1	-150.0	48.6	1.1891
800.	1281.	95.4	-283.2	-272.4	-258.9	-243.7	-227.6	65.4	1.4153
900.	1199.	131.2	-388.9	-376.9	-361.6	-344.6	-326.4	85.2	1.6578
1000.	1133.	175.4	-518.9	-505.5	-488.6	-469.6	-449.4	107.6	1.9155

BALLISTIC COEFFICIENT .54, MUZZLE VELOCITY FT/SEC 2400.

RANGE	VELOCITY	HEIGHT	100 YD	150 YD	200 YD	250 YD	300 YD	DRIFT	TIME
0.	2400.	0.0	-1.5	-1.5	-1.5	-1.5	-1.5	0.0	0.0000
100.	2243.	.1	0.0	1.2	2.7	4.4	6.3	.8	.1294
200.	2092.	2.7	-5.4	-3.1	0.0	3.4	7.1	3.1	.2677
300.	1946.	7.6	-18.8	-15.3	-10.7	-5.6	0.0	7.3	.4165
400.	1808.	15.2	-41.4	-36.7	-30.6	-23.7	-16.3	13.4	.5762
500.	1678.	26.1	-74.7	-68.8	-61.2	-52.6	-43.3	21.7	.7485
600.	1556.	41.2	-120.3	-113.3	-104.1	-93.8	-82.6	32.4	.9343
700.	1440.	61.1	-180.3	-172.0	-161.3	-149.3	-136.3	45.7	1.1348
800.	1338.	86.9	-257.0	-247.6	-235.3	-221.6	-206.8	61.8	1.3510
900.	1250.	119.5	-353.3	-342.7	-328.9	-313.4	-296.7	80.6	1.5831
1000.	1172.	160.6	-473.0	-461.2	-445.9	-428.8	-410.2	102.7	1.8336

BALLISTIC COEFFICIENT .54, MUZZLE VELOCITY FT/SEC 2500.

RANGE	VELOCITY	HEIGHT	100 YD	150 YD	200 YD	250 YD	300 YD	DRIFT	TIME
0.	2500.	0.0	-1.5	-1.5	-1.5	-1.5	-1.5	0.0	0.0000
100.	2340.	-.0	0.0	1.1	2.4	4.0	5.7	.7	.1241
200.	2186.	2.4	-4.8	-2.7	0.0	3.1	6.5	2.9	.2564
300.	2036.	6.9	-17.1	-13.9	-9.8	-5.1	0.0	6.9	.3990
400.	1893.	13.8	-37.6	-33.4	-27.9	-21.7	-14.9	12.6	.5515
500.	1758.	23.8	-67.9	-62.6	-55.8	-48.0	-39.5	20.4	.7159
600.	1631.	37.5	-109.5	-103.1	-95.0	-85.6	-75.3	30.5	.8932
700.	1511.	55.7	-164.0	-156.6	-147.1	-136.1	-124.2	43.0	1.0843
800.	1400.	79.2	-233.8	-225.3	-214.4	-201.9	-188.3	58.2	1.2906
900.	1303.	109.1	-321.5	-312.0	-299.7	-285.6	-270.3	76.2	1.5128
1000.	1219.	146.4	-429.7	-419.1	-405.5	-389.8	-372.8	97.0	1.7510

BALLISTIC COEFFICIENT .54, MUZZLE VELOCITY FT/SEC 2600.

RANGE	VELOCITY	HEIGHT	100 YD	150 YD	200 YD	250 YD	300 YD	DRIFT	TIME
0.	2600.	0.0	-1.5	-1.5	-1.5	-1.5	-1.5	0.0	0.0000
100.	2437.	-.1	0.0	1.0	2.2	3.6	5.2	.6	.1190
200.	2279.	2.2	-4.4	-2.5	0.0	2.8	5.9	2.8	.2464
300.	2127.	6.3	-15.5	-12.6	-8.9	-4.6	0.0	6.4	.3827
400.	1979.	12.7	-34.3	-30.5	-25.5	-19.8	-13.6	11.8	.5287
500.	1839.	21.9	-62.1	-57.3	-51.1	-44.0	-36.3	19.3	.6864
600.	1708.	34.4	-100.0	-94.3	-86.8	-78.3	-69.0	28.7	.8553
700.	1583.	51.0	-150.0	-143.3	-134.6	-124.7	-113.8	40.6	1.0382
800.	1466.	72.4	-213.6	-206.0	-196.0	-184.6	-172.3	54.9	1.2349
900.	1361.	99.8	-293.6	-285.0	-273.8	-261.0	-247.1	72.0	1.4476
1000.	1269.	134.1	-392.4	-382.9	-370.5	-356.3	-340.8	91.9	1.6759

BALLISTIC COEFFICIENT .54, MUZZLE VELOCITY FT/SEC 2700.

RANGE	VELOCITY	HEIGHT	100 YD	150 YD	200 YD	250 YD	300 YD	DRIFT	TIME
0.	2700.	0.0	-1.5	-1.5	-1.5	-1.5	-1.5	0.0	0.0000
100.	2533.	-.1	0.0	.8	2.0	3.3	4.7	.7	.1149
200.	2373.	1.9	-3.9	-2.3	0.0	2.6	5.4	2.6	.2372
300.	2217.	5.7	-14.0	-11.6	-8.2	-4.2	0.0	6.1	.3678
400.	2067.	11.6	-31.3	-28.0	-23.4	-18.2	-12.6	11.2	.5081
500.	1922.	20.1	-56.7	-52.7	-46.9	-40.4	-33.3	18.2	.6589
600.	1785.	31.6	-91.5	-86.6	-79.7	-71.9	-63.4	27.1	.8207
700.	1656.	46.8	-137.2	-131.5	-123.4	-114.3	-104.4	38.3	.9951
800.	1535.	66.5	-195.5	-188.9	-179.8	-169.3	-158.0	51.8	1.1832
900.	1422.	91.5	-268.4	-261.1	-250.7	-239.0	-226.3	68.0	1.3862
1000.	1322.	122.9	-358.9	-350.8	-339.3	-326.3	-312.1	87.0	1.6053

BALLISTIC COEFFICIENT .54, MUZZLE VELOCITY FT/SEC 2800.

RANGE	VELOCITY	HEIGHT	100 YD	150 YD	200 YD	250 YD	300 YD	DRIFT	TIME
0.	2800.	0.0	-1.5	-1.5	-1.5	-1.5	-1.5	0.0	0.0000
100.	2629.	-.2	0.0	.7	1.8	3.0	4.3	.5	.1102
200.	2465.	1.7	-3.5	-2.1	0.0	2.4	5.0	2.4	.2281
300.	2307.	5.3	-12.9	-10.6	-7.5	-3.9	0.0	5.7	.3540
400.	2153.	10.7	-28.7	-25.7	-21.6	-16.8	-11.6	10.6	.4886
500.	2005.	18.5	-52.1	-48.4	-43.2	-37.2	-30.6	17.1	.6330
600.	1864.	29.1	-84.0	-79.5	-73.4	-66.1	-58.3	25.5	.7879
700.	1730.	43.0	-126.0	-120.8	-113.6	-105.2	-96.0	36.1	.9551
800.	1605.	61.1	-179.7	-173.7	-165.5	-155.9	-145.4	49.0	1.1354
900.	1487.	84.1	-246.6	-239.9	-230.6	-219.8	-208.0	64.3	1.3294
1000.	1378.	113.0	-329.5	-322.1	-311.8	-299.7	-286.7	82.3	1.5393

BALLISTIC COEFFICIENT .54, MUZZLE VELOCITY FT/SEC 2900.

RANGE	VELOCITY	HEIGHT	100 YD	150 YD	200 YD	250 YD	300 YD	DRIFT	TIME
0.	2900.	0.0	-1.5	-1.5	-1.5	-1.5	-1.5	0.0	0.0000
100.	2725.	-.2	0.0	.6	1.6	2.7	3.9	.6	.1067
200.	2558.	1.6	-3.2	-1.9	0.0	2.2	4.6	2.4	.2203
300.	2396.	4.9	-11.7	-9.8	-7.0	-3.7	0.0	5.5	.3416
400.	2240.	9.9	-26.4	-23.8	-20.0	-15.6	-10.7	10.1	.4711
500.	2089.	17.1	-47.9	-44.6	-39.9	-34.4	-28.3	16.3	.6096
600.	1943.	26.9	-77.4	-73.5	-67.8	-61.3	-53.9	24.3	.7588
700.	1805.	39.8	-116.1	-111.5	-104.9	-97.2	-88.7	34.3	.9187
800.	1675.	56.4	-165.4	-160.3	-152.7	-143.9	-134.1	46.4	1.0914
900.	1553.	77.6	-227.2	-221.4	-212.8	-202.9	-192.0	61.0	1.2776
1000.	1438.	104.2	-303.3	-296.9	-287.4	-276.4	-264.2	78.2	1.4785

BALLISTIC COEFFICIENT .54, MUZZLE VELOCITY FT/SEC 3000.

RANGE	VELOCITY	HEIGHT	100 YD	150 YD	200 YD	250 YD	300 YD	DRIFT	TIME
0.	3000.	0.0	-1.5	-1.5	-1.5	-1.5	-1.5	0.0	0.0000
100.	2821.	-.2	0.0	.6	1.4	2.5	3.6	.6	.1034
200.	2650.	1.4	-2.9	-1.7	0.0	2.0	4.3	2.3	.2131
300.	2485.	4.5	-10.7	-9.0	-6.4	-3.4	0.0	5.3	.3299
400.	2326.	9.2	-24.2	-22.0	-18.5	-14.4	-9.9	9.7	.4548
500.	2172.	15.9	-44.1	-41.3	-36.9	-31.8	-26.3	15.5	.5881
600.	2023.	24.9	-71.5	-68.1	-62.8	-56.7	-50.0	23.1	.7315
700.	1880.	36.9	-107.2	-103.2	-97.1	-90.0	-82.1	32.6	.8851
800.	1746.	52.2	-152.8	-148.2	-141.3	-133.1	-124.2	44.1	1.0508
900.	1620.	71.8	-209.7	-204.6	-196.8	-187.6	-177.5	57.9	1.2291
1000.	1501.	96.2	-279.9	-274.2	-265.5	-255.3	-244.1	74.2	1.4215

BALLISTIC COEFFICIENT .54, MUZZLE VELOCITY FT/SEC 3100.

RANGE	VELOCITY	HEIGHT	100 YD	150 YD	200 YD	250 YD	300 YD	DRIFT	TIME
0.	3100.	0.0	-1.5	-1.5	-1.5	-1.5	-1.5	0.0	0.0000
100.	2917.	-.3	0.0	.5	1.3	2.2	3.3	.5	.0998
200.	2742.	1.3	-2.6	-1.6	0.0	1.9	4.0	2.1	.2056
300.	2574.	4.1	-9.8	-8.3	-6.0	-3.1	0.0	5.0	.3186
400.	2412.	8.5	-22.3	-20.4	-17.2	-13.4	-9.2	9.2	.4391
500.	2255.	14.7	-40.8	-38.4	-34.4	-29.7	-24.5	14.8	.5678
600.	2103.	23.1	-66.1	-63.2	-58.4	-52.7	-46.5	22.0	.7056
700.	1957.	34.2	-99.3	-95.8	-90.3	-83.7	-76.4	31.0	.8535
800.	1818.	48.4	-141.4	-137.4	-131.1	-123.5	-115.2	41.9	1.0122
900.	1687.	66.5	-194.1	-189.7	-182.5	-174.0	-164.7	55.0	1.1837
1000.	1564.	89.2	-259.1	-254.2	-246.2	-236.8	-226.4	70.6	1.3686

BALLISTIC COEFFICIENT .54, MUZZLE VELOCITY FT/SEC 3200.

RANGE	VELOCITY	HEIGHT	100 YD	150 YD	200 YD	250 YD	300 YD	DRIFT	TIME
0.	3200.	0.0	-1.5	-1.5	-1.5	-1.5	-1.5	0.0	0.0000
100.	3013.	-.3	0.0	.5	1.2	2.1	3.0	.5	.0965
200.	2834.	1.2	-2.3	-1.4	0.0	1.8	3.7	2.0	.1991
300.	2662.	3.8	-9.0	-7.7	-5.5	-2.9	0.0	4.8	.3085
400.	2497.	7.9	-20.7	-18.9	-16.0	-12.5	-8.6	8.8	.4250
500.	2337.	13.7	-37.9	-35.6	-32.0	-27.6	-22.8	14.2	.5492
600.	2183.	21.6	-61.4	-58.6	-54.3	-49.1	-43.3	21.0	.6817
700.	2034.	31.9	-92.3	-89.1	-84.1	-77.9	-71.2	29.6	.8245
800.	1891.	45.1	-131.4	-127.7	-122.0	-115.0	-107.3	40.0	.9772
900.	1756.	61.8	-180.3	-176.2	-169.7	-161.8	-153.1	52.5	1.1418
1000.	1629.	82.8	-240.4	-235.9	-228.7	-219.9	-210.3	67.2	1.3193

BALLISTIC COEFFICIENT .54, MUZZLE VELOCITY FT/SEC 3300.

RANGE	VELOCITY	HEIGHT	100 YD	150 YD	200 YD	250 YD	300 YD	DRIFT	TIME
0.	3300.	0.0	-1.5	-1.5	-1.5	-1.5	-1.5	0.0	0.0000
100.	3109.	-.3	0.0	.4	1.0	1.8	2.8	.5	.0940
200.	2926.	1.0	-2.1	-1.4	0.0	1.6	3.4	2.0	.1933
300.	2751.	3.5	-8.3	-7.2	-5.1	-2.7	0.0	4.6	.2990
400.	2582.	7.4	-19.1	-17.6	-14.9	-11.7	-8.0	8.4	.4116
500.	2420.	12.8	-35.1	-33.2	-29.8	-25.8	-21.3	13.6	.5316
600.	2262.	20.2	-57.0	-54.8	-50.7	-45.9	-40.5	20.1	.6599
700.	2111.	29.7	-85.7	-83.1	-78.3	-72.7	-66.4	28.3	.7971
800.	1964.	42.1	-122.2	-119.2	-113.8	-107.4	-100.1	38.2	.9446
900.	1825.	57.6	-167.6	-164.2	-158.1	-150.9	-142.8	50.1	1.1029
1000.	1694.	77.1	-223.4	-219.7	-212.9	-205.0	-195.8	64.1	1.2735

BALLISTIC COEFFICIENT .54, MUZZLE VELOCITY FT/SEC 3400.

RANGE	VELOCITY	HEIGHT	100 YD	150 YD	200 YD	250 YD	300 YD	DRIFT	TIME
0.	3400.	0.0	-1.5	-1.5	-1.5	-1.5	-1.5	0.0	0.0000
100.	3206.	-.4	0.0	.3	.9	1.7	2.5	.4	.0907
200.	3019.	.9	-1.9	-1.2	0.0	1.5	3.2	1.9	.1870
300.	2839.	3.3	-7.6	-6.6	-4.8	-2.5	0.0	4.3	.2894
400.	2667.	6.9	-17.7	-16.3	-13.9	-10.9	-7.5	8.0	.3986
500.	2502.	12.0	-32.6	-30.9	-27.9	-24.1	-20.0	13.0	.5150
600.	2342.	18.8	-53.1	-51.1	-47.5	-42.9	-38.0	19.3	.6390
700.	2187.	27.8	-79.9	-77.5	-73.3	-67.9	-62.1	27.0	.7712
800.	2038.	39.3	-114.0	-111.3	-106.4	-100.4	-93.7	36.6	.9137
900.	1895.	53.8	-156.2	-153.2	-147.8	-140.9	-133.5	47.9	1.0661
1000.	1759.	71.9	-208.2	-204.8	-198.8	-191.2	-182.9	61.2	1.2303

BALLISTIC COEFFICIENT .54, MUZZLE VELOCITY FT/SEC 3500.

RANGE	VELOCITY	HEIGHT	100 YD	150 YD	200 YD	250 YD	300 YD	DRIFT	TIME
0.	3500.	0.0	-1.5	-1.5	-1.5	-1.5	-1.5	0.0	0.0000
100.	3302.	-.4	0.0	.3	.9	1.5	2.3	.5	.0883
200.	3111.	.8	-1.7	-1.1	0.0	1.4	3.0	1.9	.1823
300.	2928.	3.1	-7.0	-6.2	-4.4	-2.4	0.0	4.3	.2815
400.	2752.	6.4	-16.4	-15.3	-13.0	-10.2	-7.0	7.8	.3872
500.	2584.	11.2	-30.4	-28.9	-26.1	-22.6	-18.7	12.5	.4997
600.	2421.	17.7	-49.6	-47.8	-44.4	-40.3	-35.5	18.5	.6197
700.	2264.	26.1	-74.7	-72.6	-68.6	-63.8	-58.3	26.0	.7479
800.	2112.	36.8	-106.4	-104.1	-99.5	-94.0	-87.7	35.1	.8849
900.	1965.	50.4	-146.0	-143.4	-138.3	-132.1	-124.9	45.9	1.0323
1000.	1826.	67.3	-194.5	-191.7	-185.9	-179.0	-171.1	58.7	1.1906

BALLISTIC COEFFICIENT .54, MUZZLE VELOCITY FT/SEC 3600.

RANGE	VELOCITY	HEIGHT	100 YD	150 YD	200 YD	250 YD	300 YD	DRIFT	TIME
0.	3600.	0.0	-1.5	-1.5	-1.5	-1.5	-1.5	0.0	0.0000
100.	3398.	-.4	0.0	.2	.8	1.4	2.1	.4	.0857
200.	3203.	.7	-1.5	-1.0	0.0	1.3	2.8	1.7	.1765
300.	3017.	2.8	-6.4	-5.7	-4.1	-2.2	0.0	4.0	.2729
400.	2837.	6.0	-15.2	-14.2	-12.1	-9.5	-6.6	7.4	.3754
500.	2665.	10.5	-28.2	-27.1	-24.5	-21.2	-17.6	12.0	.4846
600.	2500.	16.6	-46.2	-44.9	-41.7	-37.8	-33.4	17.8	.6010
700.	2340.	24.5	-69.8	-68.1	-64.5	-59.8	-54.8	25.0	.7251
800.	2186.	34.5	-99.5	-97.6	-93.4	-88.1	-82.4	33.6	.8574
900.	2036.	47.3	-136.6	-134.5	-129.8	-123.9	-117.4	44.0	1.0000
1000.	1893.	63.0	-181.9	-179.6	-174.3	-167.7	-160.5	56.2	1.1525

BALLISTIC COEFFICIENT .54, MUZZLE VELOCITY FT/SEC 3700.

RANGE	VELOCITY	HEIGHT	100 YD	150 YD	200 YD	250 YD	300 YD	DRIFT	TIME
0.	3700.	0.0	-1.5	-1.5	-1.5	-1.5	-1.5	0.0	0.0000
100.	3494.	-.4	0.0	.2	.7	1.3	2.0	.4	.0836
200.	3296.	.7	-1.4	-1.0	0.0	1.2	2.6	1.7	.1721
300.	3106.	2.6	-6.0	-5.3	-3.9	-2.1	0.0	4.0	.2661
400.	2923.	5.7	-14.2	-13.3	-11.4	-9.0	-6.2	7.3	.3656
500.	2747.	9.9	-26.4	-25.4	-23.0	-20.0	-16.5	11.6	.4713
600.	2579.	15.6	-43.3	-42.0	-39.2	-35.5	-31.4	17.2	.5841
700.	2416.	23.1	-65.4	-63.9	-60.6	-56.3	-51.5	24.1	.7043
800.	2259.	32.5	-93.4	-91.7	-87.9	-83.1	-77.5	32.4	.8328
900.	2107.	44.4	-128.1	-126.2	-122.0	-116.5	-110.2	42.3	.9702
1000.	1961.	59.2	-170.6	-168.6	-163.8	-157.7	-150.8	54.0	1.1178

BALLISTIC COEFFICIENT .54, MUZZLE VELOCITY FT/SEC 3800.

RANGE	VELOCITY	HEIGHT	100 YD	150 YD	200 YD	250 YD	300 YD	DRIFT	TIME
0.	3800.	0.0	-1.5	-1.5	-1.5	-1.5	-1.5	0.0	0.0000
100.	3590.	-.4	0.0	.2	.6	1.2	1.8	.5	.0817
200.	3389.	.6	-1.2	-.9	0.0	1.1	2.4	1.7	.1675
300.	3195.	2.5	-5.4	-5.0	-3.6	-1.9	0.0	3.8	.2586
400.	3008.	5.3	-13.1	-12.5	-10.7	-8.4	-5.8	6.9	.3553
500.	2829.	9.3	-24.6	-23.8	-21.6	-18.8	-15.5	11.2	.4582
600.	2657.	14.7	-40.4	-39.5	-36.9	-33.4	-29.6	16.6	.5677
700.	2492.	21.7	-61.2	-60.1	-57.0	-53.0	-48.5	23.2	.6844
800.	2333.	30.7	-87.5	-86.3	-82.8	-78.2	-73.1	31.2	.8088
900.	2178.	41.8	-120.1	-118.7	-114.7	-109.6	-103.8	40.7	.9416
1000.	2029.	55.7	-160.1	-158.6	-154.1	-148.4	-142.0	52.0	1.0847

BALLISTIC COEFFICIENT .54, MUZZLE VELOCITY FT/SEC 3900.

RANGE	VELOCITY	HEIGHT	100 YD	150 YD	200 YD	250 YD	300 YD	DRIFT	TIME
0.	3900.	0.0	-1.5	-1.5	-1.5	-1.5	-1.5	0.0	0.0000
100.	3687.	-.5	0.0	.1	.5	1.1	1.7	.3	.0789
200.	3481.	.5	-1.1	-.8	0.0	1.1	2.3	1.5	.1626
300.	3284.	2.3	-5.0	-4.6	-3.4	-1.8	0.0	3.7	.2515
400.	3094.	5.0	-12.2	-11.7	-10.1	-8.0	-5.5	6.7	.3456
500.	2911.	8.8	-23.0	-22.4	-20.4	-17.7	-14.7	10.7	.4456
600.	2736.	13.9	-37.9	-37.1	-34.7	-31.6	-27.9	15.9	.5517
700.	2568.	20.5	-57.5	-56.5	-53.8	-50.1	-45.7	22.3	.6649
800.	2406.	28.9	-82.3	-81.2	-78.0	-73.8	-68.9	30.0	.7858
900.	2249.	39.4	-113.0	-111.8	-108.3	-103.5	-98.0	39.2	.9148
1000.	2098.	52.5	-150.6	-149.2	-145.3	-140.0	-133.8	49.9	1.0529

BALLISTIC COEFFICIENT .54, MUZZLE VELOCITY FT/SEC 4000.

RANGE	VELOCITY	HEIGHT	100 YD	150 YD	200 YD	250 YD	300 YD	DRIFT	TIME
0.	4000.	0.0	-1.5	-1.5	-1.5	-1.5	-1.5	0.0	0.0000
100.	3783.	-.5	0.0	.1	.5	1.0	1.5	.4	.0772
200.	3574.	.5	-1.0	-.7	0.0	1.0	2.1	1.6	.1591
300.	3373.	2.1	-4.6	-4.3	-3.2	-1.7	0.0	3.6	.2453
400.	3179.	4.7	-11.3	-10.9	-9.4	-7.5	-5.2	6.5	.3368
500.	2993.	8.3	-21.5	-21.0	-19.1	-16.7	-13.8	10.4	.4341
600.	2815.	13.1	-35.6	-35.0	-32.8	-29.8	-26.4	15.4	.5377
700.	2644.	19.4	-54.0	-53.3	-50.7	-47.3	-43.3	21.6	.6475
800.	2479.	27.3	-77.4	-76.6	-73.6	-69.6	-65.1	29.0	.7646
900.	2320.	37.3	-106.4	-105.5	-102.1	-97.7	-92.6	37.8	.8898
1000.	2166.	49.5	-141.7	-140.7	-137.0	-132.1	-126.4	48.1	1.0235

BALLISTIC COEFFICIENT .55, MUZZLE VELOCITY FT/SEC 2000.

RANGE	VELOCITY	HEIGHT	100 YD	150 YD	200 YD	250 YD	300 YD	DRIFT	TIME
0.	2000.	0.0	-1.5	-1.5	-1.5	-1.5	-1.5	0.0	0.0000
100.	1861.	.4	0.0	2.0	4.3	6.9	9.6	.9	.1552
200.	1730.	4.2	-8.6	-4.6	0.0	5.1	10.7	4.0	.3225
300.	1607.	11.4	-28.9	-23.0	-16.0	-8.3	0.0	9.3	.5026
400.	1491.	22.5	-62.6	-54.7	-45.4	-35.1	-24.1	16.9	.6962
500.	1384.	38.6	-112.0	-102.2	-90.5	-77.7	-63.9	27.3	.9053
600.	1291.	60.5	-179.7	-167.8	-153.9	-138.5	-121.9	40.5	1.1300
700.	1210.	89.4	-268.1	-254.3	-238.1	-220.1	-200.8	56.4	1.3703
800.	1142.	126.1	-380.3	-364.5	-345.9	-325.3	-303.3	74.9	1.6258
900.	1090.	171.7	-519.3	-501.6	-480.7	-457.5	-432.7	96.0	1.8953
1000.	1048.	227.5	-689.7	-669.9	-646.7	-621.0	-593.4	119.8	2.1808

BALLISTIC COEFFICIENT .55, MUZZLE VELOCITY FT/SEC 2100.

RANGE	VELOCITY	HEIGHT	100 YD	150 YD	200 YD	250 YD	300 YD	DRIFT	TIME
0.	2100.	0.0	-1.5	-1.5	-1.5	-1.5	-1.5	0.0	0.0000
100.	1956.	.3	0.0	1.7	3.8	6.1	8.6	.9	.1480
200.	1820.	3.8	-7.6	-4.2	0.0	4.6	9.6	3.7	.3066
300.	1692.	10.2	-25.7	-20.6	-14.4	-7.4	0.0	8.6	.4777
400.	1571.	20.3	-56.1	-49.3	-41.0	-31.8	-21.8	16.0	.6621
500.	1456.	34.8	-100.5	-92.0	-81.6	-70.0	-57.6	25.7	.8604
600.	1354.	54.7	-161.6	-151.3	-138.8	-125.0	-110.1	38.3	1.0746
700.	1265.	80.8	-241.2	-229.2	-214.7	-198.5	-181.1	53.4	1.3034
800.	1187.	114.4	-342.8	-329.1	-312.4	-294.0	-274.1	71.4	1.5486
900.	1124.	156.3	-469.1	-453.7	-435.0	-414.2	-391.9	92.0	1.8087
1000.	1076.	207.2	-622.6	-605.5	-584.7	-561.6	-536.8	114.9	2.0816

BALLISTIC COEFFICIENT .55, MUZZLE VELOCITY FT/SEC 2200.

RANGE	VELOCITY	HEIGHT	100 YD	150 YD	200 YD	250 YD	300 YD	DRIFT	TIME
0.	2200.	0.0	-1.5	-1.5	-1.5	-1.5	-1.5	0.0	0.0000
100.	2053.	.2	0.0	1.5	3.4	5.5	7.7	.9	.1414
200.	1911.	3.4	-6.8	-3.8	0.0	4.2	8.6	3.5	.2929
300.	1778.	9.2	-23.1	-18.6	-12.9	-6.7	0.0	8.2	.4556
400.	1652.	18.3	-50.5	-44.5	-36.9	-28.6	-19.7	15.0	.6306
500.	1533.	31.5	-90.6	-83.1	-73.6	-63.2	-52.1	24.2	.8191
600.	1422.	49.4	-145.4	-136.4	-125.1	-112.6	-99.2	35.9	1.0222
700.	1324.	73.2	-217.6	-207.1	-193.9	-179.3	-163.7	50.5	1.2412
800.	1238.	103.8	-309.8	-297.8	-282.7	-266.0	-248.2	67.8	1.4761
900.	1164.	142.2	-424.3	-410.8	-393.8	-375.0	-354.9	87.8	1.7259
1000.	1108.	189.4	-564.9	-549.9	-531.0	-510.2	-487.8	110.3	1.9906

BALLISTIC COEFFICIENT .55, MUZZLE VELOCITY FT/SEC 2300.

RANGE	VELOCITY	HEIGHT	100 YD	150 YD	200 YD	250 YD	300 YD	DRIFT	TIME
0.	2300.	0.0	-1.5	-1.5	-1.5	-1.5	-1.5	0.0	0.0000
100.	2150.	.1	0.0	1.3	3.0	4.9	6.9	.8	.1349
200.	2004.	3.0	-6.1	-3.4	0.0	3.8	7.8	3.3	.2795
300.	1865.	8.3	-20.8	-16.8	-11.7	-6.0	0.0	7.6	.4343
400.	1734.	16.6	-45.6	-40.2	-33.4	-25.9	-17.9	14.0	.6013
500.	1611.	28.5	-81.9	-75.3	-66.8	-57.4	-47.3	22.7	.7809
600.	1494.	44.8	-131.6	-123.6	-113.4	-102.1	-90.0	33.7	.9741
700.	1387.	66.4	-196.9	-187.6	-175.7	-162.5	-148.5	47.5	1.1828
800.	1293.	94.3	-280.6	-270.0	-256.4	-241.3	-225.3	64.1	1.4076
900.	1212.	129.4	-384.6	-372.7	-357.4	-340.4	-322.4	83.2	1.6469
1000.	1143.	172.9	-512.6	-499.2	-482.3	-463.4	-443.3	105.2	1.9020

BALLISTIC COEFFICIENT .55, MUZZLE VELOCITY FT/SEC 2400.

RANGE	VELOCITY	HEIGHT	100 YD	150 YD	200 YD	250 YD	300 YD	DRIFT	TIME
0.	2400.	0.0	-1.5	-1.5	-1.5	-1.5	-1.5	0.0	0.0000
100.	2246.	.1	0.0	1.2	2.7	4.4	6.3	.8	.1293
200.	2098.	2.7	-5.4	-3.1	0.0	3.4	7.1	3.1	.2674
300.	1954.	7.5	-18.8	-15.3	-10.7	-5.5	0.0	7.2	.4156
400.	1818.	15.1	-41.2	-36.5	-30.4	-23.6	-16.2	13.1	.5745
500.	1690.	25.9	-74.2	-68.4	-60.7	-52.2	-43.0	21.3	.7458
600.	1569.	40.8	-119.5	-112.5	-103.3	-93.1	-82.0	31.8	.9304
700.	1455.	60.4	-178.7	-170.5	-159.8	-147.9	-134.9	44.7	1.1289
800.	1352.	85.8	-254.5	-245.1	-232.9	-219.2	-204.5	60.4	1.3430
900.	1263.	117.9	-349.3	-338.7	-324.9	-309.6	-292.9	78.7	1.5724
1000.	1186.	157.9	-465.9	-454.2	-438.9	-421.8	-403.4	100.0	1.8179

BALLISTIC COEFFICIENT .55, MUZZLE VELOCITY FT/SEC 2500.

RANGE	VELOCITY	HEIGHT	100 YD	150 YD	200 YD	250 YD	300 YD	DRIFT	TIME
0.	2500.	0.0	-1.5	-1.5	-1.5	-1.5	-1.5	0.0	0.0000
100.	2343.	-.0	0.0	1.0	2.4	4.0	5.7	.7	.1240
200.	2191.	2.4	-4.8	-2.7	0.0	3.1	6.5	2.8	.2561
300.	2044.	6.9	-17.0	-13.9	-9.8	-5.1	0.0	6.7	.3982
400.	1903.	13.8	-37.4	-33.2	-27.8	-21.6	-14.8	12.3	.5501
500.	1770.	23.7	-67.5	-62.3	-55.5	-47.7	-39.2	20.0	.7135
600.	1644.	37.2	-108.7	-102.4	-94.2	-84.9	-74.7	29.8	.8894
700.	1526.	55.1	-162.7	-155.3	-145.8	-134.9	-123.0	42.0	1.0788
800.	1415.	78.3	-231.6	-223.2	-212.3	-199.9	-186.3	56.9	1.2832
900.	1318.	107.7	-318.0	-308.5	-296.3	-282.3	-267.0	74.4	1.5029
1000.	1233.	144.4	-424.9	-414.4	-400.7	-385.2	-368.2	94.9	1.7390

BALLISTIC COEFFICIENT .55, MUZZLE VELOCITY FT/SEC 2600.

RANGE	VELOCITY	HEIGHT	100 YD	150 YD	200 YD	250 YD	300 YD	DRIFT	TIME
0.	2600.	0.0	-1.5	-1.5	-1.5	-1.5	-1.5	0.0	0.0000
100.	2440.	-.1	0.0	.9	2.2	3.6	5.2	.6	.1189
200.	2285.	2.2	-4.4	-2.5	0.0	2.8	5.9	2.7	.2462
300.	2135.	6.3	-15.5	-12.6	-8.9	-4.6	0.0	6.3	.3821
400.	1990.	12.6	-34.1	-30.3	-25.4	-19.7	-13.5	11.6	.5275
500.	1852.	21.7	-61.7	-56.9	-50.7	-43.6	-35.9	18.8	.6838
600.	1722.	34.1	-99.2	-93.5	-86.1	-77.6	-68.3	28.0	.8516
700.	1599.	50.5	-148.7	-142.1	-133.4	-123.5	-112.7	39.6	1.0329
800.	1483.	71.6	-211.5	-203.9	-194.0	-182.7	-170.3	53.6	1.2274
900.	1377.	98.5	-290.4	-281.8	-270.7	-257.9	-244.0	70.3	1.4379
1000.	1285.	132.1	-387.6	-378.1	-365.7	-351.6	-336.1	89.7	1.6634

BALLISTIC COEFFICIENT .55, MUZZLE VELOCITY FT/SEC 2700.

RANGE	VELOCITY	HEIGHT	100 YD	150 YD	200 YD	250 YD	300 YD	DRIFT	TIME
0.	2700.	0.0	-1.5	-1.5	-1.5	-1.5	-1.5	0.0	0.0000
100.	2536.	-.1	0.0	.8	2.0	3.3	4.7	.7	.1149
200.	2378.	1.9	-3.9	-2.3	0.0	2.6	5.4	2.6	.2368
300.	2225.	5.7	-14.0	-11.6	-8.1	-4.2	0.0	6.0	.3672
400.	2077.	11.6	-31.1	-27.8	-23.3	-18.1	-12.4	10.9	.5066
500.	1935.	19.9	-56.4	-52.3	-46.6	-40.1	-33.1	17.8	.6566
600.	1800.	31.3	-90.8	-85.9	-79.1	-71.3	-62.8	26.5	.8171
700.	1672.	46.3	-136.0	-130.4	-122.4	-113.3	-103.4	37.4	.9902
800.	1552.	65.7	-193.7	-187.2	-178.1	-167.6	-156.4	50.6	1.1765
900.	1439.	90.3	-265.7	-258.4	-248.1	-236.4	-223.7	66.4	1.3773
1000.	1339.	121.1	-354.6	-346.5	-335.1	-322.1	-308.0	84.9	1.5934

BALLISTIC COEFFICIENT .55, MUZZLE VELOCITY FT/SEC 2800.

RANGE	VELOCITY	HEIGHT	100 YD	150 YD	200 YD	250 YD	300 YD	DRIFT	TIME
0.	2800.	0.0	-1.5	-1.5	-1.5	-1.5	-1.5	0.0	0.0000
100.	2633.	-.2	0.0	.7	1.8	3.0	4.3	.5	.1102
200.	2471.	1.7	-3.5	-2.1	0.0	2.4	5.0	2.4	.2278
300.	2315.	5.2	-12.8	-10.6	-7.5	-3.9	0.0	5.6	.3533
400.	2164.	10.6	-28.5	-25.6	-21.5	-16.7	-11.5	10.3	.4872
500.	2018.	18.4	-51.8	-48.1	-42.9	-36.9	-30.4	16.8	.6309
600.	1879.	28.8	-83.4	-79.0	-72.8	-65.6	-57.8	25.0	.7847
700.	1747.	42.6	-125.0	-119.9	-112.7	-104.3	-95.2	35.3	.9505
800.	1623.	60.4	-177.9	-172.0	-163.8	-154.2	-143.8	47.8	1.1287
900.	1505.	83.0	-244.0	-237.3	-228.1	-217.3	-205.6	62.7	1.3206
1000.	1397.	111.3	-325.5	-318.2	-307.9	-295.9	-282.9	80.3	1.5279

BALLISTIC COEFFICIENT .55, MUZZLE VELOCITY FT/SEC 2900.

RANGE	VELOCITY	HEIGHT	100 YD	150 YD	200 YD	250 YD	300 YD	DRIFT	TIME
0.	2900.	0.0	-1.5	-1.5	-1.5	-1.5	-1.5	0.0	0.0000
100.	2728.	-.2	0.0	.6	1.6	2.7	3.9	.6	.1060
200.	2564.	1.6	-3.2	-1.9	0.0	2.2	4.6	2.3	.2200
300.	2405.	4.8	-11.7	-9.8	-6.9	-3.6	0.0	5.4	.3410
400.	2251.	9.9	-26.2	-23.7	-19.9	-15.5	-10.6	9.9	.4700
500.	2102.	17.0	-47.6	-44.4	-39.7	-34.2	-28.1	16.0	.6079
600.	1958.	26.7	-76.9	-73.0	-67.4	-60.8	-53.5	23.8	.7557
700.	1822.	39.4	-115.1	-110.6	-104.0	-96.3	-87.8	33.5	.9143
800.	1694.	55.8	-163.9	-158.7	-151.2	-142.4	-132.7	45.3	1.0851
900.	1572.	76.6	-224.9	-219.1	-210.6	-200.7	-189.8	59.5	1.2693
1000.	1458.	102.6	-299.7	-293.2	-283.8	-272.8	-260.7	76.2	1.4674

BALLISTIC COEFFICIENT .55, MUZZLE VELOCITY FT/SEC 3000.

RANGE	VELOCITY	HEIGHT	100 YD	150 YD	200 YD	250 YD	300 YD	DRIFT	TIME
0.	3000.	0.0	-1.5	-1.5	-1.5	-1.5	-1.5	0.0	0.0000
100.	2824.	-.2	0.0	.6	1.4	2.5	3.6	.6	.1033
200.	2656.	1.4	-2.9	-1.7	0.0	2.0	4.3	2.3	.2128
300.	2494.	4.5	-10.7	-9.0	-6.4	-3.3	0.0	5.2	.3295
400.	2337.	9.1	-24.2	-21.9	-18.4	-14.3	-9.9	9.5	.4538
500.	2186.	15.7	-43.9	-41.1	-36.7	-31.6	-26.0	15.2	.5862
600.	2039.	24.7	-71.1	-67.6	-62.4	-56.3	-49.6	22.7	.7288
700.	1898.	36.5	-106.4	-102.4	-96.3	-89.2	-81.4	31.9	.8810
800.	1765.	51.7	-151.4	-146.9	-139.9	-131.8	-122.9	43.1	1.0449
900.	1640.	70.8	-207.6	-202.5	-194.7	-185.6	-175.5	56.5	1.2213
1000.	1522.	94.8	-276.7	-271.0	-262.3	-252.2	-241.0	72.4	1.4113

BALLISTIC COEFFICIENT .55, MUZZLE VELOCITY FT/SEC 3100.

RANGE	VELOCITY	HEIGHT	100 YD	150 YD	200 YD	250 YD	300 YD	DRIFT	TIME
0.	3100.	0.0	-1.5	-1.5	-1.5	-1.5	-1.5	0.0	0.0000
100.	2921.	-.3	0.0	.5	1.3	2.2	3.3	.5	.0998
200.	2748.	1.3	-2.6	-1.6	0.0	1.9	3.9	2.1	.2055
300.	2583.	4.1	-9.8	-8.3	-5.9	-3.1	0.0	4.9	.3181
400.	2423.	8.5	-22.2	-20.3	-17.1	-13.3	-9.2	9.0	.4380
500.	2269.	14.6	-40.6	-38.1	-34.1	-29.5	-24.3	14.5	.5660
600.	2119.	22.9	-65.6	-62.7	-57.9	-52.3	-46.0	21.5	.7025
700.	1975.	33.9	-98.4	-95.0	-89.4	-82.9	-75.6	30.2	.8492
800.	1838.	47.9	-140.2	-136.4	-130.0	-122.5	-114.2	41.0	1.0071
900.	1708.	65.6	-192.1	-187.7	-180.5	-172.1	-162.7	53.7	1.1760
1000.	1586.	87.9	-256.1	-251.3	-243.3	-233.9	-223.5	68.8	1.3587

BALLISTIC COEFFICIENT .55, MUZZLE VELOCITY FT/SEC 3200.

RANGE	VELOCITY	HEIGHT	100 YD	150 YD	200 YD	250 YD	300 YD	DRIFT	TIME
0.	3200.	0.0	-1.5	-1.5	-1.5	-1.5	-1.5	0.0	0.0000
100.	3017.	-.3	0.0	.4	1.2	2.0	3.0	.5	.0964
200.	2841.	1.1	-2.3	-1.4	0.0	1.8	3.7	2.0	.1988
300.	2671.	3.8	-9.0	-7.7	-5.5	-2.9	0.0	4.7	.3079
400.	2509.	7.9	-20.6	-18.8	-15.9	-12.4	-8.6	8.6	.4241
500.	2352.	13.6	-37.7	-35.4	-31.9	-27.5	-22.7	13.9	.5476
600.	2199.	21.4	-61.0	-58.3	-54.0	-48.7	-43.0	20.5	.6792
700.	2052.	31.6	-91.6	-88.4	-83.4	-77.2	-70.6	28.9	.8207
800.	1911.	44.6	-130.3	-126.7	-121.0	-113.9	-106.3	39.1	.9722
900.	1777.	61.1	-178.5	-174.5	-168.1	-160.1	-151.5	51.2	1.1349
1000.	1651.	81.6	-237.7	-233.3	-226.1	-217.3	-207.7	65.6	1.3100

BALLISTIC COEFFICIENT .55, MUZZLE VELOCITY FT/SEC 3300.

RANGE	VELOCITY	HEIGHT	100 YD	150 YD	200 YD	250 YD	300 YD	DRIFT	TIME
0.	3300.	0.0	-1.5	-1.5	-1.5	-1.5	-1.5	0.0	0.0000
100.	3113.	-.3	0.0	.4	1.0	1.8	2.8	.5	.0940
200.	2933.	1.0	-2.1	-1.3	0.0	1.6	3.4	2.0	.1930
300.	2760.	3.5	-8.3	-7.1	-5.1	-2.7	0.0	4.6	.2986
400.	2594.	7.3	-19.0	-17.5	-14.8	-11.6	-8.0	8.3	.4106
500.	2434.	12.7	-34.8	-33.0	-29.6	-25.7	-21.1	13.3	.5299
600.	2279.	20.0	-56.6	-54.4	-50.4	-45.6	-40.1	19.7	.6573
700.	2130.	29.5	-85.0	-82.5	-77.8	-72.2	-65.8	27.7	.7935
800.	1985.	41.6	-121.0	-118.1	-112.7	-106.3	-99.0	37.3	.9393
900.	1847.	56.9	-166.0	-162.6	-156.6	-149.4	-141.2	48.9	1.0962
1000.	1717.	76.0	-220.8	-217.1	-210.4	-202.4	-193.3	62.5	1.2643

BALLISTIC COEFFICIENT .55, MUZZLE VELOCITY FT/SEC 3400.

RANGE	VELOCITY	HEIGHT	100 YD	150 YD	200 YD	250 YD	300 YD	DRIFT	TIME
0.	3400.	0.0	-1.5	-1.5	-1.5	-1.5	-1.5	0.0	0.0000
100.	3209.	-.4	0.0	.3	.9	1.7	2.5	.4	.0907
200.	3025.	.9	-1.9	-1.2	0.0	1.5	3.2	1.8	.1868
300.	2849.	3.3	-7.6	-6.5	-4.7	-2.5	0.0	4.3	.2889
400.	2679.	6.8	-17.6	-16.2	-13.8	-10.8	-7.5	7.9	.3976
500.	2517.	11.9	-32.5	-30.8	-27.8	-24.0	-19.9	12.7	.5135
600.	2359.	18.7	-52.8	-50.8	-47.2	-42.7	-37.7	18.9	.6366
700.	2207.	27.6	-79.3	-76.9	-72.7	-67.5	-61.6	26.4	.7679
800.	2059.	38.9	-113.0	-110.3	-105.5	-99.5	-92.8	35.7	.9088
900.	1918.	53.2	-154.8	-151.8	-146.4	-139.6	-132.1	46.8	1.0599
1000.	1783.	70.9	-206.0	-202.7	-196.7	-189.2	-180.8	59.8	1.2220

BALLISTIC COEFFICIENT .55, MUZZLE VELOCITY FT/SEC 3500.

RANGE	VELOCITY	HEIGHT	100 YD	150 YD	200 YD	250 YD	300 YD	DRIFT	TIME
0.	3500.	0.0	-1.5	-1.5	-1.5	-1.5	-1.5	0.0	0.0000
100.	3305.	-.4	0.0	.3	.9	1.5	2.3	.5	.0883
200.	3118.	.8	-1.7	-1.1	0.0	1.4	2.9	1.9	.1821
300.	2938.	3.0	-7.0	-6.1	-4.4	-2.3	0.0	4.2	.2808
400.	2765.	6.4	-16.3	-15.2	-12.9	-10.2	-7.0	7.7	.3863
500.	2599.	11.2	-30.2	-28.8	-25.9	-22.5	-18.6	12.3	.4982
600.	2438.	17.5	-49.2	-47.5	-44.0	-39.9	-35.3	18.1	.6172
700.	2283.	25.9	-74.1	-72.1	-68.1	-63.3	-57.8	25.4	.7445
800.	2134.	36.5	-105.6	-103.3	-98.7	-93.2	-87.0	34.3	.8805
900.	1989.	49.8	-144.5	-142.0	-136.8	-130.7	-123.6	44.8	1.0260
1000.	1851.	66.4	-192.4	-189.5	-183.8	-177.0	-169.2	57.3	1.1825

BALLISTIC COEFFICIENT .55, MUZZLE VELOCITY FT/SEC 3600.

RANGE	VELOCITY	HEIGHT	100 YD	150 YD	200 YD	250 YD	300 YD	DRIFT	TIME
0.	3600.	0.0	-1.5	-1.5	-1.5	-1.5	-1.5	0.0	0.0000
100.	3401.	-.4	0.0	.2	.8	1.4	2.1	.4	.0857
200.	3210.	.7	-1.5	-1.0	0.0	1.3	2.8	1.7	.1763
300.	3027.	2.8	-6.4	-5.7	-4.1	-2.2	0.0	3.9	.2724
400.	2850.	6.0	-15.1	-14.1	-12.1	-9.4	-6.6	7.2	.3745
500.	2681.	10.5	-28.1	-26.9	-24.3	-21.0	-17.4	11.7	.4831
600.	2518.	16.5	-46.0	-44.6	-41.4	-37.5	-33.2	17.4	.5990
700.	2360.	24.3	-69.2	-67.6	-64.0	-59.4	-54.3	24.4	.7220
800.	2208.	34.2	-98.7	-96.8	-92.6	-87.4	-81.6	32.8	.8532
900.	2060.	46.7	-135.3	-133.2	-128.5	-122.6	-116.1	43.0	.9940
1000.	1919.	62.2	-180.0	-177.7	-172.5	-165.9	-158.7	54.9	1.1451

BALLISTIC COEFFICIENT .55, MUZZLE VELOCITY FT/SEC 3700.

RANGE	VELOCITY	HEIGHT	100 YD	150 YD	200 YD	250 YD	300 YD	DRIFT	TIME
0.	3700.	0.0	-1.5	-1.5	-1.5	-1.5	-1.5	0.0	0.0000
100.	3498.	-.4	0.0	.2	.7	1.3	2.0	.4	.0835
200.	3303.	.7	-1.4	-.9	0.0	1.2	2.6	1.7	.1718
300.	3116.	2.6	-5.9	-5.3	-3.9	-2.1	0.0	3.9	.2657
400.	2936.	5.6	-14.0	-13.2	-11.3	-8.9	-6.1	7.1	.3645
500.	2763.	9.9	-26.3	-25.2	-22.9	-19.8	-16.4	11.4	.4701
600.	2597.	15.5	-43.0	-41.8	-38.9	-35.3	-31.1	16.8	.5820
700.	2437.	22.8	-64.8	-63.4	-60.1	-55.8	-51.0	23.5	.7011
800.	2282.	32.2	-92.5	-90.9	-87.1	-82.3	-76.7	31.6	.8285
900.	2132.	43.9	-126.9	-125.0	-120.8	-115.3	-109.1	41.3	.9646
1000.	1987.	58.4	-168.7	-166.6	-161.9	-155.9	-148.9	52.7	1.1101

BALLISTIC COEFFICIENT .55, MUZZLE VELOCITY FT/SEC 3800.

RANGE	VELOCITY	HEIGHT	100 YD	150 YD	200 YD	250 YD	300 YD	DRIFT	TIME
0.	3800.	0.0	-1.5	-1.5	-1.5	-1.5	-1.5	0.0	0.0000
100.	3594.	-.4	0.0	.2	.6	1.2	1.8	.5	.0816
200.	3396.	.6	-1.2	-.9	0.0	1.1	2.4	1.7	.1674
300.	3205.	2.5	-5.4	-5.0	-3.6	-1.9	0.0	3.8	.2583
400.	3021.	5.3	-13.0	-12.4	-10.6	-8.4	-5.8	6.8	.3545
500.	2845.	9.3	-24.4	-23.7	-21.4	-18.6	-15.4	10.9	.4568
600.	2676.	14.6	-40.2	-39.2	-36.6	-33.2	-29.3	16.2	.5656
700.	2513.	21.6	-60.8	-59.7	-56.6	-52.7	-48.2	22.7	.6817
800.	2356.	30.4	-86.8	-85.6	-82.1	-77.6	-72.4	30.5	.8050
900.	2203.	41.3	-119.0	-117.6	-113.6	-108.6	-102.8	39.8	.9364
1000.	2056.	55.0	-158.4	-156.9	-152.5	-146.9	-140.4	50.7	1.0776

BALLISTIC COEFFICIENT .55, MUZZLE VELOCITY FT/SEC 3900.

RANGE	VELOCITY	HEIGHT	100 YD	150 YD	200 YD	250 YD	300 YD	DRIFT	TIME
0.	3900.	0.0	-1.5	-1.5	-1.5	-1.5	-1.5	0.0	0.0000
100.	3690.	-.5	0.0	.1	.5	1.1	1.7	.3	.0788
200.	3489.	.5	-1.1	-.8	0.0	1.1	2.3	1.5	.1624
300.	3294.	2.3	-5.0	-4.6	-3.4	-1.8	0.0	3.6	.2509
400.	3107.	5.0	-12.2	-11.6	-10.1	-7.9	-5.5	6.6	.3449
500.	2927.	8.7	-22.9	-22.2	-20.3	-17.6	-14.6	10.5	.4443
600.	2755.	13.8	-37.7	-36.9	-34.5	-31.4	-27.7	15.6	.5500
700.	2589.	20.3	-57.1	-56.1	-53.4	-49.7	-45.4	21.8	.6622
800.	2429.	28.6	-81.6	-80.5	-77.4	-73.1	-68.3	29.3	.7818
900.	2275.	39.0	-112.0	-110.7	-107.2	-102.4	-97.0	38.2	.9095
1000.	2125.	51.7	-148.9	-147.6	-143.6	-138.3	-132.3	48.7	1.0458

BALLISTIC COEFFICIENT .55, MUZZLE VELOCITY FT/SEC 4000.

RANGE	VELOCITY	HEIGHT	100 YD	150 YD	200 YD	250 YD	300 YD	DRIFT	TIME
0.	4000.	0.0	-1.5	-1.5	-1.5	-1.5	-1.5	0.0	0.0000
100.	3787.	-.5	0.0	.1	.5	1.0	1.5	.4	.0771
200.	3581.	.5	-1.0	-.7	0.0	1.0	2.1	1.6	.1589
300.	3384.	2.1	-4.6	-4.3	-3.2	-1.7	0.0	3.5	.2449
400.	3193.	4.7	-11.3	-10.9	-9.4	-7.4	-5.2	6.4	.3362
500.	3010.	8.2	-21.4	-20.9	-19.0	-16.6	-13.7	10.2	.4328
600.	2834.	13.0	-35.4	-34.7	-32.5	-29.6	-26.2	15.1	.5355
700.	2665.	19.2	-53.6	-52.9	-50.3	-46.9	-42.9	21.1	.6448
800.	2503.	27.1	-76.8	-76.0	-73.0	-69.1	-64.6	28.4	.7612
900.	2346.	36.8	-105.5	-104.6	-101.2	-96.8	-91.7	37.0	.8850
1000.	2194.	48.9	-140.3	-139.3	-135.5	-130.7	-125.0	47.0	1.0170

BALLISTIC COEFFICIENT .56, MUZZLE VELOCITY FT/SEC 2000.

RANGE	VELOCITY	HEIGHT	100 YD	150 YD	200 YD	250 YD	300 YD	DRIFT	TIME
0.	2000.	0.0	-1.5	-1.5	-1.5	-1.5	-1.5	0.0	0.0000
100.	1864.	.4	0.0	2.0	4.3	6.8	9.6	.9	.1551
200.	1735.	4.2	-8.6	-4.6	0.0	5.1	10.6	3.9	.3221
300.	1613.	11.3	-28.7	-22.8	-15.9	-8.2	0.0	9.1	.5015
400.	1499.	22.4	-62.2	-54.4	-45.1	-34.9	-23.9	16.6	.6942
500.	1393.	38.3	-111.3	-101.5	-89.9	-77.1	-63.4	26.8	.9022
600.	1300.	60.0	-178.3	-166.6	-152.6	-137.3	-120.9	39.6	1.1251
700.	1219.	88.5	-265.9	-252.2	-235.9	-218.0	-198.9	55.2	1.3636
800.	1150.	124.8	-376.9	-361.2	-342.6	-322.2	-300.3	73.5	1.6175
900.	1097.	169.8	-514.2	-496.5	-475.6	-452.6	-428.0	94.1	1.8848
1000.	1056.	224.0	-679.5	-659.9	-636.7	-611.1	-583.7	116.8	2.1636

BALLISTIC COEFFICIENT .56, MUZZLE VELOCITY FT/SEC 2100.

RANGE	VELOCITY	HEIGHT	100 YD	150 YD	200 YD	250 YD	300 YD	DRIFT	TIME
0.	2100.	0.0	-1.5	-1.5	-1.5	-1.5	-1.5	0.0	0.0000
100.	1959.	.3	0.0	1.7	3.8	6.1	8.5	.9	.1479
200.	1825.	3.8	-7.6	-4.2	0.0	4.6	9.5	3.6	.3063
300.	1698.	10.2	-25.6	-20.5	-14.3	-7.4	0.0	8.5	.4767
400.	1579.	20.2	-55.9	-49.0	-40.7	-31.6	-21.7	15.6	.6603
500.	1466.	34.5	-99.9	-91.4	-81.0	-69.5	-57.2	25.2	.8573
600.	1364.	54.2	-160.3	-150.0	-137.6	-123.8	-109.0	37.4	1.0696
700.	1275.	80.1	-239.4	-227.5	-212.9	-196.9	-179.7	52.4	1.2975
800.	1197.	113.2	-339.8	-326.2	-309.5	-291.2	-271.5	70.0	1.5406
900.	1133.	154.4	-464.3	-449.0	-430.2	-409.6	-387.4	90.2	1.7981
1000.	1084.	204.7	-615.9	-598.9	-578.1	-555.2	-530.5	112.7	2.0690

BALLISTIC COEFFICIENT .56, MUZZLE VELOCITY FT/SEC 2200.

RANGE	VELOCITY	HEIGHT	100 YD	150 YD	200 YD	250 YD	300 YD	DRIFT	TIME
0.	2200.	0.0	-1.5	-1.5	-1.5	-1.5	-1.5	0.0	0.0000
100.	2055.	.2	0.0	1.5	3.4	5.4	7.7	.9	.1413
200.	1916.	3.4	-6.8	-3.8	0.0	4.1	8.6	3.5	.2925
300.	1785.	9.2	-23.0	-18.5	-12.9	-6.7	0.0	8.0	.4547
400.	1660.	18.2	-50.2	-44.2	-36.7	-28.5	-19.5	14.7	.6289
500.	1543.	31.2	-90.0	-82.6	-73.1	-62.8	-51.7	23.7	.8163
600.	1433.	49.0	-144.5	-135.5	-124.2	-111.8	-98.4	35.2	1.0182
700.	1335.	72.5	-216.0	-205.5	-192.3	-177.9	-162.3	49.5	1.2356
800.	1250.	102.6	-306.8	-294.8	-279.8	-263.3	-245.4	66.3	1.4676
900.	1175.	140.8	-420.7	-407.2	-390.3	-371.7	-351.6	86.2	1.7170
1000.	1117.	187.0	-558.4	-543.4	-524.6	-504.0	-481.6	108.1	1.9776

BALLISTIC COEFFICIENT .56, MUZZLE VELOCITY FT/SEC 2300.

RANGE	VELOCITY	HEIGHT	100 YD	150 YD	200 YD	250 YD	300 YD	DRIFT	TIME
0.	2300.	0.0	-1.5	-1.5	-1.5	-1.5	-1.5	0.0	0.0000
100.	2152.	.1	0.0	1.3	3.0	4.9	6.9	.8	.1348
200.	2009.	3.0	-6.0	-3.4	0.0	3.7	7.8	3.2	.2791
300.	1873.	8.3	-20.7	-16.7	-11.6	-6.0	0.0	7.4	.4335
400.	1743.	16.5	-45.4	-40.1	-33.3	-25.8	-17.8	13.7	.5998
500.	1621.	28.3	-81.4	-74.8	-66.3	-56.9	-46.9	22.2	.7781
600.	1506.	44.4	-130.7	-122.7	-112.6	-101.3	-89.3	33.0	.9701
700.	1399.	65.7	-195.3	-186.0	-174.2	-161.0	-147.0	46.4	1.1769
800.	1306.	93.2	-277.8	-267.1	-253.6	-238.6	-222.6	62.6	1.3989
900.	1224.	127.8	-380.6	-368.6	-353.4	-336.5	-318.5	81.4	1.6364
1000.	1154.	171.0	-508.0	-494.8	-477.8	-459.1	-439.0	103.4	1.8917

BALLISTIC COEFFICIENT .56, MUZZLE VELOCITY FT/SEC 2400.

RANGE	VELOCITY	HEIGHT	100 YD	150 YD	200 YD	250 YD	300 YD	DRIFT	TIME
0.	2400.	0.0	-1.5	-1.5	-1.5	-1.5	-1.5	0.0	0.0000
100.	2249.	.1	0.0	1.2	2.7	4.4	6.2	.7	.1292
200.	2103.	2.7	-5.4	-3.1	0.0	3.4	7.1	3.0	.2671
300.	1962.	7.5	-18.7	-15.2	-10.6	-5.5	0.0	7.0	.4148
400.	1828.	15.0	-41.0	-36.4	-30.2	-23.4	-16.1	12.9	.5731
500.	1701.	25.8	-73.8	-67.9	-60.3	-51.8	-42.6	20.8	.7432
600.	1581.	40.5	-118.7	-111.7	-102.5	-92.3	-81.3	31.1	.9265
700.	1468.	59.8	-177.2	-169.1	-158.3	-146.5	-133.6	43.7	1.1232
800.	1366.	84.8	-252.1	-242.8	-230.5	-216.9	-202.2	59.0	1.3353
900.	1277.	116.5	-345.8	-335.3	-321.5	-306.2	-289.7	77.0	1.5627
1000.	1199.	155.7	-460.6	-449.0	-433.6	-416.6	-398.3	97.8	1.8055

BALLISTIC COEFFICIENT .56, MUZZLE VELOCITY FT/SEC 2500.

RANGE	VELOCITY	HEIGHT	100 YD	150 YD	200 YD	250 YD	300 YD	DRIFT	TIME
0.	2500.	0.0	-1.5	-1.5	-1.5	-1.5	-1.5	0.0	0.0000
100.	2346.	-.0	0.0	1.0	2.4	4.0	5.6	.7	.1239
200.	2196.	2.4	-4.8	-2.7	0.0	3.1	6.5	2.8	.2558
300.	2052.	6.8	-16.9	-13.8	-9.7	-5.1	0.0	6.6	.3974
400.	1913.	13.7	-37.3	-33.1	-27.7	-21.5	-14.7	12.1	.5488
500.	1782.	23.5	-67.2	-61.9	-55.1	-47.4	-39.0	19.6	.7113
600.	1657.	36.9	-108.0	-101.7	-93.5	-84.3	-74.1	29.2	.8858
700.	1540.	54.6	-161.4	-154.1	-144.5	-133.7	-121.9	41.1	1.0735
800.	1430.	77.4	-229.4	-221.1	-210.2	-197.8	-184.3	55.6	1.2758
900.	1333.	106.3	-314.6	-305.2	-293.0	-279.1	-263.8	72.7	1.4932
1000.	1248.	142.2	-419.5	-409.1	-395.4	-380.0	-363.1	92.6	1.7260

BALLISTIC COEFFICIENT .56, MUZZLE VELOCITY FT/SEC 2600.

RANGE	VELOCITY	HEIGHT	100 YD	150 YD	200 YD	250 YD	300 YD	DRIFT	TIME
0.	2600.	0.0	-1.5	-1.5	-1.5	-1.5	-1.5	0.0	0.0000
100.	2443.	-.1	0.0	.9	2.2	3.6	5.1	.6	.1189
200.	2290.	2.2	-4.4	-2.5	0.0	2.8	5.9	2.7	.2459
300.	2143.	6.2	-15.4	-12.6	-8.9	-4.6	0.0	6.2	.3815
400.	2000.	12.5	-34.0	-30.2	-25.3	-19.6	-13.5	11.4	.5262
500.	1864.	21.5	-61.3	-56.5	-50.3	-43.3	-35.6	18.4	.6814
600.	1735.	33.8	-98.6	-93.0	-85.5	-77.1	-67.8	27.5	.8484
700.	1614.	49.9	-147.5	-140.9	-132.2	-122.4	-111.5	38.7	1.0277
800.	1499.	70.7	-209.5	-202.0	-192.1	-180.8	-168.5	52.3	1.2204
900.	1393.	97.2	-287.2	-278.7	-267.6	-254.9	-241.0	68.6	1.4284
1000.	1300.	130.1	-382.9	-373.5	-361.1	-347.0	-331.6	87.5	1.6512

BALLISTIC COEFFICIENT .56, MUZZLE VELOCITY FT/SEC 2700.

RANGE	VELOCITY	HEIGHT	100 YD	150 YD	200 YD	250 YD	300 YD	DRIFT	TIME
0.	2700.	0.0	-1.5	-1.5	-1.5	-1.5	-1.5	0.0	0.0000
100.	2539.	-.1	0.0	.8	1.9	3.2	4.7	.6	.1148
200.	2384.	1.9	-3.9	-2.3	0.0	2.6	5.4	2.5	.2366
300.	2233.	5.7	-14.0	-11.5	-8.1	-4.2	0.0	5.9	.3666
400.	2088.	11.5	-31.0	-27.7	-23.2	-18.0	-12.4	10.7	.5054
500.	1947.	19.8	-56.1	-52.0	-46.3	-39.8	-32.8	17.4	.6544
600.	1814.	31.0	-90.2	-85.3	-78.5	-70.7	-62.3	25.9	.8138
700.	1688.	45.9	-135.0	-129.3	-121.3	-112.3	-102.4	36.5	.9854
800.	1569.	65.0	-192.0	-185.5	-176.4	-166.1	-154.8	49.5	1.1700
900.	1457.	89.1	-262.9	-255.6	-245.4	-233.7	-221.0	64.8	1.3683
1000.	1356.	119.4	-350.6	-342.5	-331.1	-318.2	-304.1	82.9	1.5823

BALLISTIC COEFFICIENT .56, MUZZLE VELOCITY FT/SEC 2800.

RANGE	VELOCITY	HEIGHT	100 YD	150 YD	200 YD	250 YD	300 YD	DRIFT	TIME
0.	2800.	0.0	-1.5	-1.5	-1.5	-1.5	-1.5	0.0	0.0000
100.	2635.	-.2	0.0	.7	1.8	3.0	4.2	.5	.1101
200.	2477.	1.7	-3.5	-2.1	0.0	2.4	5.0	2.3	.2275
300.	2323.	5.2	-12.7	-10.5	-7.5	-3.9	0.0	5.5	.3527
400.	2175.	10.6	-28.4	-25.5	-21.4	-16.6	-11.4	10.1	.4859
500.	2031.	18.2	-51.5	-47.8	-42.7	-36.7	-30.3	16.4	.6290
600.	1893.	28.6	-82.9	-78.5	-72.4	-65.2	-57.4	24.4	.7818
700.	1763.	42.2	-124.0	-118.9	-111.7	-103.3	-94.3	34.5	.9460
800.	1640.	59.7	-176.4	-170.5	-162.3	-152.7	-142.4	46.7	1.1225
900.	1524.	81.9	-241.6	-235.0	-225.7	-214.9	-203.3	61.3	1.3124
1000.	1415.	109.7	-321.6	-314.3	-304.0	-292.1	-279.1	78.3	1.5166

BALLISTIC COEFFICIENT .56, MUZZLE VELOCITY FT/SEC 2900.

RANGE	VELOCITY	HEIGHT	100 YD	150 YD	200 YD	250 YD	300 YD	DRIFT	TIME
0.	2900.	0.0	-1.5	-1.5	-1.5	-1.5	-1.5	0.0	0.0000
100.	2731.	-.2	0.0	.6	1.6	2.7	3.9	.5	.1065
200.	2570.	1.6	-3.2	-1.9	0.0	2.2	4.6	2.3	.2198
300.	2413.	4.8	-11.7	-9.7	-6.9	-3.6	0.0	5.3	.3404
400.	2262.	9.8	-26.1	-23.5	-19.8	-15.4	-10.6	9.7	.4688
500.	2115.	16.9	-47.3	-44.1	-39.4	-33.9	-27.9	15.6	.6057
600.	1974.	26.4	-76.3	-72.5	-66.8	-60.2	-53.0	23.2	.7526
700.	1839.	39.0	-114.4	-109.9	-103.3	-95.6	-87.2	32.8	.9104
800.	1711.	55.1	-162.4	-157.3	-149.7	-140.9	-131.3	44.3	1.0790
900.	1591.	75.6	-222.6	-216.8	-208.3	-198.5	-187.7	58.1	1.2613
1000.	1478.	101.1	-296.2	-289.7	-280.3	-269.3	-257.3	74.3	1.4566

BALLISTIC COEFFICIENT .56, MUZZLE VELOCITY FT/SEC 3000.

RANGE	VELOCITY	HEIGHT	100 YD	150 YD	200 YD	250 YD	300 YD	DRIFT	TIME
0.	3000.	0.0	-1.5	-1.5	-1.5	-1.5	-1.5	0.0	0.0000
100.	2828.	-.2	0.0	.6	1.4	2.4	3.6	.6	.1032
200.	2662.	1.4	-2.9	-1.7	0.0	2.0	4.3	2.2	.2126
300.	2503.	4.4	-10.7	-9.0	-6.4	-3.3	0.0	5.1	.3290
400.	2348.	9.1	-24.1	-21.8	-18.3	-14.3	-9.8	9.3	.4528
500.	2199.	15.7	-43.7	-40.8	-36.5	-31.4	-25.9	14.9	.5846
600.	2054.	24.5	-70.6	-67.2	-62.0	-55.9	-49.2	22.2	.7260
700.	1915.	36.2	-105.7	-101.7	-95.7	-88.6	-80.8	31.2	.8773
800.	1784.	51.1	-150.2	-145.7	-138.7	-130.6	-121.7	42.2	1.0395
900.	1659.	70.0	-205.6	-200.5	-192.7	-183.6	-173.6	55.2	1.2138
1000.	1542.	93.5	-273.6	-267.9	-259.3	-249.1	-238.0	70.6	1.4013

BALLISTIC COEFFICIENT .56, MUZZLE VELOCITY FT/SEC 3100.

RANGE	VELOCITY	HEIGHT	100 YD	150 YD	200 YD	250 YD	300 YD	DRIFT	TIME
0.	3100.	0.0	-1.5	-1.5	-1.5	-1.5	-1.5	0.0	0.0000
100.	2924.	-.3	0.0	.5	1.3	2.2	3.3	.5	.0996
200.	2754.	1.3	-2.6	-1.6	0.0	1.9	3.9	2.1	.2053
300.	2592.	4.1	-9.8	-8.3	-5.9	-3.1	0.0	4.8	.3175
400.	2434.	8.4	-22.1	-20.2	-16.9	-13.2	-9.1	8.8	.4368
500.	2282.	14.5	-40.3	-37.9	-33.9	-29.3	-24.1	14.1	.5642
600.	2135.	22.8	-65.3	-62.4	-57.6	-52.0	-45.8	21.0	.7002
700.	1993.	33.6	-97.7	-94.4	-88.7	-82.3	-75.0	29.6	.8455
800.	1857.	47.4	-138.9	-135.1	-128.6	-121.2	-112.9	40.0	1.0014
900.	1728.	64.8	-190.2	-185.9	-178.6	-170.3	-161.0	52.4	1.1688
1000.	1607.	86.6	-253.2	-248.4	-240.4	-231.1	-220.7	67.1	1.3491

BALLISTIC COEFFICIENT .56, MUZZLE VELOCITY FT/SEC 3200.

RANGE	VELOCITY	HEIGHT	100 YD	150 YD	200 YD	250 YD	300 YD	DRIFT	TIME
0.	3200.	0.0	-1.5	-1.5	-1.5	-1.5	-1.5	0.0	0.0000
100.	3020.	-.3	0.0	.4	1.2	2.0	3.0	.5	.0964
200.	2847.	1.1	-2.3	-1.4	0.0	1.8	3.7	2.0	.1986
300.	2680.	3.8	-9.0	-7.6	-5.5	-2.8	0.0	4.6	.3073
400.	2520.	7.8	-20.5	-18.7	-15.9	-12.4	-8.6	8.5	.4232
500.	2365.	13.6	-37.5	-35.3	-31.7	-27.3	-22.6	13.6	.5459
600.	2215.	21.2	-60.6	-57.9	-53.6	-48.4	-42.7	20.1	.6767
700.	2070.	31.3	-90.8	-87.7	-82.7	-76.6	-69.9	28.3	.8169
800.	1931.	44.2	-129.2	-125.7	-119.9	-112.9	-105.3	38.2	.9672
900.	1798.	60.3	-176.7	-172.8	-166.3	-158.4	-149.9	50.0	1.1279
1000.	1673.	80.5	-235.1	-230.7	-223.5	-214.7	-205.3	64.0	1.3010

BALLISTIC COEFFICIENT .56, MUZZLE VELOCITY FT/SEC 3300.

RANGE	VELOCITY	HEIGHT	100 YD	150 YD	200 YD	250 YD	300 YD	DRIFT	TIME
0.	3300.	0.0	-1.5	-1.5	-1.5	-1.5	-1.5	0.0	0.0000
100.	3116.	-.3	0.0	.4	1.0	1.8	2.7	.5	.0939
200.	2939.	1.0	-2.1	-1.3	0.0	1.6	3.4	1.9	.1926
300.	2769.	3.5	-8.2	-7.1	-5.1	-2.7	0.0	4.4	.2979
400.	2606.	7.3	-18.9	-17.4	-14.8	-11.5	-7.9	8.1	.4096
500.	2448.	12.6	-34.6	-32.8	-29.5	-25.5	-21.0	13.0	.5283
600.	2296.	19.8	-56.3	-54.1	-50.1	-45.3	-39.9	19.3	.6550
700.	2148.	29.2	-84.4	-81.9	-77.2	-71.6	-65.3	27.1	.7902
800.	2005.	41.2	-120.1	-117.2	-111.8	-105.4	-98.2	36.5	.9347
900.	1869.	56.2	-164.3	-161.0	-155.0	-147.8	-139.7	47.7	1.0894
1000.	1740.	75.0	-218.5	-214.9	-208.2	-200.2	-191.2	61.1	1.2560

BALLISTIC COEFFICIENT .56, MUZZLE VELOCITY FT/SEC 3400.

RANGE	VELOCITY	HEIGHT	100 YD	150 YD	200 YD	250 YD	300 YD	DRIFT	TIME
0.	3400.	0.0	-1.5	-1.5	-1.5	-1.5	-1.5	0.0	0.0000
100.	3212.	-.4	0.0	.3	.9	1.7	2.5	.4	.0907
200.	3032.	.9	-1.9	-1.2	0.0	1.5	3.2	1.8	.1866
300.	2858.	3.2	-7.5	-6.5	-4.7	-2.5	0.0	4.2	.2885
400.	2691.	6.8	-17.5	-16.1	13.8	-10.8	-7.4	7.7	.3968
500.	2531.	11.8	-32.3	-30.6	-27.6	-23.9	-19.7	12.5	.5121
600.	2376.	18.6	-52.4	-50.4	-46.8	-42.4	-37.4	18.4	.6342
700.	2225.	27.3	-78.7	-76.4	-72.2	-67.0	-61.2	25.9	.7646
800.	2080.	38.5	-111.9	-109.2	-104.5	-98.5	-91.9	34.8	.9038
900.	1940.	52.5	-153.3	-150.3	-144.9	-138.2	-130.7	45.7	1.0536
1000.	1807.	69.9	-203.7	-200.4	-194.4	-187.0	-178.6	58.3	1.2136

BALLISTIC COEFFICIENT .56, MUZZLE VELOCITY FT/SEC 3500.

RANGE	VELOCITY	HEIGHT	100 YD	150 YD	200 YD	250 YD	300 YD	DRIFT	TIME
0.	3500.	0.0	-1.5	-1.5	-1.5	-1.5	-1.5	0.0	0.0000
100.	3309.	-.4	0.0	.3	.9	1.5	2.3	.5	.0883
200.	3124.	.8	-1.7	-1.1	0.0	1.4	2.9	1.8	.1819
300.	2947.	3.0	-6.9	-6.1	-4.4	-2.3	0.0	4.1	.2803
400.	2777.	6.4	-16.2	-15.1	-12.8	-10.1	-7.0	7.5	.3852
500.	2613.	11.1	-30.0	-28.6	-25.7	-22.3	-18.5	12.0	.4967
600.	2455.	17.4	-48.9	-47.2	-43.8	-39.7	-35.0	17.7	.6151
700.	2303.	25.6	-73.5	-71.6	-67.6	-62.8	-57.4	24.9	.7415
800.	2155.	36.1	-104.7	-102.4	-97.8	-92.4	-86.2	33.5	.8760
900.	2012.	49.2	-143.3	-140.7	-135.6	-129.4	-122.5	43.8	1.0202
1000.	1875.	65.5	-190.3	-187.4	-181.8	-174.9	-167.2	55.8	1.1745

BALLISTIC COEFFICIENT .56, MUZZLE VELOCITY FT/SEC 3600.

RANGE	VELOCITY	HEIGHT	100 YD	150 YD	200 YD	250 YD	300 YD	DRIFT	TIME
0.	3600.	0.0	-1.5	-1.5	-1.5	-1.5	-1.5	0.0	0.0000
100.	3405.	-.4	0.0	.2	.8	1.4	2.1	.4	.0856
200.	3217.	.7	-1.5	-1.0	0.0	1.3	2.7	1.7	.1762
300.	3036.	2.8	-6.4	-5.7	-4.1	-2.1	0.0	3.9	.2720
400.	2862.	6.0	-15.0	-14.1	-12.0	-9.4	-6.5	7.1	.3737
500.	2695.	10.4	-27.9	-26.8	-24.2	-20.9	-17.3	11.5	.4819
600.	2535.	16.4	-45.7	-44.3	-41.2	-37.2	-33.0	17.1	.5969
700.	2380.	24.1	-68.7	-67.1	-63.4	-58.9	-53.9	23.8	.7188
800.	2229.	33.9	-97.9	-96.0	-91.9	-86.7	-80.9	32.1	.8491
900.	2084.	46.1	-134.0	-131.9	-127.2	-121.3	-114.9	41.9	.9881
1000.	1943.	61.4	-178.1	-175.8	-170.6	-164.1	-156.9	53.5	1.1375

BALLISTIC COEFFICIENT .56, MUZZLE VELOCITY FT/SEC 3700.

RANGE	VELOCITY	HEIGHT	100 YD	150 YD	200 YD	250 YD	300 YD	DRIFT	TIME
0.	3700.	0.0	-1.5	-1.5	-1.5	-1.5	-1.5	0.0	0.0000
100.	3501.	-.4	0.0	.2	.7	1.3	2.0	.4	.0834
200.	3310.	.7	-1.4	-.9	0.0	1.2	2.6	1.7	.1717
300.	3126.	2.6	-5.9	-5.3	-3.9	-2.1	0.0	3.9	.2653
400.	2948.	5.6	-14.0	-13.1	-11.3	-8.8	-6.1	6.9	.3636
500.	2778.	9.8	-26.1	-25.1	-22.7	-19.7	-16.2	11.1	.4685
600.	2614.	15.4	-42.7	-41.5	-38.7	-35.0	-30.9	16.4	.5799
700.	2456.	22.7	-64.4	-63.0	-59.7	-55.4	-50.6	23.0	.6983
800.	2304.	31.9	-91.8	-90.2	-86.4	-81.6	-76.1	31.0	.8246
900.	2156.	43.4	-125.7	-123.9	-119.6	-114.2	-108.0	40.4	.9591
1000.	2013.	57.7	-167.1	-165.0	-160.3	-154.2	-147.3	51.5	1.1032

BALLISTIC COEFFICIENT .56, MUZZLE VELOCITY FT/SEC 3800.

RANGE	VELOCITY	HEIGHT	100 YD	150 YD	200 YD	250 YD	300 YD	DRIFT	TIME
0.	3800.	0.0	-1.5	-1.5	-1.5	-1.5	-1.5	0.0	0.0000
100.	3598.	-.4	0.0	.1	.6	1.2	1.8	.5	.0816
200.	3403.	.6	-1.2	-.9	0.0	1.1	2.4	1.7	.1673
300.	3215.	2.4	-5.4	-4.9	-3.6	-1.9	0.0	3.7	.2579
400.	3034.	5.3	-12.9	-12.3	-10.6	-8.3	-5.8	6.7	.3538
500.	2860.	9.2	-24.3	-23.5	-21.3	-18.5	-15.3	10.7	.4555
600.	2694.	14.5	-39.9	-39.0	-36.4	-33.0	-29.1	15.9	.5638
700.	2533.	21.4	-60.4	-59.3	-56.2	-52.3	-47.8	22.2	.6789
800.	2378.	30.0	-86.1	-84.9	-81.3	-76.9	-71.7	29.8	.8010
900.	2227.	40.9	-117.9	-116.6	-112.6	-107.6	-101.8	38.9	.9313
1000.	2082.	54.2	-156.7	-155.2	-150.7	-145.2	-138.7	49.4	1.0704

BALLISTIC COEFFICIENT .56, MUZZLE VELOCITY FT/SEC 3900.

RANGE	VELOCITY	HEIGHT	100 YD	150 YD	200 YD	250 YD	300 YD	DRIFT	TIME
0.	3900.	0.0	-1.5	-1.5	-1.5	-1.5	-1.5	0.0	0.0000
100.	3694.	-.5	0.0	.1	.5	1.1	1.7	.3	.0786
200.	3496.	.5	-1.1	-.8	0.0	1.1	2.3	1.5	.1621
300.	3304.	2.3	-5.0	-4.6	-3.4	-1.8	0.0	3.5	.2505
400.	3120.	4.9	-12.1	-11.6	-10.0	-7.9	-5.5	6.4	.3442
500.	2943.	8.7	-22.7	-22.1	-20.1	-17.4	-14.5	10.2	.4427
600.	2773.	13.7	-37.5	-36.6	-34.3	-31.1	-27.5	15.2	.5479
700.	2610.	20.1	-56.7	-55.7	-53.0	-49.2	-45.1	21.3	.6595
800.	2452.	28.3	-80.9	-79.8	-76.7	-72.4	-67.7	28.6	.7780
900.	2299.	38.5	-111.0	-109.8	-106.2	-101.4	-96.1	37.4	.9046
1000.	2151.	51.1	-147.5	-146.1	-142.2	-136.9	-130.9	47.6	1.0394

BALLISTIC COEFFICIENT .56, MUZZLE VELOCITY FT/SEC 4000.

RANGE	VELOCITY	HEIGHT	100 YD	150 YD	200 YD	250 YD	300 YD	DRIFT	TIME
0.	4000.	0.0	-1.5	-1.5	-1.5	-1.5	-1.5	0.0	0.0000
100.	3791.	-.5	0.0	.1	.5	1.0	1.5	.4	.0771
200.	3589.	.5	-.9	-.8	0.0	1.0	2.1	1.5	.1588
300.	3394.	2.1	-4.6	-4.3	-3.2	-1.7	0.0	3.5	.2446
400.	3206.	4.7	-11.3	-10.9	-9.4	-7.4	-5.1	6.3	.3355
500.	3026.	8.2	-21.3	-20.8	-18.9	-16.5	-13.6	10.0	.4317
600.	2853.	12.9	-35.1	-34.5	-32.3	-29.4	-26.0	14.7	.5337
700.	2686.	19.0	-53.3	-52.6	-49.9	-46.5	-42.5	20.6	.6422
800.	2526.	26.8	-76.3	-75.5	-72.5	-68.6	-64.0	27.8	.7578
900.	2371.	36.4	-104.5	-103.6	-100.3	-95.9	-90.8	36.1	.8802
1000.	2221.	48.3	-138.9	-137.9	-134.2	-129.3	-123.6	45.9	1.0108

BALLISTIC COEFFICIENT .57, MUZZLE VELOCITY FT/SEC 2000.

RANGE	VELOCITY	HEIGHT	100 YD	150 YD	200 YD	250 YD	300 YD	DRIFT	TIME
0.	2000.	0.0	-1.5	-1.5	-1.5	-1.5	-1.5	0.0	0.0000
100.	1866.	.4	0.0	2.0	4.3	6.8	9.5	.9	.1550
200.	1739.	4.2	-8.5	-4.6	0.0	5.1	10.5	3.8	.3218
300.	1620.	11.3	-28.6	-22.7	-15.8	-8.2	0.0	8.9	.5004
400.	1507.	22.3	-62.0	-54.1	-44.9	-34.7	-23.8	16.3	.6925
500.	1402.	38.0	-110.7	-100.9	-89.3	-76.6	-63.0	26.2	.8991
600.	1309.	59.5	-177.2	-165.4	-151.5	-136.3	-120.0	38.8	1.1207
700.	1229.	87.7	-263.9	-250.2	-234.0	-216.2	-197.2	54.1	1.3575
800.	1159.	123.6	-373.8	-358.2	-339.6	-319.3	-297.5	72.1	1.6097
900.	1105.	168.0	-509.4	-491.8	-470.9	-448.0	-423.6	92.4	1.8747
1000.	1062.	221.8	-673.8	-654.2	-631.0	-605.6	-578.4	115.0	2.1531

BALLISTIC COEFFICIENT .57, MUZZLE VELOCITY FT/SEC 2100.

RANGE	VELOCITY	HEIGHT	100 YD	150 YD	200 YD	250 YD	300 YD	DRIFT	TIME
0.	2100.	0.0	-1.5	-1.5	-1.5	-1.5	-1.5	0.0	0.0000
100.	1961.	.3	0.0	1.7	3.8	6.1	8.5	.9	.1478
200.	1830.	3.7	-7.6	-4.2	0.0	4.6	9.4	3.6	.3061
300.	1705.	10.1	-25.5	-20.4	-14.2	-7.3	0.0	8.3	.4757
400.	1587.	20.1	-55.6	-48.8	-40.5	-31.4	-21.6	15.3	.6585
500.	1476.	34.3	-99.3	-90.8	-80.4	-69.0	-56.8	24.6	.8542
600.	1374.	53.7	-159.3	-149.0	-136.6	-122.9	-108.2	36.7	1.0655
700.	1286.	79.2	-237.4	-225.5	-210.9	-195.0	-177.8	51.2	1.2910
800.	1208.	111.9	-336.6	-323.0	-306.4	-288.2	-268.6	68.5	1.5320
900.	1142.	152.6	-459.6	-444.3	-425.6	-405.1	-383.0	88.3	1.7876
1000.	1092.	202.3	-609.6	-592.6	-571.8	-549.0	-524.5	110.6	2.0568

BALLISTIC COEFFICIENT .57, MUZZLE VELOCITY FT/SEC 2200.

RANGE	VELOCITY	HEIGHT	100 YD	150 YD	200 YD	250 YD	300 YD	DRIFT	TIME
0.	2200.	0.0	-1.5	-1.5	-1.5	-1.5	-1.5	0.0	0.0000
100.	2058.	.2	0.0	1.5	3.4	5.4	7.6	.9	.1412
200.	1921.	3.3	-6.8	-3.8	0.0	4.1	8.5	3.4	.2922
300.	1791.	9.1	-22.9	-18.4	-12.8	-6.7	0.0	7.9	.4537
400.	1669.	18.1	-50.0	-44.0	-36.5	-28.3	-19.4	14.4	.6272
500.	1553.	31.0	-89.6	-82.1	-72.7	-62.5	-51.4	23.2	.8137
600.	1444.	48.6	-143.5	-134.5	-123.3	-111.0	-97.7	34.5	1.0141
700.	1346.	71.8	-214.2	-203.7	-190.6	-176.3	-160.7	48.4	1.2293
800.	1261.	101.5	-303.9	-292.0	-276.9	-260.6	-242.8	64.8	1.4594
900.	1186.	138.8	-415.7	-402.2	-385.3	-367.0	-346.9	84.1	1.7051
1000.	1126.	184.6	-552.3	-537.3	-518.5	-498.1	-475.8	105.9	1.9651

BALLISTIC COEFFICIENT .57, MUZZLE VELOCITY FT/SEC 2300.

RANGE	VELOCITY	HEIGHT	100 YD	150 YD	200 YD	250 YD	300 YD	DRIFT	TIME
0.	2300.	0.0	-1.5	-1.5	-1.5	-1.5	-1.5	0.0	0.0000
100.	2155.	.1	0.0	1.3	3.0	4.9	6.9	.7	.1346
200.	2014.	3.0	-6.0	-3.4	0.0	3.7	7.7	3.1	.2787
300.	1879.	8.2	-20.6	-16.6	-11.6	-6.0	0.0	7.3	.4327
400.	1752.	16.4	-45.1	-39.8	-33.1	-25.6	-17.6	13.4	.5981
500.	1632.	28.1	-81.0	-74.3	-65.9	-56.6	-46.6	21.7	.7756
600.	1518.	44.1	-129.9	-121.9	-111.8	-100.6	-88.6	32.3	.9663
700.	1412.	65.1	-193.8	-184.4	-172.7	-159.6	-145.6	45.4	1.1712
800.	1318.	92.2	-275.3	-264.7	-251.2	-236.2	-220.3	61.2	1.3914
900.	1236.	126.3	-377.1	-365.1	-349.9	-333.1	-315.2	79.7	1.6270
1000.	1165.	168.3	-501.0	-487.7	-470.8	-452.2	-432.2	100.7	1.8767

BALLISTIC COEFFICIENT .57, MUZZLE VELOCITY FT/SEC 2400.

RANGE	VELOCITY	HEIGHT	100 YD	150 YD	200 YD	250 YD	300 YD	DRIFT	TIME
0.	2400.	0.0	-1.5	-1.5	-1.5	-1.5	-1.5	0.0	0.0000
100.	2252.	.1	0.0	1.2	2.7	4.4	6.2	.7	.1291
200.	2108.	2.7	-5.4	-3.0	0.0	3.4	7.0	2.9	.2667
300.	1969.	7.5	-18.6	-15.1	-10.5	-5.4	0.0	6.8	.4139
400.	1837.	14.9	-40.9	-36.2	-30.1	-23.3	-16.1	12.7	.5719
500.	1712.	25.6	-73.3	-67.5	-59.9	-51.4	-42.3	20.4	.7406
600.	1593.	40.1	-117.9	-110.9	-101.8	-91.6	-80.7	30.4	.9228
700.	1482.	59.2	-175.8	-167.7	-157.0	-145.1	-132.4	42.7	1.1176
800.	1379.	83.9	-249.7	-240.4	-228.2	-214.6	-200.1	57.7	1.3277
900.	1290.	115.0	-342.1	-331.6	-317.9	-302.6	-286.3	75.3	1.5528
1000.	1212.	153.6	-455.3	-443.7	-428.4	-411.4	-393.3	95.6	1.7929

BALLISTIC COEFFICIENT .57, MUZZLE VELOCITY FT/SEC 2500.

RANGE	VELOCITY	HEIGHT	100 YD	150 YD	200 YD	250 YD	300 YD	DRIFT	TIME
0.	2500.	0.0	-1.5	-1.5	-1.5	-1.5	-1.5	0.0	0.0000
100.	2348.	-.0	0.0	1.0	2.4	3.9	5.6	.7	.1239
200.	2202.	2.4	-4.8	-2.7	0.0	3.1	6.4	2.7	.2556
300.	2060.	6.8	-16.9	-13.7	-9.7	-5.0	0.0	6.5	.3967
400.	1923.	13.6	-37.1	-33.0	-27.5	-21.4	-14.7	11.9	.5476
500.	1793.	23.4	-66.8	-61.6	-54.8	-47.0	-38.7	19.2	.7089
600.	1670.	36.6	-107.3	-101.0	-92.8	-83.6	-73.5	28.6	.8823
700.	1554.	54.1	-160.2	-152.9	-143.4	-132.6	-120.9	40.2	1.0686
800.	1445.	76.5	-227.5	-219.1	-208.2	-195.9	-182.5	54.4	1.2689
900.	1347.	104.9	-311.4	-302.1	-289.8	-275.9	-260.9	71.1	1.4839
1000.	1262.	140.2	-414.6	-404.2	-390.6	-375.1	-358.4	90.4	1.7138

BALLISTIC COEFFICIENT .57, MUZZLE VELOCITY FT/SEC 2600.

RANGE	VELOCITY	HEIGHT	100 YD	150 YD	200 YD	250 YD	300 YD	DRIFT	TIME
0.	2600.	0.0	-1.5	-1.5	-1.5	-1.5	-1.5	0.0	0.0000
100.	2445.	-.1	0.0	.9	2.2	3.6	5.1	.6	.1188
200.	2296.	2.1	-4.4	-2.5	0.0	2.8	5.9	2.6	.2456
300.	2150.	6.2	-15.3	-12.5	-8.8	-4.6	0.0	6.1	.3807
400.	2010.	12.5	-33.8	-30.1	-25.1	-19.5	-13.4	11.2	.5250
500.	1876.	21.4	-60.9	-56.2	-50.0	-43.0	-35.4	18.0	.6792
600.	1748.	33.5	-98.0	-92.3	-84.9	-76.5	-67.3	26.9	.8451
700.	1628.	49.5	-146.3	-139.8	-131.1	-121.3	-110.6	37.9	1.0229
800.	1515.	70.0	-207.8	-200.3	-190.4	-179.2	-166.9	51.2	1.2140
900.	1409.	96.0	-284.3	-275.9	-264.7	-252.1	-238.3	67.0	1.4194
1000.	1316.	128.4	-378.8	-369.4	-357.0	-343.0	-327.7	85.6	1.6402

BALLISTIC COEFFICIENT .57, MUZZLE VELOCITY FT/SEC 2700.

RANGE	VELOCITY	HEIGHT	100 YD	150 YD	200 YD	250 YD	300 YD	DRIFT	TIME
0.	2700.	0.0	-1.5	-1.5	-1.5	-1.5	-1.5	0.0	0.0000
100.	2542.	-.1	0.0	.8	1.9	3.2	4.6	.6	.1147
200.	2389.	1.9	-3.9	-2.3	0.0	2.6	5.4	2.5	.2363
300.	2241.	5.7	-13.9	-11.5	-8.1	-4.2	0.0	5.8	.3661
400.	2098.	11.5	-30.9	-27.6	-23.1	-18.0	-12.3	10.6	.5044
500.	1959.	19.7	-55.8	-51.7	-46.0	-39.6	-32.6	17.0	.6524
600.	1828.	30.8	-89.7	-84.8	-78.0	-70.3	-61.8	25.4	.8108
700.	1703.	45.4	-133.9	-128.3	-120.3	-111.3	-101.5	35.7	.9807
800.	1585.	64.3	-190.3	-183.9	-174.8	-164.5	-153.2	48.4	1.1638
900.	1474.	88.0	-260.2	-253.0	-242.7	-231.2	-218.5	63.3	1.3597
1000.	1373.	117.7	-346.4	-338.3	-327.0	-314.1	-300.0	80.9	1.5708

BALLISTIC COEFFICIENT .57, MUZZLE VELOCITY FT/SEC 2800.

RANGE	VELOCITY	HEIGHT	100 YD	150 YD	200 YD	250 YD	300 YD	DRIFT	TIME
0.	2800.	0.0	-1.5	-1.5	-1.5	-1.5	-1.5	0.0	0.0000
100.	2638.	-.2	0.0	.7	1.8	2.9	4.2	.5	.1101
200.	2482.	1.7	-3.5	-2.1	0.0	2.4	5.0	2.3	.2272
300.	2331.	5.2	-12.7	-10.5	-7.4	-3.9	0.0	5.4	.3522
400.	2185.	10.5	-28.3	-25.4	-21.2	-16.5	-11.3	9.9	.4848
500.	2044.	18.1	-51.2	-47.6	-42.5	-36.5	-30.1	16.1	.6272
600.	1907.	28.4	-82.4	-78.1	-71.9	-64.7	-57.0	24.0	.7790
700.	1779.	41.8	-123.2	-118.1	-110.9	-102.6	-93.5	33.8	.9419
800.	1657.	59.1	-174.9	-169.1	-160.9	-151.3	-141.0	45.7	1.1166
900.	1542.	80.9	-239.2	-232.6	-223.4	-212.7	-201.1	59.8	1.3043
1000.	1433.	108.2	-318.1	-310.9	-300.6	-288.7	-275.8	76.5	1.5063

BALLISTIC COEFFICIENT .57, MUZZLE VELOCITY FT/SEC 2900.

RANGE	VELOCITY	HEIGHT	100 YD	150 YD	200 YD	250 YD	300 YD	DRIFT	TIME
0.	2900.	0.0	-1.5	-1.5	-1.5	-1.5	-1.5	0.0	0.0000
100.	2734.	-.2	0.0	.6	1.6	2.7	3.9	.5	.1064
200.	2575.	1.6	-3.2	-1.9	0.0	2.2	4.6	2.2	.2196
300.	2421.	4.8	-11.6	-9.7	-6.9	-3.6	0.0	5.2	.3398
400.	2272.	9.7	-26.0	-23.4	-19.6	-15.3	-10.5	9.5	.4676
500.	2128.	16.8	-47.1	-43.9	-39.2	-33.7	-27.8	15.3	.6040
600.	1988.	26.2	-75.9	-72.0	-66.4	-59.8	-52.7	22.7	.7498
700.	1855.	38.6	-113.4	-108.9	-102.3	-94.7	-86.3	32.0	.9060
800.	1729.	54.6	-161.1	-155.9	-148.4	-139.6	-130.1	43.3	1.0735
900.	1610.	74.7	-220.4	-214.6	-206.1	-196.3	-185.6	56.8	1.2535
1000.	1497.	99.7	-292.9	-286.5	-277.1	-266.1	-254.2	72.5	1.4465

BALLISTIC COEFFICIENT .57, MUZZLE VELOCITY FT/SEC 3000.

RANGE	VELOCITY	HEIGHT	100 YD	150 YD	200 YD	250 YD	300 YD	DRIFT	TIME
0.	3000.	0.0	-1.5	-1.5	-1.5	-1.5	-1.5	0.0	0.0000
100.	2831.	-.2	0.0	.6	1.4	2.4	3.6	.5	.1031
200.	2668.	1.4	-2.9	-1.7	0.0	2.0	4.2	2.2	.2123
300.	2511.	4.4	-10.7	-8.9	-6.4	-3.3	0.0	5.0	.3285
400.	2359.	9.0	-24.0	-21.7	-18.2	-14.2	-9.8	9.1	.4518
500.	2212.	15.6	-43.5	-40.6	-36.3	-31.3	-25.7	14.6	.5829
600.	2069.	24.4	-70.1	-66.7	-61.6	-55.5	-48.8	21.7	.7232
700.	1932.	35.9	-104.9	-100.9	-94.9	-87.9	-80.1	30.5	.8734
800.	1802.	50.6	-148.9	-144.3	-137.5	-129.4	-120.5	41.2	1.0340
900.	1679.	69.1	-203.6	-198.5	-190.8	-181.7	-171.7	53.9	1.2065
1000.	1562.	92.3	-270.8	-265.1	-256.5	-246.4	-235.3	69.0	1.3920

BALLISTIC COEFFICIENT .57, MUZZLE VELOCITY FT/SEC 3100.

RANGE	VELOCITY	HEIGHT	100 YD	150 YD	200 YD	250 YD	300 YD	DRIFT	TIME
0.	3100.	0.0	-1.5	-1.5	-1.5	-1.5	-1.5	0.0	0.0000
100.	2927.	-.3	0.0	.5	1.3	2.2	3.2	.5	.0995
200.	2760.	1.3	-2.6	-1.6	0.0	1.8	3.9	2.0	.2051
300.	2600.	4.1	-9.7	-8.3	-5.9	-3.1	0.0	4.7	.3170
400.	2445.	8.4	-22.0	-20.1	-16.8	-13.2	-9.0	8.6	.4358
500.	2295.	14.4	-40.2	-37.8	-33.7	-29.1	-24.0	13.9	.5626
600.	2150.	22.6	-64.9	-62.0	-57.2	-51.7	-45.4	20.6	.6977
700.	2010.	33.3	-97.1	-93.7	-88.1	-81.7	-74.4	29.0	.8420
800.	1875.	46.9	-137.8	-133.9	-127.5	-120.2	-111.9	39.1	.9963
900.	1748.	64.1	-188.5	-184.2	-176.9	-168.7	-159.4	51.2	1.1621
1000.	1628.	85.4	-250.5	-245.7	-237.6	-228.5	-218.1	65.5	1.3399

BALLISTIC COEFFICIENT .57, MUZZLE VELOCITY FT/SEC 3200.

RANGE	VELOCITY	HEIGHT	100 YD	150 YD	200 YD	250 YD	300 YD	DRIFT	TIME
0.	3200.	0.0	-1.5	-1.5	-1.5	-1.5	-1.5	0.0	0.0000
100.	3023.	-.3	0.0	.4	1.2	2.0	3.0	.5	.0963
200.	2853.	1.1	-2.3	-1.4	0.0	1.7	3.6	1.9	.1984
300.	2689.	3.8	-8.9	-7.6	-5.5	-2.9	0.0	4.5	.3069
400.	2531.	7.8	-20.4	-18.7	-15.8	-12.3	-8.5	8.3	.4222
500.	2379.	13.5	-37.2	-35.1	-31.5	-27.1	-22.4	13.3	.5442
600.	2231.	21.1	-60.3	-57.6	-53.3	-48.1	-42.4	19.7	.6745
700.	2088.	31.0	-90.2	-87.1	-82.1	-76.0	-69.4	27.7	.8134
800.	1950.	43.7	-128.1	-124.7	-118.9	-112.0	-104.3	37.4	.9623
900.	1819.	59.6	-175.0	-171.1	-164.7	-156.8	-148.3	48.8	1.1213
1000.	1695.	79.4	-232.6	-228.2	-221.0	-212.3	-202.8	62.4	1.2923

BALLISTIC COEFFICIENT .57, MUZZLE VELOCITY FT/SEC 3300.

RANGE	VELOCITY	HEIGHT	100 YD	150 YD	200 YD	250 YD	300 YD	DRIFT	TIME
0.	3300.	0.0	-1.5	-1.5	-1.5	-1.5	-1.5	0.0	0.0000
100.	3119.	-.3	0.0	.4	1.0	1.8	2.7	.5	.0939
200.	2945.	1.0	-2.0	-1.3	0.0	1.6	3.4	1.9	.1923
300.	2778.	3.5	-8.1	-7.1	-5.1	-2.7	0.0	4.3	.2973
400.	2617.	7.3	-18.8	-17.3	-14.7	-11.5	-7.9	7.9	.4086
500.	2462.	12.6	-34.5	-32.7	-29.4	-25.3	-20.9	12.7	.5269
600.	2312.	19.7	-55.9	-53.7	-49.8	-45.0	-39.6	18.9	.6527
700.	2166.	29.0	-83.8	-81.2	-76.6	-71.0	-64.8	26.4	.7866
800.	2025.	40.8	-119.1	-116.2	-111.0	-104.5	-97.4	35.7	.9302
900.	1890.	55.6	-162.8	-159.5	-153.6	-146.4	-138.4	46.7	1.0833
1000.	1762.	74.0	-216.2	-212.6	-206.0	-198.0	-189.1	59.6	1.2477

BALLISTIC COEFFICIENT .57, MUZZLE VELOCITY FT/SEC 3400.

RANGE	VELOCITY	HEIGHT	100 YD	150 YD	200 YD	250 YD	300 YD	DRIFT	TIME
0.	3400.	0.0	-1.5	-1.5	-1.5	-1.5	-1.5	0.0	0.0000
100.	3216.	-.4	0.0	.3	.9	1.7	2.5	.4	.0906
200.	3038.	.9	-1.9	-1.2	0.0	1.5	3.1	1.7	.1864
300.	2867.	3.2	-7.5	-6.5	-4.7	-2.5	0.0	4.1	.2880
400.	2703.	6.8	-17.4	-16.1	-13.7	-10.7	-7.4	7.6	.3960
500.	2545.	11.8	-32.1	-30.4	-27.5	-23.8	-19.6	12.2	.5105
600.	2392.	18.4	-52.1	-50.1	-46.6	-42.1	-37.1	18.1	.6321
700.	2244.	27.1	-78.2	-75.9	-71.7	-66.5	-60.7	25.4	.7617
800.	2100.	38.1	-111.2	-108.5	-103.7	-97.8	-91.2	34.2	.8999
900.	1961.	51.9	-151.9	-148.9	-143.6	-136.9	-129.4	44.6	1.0477
1000.	1830.	69.1	-201.7	-198.4	-192.4	-185.0	-176.7	57.0	1.2060

BALLISTIC COEFFICIENT .57, MUZZLE VELOCITY FT/SEC 3500.

RANGE	VELOCITY	HEIGHT	100 YD	150 YD	200 YD	250 YD	300 YD	DRIFT	TIME
0.	3500.	0.0	-1.5	-1.5	-1.5	-1.5	-1.5	0.0	0.0000
100.	3312.	-.4	0.0	.3	.8	1.5	2.3	.5	.0883
200.	3131.	.8	-1.7	-1.1	0.0	1.4	2.9	1.8	.1817
300.	2956.	3.0	-6.9	-6.0	-4.3	-2.3	0.0	4.0	.2798
400.	2789.	6.3	-16.2	-15.0	-12.8	-10.0	-7.0	7.3	.3845
500.	2627.	11.0	-29.8	-28.4	-25.6	-22.2	-18.4	11.7	.4953
600.	2472.	17.3	-48.6	-46.9	-43.5	-39.4	-34.8	17.4	.6130
700.	2321.	25.4	-73.0	-71.0	-67.1	-62.3	-56.9	24.4	.7384
800.	2175.	35.7	-103.8	-101.5	-97.0	-91.6	-85.4	32.7	.8717
900.	2034.	48.7	-142.0	-139.5	-134.4	-128.3	-121.3	42.8	1.0148
1000.	1898.	64.6	-188.4	-185.6	-179.9	-173.2	-165.4	54.6	1.1672

BALLISTIC COEFFICIENT .57, MUZZLE VELOCITY FT/SEC 3600.

RANGE	VELOCITY	HEIGHT	100 YD	150 YD	200 YD	250 YD	300 YD	DRIFT	TIME
0.	3600.	0.0	-1.5	-1.5	-1.5	-1.5	-1.5	0.0	0.0000
100.	3408.	-.4	0.0	.2	.8	1.4	2.1	.4	.0855
200.	3224.	.7	-1.5	-1.0	0.0	1.3	2.7	1.6	.1760
300.	3046.	2.8	-6.3	-5.6	-4.1	-2.1	0.0	3.8	.2715
400.	2874.	5.9	-14.9	-14.0	-11.9	-9.3	-6.5	7.0	.3728
500.	2710.	10.3	-27.8	-26.6	-24.1	-20.8	-17.3	11.3	.4806
600.	2552.	16.2	-45.4	-44.0	-40.9	-37.0	-32.7	16.7	.5947
700.	2398.	23.9	-68.3	-66.6	-63.0	-58.5	-53.5	23.4	.7161
800.	2250.	33.5	-97.2	-95.3	-91.2	-86.0	-80.3	31.4	.8453
900.	2106.	45.6	-132.9	-130.8	-126.1	-120.3	-113.9	41.0	.9830
1000.	1968.	60.6	-176.3	-174.0	-168.8	-162.3	-155.2	52.3	1.1303

BALLISTIC COEFFICIENT .57, MUZZLE VELOCITY FT/SEC 3700.

RANGE	VELOCITY	HEIGHT	100 YD	150 YD	200 YD	250 YD	300 YD	DRIFT	TIME
0.	3700.	0.0	-1.5	-1.5	-1.5	-1.5	-1.5	0.0	0.0000
100.	3505.	-.4	0.0	.2	.7	1.3	2.0	.4	.0833
200.	3316.	.7	-1.4	-.9	0.0	1.2	2.6	1.7	.1716
300.	3135.	2.6	-5.9	-5.3	-3.9	-2.1	0.0	3.8	.2648
400.	2961.	5.6	-13.9	-13.1	-11.2	-8.8	-6.0	6.8	.3628
500.	2793.	9.7	-26.0	-25.0	-22.6	-19.6	-16.2	10.9	.4675
600.	2631.	15.3	-42.5	-41.3	-38.4	-34.8	-30.7	16.1	.5780
700.	2476.	22.5	-64.0	-62.5	-59.2	-55.0	-50.2	22.5	.6955
800.	2325.	31.6	-91.1	-89.5	-85.7	-80.9	-75.4	30.3	.8207
900.	2179.	42.9	-124.6	-122.7	-118.5	-113.0	-106.9	39.4	.9538
1000.	2037.	57.0	-165.5	-163.5	-158.7	-152.7	-145.9	50.3	1.0967

BALLISTIC COEFFICIENT .57, MUZZLE VELOCITY FT/SEC 3800.

RANGE	VELOCITY	HEIGHT	100 YD	150 YD	200 YD	250 YD	300 YD	DRIFT	TIME
0.	3800.	0.0	-1.5	-1.5	-1.5	-1.5	-1.5	0.0	0.0000
100.	3601.	-.4	0.0	.1	.6	1.1	1.8	.5	.0816
200.	3409.	.6	-1.2	-.9	0.0	1.1	2.4	1.6	.1670
300.	3225.	2.4	-5.4	-4.9	-3.6	-1.9	0.0	3.6	.2575
400.	3047.	5.2	-12.9	-12.3	-10.5	-8.3	-5.7	6.5	.3530
500.	2875.	9.2	-24.2	-23.4	-21.2	-18.4	-15.2	10.5	.4543
600.	2711.	14.4	-39.7	-38.8	-36.2	-32.8	-29.0	15.6	.5621
700.	2552.	21.2	-59.9	-58.9	-55.8	-51.9	-47.4	21.7	.6761
800.	2399.	29.8	-85.5	-84.3	-80.7	-76.3	-71.1	29.2	.7974
900.	2251.	40.5	-117.0	-115.7	-111.7	-106.6	-100.9	38.0	.9266
1000.	2107.	53.6	-155.2	-153.8	-149.4	-143.8	-137.4	48.4	1.0642

BALLISTIC COEFFICIENT .57, MUZZLE VELOCITY FT/SEC 3900.

RANGE	VELOCITY	HEIGHT	100 YD	150 YD	200 YD	250 YD	300 YD	DRIFT	TIME
0.	3900.	0.0	-1.5	-1.5	-1.5	-1.5	-1.5	0.0	0.0000
100.	3698.	-.5	0.0	.1	.5	1.1	1.7	.3	.0785
200.	3502.	.5	-1.1	-.8	0.0	1.1	2.3	1.4	.1619
300.	3314.	2.3	-5.0	-4.5	-3.4	-1.8	0.0	3.4	.2502
400.	3133.	4.9	-12.1	-11.5	-10.0	-7.8	-5.5	6.3	.3435
500.	2958.	8.6	-22.6	-21.9	-20.0	-17.3	-14.3	10.0	.4415
600.	2791.	13.6	-37.3	-36.4	-34.1	-30.9	-27.3	14.9	.5463
700.	2629.	20.0	-56.3	-55.3	-52.6	-48.9	-44.7	20.8	.6569
800.	2474.	28.0	-80.3	-79.2	-76.1	-71.9	-67.1	28.0	.7745
900.	2323.	38.1	-110.0	-108.8	-105.3	-100.5	-95.1	36.5	.8998
1000.	2177.	50.5	-146.0	-144.6	-140.8	-135.5	-129.5	46.4	1.0330

BALLISTIC COEFFICIENT .57, MUZZLE VELOCITY FT/SEC 4000.

RANGE	VELOCITY	HEIGHT	100 YD	150 YD	200 YD	250 YD	300 YD	DRIFT	TIME
0.	4000.	0.0	-1.5	-1.5	-1.5	-1.5	-1.5	0.0	0.0000
100.	3794.	-.5	0.0	.1	.5	1.0	1.5	.4	.0770
200.	3596.	.5	-.9	-.8	0.0	1.0	2.1	1.5	.1587
300.	3404.	2.1	-4.6	-4.3	-3.2	-1.7	0.0	3.4	.2443
400.	3219.	4.6	-11.2	-10.8	-9.3	-7.4	-5.1	6.1	.3349
500.	3042.	8.1	-21.2	-20.7	-18.8	-16.4	-13.6	9.8	.4305
600.	2871.	12.8	-34.9	-34.3	-32.1	-29.2	-25.8	14.4	.5320
700.	2706.	18.9	-52.9	-52.3	-49.6	-46.3	-42.3	20.2	.6399
800.	2548.	26.6	-75.6	-74.9	-71.9	-68.0	-63.4	27.1	.7542
900.	2395.	36.1	-103.6	-102.8	-99.4	-95.1	-89.9	35.3	.8757
1000.	2247.	47.7	-137.7	-136.7	-132.9	-128.2	-122.4	44.9	1.0051

BALLISTIC COEFFICIENT .58, MUZZLE VELOCITY FT/SEC 2000.

RANGE	VELOCITY	HEIGHT	100 YD	150 YD	200 YD	250 YD	300 YD	DRIFT	TIME
0.	2000.	0.0	-1.5	-1.5	-1.5	-1.5	-1.5	0.0	0.0000
100.	1868.	.4	0.0	1.9	4.3	6.8	9.5	.9	.1549
200.	1744.	4.2	-8.5	-4.6	0.0	5.1	10.5	3.8	.3214
300.	1626.	11.2	-28.5	-22.7	-15.7	-8.1	0.0	8.7	.4995
400.	1514.	22.2	-61.7	-53.9	-44.6	-34.5	-23.7	16.0	.6908
500.	1410.	37.8	-110.0	-100.3	-88.7	-76.1	-62.5	25.7	.8961
600.	1318.	59.1	-176.0	-164.3	-150.4	-135.2	-119.0	38.1	1.1164
700.	1238.	87.0	-262.1	-248.4	-232.2	-214.5	-195.6	53.1	1.3517
800.	1168.	122.3	-370.4	-354.8	-336.3	-316.1	-294.4	70.6	1.6012
900.	1112.	166.2	-504.7	-487.2	-466.3	-443.6	-419.2	90.6	1.8649
1000.	1069.	219.1	-666.5	-647.0	-623.8	-598.5	-571.4	112.7	2.1401

BALLISTIC COEFFICIENT .58, MUZZLE VELOCITY FT/SEC 2100.

RANGE	VELOCITY	HEIGHT	100 YD	150 YD	200 YD	250 YD	300 YD	DRIFT	TIME
0.	2100.	0.0	-1.5	-1.5	-1.5	-1.5	-1.5	0.0	0.0000
100.	1964.	.3	0.0	1.7	3.8	6.0	8.5	.8	.1477
200.	1834.	3.7	-7.6	-4.2	0.0	4.5	9.4	3.5	.3058
300.	1711.	10.1	-25.4	-20.3	-14.1	-7.3	0.0	8.1	.4747
400.	1595.	20.0	-55.3	-48.5	-40.2	-31.2	-21.5	15.0	.6568
500.	1485.	34.0	-98.7	-90.2	-79.8	-68.5	-56.4	24.1	.8514
600.	1384.	53.3	-158.0	-147.8	-135.4	-121.8	-107.2	35.8	1.0608
700.	1295.	78.6	-235.7	-223.8	-209.3	-193.4	-176.4	50.2	1.2854
800.	1218.	110.8	-333.7	-320.2	-303.5	-285.4	-266.0	67.1	1.5242
900.	1151.	151.0	-455.3	-440.0	-421.3	-400.9	-379.1	86.6	1.7779
1000.	1100.	199.9	-603.2	-586.2	-565.4	-542.7	-518.5	108.4	2.0447

BALLISTIC COEFFICIENT .58, MUZZLE VELOCITY FT/SEC 2200.

RANGE	VELOCITY	HEIGHT	100 YD	150 YD	200 YD	250 YD	300 YD	DRIFT	TIME
0.	2200.	0.0	-1.5	-1.5	-1.5	-1.5	-1.5	0.0	0.0000
100.	2060.	.2	0.0	1.5	3.4	5.4	7.6	.8	.1411
200.	1926.	3.3	-6.7	-3.7	0.0	4.0	8.5	3.4	.2919
300.	1798.	9.1	-22.8	-18.3	-12.7	-6.7	0.0	7.7	.4528
400.	1677.	18.0	-49.8	-43.8	-36.3	-28.2	-19.3	14.1	.6256
500.	1563.	30.8	-89.1	-81.6	-72.3	-62.2	-51.1	22.8	.8112
600.	1455.	48.2	-142.6	-133.6	-122.4	-110.2	-96.9	33.8	1.0101
700.	1357.	71.1	-212.6	-202.2	-189.1	-174.9	-159.4	47.4	1.2239
800.	1272.	100.4	-301.4	-289.4	-274.5	-258.3	-240.5	63.5	1.4519
900.	1197.	137.3	-411.9	-398.5	-381.6	-363.4	-343.5	82.5	1.6958
1000.	1135.	182.8	-547.8	-532.8	-514.1	-493.9	-471.7	104.2	1.9555

BALLISTIC COEFFICIENT .58, MUZZLE VELOCITY FT/SEC 2300.

RANGE	VELOCITY	HEIGHT	100 YD	150 YD	200 YD	250 YD	300 YD	DRIFT	TIME
0.	2300.	0.0	-1.5	-1.5	-1.5	-1.5	-1.5	0.0	0.0000
100.	2157.	.1	0.0	1.3	3.0	4.9	6.9	.7	.1345
200.	2019.	3.0	-6.0	-3.3	0.0	3.7	7.7	3.1	.2784
300.	1886.	8.2	-20.6	-16.6	-11.5	-6.0	0.0	7.2	.4320
400.	1761.	16.3	-44.9	-39.6	-32.9	-25.4	-17.5	13.2	.5965
500.	1642.	27.9	-80.6	-73.9	-65.5	-56.2	-46.3	21.3	.7732
600.	1529.	43.7	-129.1	-121.1	-111.0	-99.8	-87.9	31.7	.9626
700.	1424.	64.5	-192.3	-183.0	-171.3	-158.2	-144.3	44.5	1.1658
800.	1331.	91.2	-272.9	-262.3	-248.9	-234.0	-218.1	59.9	1.3840
900.	1249.	124.8	-373.2	-361.2	-346.1	-329.4	-311.5	78.0	1.6170
1000.	1176.	166.4	-496.3	-482.9	-466.2	-447.6	-427.7	98.8	1.8659

BALLISTIC COEFFICIENT .58, MUZZLE VELOCITY FT/SEC 2400.

RANGE	VELOCITY	HEIGHT	100 YD	150 YD	200 YD	250 YD	300 YD	DRIFT	TIME
0.	2400.	0.0	-1.5	-1.5	-1.5	-1.5	-1.5	0.0	0.0000
100.	2254.	.0	0.0	1.2	2.7	4.4	6.2	.7	.1290
200.	2113.	2.7	-5.3	-3.0	0.0	3.4	7.0	2.9	.2663
300.	1976.	7.4	-18.5	-15.0	-10.5	-5.4	0.0	6.7	.4130
400.	1846.	14.9	-40.7	-36.0	-30.0	-23.2	-16.0	12.4	.5704
500.	1722.	25.4	-72.9	-67.1	-59.6	-51.1	-42.1	19.9	.7383
600.	1605.	39.8	-117.1	-110.1	-101.1	-90.9	-80.1	29.8	.9191
700.	1495.	58.7	-174.5	-166.4	-155.8	-143.9	-131.3	41.8	1.1125
800.	1393.	83.0	-247.6	-238.3	-226.2	-212.6	-198.2	56.4	1.3207
900.	1303.	113.6	-338.7	-328.2	-314.6	-299.3	-283.1	73.6	1.5434
1000.	1225.	151.5	-450.3	-438.7	-423.5	-406.5	-388.6	93.5	1.7811

BALLISTIC COEFFICIENT .58, MUZZLE VELOCITY FT/SEC 2500.

RANGE	VELOCITY	HEIGHT	100 YD	150 YD	200 YD	250 YD	300 YD	DRIFT	TIME
0.	2500.	0.0	-1.5	-1.5	-1.5	-1.5	-1.5	0.0	0.0000
100.	2351.	-.0	0.0	1.0	2.4	3.9	5.6	.7	.1238
200.	2207.	2.4	-4.8	-2.7	0.0	3.1	6.4	2.7	.2553
300.	2067.	6.8	-16.8	-13.7	-9.6	-5.0	0.0	6.3	.3958
400.	1932.	13.6	-37.0	-32.8	-27.4	-21.2	-14.6	11.6	.5462
500.	1804.	23.2	-66.4	-61.2	-54.4	-46.7	-38.4	18.8	.7066
600.	1683.	36.3	-106.6	-100.4	-92.2	-82.9	-73.0	28.0	.8789
700.	1568.	53.6	-159.1	-151.8	-142.3	-131.5	-119.9	39.4	1.0639
800.	1460.	75.7	-225.5	-217.2	-206.3	-194.0	-180.7	53.2	1.2620
900.	1362.	103.7	-308.4	-299.1	-286.8	-272.9	-258.1	69.5	1.4750
1000.	1276.	138.5	-410.3	-400.0	-386.4	-370.9	-354.4	88.5	1.7030

BALLISTIC COEFFICIENT .58, MUZZLE VELOCITY FT/SEC 2600.

RANGE	VELOCITY	HEIGHT	100 YD	150 YD	200 YD	250 YD	300 YD	DRIFT	TIME
0.	2600.	0.0	-1.5	-1.5	-1.5	-1.5	-1.5	0.0	0.0000
100.	2448.	-.1	0.0	.9	2.2	3.6	5.1	.6	.1188
200.	2301.	2.1	-4.3	-2.5	0.0	2.8	5.8	2.6	.2454
300.	2158.	6.2	-15.2	-12.4	-8.7	-4.5	0.0	5.9	.3798
400.	2020.	12.4	-33.7	-30.0	-25.0	-19.4	-13.4	10.9	.5237
500.	1887.	21.3	-60.6	-55.9	-49.8	-42.8	-35.2	17.7	.6772
600.	1761.	33.3	-97.3	-91.7	-84.3	-75.9	-66.8	26.3	.8418
700.	1642.	49.0	-145.3	-138.8	-130.1	-120.3	-109.7	37.1	1.0184
800.	1530.	69.3	-206.0	-198.6	-188.7	-177.5	-165.4	50.1	1.2077
900.	1424.	94.8	-281.5	-273.1	-262.0	-249.4	-235.8	65.5	1.4108
1000.	1331.	126.6	-374.5	-365.2	-352.8	-338.8	-323.7	83.6	1.6290

BALLISTIC COEFFICIENT .58, MUZZLE VELOCITY FT/SEC 2700.

RANGE	VELOCITY	HEIGHT	100 YD	150 YD	200 YD	250 YD	300 YD	DRIFT	TIME
0.	2700.	0.0	−1.5	−1.5	−1.5	−1.5	−1.5	0.0	0.0000
100.	2545.	−.1	0.0	.8	1.9	3.2	4.6	.6	.1146
200.	2394.	1.9	−3.9	−2.3	0.0	2.6	5.4	2.4	.2361
300.	2249.	5.7	−13.9	−11.4	−8.0	−4.2	0.0	5.7	.3655
400.	2107.	11.4	−30.7	−27.5	−23.0	−17.9	−12.2	10.3	.5032
500.	1971.	19.5	−55.4	−51.4	−45.7	−39.3	−32.3	16.7	.6503
600.	1841.	30.6	−89.2	−84.3	−77.5	−69.9	−61.4	24.9	.8081
700.	1718.	45.0	−132.9	−127.2	−119.3	−110.4	−100.5	34.9	.9762
800.	1601.	63.6	−188.7	−182.3	−173.2	−163.0	−151.7	47.3	1.1576
900.	1491.	86.9	−257.7	−250.5	−240.3	−228.8	−216.1	61.9	1.3515
1000.	1389.	116.1	−342.6	−334.6	−323.2	−310.4	−296.4	79.0	1.5602

BALLISTIC COEFFICIENT .58, MUZZLE VELOCITY FT/SEC 2800.

RANGE	VELOCITY	HEIGHT	100 YD	150 YD	200 YD	250 YD	300 YD	DRIFT	TIME
0.	2800.	0.0	−1.5	−1.5	−1.5	−1.5	−1.5	0.0	0.0000
100.	2641.	−.2	0.0	.7	1.7	2.9	4.2	.5	.1100
200.	2488.	1.7	−3.5	−2.1	0.0	2.4	4.9	2.3	.2271
300.	2339.	5.2	−12.7	−10.5	−7.4	−3.9	0.0	5.3	.3516
400.	2195.	10.5	−28.1	−25.3	−21.1	−16.4	−11.3	9.7	.4838
500.	2056.	18.0	−50.9	−47.3	−42.2	−36.3	−29.8	15.8	.6253
600.	1921.	28.2	−82.0	−77.6	−71.5	−64.3	−56.6	23.5	.7764
700.	1794.	41.5	−122.3	−117.2	−110.0	−101.7	−92.7	33.0	.9377
800.	1673.	58.5	−173.4	−167.6	−159.4	−149.9	−139.6	44.7	1.1109
900.	1559.	80.0	−237.0	−230.5	−221.3	−210.6	−199.0	58.5	1.2968
1000.	1451.	106.7	−314.8	−307.6	−297.3	−285.4	−272.6	74.8	1.4963

BALLISTIC COEFFICIENT .58, MUZZLE VELOCITY FT/SEC 2900.

RANGE	VELOCITY	HEIGHT	100 YD	150 YD	200 YD	250 YD	300 YD	DRIFT	TIME
0.	2900.	0.0	−1.5	−1.5	−1.5	−1.5	−1.5	0.0	0.0000
100.	2737.	−.2	0.0	.6	1.6	2.7	3.9	.5	.1064
200.	2581.	1.6	−3.2	−1.9	0.0	2.2	4.5	2.2	.2194
300.	2429.	4.8	−11.6	−9.6	−6.8	−3.5	0.0	5.1	.3391
400.	2282.	9.7	−25.9	−23.3	−19.5	−15.2	−10.5	9.3	.4665
500.	2140.	16.7	−46.9	−43.7	−39.0	−33.5	−27.7	15.0	.6025
600.	2002.	26.1	−75.5	−71.6	−66.0	−59.4	−52.3	22.3	.7473
700.	1871.	38.3	−112.6	−108.1	−101.5	−93.9	−85.6	31.3	.9020
800.	1746.	54.0	−159.8	−154.7	−147.2	−138.4	−129.0	42.3	1.0682
900.	1628.	73.8	−218.3	−212.6	−204.1	−194.3	−183.7	55.5	1.2462
1000.	1516.	98.4	−289.9	−283.5	−274.1	−263.2	−251.4	70.9	1.4372

BALLISTIC COEFFICIENT .58, MUZZLE VELOCITY FT/SEC 3000.

RANGE	VELOCITY	HEIGHT	100 YD	150 YD	200 YD	250 YD	300 YD	DRIFT	TIME
0.	3000.	0.0	−1.5	−1.5	−1.5	−1.5	−1.5	0.0	0.0000
100.	2833.	−.2	0.0	.6	1.4	2.4	3.5	.5	.1030
200.	2673.	1.4	−2.8	−1.7	0.0	2.0	4.2	2.1	.2121
300.	2519.	4.4	−10.6	−8.9	−6.4	−3.3	0.0	4.9	.3281
400.	2369.	9.0	−23.8	−21.6	−18.2	−14.1	−9.7	8.9	.4507
500.	2224.	15.5	−43.3	−40.4	−36.1	−31.1	−25.5	14.3	.5813
600.	2084.	24.2	−69.7	−66.2	−61.1	−55.1	−48.4	21.2	.7205
700.	1948.	35.6	−104.2	−100.2	−94.2	−87.2	−79.4	29.9	.8697
800.	1819.	50.0	−147.7	−143.1	−136.3	−128.2	−119.3	40.3	1.0287
900.	1697.	68.3	−201.8	−196.6	−188.9	−179.9	−169.9	52.7	1.1995
1000.	1582.	91.1	−268.0	−262.3	−253.8	−243.7	−232.6	67.4	1.3831

BALLISTIC COEFFICIENT .58, MUZZLE VELOCITY FT/SEC 3100.

RANGE	VELOCITY	HEIGHT	100 YD	150 YD	200 YD	250 YD	300 YD	DRIFT	TIME
0.	3100.	0.0	−1.5	−1.5	−1.5	−1.5	−1.5	0.0	0.0000
100.	2930.	−.3	0.0	.5	1.3	2.2	3.2	.5	.0994
200.	2766.	1.3	−2.6	−1.6	0.0	1.8	3.9	2.0	.2048
300.	2608.	4.1	−9.7	−8.2	−5.8	−3.1	0.0	4.6	.3164
400.	2456.	8.3	−21.9	−20.0	−16.8	−13.1	−9.0	8.4	.4349
500.	2308.	14.4	−40.0	−37.6	−33.6	−29.0	−23.8	13.6	.5611
600.	2165.	22.4	−64.5	−61.6	−56.8	−51.2	−45.1	20.1	.6950
700.	2027.	33.0	−96.5	−93.1	−87.5	−81.1	−73.9	28.4	.8386
800.	1894.	46.4	−136.8	−132.9	−126.5	−119.2	−111.0	38.3	.9915
900.	1768.	63.3	−186.8	−182.5	−175.3	−167.0	−157.8	50.1	1.1555
1000.	1648.	84.3	−248.0	−243.2	−235.2	−226.0	−215.8	64.0	1.3314

BALLISTIC COEFFICIENT .58, MUZZLE VELOCITY FT/SEC 3200.

RANGE	VELOCITY	HEIGHT	100 YD	150 YD	200 YD	250 YD	300 YD	DRIFT	TIME
0.	3200.	0.0	-1.5	-1.5	-1.5	-1.5	-1.5	0.0	0.0000
100.	3026.	-.3	0.0	.4	1.2	2.0	3.0	.4	.0963
200.	2858.	1.1	-2.3	-1.4	0.0	1.7	3.6	1.9	.1982
300.	2697.	3.8	-8.9	-7.6	-5.4	-2.9	0.0	4.4	.3065
400.	2542.	7.8	-20.3	-18.6	-15.7	-12.3	-8.4	8.1	.4212
500.	2392.	13.4	-37.1	-34.9	-31.3	-27.0	-22.2	13.0	.5428
600.	2246.	20.9	-59.9	-57.4	-53.0	-47.9	-42.1	19.3	.6724
700.	2105.	30.8	-89.6	-86.6	-81.6	-75.5	-68.8	27.1	.8103
800.	1969.	43.3	-127.1	-123.7	-117.9	-111.0	-103.4	36.5	.9575
900.	1839.	59.0	-173.7	-169.8	-163.3	-155.5	-146.9	47.8	1.1155
1000.	1715.	78.4	-230.2	-225.9	-218.7	-210.0	-200.5	61.0	1.2839

BALLISTIC COEFFICIENT .58, MUZZLE VELOCITY FT/SEC 3300.

RANGE	VELOCITY	HEIGHT	100 YD	150 YD	200 YD	250 YD	300 YD	DRIFT	TIME
0.	3300.	0.0	-1.5	-1.5	-1.5	-1.5	-1.5	0.0	0.0000
100.	3122.	-.3	0.0	.4	1.0	1.8	2.7	.5	.0938
200.	2951.	1.0	-2.0	-1.3	0.0	1.6	3.4	1.8	.1921
300.	2786.	3.5	-8.1	-7.0	-5.1	-2.7	0.0	4.3	.2969
400.	2628.	7.2	-18.7	-17.2	-14.6	-11.4	-7.9	7.8	.4078
500.	2475.	12.5	-34.3	-32.5	-29.2	-25.2	-20.8	12.5	.5254
600.	2327.	19.6	-55.6	-53.4	-49.5	-44.7	-39.4	18.5	.6506
700.	2183.	28.7	-83.2	-80.7	-76.1	-70.5	-64.3	25.9	.7835
800.	2044.	40.4	-118.3	-115.4	-110.1	-103.7	-96.6	35.0	.9259
900.	1910.	55.0	-161.4	-158.2	-152.3	-145.1	-137.1	45.7	1.0777
1000.	1783.	73.1	-214.2	-210.6	-204.0	-196.0	-187.1	58.3	1.2402

BALLISTIC COEFFICIENT .58, MUZZLE VELOCITY FT/SEC 3400.

RANGE	VELOCITY	HEIGHT	100 YD	150 YD	200 YD	250 YD	300 YD	DRIFT	TIME
0.	3400.	0.0	-1.5	-1.5	-1.5	-1.5	-1.5	0.0	0.0000
100.	3219.	-.4	0.0	.3	.9	1.7	2.5	.4	.0906
200.	3044.	.9	-1.8	-1.2	0.0	1.5	3.1	1.7	.1862
300.	2876.	3.2	-7.5	-6.5	-4.7	-2.5	0.0	4.0	.2875
400.	2714.	6.7	-17.3	-16.0	-13.7	-10.7	-7.4	7.4	.3952
500.	2558.	11.7	-31.9	-30.3	-27.3	-23.6	-19.5	11.9	.5090
600.	2407.	18.3	-51.8	-49.8	-46.3	-41.9	-36.9	17.7	.6301
700.	2261.	26.9	-77.7	-75.4	-71.2	-66.1	-60.3	24.8	.7587
800.	2120.	37.7	-110.2	-107.6	-102.8	-96.9	-90.3	33.3	.8953
900.	1983.	51.3	-150.5	-147.5	-142.2	-135.6	-128.1	43.6	1.0417
1000.	1852.	68.2	-199.7	-196.4	-190.4	-183.1	-174.8	55.6	1.1985

BALLISTIC COEFFICIENT .58, MUZZLE VELOCITY FT/SEC 3500.

RANGE	VELOCITY	HEIGHT	100 YD	150 YD	200 YD	250 YD	300 YD	DRIFT	TIME
0.	3500.	0.0	-1.5	-1.5	-1.5	-1.5	-1.5	0.0	0.0000
100.	3315.	-.4	0.0	.3	.8	1.5	2.3	.4	.0883
200.	3137.	.8	-1.7	-1.1	0.0	1.3	2.9	1.8	.1815
300.	2965.	3.0	-6.9	-6.0	-4.3	-2.3	0.0	3.9	.2795
400.	2800.	6.3	-16.1	-15.0	-12.7	-10.0	-7.0	7.3	.3841
500.	2641.	11.0	-29.7	-28.3	-25.5	-22.1	-18.3	11.5	.4941
600.	2488.	17.2	-48.3	-46.6	-43.3	-39.2	-34.6	17.0	.6111
700.	2339.	25.2	-72.6	-70.6	-66.7	-61.9	-56.5	23.9	.7357
800.	2195.	35.4	-103.0	-100.8	-96.3	-90.9	-84.7	32.1	.8678
900.	2056.	48.2	-140.8	-138.3	-133.2	-127.1	-120.2	41.9	1.0094
1000.	1921.	63.9	-186.7	-183.9	-178.3	-171.5	-163.8	53.4	1.1605

BALLISTIC COEFFICIENT .58, MUZZLE VELOCITY FT/SEC 3600.

RANGE	VELOCITY	HEIGHT	100 YD	150 YD	200 YD	250 YD	300 YD	DRIFT	TIME
0.	3600.	0.0	-1.5	-1.5	-1.5	-1.5	-1.5	0.0	0.0000
100.	3412.	-.4	0.0	.2	.8	1.4	2.1	.4	.0854
200.	3230.	.7	-1.5	-1.0	0.0	1.3	2.7	1.6	.1758
300.	3055.	2.8	-6.3	-5.6	-4.1	-2.1	0.0	3.7	.2710
400.	2886.	5.9	-14.9	-14.0	-11.9	-9.3	-6.5	6.8	.3722
500.	2724.	10.3	-27.7	-26.5	-23.9	-20.7	-17.2	11.0	.4794
600.	2568.	16.1	-45.1	-43.7	-40.6	-36.7	-32.5	16.3	.5928
700.	2417.	23.7	-67.9	-66.2	-62.6	-58.1	-53.1	22.9	.7133
800.	2270.	33.2	-96.5	-94.6	-90.5	-85.3	-79.6	30.7	.8414
900.	2128.	45.1	-131.7	-129.6	-125.0	-119.2	-112.8	40.1	.9778
1000.	1991.	59.8	-174.6	-172.3	-167.1	-160.7	-153.6	51.1	1.1235

BALLISTIC COEFFICIENT .58, MUZZLE VELOCITY FT/SEC 3700.

RANGE	VELOCITY	HEIGHT	100 YD	150 YD	200 YD	250 YD	300 YD	DRIFT	TIME
0.	3700.	0.0	-1.5	-1.5	-1.5	-1.5	-1.5	0.0	0.0000
100.	3508.	-.4	0.0	.2	.7	1.3	2.0	.4	.0833
200.	3323.	.7	-1.4	-1.0	0.0	1.2	2.6	1.6	.1714
300.	3144.	2.6	-5.9	-5.3	-3.8	-2.0	0.0	3.7	.2644
400.	2972.	5.6	-13.9	-13.1	-11.2	-8.8	-6.0	6.7	.3623
500.	2807.	9.7	-25.9	-24.9	-22.5	-19.5	-16.1	10.7	.4664
600.	2648.	15.2	-42.2	-41.0	-38.2	-34.6	-30.5	15.8	.5763
700.	2494.	22.3	-63.6	-62.2	-58.9	-54.7	-49.9	22.1	.6931
800.	2345.	31.3	-90.5	-88.9	-85.1	-80.3	-74.9	29.7	.8173
900.	2201.	42.5	-123.6	-121.8	-117.5	-112.1	-106.0	38.6	.9491
1000.	2061.	56.3	-163.9	-161.9	-157.2	-151.2	-144.4	49.2	1.0901

BALLISTIC COEFFICIENT .58, MUZZLE VELOCITY FT/SEC 3800.

RANGE	VELOCITY	HEIGHT	100 YD	150 YD	200 YD	250 YD	300 YD	DRIFT	TIME
0.	3800.	0.0	-1.5	-1.5	-1.5	-1.5	-1.5	0.0	0.0000
100.	3604.	-.4	0.0	.1	.6	1.1	1.8	.5	.0816
200.	3416.	.6	-1.2	-.9	0.0	1.1	2.4	1.6	.1667
300.	3234.	2.4	-5.3	-4.9	-3.6	-1.9	0.0	3.6	.2572
400.	3059.	5.2	-12.8	-12.2	-10.5	-8.2	-5.7	6.4	.3522
500.	2890.	9.1	-24.1	-23.3	-21.1	-18.3	-15.1	10.3	.4533
600.	2728.	14.3	-39.5	-38.6	-36.0	-32.6	-28.8	15.2	.5603
700.	2571.	21.0	-59.5	-58.5	-55.4	-51.5	-47.0	21.3	.6736
800.	2420.	29.5	-84.8	-83.7	-80.2	-75.6	-70.6	28.6	.7939
900.	2274.	40.0	-116.0	-114.7	-110.7	-105.7	-99.9	37.2	.9218
1000.	2132.	53.0	-153.8	-152.4	-148.0	-142.3	-136.0	47.3	1.0581

BALLISTIC COEFFICIENT .58, MUZZLE VELOCITY FT/SEC 3900.

RANGE	VELOCITY	HEIGHT	100 YD	150 YD	200 YD	250 YD	300 YD	DRIFT	TIME
0.	3900.	0.0	-1.5	-1.5	-1.5	-1.5	-1.5	0.0	0.0000
100.	3701.	-.5	0.0	.1	.5	1.0	1.6	.3	.0785
200.	3509.	.5	-1.0	-.8	0.0	1.1	2.3	1.4	.1617
300.	3324.	2.2	-4.9	-4.5	-3.4	-1.8	0.0	3.4	.2498
400.	3145.	4.9	-12.0	-11.5	-9.9	-7.8	-5.4	6.2	.3428
500.	2973.	8.6	-22.5	-21.9	-19.9	-17.3	-14.3	9.9	.4407
600.	2808.	13.5	-37.1	-36.3	-34.0	-30.8	-27.2	14.6	.5447
700.	2648.	19.8	-55.9	-55.0	-52.3	-48.6	-44.4	20.4	.6546
800.	2495.	27.8	-79.8	-78.7	-75.6	-71.4	-66.6	27.4	.7713
900.	2346.	37.7	-109.2	-108.0	-104.5	-99.8	-94.4	35.8	.8955
1000.	2202.	49.9	-144.7	-143.4	-139.5	-134.3	-128.2	45.4	1.0272

BALLISTIC COEFFICIENT .58, MUZZLE VELOCITY FT/SEC 4000.

RANGE	VELOCITY	HEIGHT	100 YD	150 YD	200 YD	250 YD	300 YD	DRIFT	TIME
0.	4000.	0.0	-1.5	-1.5	-1.5	-1.5	-1.5	0.0	0.0000
100.	3798.	-.5	0.0	.1	.5	.9	1.5	.3	.0769
200.	3602.	.5	-.9	-.8	0.0	.9	2.1	1.5	.1585
300.	3414.	2.1	-4.5	-4.3	-3.1	-1.7	0.0	3.3	.2438
400.	3232.	4.6	-11.2	-10.8	-9.3	-7.4	-5.1	6.0	.3342
500.	3057.	8.1	-21.1	-20.6	-18.7	-16.3	-13.5	9.6	.4294
600.	2888.	12.8	-34.7	-34.2	-31.9	-29.1	-25.7	14.2	.5305
700.	2726.	18.8	-52.6	-51.9	-49.3	-46.0	-42.0	19.8	.6376
800.	2569.	26.3	-75.1	-74.3	-71.3	-67.5	-63.0	26.6	.7509
900.	2418.	35.7	-102.8	-102.0	-98.5	-94.3	-89.2	34.6	.8714
1000.	2272.	47.2	-136.4	-135.4	-131.7	-126.9	-121.2	43.9	.9993

BALLISTIC COEFFICIENT .59, MUZZLE VELOCITY FT/SEC 2000.

RANGE	VELOCITY	HEIGHT	100 YD	150 YD	200 YD	250 YD	300 YD	DRIFT	TIME
0.	2000.	0.0	-1.5	-1.5	-1.5	-1.5	-1.5	0.0	0.0000
100.	1870.	.4	0.0	1.9	4.2	6.8	9.5	.9	.1548
200.	1748.	4.2	-8.5	-4.6	0.0	5.0	10.4	3.7	.3209
300.	1632.	11.2	-28.4	-22.6	-15.7	-8.1	0.0	8.6	.4986
400.	1522.	22.0	-61.4	-53.7	-44.4	-34.4	-23.5	15.7	.6891
500.	1418.	37.5	-109.4	-99.7	-88.2	-75.6	-62.1	25.2	.8932
600.	1327.	58.6	-174.9	-163.2	-149.4	-134.3	-118.1	37.3	1.1122
700.	1247.	86.2	-260.1	-246.5	-230.3	-212.7	-193.8	52.0	1.3456
800.	1176.	121.3	-368.1	-352.5	-334.1	-313.9	-292.3	69.5	1.5949
900.	1120.	164.5	-500.1	-482.6	-461.9	-439.2	-414.9	88.9	1.8553
1000.	1076.	216.8	-660.1	-640.7	-617.7	-592.4	-565.4	110.7	2.1287

BALLISTIC COEFFICIENT .59, MUZZLE VELOCITY FT/SEC 2100.

RANGE	VELOCITY	HEIGHT	100 YD	150 YD	200 YD	250 YD	300 YD	DRIFT	TIME
0.	2100.	0.0	-1.5	-1.5	-1.5	-1.5	-1.5	0.0	0.0000
100.	1966.	.3	0.0	1.7	3.8	6.0	8.4	.8	.1475
200.	1838.	3.7	-7.6	-4.1	0.0	4.5	9.3	3.5	.3055
300.	1717.	10.0	-25.3	-20.2	-14.0	-7.2	0.0	8.0	.4738
400.	1603.	19.9	-55.1	-48.3	-40.0	-31.0	-21.3	14.7	.6552
500.	1494.	33.8	-98.2	-89.7	-79.3	-68.0	-56.0	23.6	.8487
600.	1394.	52.9	-157.1	-146.9	-134.4	-120.9	-106.4	35.1	1.0568
700.	1306.	77.8	-233.8	-221.9	-207.4	-191.6	-174.7	49.2	1.2793
800.	1228.	109.6	-331.0	-317.4	-300.8	-282.7	-263.5	65.8	1.5165
900.	1160.	149.3	-451.0	-435.7	-417.0	-396.7	-375.1	84.9	1.7681
1000.	1108.	197.7	-597.4	-580.4	-559.6	-537.0	-513.0	106.4	2.0333

BALLISTIC COEFFICIENT .59, MUZZLE VELOCITY FT/SEC 2200.

RANGE	VELOCITY	HEIGHT	100 YD	150 YD	200 YD	250 YD	300 YD	DRIFT	TIME
0.	2200.	0.0	-1.5	-1.5	-1.5	-1.5	-1.5	0.0	0.0000
100.	2063.	.2	0.0	1.5	3.4	5.4	7.6	.8	.1410
200.	1930.	3.3	-6.7	-3.7	0.0	4.0	8.5	3.3	.2915
300.	1804.	9.1	-22.8	-18.3	-12.7	-6.6	0.0	7.6	.4520
400.	1685.	17.9	-49.5	-43.5	-36.1	-28.0	-19.2	13.8	.6240
500.	1572.	30.6	-88.7	-81.2	-71.9	-61.8	-50.7	22.3	.8087
600.	1466.	47.8	-141.7	-132.7	-121.5	-109.4	-96.1	33.1	1.0061
700.	1368.	70.5	-211.0	-200.5	-187.5	-173.3	-157.9	46.4	1.2181
800.	1283.	99.4	-298.9	-286.9	-272.0	-255.9	-238.2	62.2	1.4446
900.	1208.	135.7	-407.9	-394.4	-377.7	-359.5	-339.6	80.7	1.6858
1000.	1144.	180.1	-540.7	-525.7	-507.1	-486.9	-464.8	101.6	1.9412

BALLISTIC COEFFICIENT .59, MUZZLE VELOCITY FT/SEC 2300.

RANGE	VELOCITY	HEIGHT	100 YD	150 YD	200 YD	250 YD	300 YD	DRIFT	TIME
0.	2300.	0.0	-1.5	-1.5	-1.5	-1.5	-1.5	0.0	0.0000
100.	2160.	.1	0.0	1.3	3.0	4.9	6.8	.7	.1344
200.	2024.	3.0	-6.0	-3.3	0.0	3.7	7.7	3.0	.2781
300.	1893.	8.2	-20.5	-16.5	-11.5	-5.9	0.0	7.0	.4313
400.	1769.	16.3	-44.8	-39.4	-32.8	-25.3	-17.4	12.9	.5953
500.	1652.	27.8	-80.2	-73.5	-65.2	-55.9	-46.0	20.9	.7709
600.	1541.	43.4	-128.3	-120.2	-110.2	-99.1	-87.2	31.0	.9589
700.	1436.	63.9	-191.0	-181.7	-170.0	-157.0	-143.2	43.6	1.1608
800.	1342.	90.3	-270.7	-260.0	-246.6	-231.8	-216.0	58.7	1.3769
900.	1260.	123.3	-369.5	-357.5	-342.5	-325.8	-308.0	76.3	1.6074
1000.	1188.	164.1	-490.5	-477.1	-460.4	-441.9	-422.1	96.6	1.8531

BALLISTIC COEFFICIENT .59, MUZZLE VELOCITY FT/SEC 2400.

RANGE	VELOCITY	HEIGHT	100 YD	150 YD	200 YD	250 YD	300 YD	DRIFT	TIME
0.	2400.	0.0	-1.5	-1.5	-1.5	-1.5	-1.5	0.0	0.0000
100.	2257.	.0	0.0	1.2	2.7	4.4	6.1	.7	.1290
200.	2117.	2.6	-5.3	-3.0	0.0	3.4	7.0	2.8	.2659
300.	1983.	7.4	-18.4	-15.0	-10.5	-5.4	0.0	6.6	.4123
400.	1854.	14.8	-40.5	-35.8	-29.8	-23.0	-15.9	12.1	.5688
500.	1732.	25.3	-72.6	-66.8	-59.3	-50.8	-41.9	19.6	.7362
600.	1617.	39.5	-116.3	-109.4	-100.4	-90.2	-79.4	29.1	.9155
700.	1508.	58.1	-173.3	-165.2	-154.6	-142.8	-130.2	40.9	1.1076
800.	1406.	82.1	-245.5	-236.2	-224.2	-210.6	-196.3	55.2	1.3138
900.	1316.	112.3	-335.6	-325.2	-311.7	-296.4	-280.3	72.1	1.5348
1000.	1237.	149.7	-445.8	-434.2	-419.2	-402.3	-384.4	91.6	1.7703

BALLISTIC COEFFICIENT .59, MUZZLE VELOCITY FT/SEC 2500.

RANGE	VELOCITY	HEIGHT	100 YD	150 YD	200 YD	250 YD	300 YD	DRIFT	TIME
0.	2500.	0.0	-1.5	-1.5	-1.5	-1.5	-1.5	0.0	0.0000
100.	2353.	-.0	0.0	1.0	2.4	3.9	5.6	.7	.1238
200.	2211.	2.4	-4.8	-2.7	0.0	3.1	6.4	2.6	.2550
300.	2074.	6.7	-16.7	-13.6	-9.5	-4.9	0.0	6.2	.3950
400.	1941.	13.5	-36.8	-32.7	-27.2	-21.1	-14.5	11.4	.5448
500.	1814.	23.1	-66.0	-60.9	-54.1	-46.4	-38.2	18.4	.7044
600.	1695.	36.0	-105.9	-99.7	-91.6	-82.3	-72.5	27.4	.8756
700.	1581.	53.1	-157.9	-150.7	-141.2	-130.4	-119.0	38.6	1.0592
800.	1474.	74.9	-223.6	-215.3	-204.5	-192.1	-179.0	52.0	1.2554
900.	1376.	102.5	-305.5	-296.2	-284.0	-270.2	-255.4	68.0	1.4665
1000.	1290.	136.6	-405.8	-395.5	-381.9	-366.5	-350.1	86.5	1.6916

BALLISTIC COEFFICIENT .59, MUZZLE VELOCITY FT/SEC 2600.

RANGE	VELOCITY	HEIGHT	100 YD	150 YD	200 YD	250 YD	300 YD	DRIFT	TIME
0.	2600.	0.0	-1.5	-1.5	-1.5	-1.5	-1.5	0.0	0.0000
100.	2451.	-.1	0.0	.9	2.2	3.6	5.1	.6	.1187
200.	2306.	2.1	-4.3	-2.5	0.0	2.8	5.8	2.5	.2451
300.	2165.	6.2	-15.2	-12.4	-8.7	-4.5	0.0	5.8	.3791
400.	2029.	12.4	-33.6	-29.8	-24.9	-19.3	-13.3	10.8	.5226
500.	1898.	21.1	-60.3	-55.7	-49.5	-42.5	-35.0	17.3	.6753
600.	1774.	33.0	-96.8	-91.2	-83.8	-75.4	-66.4	25.8	.8390
700.	1656.	48.6	-144.3	-137.8	-129.1	-119.4	-108.9	36.3	1.0140
800.	1545.	68.6	-204.4	-197.0	-187.1	-175.9	-163.9	49.0	1.2016
900.	1440.	93.7	-279.1	-270.7	-259.6	-247.0	-233.5	64.2	1.4030
1000.	1346.	125.0	-370.6	-361.3	-349.0	-335.0	-320.0	81.8	1.6185

BALLISTIC COEFFICIENT .59, MUZZLE VELOCITY FT/SEC 2700.

RANGE	VELOCITY	HEIGHT	100 YD	150 YD	200 YD	250 YD	300 YD	DRIFT	TIME
0.	2700.	0.0	-1.5	-1.5	-1.5	-1.5	-1.5	0.0	0.0000
100.	2547.	-.1	0.0	.8	1.9	3.2	4.6	.6	.1145
200.	2399.	1.9	-3.9	-2.3	0.0	2.6	5.3	2.4	.2359
300.	2256.	5.6	-13.8	-11.4	-8.0	-4.2	0.0	5.6	.3649
400.	2117.	11.3	-30.6	-27.3	-22.8	-17.7	-12.1	10.1	.5018
500.	1982.	19.4	-55.1	-51.1	-45.4	-39.0	-32.1	16.3	.6482
600.	1854.	30.3	-88.5	-83.7	-76.9	-69.3	-60.9	24.3	.8049
700.	1732.	44.6	-132.1	-126.4	-118.5	-109.6	-99.8	34.2	.9723
800.	1617.	62.9	-187.1	-180.7	-171.6	-161.4	-150.3	46.2	1.1516
900.	1507.	85.9	-255.4	-248.2	-238.0	-226.5	-214.0	60.5	1.3438
1000.	1406.	114.6	-339.1	-331.0	-319.7	-306.9	-293.0	77.3	1.5500

BALLISTIC COEFFICIENT .59, MUZZLE VELOCITY FT/SEC 2800.

RANGE	VELOCITY	HEIGHT	100 YD	150 YD	200 YD	250 YD	300 YD	DRIFT	TIME
0.	2800.	0.0	-1.5	-1.5	-1.5	-1.5	-1.5	0.0	0.0000
100.	2644.	-.2	0.0	.7	1.7	2.9	4.2	.5	.1100
200.	2493.	1.7	-3.5	-2.0	0.0	2.4	4.9	2.2	.2269
300.	2347.	5.2	-12.6	-10.5	-7.4	-3.9	0.0	5.2	.3511
400.	2205.	10.4	-28.0	-25.1	-21.0	-16.3	-11.2	9.5	.4827
500.	2067.	17.9	-50.7	-47.0	-41.9	-36.0	-29.6	15.4	.6234
600.	1935.	28.0	-81.4	-77.1	-71.0	-63.9	-56.2	23.0	.7735
700.	1808.	41.1	-121.4	-116.3	-109.2	-100.9	-91.9	32.3	.9336
800.	1689.	57.9	-172.0	-166.2	-158.0	-148.6	-138.3	43.7	1.1053
900.	1576.	79.1	-234.9	-228.4	-219.2	-208.6	-197.0	57.3	1.2896
1000.	1469.	105.3	-311.5	-304.3	-294.0	-282.3	-269.4	73.1	1.4865

BALLISTIC COEFFICIENT .59, MUZZLE VELOCITY FT/SEC 2900.

RANGE	VELOCITY	HEIGHT	100 YD	150 YD	200 YD	250 YD	300 YD	DRIFT	TIME
0.	2900.	0.0	-1.5	-1.5	-1.5	-1.5	-1.5	0.0	0.0000
100.	2740.	-.2	0.0	.6	1.6	2.7	3.8	.5	.1063
200.	2586.	1.6	-3.2	-1.9	0.0	2.2	4.5	2.2	.2191
300.	2437.	4.7	-11.5	-9.6	-6.8	-3.5	0.0	4.9	.3385
400.	2292.	9.7	-25.8	-23.2	-19.5	-15.1	-10.4	9.1	.4656
500.	2152.	16.6	-46.6	-43.4	-38.7	-33.3	-27.5	14.7	.6006
600.	2016.	25.9	-75.0	-71.2	-65.6	-59.0	-52.0	21.8	.7447
700.	1886.	38.0	-111.9	-107.4	-100.8	-93.2	-85.0	30.7	.8984
800.	1762.	53.5	-158.5	-153.4	-145.9	-137.2	-127.9	41.4	1.0629
900.	1645.	73.0	-216.4	-210.7	-202.2	-192.4	-181.9	54.3	1.2393
1000.	1535.	97.1	-287.0	-280.6	-271.3	-260.4	-248.7	69.3	1.4281

BALLISTIC COEFFICIENT .59, MUZZLE VELOCITY FT/SEC 3000.

RANGE	VELOCITY	HEIGHT	100 YD	150 YD	200 YD	250 YD	300 YD	DRIFT	TIME
0.	3000.	0.0	-1.5	-1.5	-1.5	-1.5	-1.5	0.0	0.0000
100.	2836.	-.2	0.0	.6	1.4	2.4	3.5	.5	.1029
200.	2678.	1.4	-2.8	-1.7	0.0	2.0	4.2	2.1	.2118
300.	2527.	4.4	-10.6	-8.9	-6.3	-3.3	0.0	4.9	.3276
400.	2379.	9.0	-23.7	-21.4	-18.1	-14.0	-9.6	8.7	.4497
500.	2236.	15.4	-43.1	-40.2	-36.0	-31.0	-25.4	14.1	.5799
600.	2098.	24.0	-69.4	-65.9	-60.9	-54.8	-48.2	20.8	.7184
700.	1964.	35.3	-103.5	-99.5	-93.6	-86.5	-78.8	29.2	.8662
800.	1836.	49.6	-146.7	-142.1	-135.3	-127.3	-118.4	39.5	1.0242
900.	1715.	67.5	-199.9	-194.7	-187.2	-178.1	-168.1	51.5	1.1927
1000.	1601.	89.9	-265.3	-259.6	-251.1	-241.1	-230.0	65.9	1.3743

BALLISTIC COEFFICIENT .59, MUZZLE VELOCITY FT/SEC 3100.

RANGE	VELOCITY	HEIGHT	100 YD	150 YD	200 YD	250 YD	300 YD	DRIFT	TIME
0.	3100.	0.0	-1.5	-1.5	-1.5	-1.5	-1.5	0.0	0.0000
100.	2933.	-.3	0.0	.5	1.3	2.2	3.2	.4	.0993
200.	2771.	1.3	-2.6	-1.6	0.0	1.8	3.9	1.9	.2045
300.	2616.	4.0	-9.6	-8.2	-5.8	-3.0	0.0	4.5	.3158
400.	2466.	8.3	-21.8	-19.9	-16.7	-13.0	-9.0	8.3	.4340
500.	2320.	14.3	-39.8	-37.3	-33.4	-28.8	-23.7	13.3	.5595
600.	2179.	22.3	-64.1	-61.2	-56.4	-50.9	-44.8	19.7	.6927
700.	2043.	32.7	-95.9	-92.5	-87.0	-80.5	-73.4	27.8	.8354
800.	1911.	46.0	-135.8	-132.0	-125.6	-118.2	-110.1	37.5	.9871
900.	1786.	62.7	-185.3	-180.9	-173.8	-165.5	-156.4	49.0	1.1494
1000.	1668.	83.3	-245.6	-240.8	-232.9	-223.6	-213.5	62.6	1.3232

BALLISTIC COEFFICIENT .59, MUZZLE VELOCITY FT/SEC 3200.

RANGE	VELOCITY	HEIGHT	100 YD	150 YD	200 YD	250 YD	300 YD	DRIFT	TIME
0.	3200.	0.0	-1.5	-1.5	-1.5	-1.5	-1.5	0.0	0.0000
100.	3029.	-.3	0.0	.4	1.1	2.0	3.0	.4	.0962
200.	2864.	1.1	-2.3	-1.4	0.0	1.7	3.6	1.8	.1980
300.	2705.	3.7	-8.9	-7.6	-5.4	-2.9	0.0	4.4	.3060
400.	2552.	7.7	-20.2	-18.5	-15.6	-12.2	-8.4	8.0	.4203
500.	2404.	13.3	-36.9	-34.8	-31.2	-26.9	-22.1	12.8	.5414
600.	2261.	20.8	-59.6	-57.0	-52.7	-47.6	-41.8	18.9	.6702
700.	2121.	30.5	-89.0	-86.0	-80.9	-74.9	-68.2	26.5	.8068
800.	1987.	42.8	-126.1	-122.7	-116.9	-110.1	-102.4	35.7	.9530
900.	1858.	58.3	-172.0	-168.1	-161.7	-153.9	-145.4	46.7	1.1090
1000.	1736.	77.4	-228.0	-223.8	-216.6	-208.0	-198.4	59.6	1.2763

BALLISTIC COEFFICIENT .59, MUZZLE VELOCITY FT/SEC 3300.

RANGE	VELOCITY	HEIGHT	100 YD	150 YD	200 YD	250 YD	300 YD	DRIFT	TIME
0.	3300.	0.0	-1.5	-1.5	-1.5	-1.5	-1.5	0.0	0.0000
100.	3125.	-.3	0.0	.4	1.0	1.8	2.7	.5	.0938
200.	2957.	1.0	-2.0	-1.3	0.0	1.6	3.4	1.8	.1919
300.	2795.	3.5	-8.1	-7.0	-5.1	-2.7	0.0	4.2	.2967
400.	2638.	7.2	-18.6	-17.2	-14.6	-11.4	-7.8	7.6	.4071
500.	2488.	12.4	-34.1	-32.3	-29.1	-25.1	-20.6	12.3	.5241
600.	2342.	19.4	-55.3	-53.2	-49.2	-44.5	-39.1	18.2	.6487
700.	2200.	28.5	-82.7	-80.2	-75.6	-70.0	-63.8	25.4	.7806
800.	2063.	40.0	-117.4	-114.5	-109.2	-102.9	-95.7	34.2	.9216
900.	1930.	54.4	-160.1	-156.9	-151.0	-143.8	-135.8	44.7	1.0721
1000.	1804.	72.2	-212.1	-208.5	-201.9	-194.0	-185.0	56.9	1.2326

BALLISTIC COEFFICIENT .59, MUZZLE VELOCITY FT/SEC 3400.

RANGE	VELOCITY	HEIGHT	100 YD	150 YD	200 YD	250 YD	300 YD	DRIFT	TIME
0.	3400.	0.0	-1.5	-1.5	-1.5	-1.5	-1.5	0.0	0.0000
100.	3222.	-.4	0.0	.3	.9	1.7	2.5	.4	.0905
200.	3050.	.9	-1.8	-1.2	0.0	1.5	3.1	1.7	.1860
300.	2884.	3.2	-7.4	-6.5	-4.7	-2.5	0.0	3.9	.2871
400.	2725.	6.7	-17.3	-16.0	-13.6	-10.6	-7.4	7.3	.3944
500.	2571.	11.6	-31.8	-30.1	-27.2	-23.5	-19.4	11.7	.5078
600.	2422.	18.2	-51.5	-49.5	-46.0	-41.6	-36.6	17.4	.6280
700.	2278.	26.7	-77.2	-74.8	-70.7	-65.5	-59.8	24.3	.7556
800.	2139.	37.4	-109.5	-106.9	-102.2	-96.3	-89.7	32.7	.8918
900.	2003.	50.8	-149.3	-146.4	-141.1	-134.4	-127.0	42.7	1.0366
1000.	1874.	67.4	-197.7	-194.4	-188.5	-181.1	-172.9	54.4	1.1912

BALLISTIC COEFFICIENT .59, MUZZLE VELOCITY FT/SEC 3500.

RANGE	VELOCITY	HEIGHT	100 YD	150 YD	200 YD	250 YD	300 YD	DRIFT	TIME
0.	3500.	0.0	-1.5	-1.5	-1.5	-1.5	-1.5	0.0	0.0000
100.	3318.	-.4	0.0	.3	.8	1.5	2.3	.4	.0882
200.	3143.	.8	-1.7	-1.1	0.0	1.3	2.9	1.7	.1813
300.	2974.	3.0	-6.9	-6.0	-4.3	-2.3	0.0	3.9	.2792
400.	2811.	6.3	-16.0	-14.9	-12.7	-10.0	-6.9	7.1	.3832
500.	2654.	10.9	-29.6	-28.2	-25.4	-22.0	-18.1	11.3	.4929
600.	2503.	17.1	-48.1	-46.4	-43.0	-39.0	-34.4	16.7	.6094
700.	2356.	25.0	-72.1	-70.2	-66.2	-61.5	-56.1	23.4	.7330
800.	2214.	35.1	-102.3	-100.1	-95.6	-90.2	-84.0	31.4	.8641
900.	2076.	47.6	-139.5	-137.0	-132.0	-125.9	-118.9	40.9	1.0039
1000.	1943.	63.1	-184.9	-182.1	-176.5	-169.8	-162.0	52.2	1.1536

BALLISTIC COEFFICIENT .59, MUZZLE VELOCITY FT/SEC 3600.

RANGE	VELOCITY	HEIGHT	100 YD	150 YD	200 YD	250 YD	300 YD	DRIFT	TIME
0.	3600.	0.0	-1.5	-1.5	-1.5	-1.5	-1.5	0.0	0.0000
100.	3415.	-.4	0.0	.2	.7	1.4	2.1	.3	.0853
200.	3236.	.7	-1.5	-1.0	0.0	1.3	2.7	1.6	.1756
300.	3063.	2.8	-6.3	-5.6	-4.0	-2.1	0.0	3.6	.2706
400.	2897.	5.9	-14.9	-13.9	-11.9	-9.3	-6.5	6.7	.3716
500.	2737.	10.2	-27.6	-26.4	-23.8	-20.6	-17.1	10.8	.4781
600.	2583.	16.0	-44.9	-43.5	-40.4	-36.5	-32.3	16.0	.5910
700.	2434.	23.5	-67.4	-65.7	-62.1	-57.6	-52.7	22.4	.7105
800.	2290.	32.9	-95.8	-93.9	-89.8	-84.6	-79.0	30.1	.8377
900.	2150.	44.7	-130.7	-128.6	-124.0	-118.2	-111.8	39.2	.9729
1000.	2014.	59.2	-173.1	-170.7	-165.6	-159.2	-152.1	49.9	1.1171

BALLISTIC COEFFICIENT .59, MUZZLE VELOCITY FT/SEC 3700.

RANGE	VELOCITY	HEIGHT	100 YD	150 YD	200 YD	250 YD	300 YD	DRIFT	TIME
0.	3700.	0.0	-1.5	-1.5	-1.5	-1.5	-1.5	0.0	0.0000
100.	3511.	-.4	0.0	.2	.7	1.3	1.9	.4	.0832
200.	3329.	.7	-1.3	-1.0	0.0	1.2	2.5	1.6	.1713
300.	3153.	2.6	-5.8	-5.3	-3.8	-2.0	0.0	3.6	.2640
400.	2984.	5.5	-13.8	-13.0	-11.1	-8.7	-6.0	6.6	.3617
500.	2821.	9.6	-25.7	-24.8	-22.4	-19.4	-16.0	10.5	.4651
600.	2663.	15.1	-42.0	-40.8	-38.0	-34.4	-30.3	15.5	.5745
700.	2512.	22.2	-63.2	-61.9	-58.5	-54.3	-49.6	21.7	.6907
800.	2365.	31.0	-89.9	-88.3	-84.5	-79.7	-74.3	29.1	.8138
900.	2223.	42.1	-122.6	-120.8	-116.5	-111.2	-105.1	37.8	.9444
1000.	2085.	55.6	-162.4	-160.4	-155.6	-149.7	-142.9	48.0	1.0837

BALLISTIC COEFFICIENT .59, MUZZLE VELOCITY FT/SEC 3800.

RANGE	VELOCITY	HEIGHT	100 YD	150 YD	200 YD	250 YD	300 YD	DRIFT	TIME
0.	3800.	0.0	-1.5	-1.5	-1.5	-1.5	-1.5	0.0	0.0000
100.	3608.	-.4	0.0	.1	.6	1.1	1.8	.5	.0815
200.	3422.	.6	-1.2	-.9	0.0	1.1	2.4	1.5	.1665
300.	3243.	2.4	-5.3	-4.9	-3.6	-1.9	0.0	3.5	.2567
400.	3070.	5.2	-12.8	-12.2	-10.5	-8.2	-5.7	6.3	.3517
500.	2904.	9.1	-24.0	-23.3	-21.1	-18.3	-15.1	10.2	.4524
600.	2744.	14.2	-39.2	-38.4	-35.8	-32.4	-28.6	14.9	.5586
700.	2590.	20.9	-59.1	-58.2	-55.1	-51.2	-46.7	20.9	.6712
800.	2440.	29.2	-84.1	-83.0	-79.5	-75.0	-70.0	27.9	.7903
900.	2296.	39.6	-115.0	-113.8	-109.8	-104.8	-99.1	36.4	.9173
1000.	2155.	52.4	-152.4	-151.0	-146.6	-141.0	-134.7	46.2	1.0520

BALLISTIC COEFFICIENT .59, MUZZLE VELOCITY FT/SEC 3900.

RANGE	VELOCITY	HEIGHT	100 YD	150 YD	200 YD	250 YD	300 YD	DRIFT	TIME
0.	3900.	0.0	-1.5	-1.5	-1.5	-1.5	-1.5	0.0	0.0000
100.	3704.	-.5	0.0	.1	.5	1.0	1.6	.3	.0784
200.	3515.	.5	-1.0	-.8	0.0	1.1	2.3	1.3	.1615
300.	3333.	2.2	-4.9	-4.5	-3.4	-1.8	0.0	3.3	.2495
400.	3157.	4.9	-11.9	-11.4	-9.9	-7.8	-5.4	6.0	.3420
500.	2988.	8.5	-22.4	-21.8	-19.9	-17.2	-14.2	9.7	.4396
600.	2824.	13.4	-36.8	-36.0	-33.7	-30.6	-27.0	14.3	.5429
700.	2667.	19.7	-55.6	-54.6	-52.0	-48.3	-44.1	20.0	.6521
800.	2515.	27.6	-79.3	-78.2	-75.1	-70.9	-66.1	26.9	.7683
900.	2368.	37.4	-108.3	-107.1	-103.7	-98.9	-93.5	35.0	.8911
1000.	2226.	49.3	-143.5	-142.1	-138.3	-133.0	-127.0	44.4	1.0216

BALLISTIC COEFFICIENT .59, MUZZLE VELOCITY FT/SEC 4000.

RANGE	VELOCITY	HEIGHT	100 YD	150 YD	200 YD	250 YD	300 YD	DRIFT	TIME
0.	4000.	0.0	-1.5	-1.5	-1.5	-1.5	-1.5	0.0	0.0000
100.	3801.	-.5	0.0	.1	.5	.9	1.5	.3	.0769
200.	3609.	.5	-.9	-.8	0.0	.9	2.1	1.5	.1584
300.	3423.	2.1	-4.5	-4.2	-3.1	-1.7	0.0	3.2	.2434
400.	3244.	4.6	-11.1	-10.7	-9.2	-7.3	-5.1	5.9	.3335
500.	3071.	8.1	-21.0	-20.5	-18.6	-16.3	-13.5	9.4	.4285
600.	2905.	12.7	-34.6	-34.0	-31.8	-29.0	-25.6	13.9	.5292
700.	2745.	18.6	-52.3	-51.6	-49.0	-45.7	-41.7	19.4	.6353
800.	2590.	26.1	-74.6	-73.8	-70.8	-67.0	-62.5	26.0	.7479
900.	2441.	35.3	-101.9	-101.1	-97.7	-93.4	-88.4	33.8	.8670
1000.	2296.	46.7	-135.2	-134.2	-130.5	-125.8	-120.1	42.9	.9939

BALLISTIC COEFFICIENT .60, MUZZLE VELOCITY FT/SEC 2000.

RANGE	VELOCITY	HEIGHT	100 YD	150 YD	200 YD	250 YD	300 YD	DRIFT	TIME
0.	2000.	0.0	-1.5	-1.5	-1.5	-1.5	-1.5	0.0	0.0000
100.	1873.	.4	0.0	1.9	4.2	6.7	9.4	.8	.1548
200.	1752.	4.2	-8.5	-4.6	0.0	5.0	10.4	3.6	.3205
300.	1637.	11.1	-28.3	-22.5	-15.6	-8.1	0.0	8.4	.4978
400.	1529.	21.9	-61.2	-53.4	-44.2	-34.2	-23.4	15.4	.6875
500.	1427.	37.3	-108.9	-99.2	-87.7	-75.2	-61.7	24.7	.8906
600.	1336.	58.2	-174.0	-162.3	-148.6	-133.5	-117.3	36.7	1.1085
700.	1256.	85.5	-258.3	-244.8	-228.7	-211.1	-192.3	51.1	1.3402
800.	1185.	120.0	-364.5	-349.0	-330.6	-310.5	-289.0	67.9	1.5860
900.	1128.	162.8	-495.8	-478.3	-457.7	-435.1	-410.8	87.3	1.8460
1000.	1082.	214.8	-655.0	-635.6	-612.7	-587.6	-560.6	109.0	2.1191

BALLISTIC COEFFICIENT .60, MUZZLE VELOCITY FT/SEC 2100.

RANGE	VELOCITY	HEIGHT	100 YD	150 YD	200 YD	250 YD	300 YD	DRIFT	TIME
0.	2100.	0.0	-1.5	-1.5	-1.5	-1.5	-1.5	0.0	0.0000
100.	1968.	.3	0.0	1.7	3.8	6.0	8.4	.8	.1474
200.	1842.	3.7	-7.5	-4.1	0.0	4.5	9.3	3.4	.3052
300.	1723.	10.0	-25.3	-20.1	-13.9	-7.2	0.0	7.8	.4730
400.	1610.	19.7	-54.8	-48.0	-39.8	-30.8	-21.2	14.4	.6535
500.	1503.	33.6	-97.7	-89.2	-78.9	-67.6	-55.6	23.2	.8461
600.	1403.	52.5	-156.1	-145.9	-133.5	-120.0	-105.6	34.4	1.0529
700.	1315.	77.2	-232.3	-220.4	-206.0	-190.2	-173.4	48.3	1.2742
800.	1237.	108.6	-328.6	-314.9	-298.4	-280.4	-261.2	64.6	1.5096
900.	1169.	147.7	-447.1	-431.8	-413.2	-392.9	-371.4	83.3	1.7589
1000.	1115.	195.5	-591.6	-574.5	-553.9	-531.4	-507.4	104.4	2.0219

BALLISTIC COEFFICIENT .60, MUZZLE VELOCITY FT/SEC 2200.

RANGE	VELOCITY	HEIGHT	100 YD	150 YD	200 YD	250 YD	300 YD	DRIFT	TIME
0.	2200.	0.0	-1.5	-1.5	-1.5	-1.5	-1.5	0.0	0.0000
100.	2065.	.2	0.0	1.5	3.3	5.4	7.6	.8	.1409
200.	1935.	3.3	-6.7	-3.7	0.0	4.0	8.4	3.2	.2911
300.	1810.	9.0	-22.7	-18.2	-12.6	-6.6	0.0	7.4	.4511
400.	1693.	17.9	-49.3	-43.3	-35.9	-27.9	-19.1	13.6	.6225
500.	1581.	30.5	-88.2	-80.7	-71.5	-61.4	-50.4	21.9	.8063
600.	1476.	47.5	-140.8	-131.8	-120.7	-108.6	-95.4	32.4	1.0023
700.	1379.	69.8	-209.4	-198.9	-186.0	-171.9	-156.5	45.4	1.2127
800.	1294.	98.5	-296.7	-284.7	-269.9	-253.8	-236.3	61.1	1.4380
900.	1219.	134.2	-404.1	-390.7	-374.0	-355.9	-336.1	79.1	1.6765
1000.	1153.	178.5	-536.9	-521.9	-503.5	-483.3	-461.4	100.2	1.9327

BALLISTIC COEFFICIENT .60, MUZZLE VELOCITY FT/SEC 2300.

RANGE	VELOCITY	HEIGHT	100 YD	150 YD	200 YD	250 YD	300 YD	DRIFT	TIME
0.	2300.	0.0	-1.5	-1.5	-1.5	-1.5	-1.5	0.0	0.0000
100.	2162.	.1	0.0	1.3	3.0	4.8	6.8	.7	.1343
200.	2028.	3.0	-6.0	-3.3	0.0	3.7	7.6	3.0	.2779
300.	1899.	8.1	-20.4	-16.4	-11.4	-5.9	0.0	6.9	.4305
400.	1777.	16.2	-44.6	-39.3	-32.6	-25.2	-17.4	12.7	.5940
500.	1661.	27.6	-79.8	-73.1	-64.8	-55.6	-45.7	20.5	.7686
600.	1551.	43.1	-127.5	-119.5	-109.6	-98.5	-86.7	30.4	.9556
700.	1448.	63.4	-189.7	-180.3	-168.7	-155.8	-142.0	42.7	1.1559
800.	1354.	89.4	-268.7	-258.0	-244.7	-230.0	-214.2	57.6	1.3706
900.	1272.	121.9	-366.3	-354.2	-339.3	-322.7	-305.0	74.8	1.5987
1000.	1199.	162.1	-485.6	-472.2	-455.6	-437.2	-417.5	94.6	1.8420

BALLISTIC COEFFICIENT .60, MUZZLE VELOCITY FT/SEC 2400.

RANGE	VELOCITY	HEIGHT	100 YD	150 YD	200 YD	250 YD	300 YD	DRIFT	TIME
0.	2400.	0.0	-1.5	-1.5	-1.5	-1.5	-1.5	0.0	0.0000
100.	2259.	.0	0.0	1.2	2.7	4.3	6.1	.7	.1289
200.	2122.	2.6	-5.3	-3.0	0.0	3.4	7.0	2.7	.2656
300.	1989.	7.4	-18.4	-14.9	-10.4	-5.4	0.0	6.5	.4117
400.	1863.	14.7	-40.3	-35.6	-29.7	-22.9	-15.7	11.9	.5674
500.	1742.	25.1	-72.3	-66.5	-59.0	-50.5	-41.6	19.2	.7342
600.	1628.	39.2	-115.6	-108.7	-99.7	-89.6	-78.8	28.6	.9122
700.	1520.	57.6	-172.1	-164.0	-153.5	-141.7	-129.2	40.1	1.1030
800.	1419.	81.3	-243.5	-234.2	-222.3	-208.7	-194.4	54.1	1.3072
900.	1329.	111.0	-332.5	-322.0	-308.6	-293.4	-277.3	70.6	1.5260
1000.	1250.	147.8	-441.1	-429.5	-414.6	-397.7	-379.8	89.6	1.7590

BALLISTIC COEFFICIENT .60, MUZZLE VELOCITY FT/SEC 2500.

RANGE	VELOCITY	HEIGHT	100 YD	150 YD	200 YD	250 YD	300 YD	DRIFT	TIME
0.	2500.	0.0	-1.5	-1.5	-1.5	-1.5	-1.5	0.0	0.0000
100.	2356.	-.0	0.0	1.0	2.4	3.9	5.5	.7	.1237
200.	2216.	2.4	-4.8	-2.7	0.0	3.1	6.3	2.6	.2547
300.	2081.	6.7	-16.6	-13.5	-9.5	-4.9	0.0	6.0	.3942
400.	1949.	13.4	-36.6	-32.5	-27.1	-21.0	-14.5	11.2	.5436
500.	1825.	22.9	-65.7	-60.6	-53.8	-46.2	-38.0	18.0	.7025
600.	1706.	35.8	-105.2	-99.0	-90.9	-81.8	-72.0	26.8	.8724
700.	1594.	52.6	-156.8	-149.6	-140.2	-129.5	-118.1	37.8	1.0547
800.	1488.	74.1	-221.7	-213.5	-202.7	-190.5	-177.4	50.9	1.2491
900.	1390.	101.3	-302.6	-293.3	-281.2	-267.4	-252.8	66.5	1.4580
1000.	1303.	134.9	-401.5	-391.3	-377.8	-362.4	-346.2	84.6	1.6809

BALLISTIC COEFFICIENT .60, MUZZLE VELOCITY FT/SEC 2600.

RANGE	VELOCITY	HEIGHT	100 YD	150 YD	200 YD	250 YD	300 YD	DRIFT	TIME
0.	2600.	0.0	-1.5	-1.5	-1.5	-1.5	-1.5	0.0	0.0000
100.	2453.	-.1	0.0	.9	2.2	3.6	5.0	.6	.1187
200.	2310.	2.1	-4.3	-2.5	0.0	2.8	5.8	2.5	.2448
300.	2172.	6.1	-15.1	-12.3	-8.7	-4.5	0.0	5.7	.3785
400.	2038.	12.3	-33.4	-29.7	-24.8	-19.2	-13.3	10.6	.5216
500.	1908.	21.0	-60.1	-55.4	-49.3	-42.3	-34.9	17.0	.6736
600.	1786.	32.8	-96.2	-90.7	-83.3	-74.9	-66.0	25.3	.8361
700.	1669.	48.2	-143.3	-136.8	-128.2	-118.4	-108.0	35.6	1.0098
800.	1559.	67.9	-202.8	-195.5	-185.6	-174.4	-162.5	48.0	1.1959
900.	1455.	92.7	-276.5	-268.2	-257.2	-244.6	-231.2	62.8	1.3951
1000.	1361.	123.4	-366.8	-357.6	-345.3	-331.3	-316.4	80.0	1.6084

BALLISTIC COEFFICIENT .60, MUZZLE VELOCITY FT/SEC 2700.

RANGE	VELOCITY	HEIGHT	100 YD	150 YD	200 YD	250 YD	300 YD	DRIFT	TIME
0.	2700.	0.0	-1.5	-1.5	-1.5	-1.5	-1.5	0.0	0.0000
100.	2550.	-.1	0.0	.8	1.9	3.2	4.6	.6	.1144
200.	2404.	1.9	-3.9	-2.3	0.0	2.5	5.3	2.4	.2357
300.	2263.	5.6	-13.8	-11.4	-8.0	-4.2	0.0	5.4	.3643
400.	2126.	11.3	-30.5	-27.2	-22.7	-17.6	-12.1	9.9	.5008
500.	1993.	19.3	-54.9	-50.9	-45.2	-38.8	-31.9	16.0	.6466
600.	1866.	30.1	-88.0	-83.2	-76.4	-68.7	-60.4	23.8	.8019
700.	1746.	44.3	-131.2	-125.6	-117.7	-108.7	-99.0	33.5	.9684
800.	1632.	62.3	-185.7	-179.3	-170.2	-160.1	-149.0	45.3	1.1462
900.	1523.	85.0	-253.3	-246.0	-235.8	-224.4	-211.9	59.2	1.3365
1000.	1422.	113.1	-335.6	-327.6	-316.3	-303.5	-289.7	75.5	1.5402

BALLISTIC COEFFICIENT .60, MUZZLE VELOCITY FT/SEC 2800.

RANGE	VELOCITY	HEIGHT	100 YD	150 YD	200 YD	250 YD	300 YD	DRIFT	TIME
0.	2800.	0.0	-1.5	-1.5	-1.5	-1.5	-1.5	0.0	0.0000
100.	2646.	-.2	0.0	.7	1.7	2.9	4.2	.5	.1099
200.	2498.	1.7	-3.5	-2.0	0.0	2.3	4.9	2.2	.2268
300.	2354.	5.1	-12.6	-10.4	-7.3	-3.8	0.0	5.1	.3506
400.	2214.	10.4	-27.9	-25.0	-20.9	-16.3	-11.1	9.4	.4817
500.	2079.	17.8	-50.4	-46.7	-41.6	-35.8	-29.4	15.1	.6214
600.	1948.	27.8	-81.0	-76.6	-70.5	-63.5	-55.8	22.5	.7709
700.	1823.	40.8	-120.6	-115.5	-108.4	-100.2	-91.2	31.7	.9299
800.	1705.	57.3	-170.7	-164.9	-156.7	-147.4	-137.1	42.7	1.1000
900.	1593.	78.2	-232.9	-226.4	-217.2	-206.7	-195.1	56.0	1.2825
1000.	1487.	104.0	-308.4	-301.1	-290.9	-279.2	-266.4	71.4	1.4771

BALLISTIC COEFFICIENT .60, MUZZLE VELOCITY FT/SEC 2900.

RANGE	VELOCITY	HEIGHT	100 YD	150 YD	200 YD	250 YD	300 YD	DRIFT	TIME
0.	2900.	0.0	-1.5	-1.5	-1.5	-1.5	-1.5	0.0	0.0000
100.	2743.	-.2	0.0	.6	1.6	2.7	3.8	.5	.1063
200.	2591.	1.6	-3.1	-1.9	0.0	2.2	4.5	2.1	.2189
300.	2444.	4.7	-11.5	-9.6	-6.7	-3.5	0.0	4.9	.3379
400.	2302.	9.6	-25.7	-23.1	-19.4	-15.0	-10.4	9.0	.4647
500.	2163.	16.5	-46.4	-43.2	-38.5	-33.1	-27.3	14.4	.5989
600.	2030.	25.7	-74.6	-70.8	-65.2	-58.7	-51.7	21.4	.7424
700.	1901.	37.7	-111.2	-106.7	-100.2	-92.6	-84.4	30.1	.8949
800.	1778.	53.0	-157.4	-152.3	-144.9	-136.2	-126.9	40.6	1.0582
900.	1662.	72.2	-214.6	-208.9	-200.4	-190.7	-180.2	53.1	1.2327
1000.	1553.	96.0	-284.3	-278.0	-268.6	-257.7	-246.1	67.8	1.4195

BALLISTIC COEFFICIENT .60, MUZZLE VELOCITY FT/SEC 3000.

RANGE	VELOCITY	HEIGHT	100 YD	150 YD	200 YD	250 YD	300 YD	DRIFT	TIME
0.	3000.	0.0	-1.5	-1.5	-1.5	-1.5	-1.5	0.0	0.0000
100.	2839.	-.2	0.0	.6	1.4	2.4	3.5	.5	.1028
200.	2684.	1.4	-2.8	-1.7	0.0	2.0	4.2	2.0	.2116
300.	2534.	4.4	-10.6	-8.8	-6.3	-3.3	0.0	4.8	.3271
400.	2389.	8.9	-23.7	-21.3	-18.0	-14.0	-9.6	8.6	.4488
500.	2248.	15.3	-42.9	-40.0	-35.8	-30.8	-25.3	13.8	.5784
600.	2112.	23.8	-69.0	-65.5	-60.5	-54.4	-47.8	20.4	.7159
700.	1979.	34.9	-102.8	-98.7	-92.8	-85.8	-78.1	28.6	.8624
800.	1853.	49.1	-145.5	-140.9	-134.2	-126.2	-117.4	38.6	1.0193
900.	1733.	66.8	-198.3	-193.1	-185.6	-176.6	-166.7	50.4	1.1866
1000.	1620.	88.8	-262.7	-256.9	-248.5	-238.5	-227.5	64.4	1.3656

BALLISTIC COEFFICIENT .60, MUZZLE VELOCITY FT/SEC 3100.

RANGE	VELOCITY	HEIGHT	100 YD	150 YD	200 YD	250 YD	300 YD	DRIFT	TIME
0.	3100.	0.0	-1.5	-1.5	-1.5	-1.5	-1.5	0.0	0.0000
100.	2935.	-.3	0.0	.5	1.3	2.2	3.2	.4	.0991
200.	2776.	1.3	-2.5	-1.6	0.0	1.9	3.9	1.9	.2042
300.	2624.	4.0	-9.6	-8.2	-5.8	-3.0	0.0	4.4	.3154
400.	2476.	8.3	-21.8	-19.8	-16.7	-13.0	-8.9	8.1	.4331
500.	2332.	14.2	-39.6	-37.2	-33.3	-28.6	-23.6	13.1	.5582
600.	2193.	22.1	-63.8	-60.9	-56.2	-50.6	-44.6	19.4	.6906
700.	2058.	32.5	-95.3	-91.9	-86.4	-79.9	-72.8	27.2	.8321
800.	1928.	45.6	-134.9	-131.0	-124.7	-117.3	-109.3	36.7	.9827
900.	1805.	62.0	-183.7	-179.3	-172.3	-163.9	-154.9	47.9	1.1433
1000.	1687.	82.3	-243.3	-238.4	-230.6	-221.3	-211.2	61.2	1.3152

BALLISTIC COEFFICIENT .60, MUZZLE VELOCITY FT/SEC 3200.

RANGE	VELOCITY	HEIGHT	100 YD	150 YD	200 YD	250 YD	300 YD	DRIFT	TIME
0.	3200.	0.0	-1.5	-1.5	-1.5	-1.5	-1.5	0.0	0.0000
100.	3032.	-.3	0.0	.4	1.1	2.0	2.9	.4	.0962
200.	2869.	1.1	-2.3	-1.4	0.0	1.7	3.6	1.8	.1978
300.	2713.	3.7	-8.8	-7.6	-5.4	-2.8	0.0	4.3	.3056
400.	2562.	7.7	-20.1	-18.4	-15.5	-12.1	-8.3	7.8	.4193
500.	2416.	13.2	-36.7	-34.6	-31.0	-26.7	-22.0	12.5	.5400
600.	2275.	20.7	-59.2	-56.7	-52.4	-47.2	-41.5	18.6	.6679
700.	2138.	30.3	-88.5	-85.5	-80.5	-74.5	-67.9	26.0	.8041
800.	2004.	42.5	-125.3	-121.9	-116.1	-109.2	-101.7	35.0	.9490
900.	1877.	57.7	-170.6	-166.8	-160.3	-152.5	-144.0	45.7	1.1034
1000.	1756.	76.5	-225.9	-221.6	-214.4	-205.8	-196.4	58.3	1.2687

BALLISTIC COEFFICIENT .60, MUZZLE VELOCITY FT/SEC 3300.

RANGE	VELOCITY	HEIGHT	100 YD	150 YD	200 YD	250 YD	300 YD	DRIFT	TIME
0.	3300.	0.0	-1.5	-1.5	-1.5	-1.5	-1.5	0.0	0.0000
100.	3128.	-.3	0.0	.4	1.0	1.8	2.7	.5	.0938
200.	2962.	1.0	-2.0	-1.3	0.0	1.6	3.4	1.8	.1918
300.	2803.	3.5	-8.1	-7.0	-5.1	-2.7	0.0	4.2	.2964
400.	2649.	7.2	-18.5	-17.1	-14.5	-11.3	-7.7	7.5	.4063
500.	2500.	12.4	-34.0	-32.2	-29.0	-25.0	-20.5	12.0	.5230
600.	2356.	19.3	-55.0	-52.9	-49.0	-44.2	-38.8	17.8	.6467
700.	2216.	28.3	-82.2	-79.7	-75.1	-69.6	-63.3	24.9	.7777
800.	2081.	39.6	-116.4	-113.5	-108.3	-102.0	-94.8	33.4	.9172
900.	1950.	53.9	-158.8	-155.6	-149.7	-142.6	-134.5	43.7	1.0666
1000.	1825.	71.3	-210.1	-206.5	-200.0	-192.1	-183.1	55.7	1.2255

BALLISTIC COEFFICIENT .60, MUZZLE VELOCITY FT/SEC 3400.

RANGE	VELOCITY	HEIGHT	100 YD	150 YD	200 YD	250 YD	300 YD	DRIFT	TIME
0.	3400.	0.0	-1.5	-1.5	-1.5	-1.5	-1.5	0.0	0.0000
100.	3225.	-.4	0.0	.3	.9	1.7	2.5	.4	.0905
200.	3055.	.9	-1.8	-1.2	0.0	1.5	3.1	1.6	.1857
300.	2892.	3.2	-7.4	-6.4	-4.7	-2.5	0.0	3.9	.2869
400.	2735.	6.7	-17.2	-15.9	-13.5	-10.6	-7.3	7.2	.3936
500.	2584.	11.6	-31.6	-30.0	-27.1	-23.3	-19.2	11.5	.5065
600.	2437.	18.1	-51.2	-49.2	-45.7	-41.3	-36.3	17.0	.6259
700.	2295.	26.5	-76.7	-74.4	-70.3	-65.1	-59.4	23.8	.7530
800.	2157.	37.1	-108.7	-106.0	-101.4	-95.4	-88.9	32.0	.8876
900.	2023.	50.3	-148.2	-145.2	-139.9	-133.2	-125.9	41.8	1.0315
1000.	1895.	66.6	-196.0	-192.7	-186.8	-179.4	-171.2	53.2	1.1845

BALLISTIC COEFFICIENT .60, MUZZLE VELOCITY FT/SEC 3500.

RANGE	VELOCITY	HEIGHT	100 YD	150 YD	200 YD	250 YD	300 YD	DRIFT	TIME
0.	3500.	0.0	−1.5	−1.5	−1.5	−1.5	−1.5	0.0	0.0000
100.	3321.	−.4	0.0	.3	.8	1.5	2.3	.4	.0882
200.	3149.	.8	−1.7	−1.1	0.0	1.3	2.9	1.7	.1811
300.	2982.	3.0	−6.8	−6.0	−4.3	−2.3	0.0	3.8	.2789
400.	2822.	6.3	−15.9	−14.8	−12.6	−9.9	−6.8	6.9	.3823
500.	2667.	10.8	−29.4	−28.0	−25.2	−21.9	−18.0	11.1	.4916
600.	2518.	17.0	−47.8	−46.2	−42.8	−38.8	−34.2	16.4	.6077
700.	2373.	24.8	−71.6	−69.7	−65.8	−61.1	−55.7	22.9	.7303
800.	2233.	34.8	−101.6	−99.4	−94.9	−89.6	−83.4	30.8	.8606
900.	2097.	47.2	−138.5	−136.0	−131.0	−125.0	−118.0	40.1	.9993
1000.	1965.	62.4	−183.2	−180.5	−174.9	−168.2	−160.4	51.0	1.1470

BALLISTIC COEFFICIENT .60, MUZZLE VELOCITY FT/SEC 3600.

RANGE	VELOCITY	HEIGHT	100 YD	150 YD	200 YD	250 YD	300 YD	DRIFT	TIME
0.	3600.	0.0	−1.5	−1.5	−1.5	−1.5	−1.5	0.0	0.0000
100.	3418.	−.4	0.0	.2	.7	1.4	2.1	.3	.0851
200.	3242.	.7	−1.5	−1.0	0.0	1.3	2.7	1.5	.1754
300.	3072.	2.8	−6.3	−5.6	−4.1	−2.1	0.0	3.6	.2704
400.	2908.	5.9	−14.8	−13.9	−11.8	−9.3	−6.4	6.6	.3710
500.	2750.	10.2	−27.5	−26.3	−23.7	−20.5	−17.0	10.6	.4770
600.	2598.	15.9	−44.7	−43.2	−40.2	−36.3	−32.1	15.7	.5892
700.	2451.	23.3	−67.0	−65.3	−61.8	−57.3	−52.3	21.9	.7079
800.	2309.	32.7	−95.1	−93.2	−89.1	−84.0	−78.3	29.5	.8342
900.	2170.	44.2	−129.6	−127.5	−122.9	−117.1	−110.7	38.4	.9680
1000.	2036.	58.5	−171.7	−169.3	−164.2	−157.8	−150.7	48.9	1.1111

BALLISTIC COEFFICIENT .60, MUZZLE VELOCITY FT/SEC 3700.

RANGE	VELOCITY	HEIGHT	100 YD	150 YD	200 YD	250 YD	300 YD	DRIFT	TIME
0.	3700.	0.0	−1.5	−1.5	−1.5	−1.5	−1.5	0.0	0.0000
100.	3514.	−.4	0.0	.2	.7	1.3	1.9	.4	.0831
200.	3335.	.7	−1.3	−1.0	0.0	1.2	2.5	1.6	.1712
300.	3162.	2.6	−5.8	−5.2	−3.8	−2.0	0.0	3.6	.2635
400.	2995.	5.5	−13.8	−13.0	−11.1	−8.7	−6.0	6.4	.3608
500.	2834.	9.6	−25.6	−24.6	−22.2	−19.3	−15.9	10.3	.4638
600.	2679.	15.0	−41.8	−40.6	−37.7	−34.2	−30.1	15.2	.5726
700.	2529.	22.0	−62.9	−61.5	−58.1	−54.0	−49.3	21.3	.6884
800.	2384.	30.8	−89.2	−87.6	−83.8	−79.1	−73.7	28.4	.8103
900.	2244.	41.7	−121.7	−120.0	−115.7	−110.4	−104.3	37.0	.9401
1000.	2107.	55.0	−161.0	−159.0	−154.3	−148.4	−141.6	47.0	1.0780

BALLISTIC COEFFICIENT .60, MUZZLE VELOCITY FT/SEC 3800.

RANGE	VELOCITY	HEIGHT	100 YD	150 YD	200 YD	250 YD	300 YD	DRIFT	TIME
0.	3800.	0.0	−1.5	−1.5	−1.5	−1.5	−1.5	0.0	0.0000
100.	3611.	−.4	0.0	.1	.6	1.1	1.8	.4	.0815
200.	3428.	.6	−1.2	−.9	0.0	1.1	2.4	1.5	.1664
300.	3252.	2.4	−5.3	−4.9	−3.6	−1.9	0.0	3.4	.2562
400.	3082.	5.2	−12.7	−12.2	−10.4	−8.2	−5.7	6.2	.3512
500.	2918.	9.0	−23.9	−23.2	−21.0	−18.2	−15.1	10.0	.4515
600.	2760.	14.1	−39.1	−38.2	−35.6	−32.2	−28.5	14.7	.5571
700.	2607.	20.7	−58.8	−57.8	−54.7	−50.8	−46.4	20.4	.6688
800.	2460.	29.0	−83.6	−82.5	−79.0	−74.5	−69.5	27.4	.7872
900.	2317.	39.3	−114.1	−112.9	−108.9	−103.9	−98.3	35.6	.9129
1000.	2178.	51.8	−151.0	−149.6	−145.2	−139.6	−133.4	45.2	1.0462

BALLISTIC COEFFICIENT .60, MUZZLE VELOCITY FT/SEC 3900.

RANGE	VELOCITY	HEIGHT	100 YD	150 YD	200 YD	250 YD	300 YD	DRIFT	TIME
0.	3900.	0.0	−1.5	−1.5	−1.5	−1.5	−1.5	0.0	0.0000
100.	3708.	−.5	0.0	.1	.5	1.0	1.6	.3	.0784
200.	3522.	.5	−1.0	−.8	0.0	1.1	2.2	1.3	.1613
300.	3342.	2.2	−4.9	−4.5	−3.4	−1.8	0.0	3.2	.2492
400.	3169.	4.8	−11.9	−11.3	−9.8	−7.7	−5.3	5.9	.3412
500.	3001.	8.5	−22.3	−21.6	−19.7	−17.1	−14.1	9.5	.4383
600.	2840.	13.3	−36.6	−35.8	−33.5	−30.4	−26.8	14.0	.5410
700.	2685.	19.5	−55.2	−54.3	−51.6	−47.9	−43.8	19.6	.6498
800.	2535.	27.4	−78.7	−77.6	−74.6	−70.4	−65.6	26.4	.7652
900.	2390.	37.0	−107.5	−106.2	−102.8	−98.1	−92.7	34.3	.8869
1000.	2249.	48.8	−142.3	−140.9	−137.2	−131.9	−125.9	43.5	1.0164

BALLISTIC COEFFICIENT .60, MUZZLE VELOCITY FT/SEC 4000.

RANGE	VELOCITY	HEIGHT	100 YD	150 YD	200 YD	250 YD	300 YD	DRIFT	TIME
0.	4000.	0.0	-1.5	-1.5	-1.5	-1.5	-1.5	0.0	0.0000
100.	3804.	-.5	0.0	.1	.5	.9	1.5	.3	.0769
200.	3615.	.5	-.9	-.7	0.0	.9	2.1	1.4	.1582
300.	3432.	2.1	-4.5	-4.2	-3.1	-1.7	0.0	3.2	.2431
400.	3256.	4.6	-11.0	-10.7	-9.2	-7.3	-5.0	5.8	.3327
500.	3085.	8.0	-20.9	-20.4	-18.6	-16.2	-13.4	9.3	.4277
600.	2921.	12.6	-34.4	-33.9	-31.6	-28.8	-25.4	13.7	.5278
700.	2763.	18.5	-52.0	-51.3	-48.7	-45.4	-41.5	19.1	.6333
800.	2611.	25.9	-74.0	-73.3	-70.3	-66.5	-62.0	25.5	.7448
900.	2463.	35.0	-101.1	-100.3	-97.0	-92.7	-87.7	33.1	.8631
1000.	2320.	46.2	-134.0	-133.1	-129.3	-124.6	-119.0	42.0	.9886

BALLISTIC COEFFICIENT .61, MUZZLE VELOCITY FT/SEC 2000.

RANGE	VELOCITY	HEIGHT	100 YD	150 YD	200 YD	250 YD	300 YD	DRIFT	TIME
0.	2000.	0.0	-1.5	-1.5	-1.5	-1.5	-1.5	0.0	0.0000
100.	1875.	.4	0.0	1.9	4.2	6.7	9.4	.8	.1547
200.	1756.	4.2	-8.4	-4.6	0.0	5.0	10.4	3.5	.3201
300.	1643.	11.1	-28.2	-22.4	-15.6	-8.1	0.0	8.3	.4970
400.	1536.	21.8	-60.9	-53.2	-44.0	-34.0	-23.3	15.1	.6858
500.	1435.	37.1	-108.4	-98.7	-87.3	-74.8	-61.3	24.3	.8881
600.	1344.	57.8	-172.8	-161.2	-147.5	-132.5	-116.3	36.0	1.1043
700.	1264.	84.7	-256.4	-242.9	-226.9	-209.3	-190.5	50.0	1.3343
800.	1193.	119.1	-362.3	-346.8	-328.5	-308.5	-287.0	66.9	1.5799
900.	1135.	161.7	-493.0	-475.6	-455.1	-432.5	-408.3	86.2	1.8395
1000.	1089.	212.4	-648.3	-629.1	-606.2	-581.1	-554.2	106.8	2.1071

BALLISTIC COEFFICIENT .61, MUZZLE VELOCITY FT/SEC 2100.

RANGE	VELOCITY	HEIGHT	100 YD	150 YD	200 YD	250 YD	300 YD	DRIFT	TIME
0.	2100.	0.0	-1.5	-1.5	-1.5	-1.5	-1.5	0.0	0.0000
100.	1970.	.3	0.0	1.7	3.8	6.0	8.4	.8	.1472
200.	1846.	3.7	-7.5	-4.1	0.0	4.5	9.3	3.3	.3047
300.	1729.	10.0	-25.2	-20.1	-13.9	-7.2	0.0	7.7	.4723
400.	1617.	19.6	-54.6	-47.8	-39.6	-30.6	-21.0	14.1	.6518
500.	1512.	33.4	-97.3	-88.8	-78.5	-67.3	-55.3	22.8	.8437
600.	1413.	52.1	-155.2	-144.9	-132.7	-119.2	-104.8	33.8	1.0491
700.	1325.	76.5	-230.6	-218.7	-204.4	-188.6	-171.8	47.3	1.2686
800.	1247.	107.6	-325.9	-312.3	-295.9	-277.9	-258.7	63.3	1.5023
900.	1178.	146.3	-443.5	-428.1	-409.7	-389.5	-367.9	81.8	1.7503
1000.	1123.	193.3	-586.0	-568.9	-548.4	-525.9	-502.0	102.5	2.0109

BALLISTIC COEFFICIENT .61, MUZZLE VELOCITY FT/SEC 2200.

RANGE	VELOCITY	HEIGHT	100 YD	150 YD	200 YD	250 YD	300 YD	DRIFT	TIME
0.	2200.	0.0	-1.5	-1.5	-1.5	-1.5	-1.5	0.0	0.0000
100.	2067.	.2	0.0	1.5	3.3	5.3	7.5	.8	.1408
200.	1939.	3.3	-6.7	-3.7	0.0	4.0	8.4	3.2	.2908
300.	1816.	9.0	-22.6	-18.1	-12.6	-6.5	0.0	7.3	.4503
400.	1700.	17.8	-49.1	-43.1	-35.8	-27.7	-19.0	13.3	.6210
500.	1590.	30.3	-87.8	-80.3	-71.1	-61.0	-50.1	21.5	.8039
600.	1486.	47.1	-139.9	-130.9	-119.9	-107.8	-94.7	31.8	.9986
700.	1390.	69.3	-208.0	-197.5	-184.6	-170.6	-155.3	44.5	1.2077
800.	1304.	97.5	-294.1	-282.1	-267.4	-251.3	-233.9	59.8	1.4305
900.	1229.	132.8	-400.7	-387.2	-370.7	-352.6	-333.0	77.5	1.6678
1000.	1163.	176.0	-530.1	-515.1	-496.7	-476.6	-454.8	97.7	1.9189

BALLISTIC COEFFICIENT .61, MUZZLE VELOCITY FT/SEC 2300.

RANGE	VELOCITY	HEIGHT	100 YD	150 YD	200 YD	250 YD	300 YD	DRIFT	TIME
0.	2300.	0.0	-1.5	-1.5	-1.5	-1.5	-1.5	0.0	0.0000
100.	2164.	.1	0.0	1.3	3.0	4.8	6.8	.7	.1342
200.	2032.	2.9	-6.0	-3.3	0.0	3.7	7.6	3.0	.2776
300.	1905.	8.1	-20.4	-16.4	-11.4	-5.9	0.0	6.8	.4300
400.	1785.	16.1	-44.4	-39.1	-32.5	-25.2	-17.2	12.5	.5927
500.	1670.	27.4	-79.4	-72.7	-64.4	-55.3	-45.4	20.1	.7664
600.	1562.	42.8	-126.8	-118.8	-108.9	-97.9	-86.0	29.9	.9523
700.	1459.	62.8	-188.4	-179.0	-167.4	-154.6	-140.8	41.9	1.1509
800.	1366.	88.5	-266.4	-255.7	-242.4	-227.8	-212.0	56.3	1.3636
900.	1284.	120.6	-363.0	-351.0	-336.1	-319.7	-301.9	73.3	1.5902
1000.	1211.	160.2	-480.8	-467.4	-450.9	-432.6	-412.8	92.7	1.8310

BALLISTIC COEFFICIENT .61, MUZZLE VELOCITY FT/SEC 2400.

RANGE	VELOCITY	HEIGHT	100 YD	150 YD	200 YD	250 YD	300 YD	DRIFT	TIME
0.	2400.	0.0	-1.5	-1.5	-1.5	-1.5	-1.5	0.0	0.0000
100.	2261.	.0	0.0	1.2	2.7	4.3	6.1	.7	.1288
200.	2126.	2.6	-5.3	-3.0	0.0	3.4	6.9	2.7	.2654
300.	1996.	7.4	-18.3	-14.9	-10.4	-5.4	0.0	6.4	.4111
400.	1871.	14.6	-40.1	-35.5	-29.5	-22.8	-15.6	11.6	.5661
500.	1752.	25.0	-71.9	-66.1	-58.6	-50.2	-41.3	18.8	.7320
600.	1639.	38.9	-115.0	-108.0	-99.1	-89.0	-78.3	28.0	.9091
700.	1532.	57.2	-171.0	-162.9	-152.4	-140.7	-128.1	39.3	1.0985
800.	1432.	80.5	-241.7	-232.5	-220.5	-207.1	-192.8	53.0	1.3012
900.	1341.	109.8	-329.5	-319.1	-305.7	-290.6	-274.5	69.1	1.5177
1000.	1262.	145.9	-436.6	-425.1	-410.1	-393.3	-375.5	87.7	1.7482

BALLISTIC COEFFICIENT .61, MUZZLE VELOCITY FT/SEC 2500.

RANGE	VELOCITY	HEIGHT	100 YD	150 YD	200 YD	250 YD	300 YD	DRIFT	TIME
0.	2500.	0.0	-1.5	-1.5	-1.5	-1.5	-1.5	0.0	0.0000
100.	2358.	-.0	0.0	1.0	2.4	3.9	5.5	.6	.1236
200.	2221.	2.4	-4.7	-2.7	0.0	3.0	6.3	2.5	.2544
300.	2087.	6.7	-16.6	-13.5	-9.5	-4.9	0.0	5.9	.3938
400.	1958.	13.4	-36.5	-32.4	-27.0	-20.9	-14.4	11.0	.5424
500.	1835.	22.8	-65.4	-60.3	-53.6	-46.0	-37.8	17.7	.7008
600.	1718.	35.5	-104.6	-98.4	-90.3	-81.2	-71.4	26.3	.8692
700.	1607.	52.2	-155.7	-148.5	-139.1	-128.5	-117.0	37.0	1.0503
800.	1502.	73.4	-220.0	-211.8	-201.1	-188.9	-175.8	49.8	1.2431
900.	1403.	100.2	-299.9	-290.7	-278.6	-264.9	-250.2	65.1	1.4500
1000.	1317.	133.3	-397.7	-387.5	-374.0	-358.8	-342.5	82.9	1.6711

BALLISTIC COEFFICIENT .61, MUZZLE VELOCITY FT/SEC 2600.

RANGE	VELOCITY	HEIGHT	100 YD	150 YD	200 YD	250 YD	300 YD	DRIFT	TIME
0.	2600.	0.0	-1.5	-1.5	-1.5	-1.5	-1.5	0.0	0.0000
100.	2455.	-.1	0.0	.9	2.1	3.5	5.0	.6	.1186
200.	2315.	2.1	-4.3	-2.4	0.0	2.8	5.7	2.4	.2445
300.	2179.	6.1	-15.1	-12.3	-8.6	-4.4	0.0	5.6	.3779
400.	2046.	12.3	-33.3	-29.6	-24.7	-19.1	-13.2	10.4	.5205
500.	1919.	20.9	-59.8	-55.2	-49.1	-42.1	-34.7	16.7	.6719
600.	1798.	32.6	-95.7	-90.1	-82.8	-74.4	-65.5	24.8	.8331
700.	1682.	47.8	-142.3	-135.9	-127.3	-117.5	-107.2	34.8	1.0057
800.	1573.	67.3	-201.4	-194.0	-184.2	-173.0	-161.2	47.0	1.1904
900.	1470.	91.6	-274.1	-265.8	-254.8	-242.2	-228.9	61.4	1.3874
1000.	1375.	122.0	-363.3	-354.1	-341.8	-327.9	-313.1	78.3	1.5988

BALLISTIC COEFFICIENT .61, MUZZLE VELOCITY FT/SEC 2700.

RANGE	VELOCITY	HEIGHT	100 YD	150 YD	200 YD	250 YD	300 YD	DRIFT	TIME
0.	2700.	0.0	-1.5	-1.5	-1.5	-1.5	-1.5	0.0	0.0000
100.	2552.	-.1	0.0	.8	1.9	3.2	4.6	.6	.1143
200.	2409.	1.9	-3.9	-2.3	0.0	2.5	5.3	2.3	.2354
300.	2270.	5.6	-13.7	-11.3	-7.9	-4.1	0.0	5.3	.3637
400.	2135.	11.3	-30.4	-27.2	-22.7	-17.6	-12.1	9.8	.5000
500.	2004.	19.2	-54.7	-50.6	-45.0	-38.6	-31.8	15.7	.6450
600.	1878.	29.9	-87.5	-82.7	-75.9	-68.3	-60.1	23.3	.7993
700.	1759.	43.9	-130.3	-124.7	-116.8	-107.9	-98.3	32.8	.9644
800.	1646.	61.7	-184.4	-178.0	-168.9	-158.8	-147.8	44.4	1.1409
900.	1539.	84.1	-251.1	-243.9	-233.7	-222.3	-209.9	58.0	1.3293
1000.	1438.	111.8	-332.7	-324.6	-313.3	-300.6	-286.9	74.0	1.5314

BALLISTIC COEFFICIENT .61, MUZZLE VELOCITY FT/SEC 2800.

RANGE	VELOCITY	HEIGHT	100 YD	150 YD	200 YD	250 YD	300 YD	DRIFT	TIME
0.	2800.	0.0	-1.5	-1.5	-1.5	-1.5	-1.5	0.0	0.0000
100.	2649.	-.2	0.0	.7	1.7	2.9	4.2	.5	.1099
200.	2503.	1.7	-3.5	-2.0	0.0	2.3	4.9	2.2	.2266
300.	2361.	5.1	-12.5	-10.4	-7.3	-3.8	0.0	5.0	.3501
400.	2223.	10.3	-27.8	-24.9	-20.8	-16.2	-11.1	9.2	.4808
500.	2090.	17.7	-50.2	-46.6	-41.5	-35.7	-29.3	14.8	.6200
600.	1960.	27.6	-80.5	-76.2	-70.1	-63.1	-55.4	22.1	.7684
700.	1837.	40.5	-119.9	-114.9	-107.7	-99.6	-90.6	31.1	.9266
800.	1720.	56.8	-169.3	-163.6	-155.4	-146.1	-135.9	41.8	1.0947
900.	1609.	77.4	-230.8	-224.4	-215.2	-204.7	-193.2	54.8	1.2756
1000.	1504.	102.7	-305.4	-298.2	-288.0	-276.5	-263.6	69.8	1.4683

BALLISTIC COEFFICIENT .61, MUZZLE VELOCITY FT/SEC 2900.

RANGE	VELOCITY	HEIGHT	100 YD	150 YD	200 YD	250 YD	300 YD	DRIFT	TIME
0.	2900.	0.0	-1.5	-1.5	-1.5	-1.5	-1.5	0.0	0.0000
100.	2745.	-.2	0.0	.6	1.6	2.7	3.8	.5	.1063
200.	2596.	1.5	-3.1	-1.9	0.0	2.2	4.5	2.1	.2187
300.	2451.	4.7	-11.4	-9.5	-6.7	-3.5	0.0	4.8	.3375
400.	2311.	9.6	-25.6	-23.0	-19.3	-15.0	-10.3	8.8	.4637
500.	2175.	16.4	-46.1	-43.0	-38.3	-32.9	-27.1	14.1	.5972
600.	2043.	25.5	-74.2	-70.5	-64.8	-58.3	-51.4	21.0	.7401
700.	1915.	37.4	-110.5	-106.1	-99.6	-92.0	-83.9	29.5	.8917
800.	1794.	52.5	-156.2	-151.2	-143.7	-135.0	-125.8	39.7	1.0534
900.	1679.	71.4	-212.8	-207.1	-198.7	-188.9	-178.5	51.9	1.2262
1000.	1570.	94.9	-281.7	-275.4	-266.1	-255.2	-243.6	66.3	1.4113

BALLISTIC COEFFICIENT .61, MUZZLE VELOCITY FT/SEC 3000.

RANGE	VELOCITY	HEIGHT	100 YD	150 YD	200 YD	250 YD	300 YD	DRIFT	TIME
0.	3000.	0.0	-1.5	-1.5	-1.5	-1.5	-1.5	0.0	0.0000
100.	2841.	-.2	0.0	.6	1.4	2.4	3.5	.5	.1027
200.	2689.	1.4	-2.8	-1.7	0.0	2.0	4.2	2.0	.2115
300.	2541.	4.4	-10.5	-8.8	-6.3	-3.3	0.0	4.7	.3266
400.	2398.	8.9	-23.6	-21.3	-17.9	-13.9	-9.6	8.5	.4480
500.	2259.	15.2	-42.7	-39.8	-35.6	-30.7	-25.2	13.5	.5769
600.	2125.	23.7	-68.6	-65.1	-60.1	-54.1	-47.5	20.0	.7136
700.	1994.	34.7	-102.2	-98.2	-92.3	-85.3	-77.6	28.1	.8594
800.	1869.	48.7	-144.4	-139.8	-133.1	-125.2	-116.4	37.8	1.0146
900.	1750.	66.2	-196.7	-191.6	-184.0	-175.1	-165.2	49.4	1.1806
1000.	1638.	87.8	-260.3	-254.6	-246.2	-236.2	-225.3	63.0	1.3579

BALLISTIC COEFFICIENT .61, MUZZLE VELOCITY FT/SEC 3100.

RANGE	VELOCITY	HEIGHT	100 YD	150 YD	200 YD	250 YD	300 YD	DRIFT	TIME
0.	3100.	0.0	-1.5	-1.5	-1.5	-1.5	-1.5	0.0	0.0000
100.	2938.	-.3	0.0	.5	1.3	2.2	3.2	.4	.0990
200.	2782.	1.2	-2.5	-1.6	0.0	1.9	3.9	1.8	.2039
300.	2631.	4.0	-9.6	-8.1	-5.8	-3.0	0.0	4.3	.3150
400.	2485.	8.2	-21.7	-19.7	-16.6	-12.9	-8.9	8.0	.4323
500.	2344.	14.1	-39.5	-37.0	-33.1	-28.5	-23.5	12.9	.5569
600.	2207.	22.0	-63.5	-60.5	-55.9	-50.3	-44.3	19.0	.6885
700.	2074.	32.2	-94.6	-91.2	-85.8	-79.3	-72.2	26.6	.8287
800.	1945.	45.2	-133.9	-130.0	-123.8	-116.4	-108.3	35.9	.9784
900.	1822.	61.4	-182.2	-177.8	-170.8	-162.5	-153.4	46.9	1.1374
1000.	1706.	81.3	-241.0	-236.1	-228.4	-219.1	-209.0	59.8	1.3075

BALLISTIC COEFFICIENT .61, MUZZLE VELOCITY FT/SEC 3200.

RANGE	VELOCITY	HEIGHT	100 YD	150 YD	200 YD	250 YD	300 YD	DRIFT	TIME
0.	3200.	0.0	-1.5	-1.5	-1.5	-1.5	-1.5	0.0	0.0000
100.	3034.	-.3	0.0	.4	1.1	2.0	2.9	.4	.0961
200.	2875.	1.1	-2.3	-1.4	0.0	1.7	3.6	1.8	.1976
300.	2721.	3.7	-8.8	-7.5	-5.4	-2.8	0.0	4.2	.3052
400.	2572.	7.7	-20.0	-18.4	-15.5	-12.0	-8.3	7.7	.4186
500.	2428.	13.2	-36.5	-34.4	-30.8	-26.5	-21.8	12.3	.5386
600.	2289.	20.5	-58.9	-56.4	-52.1	-46.9	-41.3	18.2	.6659
700.	2153.	30.0	-87.9	-84.9	-79.9	-73.9	-67.3	25.5	.8010
800.	2022.	42.1	-124.4	-121.0	-115.3	-108.3	-100.9	34.3	.9449
900.	1895.	57.1	-169.3	-165.5	-159.0	-151.2	-142.8	44.8	1.0980
1000.	1775.	75.7	-223.9	-219.7	-212.6	-203.9	-194.5	57.1	1.2618

BALLISTIC COEFFICIENT .61, MUZZLE VELOCITY FT/SEC 3300.

RANGE	VELOCITY	HEIGHT	100 YD	150 YD	200 YD	250 YD	300 YD	DRIFT	TIME
0.	3300.	0.0	-1.5	-1.5	-1.5	-1.5	-1.5	0.0	0.0000
100.	3131.	-.3	0.0	.4	1.0	1.8	2.7	.5	.0937
200.	2968.	1.0	-2.0	-1.3	0.0	1.6	3.4	1.7	.1917
300.	2810.	3.5	-8.1	-7.0	-5.0	-2.7	0.0	4.1	.2959
400.	2659.	7.1	-18.5	-17.1	-14.4	-11.3	-7.7	7.4	.4055
500.	2512.	12.3	-33.9	-32.1	-28.8	-24.9	-20.4	11.8	.5218
600.	2370.	19.2	-54.7	-52.6	-48.6	-43.9	-38.6	17.5	.6447
700.	2232.	28.1	-81.7	-79.3	-74.6	-69.1	-62.9	24.4	.7751
800.	2098.	39.3	-115.7	-112.9	-107.6	-101.3	-94.2	32.8	.9138
900.	1969.	53.3	-157.5	-154.3	-148.4	-141.3	-133.3	42.8	1.0612
1000.	1845.	70.6	-208.3	-204.8	-198.2	-190.3	-181.5	54.5	1.2189

BALLISTIC COEFFICIENT .61, MUZZLE VELOCITY FT/SEC 3400.

RANGE	VELOCITY	HEIGHT	100 YD	150 YD	200 YD	250 YD	300 YD	DRIFT	TIME
0.	3400.	0.0	-1.5	-1.5	-1.5	-1.5	-1.5	0.0	0.0000
100.	3227.	-.4	0.0	.3	.9	1.7	2.5	.4	.0905
200.	3061.	.9	-1.8	-1.2	0.0	1.5	3.1	1.6	.1855
300.	2900.	3.2	-7.4	-6.4	-4.7	-2.4	0.0	3.9	.2866
400.	2745.	6.7	-17.1	-15.8	-13.5	-10.5	-7.2	7.0	.3929
500.	2596.	11.5	-31.5	-29.8	-26.9	-23.2	-19.1	11.3	.5053
600.	2451.	17.9	-50.9	-49.0	-45.5	-41.0	-36.1	16.7	.6241
700.	2311.	26.3	-76.2	-73.9	-69.9	-64.6	-58.9	23.3	.7503
800.	2175.	36.7	-107.9	-105.3	-100.6	-94.6	-88.1	31.3	.8838
900.	2043.	49.8	-147.1	-144.1	-138.9	-132.2	-124.8	40.9	1.0267
1000.	1915.	65.9	-194.4	-191.1	-185.3	-177.8	-169.7	52.1	1.1783

BALLISTIC COEFFICIENT .61, MUZZLE VELOCITY FT/SEC 3500.

RANGE	VELOCITY	HEIGHT	100 YD	150 YD	200 YD	250 YD	300 YD	DRIFT	TIME
0.	3500.	0.0	-1.5	-1.5	-1.5	-1.5	-1.5	0.0	0.0000
100.	3324.	-.4	0.0	.3	.8	1.5	2.3	.4	.0882
200.	3154.	.8	-1.7	-1.1	0.0	1.3	2.9	1.7	.1809
300.	2990.	3.0	-6.8	-6.0	-4.3	-2.3	0.0	3.7	.2784
400.	2832.	6.2	-15.9	-14.8	-12.5	-9.8	-6.8	6.8	.3815
500.	2679.	10.8	-29.3	-27.9	-25.1	-21.7	-17.9	10.9	.4903
600.	2532.	16.9	-47.6	-45.9	-42.6	-38.6	-34.0	16.1	.6059
700.	2390.	24.7	-71.2	-69.3	-65.4	-60.7	-55.3	22.5	.7277
800.	2251.	34.5	-101.0	-98.8	-94.3	-88.9	-82.8	30.2	.8572
900.	2117.	46.7	-137.4	-134.9	-129.9	-123.8	-116.9	39.2	.9943
1000.	1986.	61.7	-181.6	-178.8	-173.3	-166.5	-158.9	49.9	1.1406

BALLISTIC COEFFICIENT .61, MUZZLE VELOCITY FT/SEC 3600.

RANGE	VELOCITY	HEIGHT	100 YD	150 YD	200 YD	250 YD	300 YD	DRIFT	TIME
0.	3600.	0.0	-1.5	-1.5	-1.5	-1.5	-1.5	0.0	0.0000
100.	3421.	-.4	0.0	.2	.7	1.4	2.1	.3	.0851
200.	3248.	.7	-1.5	-1.0	0.0	1.3	2.7	1.5	.1752
300.	3080.	2.8	-6.3	-5.6	-4.1	-2.1	0.0	3.5	.2701
400.	2919.	5.8	-14.8	-13.8	-11.8	-9.3	-6.4	6.5	.3705
500.	2763.	10.1	-27.3	-26.1	-23.6	-20.4	-16.9	10.4	.4760
600.	2613.	15.8	-44.4	-43.0	-40.0	-36.1	-31.8	15.4	.5874
700.	2468.	23.1	-66.6	-64.9	-61.4	-56.9	-51.9	21.5	.7055
800.	2327.	32.4	-94.5	-92.6	-88.6	-83.4	-77.7	28.9	.8309
900.	2191.	43.8	-128.7	-126.5	-122.0	-116.2	-109.8	37.6	.9635
1000.	2058.	57.9	-170.2	-167.8	-162.8	-156.4	-149.3	47.8	1.1052

BALLISTIC COEFFICIENT .61, MUZZLE VELOCITY FT/SEC 3700.

RANGE	VELOCITY	HEIGHT	100 YD	150 YD	200 YD	250 YD	300 YD	DRIFT	TIME
0.	3700.	0.0	-1.5	-1.5	-1.5	-1.5	-1.5	0.0	0.0000
100.	3517.	-.4	0.0	.2	.7	1.3	1.9	.3	.0831
200.	3341.	.7	-1.3	-.9	0.0	1.2	2.5	1.6	.1710
300.	3170.	2.6	-5.8	-5.2	-3.8	-2.0	0.0	3.5	.2631
400.	3006.	5.5	-13.7	-12.9	-11.0	-8.7	-6.0	6.3	.3601
500.	2847.	9.5	-25.5	-24.5	-22.1	-19.2	-15.8	10.1	.4626
600.	2694.	14.9	-41.6	-40.4	-37.5	-34.0	-30.0	14.9	.5712
700.	2546.	21.8	-62.5	-61.1	-57.8	-53.7	-49.0	20.8	.6859
800.	2403.	30.5	-88.7	-87.1	-83.3	-78.6	-73.2	27.9	.8072
900.	2264.	41.3	-120.8	-119.0	-114.8	-109.5	-103.5	36.3	.9358
1000.	2129.	54.4	-159.7	-157.7	-152.9	-147.1	-140.4	46.0	1.0723

BALLISTIC COEFFICIENT .61, MUZZLE VELOCITY FT/SEC 3800.

RANGE	VELOCITY	HEIGHT	100 YD	150 YD	200 YD	250 YD	300 YD	DRIFT	TIME
0.	3800.	0.0	-1.5	-1.5	-1.5	-1.5	-1.5	0.0	0.0000
100.	3614.	-.4	0.0	.1	.6	1.1	1.8	.4	.0814
200.	3434.	.6	-1.2	-.9	0.0	1.1	2.4	1.5	.1663
300.	3261.	2.4	-5.3	-4.8	-3.5	-1.9	0.0	3.3	.2558
400.	3093.	5.2	-12.7	-12.1	-10.4	-8.2	-5.7	6.1	.3507
500.	2931.	9.0	-23.7	-23.0	-20.9	-18.1	-15.0	9.8	.4502
600.	2775.	14.0	-38.8	-38.0	-35.3	-32.0	-28.3	14.4	.5552
700.	2624.	20.6	-58.4	-57.4	-54.4	-50.5	-46.1	20.0	.6664
800.	2479.	28.8	-83.0	-81.9	-78.4	-73.9	-69.0	26.8	.7840
900.	2338.	38.9	-113.3	-112.1	-108.2	-103.1	-97.6	34.9	.9090
1000.	2201.	51.3	-149.8	-148.4	-144.0	-138.5	-132.3	44.3	1.0410

BALLISTIC COEFFICIENT .61, MUZZLE VELOCITY FT/SEC 3900.

RANGE	VELOCITY	HEIGHT	100 YD	150 YD	200 YD	250 YD	300 YD	DRIFT	TIME
0.	3900.	0.0	-1.5	-1.5	-1.5	-1.5	-1.5	0.0	0.0000
100.	3711.	-.5	0.0	.1	.5	1.0	1.6	.3	.0784
200.	3528.	.5	-1.0	-.8	0.0	1.0	2.2	1.3	.1612
300.	3351.	2.2	-4.9	-4.5	-3.3	-1.8	0.0	3.2	.2488
400.	3180.	4.8	-11.8	-11.3	-9.8	-7.7	-5.3	5.8	.3405
500.	3015.	8.4	-22.2	-21.5	-19.6	-17.0	-14.1	9.3	.4374
600.	2856.	13.2	-36.4	-35.6	-33.3	-30.2	-26.6	13.7	.5396
700.	2703.	19.4	-54.9	-54.0	-51.3	-47.7	-43.5	19.2	.6478
800.	2555.	27.1	-78.1	-77.1	-74.0	-69.9	-65.1	25.8	.7620
900.	2411.	36.7	-106.7	-105.5	-102.1	-97.3	-92.0	33.6	.8830
1000.	2272.	48.3	-141.1	-139.7	-135.9	-130.7	-124.8	42.6	1.0111

BALLISTIC COEFFICIENT .61, MUZZLE VELOCITY FT/SEC 4000.

RANGE	VELOCITY	HEIGHT	100 YD	150 YD	200 YD	250 YD	300 YD	DRIFT	TIME
0.	4000.	0.0	-1.5	-1.5	-1.5	-1.5	-1.5	0.0	0.0000
100.	3807.	-.5	0.0	.1	.5	.9	1.5	.3	.0770
200.	3621.	.4	-.9	-.7	0.0	.9	2.1	1.4	.1581
300.	3441.	2.1	-4.5	-4.2	-3.1	-1.7	0.0	3.1	.2428
400.	3267.	4.5	-11.0	-10.6	-9.1	-7.3	-5.0	5.7	.3323
500.	3099.	8.0	-20.8	-20.4	-18.5	-16.2	-13.3	9.1	.4269
600.	2937.	12.5	-34.2	-33.6	-31.4	-28.6	-25.2	13.4	.5259
700.	2781.	18.4	-51.6	-50.9	-48.3	-45.0	-41.1	18.6	.6308
800.	2630.	25.7	-73.5	-72.8	-69.8	-66.0	-61.5	25.0	.7419
900.	2485.	34.7	-100.4	-99.6	-96.2	-92.0	-86.9	32.4	.8593
1000.	2343.	45.7	-132.9	-132.0	-128.3	-123.6	-118.0	41.2	.9839

BALLISTIC COEFFICIENT .62, MUZZLE VELOCITY FT/SEC 2000.

RANGE	VELOCITY	HEIGHT	100 YD	150 YD	200 YD	250 YD	300 YD	DRIFT	TIME
0.	2000.	0.0	-1.5	-1.5	-1.5	-1.5	-1.5	0.0	0.0000
100.	1877.	.4	0.0	1.9	4.2	6.7	9.4	.8	.1546
200.	1759.	4.2	-8.4	-4.6	0.0	5.0	10.4	3.5	.3197
300.	1648.	11.1	-28.1	-22.4	-15.5	-8.0	0.0	8.1	.4962
400.	1542.	21.7	-60.7	-53.0	-43.9	-33.9	-23.1	14.8	.6843
500.	1442.	36.9	-107.9	-98.3	-86.9	-74.4	-61.0	23.9	.8857
600.	1352.	57.4	-171.8	-160.3	-146.6	-131.6	-115.6	35.3	1.1006
700.	1273.	84.1	-255.0	-241.6	-225.6	-208.1	-189.4	49.3	1.3298
800.	1202.	117.9	-359.1	-343.7	-325.5	-305.5	-284.1	65.5	1.5719
900.	1143.	159.6	-487.4	-470.1	-449.6	-427.1	-403.0	84.1	1.8280
1000.	1096.	210.2	-642.6	-623.4	-600.6	-575.6	-548.9	105.0	2.0966

BALLISTIC COEFFICIENT .62, MUZZLE VELOCITY FT/SEC 2100.

RANGE	VELOCITY	HEIGHT	100 YD	150 YD	200 YD	250 YD	300 YD	DRIFT	TIME
0.	2100.	0.0	-1.5	-1.5	-1.5	-1.5	-1.5	0.0	0.0000
100.	1972.	.3	0.0	1.7	3.7	6.0	8.4	.7	.1471
200.	1850.	3.7	-7.5	-4.1	0.0	4.5	9.3	3.3	.3043
300.	1734.	9.9	-25.1	-20.0	-13.9	-7.2	0.0	7.6	.4716
400.	1624.	19.6	-54.4	-47.6	-39.4	-30.4	-20.9	13.9	.6504
500.	1520.	33.2	-96.9	-88.3	-78.2	-66.9	-55.0	22.4	.8414
600.	1422.	51.7	-154.3	-144.0	-131.8	-118.4	-104.0	33.1	1.0454
700.	1334.	76.0	-229.3	-217.4	-203.2	-187.5	-170.7	46.5	1.2641
800.	1257.	106.6	-323.5	-309.9	-293.6	-275.6	-256.5	62.1	1.4955
900.	1188.	144.8	-439.6	-424.3	-406.0	-385.8	-364.2	80.2	1.7412
1000.	1131.	191.3	-580.6	-563.6	-543.3	-520.8	-496.9	100.6	2.0002

BALLISTIC COEFFICIENT .62, MUZZLE VELOCITY FT/SEC 2200.

RANGE	VELOCITY	HEIGHT	100 YD	150 YD	200 YD	250 YD	300 YD	DRIFT	TIME
0.	2200.	0.0	-1.5	-1.5	-1.5	-1.5	-1.5	0.0	0.0000
100.	2069.	.2	0.0	1.5	3.3	5.3	7.5	.8	.1407
200.	1943.	3.3	-6.7	-3.7	0.0	4.0	8.4	3.1	.2905
300.	1822.	9.0	-22.5	-18.0	-12.5	-6.5	0.0	7.1	.4496
400.	1708.	17.7	-48.9	-42.9	-35.6	-27.6	-18.9	13.1	.6196
500.	1599.	30.1	-87.3	-79.9	-70.7	-60.7	-49.8	21.1	.8016
600.	1496.	46.8	-139.1	-130.2	-119.1	-107.1	-94.1	31.2	.9952
700.	1400.	68.7	-206.6	-196.1	-183.3	-169.3	-154.0	43.7	1.2027
800.	1315.	96.6	-292.1	-280.1	-265.4	-249.4	-232.0	58.7	1.4244
900.	1240.	131.4	-397.4	-384.0	-367.4	-349.4	-329.8	76.1	1.6594
1000.	1173.	174.5	-526.8	-511.8	-493.4	-473.4	-451.7	96.3	1.9109

BALLISTIC COEFFICIENT .62, MUZZLE VELOCITY FT/SEC 2300.

RANGE	VELOCITY	HEIGHT	100 YD	150 YD	200 YD	250 YD	300 YD	DRIFT	TIME
0.	2300.	0.0	-1.5	-1.5	-1.5	-1.5	-1.5	0.0	0.0000
100.	2166.	.1	0.0	1.3	3.0	4.8	6.8	.6	.1341
200.	2037.	2.9	-6.0	-3.3	0.0	3.6	7.6	2.9	.2774
300.	1911.	8.1	-20.3	-16.3	-11.4	-5.9	0.0	6.7	.4294
400.	1792.	16.0	-44.2	-38.9	-32.3	-25.0	-17.1	12.3	.5913
500.	1679.	27.3	-79.0	-72.3	-64.0	-55.0	-45.1	19.7	.7642
600.	1572.	42.5	-126.2	-118.1	-108.2	-97.3	-85.5	29.3	.9492
700.	1471.	62.3	-187.1	-177.7	-166.2	-153.5	-139.6	41.0	1.1462
800.	1377.	87.7	-264.4	-253.7	-240.5	-226.0	-210.1	55.2	1.3573
900.	1295.	119.4	-360.2	-348.1	-333.3	-317.0	-299.1	71.9	1.5823
1000.	1222.	158.3	-476.2	-462.8	-446.3	-428.2	-408.3	90.8	1.8204

BALLISTIC COEFFICIENT .62, MUZZLE VELOCITY FT/SEC 2400.

RANGE	VELOCITY	HEIGHT	100 YD	150 YD	200 YD	250 YD	300 YD	DRIFT	TIME
0.	2400.	0.0	-1.5	-1.5	-1.5	-1.5	-1.5	0.0	0.0000
100.	2263.	.0	0.0	1.2	2.6	4.3	6.1	.7	.1287
200.	2131.	2.6	-5.3	-3.0	0.0	3.3	6.9	2.7	.2652
300.	2002.	7.3	-18.3	-14.8	-10.4	-5.4	0.0	6.3	.4105
400.	1879.	14.6	-39.9	-35.3	-29.4	-22.7	-15.5	11.4	.5649
500.	1761.	24.8	-71.5	-65.7	-58.3	-50.0	-41.0	18.5	.7299
600.	1650.	38.7	-114.4	-107.4	-98.5	-88.5	-77.8	27.5	.9062
700.	1544.	56.7	-169.9	-161.8	-151.3	-139.7	-127.2	38.6	1.0941
800.	1444.	79.8	-240.0	-230.7	-218.8	-205.5	-191.2	52.0	1.2953
900.	1354.	108.8	-327.0	-316.6	-303.2	-288.2	-272.1	67.8	1.5104
1000.	1274.	144.4	-432.9	-421.3	-406.4	-389.8	-371.9	86.0	1.7389

BALLISTIC COEFFICIENT .62, MUZZLE VELOCITY FT/SEC 2500.

RANGE	VELOCITY	HEIGHT	100 YD	150 YD	200 YD	250 YD	300 YD	DRIFT	TIME
0.	2500.	0.0	-1.5	-1.5	-1.5	-1.5	-1.5	0.0	0.0000
100.	2360.	-.0	0.0	1.0	2.4	3.9	5.5	.6	.1236
200.	2225.	2.4	-4.7	-2.7	0.0	3.0	6.3	2.5	.2542
300.	2093.	6.7	-16.6	-13.5	-9.4	-4.9	0.0	5.9	.3933
400.	1966.	13.3	-36.3	-32.2	-26.8	-20.8	-14.2	10.8	.5412
500.	1844.	22.7	-65.2	-60.0	-53.3	-45.8	-37.6	17.4	.6989
600.	1729.	35.3	-104.0	-97.9	-89.8	-80.8	-70.9	25.8	.8665
700.	1619.	51.8	-154.7	-147.5	-138.1	-127.5	-116.0	36.2	1.0459
800.	1515.	72.8	-218.5	-210.3	-199.6	-187.5	-174.4	48.9	1.2376
900.	1417.	99.1	-297.4	-288.2	-276.1	-262.6	-247.8	63.8	1.4425
1000.	1330.	131.7	-393.9	-383.6	-370.2	-355.1	-338.7	81.2	1.6612

BALLISTIC COEFFICIENT .62, MUZZLE VELOCITY FT/SEC 2600.

RANGE	VELOCITY	HEIGHT	100 YD	150 YD	200 YD	250 YD	300 YD	DRIFT	TIME
0.	2600.	0.0	-1.5	-1.5	-1.5	-1.5	-1.5	0.0	0.0000
100.	2458.	-.1	0.0	.9	2.1	3.5	5.0	.6	.1186
200.	2319.	2.1	-4.3	-2.4	0.0	2.8	5.7	2.4	.2443
300.	2185.	6.1	-15.0	-12.3	-8.6	-4.4	0.0	5.5	.3774
400.	2055.	12.2	-33.2	-29.5	-24.6	-19.0	-13.1	10.2	.5194
500.	1929.	20.8	-59.5	-54.9	-48.8	-41.9	-34.5	16.4	.6701
600.	1809.	32.3	-95.1	-89.6	-82.3	-73.9	-65.1	24.3	.8304
700.	1695.	47.4	-141.4	-134.9	-126.4	-116.7	-106.3	34.1	1.0017
800.	1587.	65.7	-199.9	-192.5	-182.8	-171.6	-159.8	46.1	1.1850
900.	1485.	90.7	-271.8	-263.5	-252.5	-240.0	-226.7	60.1	1.3800
1000.	1390.	120.5	-359.7	-350.5	-338.3	-324.4	-309.6	76.6	1.5891

BALLISTIC COEFFICIENT .62, MUZZLE VELOCITY FT/SEC 2700.

RANGE	VELOCITY	HEIGHT	100 YD	150 YD	200 YD	250 YD	300 YD	DRIFT	TIME
0.	2700.	0.0	-1.5	-1.5	-1.5	-1.5	-1.5	0.0	0.0000
100.	2555.	-.1	0.0	.8	1.9	3.2	4.6	.6	.1143
200.	2414.	1.9	-3.9	-2.2	0.0	2.5	5.3	2.3	.2352
300.	2277.	5.6	-13.7	-11.3	-7.9	-4.1	0.0	5.2	.3631
400.	2143.	11.2	-30.3	-27.1	-22.6	-17.5	-12.0	9.6	.4991
500.	2014.	19.1	-54.4	-50.4	-44.8	-38.4	-31.6	15.4	.6433
600.	1890.	29.7	-87.1	-82.3	-75.5	-67.9	-59.7	22.9	.7969
700.	1772.	43.6	-129.6	-124.0	-116.1	-107.2	-97.7	32.2	.9609
800.	1660.	61.2	-183.1	-176.6	-167.7	-157.5	-146.6	43.5	1.1358
900.	1554.	83.2	-249.2	-241.9	-231.8	-220.4	-208.1	56.8	1.3227
1000.	1453.	110.5	-329.5	-321.5	-310.3	-297.6	-283.9	72.4	1.5224

BALLISTIC COEFFICIENT .62, MUZZLE VELOCITY FT/SEC 2800.

RANGE	VELOCITY	HEIGHT	100 YD	150 YD	200 YD	250 YD	300 YD	DRIFT	TIME
0.	2800.	0.0	-1.5	-1.5	-1.5	-1.5	-1.5	0.0	0.0000
100.	2651.	-.2	0.0	.7	1.7	2.9	4.2	.5	.1098
200.	2507.	1.7	-3.5	-2.0	0.0	2.3	4.9	2.1	.2264
300.	2368.	5.1	-12.5	-10.3	-7.3	-3.8	0.0	4.9	.3495
400.	2232.	10.3	-27.7	-24.9	-20.8	-16.2	-11.1	9.0	.4799
500.	2100.	17.6	-50.0	-46.4	-41.3	-35.6	-29.2	14.6	.6186
600.	1972.	27.4	-80.0	-75.7	-69.6	-62.7	-55.0	21.6	.7657
700.	1850.	40.1	-119.1	-114.1	-106.9	-98.9	-89.9	30.4	.9229
800.	1735.	56.3	-168.2	-162.5	-154.3	-145.1	-134.9	41.0	1.0902
900.	1624.	76.5	-228.9	-222.4	-213.3	-202.9	-191.4	53.6	1.2689
1000.	1520.	101.6	-302.8	-295.6	-285.4	-273.9	-261.1	68.4	1.4600

BALLISTIC COEFFICIENT .62, MUZZLE VELOCITY FT/SEC 2900.

RANGE	VELOCITY	HEIGHT	100 YD	150 YD	200 YD	250 YD	300 YD	DRIFT	TIME
0.	2900.	0.0	-1.5	-1.5	-1.5	-1.5	-1.5	0.0	0.0000
100.	2748.	-.2	0.0	.6	1.6	2.6	3.8	.5	.1062
200.	2600.	1.5	-3.1	-1.9	0.0	2.2	4.5	2.0	.2185
300.	2458.	4.7	-11.4	-9.5	-6.7	-3.5	0.0	4.7	.3370
400.	2320.	9.5	-25.4	-22.9	-19.2	-14.9	-10.3	8.6	.4627
500.	2185.	16.3	-45.9	-42.8	-38.1	-32.7	-27.0	13.8	.5958
600.	2055.	25.4	-73.8	-70.1	-64.5	-58.0	-51.1	20.6	.7378
700.	1929.	37.1	-109.9	-105.5	-98.9	-91.4	-83.3	28.9	.8885
800.	1809.	52.0	-155.1	-150.1	-142.6	-134.0	-124.7	38.9	1.0487
900.	1696.	70.7	-211.0	-205.4	-197.0	-187.2	-176.9	50.9	1.2200
1000.	1587.	93.8	-279.2	-272.9	-263.6	-252.8	-241.2	64.9	1.4032

BALLISTIC COEFFICIENT .62, MUZZLE VELOCITY FT/SEC 3000.

RANGE	VELOCITY	HEIGHT	100 YD	150 YD	200 YD	250 YD	300 YD	DRIFT	TIME
0.	3000.	0.0	-1.5	-1.5	-1.5	-1.5	-1.5	0.0	0.0000
100.	2844.	-.2	0.0	.6	1.4	2.4	3.5	.5	.1027
200.	2693.	1.4	-2.8	-1.7	0.0	2.0	4.2	2.0	.2113
300.	2548.	4.4	-10.5	-8.8	-6.2	-3.3	0.0	4.6	.3260
400.	2407.	8.8	-23.5	-21.2	-17.8	-13.9	-9.5	8.3	.4472
500.	2271.	15.1	-42.5	-39.7	-35.4	-30.5	-25.0	13.3	.5754
600.	2138.	23.6	-68.3	-64.9	-59.8	-53.9	-47.3	19.7	.7117
700.	2009.	34.5	-101.6	-97.6	-91.7	-84.8	-77.2	27.5	.8564
800.	1885.	48.2	-143.5	-138.9	-132.2	-124.3	-115.5	37.0	1.0103
900.	1767.	65.5	-195.2	-190.1	-182.5	-173.6	-163.8	48.4	1.1748
1000.	1656.	86.8	-258.1	-252.4	-243.9	-234.1	-223.1	61.7	1.3503

BALLISTIC COEFFICIENT .62, MUZZLE VELOCITY FT/SEC 3100.

RANGE	VELOCITY	HEIGHT	100 YD	150 YD	200 YD	250 YD	300 YD	DRIFT	TIME
0.	3100.	0.0	-1.5	-1.5	-1.5	-1.5	-1.5	0.0	0.0000
100.	2940.	-.3	0.0	.5	1.3	2.2	3.2	.4	.0989
200.	2787.	1.2	-2.5	-1.6	0.0	1.8	3.8	1.8	.2039
300.	2638.	4.0	-9.6	-8.1	-5.8	-3.0	0.0	4.3	.3146
400.	2495.	8.2	-21.7	-19.7	-16.6	-12.9	-8.9	7.8	.4316
500.	2355.	14.1	-39.3	-36.9	-33.0	-28.4	-23.4	12.6	.5556
600.	2220.	21.9	-63.2	-60.2	-55.5	-50.0	-44.0	18.6	.6864
700.	2089.	32.0	-94.1	-90.7	-85.2	-78.8	-71.8	26.1	.8258
800.	1961.	44.8	-133.1	-129.1	-122.9	-115.5	-107.5	35.2	.9743
900.	1840.	60.8	-181.0	-176.6	-169.6	-161.3	-152.2	46.0	1.1323
1000.	1724.	80.4	-238.9	-234.0	-226.2	-217.0	-207.0	58.5	1.3002

BALLISTIC COEFFICIENT .62, MUZZLE VELOCITY FT/SEC 3200.

RANGE	VELOCITY	HEIGHT	100 YD	150 YD	200 YD	250 YD	300 YD	DRIFT	TIME
0.	3200.	0.0	-1.5	-1.5	-1.5	-1.5	-1.5	0.0	0.0000
100.	3037.	-.3	0.0	.4	1.1	2.0	2.9	.4	.0961
200.	2880.	1.1	-2.3	-1.4	0.0	1.7	3.6	1.7	.1974
300.	2728.	3.7	-8.8	-7.5	-5.4	-2.8	0.0	4.1	.3048
400.	2582.	7.6	-20.0	-18.3	-15.4	-12.0	-8.3	7.5	.4179
500.	2440.	13.1	-36.3	-34.2	-30.7	-26.3	-21.7	12.0	.5371
600.	2302.	20.4	-58.6	-56.1	-51.8	-46.6	-41.1	17.9	.6640
700.	2168.	29.8	-87.4	-84.4	-79.4	-73.3	-66.9	25.0	.7980
800.	2038.	41.8	-123.6	-120.3	-114.6	-107.6	-100.2	33.7	.9412
900.	1913.	56.6	-168.1	-164.3	-157.9	-150.1	-141.7	43.9	1.0931
1000.	1794.	74.8	-222.0	-217.8	-210.7	-202.0	-192.7	55.8	1.2548

BALLISTIC COEFFICIENT .62, MUZZLE VELOCITY FT/SEC 3300.

RANGE	VELOCITY	HEIGHT	100 YD	150 YD	200 YD	250 YD	300 YD	DRIFT	TIME
0.	3300.	0.0	-1.5	-1.5	-1.5	-1.5	-1.5	0.0	0.0000
100.	3134.	-.3	0.0	.4	1.0	1.8	2.7	.5	.0937
200.	2973.	1.0	-2.0	-1.3	0.0	1.6	3.3	1.7	.1917
300.	2818.	3.4	-8.0	-7.0	-5.0	-2.6	0.0	4.0	.2955
400.	2668.	7.1	-18.4	-17.0	-14.3	-11.2	-7.7	7.2	.4047
500.	2524.	12.3	-33.7	-32.0	-28.7	-24.7	-20.3	11.6	.5207
600.	2384.	19.1	-54.4	-52.3	-48.3	-43.6	-38.4	17.1	.6427
700.	2248.	27.9	-81.3	-78.8	-74.2	-68.7	-62.6	24.0	.7725
800.	2115.	39.0	-114.9	-112.1	-106.8	-100.5	-93.5	32.1	.9098
900.	1987.	52.8	-156.3	-153.1	-147.1	-140.1	-132.2	41.9	1.0561
1000.	1864.	69.7	-206.4	-202.9	-196.3	-188.4	-179.6	53.3	1.2119

BALLISTIC COEFFICIENT .62, MUZZLE VELOCITY FT/SEC 3400.

RANGE	VELOCITY	HEIGHT	100 YD	150 YD	200 YD	250 YD	300 YD	DRIFT	TIME
0.	3400.	0.0	-1.5	-1.5	-1.5	-1.5	-1.5	0.0	0.0000
100.	3230.	-.4	0.0	.3	.9	1.6	2.5	.4	.0904
200.	3066.	.9	-1.8	-1.2	0.0	1.5	3.1	1.6	.1855
300.	2908.	3.2	-7.4	-6.4	-4.7	-2.5	0.0	3.8	.2863
400.	2755.	6.6	-17.1	-15.8	-13.4	-10.5	-7.2	6.9	.3923
500.	2608.	11.4	-31.3	-29.7	-26.8	-23.1	-19.0	11.1	.5041
600.	2465.	17.8	-50.7	-48.7	-45.2	-40.8	-35.9	16.4	.6223
700.	2327.	26.1	-75.8	-73.5	-69.4	-64.3	-58.5	22.9	.7477
800.	2192.	36.5	-107.2	-104.6	-99.9	-94.0	-87.5	30.7	.8804
900.	2062.	49.4	-146.0	-143.0	-137.8	-131.1	-123.7	40.1	1.0218
1000.	1935.	65.2	-192.8	-189.5	-183.6	-176.3	-168.1	51.0	1.1720

BALLISTIC COEFFICIENT .62, MUZZLE VELOCITY FT/SEC 3500.

RANGE	VELOCITY	HEIGHT	100 YD	150 YD	200 YD	250 YD	300 YD	DRIFT	TIME
0.	3500.	0.0	-1.5	-1.5	-1.5	-1.5	-1.5	0.0	0.0000
100.	3327.	-.4	0.0	.3	.8	1.5	2.3	.4	.0882
200.	3159.	.8	-1.7	-1.1	0.0	1.4	2.9	1.6	.1807
300.	2998.	3.0	-6.8	-6.0	-4.3	-2.3	0.0	3.7	.2779
400.	2842.	6.2	-15.8	-14.7	-12.5	-9.8	-6.8	6.6	.3806
500.	2691.	10.7	-29.1	-27.8	-25.0	-21.6	-17.9	10.7	.4893
600.	2546.	16.8	-47.3	-45.7	-42.4	-38.3	-33.8	15.8	.6041
700.	2406.	24.5	-70.8	-68.9	-65.0	-60.3	-55.0	22.1	.7254
800.	2269.	34.2	-100.3	-98.1	-93.7	-88.3	-82.2	29.6	.8537
900.	2136.	46.3	-136.5	-134.0	-129.0	-122.9	-116.2	38.5	.9901
1000.	2007.	61.1	-180.2	-177.4	-171.9	-165.1	-157.6	48.9	1.1349

BALLISTIC COEFFICIENT .62, MUZZLE VELOCITY FT/SEC 3600.

RANGE	VELOCITY	HEIGHT	100 YD	150 YD	200 YD	250 YD	300 YD	DRIFT	TIME
0.	3600.	0.0	-1.5	-1.5	-1.5	-1.5	-1.5	0.0	0.0000
100.	3424.	-.4	0.0	.2	.7	1.4	2.1	.3	.0850
200.	3253.	.7	-1.5	-1.0	0.0	1.3	2.7	1.5	.1749
300.	3088.	2.7	-6.3	-5.6	-4.1	-2.2	0.0	3.5	.2699
400.	2929.	5.8	-14.7	-13.8	-11.8	-9.2	-6.3	6.4	.3696
500.	2775.	10.1	-27.2	-26.0	-23.5	-20.3	-16.7	10.2	.4746
600.	2627.	15.7	-44.2	-42.8	-39.8	-35.9	-31.6	15.1	.5858
700.	2484.	23.0	-66.2	-64.6	-61.1	-56.6	-51.6	21.1	.7032
800.	2345.	32.1	-93.9	-92.0	-88.0	-82.9	-77.2	28.4	.8278
900.	2210.	43.4	-127.8	-125.6	-121.1	-115.4	-108.9	36.8	.9593
1000.	2079.	57.2	-168.7	-166.3	-161.3	-154.9	-147.8	46.8	1.0991

BALLISTIC COEFFICIENT .62, MUZZLE VELOCITY FT/SEC 3700.

RANGE	VELOCITY	HEIGHT	100 YD	150 YD	200 YD	250 YD	300 YD	DRIFT	TIME
0.	3700.	0.0	-1.5	-1.5	-1.5	-1.5	-1.5	0.0	0.0000
100.	3520.	-.4	0.0	.2	.7	1.3	1.9	.3	.0830
200.	3346.	.7	-1.3	-.9	0.0	1.2	2.5	1.5	.1709
300.	3178.	2.6	-5.8	-5.2	-3.7	-2.0	0.0	3.4	.2626
400.	3016.	5.5	-13.7	-12.9	-11.0	-8.6	-6.0	6.2	.3595
500.	2860.	9.5	-25.4	-24.4	-22.0	-19.1	-15.8	9.9	.4616
600.	2708.	14.8	-41.4	-40.2	-37.4	-33.9	-29.9	14.6	.5697
700.	2563.	21.7	-62.1	-60.7	-57.4	-53.3	-48.6	20.4	.6835
800.	2421.	30.3	-88.1	-86.5	-82.7	-78.1	-72.7	27.4	.8040
900.	2284.	40.9	-119.9	-118.1	-113.9	-108.7	-102.6	35.5	.9315
1000.	2151.	53.9	-158.4	-156.4	-151.7	-145.9	-139.2	45.1	1.0669

BALLISTIC COEFFICIENT .62, MUZZLE VELOCITY FT/SEC 3800.

RANGE	VELOCITY	HEIGHT	100 YD	150 YD	200 YD	250 YD	300 YD	DRIFT	TIME
0.	3800.	0.0	-1.5	-1.5	-1.5	-1.5	-1.5	0.0	0.0000
100.	3617.	-.4	0.0	.1	.6	1.1	1.8	.4	.0814
200.	3440.	.6	-1.1	-.9	0.0	1.1	2.4	1.5	.1662
300.	3269.	2.4	-5.3	-4.8	-3.5	-1.9	0.0	3.3	.2557
400.	3103.	5.1	-12.7	-12.1	-10.4	-8.2	-5.7	6.1	.3502
500.	2944.	8.9	-23.6	-22.9	-20.7	-17.9	-14.8	9.5	.4489
600.	2790.	14.0	-38.6	-37.8	-35.2	-31.9	-28.1	14.1	.5539
700.	2641.	20.4	-58.1	-57.1	-54.1	-50.2	-45.8	19.7	.6644
800.	2498.	28.6	-82.6	-81.4	-78.0	-73.5	-68.5	26.4	.7814
900.	2358.	38.6	-112.5	-111.3	-107.4	-102.4	-96.8	34.2	.9051
1000.	2223.	50.8	-148.6	-147.2	-142.9	-137.3	-131.1	43.4	1.0359

BALLISTIC COEFFICIENT .62, MUZZLE VELOCITY FT/SEC 3900.

RANGE	VELOCITY	HEIGHT	100 YD	150 YD	200 YD	250 YD	300 YD	DRIFT	TIME
0.	3900.	0.0	-1.5	-1.5	-1.5	-1.5	-1.5	0.0	0.0000
100.	3714.	-.5	0.0	.1	.5	1.0	1.6	.3	.0784
200.	3533.	.5	-1.0	-.8	0.0	1.0	2.2	1.3	.1612
300.	3359.	2.2	-4.9	-4.5	-3.3	-1.8	0.0	3.1	.2484
400.	3191.	4.8	-11.8	-11.2	-9.7	-7.6	-5.3	5.7	.3400
500.	3028.	8.4	-22.1	-21.4	-19.5	-16.9	-14.0	9.1	.4365
600.	2871.	13.1	-36.2	-35.4	-33.1	-30.0	-26.5	13.5	.5381
700.	2720.	19.3	-54.6	-53.7	-51.0	-47.4	-43.3	18.9	.6458
800.	2573.	26.9	-77.6	-76.6	-73.5	-69.4	-64.7	25.3	.7592
900.	2432.	36.3	-105.8	-104.7	-101.2	-96.6	-91.3	32.9	.8790
1000.	2294.	47.8	-139.9	-138.6	-134.8	-129.6	-123.7	41.7	1.0061

BALLISTIC COEFFICIENT .62, MUZZLE VELOCITY FT/SEC 4000.

RANGE	VELOCITY	HEIGHT	100 YD	150 YD	200 YD	250 YD	300 YD	DRIFT	TIME
0.	4000.	0.0	-1.5	-1.5	-1.5	-1.5	-1.5	0.0	0.0000
100.	3810.	-.5	0.0	.1	.5	.9	1.5	.4	.0770
200.	3627.	.4	-.9	-.7	0.0	.9	2.1	1.4	.1579
300.	3450.	2.1	-4.5	-4.2	-3.1	-1.7	0.0	3.1	.2426
400.	3278.	4.5	-11.0	-10.6	-9.1	-7.3	-5.0	5.6	.3320
500.	3113.	8.0	-20.7	-20.3	-18.4	-16.1	-13.3	9.0	.4260
600.	2953.	12.5	-33.9	-33.4	-31.2	-28.4	-25.0	13.1	.5244
700.	2798.	18.3	-51.4	-50.7	-48.2	-44.9	-41.0	18.4	.6293
800.	2649.	25.5	-73.0	-72.3	-69.4	-65.6	-61.1	24.5	.7393
900.	2506.	34.4	-99.7	-98.9	-95.6	-91.4	-86.4	31.8	.8559
1000.	2366.	45.3	-131.9	-131.0	-127.3	-122.6	-117.0	40.3	.9791

BALLISTIC COEFFICIENT .63, MUZZLE VELOCITY FT/SEC 2000.

RANGE	VELOCITY	HEIGHT	100 YD	150 YD	200 YD	250 YD	300 YD	DRIFT	TIME
0.	2000.	0.0	-1.5	-1.5	-1.5	-1.5	-1.5	0.0	0.0000
100.	1878.	.4	0.0	1.9	4.2	6.7	9.4	.8	.1545
200.	1763.	4.1	-8.4	-4.6	0.0	5.0	10.3	3.4	.3195
300.	1653.	11.0	-28.1	-22.3	-15.5	-8.0	0.0	8.0	.4954
400.	1549.	21.6	-60.5	-52.8	-43.7	-33.7	-23.1	14.6	.6829
500.	1450.	36.7	-107.4	-97.8	-86.4	-74.0	-60.6	23.4	.8832
600.	1360.	57.0	-170.9	-159.4	-145.7	-130.8	-114.8	34.6	1.0969
700.	1281.	83.4	-253.1	-239.7	-223.7	-206.4	-187.7	48.2	1.3240
800.	1211.	116.8	-356.5	-341.2	-322.9	-303.1	-281.7	64.3	1.5651
900.	1150.	158.2	-483.6	-466.4	-445.9	-423.5	-399.5	82.7	1.8197
1000.	1103.	208.2	-637.1	-618.0	-595.2	-570.4	-543.6	103.2	2.0863

BALLISTIC COEFFICIENT .63, MUZZLE VELOCITY FT/SEC 2100.

RANGE	VELOCITY	HEIGHT	100 YD	150 YD	200 YD	250 YD	300 YD	DRIFT	TIME
0.	2100.	0.0	-1.5	-1.5	-1.5	-1.5	-1.5	0.0	0.0000
100.	1974.	.3	0.0	1.7	3.7	6.0	8.4	.7	.1470
200.	1854.	3.7	-7.5	-4.0	0.0	4.5	9.3	3.2	.3038
300.	1740.	9.9	-25.1	-19.9	-13.9	-7.2	0.0	7.5	.4710
400.	1631.	19.5	-54.2	-47.4	-39.3	-30.3	-20.8	13.7	.6490
500.	1528.	33.0	-96.4	-87.9	-77.8	-66.6	-54.6	22.0	.8391
600.	1431.	51.4	-153.5	-143.3	-131.2	-117.7	-103.4	32.6	1.0421
700.	1343.	75.3	-227.7	-215.7	-201.6	-185.9	-169.2	45.5	1.2587
800.	1266.	105.6	-321.0	-307.4	-291.2	-273.3	-254.2	60.9	1.4887
900.	1197.	143.4	-436.4	-421.0	-402.9	-382.7	-361.2	78.8	1.7332
1000.	1139.	189.3	-575.5	-558.4	-538.3	-515.9	-492.0	98.8	1.9900

BALLISTIC COEFFICIENT .63, MUZZLE VELOCITY FT/SEC 2200.

RANGE	VELOCITY	HEIGHT	100 YD	150 YD	200 YD	250 YD	300 YD	DRIFT	TIME
0.	2200.	0.0	-1.5	-1.5	-1.5	-1.5	-1.5	0.0	0.0000
100.	2071.	.2	0.0	1.5	3.3	5.3	7.5	.7	.1406
200.	1947.	3.3	-6.7	-3.7	0.0	4.0	8.3	3.1	.2902
300.	1828.	8.9	-22.5	-18.0	-12.5	-6.5	0.0	7.0	.4491
400.	1715.	17.6	-48.7	-42.7	-35.4	-27.4	-18.7	12.8	.6182
500.	1607.	29.9	-86.9	-79.4	-70.3	-60.3	-49.4	20.7	.7993
600.	1506.	46.5	-138.4	-129.4	-118.4	-106.5	-93.4	30.6	.9920
700.	1410.	68.1	-205.3	-194.8	-182.0	-168.0	-152.8	42.8	1.1980
800.	1325.	95.7	-289.8	-277.9	-263.2	-247.2	-229.9	57.5	1.4177
900.	1250.	130.1	-394.2	-380.7	-364.2	-346.2	-326.7	74.6	1.6510
1000.	1183.	172.1	-520.5	-505.5	-487.2	-467.2	-445.5	94.0	1.8979

BALLISTIC COEFFICIENT .63, MUZZLE VELOCITY FT/SEC 2300.

RANGE	VELOCITY	HEIGHT	100 YD	150 YD	200 YD	250 YD	300 YD	DRIFT	TIME
0.	2300.	0.0	-1.5	-1.5	-1.5	-1.5	-1.5	0.0	0.0000
100.	2168.	.1	0.0	1.3	3.0	4.8	6.8	.6	.1340
200.	2041.	2.9	-6.0	-3.3	0.0	3.6	7.6	2.9	.2771
300.	1917.	8.1	-20.3	-16.3	-11.3	-6.0	0.0	6.6	.4288
400.	1800.	16.0	-44.1	-38.7	-32.1	-24.9	-17.0	12.0	.5900
500.	1688.	27.1	-78.6	-71.9	-63.7	-54.7	-44.8	19.3	.7621
600.	1582.	42.2	-125.5	-117.5	-107.6	-96.8	-84.9	28.8	.9461
700.	1482.	61.8	-185.9	-176.5	-165.0	-152.4	-138.5	40.2	1.1415
800.	1388.	86.9	-262.4	-251.7	-238.5	-224.1	-208.2	54.1	1.3510
900.	1306.	118.1	-356.9	-344.9	-330.1	-313.9	-296.0	70.4	1.5738
1000.	1233.	156.8	-472.5	-459.1	-442.6	-424.7	-404.8	89.3	1.8115

BALLISTIC COEFFICIENT .63, MUZZLE VELOCITY FT/SEC 2400.

RANGE	VELOCITY	HEIGHT	100 YD	150 YD	200 YD	250 YD	300 YD	DRIFT	TIME
0.	2400.	0.0	-1.5	-1.5	-1.5	-1.5	-1.5	0.0	0.0000
100.	2265.	.0	0.0	1.2	2.6	4.3	6.1	.6	.1287
200.	2135.	2.6	-5.3	-3.0	0.0	3.3	6.9	2.7	.2651
300.	2008.	7.3	-18.2	-14.8	-10.3	-5.4	0.0	6.1	.4099
400.	1886.	14.5	-39.8	-35.2	-29.2	-22.6	-15.5	11.2	.5638
500.	1770.	24.7	-71.2	-65.5	-58.0	-49.8	-40.8	18.2	.7282
600.	1660.	38.4	-113.7	-106.8	-97.9	-88.0	-77.3	27.0	.9032
700.	1556.	56.3	-168.9	-160.8	-150.3	-138.8	-126.3	37.9	1.0901
800.	1456.	79.0	-238.2	-229.0	-217.0	-203.9	-189.6	50.9	1.2894
900.	1366.	107.6	-324.1	-313.7	-300.2	-285.4	-269.3	66.4	1.5022
1000.	1286.	142.7	-428.6	-417.1	-402.1	-385.7	-367.8	84.2	1.7285

BALLISTIC COEFFICIENT .63, MUZZLE VELOCITY FT/SEC 2500.

RANGE	VELOCITY	HEIGHT	100 YD	150 YD	200 YD	250 YD	300 YD	DRIFT	TIME
0.	2500.	0.0	-1.5	-1.5	-1.5	-1.5	-1.5	0.0	0.0000
100.	2363.	-.0	0.0	1.0	2.4	3.9	5.5	.6	.1235
200.	2229.	2.3	-4.7	-2.7	0.0	3.0	6.3	2.5	.2541
300.	2100.	6.7	-16.5	-13.4	-9.4	-4.9	0.0	5.8	.3929
400.	1974.	13.2	-36.2	-32.1	-26.7	-20.7	-14.1	10.5	.5399
500.	1854.	22.6	-64.8	-59.7	-52.9	-45.5	-37.2	17.0	.6968
600.	1739.	35.1	-103.5	-97.4	-89.3	-80.4	-70.5	25.3	.8639
700.	1631.	51.4	-153.7	-146.6	-137.2	-126.7	-115.2	35.6	1.0421
800.	1528.	72.1	-217.0	-208.8	-198.1	-186.1	-172.9	47.9	1.2322
900.	1430.	98.1	-295.1	-285.8	-273.8	-260.3	-245.5	62.5	1.4352
1000.	1343.	130.2	-390.3	-380.1	-366.6	-351.7	-335.2	79.5	1.6518

BALLISTIC COEFFICIENT .63, MUZZLE VELOCITY FT/SEC 2600.

RANGE	VELOCITY	HEIGHT	100 YD	150 YD	200 YD	250 YD	300 YD	DRIFT	TIME
0.	2600.	0.0	-1.5	-1.5	-1.5	-1.5	-1.5	0.0	0.0000
100.	2460.	-.1	0.0	.9	2.1	3.5	5.0	.6	.1185
200.	2324.	2.1	-4.3	-2.4	0.0	2.8	5.7	2.3	.2441
300.	2191.	6.1	-15.0	-12.2	-8.6	-4.4	0.0	5.4	.3769
400.	2063.	12.1	-33.0	-29.3	-24.5	-18.9	-13.1	10.0	.5183
500.	1939.	20.7	-59.2	-54.6	-48.5	-41.6	-34.3	16.1	.6683
600.	1820.	32.1	-94.6	-89.1	-81.8	-73.4	-64.6	23.8	.8276
700.	1708.	47.0	-140.5	-134.0	-125.5	-115.8	-105.5	33.5	.9978
800.	1601.	66.1	-198.4	-191.1	-181.4	-170.3	-158.5	45.2	1.1797
900.	1499.	89.7	-269.6	-261.3	-250.3	-237.9	-224.6	58.9	1.3730
1000.	1404.	119.1	-356.4	-347.2	-335.1	-321.2	-306.5	75.0	1.5800

BALLISTIC COEFFICIENT .63, MUZZLE VELOCITY FT/SEC 2700.

RANGE	VELOCITY	HEIGHT	100 YD	150 YD	200 YD	250 YD	300 YD	DRIFT	TIME
0.	2700.	0.0	-1.5	-1.5	-1.5	-1.5	-1.5	0.0	0.0000
100.	2557.	-.1	0.0	.8	1.9	3.2	4.5	.5	.1142
200.	2418.	1.9	-3.8	-2.2	0.0	2.5	5.2	2.2	.2350
300.	2283.	5.6	-13.6	-11.2	-7.9	-4.1	0.0	5.1	.3625
400.	2152.	11.2	-30.1	-26.9	-22.5	-17.4	-12.0	9.4	.4979
500.	2024.	19.0	-54.2	-50.2	-44.6	-38.2	-31.5	15.2	.6418
600.	1902.	29.6	-86.7	-81.8	-75.1	-67.5	-59.4	22.5	.7945
700.	1785.	43.3	-128.8	-123.2	-115.4	-106.5	-97.0	31.6	.9575
800.	1674.	60.6	-181.8	-175.4	-166.4	-156.3	-145.4	42.6	1.1309
900.	1569.	82.4	-247.3	-240.0	-230.0	-218.6	-206.4	55.7	1.3163
1000.	1469.	109.2	-326.5	-318.5	-307.3	-294.6	-281.1	70.8	1.5136

BALLISTIC COEFFICIENT .63, MUZZLE VELOCITY FT/SEC 2800.

RANGE	VELOCITY	HEIGHT	100 YD	150 YD	200 YD	250 YD	300 YD	DRIFT	TIME
0.	2800.	0.0	-1.5	-1.5	-1.5	-1.5	-1.5	0.0	0.0000
100.	2653.	-.2	0.0	.7	1.7	2.9	4.1	.5	.1098
200.	2512.	1.7	-3.5	-2.0	0.0	2.3	4.8	2.1	.2262
300.	2374.	5.1	-12.4	-10.3	-7.2	-3.8	0.0	4.8	.3489
400.	2240.	10.3	-27.6	-24.8	-20.7	-16.1	-11.0	8.9	.4791
500.	2110.	17.5	-49.7	-46.1	-41.0	-35.3	-29.0	14.3	.6169
600.	1984.	27.2	-79.6	-75.3	-69.2	-62.3	-54.7	21.2	.7634
700.	1864.	39.8	-118.3	-113.3	-106.2	-98.2	-89.3	29.8	.9193
800.	1749.	55.8	-167.1	-161.4	-153.2	-144.1	-133.9	40.2	1.0856
900.	1640.	75.8	-227.2	-220.7	-211.6	-201.3	-189.8	52.5	1.2628
1000.	1536.	100.4	-300.1	-292.9	-282.7	-271.3	-258.6	67.0	1.4519

BALLISTIC COEFFICIENT .63, MUZZLE VELOCITY FT/SEC 2900.

RANGE	VELOCITY	HEIGHT	100 YD	150 YD	200 YD	250 YD	300 YD	DRIFT	TIME
0.	2900.	0.0	-1.5	-1.5	-1.5	-1.5	-1.5	0.0	0.0000
100.	2750.	-.2	0.0	.6	1.6	2.6	3.8	.5	.1062
200.	2605.	1.5	-3.1	-1.9	0.0	2.2	4.5	2.0	.2182
300.	2465.	4.7	-11.3	-9.5	-6.7	-3.5	0.0	4.6	.3366
400.	2328.	9.5	-25.4	-22.9	-19.2	-14.8	-10.2	8.5	.4620
500.	2196.	16.2	-45.7	-42.6	-38.0	-32.6	-26.8	13.6	.5945
600.	2067.	25.2	-73.4	-69.7	-64.1	-57.6	-50.7	20.2	.7354
700.	1943.	36.9	-109.2	-104.8	-98.3	-90.8	-82.7	28.4	.8852
800.	1824.	51.6	-154.1	-149.1	-141.6	-133.0	-123.8	38.2	1.0444
900.	1711.	70.0	-209.3	-203.7	-195.4	-185.6	-175.3	49.8	1.2139
1000.	1604.	92.7	-276.7	-270.5	-261.2	-250.4	-238.9	63.5	1.3954

BALLISTIC COEFFICIENT .63, MUZZLE VELOCITY FT/SEC 3000.

RANGE	VELOCITY	HEIGHT	100 YD	150 YD	200 YD	250 YD	300 YD	DRIFT	TIME
0.	3000.	0.0	-1.5	-1.5	-1.5	-1.5	-1.5	0.0	0.0000
100.	2846.	-.2	0.0	.6	1.4	2.4	3.5	.5	.1027
200.	2698.	1.4	-2.8	-1.7	0.0	2.0	4.1	2.0	.2112
300.	2555.	4.3	-10.4	-8.7	-6.2	-3.3	0.0	4.5	.3255
400.	2416.	8.8	-23.4	-21.1	-17.8	-13.8	-9.5	8.2	.4463
500.	2281.	15.1	-42.3	-39.5	-35.2	-30.3	-24.9	13.0	.5740
600.	2150.	23.4	-67.9	-64.5	-59.4	-53.6	-47.0	19.3	.7095
700.	2023.	34.2	-101.0	-97.1	-91.1	-84.3	-76.7	27.0	.8534
800.	1900.	47.9	-142.6	-138.0	-131.3	-123.4	-114.7	36.3	1.0063
900.	1784.	64.9	-193.8	-188.7	-181.1	-172.3	-162.5	47.4	1.1695
1000.	1673.	85.8	-255.8	-250.2	-241.7	-231.9	-221.0	60.4	1.3430

BALLISTIC COEFFICIENT .63, MUZZLE VELOCITY FT/SEC 3100.

RANGE	VELOCITY	HEIGHT	100 YD	150 YD	200 YD	250 YD	300 YD	DRIFT	TIME
0.	3100.	0.0	-1.5	-1.5	-1.5	-1.5	-1.5	0.0	0.0000
100.	2943.	-.3	0.0	.5	1.3	2.2	3.2	.4	.0988
200.	2791.	1.2	-2.5	-1.6	0.0	1.8	3.8	1.8	.2038
300.	2645.	4.0	-9.6	-8.1	-5.7	-3.0	0.0	4.2	.3142
400.	2504.	8.2	-21.6	-19.6	-16.5	-12.8	-8.8	7.7	.4310
500.	2366.	14.0	-39.1	-36.7	-32.8	-28.2	-23.2	12.4	.5542
600.	2233.	21.8	-62.9	-60.0	-55.3	-49.8	-43.8	18.3	.6847
700.	2103.	31.8	-93.6	-90.2	-84.7	-78.3	-71.3	25.7	.8232
800.	1977.	44.4	-132.1	-128.2	-121.9	-114.6	-106.6	34.5	.9699
900.	1857.	60.2	-179.5	-175.1	-168.0	-159.8	-150.8	45.0	1.1265
1000.	1742.	79.6	-237.0	-232.2	-224.3	-215.2	-205.2	57.3	1.2935

BALLISTIC COEFFICIENT .63, MUZZLE VELOCITY FT/SEC 3200.

RANGE	VELOCITY	HEIGHT	100 YD	150 YD	200 YD	250 YD	300 YD	DRIFT	TIME
0.	3200.	0.0	-1.5	-1.5	-1.5	-1.5	-1.5	0.0	0.0000
100.	3040.	-.3	0.0	.4	1.1	2.0	2.9	.4	.0960
200.	2885.	1.1	-2.3	-1.4	0.0	1.7	3.6	1.7	.1973
300.	2735.	3.7	-8.8	-7.5	-5.4	-2.8	0.0	4.1	.3043
400.	2591.	7.6	-19.9	-18.2	-15.4	-11.9	-8.2	7.4	.4171
500.	2451.	13.0	-36.2	-34.1	-30.5	-26.2	-21.6	11.8	.5360
600.	2315.	20.3	-58.3	-55.8	-51.5	-46.4	-40.8	17.5	.6621
700.	2183.	29.6	-86.9	-83.9	-78.9	-72.9	-66.4	24.5	.7953
800.	2055.	41.4	-122.8	-119.5	-113.8	-106.9	-99.5	33.0	.9374
900.	1931.	56.1	-166.9	-163.1	-156.7	-148.9	-140.6	43.0	1.0881
1000.	1813.	74.0	-220.1	-215.9	-208.7	-200.1	-190.9	54.7	1.2481

BALLISTIC COEFFICIENT .63, MUZZLE VELOCITY FT/SEC 3300.

RANGE	VELOCITY	HEIGHT	100 YD	150 YD	200 YD	250 YD	300 YD	DRIFT	TIME
0.	3300.	0.0	-1.5	-1.5	-1.5	-1.5	-1.5	0.0	0.0000
100.	3136.	-.3	0.0	.4	1.0	1.8	2.7	.5	.0936
200.	2978.	1.0	-2.0	-1.3	0.0	1.6	3.3	1.7	.1916
300.	2825.	3.4	-8.0	-6.9	-4.9	-2.6	0.0	3.9	.2950
400.	2678.	7.1	-18.3	-16.9	-14.3	-11.1	-7.7	7.1	.4039
500.	2535.	12.2	-33.6	-31.8	-28.5	-24.6	-20.3	11.4	.5195
600.	2397.	19.0	-54.2	-52.1	-48.1	-43.4	-38.2	16.8	.6411
700.	2262.	27.7	-80.8	-78.4	-73.7	-68.2	-62.2	23.5	.7699
800.	2132.	38.7	-114.2	-111.4	-106.1	-99.8	-92.9	31.5	.9064
900.	2005.	52.3	-155.2	-152.1	-146.1	-139.0	-131.2	41.1	1.0515
1000.	1883.	69.0	-204.7	-201.2	-194.6	-186.7	-178.1	52.2	1.2057

BALLISTIC COEFFICIENT .63, MUZZLE VELOCITY FT/SEC 3400.

RANGE	VELOCITY	HEIGHT	100 YD	150 YD	200 YD	250 YD	300 YD	DRIFT	TIME
0.	3400.	0.0	-1.5	-1.5	-1.5	-1.5	-1.5	0.0	0.0000
100.	3233.	-.4	0.0	.3	.9	1.6	2.5	.4	.0904
200.	3071.	.9	-1.8	-1.2	0.0	1.5	3.1	1.6	.1854
300.	2915.	3.2	-7.4	-6.4	-4.7	-2.5	0.0	3.8	.2860
400.	2765.	6.6	-17.0	-15.7	-13.4	-10.5	-7.2	6.8	.3916
500.	2619.	11.4	-31.2	-29.6	-26.6	-23.0	-18.9	10.8	.5028
600.	2479.	17.7	-50.4	-48.5	-45.0	-40.6	-35.6	16.0	.6206
700.	2342.	25.9	-75.4	-73.2	-69.0	-63.9	-58.2	22.5	.7455
800.	2209.	36.2	-106.6	-104.0	-99.3	-93.4	-86.8	30.1	.8771
900.	2080.	48.9	-144.8	-141.9	-136.6	-130.0	-122.6	39.2	1.0168
1000.	1955.	64.5	-191.2	-188.0	-182.1	-174.8	-166.6	49.9	1.1661

BALLISTIC COEFFICIENT .63, MUZZLE VELOCITY FT/SEC 3500.

RANGE	VELOCITY	HEIGHT	100 YD	150 YD	200 YD	250 YD	300 YD	DRIFT	TIME
0.	3500.	0.0	-1.5	-1.5	-1.5	-1.5	-1.5	0.0	0.0000
100.	3330.	-.4	0.0	.3	.8	1.5	2.2	.4	.0881
200.	3165.	.8	-1.6	-1.1	0.0	1.4	2.9	1.6	.1805
300.	3005.	2.9	-6.7	-5.9	-4.3	-2.2	0.0	3.6	.2775
400.	2852.	6.2	-15.7	-14.6	-12.4	-9.7	-6.7	6.5	.3800
500.	2703.	10.7	-29.0	-27.7	-24.9	-21.5	-17.8	10.5	.4883
600.	2560.	16.7	-47.1	-45.4	-42.1	-38.1	-33.6	15.5	.6023
700.	2421.	24.3	-70.4	-68.5	-64.7	-59.9	-54.7	21.6	.7230
800.	2286.	34.0	-99.6	-97.5	-93.0	-87.6	-81.6	29.0	.8504
900.	2155.	45.9	-135.4	-133.0	-128.0	-121.9	-115.2	37.7	.9855
1000.	2027.	60.5	-178.8	-176.1	-170.5	-163.7	-156.3	47.9	1.1293

BALLISTIC COEFFICIENT .63, MUZZLE VELOCITY FT/SEC 3600.

RANGE	VELOCITY	HEIGHT	100 YD	150 YD	200 YD	250 YD	300 YD	DRIFT	TIME
0.	3600.	0.0	-1.5	-1.5	-1.5	-1.5	-1.5	0.0	0.0000
100.	3426.	-.4	0.0	.2	.7	1.4	2.1	.3	.0850
200.	3259.	.7	-1.5	-1.0	0.0	1.3	2.7	1.4	.1747
300.	3096.	2.7	-6.3	-5.5	-4.1	-2.2	0.0	3.5	.2696
400.	2939.	5.8	-14.6	-13.7	-11.7	-9.1	-6.3	6.2	.3687
500.	2787.	10.0	-27.1	-25.9	-23.4	-20.2	-16.6	10.0	.4737
600.	2641.	15.6	-44.0	-42.6	-39.6	-35.8	-31.5	14.8	.5843
700.	2500.	22.9	-65.9	-64.3	-60.8	-56.4	-51.3	20.7	.7012
800.	2363.	31.9	-93.3	-91.5	-87.5	-82.4	-76.7	27.8	.8247
900.	2229.	43.1	-126.9	-124.8	-120.3	-114.6	-108.1	36.1	.9553
1000.	2100.	56.7	-167.5	-165.1	-160.2	-153.8	-146.6	45.9	1.0941

BALLISTIC COEFFICIENT .63, MUZZLE VELOCITY FT/SEC 3700.

RANGE	VELOCITY	HEIGHT	100 YD	150 YD	200 YD	250 YD	300 YD	DRIFT	TIME
0.	3700.	0.0	-1.5	-1.5	-1.5	-1.5	-1.5	0.0	0.0000
100.	3523.	-.4	0.0	.2	.7	1.3	1.9	.3	.0830
200.	3352.	.6	-1.3	-.9	0.0	1.2	2.5	1.5	.1707
300.	3186.	2.6	-5.7	-5.1	-3.7	-2.0	0.0	3.4	.2623
400.	3026.	5.4	-13.6	-12.8	-10.9	-8.6	-5.9	6.1	.3589
500.	2872.	9.4	-25.2	-24.2	-21.9	-19.0	-15.7	9.7	.4606
600.	2723.	14.7	-41.2	-40.0	-37.2	-33.7	-29.7	14.4	.5682
700.	2579.	21.5	-61.8	-60.4	-57.1	-53.0	-48.4	20.0	.6815
800.	2439.	30.0	-87.5	-85.9	-82.1	-77.4	-72.1	26.8	.8008
900.	2304.	40.6	-119.1	-117.3	-113.1	-107.8	-101.9	34.8	.9277
1000.	2172.	53.3	-157.1	-155.1	-150.4	-144.5	-137.9	44.1	1.0615

BALLISTIC COEFFICIENT .63, MUZZLE VELOCITY FT/SEC 3800.

RANGE	VELOCITY	HEIGHT	100 YD	150 YD	200 YD	250 YD	300 YD	DRIFT	TIME
0.	3800.	0.0	-1.5	-1.5	-1.5	-1.5	-1.5	0.0	0.0000
100.	3620.	-.4	0.0	.1	.6	1.1	1.8	.4	.0814
200.	3445.	.6	-1.1	-.9	0.0	1.1	2.4	1.4	.1661
300.	3277.	2.4	-5.3	-4.8	-3.5	-1.9	0.0	3.3	.2556
400.	3114.	5.1	-12.6	-12.1	-10.3	-8.1	-5.6	6.0	.3496
500.	2956.	8.9	-23.5	-22.8	-20.6	-17.9	-14.7	9.4	.4479
600.	2804.	13.9	-38.5	-37.6	-35.1	-31.8	-28.0	13.9	.5527
700.	2657.	20.3	-57.8	-56.8	-53.8	-49.9	-45.5	19.3	.6624
800.	2516.	28.4	-82.1	-81.0	-77.5	-73.1	-68.1	25.9	.7787
900.	2378.	38.2	-111.7	-110.4	-106.5	-101.6	-95.9	33.5	.9011
1000.	2244.	50.3	-147.5	-146.1	-141.8	-136.3	-130.0	42.5	1.0311

BALLISTIC COEFFICIENT .63, MUZZLE VELOCITY FT/SEC 3900.

RANGE	VELOCITY	HEIGHT	100 YD	150 YD	200 YD	250 YD	300 YD	DRIFT	TIME
0.	3900.	0.0	-1.5	-1.5	-1.5	-1.5	-1.5	0.0	0.0000
100.	3717.	-.5	0.0	.1	.5	1.0	1.6	.3	.0784
200.	3539.	.5	-1.0	-.8	0.0	1.0	2.2	1.3	.1611
300.	3367.	2.2	-4.8	-4.5	-3.3	-1.8	0.0	3.0	.2481
400.	3201.	4.8	-11.7	-11.2	-9.7	-7.6	-5.3	5.6	.3395
500.	3041.	8.4	-22.0	-21.4	-19.4	-16.9	-13.9	9.0	.4355
600.	2886.	13.1	-36.0	-35.3	-33.0	-29.9	-26.4	13.2	.5368
700.	2736.	19.1	-54.3	-53.4	-50.7	-47.1	-43.0	18.5	.6437
800.	2592.	26.7	-77.2	-76.1	-73.1	-68.9	-64.3	24.8	.7565
900.	2452.	36.0	-105.1	-103.9	-100.5	-95.8	-90.6	32.2	.8753
1000.	2316.	47.4	-138.8	-137.6	-133.7	-128.6	-122.7	40.8	1.0013

BALLISTIC COEFFICIENT .63, MUZZLE VELOCITY FT/SEC 4000.

RANGE	VELOCITY	HEIGHT	100 YD	150 YD	200 YD	250 YD	300 YD	DRIFT	TIME
0.	4000.	0.0	-1.5	-1.5	-1.5	-1.5	-1.5	0.0	0.0000
100.	3813.	-.5	0.0	.1	.5	.9	1.5	.4	.0770
200.	3633.	.4	-.9	-.7	0.0	.9	2.1	1.4	.1577
300.	3458.	2.1	-4.4	-4.2	-3.1	-1.7	0.0	3.0	.2423
400.	3289.	4.5	-10.9	-10.6	-9.1	-7.2	-5.0	5.5	.3314
500.	3125.	7.9	-20.6	-20.2	-18.4	-16.0	-13.2	8.8	.4251
600.	2968.	12.4	-33.8	-33.3	-31.1	-28.2	-24.9	12.9	.5232
700.	2815.	18.1	-51.0	-50.4	-47.9	-44.6	-40.7	18.0	.6273
800.	2668.	25.3	-72.5	-71.8	-68.9	-65.1	-60.7	24.0	.7365
900.	2526.	34.1	-99.1	-98.3	-95.0	-90.7	-85.7	31.2	.8525
1000.	2388.	44.8	-130.8	-129.9	-126.3	-121.5	-116.0	39.5	.9744

BALLISTIC COEFFICIENT .64, MUZZLE VELOCITY FT/SEC 2000.

RANGE	VELOCITY	HEIGHT	100 YD	150 YD	200 YD	250 YD	300 YD	DRIFT	TIME
0.	2000.	0.0	-1.5	-1.5	-1.5	-1.5	-1.5	0.0	0.0000
100.	1880.	.4	0.0	1.9	4.2	6.7	9.3	.8	.1544
200.	1767.	4.1	-8.4	-4.5	0.0	4.9	10.3	3.4	.3192
300.	1658.	11.0	-28.0	-22.2	-15.4	-8.0	0.0	7.8	.4946
400.	1555.	21.5	-60.2	-52.6	-43.5	-33.6	-23.0	14.4	.6816
500.	1458.	36.5	-106.9	-97.3	-85.9	-73.6	-60.3	23.0	.8809
600.	1368.	56.6	-169.9	-158.4	-144.8	-130.0	-114.0	34.0	1.0933
700.	1290.	82.8	-251.6	-238.1	-222.2	-205.0	-186.3	47.4	1.3191
800.	1219.	115.8	-354.0	-338.6	-320.4	-300.7	-279.4	63.1	1.5584
900.	1158.	156.9	-480.4	-463.2	-442.7	-420.5	-396.5	81.4	1.8124
1000.	1109.	206.2	-632.0	-612.8	-590.1	-565.4	-538.8	101.5	2.0765

BALLISTIC COEFFICIENT .64, MUZZLE VELOCITY FT/SEC 2100.

RANGE	VELOCITY	HEIGHT	100 YD	150 YD	200 YD	250 YD	300 YD	DRIFT	TIME
0.	2100.	0.0	-1.5	-1.5	-1.5	-1.5	-1.5	0.0	0.0000
100.	1976.	.3	0.0	1.7	3.7	6.0	8.3	.7	.1469
200.	1858.	3.7	-7.4	-4.0	0.0	4.5	9.2	3.1	.3034
300.	1745.	9.9	-25.0	-19.9	-13.9	-7.2	0.0	7.3	.4702
400.	1638.	19.4	-54.0	-47.2	-39.2	-30.2	-20.7	13.4	.6477
500.	1536.	32.9	-96.0	-87.4	-77.4	-66.2	-54.3	21.6	.8369
600.	1439.	51.1	-152.8	-142.5	-130.5	-117.1	-102.8	32.0	1.0389
700.	1352.	74.7	-226.3	-214.4	-200.4	-184.7	-168.0	44.7	1.2540
800.	1275.	104.8	-319.1	-305.4	-289.4	-271.5	-252.4	59.8	1.4829
900.	1206.	142.0	-432.8	-417.4	-399.4	-379.2	-357.8	77.2	1.7245
1000.	1148.	187.4	-570.5	-553.4	-533.4	-511.0	-487.2	97.0	1.9797

BALLISTIC COEFFICIENT .64, MUZZLE VELOCITY FT/SEC 2200.

RANGE	VELOCITY	HEIGHT	100 YD	150 YD	200 YD	250 YD	300 YD	DRIFT	TIME
0.	2200.	0.0	-1.5	-1.5	-1.5	-1.5	-1.5	0.0	0.0000
100.	2073.	.2	0.0	1.5	3.3	5.3	7.5	.7	.1404
200.	1950.	3.3	-6.6	-3.7	0.0	4.0	8.3	3.0	.2899
300.	1833.	8.9	-22.5	-18.0	-12.5	-6.5	0.0	6.9	.4486
400.	1722.	17.5	-48.5	-42.5	-35.2	-27.3	-18.6	12.6	.6169
500.	1616.	29.7	-86.5	-79.0	-69.9	-59.9	-49.1	20.3	.7971
600.	1515.	46.2	-137.7	-128.8	-117.8	-105.8	-92.8	30.0	.9889
700.	1420.	67.6	-204.0	-193.5	-180.7	-166.8	-151.6	42.0	1.1933
800.	1335.	95.0	-288.0	-276.1	-261.4	-245.5	-228.1	56.5	1.4121
900.	1260.	128.8	-390.9	-377.4	-361.0	-343.0	-323.5	73.1	1.6427
1000.	1192.	170.6	-516.8	-501.8	-483.5	-463.6	-441.9	92.5	1.8893

BALLISTIC COEFFICIENT .64, MUZZLE VELOCITY FT/SEC 2300.

RANGE	VELOCITY	HEIGHT	100 YD	150 YD	200 YD	250 YD	300 YD	DRIFT	TIME
0.	2300.	0.0	-1.5	-1.5	-1.5	-1.5	-1.5	0.0	0.0000
100.	2170.	.1	0.0	1.3	3.0	4.8	6.7	.6	.1340
200.	2045.	2.9	-6.0	-3.3	0.0	3.6	7.5	2.8	.2769
300.	1923.	8.1	-20.2	-16.3	-11.3	-5.9	0.0	6.5	.4282
400.	1807.	15.9	-43.9	-38.6	-32.0	-24.8	-16.9	11.8	.5888
500.	1697.	27.0	-78.2	-71.6	-63.3	-54.4	-44.5	19.0	.7601
600.	1592.	41.9	-124.8	-116.8	-106.9	-96.2	-84.3	28.2	.9430
700.	1492.	61.3	-184.8	-175.5	-163.9	-151.4	-137.5	39.5	1.1372
800.	1399.	86.1	-260.5	-249.9	-236.7	-222.3	-206.5	53.1	1.3451
900.	1317.	117.0	-354.2	-342.3	-327.4	-311.3	-293.5	69.1	1.5664
1000.	1244.	154.9	-467.9	-454.6	-438.1	-420.2	-400.4	87.4	1.8009

BALLISTIC COEFFICIENT .64, MUZZLE VELOCITY FT/SEC 2400.

RANGE	VELOCITY	HEIGHT	100 YD	150 YD	200 YD	250 YD	300 YD	DRIFT	TIME
0.	2400.	0.0	-1.5	-1.5	-1.5	-1.5	-1.5	0.0	0.0000
100.	2268.	.0	0.0	1.2	2.6	4.3	6.1	.6	.1286
200.	2139.	2.6	-5.3	-3.0	0.0	3.3	6.8	2.6	.2649
300.	2014.	7.3	-18.2	-14.7	-10.3	-5.4	0.0	6.0	.4093
400.	1894.	14.5	-39.7	-35.1	-29.1	-22.6	-15.4	11.1	.5628
500.	1779.	24.6	-71.0	-65.2	-57.7	-49.6	-40.6	17.9	.7265
600.	1670.	38.2	-113.2	-106.2	-97.3	-87.5	-76.8	26.5	.9004
700.	1567.	55.9	-167.9	-159.8	-149.4	-138.0	-125.5	37.2	1.0861
800.	1468.	78.3	-236.5	-227.3	-215.4	-202.3	-188.0	49.9	1.2837
900.	1378.	106.5	-321.5	-311.1	-297.6	-283.0	-266.9	65.1	1.4947
1000.	1298.	141.1	-424.8	-413.3	-398.4	-382.0	-364.2	82.6	1.7191

BALLISTIC COEFFICIENT .64, MUZZLE VELOCITY FT/SEC 2500.

RANGE	VELOCITY	HEIGHT	100 YD	150 YD	200 YD	250 YD	300 YD	DRIFT	TIME
0.	2500.	0.0	-1.5	-1.5	-1.5	-1.5	-1.5	0.0	0.0000
100.	2365.	-.0	0.0	1.0	2.4	3.9	5.5	.6	.1234
200.	2233.	2.3	-4.7	-2.7	0.0	3.0	6.2	2.4	.2539
300.	2106.	6.6	-16.5	-13.4	-9.4	-4.9	0.0	5.7	.3922
400.	1982.	13.2	-36.0	-31.9	-26.5	-20.6	-14.1	10.3	.5388
500.	1863.	22.4	-64.5	-59.3	-52.6	-45.2	-37.0	16.7	.6949
600.	1750.	34.9	-103.0	-96.9	-88.8	-79.9	-70.1	24.9	.8612
700.	1642.	51.0	-152.9	-145.7	-136.3	-125.9	-114.5	34.9	1.0383
800.	1540.	71.5	-215.5	-207.3	-196.6	-184.7	-171.6	47.0	1.2268
900.	1444.	97.2	-292.8	-283.6	-271.6	-258.2	-243.5	61.3	1.4283
1000.	1356.	128.8	-387.1	-376.8	-363.4	-348.5	-332.2	78.0	1.6432

BALLISTIC COEFFICIENT .64, MUZZLE VELOCITY FT/SEC 2600.

RANGE	VELOCITY	HEIGHT	100 YD	150 YD	200 YD	250 YD	300 YD	DRIFT	TIME
0.	2600.	0.0	-1.5	-1.5	-1.5	-1.5	-1.5	0.0	0.0000
100.	2462.	-.1	0.0	.9	2.1	3.5	5.0	.5	.1185
200.	2328.	2.1	-4.3	-2.4	0.0	2.8	5.7	2.3	.2439
300.	2197.	6.1	-14.9	-12.2	-8.5	-4.4	0.0	5.3	.3764
400.	2071.	12.1	-32.9	-29.2	-24.3	-18.8	-12.9	9.8	.5171
500.	1948.	20.6	-59.0	-54.4	-48.3	-41.4	-34.1	15.8	.6667
600.	1831.	32.0	-94.2	-88.7	-81.4	-73.1	-64.3	23.4	.8255
700.	1720.	46.7	-139.6	-133.2	-124.7	-115.0	-104.7	32.8	.9940
800.	1614.	65.5	-197.0	-189.7	-180.0	-168.9	-157.2	44.3	1.1745
900.	1513.	88.9	-267.6	-259.3	-248.4	-236.0	-222.8	57.7	1.3665
1000.	1418.	117.8	-353.3	-344.1	-332.0	-318.2	-303.5	73.5	1.5714

BALLISTIC COEFFICIENT .64, MUZZLE VELOCITY FT/SEC 2700.

RANGE	VELOCITY	HEIGHT	100 YD	150 YD	200 YD	250 YD	300 YD	DRIFT	TIME
0.	2700.	0.0	-1.5	-1.5	-1.5	-1.5	-1.5	0.0	0.0000
100.	2559.	-.1	0.0	.8	1.9	3.2	4.5	.5	.1141
200.	2422.	1.9	-3.8	-2.2	0.0	2.5	5.2	2.2	.2347
300.	2289.	5.5	-13.6	-11.2	-7.9	-4.1	0.0	5.1	.3621
400.	2160.	11.1	-30.0	-26.8	-22.3	-17.3	-11.9	9.2	.4968
500.	2034.	18.9	-54.0	-50.0	-44.4	-38.1	-31.3	14.9	.6403
600.	1913.	29.4	-86.3	-81.5	-74.8	-67.2	-59.1	22.1	.7924
700.	1798.	42.9	-128.1	-122.4	-114.6	-105.8	-96.3	31.0	.9539
800.	1688.	60.1	-180.6	-174.1	-165.2	-155.1	-144.3	41.8	1.1261
900.	1583.	81.6	-245.4	-238.1	-228.2	-216.8	-204.6	54.6	1.3100
1000.	1484.	108.0	-323.7	-315.6	-304.5	-291.8	-278.3	69.4	1.5052

BALLISTIC COEFFICIENT .64, MUZZLE VELOCITY FT/SEC 2800.

RANGE	VELOCITY	HEIGHT	100 YD	150 YD	200 YD	250 YD	300 YD	DRIFT	TIME
0.	2800.	0.0	-1.5	-1.5	-1.5	-1.5	-1.5	0.0	0.0000
100.	2656.	-.2	0.0	.7	1.7	2.9	4.1	.5	.1097
200.	2516.	1.7	-3.5	-2.0	0.0	2.3	4.8	2.1	.2261
300.	2380.	5.1	-12.4	-10.3	-7.2	-3.8	0.0	4.7	.3484
400.	2249.	10.2	-27.6	-24.7	-20.6	-16.1	-11.0	8.7	.4783
500.	2120.	17.4	-49.5	-45.9	-40.8	-35.1	-28.8	14.0	.6152
600.	1996.	27.1	-79.3	-75.0	-68.9	-62.0	-54.5	20.8	.7613
700.	1877.	39.5	-117.6	-112.6	-105.5	-97.5	-88.7	29.2	.9160
800.	1763.	55.4	-166.0	-160.2	-152.1	-143.0	-132.9	39.4	1.0811
900.	1655.	75.1	-225.5	-219.1	-209.9	-199.7	-188.3	51.5	1.2569
1000.	1552.	99.4	-297.7	-290.5	-280.3	-268.9	-256.3	65.6	1.4442

BALLISTIC COEFFICIENT .64, MUZZLE VELOCITY FT/SEC 2900.

RANGE	VELOCITY	HEIGHT	100 YD	150 YD	200 YD	250 YD	300 YD	DRIFT	TIME
0.	2900.	0.0	-1.5	-1.5	-1.5	-1.5	-1.5	0.0	0.0000
100.	2752.	-.2	0.0	.6	1.5	2.6	3.8	.5	.1062
200.	2610.	1.5	-3.1	-1.9	0.0	2.2	4.4	2.0	.2180
300.	2471.	4.7	-11.3	-9.5	-6.7	-3.4	0.0	4.5	.3361
400.	2337.	9.5	-25.3	-22.8	-19.1	-14.8	-10.2	8.3	.4612
500.	2206.	16.1	-45.6	-42.5	-37.8	-32.5	-26.7	13.4	.5931
600.	2079.	25.0	-73.0	-69.3	-63.7	-57.2	-50.3	19.8	.7330
700.	1956.	36.6	-108.6	-104.2	-97.7	-90.2	-82.2	27.8	.8822
800.	1839.	51.2	-153.2	-148.2	-140.8	-132.2	-123.0	37.5	1.0405
900.	1727.	69.3	-207.8	-202.2	-193.9	-184.2	-173.9	48.8	1.2083
1000.	1621.	91.7	-274.2	-268.1	-258.8	-248.0	-236.5	62.2	1.3877

BALLISTIC COEFFICIENT .64, MUZZLE VELOCITY FT/SEC 3000.

RANGE	VELOCITY	HEIGHT	100 YD	150 YD	200 YD	250 YD	300 YD	DRIFT	TIME
0.	3000.	0.0	-1.5	-1.5	-1.5	-1.5	-1.5	0.0	0.0000
100.	2849.	-.2	0.0	.6	1.4	2.4	3.5	.5	.1026
200.	2703.	1.4	-2.8	-1.7	0.0	2.0	4.1	1.9	.2111
300.	2562.	4.3	-10.4	-8.7	-6.2	-3.2	0.0	4.4	.3250
400.	2425.	8.8	-23.3	-21.1	-17.7	-13.8	-9.4	8.0	.4455
500.	2292.	15.0	-42.1	-39.3	-35.1	-30.2	-24.8	12.8	.5728
600.	2162.	23.3	-67.5	-64.2	-59.1	-53.2	-46.7	18.9	.7073
700.	2037.	34.0	-100.5	-96.6	-90.6	-83.8	-76.3	26.5	.8507
800.	1915.	47.5	-141.7	-137.3	-130.5	-122.6	-114.0	35.7	1.0026
900.	1800.	64.3	-192.3	-187.3	-179.6	-170.8	-161.1	46.4	1.1639
1000.	1690.	84.9	-253.7	-248.1	-239.6	-229.8	-219.0	59.1	1.3359

BALLISTIC COEFFICIENT .64, MUZZLE VELOCITY FT/SEC 3100.

RANGE	VELOCITY	HEIGHT	100 YD	150 YD	200 YD	250 YD	300 YD	DRIFT	TIME
0.	3100.	0.0	-1.5	-1.5	-1.5	-1.5	-1.5	0.0	0.0000
100.	2945.	-.3	0.0	.5	1.3	2.2	3.2	.4	.0988
200.	2796.	1.2	-2.6	-1.6	0.0	1.8	3.8	1.8	.2038
300.	2652.	4.0	-9.5	-8.1	-5.7	-3.0	0.0	4.1	.3138
400.	2512.	8.1	-21.5	-19.6	-16.4	-12.8	-8.8	7.6	.4303
500.	2377.	13.9	-39.0	-36.5	-32.6	-28.1	-23.1	12.1	.5528
600.	2245.	21.6	-62.6	-59.7	-55.0	-49.5	-43.6	18.0	.6829
700.	2117.	31.5	-93.0	-89.6	-84.1	-77.8	-70.8	25.1	.8201
800.	1993.	44.1	-131.3	-127.4	-121.1	-113.9	-105.9	33.8	.9663
900.	1873.	59.6	-178.2	-173.8	-166.7	-158.6	-149.6	44.1	1.1214
1000.	1760.	78.7	-235.0	-230.1	-222.2	-213.2	-203.2	56.1	1.2866

BALLISTIC COEFFICIENT .64, MUZZLE VELOCITY FT/SEC 3200.

RANGE	VELOCITY	HEIGHT	100 YD	150 YD	200 YD	250 YD	300 YD	DRIFT	TIME
0.	3200.	0.0	-1.5	-1.5	-1.5	-1.5	-1.5	0.0	0.0000
100.	3042.	-.3	0.0	.4	1.1	2.0	2.9	.4	.0960
200.	2889.	1.1	-2.3	-1.4	0.0	1.7	3.5	1.7	.1972
300.	2742.	3.7	-8.7	-7.5	-5.3	-2.8	0.0	4.0	.3039
400.	2600.	7.6	-19.8	-18.1	-15.3	-11.9	-8.2	7.3	.4164
500.	2462.	13.0	-36.1	-33.9	-30.4	-26.1	-21.5	11.6	.5348
600.	2327.	20.2	-58.1	-55.5	-51.3	-46.2	-40.6	17.2	.6603
700.	2197.	29.4	-86.4	-83.5	-78.5	-72.5	-66.1	24.0	.7928
800.	2070.	41.1	-122.0	-118.6	-112.9	-106.1	-98.7	32.3	.9336
900.	1948.	55.6	-165.7	-161.9	-155.5	-147.8	-139.5	42.1	1.0832
1000.	1831.	73.3	-218.4	-214.2	-207.1	-198.5	-189.3	53.6	1.2420

BALLISTIC COEFFICIENT .64, MUZZLE VELOCITY FT/SEC 3300.

RANGE	VELOCITY	HEIGHT	100 YD	150 YD	200 YD	250 YD	300 YD	DRIFT	TIME
0.	3300.	0.0	-1.5	-1.5	-1.5	-1.5	-1.5	0.0	0.0000
100.	3139.	-.3	0.0	.3	1.0	1.8	2.7	.5	.0936
200.	2983.	1.0	-2.0	-1.3	0.0	1.6	3.3	1.7	.1914
300.	2832.	3.4	-8.0	-6.9	-4.9	-2.6	0.0	3.8	.2945
400.	2687.	7.1	-18.3	-16.9	-14.2	-11.1	-7.6	7.0	.4032
500.	2546.	12.1	-33.4	-31.7	-28.4	-24.4	-20.2	11.2	.5182
600.	2410.	18.9	-53.9	-51.8	-47.9	-43.1	-38.0	16.5	.6393
700.	2277.	27.5	-80.4	-77.9	-73.3	-67.8	-61.8	23.0	.7673
800.	2148.	38.4	-113.5	-110.7	-105.4	-99.1	-92.3	30.9	.9030
900.	2023.	51.9	-154.1	-151.0	-145.0	-138.0	-130.3	40.3	1.0470
1000.	1902.	68.4	-203.2	-199.7	-193.1	-185.2	-176.7	51.2	1.1998

BALLISTIC COEFFICIENT .64, MUZZLE VELOCITY FT/SEC 3400.

RANGE	VELOCITY	HEIGHT	100 YD	150 YD	200 YD	250 YD	300 YD	DRIFT	TIME
0.	3400.	0.0	-1.5	-1.5	-1.5	-1.5	-1.5	0.0	0.0000
100.	3235.	-.4	0.0	.3	.9	1.6	2.5	.4	.0903
200.	3076.	.9	-1.8	-1.2	0.0	1.4	3.1	1.6	.1853
300.	2922.	3.2	-7.4	-6.4	-4.6	-2.5	0.0	3.7	.2857
400.	2774.	6.6	-16.9	-15.6	-13.3	-10.4	-7.1	6.6	.3907
500.	2630.	11.3	-31.1	-29.5	-26.5	-22.9	-18.8	10.7	.5018
600.	2492.	17.6	-50.2	-48.3	-44.8	-40.5	-35.5	15.8	.6191
700.	2357.	25.8	-75.0	-72.8	-68.6	-63.6	-57.8	22.1	.7431
800.	2225.	35.9	-105.9	-103.4	-98.6	-92.9	-86.3	29.6	.8739
900.	2098.	48.5	-143.9	-141.0	-135.7	-129.2	-121.8	38.5	1.0129
1000.	1974.	63.8	-189.6	-186.4	-180.5	-173.3	-165.0	48.9	1.1600

BALLISTIC COEFFICIENT .64, MUZZLE VELOCITY FT/SEC 3500.

RANGE	VELOCITY	HEIGHT	100 YD	150 YD	200 YD	250 YD	300 YD	DRIFT	TIME
0.	3500.	0.0	-1.5	-1.5	-1.5	-1.5	-1.5	0.0	0.0000
100.	3332.	-.4	0.0	.3	.8	1.5	2.2	.4	.0881
200.	3170.	.8	-1.6	-1.1	0.0	1.4	2.9	1.6	.1803
300.	3013.	2.9	-6.7	-5.9	-4.3	-2.2	0.0	3.5	.2772
400.	2861.	6.2	-15.7	-14.6	-12.4	-9.7	-6.7	6.4	.3794
500.	2715.	10.6	-28.9	-27.6	-24.8	-21.4	-17.7	10.3	.4873
600.	2573.	16.6	-46.9	-45.3	-42.0	-37.9	-33.4	15.2	.6008
700.	2436.	24.2	-70.0	-68.1	-64.3	-59.5	-54.3	21.2	.7205
800.	2303.	33.7	-99.0	-96.9	-92.5	-87.0	-81.1	28.5	.8474
900.	2173.	45.5	-134.4	-132.1	-127.1	-121.0	-114.3	36.9	.9812
1000.	2047.	59.9	-177.4	-174.8	-169.3	-162.4	-155.0	47.0	1.1239

BALLISTIC COEFFICIENT .64, MUZZLE VELOCITY FT/SEC 3600.

RANGE	VELOCITY	HEIGHT	100 YD	150 YD	200 YD	250 YD	300 YD	DRIFT	TIME
0.	3600.	0.0	-1.5	-1.5	-1.5	-1.5	-1.5	0.0	0.0000
100.	3429.	-.4	0.0	.2	.7	1.4	2.1	.3	.0850
200.	3264.	.7	-1.5	-1.0	0.0	1.3	2.7	1.4	.1746
300.	3103.	2.7	-6.2	-5.5	-4.1	-2.2	0.0	3.4	.2693
400.	2949.	5.8	-14.5	-13.6	-11.6	-9.1	-6.2	6.1	.3680
500.	2799.	10.0	-27.0	-25.8	-23.4	-20.2	-16.6	9.9	.4730
600.	2655.	15.6	-43.8	-42.4	-39.4	-35.6	-31.3	14.6	.5828
700.	2515.	22.7	-65.6	-63.9	-60.5	-56.1	-51.0	20.4	.6992
800.	2380.	31.6	-92.7	-90.8	-86.9	-81.8	-76.1	27.3	.8215
900.	2248.	42.7	-126.1	-124.0	-119.5	-113.8	-107.4	35.5	.9514
1000.	2120.	56.1	-166.1	-163.7	-158.8	-152.5	-145.2	44.9	1.0884

BALLISTIC COEFFICIENT .64, MUZZLE VELOCITY FT/SEC 3700.

RANGE	VELOCITY	HEIGHT	100 YD	150 YD	200 YD	250 YD	300 YD	DRIFT	TIME
0.	3700.	0.0	-1.5	-1.5	-1.5	-1.5	-1.5	0.0	0.0000
100.	3526.	-.4	0.0	.2	.7	1.3	1.9	.3	.0830
200.	3357.	.6	-1.3	-.9	0.0	1.2	2.5	1.5	.1706
300.	3194.	2.5	-5.7	-5.1	-3.7	-2.0	0.0	3.3	.2621
400.	3036.	5.4	-13.5	-12.8	-10.9	-8.5	-5.9	6.0	.3583
500.	2884.	9.4	-25.1	-24.1	-21.8	-18.9	-15.6	9.5	.4597
600.	2736.	14.7	-41.0	-39.8	-37.0	-33.4	-29.5	14.1	.5667
700.	2594.	21.4	-61.4	-60.0	-56.8	-52.6	-48.1	19.7	.6794
800.	2456.	29.8	-87.0	-85.4	-81.7	-76.9	-71.7	26.3	.7981
900.	2322.	40.2	-118.3	-116.5	-112.3	-107.0	-101.1	34.1	.9238
1000.	2192.	52.8	-155.9	-153.9	-149.3	-143.4	-136.8	43.3	1.0566

BALLISTIC COEFFICIENT .64, MUZZLE VELOCITY FT/SEC 3800.

RANGE	VELOCITY	HEIGHT	100 YD	150 YD	200 YD	250 YD	300 YD	DRIFT	TIME
0.	3800.	0.0	-1.5	-1.5	-1.5	-1.5	-1.5	0.0	0.0000
100.	3623.	-.4	0.0	.1	.6	1.1	1.7	.4	.0813
200.	3451.	.6	-1.1	-.9	0.0	1.1	2.4	1.4	.1660
300.	3285.	2.4	-5.2	-4.8	-3.5	-1.9	0.0	3.3	.2553
400.	3124.	5.1	-12.6	-12.0	-10.3	-8.1	-5.6	5.9	.3491
500.	2968.	8.8	-23.4	-22.7	-20.5	-17.8	-14.7	9.2	.4472
600.	2818.	13.8	-38.3	-37.4	-34.9	-31.6	-27.8	13.6	.5512
700.	2673.	20.2	-57.4	-56.4	-53.4	-49.6	-45.2	18.9	.6602
800.	2533.	28.1	-81.6	-80.5	-77.0	-72.6	-67.6	25.4	.7759
900.	2397.	37.9	-111.0	-109.7	-105.8	-100.9	-95.3	32.9	.8975
1000.	2265.	49.8	-146.4	-145.0	-140.7	-135.2	-129.0	41.7	1.0263

BALLISTIC COEFFICIENT .64, MUZZLE VELOCITY FT/SEC 3900.

RANGE	VELOCITY	HEIGHT	100 YD	150 YD	200 YD	250 YD	300 YD	DRIFT	TIME
0.	3900.	0.0	-1.5	-1.5	-1.5	-1.5	-1.5	0.0	0.0000
100.	3719.	-.5	0.0	.1	.5	1.0	1.6	.2	.0783
200.	3545.	.5	-1.0	-.8	0.0	1.0	2.2	1.3	.1610
300.	3375.	2.2	-4.8	-4.4	-3.3	-1.7	0.0	3.0	.2477
400.	3211.	4.8	-11.7	-11.2	-9.7	-7.6	-5.3	5.5	.3390
500.	3053.	8.3	-21.9	-21.3	-19.3	-16.8	-13.9	8.8	.4345
600.	2900.	13.0	-35.9	-35.2	-32.9	-29.8	-26.3	13.0	.5357
700.	2752.	19.0	-54.0	-53.1	-50.5	-46.9	-42.8	18.2	.6419
800.	2610.	26.5	-76.7	-75.7	-72.6	-68.5	-63.8	24.3	.7537
900.	2471.	35.7	-104.4	-103.3	-99.8	-95.2	-89.9	31.6	.8718
1000.	2337.	47.0	-137.9	-136.6	-132.8	-127.6	-121.8	40.1	.9969

BALLISTIC COEFFICIENT .64, MUZZLE VELOCITY FT/SEC 4000.

RANGE	VELOCITY	HEIGHT	100 YD	150 YD	200 YD	250 YD	300 YD	DRIFT	TIME
0.	4000.	0.0	-1.5	-1.5	-1.5	-1.5	-1.5	0.0	0.0000
100.	3816.	-.5	0.0	.1	.4	.9	1.5	.4	.0771
200.	3638.	.4	-.9	-.7	0.0	1.0	2.1	1.3	.1576
300.	3466.	2.1	-4.4	-4.2	-3.1	-1.6	0.0	3.0	.2420
400.	3299.	4.5	-10.8	-10.5	-9.1	-7.1	-4.9	5.4	.3306
500.	3138.	7.9	-20.5	-20.1	-18.3	-15.9	-13.2	8.7	.4242
600.	2982.	12.3	-33.7	-33.2	-31.0	-28.1	-24.8	12.7	.5221
700.	2831.	18.0	-50.7	-50.1	-47.6	-44.2	-40.4	17.6	.6252
800.	2686.	25.1	-72.1	-71.4	-68.5	-64.7	-60.3	23.6	.7339
900.	2546.	33.8	-98.3	-97.6	-94.3	-90.0	-85.1	30.6	.8490
1000.	2409.	44.4	-129.9	-129.0	-125.4	-120.6	-115.1	38.7	.9701

BALLISTIC COEFFICIENT .65, MUZZLE VELOCITY FT/SEC 2000.

RANGE	VELOCITY	HEIGHT	100 YD	150 YD	200 YD	250 YD	300 YD	DRIFT	TIME
0.	2000.	0.0	-1.5	-1.5	-1.5	-1.5	-1.5	0.0	0.0000
100.	1882.	.4	0.0	1.9	4.2	6.6	9.3	.8	.1544
200.	1770.	4.1	-8.4	-4.5	0.0	4.9	10.2	3.3	.3189
300.	1663.	11.0	-27.9	-22.1	-15.3	-8.0	0.0	7.7	.4939
400.	1562.	21.5	-60.0	-52.4	-43.3	-33.5	-22.9	14.1	.6803
500.	1465.	36.3	-106.4	-96.8	-85.5	-73.2	-59.9	22.6	.8784
600.	1376.	56.3	-169.1	-157.6	-144.0	-129.3	-113.3	33.4	1.0900
700.	1297.	82.2	-250.1	-236.6	-220.8	-203.6	-185.0	46.5	1.3145
800.	1227.	114.9	-351.7	-336.3	-318.2	-298.6	-277.3	62.0	1.5522
900.	1166.	155.3	-476.3	-459.0	-438.6	-416.6	-392.6	79.8	1.8033
1000.	1116.	204.3	-626.7	-607.6	-584.9	-560.4	-533.8	99.7	2.0667

BALLISTIC COEFFICIENT .65, MUZZLE VELOCITY FT/SEC 2100.

RANGE	VELOCITY	HEIGHT	100 YD	150 YD	200 YD	250 YD	300 YD	DRIFT	TIME
0.	2100.	0.0	-1.5	-1.5	-1.5	-1.5	-1.5	0.0	0.0000
100.	1978.	.3	0.0	1.7	3.7	5.9	8.3	.7	.1467
200.	1861.	3.7	-7.4	-4.0	0.0	4.5	9.2	3.0	.3030
300.	1750.	9.8	-24.9	-19.8	-13.8	-7.1	0.0	7.2	.4695
400.	1644.	19.3	-53.8	-47.0	-39.0	-30.1	-20.6	13.2	.6464
500.	1544.	32.7	-95.6	-87.0	-77.1	-65.9	-54.1	21.2	.8348
600.	1448.	50.7	-152.0	-141.7	-129.8	-116.4	-102.2	31.4	1.0357
700.	1361.	74.2	-225.0	-213.0	-199.0	-183.4	-166.8	43.9	1.2494
800.	1284.	103.9	-316.7	-303.0	-287.1	-269.2	-250.2	58.7	1.4763
900.	1215.	140.8	-429.7	-414.3	-396.4	-376.3	-354.9	75.9	1.7170
1000.	1155.	185.9	-567.0	-549.9	-530.0	-507.7	-483.9	95.7	1.9721

BALLISTIC COEFFICIENT .65, MUZZLE VELOCITY FT/SEC 2200.

RANGE	VELOCITY	HEIGHT	100 YD	150 YD	200 YD	250 YD	300 YD	DRIFT	TIME
0.	2200.	0.0	-1.5	-1.5	-1.5	-1.5	-1.5	0.0	0.0000
100.	2075.	.2	0.0	1.5	3.3	5.3	7.5	.7	.1403
200.	1954.	3.3	-6.6	-3.6	0.0	4.0	8.3	3.0	.2896
300.	1839.	8.9	-22.4	-17.9	-12.5	-6.5	0.0	6.9	.4480
400.	1728.	17.5	-48.4	-42.4	-35.1	-27.2	-18.5	12.4	.6158
500.	1624.	29.6	-86.1	-78.7	-69.5	-59.6	-48.8	19.9	.7950
600.	1524.	45.9	-137.1	-128.1	-117.2	-105.2	-92.2	29.5	.9859
700.	1430.	67.1	-202.9	-192.5	-179.7	-165.8	-150.6	41.3	1.1892
800.	1345.	94.1	-285.9	-273.9	-259.4	-243.4	-226.1	55.4	1.4058
900.	1270.	127.6	-388.0	-374.6	-358.2	-340.2	-320.8	71.8	1.6353
1000.	1203.	168.6	-511.7	-496.8	-478.5	-458.6	-437.0	90.6	1.8784

BALLISTIC COEFFICIENT .65, MUZZLE VELOCITY FT/SEC 2300.

RANGE	VELOCITY	HEIGHT	100 YD	150 YD	200 YD	250 YD	300 YD	DRIFT	TIME
0.	2300.	0.0	-1.5	-1.5	-1.5	-1.5	-1.5	0.0	0.0000
100.	2172.	.1	0.0	1.3	3.0	4.8	6.7	.6	.1339
200.	2048.	2.9	-5.9	-3.3	0.0	3.6	7.5	2.8	.2766
300.	1928.	8.0	-20.2	-16.2	-11.3	-5.9	0.0	6.4	.4276
400.	1814.	15.8	-43.7	-38.4	-31.8	-24.7	-16.8	11.6	.5875
500.	1705.	26.8	-77.9	-71.3	-63.0	-54.1	-44.3	18.6	.7581
600.	1601.	41.7	-124.2	-116.2	-106.3	-95.6	-83.8	27.7	.9401
700.	1503.	60.9	-183.7	-174.4	-162.9	-150.3	-136.6	38.7	1.1331
800.	1410.	85.4	-258.7	-248.1	-235.0	-220.6	-204.9	52.1	1.3393
900.	1328.	115.9	-351.4	-339.5	-324.7	-308.6	-290.9	67.7	1.5588
1000.	1254.	153.4	-464.1	-450.9	-434.4	-416.5	-396.8	85.8	1.7919

BALLISTIC COEFFICIENT .65, MUZZLE VELOCITY FT/SEC 2400.

RANGE	VELOCITY	HEIGHT	100 YD	150 YD	200 YD	250 YD	300 YD	DRIFT	TIME
0.	2400.	0.0	-1.5	-1.5	-1.5	-1.5	-1.5	0.0	0.0000
100.	2270.	.0	0.0	1.2	2.6	4.3	6.0	.6	.1285
200.	2143.	2.6	-5.3	-3.0	0.0	3.2	6.8	2.6	.2647
300.	2020.	7.3	-18.1	-14.7	-10.2	-5.3	0.0	5.9	.4087
400.	1901.	14.4	-39.6	-34.9	-29.0	-22.5	-15.4	10.9	.5618
500.	1788.	24.5	-70.7	-64.9	-57.5	-49.3	-40.4	17.5	.7246
600.	1680.	37.9	-112.6	-105.7	-96.7	-87.0	-76.3	26.0	.8976
700.	1578.	55.5	-166.9	-158.9	-148.4	-137.1	-124.6	36.5	1.0822
800.	1480.	77.7	-234.9	-225.6	-213.7	-200.8	-186.5	48.9	1.2780
900.	1390.	105.5	-319.0	-308.6	-295.2	-280.6	-264.5	63.8	1.4875
1000.	1310.	139.6	-421.1	-409.6	-394.7	-378.5	-360.7	80.9	1.7099

BALLISTIC COEFFICIENT .65, MUZZLE VELOCITY FT/SEC 2500.

RANGE	VELOCITY	HEIGHT	100 YD	150 YD	200 YD	250 YD	300 YD	DRIFT	TIME
0.	2500.	0.0	-1.5	-1.5	-1.5	-1.5	-1.5	0.0	0.0000
100.	2367.	-.0	0.0	1.0	2.4	3.8	5.5	.6	.1233
200.	2237.	2.3	-4.7	-2.7	0.0	3.0	6.2	2.4	.2537
300.	2111.	6.6	-16.4	-13.3	-9.3	-4.9	0.0	5.6	.3916
400.	1989.	13.1	-35.9	-31.8	-26.5	-20.5	-14.0	10.2	.5379
500.	1872.	22.3	-64.2	-59.1	-52.4	-45.0	-36.9	16.4	.6932
600.	1760.	34.6	-102.4	-96.3	-88.3	-79.4	-69.6	24.4	.8585
700.	1654.	50.6	-152.0	-144.8	-135.5	-125.1	-113.7	34.3	1.0346
800.	1553.	70.9	-214.2	-206.0	-195.3	-183.4	-170.4	46.1	1.2220
900.	1457.	96.2	-290.6	-281.4	-269.4	-256.0	-241.4	60.1	1.4215
1000.	1369.	127.4	-383.6	-373.3	-359.9	-345.1	-328.9	76.4	1.6340

BALLISTIC COEFFICIENT .65, MUZZLE VELOCITY FT/SEC 2600.

RANGE	VELOCITY	HEIGHT	100 YD	150 YD	200 YD	250 YD	300 YD	DRIFT	TIME
0.	2600.	0.0	-1.5	-1.5	-1.5	-1.5	-1.5	0.0	0.0000
100.	2464.	-.1	0.0	.9	2.1	3.5	5.0	.5	.1184
200.	2332.	2.1	-4.3	-2.4	0.0	2.7	5.7	2.3	.2438
300.	2203.	6.0	-14.9	-12.2	-8.5	-4.4	0.0	5.2	.3759
400	2079.	12.0	-32.7	-29.1	-24.2	-18.7	-12.8	9.6	.5159
500.	1957.	20.5	-58.7	-54.1	-48.1	-41.2	-33.9	15.5	.6651
600.	1842.	31.8	-93.8	-88.3	-81.0	-72.8	-64.0	23.0	.8232
700.	1731.	46.4	-138.9	-132.5	-124.0	-114.4	-104.1	32.2	.9908
800.	1626.	64.9	-195.7	-188.4	-178.7	-167.7	-156.0	43.4	1.1697
900.	1527.	88.1	-265.7	-257.4	-246.5	-234.2	-220.9	56.6	1.3602
1000.	1432.	116.5	-350.4	-341.2	-329.1	-315.4	-300.7	72.0	1.5632

BALLISTIC COEFFICIENT .65, MUZZLE VELOCITY FT/SEC 2700.

RANGE	VELOCITY	HEIGHT	100 YD	150 YD	200 YD	250 YD	300 YD	DRIFT	TIME
0.	2700.	0.0	-1.5	-1.5	-1.5	-1.5	-1.5	0.0	0.0000
100.	2561.	-.1	0.0	.8	1.9	3.2	4.5	.5	.1140
200.	2426.	1.9	-3.8	-2.2	0.0	2.5	5.2	2.2	.2345
300.	2295.	5.5	-13.6	-11.1	-7.8	-4.1	0.0	5.0	.3616
400.	2168.	11.1	-29.9	-26.7	-22.3	-17.2	-11.8	9.1	.4959
500.	2044.	18.8	-53.8	-49.7	-44.2	-37.9	-31.2	14.7	.6389
600.	1924.	29.2	-85.9	-81.1	-74.5	-66.9	-58.8	21.7	.7902
700.	1810.	42.6	-127.3	-121.6	-113.9	-105.1	-95.6	30.4	.9505
800.	1701.	59.6	-179.4	-172.9	-164.1	-154.0	-143.2	40.9	1.1215
900.	1598.	80.8	-243.6	-236.3	-226.4	-215.0	-202.8	53.5	1.3038
1000.	1499.	106.8	-321.0	-312.9	-301.9	-289.2	-275.7	68.0	1.4972

BALLISTIC COEFFICIENT .65, MUZZLE VELOCITY FT/SEC 2800.

RANGE	VELOCITY	HEIGHT	100 YD	150 YD	200 YD	250 YD	300 YD	DRIFT	TIME
0.	2800.	0.0	-1.5	-1.5	-1.5	-1.5	-1.5	0.0	0.0000
100.	2658.	-.2	0.0	.7	1.7	2.9	4.1	.4	.1097
200.	2520.	1.7	-3.5	-2.0	0.0	2.3	4.8	2.0	.2259
300.	2387.	5.1	-12.4	-10.2	-7.2	-3.8	0.0	4.7	.3480
400.	2257.	10.2	-27.5	-24.6	-20.5	-16.0	-11.0	8.6	.4774
500.	2130.	17.3	-49.3	-45.8	-40.7	-35.0	-28.7	13.8	.6141
600.	2007.	26.9	-78.9	-74.6	-68.5	-61.7	-54.2	20.5	.7592
700.	1889.	39.3	-117.0	-112.0	-104.9	-97.0	-88.1	28.7	.9130
800.	1777.	54.9	-165.0	-159.3	-151.2	-142.1	-132.0	38.7	1.0770
900.	1669.	74.4	-223.9	-217.4	-208.3	-198.1	-186.7	50.5	1.2512
1000.	1568.	98.4	-295.3	-288.1	-278.0	-266.6	-254.1	64.3	1.4369

BALLISTIC COEFFICIENT .65, MUZZLE VELOCITY FT/SEC 2900.

RANGE	VELOCITY	HEIGHT	100 YD	150 YD	200 YD	250 YD	300 YD	DRIFT	TIME
0.	2900.	0.0	-1.5	-1.5	-1.5	-1.5	-1.5	0.0	0.0000
100.	2754.	-.2	0.0	.6	1.5	2.6	3.8	.5	.1062
200.	2614.	1.5	-3.1	-1.8	0.0	2.1	4.4	1.9	.2178
300.	2478.	4.7	-11.3	-9.4	-6.7	-3.4	0.0	4.4	.3356
400.	2345.	9.4	-25.2	-22.7	-19.1	-14.8	-10.2	8.2	.4605
500.	2216.	16.1	-45.4	-42.3	-37.7	-32.3	-26.6	13.1	.5918
600.	2091.	24.9	-72.7	-69.0	-63.5	-57.0	-50.2	19.5	.7312
700.	1969.	36.3	-107.9	-103.6	-97.1	-89.6	-81.6	27.3	.8790
800.	1853.	50.8	-152.1	-147.1	-139.8	-131.2	-122.0	36.7	1.0361
900.	1742.	68.7	-206.4	-200.8	-192.5	-182.9	-172.6	47.9	1.2031
1000.	1637.	90.8	-272.1	-265.9	-256.7	-246.0	-234.5	60.9	1.3808

BALLISTIC COEFFICIENT .65, MUZZLE VELOCITY FT/SEC 3000.

RANGE	VELOCITY	HEIGHT	100 YD	150 YD	200 YD	250 YD	300 YD	DRIFT	TIME
0.	3000.	0.0	-1.5	-1.5	-1.5	-1.5	-1.5	0.0	0.0000
100.	2851.	-.2	0.0	.6	1.4	2.4	3.5	.5	.1026
200.	2707.	1.4	-2.8	-1.7	0.0	2.0	4.1	1.9	.2109
300.	2568.	4.3	-10.4	-8.7	-6.2	-3.2	0.0	4.3	.3247
400.	2433.	8.7	-23.2	-21.0	-17.6	-13.7	-9.4	7.8	.4446
500.	2302.	14.9	-42.0	-39.2	-35.0	-30.1	-24.7	12.6	.5716
600.	2174.	23.1	-67.2	-63.9	-58.8	-52.9	-46.5	18.5	.7053
700.	2050.	33.8	-100.0	-96.1	-90.1	-83.3	-75.8	26.0	.8479
800.	1930.	47.1	-140.9	-136.4	-129.6	-121.8	-113.2	35.0	.9988
900.	1815.	63.7	-190.9	-185.9	-178.3	-169.4	-159.8	45.5	1.1586
1000.	1706.	84.0	-251.6	-246.1	-237.5	-227.7	-217.0	57.9	1.3290

BALLISTIC COEFFICIENT .65, MUZZLE VELOCITY FT/SEC 3100.

RANGE	VELOCITY	HEIGHT	100 YD	150 YD	200 YD	250 YD	300 YD	DRIFT	TIME
0.	3100.	0.0	-1.5	-1.5	-1.5	-1.5	-1.5	0.0	0.0000
100.	2948.	-.3	0.0	.5	1.3	2.2	3.2	.3	.0987
200.	2800.	1.2	-2.6	-1.6	0.0	1.8	3.8	1.8	.2037
300.	2658.	4.0	-9.5	-8.0	-5.7	-3.0	0.0	4.1	.3134
400.	2521.	8.1	-21.5	-19.5	-16.3	-12.8	-8.8	7.5	.4296
500.	2387.	13.9	-38.8	-36.4	-32.4	-27.9	-23.0	11.9	.5517
600.	2257.	21.5	-62.3	-59.4	-54.7	-49.3	-43.3	17.7	.6810
700.	2131.	31.3	-92.6	-89.2	-83.7	-77.4	-70.4	24.7	.8177
800.	2008.	43.7	-130.5	-126.7	-120.3	-113.1	-105.2	33.2	.9628
900.	1890.	59.1	-177.0	-172.6	-165.5	-157.4	-148.5	43.2	1.1166
1000.	1777.	78.0	-233.3	-228.5	-220.5	-211.6	-201.6	55.1	1.2806

BALLISTIC COEFFICIENT .65, MUZZLE VELOCITY FT/SEC 3200.

RANGE	VELOCITY	HEIGHT	100 YD	150 YD	200 YD	250 YD	300 YD	DRIFT	TIME
0.	3200.	0.0	-1.5	-1.5	-1.5	-1.5	-1.5	0.0	0.0000
100.	3044.	-.3	0.0	.4	1.1	2.0	2.9	.4	.0959
200.	2894.	1.1	-2.3	-1.4	0.0	1.7	3.5	1.7	.1972
300.	2749.	3.7	-8.7	-7.4	-5.3	-2.8	0.0	3.9	.3036
400.	2608.	7.5	-19.8	-18.1	-15.2	-11.8	-8.1	7.1	.4156
500.	2472.	12.9	-35.9	-33.8	-30.2	-26.0	-21.4	11.4	.5337
600.	2340.	20.1	-57.8	-55.3	-51.0	-46.0	-40.4	16.9	.6587
700.	2211.	29.2	-86.0	-83.0	-78.0	-72.1	-65.6	23.6	.7904
800.	2086.	40.8	-121.2	-117.8	-112.1	-105.4	-98.0	31.7	.9300
900.	1964.	55.1	-164.5	-160.7	-154.3	-146.7	-138.4	41.3	1.0785
1000.	1848.	72.6	-216.7	-212.5	-205.4	-196.9	-187.7	52.5	1.2359

BALLISTIC COEFFICIENT .65, MUZZLE VELOCITY FT/SEC 3300.

RANGE	VELOCITY	HEIGHT	100 YD	150 YD	200 YD	250 YD	300 YD	DRIFT	TIME
0.	3300.	0.0	-1.5	-1.5	-1.5	-1.5	-1.5	0.0	0.0000
100.	3141.	-.3	0.0	.3	1.0	1.8	2.6	.5	.0935
200.	2987.	1.0	-2.0	-1.3	0.0	1.6	3.3	1.6	.1912
300.	2839.	3.4	-7.9	-6.9	-4.9	-2.5	0.0	3.7	.2940
400.	2695.	7.0	-18.2	-16.8	-14.2	-11.0	-7.6	6.9	.4027
500.	2557.	12.1	-33.3	-31.5	-28.2	-24.3	-20.1	11.0	.5170
600.	2422.	18.8	-53.7	-51.6	-47.6	-42.9	-37.8	16.2	.6376
700.	2291.	27.3	-79.9	-77.5	-72.9	-67.4	-61.5	22.6	.7650
800.	2164.	38.1	-112.8	-110.0	-104.7	-98.4	-91.6	30.3	.8995
900.	2040.	51.4	-153.2	-150.0	-144.1	-137.0	-129.4	39.5	1.0427
1000.	1920.	67.7	-201.8	-198.3	-191.7	-183.8	-175.4	50.2	1.1944

BALLISTIC COEFFICIENT .65, MUZZLE VELOCITY FT/SEC 3400.

RANGE	VELOCITY	HEIGHT	100 YD	150 YD	200 YD	250 YD	300 YD	DRIFT	TIME
0.	3400.	0.0	-1.5	-1.5	-1.5	-1.5	-1.5	0.0	0.0000
100.	3238.	-.4	0.0	.3	.9	1.6	2.4	.4	.0903
200.	3081.	.9	-1.8	-1.2	0.0	1.4	3.1	1.5	.1852
300.	2929.	3.2	-7.3	-6.4	-4.6	-2.4	0.0	3.6	.2851
400.	2783.	6.5	-16.9	-15.6	-13.2	-10.4	-7.1	6.5	.3900
500.	2641.	11.3	-30.9	-29.4	-26.4	-22.8	-18.7	10.5	.5008
600.	2504.	17.6	-50.0	-48.1	-44.6	-40.3	-35.4	15.5	.6177
700.	2371.	25.6	-74.6	-72.4	-68.2	-63.2	-57.5	21.7	.7407
800.	2241.	35.7	-105.4	-102.8	-98.1	-92.4	-85.8	29.1	.8710
900.	2115.	48.1	-142.9	-140.0	-134.7	-128.2	-120.9	37.7	1.0084
1000.	1993.	63.2	-188.2	-185.0	-179.1	-172.0	-163.8	47.9	1.1546

BALLISTIC COEFFICIENT .65, MUZZLE VELOCITY FT/SEC 3500.

RANGE	VELOCITY	HEIGHT	100 YD	150 YD	200 YD	250 YD	300 YD	DRIFT	TIME
0.	3500.	0.0	-1.5	-1.5	-1.5	-1.5	-1.5	0.0	0.0000
100.	3335.	-.4	0.0	.3	.8	1.5	2.2	.4	.0881
200.	3175.	.8	-1.6	-1.1	0.0	1.4	2.9	1.5	.1801
300.	3020.	2.9	-6.7	-5.9	-4.3	-2.2	0.0	3.5	.2769
400.	2870.	6.1	-15.6	-14.6	-12.4	-9.6	-6.7	6.3	.3787
500.	2726.	10.6	-28.8	-27.5	-24.8	-21.3	-17.6	10.2	.4863
600.	2586.	16.5	-46.6	-45.1	-41.8	-37.7	-33.2	15.0	.5994
700.	2451.	24.0	-69.6	-67.8	-63.9	-59.1	-54.0	20.8	.7183
800.	2319.	33.5	-98.4	-96.3	-91.9	-86.4	-80.5	27.9	.8443
900.	2191.	45.1	-133.6	-131.2	-126.3	-120.1	-113.4	36.2	.9773
1000.	2066.	59.3	-176.0	-173.4	-167.9	-161.1	-153.7	46.0	1.1185

BALLISTIC COEFFICIENT .65, MUZZLE VELOCITY FT/SEC 3600.

RANGE	VELOCITY	HEIGHT	100 YD	150 YD	200 YD	250 YD	300 YD	DRIFT	TIME
0.	3600.	0.0	-1.5	-1.5	-1.5	-1.5	-1.5	0.0	0.0000
100.	3432.	-.4	0.0	.2	.7	1.4	2.1	.3	.0850
200.	3269.	.7	-1.5	-1.0	0.0	1.3	2.7	1.4	.1746
300.	3111.	2.7	-6.2	-5.5	-4.0	-2.2	0.0	3.4	.2691
400.	2958.	5.7	-14.5	-13.6	-11.6	-9.1	-6.2	6.0	.3674
500.	2810.	9.9	-26.9	-25.7	-23.2	-20.1	-16.5	9.7	.4719
600.	2668.	15.5	-43.6	-42.2	-39.2	-35.4	-31.1	14.3	.5812
700.	2530.	22.6	-65.2	-63.6	-60.1	-55.7	-50.7	20.0	.6971
800.	2396.	31.4	-92.2	-90.3	-86.4	-81.4	-75.6	26.8	.8188
900.	2266.	42.3	-125.2	-123.1	-118.6	-113.0	-106.6	34.8	.9475
1000.	2139.	55.6	-165.0	-162.7	-157.7	-151.4	-144.3	44.1	1.0839

BALLISTIC COEFFICIENT .65, MUZZLE VELOCITY FT/SEC 3700.

RANGE	VELOCITY	HEIGHT	100 YD	150 YD	200 YD	250 YD	300 YD	DRIFT	TIME
0.	3700.	0.0	-1.5	-1.5	-1.5	-1.5	-1.5	0.0	0.0000
100.	3528.	-.4	0.0	.2	.7	1.3	1.9	.3	.0830
200.	3362.	.6	-1.3	-.9	0.0	1.2	2.5	1.4	.1704
300.	3201.	2.5	-5.7	-5.1	-3.7	-2.0	0.0	3.3	.2619
400.	3046.	5.4	-13.5	-12.7	-10.9	-8.5	-5.9	5.9	.3577
500.	2895.	9.4	-25.1	-24.1	-21.8	-18.8	-15.5	9.4	.4590
600.	2750.	14.6	-40.8	-39.6	-36.8	-33.3	-29.4	13.9	.5654
700.	2609.	21.3	-61.1	-59.7	-56.5	-52.3	-47.7	19.3	.6773
800.	2473.	29.6	-86.4	-84.9	-81.2	-76.4	-71.2	25.8	.7953
900.	2341.	39.9	-117.6	-115.8	-111.6	-106.3	-100.4	33.5	.9203
1000.	2212.	52.4	-154.8	-152.8	-148.2	-142.3	-135.7	42.4	1.0519

BALLISTIC COEFFICIENT .65, MUZZLE VELOCITY FT/SEC 3800.

RANGE	VELOCITY	HEIGHT	100 YD	150 YD	200 YD	250 YD	300 YD	DRIFT	TIME
0.	3800.	0.0	-1.5	-1.5	-1.5	-1.5	-1.5	0.0	0.0000
100.	3625.	-.4	0.0	.1	.6	1.1	1.7	.4	.0813
200.	3456.	.6	-1.1	-.9	0.0	1.1	2.3	1.4	.1659
300.	3292.	2.4	-5.2	-4.8	-3.5	-1.9	0.0	3.2	.2549
400.	3133.	5.1	-12.5	-12.0	-10.3	-8.1	-5.6	5.8	.3485
500.	2980.	8.8	-23.4	-22.6	-20.5	-17.8	-14.7	9.1	.4465
600.	2832.	13.7	-38.1	-37.2	-34.6	-31.4	-27.6	13.4	.5496
700.	2688.	20.1	-57.2	-56.1	-53.2	-49.4	-45.0	18.6	.6583
800.	2550.	27.9	-81.1	-79.9	-76.5	-72.2	-67.2	24.9	.7731
900.	2416.	37.6	-110.3	-109.0	-105.1	-100.3	-94.6	32.3	.8940
1000.	2285.	49.3	-145.3	-143.9	-139.6	-134.2	-128.0	40.9	1.0216

BALLISTIC COEFFICIENT .65, MUZZLE VELOCITY FT/SEC 3900.

RANGE	VELOCITY	HEIGHT	100 YD	150 YD	200 YD	250 YD	300 YD	DRIFT	TIME
0.	3900.	0.0	-1.5	-1.5	-1.5	-1.5	-1.5	0.0	0.0000
100.	3722.	-.5	0.0	.1	.5	1.0	1.6	.2	.0783
200.	3550.	.5	-1.0	-.8	0.0	1.0	2.2	1.3	.1610
300.	3383.	2.2	-4.8	-4.4	-3.3	-1.7	0.0	2.9	.2474
400.	3221.	4.8	-11.7	-11.2	-9.6	-7.6	-5.3	5.4	.3385
500.	3065.	8.3	-21.8	-21.2	-19.3	-16.7	-13.8	8.6	.4337
600.	2914.	13.0	-35.8	-35.0	-32.7	-29.7	-26.2	12.8	.5345
700.	2768.	18.9	-53.7	-52.9	-50.2	-46.6	-42.5	17.9	.6400
800.	2627.	26.3	-76.2	-75.2	-72.1	-68.0	-63.4	23.9	.7511
900.	2490.	35.5	-103.8	-102.6	-99.2	-94.6	-89.4	31.0	.8685
1000.	2357.	46.5	-136.9	-135.7	-131.8	-126.7	-120.9	39.3	.9926

BALLISTIC COEFFICIENT .65, MUZZLE VELOCITY FT/SEC 4000.

RANGE	VELOCITY	HEIGHT	100 YD	150 YD	200 YD	250 YD	300 YD	DRIFT	TIME
0.	4000.	0.0	-1.5	-1.5	-1.5	-1.5	-1.5	0.0	0.0000
100.	3819.	-.5	0.0	.1	.4	.9	1.5	.4	.0771
200.	3644.	.4	-.9	-.7	0.0	1.0	2.0	1.3	.1574
300.	3474.	2.1	-4.4	-4.1	-3.1	-1.6	0.0	2.9	.2417
400.	3309.	4.5	-10.8	-10.5	-9.0	-7.1	-4.9	5.3	.3302
500.	3150.	7.9	-20.4	-20.0	-18.2	-15.8	-13.1	8.5	.4233
600.	2996.	12.3	-33.5	-33.0	-30.8	-27.9	-24.7	12.4	.5206
700.	2847.	17.9	-50.4	-49.8	-47.3	-43.9	-40.1	17.3	.6232
800.	2704.	24.9	-71.7	-71.0	-68.2	-64.3	-60.0	23.2	.7317
900.	2565.	33.6	-97.7	-96.9	-93.7	-89.4	-84.5	30.0	.8455
1000.	2430.	44.0	-128.9	-128.0	-124.4	-119.6	-114.2	38.0	.9657

BALLISTIC COEFFICIENT .66, MUZZLE VELOCITY FT/SEC 2000.

RANGE	VELOCITY	HEIGHT	100 YD	150 YD	200 YD	250 YD	300 YD	DRIFT	TIME
0.	2000.	0.0	-1.5	-1.5	-1.5	-1.5	-1.5	0.0	0.0000
100.	1884.	.4	0.0	1.9	4.2	6.6	9.3	.8	.1544
200.	1773.	4.1	-8.3	-4.5	0.0	4.9	10.2	3.3	.3187
300.	1668.	10.9	-27.8	-22.0	-15.3	-8.0	0.0	7.6	.4931
400.	1568.	21.4	-59.8	-52.2	-43.1	-33.4	-22.8	13.9	.6790
500.	1472.	36.1	-105.9	-96.3	-85.0	-72.8	-59.6	22.2	.8761
600.	1384.	55.9	-168.1	-156.6	-143.0	-128.4	-112.5	32.8	1.0864
700.	1306.	81.6	-248.5	-235.1	-219.3	-202.2	-183.7	45.7	1.3097
800.	1235.	114.1	-349.7	-334.4	-316.3	-296.8	-275.6	61.0	1.5468
900.	1173.	154.5	-474.3	-457.1	-436.8	-414.9	-391.0	78.9	1.7983
1000.	1123.	202.4	-621.7	-602.6	-580.0	-555.6	-529.1	98.1	2.0572

BALLISTIC COEFFICIENT .66, MUZZLE VELOCITY FT/SEC 2100.

RANGE	VELOCITY	HEIGHT	100 YD	150 YD	200 YD	250 YD	300 YD	DRIFT	TIME
0.	2100.	0.0	-1.5	-1.5	-1.5	-1.5	-1.5	0.0	0.0000
100.	1980.	.3	0.0	1.7	3.7	5.9	8.3	.7	.1466
200.	1865.	3.7	-7.4	-4.0	0.0	4.4	9.2	3.0	.3028
300.	1755.	9.8	-24.9	-19.7	-13.7	-7.1	0.0	7.1	.4687
400.	1651.	19.2	-53.7	-46.8	-38.9	-30.0	-20.5	13.0	.6452
500.	1551.	32.5	-95.3	-86.7	-76.7	-65.6	-53.8	20.9	.8328
600.	1456.	50.4	-151.2	-141.0	-129.0	-115.7	-101.5	30.9	1.0324
700.	1370.	73.6	-223.7	-211.7	-197.7	-182.2	-165.7	43.1	1.2449
800.	1293.	103.1	-314.9	-301.2	-285.3	-267.6	-248.7	57.7	1.4710
900.	1224.	139.4	-426.3	-410.9	-393.0	-373.0	-351.7	74.5	1.7088
1000.	1164.	183.7	-561.2	-544.0	-524.1	-501.9	-478.3	93.6	1.9604

BALLISTIC COEFFICIENT .66, MUZZLE VELOCITY FT/SEC 2200.

RANGE	VELOCITY	HEIGHT	100 YD	150 YD	200 YD	250 YD	300 YD	DRIFT	TIME
0.	2200.	0.0	-1.5	-1.5	-1.5	-1.5	-1.5	0.0	0.0000
100.	2077.	.2	0.0	1.5	3.3	5.3	7.5	.7	.1402
200.	1958.	3.3	-6.6	-3.6	0.0	4.0	8.3	2.9	.2894
300.	1844.	8.9	-22.4	-17.9	-12.4	-6.5	0.0	6.7	.4474
400.	1735.	17.4	-48.3	-42.3	-35.0	-27.1	-18.5	12.2	.6148
500.	1631.	29.4	-85.8	-78.3	-69.2	-59.3	-48.5	19.6	.7932
600.	1533.	45.6	-136.4	-127.4	-116.5	-104.6	-91.7	29.0	.9829
700.	1439.	66.7	-201.9	-191.4	-178.7	-164.8	-149.7	40.6	1.1852
800.	1355.	93.4	-284.2	-272.3	-257.7	-241.9	-224.6	54.5	1.4005
900.	1280.	126.5	-385.2	-371.7	-355.4	-337.5	-318.1	70.5	1.6279
1000.	1212.	166.9	-507.5	-492.6	-474.4	-454.5	-433.0	88.9	1.8689

BALLISTIC COEFFICIENT .66, MUZZLE VELOCITY FT/SEC 2300.

RANGE	VELOCITY	HEIGHT	100 YD	150 YD	200 YD	250 YD	300 YD	DRIFT	TIME
0.	2300.	0.0	-1.5	-1.5	-1.5	-1.5	-1.5	0.0	0.0000
100.	2174.	.1	0.0	1.3	3.0	4.8	6.7	.6	.1338
200.	2052.	2.9	-5.9	-3.3	0.0	3.6	7.5	2.7	.2763
300.	1934.	8.0	-20.1	-16.2	-11.2	-5.9	0.0	6.3	.4269
400.	1821.	15.8	-43.6	-38.3	-31.7	-24.5	-16.7	11.4	.5864
500.	1713.	26.7	-77.5	-71.0	-62.7	-53.8	-44.0	18.3	.7562
600.	1611.	41.4	-123.5	-115.6	-105.7	-95.0	-83.3	27.2	.9371
700.	1513.	60.5	-182.7	-173.5	-162.0	-149.4	-135.8	38.1	1.1292
800.	1421.	84.6	-257.0	-246.4	-233.3	-218.9	-203.3	51.1	1.3337
900.	1338.	114.8	-348.9	-337.0	-322.2	-306.1	-288.5	66.5	1.5518
1000.	1265.	151.7	-459.9	-446.7	-430.3	-412.4	-392.8	84.1	1.7822

BALLISTIC COEFFICIENT .66, MUZZLE VELOCITY FT/SEC 2400.

RANGE	VELOCITY	HEIGHT	100 YD	150 YD	200 YD	250 YD	300 YD	DRIFT	TIME
0.	2400.	0.0	-1.5	-1.5	-1.5	-1.5	-1.5	0.0	0.0000
100.	2272.	.0	0.0	1.1	2.6	4.3	6.0	.6	.1284
200.	2147.	2.6	-5.3	-3.0	0.0	3.3	6.8	2.5	.2643
300.	2025.	7.2	-18.1	-14.7	-10.2	-5.3	0.0	5.8	.4082
400.	1908.	14.3	-39.5	-34.9	-28.9	-22.4	-15.3	10.7	.5609
500.	1796.	24.3	-70.3	-64.6	-57.2	-49.0	-40.2	17.2	.7228
600.	1690.	37.7	-112.0	-105.1	-96.2	-86.5	-75.8	25.5	.8949
700.	1588.	55.1	-166.0	-157.9	-147.6	-136.2	-123.8	35.8	1.0785
800.	1492.	77.0	-233.4	-224.2	-212.3	-199.3	-185.1	48.0	1.2729
900.	1402.	104.5	-316.6	-306.2	-292.9	-278.2	-262.3	62.6	1.4805
1000.	1321.	138.1	-417.6	-406.1	-391.3	-375.0	-357.3	79.4	1.7011

BALLISTIC COEFFICIENT .66, MUZZLE VELOCITY FT/SEC 2500.

RANGE	VELOCITY	HEIGHT	100 YD	150 YD	200 YD	250 YD	300 YD	DRIFT	TIME
0.	2500.	0.0	-1.5	-1.5	-1.5	-1.5	-1.5	0.0	0.0000
100.	2369.	-.0	0.0	1.0	2.4	3.8	5.4	.6	.1233
200.	2241.	2.3	-4.7	-2.7	0.0	2.9	6.2	2.4	.2535
300.	2117.	6.6	-16.3	-13.3	-9.3	-4.8	0.0	5.4	.3909
400.	1997.	13.1	-35.8	-31.7	-26.4	-20.5	-14.0	10.0	.5370
500.	1881.	22.2	-63.9	-58.8	-52.1	-44.8	-36.7	16.1	.6916
600.	1770.	34.4	-102.0	-95.9	-87.9	-79.0	-69.4	24.0	.8562
700.	1665.	50.3	-151.1	-144.0	-134.6	-124.3	-113.0	33.6	1.0311
800.	1565.	70.4	-212.8	-204.7	-194.0	-182.2	-169.3	45.3	1.2172
900.	1469.	95.3	-288.4	-279.3	-267.2	-254.0	-239.4	58.9	1.4148
1000.	1381.	126.1	-380.3	-370.1	-356.7	-342.0	-325.8	74.9	1.6253

BALLISTIC COEFFICIENT .66, MUZZLE VELOCITY FT/SEC 2600.

RANGE	VELOCITY	HEIGHT	100 YD	150 YD	200 YD	250 YD	300 YD	DRIFT	TIME
0.	2600.	0.0	-1.5	-1.5	-1.5	-1.5	-1.5	0.0	0.0000
100.	2466.	-.1	0.0	.9	2.1	3.5	5.0	.5	.1183
200.	2336.	2.1	-4.3	-2.4	0.0	2.7	5.7	2.3	.2436
300.	2209.	6.0	-14.9	-12.1	-8.5	-4.4	0.0	5.2	.3754
400.	2086.	12.0	-32.6	-29.0	-24.1	-18.7	-12.8	9.4	.5151
500.	1966.	20.4	-58.5	-53.9	-47.8	-41.0	-33.7	15.2	.6634
600.	1852.	31.6	-93.2	-87.7	-80.5	-72.3	-63.5	22.6	.8206
700.	1743.	46.1	-138.2	-131.8	-123.3	-113.8	-103.5	31.7	.9876
800.	1639.	64.4	-194.5	-187.2	-177.5	-166.6	-154.9	42.6	1.1651
900.	1540.	87.2	-263.7	-255.4	-244.6	-232.3	-219.1	55.5	1.3539
1000.	1446.	115.4	-347.7	-338.5	-326.4	-312.8	-298.1	70.7	1.5553

BALLISTIC COEFFICIENT .66, MUZZLE VELOCITY FT/SEC 2700.

RANGE	VELOCITY	HEIGHT	100 YD	150 YD	200 YD	250 YD	300 YD	DRIFT	TIME
0.	2700.	0.0	-1.5	-1.5	-1.5	-1.5	-1.5	0.0	0.0000
100.	2563.	-.1	0.0	.8	1.9	3.2	4.5	.5	.1140
200.	2430.	1.9	-3.8	-2.2	0.0	2.5	5.2	2.1	.2342
300.	2301.	5.5	-13.5	-11.1	-7.8	-4.0	0.0	4.9	.3612
400.	2175.	11.0	-29.8	-26.6	-22.2	-17.1	-11.7	8.9	.4949
500.	2053.	18.8	-53.6	-49.5	-44.0	-37.7	-31.0	14.4	.6374
600.	1935.	29.0	-85.5	-80.6	-74.1	-66.5	-58.4	21.3	.7879
700.	1822.	42.3	-126.6	-120.9	-113.3	-104.4	-95.0	29.8	.9473
800.	1714.	59.1	-178.2	-171.7	-163.0	-152.9	-142.1	40.1	1.1170
900.	1611.	80.1	-241.8	-234.5	-224.6	-213.3	-201.1	52.4	1.2978
1000.	1514.	105.8	-318.5	-310.4	-299.5	-286.9	-273.4	66.7	1.4899

BALLISTIC COEFFICIENT .66, MUZZLE VELOCITY FT/SEC 2800.

RANGE	VELOCITY	HEIGHT	100 YD	150 YD	200 YD	250 YD	300 YD	DRIFT	TIME
0.	2800.	0.0	-1.5	-1.5	-1.5	-1.5	-1.5	0.0	0.0000
100.	2660.	-.2	0.0	.7	1.7	2.9	4.1	.4	.1096
200.	2524.	1.7	-3.5	-2.0	0.0	2.3	4.8	2.0	.2257
300.	2393.	5.0	-12.3	-10.2	-7.2	-3.8	0.0	4.6	.3476
400.	2264.	10.1	-27.4	-24.5	-20.5	-15.9	-10.9	8.4	.4765
500.	2140.	17.3	-49.2	-45.6	-40.5	-34.9	-28.6	13.6	.6129
600.	2018.	26.8	-78.5	-74.2	-68.2	-61.4	-53.8	20.1	.7571
700.	1901.	39.0	-116.4	-111.4	-104.3	-96.4	-87.6	28.2	.9102
800.	1790.	54.5	-164.0	-158.3	-150.2	-141.1	-131.1	38.0	1.0729
900.	1684.	73.7	-222.3	-215.9	-206.8	-196.6	-185.2	49.5	1.2456
1000.	1583.	97.4	-293.0	-285.8	-275.7	-264.4	-251.8	63.1	1.4297

BALLISTIC COEFFICIENT .66, MUZZLE VELOCITY FT/SEC 2900.

RANGE	VELOCITY	HEIGHT	100 YD	150 YD	200 YD	250 YD	300 YD	DRIFT	TIME
0.	2900.	0.0	-1.5	-1.5	-1.5	-1.5	-1.5	0.0	0.0000
100.	2757.	-.2	0.0	.6	1.5	2.6	3.7	.5	.1062
200.	2618.	1.5	-3.1	-1.8	0.0	2.1	4.4	1.9	.2175
300.	2484.	4.6	-11.2	-9.4	-6.6	-3.4	0.0	4.4	.3353
400.	2353.	9.4	-25.1	-22.7	-19.0	-14.7	-10.1	8.1	.4597
500.	2226.	16.0	-45.2	-42.1	-37.5	-32.2	-26.5	12.9	.5906
600.	2102.	24.8	-72.4	-68.7	-63.2	-56.8	-49.9	19.1	.7295
700.	1982.	36.1	-107.3	-103.0	-96.5	-89.1	-81.0	26.7	.8760
800.	1867.	50.4	-151.1	-146.1	-138.8	-130.3	-121.1	36.0	1.0319
900.	1757.	68.1	-204.9	-199.3	-191.1	-181.5	-171.2	46.9	1.1977
1000.	1652.	89.9	-270.0	-263.8	-254.7	-244.0	-232.5	59.8	1.3740

BALLISTIC COEFFICIENT .66, MUZZLE VELOCITY FT/SEC 3000.

RANGE	VELOCITY	HEIGHT	100 YD	150 YD	200 YD	250 YD	300 YD	DRIFT	TIME
0.	3000.	0.0	-1.5	-1.5	-1.5	-1.5	-1.5	0.0	0.0000
100.	2853.	-.2	0.0	.6	1.4	2.4	3.4	.5	.1026
200.	2711.	1.4	-2.8	-1.7	0.0	2.0	4.1	1.9	.2108
300.	2574.	4.3	-10.3	-8.7	-6.1	-3.2	0.0	4.3	.3243
400.	2441.	8.7	-23.1	-20.9	-17.5	-13.6	-9.3	7.7	.4437
500.	2312.	14.9	-41.8	-39.0	-34.8	-29.9	-24.6	12.4	.5703
600.	2186.	23.0	-66.9	-63.6	-58.5	-52.6	-46.2	18.2	.7035
700.	2063.	33.5	-99.4	-95.5	-89.6	-82.8	-75.3	25.5	.8451
800.	1944.	46.8	-140.0	-135.6	-128.8	-121.0	-112.4	34.3	.9950
900.	1831.	63.2	-189.7	-184.7	-177.1	-168.3	-158.7	44.7	1.1539
1000.	1723.	83.2	-249.6	-244.1	-235.6	-225.8	-215.1	56.7	1.3224

BALLISTIC COEFFICIENT .66, MUZZLE VELOCITY FT/SEC 3100.

RANGE	VELOCITY	HEIGHT	100 YD	150 YD	200 YD	250 YD	300 YD	DRIFT	TIME
0.	3100.	0.0	-1.5	-1.5	-1.5	-1.5	-1.5	0.0	0.0000
100.	2950.	-.3	0.0	.5	1.3	2.2	3.2	.3	.0987
200.	2805.	1.2	-2.5	-1.6	0.0	1.8	3.8	1.8	.2035
300.	2665.	4.0	-9.5	-8.0	-5.7	-3.0	0.0	4.0	.3130
400.	2529.	8.1	-21.4	-19.5	-16.3	-12.7	-8.8	7.4	.4289
500.	2397.	13.8	-38.7	-36.3	-32.3	-27.9	-22.9	11.7	.5506
600.	2269.	21.4	-62.0	-59.1	-54.4	-49.0	-43.1	17.3	.6792
700.	2144.	31.2	-92.2	-88.8	-83.3	-77.0	-70.1	24.3	.8153
800.	2022.	43.4	-129.8	-125.9	-119.6	-112.4	-104.5	32.6	.9593
900.	1905.	58.6	-175.9	-171.5	-164.4	-156.4	-147.5	42.4	1.1121
1000.	1794.	77.2	-231.5	-226.7	-218.8	-209.9	-199.9	54.0	1.2744

BALLISTIC COEFFICIENT .66, MUZZLE VELOCITY FT/SEC 3200.

RANGE	VELOCITY	HEIGHT	100 YD	150 YD	200 YD	250 YD	300 YD	DRIFT	TIME
0.	3200.	0.0	-1.5	-1.5	-1.5	-1.5	-1.5	0.0	0.0000
100.	3047.	-.3	0.0	.4	1.1	2.0	2.9	.4	.0959
200.	2898.	1.1	-2.3	-1.4	0.0	1.7	3.5	1.7	.1971
300.	2755.	3.7	-8.7	-7.4	-5.3	-2.8	0.0	3.9	.3034
400.	2617.	7.5	-19.7	-18.0	-15.1	-11.8	-8.1	7.0	.4148
500.	2482.	12.9	-35.8	-33.6	-30.1	-25.9	-21.3	11.2	.5326
600.	2352.	20.0	-57.6	-55.0	-50.8	-45.8	-40.2	16.6	.6571
700.	2224.	29.1	-85.6	-82.6	-77.6	-71.7	-65.2	23.2	.7880
800.	2101.	40.5	-120.6	-117.2	-111.5	-104.9	-97.4	31.2	.9270
900.	1981.	54.6	-163.3	-159.5	-153.1	-145.6	-137.2	40.5	1.0736
1000.	1866.	71.8	-215.0	-210.7	-203.6	-195.3	-186.0	51.4	1.2297

BALLISTIC COEFFICIENT .66, MUZZLE VELOCITY FT/SEC 3300.

RANGE	VELOCITY	HEIGHT	100 YD	150 YD	200 YD	250 YD	300 YD	DRIFT	TIME
0.	3300.	0.0	-1.5	-1.5	-1.5	-1.5	-1.5	0.0	0.0000
100.	3144.	-.3	0.0	.4	1.0	1.8	2.6	.5	.0935
200.	2992.	1.0	-2.0	-1.3	0.0	1.6	3.3	1.6	.1910
300.	2846.	3.4	-7.9	-6.8	-4.9	-2.5	0.0	3.7	.2936
400.	2704.	7.0	-18.2	-16.8	-14.2	-11.0	-7.6	6.8	.4021
500.	2567.	12.0	-33.1	-31.4	-28.1	-24.2	-20.0	10.8	.5159
600.	2434.	18.7	-53.4	-51.3	-47.4	-42.6	-37.6	15.9	.6358
700.	2305.	27.2	-79.6	-77.1	-72.5	-67.0	-61.1	22.2	.7627
800.	2179.	37.8	-112.1	-109.3	-104.1	-97.7	-91.0	29.7	.8962
900.	2057.	51.0	-152.1	-149.0	-143.1	-136.0	-128.5	38.8	1.0384
1000.	1938.	67.1	-200.2	-196.7	-190.2	-182.3	-173.9	49.2	1.1887

BALLISTIC COEFFICIENT .66, MUZZLE VELOCITY FT/SEC 3400.

RANGE	VELOCITY	HEIGHT	100 YD	150 YD	200 YD	250 YD	300 YD	DRIFT	TIME
0.	3400.	0.0	-1.5	-1.5	-1.5	-1.5	-1.5	0.0	0.0000
100.	3240.	-.4	0.0	.3	.9	1.6	2.4	.4	.0902
200.	3086.	.9	-1.8	-1.2	0.0	1.4	3.0	1.5	.1852
300.	2936.	3.1	-7.3	-6.3	-4.6	-2.4	0.0	3.5	.2846
400.	2792.	6.5	-16.8	-15.6	-13.2	-10.3	-7.1	6.5	.3896
500.	2652.	11.2	-30.8	-29.2	-26.3	-22.7	-18.7	10.3	.4998
600.	2517.	17.5	-49.8	-47.9	-44.4	-40.1	-35.3	15.3	.6162
700.	2385.	25.4	-74.2	-72.0	-67.8	-62.9	-57.2	21.3	.7385
800.	2257.	35.4	-104.8	-102.2	-97.5	-91.8	-85.3	28.5	.8679
900.	2132.	47.7	-142.0	-139.2	-133.8	-127.4	-120.2	37.0	1.0046
1000.	2011.	62.7	-186.9	-183.7	-177.8	-170.7	-162.6	47.0	1.1494

BALLISTIC COEFFICIENT .66, MUZZLE VELOCITY FT/SEC 3500.

RANGE	VELOCITY	HEIGHT	100 YD	150 YD	200 YD	250 YD	300 YD	DRIFT	TIME
0.	3500.	0.0	-1.5	-1.5	-1.5	-1.5	-1.5	0.0	0.0000
100.	3337.	-.4	0.0	.3	.8	1.5	2.2	.4	.0880
200.	3179.	.8	-1.6	-1.1	0.0	1.4	2.8	1.5	.1799
300.	3027.	2.9	-6.7	-5.9	-4.3	-2.2	0.0	3.4	.2766
400.	2879.	6.1	-15.5	-14.5	-12.3	-9.6	-6.6	6.2	.3781
500.	2736.	10.6	-28.7	-27.4	-24.7	-21.2	-17.5	10.0	.4853
600.	2599.	16.4	-46.4	-44.9	-41.6	-37.5	-33.1	14.7	.5979
700.	2465.	23.9	-69.3	-67.5	-63.6	-58.8	-53.7	20.5	.7163
800.	2335.	33.2	-97.9	-95.8	-91.5	-86.0	-80.1	27.4	.8416
900.	2208.	44.7	-132.7	-130.4	-125.5	-119.3	-112.7	35.6	.9735
1000.	2085.	58.7	-174.7	-172.1	-166.7	-159.8	-152.4	45.1	1.1133

BALLISTIC COEFFICIENT .66, MUZZLE VELOCITY FT/SEC 3600.

RANGE	VELOCITY	HEIGHT	100 YD	150 YD	200 YD	250 YD	300 YD	DRIFT	TIME
0.	3600.	0.0	-1.5	-1.5	-1.5	-1.5	-1.5	0.0	0.0000
100.	3434.	-.4	0.0	.2	.7	1.4	2.1	.3	.0850
200.	3273.	.7	-1.5	-1.0	0.0	1.2	2.7	1.4	.1746
300.	3118.	2.7	-6.2	-5.5	-4.0	-2.1	0.0	3.3	.2688
400.	2967.	5.7	-14.5	-13.5	-11.5	-9.0	-6.2	5.9	.3669
500.	2821.	9.9	-26.8	-25.6	-23.1	-20.0	-16.4	9.5	.4709
600.	2681.	15.4	-43.3	-41.9	-38.9	-35.2	-30.9	14.0	.5797
700.	2544.	22.4	-64.9	-63.3	-59.7	-55.4	-50.4	19.6	.6949
800.	2412.	31.2	-91.7	-89.8	-85.8	-80.9	-75.1	26.3	.8160
900.	2283.	42.0	-124.4	-122.3	-117.8	-112.2	-105.8	34.1	.9437
1000.	2158.	55.1	-163.7	-161.4	-156.4	-150.2	-143.0	43.2	1.0787

BALLISTIC COEFFICIENT .66, MUZZLE VELOCITY FT/SEC 3700.

RANGE	VELOCITY	HEIGHT	100 YD	150 YD	200 YD	250 YD	300 YD	DRIFT	TIME
0.	3700.	0.0	-1.5	-1.5	-1.5	-1.5	-1.5	0.0	0.0000
100.	3531.	-.4	0.0	.2	.7	1.2	1.9	.3	.0830
200.	3367.	.6	-1.3	-.9	0.0	1.2	2.5	1.4	.1702
300.	3208.	2.5	-5.7	-5.1	-3.7	-2.0	0.0	3.2	.2616
400.	3055.	5.4	-13.4	-12.6	-10.8	-8.5	-5.8	5.8	.3571
500.	2906.	9.3	-25.0	-24.0	-21.7	-18.8	-15.5	9.3	.4582
600.	2763.	14.5	-40.6	-39.5	-36.7	-33.2	-29.2	13.7	.5641
700.	2624.	21.1	-60.7	-59.4	-56.2	-52.0	-47.4	18.9	.6752
800.	2489.	29.4	-86.0	-84.4	-80.7	-76.0	-70.8	25.4	.7928
900.	2359.	39.6	-116.8	-115.1	-110.9	-105.6	-99.7	32.9	.9168
1000.	2231.	51.9	-153.7	-151.8	-147.2	-141.3	-134.8	41.6	1.0474

BALLISTIC COEFFICIENT .66, MUZZLE VELOCITY FT/SEC 3800.

RANGE	VELOCITY	HEIGHT	100 YD	150 YD	200 YD	250 YD	300 YD	DRIFT	TIME
0.	3800.	0.0	-1.5	-1.5	-1.5	-1.5	-1.5	0.0	0.0000
100.	3628.	-.4	0.0	.1	.6	1.1	1.7	.4	.0813
200.	3461.	.6	-1.1	-.8	0.0	1.1	2.3	1.4	.1658
300.	3299.	2.4	-5.2	-4.8	-3.5	-1.9	0.0	3.1	.2545
400.	3143.	5.1	-12.5	-11.9	-10.2	-8.1	-5.6	5.7	.3480
500.	2991.	8.8	-23.2	-22.5	-20.4	-17.7	-14.6	8.9	.4455
600.	2845.	13.7	-37.9	-37.0	-34.5	-31.2	-27.5	13.1	.5481
700.	2703.	19.9	-56.9	-55.9	-52.9	-49.1	-44.8	18.3	.6566
800.	2567.	27.7	-80.6	-79.4	-76.0	-71.7	-66.8	24.4	.7705
900.	2434.	37.3	-109.5	-108.2	-104.4	-99.5	-94.0	31.7	.8904
1000.	2304.	48.9	-144.4	-142.9	-138.7	-133.3	-127.1	40.1	1.0173

BALLISTIC COEFFICIENT .66, MUZZLE VELOCITY FT/SEC 3900.

RANGE	VELOCITY	HEIGHT	100 YD	150 YD	200 YD	250 YD	300 YD	DRIFT	TIME
0.	3900.	0.0	-1.5	-1.5	-1.5	-1.5	-1.5	0.0	0.0000
100.	3725.	-.5	0.0	.1	.5	1.0	1.6	.2	.0783
200.	3555.	.5	-1.0	-.8	0.0	1.0	2.2	1.2	.1609
300.	3390.	2.2	-4.8	-4.4	-3.3	-1.7	0.0	2.9	.2473
400.	3231.	4.7	-11.6	-11.1	-9.6	-7.5	-5.2	5.3	.3380
500.	3077.	8.3	-21.7	-21.1	-19.2	-16.7	-13.8	8.5	.4330
600.	2927.	12.9	-35.6	-34.9	-32.5	-29.5	-26.0	12.6	.5331
700.	2783.	18.8	-53.4	-52.6	-49.8	-46.3	-42.2	17.5	.6379
800.	2644.	26.2	-75.8	-74.8	-71.7	-67.7	-63.0	23.5	.7487
900.	2509.	35.2	-103.2	-102.1	-98.6	-94.0	-88.8	30.5	.8655
1000.	2377.	46.1	-135.9	-134.7	-130.8	-125.7	-119.9	38.5	.9881

BALLISTIC COEFFICIENT .66, MUZZLE VELOCITY FT/SEC 4000.

RANGE	VELOCITY	HEIGHT	100 YD	150 YD	200 YD	250 YD	300 YD	DRIFT	TIME
0.	4000.	0.0	-1.5	-1.5	-1.5	-1.5	-1.5	0.0	0.0000
100.	3822.	-.5	0.0	.1	.4	.9	1.5	.4	.0771
200.	3649.	.4	-.9	-.7	0.0	1.0	2.0	1.3	.1572
300.	3481.	2.0	-4.4	-4.1	-3.1	-1.6	0.0	2.9	.2414
400.	3319.	4.5	-10.8	-10.4	-9.0	-7.1	-4.9	5.2	.3298
500.	3162.	7.8	-20.3	-19.9	-18.1	-15.7	-13.0	8.3	.4224
600.	3010.	12.2	-33.3	-32.8	-30.7	-27.8	-24.5	12.2	.5194
700.	2863.	17.8	-50.1	-49.6	-47.1	-43.7	-39.9	17.0	.6215
800.	2721.	24.8	-71.3	-70.6	-67.8	-63.9	-59.6	22.8	.7293
900.	2583.	33.3	-97.1	-96.3	-93.1	-88.8	-83.9	29.5	.8425
1000.	2450.	43.6	-127.9	-127.1	-123.6	-118.7	-113.3	37.2	.9616

BALLISTIC COEFFICIENT .67, MUZZLE VELOCITY FT/SEC 2000.

RANGE	VELOCITY	HEIGHT	100 YD	150 YD	200 YD	250 YD	300 YD	DRIFT	TIME
0.	2000.	0.0	-1.5	-1.5	-1.5	-1.5	-1.5	0.0	0.0000
100.	1886.	.4	0.0	1.9	4.2	6.6	9.2	.8	.1543
200.	1776.	4.1	-8.3	-4.5	0.0	4.9	10.1	3.2	.3185
300.	1672.	10.9	-27.7	-22.0	-15.2	-7.9	0.0	7.5	.4924
400.	1573.	21.3	-59.6	-52.0	-43.0	-33.2	-22.7	13.7	.6777
500.	1479.	35.9	-105.4	-95.8	-84.5	-72.4	-59.2	21.8	.8739
600.	1391.	55.6	-167.3	-155.8	-142.3	-127.7	-111.9	32.2	1.0832
700.	1313.	81.1	-247.4	-234.0	-218.2	-201.2	-182.7	45.0	1.3058
800.	1244.	113.1	-347.2	-331.9	-313.9	-294.4	-273.3	59.9	1.5404
900.	1181.	152.7	-469.4	-452.1	-431.9	-410.0	-386.3	77.1	1.7878
1000.	1130.	200.6	-617.0	-597.8	-575.3	-551.0	-524.7	96.5	2.0481

BALLISTIC COEFFICIENT .67, MUZZLE VELOCITY FT/SEC 2100.

RANGE	VELOCITY	HEIGHT	100 YD	150 YD	200 YD	250 YD	300 YD	DRIFT	TIME
0.	2100.	0.0	-1.5	-1.5	-1.5	-1.5	-1.5	0.0	0.0000
100.	1982.	.3	0.0	1.7	3.7	5.9	8.3	.7	.1466
200.	1868.	3.6	-7.4	-4.0	0.0	4.4	9.1	3.0	.3025
300.	1760.	9.8	-24.8	-19.6	-13.7	-7.1	0.0	6.9	.4680
400.	1657.	19.2	-53.5	-46.6	-38.7	-29.9	-20.5	12.8	.6440
500.	1558.	32.4	-94.9	-86.3	-76.4	-65.4	-53.6	20.5	.8309
600.	1465.	50.1	-150.5	-140.2	-128.3	-115.0	-100.9	30.3	1.0293
700.	1378.	73.1	-222.3	-210.4	-196.5	-181.0	-164.6	42.3	1.2405
800.	1302.	102.2	-312.5	-298.8	-283.0	-265.3	-246.5	56.6	1.4644
900.	1233.	138.4	-423.8	-408.4	-390.6	-370.7	-349.5	73.3	1.7024
1000.	1172.	182.0	-556.8	-539.7	-519.9	-497.7	-474.2	92.0	1.9513

BALLISTIC COEFFICIENT .67, MUZZLE VELOCITY FT/SEC 2200.

RANGE	VELOCITY	HEIGHT	100 YD	150 YD	200 YD	250 YD	300 YD	DRIFT	TIME
0.	2200.	0.0	-1.5	-1.5	-1.5	-1.5	-1.5	0.0	0.0000
100.	2079.	.2	0.0	1.5	3.3	5.3	7.4	.7	.1401
200.	1961.	3.3	-6.6	-3.6	0.0	4.0	8.2	2.9	.2891
300.	1849.	8.8	-22.3	-17.8	-12.3	-6.4	0.0	6.6	.4466
400.	1741.	17.3	-48.1	-42.2	-34.9	-27.0	-18.4	12.0	.6137
500.	1639.	29.3	-85.5	-78.0	-68.9	-59.0	-48.4	19.3	.7913
600.	1541.	45.4	-135.8	-126.8	-115.9	-104.0	-91.2	28.5	.9800
700.	1449.	66.2	-200.7	-190.3	-177.6	-163.7	-148.8	39.9	1.1811
800.	1364.	92.6	-282.2	-270.3	-255.8	-239.9	-222.8	53.4	1.3945
900.	1289.	125.3	-382.4	-369.0	-352.6	-334.8	-315.6	69.2	1.6207
1000.	1222.	165.2	-503.4	-488.4	-470.3	-450.4	-429.1	87.3	1.8596

BALLISTIC COEFFICIENT .67, MUZZLE VELOCITY FT/SEC 2300.

RANGE	VELOCITY	HEIGHT	100 YD	150 YD	200 YD	250 YD	300 YD	DRIFT	TIME
0.	2300.	0.0	-1.5	-1.5	-1.5	-1.5	-1.5	0.0	0.0000
100.	2176.	.1	0.0	1.3	3.0	4.7	6.7	.6	.1337
200.	2056.	2.9	-5.9	-3.3	0.0	3.6	7.5	2.7	.2761
300.	1939.	8.0	-20.1	-16.1	-11.2	-5.8	0.0	6.2	.4263
400.	1827.	15.7	-43.5	-38.2	-31.6	-24.5	-16.7	11.2	.5855
500.	1721.	26.6	-77.2	-70.6	-62.4	-53.5	-43.8	18.0	.7543
600.	1620.	41.1	-122.9	-115.0	-105.1	-94.4	-82.7	26.7	.9342
700.	1523.	60.0	-181.7	-172.5	-161.0	-148.5	-134.9	37.4	1.1255
800.	1431.	84.0	-255.5	-245.0	-231.8	-217.5	-202.0	50.2	1.3288
900.	1349.	113.8	-346.4	-334.6	-319.8	-303.7	-286.2	65.3	1.5449
1000.	1275.	150.3	-456.6	-443.4	-427.0	-409.1	-389.7	82.7	1.7740

BALLISTIC COEFFICIENT .67, MUZZLE VELOCITY FT/SEC 2400.

RANGE	VELOCITY	HEIGHT	100 YD	150 YD	200 YD	250 YD	300 YD	DRIFT	TIME
0.	2400.	0.0	-1.5	-1.5	-1.5	-1.5	-1.5	0.0	0.0000
100.	2273.	.0	0.0	1.1	2.6	4.3	6.0	.6	.1283
200.	2150.	2.6	-5.2	-3.0	0.0	3.3	6.8	2.5	.2640
300.	2030.	7.2	-18.1	-14.6	-10.2	-5.3	0.0	5.8	.4078
400.	1915.	14.3	-39.4	-34.8	-28.9	-22.3	-15.3	10.6	.5600
500.	1804.	24.2	-70.1	-64.3	-56.9	-48.8	-39.9	16.9	.7211
600.	1699.	37.5	-111.5	-104.6	-95.7	-85.9	-75.3	25.0	.8923
700.	1599.	54.7	-165.1	-157.0	-146.7	-135.3	-122.9	35.2	1.0747
800.	1503.	76.4	-231.9	-222.7	-210.9	-197.9	-183.7	47.1	1.2679
900.	1413.	103.5	-314.3	-304.0	-290.7	-276.0	-260.1	61.4	1.4739
1000.	1333.	136.8	-414.4	-402.9	-388.1	-371.8	-354.1	77.9	1.6928

BALLISTIC COEFFICIENT .67, MUZZLE VELOCITY FT/SEC 2500.

RANGE	VELOCITY	HEIGHT	100 YD	150 YD	200 YD	250 YD	300 YD	DRIFT	TIME
0.	2500.	0.0	-1.5	-1.5	-1.5	-1.5	-1.5	0.0	0.0000
100.	2371.	-.0	0.0	1.0	2.4	3.8	5.4	.6	.1232
200.	2245.	2.3	-4.7	-2.7	0.0	2.9	6.1	2.3	.2533
300.	2123.	6.6	-16.3	-13.2	-9.2	-4.8	0.0	5.3	.3904
400.	2004.	13.0	-35.7	-31.6	-26.3	-20.4	-14.0	9.9	.5361
500.	1889.	22.1	-63.7	-58.6	-51.9	-44.6	-36.6	15.9	.6901
600.	1780.	34.3	-101.6	-95.5	-87.5	-78.6	-69.0	23.6	.8540
700.	1676.	49.9	-150.3	-143.2	-133.8	-123.5	-112.3	33.0	1.0276
800.	1576.	69.8	-211.6	-203.4	-192.7	-181.0	-168.1	44.4	1.2125
900.	1482.	94.4	-286.3	-277.2	-265.1	-251.9	-237.5	57.8	1.4083
1000.	1394.	124.8	-377.3	-367.1	-353.7	-339.0	-323.0	73.4	1.6173

BALLISTIC COEFFICIENT .67, MUZZLE VELOCITY FT/SEC 2600.

RANGE	VELOCITY	HEIGHT	100 YD	150 YD	200 YD	250 YD	300 YD	DRIFT	TIME
0.	2600.	0.0	-1.5	-1.5	-1.5	-1.5	-1.5	0.0	0.0000
100.	2468.	-.1	0.0	.9	2.1	3.5	4.9	.5	.1183
200.	2340.	2.1	-4.3	-2.4	0.0	2.7	5.6	2.2	.2434
300.	2215.	6.0	-14.8	-12.1	-8.4	-4.4	0.0	5.1	.3749
400.	2093.	12.0	-32.6	-28.9	-24.1	-18.7	-12.8	9.3	.5144
500.	1975.	20.3	-58.2	-53.6	-47.6	-40.8	-33.5	14.9	.6617
600.	1862.	31.4	-92.7	-87.2	-80.0	-71.9	-63.1	22.1	.8180
700.	1754.	45.7	-137.4	-131.0	-122.5	-113.1	-102.8	31.1	.9842
800.	1651.	63.9	-193.4	-186.0	-176.4	-165.5	-153.8	41.8	1.1607
900.	1553.	86.5	-262.0	-253.7	-242.9	-230.7	-217.5	54.5	1.3482
1000.	1460.	114.2	-344.9	-335.7	-323.6	-310.1	-295.5	69.3	1.5475

BALLISTIC COEFFICIENT .67, MUZZLE VELOCITY FT/SEC 2700.

RANGE	VELOCITY	HEIGHT	100 YD	150 YD	200 YD	250 YD	300 YD	DRIFT	TIME
0.	2700.	0.0	-1.5	-1.5	-1.5	-1.5	-1.5	0.0	0.0000
100.	2565.	-.1	0.0	.8	1.9	3.2	4.5	.5	.1140
200.	2434.	1.9	-3.8	-2.2	0.0	2.5	5.2	2.1	.2339
300.	2307.	5.5	-13.5	-11.1	-7.8	-4.0	0.0	4.8	.3607
400.	2183.	11.0	-29.7	-26.5	-22.1	-17.1	-11.7	8.7	.4941
500.	2062.	18.7	-53.3	-49.3	-43.9	-37.6	-30.8	14.1	.6359
600.	1945.	28.9	-85.1	-80.2	-73.7	-66.2	-58.1	21.0	.7857
700.	1833.	42.1	-126.0	-120.4	-112.8	-104.0	-94.5	29.4	.9446
800.	1727.	58.7	-177.1	-170.7	-162.0	-151.9	-141.1	39.4	1.1128
900.	1625.	79.4	-240.1	-232.8	-223.0	-211.7	-199.6	51.4	1.2921
1000.	1528.	104.8	-316.1	-308.0	-297.1	-284.6	-271.1	65.4	1.4826

BALLISTIC COEFFICIENT .67, MUZZLE VELOCITY FT/SEC 2800.

RANGE	VELOCITY	HEIGHT	100 YD	150 YD	200 YD	250 YD	300 YD	DRIFT	TIME
0.	2800.	0.0	-1.5	-1.5	-1.5	-1.5	-1.5	0.0	0.0000
100.	2662.	-.2	0.0	.7	1.7	2.9	4.1	.4	.1095
200.	2528.	1.7	-3.4	-2.0	0.0	2.3	4.8	2.0	.2255
300.	2398.	5.0	-12.3	-10.2	-7.2	-3.8	0.0	4.5	.3472
400.	2272.	10.1	-27.3	-24.4	-20.4	-15.8	-10.8	8.3	.4756
500.	2149.	17.2	-49.0	-45.4	-40.3	-34.7	-28.4	13.3	.6114
600.	2029.	26.6	-78.2	-73.9	-67.9	-61.1	-53.6	19.8	.7552
700.	1913.	38.8	-115.9	-110.9	-103.8	-95.9	-87.2	27.7	.9075
800.	1803.	54.1	-163.0	-157.3	-149.2	-140.1	-130.1	37.2	1.0688
900.	1698.	73.1	-220.7	-214.3	-205.2	-195.0	-183.8	48.5	1.2401
1000.	1597.	96.4	-290.7	-283.6	-273.5	-262.2	-249.6	61.8	1.4227

BALLISTIC COEFFICIENT .67, MUZZLE VELOCITY FT/SEC 2900.

RANGE	VELOCITY	HEIGHT	100 YD	150 YD	200 YD	250 YD	300 YD	DRIFT	TIME
0.	2900.	0.0	-1.5	-1.5	-1.5	-1.5	-1.5	0.0	0.0000
100.	2759.	-.2	0.0	.6	1.5	2.6	3.7	.5	.1062
200.	2622.	1.5	-3.0	-1.8	0.0	2.1	4.4	1.8	.2174
300.	2490.	4.6	-11.2	-9.4	-6.6	-3.4	0.0	4.3	.3350
400.	2361.	9.4	-25.0	-22.6	-18.9	-14.7	-10.1	7.9	.4589
500.	2235.	15.9	-45.0	-42.0	-37.4	-32.1	-26.4	12.7	.5895
600.	2113.	24.6	-72.0	-68.3	-62.8	-56.5	-49.6	18.8	.7273
700.	1994.	35.9	-106.8	-102.5	-96.1	-88.6	-80.6	26.3	.8735
800.	1880.	50.0	-150.2	-145.3	-138.0	-129.5	-120.3	35.3	1.0282
900.	1771.	67.5	-203.6	-198.0	-189.8	-180.3	-169.9	46.1	1.1929
1000.	1668.	89.0	-267.9	-261.8	-252.7	-242.0	-230.5	58.6	1.3674

BALLISTIC COEFFICIENT .67, MUZZLE VELOCITY FT/SEC 3000.

RANGE	VELOCITY	HEIGHT	100 YD	150 YD	200 YD	250 YD	300 YD	DRIFT	TIME
0.	3000.	0.0	-1.5	-1.5	-1.5	-1.5	-1.5	0.0	0.0000
100.	2855.	-.2	0.0	.6	1.4	2.4	3.4	.4	.1025
200.	2716.	1.4	-2.8	-1.7	0.0	1.9	4.1	1.9	.2106
300.	2580.	4.3	-10.3	-8.7	-6.1	-3.2	0.0	4.2	.3240
400.	2449.	8.7	-23.1	-20.8	-17.5	-13.6	-9.3	7.6	.4431
500.	2321.	14.8	-41.6	-38.9	-34.6	-29.8	-24.4	12.2	.5690
600.	2197.	22.9	-66.6	-63.3	-58.2	-52.4	-46.0	17.9	.7018
700.	2076.	33.3	-98.8	-94.9	-89.0	-82.2	-74.7	25.0	.8423
800.	1958.	46.4	-139.2	-134.7	-128.0	-120.2	-111.7	33.7	.9914
900.	1846.	62.6	-188.5	-183.5	-175.9	-167.1	-157.5	43.9	1.1492
1000.	1738.	82.4	-247.9	-242.3	-233.9	-224.2	-213.5	55.7	1.3165

BALLISTIC COEFFICIENT .67, MUZZLE VELOCITY FT/SEC 3100.

RANGE	VELOCITY	HEIGHT	100 YD	150 YD	200 YD	250 YD	300 YD	DRIFT	TIME
0.	3100.	0.0	-1.5	-1.5	-1.5	-1.5	-1.5	0.0	0.0000
100.	2952.	-.3	0.0	.5	1.3	2.2	3.1	.3	.0986
200.	2809.	1.2	-2.5	-1.6	0.0	1.8	3.8	1.7	.2033
300.	2671.	3.9	-9.4	-8.0	-5.6	-3.0	0.0	3.9	.3125
400.	2537.	8.0	-21.3	-19.4	-16.2	-12.7	-8.7	7.2	.4281
500.	2407.	13.7	-38.5	-36.1	-32.2	-27.7	-22.8	11.5	.5495
600.	2280.	21.3	-61.7	-58.8	-54.1	-48.8	-42.9	17.0	.6773
700.	2157.	30.9	-91.6	-88.3	-82.7	-76.5	-69.6	23.8	.8126
800.	2037.	43.1	-129.1	-125.2	-118.9	-111.8	-103.9	32.0	.9561
900.	1921.	58.2	-174.8	-170.5	-163.4	-155.4	-146.5	41.7	1.1078
1000.	1810.	76.5	-229.8	-225.0	-217.1	-208.2	-198.3	52.9	1.2683

BALLISTIC COEFFICIENT .67, MUZZLE VELOCITY FT/SEC 3200.

RANGE	VELOCITY	HEIGHT	100 YD	150 YD	200 YD	250 YD	300 YD	DRIFT	TIME
0.	3200.	0.0	-1.5	-1.5	-1.5	-1.5	-1.5	0.0	0.0000
100.	3049.	-.3	0.0	.4	1.1	2.0	2.9	.4	.0958
200.	2903.	1.1	-2.3	-1.4	0.0	1.7	3.5	1.7	.1970
300.	2761.	3.7	-8.7	-7.4	-5.3	-2.8	0.0	3.8	.3031
400.	2625.	7.5	-19.6	-17.9	-15.1	-11.8	-8.0	6.9	.4142
500.	2492.	12.8	-35.7	-33.5	-30.0	-25.9	-21.2	11.1	.5317
600.	2363.	19.9	-57.4	-54.8	-50.5	-45.6	-40.0	16.4	.6555
700.	2238.	28.9	-85.2	-82.2	-77.2	-71.4	-64.9	22.8	.7859
800.	2115.	40.2	-119.9	-116.4	-110.8	-104.2	-96.7	30.5	.9235
900.	1997.	54.2	-162.4	-158.5	-152.1	-144.7	-136.3	39.7	1.0696
1000.	1882.	71.2	-213.5	-209.2	-202.1	-193.8	-184.5	50.5	1.2242

BALLISTIC COEFFICIENT .67, MUZZLE VELOCITY FT/SEC 3300.

RANGE	VELOCITY	HEIGHT	100 YD	150 YD	200 YD	250 YD	300 YD	DRIFT	TIME
0.	3300.	0.0	-1.5	-1.5	-1.5	-1.5	-1.5	0.0	0.0000
100.	3146.	-.3	0.0	.4	1.0	1.8	2.6	.4	.0935
200.	2997.	1.0	-2.0	-1.3	0.0	1.6	3.3	1.6	.1907
300.	2852.	3.4	-7.9	-6.8	-4.9	-2.5	0.0	3.6	.2933
400.	2712.	7.0	-18.1	-16.7	-14.1	-11.0	-7.6	6.7	.4015
500.	2577.	12.0	-33.0	-31.3	-28.1	-24.1	-19.9	10.6	.5150
600.	2446.	18.6	-53.1	-51.0	-47.2	-42.4	-37.4	15.6	.6342
700.	2318.	27.0	-79.1	-76.7	-72.2	-66.6	-60.8	21.8	.7604
800.	2194.	37.6	-111.5	-108.7	-103.5	-97.2	-90.5	29.2	.8933
900.	2073.	50.6	-151.1	-147.9	-142.1	-135.0	-127.5	38.0	1.0340
1000.	1955.	66.5	-198.8	-195.3	-188.9	-181.0	-172.6	48.3	1.1833

BALLISTIC COEFFICIENT .67, MUZZLE VELOCITY FT/SEC 3400.

RANGE	VELOCITY	HEIGHT	100 YD	150 YD	200 YD	250 YD	300 YD	DRIFT	TIME
0.	3400.	0.0	-1.5	-1.5	-1.5	-1.5	-1.5	0.0	0.0000
100.	3243.	-.4	0.0	.3	.9	1.6	2.4	.3	.0902
200.	3090.	.9	-1.8	-1.2	0.0	1.4	3.0	1.5	.1851
300.	2943.	3.1	-7.3	-6.3	-4.5	-2.4	0.0	3.4	.2841
400.	2800.	6.5	-16.8	-15.6	-13.2	-10.3	-7.2	6.4	.3893
500.	2662.	11.2	-30.7	-29.1	-26.2	-22.6	-18.6	10.1	.4988
600.	2528.	17.4	-49.6	-47.7	-44.2	-39.9	-35.1	15.0	.6148
700.	2399.	25.3	-73.9	-71.7	-67.5	-62.5	-57.0	20.9	.7364
800.	2272.	35.2	-104.2	-101.6	-96.9	-91.2	-84.8	28.0	.8649
900.	2149.	47.3	-141.1	-138.3	-132.9	-126.6	-119.4	36.4	1.0007
1000.	2029.	62.1	-185.6	-182.5	-176.5	-169.4	-161.5	46.1	1.1445

BALLISTIC COEFFICIENT .67, MUZZLE VELOCITY FT/SEC 3500.

RANGE	VELOCITY	HEIGHT	100 YD	150 YD	200 YD	250 YD	300 YD	DRIFT	TIME
0.	3500.	0.0	-1.5	-1.5	-1.5	-1.5	-1.5	0.0	0.0000
100.	3340.	-.4	0.0	.3	.8	1.5	2.2	.4	.0880
200.	3184.	.8	-1.6	-1.1	0.0	1.4	2.8	1.5	.1798
300.	3034.	2.9	-6.7	-5.9	-4.3	-2.2	0.0	3.4	.2763
400.	2888.	6.1	-15.5	-14.5	-12.3	-9.6	-6.6	6.1	.3777
500.	2747.	10.5	-28.6	-27.3	-24.6	-21.1	-17.5	9.8	.4844
600.	2611.	16.3	-46.2	-44.7	-41.4	-37.3	-32.9	14.4	.5963
700.	2479.	23.7	-68.9	-67.1	-63.3	-58.5	-53.4	20.1	.7142
800.	2350.	33.0	-97.4	-95.3	-91.0	-85.5	-79.6	27.0	.8389
900.	2225.	44.4	-131.9	-129.6	-124.7	-118.5	-111.9	34.9	.9698
1000.	2103.	58.3	-173.6	-171.0	-165.6	-158.7	-151.4	44.3	1.1088

BALLISTIC COEFFICIENT .67, MUZZLE VELOCITY FT/SEC 3600.

RANGE	VELOCITY	HEIGHT	100 YD	150 YD	200 YD	250 YD	300 YD	DRIFT	TIME
0.	3600.	0.0	-1.5	-1.5	-1.5	-1.5	-1.5	0.0	0.0000
100.	3436.	-.4	0.0	.2	.7	1.4	2.1	.3	.0850
200.	3278.	.7	-1.5	-1.0	0.0	1.2	2.7	1.4	.1746
300.	3124.	2.7	-6.2	-5.5	-4.0	-2.1	0.0	3.3	.2685
400.	2976.	5.7	-14.4	-13.5	-11.5	-9.0	-6.2	5.8	.3666
500.	2832.	9.8	-26.6	-25.5	-22.9	-19.9	-16.3	9.3	.4697
600.	2693.	15.3	-43.2	-41.8	-38.8	-35.1	-30.8	13.8	.5785
700.	2559.	22.3	-64.5	-62.9	-59.4	-55.1	-50.1	19.3	.6928
800.	2428.	31.0	-91.1	-89.3	-85.3	-80.3	-74.6	25.8	.8132
900.	2300.	41.7	-123.7	-121.6	-117.0	-111.5	-105.1	33.5	.9403
1000.	2177.	54.6	-162.5	-160.2	-155.2	-149.0	-141.9	42.4	1.0740

BALLISTIC COEFFICIENT .67, MUZZLE VELOCITY FT/SEC 3700.

RANGE	VELOCITY	HEIGHT	100 YD	150 YD	200 YD	250 YD	300 YD	DRIFT	TIME
0.	3700.	0.0	-1.5	-1.5	-1.5	-1.5	-1.5	0.0	0.0000
100.	3533.	-.4	0.0	.2	.6	1.2	1.9	.3	.0830
200.	3372.	.6	-1.3	-.9	0.0	1.2	2.5	1.4	.1701
300.	3215.	2.5	-5.7	-5.1	-3.7	-2.0	0.0	3.2	.2614
400.	3064.	5.4	-13.4	-12.6	-10.8	-8.4	-5.8	5.7	.3566
500.	2917.	9.3	-24.9	-24.0	-21.7	-18.7	-15.5	9.2	.4575
600.	2775.	14.4	-40.4	-39.2	-36.5	-33.0	-29.0	13.4	.5625
700.	2638.	21.0	-60.5	-59.1	-55.9	-51.8	-47.2	18.6	.6735
800.	2505.	29.2	-85.6	-84.0	-80.4	-75.7	-70.4	25.0	.7904
900.	2376.	39.3	-116.0	-114.3	-110.2	-104.9	-99.0	32.3	.9132
1000.	2250.	51.5	-152.7	-150.8	-146.2	-140.4	-133.8	40.9	1.0432

BALLISTIC COEFFICIENT .67, MUZZLE VELOCITY FT/SEC 3800.

RANGE	VELOCITY	HEIGHT	100 YD	150 YD	200 YD	250 YD	300 YD	DRIFT	TIME
0.	3800.	0.0	-1.5	-1.5	-1.5	-1.5	-1.5	0.0	0.0000
100.	3630.	-.4	0.0	.1	.6	1.1	1.7	.4	.0812
200.	3466.	.6	-1.1	-.8	0.0	1.1	2.3	1.4	.1657
300.	3307.	2.4	-5.2	-4.7	-3.5	-1.8	0.0	3.1	.2542
400.	3152.	5.0	-12.4	-11.9	-10.2	-8.0	-5.5	5.6	.3474
500.	3003.	8.7	-23.1	-22.4	-20.3	-17.6	-14.5	8.8	.4445
600.	2858.	13.6	-37.7	-36.8	-34.3	-31.1	-27.4	12.9	.5469
700.	2718.	19.8	-56.6	-55.6	-52.7	-48.9	-44.6	18.0	.6549
800.	2583.	27.6	-80.2	-79.0	-75.6	-71.3	-66.4	24.0	.7681
900.	2451.	37.0	-108.9	-107.6	-103.8	-98.9	-93.3	31.1	.8871
1000.	2323.	48.5	-143.4	-141.9	-137.7	-132.3	-126.1	39.3	1.0130

BALLISTIC COEFFICIENT .67, MUZZLE VELOCITY FT/SEC 3900.

RANGE	VELOCITY	HEIGHT	100 YD	150 YD	200 YD	250 YD	300 YD	DRIFT	TIME
0.	3900.	0.0	-1.5	-1.5	-1.5	-1.5	-1.5	0.0	0.0000
100.	3727.	-.5	0.0	.1	.5	1.0	1.6	.2	.0783
200.	3560.	.5	-1.0	-.8	0.0	1.0	2.2	1.2	.1608
300.	3398.	2.2	-4.8	-4.4	-3.3	-1.8	0.0	2.9	.2471
400.	3240.	4.7	-11.6	-11.1	-9.5	-7.5	-5.2	5.2	.3375
500.	3088.	8.2	-21.7	-21.1	-19.1	-16.6	-13.7	8.4	.4324
600.	2941.	12.8	-35.4	-34.7	-32.3	-29.3	-25.8	12.3	.5315
700.	2798.	18.7	-53.3	-52.4	-49.7	-46.2	-42.1	17.3	.6367
800.	2660.	26.0	-75.4	-74.4	-71.3	-67.3	-62.6	23.1	.7464
900.	2527.	34.9	-102.6	-101.5	-98.0	-93.4	-88.2	29.9	.8624
1000.	2397.	45.7	-135.0	-133.8	-129.9	-124.9	-119.0	37.8	.9842

BALLISTIC COEFFICIENT .67, MUZZLE VELOCITY FT/SEC 4000.

RANGE	VELOCITY	HEIGHT	100 YD	150 YD	200 YD	250 YD	300 YD	DRIFT	TIME
0.	4000.	0.0	-1.5	-1.5	-1.5	-1.5	-1.5	0.0	0.0000
100.	3824.	-.5	0.0	.1	.4	.9	1.5	.4	.0770
200.	3654.	.4	-.9	-.7	0.0	1.0	2.0	1.2	.1571
300.	3489.	2.0	-4.4	-4.1	-3.1	-1.6	0.0	2.8	.2411
400.	3329.	4.5	-10.7	-10.4	-9.0	-7.1	-4.9	5.2	.3293
500.	3174.	7.8	-20.2	-19.8	-18.0	-15.6	-12.9	8.2	.4214
600.	3023.	12.1	-33.2	-32.7	-30.6	-27.7	-24.4	12.0	.5183
700.	2878.	17.7	-49.9	-49.3	-46.8	-43.5	-39.7	16.7	.6198
800.	2737.	24.6	-70.9	-70.2	-67.4	-63.6	-59.3	22.3	.7270
900.	2602.	33.1	-96.5	-95.7	-92.6	-88.2	-83.4	28.9	.8395
1000.	2470.	43.3	-127.1	-126.3	-122.7	-117.9	-112.6	36.6	.9577

BALLISTIC COEFFICIENT .68, MUZZLE VELOCITY FT/SEC 2000.

RANGE	VELOCITY	HEIGHT	100 YD	150 YD	200 YD	250 YD	300 YD	DRIFT	TIME
0.	2000.	0.0	-1.5	-1.5	-1.5	-1.5	-1.5	0.0	0.0000
100.	1887.	.4	0.0	1.9	4.2	6.6	9.2	.7	.1542
200.	1780.	4.1	-8.3	-4.5	0.0	4.9	10.1	3.2	.3182
300.	1677.	10.9	-27.6	-21.9	-15.1	-7.9	0.0	7.3	.4917
400.	1579.	21.2	-59.4	-51.8	-42.8	-33.1	-22.6	13.5	.6764
500.	1486.	35.7	-105.0	-95.4	-84.2	-72.0	-58.9	21.4	.8718
600.	1399.	55.3	-166.5	-155.0	-141.5	-127.0	-111.2	31.7	1.0802
700.	1321.	80.5	-245.7	-232.3	-216.6	-199.6	-181.3	44.2	1.3009
800.	1251.	112.3	-345.0	-329.7	-311.7	-292.3	-271.4	58.9	1.5345
900.	1189.	151.5	-466.4	-449.1	-428.9	-407.1	-383.5	75.8	1.7807
1000.	1136.	199.2	-613.6	-594.4	-571.9	-547.7	-521.5	95.2	2.0410

BALLISTIC COEFFICIENT .68, MUZZLE VELOCITY FT/SEC 2100.

RANGE	VELOCITY	HEIGHT	100 YD	150 YD	200 YD	250 YD	300 YD	DRIFT	TIME
0.	2100.	0.0	-1.5	-1.5	-1.5	-1.5	-1.5	0.0	0.0000
100.	1983.	.3	0.0	1.7	3.7	5.9	8.2	.7	.1466
200.	1871.	3.6	-7.4	-4.0	0.0	4.4	9.1	2.9	.3022
300.	1764.	9.7	-24.7	-19.6	-13.7	-7.1	0.0	6.9	.4675
400.	1663.	19.1	-53.3	-46.5	-38.6	-29.8	-20.3	12.6	.6428
500.	1565.	32.2	-94.5	-86.0	-76.1	-65.1	-53.3	20.2	.8290
600.	1473.	49.8	-149.7	-139.5	-127.6	-114.4	-100.3	29.8	1.0263
700.	1387.	72.6	-221.1	-209.2	-195.3	-179.9	-163.4	41.6	1.2363
800.	1310.	101.4	-310.6	-297.0	-281.2	-263.5	-244.7	55.6	1.4589
900.	1242.	137.1	-420.4	-405.1	-387.3	-367.5	-346.3	71.9	1.6945
1000.	1180.	180.3	-552.3	-535.3	-515.5	-493.5	-469.9	90.4	1.9421

BALLISTIC COEFFICIENT .68, MUZZLE VELOCITY FT/SEC 2200.

RANGE	VELOCITY	HEIGHT	100 YD	150 YD	200 YD	250 YD	300 YD	DRIFT	TIME
0.	2200.	0.0	-1.5	-1.5	-1.5	-1.5	-1.5	0.0	0.0000
100.	2081.	.2	0.0	1.5	3.3	5.3	7.4	.6	.1400
200.	1965.	3.3	-6.6	-3.6	0.0	4.0	8.2	2.8	.2888
300.	1854.	8.8	-22.2	-17.7	-12.3	-6.3	0.0	6.5	.4458
400.	1747.	17.3	-48.0	-42.0	-34.8	-26.8	-18.4	11.8	.6125
500.	1646.	29.2	-85.2	-77.7	-68.7	-58.7	-48.2	19.0	.7895
600.	1550.	45.1	-135.2	-126.2	-115.4	-103.5	-90.8	28.0	.9775
700.	1458.	65.8	-199.7	-189.2	-176.5	-162.6	-147.9	39.2	1.1771
800.	1374.	91.9	-280.4	-268.5	-254.0	-238.1	-221.2	52.5	1.3890
900.	1299.	124.2	-379.7	-366.3	-350.0	-332.1	-313.2	68.0	1.6137
1000.	1231.	163.7	-499.7	-484.8	-466.7	-446.9	-425.8	85.8	1.8512

BALLISTIC COEFFICIENT .68, MUZZLE VELOCITY FT/SEC 2300.

RANGE	VELOCITY	HEIGHT	100 YD	150 YD	200 YD	250 YD	300 YD	DRIFT	TIME
0.	2300.	0.0	-1.5	-1.5	-1.5	-1.5	-1.5	0.0	0.0000
100.	2178.	.1	0.0	1.3	3.0	4.7	6.7	.6	.1337
200.	2059.	2.9	-5.9	-3.3	0.0	3.6	7.4	2.6	.2758
300.	1944.	8.0	-20.0	-16.1	-11.2	-5.8	0.0	6.1	.4258
400.	1834.	15.7	-43.4	-38.1	-31.6	-24.4	-16.7	11.1	.5846
500.	1729.	26.4	-77.0	-70.4	-62.2	-53.3	-43.6	17.7	.7528
600.	1628.	40.9	-122.3	-114.5	-104.6	-93.9	-82.3	26.2	.9317
700.	1533.	59.6	-180.8	-171.6	-160.1	-147.6	-134.1	36.7	1.1217
800.	1442.	83.4	-254.0	-243.6	-230.4	-216.1	-200.6	49.3	1.3238
900.	1359.	112.8	-344.0	-332.2	-317.4	-301.3	-283.9	64.1	1.5382
1000.	1286.	148.8	-452.7	-439.7	-423.2	-405.3	-386.0	81.1	1.7650

BALLISTIC COEFFICIENT .68, MUZZLE VELOCITY FT/SEC 2400.

RANGE	VELOCITY	HEIGHT	100 YD	150 YD	200 YD	250 YD	300 YD	DRIFT	TIME
0.	2400.	0.0	-1.5	-1.5	-1.5	-1.5	-1.5	0.0	0.0000
100.	2275.	.0	0.0	1.1	2.6	4.3	6.0	.6	.1283
200.	2154.	2.6	-5.2	-2.9	0.0	3.3	6.8	2.4	.2637
300.	2036.	7.2	-18.0	-14.6	-10.2	-5.3	0.0	5.7	.4073
400.	1921.	14.3	-39.2	-34.7	-28.8	-22.2	-15.2	10.4	.5591
500.	1812.	24.1	-69.8	-64.0	-56.7	-48.5	-39.7	16.6	.7194
600.	1708.	37.3	-110.9	-104.1	-95.2	-85.4	-74.9	24.6	.8898
700.	1609.	54.3	-164.2	-156.1	-145.8	-134.4	-122.1	34.5	1.0711
800.	1514.	75.8	-230.6	-221.4	-209.7	-196.6	-182.5	46.3	1.2632
900.	1424.	102.6	-312.1	-301.8	-288.6	-273.9	-258.0	60.3	1.4675
1000.	1344.	135.5	-411.2	-399.7	-385.0	-368.6	-351.0	76.5	1.6846

BALLISTIC COEFFICIENT .68, MUZZLE VELOCITY FT/SEC 2500.

RANGE	VELOCITY	HEIGHT	100 YD	150 YD	200 YD	250 YD	300 YD	DRIFT	TIME
0.	2500.	0.0	-1.5	-1.5	-1.5	-1.5	-1.5	0.0	0.0000
100.	2373.	-.0	0.0	1.0	2.4	3.8	5.4	.5	.1231
200.	2249.	2.3	-4.7	-2.7	0.0	2.9	6.1	2.3	.2531
300.	2128.	6.6	-16.3	-13.2	-9.2	-4.8	0.0	5.3	.3900
400.	2011.	13.0	-35.6	-31.5	-26.2	-20.3	-13.9	9.7	.5352
500.	1897.	22.0	-63.5	-58.4	-51.7	-44.4	-36.4	15.6	.6886
600.	1789.	34.1	-101.1	-95.0	-87.0	-78.2	-68.6	23.2	.8516
700.	1686.	49.6	-149.5	-142.4	-133.1	-122.8	-111.5	32.4	1.0242
800.	1588.	69.3	-210.3	-202.2	-191.5	-179.7	-166.9	43.6	1.2080
900.	1494.	93.6	-284.4	-275.2	-263.2	-250.0	-235.6	56.7	1.4022
1000.	1406.	123.6	-374.4	-364.2	-350.8	-336.2	-320.1	72.1	1.6094

BALLISTIC COEFFICIENT .68, MUZZLE VELOCITY FT/SEC 2600.

RANGE	VELOCITY	HEIGHT	100 YD	150 YD	200 YD	250 YD	300 YD	DRIFT	TIME
0.	2600.	0.0	-1.5	-1.5	-1.5	-1.5	-1.5	0.0	0.0000
100.	2470.	-.1	0.0	.9	2.1	3.5	4.9	.5	.1182
200.	2343.	2.1	-4.2	-2.4	0.0	2.7	5.6	2.2	.2433
300.	2220.	6.0	-14.8	-12.0	-8.4	-4.4	0.0	5.0	.3744
400.	2100.	11.9	-32.5	-28.8	-24.0	-18.6	-12.8	9.2	.5137
500.	1984.	20.2	-58.0	-53.4	-47.4	-40.6	-33.3	14.7	.6603
600.	1872.	31.2	-92.3	-86.9	-79.6	-71.5	-62.8	21.8	.8159
700.	1765.	45.5	-136.7	-130.3	-121.8	-112.4	-102.2	30.5	.9811
800.	1663.	63.4	-192.2	-184.9	-175.2	-164.5	-152.8	41.1	1.1565
900.	1566.	85.8	-260.3	-252.1	-241.2	-229.1	-215.9	53.5	1.3426
1000.	1473.	113.1	-342.2	-333.1	-321.0	-307.5	-292.9	67.9	1.5398

BALLISTIC COEFFICIENT .68, MUZZLE VELOCITY FT/SEC 2700.

RANGE	VELOCITY	HEIGHT	100 YD	150 YD	200 YD	250 YD	300 YD	DRIFT	TIME
0.	2700.	0.0	-1.5	-1.5	-1.5	-1.5	-1.5	0.0	0.0000
100.	2567.	-.1	0.0	.8	1.9	3.1	4.5	.5	.1140
200.	2438.	1.9	-3.8	-2.2	0.0	2.5	5.2	2.0	.2337
300.	2312.	5.5	-13.5	-11.0	-7.8	-4.0	0.0	4.7	.3603
400.	2190.	10.9	-29.6	-26.4	-22.1	-17.0	-11.7	8.6	.4934
500.	2071.	18.6	-53.1	-49.0	-43.6	-37.4	-30.7	13.9	.6344
600.	1955.	28.7	-84.7	-79.9	-73.4	-65.9	-57.8	20.6	.7837
700.	1844.	41.8	-125.4	-119.8	-112.2	-103.4	-94.0	28.9	.9418
800.	1739.	58.3	-176.2	-169.7	-161.1	-151.0	-140.3	38.7	1.1090
900.	1638.	78.7	-238.5	-231.2	-221.5	-210.2	-198.1	50.5	1.2868
1000.	1542.	103.8	-313.7	-305.6	-294.8	-282.3	-268.8	64.1	1.4755

BALLISTIC COEFFICIENT .68, MUZZLE VELOCITY FT/SEC 2800.

RANGE	VELOCITY	HEIGHT	100 YD	150 YD	200 YD	250 YD	300 YD	DRIFT	TIME
0.	2800.	0.0	-1.5	-1.5	-1.5	-1.5	-1.5	0.0	0.0000
100.	2664.	-.2	0.0	.7	1.7	2.8	4.1	.4	.1095
200.	2532.	1.7	-3.4	-2.0	0.0	2.3	4.8	1.9	.2253
300.	2404.	5.0	-12.3	-10.2	-7.1	-3.8	0.0	4.5	.3468
400.	2279.	10.1	-27.2	-24.3	-20.3	-15.8	-10.8	8.1	.4747
500.	2158.	17.1	-48.7	-45.2	-40.1	-34.5	-28.3	13.1	.6099
600.	2040.	26.5	-77.9	-73.6	-67.6	-60.8	-53.3	19.5	.7534
700.	1925.	38.5	-115.3	-110.4	-103.3	-95.4	-86.7	27.2	.9048
800.	1816.	53.7	-162.0	-156.3	-148.2	-139.2	-129.2	36.5	1.0648
900.	1711.	72.4	-219.2	-212.8	-203.8	-193.6	-182.4	47.6	1.2349
1000.	1612.	95.5	-288.5	-281.4	-271.3	-260.0	-247.5	60.6	1.4158

BALLISTIC COEFFICIENT .68, MUZZLE VELOCITY FT/SEC 2900.

RANGE	VELOCITY	HEIGHT	100 YD	150 YD	200 YD	250 YD	300 YD	DRIFT	TIME
0.	2900.	0.0	-1.5	-1.5	-1.5	-1.5	-1.5	0.0	0.0000
100.	2761.	-.2	0.0	.6	1.5	2.6	3.7	.5	.1061
200.	2626.	1.5	-3.0	-1.8	0.0	2.1	4.4	1.8	.2172
300.	2495.	4.6	-11.2	-9.4	-6.6	-3.5	0.0	4.3	.3347
400.	2368.	9.3	-24.9	-22.5	-18.8	-14.6	-10.0	7.8	.4581
500.	2244.	15.9	-44.9	-41.8	-37.3	-32.0	-26.2	12.5	.5883
600.	2124.	24.5	-71.7	-68.0	-62.5	-56.2	-49.3	18.4	.7254
700.	2006.	35.7	-106.3	-102.0	-95.6	-88.2	-80.1	25.8	.8710
800.	1893.	49.7	-149.4	-144.5	-137.2	-128.7	-119.5	34.7	1.0247
900.	1785.	67.0	-202.3	-196.8	-188.6	-179.1	-168.7	45.2	1.1881
1000.	1682.	88.2	-265.9	-259.8	-250.7	-240.2	-228.6	57.5	1.3611

BALLISTIC COEFFICIENT .68, MUZZLE VELOCITY FT/SEC 3000.

RANGE	VELOCITY	HEIGHT	100 YD	150 YD	200 YD	250 YD	300 YD	DRIFT	TIME
0.	3000.	0.0	-1.5	-1.5	-1.5	-1.5	-1.5	0.0	0.0000
100.	2857.	-.2	0.0	.6	1.4	2.4	3.4	.4	.1025
200.	2720.	1.4	-2.8	-1.7	0.0	1.9	4.1	1.8	.2105
300.	2586.	4.3	-10.3	-8.6	-6.1	-3.2	0.0	4.2	.3236
400.	2457.	8.6	-23.0	-20.8	-17.4	-13.5	-9.3	7.5	.4425
500.	2330.	14.7	-41.5	-38.7	-34.5	-29.7	-24.3	12.0	.5680
600.	2208.	22.8	-66.4	-63.0	-58.0	-52.2	-45.8	17.6	.7001
700.	2088.	33.1	-98.4	-94.5	-88.6	-81.8	-74.3	24.6	.8399
800.	1972.	46.1	-138.3	-133.9	-127.1	-119.4	-110.9	33.0	.9877
900.	1860.	62.1	-187.1	-182.1	-174.5	-165.8	-156.2	43.0	1.1441
1000.	1754.	81.7	-246.1	-240.5	-232.1	-222.4	-211.7	54.6	1.3104

BALLISTIC COEFFICIENT .68, MUZZLE VELOCITY FT/SEC 3100.

RANGE	VELOCITY	HEIGHT	100 YD	150 YD	200 YD	250 YD	300 YD	DRIFT	TIME
0.	3100.	0.0	-1.5	-1.5	-1.5	-1.5	-1.5	0.0	0.0000
100.	2954.	-.3	0.0	.5	1.3	2.2	3.1	.3	.0986
200.	2813.	1.2	-2.5	-1.6	0.0	1.8	3.7	1.7	.2031
300.	2677.	3.9	-9.4	-8.0	-5.6	-2.9	0.0	3.8	.3121
400.	2545.	8.0	-21.2	-19.3	-16.2	-12.6	-8.7	7.1	.4274
500.	2416.	13.7	-38.4	-36.0	-32.1	-27.6	-22.7	11.3	.5484
600.	2291.	21.2	-61.5	-58.6	-53.9	-48.6	-42.7	16.7	.6758
700.	2169.	30.8	-91.2	-87.8	-82.3	-76.1	-69.2	23.4	.8101
800.	2051.	42.8	-128.4	-124.5	-118.3	-111.1	-103.3	31.4	.9528
900.	1936.	57.7	-173.7	-169.3	-162.3	-154.3	-145.5	40.9	1.1033
1000.	1826.	75.8	-228.2	-223.3	-215.5	-206.6	-196.8	51.9	1.2627

BALLISTIC COEFFICIENT .68, MUZZLE VELOCITY FT/SEC 3200.

RANGE	VELOCITY	HEIGHT	100 YD	150 YD	200 YD	250 YD	300 YD	DRIFT	TIME
0.	3200.	0.0	-1.5	-1.5	-1.5	-1.5	-1.5	0.0	0.0000
100.	3051.	-.3	0.0	.4	1.1	2.0	2.9	.4	.0958
200.	2907.	1.1	-2.3	-1.4	0.0	1.6	3.5	1.7	.1969
300.	2768.	3.6	-8.6	-7.3	-5.2	-2.8	0.0	3.8	.3026
400.	2633.	7.5	-19.6	-17.8	-15.0	-11.7	-8.0	6.8	.4136
500.	2502.	12.8	-35.6	-33.4	-29.9	-25.8	-21.2	10.9	.5308
600.	2374.	19.8	-57.1	-54.5	-50.3	-45.4	-39.8	16.1	.6538
700.	2250.	28.7	-84.8	-81.8	-76.8	-71.1	-64.6	22.4	.7837
800.	2130.	39.9	-119.3	-115.8	-110.2	-103.6	-96.2	30.0	.9205
900.	2012.	53.7	-161.4	-157.5	-151.2	-143.8	-135.5	39.0	1.0655
1000.	1899.	70.6	-212.0	-207.7	-200.7	-192.5	-183.2	49.5	1.2189

BALLISTIC COEFFICIENT .68, MUZZLE VELOCITY FT/SEC 3300.

RANGE	VELOCITY	HEIGHT	100 YD	150 YD	200 YD	250 YD	300 YD	DRIFT	TIME
0.	3300.	0.0	-1.5	-1.5	-1.5	-1.5	-1.5	0.0	0.0000
100.	3148.	-.3	0.0	.4	1.0	1.8	2.6	.4	.0934
200.	3001.	1.0	-2.0	-1.3	0.0	1.6	3.3	1.5	.1905
300.	2858.	3.4	-7.9	-6.8	-4.9	-2.5	0.0	3.6	.2930
400.	2720.	7.0	-18.1	-16.6	-14.1	-10.9	-7.6	6.6	.4009
500.	2587.	11.9	-32.9	-31.2	-28.0	-24.0	-19.8	10.5	.5140
600.	2457.	18.5	-52.9	-50.8	-47.0	-42.3	-37.2	15.4	.6328
700.	2331.	26.9	-78.8	-76.3	-71.9	-66.4	-60.5	21.5	.7584
800.	2208.	37.3	-110.9	-108.1	-103.0	-96.7	-90.0	28.7	.8904
900.	2089.	50.2	-150.2	-147.0	-141.3	-134.2	-126.6	37.3	1.0302
1000.	1972.	65.8	-197.4	-193.8	-187.5	-179.6	-171.2	47.3	1.1779

BALLISTIC COEFFICIENT .68, MUZZLE VELOCITY FT/SEC 3400.

RANGE	VELOCITY	HEIGHT	100 YD	150 YD	200 YD	250 YD	300 YD	DRIFT	TIME
0.	3400.	0.0	-1.5	-1.5	-1.5	-1.5	-1.5	0.0	0.0000
100.	3245.	-.4	0.0	.3	.9	1.6	2.4	.3	.0901
200.	3095.	.9	-1.8	-1.2	0.0	1.4	3.0	1.5	.1850
300.	2949.	3.1	-7.2	-6.3	-4.5	-2.4	0.0	3.4	.2838
400.	2808.	6.5	-16.8	-15.5	-13.1	-10.3	-7.1	6.3	.3886
500.	2672.	11.1	-30.6	-29.0	-26.0	-22.5	-18.5	10.0	.4977
600.	2540.	17.3	-49.4	-47.5	-44.0	-39.7	-35.0	14.8	.6133
700.	2412.	25.1	-73.5	-71.4	-67.2	-62.2	-56.7	20.5	.7344
800.	2287.	34.9	-103.6	-101.1	-96.3	-90.6	-84.3	27.5	.8620
900.	2165.	46.9	-140.2	-137.4	-132.0	-125.6	-118.5	35.6	.9966
1000.	2046.	61.6	-184.4	-181.3	-175.3	-168.3	-160.3	45.3	1.1397

BALLISTIC COEFFICIENT .68, MUZZLE VELOCITY FT/SEC 3500.

RANGE	VELOCITY	HEIGHT	100 YD	150 YD	200 YD	250 YD	300 YD	DRIFT	TIME
0.	3500.	0.0	-1.5	-1.5	-1.5	-1.5	-1.5	0.0	0.0000
100.	3342.	-.4	0.0	.3	.8	1.5	2.2	.4	.0880
200.	3189.	.8	-1.6	-1.1	0.0	1.4	2.8	1.5	.1797
300.	3040.	2.9	-6.6	-5.9	-4.2	-2.2	0.0	3.3	.2760
400.	2896.	6.1	-15.5	-14.5	-12.3	-9.5	-6.6	6.1	.3773
500.	2757.	10.5	-28.5	-27.2	-24.5	-21.1	-17.4	9.7	.4836
600.	2623.	16.2	-46.0	-44.4	-41.2	-37.1	-32.7	14.2	.5949
700.	2492.	23.6	-68.6	-66.8	-63.0	-58.2	-53.1	19.8	.7124
800.	2365.	32.8	-96.9	-94.8	-90.4	-85.0	-79.1	26.5	.8362
900.	2241.	44.1	-131.2	-128.9	-124.0	-117.8	-111.3	34.3	.9665
1000.	2121.	57.7	-172.3	-169.7	-164.3	-157.4	-150.1	43.4	1.1037

BALLISTIC COEFFICIENT .68, MUZZLE VELOCITY FT/SEC 3600.

RANGE	VELOCITY	HEIGHT	100 YD	150 YD	200 YD	250 YD	300 YD	DRIFT	TIME
0.	3600.	0.0	-1.5	-1.5	-1.5	-1.5	-1.5	0.0	0.0000
100.	3439.	-.4	0.0	.2	.7	1.3	2.1	.3	.0849
200.	3283.	.7	-1.5	-1.0	0.0	1.2	2.6	1.4	.1745
300.	3131.	2.7	-6.2	-5.5	-4.0	-2.1	0.0	3.2	.2682
400.	2984.	5.7	-14.4	-13.5	-11.5	-9.0	-6.2	5.8	.3660
500.	2842.	9.8	-26.5	-25.4	-22.8	-19.7	-16.2	9.2	.4687
600.	2705.	15.2	-43.0	-41.6	-38.6	-34.9	-30.7	13.6	.5772
700.	2572.	22.2	-64.2	-62.6	-59.1	-54.8	-49.8	18.9	.6910
800.	2443.	30.8	-90.6	-88.8	-84.7	-79.8	-74.2	25.3	.8104
900.	2317.	41.4	-122.9	-120.8	-116.3	-110.7	-104.4	32.9	.9367
1000.	2195.	54.2	-161.5	-159.2	-154.2	-148.0	-140.9	41.6	1.0697

BALLISTIC COEFFICIENT .68, MUZZLE VELOCITY FT/SEC 3700.

RANGE	VELOCITY	HEIGHT	100 YD	150 YD	200 YD	250 YD	300 YD	DRIFT	TIME
0.	3700.	0.0	-1.5	-1.5	-1.5	-1.5	-1.5	0.0	0.0000
100.	3536.	-.4	0.0	.2	.6	1.2	1.9	.3	.0830
200.	3377.	.6	-1.3	-.9	0.0	1.2	2.5	1.4	.1699
300.	3222.	2.5	-5.7	-5.1	-3.7	-2.0	0.0	3.1	.2611
400.	3073.	5.3	-13.4	-12.6	-10.8	-8.4	-5.8	5.6	.3562
500.	2928.	9.3	-24.8	-23.8	-21.6	-18.7	-15.4	9.0	.4565
600.	2788.	14.4	-40.2	-39.1	-36.4	-32.9	-28.9	13.2	.5613
700.	2652.	20.9	-60.2	-58.8	-55.7	-51.6	-47.0	18.3	.6717
800.	2521.	29.1	-85.1	-83.6	-80.0	-75.3	-70.0	24.5	.7881
900.	2393.	39.0	-115.4	-113.6	-109.6	-104.3	-98.4	31.7	.9100
1000.	2268.	51.0	-151.7	-149.8	-145.2	-139.4	-132.8	40.1	1.0388

BALLISTIC COEFFICIENT .68, MUZZLE VELOCITY FT/SEC 3800.

RANGE	VELOCITY	HEIGHT	100 YD	150 YD	200 YD	250 YD	300 YD	DRIFT	TIME
0.	3800.	0.0	-1.5	-1.5	-1.5	-1.5	-1.5	0.0	0.0000
100.	3633.	-.4	0.0	.1	.6	1.1	1.7	.4	.0812
200.	3471.	.6	-1.1	-.8	0.0	1.1	2.3	1.3	.1655
300.	3313.	2.3	-5.2	-4.7	-3.5	-1.8	0.0	3.0	.2540
400.	3161.	5.0	-12.4	-11.8	-10.1	-7.9	-5.5	5.5	.3468
500.	3013.	8.7	-23.0	-22.3	-20.2	-17.5	-14.4	8.6	.4437
600.	2871.	13.5	-37.6	-36.7	-34.2	-30.9	-27.2	12.7	.5456
700.	2732.	19.7	-56.4	-55.3	-52.4	-48.6	-44.3	17.7	.6531
800.	2598.	27.4	-79.8	-78.6	-75.2	-70.9	-66.0	23.6	.7657
900.	2469.	36.8	-108.2	-106.9	-103.1	-98.2	-92.7	30.5	.8840
1000.	2342.	48.1	-142.5	-141.0	-136.8	-131.4	-125.3	38.7	1.0091

BALLISTIC COEFFICIENT .68, MUZZLE VELOCITY FT/SEC 3900.

RANGE	VELOCITY	HEIGHT	100 YD	150 YD	200 YD	250 YD	300 YD	DRIFT	TIME
0.	3900.	0.0	-1.5	-1.5	-1.5	-1.5	-1.5	0.0	0.0000
100.	3730.	-.5	0.0	.1	.5	1.0	1.6	.2	.0783
200.	3565.	.5	-1.0	-.8	0.0	1.0	2.2	1.2	.1607
300.	3405.	2.2	-4.8	-4.4	-3.2	-1.7	0.0	2.8	.2468
400.	3250.	4.7	-11.5	-11.0	-9.5	-7.5	-5.1	5.1	.3368
500.	3099.	8.2	-21.6	-21.0	-19.1	-16.6	-13.7	8.3	.4317
600.	2953.	12.7	-35.2	-34.5	-32.2	-29.2	-25.7	12.1	.5303
700.	2812.	18.6	-53.0	-52.2	-49.4	-45.9	-41.9	17.0	.6349
800.	2676.	25.8	-74.9	-74.0	-70.9	-66.9	-62.2	22.6	.7439
900.	2544.	34.7	-101.9	-100.9	-97.4	-92.9	-87.6	29.4	.8593
1000.	2416.	45.4	-134.1	-132.9	-129.0	-124.0	-118.2	37.1	.9802

BALLISTIC COEFFICIENT .68, MUZZLE VELOCITY FT/SEC 4000.

RANGE	VELOCITY	HEIGHT	100 YD	150 YD	200 YD	250 YD	300 YD	DRIFT	TIME
0.	4000.	0.0	-1.5	-1.5	-1.5	-1.5	-1.5	0.0	0.0000
100.	3827.	-.5	0.0	.1	.4	.9	1.4	.3	.0769
200.	3659.	.4	-.9	-.7	0.0	1.0	2.0	1.2	.1569
300.	3496.	2.0	-4.3	-4.1	-3.0	-1.6	0.0	2.8	.2408
400.	3338.	4.4	-10.7	-10.4	-9.0	-7.0	-4.9	5.1	.3289
500.	3185.	7.7	-20.1	-19.7	-18.0	-15.6	-12.9	8.0	.4206
600.	3037.	12.1	-33.0	-32.5	-30.4	-27.5	-24.3	11.8	.5171
700.	2893.	17.6	-49.7	-49.1	-46.7	-43.3	-39.6	16.4	.6185
800.	2754.	24.5	-70.6	-69.9	-67.1	-63.2	-59.0	22.0	.7249
900.	2619.	32.8	-95.9	-95.1	-92.0	-87.6	-82.8	28.4	.8363
1000.	2489.	42.9	-126.3	-125.5	-122.0	-117.1	-111.8	35.9	.9540

BALLISTIC COEFFICIENT .69, MUZZLE VELOCITY FT/SEC 2000.

RANGE	VELOCITY	HEIGHT	100 YD	150 YD	200 YD	250 YD	300 YD	DRIFT	TIME
0.	2000.	0.0	-1.5	-1.5	-1.5	-1.5	-1.5	0.0	0.0000
100.	1889.	.4	0.0	1.9	4.2	6.6	9.2	.7	.1542
200.	1783.	4.1	-8.3	-4.5	0.0	4.8	10.1	3.2	.3180
300.	1681.	10.8	-27.5	-21.8	-15.1	-7.8	0.0	7.2	.4911
400.	1585.	21.1	-59.2	-51.6	-42.6	-32.9	-22.5	13.2	.6752
500.	1492.	35.6	-104.6	-95.0	-83.8	-71.7	-58.6	21.1	.8698
600.	1406.	54.9	-165.7	-154.2	-140.8	-126.2	-110.6	31.2	1.0771
700.	1328.	80.0	-244.5	-231.1	-215.4	-198.5	-180.2	43.4	1.2969
800.	1259.	111.4	-342.9	-327.6	-309.6	-290.3	-269.4	57.9	1.5288
900.	1196.	150.4	-463.6	-446.4	-426.2	-404.4	-381.0	74.7	1.7743
1000.	1143.	197.1	-607.8	-588.6	-566.2	-542.0	-515.9	93.3	2.0302

BALLISTIC COEFFICIENT .69, MUZZLE VELOCITY FT/SEC 2100.

RANGE	VELOCITY	HEIGHT	100 YD	150 YD	200 YD	250 YD	300 YD	DRIFT	TIME
0.	2100.	0.0	-1.5	-1.5	-1.5	-1.5	-1.5	0.0	0.0000
100.	1985.	.3	0.0	1.7	3.7	5.9	8.2	.6	.1465
200.	1875.	3.6	-7.3	-4.0	0.0	4.4	9.1	2.9	.3020
300.	1769.	9.7	-24.7	-19.6	-13.6	-7.0	0.0	6.8	.4669
400.	1668.	19.0	-53.1	-46.3	-38.4	-29.6	-20.2	12.4	.6417
500.	1572.	32.1	-94.1	-85.7	-75.8	-64.8	-53.0	19.9	.8272
600.	1481.	49.5	-149.0	-138.8	-126.9	-113.7	-99.6	29.2	1.0233
700.	1395.	72.2	-220.0	-208.2	-194.3	-179.0	-162.5	40.9	1.2326
800.	1319.	100.7	-308.7	-295.2	-279.4	-261.8	-243.0	54.7	1.4536
900.	1250.	135.9	-417.6	-402.3	-384.5	-364.7	-343.6	70.7	1.6874
1000.	1189.	178.7	-548.5	-531.5	-511.7	-489.8	-466.3	88.9	1.9337

BALLISTIC COEFFICIENT .69, MUZZLE VELOCITY FT/SEC 2200.

RANGE	VELOCITY	HEIGHT	100 YD	150 YD	200 YD	250 YD	300 YD	DRIFT	TIME
0.	2200.	0.0	-1.5	-1.5	-1.5	-1.5	-1.5	0.0	0.0000
100.	2082.	.2	0.0	1.5	3.3	5.3	7.4	.6	.1400
200.	1968.	3.2	-6.6	-3.6	0.0	4.0	8.2	2.8	.2885
300.	1858.	8.8	-22.1	-17.6	-12.2	-6.3	0.0	6.3	.4451
400.	1753.	17.2	-47.8	-41.8	-34.6	-26.7	-18.3	11.6	.6114
500.	1653.	29.0	-84.8	-77.4	-68.4	-58.5	-48.0	18.7	.7878
600.	1558.	44.9	-134.6	-125.6	-114.9	-102.9	-90.4	27.6	.9749
700.	1467.	65.3	-198.5	-188.1	-175.5	-161.6	-147.0	38.5	1.1731
800.	1383.	91.2	-278.6	-266.7	-252.4	-236.4	-219.7	51.5	1.3837
900.	1308.	123.2	-377.1	-363.7	-347.5	-329.6	-310.8	66.8	1.6069
1000.	1241.	162.3	-496.0	-481.1	-463.1	-443.2	-422.3	84.3	1.8428

BALLISTIC COEFFICIENT .69, MUZZLE VELOCITY FT/SEC 2300.

RANGE	VELOCITY	HEIGHT	100 YD	150 YD	200 YD	250 YD	300 YD	DRIFT	TIME
0.	2300.	0.0	-1.5	-1.5	-1.5	-1.5	-1.5	0.0	0.0000
100.	2180.	.1	0.0	1.3	2.9	4.7	6.7	.6	.1336
200.	2063.	2.9	-5.9	-3.3	0.0	3.6	7.4	2.6	.2755
300.	1949.	7.9	-20.0	-16.1	-11.1	-5.8	0.0	6.0	.4252
400.	1840.	15.6	-43.3	-38.0	-31.5	-24.3	-16.6	10.9	.5838
500.	1736.	26.3	-76.7	-70.2	-62.0	-53.1	-43.4	17.4	.7512
600.	1637.	40.7	-121.8	-114.0	-104.1	-93.4	-81.9	25.8	.9293
700.	1542.	59.3	-179.8	-170.7	-159.2	-146.7	-133.2	36.1	1.1181
800.	1452.	82.7	-252.5	-242.1	-229.0	-214.6	-199.3	48.5	1.3189
900.	1370.	111.8	-341.6	-329.9	-315.1	-299.0	-281.7	62.9	1.5316
1000.	1296.	147.4	-449.5	-436.5	-420.1	-402.2	-383.0	79.7	1.7571

BALLISTIC COEFFICIENT .69, MUZZLE VELOCITY FT/SEC 2400.

RANGE	VELOCITY	HEIGHT	100 YD	150 YD	200 YD	250 YD	300 YD	DRIFT	TIME
0.	2400.	0.0	-1.5	-1.5	-1.5	-1.5	-1.5	0.0	0.0000
100.	2277.	.0	0.0	1.1	2.6	4.2	6.0	.6	.1282
200.	2157.	2.6	-5.2	-2.9	0.0	3.3	6.8	2.4	.2634
300.	2041.	7.2	-18.0	-14.6	-10.2	-5.2	0.0	5.6	.4068
400.	1928.	14.2	-39.1	-34.5	-28.7	-22.1	-15.1	10.2	.5581
500.	1820.	24.0	-69.5	-63.8	-56.4	-48.2	-39.5	16.3	.7177
600.	1717.	37.0	-110.4	-103.5	-94.8	-84.9	-74.4	24.2	.8873
700.	1619.	53.9	-163.2	-155.2	-145.0	-133.5	-121.2	33.9	1.0675
800.	1525.	75.3	-229.3	-220.1	-208.4	-195.3	-181.3	45.5	1.2586
900.	1436.	101.8	-310.2	-299.9	-286.8	-272.0	-256.2	59.2	1.4616
1000.	1355.	134.3	-408.3	-396.9	-382.3	-365.8	-348.3	75.2	1.6772

BALLISTIC COEFFICIENT .69, MUZZLE VELOCITY FT/SEC 2500.

RANGE	VELOCITY	HEIGHT	100 YD	150 YD	200 YD	250 YD	300 YD	DRIFT	TIME
0.	2500.	0.0	-1.5	-1.5	-1.5	-1.5	-1.5	0.0	0.0000
100.	2374.	-.0	0.0	1.0	2.3	3.8	5.4	.5	.1230
200.	2252.	2.3	-4.7	-2.7	0.0	2.9	6.1	2.3	.2529
300.	2133.	6.5	-16.3	-13.2	-9.2	-4.8	0.0	5.2	.3897
400.	2017.	13.0	-35.5	-31.4	-26.1	-20.2	-13.8	9.6	.5343
500.	1905.	21.9	-63.3	-58.2	-51.5	-44.2	-36.2	15.4	.6873
600.	1798.	33.9	-100.6	-94.6	-86.6	-77.8	-68.1	22.8	.8493
700.	1696.	49.3	-148.7	-141.6	-132.3	-122.0	-110.8	31.8	1.0209
800.	1599.	68.8	-209.0	-200.9	-190.2	-178.5	-165.7	42.8	1.2035
900.	1506.	92.9	-282.6	-273.4	-261.4	-248.3	-233.8	55.7	1.3964
1000.	1418.	122.4	-371.6	-361.4	-348.1	-333.4	-317.4	70.7	1.6018

BALLISTIC COEFFICIENT .69, MUZZLE VELOCITY FT/SEC 2600.

RANGE	VELOCITY	HEIGHT	100 YD	150 YD	200 YD	250 YD	300 YD	DRIFT	TIME
0.	2600.	0.0	-1.5	-1.5	-1.5	-1.5	-1.5	0.0	0.0000
100.	2472.	-.1	0.0	.9	2.1	3.5	4.9	.5	.1182
200.	2347.	2.1	-4.2	-2.4	0.0	2.7	5.6	2.2	.2431
300.	2226.	6.0	-14.8	-12.0	-8.4	-4.4	0.0	4.9	.3741
400.	2107.	11.9	-32.4	-28.7	-23.9	-18.5	-12.7	9.0	.5127
500.	1992.	20.1	-57.8	-53.2	-47.2	-40.5	-33.2	14.5	.6590
600.	1881.	31.0	-92.0	-86.5	-79.2	-71.2	-62.4	21.4	.8139
700.	1775.	45.2	-136.1	-129.7	-121.2	-111.8	-101.6	30.0	.9783
800.	1674.	63.0	-191.1	-183.8	-174.1	-163.4	-151.7	40.3	1.1522
900.	1578.	85.1	-258.6	-250.4	-239.5	-227.5	-214.3	52.6	1.3371
1000.	1486.	112.0	-339.6	-330.5	-318.4	-305.0	-290.4	66.6	1.5325

BALLISTIC COEFFICIENT .69, MUZZLE VELOCITY FT/SEC 2700.

RANGE	VELOCITY	HEIGHT	100 YD	150 YD	200 YD	250 YD	300 YD	DRIFT	TIME
0.	2700.	0.0	-1.5	-1.5	-1.5	-1.5	-1.5	0.0	0.0000
100.	2569.	-.1	0.0	.8	1.9	3.1	4.5	.5	.1139
200.	2442.	1.9	-3.8	-2.1	0.0	2.5	5.2	2.0	.2335
300.	2318.	5.5	-13.4	-11.0	-7.8	-4.0	0.0	4.7	.3598
400.	2197.	10.9	-29.5	-26.3	-22.0	-17.0	-11.6	8.5	.4927
500.	2079.	18.5	-52.8	-48.8	-43.4	-37.2	-30.5	13.6	.6328
600.	1965.	28.6	-84.3	-79.5	-73.0	-65.5	-57.5	20.2	.7816
700.	1856.	41.5	-124.6	-119.0	-111.5	-102.7	-93.3	28.3	.9385
800.	1751.	57.9	-175.1	-168.7	-160.1	-150.1	-139.4	38.0	1.1050
900.	1651.	78.1	-237.0	-229.7	-220.1	-208.8	-196.8	49.6	1.2816
1000.	1556.	102.8	-311.5	-303.5	-292.7	-280.2	-266.8	63.0	1.4690

BALLISTIC COEFFICIENT .69, MUZZLE VELOCITY FT/SEC 2800.

RANGE	VELOCITY	HEIGHT	100 YD	150 YD	200 YD	250 YD	300 YD	DRIFT	TIME
0.	2800.	0.0	-1.5	-1.5	-1.5	-1.5	-1.5	0.0	0.0000
100.	2666.	-.2	0.0	.7	1.7	2.8	4.1	.4	.1094
200.	2536.	1.7	-3.4	-2.0	0.0	2.2	4.7	1.9	.2251
300.	2410.	5.0	-12.3	-10.1	-7.1	-3.7	0.0	4.4	.3464
400.	2286.	10.0	-27.1	-24.2	-20.2	-15.7	-10.7	8.0	.4741
500.	2166.	17.0	-48.5	-45.0	-40.0	-34.3	-28.1	12.8	.6086
600.	2050.	26.4	-77.6	-73.3	-67.3	-60.5	-53.1	19.1	.7515
700.	1936.	38.3	-114.8	-109.8	-102.7	-94.9	-86.1	26.8	.9020
800.	1828.	53.3	-161.2	-155.5	-147.4	-138.4	-128.5	35.9	1.0613
900.	1725.	71.9	-217.8	-211.5	-202.4	-192.3	-181.1	46.7	1.2299
1000.	1626.	94.6	-286.4	-279.3	-269.2	-258.0	-245.5	59.5	1.4093

BALLISTIC COEFFICIENT .69, MUZZLE VELOCITY FT/SEC 2900.

RANGE	VELOCITY	HEIGHT	100 YD	150 YD	200 YD	250 YD	300 YD	DRIFT	TIME
0.	2900.	0.0	-1.5	-1.5	-1.5	-1.5	-1.5	0.0	0.0000
100.	2763.	-.2	0.0	.6	1.5	2.6	3.7	.5	.1061
200.	2630.	1.5	-3.0	-1.8	0.0	2.1	4.4	1.8	.2171
300.	2501.	4.6	-11.2	-9.3	-6.6	-3.5	0.0	4.2	.3344
400.	2375.	9.3	-24.9	-22.4	-18.8	-14.5	-9.9	7.7	.4573
500.	2253.	15.8	-44.7	-41.7	-37.1	-31.9	-26.1	12.3	.5872
600.	2134.	24.4	-71.4	-67.8	-62.3	-56.0	-49.1	18.2	.7239
700.	2018.	35.4	-105.8	-101.5	-95.1	-87.7	-79.7	25.4	.8684
800.	1906.	49.3	-148.6	-143.7	-136.4	-128.0	-118.8	34.1	1.0214
900.	1799.	66.5	-201.0	-195.4	-187.3	-177.8	-167.4	44.4	1.1832
1000.	1697.	87.4	-264.0	-257.9	-248.8	-238.3	-226.8	56.4	1.3548

BALLISTIC COEFFICIENT .69, MUZZLE VELOCITY FT/SEC 3000.

RANGE	VELOCITY	HEIGHT	100 YD	150 YD	200 YD	250 YD	300 YD	DRIFT	TIME
0.	3000.	0.0	-1.5	-1.5	-1.5	-1.5	-1.5	0.0	0.0000
100.	2860.	-.2	0.0	.6	1.4	2.4	3.4	.4	.1025
200.	2724.	1.4	-2.8	-1.7	0.0	1.9	4.1	1.8	.2103
300.	2592.	4.3	-10.3	-8.6	-6.1	-3.2	0.0	4.1	.3232
400.	2464.	8.6	-22.9	-20.7	-17.3	-13.5	-9.2	7.4	.4418
500.	2339.	14.7	-41.4	-38.6	-34.4	-29.6	-24.3	11.8	.5670
600.	2218.	22.7	-66.1	-62.7	-57.7	-51.9	-45.5	17.3	.6984
700.	2100.	32.9	-98.0	-94.1	-88.2	-81.5	-74.0	24.2	.8378
800.	1985.	45.8	-137.6	-133.1	-126.4	-118.7	-110.2	32.4	.9844
900.	1874.	61.6	-186.0	-180.9	-173.4	-164.7	-155.2	42.2	1.1398
1000.	1769.	81.0	-244.4	-238.8	-230.5	-220.8	-210.2	53.6	1.3047

BALLISTIC COEFFICIENT .69, MUZZLE VELOCITY FT/SEC 3100.

RANGE	VELOCITY	HEIGHT	100 YD	150 YD	200 YD	250 YD	300 YD	DRIFT	TIME
0.	3100.	0.0	-1.5	-1.5	-1.5	-1.5	-1.5	0.0	0.0000
100.	2956.	-.3	0.0	.5	1.3	2.2	3.1	.3	.0985
200.	2817.	1.2	-2.5	-1.6	0.0	1.8	3.7	1.7	.2029
300.	2683.	3.9	-9.4	-7.9	-5.6	-2.9	0.0	3.8	.3117
400.	2552.	8.0	-21.2	-19.2	-16.1	-12.5	-8.6	7.0	.4267
500.	2425.	13.6	-38.2	-35.8	-31.9	-27.5	-22.6	11.1	.5472
600.	2302.	21.1	-61.3	-58.4	-53.7	-48.3	-42.5	16.5	.6743
700.	2181.	30.6	-90.7	-87.3	-81.9	-75.6	-68.8	22.9	.8078
800.	2064.	42.5	-127.7	-123.8	-117.6	-110.4	-102.6	30.9	.9495
900.	1951.	57.2	-172.6	-168.3	-161.3	-153.2	-144.5	40.2	1.0991
1000.	1842.	75.2	-226.8	-221.9	-214.2	-205.2	-195.5	51.0	1.2575

BALLISTIC COEFFICIENT .69, MUZZLE VELOCITY FT/SEC 3200.

RANGE	VELOCITY	HEIGHT	100 YD	150 YD	200 YD	250 YD	300 YD	DRIFT	TIME
0.	3200.	0.0	-1.5	-1.5	-1.5	-1.5	-1.5	0.0	0.0000
100.	3053.	-.3	0.0	.4	1.1	1.9	2.9	.3	.0957
200.	2911.	1.1	-2.3	-1.4	0.0	1.6	3.5	1.6	.1968
300.	2774.	3.6	-8.6	-7.3	-5.2	-2.8	0.0	3.7	.3021
400.	2640.	7.4	-19.5	-17.8	-15.0	-11.7	-8.0	6.7	.4131
500.	2511.	12.7	-35.5	-33.3	-29.8	-25.7	-21.1	10.8	.5299
600.	2385.	19.7	-56.9	-54.3	-50.0	-45.2	-39.6	15.8	.6522
700.	2263.	28.6	-84.4	-81.4	-76.4	-70.8	-64.3	22.0	.7815
800.	2143.	39.7	-118.7	-115.2	-109.6	-103.1	-95.7	29.5	.9178
900.	2027.	53.3	-160.5	-156.6	-150.2	-143.0	-134.7	38.3	1.0616
1000.	1915.	70.0	-210.8	-206.4	-199.4	-191.3	-182.1	48.7	1.2140

BALLISTIC COEFFICIENT .69, MUZZLE VELOCITY FT/SEC 3300.

RANGE	VELOCITY	HEIGHT	100 YD	150 YD	200 YD	250 YD	300 YD	DRIFT	TIME
0.	3300.	0.0	-1.5	-1.5	-1.5	-1.5	-1.5	0.0	0.0000
100.	3150.	-.3	0.0	.4	1.0	1.8	2.6	.4	.0934
200.	3005.	1.0	-2.0	-1.3	0.0	1.6	3.2	1.5	.1904
300.	2864.	3.4	-7.8	-6.8	-4.9	-2.5	0.0	3.5	.2927
400.	2728.	6.9	-18.0	-16.6	-14.0	-10.9	-7.5	6.4	.4003
500.	2597.	11.9	-32.8	-31.0	-27.9	-24.0	-19.7	10.3	.5130
600.	2469.	18.4	-52.7	-50.6	-46.8	-42.1	-37.1	15.1	.6314
700.	2344.	26.7	-78.5	-76.0	-71.5	-66.1	-60.2	21.1	.7564
800.	2222.	37.1	-110.4	-107.5	-102.4	-96.2	-89.5	28.2	.8876
900.	2104.	49.8	-149.4	-146.2	-140.5	-133.5	-125.9	36.7	1.0266
1000.	1989.	65.3	-196.0	-192.5	-186.2	-178.4	-169.9	46.4	1.1729

BALLISTIC COEFFICIENT .69, MUZZLE VELOCITY FT/SEC 3400.

RANGE	VELOCITY	HEIGHT	100 YD	150 YD	200 YD	250 YD	300 YD	DRIFT	TIME
0.	3400.	0.0	-1.5	-1.5	-1.5	-1.5	-1.5	0.0	0.0000
100.	3247.	-.4	0.0	.3	.9	1.6	2.4	.3	.0900
200.	3099.	.9	-1.8	-1.2	0.0	1.4	3.0	1.5	.1849
300.	2955.	3.1	-7.2	-6.3	-4.5	-2.4	0.0	3.3	.2835
400.	2816.	6.5	-16.7	-15.5	-13.1	-10.2	-7.1	6.2	.3879
500.	2682.	11.1	-30.5	-28.9	-25.9	-22.4	-18.5	9.8	.4967
600.	2551.	17.2	-49.2	-47.3	-43.7	-39.5	-34.8	14.5	.6117
700.	2425.	25.0	-73.2	-71.0	-66.8	-61.9	-56.4	20.2	.7323
800.	2301.	34.7	-103.1	-100.6	-95.8	-90.2	-83.9	27.0	.8594
900.	2181.	46.6	-139.4	-136.6	-131.2	-124.8	-117.7	35.0	.9929
1000.	2063.	61.1	-183.2	-180.1	-174.1	-167.0	-159.1	44.4	1.1348

BALLISTIC COEFFICIENT .69, MUZZLE VELOCITY FT/SEC 3500.

RANGE	VELOCITY	HEIGHT	100 YD	150 YD	200 YD	250 YD	300 YD	DRIFT	TIME
0.	3500.	0.0	-1.5	-1.5	-1.5	-1.5	-1.5	0.0	0.0000
100.	3344.	-.4	0.0	.3	.8	1.5	2.2	.4	.0880
200.	3193.	.8	-1.6	-1.1	0.0	1.4	2.8	1.4	.1796
300.	3046.	2.9	-6.6	-5.8	-4.2	-2.2	0.0	3.3	.2757
400.	2905.	6.1	-15.5	-14.4	-12.3	-9.5	-6.6	6.0	.3769
500.	2767.	10.4	-28.4	-27.1	-24.4	-21.0	-17.3	9.5	.4826
600.	2634.	16.2	-45.8	-44.3	-41.0	-36.9	-32.6	14.0	.5937
700.	2505.	23.5	-68.4	-66.5	-62.8	-58.0	-52.9	19.5	.7107
800.	2379.	32.6	-96.3	-94.2	-89.9	-84.4	-78.6	26.0	.8334
900.	2257.	43.8	-130.5	-128.1	-123.2	-117.1	-110.6	33.7	.9631
1000.	2138.	57.3	-171.4	-168.7	-163.3	-156.5	-149.3	42.7	1.0996

BALLISTIC COEFFICIENT .69, MUZZLE VELOCITY FT/SEC 3600.

RANGE	VELOCITY	HEIGHT	100 YD	150 YD	200 YD	250 YD	300 YD	DRIFT	TIME
0.	3600.	0.0	-1.5	-1.5	-1.5	-1.5	-1.5	0.0	0.0000
100.	3441.	-.4	0.0	.2	.7	1.3	2.0	.3	.0849
200.	3287.	.7	-1.5	-1.0	0.0	1.2	2.6	1.3	.1743
300.	3138.	2.7	-6.1	-5.5	-4.0	-2.1	0.0	3.1	.2679
400.	2993.	5.7	-14.3	-13.4	-11.4	-8.9	-6.1	5.6	.3653
500.	2853.	9.8	-26.4	-25.3	-22.8	-19.7	-16.2	9.0	.4679
600.	2717.	15.2	-42.8	-41.5	-38.5	-34.8	-30.6	13.4	.5760
700.	2585.	22.1	-63.9	-62.4	-58.8	-54.5	-49.6	18.6	.6892
800.	2458.	30.6	-90.2	-88.4	-84.3	-79.4	-73.8	24.9	.8081
900.	2333.	41.1	-122.2	-120.2	-115.7	-110.1	-103.8	32.3	.9336
1000.	2212.	53.7	-160.5	-158.2	-153.2	-147.0	-140.0	40.8	1.0654

BALLISTIC COEFFICIENT .69, MUZZLE VELOCITY FT/SEC 3700.

RANGE	VELOCITY	HEIGHT	100 YD	150 YD	200 YD	250 YD	300 YD	DRIFT	TIME
0.	3700.	0.0	-1.5	-1.5	-1.5	-1.5	-1.5	0.0	0.0000
100.	3538.	-.4	0.0	.2	.6	1.2	1.9	.3	.0831
200.	3381.	.6	-1.3	-.9	0.0	1.2	2.5	1.3	.1698
300.	3229.	2.5	-5.6	-5.1	-3.7	-2.0	0.0	3.1	.2608
400.	3081.	5.3	-13.3	-12.6	-10.8	-8.4	-5.8	5.6	.3559
500.	2938.	9.2	-24.6	-23.7	-21.4	-18.5	-15.2	8.8	.4554
600.	2800.	14.3	-40.1	-39.0	-36.3	-32.8	-28.9	13.0	.5605
700.	2666.	20.8	-59.9	-58.5	-55.4	-51.3	-46.7	18.0	.6699
800.	2536.	28.9	-84.7	-83.1	-79.5	-74.9	-69.6	24.1	.7856
900.	2409.	38.7	-114.7	-113.0	-108.9	-103.7	-97.8	31.2	.9069
1000.	2286.	50.6	-150.7	-148.8	-144.3	-138.5	-131.9	39.4	1.0346

BALLISTIC COEFFICIENT .69, MUZZLE VELOCITY FT/SEC 3800.

RANGE	VELOCITY	HEIGHT	100 YD	150 YD	200 YD	250 YD	300 YD	DRIFT	TIME
0.	3800.	0.0	-1.5	-1.5	-1.5	-1.5	-1.5	0.0	0.0000
100.	3635.	-.4	0.0	.1	.6	1.1	1.7	.4	.0811
200.	3475.	.6	-1.1	-.8	0.0	1.1	2.3	1.3	.1654
300.	3320.	2.3	-5.2	-4.7	-3.5	-1.8	0.0	3.0	.2538
400.	3170.	5.0	-12.3	-11.7	-10.1	-7.9	-5.5	5.3	.3462
500.	3024.	8.7	-23.0	-22.2	-20.1	-17.4	-14.4	8.5	.4430
600.	2883.	13.5	-37.4	-36.5	-34.0	-30.7	-27.1	12.5	.5445
700.	2746.	19.6	-56.1	-55.1	-52.2	-48.3	-44.1	17.4	.6514
800.	2614.	27.2	-79.3	-78.1	-74.8	-70.4	-65.6	23.2	.7633
900.	2485.	36.5	-107.6	-106.3	-102.6	-97.6	-92.2	30.0	.8810
1000.	2360.	47.8	-141.6	-140.1	-136.0	-130.5	-124.4	38.0	1.0052

BALLISTIC COEFFICIENT .69, MUZZLE VELOCITY FT/SEC 3900.

RANGE	VELOCITY	HEIGHT	100 YD	150 YD	200 YD	250 YD	300 YD	DRIFT	TIME
0.	3900.	0.0	-1.5	-1.5	-1.5	-1.5	-1.5	0.0	0.0000
100.	3732.	-.5	0.0	.1	.5	1.0	1.6	.2	.0782
200.	3570.	.5	-1.0	-.8	0.0	1.0	2.1	1.2	.1607
300.	3412.	2.2	-4.7	-4.4	-3.2	-1.7	0.0	2.8	.2464
400.	3259.	4.7	-11.4	-11.0	-9.4	-7.4	-5.1	5.0	.3362
500.	3110.	8.2	-21.6	-21.0	-19.0	-16.5	-13.6	8.2	.4310
600.	2966.	12.7	-35.1	-34.4	-32.0	-29.1	-25.6	11.9	.5293
700.	2827.	18.5	-52.7	-51.9	-49.1	-45.7	-41.6	16.6	.6331
800.	2692.	25.7	-74.6	-73.7	-70.5	-66.6	-61.9	22.2	.7418
900.	2561.	34.4	-101.3	-100.2	-96.7	-92.2	-87.0	28.8	.8561
1000.	2434.	45.0	-133.2	-132.1	-128.1	-123.2	-117.4	36.4	.9762

BALLISTIC COEFFICIENT .69, MUZZLE VELOCITY FT/SEC 4000.

RANGE	VELOCITY	HEIGHT	100 YD	150 YD	200 YD	250 YD	300 YD	DRIFT	TIME
0.	4000.	0.0	-1.5	-1.5	-1.5	-1.5	-1.5	0.0	0.0000
100.	3829.	-.5	0.0	.1	.4	.9	1.4	.3	.0769
200.	3664.	.4	-.9	-.7	0.0	1.0	2.0	1.2	.1569
300.	3503.	2.0	-4.3	-4.1	-3.0	-1.6	0.0	2.7	.2405
400.	3347.	4.4	-10.7	-10.3	-8.9	-7.0	-4.9	5.0	.3284
500.	3196.	7.7	-20.1	-19.7	-17.9	-15.5	-12.9	7.9	.4200
600.	3049.	12.0	-32.9	-32.4	-30.3	-27.4	-24.2	11.6	.5159
700.	2907.	17.5	-49.6	-49.0	-46.5	-43.1	-39.5	16.2	.6171
800.	2770.	24.3	-70.2	-69.5	-66.7	-62.9	-58.6	21.6	.7227
900.	2637.	32.6	-95.4	-94.6	-91.5	-87.1	-82.4	27.9	.8337
1000.	2508.	42.6	-125.6	-124.8	-121.3	-116.5	-111.2	35.3	.9506

BALLISTIC COEFFICIENT .70, MUZZLE VELOCITY FT/SEC 2000.

RANGE	VELOCITY	HEIGHT	100 YD	150 YD	200 YD	250 YD	300 YD	DRIFT	TIME
0.	2000.	0.0	-1.5	-1.5	-1.5	-1.5	-1.5	0.0	0.0000
100.	1890.	.4	0.0	1.9	4.1	6.6	9.2	.7	.1541
200.	1786.	4.1	-8.3	-4.5	0.0	4.8	10.0	3.1	.3177
300.	1686.	10.8	-27.5	-21.7	-15.0	-7.8	0.0	7.1	.4904
400.	1590.	21.1	-59.0	-51.4	-42.5	-32.8	-22.4	13.0	.6740
500.	1499.	35.4	-104.1	-94.6	-83.4	-71.3	-58.3	20.7	.8678
600.	1413.	54.6	-164.9	-153.4	-140.1	-125.5	-110.0	30.6	1.0741
700.	1336.	79.5	-243.4	-230.0	-214.4	-197.5	-179.3	42.8	1.2933
800.	1267.	110.6	-340.9	-325.6	-307.8	-288.4	-267.7	56.9	1.5234
900.	1204.	149.1	-460.3	-443.1	-423.0	-401.2	-377.9	73.3	1.7667
1000.	1150.	195.5	-603.6	-584.4	-562.1	-537.9	-512.0	91.8	2.0218

BALLISTIC COEFFICIENT .70, MUZZLE VELOCITY FT/SEC 2100.

RANGE	VELOCITY	HEIGHT	100 YD	150 YD	200 YD	250 YD	300 YD	DRIFT	TIME
0.	2100.	0.0	-1.5	-1.5	-1.5	-1.5	-1.5	0.0	0.0000
100.	1987.	.3	0.0	1.7	3.7	5.9	8.2	.6	.1465
200.	1878.	3.6	-7.3	-3.9	0.0	4.4	9.1	2.8	.3017
300.	1773.	9.7	-24.6	-19.5	-13.6	-7.0	0.0	6.7	.4664
400.	1674.	18.9	-52.9	-46.2	-38.3	-29.5	-20.1	12.2	.6405
500.	1579.	31.9	-93.8	-85.3	-75.5	-64.5	-52.8	19.5	.8254
600.	1488.	49.2	-148.3	-138.2	-126.3	-113.1	-99.1	28.8	1.0206
700.	1403.	71.7	-218.7	-206.9	-193.1	-177.7	-161.3	40.2	1.2283
800.	1327.	99.9	-306.9	-293.4	-277.6	-260.0	-241.3	53.8	1.4484
900.	1259.	134.8	-414.7	-399.5	-381.8	-362.0	-340.9	69.5	1.6805
1000.	1197.	177.3	-544.8	-527.9	-508.2	-486.2	-462.8	87.5	1.9258

BALLISTIC COEFFICIENT .70, MUZZLE VELOCITY FT/SEC 2200.

RANGE	VELOCITY	HEIGHT	100 YD	150 YD	200 YD	250 YD	300 YD	DRIFT	TIME
0.	2200.	0.0	-1.5	-1.5	-1.5	-1.5	-1.5	0.0	0.0000
100.	2084.	.2	0.0	1.5	3.3	5.3	7.3	.6	.1400
200.	1971.	3.2	-6.5	-3.6	0.0	4.0	8.1	2.7	.2882
300.	1863.	8.7	-22.0	-17.6	-12.2	-6.2	0.0	6.2	.4445
400.	1759.	17.1	-47.6	-41.6	-34.5	-26.5	-18.2	11.4	.6102
500.	1660.	28.9	-84.5	-77.1	-68.1	-58.2	-47.8	18.3	.7861
600.	1566.	44.6	-134.0	-125.1	-114.4	-102.4	-90.0	27.1	.9724
700.	1476.	64.9	-197.4	-187.0	-174.5	-160.6	-146.0	37.8	1.1693
800.	1392.	90.5	-277.0	-265.1	-250.9	-234.9	-218.3	50.7	1.3788
900.	1317.	122.2	-374.7	-361.3	-345.3	-327.3	-308.6	65.7	1.6006
1000.	1250.	160.8	-492.3	-477.4	-459.6	-439.7	-418.9	82.9	1.8345

BALLISTIC COEFFICIENT .70, MUZZLE VELOCITY FT/SEC 2300.

RANGE	VELOCITY	HEIGHT	100 YD	150 YD	200 YD	250 YD	300 YD	DRIFT	TIME
0.	2300.	0.0	-1.5	-1.5	-1.5	-1.5	-1.5	0.0	0.0000
100.	2181.	.1	0.0	1.3	2.9	4.7	6.6	.5	.1335
200.	2066.	2.9	-5.9	-3.3	0.0	3.6	7.4	2.5	.2752
300.	1954.	7.9	-19.9	-16.0	-11.1	-5.8	0.0	5.9	.4247
400.	1846.	15.5	-43.1	-37.9	-31.4	-24.2	-16.5	10.7	.5826
500.	1743.	26.2	-76.4	-69.9	-61.8	-52.9	-43.3	17.2	.7497
600.	1645.	40.5	-121.3	-113.5	-103.7	-93.0	-81.5	25.4	.9269
700.	1551.	58.9	-179.0	-169.9	-158.5	-146.0	-132.5	35.5	1.1148
800.	1462.	82.1	-251.0	-240.5	-227.5	-213.2	-197.9	47.6	1.3139
900.	1380.	110.9	-339.2	-327.5	-312.8	-296.8	-279.5	61.8	1.5251
1000.	1306.	146.0	-446.0	-432.9	-416.6	-398.8	-379.6	78.2	1.7487

BALLISTIC COEFFICIENT .70, MUZZLE VELOCITY FT/SEC 2400.

RANGE	VELOCITY	HEIGHT	100 YD	150 YD	200 YD	250 YD	300 YD	DRIFT	TIME
0.	2400.	0.0	-1.5	-1.5	-1.5	-1.5	-1.5	0.0	0.0000
100.	2279.	.0	0.0	1.1	2.6	4.2	6.0	.5	.1281
200.	2161.	2.6	-5.2	-2.9	0.0	3.3	6.8	2.3	.2631
300.	2046.	7.2	-18.0	-14.5	-10.2	-5.2	0.0	5.5	.4063
400.	1934.	14.1	-39.0	-34.4	-28.6	-22.0	-15.1	10.0	.5571
500.	1828.	23.9	-69.3	-63.6	-56.3	-48.1	-39.4	16.1	.7164
600.	1726.	36.9	-110.0	-103.1	-94.4	-84.5	-74.1	23.8	.8851
700.	1628.	53.6	-162.5	-154.5	-144.3	-132.8	-120.6	33.3	1.0643
800.	1535.	74.7	-228.0	-218.8	-207.2	-194.0	-180.1	44.7	1.2541
900.	1447.	101.0	-308.3	-298.0	-284.9	-270.1	-254.4	58.2	1.4557
1000.	1366.	133.0	-405.1	-393.7	-379.1	-362.7	-345.3	73.8	1.6691

BALLISTIC COEFFICIENT .70, MUZZLE VELOCITY FT/SEC 2500.

RANGE	VELOCITY	HEIGHT	100 YD	150 YD	200 YD	250 YD	300 YD	DRIFT	TIME
0.	2500.	0.0	-1.5	-1.5	-1.5	-1.5	-1.5	0.0	0.0000
100.	2376.	-.0	0.0	1.0	2.3	3.8	5.4	.5	.1230
200.	2256.	2.3	-4.7	-2.7	0.0	2.9	6.1	2.2	.2527
300.	2138.	6.5	-16.2	-13.2	-9.2	-4.8	0.0	5.2	.3894
400.	2024.	12.9	-35.4	-31.3	-26.0	-20.2	-13.7	9.4	.5335
500.	1913.	21.8	-63.1	-58.0	-51.4	-44.1	-36.0	15.1	.6861
600.	1807.	33.7	-100.2	-94.1	-86.1	-77.4	-67.7	22.4	.8471
700.	1706.	49.0	-148.0	-140.9	-131.6	-121.3	-110.1	31.3	1.0177
800.	1610.	68.3	-207.8	-199.7	-189.1	-177.4	-164.5	42.1	1.1991
900.	1518.	92.1	-280.9	-271.7	-259.8	-246.6	-232.2	54.7	1.3909
1000.	1430.	121.3	-369.0	-358.9	-345.5	-330.9	-314.9	69.5	1.5947

BALLISTIC COEFFICIENT .70, MUZZLE VELOCITY FT/SEC 2600.

RANGE	VELOCITY	HEIGHT	100 YD	150 YD	200 YD	250 YD	300 YD	DRIFT	TIME
0.	2600.	0.0	-1.5	-1.5	-1.5	-1.5	-1.5	0.0	0.0000
100.	2474.	-.1	0.0	.9	2.1	3.5	4.9	.5	.1181
200.	2351.	2.1	-4.2	-2.4	0.0	2.7	5.6	2.1	.2429
300.	2231.	6.0	-14.7	-12.0	-8.4	-4.4	0.0	4.9	.3738
400.	2114.	11.8	-32.2	-28.6	-23.8	-18.4	-12.6	8.8	.5118
500.	2000.	20.0	-57.6	-53.1	-47.0	-40.3	-33.0	14.2	.6578
600.	1891.	30.9	-91.6	-86.1	-78.9	-70.9	-62.1	21.1	.8119
700.	1786.	44.9	-135.4	-129.1	-120.6	-111.2	-101.0	29.5	.9754
800.	1686.	62.5	-190.0	-182.7	-173.1	-162.4	-150.7	39.6	1.1482
900.	1590.	84.4	-257.0	-248.8	-238.0	-225.9	-212.8	51.6	1.3318
1000.	1499.	111.0	-337.2	-328.1	-316.0	-302.6	-288.1	65.4	1.5255

BALLISTIC COEFFICIENT .70, MUZZLE VELOCITY FT/SEC 2700.

RANGE	VELOCITY	HEIGHT	100 YD	150 YD	200 YD	250 YD	300 YD	DRIFT	TIME
0.	2700.	0.0	-1.5	-1.5	-1.5	-1.5	-1.5	0.0	0.0000
100.	2571.	-.1	0.0	.8	1.9	3.1	4.5	.5	.1139
200.	2445.	1.9	-3.8	-2.1	0.0	2.5	5.2	2.0	.2334
300.	2323.	5.5	-13.4	-11.0	-7.7	-4.0	0.0	4.6	.3594
400.	2204.	10.9	-29.4	-26.2	-21.9	-16.9	-11.6	8.4	.4919
500.	2088.	18.4	-52.7	-48.7	-43.3	-37.1	-30.4	13.4	.6318
600.	1975.	28.4	-83.9	-79.1	-72.6	-65.1	-57.1	19.9	.7795
700.	1866.	41.3	-124.0	-118.4	-110.9	-102.1	-92.8	27.8	.9356
800.	1763.	57.5	-174.2	-167.7	-159.1	-149.2	-138.5	37.4	1.1012
900.	1663.	77.5	-235.5	-228.3	-218.6	-207.4	-195.4	48.7	1.2766
1000.	1569.	101.9	-309.4	-301.4	-290.6	-278.2	-264.8	61.9	1.4626

BALLISTIC COEFFICIENT .70, MUZZLE VELOCITY FT/SEC 2800.

RANGE	VELOCITY	HEIGHT	100 YD	150 YD	200 YD	250 YD	300 YD	DRIFT	TIME
0.	2800.	0.0	-1.5	-1.5	-1.5	-1.5	-1.5	0.0	0.0000
100.	2668.	-.2	0.0	.7	1.7	2.8	4.1	.4	.1093
200.	2540.	1.7	-3.4	-2.0	0.0	2.2	4.7	1.9	.2249
300.	2415.	5.0	-12.2	-10.1	-7.1	-3.7	0.0	4.3	.3460
400.	2293.	10.0	-27.0	-24.2	-20.2	-15.7	-10.7	7.9	.4734
500.	2175.	17.0	-48.4	-44.8	-39.8	-34.2	-28.0	12.6	.6074
600.	2060.	26.2	-77.3	-73.0	-67.0	-60.3	-52.8	18.8	.7496
700.	1948.	38.1	-114.2	-109.3	-102.2	-94.4	-85.7	26.3	.8994
800.	1840.	53.0	-160.4	-154.7	-146.7	-137.7	-127.8	35.4	1.0580
900.	1738.	71.3	-216.6	-210.3	-201.2	-191.1	-179.9	46.0	1.2254
1000.	1640.	93.8	-284.5	-277.4	-267.3	-256.1	-243.7	58.4	1.4031

BALLISTIC COEFFICIENT .70, MUZZLE VELOCITY FT/SEC 2900.

RANGE	VELOCITY	HEIGHT	100 YD	150 YD	200 YD	250 YD	300 YD	DRIFT	TIME
0.	2900.	0.0	-1.5	-1.5	-1.5	-1.5	-1.5	0.0	0.0000
100.	2765.	-.2	0.0	.6	1.5	2.6	3.7	.4	.1060
200.	2634.	1.5	-3.0	-1.8	0.0	2.1	4.4	1.8	.2170
300.	2506.	4.6	-11.2	-9.3	-6.6	-3.4	0.0	4.2	.3341
400.	2383.	9.3	-24.8	-22.3	-18.7	-14.5	-9.9	7.5	.4566
500.	2262.	15.7	-44.6	-41.5	-37.0	-31.7	-26.0	12.1	.5860
600.	2144.	24.3	-71.2	-67.5	-62.1	-55.7	-48.9	17.9	.7223
700.	2030.	35.3	-105.3	-101.0	-94.7	-87.3	-79.3	25.0	.8661
800.	1919.	49.0	-147.9	-143.0	-135.7	-127.3	-118.1	33.6	1.0182
900.	1813.	65.9	-199.7	-194.2	-186.0	-176.6	-166.2	43.6	1.1786
1000.	1711.	86.6	-262.2	-256.0	-247.0	-236.4	-225.0	55.3	1.3488

BALLISTIC COEFFICIENT .70, MUZZLE VELOCITY FT/SEC 3000.

RANGE	VELOCITY	HEIGHT	100 YD	150 YD	200 YD	250 YD	300 YD	DRIFT	TIME
0.	3000.	0.0	-1.5	-1.5	-1.5	-1.5	-1.5	0.0	0.0000
100.	2862.	-.2	0.0	.6	1.4	2.4	3.4	.4	.1024
200.	2727.	1.4	-2.8	-1.7	0.0	1.9	4.0	1.8	.2101
300.	2597.	4.3	-10.2	-8.6	-6.1	-3.2	0.0	4.0	.3229
400.	2471.	8.6	-22.8	-20.6	-17.3	-13.4	-9.2	7.2	.4411
500.	2348.	14.6	-41.3	-38.4	-34.3	-29.5	-24.2	11.6	.5660
600.	2228.	22.6	-65.9	-62.5	-57.5	-51.7	-45.4	17.1	.6970
700.	2112.	32.7	-97.5	-93.5	-87.7	-81.0	-73.6	23.8	.8352
800.	1998.	45.5	-136.9	-132.4	-125.8	-118.1	-109.6	31.9	.9813
900.	1888.	61.2	-184.9	-179.9	-172.4	-163.7	-154.2	41.5	1.1356
1000.	1784.	80.3	-242.9	-237.2	-229.0	-219.3	-208.7	52.7	1.2994

BALLISTIC COEFFICIENT .70, MUZZLE VELOCITY FT/SEC 3100.

RANGE	VELOCITY	HEIGHT	100 YD	150 YD	200 YD	250 YD	300 YD	DRIFT	TIME
0.	3100.	0.0	-1.5	-1.5	-1.5	-1.5	-1.5	0.0	0.0000
100.	2958.	-.3	0.0	.5	1.3	2.2	3.1	.3	.0985
200.	2821.	1.2	-2.5	-1.5	0.0	1.8	3.7	1.6	.2027
300.	2688.	3.9	-9.4	-7.9	-5.6	-2.9	0.0	3.7	.3115
400.	2560.	8.0	-21.1	-19.1	-16.1	-12.5	-8.6	6.8	.4259
500.	2434.	13.6	-38.1	-35.7	-31.8	-27.3	-22.5	10.9	.5460
600.	2312.	21.0	-61.0	-58.1	-53.5	-48.1	-42.3	16.2	.6726
700.	2193.	30.4	-90.4	-86.9	-81.6	-75.3	-68.5	22.6	.8057
800.	2078.	42.2	-126.9	-123.0	-116.9	-109.7	-101.9	30.3	.9461
900.	1965.	56.8	-171.6	-167.2	-160.3	-152.2	-143.5	39.4	1.0950
1000.	1857.	74.5	-225.1	-220.2	-212.5	-203.5	-193.8	50.0	1.2517

BALLISTIC COEFFICIENT .70, MUZZLE VELOCITY FT/SEC 3200.

RANGE	VELOCITY	HEIGHT	100 YD	150 YD	200 YD	250 YD	300 YD	DRIFT	TIME
0.	3200.	0.0	-1.5	-1.5	-1.5	-1.5	-1.5	0.0	0.0000
100.	3055.	-.3	0.0	.4	1.1	1.9	2.9	.3	.0956
200.	2915.	1.1	-2.3	-1.4	0.0	1.6	3.4	1.6	.1968
300.	2779.	3.6	-8.6	-7.3	-5.2	-2.7	0.0	3.6	.3016
400.	2648.	7.4	-19.5	-17.8	-14.9	-11.7	-8.0	6.6	.4125
500.	2520.	12.7	-35.4	-33.2	-29.7	-25.6	-21.1	10.6	.5289
600.	2396.	19.6	-56.7	-54.1	-49.8	-45.0	-39.5	15.5	.6508
700.	2275.	28.4	-84.0	-81.0	-76.0	-70.4	-64.0	21.6	.7792
800.	2157.	39.4	-118.0	-114.6	-108.9	-102.4	-95.1	29.0	.9145
900.	2042.	53.0	-159.6	-155.8	-149.4	-142.1	-133.9	37.7	1.0579
1000.	1931.	69.4	-209.4	-205.1	-198.0	-189.9	-180.8	47.8	1.2090

BALLISTIC COEFFICIENT .70, MUZZLE VELOCITY FT/SEC 3300.

RANGE	VELOCITY	HEIGHT	100 YD	150 YD	200 YD	250 YD	300 YD	DRIFT	TIME
0.	3300.	0.0	-1.5	-1.5	-1.5	-1.5	-1.5	0.0	0.0000
100.	3152.	-.3	0.0	.4	1.0	1.8	2.6	.4	.0933
200.	3009.	1.0	-2.0	-1.3	0.0	1.5	3.2	1.5	.1903
300.	2870.	3.4	-7.8	-6.7	-4.9	-2.5	0.0	3.5	.2923
400.	2736.	6.9	-17.9	-16.5	-14.0	-10.9	-7.5	6.3	.3997
500.	2606.	11.8	-32.7	-30.9	-27.7	-23.9	-19.7	10.1	.5120
600.	2479.	18.3	-52.5	-50.4	-46.6	-42.0	-36.9	14.9	.6299
700.	2356.	26.6	-78.1	-75.6	-71.2	-65.8	-59.9	20.8	.7545
800.	2236.	36.9	-109.9	-107.0	-102.0	-95.8	-89.0	27.8	.8851
900.	2119.	49.4	-148.4	-145.2	-139.5	-132.6	-124.9	36.0	1.0225
1000.	2005.	64.8	-194.9	-191.3	-185.0	-177.3	-168.8	45.6	1.1684

BALLISTIC COEFFICIENT .70, MUZZLE VELOCITY FT/SEC 3400.

RANGE	VELOCITY	HEIGHT	100 YD	150 YD	200 YD	250 YD	300 YD	DRIFT	TIME
0.	3400.	0.0	-1.5	-1.5	-1.5	-1.5	-1.5	0.0	0.0000
100.	3249.	-.4	0.0	.3	.9	1.6	2.4	.3	.0899
200.	3103.	.9	-1.8	-1.2	0.0	1.4	3.0	1.5	.1848
300.	2961.	3.1	-7.2	-6.3	-4.5	-2.4	0.0	3.3	.2832
400.	2824.	6.4	-16.6	-15.4	-13.0	-10.2	-7.0	6.0	.3872
500.	2691.	11.1	-30.4	-28.8	-25.8	-22.3	-18.4	9.6	.4960
600.	2562.	17.1	-49.0	-47.1	-43.5	-39.3	-34.6	14.2	.6103
700.	2437.	24.8	-72.8	-70.6	-66.4	-61.5	-56.0	19.8	.7302
800.	2315.	34.5	-102.6	-100.1	-95.3	-89.7	-83.4	26.5	.8567
900.	2196.	46.3	-138.7	-135.9	-130.4	-124.1	-117.1	34.4	.9896
1000.	2080.	60.5	-181.9	-178.8	-172.8	-165.8	-157.9	43.6	1.1298

BALLISTIC COEFFICIENT .70, MUZZLE VELOCITY FT/SEC 3500.

RANGE	VELOCITY	HEIGHT	100 YD	150 YD	200 YD	250 YD	300 YD	DRIFT	TIME
0.	3500.	0.0	-1.5	-1.5	-1.5	-1.5	-1.5	0.0	0.0000
100.	3346.	-.4	0.0	.3	.8	1.5	2.2	.4	.0879
200.	3197.	.8	-1.6	-1.1	0.0	1.4	2.8	1.4	.1796
300.	3053.	2.9	-6.6	-5.8	-4.2	-2.2	0.0	3.2	.2753
400.	2913.	6.1	-15.4	-14.4	-12.2	-9.5	-6.6	5.9	.3765
500.	2777.	10.4	-28.2	-26.9	-24.2	-20.8	-17.2	9.3	.4815
600.	2645.	16.1	-45.7	-44.1	-40.9	-36.8	-32.5	13.8	.5924
700.	2518.	23.4	-68.1	-66.2	-62.5	-57.7	-52.7	19.2	.7090
800.	2394.	32.4	-95.9	-93.8	-89.4	-84.0	-78.3	25.6	.8310
900.	2273.	43.4	-129.7	-127.3	-122.5	-116.4	-109.9	33.1	.9596
1000.	2155.	56.8	-170.2	-167.5	-162.2	-155.4	-148.2	41.9	1.0950

BALLISTIC COEFFICIENT .70, MUZZLE VELOCITY FT/SEC 3600.

RANGE	VELOCITY	HEIGHT	100 YD	150 YD	200 YD	250 YD	300 YD	DRIFT	TIME
0.	3600.	0.0	-1.5	-1.5	-1.5	-1.5	-1.5	0.0	0.0000
100.	3443.	-.4	0.0	.2	.7	1.3	2.0	.3	.0849
200.	3291.	.7	-1.4	-1.0	0.0	1.2	2.6	1.3	.1741
300.	3144.	2.7	-6.1	-5.5	-4.0	-2.1	0.0	3.1	.2676
400.	3001.	5.6	-14.2	-13.4	-11.4	-8.9	-6.1	5.5	.3647
500.	2862.	9.7	-26.3	-25.2	-22.7	-19.6	-16.1	8.9	.4671
600.	2728.	15.1	-42.7	-41.3	-38.3	-34.6	-30.4	13.2	.5747
700.	2598.	21.9	-63.6	-62.1	-58.6	-54.3	-49.4	18.3	.6874
800.	2472.	30.4	-89.7	-87.9	-83.9	-79.0	-73.4	24.5	.8056
900.	2349.	40.8	-121.6	-119.6	-115.1	-109.5	-103.2	31.8	.9305
1000.	2229.	53.3	-159.5	-157.3	-152.3	-146.1	-139.1	40.1	1.0614

BALLISTIC COEFFICIENT .70, MUZZLE VELOCITY FT/SEC 3700.

RANGE	VELOCITY	HEIGHT	100 YD	150 YD	200 YD	250 YD	300 YD	DRIFT	TIME
0.	3700.	0.0	-1.5	-1.5	-1.5	-1.5	-1.5	0.0	0.0000
100.	3540.	-.4	0.0	.2	.6	1.2	1.9	.3	.0831
200.	3386.	.6	-1.3	-.9	0.0	1.2	2.5	1.3	.1698
300.	3235.	2.5	-5.6	-5.1	-3.7	-2.0	0.0	3.0	.2606
400.	3090.	5.3	-13.3	-12.5	-10.7	-8.4	-5.8	5.5	.3555
500.	2948.	9.2	-24.5	-23.6	-21.3	-18.4	-15.2	8.6	.4545
600.	2811.	14.3	-39.9	-38.8	-36.1	-32.6	-28.7	12.8	.5592
700.	2679.	20.7	-59.6	-58.2	-55.1	-51.0	-46.5	17.7	.6681
800.	2550.	28.7	-84.2	-82.7	-79.1	-74.4	-69.2	23.7	.7832
900.	2425.	38.5	-114.0	-112.3	-108.3	-103.1	-97.2	30.6	.9037
1000.	2304.	50.2	-149.8	-147.9	-143.4	-137.6	-131.0	38.7	1.0307

BALLISTIC COEFFICIENT .70, MUZZLE VELOCITY FT/SEC 3800.

RANGE	VELOCITY	HEIGHT	100 YD	150 YD	200 YD	250 YD	300 YD	DRIFT	TIME
0.	3800.	0.0	-1.5	-1.5	-1.5	-1.5	-1.5	0.0	0.0000
100.	3638.	-.4	0.0	.1	.6	1.1	1.7	.4	.0811
200.	3480.	.6	-1.1	-.8	0.0	1.1	2.3	1.3	.1653
300.	3327.	2.3	-5.1	-4.7	-3.5	-1.8	0.0	2.9	.2536
400.	3178.	5.0	-12.3	-11.7	-10.0	-7.8	-5.4	5.2	.3456
500.	3034.	8.6	-22.9	-22.2	-20.1	-17.4	-14.3	8.4	.4422
600.	2895.	13.4	-37.3	-36.4	-33.9	-30.7	-27.0	12.3	.5436
700.	2760.	19.5	-55.9	-54.9	-52.0	-48.2	-43.9	17.1	.6500
800.	2629.	27.1	-78.9	-77.7	-74.4	-70.1	-65.2	22.8	.7611
900.	2502.	36.3	-107.1	-105.8	-102.1	-97.2	-91.7	29.5	.8784
1000.	2378.	47.4	-140.7	-139.2	-135.1	-129.6	-123.5	37.3	1.0012

BALLISTIC COEFFICIENT .70, MUZZLE VELOCITY FT/SEC 3900.

RANGE	VELOCITY	HEIGHT	100 YD	150 YD	200 YD	250 YD	300 YD	DRIFT	TIME
0.	3900.	0.0	-1.5	-1.5	-1.5	-1.5	-1.5	0.0	0.0000
100.	3735.	-.5	0.0	.1	.5	1.0	1.6	.2	.0782
200.	3574.	.5	-1.0	-.8	0.0	1.0	2.1	1.2	.1606
300.	3418.	2.2	-4.7	-4.4	-3.2	-1.7	0.0	2.7	.2460
400.	3267.	4.7	-11.4	-11.0	-9.4	-7.4	-5.1	5.0	.3359
500.	3120.	8.1	-21.5	-20.9	-19.0	-16.5	-13.6	8.0	.4303
600.	2978.	12.7	-35.0	-34.3	-32.0	-29.0	-25.6	11.8	.5285
700.	2840.	18.4	-52.4	-51.6	-48.8	-45.4	-41.4	16.3	.6312
800.	2707.	25.5	-74.2	-73.3	-70.2	-66.3	-61.7	21.9	.7398
900.	2578.	34.2	-100.8	-99.7	-96.2	-91.8	-86.6	28.4	.8534
1000.	2452.	44.7	-132.4	-131.3	-127.3	-122.4	-116.7	35.8	.9725

BALLISTIC COEFFICIENT .70, MUZZLE VELOCITY FT/SEC 4000.

RANGE	VELOCITY	HEIGHT	100 YD	150 YD	200 YD	250 YD	300 YD	DRIFT	TIME
0.	4000.	0.0	-1.5	-1.5	-1.5	-1.5	-1.5	0.0	0.0000
100.	3832.	-.5	0.0	.1	.4	.9	1.4	.3	.0768
200.	3668.	.4	-.9	-.7	0.0	1.0	2.0	1.2	.1568
300.	3510.	2.0	-4.3	-4.1	-3.0	-1.6	0.0	2.7	.2402
400.	3356.	4.4	-10.6	-10.3	-8.9	-7.0	-4.9	4.9	.3280
500.	3206.	7.7	-20.0	-19.6	-17.9	-15.5	-12.9	7.8	.4194
600.	3062.	12.0	-32.8	-32.3	-30.1	-27.3	-24.1	11.4	.5148
700.	2921.	17.4	-49.4	-48.8	-46.3	-43.0	-39.3	16.0	.6158
800.	2785.	24.2	-69.8	-69.1	-66.3	-62.5	-58.3	21.2	.7205
900.	2653.	32.4	-94.9	-94.1	-91.0	-86.6	-81.9	27.5	.8310
1000.	2526.	42.3	-124.9	-124.1	-120.5	-115.7	-110.5	34.7	.9472